Emerging Paradigms for Antibiotic-Resistant Infections: Beyond the Pill

Mayank Gangwar • Gopal Nath
Editors

Emerging Paradigms for Antibiotic-Resistant Infections: Beyond the Pill

Springer

Editors
Mayank Gangwar
Faculty of Medicine, Viral Research and Diagnostic Laboratory, Department of Microbiology, Institute of Medical Sciences, Banaras Hindu University, Varanasi, Uttar Pradesh, India

Clinical Studies and Trials Unit, Division of Development Research, Indian Council of Medical Research, Department of Health Research, Ministry of Health and Family Welfare, Govt. of India, V. Ramalingaswami Bhawan, New Delhi, India

Gopal Nath
Department of Microbiology
Banaras Hindu University
Varanasi, Uttar Pradesh, India

ISBN 978-981-97-5271-3 ISBN 978-981-97-5272-0 (eBook)
https://doi.org/10.1007/978-981-97-5272-0

This Springer imprint is published by the registered company Springer Nature Singapore Pte Ltd.
The registered company address is: 152 Beach Road, #21-01/04 Gateway East, Singapore 189721, Singapore

This book is dedicated to Pandit Madan Mohan Malaviya, founder of Banaras Hindu University.

In honor of Pandit Madan Mohan Malaviya,
Visionary architect of knowledge,
Whose fervent dedication forged
The grandeur of Banaras Hindu University.

To the luminary whose inspired vision
Gave birth to Asia's sprawling sanctuary of learning,
A beacon of enlightenment, a bastion of tradition,
Guiding countless seekers toward wisdom's yearning.

In reverence, we dedicate these pages,
To the stalwart champion of education's cause, whose legacy continues to inspire,
Echoing through the hallowed halls without pause.

To Pandit Madan Mohan Malaviya,
Whose indomitable spirit and tireless endeavor,
Uplifted minds, nurtured hearts, and
Enshrined the pursuit of knowledge forever.

Foreword

It is a great pleasure and enthusiasm for me to write the foreword of the latest creation *Emerging Paradigms for Antibiotic-Resistant Infections: Beyond the Pill*, published by Springer, Nature. As the Director of the Institute of Medical Sciences, I am immensely proud of the work put forth by our esteemed editors, *Prof. Gopal Nath* and *Dr. Mayank Gangwar*, who have meticulously curated a collection of insights and perspectives that transcend conventional approaches to combatting antibiotic-resistant infections.

In recent years, the emergence of antibiotic-resistant bacteria has posed a great threat to global public health, challenging the efficacy of traditional antibiotic therapies. In response to this pressing issue, ***Emerging Paradigms for Antibiotic-Resistant Infections: Beyond the Pill*** delves into innovative strategies and emerging paradigms that extend beyond conventional pharmaceutical interventions. Through a multidisciplinary approach, this book explores the intersections of Microbiology, Ethnopharmacology, Immunology, and Public Health Policy, or challenges to offer novel perspectives and solutions in the battle against antibiotic resistance.

Prof. Nath and *Dr. Gangwar* have been able to attract a distinguished group of contributors, comprising leading researchers, clinicians, and policymakers from around the world, whose collective expertise and diverse perspectives enrich the discourse on antibiotic resistance. From the exploration of alternative antimicrobial agents to the examination of host–pathogen interactions and the role of environmental factors, each chapter presents cutting-edge research and practical insights that promise to reshape our understanding and approach to combating antibiotic-resistant infections.

As we navigate the complexities of antibiotic resistance in the twenty-first century, it is imperative that we embrace a holistic and forward-thinking approach. *Emerging Paradigms for Antibiotic-Resistant Infections: Beyond the Pill* serves as a beacon of knowledge and innovation, guiding us towards a future where effective strategies for infection control extend beyond the confines of traditional pharmaceutical solutions.

I appreciate their organized way of writing that focuses on major areas of antimicrobial alternatives. Both national and international researchers have contributed to this book. Chapters of this book are divided into eight major parts, viz. *I-The Rise of Antibiotic Resistance; II-Understanding Non-antibiotic Advanced Therapies; III-Phage Therapy: Viral Warriors Against Bacterial Invaders; IV-Immunotherapies:*

Boosting the Body's Defense; V-Natural Antimicrobial Solutions: Nature's Weapon, VI-Bacteriophage-Derived Enzymes: Enlisting Nature's Allies; VII-Overcoming Challenges and Regulatory Hurdles; and *VIII-The Future of Non-Antibiotic Therapies*.

In short, *Prof. Nath's* and *Dr. Gangwar's* book is unique and covers almost all the dimensions of alternatives to antibiotics. I am sure it will prove to be a treasure for anyone who is interested in recent innovations, research, and development towards antimicrobials. It is my pleasure to endorse *Prof. Nath* experience in Microbiology and his commendable efficiency of teamwork with world leaders in this particular field. As a result of this effort as a book, both editors and the institution are enriched and privileged to have an opportunity of working with such international laureates. My heartiest congratulations and wishes are always with *Prof. Nath* and his team for producing such a nice piece of work that will guide young minds for long.

I extend my sincerest gratitude to *Prof. Gopal Nath, Dr. Mayank Gangwar*, and all the contributing authors for their dedication and expertise in bringing this seminal work to fruition. It is my hope that this book will serve as a catalyst for continued dialog, collaboration, and action in the global fight against antibiotic resistance.

Faculty of Medicine, Institute of Medical Sciences S. N. Sankhwar
Banaras Hindu University
Varanasi, India

Preface

In an age where microbial adversaries are rapidly evolving, the battle against antibiotic-resistant infections has become one of the most pressing challenges of our time. The rise of antibiotic resistance threatens not only individual health but also the very foundation of modern medicine. As traditional antibiotics struggle to keep pace with the mutating pathogens, the need for innovative solutions has never been more urgent.

This book, *Emerging Paradigms for Antibiotic-Resistant Infections: Beyond the Pill*, stands as a testament to the collaborative efforts of national and international researchers who have dedicated their expertise to confronting this formidable threat. Within its pages lie a collection of insights, discoveries, and strategies aimed at transcending conventional approaches to combat antibiotic resistance.

The chapters of this book are thoughtfully organized into eight major parts, each delving into a distinct aspect of the multifaceted challenge posed by antibiotic-resistant infections: Part I includes the The Rise of Antibiotic Resistance. In this foundational part, we embark on a journey through the historical and epidemiological landscape of antibiotic resistance. We explore the factors driving the emergence and spread of resistant microbes, laying the groundwork for understanding the gravity of the current crisis. Part II includes the Understanding Non-antibiotic Advanced Therapies. Moving beyond traditional antibiotic treatments, this part explores a diverse array of advanced therapies poised to revolutionize the field. From microbiome modulation to host-directed therapies, these innovative approaches offer promising avenues for combating resistant infections. Part III includes the Phage Therapy: Viral Warriors Against Bacterial Invaders. Phage therapy, once overshadowed by the advent of antibiotics, has experienced a resurgence as a potent weapon against antibiotic-resistant bacteria. Here, we delve into the mechanisms, applications, and challenges of harnessing bacteriophages as precision antimicrobial agents. Part IV includes chapters related to Immunotherapies: Boosting the Body's Defense. Harnessing the power of the immune system, immunotherapies represent a paradigm shift in the fight against resistant infections. Through vaccines, monoclonal antibodies, and immune modulators, researchers are unlocking new avenues to bolster the body's natural defenses. Part V of the book includes the Natural Antimicrobial Solutions: Nature's Weapon. Nature has long provided a rich source of antimicrobial compounds, offering potential alternatives to synthetic antibiotics. This part explores the diverse arsenal of natural antimicrobials, from plant-derived compounds to antimicrobial peptides, and their prospects in combating resistant pathogens. Part VI of this book includes the Bacteriophage-Derived Enzymes: Enlisting Nature's Allies. Beyond whole phage therapy, bacteriophage-derived enzymes have emerged as versatile tools for targeting specific bacterial pathogens. Here, we explore the therapeutic potential of these enzymes and their applications in overcoming antibiotic resistance. Part VII includes the Overcoming Challenges and Regulatory Hurdles. The translation of novel antimicrobial therapies from the laboratory to the clinic presents a myriad of challenges and regulatory hurdles. This part examines the obstacles impeding the adoption of non-antibiotic treatments and explores strategies to navigate the complex landscape of drug development and regulation. The last part VIII includes the Future of Non-antibiotic Therapies. As we peer into the future, the horizon is marked by both promise and uncertainty. In this final part, we contemplate the evolving landscape of non-antibiotic therapies and envision the transformative impact they may have on the global fight against antibiotic-resistant infections.

As editors of this volume, we are immensely grateful to the contributors whose expertise and dedication have enriched its pages. Their collective insights and tireless efforts have illuminated new pathways toward a future where antibiotic-resistant infections no longer pose a threat to human health. It is our hope that this book will serve as a beacon of inspiration and knowledge for researchers, clinicians, policymakers, and all those vested in the pursuit of a world free from the shadow of antibiotic resistance.

Together, let us embark on this journey Beyond the Pill, where innovation, collaboration, and determination converge to conquer one of the greatest challenges of our time. Therefore, it is a crucial and vital need for continuous exchange of shared knowledge among research contributors throughout the world. This book provides comprehensive information on the effective utilization of different alternatives for antimicrobials with existing mechanism. This also sheds some light on unknown and recent technologies about current research, which may further be advantageous to all the microbiologists.

Great efforts have been made by all contributors in preparation of these chapters in this book to improve our lives against antibiotic-resistant infections. This book will be valuable for students, teachers, academicians, and researchers related to microbiology, Ayurveda, natural product chemistry, medicinal chemistry, pharmaceutical sciences, biotechnology, and molecular biology.

Varanasi, Uttar Pradesh, India Mayank Gangwar
Varanasi, Uttar Pradesh, India Gopal Nath

Take up one idea. Make that one idea your life—think of it, dream of it, live on that idea. Let the brain, muscles, nerves, every part of your body, be full of that idea, and just leave every other idea alone. This is the way to success that is the way great spiritual giants are produced.

–Swami Vivekananda, World's most respected Vedanta thinker (1863–1902),
Vedanta Philosophy: Lectures by Swami Vivekananda,
Kessinger Publishing, USA, 1996, Page 70

Contents

Part I The Rise of Antibiotic Resistance

1 **CRISPR–Cas Against Antimicrobial-Resistant Pathogens** 3
Siddhant Tripathi, Yashika Sharma, Manish Kumar Gautam, and Dileep Kumar

2 **Upsurge in Antibiotic-Resistant Infections** 25
Sindhuprava Rana, Vibhor Joshi, Ganesh Chandra Sahoo, Maneesh Kumar, and Krishna Pandey

3 **Navigating the Challenges of Antibiotic Resistance: Origins, Mechanisms, and Global Responses** 43
Lakshmi Jyothi Tadi and Shazia Naaz

4 **The Rise of Antibiotic Resistance: A "Silent Killer"** 53
Kulsoom and Amit Lather

Part II Understanding Non-antibiotic Advanced Therapies

5 **Alternative Approaches to Counter Multidrug-Resistant Bacterial Pathogens** 69
Deepanshi Saxena, Rahul Maitra, Arunava Dasgupta, and Sidharth Chopra

6 **Anti-Quorum-Sensing Techniques: A Potential Strategy to Manage Antibiotic Resistance** 123
Srishti Singh, Alok Kumar Singh, Mayank Gangwar, and Gopal Nath

7 **Revolutionizing Healthcare: Unleashing the Potential of Immunotherapies to Strengthen the Body's Immune Defense Against Antibiotic-Resistant Pathogen** 147
Anasuya Patil, Vaishali P. Nagulwar, Sandeep Kumar Yadav, Fauzia Tabassum, and Singh Shekhar Gautam

8 Beyond Antibiotics: Pioneering Strategies in Infection Control to Counter Antibiotic Resistance's Rising Tide 173
Neha Dhami, Mayank Gangwar, Dileep Kumar, Avinash Kumar Rao, and Sandeep Kumar

9 Revolutionizing Infection Control: Harnessing the Power of Stem Cells for Precision Antimicrobial Therapy 197
Arup Dey, Pankaj Kumar Sharma, Ashutosh Pathak, Sanjay Kumar Yadav, Jashanjit Singh, and Sourav Ghosh

10 Combatting Antibiotic Resistance: Leveraging Fecal Microbial Transplantation for Gut Health 211
Prashasti Upreti, Lalchand D. Devhare, Lara Hashim Abdulmageed, Yamjala Ganesh Kumar, Rajesh Kumar, and G. Dharmamoorthy

11 A Journey Beyond Resistance: Discovering the Promise of Non-Antibiotic Therapies 233
Nitish Kumar Singh, Aprajita Singh, and Raghvendra Dubey

12 The Rise of Antibiotic Resistance: A Global Threat, Origin, and Evolution of Antibiotic Resistance: Current Scenario and Future Prospective 261
Vishal Bhojyawal, Mansi Kesarwani, and Sweety Gupta

13 Functional Nanomaterials in Antimicrobial Therapy for Antibiotic-Resistant Bacterial Infections 277
Gudikandula Krishna, Dasari Thrimothi, and V. Srilekha

14 Association and Contribution of Dentistry in Creating and Sharing the Burden of Antibiotic-Resistant Microbial World ... 295
Alka Shukla, Sukesh Sharma, Mayank Gangwar, Gaurav Sharma, and B. Apurva Rao

Part III Phage Therapy: Viral Warrior Against Bacterial Invaders

15 Bacteriophages: The Natural Combatants to Fight AMR 315
Aiswarya Mohan, Daman Saluja, and Urmi Bajpai

16 Reviving Phage Therapy: A Novel Approach to Treat Respiratory-Associated Infections 341
Ramesh Nachimuthu, Ramya Juliet, S. Bershiyal, and Oishi Mitra

17 The Art and Science of Defeating Superbugs: Present and Future of Phage Therapy 361
Sunil Kumar Raman, Chunna Yadav, Himanshu Bansode, Shilpi Mishra, Unnati Sharma, and Niharika Mishra

18 Phage Therapy: Challenges in Its Development and Clinical Implementation 387
Vijaya Lakshmi Nag, Yashik Bansal, and Mahesh Jansari

19 Breakthroughs in the Non-Antibiotic Therapies for Infectious Disease Treatment from Phages to Peptides and Beyond 409
Lokesh and Sruti Patel

20 Synergistic Effects of Phage and Antibiotic Combinations Against Pathogenic Bacteria 435
Minakshi Sahu, Ranjeet Kumar Vishwakarma, Deepak Kumar, Mayank Gangwar, Krupa Parmar, and Gopal Nath

21 Phage Therapy Reimagined: Viral Solutions for Antibiotic Resistant Challenges 449
Babita, Seema Sangwan, and Nishu Sehrawat

Part IV Immunotherapies: Boosting the Body's Defense

22 Diverse Properties of *Tinospora cordifolia* (Giloy, Heart Leaved Moonseed) World Wild Use for Immunotherapies; Boosting the Body's Defense and Immune Support 471
Preeti Semwal, Biresh Kumar Sarkar, Kumari Sonu, Garima Chand, Lalchand D. Devhare, and Tanuj Hooda

Part V Natural Antimicrobial Solutions: Nature's Weapon

23 Antimicrobial Efficacy of Rutin: Bridging the Gap Between Traditional Medicine and Modern Science 491
Divya Gupta, Raghvendra Raman Mishra, Pragya Mishra, Shubham Singh, Sadhana Shrivastava, Sangeeta Shukla, and Subodh Kumar

24 Beyond Pharmaceuticals: Tapping into Nature's Weapons Against Pathogenic Resilience 511
Ritu Raj Patel, Aradhana Mishra, Sudhir Kumar Singh, and Meenakshi Singh

25 Exploring the Potential of Phytochemicals: A Herbal Boon Against the MDR Bacteria Associated with the Chronic Wounds 539
Poonam Pal, M. Sree Vidhya, E. Karthikeyan, N. S. Disha, Niravsinh M. Rathod, and Sanjeev Kumar

26 Harnessing Nature's Pharmacy: Medicinal Plants Combatting Bacterial Infections in the Aquatic Environment 555
Ruksar, Tejpal Dahiya, Tannu Goyal, Abhimanyu, and Priyanka Priyanka Joshi

27 Essential Oil- Natural Weapons to Overcome Antimicrobial Resistance 589
Surabhi Shakya, Harikumar Pallathadka, Nangthombam Grihalaxmi Devi, and Amit Kumar Verma

28 Antimicrobial Activity of Plant and Cyanobacterial Secondary Metabolites for Therapeutic Advancements in Modern Medicine 607
Rupanshee Srivastava, Nidhi Singh, Tripti Kanda, Sadhana Yadav, Shivam Yadav, and Neelam Atri

29 Antimicrobial Aspects of Brazilian Red Propolis and Its Compounds 621
Nagela Bernadelli Sousa Silva, Daniela Silva Gonçalves, Gabriel Guimarães Calefi, Diego Godina Prado, Luís Carlos Scalon Cunha, and Carlos Henrique Gomes Martins

30 Phytochemicals as Source of Novel Antimicrobial Agents Against the Rapidly Evolving Antimicrobial Resistance 641
Ashapurna Sinha, Anupama Rani, Vaishali Mishra, Shiwa Chaubey, and Farina Mujeeb

31 Natural Antimicrobial Compounds Are an Alternative to Synthetic Antibiotics 671
Varsha Rini Guria, Pavan Kumar Padarthi, P. Gayatri Devi, Vishal Chauhan, and K. S. Sridevi Sangeetha

Part VI Bacteriophage-Derived Enzymes: Enlisting Nature's Allies

32 Phage Power: Harnessing Nature's Arsenal in the Battle Against Microbial Threats for Sustainable Healthcare 685
Jyotirmayee, Reshma Khanda, and Malkhey Verma

33 Bacteriophage-Derived Enzymes as Potential Alternatives for the Treatment of MDR Bacterial Infections 717
Subhash Lal Karn, Satyanam Kumar Bhartiya, Sandeep Pokhrel, Mayank Gangwar, and Gopal Nath

34 Nature's Allies: Bacteriophage-Derived Enzymes to Combat Resistant Bacteria 729
Shraddha Manish Gupta, Dinesh D. Rishipathak, Hemant U. Chikhale, and Iram Khan

35 Advancing Beyond Antibiotics: The Pioneering Use of Bacteriophage Enzymes 743
Nishu Sehrawat, Seema Sangwan, Babita, and Renu Lamba

Part VII Overcoming Challenges and Regulatory Hurdles

36 Overcoming Challenges and Regulatory Hurdles for Sustainable Bacteriophage Therapy Products 771
Rajal Dave and Sangeeta Ahiwale

37 Non-Traditional Antibacterial Therapeutic: Focus on Safety, Manufacturing, and Regulatory Dynamics 787
Deepa Dehari, Aiswarya Chaudhuri, Dulla Naveen Kumar, and Ashish Kumar Agrawal

Part VIII Future of Non-antibiotics Therapies

38 Harnessing Artificial Intelligence and Machine Learning in the Battle Against Antimicrobial-Resistant Infections 813
Arup Acharjee, Sauparnika Prasanna Kumar, Papia Acharjee, and Graham Ball

Correction to: The Art and Science of Defeating Superbugs: Present and Future of Phage Therapy C1
Sunil Kumar Raman, Chunna Yadav, Himanshu Bansode, Shilpi Mishra, Unnati Sharma, and Niharika Mishra

Editors and Contributors

About the Editors

Gopal Nath is widely known for his pioneering research on Bacteriophage Therapy in India, a potential alternative to antibiotics in the era of emerging antibiotic resistance worldwide. For the first time, he has conducted clinical phage therapy trials in India for acute and chronic wound infections. He has carried out preclinical work to discover the PK and PD of phages for life-threatening septicemia, urinary tract infections, and recalcitrant bone infections. For the first time, he established that *Salmonella Typhi* (The second bacteria implicated in carcinogenesis) induces gallbladder cancer. He has published his novel findings in high-impact journals with good citations. His research attributes are H factor 57, i10-index 173 with total citations of 12,463. He is a popular teacher for >38 years. As per ADI Scientific Index 2023, Dr. Nath is in the top 3% of Indian scientists. He completed his MBBS (1985) from Allahabad University; MD (1988), and PhD in Microbiology (2002) from Banaras Hindu University. He got various extramural research grants from several national and state agencies in the field of medicine, *viz.*, ICMR, DBT, DST, UGC, MHRD, CST-UP, NMCG, DHR, and IOE-BHU. He has received the Ernest Hanbury Hankin Lifetime Achievement Award from the Society for Bacteriophage Research and Therapy (2021), Dr. P.N. Chuttani Oration Award by the National Academy of Medical Sciences (2019), Dr. U.C. Chaturvedi Oration Award by IAMM UP-UK Chapter (2017), Fellow of the National Academy of Medical Sciences (2016), and many more. He has been widely applauded for delivering didactic lectures (>40)

nationwide on various platforms. He visited Sidney Kimmel Cancer Center, San Diego, CA, USA, as visiting Professor. Dr. Nath had been a member of the selection committee of the prestigious International King Faisal Award in Biology Saudi Arabia in 2015. Currently, he is working as Dean (Research), Institute of Medical Sciences, Banaras Hindu University, Varanasi.

Mayank Gangwar is currently serving as Scientist-D in Clinical Studies and Trials Unit, INTENT, Division of Development Research, ICMR, Headquarters, New Delhi. Dr. Gangwar, oversees clinical trials, regulatory compliance, stakeholder collaborations, clinical trial design and management, regulatory guidance and compliance, medical device, and diagnostic validation. His extraordinary research odyssey commenced in 2011 as a DST-JRF, where he embarked on a relentless pursuit of understanding infectious microorganisms. With an unwavering passion for science, he not only mastered the intricacies of molecular assays but also showcased his brilliance on national and international scientific stages through impactful presentations and talks. His academic journey is a testament to his unwavering commitment, culminating in a Doctorate in Pharmacology from Banaras Hindu University, complemented by a Master's and Bachelor of Pharmacy. Beyond academia, Dr. Gangwar's imprint on the scientific community is indelible. His prolific contributions to research are encapsulated in over 50 publications, bearing witness to his profound influence with an impressive citation count of 1635, an h-index of 21, and an i10-index of 40. Recognition and accolades have been his companions, with the CSIR-Direct SRF, DBT-CTEP and ICMR International Travel Grant among his crowning achievements. He served 5 years as Scientist-C in VRDL, IMS, BHU towards research management and quality standards, augmented by ISO and NABL accreditation training by ICMR, demonstrating his dedication towards diagnostic and research approach. His impact extends beyond the laboratory, as he has contributed as a board editor and held editorial roles in esteemed national and international journals, including a tenure as a Research Topic Editor in the *Frontiers in Pharmacology*.

Additionally, his commitment to excellence is underscored by his certification as a Registered Pharmacist with the Pharmacy Council of India, and his status as a Life Member of the Medicinal and Aromatic Plant Association of India (MAPAI). Dr. Gangwar's multifaceted contributions reach their zenith with his involvement in the Society for Bacteriophage Research and Therapy, where he has held office, further solidifying his role in shaping the future of research. Dr. Gangwar's journey is an awe-inspiring testament to the limitless horizons that await those who dare to explore the frontiers of science, leaving an indelible legacy that energizes and inspires the scientific community to push the boundaries of knowledge and discovery.

Contributors

Lara Hashim Abdulmageed University of Anbar, Ramadi, Anbar, Iraq

Abhimanyu Department of Zoology, College of Basic Sciences and Humanities, Chaudhary Charan Singh Haryana Agricultural University, Hisar, Haryana, India

Arup Acharjee Department of Zoology, University of Allahabad, Prayagraj, India

Papia Acharjee Department of Zoology, Banaras Hindu University, Varanasi, India

Ashish Kumar Agrawal Department of Pharmaceutical Engineering and Technology, Indian Institute of Technology (BHU), Varanasi, Uttar Pradesh, India

Sangeeta Ahiwale Department of Microbiology, Mahatma Phule Mahavidyalaya, Pimpri, Pune, Maharashtra, India

Neelam Atri Department of Botany, M.M.V., Banaras Hindu University, Varanasi, Uttar Pradesh, India

Babita Department of Microbiology, CCS Haryana Agricultural University, Hisar, Haryana, India

Urmi Bajpai Department of Biomedical Science, Acharya Narendra Dev College, University of Delhi, Delhi, India

Graham Ball Medical Technology Research Centre, Anglia Ruskin University, Cambridge, UK

Yashik Bansal Department of Microbiology, Employees State Insurance Corporation Medical College Hospital, Alwar, Rajasthan, India

Himanshu Bansode CSIR-Central Drug Research Institute, Jankipuram Extension, Lucknow, Uttar Pradesh, India

S. Bershiyal School of Biosciences and Technology, Vellore Institute of Technology, Vellore, Tamil Nadu, India

Satyanam Kumar Bhartiya Department of General Surgery, Institute of Medical Sciences, Banaras Hindu University, Varanasi, India

Vishal Bhojyawal Department of Zoology, T. D. P. G. College, V. B. S. Purvanchal University, Jaunpur, Uttar Pradesh, India

Gabriel Guimarães Calefi Institute of Biomedical Sciences (ICBIM), Federal University of Uberlândia, Uberlândia, MG, Brazil

Garima Chand College of Pharmacy, Graphic Era Hill University, Bhimtal, Uttarakhand, India

Shiwa Chaubey Phytochemistry Division, CSIR-National Botanical Research Institute, Lucknow, Uttar Pradesh, India

Aiswarya Chaudhuri Department of Pharmaceutical Engineering and Technology, Indian Institute of Technology (BHU), Varanasi, Uttar Pradesh, India

Vishal Chauhan MAHSA University, Jenjarom, Selangor, Malaysia

Hemant U. Chikhale Department of Pharmaceutical Chemistry, Sandip Institute of Pharmaceutical Sciences, Nashik, Maharashtra, India

Sidharth Chopra Department of Molecular Microbiology and Immunology, CSIR-Central Drug Research Institute, Lucknow, Uttar Pradesh, India

AcSIR: Academy of Scientific and Innovative Research (AcSIR), Ghaziabad, India

Luís Carlos Scalon Cunha Chemistry Department, Federal Institute of Triângulo Mineiro, Uberaba, MG, Brazil

Tejpal Dahiya Department of Zoology, College of Basic Sciences and Humanities, Chaudhary Charan Singh Haryana Agricultural University, Hisar, Haryana, India

Arunava Dasgupta Department of Molecular Microbiology and Immunology, CSIR-Central Drug Research Institute, Lucknow, Uttar Pradesh, India

AcSIR: Academy of Scientific and Innovative Research (AcSIR), Ghaziabad, India

Rajal Dave Department of Microbiology, Faculty of Science, Atmiya University, "Yogidham Gurukul", Rajkot, Gujarat, India

Deepa Dehari Department of Pharmaceutical Engineering and Technology, Indian Institute of Technology (BHU), Varanasi, Uttar Pradesh, India

Lalchand D. Devhare Manwatkar College of Pharmacy, Ghodpeth, Maharashtra, India

Nangthombam Grihalaxmi Devi Manipur International University, Imphal, Manipur, India

P. Gayatri Devi NRI College of Pharmacy, Vijayawada, Andhra Pradesh, India

Arup Dey Department of Pharmaceutical Technology, School of Health and Medical Science, Adamas University, Kolkata, West Bengal, India

Neha Dhami Department of Social Work, Uttarakhand Open University, Haldwani, India

Department of Education, Kumaun University, Nainital, Uttarakhand, India

G. Dharmamoorthy Department of Pharmaceutical Analysis, MB School of Pharmaceutical Sciences (Erstwhile Sree Vidyanikethan College of Pharmacy), Mohan Babu University, Tirupati, Andhra Pradesh, India

N. . S. Disha Department of Analysis, Saveetha Institute of Medical and Technical Sciences, Chennai, Tamil Nadu, India

Raghvendra Dubey Department of Life and Basic Sciences, Jaipur National University, Jaipur, India

Mayank Gangwar Faculty of Medicine, Viral Research and Diagnostic Laboratory, Department of Microbiology, Institute of Medical Sciences, Banaras Hindu University, Varanasi, Uttar Pradesh, India

Clinical Studies and Trials Unit, Division of Development Research, Indian Council of Medical Research, Department of Health Research, Ministry of Health and Family Welfare, Govt. of India, V. Ramalingaswami Bhawan, New Delhi, India

Manish Kumar Gautam Faculty of Pharmaceutical Science, Assam Down Town University, Guwahati, Assam, India

Singh Shekhar Gautam Educational Institute of Pharmacy, Prayagraj, Uttar Pradesh, India

Sourav Ghosh Department of Pharmaceutical Technology, School of Health and Medical Science, Adamas University, Kolkata, West Bengal, India

Daniela Silva Gonçalves Institute of Biomedical Sciences (ICBIM), Federal University of Uberlândia, Uberlândia, MG, Brazil

Tannu Goyal Department of Zoology, Jan Nayak Chaudhary Devi Lal Memorial College, Chaudhary Devi Lal University, Sirsa, Haryana, India

Divya Gupta Reproductive Biology and Toxicology Lab, School of Studies in Zoology, Jiwaji University, Gwalior, Madhya Pradesh, India

Shraddha Manish Gupta Department of Pharmaceutical Sciences, School of Health Sciences and Technology, UPES, Dehradun, Uttarakhand, India

Sweety Gupta Department of Zoology, T. D. P. G. College, V. B. S. Purvanchal University, Jaunpur, Uttar Pradesh, India

Varsha Rini Guria Special Centre for Molecular Medicine, Jawaharlal Nehru University, New Delhi, India

KIIT School of Biotechnology, KIIT University, Bhubaneshwar, Odisha, India

Tanuj Hooda MM College of Pharmacy, Maharishi Markandeshwar (deemed to be) University, Mullana, Ambala, Haryana, India

Mahesh Jansari Department of Pulmonary Medicine, All India Institute of Medical Sciences, Kalyani, West Bengal, India

Priyanka Joshi Department of Zoology, College of Basic Sciences and Humanities, Chaudhary Charan Singh Haryana Agricultural University, Hisar, Haryana, India

Vibhor Joshi Environmental Biotechnology Genetics and Molecular Biology Department, ICMR-National Institute for Research in Environmental Health, Bhopal, Madhya Pradesh, India

Ramya Juliet School of Biosciences and Technology, Vellore Institute of Technology, Vellore, Tamil Nadu, India

Jyotirmayee School of Biotechnology, Institute of Science, Banaras Hindu University, Varanasi, Uttar Pradesh, India

Tripti Kanda Department of Botany, M.M.V., Banaras Hindu University, Varanasi, Uttar Pradesh, India

Subhash Lal Karn Department of Microbiology, Nepalgunj Medical College, Kohalpur, Banke, Nepal

E. Karthikeyan Department of Pharmaceutical Chemistry, Saveetha Institute of Medical and Technical Sciences, Chennai, Tamil Nadu, India

Mansi Kesarwani Department of Biochemistry, University of Allahabad, Prayagraj, Uttar Pradesh, India

Reshma Khanda School of Life Sciences, Sambalpur University, Jyoti Vihar, Burla, Sambalpur, Odisha, India

Iram Khan Department of Pharmaceutical Sciences, School of Health Sciences and Technology, UPES, Dehradun, Uttarakhand, India

Gudikandula Krishna Department of Microbiology, Kakatiya University, Warangal, Telangana, India

Kulsoom School of Life Science and Biotechnology, Chhatrapati Shahu Ji Maharaj University, Kalyanpur, Kanpur, Uttar Pradesh, India

Deepak Kumar Department of Microbiology, Institute of Medical Sciences, Banaras Hindu University, Varanasi, Uttar Pradesh, India

Dileep Kumar Poona College of Pharmacy, Bharati Vidyapeeth (Deemed to be) University, Pune, Maharashtra, India

Dulla Naveen Kumar Department of Pharmaceutical Engineering and Technology, Indian Institute of Technology (BHU), Varanasi, Uttar Pradesh, India

Maneesh Kumar State VRDL, All India Institute of Medical Sciences, Deoghar, Jharkhand, India

Rajesh Kumar Maharaja Agrasen School of Pharmacy, Maharaja Agrasen University, Solan, Himachal Pradesh, India

Sandeep Kumar MM College of Pharmacy, Maharishi Markandeshwar (Deemed to be) University, Ambala, Haryana, India

Sanjeev Kumar Department of Dravyaguna, Faculty of Ayurveda, Institute of Medical Sciences, Banaras Hindu University, Varanasi, Uttar Pradesh, India

Sauparnika Prasanna Kumar SIES College of Arts Science and Commerce, Mumbai, India

Subodh Kumar Department of Medical Laboratory Technology, School of Allied Health Sciences and Management, Delhi Pharmaceutical Sciences and Research University, New Delhi, India

Yamjala Ganesh Kumar Department of Pharmacognosy and Phytochemistry, Sree Dattha Institute of Pharmacy, Hyderabad, Telangana, India

Renu Lamba Department of Botany and Plant Physiology, CCS Haryana Agricultural University, Hisar, Haryana, India

Amit Lather MM College of Pharmacy, Maharishi Markandeshwar (Deemed to be) University, Ambala, Haryana, India

Lokesh Department of Biotechnology, Shri Alpesh N. Patel Post Graduate Institute of Science and Research, Affiliated with Sardar Patel University, Anand, Gujarat, India

Rahul Maitra Department of Molecular Microbiology and Immunology, CSIR-Central Drug Research Institute, Lucknow, Uttar Pradesh, India

AcSIR: Academy of Scientific and Innovative Research (AcSIR), Ghaziabad, India

Carlos Henrique Gomes Martins Institute of Biomedical Sciences (ICBIM), Federal University of Uberlândia, Uberlândia, MG, Brazil

Aradhana Mishra Department of Medicinal Chemistry, Faculty of Ayurveda, Institute of Medical Sciences, Banaras Hindu University, Varanasi, Uttar Pradesh, India

Niharika Mishra Krishna Institute of Pharmacy and Sciences, Kanpur, Uttar Pradesh, India

Pragya Mishra Food Processing and Management, DDU KAUSHAL KENDRA, RGSC, Banaras Hindu University, Varanasi, Uttar Pradesh, India

Raghvendra Raman Mishra Medical Laboratory Technology, DDU KAUSHAL KENDRA, RGSC, Banaras Hindu University, Varanasi, Uttar Pradesh, India

Vaishali Mishra Phytochemistry Division, CSIR-National Botanical Research Institute, Lucknow, Uttar Pradesh, India

Shilpi Mishra Institute of Pharmaceutical Sciences, Department of Pharmacology, University of Lucknow, Lucknow, India

Oishi Mitra School of Biosciences and Technology, Vellore Institute of Technology, Vellore, Tamil Nadu, India

Aiswarya Mohan Department of Biomedical Science, Acharya Narendra Dev College, University of Delhi, New Delhi, India

Department of Epidemiology and Public Health, Central University of Tamil Nadu, Nilakkudi, Tamil Nadu, India

Farina Mujeeb Department of Biosciences, Integral University, Lucknow, Uttar Pradesh, India

Shazia Naaz Mamata Academy of Medical Science, Hyderabad, Telangana, India

Ramesh Nachimuthu School of Biosciences and Technology, Vellore Institute of Technology, Vellore, Tamil Nadu, India

Vaishali P. Nagulwar Government College of Pharmacy, Chhatrapati Sambhajinagar, Chhatrapati Sambhajinagar, India

Vijaya Lakshmi Nag Department of Microbiology, All India Institute of Medical Sciences, Jodhpur, Rajasthan, India

Gopal Nath Department of Microbiology, Institute of Medical Sciences, Banaras Hindu University, Varanasi, Uttar Pradesh, India

Pavan Kumar Padarthi Chebrolu Hanumaiah Institute of Pharmaceutical Sciences, Chowdavaram, Guntur, Andhra Pradesh, India

Harikumar Pallathadka Manipur International University, Imphal, Manipur, India

Poonam Pal Department of Dravyaguna, Faculty of Ayurveda, Institute of Medical Sciences, Banaras Hindu University, Varanasi, Uttar Pradesh, India

Krishna Pandey Department of Clinical Medicine, ICMR-Rajendra Memorial Research Institute of Medical Sciences, Patna, Bihar, India

Krupa Parmar Division of Clinical Microbiology, Department of Laboratory Medicine and Pathology, Mayo Clinic, Rochester, MN, USA

Ritu Raj Patel Department of Medicinal Chemistry, Faculty of Ayurveda, Institute of Medical Sciences, Banaras Hindu University, Varanasi, Uttar Pradesh, India

Sruti Patel Dr. Vikram Sarabhai Institute of Cell and Molecular Biology, The Maharaja Sayajirao University of Baroda, Vadodara, Gujarat, India

Ashutosh Pathak Dr. Shakuntala Mishra National Rehabilitation University, Lucknow, Uttar Pradesh, India

Anasuya Patil KLE College of Pharmacy, Bengaluru, Karnataka, India

Sandeep Pokhrel Department of Microbiology, Nepalgunj Medical College, Kohalpur, Banke, Nepal

Diego Godina Prado Institute of Chemistry (IQ), Federal University of Uberlândia, Uberlândia, MG, Brazil

Sunil Kumar Raman CSIR-Central Drug Research Institute, Jankipuram Extension, Lucknow, Uttar Pradesh, India

Sindhuprava Rana Biostatistics and Bioinformatics Department, ICMR-National Institute for Research in Environmental Health, Bhopal, Madhya Pradesh, India

Anupama Rani Department of Biosciences, Integral University, Lucknow, Uttar Pradesh, India

Avinash Kumar Rao Patel College of Pharmacy, Madhyanchal Professional University (MPU), Bhopal, Madhya Pradesh, India

B. Apurva Rao Department of Paediatric Dentistry, SDM College, Dharwad, Karnataka, India

Niravsinh M. Rathod Parul Institute of Pharmacy, Faculty of Pharmacy, Parul University, Waghodia, Vadodara, Gujarat, India

Dinesh D. Rishipathak Department of Pharmaceutical Chemistry, MET's Institute of Pharmacy, Nashik, Maharashtra, India

Ruksar Department of Zoology, College of Basic Sciences and Humanities, Chaudhary Charan Singh Haryana Agricultural University, Hisar, Haryana, India

Ganesh Chandra Sahoo Department of Virology, ICMR-Rajendra Memorial Research Institute of Medical Sciences, Patna, Bihar, India

Minakshi Sahu Department of Microbiology, Institute of Medical Sciences, Banaras Hindu University, Varanasi, Uttar Pradesh, India

Daman Saluja Dr B R Ambedkar Centre of Biomedical Research, University of Delhi, Delhi, India

K. S. Sridevi Sangeetha Faculty of Allied Health Sciences, Meenakshi Academy of Higher Education and Research, Chennai, Tamil Nadu, India

Seema Sangwan Department of Microbiology, CCS Haryana Agricultural University, Hisar, Haryana, India

Biresh Kumar Sarkar Central Ayurveda Research Institute, CCRAS, Ministry of AYUSH, New Delhi, India

Deepanshi Saxena Department of Molecular Microbiology and Immunology, CSIR-Central Drug Research Institute, Lucknow, Uttar Pradesh, India

Nishu Sehrawat Department of Microbiology, CCS Haryana Agricultural University, Hisar, Haryana, India

Preeti Semwal Shri Guru Ram Rai University, Dehradun, Uttarakhand, India

Surabhi Shakya Department of Pharmacy, MJP Rohilkhand University, Bareilly, Uttar Pradesh, India

Gaurav Sharma Department of Public Health Dentistry, SCB Dental College and Hospital, Cuttack, Odisha, India

Pankaj Kumar Sharma Apex Institute of Pharmaceutical Sciences, Jaipur, Rajasthan, India

Sukesh Sharma Department of Paediatrics, Motilal Nehru Medical College, Prayagraj, Uttar Pradesh, India

Unnati Sharma School of Pharmaceutical Sciences, Chhatrapati Shahuji Maharaj University, Kanpur, Uttar Pradesh, India

Yashika Sharma Poona College of Pharmacy, Bharati Vidyapeeth (Deemed to be) University, Pune, Maharashtra, India

Sadhana Shrivastava Reproductive Biology and Toxicology Lab, School of Studies in Zoology, Jiwaji University, Gwalior, Madhya Pradesh, India

Alka Shukla Viral Research and Diagnostic Laboratory, Department of Microbiology, Faculty of Medicine, Institute of Medical Sciences, Banaras Hindu University, Varanasi, Uttar Pradesh, India

Sangeeta Shukla Reproductive Biology and Toxicology Lab, School of Studies in Zoology, Jiwaji University, Gwalior, Madhya Pradesh, India

Nagela Bernadelli Sousa Silva Institute of Biomedical Sciences (ICBIM), Federal University of Uberlândia, Uberlândia, MG, Brazil

Alok Kumar Singh Department of Botany, CMP Degree College, University of Allahabad, Prayagraj, Uttar Pradesh, India

Aprajita Singh Department of Microbiology, Institute of Medical Sciences, Banaras Hindu University, Varanasi, Uttar Pradesh, India

Jashanjit Singh Gurukul Institute of Pharmaceutical Sciences, Fatehgarh Sahib, Punjab, India

Meenakshi Singh Department of Medicinal Chemistry, Faculty of Ayurveda, Institute of Medical Sciences, Banaras Hindu University, Varanasi, Uttar Pradesh, India

Nidhi Singh Department of Botany, M.M.V., Banaras Hindu University, Varanasi, Uttar Pradesh, India

Nitish Kumar Singh Department of Life and Basic Sciences, Jaipur National University, Jaipur, India

Shubham Singh Department of Zoology, Guru Ghasidas Vishwavidyalaya, Bilaspur, Chhattisgarh, India

Srishti Singh Department of Botany, CMP Degree College, University of Allahabad, Prayagraj, Uttar Pradesh, India

Sudhir Kumar Singh Department of Microbiology, Institute of Medical Sciences, Banaras Hindu University, Varanasi, Uttar Pradesh, India

Ashapurna Sinha Department of Biosciences, Integral University, Lucknow, Uttar Pradesh, India

Kumari Sonu IIMT College of Pharmacy, Knowledge Park III, Greater Noida, Uttar Pradesh, India

V. Srilekha Department of Biotechnology, Chaitanya (Deemed to be University), Warangal, Telangana, India

Rupanshee Srivastava Department of Botany, M.M.V., Banaras Hindu University, Varanasi, Uttar Pradesh, India

Fauzia Tabassum Department of Pharmacology, Vision College, Riyadh, Kingdom of Saudi Arabia

Lakshmi Jyothi Tadi Department of Microbiology, AIIMS Bibinagar, Hyderabad, Telangana, India

Dasari Thrimothi Department of Microbiology, Kakatiya University, Warangal, Telangana, India

Siddhant Tripathi Poona College of Pharmacy, Bharati Vidyapeeth (Deemed to be) University, Pune, Maharashtra, India

Prashasti Upreti Department of Microbiology, Graphic Era Deemed to be University, Dehradun, Uttarakhand, India

Amit Kumar Verma Department of Pharmacy, MJP Rohilkhand University, Bareilly, Uttar Pradesh, India

Manipur International University, Imphal, Manipur, India

Malkhey Verma School of Biotechnology, Institute of Science, Banaras Hindu University, Varanasi, Uttar Pradesh, India

M. Sree Vidhya Department of Pharmacology, Saveetha Institute of Medical and Technical Sciences, Chennai, Tamil Nadu, India

Ranjeet Kumar Vishwakarma Department of Physiology, Institute of Medical Sciences, Banaras Hindu University, Varanasi, Uttar Pradesh, India

Chunna Yadav CSIR-Central Drug Research Institute, Jankipuram Extension, Lucknow, Uttar Pradesh, India

Sadhana Yadav Department of Botany, M.M.V., Banaras Hindu University, Varanasi, Uttar Pradesh, India

Sandeep Kumar Yadav Department of Pharmacology, SGT College of Pharmacy, Shree Guru Gobind Singh Tricentenary (SGT) University, Gurugram, Haryana, India

Sanjay Kumar Yadav Dr. Shakuntala Mishra National Rehabilitation University, Lucknow, Uttar Pradesh, India

Shivam Yadav Department of Botany, University of Allahabad, Prayagraj, Uttar Pradesh, India

Part I

The Rise of Antibiotic Resistance

CRISPR–Cas Against Antimicrobial-Resistant Pathogens

1

Siddhant Tripathi, Yashika Sharma, Manish Kumar Gautam, and Dileep Kumar

Abstract

Our ability to treat infectious diseases and create new antibiotics is hampered by the rapid emergence of bacteria resistant to antibiotics. One of the innovative methods for handling antibiotic-resistant strains is the utilization of the clustered regularly interspaced short palindromic repeat-associated system (CRISPR–Cas), which serves as an adaptive immune system for bacteria. When applied to bacterial genomic regions, the programmable Cas nuclease in this technology has the potential to be extremely lethal toward or beneficial in decreasing antibiotic resistance. The goal of this book chapter is to examine the process of promoting bacterial antibiotic sensitization through the CRISPR–Cas system. The use of CRISPR–Cas methods could lead to the development of intelligent antibiotics that can identify harmful benign microorganisms and end multidrug-resistant (MDR) infections. These systems offer opportunities for studying the microbial consortia, controlling industrial fermentation, and treating infections resistant to various medications. They possess the capacity to eliminate particular bacterial strains in a sequence-specific manner, both quantitatively and selectively.

Keywords

Antibiotic-resistant bacteria · CRISPR–Cas system · Sequence-specific manner · Resensitization · Genome editing

S. Tripathi · Y. Sharma · D. Kumar (✉)
Poona College of Pharmacy, Bharati Vidyapeeth (Deemed to be) University, Pune, Maharashtra, India

M. K. Gautam
Faculty of Pharmaceutical Science, Assam Down Town University, Guwahati, Assam, India

M. Gangwar, G. Nath (eds.), *Emerging Paradigms for Antibiotic-Resistant Infections: Beyond the Pill*, https://doi.org/10.1007/978-981-97-5272-0_1

1.1 Introduction

Since their initial discovery about 90 years ago, antibiotic compounds have had a tremendous influence on modern medicine. Due to the large amount of antibiotics released into the environment, their overuse in veterinary and agricultural contexts raises serious concerns, although they can treat diseases that were previously fatal (Kümmerer 2003). Numerous antibiotic exposures result in strong selection pressures, which promote the evolution and spread of genes encoding antimicrobial resistance in both commensal and pathogenic bacteria (World Health Organization 2014; Wright 2007). Using a number of strategies and mechanisms, such as efflux pumping, enzymatic deactivation of antibiotic molecules, and chemical modifications (of the ribosome and the cell wall), these antibiotic-resistant genes allow bacteria to withstand antibiotics by safeguarding their cellular targets (Alekshun and Levy 2007). When combined, these resistance mechanisms pose a threat to the antibiotics' ability to treat patients (World Health Organization 2014). As stated in a Centers for Disease Control and Prevention report from 2013, antibiotic-resistant bacteria (AMR) infect more than two million people each year, resulting in 23,000 deaths (Frieden 2013). Another forecast states that ten million deaths per year will result from drug-resistant infections by 2050. Drug-resistant pathogens will therefore be the leading cause of death, surpassing deaths by diabetes and cancer and even automobile-related accidents (Erickson et al. 2017). One factor contributing to the emergence of antibiotic resistance in bacteria is their ability to acquire remarkable phenotypic and genotypic heterogeneity for survival in hostile conditions (Erickson et al. 2017).

There are currently just 37 antibiotic drugs undergoing phase II or III clinical trials, and the primary cause of AMR is the decrease in the development of new antibiotics. For Gram-negative bacterial infections, this has resulted in zero approval of a novel antibiotic class in more than 45 years (de la Fuente-Nunez et al. 2017). Furthermore, developing, screening, and testing antibiotics is expensive and requires significant resources (Greene 2018). These elements compelled us to look for different ways to combat AMR pathogens, which included creating a special antibacterial toolkit with precise target capabilities. To do this, scientists have created a variety of antibacterial treatments, including bacteriophage therapies, bacteriocins, antibodies, and anti-virulence compounds, based on unique peptide and nucleic acid bases (de la Fuente-Nunez et al. 2017). We discuss about the bacterial adaptive immune system in our review as well as about the role of the clustered regularly interspaced short palindromic repeat-associated system (CRISPR–Cas) in combating the escalating threats posed by AMR.

1.2 Three Different CRISPR–Cas System Types

The discovery of a set of 29 nucleotide (nt) repeats in *Escherichia coli* by Nakata and associates, while researching the *iap* gene in 1987, marked the beginning of the CRISPR–Cas system's history (Ishino et al. 1987). Over the next 10 years, sequencing of a large number of microbial genomes revealed additional repeat elements from the genomes of different strains of bacteria and archaea. Eventually, this distinct group of repeated sequences spaced apart came to be known as clustered repeat elements (Mojica et al. 2000). Mojica and Jansen used the term "CRISPR" in 2002 (Hsu et al. 2014). The separation of spacer sequences from direct repeats in 2005, which suggested their phage association or extrachromosomal origins, was a significant advance (Mojica et al. 2005; Pourcel et al. 2005). The year 2010 saw the fundamental operation and mechanism of the CRISPR–Cas system come to light. This system consists of adjacent 6–20 genes encoding CRISPR-associated (Cas) proteins and a genetic locus with non-repetitive spacer sequences (Shabbir et al. 2016a, b). Several scientists have started utilizing the CRISPR–Cas system to create phage-resistant dairy cultures and for biotechnological purposes (Hsu et al. 2014).

Bacteriophages, phages, and mobile genetic elements (MGEs) are among the invaders that the CRISPR–Cas system, an adaptive immune system of bacteria and archaea, defends against (Barrangou et al. 2007). The CRISPR–Cas system breaks down foreign genetic elements in three stages (Fig. 1.1). Adaptation or spacer acquisition is the process of incorporating the spacer sequence into the CRISPR array following recognition (Garneau et al. 2010). RNA polymerase (RNAP) is responsible for the transcription of pre-CRISPR RNA (pre-crRNA) during the second stage of biogenesis, also known as the expression of CRISPR RNA (crRNA). Specific endoribonucleases then cleave these pre-crRNAs to produce small crRNAs, which are also known as guide RNAs based on their function (Brouns et al. 2008; Carte et al. 2008). When crRNAs identify foreign DNA or RNA and also produce base pairs that are nearly perfectly complementary to it, then this process is known as interference (Deveau et al. 2010; Shabbir et al. 2016a, b). Cleavage of the foreign nucleic acid complex containing crRNAs is the outcome. However, cleavage does not occur as well and the host is vulnerable to infection if there is a mutation in the protospacer adjacent motif (PAM) or incompatibility between the spacer and the invader's DNA (Bhaya et al. 2011).

There are three subtypes of the CRISPR–Cas system: I, II, and III. The signature genes found in each type serve as the basis for this classification. For example, Cas I carries the *cas3* gene, Cas II the *cas9* gene, and Cas III the *cas10* gene. The presence of cas1 and cas2 proteins in all CRISPR system types and subtypes is significant, as these two proteins are essential for spacers (Shabbir et al. 2016a, b). Table 1.1 lists the various CRISPR–Cas system types found in relevant bacterial species.

1.2.1 Type I System

Most bacteria and archaea contain the CRISPR–Cas type I system (Makarova et al. 2011). Six subtypes (A–F) encoding the *cas3* gene further subdivide this system. The multi-domain protein Cas3 possesses both nuclease and helicase functions (Sinkunas et al. 2011). Two domains make up the Cas3 protein: the C-terminal DExH helicase domain, which unwinds double-stranded DNA, and the N-terminal

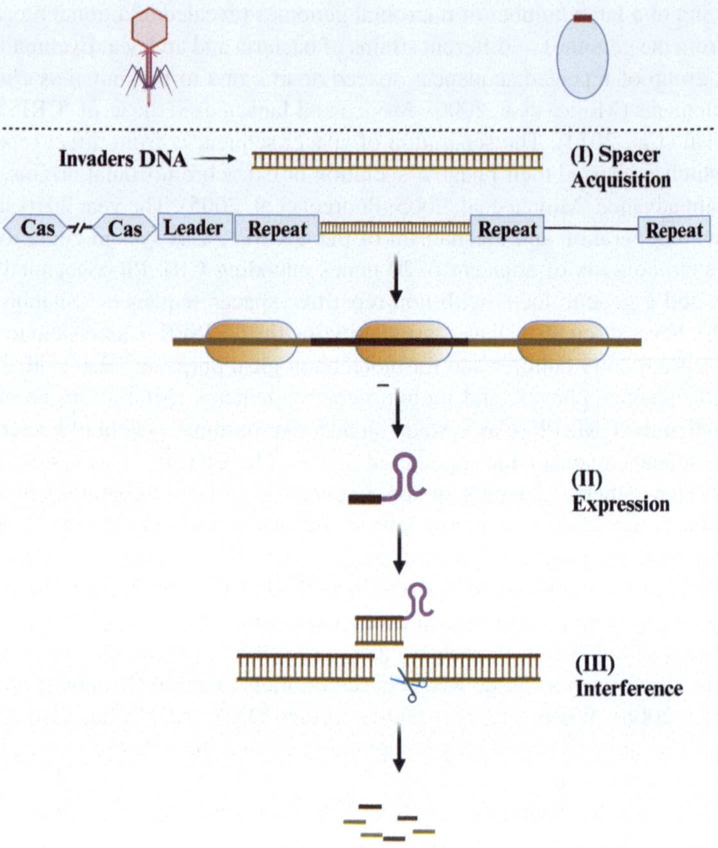

Fig. 1.1 The CRISPR–Cas immunity mechanism consists of three phases. Phase (I): Acquiring spacer acquisition. The CRISPR array's leader end integrates targeted viral or plasmid double-strand breaks with the host DNA during the initial stage. Spacers that are distinct from repeats make up a CRISPR array. When cas1 and cas2 proteins are present in close proximity to a CRISPR array, spacer acquisition takes place. Phase (II): crRNA biogenesis. During this phase, the RNA polymerase toward the leader end helps transcribe pre-CRISPR RNA (pre-crRNA) into mature crRNA. Phase (III): Interference. The cleavage of foreign genetic elements occurs in the final stage due to a specific match between the target sequence and the crRNA spacer (Shabbir et al. 2016a, b)

HD phosphohydrolase domain, which cleaves DNA (Shabbir et al. 2016a, b). Together, these two domains break down the DNA of the invader. However, cas3 by itself is unable to recognize the DNA of an intruder and shield cells from infection (Brouns et al. 2008; Jore et al. 2011). Together, all subtypes of the Cas proteins of the type I immune system form the CRISPR-associated complex for antiviral defense also referred to as the crRNA-guided surveillance complex (CASCADE). Target sequences complementary to the crRNA spacer are recognized by these complexes and subsequently bound to them. *E. coli* K12 (type I-E) provided the initial description of the crRNA-guided surveillance complex (Brouns et al. 2008; Jore et al. 2011). Together, five distinct Cas proteins make up the complex. Cas6e,

Table 1.1 Different CRISPR–Cas system types present in different bacterial species

CRISPR–Cas system	Species of bacteria	References
Type I	*Yersinia pestis*, *Campylobacter concisus*, *C. curves*, *Campylobacter fetus*, *Cryptosporidium hominis*, *Campylobacter rectus*, *E. coli*, *P. aeruginosa*, *Myxococcus xanthus*, *Bacillus halodurans*, *Salmonella* species, *Propionibacterium acnes*, and *Erwinia amylovora*	Brouns et al. (2008); Fabre et al. (2012); Haurwitz et al. (2010); Makarova et al. (2011); Marinelli et al. (2012); Vergnaud et al. (2007); Viswanathan et al. (2007)
Type II	*F. novicida*, *Streptococcus pyogenes*, *S. aureus*, *Mycoplasma gallisepticum*, *E. faecalis*, *Legionella pneumophila*, *Listeria monocytogenes*, *Streptococcus thermophilus*, *Streptococcus mutants*, *N. meningitides*, *C. jejuni*	Bhaya et al. (2011); Bourgogne et al. (2008); Delaney et al. (2012); Frieden (2013); Gunderson and Cianciotto (2013); Kinnevey et al. (2013); Louwen et al. (2013); Sampson et al. (2013); Sampson and Weiss (2014)
Type III	*S. epidermidis*, *Mycobacterium tuberculosis*, and *P. furiosus*	Abadia et al. (2010); Hale et al. (2009); Marraffini and Sontheimer (2008)

formerly called Cse3 or CasE, facilitates the maturation of crRNA. Mature crRNA contributes to the identification and cleavage of invader DNA while staying bound to the CASCADE complex (Shabbir et al. 2016a, b). Another report of a similar kind of complex involved *Saccharolobus solfataricus* (Lintner et al. 2011). Further research has revealed the presence of crRNA-guided surveillance complexes in *Bacillus halodurans* (type I-C) (Nam et al. 2012) and *Pseudomonas aeruginosa* (type I-F) (Wiedenheft et al. 2011).

1.2.2 Type II System

The type II system is unique to bacteria (Kira S Makarova et al. 2011). Of the CRISPR–Cas variants, the type II system is the most simple (Bhaya et al. 2011). There are four genes in the CRISPR–Cas type II system: *cas1*, *cas2*, *cas9*, and *cas4* in type II-B cases and *csn2* in type II-A cases. Cas9 is the distinctive protein of the type II system and plays a part in invader DNA cleavage and crRNA biogenesis (Deltcheva et al. 2011). The two domains that constitute the *cas9* gene are the HNH and RuvC domains (Jinek et al. 2012). While the RuvC domain cleaves non-complementary strands, the HNH domain helps cleave complementary DNA to the crRNA guide (Jinek et al. 2012). In the type II system, transactivating crRNA, or tracrRNA, is necessary for the biogenesis of crRNA. At the opposing strand of the CRISPR–Cas locus, *Streptococcus pyogenes* encodes tracrRNA (Shabbir et al. 2016a, b). The production of double-stranded RNA (dsRNA) is the result of hybridization between tracrRNA and crRNA repeats. Cellular non-Cas RNase III enzyme recognizes and cleaves dsRNA. The deletion of the *cas9* gene inhibits the biogenesis of crRNA, though its exact function in crRNA biogenesis is unknown (Deltcheva et al. 2011). Jinek et al. (2012) showed that in order for the cas9 enzyme to cleave target DNA, it needs both tracrRNA and crRNA. Since it consists of a single protein that has all the domains needed for DNA cleavage, the CRISPR–Cas type II system is an excellent choice for modifying genomes (cas9) (Cong et al. 2013).

1.2.3 Type III System

The two subtypes that comprise the type III system are type III-A and type III-B (Makarova et al. 2011). The type III-B system has only been observed alongside different CRISPR types; despite this, it is most frequently found in archaea. Encoding both *cas6* and *cas10* genes is possible with the CRISPR–Cas type III system. Cas10, also referred to as a repeat-associated mysterious protein (RAMP), may play a role in DNA cleavage and crRNA maturation (Anantharaman et al. 2010). Cas6 is an endoribonuclease that functions independently of the CASCADE complex (Zhang et al. 2012). Mature crRNA binds to the CASCADE complex of the type III system, which then cleaves foreign RNA (Wang et al. 2011). Furthermore, archaea possessing a type III system and a CRISPR–Cas type I-A or I-B may share cas6 (Deng et al. 2013).

Despite their similarities, the type III CRISPR–Cas system's two subtypes appear to target different substrates chemically. For instance, the *Staphylococcus epidermidis* type III-A system targets DNA, whereas the *S. solfataricus* and *Pyrococcus furiosus* type III-B system cleaves RNA (Shabbir et al. 2016a, b). This demonstrates the variety of CRISPR systems found in the type III system.

1.3 The Role of the CRISPR–Cas System in Combating Bacterial Virulence

Numerous investigations have demonstrated that the CRISPR–Cas system serves purposes other than protecting bacteria from outside threats. This system is involved in the regulation of both bacterial pathogenicity and endogenous transcription. For example, *Francisella novicida*, the bacterium that causes human disease, replicates intracellularly by eluding the host immune system. This bacterium uses a number of strategies to interfere with the actions of host macrophages and other immune cells. Following macrophage engulfment, *F. novicida* penetrates the phagosome, a space that houses antimicrobials and several immune recognition receptors (Jones et al. 2012). Among the receptors that are capable of identifying bacterial lipoproteins (BLPs) is toll-like receptor 2 (TLR2) (Sampson and Weiss 2014). By recruiting and activating immune cells, TLR2 activation triggers a pro-inflammatory response that ultimately aids in the removal of the bacterial pathogen. *F. novicida* employs the regulators cas9, sacRNA (small, CRISPR-associated RNA), and tracrRNA to inhibit the expression of BLPs (Sampson et al. 2013). Thus, by inhibiting TLR2 activation, this pathogen can persist within the host. According to reports, sacRNA, tracrRNA, and cas9 deletion mutants significantly increase the inflammatory response compared to the wild-type ones, indicating that *F. novicida* nevertheless causes a significant inflammatory response in the absence of these regulators (Sampson et al. 2013). Further highlighting the significance of the CRISPR–Cas system as a virulence regulator in *F. novicida* is the report that these mutants were unable to successfully infect mice (Sampson et al. 2013).

According to another report, *Neisseria meningitides* attach themselves to the host cell surface through cas9 and replicate intracellularly (Sampson et al. 2013). Furthermore, reports have indicated that *Campylobacter jejuni* utilizes cas9 in epithelial cells for both attachment and invasion (Louwen et al. 2013). Our theory is that the CRISPR–Cas system shields *C. jejuni* from the host's innate complement system while simultaneously facilitating the bacterium's attachment to host cells. Furthermore, in *C. jejuni*, which is cst-II positive, deletion of the *cas9* gene leads to an almost total loss of virulence (Louwen et al. 2013). Although the precise method by which the *cas9* gene controls the virulence of these microbes is unknown, it is believed that cas9 is not the only factor that affects virulence. According to a recent study, the CRISPR–*cas9* gene has been able to increase the virulence of *C. jejuni* and regulate several virulence-associated genes.

1.4 Why Is the CRISPR–Cas System Used to Combat the Threat of Antibiotic Resistance?

The engineering of four major classes of DNA-binding proteins has allowed for the successful editing of genomes: zinc-finger nucleases (ZFNs) from eukaryotic transcription factors (Hsu et al. 2014), transcription activator-like effectors (TALEs) from bacteria (*Xanthomonas*) (Miller et al. 2011), meganucleases from microbial MGEs (Smith et al. 2006), and, lastly, cas9 RNA-guided DNA endonucleases from the bacterial CRISPR–Cas type II system (Cong et al. 2013).

The low sequence specificity of meganucleases for target DNA makes them unsuitable for widespread use in genome editing (Hsu et al. 2014). Due to their challenging design for binding to a desired sequence, ZFNs have additional drawbacks. They also have a restricted target site selection. Because transcription activator-like effector nucleases (TALENs) can have longer DNA-binding protein domains, which enable highly specific targeting, they are simple to design. However, because TALENs are substantially bigger than ZFNs, delivering them into cells is complicated (Strong and Musunuru 2017).

By Watson–Crick base pairing, the Cas9 nuclease of the CRISPR–Cas type II system identifies target DNA with the aid of a guide RNA. In order to retarget the CRISPR–cas9 nuclease, we can simply replace the invader sequence that is present in CRISPR guide RNAs with the desired sequence (Hsu et al. 2014).

Through a thorough analysis of the CRISPR–Cas system, scientists have been able to insert, delete, and mutate desired genes in nearly every species. It is even possible to use this technology to treat genetic diseases in living animals (Yosef et al. 2015). Additionally, this system is used in specific antibacterial preparations that target AMR pathogens within complex bacterial populations, target pathogenic bacteria with antibiotics, and, in some cases, target host cells infected with pathogenic bacteria with treatments. Because CRISPR–Cas targets specific sequences, it can differentiate between pathogenic and commensal bacterial species. This chapter emphasizes the CRISPR–Cas system's ability to combat AMR infections.

1.5 The Role of CRISPR–Cas in Combating Antibiotic-Resistant Bacteria

Studies have demonstrated a strong inverse relationship between antibiotic resistance in species like *enterococci* and the CRISPR–Cas system (Burley and Sedgley 2012); however, other studies, such as those on *E. coli*, did not find a significant relationship (Touchon et al. 2012). Three CRISPR loci are found in *enterococci*: CRISPR1–Cas, orphan CRISPR2 with lack of *cas* genes, and CRISPR3–Cas. The results show that although not all *Enterococcus faecalis* strains contain CRISPR1–Cas or CRISPR3–Cas, orphan CRISPR2 does. Furthermore, CRISPR1–Cas and CRISPR2 are functionally related to one another, according to Palmer and Gilmore. The examination of CRISPR spacers revealed that pheromone-responsive plasmids are crucial for the genome plasticity of *enterococci*, as they can mobilize virulence factors encoded in chromosomes and resistance to antibiotics. Moreover, they can displace CRISPR–Cas and encourage their transfer (Manson et al. 2010). Although no CRISPR spacers are currently known against Tn916 as vectors of antibiotic resistance in *enterococci*, the distribution of these pheromone-responsive plasmid spacers revealed that specific elements show a tendency to be embedded in CRISPR loci as spacers or are frequently encountered by *E. faecalis* (Gilmore et al. 2002; Roberts and Mullany 2009). The discovery that the *tetM* gene is present in *E. faecalis* carrying CRISPR loci, which Tn916 and other conjugative transposons commonly spread (Gilmore et al. 2002; Roberts and Mullany 2009), could suggest that conjugative transposons get past this barrier. Furthermore, the Inc18 plasmid family lacks a spacer, which is responsible for the transfer of *enterococci*'s vancomycin-resistant genes to methicillin-resistant *Staphylococcus aureus* (MRSA) (Flannagan et al. 2003; Zhu et al. 2008). Both the relative uncommonness of interspecies transfer and the comparatively poor efficacy of element transfer, which lacks mechanisms for successful couple formation, could explain these results. Furthermore, CRISPR-carrying strains of *Streptococcus* thermophiles have proved their effectiveness by acquiring novel spacers derived from the virus and becoming resistant to phage infection (Barrangou et al. 2007).

E. coli K12 strains that were carrying an artificial CRISPR–Cas system with spacers targeting lambda phage genes demonstrated decreased sensitivity to the lambda phage, as per a report by Brouns et al. (2008). Moreover, Marraffini and Sontheimer (2008) demonstrated that the CRISPR–Cas system may have limited *S. epidermidis* plasmid conjugation, suggesting a more extensive and important role for the system in preventing horizontal gene transfer (HGT). Four CRISPR loci have been identified in *E. coli*: CRISPR1, CRISPR2, CRISPR3, and CRISPR4. Each of these loci has a distinct kind of *cas* gene (Díez-Villaseñor et al. 2010; Touchon and Rocha 2010). It is interesting to note that every *Escherichia* strain has a maximum of three CRISPRs in its genome because every genome that carries CRISPR1 also lacks CRISPR4 (Touchon and Rocha 2010). Touchon et al. (2012) showed that there is no correlation between integrons, plasmids, or antibiotic resistance and the existence of *E. coli* CRISPR loci and the total number of repeats. Furthermore, the researchers evaluated the sequence similarity between the spacers and the available plasmid sequences. They did not discover any indication that the spacers matched elements related to the mobilization of antibiotic-resistant genes, such as plasmid sequences, extended-spectrum β-lactamase (ESBL) synthesis or any kind of ESBL or replicon, Tn3, intI, or ISEcp1 (Touchon et al. 2012). Among the *E. coli* strains that they analyzed, they discovered a comparatively small number of spacers that matched the plasmid genes. Each CRISPR group had its own set of spacers; 1 strain had 15, whereas the other strains had 1, 2, 3, or 4 spacers in 77% of cases, 13.2% of cases, and 7.5% of cases. Furthermore, Touchon et al. (2012) discovered that CRISPR had no impact on the transmission of antibiotic-resistant genes or on the epidemiology of plasmids in *E. coli*. These outcomes conflict with those of *enterococci*, where Palmer and Gilmore discovered an inverse relationship between antibiotic resistance and CRISPRs. These data suggest that CRISPR–Cas systems may differ in terms of their impact on antibiotic resistance in different species. These variations may result from the system's distinct evolutionary history, the use of restriction and modification systems to encourage competition between plasmids, the development of CRISPR loci through partial or total deletions in the *cas* gene cluster, or the presence of anti-CRISPR proteins similar to toxins–antitoxins (Pawluk et al. 2016; Touchon and Rocha 2010). A high number of repeats indicates a complete Cas system; an intermediate number indicates a recent erosion of the Cas system; and a low number indicates the presence of only remnants of the system (Touchon and Rocha 2010). It is interesting to note that the genomes of three closely related *E. coli* strains, namely, BL21, BL21-DE3, and B-REL606, contain several repeats without any *cas* genes, but the *E. coli* strain SMS35 has no CRISPR2 at all because of a recent sucrose operon insertion (Touchon and Rocha 2010). In general, the quantity of repeats provides useful information about the potential integrity and functionality of the system. In addition, if spacers in the host match the chromosome in the CRISPR locus, then the acquisition of a mobile element could potentially pose a significant risk. Whenever the CRISPR–Cas system degrades DNA plasmids, which typically contain spacers like chromosomal DNA, chromosomal degradation may result. Furthermore, if the CRISPR–Cas system only interferes with gene expression, then this might still enable the outside force to control the host. Therefore, it is possible that bacterial genomes have evolved a defense mechanism or a workaround for the incoming CRISPR–Cas system. This mechanism may involve native CRISPR functioning as anti-CRISPR, as seen in *P. aeruginosa* and *E. coli* (Pawluk et al. 2016; Touchon and Rocha 2010).

1.6 The Irony: Bacterial Defense Mechanisms Against Other Bacteria

Because CRISPR–Cas targets specific sequences, it can distinguish between pathogenic and commensal species. CRISPR guide RNAs can be engineered to specifically target pathogen-specific virulence and chromosomal genes, allowing the CRISPR–Cas system to be repurposed against bacteria rather than foreign invaders (Cui and Bikard 2016). AMR pathogens alone can be targeted by sequence-specific antibiotics made possible by CRISPR–cas9 technology (Kim et al. 2016). It is possible to program or utilize the double-stranded DNA nuclease Cas9 to cleave any sequence of DNA (Kim et al. 2016). In the past, researchers used a plasmid encoding cas9-guided RNAs to precisely degrade antibiotic-resistant genes in *E. coli* and *Staphylococcus aureus* (Jinek et al. 2012; van der Oost et al. 2014). Target sequence-specific Cas9 programming can increase the cytotoxicity of resistant cells. This means that with the aid of the CRISPR–cas9 system, AMR pathogens can be precisely cleaved to produce antibiotic-sensitive cells.

The main challenge in the delivery of the CRISPR–cas9 antibacterial complex is getting the 160-kDa protein RNA through bacterial membranes. Furthermore, what is the method for delivering this complex to bacteria that are Gram-positive and Gram-negative? Numerous researchers have employed species-specific phages such as CRISPR–Cas delivery vehicles to address this issue. Phages are bacterial predators that naturally feed on bacteria by injecting their DNA into them. It was reported in 2014 that by genetically encoding the phagemid, CRISPR-cas9, which is capable of being encapsulated inside the capsids (protein coat) of inactive phages and is meant to target specific chromosomal genes of bacteria. A plasmid called phagemid is intended to be enclosed within phage capsids (Citorik et al. 2014). According to a different study, CRISPR–cas9-containing genetically modified phages can target *S. aureus* antibiotic resistance (Bikard et al. 2014). All things considered, these findings showed that CRISPR–cas9 antibacterial tools show a high degree of specificity toward pathogenic bacteria while sparing nonpathogenic species, an essential prerequisite for the development of novel, unique antibiotics. These in vitro results emphasize how appealing it is to use phages to deliver CRISPR–cas9 and rapidly eliminate pathogens that have become resistant.

Furthermore, other teams have investigated the possibility of using CRISPR–cas9 to eliminate resistant bacteria from intricate bacterial populations (Gomaa et al. 2014). Other researchers have investigated gene editing techniques to resensitize bacteria to antibiotics (Yosef et al. 2015), and additional research has created phage scaffolds to expand the host range (Ando et al. 2015). These findings back up the idea of utilizing the CRISPR–cas9 machinery to combat AMR infections and newly emerging bacterial strains. By altering the guide RNA sequence, CRISPR–cas9 can exhibit remarkable programmability and adaptability. However, due to poor loading and packaging efficiencies, the techniques used to encapsulate both sgRNA and protein limit their practical use. Furthermore, toxicity issues could arise from a high administration dosage requirement (Bozzuto and Molinari 2015; Taranejoo et al. 2015).

In order to get around these issues, scientists have created a nonviral genome editing technique that successfully targets the *mecA* gene using a nano-sized CRISPR complex. They used cas9 protein, a polymer-derived protein whose cationic polymer is covalently modified. They claimed that the formation of CRISPR-sized complexes, or Cr-nanocomplexes, was successful despite affecting CRISPR–cas9's ability to cleave DNA. The purpose of the Cr-nanocomplex was to specifically target the *mecA* gene. It is possible to effectively deliver this methicillin-resistant gene to methicillin-resistant *S. aureus* (MRSA) and enable bacterial genome editing, as opposed to using the native cas9 complex or the traditional lipid-based formulation (Kang et al. 2017). The results of this study strongly imply that CRISPR–cas9 can be modified for use against AMR infections.

These techniques, however, only deal with infections that are external and topically treated, such as MRSA. Therefore, in order to treat infections specific to particular tissues and organs, systemic intracellular pathogens require alternative approaches.

1.7 Intracellular Delivery of CRISPR–cas9 Antibacterial Tools

It is possible to introduce CRISPR–cas9 antibacterial tools into bacteria by means of genetically encoded phage genomes, as previously noted. This process is more difficult when pathogenic bacteria are intracellular. In this case, it is necessary to precisely target infected cells with CRISPR–cas9 encoded in the phage prior to releasing it into the bacteria that reside inside the cells. As seen in Fig. 1.2, the specificity for both layers (host cell and intracellular bacteria) complicates this delivery process (Greene 2018).

The biggest challenge in developing new drugs is determining how effective they are in creating matrices for effective delivery and therapy. Drug delivery through molecular property modification lowers treatment costs while providing other advantages like avoiding healthy cells and requiring less dosage. There are several methods for editing the human host cell using CRISPR–cas9 (Wang et al. 2017). These tactics, however, only target human target cells; they do not address the difficulties in delivering CRISPR–cas9 encoded in phages to fight intracellular infections. It is also important to consider how different phages differ in terms of size and structure. CRISPR–cas9 provides an approach to solve these problems by enabling the genetic modification of a phage genome to create programmable antibacterial tools. Moreover, the use of encoded phages to transfer CRISPR–cas9 into infected host cells represents a breakthrough over the existing delivery methods.

Because of the known structural diversity of phages, conventional tactics like loading them with nanoparticles are impractical. Drug delivery methods employ a variety of porous nanoparticles as cargo; however, due to pore size limitations, these approaches are ineffective when large and nonsymmetrical phages are employed. Previous studies have shown that phage encapsulation (bacterial cargoes) can be directly induced to self-assemble into lipid and silica-based particle structures

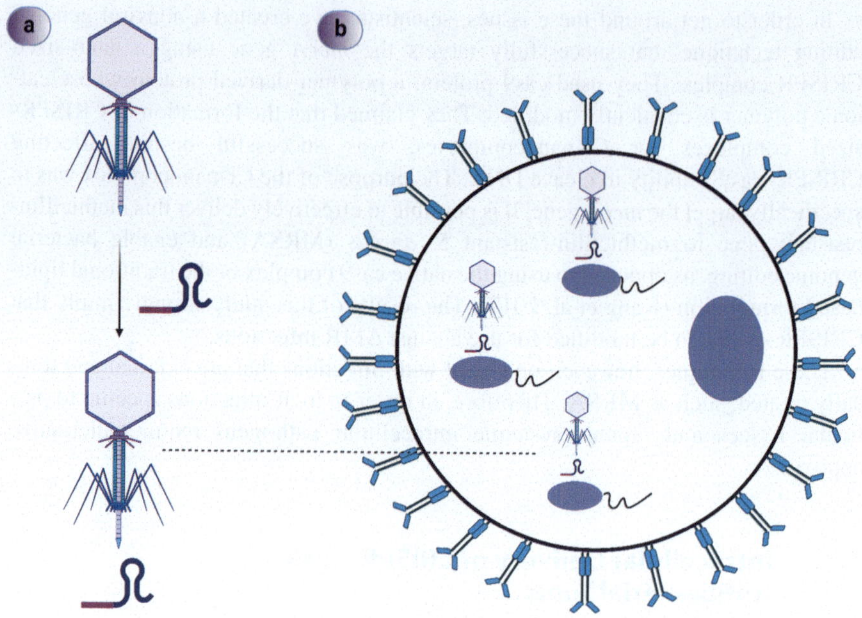

Fig. 1.2 Delivery of CRISPR–cas9 antibacterial tools into an infected cell (**a**, **b**). (**a**) Bacteriophages contain an encoded CRISPR–cas9 antibacterial tool. (**b**) Phages encoding CRISPR–cas9 were introduced within the infected host cells to combat AMR infections

(Malik et al. 2017). This work is necessary to address these problems because stabilizing bacteriophages for therapeutic purposes requires encapsulation. The use of various biological components can alter chemical formulations for encapsulation, such as silica doping and stabilizing protein applications. Encapsulated phages based on silica can avoid detection by the immune system while maintaining their normal biological function. In order to treat intracellular bacterial infections like *Burkholderia pseudomallei*, this strategy can assist in overcoming the twofold barrier problem.

Phages that encode CRISPR–cas9 offer the possibility of delivering antibacterial agents to particular species. Moreover, the versatility of CRISPR–cas9 enables the quick development of biologics to combat AMR pathogen infections. Building a library of CRISPR–cas9 guide RNAs is now feasible to combat AMR pathogens that are evolving quickly (Greene 2018).

1.8 Neutralizing Antibiotic-Resistant Genes with the CRISPR–Cas System

Generally speaking, partial repeats surrounding RNA-based spacers instruct Cas proteins to specifically target and cleave DNA in order to encode corresponding protospacers. Because the CRISPR array contains information that can be specifically targeted and used to cleave any DNA for in vivo purposes, it is possible to target the population of bacteria that possess specific genes that encode antibiotic resistance using the CRISPR–Cas system (Bikard et al. 2014; Hale et al. 2012; Yosef et al. 2015).

Recent studies have shown that when the CRISPR–Cas system targets the bacterial genome, it can cause irreversible chromosomal lesions that can lead to cell death. This can happen either intentionally or accidentally (Citorik et al. 2014; Vercoe et al. 2013). Due to the widespread conservation of the CRISPR–Cas system in bacteria and archaea, it will be necessary to isolate, optimize, and develop delivery carriers and vectors of the CRISPR–Cas system in order to create RNA-guided nucleases that can target additional strains, such as important members of the endogenous microbiota and multidrug-resistant (MDR) pathogens (Citorik et al. 2014). RNA-guided nucleases have a delivery mechanism that functions in more advanced organisms, which may also enable researchers to regulate the frequency of particular genes in wild-type populations, such as virulence determinants and antibiotic-resistant genes (Esvelt et al. 2014), transported by polymer-derivative CRISPR nanocomplexes to microbial populations (Kang et al. 2017), bacteria that carry conjugation-transmittable plasmids (Citorik et al. 2014; Kim et al. 2016), and/or bacteriophages (Bikard et al. 2014; Citorik et al. 2014). The Cas9 nuclease specifically targets antibiotic-resistant genes and certain DNA sequences found in bacterial pathogens. Gomaa et al. (2014) used the type I-E CRISPR–Cas system of *E. coli* to encode six *cas* genes within two operons: casABCDE and cas3(S. J. Brouns et al. 2008), to particular and significant sequences that are present in the genomes of various sequences. Researchers concluded that targeting multiple sites across the genome, including ftsA, nusB, msbA, and asd, can achieve extensive elimination. They also observed that the levels of elimination demonstrated by simultaneously focusing on several locations were comparable to those obtained by focusing on just one location (Gomaa et al. 2014). Even though they used a powerful technique to transfer a type I CRISPR–Cas system into bacteria, the researchers in this study used transformation to target resistance genes on extrachromosomal elements as well as chromosomal genes that are unique to a subset of the heterogeneous population. The Cas9 phagemid, a plasmid intended to be packaged in phage capsids (Melnikov et al. 1984), was used by Bikard et al. (2014) to remove MRSA strains from a bacterial population that was mixed. They also created a CRISPR–Cas9 plasmid that specifically targets plasmids like pUSA01 and pUSA02, which are resistant to tetracycline. To treat the clinical isolates of *S. aureus* USA300Φ, they constructed cas9 along with the crRNA of the methicillin-resistant gene *mecA* in the phagemid (pDB121::mecA) (Ubukata et al. 1989). None of the Cas9 plasmids that target tetracycline-resistant plasmids showed signs of cell death. On the other hand, more than 99.99% of the cells showed signs of tetracycline sensitivity, and, after receiving pDB121:mecAphagemid, the proportion of *S. aureus* USA300Φ was found to decrease from 50% before treatment to 0.4%. Furthermore, the researchers incorporated the enterotoxin *sek* gene into the phagemid and observed that it exhibited comparable efficacy in eliminating all strains of *S. aureus* (Isnard et al. 2013). In a different work, Citorik et al. employed two strategies, an M13-based phagemid and a conjugative plasmid to target the genes *blaNDM-1* and *blaSHV-18*, which encode antibiotics with pan- and extended-spectrum resistance and extended-spectrum resistance, respectively, by introducing the Cas9 nuclease into bacteria (Rasheed et al. 2000; Rasheed et al. 2013). The phages treated the resistant strains, reducing the number of viable cells by two to three logarithmic factors. They also demonstrated the ability of the CRISPR–Cas system to discriminate between vulnerable and resistant strains. Researchers discovered that only chromosomal gyrA mutations that result in quinolone-resistant *E. coli* strains were susceptible to the cytotoxic effects of the programmed gyrA phagemid. This was not the case for isogenic strains harboring the wild-type *gyrA* gene (Citorik et al. 2014; Jacoby 2005).

In a different work, Kang et al. (2017) suggested a nonviral delivery strategy using a Cas9 nanocomplex that included sgRNA targeting mecA, Cas9, and cationic polymer branched polyethyleneimine (bPEI) that is one of the most widely used carriers for gene delivery (Chung et al. 2011; Patnaik and Gupta 2013). This carrier was used to package sgRNA, which can improve the delivery of Cas9 into MRSA strains (Kang et al. 2017). In contrast to native Cas9 combined with lipofectamine, Cas9 conjugated with bPEI was able to absorb more into the bacteria, according to Kang et al., which is a carrier for gene delivery in mammalian cells (Zuris et al. 2015), or native Cas9 simply mixed with bPEI (Kang et al. 2017). They proposed that either the highly cationic property of the bPEI polymer or the consequent enhancement of the protein's polarity could be the cause of this phenomenon (Kang et al. 2017). Furthermore, unlike the untreated strains, the cultured MRSA strains treated with Cas9–bPEI could not develop in agar media containing 6 μg/mL of oxacillin. Additionally, compared to treating with the Cas9–bPEI complex lacking sgRNA as the control, therapy with the Cas9–bPEI complex was able to substantially lower the growth (a 32% reduction), whereas native Cas9/sgRNA complexed using a lipofectamine carrier exhibited no reduction in growth (Kang et al. 2017). Given that vector-free delivery of CRISPR can simply induce phenotypic changes and edit the bacterial genome while avoiding immunogenicity concerns and off-target effects, this finding may represent a step forward in the design of CRISPR-based antimicrobial drugs.

The low frequency of conjugation is one of the factors that restrict CRISPR-mediated killing by plasmids; much more effective delivery methods can be developed using phagemids or bacteriophages. Nonetheless, because conjugative plasmids have a broad host range, CRISPR nuclease delivery via plasmids remains a desirable alternative (Jain and Srivastava 2013), is simple to engineer, has a large coding capacity (Smillie et al. 2010), is resistant to restriction modification systems (Oliveira et al. 2014), and does not need a biological receptor (Pérez-Mendoza and de la Cruz 2009). Conjugative plasmids that both encode and stimulate the formation of biofilms may accelerate conjugative plasmid transfer because of the increased cell-to-cell contact they provide (Hausner and Wuertz 1999). This makes them potentially useful for delivering molecular tools like CRISPR nucleases that alter the makeup of microbial communities, indicating that a large number of them are biofilms (López-Igual et al. 2019; Peters et al. 2019). Hamilton et al. (2019) study created a cis-conjugative system in which the plasmid encodes conjugative machinery and CRISPR nucleases. They created 65 total sgRNAs that target nonessential genes, genes with unclear phenotypes, and 38, 23, and 4 essential genes. They discovered that conjugative transfer from *E. coli* to *Salmonella enterica* occurs more frequently in plasmids carrying the CRISPR nuclease and conjugative machinery when the conditions are favorable for cell-to-cell contact. Their results show that single or multiplexed sgRNAs that target putative lipoprotein (yghJ), putative oxidoreductase (aegA), glutamate/aspartate transporter (gltJ), and other nonessential genes more effectively kill *S. enterica* than do essential genes. Additionally, they showed how the cis-conjugative plasmid recipients can potentially donate to further conjugation cycles, possibly resulting in an exponential rise in the number of conjugative donor bacteria in the population. They proposed that this delivery mechanism in conjunction with CRISPR nucleases might be a useful tool for changing the microbiome.

Using phagemid targeting intimin, a chromosomally encoded virulence factor of *E. coli* O157:H7 that is essential for intestinal pathology and colonization (Kaper et al. 2004), According to Citorik and colleagues, treating entero hemorrhagic *E. coli* (EHEC) would result in a twenty times decrease in the number of viable cells (Kang et al. 2017). These results demonstrated that the activity of the CRISPR–Cas system can specifically kill bacteria with specific genomes, potentially reducing the frequency of unwanted genes in the bacterial population without endangering nearby organisms, such as virulence loci, antibiotic resistance, or metabolic pathways. Furthermore, complex bacterial communities are able to have their composition changed using the CRISPR–Cas system. While prebiotics and probiotics, which are drugs that alter the human microbiota, have the potential to lessen suffering from a variety of diseases, their precise mechanisms of action are still poorly understood (Sonnenburg and Fischbach 2011).

Researchers tested the Cas9 phagemid's potency against infections using an in vivo mouse model (Bikard et al. 2014). Topical application of the CRISPR–Cas9 antimicrobial pDB121::aphphagemid (against the kanamycin-resistant gene *aph* (Gray and Fitch 1983)) reduced the proportion of RNKΦ cells in the back of infected CD1 mice, which was substantially distinct from treatments with 200 mg/mouse of streptomycin or 2% mupirocin ($P < 0.0001$). Streptomycin treatment wiped out the infection, whereas mupirocin did not (Bikard et al. 2014). Moreover, Citorik et al. (2014) showed that EHEC treated with the cas9-eae phagemid in *Galleria mellonella* larvae had a significantly higher survival rate compared to the group that received no treatment. Additionally, it proved much less effective than treating the strain with carbenicillin, to which it was susceptible, and much more effective than treating the strain with chloramphenicol, to which it was resistant. To assess the effectiveness of both phage therapy and antibiotics against different fungal, Gram-positive, and Gram-negative pathogens, *G. mellonella* was employed as an infection model (Desbois and Coote 2012). A Cas13a-based phage was developed by further research, similar to that of Kiga et al. (2020), to specifically target methicillin-resistant *S. aureus* and carbapenem-resistant *E. coli*.. The above findings support the use of phagemid and the CRISPR–Cas system as viable alternatives when bacterial strains show high resistance to currently available antimicrobial agents.

However, there are two significant disadvantages to phagemid use. First off, the total amount of phagemid required for treatments is far more than the size of the target population since the phagemid fails to generate additional phages following infections. Second, the enormous number of phagemids and their restricted host range could prevent their general use.

Utilizing a nuclease having two or more crRNA guides to start distinct plasmid and chromosomal sequences has the advantage of allowing for targeted cell range extension and a decrease in resistant clones that create target mutations to avoid phagemid treatment. By giving nonpathogenic bacterial strains the sequence-specific Cas9 nuclease and by quickly reprogramming it to target distinct sequences, it is also feasible to immunize them against the spread of virulence and/or antibiotic-resistant plasmids. In a bacterial population, this lowers the plasmid content without resulting in cell death.

Most studies focused on the target genes with the highest killing efficiency and phage, conjugative, and polymeric nanoparticle-based CRISPR delivery systems as possible future directions, depending on the type of dysbiosis and microbiome. According to the majority of research studies, planktonic and biofilm conditions might benefit from a combination of delivery methods based on phage, conjugative, and polymeric nanoparticles. In addition, total eradication of the target organism or organisms and a dearth of specialized instruments to manage pathogenic species or change the makeup of microbial communities represent major challenges in microbiome manipulation, microbiology, and infectious disease control (Roach et al. 2017). The potential application of CRISPR-based nucleases as sequence-specific antimicrobial agents could lessen the target's and infectious bacteria's proportion but still necessitates the creation of a stable and widely used delivery strategy. Therefore, the challenges that CRISPR–Cas-based antibacterial tools face could include the architecture of CRISPR–Cas delivery vectors or vehicles, intricate bacterial communities, and various ways in which an organism resists a single antibacterial agent, potential for target gene mutations, regulatory frameworks, and the social obligations associated with CRISPR–Cas-based antibacterial tools.

1.9 Conclusions and Future Perspectives

Because of their clever design, CRISPR–Cas-based antibiotics and small molecules may pave the way for the discovery of new antimicrobials, sequence-informed methodologies, and ease of use in a field where time- and money-consuming broad-spectrum screening has traditionally dominated practice. Modern antibiotic therapy now targets the simple acquisition of genes conferring antibiotic resistance, genome stability manipulation, and facilitating the colonization of novel habitats like the antibiotic-filled hospital environment among other things.

We hope that CRISPR–Cas methods will pave the way for the creation of intelligent antibiotics that can distinguish between pathogenic and nonpathogenic microorganisms and eradicate multidrug-resistant pathogens. These systems offer opportunities in the treatment of multidrug-resistant infections, the investigation of microbial consortia, and the regulation of industrial fermentations by allowing for the quantitative and selective elimination of individual bacterial strains according to a sequence-specific manner.

References

Abadia E, Zhang J, dos Vultos T, Ritacco V, Kremer K, Aktas E, Matsumoto T, Refregier G, van Soolingen D, Gicquel B, Sola C (2010) Resolving lineage assignation on Mycobacterium tuberculosis clinical isolates classified by spoligotyping with a new high-throughput 3R SNPs based method. Infect Genet Evol 10(7):1066–1074. https://doi.org/10.1016/j.meegid.2010.07.006

Alekshun MN, Levy SB (2007) Molecular mechanisms of antibacterial multidrug resistance. Cell 128(6):1037–1050

Anantharaman V, Iyer LM, Aravind L (2010) Presence of a classical RRM-fold palm domain in Thg1-type 3′-5′ nucleic acid polymerases and the origin of the GGDEF and CRISPR polymerase domains. Biol Direct 5:1–9

Ando H, Lemire S, Pires DP, Lu TK (2015) Engineering modular viral scaffolds for targeted bacterial population editing. Cell Syst 1(3):187–196. https://doi.org/10.1016/j.cels.2015.08.013

Barrangou R, Fremaux C, Deveau H, Richards M, Boyaval P, Moineau S, Romero DA, Horvath P (2007) CRISPR provides acquired resistance against viruses in prokaryotes. Science 315(5819):1709–1712

Bhaya D, Davison M, Barrangou R (2011) CRISPR-Cas systems in bacteria and archaea: versatile small RNAs for adaptive defense and regulation. Annu Rev Genet 45:273–297

Bikard D, Euler CW, Jiang W, Nussenzweig PM, Goldberg GW, Duportet X, Fischetti VA, Marraffini LA (2014) Exploiting CRISPR-Cas nucleases to produce sequence-specific antimicrobials. Nat Biotechnol 32(11):1146–1150. https://doi.org/10.1038/nbt.3043

Bourgogne A, Garsin DA, Qin X, Singh KV, Sillanpaa J, Yerrapragada S, Ding Y, Dugan-Rocha S, Buhay C, Shen H, Chen G, Williams G, Muzny D, Maadani A, Fox KA, Gioia J, Chen L, Shang Y, Arias CA, Nallapareddy SR, Zhao M, Prakash VP, Chowdhury S, Jiang H, Gibbs RA, Murray BE, Highlander SK, Weinstock GM (2008) Large scale variation in *Enterococcus faecalis* illustrated by the genome analysis of strain OG1RF. Genome Biol 9(7):R110. https://doi.org/10.1186/gb-2008-9-7-r110

Bozzuto G, Molinari A (2015) Liposomes as nanomedical devices. Int J Nanomedicine 10:975–999. https://doi.org/10.2147/ijn.S68861

Brouns SJ, Jore MM, Lundgren M, Westra ER, Slijkhuis RJ, Snijders AP, Dickman MJ, Makarova KS, Koonin EV, Van Der Oost J (2008) Small CRISPR RNAs guide antiviral defense in prokaryotes. Science 321(5891):960–964

Burley KM, Sedgley CM (2012) CRISPR-Cas, a prokaryotic adaptive immune system, in endodontic, oral, and multidrug-resistant hospital-acquired *Enterococcus faecalis*. J Endod 38(11):1511–1515

Carte J, Wang R, Li H, Terns RM, Terns MP (2008) Cas6 is an endoribonuclease that generates guide RNAs for invader defense in prokaryotes. Genes Dev 22(24):3489–3496

Chung HJ, Hong CA, Lee SH, Jo SD, Park TG (2011) Reducible siRNA dimeric conjugates for efficient cellular uptake and gene silencing. Bioconjug Chem 22(2):299–306. https://doi.org/10.1021/bc100438m

Citorik RJ, Mimee M, Lu TK (2014) Sequence-specific antimicrobials using efficiently delivered RNA-guided nucleases. Nat Biotechnol 32(11):1141–1145. https://doi.org/10.1038/nbt.3011

Cong L, Ran FA, Cox D, Lin S, Barretto R, Habib N, Hsu PD, Wu X, Jiang W, Marraffini LA (2013) Multiplex genome engineering using CRISPR/Cas systems. Science 339(6121):819–823

Cui L, Bikard D (2016) Consequences of Cas9 cleavage in the chromosome of Escherichia coli. Nucleic Acids Res 44(9):4243–4251 de la Fuente-Nunez C, Torres MD, Mojica FJ, Lu TK (2017) Next-generation precision antimicrobials: towards personalized treatment of infectious diseases. Curr Opin Microbiol 37:95–102

Delaney NF, Balenger S, Bonneaud C, Marx CJ, Hill GE, Ferguson-Noel N, Tsai P, Rodrigo A, Edwards SV (2012) Ultrafast evolution and loss of CRISPRs following a host shift in a novel wildlife pathogen, Mycoplasma gallisepticum. PLoS Genet 8(2):e1002511. https://doi.org/10.1371/journal.pgen.1002511

Deltcheva E, Chylinski K, Sharma CM, Gonzales K, Chao Y, Pirzada ZA, Eckert MR, Vogel J, Charpentier E (2011) CRISPR RNA maturation by trans-encoded small RNA and host factor RNase III. Nature 471(7340):602–607

Deng L, Garrett RA, Shah SA, Peng X, She Q (2013) A novel interference mechanism by a type IIIB CRISPR-Cmr module in Sulfolobus. Mol Microbiol 87(5):1088–1099

Desbois AP, Coote PJ (2012) Utility of greater wax moth larva (Galleria mellonella) for evaluating the toxicity and efficacy of new antimicrobial agents. Adv Appl Microbiol 78:25–53. https://doi.org/10.1016/b978-0-12-394805-2.00002-6

Deveau H, Garneau JE, Moineau S (2010) CRISPR/Cas system and its role in phage-bacteria interactions. Ann Rev Microbiol 64:475–493

Díez-Villaseñor C, Almendros C, García-Martínez J, Mojica FJ (2010) Diversity of CRISPR loci in Escherichia coli. Microbiology 156(5):1351–1361

Erickson KE, Otoupal PB, Chatterjee A (2017) Transcriptome-level signatures in gene expression and gene expression variability during bacterial adaptive evolution. mSphere 2(1):e00009–e00017

Esvelt KM, Smidler AL, Catteruccia F, Church GM (2014) Concerning RNA-guided gene drives for the alteration of wild populations. Elife 3:e03401. https://doi.org/10.7554/eLife.03401

Fabre L, Zhang J, Guigon G, Le Hello S, Guibert V, Accou-Demartin M, de Romans S, Lim C, Roux C, Passet V, Diancourt L, Guibourdenche M, Issenhuth-Jeanjean S, Achtman M, Brisse S, Sola C, Weill F-X (2012) CRISPR typing and subtyping for improved laboratory surveillance of salmonella infections. PLoS One 7(5):e36995. https://doi.org/10.1371/journal.pone.0036995

Flannagan SE, Chow JW, Donabedian SM, Brown WJ, Perri MB, Zervos MJ, Ozawa Y, Clewell DB (2003) Plasmid content of a vancomycin-resistant *Enterococcus faecalis* isolate from a patient also colonized by Staphylococcus aureus with a VanA phenotype. Antimicrob Agents Chemother 47(12):3954–3959

Frieden T (2013) Antibiotic resistance threats in the United States, 2013 | antibiotic/antimicrobial resistance report. U.S. Department of Health and Human Services, Centers for Disease Control and Prevention

Garneau JE, Dupuis M-È, Villion M, Romero DA, Barrangou R, Boyaval P, Fremaux C, Horvath P, Magadán AH, Moineau S (2010) The CRISPR/Cas bacterial immune system cleaves bacteriophage and plasmid DNA. Nature 468(7320):67–71

Gilmore MS, Clewell DB, Courvalin P, Dunny GM, Murray BE, Rice LB (2002) The enterococci: pathogenesis, molecular biology, and antibiotic resistance, vol 10. ASM Press, Washington

Gomaa AA, Klumpe HE, Luo ML, Selle K, Barrangou R, Beisel CL (2014) Programmable removal of bacterial strains by use of genome-targeting CRISPR-Cas systems. mBio 5(1):e00928–e00913. https://doi.org/10.1128/mBio.00928-13

Gray GS, Fitch WM (1983) Evolution of antibiotic resistance genes: the DNA sequence of a kanamycin resistance gene from Staphylococcus aureus. Mol Biol Evol 1(1):57–66. https://doi.org/10.1093/oxfordjournals.molbev.a040298

Greene AC (2018) CRISPR-based antibacterials: transforming bacterial defense into offense. Trends Biotechnol 36(2):127–130. https://doi.org/10.1016/j.tibtech.2017.10.021

Gunderson FF, Cianciotto NP (2013) The CRISPR-associated gene cas2 of Legionella pneumophila is required for intracellular infection of amoebae. mBio 4(2):e00074–e00013. https://doi.org/10.1128/mBio.00074-13

Hale CR, Zhao P, Olson S, Duff MO, Graveley BR, Wells L, Terns RM, Terns MP (2009) RNA-guided RNA cleavage by a CRISPR RNA-Cas protein complex. Cell 139(5):945–956. https://doi.org/10.1016/j.cell.2009.07.040

Hale CR, Majumdar S, Elmore J, Pfister N, Compton M, Olson S, Resch AM, Glover CV 3rd, Graveley BR, Terns RM, Terns MP (2012) Essential features and rational design of CRISPR RNAs that function with the Cas RAMP module complex to cleave RNAs. Mol Cell 45(3):292–302. https://doi.org/10.1016/j.molcel.2011.10.023

Hamilton TA, Pellegrino GM, Therrien JA, Ham DT, Bartlett PC, Karas BJ, Gloor GB, Edgell DR (2019) Efficient inter-species conjugative transfer of a CRISPR nuclease for targeted bacterial killing. Nat Commun 10(1):4544. https://doi.org/10.1038/s41467-019-12448-3

Haurwitz RE, Jinek M, Wiedenheft B, Zhou K, Doudna JA (2010) Sequence- and structure-specific RNA processing by a CRISPR endonuclease. Science 329(5997):1355–1358. https://doi.org/10.1126/science.1192272

Hausner M, Wuertz S (1999) High rates of conjugation in bacterial biofilms as determined by quantitative in situ analysis. Appl Environ Microbiol 65(8):3710–3713. https://doi.org/10.1128/aem.65.8.3710-3713.1999

Hsu PD, Lander ES, Zhang F (2014) Development and applications of CRISPR-Cas9 for genome engineering. Cell 157(6):1262–1278

Ishino Y, Shinagawa H, Makino K, Amemura M, Nakata A (1987) Nucleotide sequence of the iap gene, responsible for alkaline phosphatase isozyme conversion in Escherichia coli, and identification of the gene product. J Bacteriol 169(12):5429–5433

Isnard C, Malbruny B, Leclercq R, Cattoir V (2013) Genetic basis for in vitro and in vivo resistance to lincosamides, streptogramins A, and pleuromutilins (LSAP phenotype) in Enterococcus faecium. Antimicrob Agents Chemother 57(9):4463–4469. https://doi.org/10.1128/aac.01030-13

Jacoby GA (2005) Mechanisms of resistance to quinolones. Clin Infect Dis 41(Suppl 2):S120–S126. https://doi.org/10.1086/428052

Jain A, Srivastava P (2013) Broad host range plasmids. FEMS Microbiol Lett 348(2):87–96. https://doi.org/10.1111/1574-6968.12241

Jinek M, Chylinski K, Fonfara I, Hauer M, Doudna JA, Charpentier E (2012) A programmable dual-RNA–guided DNA endonuclease in adaptive bacterial immunity. Science 337(6096):816–821

Jones CL, Sampson TR, Nakaya HI, Pulendran B, Weiss DS (2012) Repression of bacterial lipoprotein production by Francisella novicida facilitates evasion of innate immune recognition. Cell Microbiol 14(10):1531–1543

Jore MM, Lundgren M, Van Duijn E, Bultema JB, Westra ER, Waghmare SP, Wiedenheft B, Pul Ü, Wurm R, Wagner R (2011) Structural basis for CRISPR RNA-guided DNA recognition by Cascade. Nat Struct Mol Biol 18(5):529–536

Kang YK, Kwon K, Ryu JS, Lee HN, Park C, Chung HJ (2017) Nonviral genome editing based on a polymer-derivatized CRISPR nanocomplex for targeting bacterial pathogens and antibiotic resistance. Bioconjug Chem 28(4):957–967. https://doi.org/10.1021/acs.bioconjchem.6b00676

Kaper JB, Nataro JP, Mobley HL (2004) Pathogenic Escherichia coli. Nat Rev Microbiol 2(2):123–140. https://doi.org/10.1038/nrmicro818

Kiga K, Tan XE, Ibarra-Chávez R, Watanabe S, Aiba Y, Sato'o Y, Li FY, Sasahara T, Cui B, Kawauchi M, Boonsiri T, Thitiananpakorn K, Taki Y, Azam AH, Suzuki M, Penadés JR, Cui L (2020) Development of CRISPR-Cas13a-based antimicrobials capable of sequence-specific killing of target bacteria. Nat Commun 11(1):2934. https://doi.org/10.1038/s41467-020-16731-6

Kim JS, Cho DH, Park M, Chung WJ, Shin D, Ko KS, Kweon DH (2016) CRISPR/Cas9-mediated re-sensitization of antibiotic-resistant Escherichia coli harboring extended-Spectrum β-lactamases. J Microbiol Biotechnol 26(2):394–401. https://doi.org/10.4014/jmb.1508.08080

Kinnevey PM, Shore AC, Brennan GI, Sullivan DJ, Ehricht R, Monecke S, Slickers P, Coleman DC (2013) Emergence of sequence type 779 methicillin-resistant Staphylococcus aureus harboring a novel pseudo staphylococcal cassette chromosome mec (SCCmec)-SCC-SCCCRISPR composite element in Irish hospitals. Antimicrob Agents Chemother 57(1):524–531. https://doi.org/10.1128/aac.01689-12

Kümmerer K (2003) Significance of antibiotics in the environment. J Antimicrob Chemother 52(1):5–7

Lintner NG, Kerou M, Brumfield SK, Graham S, Liu H, Naismith JH, Sdano M, Peng N, She Q, Copié V (2011) Structural and functional characterization of an archaeal clustered regularly interspaced short palindromic repeat (CRISPR)-associated complex for antiviral defense (CASCADE). J Biol Chem 286(24):21643–21656

López-Igual R, Bernal-Bayard J, Rodríguez-Patón A, Ghigo JM, Mazel D (2019) Engineered toxin-intein antimicrobials can selectively target and kill antibiotic-resistant bacteria in mixed populations. Nat Biotechnol 37(7):755–760. https://doi.org/10.1038/s41587-019-0105-3

Louwen R, Horst-Kreft D, De Boer A, Van Der Graaf L, de Knegt G, Hamersma M, Heikema A, Timms A, Jacobs B, Wagenaar J (2013) A novel link between *Campylobacter jejuni* bacteriophage defence, virulence and Guillain–Barré syndrome. Eur J Clin Microbiol Infect Dis 32:207–226

Makarova KS, Haft DH, Barrangou R, Brouns SJ, Charpentier E, Horvath P, Moineau S, Mojica FJ, Wolf YI, Yakunin AF (2011) Evolution and classification of the CRISPR–Cas systems. Nat Rev Microbiol 9(6):467–477

Malik DJ, Sokolov IJ, Vinner GK, Mancuso F, Cinquerrui S, Vladisavljevic GT, Clokie MRJ, Garton NJ, Stapley AGF, Kirpichnikova A (2017) Formulation, stabilisation and encapsulation of bacteriophage for phage therapy. Adv Colloid Interface Sci 249:100–133. https://doi.org/10.1016/j.cis.2017.05.014

Manson JM, Hancock LE, Gilmore MS (2010) Mechanism of chromosomal transfer of *Enterococcus faecalis* pathogenicity island, capsule, antimicrobial resistance, and other traits. Proc Natl Acad Sci U S A 107(27):12269–12274

Marinelli LJ, Fitz-Gibbon S, Hayes C, Bowman C, Inkeles M, Loncaric A, Russell DA, Jacobs-Sera D, Cokus S, Pellegrini M, Kim J, Miller JF, Hatfull GF, Modlin RL (2012) Propionibacterium acnes bacteriophages display limited genetic diversity and broad killing activity against bacterial skin isolates. mBio 3(5):e00279–e00212. https://doi.org/10.1128/mBio.00279-12

Marraffini LA, Sontheimer EJ (2008) CRISPR interference limits horizontal gene transfer in staphylococci by targeting DNA. Science 322(5909):1843–1845

Melnikov AA, Tchernov AP, Fodor I, Bayev AA (1984) Lambda phagemids and their transducing properties. Gene 28(1):29–35. https://doi.org/10.1016/0378-1119(84)90084-2

Miller JC, Tan S, Qiao G, Barlow KA, Wang J, Xia DF, Meng X, Paschon DE, Leung E, Hinkley SJ, Dulay GP, Hua KL, Ankoudinova I, Cost GJ, Urnov FD, Zhang HS, Holmes MC, Zhang L, Gregory PD, Rebar EJ (2011) A TALE nuclease architecture for efficient genome editing. Nat Biotechnol 29(2):143–148. https://doi.org/10.1038/nbt.1755

Mojica FJ, Díez-Villaseñor C, Soria E, Juez G (2000) Biological significance of a family of regularly spaced repeats in the genomes of archaea, bacteria and mitochondria. Mol Microbiol 36(1):244–246

Mojica FJ, Díez-Villaseñor CS, García-Martínez J, Soria E (2005) Intervening sequences of regularly spaced prokaryotic repeats derive from foreign genetic elements. J Mol Evol 60:174–182

Nam KH, Haitjema C, Liu X, Ding F, Wang H, DeLisa MP, Ke A (2012) Cas5d protein processes pre-crRNA and assembles into a cascade-like interference complex in subtype IC/Dvulg CRISPR-Cas system. Structure 20(9):1574–1584

Oliveira PH, Touchon M, Rocha EP (2014) The interplay of restriction-modification systems with mobile genetic elements and their prokaryotic hosts. Nucleic Acids Res 42(16):10618–10631. https://doi.org/10.1093/nar/gku734

Patnaik S, Gupta KC (2013) Novel polyethylenimine-derived nanoparticles for in vivo gene delivery. Expert Opin Drug Deliv 10(2):215–228. https://doi.org/10.1517/17425247.2013.744964

Pawluk A, Staals RH, Taylor C, Watson BN, Saha S, Fineran PC, Maxwell KL, Davidson AR (2016) Inactivation of CRISPR-Cas systems by anti-CRISPR proteins in diverse bacterial species. Nat Microbiol 1(8):1–6

Pérez-Mendoza D, de la Cruz F (2009) Escherichia coli genes affecting recipient ability in plasmid conjugation: are there any? BMC Genomics 10:71. https://doi.org/10.1186/1471-2164-10-71

Peters JM, Koo BM, Patino R, Heussler GE, Hearne CC, Qu J, Inclan YF, Hawkins JS, Lu CHS, Silvis MR, Harden MM, Osadnik H, Peters JE, Engel JN, Dutton RJ, Grossman AD, Gross CA, Rosenberg OS (2019) Enabling genetic analysis of diverse bacteria with mobile-CRISPRi. Nat Microbiol 4(2):244–250. https://doi.org/10.1038/s41564-018-0327-z

Pourcel C, Salvignol G, Vergnaud G (2005) CRISPR elements in Yersinia pestis acquire new repeats by preferential uptake of bacteriophage DNA, and provide additional tools for evolutionary studies. Microbiology 151(3):653–663

Rasheed JK, Anderson GJ, Yigit H, Queenan AM, Doménech-Sánchez A, Swenson JM, Biddle JW, Ferraro MJ, Jacoby GA, Tenover FC (2000) Characterization of the extended-spectrum beta-lactamase reference strain, Klebsiella pneumoniae K6 (ATCC 700603), which produces the novel enzyme SHV-18. Antimicrob Agents Chemother 44(9):2382–2388. https://doi.org/10.1128/aac.44.9.2382-2388.2000

Rasheed JK, Kitchel B, Zhu W, Anderson KF, Clark NC, Ferraro MJ, Savard P, Humphries RM, Kallen AJ, Limbago BM (2013) New Delhi metallo-β-lactamase-producing Enterobacteriaceae, United States. Emerg Infect Dis 19(6):870–878. https://doi.org/10.3201/eid1906.121515

Roach DR, Leung CY, Henry M, Morello E, Singh D, Di Santo JP, Weitz JS, Debarbieux L (2017) Synergy between the host immune system and bacteriophage is essential for successful phage therapy against an acute respiratory pathogen. Cell Host Microbe 22(1):38–47.e34. https://doi.org/10.1016/j.chom.2017.06.018

Roberts AP, Mullany P (2009) A modular master on the move: the Tn916 family of mobile genetic elements. Trends Microbiol 17(6):251–258

Sampson TR, Weiss DS (2014) CRISPR-Cas systems: new players in gene regulation and bacterial physiology. Front Cell Infect Microbiol 4:37

Sampson TR, Saroj SD, Llewellyn AC, Tzeng Y-L, Weiss DS (2013) A CRISPR/Cas system mediates bacterial innate immune evasion and virulence. Nature 497(7448):254–257

Shabbir MA, Hao H, Shabbir MZ, Wu Q, Sattar A, Yuan Z (2016a) Bacteria vs. bacteriophages: parallel evolution of immune arsenals. Front Microbiol 7:1292

Shabbir MAB, Hao H, Shabbir MZ, Hussain HI, Iqbal Z, Ahmed S, Sattar A, Iqbal M, Li J, Yuan Z (2016b) Survival and evolution of CRiSPR–Cas system in prokaryotes and its applications. Front Immunol 7:375

Sinkunas T, Gasiunas G, Fremaux C, Barrangou R, Horvath P, Siksnys V (2011) Cas3 is a single-stranded DNA nuclease and ATP-dependent helicase in the CRISPR/Cas immune system. EMBO J 30(7):1335–1342

Smillie C, Garcillán-Barcia MP, Francia MV, Rocha EP, de la Cruz F (2010) Mobility of plasmids. Microbiol Mol Biol Rev 74(3):434–452. https://doi.org/10.1128/mmbr.00020-10

Smith J, Grizot S, Arnould S, Duclert A, Epinat JC, Chames P, Prieto J, Redondo P, Blanco FJ, Bravo J, Montoya G, Pâques F, Duchateau P (2006) A combinatorial approach to create artificial homing endonucleases cleaving chosen sequences. Nucleic Acids Res 34(22):e149. https://doi.org/10.1093/nar/gkl720

Sonnenburg JL, Fischbach MA (2011) Community health care: therapeutic opportunities in the human microbiome. Sci Transl Med 3(78):78ps12. https://doi.org/10.1126/scitranslmed.3001626

Strong A, Musunuru K (2017) Genome editing in cardiovascular diseases. Nat Rev Cardiol 14(1):11–20. https://doi.org/10.1038/nrcardio.2016.139

Taranejoo S, Liu J, Verma P, Hourigan K (2015) A review of the developments of characteristics of PEI derivatives for gene delivery applications. J Appl Polym Sci 132. https://doi.org/10.1002/app.42096

Touchon M, Rocha EP (2010) The small, slow and specialized CRISPR and anti-CRISPR of Escherichia and Salmonella. PLoS One 5(6):e11126

Touchon M, Charpentier S, Pognard D, Picard B, Arlet G, Rocha EP, Denamur E, Branger C (2012) Antibiotic resistance plasmids spread among natural isolates of Escherichia coli in spite of CRISPR elements. Microbiology 158(12):2997–3004

Ubukata K, Nonoguchi R, Matsuhashi M, Konno M (1989) Expression and inducibility in Staphylococcus aureus of the mecA gene, which encodes a methicillin-resistant *S. aureus*-specific penicillin-binding protein. J Bacteriol 171(5):2882–2885. https://doi.org/10.1128/jb.171.5.2882-2885.1989

van der Oost J, Westra ER, Jackson RN, Wiedenheft B (2014) Unravelling the structural and mechanistic basis of CRISPR-Cas systems. Nat Rev Microbiol 12(7):479–492. https://doi.org/10.1038/nrmicro3279

Vercoe RB, Chang JT, Dy RL, Taylor C, Gristwood T, Clulow JS, Richter C, Przybilski R, Pitman AR, Fineran PC (2013) Cytotoxic chromosomal targeting by CRISPR/Cas systems can reshape bacterial genomes and expel or remodel pathogenicity islands. PLoS Genet 9(4):e1003454. https://doi.org/10.1371/journal.pgen.1003454

Vergnaud G, Li Y, Gorgé O, Cui Y, Song Y, Zhou D, Grissa I, Dentovskaya SV, Platonov ME, Rakin A, Balakhonov SV, Neubauer H, Pourcel C, Anisimov AP, Yang R (2007) Analysis of the three Yersinia pestis CRISPR loci provides new tools for phylogenetic studies and possibly for the investigation of ancient DNA. Adv Exp Med Biol 603:327–338. https://doi.org/10.1007/978-0-387-72124-8_30

Viswanathan P, Murphy K, Julien B, Garza AG, Kroos L (2007) Regulation of dev, an operon that includes genes essential for Myxococcus xanthus development and CRISPR-associated genes and repeats. J Bacteriol 189(10):3738–3750. https://doi.org/10.1128/jb.00187-07

Wang R, Preamplume G, Terns MP, Terns RM, Li H (2011) Interaction of the Cas6 riboendonuclease with CRISPR RNAs: recognition and cleavage. Structure 19(2):257–264

Wang HX, Li M, Lee CM, Chakraborty S, Kim HW, Bao G, Leong KW (2017) CRISPR/Cas9-based genome editing for disease modeling and therapy: challenges and opportunities for non-viral delivery. Chem Rev 117(15):9874–9906. https://doi.org/10.1021/acs.chemrev.6b00799

Wiedenheft B, van Duijn E, Bultema JB, Waghmare SP, Zhou K, Barendregt A, Westphal W, Heck AJ, Boekema EJ, Dickman MJ (2011) RNA-guided complex from a bacterial immune system enhances target recognition through seed sequence interactions. Proc Natl Acad Sci U S A 108(25):10092–10097

World Health Organization (2014) Antimicrobial resistance: global report on surveillance. World Health Organization [Record #1420 is using a reference type undefined in this output style]

Wright GD (2007) The antibiotic resistome: the nexus of chemical and genetic diversity. Nat Rev Microbiol 5(3):175–186

Yosef I, Manor M, Kiro R, Qimron U (2015) Temperate and lytic bacteriophages programmed to sensitize and kill antibiotic-resistant bacteria. Proc Natl Acad Sci U S A 112(23):7267–7272. https://doi.org/10.1073/pnas.1500107112

Zhang J, Rouillon C, Kerou M, Reeks J, Brugger K, Graham S, Reimann J, Cannone G, Liu H, Albers S-V (2012) Structure and mechanism of the CMR complex for CRISPR-mediated antiviral immunity. Mol Cell 45(3):303–313

Zhu W, Clark NC, McDougal LK, Hageman J, McDonald LC, Patel JB (2008) Vancomycin-resistant Staphylococcus aureus isolates associated with Inc18-like vanA plasmids in Michigan. Antimicrob Agents Chemother 52(2):452–457

Zuris JA, Thompson DB, Shu Y, Guilinger JP, Bessen JL, Hu JH, Maeder ML, Joung JK, Chen ZY, Liu DR (2015) Cationic lipid-mediated delivery of proteins enables efficient protein-based genome editing in vitro and in vivo. Nat Biotechnol 33(1):73–80. https://doi.org/10.1038/nbt.3081

Upsurge in Antibiotic-Resistant Infections

2

Sindhuprava Rana, Vibhor Joshi, Ganesh Chandra Sahoo, Maneesh Kumar, and Krishna Pandey

Abstract

Penicillin was the first antibiotic to be discovered in 1929. Over time, various antibiotics have been discovered to combat the threat of diseases caused by different microorganisms. The extensive use of the newly discovered antibiotics triggered biological resistance mechanisms that eventually led to the development of new strains of microbes. Hospital-acquired infections are the most common cases of antibiotic-resistant infections and pose a major challenge in their treatment. Therefore, there is a constant need to find new defense mechanisms against the constantly evolving microorganisms. The aim of this chapter is to compile all known molecular mechanisms that make microbes resistant to antibiotics. It focuses on the mechanisms of drug resistance such as changes in target sites, modifications of the antibiotic molecule, reduced antibiotic penetration and efflux resistance due to global cell adaptations, and how genetic mutations and

S. Rana
Biostatistics and Bioinformatics Department, ICMR-National Institute for Research in Environmental Health, Bhopal, Madhya Pradesh, India

V. Joshi (✉)
Environmental Biotechnology Genetics and Molecular Biology Department, ICMR-National Institute for Research in Environmental Health, Bhopal, Madhya Pradesh, India
e-mail: vibhor.joshi@icmr.gov.in

G. C. Sahoo
Department of Virology, ICMR-Rajendra Memorial Research Institute of Medical Sciences, Patna, Bihar, India

M. Kumar
State VRDL, All India Institute of Medical Sciences, Deoghar, Jharkhand, India

K. Pandey
Department of Clinical Medicine, ICMR-Rajendra Memorial Research Institute of Medical Sciences, Patna, Bihar, India

M. Gangwar, G. Nath (eds.), *Emerging Paradigms for Antibiotic-Resistant Infections: Beyond the Pill*, https://doi.org/10.1007/978-981-97-5272-0_2

mobile genetic parts contribute to the spread of resistance genes in microbial populations. Biological processes of antibiotic resistance, such as horizontal gene transfer, efflux pumping, and biofilm formation, are becoming increasingly common. This means that conventional antibiotic treatments are losing their effectiveness against more and more bacterial infections. Bringing together all known mechanisms of drug resistance under one umbrella will help the research community develop new ideas to overcome the challenges that arise in the fight for survival against the never-ending phenomenon of evolution. The focus of this chapter is therefore on the impact of antimicrobial resistance (AMR) on health-care, patient outcomes, and economic burdens. This underscores the importance of collaboration between policymakers, healthcare providers, researchers, and the community to mitigate the impact of this burgeoning crisis.

Keywords
Antibiotics · Evolution · Molecular mechanism · Antimicrobial resistance · Healthcare

2.1 Introduction

The spread of antibiotic-resistant infections in India poses a significant public health dilemma and underscores the urgent need for a comprehensive approach to combat antimicrobial resistance (AMR). Due to the prevalence of infectious diseases, high population density, and a heterogeneous healthcare system, India is particularly vulnerable to the emergence and spread of resistant pathogens. The widespread misuse and overuse of antibiotics in agriculture, animal husbandry, and healthcare, all of which contribute to this problem, have facilitated the spread of antibiotic-resistant bacteria (Poudel et al. 2023). Several reasons exacerbate the problem in India, including easy access to antibiotics without prescription, inadequate control of their use, and poor knowledge among healthcare providers and the population about the optimal use of antibiotics. Improved infection control measures, inadequate sanitation, and overcrowding in healthcare facilities also contribute to the spread of resistant bacteria (Cowen et al. 2015; Harkins et al. 2017). Antibiotic resistance has far-reaching consequences that affect not only the well-being of patients but also place a financial burden on healthcare systems and society in general. Antibiotic-resistant infections lead to prolonged periods of illness, increased mortality and morbidity, and an increase in healthcare expenditure due to the need for costlier and protracted treatment methods. In addition, the spread of resistant pathogens poses a threat to vital medical procedures such as organ transplants, chemotherapy, and surgery, hindering progress in healthcare and jeopardizing patient safety (Schenk et al. 2013). To address the growing incidence of antibiotic-resistant infections, a comprehensive strategy should be implemented that includes regulatory reforms, investment in healthcare infrastructure, antimicrobial stewardship initiatives, and public awareness campaigns. The key components of this comprehensive approach include tightening prescriptions of antibiotics, advocating for prudent antibiotic practices, strengthening infection prevention measures, and facilitating access to diagnostics and alternative treatments. To effectively combat antimicrobial resistance (AMR) and protect public health in India, government agencies, healthcare providers, pharmaceutical companies, research institutions, and civil society organizations need to work together (Hatfull et al. 2022). This chapter summarizes the known molecular processes that cause antibiotic resistance in microbes. These include changes in the antibiotic molecules, changes in the target sites, reduced penetration of the antibiotic, and efflux resistance, which is due to changes in the cells as a whole. This study also investigates the functions of mobile genetic elements and genetic mutations that promote the spread of resistance genes in microbial populations. This chapter provides valuable insights into the complex ways in which microbes circumvent the effects of antibiotics. This emphasizes the multifaceted nature of antimicrobial resistance and highlights the importance of understanding these mechanisms in order to develop effective strategies to combat resistance.

2.2 Current State of Antibiotic Resistance

Antimicrobial agents are medical substances that are used both to prevent and treat infections. Antibiotics, antivirals, antifungals, antiseptics, and antiparasitic drugs are just some of them. Antimicrobials are used as disinfectants to sterilize nonliving surfaces. By targeting the key stages of cellular metabolism such as the production of macromolecules and the function of enzymes in cells or cellular structures such as the cell wall and cell membranes, antimicrobials can eradicate microbes and/or prevent their growth (Reygaert 2018; Di Martino 2021; Purssell 2020; Romani et al. 2022). Antimicrobials, whether man-made or natural, always have an ecological effect when present in an environment. The development of antimicrobial resistance is one of these effects (Di Martino 2022). Antimicrobial resistance (AMR) develops in bacteria, fungi, parasites, and viruses (microorganisms) in response to drugs such as antibiotics that are commonly used to fight the infections caused by them. AMR is often called the "silent pandemic" and is a pressing issue that demands immediate action rather than being viewed as a problem for the future (Dadgostar 2019).

AMR can develop in any organism, but it can also maintain susceptibility to a variety of other substances, allowing for effective therapy in the clinical setting. Recently, the acronym "ESKAPE" has been used to refer to a specific group of bacterial pathogens responsible for the majority of hospital-acquired infections. "ESKAPE" is the abbreviation for *Enterococcus faecium*, *Staphylococcus aureus*, *Klebsiella pneumoniae*, *Acinetobacter baumannii*, *Pseudomonas Aeruginosa*, and *Enterobacter* spp. (Panda et al. 2022). Nevertheless, some critics have raised concerns about the inclusion of these bacteria and the acronym itself, arguing that it excludes other enteric pathogens such as *Escherichia coli* (Gram-negative), which is a common bacterial pathogen that causes severe infections with significant multidrug resistance (MDR). Therefore, "ESCAPE" is a more appropriate term, where "C" stands for a nosocomial pathogen, *Clostridium difficile*, which is known to readily develop MDR properties, and all intestinal bacteria of the family

Enterobacteriaceae are included under the term "E" such as *E. coli*, *K. pneumoniae*, *Proteus* spp., and *Enterobacter* spp. (Roy Chowdhury et al. 2016).

AMR is listed among the top ten risks to global health by the World Health Organization (WHO). Estimates suggest that bacterial AMR led to 4.95 million deaths globally in 2019, with 1.27 million directly attributed to this phenomenon. Tuberculosis, malaria, and human immunodeficiency virus (HIV)/acquired immunodeficiency syndrome (AIDS) accounted for 1.5 million, 643,000, and 864,000 annual global deaths due to AMR, respectively. Without intervention, it is projected that the global deaths attributed to AMR could escalate to ten million annually by 2050 (Walsh et al. 2023). In 2019, *Staphylococcus aureus*, *E. coli*, *K. pneumoniae*, *Acinetobacter baumannii*, *S. pneumoniae*, and *Pseudomonas aeruginosa* each caused more than 250,000 deaths due to AMR. Together, these 6 pathogens accounted for 929,000 out of the 1.27 million deaths attributed to AMR and 3.57 million out of the 4.95 million deaths associated with AMR worldwide. Between 100,000 and 250,000 fatalities linked to AMR were attributed to 6 other infections in 2019. According to Murray et al. (2022), *E. coli*, *K. pneumoniae*, *S. pneumoniae*, *A. baumannii*, *S. aureus*, and *M. tuberculosis* were the most frequent causes of AMR-related mortality in 2019 (Fig. 2.1).

Apart from serious consequences such as death and disability, antimicrobial resistance (AMR) also represents a considerable economic burden. The World Bank estimates that antimicrobial resistance could lead to an additional healthcare expenditure of 1 trillion US dollars by 2050 and annual gross domestic product (GDP) losses of 1–3.4 trillion US dollars by 2030. Particularly alarming is the fact that the death rates due to AMR are the highest in low- and middle-income countries (LMICs), highlighting the health crisis caused by AMR and the urgent impact on some of the world's poorest countries (Poudel et al. 2023).

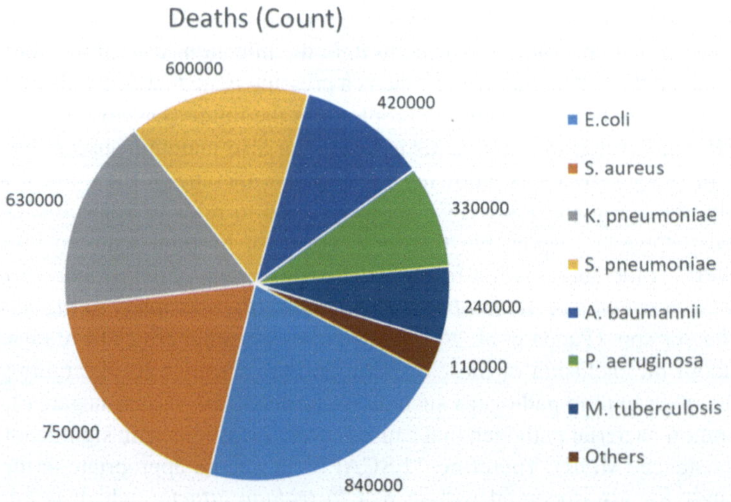

Fig. 2.1 Pie chart representing the number of deaths caused by various pathogens due to AMR

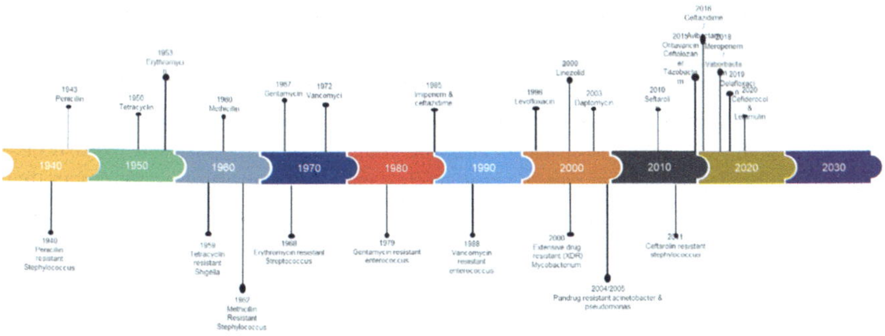

Fig. 2.2 Schematic diagram showing the timeline of antibiotic discoveries and simultaneous development of resistant species

Drugs that were once life-saving are now ineffective. Chloramphenicol, once the first choice in the treatment of typhoid fever, is no longer effective in many regions. Drug-resistant strains of tuberculosis, methicillin-resistant *Staphylococcus aureus* (MRSA), *E. coli*, and *Klebsiella pneumoniae* pose a serious threat to public health. *Plasmodium falciparum*, the parasite responsible for severe malaria, is becoming resistant to all known antimalarial drugs. HIV is increasingly resistant to common antiviral drugs. Resistance to antibiotic (Fig. 2.2), antiviral, antiparasitic, and antifungal drugs is on the rise and impairs their effectiveness (Prestinaci et al. 2015).

WHO and its affiliated networks are fighting the AMR epidemic in various ways. Working at local, regional, national, and international levels, "One Health" is a collaborative, intersectoral, and transdisciplinary strategy that aims to achieve positive health outcomes while recognizing the interdependence of humans, animals, plants, and other life-forms in the common environment (Velazquez-Meza et al. 2022). WHO started the Global Antimicrobial Resistance and Use Surveillance System (GLASS) on October 22, 2015, marking the commencement of the premier global collaborative endeavor for the standardization of AMR surveillance. Additionally, epidemiological surveillance networks, including the European Antimicrobial Resistance Surveillance Network (EARS-Net), the Central Asian and European Surveillance of Antimicrobial Resistance (CAESAR), and the Center for Disease Control and Prevention (CDC)s National Healthcare Safety Network (NHSN) in the United States, were actively involved in surveillance initiatives (Ajulo and Awosile 2024).

2.3 Historical Context

The frightening history of antibiotic resistance is an example of the struggle between human innovation and microbial evolution. As one of the most important classes of drugs, antibiotics have dominated twentieth century medicine. With their countless life-saving actions to combat bacterial infections, antibiotics have inadvertently led to resistance that is undeniably dangerous to the human civilization. Over the last century, antibiotic resistance in humans has gone through several phases. In the early research on penicillin, discovered in the 1920s, no resistance was found (Nadeem et al. 2020). From the 1950s to the 1970s, the use and overuse of antibiotics led to widespread resistance. The sharp increase in microbes resistant to many drugs between the 1980s and 2000s showed that resistance increased even more. Such resistant strains spread rapidly due to the overuse of antibiotics. Bacteria that could resist treatment multiplied rapidly in the twentieth century, posing a growing threat in both medical facilities and private homes (Sarfraz et al. 2020; Mancuso et al. 2021; Murugaiyan et al. 2022). Later, antibiotic resistance worsened between 2020 and 2022. Due to the impact of the COVID-19 pandemic on healthcare systems, antibiotics were widely used to treat secondary bacterial infections and as empirical treatment for severe cases. This increase in antibiotic use reinforced earlier concerns about antimicrobial resistance (AMR), as misuse and overuse accelerate the development of resistant strains. During the pandemic, healthcare services and supply networks were disrupted, complicating antibiotic management and infection prevention (Ayobami et al. 2022; Hawkins et al. 2022; Antuori et al. 2023; Rayan 2023). At the same time, reports of new and emerging resistant diseases emerged, emphasizing the constant adaptability of bacteria and the importance of surveillance and response measures. The combination of the COVID-19 pandemic and the current antibiotic resistance epidemic emphasized the importance of coordinated, cross-sectoral measures to combat antibiotic resistance holistically. As the globe faced the dual challenges of infectious illnesses and antimicrobial resistance, prioritizing sensible antibiotic use, investing in novel antibiotic research and development, and upgrading healthcare infrastructure remained critical for public health protection (Hawkins et al. 2022; Muteeb et al. 2023). Moreover, nature has to respond efficiently to the new antibiotics by producing more dangerous antagonists that could lead to the emergence of incurable diseases. To counter this threat to global health, research and stewardship efforts in the fight against antibiotic resistance are crucial (Table 2.1).

Modern medicine relies primarily on the use of antibiotics to fight microbial infections. Unfortunately, the overuse and misadministration of these life-saving drugs have accelerated the development of resistance in microbes (Serwecińska 2020; Muteeb et al. 2023). Experts predict a bleak future: by 2050, around ten million people will die each year from antibiotic-resistant diseases. This development emphasizes the urgent need for prudent use of antibiotics and new therapeutic approaches to combat antibiotic resistance (McLean 2023). If we do not tackle this problem comprehensively, we will jeopardize the effectiveness of important medical interventions and make previously curable infections potentially fatal. To protect public health and preserve the advances of modern medicine, concerted efforts must be taken to maintain the effectiveness of antibiotics. Researchers have observed a significant increase in antibiotic resistance over the last two decades. This worrying trend points to a turning point in the fight against infectious diseases as treatments designed to kill germs become increasingly ineffective. Antimicrobial resistance poses a global threat to patients' lives and the effectiveness of medical therapies. It also highlights the importance of working creatively together to combat the spread

Table 2.1 List of antibiotics that become resistant

Antibiotics	Resistant against bacteria	Affected role/developed behavior	References
Aminoglycosides	*Escherichia coli*, *Pseudomonas aeruginosa*, *Enterococcus faecium*	Enzymatic alteration of the antibiotic or reduced absorption into the bacterial cell	Wang et al. (2022)
Carbapenems	Carbapenem-resistant *Enterobacteriaceae* (CRE) like *Escherichia coli*, *Klebsiella pneumoniae*, *Acinetobacter baumannii*	Developed resistance to carbapenem antibiotics	Sheu et al. (2019)
Cephalosporins	*Escherichia coli*, *Klebsiella pneumoniae*	Bacteria producing extended-spectrum beta-lactamases (ESBLs)	Kumari and Sharma (2017)
Clindamycin	*Staphylococcus aureus*, *Streptococcus pneumoniae*	Used to treat anaerobic bacterial infections, skin infections, and respiratory tract infections	Brook (2012)
Ciprofloxacin	*Escherichia coli*, *Pseudomonas aeruginosa*	Used to treat a variety of bacterial infections and later become resistant	Rehman et al. (2019)
Erythromycin	*Streptococcus pneumoniae*, *Staphylococcus aureus*	Used to treat respiratory tract infections, skin infections, and sexually transmitted infections	Shafia et al. (2016)
Fluoroquinolones	*Escherichia coli*, *Pseudomonas aeruginosa*, *Neisseria gonorrhoeae*	Cause mutations in genes encoding DNA gyrase and topoisomerase IV	Redgrave et al. (2014)
Macrolides	*Streptococcus pneumoniae*, *Staphylococcus aureus*, and *Mycoplasma pneumoniae*	Developed resistance through mechanisms such as ribosomal modification or efflux pumps	Zahari et al. (2023)
Methicillin	*Staphylococcus aureus*	A type of penicillin, leading to the emergence of methicillin-resistant *Staphylococcus aureus* (MRSA)	Gajdács (2019)
Penicillin	*Staphylococcus aureus*, *Streptococcus pneumoniae*	Developed resistance to penicillin through the production of beta-lactamase enzymes or by altering their cell wall structure	Reygaert (2018)
Sulfonamides	*Escherichia coli*, *Enterobacter* species, *Staphylococcus aureus*	Developed mutations in enzymes involved in folate synthesis or through acquisition of resistant genes	Urban-Chmiel et al. (2022)

(continued)

Table 2.1 (continued)

Antibiotics	Resistant against bacteria	Affected role/developed behavior	References
Tetracyclines	*Staphylococcus aureus*, *Enterococcus faecalis*, and various Gram-negative bacterial species	Developed resistance through the acquisition of resistance genes or efflux pumps	Yılmaz and Özcengiz (2017)
Vancomycin	*Enterococcus faecium*, *Staphylococcus aureus*	Developed vancomycin-resistant *Enterococcus* (VRE) and vancomycin-resistant *Staphylococcus aureus* (VRSA) strains	Cong et al. (2020)

of drug-resistant diseases (Ayobami et al. 2022; Mancuso et al. 2021). Treating infections could become more difficult if clinicians do not address the root causes of antimicrobial resistance, which include misuse of antibiotics, lack of infection prevention measures, and lack of new antimicrobials. Therefore, this important public health problem requires the combined efforts of scientists, legislators, and clinicians (Majumder et al. 2020). Jackson et al. expressed serious concerns about the increasing incidence of drug-resistant diseases. They emphasized that this worrying trend points to a major public health problem as viruses evolve to defy established treatments. Dr. Frazee stressed that collaboration between healthcare professionals, governments, and the public is urgently needed to tackle the root causes of antimicrobial resistance. If we fail to act decisively, then we risk jeopardizing patient outcomes and putting a strain on healthcare systems (Jackson et al. 2022).

Penicillin, the first widely used antibiotic, encountered difficulties as bacteria, particularly *Staphylococcus aureus*, developed resistance by producing beta-lactamase enzymes that break down the antibiotic. Methicillin, which was developed to combat penicillin resistance, was put under pressure by the emergence of methicillin-resistant *Staphylococcus aureus* (MRSA), a powerful bacterium that causes difficult-to-treat infections. Vancomycin, which is often used as a last-resort antibiotic, has developed resistance to bacterial strains such as *Enterococcus* and *Staphylococcus aureus*, leading to vancomycin-resistant *Enterococcus* (VRE) and vancomycin-resistant *Staphylococcus aureus* (VRSA). Cephalosporins, which are broad-spectrum antibiotics, are less effective at treating infections because bacteria that produce extended-spectrum beta-lactamases (ESBLs) attack them. Carbapenems, another effective antibiotic used to treat serious infections, are becoming resistant due to carbapenem-resistant *Enterobacteriaceae* (CRE) and pose a significant threat to public health due to limited treatment options. These cases highlight the importance of implementing antibiotic stewardship programs, infection control measures, and research into new antimicrobial agents to effectively combat antibiotic resistance and ensure the efficacy of these vital drugs for future generations.

2.4 Mechanisms of Resistance Development

There are several different types of antimicrobial agents, each with its own mechanism of action against microbes. Agents that prevent the production of cell walls, damage the cell membrane, prevent the synthesis of proteins and nucleic acids, and clog metabolic pathways are the main categories. These mechanisms may originate from the microorganisms themselves or may be acquired from external sources. Enhanced comprehension of these mechanisms is expected to result in improved treatment strategies for infectious diseases and the advancement of antimicrobial agents against microorganisms like bacteria, fungi, viruses, and *Plasmodium* (Reygaert 2018).

2.4.1 Antibiotic Resistance by Bacteria can have Different Mechanisms

2.4.1.1 Antibiotic Destruction

The degradation of antibiotic substances by beta-lactamases is an example of antibiotic resistance. Kapoor et al. (2017) described how these enzymes neutralize antimicrobial agents by binding to the amide group of the lactam ring. Furthermore, modifications to antibiotic-activating enzymes serve as an additional illustration of a resistance mechanism. Changes in the nitroreductase genes (*nfsA* and *nfsB*) stop nitrofurantoin from turning into dangerous intermediates by stopping the enzyme from working normally. This alteration diminishes the effectiveness of nitrofurantoin, thereby reducing its capacity to combat bacterial infections. The aforementioned instances illuminate the complex processes through which bacteria acquire resistance to antibiotics, thereby emphasizing the perpetual competition between microbial innovation and antimicrobial treatments. It is extremely important to understand these mechanisms in order to come up with new ways to fight antibiotic resistance and make sure that antimicrobial therapies work in the fight against infectious diseases (Le et al. 2019).

2.4.1.2 Antibiotic Modification

Clinically significant resistance to aminoglycosides is primarily caused by the enzymatic modification of the antibiotic molecule. Adenylation, acetylation, or phosphorylation of aminoglycosides results in modified antibiotics with reduced affinity for their target. The enzyme that enables these reactions is called an aminoglycoside-modifying enzyme (AME). Through its enzymatic activity, AME promotes the chemical modifications that confer resistance on aminoglycosides and thus reduces their effectiveness in the fight against bacterial infections. This mechanism is a prime example of the complicated relationship between antimicrobial agents and microbial adaptation and highlights the ongoing problem of antibiotic resistance in the clinical setting. Gaining an insight into the processes that contribute to aminoglycoside resistance, including the enzymatic modification facilitated by AME, is critical to developing effective countermeasures and maintaining the clinical utility of these vital antibiotics against bacterial infections (Labby and Garneau-Tsodikova 2013).

2.4.1.3 Target Replacement/Bypass

In *Streptococcus pneumoniae* and *S. aureus*, β-lactam resistance arises from the substitution of bacterial proteins involved in penicillin binding. Similarly, methicillin resistance in *Staphylococcus aureus* demonstrates target replacement. This mechanism involves the substitution of the target protein that methicillin binds to, rendering the antibiotic ineffective. These examples highlight how bacterial pathogens employ sophisticated strategies to evade the effects of antibiotics, underscoring the ongoing challenge of antimicrobial resistance and the need for innovative approaches to combat it effectively (Stapleton and Taylor 2002).

2.4.1.4 Alteration in the Target Site

Modifications of domain V of 23S rRNA are the primary mechanisms promoting linezolid resistance. Bacteria usually possess more than one copy of the genes encoding 23S rRNA. Consequently, there is a direct correlation between the minimum inhibitory concentration (MIC) and the number of mutant alleles. Essentially, an increase in the number of mutated alleles means a corresponding increase in the resistance of the microorganisms. This shows how important it is to understand the genetic changes that cause linezolid resistance. This information will help us find ways to stop resistance and make this antibiotic more effective against bacterial infections (Long and Vester 2012).

2.4.1.5 Protection of the Target Site

Ribosomal protection proteins (RPPs), which secure the target site, demonstrate microbial resistance. This phenomenon has been documented in bacteria of all types, including both Gram-negative and Gram-positive strains. By binding to the ribosome, which is the target of numerous antibiotics, these proteins protect the ribosome from the enzymatic activity of antimicrobial agents. Antibiotics enable the continued synthesis of proteins by bacteria, which confers resistance. The identification of RPPs in a wide variety of bacterial species is remarkable and demonstrates their pervasive significance in mediating resistance mechanisms. The identification of RPPs in Gram-negative and Gram-positive bacteria underscores the adaptive capacity of microbial defense mechanisms and stresses the persistent difficulty of effectively combating antibiotic resistance among a wide range of bacterial pathogens (Connell et al. 2003).

2.4.1.6 Efflux Pump

Bacterial membrane transport proteins, which are located on the cytoplasmic membrane, utilize energy to expel toxic molecules from the cell. The initial documentation of an efflux pump, which is accountable for the expulsion of tetracycline from bacterial cells, appeared in 1980 and was observed in *Escherichia coli*. The genetic blueprint of this pump was discovered to be encoded on a plasmid, which underscores its portability and ability to be transferred. This finding represented a critical juncture in the comprehension of bacterial resistance mechanisms; it illuminated the proactive measures taken by bacteria to counteract the impact of antibiotics through the utilization of complex molecular apparatuses that expel these compounds from their intracellular milieu (Sharma et al. 2019).

2.4.1.7 Target Overproduction

Excessive production of the antibiotic target can lead to resistance by exceeding the number of antibiotic molecules that are readily available. For instance, heightened levels of dihydrofolate reductase (DHFR) have been associated with trimethoprim resistance in *Escherichia coli*. This happens when too much DHFR masks the antimicrobial effect of trimethoprim, letting microbes survive antibiotic exposure. This mechanism highlights the adaptive ability of bacteria to withstand the effects of antibiotics through the enhancement of target enzyme synthesis. Understanding these mechanisms is extremely important for coming up with effective ways to fight antibiotic resistance. This highlights the need for multifaceted approaches to protect the effectiveness of antimicrobial therapies (Kim and Ahn 2022).

2.4.1.8 Modulation of Bacterial Outer Membrane Permeability

The outer membrane of Gram-negative bacteria functions as a barrier, restricting the ingress of substances, such as antibiotics. The reduction in susceptibility to specific antibiotic agents represents a substantial component of intrinsic resistance. Additionally, modifications to the outer membrane's composition may contribute to the development of acquired resistance. Although these changes make the bacteria more resistant to antimicrobials, they also show how complicated and varied the processes are that make Gram-negative bacteria resistant. Understanding these mechanisms is of utmost importance in order to develop efficacious approaches to address antibiotic resistance in clinical environments (Delcour 2009).

2.4.1.9 Epistasis

Bacterial cells can have numerous resistance mutations. Their effects on fitness and other traits, however, defy simple additive modeling. Instead, they are epistasis-shaped, which means that interactions between mutations have an impact on them. Antibiotic resistance can take many different and frequently unexpected forms, due to the fact that these interactions can either increase or decrease the impact of specific mutations. In order to combat antimicrobial resistance in bacterial populations, it is essential to understand the complexities of epistasis and the methods by which bacteria resist antibiotics (Schenk et al. 2013).

2.5 Antibiotic Resistance Exhibited by Fungi

One of the main ways in which fungi develop resistance to antifungal agents is through alterations in the ergosterol biosynthetic enzyme, specifically Erg11 (Cyp51A) substitutions. Certain antifungal agents rely on specific permeases to enter fungal cells. There are two drug efflux systems in fungi: the ATP-binding cassette (ABC) superfamily and the major facilitator superfamily (MFS) that play a role in modulating resistance to azoles. Azole resistance can also be linked to various genomic alterations, such as loss of heterozygosity (LOH), increased chromosome copy number, and segmental or chromosomal aneuploidies. Enhanced mRNA stability is another mechanism that contributes to azole resistance, particularly in the context of the expression of *CDR1*, a drug resistance gene from *Candida albicans*, which encodes a multidrug efflux transporter (Cowen et al. 2015).

Mitochondrial dysfunction has also been associated with azole resistance, as evidenced by the ability of *Saccharomyces cerevisiae* and *Candida glabrata* to survive without complete mitochondrial DNA (formation of "petite" mutants). The *Candida* species are commonly responsible for biofilm infections, with biofilms being a prevalent form of microbial growth in nature. The molecular chaperone heat shock protein 90 (Hsp90) is a crucial cellular regulator that plays a key role in stress responses related to azole tolerance and resistance (Pais et al. 2019).

2.6 Antibiotic Resistance Exhibited by Viruses

In the realm of virology, the development of resistance to antiviral drugs and vaccines typically adheres to a shared mechanism characterized by the selection of random mutations. This process mirrors Darwin's theory of evolution, wherein spontaneous alterations arise and only the most adaptable variants persist—a phenomenon particularly pronounced in virology. When confronted with antiviral treatments during a natural infection, the prevalence of the wild-type virus notably diminishes, whereas the resistant variant—whether originating from a preexisting minor variant or a novel mutation—ascends to predominance. This transition underscores the dynamic interplay between viral populations and selective pressures, wherein the evolutionary advantage conferred by resistance mutations ensures the continued survival and proliferation of the virus, presenting a formidable challenge in the development of effective antiviral therapies and vaccines (Vere Hodge and Field 2011).

2.7 Antibiotic Resistance Exhibited by *Plasmodium*

Parasites that are resistant to chloroquine have a reduced amount of chloroquine accumulation in their digestive vacuole (DV) due to a mechanism that rapidly eliminates the drug through energy-dependent processes. Genetic analysis of crosses between chloroquine-sensitive and chloroquine-resistant strains revealed mutations in the chloroquine resistance transporter gene (*pfcrt*) as the main reason for this resistance. The presence of the *P. falciparum* multidrug resistance 1 transporter gene (*PfMDR1*) contributes to resistance to antimalarial agents that target heme (Chinappi et al. 2010).

Genome-wide association studies (GWASs) have pointed out multiple potential molecular markers associated with piperaquine resistance, such as amplification of the plasmepsin 2 and 3 genes (pm2–3) that play a role in hemoglobin degradation. It has been suggested that increased production of these plasmepsins could impede piperaquine's ability to inhibit heme detoxification processes in the DV. Additionally, K13 mutations are believed to contribute to artemisinin resistance by reducing the activation of artemisinin drugs and/or enhancing the parasite's ability to eliminate damaged proteins (Amato et al. 2017).

2.8 Alarming Consequences of Widespread Antibiotic Misuse

In his Nobel Prize speech, Fleming issued a warning about the potential rise of drug-resistant bacteria, particularly when exposed to suboptimal doses of antibiotics. Disregard for his cautionary words has led to the current crisis in combating infectious diseases, as many once-treatable infections have become untreatable again, regaining their status as the leading causes of human mortality and morbidity. Following Fleming's breakthrough, new antibiotics such as cephalosporin, penicillin, aminoglycoside, tetracycline, and glycopeptides were swiftly developed and incorporated into medical practices (Rosenblatt-Farrell 2009).

Globally, inappropriate or excessive use of antimicrobials plays a significant role in the rise of AMR. Various issues within the healthcare system contribute to this misuse, including lack of updated knowledge on appropriate antimicrobial prescription, the absence of treatment guidelines, insufficient laboratory capabilities for identifying pathogens and their susceptibility to antimicrobials, and inadequate surveillance data on AMR and antimicrobial consumption. Factors like over-the-counter access, lack of effective antimicrobial stewardship programs, and financial incentives for prescribing antibiotics can drive the inappropriate use of these medications. Furthermore, patient and public expectations for antibiotic prescriptions, as well as suboptimal infection prevention practices in healthcare facilities and poor community hygiene and sanitation, contribute to the spread of infections and the escalating use of antimicrobials (Salam et al. 2023).

2.9 Need for Alternative Therapies

The increasing threat of antibiotic resistance highlights the importance of developing alternative treatments for microbial infections in the human body. Antibiotics have been indispensable for the treatment of bacterial infections for several decades. However, the emergence of resistant strains has brought the limitations of conventional antibiotic therapy to the fore. To address this obstacle, scientists are exploring a variety of alternative methods to treat microbial infections. One potentially fruitful direction is the advancement of phage therapy, an approach in which bacteriophages—viral agents that infect and eliminate bacteria—are used as a targeted antimicrobial therapy. Phage therapy offers numerous advantages over conventional antibiotics. These include the ability to selectively target bacteria, bypass mechanisms that lead to resistance to antibiotics, and minimally disrupt the body's beneficial microbiota (Dutta et al. 2021; Hatfull et al. 2022). Based on its efficacy in clinical trials and case studies, phage therapy has shown promise for its potential clinical application in the treatment of infections caused by multidrug-resistant bacteria (Holger et al. 2021). Another method is the use of bacteriocins, which are antimicrobial peptides produced by bacteria to stop the growth of competing microorganisms. Bacteriocins have antimicrobial properties but are only effective against certain types of bacteria. This means that they could be used instead of other antibiotics (Meade et al. 2020). In addition, research into antimicrobial peptides derived from organic sources, including animals and plants, has yielded innovative candidates for the further development of therapeutics (Moretta et al. 2021).

Another interesting approach is immunotherapies, which attempt to strengthen the immune system's ability to fight infections caused by microbes. An example of this is the ability of monoclonal antibodies to directly neutralize pathogens or support immune-mediated clearance by targeting specific microbial antigens. Cytokines and agonists of toll-like receptors are immunomodulatory substances that have the potential to strengthen the host's defenses against infection (Federico et al. 2020).

The field of nanotechnology has changed significantly in its approach to fighting microbial infections, introducing new strategies characterized by exceptional accuracy and efficacy. Nanoparticles, which generally range in size from 1 to 100 nm, have special properties that make them remarkably effective against microbial pathogens. Virulence factors or certain bacterial strains can be selectively combated by functionalized nanoparticles, which achieve this by destroying microbial cell membranes, preventing the formation of biofilms, and disrupting bacterial signaling pathways. In addition, the use of nanomaterials has the potential to optimize the transport of antimicrobial agents to the specific sites of infection, thereby increasing therapeutic efficacy while reducing unintended consequences and adverse effects. Nanotechnology facilitates the production of antimicrobial coatings and surfaces for textiles, medical devices, and implants, thereby reducing the likelihood of healthcare-associated infections. In general, the use of nanotechnology to fight microbial infections is a game changer as it offers unprecedented accuracy, adaptability, and efficacy in the fight against antibiotic-resistant pathogens. Antimicrobial-functionalized nanoparticles are able to selectively destroy the membranes of bacterial cells and thereby prevent the formation of biofilms and improve the transportation of antimicrobial agents to the sites of infection (Sahoo et al. 2022).

2.10 Conclusions

A crucial requirement exists to enable efficient therapy against microbial infections in people, as the prevalence of infections that are resistant to antibiotics continues to rise. Traditional antimicrobial medicines are becoming increasingly ineffective as antibiotic resistance rises worldwide. This threatens our capacity to successfully fight infectious diseases. Comprehensive measures need to be put in place at various levels to tackle this urgent matter. The development of novel antimicrobial drugs and alternative therapies requires substantial investment in R&D, strict legislation to limit antibiotic use and misuse, and reliable monitoring systems to track the emergence of resistant microorganisms. To further reduce the development and spread of antibiotic-resistant strains, it is essential to raise public awareness and educate people on how to responsibly use antibiotics. The development and implementation of effective solutions also depend on healthcare providers, lawmakers, researchers, pharmaceutical companies, and the general public working together. Protecting public health in the present and future will require a comprehensive strategy that includes innovation, stewardship, prevention, and surveillance to overcome the obstacles presented by antibiotic resistance and guarantee that people get the treatment they need for microbial infections on time.

References

Ajulo S, Awosile B (2024) Global antimicrobial resistance and use surveillance system (GLASS 2022): investigating the relationship between antimicrobial resistance and antimicrobial consumption data across the participating countries. PLoS One 19(2):e0297921. https://doi.org/10.1371/journal.pone.0297921

Amato R, Lim P, Miotto O, Amaratunga C, Dek D, Pearson RD, Almagro-Garcia J, Neal AT, Sreng S, Suon S, Drury E, Jyothi D, Stalker J, Kwiatkowski DP, Fairhurst RM (2017) Genetic markers associated with dihydroartemisinin–piperaquine failure in Plasmodium falciparum malaria in Cambodia: a genotype–phenotype association study. Lancet Infect Dis 17(2):164–173. https://doi.org/10.1016/S1473-3099(16)30409-1

Antuori A, Giménez M, Linares G, Cardona PJ (2023) Characterization of respiratory bacterial co-infection and assessment of empirical antibiotic treatment in patients with COVID-19 at hospital admission. Sci Rep 13(1):19302

Ayobami O, Brinkwirth S, Eckmanns T, Markwart R (2022) Antibiotic resistance in hospital-acquired ESKAPE-E infections in low-and lower-middle-income countries: a systematic review and meta-analysis. Emerg Microbes Infect 11(1):443–451

Brook I (2012) Anaerobic bacteria in upper respiratory tract and head and neck infections: microbiology and treatment. Anaerobe 18(2):214–220

Chinappi M, Via A, Marcatili P, Tramontano A (2010) On the mechanism of chloroquine resistance in Plasmodium falciparum. PLoS One 5(11):e14064. https://doi.org/10.1371/journal.pone.0014064

Cong Y, Yang S, Rao X (2020) Vancomycin resistant Staphylococcus aureus infections: a review of case updating and clinical features. J Adv Res 21:169–176

Connell SR, Tracz DM, Nierhaus KH, Taylor DE (2003) Ribosomal protection proteins and their mechanism of tetracycline resistance. Antimicrob Agents Chemother 47(12):3675–3681. https://doi.org/10.1128/AAC.47.12.3675-3681.2003

Cowen LE, Sanglard D, Howard SJ, Rogers PD, Perlin DS (2015) Mechanisms of antifungal drug resistance. Cold Spring Harb Perspect Med 5(7):a019752. https://doi.org/10.1101/cshperspect.a019752

Dadgostar P (2019) Antimicrobial resistance: implications and costs. Infect Drug Resist 12:3903–3910. https://doi.org/10.2147/IDR.S234610

Delcour AH (2009) Outer membrane permeability and antibiotic resistance. Biochim Biophys Acta 1794(5):808–816. https://doi.org/10.1016/j.bbapap.2008.11.005

Di Martino P (2021) Ways to improve biocides for metalworking fluid. AIMS Microbiol 7(1):13–27. https://doi.org/10.3934/microbiol.2021002

Di Martino P (2022) Antimicrobial agents and microbial ecology. AIMS Microbiol 8(1):1
Dutta S, Sarkar R, Bordoloi KP, Deka C, Sonowal PJ (2021) Bacteriophage therapy to combat antibiotic resistance: a brief review. Pharma Innov 10(5):389–394
Federico S, Pozzetti L, Papa A, Carullo G, Gemma S, Butini S, Campiani G, Relitti N (2020) Modulation of the innate immune response by targeting toll-like receptors: a perspective on their agonists and antagonists. J Med Chem 63(22):13466–13513
Gajdács M (2019) The continuing threat of methicillin-resistant Staphylococcus aureus. Antibiotics (Basel) 8(2):52
Harkins CP, Pichon B, Doumith M, Parkhill J, Westh H, Tomasz A, de Lencastre H, Bentley SD, Kearns AM, Holden MT (2017) Methicillin-resistant Staphylococcus aureus emerged long before the introduction of methicillin into clinical practice. Genome Biol 18(1):130
Hatfull GF, Dedrick RM, Schooley RT (2022) Phage therapy for antibiotic-resistant bacterial infections. Annu Rev Med 73:197–211
Hawkins O, Scott AM, Montgomery A, Nicholas B, Mullan J, van Oijen A, Degeling C (2022) Comparing public attitudes, knowledge, beliefs and behaviours towards antibiotics and antimicrobial resistance in Australia, United Kingdom, and Sweden (2010-2021): a systematic review, meta-analysis, and comparative policy analysis. PLoS One 17(1):e0261917
Holger D, Kebriaei R, Morrisette T, Lev K, Alexander J, Rybak M (2021) Clinical pharmacology of bacteriophage therapy: a focus on multidrug-resistant pseudomonas aeruginosa infections. Antibiotics (Basel) 10(5):556
Jackson N, Belmont CR, Tarlton NJ, Allegretti YH, Adams-Sapper S, Huang YY, Borges CA, Frazee BW, Florence-Petrovic D, Hufana C, Parker A (2022) Genetic predictive factors for nonsusceptible phenotypes and multidrug resistance in expanded-Spectrum cephalosporin-resistant uropathogenic Escherichia coli from a multicenter cohort: insights into the phenotypic and genetic basis of coresistance. mSphere 7(6):e00471–e00422
Kapoor G, Saigal S, Elongavan A (2017) Action and resistance mechanisms of antibiotics: a guide for clinicians. J Anaesthesiol Clin Pharmacol 33(3):300. https://doi.org/10.4103/joacp.JOACP_349_15
Kim J, Ahn J (2022) Emergence and spread of antibiotic-resistant foodborne pathogens from farm to table. Food Sci Biotechnol 31(12):1481–1499. https://doi.org/10.1007/s10068-022-01157-1
Kumari S, Sharma S (2017) Extended-spectrum beta-lactamases in urinary isolates of Escherichia coli, Klebsiella pneumoniae and other gram-negative bacteria in tertiary care. Int J Contemp Microbiol 3(2):5
Labby KJ, Garneau-Tsodikova S (2013) Strategies to overcome the action of aminoglycoside-modifying enzymes for treating resistant bacterial infections. Future Med Chem 5(11):1285–1309. https://doi.org/10.4155/fmc.13.80
Le VVH, Davies IG, Moon CD, Wheeler D, Biggs PJ, Rakonjac J (2019) Novel 5-Nitrofuran-activating reductase in Escherichia coli. Antimicrob Agents Chemother 63(11):e00868–e00819. https://doi.org/10.1128/AAC.00868-19
Long KS, Vester B (2012) Resistance to linezolid caused by modifications at its binding site on the ribosome. Antimicrob Agents Chemother 56(2):603–612. https://doi.org/10.1128/AAC.05702-11
Majumder MA, Rahman S, Cohall D, Bharatha A, Singh K, Haque M, Gittens-St Hilaire M (2020) Antimicrobial stewardship: fighting antimicrobial resistance and protecting global public health. Infect Drug Resist 13:4713–4738
Mancuso G, Midiri A, Gerace E, Biondo C (2021) Bacterial antibiotic resistance: the most critical pathogens. Pathogens 10(10):1310
McLean S (2023) What is antimicrobial resistance and why is it a problem? In: Pre and probiotics for poultry gut health. CABI, pp 1–17
Meade E, Slattery MA, Garvey M (2020) Bacteriocins, potent antimicrobial peptides and the fight against multi drug resistant species: resistance is futile? Antibiotics (Basel) 9(1):32
Moretta A, Scieuzo C, Petrone AM, Salvia R, Manniello MD, Franco A, Lucchetti D, Vassallo A, Vogel H, Sgambato A, Falabella P (2021) Antimicrobial peptides: a new hope in biomedical and pharmaceutical fields. Front Cell Infect Microbiol 11:668632

Murray CJL, Ikuta KS, Sharara F, Swetschinski L, Robles Aguilar G, Gray A, Han C, Bisignano C, Rao P, Wool E, Johnson SC, Browne AJ, Chipeta MG, Fell F, Hackett S, Haines-Woodhouse G, KashefHamadani BH, Kumaran EAP, McManigal B et al (2022) Global burden of bacterial antimicrobial resistance in 2019: a systematic analysis. Lancet 399(10325):629–655. https://doi.org/10.1016/S0140-6736(21)02724-0

Murugaiyan J, Kumar PA, Rao GS, Iskandar K, Hawser S, Hays JP, Mohsen Y, Adukkadukkam S, Awuah WA, Jose RA, Sylvia N (2022) Progress in alternative strategies to combat antimicrobial resistance: focus on antibiotics. Antibiotics (Basel) 11(2):200

Muteeb G, Rehman MT, Shahwan M, Aatif M (2023) Origin of antibiotics and antibiotic resistance, and their impacts on drug development: a narrative review. Pharmaceuticals (Basel) 16(11):1615

Nadeem SF, Gohar UF, Tahir SF, Mukhtar H, Pornpukdeewattana S, Nukthamna P, Moula Ali AM, Bavisetty SC, Massa S (2020) Antimicrobial resistance: more than 70 years of war between humans and bacteria. Crit Rev Microbiol 46(5):578–599

Pais P, Galocha M, Viana R, Cavalheiro M, Pereira D, Teixeira MC (2019) Microevolution of the pathogenic yeasts Candida albicans and Candida glabrata during antifungal therapy and host infection. Microb Cell 6(3):142–159. https://doi.org/10.15698/mic2019.03.670

Panda SK, Buroni S, Swain SS, Bonacorsi A, da Fonseca Amorim EA, Kulshrestha M, da Silva LCN, Tiwari V (2022) Recent advances to combat ESKAPE pathogens with special reference to essential oils. Front Microbiol 13:1029098. https://doi.org/10.3389/FMICB.2022.1029098/BIBTEX

Poudel AN, Zhu S, Cooper N, Little P, Tarrant C, Hickman M, Yao G (2023) The economic burden of antibiotic resistance: a systematic review and meta-analysis. PLoS One 18(5):e0285170. https://doi.org/10.1371/journal.pone.0285170

Prestinaci F, Pezzotti P, Pantosti A (2015) Antimicrobial resistance: a global multifaceted phenomenon. Pathog Glob Health 109(7):309. https://doi.org/10.1179/2047773215Y.0000000030

Purssell E (2020) Understanding pharmacology in nursing practice. Springer International Publishing

Rayan RA (2023) Pharmaceutical effluent evokes superbugs in the environment: a call to action. Biosaf Health 5(6):363–371

Redgrave LS, Sutton SB, Webber MA, Piddock LJ (2014) Fluoroquinolone resistance: mechanisms, impact on bacteria, and role in evolutionary success. Trends Microbiol 22(8):438–445

Rehman A, Patrick WM, Lamont IL (2019) Mechanisms of ciprofloxacin resistance in Pseudomonas aeruginosa: new approaches to an old problem. J Med Microbiol 68(1):1–10

Reygaert WC (2018) An overview of the antimicrobial resistance mechanisms of bacteria. AIMS Microbiol 4(3):482–501. https://doi.org/10.3934/microbiol.2018.3.482

Romani M, Warscheid T, Nicole L, Marcon L, Di Martino P, Suzuki MT, Lebaron P, Lami R (2022) Current and future chemical treatments to fight biodeterioration of outdoor building materials and associated biofilms: moving away from ecotoxic and towards efficient, sustainable solutions. Sci Total Environ 802:149846. https://doi.org/10.1016/j.scitotenv.2021.149846

Rosenblatt-Farrell N (2009) The landscape of antibiotic resistance. Environ Health Perspect 117(6):A244. https://doi.org/10.1289/ehp.117-a244

Roy Chowdhury P, Demaere M, Chapman T, Worden P, Charles IG, Darling AE, Djordjevic SP (2016) Comparative genomic analysis of toxin-negative strains of Clostridium difficile from humans and animals with symptoms of gastrointestinal disease. BMC Microbiol 16(1):1–13. https://doi.org/10.1186/S12866-016-0653-3/TABLES/3

Sahoo J, Sarkhel S, Mukherjee N, Jaiswal A (2022) Nanomaterial-based antimicrobial coating for biomedical implants: new age solution for biofilm-associated infections. ACS Omega 7(50):45962–45980

Salam MA, Al-Amin MY, Salam MT, Pawar JS, Akhter N, Rabaan AA, Alqumber MAA (2023) Antimicrobial resistance: a growing serious threat for global public health. Healthcare 11(13):1946. https://doi.org/10.3390/healthcare11131946

Sarfraz I, Rasul A, Hussain G, Hussain SM, Samiullah K, Rasool B, Riaz A, Asrar M, Selamoglu Z (2020) Global and temporal trends in the use of antibiotics and spread of antimicrobial resis-

tance. In: Antibiotics and antimicrobial resistance genes: environmental occurrence and treatment technologies, pp 81–94
Schenk MF, Szendro IG, Salverda MLM, Krug J, de Visser JAGM (2013) Patterns of epistasis between beneficial mutations in an antibiotic resistance gene. Mol Biol Evol 30(8):1779–1787. https://doi.org/10.1093/molbev/mst096
Serwecińska L (2020) Antimicrobials and antibiotic-resistant bacteria: a risk to the environment and to public health. Water 12(12):3313
Shafia S, Chandluri P, Ganpisetti R, Lakshmi BV, Swami PA (2016) Erythromycin use as broad spectrum anti biotic. World J Phram Med Res 2:23–26
Sharma A, Gupta V, Pathania R (2019) Efflux pump inhibitors for bacterial pathogens: from bench to bedside. Indian J Med Res 149(2):129. https://doi.org/10.4103/ijmr.IJMR_2079_17
Sheu CC, Chang YT, Lin SY, Chen YH, Hsueh PR (2019) Infections caused by carbapenem-resistant Enterobacteriaceae: an update on therapeutic options. Front Microbiol 10:80
Stapleton PD, Taylor PW (2002) Methicillin resistance in Staphylococcus aureus: mechanisms and modulation. Sci Prog 85(1):57–72. https://doi.org/10.3184/003685002783238870
Urban-Chmiel R, Marek A, Stępień-Pyśniak D, Wieczorek K, Dec M, Nowaczek A, Osek J (2022) Antibiotic resistance in bacteria—a review. Antibiotics (Basel) 11(8):1079
Velazquez-Meza ME, Galarde-López M, Carrillo-Quiróz B, Alpuche-Aranda CM (2022) Antimicrobial resistance: one health approach. Vet World 15(3):743. https://doi.org/10.14202/VETWORLD.2022.743-749
Vere Hodge A, Field HJ (2011) General mechanisms of antiviral resistance. In: Genetics and evolution of infectious disease. Elsevier, pp 339–362. https://doi.org/10.1016/B978-0-12-384890-1.00013-3
Walsh TR, Gales AC, Laxminarayan R, Dodd PC (2023) Antimicrobial resistance: addressing a global threat to humanity. PLoS Med 20(7):e1004264. https://doi.org/10.1371/JOURNAL.PMED.1004264
Wang N, Luo J, Deng F, Huang Y, Zhou H (2022) Antibiotic combination therapy: a strategy to overcome bacterial resistance to aminoglycoside antibiotics. Front Pharmacol 13:839808
Yılmaz Ç, Özcengiz G (2017) Antibiotics: pharmacokinetics, toxicity, resistance and multidrug efflux pumps. Biochem Pharmacol 133:43–62
Zahari NI, Engku Abd Rahman EN, Irekeola AA, Ahmed N, Rabaan AA, Alotaibi J, Alqahtani SA, Halawi MY, Alamri IA, Almogbel MS, Alfaraj AH (2023) A review of the resistance mechanisms for β-lactams, macrolides and fluoroquinolones among Streptococcus pneumoniae. Medicina (Kaunas) 59(11):1927

Navigating the Challenges of Antibiotic Resistance: Origins, Mechanisms, and Global Responses

3

Lakshmi Jyothi Tadi and Shazia Naaz

Abstract

Antibiotic resistance is a global health issue, which undermines the efficacy of antibiotics and poses a challenge to healthcare systems worldwide. This abstract provides a brief overview of the origins, mechanisms, and evolution of antibiotic resistance as well as its multidimensional contributors, along with a snippet of information on strategies for its alleviation. The rise in antibiotic resistance is attributed to the rise and spread of antibiotic-resistant strains, mainly due to the selective pressure exerted by global antibiotic misuse. Furthermore, other mechanisms like horizontal gene transfer, mutations, and the presence of as environmental hotspots of resistance genes also contribute to the spread of resistance. The primary causes for the spread of antibiotic resistance include overprescription of antibiotics, lack of surveillance tools in healthcare systems, inadequacies in infection control practices, and antibiotic use in agriculture, livestock, and fisheries. Moreover, the fact that the pipeline for new antibiotics is practically dry further limits the available treatment options for multidrug-resistant infections. A multifaceted approach is needed for combatting antimicrobial resistance (AMR), which encompasses meticulous antibiotic stewardship, a dedicated infection control team, and one healthcare approach with an emphasis on reducing environmental misuse of antibiotics. Collaboration among healthcare providers, policymakers, researchers, and the public is essential to combatting antibiotic resistance and preserving the efficacy of antibiotics for future generations. Hence, it can be concluded that addressing antibiotic resistance requires focused

L. J. Tadi (✉)
Department of Microbiology, AIIMS Bibinagar, Hyderabad, Telangana, India

S. Naaz
Mamata Academy of Medical Science, Hyderabad, Telangana, India

M. Gangwar, G. Nath (eds.), *Emerging Paradigms for Antibiotic-Resistant Infections: Beyond the Pill*, https://doi.org/10.1007/978-981-97-5272-0_3

global action and a commitment to sustainable practices, to safeguard public health.

Keywords
Antibiotic resistance · Mechanisms of resistance · AMR drivers · Global initiatives

3.1 Introduction

The escalating phenomenon of antibiotic resistance presents a daunting challenge to global public health, posing a significant threat to the efficacy of once-potent treatments against bacterial infections. It is imperative to grasp the multifaceted causes and ramifications of this crisis to formulate effective strategies for its containment. This chapter delves into the origins, evolution, and global factors contributing to antibiotic resistance while also examining the evolving responses to this urgent issue.

3.1.1 Origins and Evolution of Antibiotic Resistance

One of the greatest discoveries of all times has long been thought to be the discovery of antibiotics (Davies and Davies 2010). Antibiotic resistance is not a new phenomenon but has been long-standing since as early as the 1930s (Davies and Davies 2010). History shows that bacteria have been an important causal factor in causing diseases and mortality (MacLean and San Millan 2019). Multiple resistance mechanisms amongst bacteria pose a major challenge to treatment by endangering the capacity of antibiotics. By 1945, Sir Alexander Flemming suggested that there will be an era of demand and abuse of antibiotics (Ventola 2015). Bacteria, in their constant struggle for survival, have developed intrinsic mechanisms to evade the effects of antibiotics. One of the key factors of antibiotic resistance is horizontal gene transfer, where resistant genes are transmitted between bacteria, which causes rapid dissemination of resistance traits within populations (Tao et al. 2022).

Moreover, any change in gene expression can lead to increased virulence in bacterial genomes, whereas horizontal gene transfer and mutations have a greater impact by spreading these resistant genes. The presence of an environmental pool of resistance genes further exacerbates the spread of resistance, as bacteria can acquire these genes through various mechanisms (Ventola 2015).

Determining and perceiving the mechanisms of evolution of antibiotic resistance is crucial for developing effective strategies to combat this global threat. By unfolding the mechanisms underlying resistance, emergence, and dissemination, researchers can identify targets and design novel approaches to curb the impact of antibiotic resistance on public health.

3.2 Mechanisms of Antibiotic Resistance

3.2.1 Enzymatic Mechanisms

3.2.1.1 Enzymatic Inactivation

This was one of the first mechanisms to be discovered. Some bacteria produce enzymes that either modify or destroy antibiotics, rendering them ineffective. For example, β-lactamases hydrolyze the β-lactam ring in antibiotics like penicillins and cephalosporins (Egorov et al. 2018).

3.2.1.2 Enzymatic Modification

Over time, bacteria may acquire enzymes that modify the target of the antibiotic like the *erm* gene that codes for methylation of part of a chromosome, which serves as an attachment site for erythromycin. Moreover, the aminoglycoside-modifying enzyme confers high resistance on aminoglycosides (Munita and Arias 2016).

3.2.2 Alteration of Target Sites

A common mechanism that bacteria use is to modify the target sites that antibiotics bind to, making them less susceptible to the antibiotic's effects. This alteration can occur through genetic mutations encoding target proteins or by acquiring resistant genes from other organisms, e.g., the altered penicillin-binding site in methicillin-resistant *Staphylococcus aureus* (MRSA) and *Streptococcus pneumoniae* (Munita and Arias 2016).

3.2.3 Efflux Pumps and Decreased Permeability

Bacteria may develop mechanisms to reduce the uptake of antibiotics into their cells. This can involve changes in membrane permeability or the expression of efflux pumps that actively remove antibiotics from the bacterial cell. Using both mechanisms, bacteria reduce the accumulation of antibiotics and lower the intracellular antibiotic concentration.

3.2.4 Modification of Metabolic Pathways

Some bacteria can alter the metabolic pathways affected by antibiotics. By adapting to this mechanism, the resistance that arises due to acquired mutational changes is varied and complex in nature (Munita and Arias 2016).

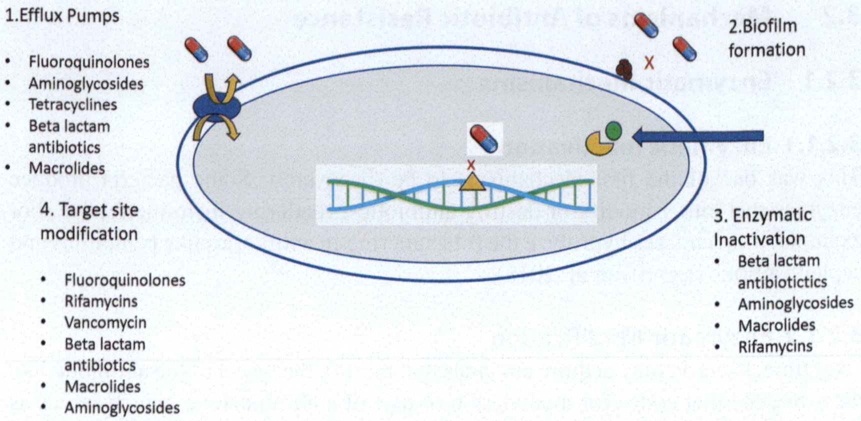

Fig. 3.1 Illustration of the important mechanisms of antimicrobial resistance

3.2.5 Biofilm Formation

Biofilms contain a sessile population embedded in a matrix that makes the colonies highly resistant to antibiotics. The protective matrix of biofilms can prevent antibiotics from reaching their target sites effectively (Roy et al. 2018).

3.2.6 Horizontal Gene Transfer

Bacteria can acquire resistance genes from other bacteria through processes such as conjugation, transformation, or transduction. This horizontal transfer of genetic material allows for rapid spread of antibiotic resistance within bacterial populations (Ventola 2015).

Basic knowledge about the mechanisms of antibiotic resistance is imperative for developing strategies to combat it and preserve the effectiveness of antibiotics for future generations (Fig. 3.1).

3.3 Global Contributors to Antibiotic Resistance

Antibiotic resistance initially emerged in hospitals where most of the antibiotics were being used (Levy and Marshall 2004). The overprescription of antibiotics, their over-the-counter availability, agricultural use of antibiotics, and poor sanitation have all accelerated the spread of resistant bacteria (Levy and Marshall 2004). This multidimensional scenario is further worsened by the dire scarcity of newer treatment options for multidrug-resistant organisms (MDROs) (Chinemerem Nwobodo et al. 2022).

Antimicrobial resistance has multifactorial causes and is influenced by several variables like increased transfer of patients in facilities, suboptimal infection control practices (poor adherence to hand hygiene measures, lack of adherence to disinfection policies), increased international trade and travel, overuse of broad-spectrum antibiotics, etc. (Chinemerem Nwobodo et al. 2022). The perils of antibiotic resistance disperse far beyond the confines of healthcare facilities and spill over into broader societal dimensions. The financial burden incurred by escalating healthcare expenditures in the wake of antibiotic resistance poses significant challenges to healthcare systems worldwide.

A multidisciplinary approach that incorporates insights from diverse fields, including public health, veterinary medicine, and environmental science, is essential to effectively combat the global impact of antibiotic resistance. By recognizing and addressing the global contributors to antibiotic resistance, stakeholders can pave the way for collaborative solutions that preserve the efficacy of antibiotics and safeguard public health on a global scale.

Moreover, understanding the varied spectrum of antibiotic resistance helps frame comprehensive strategies that encompass judicious antibiotic stewardship, enhanced infection control measures, and sustainable agricultural practices (Fig. 3.2). It is only through concerted global effort and firm commitment that we can hope to stop the tide of antibiotic resistance and safeguard the health and well-being of present and future generations.

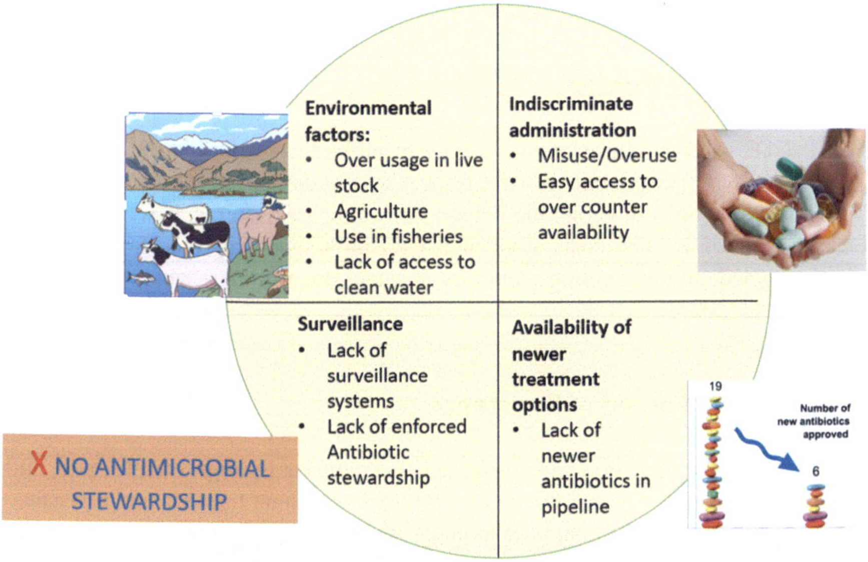

Fig. 3.2 Global contributors to antibiotic resistance

3.4 Impact of Antibiotic Resistance

Antibiotic resistance has affected both morbidity and mortality, and the financial burden is also quite difficult to estimate. Every year, two million people are affected by antibiotic-resistant infections as per the estimates by the US Centers for Disease Control and Prevention (CDC), with at least 23,000 dying due to such infections (Prestinaci et al. 2015). It is stated by multiple studies that the following has largely been contributed by antibiotic resistance:

3.4.1 Increased Morbidity and Mortality

The rise of antibiotic-resistant strains in patients undergoing chemotherapy, various invasive surgeries like hip replacement, transplants, intensive care unit (ICU)-admitted preterm neonates, etc. has led to an increase in morbidity and mortality in such cases. This entails more severe infections, prolonged hospital stays, and increased healthcare costs.

3.4.2 Limited Treatment Options

Antibiotic resistance has limited the number of effective treatment options, leading to therapy using a combination of antibiotics and broad-spectrum antibiotics; these antibiotics may be difficult to tolerate, may require adjustment of both hepatic and renal parameters, and still may not be completely effective, further contributing to the development of resistance (Alaoui Mdarhri et al. 2022).

3.4.3 Financial Burden

The US CDC estimates the economic burden of antibiotic resistance to be US$2371.4 to +US$29,289.1 per patient episode. Hence, we see that antibiotic-resistant infections place a significant economic burden on healthcare systems as well as on the society as a whole. The costs associated with prolonged length of hospital stay, additional medical interventions, and loss of wages due to illness can be substantial (Poudel et al. 2023).

3.4.4 Global Spread of Resistance

Antibiotic-resistant bacteria can spread across countries through travel, trade, and migration, making it a global public health threat. Resistant bacteria and resistance genes can transfer between humans, animals, and the environment, exacerbating the spread of resistance (Laxminarayan and Chaudhury 2016).

3.4.5 The Undermining of Progress in Medicine

Decades of progress in the field of medicine, including the prevention and treatment of infectious diseases as well as increases in life expectancy, are threatened by the advent of antibiotic resistance. Regular medical operations and common diseases become riskier and potentially fatal without adequate antibiotics (Chinemerem Nwobodo et al. 2022).

There lies a need to address the pressing impact of antibiotic resistance, which requires coordinated efforts at the local, national, and international levels, including promoting antibiotic stewardship, investing in the research and development of new antibiotics, improving infection prevention and control measures, and raising awareness among healthcare providers and the general public.

3.5 Global Initiatives for AMR

Global initiatives have been adopted to tackle the rising AMR. A few important initiatives are listed below:

3.5.1 World Health Organization (WHO) Global Antimicrobial Resistance and Use Surveillance System (GLASS)

The WHO member states approved the Global Action Plan (GAP) for AMR in 2015, to combat the rising antimicrobial resistance. The goal of GAP-AMR was to prevent infections and ensure safe and effective treatment. Furthermore, as surveillance is an essential tool for monitoring the aforesaid task, GLASS was launched to strengthen surveillance and research and optimize the use of antibiotics in human and animal health and environment (WHO, GLASS Report 2021).

3.5.2 World Antibiotic Awareness Week

This is an initiative taken by WHO. It is an annual campaign that aims to create and increase awareness of antibiotic resistance and promote best practices for antibiotic use across various sectors: healthcare providers, policymakers, the general public, and animal health environment. It is celebrated each year between the 18th and the 24th of November (World AMR awareness week 2023).

3.5.3 Global Antibiotic Resistance Partnership (GARP)

This is a collaborative partnership between the Center for Disease Dynamics, Economics & Policy (CDDEP) and partners worldwide. These partners are actively working through GARP to develop country-specific evidence to guide policies and interventions to tackle antibiotic resistance across low/middle-income countries (GARP, One Health Trust 2023).

3.5.4 The Global Health Security Agenda (GHSA)

GHSA is a partnership of nations, international organizations, and other stakeholders working to strengthen global health security and preparedness. More than 70 countries have signed this partnership. One of the key objectives is to combat antibiotic resistance (CDC and the Global Health Security Agenda 2021).

3.5.5 The United Nations (UN) Interagency Coordination Group (IACG) on Antimicrobial Resistance

Established in 2016, IACG brings together high-level representatives from UN agencies, international organizations, and other stakeholders (experts across different sectors like animal health, agriculture, and environment) to provide guidance and recommendations on antimicrobial resistance (Interagency Coordination Group on Antimicrobial Resistance secretary-general 2017).

3.5.6 Combating Antibiotic-Resistant Bacteria (CARB) Task Force (USA)

Its inception saw the establishment of coordinated efforts across government agencies to address antibiotic resistance through research, surveillance, and prevention strategies. It has five goals, namely, reducing the spread of resistant organisms, strengthening National One Health Surveillance Efforts to combat resistance, advancing development and using rapid tests for disease diagnosis, accelerating research and development for initiating appropriate therapy, and improving international collaboration (U.S. National Action Plan for Combating antibiotic-resistant Bacteria 2021).

3.5.7 European Antibiotic Awareness Day (EAAD)

This is a European health initiative coordinated by the European Centre for Disease Prevention and Control (ECDC). It is an annual initiative similar to World Antibiotic Awareness Week, which was started to raise awareness of antibiotic resistance and promote prudent antibiotic use across Europe. This year, one health approach was taken up (European Antibiotic Awareness Day (EAAD) 2023).

Hence, such collaborative efforts & its strengthening, support from the government, and coming together of various realms of working classes is required to deal with and reduce the burden of antibiotic resistance.

3.6 Conclusions

To summarize, the emergence of antibiotic resistance poses an immense global health challenge, demanding coordinated action on a worldwide scale. By comprehending the underlying causes, evolutionary causes, and global drivers of antibiotic resistance, various nations can create evidence-based strategies to preserve the efficacy of antibiotics and hence preserve the well-being of our society.

References

Alaoui Mdarhri H, Benmessaoud R, Yacoubi H, Seffar L, Guennouni Assimi H, Hamam M, Boussettine R, Filali-Ansari N, Lahlou FA, Diawara I, Ennaji MM, Kettani-Halabi M (2022) Alternatives therapeutic approaches to conventional antibiotics: advantages, limitations and potential application in medicine. Antibiotics (Basel) 11(12):1826. https://doi.org/10.3390/antibiotics11121826

Centers for Disease Control and Prevention (2021) CDC and the global health security agenda. Centers for Disease Control and Prevention. https://www.cdc.gov/globalhealth/security/index.htm

Chinemerem Nwobodo D, Ugwu MC, OliselokeAnie C, Al-Ouqaili MTS, ChineduIkem J, Victor Chigozie U, Saki M (2022) Antibiotic resistance: the challenges and some emerging strategies for tackling a global menace. J Clin Lab Anal 36(9):e24655. https://doi.org/10.1002/jcla.24655

Davies J, Davies D (2010) Origins and evolution of antibiotic resistance. Microbiol Mol Biol Rev 74(3):417–433. https://doi.org/10.1128/mmbr.00016-10

Egorov AM, Ulyashova MM, Rubtsova MY (2018) Bacterial enzymes and antibiotic resistance. Acta Naturae 10(4):33–48. https://doi.org/10.32607/20758251-2018-10-4-33-48

European Antibiotic Awareness Day (EAAD) (2023). https://www.ecdc.europa.eu/en/news-events/european-antibiotic-awareness-day-eaad-2023

Global antibiotic resistance partnership. One Health Trust (2023). https://onehealthtrust.org/projects/global-antibiotic-resistance-partnership/

Interagency Coordination Group on Antimicrobial Resistance secretary-general (2017). https://www.un.org/sg/en/content/sg/personnel-appointments/2017-03-17/interagency-coordination-group-antimicrobial-resistance

Laxminarayan R, Chaudhury RR (2016) Antibiotic resistance in India: drivers and opportunities for action. PLoS Med 13(3):e1001974. https://doi.org/10.1371/journal.pmed.1001974

Levy SB, Marshall B (2004) Antibacterial resistance worldwide: causes, challenges and responses. Nat Med 10(12 Suppl):S122–S129. https://doi.org/10.1038/nm1145

MacLean RC, San Millan A (2019) The evolution of antibiotic resistance. Science 365(6458):1082–1083. https://doi.org/10.1126/science.aax3879

Munita JM, Arias CA (2016) Mechanisms of antibiotic resistance. Microbiol Spectr 4(2). https://doi.org/10.1128/microbiolspec.VMBF-0016-2015

Poudel AN, Zhu S, Cooper N, Little P, Tarrant C, Hickman M, Yao G (2023) The economic burden of antibiotic resistance: a systematic review and meta-analysis. PLoS One 18(5):e0285170. https://doi.org/10.1371/journal.pone.0285170

Prestinaci F, Pezzotti P, Pantosti A (2015) Antimicrobial resistance: a global multifaceted phenomenon. Pathog Glob Health 109(7):309–318. https://doi.org/10.1179/2047773215Y.0000000030

Roy R, Tiwari M, Donelli G, Tiwari V (2018) Strategies for combating bacterial biofilms: a focus on anti-biofilm agents and their mechanisms of action. Virulence 9(1):522–554. https://doi.org/10.1080/21505594.2017.1313372

Tao S, Chen H, Li N, Wang T, Liang W (2022) The spread of antibiotic resistance genes in vivo model. Can J Infect Dis Med Microbiol 2022:3348695. https://doi.org/10.1155/2022/3348695

U.S. National Action Plan for combating antibiotic-resistant bacteria (2021). https://www.cdc.gov/drugresistance/us-activities/national-action-plan.html

Ventola CL (2015) The antibiotic resistance crisis: part 1: causes and threats. P T 40(4):277–283

World AMR Awareness Week (2023). https://www.who.int/news-room/events/detail/2023/11/18/default-calendar/world-amr-awareness-week-2023

World Health Organization (2021) Global Antimicrobial Resistance and Use Surveillance System (GLASS) report: 2021. https://www.who.int/publications/i/item/9789240027336

The Rise of Antibiotic Resistance: A "Silent Killer"

4

Kulsoom and Amit Lather

Abstract

Antibiotics have been a transformative discovery in medicine, significantly reducing mortality and morbidity from infectious diseases. Before antibiotics, infectious disease was among the leading cause of death. In 1928, Alexander Fleming, a professor of bacteriology, discovered penicillin, the first true antibiotic. However, it took over a decade for penicillin to be introduced as a treatment for bacterial infections. The period from 1940 to 1962 is often referred to as the "golden era" of antibiotics. For many years, antibiotics effectively saved countless lives.

Unfortunately, the future of antibiotics is now uncertain due to emergence of drug-resistant bacteria, which pose substantial challenges to human health. These resistance bacteria can cause life-threatening blood infections and complicate the treatment of various common infections. According to the World Health Organization (WHO), an estimated 5.2 million people in the Western Pacific region will die due to drug-resistant bacterial infections. This increase in resistance is attributed to changes in bacteria that render antibiotics less effective. Resistance is rising for both first-line drugs (such as ciprofloxacin, ampicillin, and co-trimoxazole) and second-line treatments (including fluoroquinolones). Antimicrobial resistance (AMR) now kills more people than cancers and road accidents combined. Given this context, antibiotic resistance can be considered a "silent killer."

Kulsoom (✉)
School of Life science and Biotechnology, Chhatrapati Shahu Ji Maharaj University, Kalyanpur, Kanpur, Uttar Pradesh, India

A. Lather
MM College of Pharmacy, Maharishi Markandeshwar (deemed to be) University, Ambala, Haryana, India

M. Gangwar, G. Nath (eds.), *Emerging Paradigms for Antibiotic-Resistant Infections: Beyond the Pill*, https://doi.org/10.1007/978-981-97-5272-0_4

Keywords

Antibiotic resistance · Superbug · Super resistivity · Antibiotic era · Strategies of bacterial resistivity

4.1 Introduction

Antibiotic resistance is a unique characteristic of bacteria that allows them to overcome the mechanisms of drugs designed to kill them. The discovery of antibiotics has transformed both human and animal health. Antibiotics, which are chemical substances produced primarily by living microorganism, are used to treat bacterial infections. They are the most momentous class of pharmaceuticals and one of the most influential medical inventions of the twentieth century. Over the years, antibiotics have undeniably saved millions of lives, proving invaluable in the fight against bacterial infections. They represent the most significant class of pharmaceuticals and one of the most influential medical inventions of the twentieth century. They function by either killing bacteria or preventing them from reproducing and spreading (Halawa et al. 2024).

Antibiotics can treat a wide range of health conditions, from minor to severe. In our daily lives, when we fall ill and seek medical advice, we are often prescribed antibiotics to aid our recovery. More than 100 different antibiotics are available to cure numerous diseases, both minor and life-threatening. Despite over a century of antibiotics research and development, the world is now a rapidly running out of these lifesaving drugs.

Bacteria, having evolved for more than three billon years, possess the ability to adapt and modify themselves for survival. This process leads to the development of antibiotic resistance, where bacteria that were previously sensitive to a particular drug become resistant to its effects. This phenomenon poses a significant challenge to treat infections and underscore the urgent need for new antibiotics and strategies to combat resistance.

Antimicrobial resistance is best described as the acquired ability of bacteria to survive or resist the effects of antibiotics that previously inhibited their growth or killed them. This resistance mechanism arises due to changes in the bacterial genome, allowing bacteria to combat the effects of antibiotics.

Antibiotic resistance poses significant challenges, leading to treatment failures, persistent disease symptoms, and increased mortality rates, all of which contribute to higher medical costs. This chapter addresses the mechanisms of resistance, surveillance of bacterial resistance, the current situation, and future approaches for antibiotics. According to the World Health Organization (WHO), antibiotic resistance is a growing global threat, leaving communities with infections that are increasingly difficult or impossible to treat.

A 2022 structured analysis on antimicrobial resistance (AMR) estimated that out of 4.95 million deaths in 204 countries, 1.27 million were attributable to AMR. The WHO has emphasized the urgent need for action to prevent a post-antibiotic era.

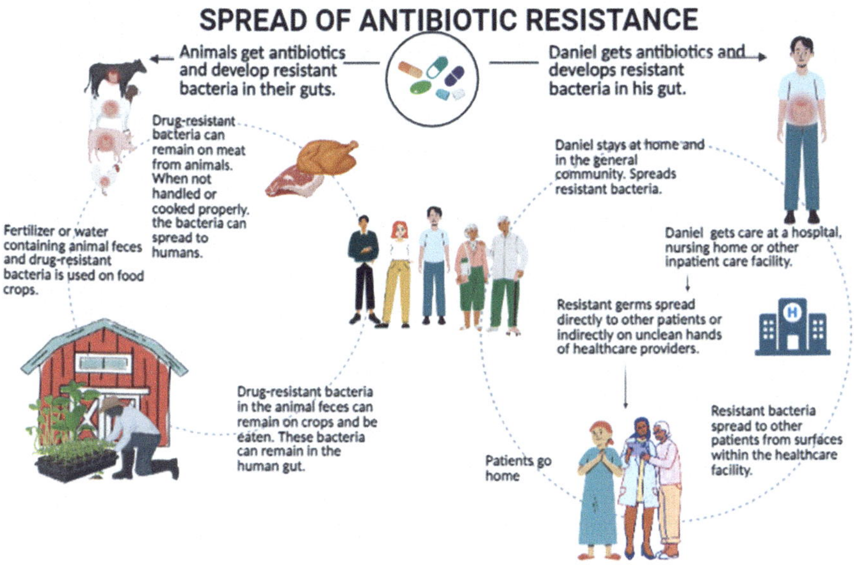

Fig. 4.1 A web of antiobic-resistant development and spread

In the United States, several entities, including centers within the Food and Drug Administration (FAD), are developing strategies to combat antimicrobial resistance. Many researchers refer to resistant bacteria as "superbugs" due to their ability to adapt and evolve, showing resistance to multiple drugs and addressing antibiotic resistance already cost millions of dollars.

The World Bank estimates that antimicrobial resistance could result in medical costs of 1 trillion US dollars by 2050 and a loss of 1 to 3.4 trillion US dollars in Gross Domestic product (GDP) per year by 2030. A 2019 AMR survey found that drug resistance was directly responsible for 1.27 million deaths and contributed to 4.95 million deaths globally (Poudel et al. 2023).

In 2019, Dr. Kamini Walia, a senior scientist at the Indian Council of Medical Research (ICMR), warned that antimicrobial resistance could trigger the next pandemic in the near future. The current global scenario points toward a post-antibiotic era, where common infections, minor injuries, and even scratches could potentially lead to deaths. Antimicrobial resistance undermines modern healthcare and poses a severe threat to millions of live (Fig. 4.1).

4.2 Accidental Discovery of Antibiotics

The first antibiotic, penicillin, was discovered accidentally. In 1928, Alexander Fleming returned from holiday to find mold growing on a petri dish containing colonies of Staphylococcus bacteria. He realized that the mold was producing a self-defense chemical, which he identified as penicillin. This marked the discovery of the first true antibiotic (Ribeiro da Cunha et al. 2019).

4.3 Historical Attributes of Antibiotic

Before the discovery of antibiotics, knowledge about microbes and infectious diseases was limited. Illnesses often led to pandemics and the deaths of millions of people. In 1676, Antoine van Leeuwenhoek discovered microscopic organisms, laying the groundwork for antibiotic production. In 1871, Joseph Lister revealed that Penicillium glaucum could inhibit bacterial growth.

In the late 1800s, French scientist Louis Pasteur worked on Bacillus anthracis, and German doctor Robert Koch focused on Mycobacterium tuberculosis. Their research established crucial links between specific bacteria and diseases, forming the basis for microbiology and antibiotics.

The plague, caused by Yersinia pestis and transmitted through infected fleas, led to the "Black Death" pandemic in the fourteenth century, resulting in over 50 million deaths in Europe (Uddin et al. 2021).

4.4 Mechanism of Antibiotic Resistant

Antibiotics are potent medications that function effectively against specific types of illnesses. Antimicrobial resistance occurs when bacteria can no longer be controlled or killed by certain antibiotics. This resistance is influenced by genetic and environmental factors, leading bacteria to develop mechanisms against drugs. Antibiotic resistance can be categorized into two types: natural and acquired resistance.

Natural resistance is innate and occurs naturally in organisms. Mediated resistance, on the other hand, results from the activation of resistance genes in bacteria due to antibiotic exposure.

Acquired resistance can result from genetic changes in bacteria through mechanisms such as transposition, congugation, transformation, or mutation in their chromosomal DNA (Moo et al. 2020).

The process of antimicrobial resistance can be categorized into four mechanisms.

1. Therapeutic absorption restriction: Limiting the drug's ability to enter the bacterial cell.
2. Target modification: Altering the drug's target site within the bacteria.
3. Drug efflux: Actively pumping the drug out of the bacterial cell.
4. Drug inactivation: Producing enzymes that chemically modify and inactivate the drug.

There are structural differences between Gram-negative and Gram-positive bacteria. The outer membrane of Gram-negative bacteria contains lipopolysaccharides (LPS), making them more harmful due to their protective outer shell. The outer shell enhances resistance to antibiotics. The latest WHO list shows that most antibiotic-resistant bacteria are Gram-negative. Gram-positive bacteria, lacking this outer membrane, are less likely to exhibit drug uptake limitation and drug efflux mechanism (Fig. 4.2).

Antibiotic resistance is the capability of microbes, such as viruses, bacteria, and archea to evolve over time and according to environmental factors, the primary intent of the bacteria reproduces, survives, and proliferates rapidly. When bacteria encounter obstacles like antibiotic that obstructed the growth of bacteria, they can induce genetic changes that make them unsusceptible to the medication, ensuring their survival.

Several factors contribute to the complexity of antibiotic resistance: overuse and misuse of antibiotics, incorrect medication and prescription, reduced patient sensitivity, self-medication, improper healthcare supplies, inadequate hygiene practices, and extensive use of antibiotics in agriculture.

The augmentation in resistance is propelled by blending of microorganisms exposed to antibiotics. The spread of these microorganisms can continuously expose to the environment or bacterial hosts (e.g., humans, animals, and plants) (Reygaert 2018).

Prescription guidelines for antibiotics change over time based on the latest research and updates on resistance and the nature of microorganisms. Almost 30 years ago, antibiotics were prescribed for 10 days. Further research revealed no substantial benefit for longer use over shorter periods. About 15 years ago, the

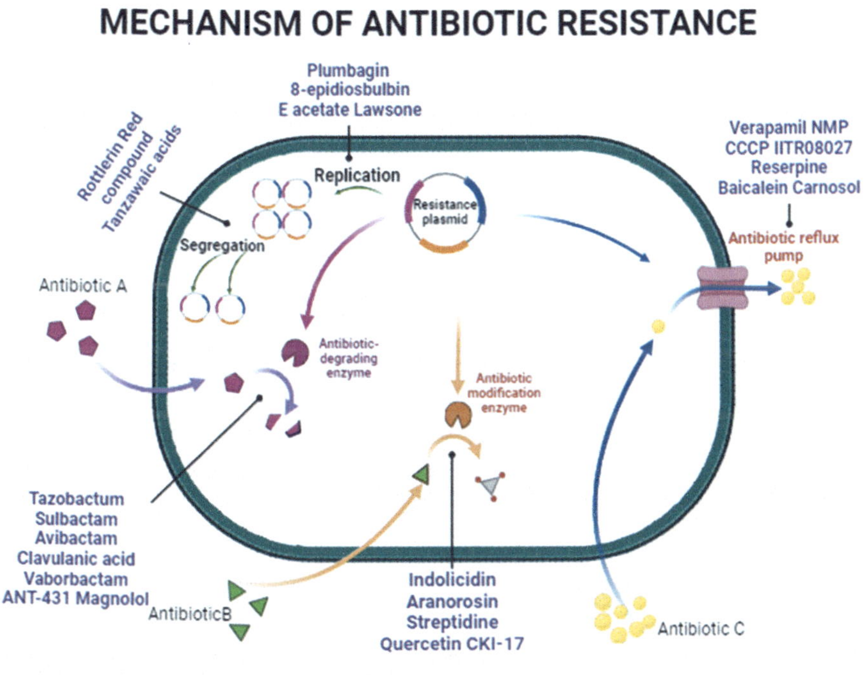

Fig. 4.2 Antibiotic-resistant development mechanism (A schematic representation)

recommended duration was reduced to 5 days, which is still followed in most countries. Currently, the guidelines generally suggest 2–3 days of antibiotic treatment.

Therefore, antibiotics for medical care in animal, human health, and agriculture should be used only when necessary and for the required period. This ensures the effective use of antibiotics and helps prevent the development of resistance.

4.5 Causes of Antimicrobial Resistant

Antimicrobial resistance in microorganisms is not caused by a single factor; rather, it is a multifaceted issue with various interconnected factors.

4.5.1 Genetic Variation or Mutation

Minor alteration in base pair such as point mutations can lead to the replacement of one or more amino acid that could be the target for enzyme, cell structure, or the cell wall of microorganisms. These mutations regulate gene and chromosomal structure, leading to the emergence of new strains. The newly formed defenses can effectively inhibit previously effective antibiotics (Uddin et al. 2021).

4.5.2 Replication and Genetic Exchange

Previously susceptible strains can acquire resistance from other species or genera. Resistance genes carried by the mobile elements like transposons, bacteriophages, and plasmids allow transmission to different genera and species of microorganisms. Resistant bacteria can exchange genetic material, enabling nonresistant bacteria to transform into resistant ones (Uddin et al. 2021).

4.5.3 Selection Pressure of Environment

Environmental conditions provoke bacteria to survive and continuously evolve through mutation or newly acquired genes. When microbes are exposed to an antimicrobial environment, those without resistant genes eradicated, leaving behind those that have acquired resistance. The newly developed resistant bacteria become dominant in the population (Uddin et al. 2021).

4.5.4 Inaccurate Diagnosis

Inaccurate diagnosis of infections can lead to the use of broad-spectrum antibiotics instead of narrow-spectrum antibiotics, which would be more appropriate. This creates selective pressure and accelerates antimicrobial resistance.

4.5.5 Self-Administration

In many parts of world, antibiotics are used without proper prescriptions. Self-medication of Antibiotics (SMA) is prevalent in Southeast Asia, where antibiotics are frequently used without proper medical guidance. This practice puts patients at higher risk and fosters to antimicrobial resistance (Uddin et al. 2021).

4.5.6 Poor Medicare Facilities

Hospitals see thousands of individual dailies, each with their own microbiome without proper protocol, and bacteria can easily transmit between patients, leading to antimicrobial resistance.

4.5.7 Super Resistivity of Superbugs

A superbug is a strain of bacteria that has become resistant to multiple antimicrobial agents, including antibiotics. These bacterial resistances often arise due to the acquisition of resistance gene through horizontal gene transfer. Superbugs are difficult to treat, can cause serious infection, and are impossible to treat with standard antibiotics. They pose a significant challenge and increase healthcare cost, morbidity, and mortality.

Some well-known superbugs include multidrug-resistant (MDR), *Mycobacterium tuberculosis*, *Acinetobacter baumannii*, *Enterococcus faecium*, *Staphylococcus aureus*, *Pseudomonas aeruginosa*, *Salmonella* spp., *Klebsiella pneumoniae*, *Haemophilus influenzae*, *Staphylococcus epidermidis*, *Stenotrophomonas maltophilia*, and *Klebsiella pneumoniae*. Tuberculosis (TB), caused by *Mycobacterium tuberculosis*, infects one-third of world's population and has evolved alongside humans. Despite the revolutionary discoveries of streptomycin and isoniazid, resistance developed rapidly. Multidrug-resistant (MDR) and extremely drug-resistant (XDR) strain of TB have emerged, making treatment more challenging. The introduction of antibiotics initially provided essential treatment, but the continuous development of resistance now compromises TB therapy globally. Horizontal gene transfer plays a significant role in the misuse and overuse of antibiotics in humans and animals, and the environment have contributed to the rise of multidrug-resistant forms of pathogens, complicating treatment and control of infectious diseases.

4.6 Antibiotic-Resistant Response and Its Current Global Evaluation

In 2013, the Centre for Disease Control and Prevention (CDC) published first antimicrobial threat report highlighting the most alarming antibiotic-resistant germ in the United States.

By 2017, the CDC reported that more than 2.8 million infections and 35,000 deaths annually in the United States were attributed to antibiotic spread of resistance among gram-negative pathogens like *Salmonella enterica* and *Escherichia coli*. Beta-lactam antibiotics, in particular, face resistance from beta-lactamase enzymes, with over 1000 resistance-related genes reported (Centre for Disease Control and Prevention 2019). The 2019 CDC report showed gaps in the progress against antibiotic resistance, listing 18 germs as significant threats. Some of these include carbapenem-resistant *Acinetobacter*, *Candida auris*, and *Clostridioides difficile*. Other serious threats include drug-resistant Campylobacter, extended-spectrum beta-lactamase (ESBL)-producing *Enterobacteriaceae*, Vancomycin-resistant *Enterococci* (VRE), multidrug-resistant *Pseudomonas aeruginosa* (*P. aeruginosa*), drug-resistant *Streptococcus pneumoniae* (*S. pneumoniae*), and additionally, erythromycin-resistant group *A. Streptococcus* (Centre for Disease Control and Prevention 2019) (Fig. 4.3).

Addressing antibiotic resistance requires awareness programs, restrictions on self-medication, and efforts to improve the effectiveness of antibiotics.

Tuberculosis (TB), a common example of antimicrobial resistance, shows multidrug resistant (MDR). Multidrug-resistant tuberculosis (MDR-TB) is caused by *Mycobacterium tuberculosis,* which no longer response to the first-line drugs rifampicin and isoniazid. Malaria, caused by the protozoa Plasmodium, is another significant health concern. Artemisinin-based combination therapies (ACTs) are the

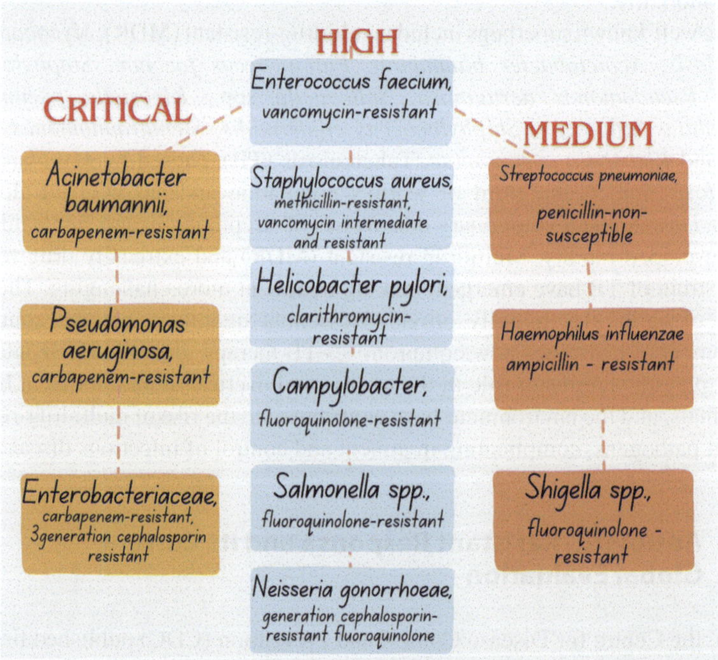

Fig. 4.3 List of prioritized pathogens for research and development by World Health Organization

first-line treatment in many countries. However, partial resistance for the artemisinin-based combination therapies (ACTs) is confirmed in the Eastern Mediterranean Region and several African countries, where genetic mutations linked to artemisinin partial resistance have been identified (World Health Organization 2023).

Data from 2019 show that antimicrobial resistance (AMR) results in approximately 250,000 deaths associated with *Staphylococcus aureus*, *K. pneumoniae*, *Acinetobacter baumannii*, *Pseudomonas aeruginosa*, and with *Escherichia coli* being the most responsible for morbidity. Additionally, there were between 100,000 and 250,000 deaths related to AMR caused by these pathogens (Fig. 4.4). Other pathogens contributing to AMR-related deaths include *Enterococcus faecium*, *Salmonella typhi*, *Streptococcus agalactiae*, *Mycobacterium tuberculosis*, *Enterococcus faecalis*, *Enterobacter* spp., and *Streptococcus*.

In 2021, research conducted on 157 patients in Chitwan Medical College Teaching Hospital in Nepal analyzed antibiotic sensitivity, utilization, and cost. The study found an average of three prescriptions per patient, with piperacillin/tazobactam and meropenem the most commonly prescribed expensive antibiotic. The total cost for antibiotic was $4440.70 with an average cost of $47.67 per patient. *Escherichia coli*, *Acinetobacter*, and *Pseudomonas* sp. were most common pathogens and were found to be resistant to multiple commonly used antibiotics (Fig. 4.5) (Marasine et al. 2021).

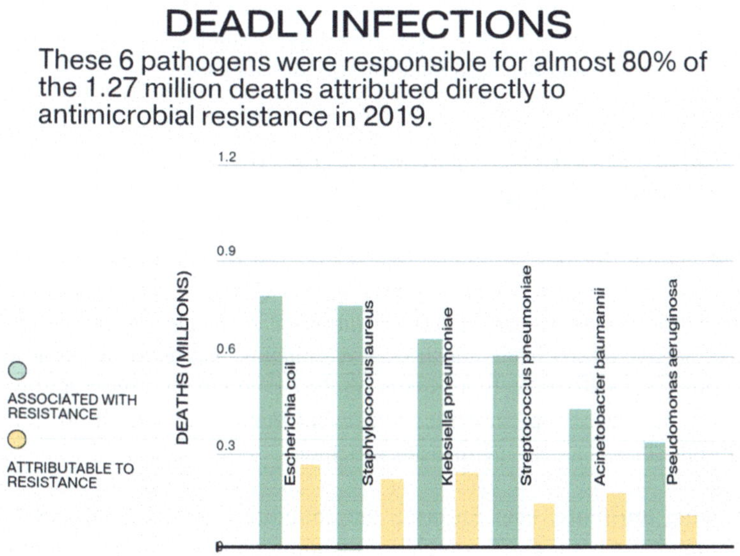

Fig. 4.4 List of six dominant pathogen which responsible for millions of deaths

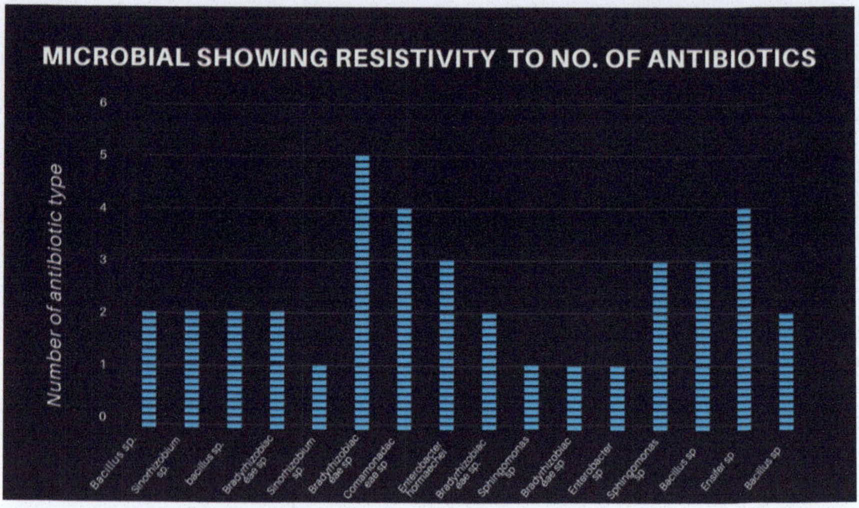

Fig. 4.5 Microbial pathogen showing resistant to no of antibiotics: *Bradyrhizobiaceae* sp. resistant to 5 antibiotics, *Comamonadaceae*, Ensifer sp. for 4 antibiotics, *Enterobacter hormaechei*, *Sphingomonas*, *Bacillus* sp. resistant to 3 antibiotics, *Bacillus*, *Sinorhizobium*, *Bacillus*, *Bradyrhizobiaceae*, *Bradyrhizobiaceae Bacillus* sp., resistant to 2 antibiotics, *Sinorhizobium*, *sphingomonas*, *Bradyrhizobiaceae*, *Enterobacter* resistant to 1 antibiotic

Addressing the issue of antibiotic resistance requires comprehensive strategies, including better diagnostic practices, controlled use of antibiotics, and global cooperation in monitoring and managing antimicrobial resistance.

4.7 Safety Measures to Minimize the Effect of Antibiotic Resistance

Instead of waiting for post-antibiotic era, it's crucial to take immediate action to overcome Antimicrobial Resistant (AMR). At the G20 summit, leaders signed a declaration on Global health, including antibiotic resistance. Combating AMR requires collective efforts from industries, governments, and public awareness campaigns (Annunziato 2019). Relying solely on the development of new antibiotics is not enough; we must adopt strategies to prevent microorganisms from becoming resistant in the first place. Some measures include:

- **Improving antibiotic use**: Ensuring the antibiotics are used appropriately in both animal and humans. This includes avoiding unnecessary prescriptions and following the prescribed dosage and duration.
- **Public awareness**: Educating the public about the dangers of antibiotic misuse and the importance of completing prescribed antibiotic courses.
- **Hygiene and infection control**: Promoting good hygiene practices and infection control measures to reduce the spread of resistant bacteria.

- **Surveillance and monitoring**: Enhancing surveillance systems to track antibiotic resistance patterns and outbreaks.
- **Research and development**: Investing in research to develop new antibiotics, alternative treatments, and rapid diagnostic tests.
- **Regulation and policy**: Implementing policies to restrict over the counter sales of antibiotics and regulating their use in agriculture and livestock.

4.8 Future Aspects and Strategies

Researchers are continuously exploring new strategies and techniques to combat Antimicrobial Resistant (AMR). Some promising approaches include:

- **Quorum-Quenching Technique (QQ)**: This method disrupts the communication between bacteria, preventing them from coordinating actions that lead to infection and resistance.
- **Whole Genome Sequencing (WGS)**: WGS helps identify resistant genes and track the spread of resistance, aiding in the development of targeted treatments.
- **Bacteriophage Therapy**: Using viruses that specifically target bacteria, known as bacteriophages, to treat bacterial infections. This approach is currently being used in Georgia and Russia with promising results.
- **Humanized Monoclonal antibodies**: These antibodies are engineered to target specific bacterial components and are currently undergoing test and trials.
- **3D Structure Approach and X-ray Crystallography**: These techniques help in creating sustainable and friendly strains of bacteria and developing specific target medication and drugs (Uddin et al. 2021).

By integrating these strategies, we can better manage and mitigate the impact of antibiotic resistance, ensuring the effectiveness of antibiotics for future generations.

4.9 Summary

Antibiotics have been miraculous drugs, saving millions of lives over the past 70 years. However, the increasing loss of their effectiveness poses a significant health risk, potentially leading to a scenario where there are "No VACCINE and No CURE." Antibiotic resistance, if it continues unchecked, acts like a slow poison due to its subtle and pervasive nature. This resistance could result in global health crises comparable to pandemics or natural disasters, with severe consequences.

Antibiotic resistance is a worldwide health issue that impacts medical facilities, hygiene practices, and economic stability. Recent reports indicate that it could directly or indirectly damage global economies and healthcare infrastructures. To combat this issue, awareness programs are essential to minimize the misuse of antibiotics and improve their effectiveness.

Tuberculosis (TB) is a common example of antimicrobial resistance, showing multidrug resistance. Multidrug-resistant tuberculosis (MDR-TB) is caused by Mycobacterium tuberculosis, no longer respond to the two most potent first-line drugs, rifampicin and isoniazid. Treating MDR-TB with second-line drugs is often toxic, expensive, and limited in effectiveness. Data from 2022 indicate that only 2 out of 5 patients respond to drug-resistant TB treatments (World Health Organization 2023).

Malaria, caused by the protozoan Plasmodium, can lead to severe health problems such as brain damage, respiratory issues, organ failure, and death. Artemisinin-based combination therapies (ACTs) are the first-line treatment for malaria. However, partial resistance to ACTs has been confirmed in many countries, particularly in the Eastern Mediterranean region and Africa, where genetic mutations are linked to artemisinin partial resistance.

A UK government-commissioned review on antimicrobial resistance estimated that AMR could cause ten million death per year by 2050 (Smith 2022). While some have criticized this projection, the World Health Organization (WHO) and other organizations agreed on the urgent need for action to address AMR.

India is currently the largest consumer of antibiotics globally, and China exhibits highest level and distinctive patterns of antibiotic resistance.

This summary emphasizes the importance of coordinated global efforts, including better regulation of antibiotic use, and enhances public awareness and continued research to develop new treatment strategies to mitigate the threat of antibiotic resistance.

References

Annunziato G (2019) Strategies to overcome antimicrobial resistance (AMR) making use of non-essential target inhibitors: a review. Int J Mol Sci 20(23):5844. https://doi.org/10.3390/ijms20235844

Centre for Disease Control and Prevention (2019) Antibiotic resistance threats in the United States, 2019. Centre for Disease Control and Prevention. https://www.cdc.gov/drugresistance/pdf/threats-report/2019-ar-threats-report-508.pdf

Halawa EM, Fadel M, Al-Rabia MW, Behairy A, Nouh NA, Abdo M, Olga R, Fericean L, Atwa AM, El-Nablaway M, Abdeen A (2024) Antibiotic action and resistance: updated review of mechanisms, spread, influencing factors, and alternative approaches for combating resistance. Front Pharmacol 14:1305294. https://doi.org/10.3389/fphar.2023.1305294

Marasine NR et al (2021) Antibiotic utilization, sensitivity, and cost in the medical intensive care unit of a tertiary care teaching hospital in Nepal. SAGE Open Med 9:20503121211043710. https://doi.org/10.1177/20503121211043710

Moo CL, Yang SK, Yusoff K, Ajat M, Thomas W, Abushelaibi A, Lim SH, Lai KS (2020) Mechanisms of antimicrobial resistance (AMR) and alternative approaches to overcome AMR. Curr Drug Discov Technol 17(4):430–447. https://doi.org/10.2174/1570163816666190304122219

Poudel AN, Zhu S, Cooper N, Little P, Tarrant C, Hickman M, Yao G (2023) The economic burden of antibiotic resistance: a systematic review and meta-analysis. PLoS One 18(5):e0285170. https://doi.org/10.1371/journal.pone.0285170

Reygaert WC (2018) An overview of the antimicrobial resistance mechanisms of bacteria. AIMS Microbiol 4(3):482–501. https://doi.org/10.3934/microbiol.2018.3.482

Ribeiro da Cunha B, Fonseca LP, Calado CRC (2019) Antibiotic discovery: where have we come from, where do we go? Antibiotics (Basel) 8(2):45. https://doi.org/10.3390/antibiotics8020045

Smith J (2022) Global burden of bacterial antimicrobial resistance in 2019: a systematic analysis. Lancet 399(10325):123–135. https://doi.org/10.1016/S0140-6736(21)02724-0

Uddin TM et al (2021) Antibiotic resistance in microbes: history, mechanisms, therapeutic strategies and future prospects. J Infect Public Health 14(12):1750–1766. https://doi.org/10.1016/j.jiph.2021.10.020

World Health Organization (2023) Antimicrobial resistance. World Health Organization. https://www.who.int/news-room/fact-sheets/detail/antimicrobial-resistance

Part II

Understanding Non-antibiotic Advanced Therapies

5 Alternative Approaches to Counter Multidrug-Resistant Bacterial Pathogens

Deepanshi Saxena, Rahul Maitra, Arunava Dasgupta, and Sidharth Chopra

Abstract

The discovery of antimicrobials has had a significant impact on the reduction of deaths related to infectious diseases. However, this advancement was no sooner followed by the emergence of resistance against these antimicrobials. Overuse and misuse of these agents have brought us to a pre-antibiotic world, with extremely limited agents targeting quickly evolving multidrug-resistant (MDR) bacterial pathogens. This scenario highlights the need to develop strategies to discover novel antibiotics that can overcome at least the presently known resistance mechanism. This review covers the newly emerging strategies that not only bypass the bacterial resistance mechanism but also potentiate available antibiotics against MDR bacterial pathogens. These nonantibiotic approaches include targeted antibacterial photodynamic therapy (APT) and phage therapy, bacterial protein degradation via BacPROTAC, genome editing utilizing clustered regularly interspaced short palindromic repeats (CRISPR)–CRISPR-associated protein (Cas), reduction of bacterial pathogenesis by anti-virulence strategies such as targeting specialized secretion systems, regulating virulence gene expression and toxin production, and reducing drug-associated toxicity by utilizing biomimetic assemblies and surfaces, using immunotherapies and vaccines, and modulating the gut microbiome via fecal microbiota transplantation (FMT), prebiotics, probiotics, postbiotics, and hemofiltration devices. Additionally, this chapter

D. Saxena
Department of Molecular Microbiology and Immunology, CSIR-Central Drug Research Institute, Lucknow, Uttar Pradesh, India

R. Maitra · A. Dasgupta · S. Chopra (✉)
Department of Molecular Microbiology and Immunology, CSIR-Central Drug Research Institute, Lucknow, Uttar Pradesh, India

AcSIR: Academy of Scientific and Innovative Research (AcSIR), Ghaziabad, India

M. Gangwar, G. Nath (eds.), *Emerging Paradigms for Antibiotic-Resistant Infections: Beyond the Pill*, https://doi.org/10.1007/978-981-97-5272-0_5

explores the exciting advancements on the horizon, such as personalized medicine, nanotechnology-based therapies, and genome mining and artificial intelligence (AI)-driven approaches. In addition, this chapter discusses the potential of these innovations on the treatment landscape and the hope that they bring by overcoming antibiotic resistance associated with MDR bacterial pathogens.

Keywords

CRISPR–Cas · Anti-virulence strategies · Biomimetics · Gut microbiome modulation · Phage therapy · Artificial intelligence

5.1 Introduction

The discovery and development of antimicrobials is a cornerstone of modern medicine, which has drastically helped reduce the mortality rate for infectious diseases (Gould 2016). However, the emergence of multidrug-resistant (MDR) microbes has left us with extremely limited treatment options, leading to an increased risk of disease spread, disability, and, finally, mortality (Report n.d.). The evolution of microbes with antimicrobial resistance (AMR) is a natural process, occurring over time, under pressure of antimicrobial agents, and is accelerated by human activities, majorly the overuse and misuse of antimicrobials. Drug-resistant infections pose a serious threat to our economic future, affecting both healthcare systems and national economies overall (2015). In its 2022 report, the Global Antimicrobial Resistance and Use Surveillance System (GLASS) highlighted the alarming prevalence of resistance among bacterial pathogens. The emergence of resistance to colistin and carbapenems, the reported last-resort antibiotics, has left us with no treatment options to treat the infections associated with these pathogens (World Health Organization 2022).

In 2019, Murray et al. conducted a worldwide survey involving 204 countries and territories and reported the death of ~4.95 million people associated with bacterial antimicrobial resistance (AMR). The major MDR pathogens responsible for these deaths were *Escherichia coli*, *Staphylococcus aureus (SA)*, *Klebsiella pneumoniae*, *Streptococcus pneumoniae*, *Acinetobacter baumannii*, and *Pseudomonas aeruginosa. These pathogens alone were responsible for* ~3.57 million deaths in 2019. Moreover, methicillin-resistant *S. aureus (MRSA) was attributed to* >1,00,000 deaths, whereas MDR tuberculosis, third-generation cephalosporin-resistant *E. coli*, carbapenem-resistant *A. baumannii*, fluoroquinolone-resistant *E. coli*, carbapenem-resistant *K. pneumoniae*, and third-generation cephalosporin-resistant *K. pneumoniae were each found to be responsible for around* 50,000–1,00,000 deaths (Murray et al. 2022).

A similar association between MDR bacterial strains and infectious diseases and mortality was observed and reported in Europe (Cassini et al. 2019) and USA (Roberts et al. 2009), where these strains were found to be responsible for ~25,000 deaths in 2019. Considering this scenario, the World Health Organization (WHO) has categorized AMR amongst the top three global public health threats (Annunziato

2019). The Organization for Economic Cooperation and Development (OECD) has anticipated that compared to 2005, the resistance against last-resort antibiotics will be doubled by 2035. Therefore, there is an urgent need to discover and/or develop novel antibacterial agents or strategies to target these emerging MDR bacterial strains. To tackle the emerging resistance, countries have adopted the Global Action Plan (GAP), bringing multiple sectors together for the "One Health" approach. The World AMR Awareness Week (WAAW), initiated by WHO, is observed from 18 to 24 November every year to raise awareness regarding AMR (WHO 2015).

Till date, bucketfuls of antibiotics have been developed to target the different aspects of bacteria. The list starts with antibiotics interfering with cell wall synthesis (e.g., penicillin, glycopeptides, cycloserine) to outer membrane and cell membrane disruption (e.g., polymyxin, colistin), protein synthesis inhibitors (e.g., aminoglycosides, tetracyclines, macrolides), nucleic acid inhibitors (e.g., quinolones, ansamycins), and metabolic pathway inhibitors (e.g., sulfonamides, salicyclates) (Hutchings et al. 2019). However, MDR bacteria have evolved to survive in the presence of almost all the available drugs by utilizing different mechanisms such as enzymatic inactivation (e.g., β-lactamases inactivating penicillins, ribosomal modification by rRNA methyltransferases), drug extrusion utilizing efflux pumps, decreased drug uptake by regulation of porins and outer membrane permeability, or by modifying the drug target (Fig. 5.1) (Annunziato 2019).

Thus, there is an urgent requirement to discover and/or develop novel strategies to deal with the difficult-to-treat infections associated with MDR bacteria. This review deals with newly emerging antibacterial strategies, apart from antibiotics, which can bypass the known resistance mechanism and/or can aid in potentiating the available antibiotics against MDR pathogenic bacteria.

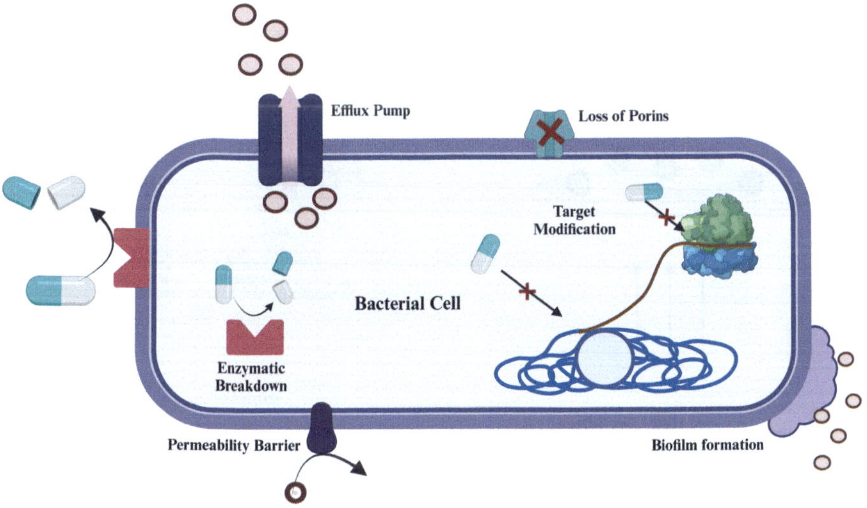

Fig. 5.1 Mechanism of resistance against currently used antibiotics in bacteria

5.2 Antibacterial Photodynamic Therapy (APT)

The concept of light-induced cell death was first reported by Oscar Raab, who observed the lethal effect of acridine red and light on *Paramecium*, a type of infusoria (Rajesh et al. 2011). Later, Von Tappeiner introduced the term "photodynamic action" for the phenomenon and explained the role of oxygen in the therapy (Rajesh et al. 2011). Photodynamic therapy (PDT) is the first Food and Drug Administration (FDA)-approved drug–device combination therapy that has the potential to meet unmet medical needs (Agostinis et al. 2011). This therapy is now used to treat a number of diseases and disorders, including different types of cancers, dental biofilms, leg ulcers, cellulitis, acne, warts, cutaneous leishmaniasis, onychomycosis, genital herpes, recurrent respiratory papillomatosis, and epidermodysplasia verruciformis (Kharkwal et al. 2011). Recently, studies have been conducted to utilize this therapy for targeting MDR bacterial pathogens and hence the name "antibacterial photodynamic therapy."

The process involves a light-absorbing compound (a photosensitizer (PS)), which, on illumination with a specific wavelength of light, initiates a photochemical reaction and produces reactive oxygen species/reactive molecular species (ROS/RMS) in the presence of oxygen (Fig. 5.2) (Feng et al. 2021). This photochemical reaction involves transition of electrons from the low-energy ground state to the excited, high-energy singlet state. This transition can be followed by decay back of the photosensitizer to its ground state with fluorescence emission or can undergo a transition to a higher-energy triplet state, where, on reaction with an organic

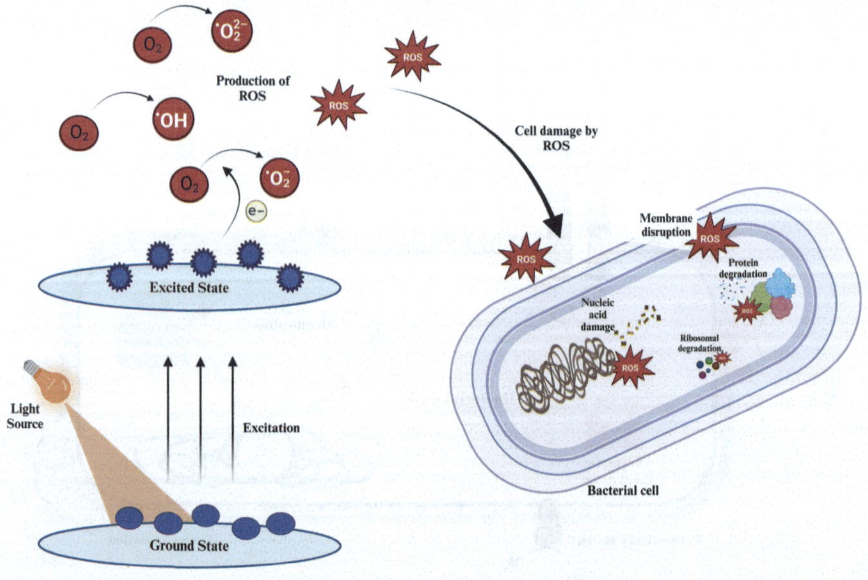

Fig. 5.2 A diagram illustrating antibacterial photodynamic therapy targeting the bacterial cell via ROS production

substrate, it can produce a free radical or can react with oxygen to produce highly reactive oxygen species, e.g., superoxide, hydroxyl radicals, and hydrogen peroxide (Rajesh et al. 2011). The produced RMS induce oxidative damage in the bacterial cell wall, cell membrane, and other biomolecules such as DNA and proteins and activate pro-cell death factors (Dosselli et al. 2012). This multi-targeted action of APT damages the bacterial cell without allowing induction of the resistance mechanism against the therapy (Kashef and Hamblin 2017; Al-Mutairi et al. 2018).

Some of the commonly utilized photosensitizers are tetrapyrrole (e.g., porphyrin and its derivatives, chlorin e6 (Ce6)), phenothiazinium (e.g., methylene blue (MB), toluidine blue, and rose bengal), natural PSs (e.g., curcumin), nanostructures (e.g., the C60 fullerene), tetracycline, and pheophorbide a (e.g., Pheo A) (Feng et al. 2021). Most of the photosensitizers used in photodynamic therapy are activated in the infrared region of the spectrum at a wavelength between 630 and 700 nm. Some of the commonly used light sources include neon lasers (633 nm), gallium–aluminum–arsenide diode lasers (630–690, 830, or 906 nm), and argon lasers (488–514 nm) (Rajesh et al. 2011). APT is advantageous over traditional methods because of its quick action, broad-spectrum target, ability to bypass the bacterial resistance mechanism, minimal cytotoxicity, and minimized off-target effects (Feng et al. 2021). The RMS generated in the therapy also have the potential to disperse pre-formed bacterial biofilms and kill bacterial cells, which would otherwise require higher antibiotic doses (Feng et al. 2021; Zhang et al. 2017b). The therapy has been successfully implemented in treating dental biofilms (Fontana et al. 2009; Wood et al. 2006), leg ulcers (Morley et al. 2013), acne (Wong et al. 2019), and cellulitis (Ibbotson 2002).

Zhu et al. designed a novel polymer, $P\alpha Gal_{50}$-*b*-$PGRB_{20}$, which selectively targets MDR *P. aeruginosa*, an important causative agent of bacterial keratitis. While exerting a safe cytotoxic effect against mammalian cells, the polymer effectively dispersed bacterial biofilms and displayed its antibacterial effect by binding to the Lec A protein that was specifically present on the membrane of MDR *P. aeruginosa* and in the extracellular polymeric substances. Upon illumination with light of 578 ± 10 nm wavelength for 30 min, the polymer was found to disrupt the bacterial membrane (BM), with significant reduction in the nucleic acid and protein content of the cell. The potency of the compound was assessed in an MDR *P. aeruginosa*-infected keratitis model, in which the polymer displayed better corneal recovery compared to the control group. Thus, the selective and effective antibacterial potency of $P\alpha Gal_{50}$-*b*-$PGRB_{20}$ to preferentially target MDR *P. aeruginosa* makes it an ideal candidate to be studied further as a novel antibacterial photodynamic therapy against this multidrug-resistant pathogen (Zhu et al. 2022).

Mitsunaga et al. reported the photoimmuno-antimicrobial strategy (PIAS) against *S. aureus*, including MRSA, where they designed an SA-targeting conjugate (SA mAb–IR700 probe conjugate; SA–IR700) with mAb targeting the peptidoglycan layer of *S. aureus*. PIAS was found to display equipotent antibacterial activity against *S. aureus* under variable conditions, including different culture media, on stimulation with a near-infrared light of wavelength 670–710 nm and an energy density of 30 J/cm^2. Additionally, oxygen toxicity and nonspecific conjugation were not found to have any effect on the bactericidal activity of PIAS on *S. aureus*. In a mouse model of MRSA intraperitoneal infection, PIAS was found to potentially eliminate 100% of MRSA without affecting the normal intestinal microflora, thus highlighting the specificity of the therapy (Mitsunaga et al. 2022).

Howlin et al. explained the potency of ultrasonically activated stream (UAS) to remove biofilms of the cariogenic pathogen *Streptococcus mutans* UA159, *Actinomyces naeslundii* ATCC 12104, and *Streptococcus oralis* ATCC 9811, grown on machine-etched glass slides and artificial teeth, thus depicting an effective treatment option to improve oral hygiene (Howlin et al. 2015).

Park et al. highlighted the role of highly pure chlorin e(6) (Ce(6)), a commonly used second-generation photosensitizer, in potentially inhibiting *Staphylococcus aureus* Xen29 in a mouse superficial wound infection model in which the mice treated with Ce(6)-mediated PDT displayed lesser bioluminescent signals compared to the controls, thus suggesting the potency of Ce(6) to control the growth of bacteria in a wound infection (Park et al. 2012).

Bisland et al. investigated the role of PDT as an alternative treatment for osteomyelitis. They exposed a bioluminescent strain of biofilm-producing *S. aureus* grown onto Kirschner wires (K-wire) to methylene blue (MB) or 5-aminolevulinic acid (ALA). The efficacy of PDT was assessed under both in vitro and in vivo conditions by implanting K-wires into the tibial medullary cavity of Sprague Dawley rats. The progression of the *S. aureus* biofilm was monitored noninvasively using bioluminescence and expressed as a percentage of the signal for each sample immediately prior to treatment. Treatment with 300 mg kg^{-1} of ALA, on illumination with light of wavelength 635 ± 10 nm and an energy density of 75 J cm^{-2}, was found to display a $\geq 4 \log_{10}$ bacterial reduction with a significant delay in bacterial growth. The results thus establish the potency of PDT for the treatment of osteomyelitis (Bisland et al. 2006).

In recent years, PDT has been used in combination with antibiotics to target emerging MDR bacterial pathogens. Feng et al. (2021) reviewed the therapeutic effects of PDT/antibiotic combinations, including PDT in combination with bacterial cell membrane-targeting antibiotics such as polymyxins (Le Guern et al. 2017), vancomycin (Huang et al. 2018), antimicrobial peptides (AMPs), and chitosan (Feng et al. 2021); PDT in combination with bacterial protein synthesis-inhibiting antibiotics (e.g., erythromycin-loaded liposomes in combination with pullulan-pheophorbide a (PU-Pheo A) (Jeong et al. 2017), kanamycin with rose bengal (Cahan et al. 2010), and gentamycin with 9-ITMPo (Rouzes et al. 2000)); and PDT in combination with ofloxacin (Oppezzo and Forte Giacobone 2018), ciprofloxacin (Ronqui et al. 2016), quinolone (Batalha et al. 2015), and rifampicin (Iluz et al. 2018).

Additionally, combinations of PDT with two or more antibiotics have also been studied to target *Mycobacterium fortuitum* infection where amikacin, clarithromycin, ethambutol, levofloxacin, and rifampicin, when administered in combination with the photosensitizer aminolevulinic acid, displayed better improvement (78.9%) compared to multiple antibiotics alone (45.8%), with no relapse of infection even after 3 months of treatment (Gong et al. 2016).

APT has proven to be an effective MDR bacteria-targeting therapy. However, its implementation has certain limitations. First, its efficacy is restricted to light-approachable areas such as the skin and just beneath the skin, lining of organs, exposed eye surface, and oral surfaces. The inefficacy of light to penetrate deeper into the tissues limits the spectrum of APT. Additionally, the short lifespan (~3.5 μs) and the short diffusion distance (~100 nm in aqueous solution) of RMS and the requirement of a continuous supply of oxygen call for further optimization of photosensitizers and the therapy to expand its applicability based on the target species and the host environment (Bartusik et al. 2012; Feng et al. 2021).

5.3 Clustered Regularly Interspaced Short Palindromic Repeats (CRISPR)–CRISPR-Associated Protein (Cas)

Amongst microbes, AMR genes are majorly shared by horizontal gene transfer (HGT) on plasmids and transposons (Christaki et al. 2020). The adaptive immune system of prokaryotes includes an RNA-guided nucleic acid degradation machinery known as clustered regularly interspaced short palindromic repeats (CRISPR) and CRISPR-associated protein (Cas), which recognizes and provides defense against invading exogenous genetic materials such as phages and plasmids (Liu et al. 2017b). This prokaryotic CRISPR–Cas system is also involved in maintaining the integrity of the cell envelope by regulating bacterial lipoproteins (Wan et al. 2021), biofilm formation, and host colonization (Maikova et al. 2018).

The CRISPR–Cas system comprises short repeat sequences (a CRISPR array), separated by spacer sequences and a set of CRISPR-associated (*Cas*) genes (Rath et al. 2015). An A-T-rich leader sequence is found downstream of the CRISPR array, which contains a promoter for initiating the transcription of the array and spacer sequences (Alkhnbashi et al. 2016).

The working mechanism of the CRISPR–Cas system in bacteria has three stages (Hille and Charpentier 2016): (1) the **adaptation stage** involving spacer sequence acquisition (Nuñez et al. 2014), (2) the **expression stage** involving CRISPR RNA (crRNA) and Cas protein generation (Makarova et al. 2015), and (3) the **interference stage** involving targeted nucleic acid cleavage by crRNA (Plagens et al. 2015; Charpentier et al. 2015). The crRNA–Cas protein complex recognizes and cleaves the nucleic acid that shares sequence complementarity to crRNA. This specificity of the CRISPR–Cas system is ingeniously utilized to target genes responsible for drug resistance, thus restoring their antibiotic susceptibility in bacteria (Fig. 5.3) (Teng et al. 2021; Lino et al. 2018).

An efficient delivery method is required to ensure successful CRISPR–Cas-mediated cleavage of drug-resistant genes or resistant bacteria (Lino et al. 2018). These delivery methods are as follows:

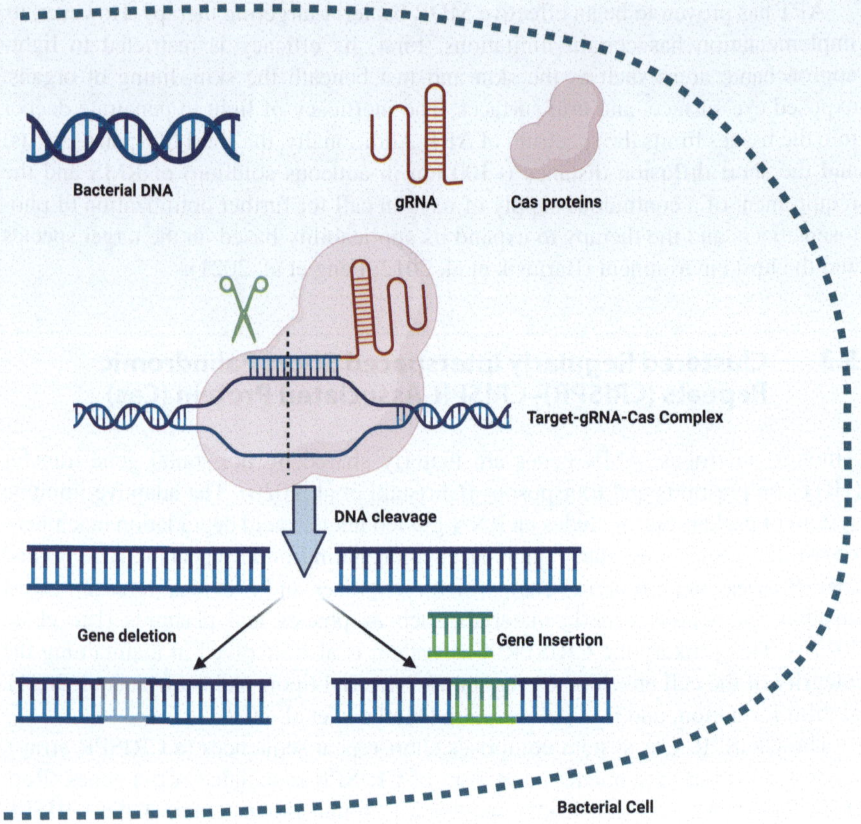

Fig. 5.3 Diagram illustrating gene editing by CRISPR–Cas

5.3.1 Plasmid Vector-Based

Use of a conjugative plasmid as a vehicle to deliver CRISPR assembly offers advantages such as broad host ranges, resistance to restrictive modification systems, and no cellular receptors. The examples of such delivery systems include the pCasCure system to target carbapenemase resistance genes such as *blaNDM* and *blaKPC* in carbapenem-resistant *Enterobacteriaceae* (Kang et al. 2017b) and the recombinant pMBLcas9-sgRNA plasmid to target the mobile colistin resistance 1 (mcr-1) plasmid in the clinical isolates of *E. coli* (Dong et al. 2019)-targeted antibacterial plasmids (TAPs) to target *E. coli* and other related members of Enterobacteriaceae (Reuter et al. 2021).

5.3.2 Phage Vector-Based

This strategy utilizes the ability of a temperate phage to integrate its genome into the bacterial cell. Examples of such a phage–CRISPR–Cas system *include* phagemids carrying the CRISPER–Cas system to target the genes responsible for drug resistance in *S. aureus* (Bikard et al. 2014) and phages carrying sgRNA and Cas9 to target more than 200 pathogenic bacteria expressing the β-lactamase enzyme (Kim et al. 2016). However, the application of the phage-based delivery method is limited by its small host range and the host's metabolic and immune pathway-mediated elimination. To broaden the host range of ϕSaBov, Yosef and colleagues (2015) designed T7-derived defective phages by mutating the tail fiber gene of the phage plasmid, whereas Park et al. (2017) utilized the tail fiber protein gene of ϕ11 (orf50). Additionally, to facilitate host-mediated elimination and to ensure its long-term efficacy, phages are encapsulated with fibers, liposomes, hydrogels, or nanoemulsions (Wu et al. 2021b; Wang et al. 2017a; Díaz et al. 2018; Chadha et al. 2017; Abdelsattar et al. 2019).

5.3.3 Extracellular Vesicle (EV) Vector-Based

Tailoring the CRISPR–Cas system inside the extracellular vesicles has proved to be an efficient delivery method with low immunogenicity and intrinsic targeting properties (Lu et al. 2017). Examples of extracellular vesicle vector-based CRISPR delivery include outer membrane vesicle (OMV)-based delivery of Cas9 ribonucleoprotein (RNP) for gene targeting (Yao et al. 2021). Although effective, this technique requires further standardization for a robust yield, scalable production, procedures for EV storage, efficient encapsulation of the CRISPR system without damaging EV integrity, and understanding in vivo trafficking (Lu et al. 2017).

5.3.4 Nanoparticle-Based

The flexible size of nanoparticle-based vectors to deliver the CRISPR–Cas9 system has emerged as a successful advancement in this field (Givens et al. 2018). For example, polyethyleneimine nanomaterials carrying assembled Cas effector proteins guide RNAs into methicillin-resistant *Staphylococcus aureus* (MRSA) to target the *mecA* gene (Kang et al. 2017b). Adding on to these advantages, the properties of nanoparticle-based delivery systems, such as excellent biocompatibility, lesser immunogenicity, good surface fictionalization, and higher safety, compared to those of virus vector-based systems have allowed the former's use in multiple areas (Riley and Vermerris 2017; Shi and Huang 2009).

Aydin et al. studied and reported the potency of the type I CRISPR system in *E. coli* to interfere with acquisition of drug-resistant plasmids. This group investigated the presence of type I-F CRISPR systems in 82 antimicrobial-susceptible and 96 MDR clinical *E. coli* isolates utilizing polymerase chain reaction (PCR) and

DNA sequencing methods. The results revealed the overrepresentation of type I-F CRISPR among susceptible strains (22.0%) compared to the MDR strains (2.1%) (Aydin et al. 2017).

Price et al. reported the potency of CRISPR–Cas, isolated from mammalian intestinal flora, to prevent the spread of antibiotic resistance plasmids, in a mouse intestinal colonization model (Price et al. 2019).

Kim et al. (2016) studied the potency of the CRISPR–Cas9 system to target and eliminate extended-spectrum β-lactamase (ESBL)-resistant plasmids, including TEM- and SHV-type ESBLs, in >1000 ESBL mutants, to restore the susceptibility of antibiotics.

Wan et al. reported the knockout of colistin resistance genes on mcr-1 plasmids by the CRISPR–Cas9 system to sensitize bacteria to colistin by curing the *mcr-1*-harboring plasmid in *E. coli.* Two *mcr-1* gene-specific sgRNAs were cloned into the pCas9 plasmid, and this recombinant plasmid was transformed into *E. coli* carrying pUC19-*mcr-1* or pHNSHP45, separately. The PCR and quantitative PCR (qPCR) results showed the successful elimination of *mcr-1*-harboring plasmids using the recombinant plasmid pCas9-*mcr. Additionally*, conjugation assays showed the potency of the engineered CRISPR–Cas9 system to prevent transfer of plasmid-borne *mcr-1* to the recipient cell (Wan et al. 2020).

Adding on to the advantages, Vad-Nielsen et al. identified the potency of CRISPR–Cas9 to target multiple genes in a single cell. An assembly of the CRISPR gRNA expression array, containing 10 individual expression cassettes (M10-CRISPRi), was able to target 10 genomic loci and inhibit multiple endogenous genes. This strategy paves the way to tackle multigene disorders such as neurodegenerative diseases (Vad-Nielsen et al. 2016). In a similar study, Wu et al. reviewed the use of the CRISPR–Cas9 system to target two super drug-resistant genes (*mcr-1* and *blaNDM-1*) simultaneously and thus restore the potency of antibiotics like carbapenem and colistin (Wu et al. 2021b).

Naturally, several phages and mobile genetic elements (MGEs) have evolved to express anti-CRISPR proteins (Acres) to tackle the prokaryotic CRISPR–Cas system (Borges et al. 2017). Examples of these Acres include AcrIF1 and AcrIIA, which act by blocking the DNA-binding sites of the CRISPR RNA (crRNA)-guided surveillance complex (Csy complex) of *Pseudomonas aeruginosa and thus deteriorate the* CRISPR-mediated adaptive immune system of the bacteria (Basgall et al. 2018; Chowdhury et al. 2017). Similarly, AcrIIC1 acts as a broad-spectrum Cas9 inhibitor by binding Cas9 to its conserved HNH catalytic domain and thus preventing Cas9-mediated attachment of foreign DNA. An another Cas9 inhibitor, AcrIIC3 induces Cas9 dimerization, thus preventing its binding to the target DNA (Harrington et al. 2017). A phage-encoded anti-CRISPR protein, AcrE1, inhibits bacterial Cas protein by binding to the C-terminal region of Cas3 and inactivating CRISPR–Cas by enzymatic action (Pawluk et al. 2017). Meeske et al. identified listeria phages (ϕLS46) that expresses an anti-CRISPR protein (AcrVIA1), which inactivates the type VI-A CRISPR system by inhibiting Cas13a nuclease (Meeske et al. 2020; Liu et al. 2017b).

Thus, the CRISPR–Cas system has proven to be a powerful gene editing tool to eliminate MDR genes and to interfere with horizontal gene transfer amongst bacteria. However, its clinical application needs further optimization of its delivery methods and target-specific action, under both in vitro and in vivo conditions.

5.4 BacPROTAC

Degradation of the targeted protein in both eukaryotes and prokaryotes is a highly regulated process under the control of energy-dependent proteases (Goldberg 1992). Only the tagged proteins are allowed to enter the protease cage (Elsasser and Finley 2005). Amongst eukaryotes, the protein destined for degradation is recognized by ubiquitin ligases that mark the target with a polyubiquitin tag. The marked protein is then recognized and degraded by the AAA regulatory particle of the 26S proteasome (Kerscher et al. 2006; Finley 2009). A similar protein degradation machinery is found amongst prokaryotes, where the specific degradation motifs on the target protein, called degrons, are identified by Clp complexes, which finally result in target protein degradation (Battesti and Gottesman 2013; Sauer and Baker 2011).

The ClpC–ClpP (ClpCP) protease system is well studied in *Bacillus subtilis* and other Gram-positive bacteria, where the AAA unfolds aseClpC and the protease ClpP plays an important role in eliminating unfolded and aggregated proteins. This ClpCP protease system is regulated by the protein kinase McsB that targets arginine residues (Fuhrmann et al. 2009; Kirstein et al. 2007).

Based on these naturally found degradation machineries, a novel approach to artificially target the desired protein and promote its degradation was introduced and this is known as proteolysis targeting chimeras (PROTACs), which are hetero-bifunctional molecules consisting of two arms connected via a linker molecule. At one arm of the PROTAC is a ligand that will bind to the target protein, whereas the other arm carries the E3 ubiquitin-recruiting ligand. These PROTACs target the protein of interest (POI), bringing it close to the ubiquitin-recruiting ligand and finally promote its ubiquitin-mediated degradation (Gao et al. 2020). In this way, these catalytic PROTACs regulate the related signaling pathways. Degradation of the targeted proteins offers certain advantages over traditional methods such as specificity and more efficiency due to their catalytic activity (Churcher 2018; Hanzl and Winter 2020; Schapira et al. 2019). Sakamoto et al. (2001) first reported PROTAC molecules against cancer cells by targeting methionine aminopeptidase-2 (MetAP-2), and, since then, PROTACs have been employed for the treatment of several diseases and disorders such as cancer, viral infection, immune disorders, neurodegenerative diseases, etc. Some examples include PROTACs targeting B-cell lymphoma 6 (BCL6) (McCoull et al. 2018), P300/CBP-associated factor/general control nonderepressible 5 (PCAF/GCN5) and interleukin-1 receptor-associated kinase 4 (IRAK4) (Bassi et al. 2018) (Nunes et al. 2019), Bruton's tyrosine kinase (BTK) (Zorba et al. 2018), focal adhesion kinase (FAK) (Popow et al. 2019), and ARV-110 (Neklesa et al. 2017). Recently, the approach has caught the attention of academics and several pharmaceutical companies.

Trentini et al. studied the degradation mechanism of phosphoarginine (pArg) proteins by expressing an inactive trapping variant of the ClpP protease (Ser98Ala, $ClpP^{TRAP}$) under in vivo conditions. The group characterized pArg as a degradation marker that recruits the ClpCP proteolytic machinery to initiate targeted protein degradation. Additionally, they explained the similarity between pArg and polyubiquitin-mediated degradation (Trentini et al. 2016). This ClpP, ClpC protease system is only found in Gram-positive bacteria and *Mycobacterium* sp. Morreale et al. worked on reprograming this ClpP, ClpC protease system to target microbial infections. The group designed the first bacterial PROTAC (BacPROTAC) that promotes the binding of a substrate to the ClpP, ClpC protease, thereby promoting its degradation. Structurally, BacPROTAC, like other PROTACs, possesses a POI-targeting ligand (biotin with a phosphorylated arginine residue: a high-affinity mercaptosuccinic acid (mSA) ligand) at one arm and a ClpCNTD anchor at the other. ClpCNTD is the amino terminal domain of the ClpC protomer, which works as a binding or docking site for POI. Biotin with a phosphorylated arginine residue targets monomeric streptavidin (mSA, a model protein), bringing it in close proximity to the ClpP:ClpC protease. The designed BacPROTAC at 100 μM concentration effectively degraded streptavidin via the ClpA, ClpP degradation mechanism in *Bacillus subtilis.* The efficacy of the designed BacPROTAC was also analyzed for various mSA fusion proteins, thus highlighting the need for an unstructured region in the target protein for efficient degradation. Moreover, the degradation efficacy of the system was not influenced by changing the length of the linker. This protein-targeting machinery was further studied in *Mycobacterium* where the ternary complex of BacPROTAC-1 with mSA and $ClpC1_{NTD}$ potentially targeted and degraded the POI–BRDTBD1 fusion protein, as shown in Fig. 5.4 (Morreale et al. 2022).

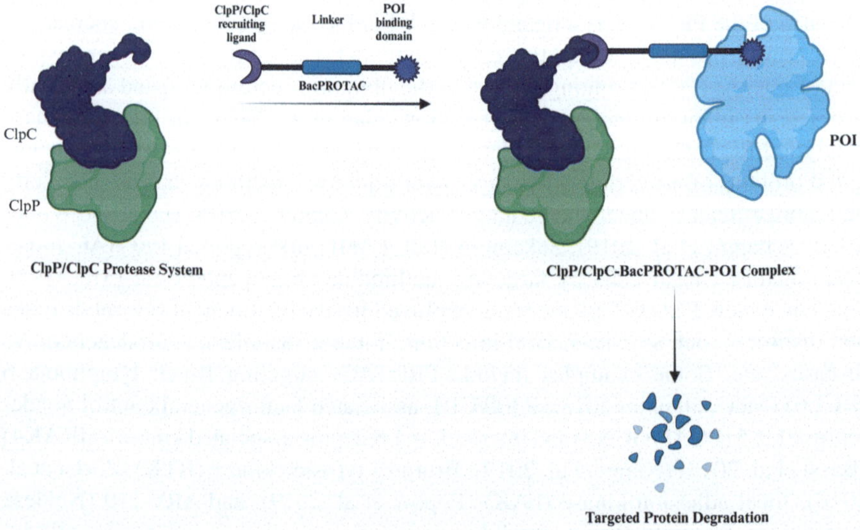

Fig. 5.4 A diagram illustrating targeted protein degradation by BacPROTAC

Although BacPROTAC is an innovative and newly emerging strategy for targeting the selective protein of interest, including even the undruggable protein targets, and for preventing emergence of resistance, it has certain limitations too. The additional outer layer of the Gram-negative and waxy protective shield of lipids in the mycobacterial cell wall limits the access of the molecule into the cell. Additionally, in vivo degradation of the molecule needs close monitoring at high cell intensity (Morreale et al. 2022). Additionally, the poor pharmacokinetic profile and the chemical instability of the phospho-guanidinium group of pArg-based PROTACs were overcome by utilizing cyclic peptides that deregulate the ClpC1P1P2 complex (Lee and Suh 2016; Morreale et al. 2022). Considering this, Morreale et al. (2022) explained the efficacy of the $ClpC1_{NTD}$-directed antibiotic CymA to target mycobacterial proteins.

5.5 Anti-Virulence Strategies

Once inside the host, bacteria express and release a number of virulent factors such as adhesions, toxins, and effector molecules to colonize and infect their host. For doing so, bacteria activate specialized secretion systems to release a variety of signaling molecules or toxins into the host environment (Rasko et al. 2008; Rumbaugh 2007). Anti-virulence therapies utilize and target bacterial virulence gene expression, toxin secretion, cell-to-cell signaling pathways, and cell adhesion and cell invasion to prevent bacterial pathogenesis. Interfering with the bacterial ability to recognize the host environment helps prevent the activation of virulence factors, thus reducing the chances of bacterial pathogenesis (Fig. 5.5). This strategy simply prevents bacterial colonization without actually killing the cell, thus offering low pressure to mutate (Paul and Leibovici 2009) (Cegelski et al. 2008).

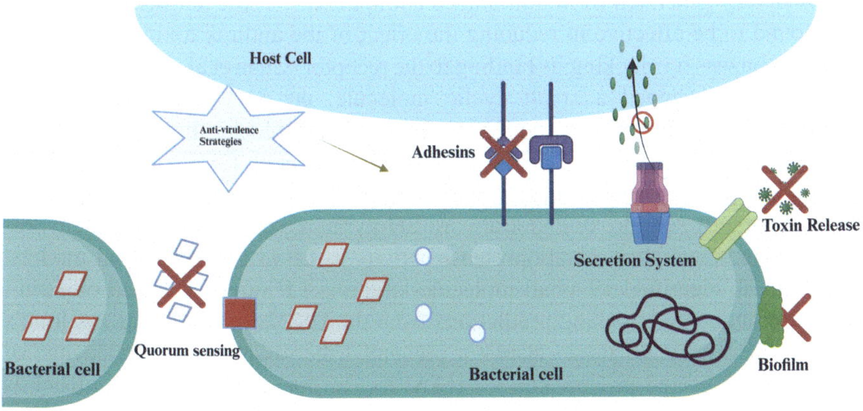

Fig. 5.5 An image illustrating the mechanism of action of anti-virulence strategies to target bacterial pathogenesis

5.5.1 Toxin Secretion

Bacteria, on recognizing and binding to the infection site, initiate the production of toxins, which exert deleterious effects on the host cells. These toxins are bacterial proteins expressed during post-exponential and early stationary phases. These toxins can be classified under different classes based on their action: exfoliative toxins (Bukowski et al. 2010), hemolysins (Powers et al. 2015), leukotoxins (Zivkovic et al. 2011), endotoxins (Bukowski et al. 2010), and others. These proteins cause physical and biochemical damage to the host cells, resulting in cell-to-cell signaling interruption and thus helping bacteria evade their host's immune system and establish infection (Wei et al. 2017). A wide variety of toxins produced by bacteria have been studied till date. For example, the pertussis toxin from *Bordetella pertussis* (Shrivastava and Miller 2009), the Shiga toxin from *Escherichia coli* (O'Brien et al. 1979), the anthrax toxin from *Bacillus anthracis* (Abrami et al. 2003), and α-hemolysin (α-toxin (AT)) from *S. aureus* (Surewaard et al. 2018).

Anti-virulence strategies targeting the toxin secretion system of bacteria act by blocking the binding of toxins to host cell receptors, or by blocking the pores generated by the toxin subunits, and/or by targeting and inhibiting the toxin subunits. Binding of a toxin to its cognate receptor can be prevented by introducing soluble analogues to the receptor. For example, Scobie et al. reported the soluble anthrax toxin receptor (sATR), tumor endothelial marker 8 (TEM8), and capillary morphogenesis protein 2 (sCMG2) that binds to the PA subunit of the anthrax toxin and inhibits its binding with cellular receptors. In a tissue culture model of intoxication, sCMG2 was found to exhibit 11.4-fold more efficacy than sATR/TEM8, with an ~1000-fold higher binding to the PA subunit. The efficacy of sCMG2 was also studied in vivo, where at 480 pmol concentration, it was found to be effective in protecting rats against lethal toxins (Scobie et al. 2005).

Chen et al. identified four single-chain variable fragments (scFvs) (targeting the PA subunit of anthrax toxin) and two scFvs (targeting the LF subunit) from a phage display library generated from immunized chimpanzees. These antibody fragments were found to be effective in reducing the effect of the anthrax toxin in F344 rats and chimpanzees by blocking its binding to the receptor (Chen et al. 2006). Karginov et al. (2005) designed a small cyclic molecule, per-6-(3-aminopropylthio)-β-cyclodextrin, which binds to the PA pore lumen and blocks the entry of lethal factor (LF) and edema factor (EF) into the cell. Shoop et al. (2005) reported an LF subunit of a hydroxamate-based inhibitor, (2R)-2-[(4-fluoro-3-methylphenyl) sulphonylamino]-*N*-hydroxy-2-(tetrahydro-2*H*-pyran-4-yl) acetamide, which prevents the zinc-dependent metalloprotease activity of LF in various animal models. Panchal et al. identified 19 small molecule-targeting LF subunits, based on structural information and docking predictions from the US National Cancer Institute's chemical repository. Of these, cisplatin prevented the toxin-mediated cleavage of the mitogen-activated protein kinases (MAPKs) in macrophages and neutrophils (Panchal et al. 2004; Turk 2007).

Le et al. (2016) studied the anti-AT human monoclonal antibody (mAb), MEDI4893*, for neutralizing the AT and thus significantly decreasing the severity of *S. aureus* infection. Later, Ortines et al. further expanded the efficacy of MEDI4893* in significantly decreasing the wound size and bacterial burden of *S. aureus*-infected nondiabetic and diabetic mice. Moreover, the mAb was also effective in regulating the host immune response by decreasing the neutrophil count and increasing the monocyte and macrophage infiltrates in nondiabetic mice while decreasing neutrophil extracellular traps (NETs) in diabetic mice (Ortines et al. 2018).

Similarly, Andersen et al. utilized an antibody-based approach to neutralize toxin A (TcdA) and toxin B (TcdB) of *Clostridium difficile*, which damage the gastrointestinal epithelium along with causing colonic inflammation in the host. These antitoxin antibodies were effective in improving the animal survival rates with no damage or inflammation observed in the gastrointestinal (GI) tract of treated hamsters for even up to 4 days (Andersen et al. 2016). In addition to monoclonal antibodies, Henry et al. designed artificial liposomes made of sphingomyelin and cholesterol (66 mol/% cholesterol). These liposomes displayed potential binding with cholesterol-dependent cytolysins (CDCs), phospholipase C, and α-hemolysin, thus competing with toxins to bind to the host cells. Additionally, the mixture of cholesterol and sphingomyelin liposomes (66 mol/% cholesterol) with sphingomyelin liposomes (sphingomyelin-only liposomes) was effective in sequestering the whole palette of toxins produced by both staphylococcal and streptococcal pathogens under in vivo conditions, thus establishing the role of therapy alone or in combination with antibiotics in combating fatal invasive staphylococcal and streptococcal diseases (Henry et al. 2015).

5.5.2 Adhesins

Adhesion to the host cell surface is an important step in colonizing and establishing an infection. Bacterial appendages, including the pili, fimbriae, and sometimes flagella, aid in adhering to the host cell surface via proteins known as adhesins (Proft and Baker 2009). These surface adhesion proteins are termed "MSCRAMMs" (microbial surface component-recognizing adhesive matrix molecules) (Dehbanipour and Ghalavand 2022). Uropathogenic *E. coli* (UPEC) encodes the type I pili and the P pili (Wright and Hultgren 2006), which helps the bacteria adhere to the urogenital epithelia and thus prevents bacterial elimination through urine (Lane and Mobley 2007). Pilicides, the mimetic of normal pilin subunits, are found to be an effective treatment to prevent pilus formation and elongation on getting incorporated into the growing pilus. Small molecules of the bicyclic 2-pyridones class were found to target P pilus biogenesis (Berg et al. 2006). Pinker et al. designed a library of bicyclic 2-pyridones that inhibits the biogenesis of the P pilus of UPEC. Inhibition of pilus formation drastically decreases the adherence of cells to bladder cells and biofilm formation by approximately 90% in laboratory and clinical *E. coli* strains (Pinkner et al. 2006). Han et al. (2010) identified biphenyl mannosides as an effective treatment to reduce bacterial colonization (Hartmann et al.

2012). Thus, inhibiting pilus biogenesis can be an effective way to prevent bacterial adhesion and biofilm formation.

5.5.3 Specialized Secretion Systems

Three distinct secretory systems are studied in bacterial pathogens to release effector molecules like toxins directly into the inclusion membrane and host cell cytoplasm. These effector molecules compete with the host cell proteins, thereby altering the mammalian signaling pathways (Galán and Cossart 2005). These specialized secretory systems are of three types: III, IV, and VI (Filloux et al. 2008). The type III secretion system (TTSS) injects effectors directly into the eukaryotic cell cytoplasm by forming a needle-type structure composed of several functional proteins. These effectors interfere and modulate signal transduction in the host cell by causing inflammation, opening of tight junctions (as in the case of enteropathogenic *E. coli*), and facilitating electrolyte secretion and/or alterations in the cytoskeleton (as in the cases of the *Shigella* and *Salmonella* species for bacterial invasion and in enteropathogenic *E. coli* for bacterial adhesion) (Keyser et al. 2008; Stavrinides et al. 2008).

The type IV bacterial secretory system forms a long pilus-like structure, which aids in transferring either the bacterial DNA (e.g., by *Agrobacteria tumefaciens*) or bacterial effector proteins (e.g., *Helicobacter pylori*) into mammalian cells (Filloux et al. 2008). The type VI bacterial secretion system forms a tubular structure to inject effector molecules into the host cells; however, the exact mechanism is largely unknown (Filloux 2009). Several studies have been conducted to design compounds targeting these secretion systems to prevent the toxic effect of the effectors. Bailey et al. identified small molecules belonging to the acylated hydrazones of salicylaldehyde that were found to specifically block the type III secretion system (TTSS) of *Chlamydia*. Moreover, these compounds were also found to impair the developmental stages of *Chlamydia* spp., thus preventing further progression of the infection (Bailey et al. 2007). Muschiol et al. (2006) studied and reported a class of small molecules, including INP0400, which inhibits the intracellular replication and infectivity of *Chlamydia trachomatis* and *Yersinia pseudotuberculosis* by blocking TTSS at defined stages. Kauppi et al. (2003) screened a chemical library of ~9400 compounds using a luciferase reporter gene assay, involving *Yersinia* outer membrane proteins (Yops) conjugated to *luxAB* to identify *Yersinia* TTSS-targeting compounds. Several other studies have been conducted to identify small molecules targeting TTSS, e.g., salicylideneaniline-containing compounds targeting enteropathogenic *E. coli* (Gauthier et al. 2005), INP0400 and INP0406 inhibiting *Shigella* spp. (Veenendaal et al. 2009), and 2-imino-5-arylidene thiazolidinone targeting *Salmonella enterica* serovar Typhimurium, *Pseudomonas syringae*, and *Yersinia enterocolytica* (Felise et al. 2008; Kline et al. 2008).

Sharma et al. identified a T3SS inhibitor, INP0341, which, at 100 μM, was found to be effective in protecting human corneal epithelial cells (HCECs) from *P. aeruginosa*-mediated cytotoxicity. INP0341 potentially increased the expression of antimicrobial peptides and ROS generation in the cells exposed to *P. aeruginosa*. In an in vivo environment, the compound proved its efficacy to efficiently attenuate the corneal infection caused by *P. aeruginosa* in a murine keratitis model as calculated from corneal opacity, clinical score, and bacterial load (Sharma et al. 2020).

5.5.4 Regulation of Virulence Gene Expression

Expression of virulence traits is highly regulated in a bacterial cell. The most studied mechanism is the one that senses environmental cues (for example, temperature and osmolarity) or chemicals produced by the host or other microbes in the host (Paul and Leibovici 2009). This cell-to-cell signaling mechanism in bacteria is referred to as a quorum-sensing system, and this system is tightly regulated and highly conserved among bacterial species (Fuqua et al. 1994). For example, *P. aeruginosa* produces a variety of virulence factors such as TTSS, proteases, and adhesins, and both their production and release are tightly regulated by quorum sensing (Fuqua et al. 1994; Gooderham and Hancock 2009). Chemical signals such acyl homoserine lactones (AHLs), including *N*-3-oxo-dodecanoyl-L-homoserine lactone (C12-HSL) and *N*-butanoyl-L-homoserine lactone (C4-HSL), are sensed by LasR–LasI and RhlR–RhlI quorum-sensing systems, whereas 4-quinolones are sensed by PqsR–pQSABCDE (Pearson et al. 1994, 1997; Brint and Ohman 1995; Pesci et al. 1999; Wilson et al. 1988; Wade et al. 2005). Once emitted, the signal binds to its specific transcription regulators and initiates virulence gene expression. Thus, targeting this signaling process of bacteria using the analogues of an autoinducer molecule can be an effective approach to bacterial targeting. Amongst bacteria, histidine kinases have been studied as the major sensory proteins for environmental cues. Histidine sensor kinase, being present on the membrane of bacteria, is considered a conserved therapeutic target for targeting pathogenic bacteria (Mitrophanov and Groisman 2008). Rasko et al. (2008)studied bacterial QseC histidine sensor kinase that recognizes bacterial quorum-sensing signal autoinducer 3 (AI-3) and/or the host signals adrenaline and/or noradrenaline (Clarke et al. 2006) and initiates a complex phosphorylation cascade that regulates virulence gene expression. Homologues of QseC are found in a variety of human and plant pathogens and thus can be utilized for multi-pathogen targeting (Rasko et al. 2008). Studies have revealed that the mutation in *qseC* results in decreased virulence in animal models infected with enterohemorrhagic *E. coli* (EHEC) (Clarke et al. 2006), *Salmonella typhimurium* (Bearson and Bearson 2008), and *Francisella tularensis* (Weiss et al. 2007).

Similarly, cholera toxin and the toxin co-regulated pilus (TCP) expression in *V. cholerae* is regulated by the transcription factor ToxT under the controlled regulation of membrane-associated ToxRS and TcpPH (Matson et al. 2007). Shakhnovich et al. screened a library of 50,000 small molecules, of which virstatin (4-[*N*-(1,8-naphthalimide)]-n-butyric acid) was found to inhibit ToxT dimerization, subsequently regulating the expression of the cholera toxin and TCP. Virstatin, at 50 μM concentration, was found to display a significant reduction in cholerae infection in an infant mouse model of infection (Shakhnovich et al. 2007).

Additionally, a halogenated furanone produced by the marine macroalga *Delisea pulchra* was found to be a competitive inhibitor with a bacterial AHL-based quorum-sensing system (Rasmussen et al. 2000). Manefield et al. (2002) identified a furanone derivative, C-30, which potentially targets the *Pseudomonas* quorum-sensing system and sensitizes bacterial biofilms to antibiotics and detergents. There are several studies illustrating the efficacy of AHL analogues with modified acyl side chains to target the AHL-based quorum-sensing system (Kline et al. 1999; Smith et al. 2003). Müh et al. (2006) identified PD12 (a tetrazole with a 12-carbon alkyl tail) and V-06-018 (a phenyl ring with a 12-carbon alkyl tail) as LasRI-based quorum-sensing inhibitors, in a high-throughput screening (HTS) of a library of 200,000 small compounds. Rasko et al. performed a high-throughput screening of a library of 150,000 small organic compounds. Based on the screening results, LED209 (*N*-phenyl-4-[[(phenylamino)thioxomethyl]amino]-benzenesulphonamide) was identified as the most potent compound inhibiting QseC-mediated virulence gene expression. The compound prevented the attachment of signaling molecules (such as AI-3, adrenaline, and noradrenaline) to QseC in several pathogenic bacterial models (Rasko et al. 2008). Similarly, *S. aureus* possesses an Agr quorum-sensing system to regulate virulence gene expression through the binding of an autoinducing peptide (AIP) to AgrC, a membrane-bound histidine sensor kinase and then activating it. This AIP-mediated cell-to-cell signaling can be targeted to tackle emerging MDR *S. aureus* infections (Novick and Geisinger 2008; George and Muir 2007; Wright and Hultgren 2006).

Ji et al. reported sulfonyladenosine compounds (anthranilyl-AMS and anthranilyl-AMSN) as *Pseudomonas* quinolone system (PQS) inhibitors by mimicking the anthranilyl-AMP intermediate involved in the production of PQS (3,4-dihydroxy-2-heptylquinoline, *Pseudomonas* quinolone signal) and HHQ (2-heptyl-4-hydroxyquinoline). The reduced PQS and HHQ production was found to be associated with a drastic decrease in the *P. aeruginosa* PA14 strain virulence (Ji et al. 2016). Allegretta et al. (2017) reported the potency of some catechol-derived compounds to prevent the attachment of *P. aeruginosa* onto its substrate by blocking the active site of PqsD, thus reducing both bacterial virulence and production of HHQ. In a similar study, Drees et al. (2016) reported 2-aminoacetophenone (2-AA) as the competitive inhibitor of PqsBC, an enzyme catalyzing octanoyl-coenzyme A (CoA) and 2-aminobenzoylacetate condensation to HHQ in *Pseudomonas putida* KT2440. Apart from these strategies, several quorum-quenching enzymes that target quorum signals after their release, but do not bind to the receptor and thus interrupt the quorum signaling in the bacteria, are studied. Such enzymes targeting the Quorum Sensing (QS) signal include acyl-homoserine lactone (acyl-HSL) lactonases, acyl-HSL acylases, and acyl-HSL oxidoreductases. Fetzner reported the potency of acyl-HSL lactonases and acyl-HSL acylases to degrade acyl-HSL via hydrolysis of the homoserine lactone ring/amide bond between the acyl tail and the homoserine lactone ring. Utari et al. reported the role of PvdQ acylase in hindering the virulence of *P. aeruginosa* in models of pulmonary

Table 5.1 List of advances in the way toward targeting bacterial virulence (Fleitas Martínez et al. 2019)

Inhibitory activity	Compounds/strategies studied	Targeted microbe	References
Compounds targeting the functional membrane microdomains (FMMs) of bacteria			
Oligomerization of RNase Rny	Zaragozic acid	*S. aureus*	García-Fernández et al. (2017)
	Miltefosine	*S. aureus*	Koch et al. (2017)
	5-doxyl-stearic acid (DSA)	*S. aureus*	Koch et al. (2017); Mielich-Süss et al. (2017)
Targeting T7SS system assembly	Zaragozic acid	*S. aureus*	García-Fernández et al. (2017)
	Simvastatin	*S. aureus*	Mielich-Süss et al. (2017)
Compounds targeting the bacterial quorum-sensing machinery			
Downregulating *mqsR*, *pgaB*, *pgaC*, *csgE*, and *csgF* gene expression	CRISPR–Cas9	*E. coli* SE15	Kang et al. (2017a)
Reducing pyocyanin and pyoverdine production	2-(Methylsulfonyl)-4-(1H-tetrazol-1-yl) pyrimidine	*P. aeruginosa*	Thomann et al. (2016)
Reducing protease and siderophore production	Diketopiperazine	*Burkholderia cenocepacia*	Scoffone et al. (2016)
Reducing alginate and pyocyanin production	Pyridoxal lactohydrazone	*P. aeruginosa*	Heidari et al. (2017)
Downregulating *lasI*, *rhlI*, *rhlR*, *pqsB*, *pqsC*, *pqsH*, ambBCDE gene expression	Coumarin	*P. aeruginosa* POA1, including clinical isolates	Zhang et al. (2018)
Reducing pyocyanin and elastase production	AHL–nitric oxide hybrids	*P. aeruginosa* PA14 and *P. aeruginosa* PAO1	Kutty et al. (2015)
Reduction in the swarming motility	Flavonoids	*P. aeruginosa* PA14	Paczkowski et al. (2017)
	Fluoro-substituted Isothiocyanates	*P. aeruginosa*	Amara et al. (2016)

(continued)

Table 5.1 (continued)

Inhibitory activity	Compounds/strategies studied	Targeted microbe	References
Reduction in the production of pyocyanin, hemolysin, elastase, proteases, rhamnolipid	Zingerone	*P. aeruginosa* PAO1 and clinical strains	Kumar et al. (2015)
Compounds targeting bacterial biofilms			
Reduction in biofilm formation	T315 compound	*S. enterica* serovar Typhimurium, *S. enterica* serovar Typhi, and *A. baumannii*	Moshiri et al. (2018)
	2-aminobenzimidazole derivatives	*S. enterica* serovar Typhimurium	Moshiri et al. (2018)
	N-(4-{fluoroanilno}-butanoyl)-Lhomoserine lactone N-(4-{chlororoanilno} butanoyl)-L-homoserine lactone	*P. aeruginosa* PA330 and *P. aeruginosa* PA282	Kalaiarasan et al. (2017)
Dispersion of pre-formed biofilms	3F1 compound	*S. mutans*	García-Fernández et al. (2017)
	p-tolyl (3-phenylpropyl) carbamate	*S. aureus*	Stephens et al. (2016)
	Zosteric acid derivatives	*E. coli*	Cattò et al. (2015)
Biofilm inhibition	Peptidomimetic compounds	*Porphyromonas gingivalis*	Tan et al. (2018)
Decreased fibronectin adherence	Chalcone	*S. aureus*	Zhang et al. (2017a)
Compounds targeting bacterial toxins			
Downregulation of *RNAIII*, *psmα*, and *hla* gene expression	ω-Hydroxyemodin	*S. aureus*	Daly et al. (2015)
	Biaryl hydroxyketones		Greenberg et al. (2018)
	Solonamide analogues		Baldry et al. (2016)

(continued)

Table 5.1 (continued)

Inhibitory activity	Compounds/strategies studied	Targeted microbe	References
Decreased production of α-hemolysin	Resveratrol	*S. aureus*	Duan et al. (2018); Tang et al. (2019)
	Lysionotin		Tang et al. (2015)
	Prim-*O*-glucosylcimifugin		Ping et al. (2018)
	Sclareol		Ping et al. (2017)
LtxA-mediated cytotoxicity inhibition	Galloylated catechins	*Aggregatibacter actinomycetemcomitans*	Krueger et al. (2018)
	Peptides		

P. aeruginosa infection. Administration of PvdQ to *P. aeruginosa*-treated mice was not only effective in reducing the bacterial load by fivefold but also in increasing the survival rate along with decreasing the lung inflammation compared to the control group (Utari et al. 2018).

Thus, anti-virulence therapies for targeting bacterial toxins, quorum-sensing inhibition, surface adhesion inhibition, and inhibiting the bacterial secretion systems have come up as beneficial approaches to target bacteria in such a way as to prevent bacterial pathogenesis (Table 5.1).

5.6 Biomimetics

Another emerging way of developing novel antibacterials is by making use of biomimetic materials to reduce the cytotoxicity associated with nanometals and antimicrobial peptides and improve both drug delivery profiles and the antibacterial efficacy of biopolymers and other antimicrobial substances such as essential oil-incorporated materials and platelet-rich plasma (PRP) (Chee and Brown 2020). This versatile strategy makes use of methods like surface mimicry, biomimetic fictionalization, and biomimetic assemblies to address the development of resistance in bacteria in a biocompatible manner. These strategies have been beautifully reviewed by Chee and Brown (2020). The upcoming section deals with a brief description of these strategies to target bacterial cells.

5.6.1 Surface Mimicry

Surface mimicry is based on the physical disruption of bacterial cells attached to material surfaces that mimic the naturally known antimicrobial surfaces such as insect wings (particularly that of cicadae, damselflies, and dragonflies), shark skin, and the carapace of *Cancer pagurus* (Elbourne et al. 2017; Ivanova et al. 2012; Michalska et al. 2018; Sakamoto et al. 2014).

Insect wings have nanopillar geometry with a thick and blunt surface. When bacteria get attached to the space in between the nanopillars, the bacterial cell wall and membrane are disrupted by the mechanical stress induced by these structures (Bazaka et al. 2015). These structures not only disrupt the bacterial physical attachment but are also able to prevent bacterial biofilm formation (Table 5.1) (Ivanova et al. 2012; Pogodin et al. 2013). However, the antimicrobial activity of insect wings is largely affected by the length and density of these nanopillars.

In contrast to the nanopillar geometry of insect wings, the surfaces of the shark skin and carapace of *Cancer pagurus* have nanogrooves and nanowrinkles, respectively. Nanogrooves of the shark skin have a ridged surface, whereas nanowrinkles are ridge-like protrusions useful for disrupting the physical attachment of bacterial cells onto the surface of a material (Sakamoto et al. 2014; Papi et al. 2016; Reddy et al. 2011; Chung et al. 2007). Papi et al. designed graphene oxide-based hydrogels with the geometry resembling the surface of a crab's carapace. The formed hydrogels displayed improved antibacterial activity against several bacterial species, including *S. aureus* and *E. coli* (Papi et al. 2016). Although surface mimicry displays a significant antibacterial effect, it cannot be translated for applications in an in vivo environment.

5.6.2 Biomimetic Functionalization

Biomimetic functionalization involves use of biomimetic materials to enhance the antimicrobial activity of the existing antimicrobial strategies, reduce associated cytotoxicity, and improve delivery profiles. For example, the cytotoxicity associated with nanoparticles can be overcome by incorporating reduction reactions that mimic chemical processes in an in vivo environment, in microorganisms and plants, and limitations of delivering an antimicrobial peptide can be improved by modifying the peptide's structure. For example, Wu et al. (2015) designed dopamine-coated silver nanoparticles (AgNPs) on a central venous catheter (CVC) surface that displayed dose-dependent antibacterial activity. Kung et al. (2016) designed a dopamine-functionalized silver nanocomposite that selectively targets bacteria while being nontoxic to endothelial cells. Ghavaminejad et al. highlighted the wound-healing property of silver nanoparticles under both in vitro and in vivo conditions. These nanoparticles were designed based on a hydrophobic dopamine methacrylamide (DMA) monomer (Ghavaminejad et al. 2016). Wang et al. (2017b) designed a biomimetic dopamine-modified ε-poly-L-lysine-polyethylene glycol-based hydrogel (PPD hydrogel) wound dressing with excellent hemostatic property and accelerated wound repair process. Meng et al. illustrated the antipathogenic property of hydrogen peroxide-releasing microgels containing catechol as the adhesive moiety. These microgels displayed excellent antibacterial activity against both Gram-positive (*S. epidermidis*) and Gram-negative (*E. coli*) bacteria (Meng et al. 2019).

5.6.3 Biomimetic Assemblies

Biomimetic assemblies include antimicrobial materials designed by assembling different components, e.g., a cell membrane-like structure or cell-mimetic morphologies to improve the target-directed delivery of an antimicrobial agent. These assembles include bioactive supramolecular assemblies, antimicrobial peptide-based particles, and cell-mimetic particles.

5.6.3.1 Biomimetic Assembled Particles

Biomimetic assembled materials make use of charged materials such as nanoparticles coated with charged lipid layers or polymeric coatings for attaching to the bacterial cell wall. The particles will then burst the bacterial cell by mechanical disruption in a way similar to surface mimicry (de Melo Carrasco et al. 2015). Xavier et al. designed a biomimetic assembled particle by incorporating an antimicrobial peptide, gramicidin, into a cationic bilayer with a polystyrene core. The assembly mimics histones with DNA compaction and was found to enhance the antibacterial activity of gramicidin against *S. aureus* (Xavier and Carmona-Ribeiro 2017). Similarly, biomimetic assembled particles utilizing cationic polymers poly(diallyldimethylammonium chloride) (PDDA) and poly(methyl methacrylate) (PMMA) were designed by Sanches et al. (2015) and were found to possess significant antibacterial efficacy.

de Melo Carrasco et al. designed cationic nanostructured particles with the poly(diallyldimethylammonium) chloride (PDDA) polymer as the outer layer, and these particles displayed significant antibacterial activity against MDR Gram-positive and Gram-negative bacterial strains via membrane disruption (de Melo Carrasco et al. 2015). Rosa et al. designed cationic biomimetic particles by enclosing polystyrene sulfate (PSS) particles into a dioctadecyldimethylammonium bromide (DODAB) bilayer (PSS/DODAB). These particles displayed significant antibacterial activity against *E. coli* (Rosa et al. 2008).

5.6.3.2 Antimicrobial Peptide-Based Particles

Knowledge of antimicrobial peptide-based particles can help in designing particles mimicking the antimicrobial peptide to improve the site-directed delivery of an antimicrobial agent to enhance its efficacy, and limit its degradation and off-target effects. Additionally, enclosing antimicrobial peptides within a membrane allows their delivery at difficult-to-reach targets and protects them from premature degradation (Azmi et al. 2016; Xiong et al. 2015; Emberger et al. 2011). Xi et al. designed a hybrid polypeptide-based nanofibrous matrix composed of poly(citrate)-ε-poly lysine (PCE) and polycaprolactone (PCL). The PCL–PCE hybrid nanofibrous matrix potentially inhibited the growth of MDR bacteria in a mouse wound infection model and was also found to enhance wound healing and the skin regeneration process in mice (Xi et al. 2018).

Ryge et al. (2008) designed 20 lysine–peptoid hybrids that displayed activity against both Gram-positive and Gram-negative bacteria, including the MDR strains (panel includes methicillin-susceptible *S. aureus* ATCC 29213, MRSA ATCC

33591, VISA ATCC 700699 MU50, VRE ATCC 700221, *P. aeruginosa* ATCC 27853, and amphotericin-B-resistant *C. albicans* ATCC 200955 and the clinical isolates of *Salmonella typhimurium* and *Cryptococcus neoformans*). Additionally, Molchanova et al. (2017) designed and synthesized lysine-based α-peptide/β--peptoid peptidomimetics oligomers that displayed significant antibacterial activity against MDR Gram-negative bacteria. Delivering peptides within the membrane makes their delivery feasible for difficult-to-reach targets, for example, sites beyond the blood–brain barrier, while retaining their antimicrobial property. Additionally, the membrane protects the peptides from their premature degradation. Thus, biomimetic assembled materials are found to possess great potential for improving the delivery of a variety of antimicrobial materials.

5.6.3.3 Cell-Mimetic Materials

Immune cells, e.g., leukocytes and platelets, have the intricate mechanisms to fight against foreign pathogenic organisms. The cell-mimetic strategy involves designing antimicrobial particles that mimic various immune cell functions. Molinaro et al. designed leukocyte-mimetic materials by enclosing polymeric lipid nanoparticles into leukocyte membranes. The assembly was termed “leukosomes,” and these leukosomes could deliver drugs to the targeted inflamed vasculature (Molinaro et al. 2016). Hammer et al. used polymersome tunability to design leuko-polymersome, consisting of binding motif-coated polymeric vesicles that mimic leukocyte interactions in the physiological environment. The leuko-polymersome was studied as a potential carrier in targeted drug discovery (Hammer et al. 2008). Parodi et al. (2013) designed nanoporous silicon particles coated with the leukocyte membrane (termed “leuko-like vectors” (LLVs)) toward a targeted drug delivery system. Besides targeted drug delivery, these cell mimetics have also shown to reduce the effect of bacterial toxins. Fang and colleagues (2019)designed nanosponges by enclosing polymeric nanoparticles with the red blood cell (RBC) membrane, which mimics nonspecific binding of bacterial toxins to the cell membrane components (Table 5.1). Sproul et al. designed nanogold composite deformable hydrogels that mimic platelets. These platelet-mimetic particles were effective in reducing bacterial growth (Sproul et al. 2020). Besides mimicking human cells, several bacterial-derived nanoparticles have also been studied as nano-vehicles for targeted drug delivery (Watanabe 2016; Liu et al. 2017a, 2018; Schulz et al. 2018; Wang et al. 2018, 2019; Yu et al. 2018). These nano-vehicles damage bacterial cells by nitrogen captivation (Wang et al. 2018). Gao et al. (2015) designed biomimetic nanovaccines (bacterial membrane-coated gold nanoparticles (BM-AuNPs)) by enclosing gold nanoparticles with *E. coli* outer membrane vesicles (OMVs). The designed nanovaccines were found to elicit bacteria-specific antibody responses, which were significantly higher when compared to those of OMVs alone. Additionally, the vaccines were observed to induce strong T-helper (Th)1- and Th17-biased cell responses against *E. coli* (Hu et al. 2013; Yang et al. 2019a). Although biomimetic strategies have proven to be effective in targeted drug discovery and in overcoming the cytotoxicity of the known antimicrobial agents, they need further optimization to be taken forward for clinical applications (Chee and Brown 2020).

5.7 Modulation of the Gut Microbiota

The human body harbors a complex ecosystem of microbial communities. Of these, maximum diversity is found in the large intestine and is collectively referred to as the gut microbiome (Lynch and Pedersen 2016). After the characterization of the human microbiome by the Human Microbiome Project (Peterson et al. 2009), high-throughput sequencing revealed the presence of around 10^{11-12} microorganisms per gram of content in the human intestine (Araos and D'Agata 2019; Lagier et al. 2012). The gut microbiome has been studied to play an important role in maintaining human health by regulating host metabolism, influencing the immune response, and checking disease progression. The gut microbiome is largely affected by factors like age, diet, lifestyle, and geographic distribution (Xu et al. 2019). Adding on to these factors, consumption of antibiotics, especially broad-spectrum antibiotics, drastically reduces the diversity of the natural microbiome. Studies have revealed that excessive or prolonged use of antibiotics (Hecker et al. 2003; Teo et al. 2012; Versporten et al. 2018) lead to the overgrowth of difficult-to-treat multidrug-resistant organisms (MDROs) (Nord et al. 1984; Jakobsson et al. 2010; Tan et al. 2020). Gut microbiota dysbiosis may result in complex gut disorders such as diarrhea (Li et al. 2021) and chronic inflammatory bowel disease (IBD) (Bistoletti et al. 2019; Baj et al. 2019), systemic metabolic diseases (Wu et al. 2021a), and neurological disorders (Gebrayel et al. 2022).

Several strategies have been studied and utilized to revert microbiome dysbiosis. These strategies include use of prebiotics, probiotics, postbiotics, metabolites, selective oropharyngeal decontamination (SOD), selective digestive decontamination (SDD), microRNA (miRNA), and fecal microbiota transplantation (FMT) (Tan et al. 2020; Gebrayel et al. 2022; Mindt and DiGiandomenico 2022). SOD and SDD implements use antimicrobial agents that selectively target pathogens in the oropharynx and gut, respectively, instead of being absorbed in the systemic circulation, thus having minimal impact on the nontarget microbiome (Wittekamp et al. 2018; Huttner et al. 2019; Tacconelli et al. 2019; Johanson et al. 1969; Stoutenbeek et al. 1984).

5.7.1 Prebiotics

Prebiotics are a group of nonviable substrates that selectively promote the growth and/or activity of beneficial microorganisms, including both indigenous and exogenously administered strains, and hence confers health benefits on the host (Hill et al. 2014). Some of the commonly used prebiotics include fructo-oligosaccharides (FOS), galacto-oligosaccharides (GOS), trans-galacto-oligosaccharides (TOS), and pectic oligosaccharide (POS) (Davani-Davari et al. 2019). These prebiotics are naturally present in dietary food products such as asparagus, sugar beet, garlic, chicory, onion, Jerusalem artichoke, wheat, honey, banana, barley, tomato, rye, soybean, human and cow milk, peas, beans, etc. and, recently, seaweeds and microalgae (Ali and Nizar 2018). Prebiotics confer their health benefits on being digested by the gut microbiota into short-chain fatty acids ((SCFAs), including lactic acid, butyric acid, and propionic acid) and peptidoglycan. These fermented products modulate the gut microbiome in their own specific manner, for example, propionate regulate T helper 2 (Th2 cells), macrophages, and bone marrow dendritic cells (Stinson et al. 2017). SCFAs act by decreasing the pH of the colon and can also affect the distant-site organs via diffusing in the blood (Hernot et al. 2009). Studies have reported the role of prebiotics in irritable bowel syndrome (IBS) and Crohn's disease associated with *Bifidobacteria*, *Faecalibacterium prausnitzii*, and/or a decreased Bacteroides-to-Firmicutes ratio (Whelan 2013). In two randomized, double-blind clinical trials, FOS consumed at a concentration of 5 g/day for 6 weeks (Paineau et al. 2008) or GOS at a concentration of 3.5 g/day for 12 weeks (Silk et al. 2009) were found to be effective in improving IBS symptoms. In an another study, FOS at 15 g/day administered for 3 weeks to a population with Crohn's disease reported elevated levels of *Bifidobacteria* in the collected fecal samples (Lindsay et al. 2006). In a similar study, prebiotics such as GOS and FOS were found to be effective in stimulating the growth of the gut microbiota (e.g., *Bifidobacteria*), which reduced the count of pathogenic bacteria in preterm infants with necrotizing enterocolitis (NEC) (Knol et al. 2005).

Probiotics are living organisms that confer health benefits on the host when administered in adequate amounts (Reid 2016). These probiotics are available as food, pills, or powders. The commonly used probiotics for gastrointestinal microflora include lactic acid bacteria (e.g., *Lactobacillus*) and yeast-based *Saccharomyces boulardii* (*cerevisiae*) (Eggers et al. 2018; Tannock et al. 2011; Salomão et al. 2016). Probiotics play an effective role in modulating the gut microbiota in several ways, including bacteriocin and short-chain fatty acid production, in lowering the pH of the gut, and in competing with other bacteria for available nutrients. These, in turn, stimulate mucosal barrier function and immunomodulation (Kechagia et al. 2013). Probiotics are also found to influence the acquired and innate immune responses of the host by inducing phagocytosis and immunoglobulin (Ig)A secretion, regulating T-cell responses, enhancing Th1 responses, and attenuating Th2 responses (Guarner and Malagelada 2003). The next-generation probiotic *Faecalibacterium prausnitzii* has been reported to improve gut dysbiosis by stimulating host immune responses and exhibits anti-inflammatory action (He et al. 2021; Lapiere et al. 2020). These probiotics have also been studied to control the transmission of healthcare-associated infections (HAIs). Caselli et al. (2018, 2019)studied the role of probiotics in significantly decreasing the risk of contracting an HAI during hospitalization in a multicenter, pre-post interventional study. Probiotics including *Lacticasei bacillus rhamnosus* GG, *Limosilacto bacillus reuteri*, *Lactobacillus casei* Shirota, and *Bifidobacterium animalis* Bb12 were found to be effective in reducing the duration of acute rotavirus diarrhea, majorly associated with infant mortality (Szajewska and Mrukowicz 2001). Additionally, probiotics, including *L. rhamnosus* GG, *S. boulardii*, *Lactobacillus acidophilus*, and *Bifidobacterium bifidum*, were found to be beneficial in preventing travelers' diarrhea in adults (McFarland 2007). Expanding the benefits of probiotics, these organisms are found to target several enteropathogens via production of bacteriocins and thereby protecting the host organism (Moslehi-Jenabian et al. 2011).

Postbiotics are defined as "preparations of inanimate microorganisms and/or their components that confers health benefits in the host" (Salminen et al. 2021). They are generally classified as carbohydrates (teichoic acids and galactose-rich polysaccharides), proteins (p40, p75 molecules, lactocepin), lipids (butyrate, acetate, propionate, lactate, dimethyl acetyl-derived plasmalogen), vitamins (B-group vitamins), organic acids (3-phenyllactic acid and propionic), and other complex molecules (lipoteichoic acids, peptidoglycan-derived muropeptides) (Thorakkattu et al. 2022). Postbiotics such as Broncho-Vaxom or Ismigen (polyvalent bacterial lysates) are used for prophylactic purposes in respiratory tract infections and chronic obstructive pulmonary disease (COPD) exacerbations and in treating gastroenteritis, enteric infections, and gut barrier dysfunction. Morales-Ferré et al. (2022) reported the role of postbiotic Lactofidus™ and short-chain galacto-oligosaccharides (scGOS)/long-chain fructo-oligosaccharides (lcFOS) in treating rats with early-life rotavirus induced-diarrhea. Corsello et al. (2017) reported the role of heat-killed *Lactobacillus paracasei* CBA L74 in avoiding diarrhea in 537 healthy Italian kids.

Gut microbiota-associated metabolites also play an important role in regulating the development and progression of several human diseases. This property of microbe-associated metabolites have been explored for the treatment local inflammation, cardiometabolic and neurological disorders, and cancers (Descamps et al. 2019; Yang et al. 2019b). For example, the gut microbe-dependent metabolite trimethylamine *N*-oxide (TMAO) was found to regulate atherosclerotic cardiovascular disease (CVD) (Yang et al. 2019b). miRNAs, short non-coding RNA molecules, are also found to be important regulators of microbiome–host interaction (Liu and Weiner 2016; Gebrayel et al. 2022).

5.7.2 Fecal Microbiota Transplantation (FMT)

Fecal microbiota transplantation (FMT) is an emerging therapeutic option for regulation of the gut microbiome. This therapy involves administration of liquefied or encapsulated preprocessed stool samples taken from a healthy donor into the colon of the patient via endoscopy or oral capsule preparation. This results in repopulation of the recipient's gut microflora, thus restoring the gut microbiome (Tan et al. 2020). The potency of FMT as a microbiome regulator was first reported in 1958, when Eiseman et al. (1958) reported the efficacy of fecal enemas in patients with pseudomembranous colitis. FMT is now an approved therapeutic option to tackle recurrent colitis, caused by *Clostridium difficile*, with cure rates of around 90% (McDonald et al. 2018; van Nood et al. 2013; Lee et al. 2016; Hvas et al. 2019). Biliński et al. 2016; Bilinski et al. 2017) studied and reported the efficacy of FMT in decolonizing ~60–100% of MDR bacteria after 1 month of treatment. Later, they also identified FMT to play an immunomodulatory role in the treatment of acute gastrointestinal graft-versus-host disease (Biliński et al. 2021).

The technique was also found to be effective in eliminating carbapenemase-producing *Enterobacterales* (Lagier et al. 2015; Bilinski et al. 2017), vancomycin-resistant *Enterococci* (VRE) (Dinh et al. 2018), and extended-spectrum β-lactamase-producing *Enterobacterales* in both immunocompetent and immunocompromised patients (Singh et al. 2018). Apart from bacteria-associated infections, FMT is now also being studied for other disorders like obesity, diabetes (Napolitano and Covasa 2020), and metabolic syndrome (Leshem et al. 2019). Although FMT has proven to be an effective therapeutic option to restore microbiome dysbiosis (Fig. 5.6), its specificity and heterogeneity both need to be evaluated further for microbe- and disease-specific optimization. Moreover, the therapy faces the limitations of manufacturing, storing, and delivery of fecal inoculum. Researchers are trying to overcome these limitations using frozen or freeze-dried stool capsules, for making them orally available in clinical settings (Cheminet et al. 2018) (Patent no. WO2017103225A1 2021). Thus, all these strategies to modulate the gut microbiome are emerging as an effective treatment option to combat emerging drug-resistant pathogens.

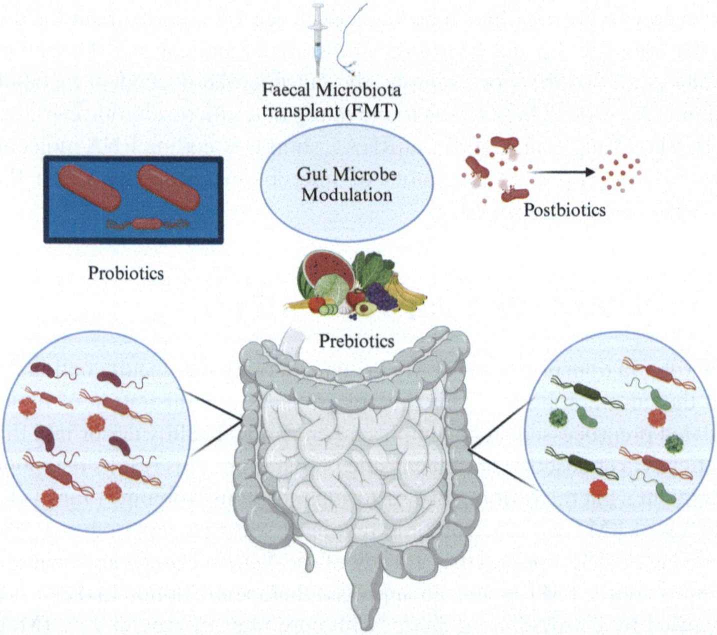

Fig. 5.6 Gut microbiome modulation as an effective way of eliminating drug-resistant pathogenic bacteria

5.8 Drug Discovery Through AI/Machine Learning (ML)

Drug discovery through artificial intelligence has been an extremely useful tool in recent times. The pharmaceutical sector generally uses AI/ML (artificial intelligence/machine learning) to control huge amounts of data and also to analyze the data to get an output. AI/ML mainly uses a set of data and builds a pattern that helps screen large amounts of data. In recent times, the medicinal chemistry field has adopted this path to choose a large number of drug-like molecules and omit the rest of the molecules in an extremely quick manner. Despite the opportunity, AI/ML has some drawbacks, including data uncertainty, handling a large set of data, diversity inside the dataset, etc. A huge amount of money (approximately US$2.8 billion) is required to develop and process a drug molecule through the drug discovery steps, but 90% molecules fail phase II clinical trials. Some algorithms (heart of ML), such as nearest-neighbor classifiers, random forest (RF), extreme learning machines, support vector machines (SVMs), and deep neural networks (DNNs), are used to predict in vivo activity and toxicity. This is why biological and pharmaceutical companies are now interested in collaborating with information technology (IT) to create a stage that can help discover different therapies. AI can also predict the affinity of a drug molecule toward the target site. If a drug molecule binds to a target different from its actual target, then this may cause a toxicity issue. AI helps solve this problem by considering the structural similarity of the drug and its target. More or less all diseases are related to some proteins that are overexpressed in certain diseases. That is why to discover a new drug, it is extremely essential to know the protein structure that is related to this disease. AI helps structure-aided drug discovery by predicting the three-dimensional (3D) structure of the protein. AlphaFold is an AI device that can do the task. Structural prediction of proteins can also help in drug repurposing (Fig. 5.7) (Paul et al. 2020; Beneke and Mackenrodt 2019; King et al. 1995).

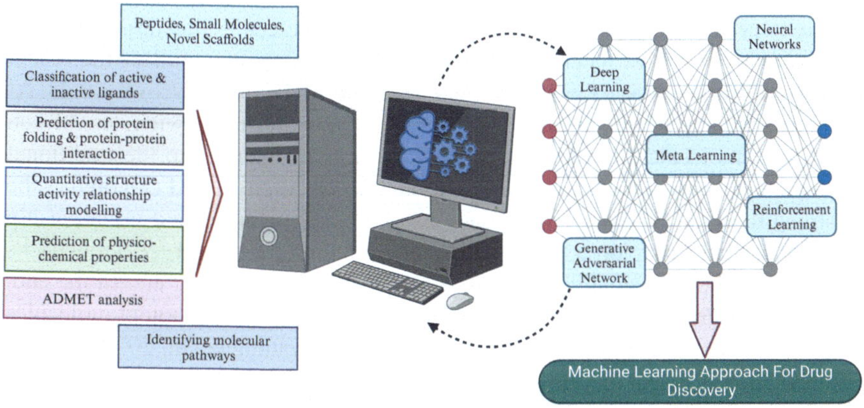

Fig. 5.7 Machine learning approach toward drug discovery

Abaucin, a new molecule found by Gary et al., is a potent antibiotic against the Gram-negative nosocomial pathogen *A. baumannii*. The authors first screened 7684 molecules through the conventional screening method and selected 480 molecules. After that, these data were used to instruct a binary classifier to predict whether the upcoming molecules are active or not. In this way, they found abaucin to be a potent target for *A. baumannii*. They tested abaucin against 41 strains of *A. baumannii*, and abaucin shows potent activity at different concentrations. Surprisingly, abaucin did not show any growth inhibitory activity against 24 carbapenem-resistant Enterobacteriaceae strains, 24 *Pseudomonas aeruginosa* strains, and 14 *Staphylococcus aureus* strains at different minimum inhibitory concentrations (MICs). To identify the target of abaucin, authors generated resistant mutant to abaucin and subjected it to whole genome sequencing, where they found mutation in lolE gene use for lipoprotein trafficking in bacteria. In an in vivo efficacy study, abaucin was tested in wound infection in mice. Mice were treated with *A. baumannii* ATCC 17978 at about 6.5×10^6cfu/mL after 1 h of infection and abaucin was treated at 2, 3, 4, 6, 10, 21, and 24 h along with vehicle control. After 25 h, all the mice were killed, and it was found that vehicle-treated mice showed a bacterial count of about 6.9×10^8cfu/g and abaucin-treated mice showed a bacterial count of about 4.0×10^7cfu/g. From these data, it is clear that abaucin shows significant activity in an in vivo wound infection model. At last, we can say that abaucin is an excellent molecule to treat *A. baumannii* infection (Liu et al. 2023).

In recent times, different AI-based drug discovery companies have started to discover drugs and some molecules have also reached the clinical trial phase of the drug discovery pipeline. The process is both cost-effective and challenging. Madura et al. studied about 24 AI-based drug discovery companies from 2010 to 2021 at a time when drug discovery through AI was on the rise. TYK2 is a protein that is a member of the Janus kinase (JAK) family of proteins. Some inhibitors of these proteins are available in the market, but they have issues with specificity, meaning that they can bind multiple isoforms of the same family. AI can find molecules that can bind more specifically to a single JAK isoform than several market-approved drugs. The inhibitor of the serotonin receptor can also be found using AI. Both 5-hydroxytryptamine (5-HT)1A and 5-HT2A are two molecules that target the serotonin receptor and have recently entered the clinic. From seeing all of the recent advances, it is clear that AI can open up a new path in the drug discovery field. AI has both advantages and drawbacks. Advantages include timesaving, as AI can screen $>10^6$ molecules in a limited amount of time, as opposed to the traditional technique that is more time-consuming. Another is specificity, as AI can predict the 3D structure of the target and after that Quantitative Structure activity relationship (QSAR) can help develop a proper molecule/inhibitor for the target. Drawbacks include limited data availability, maintenance of large sets of data, etc. In the end, one can conclude that AI can help drug discovery by finding new potent molecules against different proteins with maximum efficiency (Jayatunga et al. 2022).

5.9 Phage Therapy

A phage or a bacteriophage is a type of virus that infects bacteria. Different types of phages include lytic phages, lysogenic phages, and chronic phages. Lytic phages generally attack bacterial cells, replicate inside them, and then lyse them to release progeny phages. Lysogenic phages infect bacteria in the same way as lytic phages do, but, instead of cell lysis, these types of phages integrate their genome inside the bacterial genome. Chronic phages infect bacteria but neither lyse the bacterial cell nor integrate their genome inside the host genome. Filamentous phages are continuously released into the environment by bacteria (Housby and Mann 2009). The first phage therapy was reported in 1919 by d'Hérelle and co-workers. After d'Hérelle's success, Lang et al. used phage therapy to treat chronic orthopedic conditions (Abedon et al. 2011).

Rosanna et al. used phages to treat *S. aureus* infection in mice, including the MRSA strain. They used *S. aureus* A170 at a concentration of 10^8cfu/mouse and a phage concentration of 10^9 PFU and rescued 97% mice. Even when used at a nonlethal dose (5×10^6cfu/mouse) following a 10-day infection, phages totally reduced the bacterial burden (Capparelli et al. 2007). Phage endolysin is also a promising antibacterial target. Endolysins are mainly divided into three types: (1) glycosidases, (2) amidases, and (3) endopeptidases. Endolysins are mainly used to release progeny phages into the environment from bacterial cells. Endolysins along with holin are used to create pores in the inner membrane and peptidoglycan (Fig. 5.8).

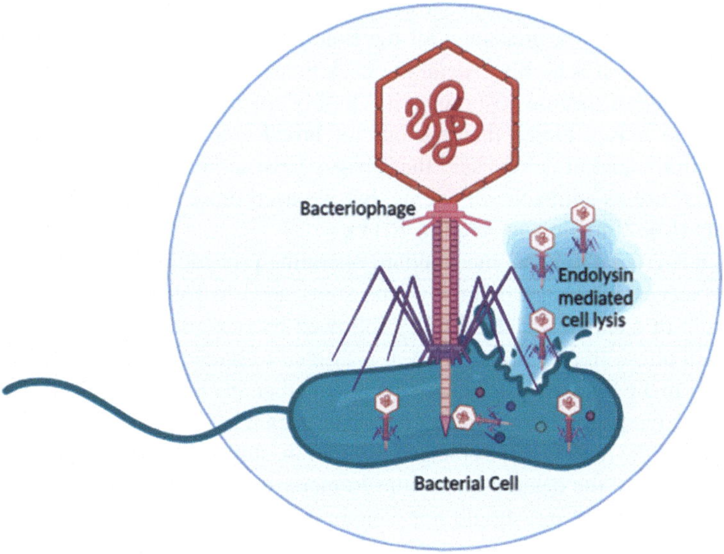

Fig. 5.8 A diagram illustrating the release of progeny phages via endolysin-mediated host cell lysis

In Gram-negative bacteria, this can be achieved by fusion of the outer membrane with the inner membrane (Gordillo Altamirano and Barr 2019).

Artilysin, combining a polycationic nonapeptide and a modular endolysin, can eliminate MDR Gram-negative strains (*A. baumannii*, *P. aeruginosa*) from 4 to 5 log reduction within 30 min (Briers et al. 2014). The phage ϕMR11-coded endolysin MV-L can successfully eliminate the MDR *S. aureus* strain, including MRSA (Rashel et al. 2007). One chimeric lysine by Daniel et al. is active against MRSA. One chimeric lysin, named ClyS, by Daniel et al. is active against MRSA. ClyS is made by fusing N-terminal catalytic domain of *S. aureus* Twort phage lysin with C-terminal cell wall-targeting domain from another *S. aureus* phage lysin (phiNM3). When tested in in vivo mouse nasal decolonization model, ClyS shows more or less 2 log reductions in 1 h with a single dose. This lysine also shows significant interaction with doxycycline and vancomycin to treat MRSA (Anu et al. 2010).

Like all therapies, phage therapy also has some pros and cons. A major advantage is its lack of resistance to antibiotics. Lytic bacteriophages generally attack bacterial cells and kill them, so bacteria do not have a chance to show resistance; however, this is not true for static antibiotics. To avoid multiple dosing, obligate lytic phages generally infect one bacterium, kill it, and produce progeny phages so that no further dosing is required. The third advantage is less toxicity; phages are less toxic than antibiotics because they only contain proteins and nucleic acids, despite having some allergic responses. The fourth advantage is specificity; specific phages are used for a particular type of bacteria, which is why other normal microflorae of the body are not harmed. The fifth advantage is different formulations and combinations; phages can be used in combination with antibiotics and different formulations can be prepared for use. The major disadvantages of phage therapy include specific environment; phages require some specific environments to replicate, and this cannot be managed all the time inside the body. Some phages also contain toxins, so it is highly crucial to check their full genome. The second disadvantage includes a specific host range; it is highly crucial to know the specific bacterial infection before phage therapy, but for broad-spectrum antibiotics, it is not required. In the end, it can be said that phages have some excellent properties that enable their use as antibacterial agents, but we need more knowledge to use them efficiently (Loc-Carrillo and Abedon 2011).

Now, it is crucial to know the standard operating procedure (SOP) of phage therapy. For successful phage therapy, some steps need to followed; the first one is preparation of a suitable phage library followed by recognition of phage-sensitive bacteria; in phage library preparation, it is crucial to screen the phage before accumulating it in the library by its genetic level (whether any toxin or undesirable genes are present); the phage library must target pathogenic bacteria like ESKAPE pathogens. Second, the SOP includes phage preparation, storage, and transport. For these, one has to choose the desired bacterium to increase the phage progeny; the bacterium must be specific and should not contain undesired genes. Phage formulations come in different forms such as creams, injections, and powders. For example, if anyone wants to treat their respiratory infection, then he/she can choose an aerosolized phage formulation. The proper pH value required to maintain phage preparation ranges between 7.35 and 7.45. The phage preparation should be maintained in cold temperature. The third step is determining an accurate route of administration to direct the interaction between the phage and bacteria. For any blood infection, phage infusion can be a good choice, and for a GI tract infection, the phage should be inside a protective particle to avoid inaction by stomach acids (Cui et al. 2019).

5.10 Nanoparticles

Antibiotic-loaded nanoparticles are now a highly efficient treatment option for various antibacterial infections. Antibiotics can cure such infections but there have some issues like bioavailability, stability, etc. Nanoparticles carry drugs to the targeted site and have some major advantages like excellent solubility, slow release of antibiotics into the environment, increased antibiotic half-life, etc. Nanoparticles can themselves act as antibacterial agents. Nano-particle made up with metals, chitosan and surfactants which impart net positive charge on the surface can be attracted with overall negative charge on bacterial cell membrane, thus destroying the membrane as like antimicrobial peptides. Antibiotics encapsulated inside nanoparticles are slowly released into the environment, which is the reason behind their increased bioavailability and stability. Nanoparticles can be engineered to release antibiotics in a specific environment (pH, specific enzyme concentration, some chemicals), which is present at the diseased site. They can also get rid of bacterial biofilm formation. Biofilms of bacteria are mainly made up of extracellular polymeric substances (EPSs) that can lead to medical complications. Generally, the bacteria present in the nucleus of a biofilm are more difficult to treat. Some nanoparticles can penetrate into biofilms and destroy bacteria. Au-based nanoparticles are mainly used as drug carriers. Drug molecules adsorb at the surface of the nanoparticle, leading to sustained and controlled release of the drug into the environment. Ag nanoparticles can themselves be used as antibacterial agents. Their mode of action is leakage of cytoplasmic components due to cell wall disintegration. Super paramagnetic iron oxide nanoparticles (SPIONs) are combined with antibodies, antimicrobial peptides, and aptamers to get activated. They are also useful for deep penetration such as to eradicate bacteria from biofilms. In case of natural or synthetic polymers, antibiotics may either attach to the surface of the nanoparticle covalently or may be incorporated physically inside the nanoparticle. Alginate is an anionic polysaccharide found on the cell wall of algae, which can be used as a matrix to carry drugs. An alginate matrix-based drug delivery system is mainly used to treat tuberculosis and fungal infections (Yeh et al. 2020).

Vaccination by virus-like particles (VLPs) provides a new horizon. These particles do not contain any genetic elements and are thus not harmful. They can generate immune responses that can eliminate the bacterial burden inside the body in a diseased condition. However, this technique still faces obstacles, which include choosing a suitable bacterial marker. Thus, judging from all sides, it will not be wrong to say that this method will open new avenues to combat AMR.

5.11 Genomic Mining

It has been observed by a group of researchers that in actinomycetes, a large number of bioactive molecules produce silent genes for some unknown reason (Nett et al. 2009; Bentley et al. 2002; Oliynyk et al. 2007). The biosynthetic gene clusters (BGCs) inside the bacterium are not expressed under laboratory conditions. A complete genome sequence of *Streptomyces coelicolor* revealed that 16 additional gene clusters are present inside the genome, which give rise to different bioactive molecules, including non-ribosomal peptide synthetases (NRPSs), polyketide synthases (PKSs), terpene synthases, and NRPS-independent siderophore synthetases (Challis 2014). After that BCGs were also found in other microorganisms and give birth to a new era to control AMR. Next-generation sequencing is mainly used to identify BGCs. One widely used in silico tool is antiSMASH (antibiotics and secondary metabolite analysis shell) (Medema et al. 2011). So now, the question is why are BCGs not upregulated in the lab environment and how can we do this? It is believed that various types of activation signals are absent in the in vitro growth culture, which is why BGCs are downregulated in the lab environment. One approach to express BCGs is through their pleiotropic expression, wherein more than one gene is simultaneously expressed and there is no induction of any other gene expression. However, one drawback is that since various metabolites are found in a mixture, downstream purification is difficult. Change in the growth condition may be induced by some BGCs, and this may be a good approach to express these types of downregulated genes. Some scientists tried to grow *Aspergillus nidulans* under several different growth conditions and obtained 45 different types of extracts from which aspoquinolones A–D and aspernidines A and B were identified (Scherlach and Hertweck 2006; Scherlach et al. 2010). Induction of BGCs by incorporation of a suitable gene adjacent to BGC that can be up-regulated by a specific signal is a possibility. The PKS–NRPS gene cluster in *A. nidulans* can be induced by incorporating a gene adjacent to the gene cluster, which is a pathway-specific activator. Now, these hybrid genes, when induced, cause upregulation of the PKS–NRPS gene cluster, from which aspyridones A and B were identified (Bergmann et al. 2007). In a well-studied strain of bacteria, it has been observed that a large number of genes are not expressed in normal laboratory conditions, thus demonstrating potent biological activity. Looking at this matter led to the idea of gene mining.

AntiSmash is a tool that can be used for genome mining. This tool is generally used for identifying and analyzing gene clusters. In recent times, this technology has become more superior. In 2019, Blin et al. presented version 5 of AntiSmash, a more developed and advanced version. This version was redesigned with the help of Python version 3 and the code also restructured. It includes rules of 52 different BGC types, which can detect *N*-acyl amino acids, β-lactones, polybrominated diphenyl ethers, C-nucleosides, pseudopyronines, fungal ribosomally synthesized and post-translationally modified peptides (RiPPs), and RaS-RiPPs. In this recent version, the type II PKS prediction system has been improved by many virtues (Blin et al. 2019).

Busi et al. found six different classes of BGCs from glacier-fed stream epilithic biofilms. They hypothesized that close to each BGC, there is a gene that can inhibit the effect of the product produced by those BGCs (mainly antimicrobial agents) (Busi et al. 2023).

5.12 Hemofiltration Devices

Apart from these strategies, several hemofiltration devices are now being introduced to remove pathogens or pathogen-associated toxins from the blood. Monard et al. (2023) published the details of the available hemofiltration devices to target bacterial pathogens in sepsis and to modulate the host cell immune system in response to inflammatory mediators (cytokines) or pathogen-associated molecular patterns (PAMPs). These hemofilters include the **Seraph® 100 Microbind® Affinity Blood Filter** (utilizes heparin to adsorb pathogens, including viruses, e.g., Zika virus, cytomegalovirus, adenovirus, Gram-negative and Gram-positive bacteria, and positively charged cytokines) (Li and Kusche-Gullberg 2016; McCrea et al. 2014; Seffer et al. 2021; Axelsson et al. 2010), the **GARNET® Hemofilter** (utilizes mannose-binding lectin (MBL) to initiate bacterial opsonization) (Dommett et al. 2006; Kang et al. 2014; Didar et al. 2015), and the **CytoSorb®** (utilizes polymyxin-B-immobilized columns inside a hemoperfusion cartridge that adsorbs bacteria-associated endotoxins, preventing early phase of bacteria-associated septic shock) (Srisawat et al. 2018; Dellinger et al. 2018; Payen et al. 2015). Liu et al. (2024) reported the efficacy of macrophage membrane-coated silicon nanowired microfluidic devices to remove PAMPs and cytokines from blood in rats with lipopolysaccharide (LPS)-induced sepsis.

5.13 Conclusions

Emerging MDR bacterial resistance to all the currently available drugs highlights the urgent need to develop newer antibiotics with novel mechanisms of action. In this scenario, the above-discussed newly emerging antibacterial therapy has come up with a solution to deal with these quickly evolving MDR bacterial pathogens. These strategies not only have the potency to bypass the resistance mechanism of MDR pathogenic bacteria but also have the ability to potentiate antibiotics against these strains. For example, the specific and multitargeted action of antibacterial photodynamic therapy does not allow bacteria to generate resistance. Targeting virulent genes by CRISPR–Cas and genome editing aids in reducing bacterial pathogenesis. Utilizing biomimetic assemblies and gut microbiome modulation, one can naturally eliminate resistant bacteria along with eliminating the risk of toxicity associated with certain antibiotics. Utilizing artificial intelligence in the drug discovery process quickens the process of identifying novel antibacterial agents and novel bacterial targets. Although there is a need to further explore the potency of this therapy, the combination of antibacterial strategies and antibiotics brings a ray of hope to treat infections caused by MDR bacterial pathogens.

Acknowledgments DS thanks UGC and RM thanks DST INSPIRE for their fellowship. This chapter bears the CSIR–CDRI communication number 10758.

References

Abdelsattar AS, Abdelrahman F, Dawoud A et al (2019) Encapsulation of E. coli phage ZCEC5 in chitosan-alginate beads as a delivery system in phage therapy. AMB Express 9:87. https://doi.org/10.1186/s13568-019-0810-9

Abedon ST, Kuhl SJ, Blasdel BG, Kutter EM (2011) Phage treatment of human infections, vol 1. pp 66–85. https://doi.org/10.4161/bact.1.2.15845

Abrami L, Liu S, Cosson P et al (2003) Anthrax toxin triggers endocytosis of its receptor via a lipid raft-mediated clathrin-dependent process. J Cell Biol 160:321–328. https://doi.org/10.1083/jcb.200211018

Agostinis P, Berg K, Cengel KA et al (2011) Photodynamic therapy of cancer: an update. CA Cancer J Clin 61:250–281. https://doi.org/10.3322/caac.20114

Ali E, Nizar NNA (2018) Preparation and processing of religious and cultural foods. Woodhead Publishing, an imprint of Elsevier, Duxford

Alkhnbashi OS, Shah SA, Garrett RA et al (2016) Characterizing leader sequences of CRISPR loci. Bioinformatics 32:i576–i585. https://doi.org/10.1093/bioinformatics/btw454

Allegretta G, Maurer CK, Eberhard J et al (2017) In-depth profiling of MvfR-regulated small molecules in Pseudomonas aeruginosa after quorum sensing inhibitor treatment. Front Microbiol 8:924. https://doi.org/10.3389/fmicb.2017.00924

Al-Mutairi R, Tovmasyan A, Batinic-Haberle I, Benov L (2018) Sublethal photodynamic treatment does not lead to development of resistance. Front Microbiol 9:1699. https://doi.org/10.3389/fmicb.2018.01699

Amara N, Gregor R, Rayo J et al (2016) Fine-tuning covalent inhibition of bacterial quorum sensing. Chembiochem 17:825–835. https://doi.org/10.1002/cbic.201500676

Andersen KK, Strokappe NM, Hultberg A et al (2016) Neutralization of Clostridium difficile toxin B mediated by engineered Lactobacilli that produce single-domain antibodies. Infect Immun 84:395–406. https://doi.org/10.1128/IAI.00870-15

Annunziato G (2019) Strategies to overcome antimicrobial resistance (AMR) making use of non-essential target inhibitors: a review. Int J Mol Sci 20:5844. https://doi.org/10.3390/ijms20235844

Anu D, Chad E, Mattias C et al (2010) Synergism between a novel chimeric lysin and oxacillin protects against infection by methicillin-resistant Staphylococcus aureus. Antimicrob Agents Chemother 54:1603–1612. https://doi.org/10.1128/aac.01625-09

Araos R, D'Agata EMC (2019) The human microbiota and infection prevention. Infect Control Hosp Epidemiol 40:585–589. https://doi.org/10.1017/ice.2019.28

Axelsson J, Ferreira M, Adolfsson L et al (2010) Cytokines in blood from septic patients interact with surface-immobilized heparin. ASAIO J 56:48–51. https://doi.org/10.1097/MAT.0b013e3181c3fec8

Aydin S, Personne Y, Newire E et al (2017) Presence of type I-F CRISPR/Cas systems is associated with antimicrobial susceptibility in Escherichia coli. J Antimicrob Chemother 72:2213–2218. https://doi.org/10.1093/jac/dkx137

Azmi F, Skwarczynski M, Toth I (2016) Towards the development of synthetic antibiotics: designs inspired by natural antimicrobial peptides. Curr Med Chem 23:4610–4624. https://doi.org/10.2174/0929867323666160825162435

Bailey L, Gylfe A, Sundin C et al (2007) Small molecule inhibitors of type III secretion in Yersinia block the Chlamydia pneumoniae infection cycle. FEBS Lett 581:587–595. https://doi.org/10.1016/j.febslet.2007.01.013

Baj A, Moro E, Bistoletti M et al (2019) Glutamatergic signaling along the microbiota-gut-brain axis. Int J Mol Sci 20:1482. https://doi.org/10.3390/ijms20061482

Baldry M, Kitir B, Frøkiær H et al (2016) The agr inhibitors Solonamide B and analogues alter immune responses to Staphylococccus aureus but do not exhibit adverse effects on immune cell functions. PLoS One 11:e0145618. https://doi.org/10.1371/journal.pone.0145618

Bartusik D, Aebisher D, Lyons AM, Greer A (2012) Bacterial inactivation by a singlet oxygen bubbler: identifying factors controlling the toxicity of (1)O2 bubbles. Environ Sci Technol 46:12098–12104. https://doi.org/10.1021/es303645n

Basgall EM, Goetting SC, Goeckel ME et al (2018) Gene drive inhibition by the anti-CRISPR proteins AcrIIA2 and AcrIIA4 in Saccharomyces cerevisiae. Microbiology 164:464–474. https://doi.org/10.1099/mic.0.000635

Bassi ZI, Fillmore MC, Miah AH et al (2018) Modulating PCAF/GCN5 immune cell function through a PROTAC approach. ACS Chem Biol 13:2862–2867. https://doi.org/10.1021/acschembio.8b00705

Batalha PN, Gomes ATPC, Forezi LSM et al (2015) Synthesis of new porphyrin/4-quinolone conjugates and evaluation of their efficiency in the photoinactivation of Staphylococcus aureus. RSC Adv 5:71228–71239. https://doi.org/10.1039/c5ra11070j

Battesti A, Gottesman S (2013) Roles of adaptor proteins in regulation of bacterial proteolysis. Curr Opin Microbiol 16:140–147. https://doi.org/10.1016/j.mib.2013.01.002

Bazaka K, Jacob MV, Chrzanowski W, Ostrikov K (2015) Anti-bacterial surfaces: natural agents, mechanisms of action, and plasma surface modification. RSC Adv 5:48739–48759. https://doi.org/10.1039/c4ra17244b

Bearson BL, Bearson SMD (2008) The role of the QseC quorum-sensing sensor kinase in colonization and norepinephrine-enhanced motility of Salmonella enterica serovar Typhimurium. Microb Pathog 44:271–278. https://doi.org/10.1016/j.micpath.2007.10.001

Beneke F, Mackenrodt M-O (2019) Artificial intelligence and collusion. IIC—Int Rev Intellect Prop Compet Law 50:109–134. https://doi.org/10.1007/s40319-018-00773-x

Bentley SD, Chater KF, Cerdeño-Tárraga A-M et al (2002) Complete genome sequence of the model actinomycete Streptomyces coelicolor A3(2). Nature 417:141–147. https://doi.org/10.1038/417141a

Berg V, Sellstedt M, Hedenström M et al (2006) Design, synthesis and evaluation of peptidomimetics based on substituted bicyclic 2-pyridones-targeting virulence of uropathogenic E. coli. Bioorg Med Chem 14:7563–7581. https://doi.org/10.1016/j.bmc.2006.07.017

Bergmann S, Schümann J, Scherlach K et al (2007) Genomics-driven discovery of PKS-NRPS hybrid metabolites from Aspergillus nidulans. Nat Chem Biol 3:213–217. https://doi.org/10.1038/nchembio869

Bikard D, Euler CW, Jiang W et al (2014) Exploiting CRISPR-Cas nucleases to produce sequence-specific antimicrobials. Nat Biotechnol 32:1146–1150. https://doi.org/10.1038/nbt.3043

Biliński J, Grzesiowski P, Muszyński J et al (2016) Fecal microbiota transplantation inhibits multidrug-resistant gut pathogens: preliminary report performed in an immunocompromised host. Arch Immunol Ther Exp 64:255–258

Bilinski J, Grzesiowski P, Sorensen N et al (2017) Fecal microbiota transplantation in patients with blood disorders inhibits gut colonization with antibiotic-resistant bacteria: results of a prospective, single-center study. Clin Infect Dis 65:364–370. https://doi.org/10.1093/cid/cix252

Biliński J, Jasiński M, Tomaszewska A et al (2021) Fecal microbiota transplantation with ruxolitinib as a treatment modality for steroid-refractory/dependent acute, gastrointestinal graft-versus-host disease: a case series. Am J Hematol 96:E461–E463

Bisland SK, Chien C, Wilson BC, Burch S (2006) Pre-clinical in vitro and in vivo studies to examine the potential use of photodynamic therapy in the treatment of osteomyelitis. Photochem Photobiol Sci 5:31–38. https://doi.org/10.1039/b507082a

Bistoletti M, Caputi V, Baranzini N et al (2019) Antibiotic treatment-induced dysbiosis differently affects BDNF and TrkB expression in the brain and in the gut of juvenile mice. PLoS One 14:e0212856. https://doi.org/10.1371/journal.pone.0212856

Blin K, Shaw S, Steinke K et al (2019) antiSMASH 5.0: updates to the secondary metabolite genome mining pipeline. Nucleic Acids Res 47:W81–W87. https://doi.org/10.1093/nar/gkz310

Borges AL, Davidson AR, Bondy-Denomy J (2017) The discovery, mechanisms, and evolutionary impact of anti-CRISPRs. Annu Rev Virol 4:37–59. https://doi.org/10.1146/annurev-virology-101416-041616

Briers Y, Walmagh M, Van Puyenbroeck V et al (2014) Engineered endolysin-based "Artilysins" to combat multidrug-resistant gram-negative pathogens. mBio 5(4):e01379–e01314. https://doi.org/10.1128/mBio.01379-14

Brint JM, Ohman DE (1995) Synthesis of multiple exoproducts in Pseudomonas aeruginosa is under the control of RhlR-RhlI, another set of regulators in strain PAO1 with homology to the autoinducer-responsive LuxR-LuxI family. J Bacteriol 177:7155–7163. https://doi.org/10.1128/jb.177.24.7155-7163.1995

Bukowski M, Wladyka B, Dubin G (2010) Exfoliative toxins of Staphylococcus aureus. Toxins (Basel) 2:1148–1165. https://doi.org/10.3390/toxins2051148

Busi SB, de Nies L, Pramateftaki P et al (2023) Glacier-fed stream biofilms harbor diverse resistomes and biosynthetic gene clusters. Microbiol Spectr 11:e04069–e04022

Cahan R, Swissa N, Gellerman G, Nitzan Y (2010) Photosensitizer-antibiotic conjugates: a novel class of antibacterial molecules. Photochem Photobiol 86:418–425. https://doi.org/10.1111/j.1751-1097.2009.00674.x

Capparelli R, Parlato M, Borriello G et al (2007) Experimental phage therapy against Staphylococcus aureus in mice. Antimicrob Agents Chemother 51:2765–2773. https://doi.org/10.1128/AAC.01513-06

Caselli E, Brusaferro S, Coccagna M et al (2018) Reducing healthcare-associated infections incidence by a probiotic-based sanitation system: a multicentre, prospective, intervention study. PLoS One 13:e0199616. https://doi.org/10.1371/journal.pone.0199616

Caselli E, Arnoldo L, Rognoni C et al (2019) Impact of a probiotic-based hospital sanitation on antimicrobial resistance and HAI-associated antimicrobial consumption and costs: a multicenter study. Infect Drug Resist 12:501–510. https://doi.org/10.2147/IDR.S194670

Cassini A, Högberg LD, Plachouras D et al (2019) Attributable deaths and disability-adjusted life-years caused by infections with antibiotic-resistant bacteria in the EU and the European economic area in 2015: a population-level modelling analysis. Lancet Infect Dis 19:56–66. https://doi.org/10.1016/S1473-3099(18)30605-4

Cattò C, Dell'Orto S, Villa F et al (2015) Unravelling the structural and molecular basis responsible for the anti-biofilm activity of zosteric acid. PLoS One 10:e0131519. https://doi.org/10.1371/journal.pone.0131519

Cegelski L, Marshall GR, Eldridge GR, Hultgren SJ (2008) The biology and future prospects of antivirulence therapies. Nat Rev Microbiol 6:17–27. https://doi.org/10.1038/nrmicro1818

Chadha P, Katare OP, Chhibber S (2017) Liposome loaded phage cocktail: enhanced therapeutic potential in resolving Klebsiella pneumoniae mediated burn wound infections. Burns 43:1532–1543. https://doi.org/10.1016/j.burns.2017.03.029

Challis GL (2014) Exploitation of the Streptomyces coelicolor A3(2) genome sequence for discovery of new natural products and biosynthetic pathways. J Ind Microbiol Biotechnol 41:219–232. https://doi.org/10.1007/s10295-013-1383-2

Charpentier E, Richter H, van der Oost J, White MF (2015) Biogenesis pathways of RNA guides in archaeal and bacterial CRISPR-Cas adaptive immunity. FEMS Microbiol Rev 39:428–441. https://doi.org/10.1093/femsre/fuv023

Chee E, Brown AC (2020) Biomimetic antimicrobial material strategies for combating antibiotic resistant bacteria. Biomater Sci 8:1089–1100. https://doi.org/10.1039/c9bm01393h

Cheminet G, Kapel N, Bleibtreu A et al (2018) Faecal microbiota transplantation with frozen capsules for relapsing Clostridium difficile infections: the first experience from 15 consecutive patients in France. J Hosp Infect 100:148–151. https://doi.org/10.1016/j.jhin.2018.07.005

Chen Z, Moayeri M, Zhou Y-H et al (2006) Efficient neutralization of anthrax toxin by chimpanzee monoclonal antibodies against protective antigen. J Infect Dis 193:625–633. https://doi.org/10.1086/500148
Chowdhury S, Carter J, Rollins MF et al (2017) Structure reveals mechanisms of viral suppressors that intercept a CRISPR RNA-guided surveillance complex. Cell 169:47–57.e11. https://doi.org/10.1016/j.cell.2017.03.012
Christaki E, Marcou M, Tofarides A (2020) Antimicrobial resistance in bacteria: mechanisms, evolution, and persistence. J Mol Evol 88:26–40. https://doi.org/10.1007/s00239-019-09914-3
Chung KK, Schumacher JF, Sampson EM et al (2007) Impact of engineered surface microtopography on biofilm formation of Staphylococcus aureus. Biointerphases 2:89–94. https://doi.org/10.1116/1.2751405
Churcher I (2018) Protac-induced protein degradation in drug discovery: breaking the rules or just making new ones? J Med Chem 61:444–452. https://doi.org/10.1021/acs.jmedchem.7b01272
Clarke MB, Hughes DT, Zhu C et al (2006) The QseC sensor kinase: a bacterial adrenergic receptor. Proc Natl Acad Sci U S A 103:10420–10425. https://doi.org/10.1073/pnas.0604343103
Corsello G, Carta M, Marinello R et al (2017) Preventive effect of cow's milk fermented with Lactobacillus paracasei CBA L74 on common infectious diseases in children: a multicenter randomized controlled trial. Nutrients 9:669. https://doi.org/10.3390/nu9070669
Cui Z, Guo X, Feng T, Li L (2019) Exploring the whole standard operating procedure for phage therapy in clinical practice. J Transl Med 17:1–7. https://doi.org/10.1186/s12967-019-2120-z
Daly SM, Elmore BO, Kavanaugh JS et al (2015) ω-Hydroxyemodin limits staphylococcus aureus quorum sensing-mediated pathogenesis and inflammation. Antimicrob Agents Chemother 59:2223–2235. https://doi.org/10.1128/AAC.04564-14
Davani-Davari D, Negahdaripour M, Karimzadeh I et al (2019) Prebiotics: definition, types, sources, mechanisms, and clinical applications. Foods 8:92. https://doi.org/10.3390/foods8030092
de Melo Carrasco LD, Sampaio JLM, Carmona-Ribeiro AM (2015) Supramolecular cationic assemblies against multidrug-resistant microorganisms: activity and mechanism of action. Int J Mol Sci 16:6337–6352. https://doi.org/10.3390/ijms16036337
Dehbanipour R, Ghalavand Z (2022) Anti-virulence therapeutic strategies against bacterial infections: recent advances. Germs 12:262–275. https://doi.org/10.18683/germs.2022.1328
Dellinger RP, Bagshaw SM, Antonelli M et al (2018) Effect of targeted polymyxin B hemoperfusion on 28-day mortality in patients with septic shock and elevated endotoxin level: the EUPHRATES randomized clinical trial. JAMA 320:1455–1463. https://doi.org/10.1001/jama.2018.14618
Descamps HC, Herrmann B, Wiredu D, Thaiss CA (2019) The path toward using microbial metabolites as therapies. EBioMedicine 44:747–754. https://doi.org/10.1016/j.ebiom.2019.05.063
Díaz A, del Valle LJ, Rodrigo N et al (2018) Antimicrobial activity of poly(ester urea) electrospun fibers loaded with bacteriophages. Fibers 6:33. https://doi.org/10.3390/fib6020033
Didar TF, Cartwright MJ, Rottman M et al (2015) Improved treatment of systemic blood infections using antibiotics with extracorporeal opsonin hemoadsorption. Biomaterials 67:382–392. https://doi.org/10.1016/j.biomaterials.2015.07.046
Dinh A, Fessi H, Duran C et al (2018) Clearance of carbapenem-resistant Enterobacteriaceae vs vancomycin-resistant enterococci carriage after faecal microbiota transplant: a prospective comparative study. J Hosp Infect 99:481–486. https://doi.org/10.1016/j.jhin.2018.02.018
Dommett RM, Klein N, Turner MW (2006) Mannose-binding lectin in innate immunity: past, present and future. Tissue Antigens 68:193–209. https://doi.org/10.1111/j.1399-0039.2006.00649.x
Dong H, Xiang H, Mu D et al (2019) Exploiting a conjugative CRISPR/Cas9 system to eliminate plasmid harbouring the mcr-1 gene from Escherichia coli. Int J Antimicrob Agents 53:1–8. https://doi.org/10.1016/j.ijantimicag.2018.09.017
Dosselli R, Millioni R, Puricelli L et al (2012) Molecular targets of antimicrobial photodynamic therapy identified by a proteomic approach. J Proteomics 77:329–343. https://doi.org/10.1016/j.jprot.2012.09.007

Drees SL, Li C, Prasetya F et al (2016) PqsBC, a condensing enzyme in the biosynthesis of the Pseudomonas aeruginosa quinolone signal: crystal structure, inhibition, and reaction mechanism. J Biol Chem 291:6610–6624. https://doi.org/10.1074/jbc.M115.708453
Duan J, Li M, Hao Z et al (2018) Subinhibitory concentrations of resveratrol reduce alpha-hemolysin production in Staphylococcus aureus isolates by downregulating saeRS. Emerg Microbes Infect 7:136. https://doi.org/10.1038/s41426-018-0142-x
Eggers S, Barker AK, Valentine S et al (2018) Effect of Lactobacillus rhamnosus HN001 on carriage of Staphylococcus aureus: results of the impact of probiotics for reducing infections in veterans (IMPROVE) study. BMC Infect Dis 18:129. https://doi.org/10.1186/s12879-018-3028-6
Eiseman B, Silen W, Bascom GS, Kauvar AJ (1958) Fecal enema as an adjunct in the treatment of pseudomembranous enterocolitis. Surgery 44:854–859
Elbourne A, Crawford RJ, Ivanova EP (2017) Nano-structured antimicrobial surfaces: from nature to synthetic analogues. J Colloid Interface Sci 508:603–616. https://doi.org/10.1016/j.jcis.2017.07.021
Elsasser S, Finley D (2005) Delivery of ubiquitinated substrates to protein-unfolding machines. Nat Cell Biol 7:742–749. https://doi.org/10.1038/ncb0805-742
Emberger M, Koller J, Laimer M et al (2011) Nosocomial staphylococcal scalded skin syndrome caused by intra-articular injection. J Eur Acad Dermatol Venereol 25:227–231. https://doi.org/10.1111/j.1468-3083.2010.03766.x
Felise HB, Nguyen HV, Pfuetzner RA et al (2008) An inhibitor of gram-negative bacterial virulence protein secretion. Cell Host Microbe 4:325–336. https://doi.org/10.1016/j.chom.2008.08.001
Feng Y, Coradi Tonon C, Ashraf S, Hasan T (2021) Photodynamic and antibiotic therapy in combination against bacterial infections: efficacy, determinants, mechanisms, and future perspectives. Adv Drug Deliv Rev 177:113941. https://doi.org/10.1016/j.addr.2021.113941
Filloux A (2009) The type VI secretion system: a tubular story. EMBO J 28:309–310. https://doi.org/10.1038/emboj.2008.301
Filloux A, Hachani A, Bleves S (2008) The bacterial type VI secretion machine: yet another player for protein transport across membranes. Microbiology 154:1570–1583. https://doi.org/10.1099/mic.0.2008/016840-0
Finley D (2009) Recognition and processing of ubiquitin-protein conjugates by the proteasome. Annu Rev Biochem 78:477–513. https://doi.org/10.1146/annurev.biochem.78.081507.101607
Fleitas Martínez O, Cardoso MH, Ribeiro SM, Franco OL (2019) Recent advances in anti-virulence therapeutic strategies with a focus on dismantling bacterial membrane microdomains, toxin neutralization, quorum-sensing interference and biofilm inhibition. Front Cell Infect Microbiol 9:74. https://doi.org/10.3389/fcimb.2019.00074
Fontana CR, Abernethy AD, Som S et al (2009) The antibacterial effect of photodynamic therapy in dental plaque-derived biofilms. J Periodontal Res 44:751–759. https://doi.org/10.1111/j.1600-0765.2008.01187.x
Fuhrmann J, Schmidt A, Spiess S et al (2009) McsB is a protein arginine kinase that phosphorylates and inhibits the heat-shock regulator CtsR. Science 324:1323–1327. https://doi.org/10.1126/science.1170088
Fuqua WC, Winans SC, Greenberg EP (1994) Quorum sensing in bacteria: the LuxR-LuxI family of cell density-responsive transcriptional regulators. J Bacteriol 176:269–275. https://doi.org/10.1128/jb.176.2.269-275.1994
Galán JE, Cossart P (2005) Host-pathogen interactions: a diversity of themes, a variety of molecular machines. Curr Opin Microbiol 8:1–3
Gao W, Fang RH, Thamphiwatana S et al (2015) Modulating antibacterial immunity via bacterial membrane-coated nanoparticles. Nano Lett 15:1403–1409. https://doi.org/10.1021/nl504798g
Gao H, Sun X, Rao Y (2020) PROTAC technology: opportunities and challenges. ACS Med Chem Lett 11:237–240. https://doi.org/10.1021/acsmedchemlett.9b00597
García-Fernández E, Koch G, Wagner RM et al (2017) Membrane microdomain disassembly inhibits MRSA antibiotic resistance. Cell 171:1354–1367.e20. https://doi.org/10.1016/j.cell.2017.10.012

Gauthier A, Robertson ML, Lowden M et al (2005) Transcriptional inhibitor of virulence factors in enteropathogenic Escherichia coli. Antimicrob Agents Chemother 49:4101–4109. https://doi.org/10.1128/AAC.49.10.4101-4109.2005

Gebrayel P, Nicco C, Al Khodor S et al (2022) Microbiota medicine: towards clinical revolution. J Transl Med 20:111. https://doi.org/10.1186/s12967-022-03296-9

George EA, Muir TW (2007) Molecular mechanisms of agr quorum sensing in virulent staphylococci. Chembiochem 8:847–855. https://doi.org/10.1002/cbic.200700023

Ghavaminejad A, Park CH, Kim CS (2016) In situ synthesis of antimicrobial silver nanoparticles within antifouling Zwitterionic hydrogels by Catecholic redox chemistry for wound healing application. Biomacromolecules 17:1213–1223. https://doi.org/10.1021/acs.biomac.6b00039

Givens BE, Naguib YW, Geary SM et al (2018) Nanoparticle-based delivery of CRISPR/Cas9 genome-editing therapeutics. AAPS J 20:108. https://doi.org/10.1208/s12248-018-0267-9

Goldberg AL (1992) The mechanism and functions of ATP-dependent proteases in bacterial and animal cells. Eur J Biochem 203:9–23. https://doi.org/10.1111/j.1432-1033.1992.tb19822.x

Gong N, Tan Y, Li M et al (2016) ALA-PDT combined with antibiotics for the treatment of multiple skin abscesses caused by Mycobacterium fortuitum. Photodiagn Photodyn Ther 15:70–72. https://doi.org/10.1016/j.pdpdt.2016.05.009

Gooderham WJ, Hancock REW (2009) Regulation of virulence and antibiotic resistance by two-component regulatory systems in Pseudomonas aeruginosa. FEMS Microbiol Rev 33:279–294. https://doi.org/10.1111/j.1574-6976.2008.00135.x

Gordillo Altamirano FL, Barr JJ (2019) Phage therapy in the postantibiotic era. Clin Microbiol Rev 32:e00066–e00018. https://doi.org/10.1128/CMR.00066-18

Gould K (2016) Antibiotics: from prehistory to the present day. J Antimicrob Chemother 71:572–575. https://doi.org/10.1093/jac/dkv484

Greenberg M, Kuo D, Jankowsky E et al (2018) Small-molecule AgrA inhibitors F12 and F19 act as antivirulence agents against gram-positive pathogens. Sci Rep 8:14578. https://doi.org/10.1038/s41598-018-32829-w

Guarner F, Malagelada J-R (2003) Gut flora in health and disease. Lancet 361:512–519. https://doi.org/10.1016/S0140-6736(03)12489-0

Hammer DA, Robbins GP, Haun JB et al (2008) Leuko-polymersomes. Faraday Discuss 139(128–129):419–420. https://doi.org/10.1039/b717821b

Han Z, Pinkner JS, Ford B et al (2010) Structure-based drug design and optimization of mannoside bacterial FimH antagonists. J Med Chem 53:4779–4792. https://doi.org/10.1021/jm100438s

Hanzl A, Winter GE (2020) Targeted protein degradation: current and future challenges. Curr Opin Chem Biol 56:35–41. https://doi.org/10.1016/j.cbpa.2019.11.012

Harrington LB, Doxzen KW, Ma E et al (2017) A broad-Spectrum inhibitor of CRISPR-Cas9. Cell 170:1224–1233.e15. https://doi.org/10.1016/j.cell.2017.07.037

Hartmann M, Papavlassopoulos H, Chandrasekaran V et al (2012) Inhibition of bacterial adhesion to live human cells: activity and cytotoxicity of synthetic mannosides. FEBS Lett 586:1459–1465. https://doi.org/10.1016/j.febslet.2012.03.059

He X, Zhao S, Li Y (2021) Faecalibacterium prausnitzii: a next-generation probiotic in gut disease improvement. Can J Infect Dis Med Microbiol 2021:1. https://doi.org/10.1155/2021/6666114

Hecker MT, Aron DC, Patel NP et al (2003) Unnecessary use of antimicrobials in hospitalized patients: current patterns of misuse with an emphasis on the antianaerobic spectrum of activity. Arch Intern Med 163:972–978. https://doi.org/10.1001/archinte.163.8.972

Heidari A, Noshiranzadeh N, Haghi F, Bikas R (2017) Inhibition of quorum sensing related virulence factors of Pseudomonas aeruginosa by pyridoxal lactohydrazone. Microb Pathog 112:103–110. https://doi.org/10.1016/j.micpath.2017.09.043

Henry BD, Neill DR, Becker KA et al (2015) Engineered liposomes sequester bacterial exotoxins and protect from severe invasive infections in mice. Nat Biotechnol 33:81–88. https://doi.org/10.1038/nbt.3037

Hernot DC, Boileau TW, Bauer LL et al (2009) In vitro fermentation profiles, gas production rates, and microbiota modulation as affected by certain fructans, galactooligosaccharides, and polydextrose. J Agric Food Chem 57:1354–1361. https://doi.org/10.1021/jf802484j

Hill C, Guarner F, Reid G et al (2014) Expert consensus document. The International Scientific Association for Probiotics and Prebiotics consensus statement on the scope and appropriate use of the term probiotic. Nat Rev Gastroenterol Hepatol 11:506–514. https://doi.org/10.1038/nrgastro.2014.66

Hille F, Charpentier E (2016) CRISPR-Cas: biology, mechanisms and relevance. Philos Trans R Soc Lond B Biol Sci 371:20150496. https://doi.org/10.1098/rstb.2015.0496

Housby JN, Mann NH (2009) Phage therapy. Drug Discov Today 14:536–540. https://doi.org/10.1016/j.drudis.2009.03.006

Howlin RP, Fabbri S, Offin DG et al (2015) Removal of dental biofilms with an ultrasonically activated water stream. J Dent Res 94:1303–1309. https://doi.org/10.1177/0022034515589284

Hu CMJ, Fang RH, Luk BT, Zhang L (2013) Nanoparticle-detained toxins for safe and effective vaccination. Nat Nanotechnol 8:933–938

Huang L, Wang M, Huang Y-Y et al (2018) Progressive cationic functionalization of chlorin derivatives for antimicrobial photodynamic inactivation and related vancomycin conjugates. Photochem Photobiol Sci 17:638–651. https://doi.org/10.1039/c7pp00389g

Hutchings MI, Truman AW, Wilkinson B (2019) Antibiotics: past, present and future. Curr Opin Microbiol 51:72–80. https://doi.org/10.1016/j.mib.2019.10.008

Huttner BD, de Lastours V, Wassenberg M et al (2019) A 5-day course of oral antibiotics followed by faecal transplantation to eradicate carriage of multidrug-resistant Enterobacteriaceae: a randomized clinical trial. Clin Microbiol Infect 25:830–838. https://doi.org/10.1016/j.cmi.2018.12.009

Hvas CL, Dahl Jørgensen SM, Jørgensen SP et al (2019) Fecal microbiota transplantation is superior to Fidaxomicin for treatment of recurrent Clostridium difficile infection. Gastroenterology 156:1324–1332.e3. https://doi.org/10.1053/j.gastro.2018.12.019

Ibbotson SH (2002) Topical 5-aminolaevulinic acid photodynamic therapy for the treatment of skin conditions other than non-melanoma skin cancer. Br J Dermatol 146:178–188. https://doi.org/10.1046/j.0007-0963.2001.04689.x

Iluz N, Maor Y, Keller N, Malik Z (2018) The synergistic effect of PDT and oxacillin on clinical isolates of Staphylococcus aureus. Lasers Surg Med 50:535–551. https://doi.org/10.1002/lsm.22785

Ivanova EP, Hasan J, Webb HK et al (2012) Natural bactericidal surfaces: mechanical rupture of Pseudomonas aeruginosa cells by cicada wings. Small 8:2489–2494. https://doi.org/10.1002/smll.201200528

Jakobsson HE, Jernberg C, Andersson AF et al (2010) Short-term antibiotic treatment has differing long-term impacts on the human throat and gut microbiome. PLoS One 5:e9836. https://doi.org/10.1371/journal.pone.0009836

Jayatunga MKP, Xie W, Ruder L et al (2022) From the analyst's couch AI in small-molecule drug discovery: a coming wave? Nat Rev Drug Discov 21(3):175–176. https://doi.org/10.1038/d41573-022-00025-1

Jeong S, Lee J, Im BN et al (2017) Combined photodynamic and antibiotic therapy for skin disorder via lipase-sensitive liposomes with enhanced antimicrobial performance. Biomaterials 141:243–250. https://doi.org/10.1016/j.biomaterials.2017.07.009

Ji C, Sharma I, Pratihar D et al (2016) Designed small-molecule inhibitors of the anthranilyl-CoA synthetase PqsA block quinolone biosynthesis in Pseudomonas aeruginosa. ACS Chem Biol 11:3061–3067. https://doi.org/10.1021/acschembio.6b00575

Johanson WG, Pierce AK, Sanford JP (1969) Changing pharyngeal bacterial flora of hospitalized patients. Emergence of gram-negative bacilli. N Engl J Med 281:1137–1140. https://doi.org/10.1056/NEJM196911202812101

Kalaiarasan E, Thirumalaswamy K, Harish BN et al (2017) Inhibition of quorum sensing-controlled biofilm formation in Pseudomonas aeruginosa by quorum-sensing inhibitors. Microb Pathog 111:99–107. https://doi.org/10.1016/j.micpath.2017.08.017

Kang JH, Super M, Yung CW et al (2014) An extracorporeal blood-cleansing device for sepsis therapy. Nat Med 20:1211–1216. https://doi.org/10.1038/nm.3640

Kang S, Kim J, Hur JK, Lee S-S (2017a) CRISPR-based genome editing of clinically important Escherichia coli SE15 isolated from indwelling urinary catheters of patients. J Med Microbiol 66:18–25. https://doi.org/10.1099/jmm.0.000406

Kang YK, Kwon K, Ryu JS et al (2017b) Nonviral genome editing based on a polymer-derivatized CRISPR nanocomplex for targeting bacterial pathogens and antibiotic resistance. Bioconjug Chem 28:957–967. https://doi.org/10.1021/acs.bioconjchem.6b00676

Karginov VA, Nestorovich EM, Moayeri M et al (2005) Blocking anthrax lethal toxin at the protective antigen channel by using structure-inspired drug design. Proc Natl Acad Sci U S A 102:15075–15080. https://doi.org/10.1073/pnas.0507488102

Kashef N, Hamblin MR (2017) Can microbial cells develop resistance to oxidative stress in antimicrobial photodynamic inactivation? Drug Resist Updat 31:31–42. https://doi.org/10.1016/j.drup.2017.07.003

Kauppi AM, Nordfelth R, Uvell H et al (2003) Targeting bacterial virulence: inhibitors of type III secretion in Yersinia. Chem Biol 10:241–249. https://doi.org/10.1016/s1074-5521(03)00046-2

Kechagia M, Basoulis D, Konstantopoulou S et al (2013) Health benefits of probiotics: a review. ISRN Nutr 2013:481651. https://doi.org/10.5402/2013/481651

Kerscher O, Felberbaum R, Hochstrasser M (2006) Modification of proteins by ubiquitin and ubiquitin-like proteins. Annu Rev Cell Dev Biol 22:159–180. https://doi.org/10.1146/annurev.cellbio.22.010605.093503

Keyser P, Elofsson M, Rosell S, Wolf-Watz H (2008) Virulence blockers as alternatives to antibiotics: type III secretion inhibitors against gram-negative bacteria. J Intern Med 264:17–29. https://doi.org/10.1111/j.1365-2796.2008.01941.x

Kharkwal GB, Sharma SK, Huang Y-Y et al (2011) Photodynamic therapy for infections: clinical applications. Lasers Surg Med 43:755–767. https://doi.org/10.1002/lsm.21080

Kim J-S, Cho D-H, Park M et al (2016) CRISPR/Cas9-mediated re-sensitization of antibiotic-resistant Escherichia coli harboring extended-spectrum β-lactamases. J Microbiol Biotechnol 26:394–401. https://doi.org/10.4014/jmb.1508.08080

King RD, Hirst JD, Sternberg MJE (1995) Comparison of artificial intelligence methods for modeling pharmaceutical QSARS. Appl Artif Intell 9:213–233. https://doi.org/10.1080/08839519508945474

Kirstein J, Dougan DA, Gerth U et al (2007) The tyrosine kinase McsB is a regulated adaptor protein for ClpCP. EMBO J 26:2061–2070. https://doi.org/10.1038/sj.emboj.7601655

Kline T, Bowman J, Iglewski BH et al (1999) Novel synthetic analogs of the Pseudomonas autoinducer. Bioorg Med Chem Lett 9:3447–3452. https://doi.org/10.1016/s0960-894x(99)00626-5

Kline T, Felise HB, Barry KC et al (2008) Substituted 2-imino-5-arylidenethiazolidin-4-one inhibitors of bacterial type III secretion. J Med Chem 51:7065–7074. https://doi.org/10.1021/jm8004515

Knol J, Boehm G, Lidestri M et al (2005) Increase of faecal bifidobacteria due to dietary oligosaccharides induces a reduction of clinically relevant pathogen germs in the faeces of formula-fed preterm infants. Acta Paediatr Suppl 94:31–33. https://doi.org/10.1111/j.1651-2227.2005.tb02152.x

Koch G, Wermser C, Acosta IC et al (2017) Attenuating Staphylococcus aureus virulence by targeting Flotillin protein scaffold activity. Cell Chem Biol 24:845–857.e6. https://doi.org/10.1016/j.chembiol.2017.05.027

Krueger E, Hayes S, Chang EH et al (2018) Receptor-based peptides for inhibition of leukotoxin activity. ACS Infect Dis 4:1073–1081. https://doi.org/10.1021/acsinfecdis.7b00230

Kumar L, Chhibber S, Kumar R et al (2015) Zingerone silences quorum sensing and attenuates virulence of Pseudomonas aeruginosa. Fitoterapia 102:84–95. https://doi.org/10.1016/j.fitote.2015.02.002

Kung ML, Lin PY, Peng SW et al (2016) Biomimetic polymer-based Ag nanocomposites as a antimicrobial platform. Appl Mater Today 4:31–39

Kutty SK, Barraud N, Ho KKK et al (2015) Hybrids of acylated homoserine lactone and nitric oxide donors as inhibitors of quorum sensing and virulence factors in Pseudomonas aeruginosa. Org Biomol Chem 13:9850–9861. https://doi.org/10.1039/c5ob01373a

Lagier J-C, Million M, Hugon P et al (2012) Human gut microbiota: repertoire and variations. Front Cell Infect Microbiol 2:136. https://doi.org/10.3389/fcimb.2012.00136

Lagier JC, Million M, Fournier PE et al (2015) Faecal microbiota transplantation for stool decolonization of OXA-48 carbapenemase-producing Klebsiella pneumoniae. J Hosp Infect 90:173–174

Lane MC, Mobley HLT (2007) Role of P-fimbrial-mediated adherence in pyelonephritis and persistence of uropathogenic Escherichia coli (UPEC) in the mammalian kidney. Kidney Int 72:19–25. https://doi.org/10.1038/sj.ki.5002230

Lapiere A, Geiger M, Robert V et al (2020) Prophylactic Faecalibacterium prausnitzii treatment prevents the acute breakdown of colonic epithelial barrier in a preclinical model of pelvic radiation disease. Gut Microbes 12:1–15. https://doi.org/10.1080/19490976.2020.1812867

Le Guern F, Sol V, Ouk C et al (2017) Enhanced photobactericidal and targeting properties of a cationic porphyrin following the attachment of polymyxin B. Bioconjug Chem 28:2493–2506. https://doi.org/10.1021/acs.bioconjchem.7b00516

Le VTM, Tkaczyk C, Chau S et al (2016) Critical role of alpha-toxin and protective effects of its neutralization by a human antibody in acute bacterial skin and skin structure infections. Antimicrob Agents Chemother 60:5640–5648. https://doi.org/10.1128/AAC.00710-16

Lee H, Suh J-W (2016) Anti-tuberculosis lead molecules from natural products targeting Mycobacterium tuberculosis ClpC1. J Ind Microbiol Biotechnol 43:205–212. https://doi.org/10.1007/s10295-015-1709-3

Lee CH, Steiner T, Petrof EO et al (2016) Frozen vs fresh fecal microbiota transplantation and clinical resolution of diarrhea in patients with recurrent Clostridium difficile infection: a randomized clinical trial. JAMA 315:142–149. https://doi.org/10.1001/jama.2015.18098

Leshem A, Horesh N, Elinav E (2019) Fecal microbial transplantation and its potential application in cardiometabolic syndrome. Front Immunol 10:1341. https://doi.org/10.3389/fimmu.2019.01341

Li J-P, Kusche-Gullberg M (2016) Heparan sulfate: biosynthesis, structure, and function. Int Rev Cell Mol Biol 325:215–273. https://doi.org/10.1016/bs.ircmb.2016.02.009

Li Y, Xia S, Jiang X et al (2021) Gut microbiota and diarrhea: an updated review. Front Cell Infect Microbiol 11:625210. https://doi.org/10.3389/fcimb.2021.625210

Lindsay JO, Whelan K, Stagg AJ et al (2006) Clinical, microbiological, and immunological effects of fructo-oligosaccharide in patients with Crohn's disease. Gut 55:348–355. https://doi.org/10.1136/gut.2005.074971

Lino CA, Harper JC, Carney JP, Timlin JA (2018) Delivering CRISPR: a review of the challenges and approaches. Drug Deliv 25:1234–1257. https://doi.org/10.1080/10717544.2018.1474964

Liu S, Weiner HL (2016) Control of the gut microbiome by fecal microRNA. Microb Cell 3:176–177. https://doi.org/10.15698/mic2016.04.492

Liu C, Gao H, Lv P et al (2017a) Extracellular vesicles as an efficient nanoplatform for the delivery of therapeutics. Hum Vaccin Immunother 13:2678–2687. https://doi.org/10.1080/21645515.2017.1363935

Liu T, Liu Z, Ye Q et al (2017b) Coupling transcriptional activation of CRISPR-Cas system and DNA repair genes by Csa3a in Sulfolobus islandicus. Nucleic Acids Res 45:8978–8992. https://doi.org/10.1093/nar/gkx612

Liu Y, Defourny KAY, Smid EJ, Abee T (2018) Gram-positive bacterial extracellular vesicles and their impact on health and disease. Front Microbiol 9:1502. https://doi.org/10.3389/fmicb.2018.01502

Liu G, Catacutan DB, Rathod K et al (2023) Deep learning-guided discovery of an antibiotic targeting Acinetobacter baumannii. Nat Chem Biol 19:1342. https://doi.org/10.1038/s41589-023-01349-8

Liu S, van Beuningen FE, Xiao X et al (2024) Macrophage membrane-coated, nanostructured adsorbent surfaces in a microfluidic device for extracorporeal blood cleansing in bacterially induced sepsis. Adv Funct Mater 34:2305913. https://doi.org/10.1002/adfm.202305913

Loc-Carrillo C, Abedon ST (2011) Pros and cons of phage therapy. Bacteriophage 1:111–114. https://doi.org/10.4161/bact.1.2.14590

Lu M, Xing H, Yang Z et al (2017) Recent advances on extracellular vesicles in therapeutic delivery: challenges, solutions, and opportunities. Eur J Pharm Biopharm 119:381–395. https://doi.org/10.1016/j.ejpb.2017.07.010

Lynch SV, Pedersen O (2016) The human intestinal microbiome in health and disease. N Engl J Med 375:2369–2379. https://doi.org/10.1056/NEJMra1600266

Maikova A, Severinov K, Soutourina O (2018) New insights into functions and possible applications of Clostridium difficile CRISPR-Cas system. Front Microbiol 9:1740. https://doi.org/10.3389/fmicb.2018.01740

Makarova KS, Wolf YI, Alkhnbashi OS et al (2015) An updated evolutionary classification of CRISPR-Cas systems. Nat Rev Microbiol 13:722–736

Manefield M, Rasmussen TB, Henzter M et al (2002) Halogenated furanones inhibit quorum sensing through accelerated LuxR turnover. Microbiology 148:1119–1127. https://doi.org/10.1099/00221287-148-4-1119

Matson JS, Withey JH, DiRita VJ (2007) Regulatory networks controlling Vibrio cholerae virulence gene expression. Infect Immun 75:5542–5549. https://doi.org/10.1128/IAI.01094-07

McCoull W, Cheung T, Anderson E et al (2018) Development of a novel B-cell lymphoma 6 (BCL6) PROTAC to provide insight into small molecule targeting of BCL6. ACS Chem Biol 13:3131–3141. https://doi.org/10.1021/acschembio.8b00698

McCrea K, Ward R, LaRosa SP (2014) Removal of Carbapenem-resistant Enterobacteriaceae (CRE) from blood by heparin-functional hemoperfusion media. PLoS One 9:e114242. https://doi.org/10.1371/journal.pone.0114242

McDonald LC, Gerding DN, Johnson S et al (2018) Clinical practice guidelines for Clostridium difficile infection in adults and children: 2017 update by the Infectious Diseases Society of America (IDSA) and Society for Healthcare Epidemiology of America (SHEA). Clin Infect Dis 66:e1–e48. https://doi.org/10.1093/cid/cix1085

McFarland LV (2007) Meta-analysis of probiotics for the prevention of traveler's diarrhea. Travel Med Infect Dis 5:97–105. https://doi.org/10.1016/j.tmaid.2005.10.003

Medema MH, Blin K, Cimermancic P et al (2011) antiSMASH: rapid identification, annotation and analysis of secondary metabolite biosynthesis gene clusters in bacterial and fungal genome sequences. Nucleic Acids Res 39:W339–W346. https://doi.org/10.1093/nar/gkr466

Meeske AJ, Jia N, Cassel AK et al (2020) A phage-encoded anti-CRISPR enables complete evasion of type VI-A CRISPR-Cas immunity. Science 369:54–59. https://doi.org/10.1126/science.abb6151

Meng H, Forooshani PK, Joshi PU et al (2019) Biomimetic recyclable microgels for on-demand generation of hydrogen peroxide and antipathogenic application. Acta Biomater 83:109–118. https://doi.org/10.1016/j.actbio.2018.10.037

Michalska M, Gambacorta F, Divan R et al (2018) Tuning antimicrobial properties of biomimetic nanopatterned surfaces. Nanoscale 10:6639–6650. https://doi.org/10.1039/c8nr00439k

Mielich-Süss B, Wagner RM, Mietrach N et al (2017) Flotillin scaffold activity contributes to type VII secretion system assembly in Staphylococcus aureus. PLoS Pathog 13:e1006728. https://doi.org/10.1371/journal.ppat.1006728

Mindt BC, DiGiandomenico A (2022) Microbiome modulation as a novel strategy to treat and prevent respiratory infections. Antibiotics (Basel) 11:474. https://doi.org/10.3390/antibiotics11040474

Mitrophanov AY, Groisman EA (2008) Signal integration in bacterial two-component regulatory systems. Genes Dev 22:2601–2611. https://doi.org/10.1101/gad.1700308

Mitsunaga M, Ito K, Nishimura T et al (2022) Antimicrobial strategy for targeted elimination of different microbes, including bacterial, fungal and viral pathogens. Commun Biol 5:647. https://doi.org/10.1038/s42003-022-03586-4

Molchanova N, Hansen PR, Damborg P et al (2017) Lysine-based α-peptide/β-peptoid peptidomimetics: influence of hydrophobicity, fluorination, and distribution of cationic charge on antimicrobial activity and cytotoxicity. ChemMedChem 12:312–318. https://doi.org/10.1002/cmdc.201600553

Molinaro R, Corbo C, Martinez JO et al (2016) Biomimetic proteolipid vesicles for targeting inflamed tissues. Nat Mater 15:1037–1046. https://doi.org/10.1038/nmat4644

Monard C, Abraham P, Schneider A, Rimmelé T (2023) New targets for extracorporeal blood purification therapies in sepsis. Blood Purif 52:1–7. https://doi.org/10.1159/000524973

Morales-Ferré C, Azagra-Boronat I, Massot-Cladera M et al (2022) Preventive effect of a postbiotic and prebiotic mixture in a rat model of early life rotavirus induced-diarrhea. Nutrients 14:1163. https://doi.org/10.3390/nu14061163

Morley S, Griffiths J, Philips G et al (2013) Phase IIa randomized, placebo-controlled study of antimicrobial photodynamic therapy in bacterially colonized, chronic leg ulcers and diabetic foot ulcers: a new approach to antimicrobial therapy. Br J Dermatol 168:617–624. https://doi.org/10.1111/bjd.12098

Morreale FE, Kleine S, Leodolter J et al (2022) BacPROTACs mediate targeted protein degradation in bacteria. Cell 185:2338–2353.e18. https://doi.org/10.1016/j.cell.2022.05.009

Moshiri J, Kaur D, Hambira CM et al (2018) Identification of a small molecule anti-biofilm agent against Salmonella enterica. Front Microbiol 9:2804. https://doi.org/10.3389/fmicb.2018.02804

Moslehi-Jenabian S, Nielsen DS, Jespersen L (2011) Application of molecular biology and genomics of probiotics for enteric cytoprotection. In: Malago JJ, Koninkx JFJG, Marinsek-Logar R (eds) Probiotic bacteria and enteric infections: cytoprotection by probiotic bacteria. Springer, Dordrecht, pp 133–153

Müh U, Schuster M, Heim R et al (2006) Novel Pseudomonas aeruginosa quorum-sensing inhibitors identified in an ultra-high-throughput screen. Antimicrob Agents Chemother 50:3674–3679. https://doi.org/10.1128/AAC.00665-06

Murray CJ, Ikuta KS, Sharara F et al (2022) Global burden of bacterial antimicrobial resistance in 2019: a systematic analysis. Lancet 399:629–655. https://doi.org/10.1016/S0140-6736(21)02724-0

Muschiol S, Bailey L, Gylfe A et al (2006) A small-molecule inhibitor of type III secretion inhibits different stages of the infectious cycle of Chlamydia trachomatis. Proc Natl Acad Sci U S A 103:14566–14571. https://doi.org/10.1073/pnas.0606412103

Napolitano M, Covasa M (2020) Microbiota transplant in the treatment of obesity and diabetes: current and future perspectives. Front Microbiol 11:590370. https://doi.org/10.3389/fmicb.2020.590370

Neklesa TK, Snyder LB, Bookbinder M et al (2017) An oral androgen receptor PROTAC degrader for prostate cancer. J Clin Oncol 35:273

Nett M, Ikeda H, Moore BS (2009) Genomic basis for natural product biosynthetic diversity in the actinomycetes. Nat Prod Rep 26:1362–1384. https://doi.org/10.1039/b817069j

Nord CE, Heimdahl A, Kager L, Malmborg AS (1984) The impact of different antimicrobial agents on the normal gastrointestinal microflora of humans. Rev Infect Dis 6(Suppl 1):S270–S275. https://doi.org/10.1093/clinids/6.supplement_1.s270

Novick RP, Geisinger E (2008) Quorum sensing in staphylococci. Annu Rev Genet 42:541–564. https://doi.org/10.1146/annurev.genet.42.110807.091640

Nunes J, McGonagle GA, Eden J et al (2019) Targeting IRAK4 for degradation with PROTACs. ACS Med Chem Lett 10:1081–1085. https://doi.org/10.1021/acsmedchemlett.9b00219

Nuñez JK, Kranzusch PJ, Noeske J et al (2014) Cas1–Cas2 complex formation mediates spacer acquisition during CRISPR–Cas adaptive immunity. Nat Struct Mol Biol 21:528–534. https://doi.org/10.1038/nsmb.2820

O'Brien AD, Gentry MK, Thompson MR et al (1979) Shigellosis and Escherichia coli diarrhea: relative importance of invasive and toxigenic mechanisms. Am J Clin Nutr 32:229–233. https://doi.org/10.1093/ajcn/32.1.229

Oliynyk M, Samborskyy M, Lester JB et al (2007) Complete genome sequence of the erythromycin-producing bacterium Saccharopolyspora erythraea NRRL23338. Nat Biotechnol 25:447–453. https://doi.org/10.1038/nbt1297

Oppezzo OJ, Forte Giacobone AF (2018) Lethal effect of photodynamic treatment on Persister bacteria. Photochem Photobiol 94:186–189. https://doi.org/10.1111/php.12843

Ortines RV, Liu H, Cheng LI et al (2018) Neutralizing alpha-toxin accelerates healing of Staphylococcus aureus-infected wounds in nondiabetic and diabetic mice. Antimicrob Agents Chemother 62:e02288–e02217. https://doi.org/10.1128/AAC.02288-17
Paczkowski JE, Mukherjee S, McCready AR et al (2017) Flavonoids suppress Pseudomonas aeruginosa virulence through allosteric inhibition of quorum-sensing receptors. J Biol Chem 292:4064–4076. https://doi.org/10.1074/jbc.M116.770552
Paineau D, Payen F, Panserieu S et al (2008) The effects of regular consumption of short-chain fructo-oligosaccharides on digestive comfort of subjects with minor functional bowel disorders. Br J Nutr 99:311–318. https://doi.org/10.1017/S000711450779894X
Panchal RG, Hermone AR, Nguyen TL et al (2004) Identification of small molecule inhibitors of anthrax lethal factor. Nat Struct Mol Biol 11:67–72. https://doi.org/10.1038/nsmb711
Papi M, Palmieri V, Bugli F et al (2016) Biomimetic antimicrobial cloak by graphene-oxide agar hydrogel. Sci Rep 6:1–7. https://doi.org/10.1038/s41598-016-0010-7
Park J-H, Ahn M-Y, Kim Y-C et al (2012) In vitro and in vivo antimicrobial effect of photodynamic therapy using a highly pure chlorin e6 against Staphylococcus aureus Xen29. Biol Pharm Bull 35:509–514. https://doi.org/10.1248/bpb.35.509
Park JY, Moon BY, Park JW et al (2017) Genetic engineering of a temperate phage-based delivery system for CRISPR/Cas9 antimicrobials against Staphylococcus aureus. Sci Rep 7:44929. https://doi.org/10.1038/srep44929
Parodi A, Quattrocchi N, van de Ven AL et al (2013) Synthetic nanoparticles functionalized with biomimetic leukocyte membranes possess cell-like functions. Nat Nanotechnol 8:61–68. https://doi.org/10.1038/nnano.2012.212
Paul M, Leibovici L (2009) Combination antimicrobial treatment versus monotherapy: the contribution of meta-analyses. Infect Dis Clin North Am 23:277–293. https://doi.org/10.1016/j.idc.2009.01.004
Paul D, Sanap G, Shenoy S (2020) Artificial intelligence in drug discovery and development
Pawluk A, Shah M, Mejdani M et al (2017) Disabling a type I-E CRISPR-Cas nuclease with a bacteriophage-encoded anti-CRISPR protein. mBio 8:e01751–e01717. https://doi.org/10.1128/mBio.01751-17
Payen DM, Guilhot J, Launey Y et al (2015) Early use of polymyxin B hemoperfusion in patients with septic shock due to peritonitis: a multicenter randomized control trial. Intensive Care Med 41:975–984. https://doi.org/10.1007/s00134-015-3751-z
Pearson JP, Gray KM, Passador L et al (1994) Structure of the autoinducer required for expression of Pseudomonas aeruginosa virulence genes. Proc Natl Acad Sci U S A 91:197–201. https://doi.org/10.1073/pnas.91.1.197
Pearson JP, Pesci EC, Iglewski BH (1997) Roles of Pseudomonas aeruginosa las and rhl quorum-sensing systems in control of elastase and rhamnolipid biosynthesis genes. J Bacteriol 179:5756–5767. https://doi.org/10.1128/jb.179.18.5756-5767.1997
Pesci EC, Milbank JB, Pearson JP et al (1999) Quinolone signaling in the cell-to-cell communication system of Pseudomonas aeruginosa. Proc Natl Acad Sci U S A 96:11229–11234. https://doi.org/10.1073/pnas.96.20.11229
Peterson J, Garges S, Giovanni M et al (2009) The NIH human microbiome project. Genome Res 19:2317–2323. https://doi.org/10.1101/gr.096651.109
Ping O, Mao S, Xuewen H et al (2017) Sclareol protects Staphylococcus aureus-induced lung cell injury via inhibiting alpha-hemolysin expression. J Microbiol Biotechnol 27:19–25. https://doi.org/10.4014/jmb.1606.06039
Ping O, Ruixue Y, Jiaqiang D et al (2018) Subinhibitory concentrations of prim-O-glucosylcimifugin decrease the expression of alpha-hemolysin in Staphylococcus aureus (USA300). Evid Based Complement Alternat Med 2018:7579808. https://doi.org/10.1155/2018/7579808
Pinkner JS, Remaut H, Buelens F et al (2006) Rationally designed small compounds inhibit pilus biogenesis in uropathogenic bacteria. Proc Natl Acad Sci U S A 103:17897–17902. https://doi.org/10.1073/pnas.0606795103

Plagens A, Richter H, Charpentier E, Randau L (2015) DNA and RNA interference mechanisms by CRISPR-Cas surveillance complexes. FEMS Microbiol Rev 39:442–463. https://doi.org/10.1093/femsre/fuv019

Pogodin S, Hasan J, Baulin VA et al (2013) Biophysical model of bacterial cell interactions with nanopatterned cicada wing surfaces. Biophys J 104:835–840. https://doi.org/10.1016/j.bpj.2012.12.046

Popow O, Paulo JA, Tatham MH et al (2019) Identification of endogenous adenomatous polyposis coli interaction partners and β-catenin-independent targets by proteomics. Mol Cancer Res 17:1828–1841. https://doi.org/10.1158/1541-7786.MCR-18-1154

Powers ME, Becker REN, Sailer A et al (2015) Synergistic action of Staphylococcus aureus α-toxin on platelets and myeloid lineage cells contributes to lethal sepsis. Cell Host Microbe 17:775–787. https://doi.org/10.1016/j.chom.2015.05.011

Price VJ, McBride SW, Hullahalli K et al (2019) Enterococcus faecalis CRISPR-Cas is a robust barrier to conjugative antibiotic resistance dissemination in the murine intestine. mSphere 4:e00464–e00419. https://doi.org/10.1128/mSphere.00464-19

Proft T, Baker EN (2009) Pili in gram-negative and gram-positive bacteria—structure, assembly and their role in disease. Cell Mol Life Sci 66:613–635. https://doi.org/10.1007/s00018-008-8477-4

Rajesh S, Koshi E, Philip K, Mohan A (2011) Antimicrobial photodynamic therapy: an overview. J Indian Soc Periodontol 15:323–327. https://doi.org/10.4103/0972-124X.92563

Rashel M, Uchiyama J, Ujihara T et al (2007) Efficient elimination of multidrug-resistant Staphylococcus aureus by cloned lysin derived from bacteriophage ϕMR11. J Infect Dis 196:1237–1247. https://doi.org/10.1086/521305

Rasko DA, Moreira CG, Li DR et al (2008) Targeting QseC signaling and virulence for antibiotic development. Science 321:1078–1080. https://doi.org/10.1126/science.1160354

Rasmussen TB, Manefield M, Andersen JB et al (2000) How Delisea pulchra furanones affect quorum sensing and swarming motility in Serratia liquefaciens MG1. Microbiology 146(Pt 12):3237–3244. https://doi.org/10.1099/00221287-146-12-3237

Rath D, Amlinger L, Hoekzema M et al (2015) Efficient programmable gene silencing by cascade. Nucleic Acids Res 43:237–246. https://doi.org/10.1093/nar/gku1257

Reddy ST, Chung KK, McDaniel CJ et al (2011) Micropatterned surfaces for reducing the risk of catheter-associated urinary tract infection: an in vitro study on the effect of sharklet micropatterned surfaces to inhibit bacterial colonization and migration of uropathogenic Escherichia coli. J Endourol 25:1547–1552. https://doi.org/10.1089/end.2010.0611

Reid G (2016) Probiotics: definition, scope and mechanisms of action. Best Pract Res Clin Gastroenterol 30:17–25. https://doi.org/10.1016/j.bpg.2015.12.001

Report T The bacterial challenge: time to react

Reuter A, Hilpert C, Dedieu-Berne A et al (2021) Targeted-antibacterial-plasmids (TAPs) combining conjugation and CRISPR/Cas systems achieve strain-specific antibacterial activity. Nucleic Acids Res 49:3584–3598. https://doi.org/10.1093/nar/gkab126

Riley MK, Vermerris W (2017) Recent advances in nanomaterials for gene delivery—a review. Nanomaterials (Basel) 7:94. https://doi.org/10.3390/nano7050094

Roberts RR, Hota B, Ahmad I et al (2009) Hospital and societal costs of antimicrobial-resistant infections in a Chicago teaching hospital: implications for antibiotic stewardship. Clin Infect Dis 49:1175–1184. https://doi.org/10.1086/605630

Ronqui MR, de Aguiar CTMSF, de Freitas LM et al (2016) Synergistic antimicrobial effect of photodynamic therapy and ciprofloxacin. J Photochem Photobiol B 158:122–129. https://doi.org/10.1016/j.jphotobiol.2016.02.036

Rosa H, Petri DFS, Carmona-Ribeiro AM (2008) Interactions between bacteriophage DNA and cationic biomimetic particles. J Phys Chem B 112:16422–16430. https://doi.org/10.1021/jp806992f

Rouzes C, Gref R, Leonard M et al (2000) Surface modification of poly(lactic acid) nanospheres using hydrophobically modified dextrans as stabilizers in an o/w emulsion/evaporation technique. J Biomed Mater Res 50:557–565. https://doi.org/10.1002/(sici)1097-4636(20000615)50:4<557::aid-jbm11>3.0.co;2-r

Rumbaugh KP (2007) Convergence of hormones and autoinducers at the host/pathogen interface. Anal Bioanal Chem 387:425–435. https://doi.org/10.1007/s00216-006-0694-9

Ryge TS, Frimodt-Møller N, Hansen PR (2008) Antimicrobial activities of twenty lysine-peptoid hybrids against clinically relevant bacteria and fungi. Chemotherapy 54:152–156. https://doi.org/10.1159/000119707

Sakamoto KM, Kim KB, Kumagai A et al (2001) Protacs: chimeric molecules that target proteins to the Skp1-Cullin-F box complex for ubiquitination and degradation. Proc Natl Acad Sci U S A 98:8554–8559. https://doi.org/10.1073/pnas.141230798

Sakamoto A, Terui Y, Horie C et al (2014) Antibacterial effects of protruding and recessed shark skin micropatterned surfaces of polyacrylate plate with a shallow groove. FEMS Microbiol Lett 361:10–16. https://doi.org/10.1111/1574-6968.12604

Salminen S, Collado MC, Endo A et al (2021) The International Scientific Association of Probiotics and Prebiotics (ISAPP) consensus statement on the definition and scope of postbiotics. Nat Rev Gastroenterol Hepatol 18:649–667. https://doi.org/10.1038/s41575-021-00440-6

Salomão MCC, Heluany-Filho MA, Menegueti MG et al (2016) A randomized clinical trial on the effectiveness of a symbiotic product to decolonize patients harboring multidrug-resistant gram-negative bacilli. Rev Soc Bras Med Trop 49:559–566. https://doi.org/10.1590/0037-8682-0233-2016

Sanches LM, Petri DFS, Melo Carrasco LD, Carmona-Ribeiro AM (2015) The antimicrobial activity of free and immobilized poly (diallyldimethylammonium) chloride in nanoparticles of poly (methylmethacrylate). J Nanobiotechnology 13:58. https://doi.org/10.1186/s12951-015-0123-3

Sauer RT, Baker TA (2011) AAA+ proteases: ATP-fueled machines of protein destruction. Annu Rev Biochem 80:587–612. https://doi.org/10.1146/annurev-biochem-060408-172623

Schapira M, Calabrese MF, Bullock AN, Crews CM (2019) Targeted protein degradation: expanding the toolbox. Nat Rev Drug Discov 18:949–963. https://doi.org/10.1038/s41573-019-0047-y

Scherlach K, Hertweck C (2006) Discovery of aspoquinolones A-D, prenylated quinoline-2-one alkaloids from Aspergillus nidulans, motivated by genome mining. Org Biomol Chem 4:3517–3520. https://doi.org/10.1039/b607011f

Scherlach K, Schuemann J, Dahse HM, Hertweck C (2010) Aspernidine A and B, prenylated isoindolinone alkaloids from the model fungus Aspergillus nidulans. J Antibiot (Tokyo) 63:375–377. https://doi.org/10.1038/ja.2010.46

Schulz E, Goes A, Garcia R et al (2018) Biocompatible bacteria-derived vesicles show inherent antimicrobial activity. J Control Release 290:46–55. https://doi.org/10.1016/j.jconrel.2018.09.030

Scobie HM, Thomas D, Marlett JM et al (2005) A soluble receptor decoy protects rats against anthrax lethal toxin challenge. J Infect Dis 192:1047–1051. https://doi.org/10.1086/432731

Scoffone VC, Chiarelli LR, Makarov V et al (2016) Discovery of new diketopiperazines inhibiting Burkholderia cenocepacia quorum sensing in vitro and in vivo. Sci Rep 6:32487. https://doi.org/10.1038/srep32487

Seffer MT, Cottam D, Forni LG, Kielstein JT (2021) Heparin 2.0: a new approach to the infection crisis. Blood Purif 50:28–34. https://doi.org/10.1159/000508647

Shakhnovich EA, Hung DT, Pierson E et al (2007) Virstatin inhibits dimerization of the transcriptional activator ToxT. Proc Natl Acad Sci U S A 104:2372–2377. https://doi.org/10.1073/pnas.0611643104

Sharma P, Elofsson M, Roy S (2020) Attenuation of Pseudomonas aeruginosa infection by INP0341, a salicylidene acylhydrazide, in a murine model of keratitis. Virulence 11:795–804. https://doi.org/10.1080/21505594.2020.1776979

Shi Y, Huang G (2009) Recent developments of biodegradable and biocompatible materials based micro/nanoparticles for delivering macromolecular therapeutics. Crit Rev Ther Drug Carrier Syst 26:29–84. https://doi.org/10.1615/critrevtherdrugcarriersyst.v26.i1.20

Shoop WL, Xiong Y, Wiltsie J et al (2005) Anthrax lethal factor inhibition. Proc Natl Acad Sci U S A 102:7958–7963. https://doi.org/10.1073/pnas.0502159102

Shrivastava R, Miller JF (2009) Virulence factor secretion and translocation by Bordetella species. Curr Opin Microbiol 12:88–93. https://doi.org/10.1016/j.mib.2009.01.001

Silk DBA, Davis A, Vulevic J et al (2009) Clinical trial: the effects of a trans-galactooligosaccharide prebiotic on faecal microbiota and symptoms in irritable bowel syndrome. Aliment Pharmacol Ther 29:508–518. https://doi.org/10.1111/j.1365-2036.2008.03911.x

Singh R, de Groot PF, Geerlings SE et al (2018) Fecal microbiota transplantation against intestinal colonization by extended spectrum beta-lactamase producing Enterobacteriaceae: a proof of principle study. BMC Res Notes 11:190. https://doi.org/10.1186/s13104-018-3293-x

Smith KM, Bu Y, Suga H (2003) Library screening for synthetic agonists and antagonists of a Pseudomonas aeruginosa autoinducer. Chem Biol 10:563–571. https://doi.org/10.1016/s1074-5521(03)00107-8

Sproul EP, Nandi S, Chee E et al (2020) Development of biomimetic antimicrobial platelet-like particles comprised of microgel nanogold composites. Regen Eng Transl Med 6:299–309. https://doi.org/10.1007/s40883-019-00121-6

Srisawat N, Tungsanga S, Lumlertgul N et al (2018) The effect of polymyxin B hemoperfusion on modulation of human leukocyte antigen DR in severe sepsis patients. Crit Care 22:279. https://doi.org/10.1186/s13054-018-2077-y

Stavrinides J, McCann HC, Guttman DS (2008) Host-pathogen interplay and the evolution of bacterial effectors. Cell Microbiol 10:285–292. https://doi.org/10.1111/j.1462-5822.2007.01078.x

Stephens MD, Yodsanit N, Melander C (2016) Evaluation of ethyl N-(2-phenethyl) carbamate analogues as biofilm inhibitors of methicillin resistant Staphylococcus aureus. Org Biomol Chem 14:6853–6856. https://doi.org/10.1039/c6ob00706f

Stinson LF, Payne MS, Keelan JA (2017) Planting the seed: origins, composition, and postnatal health significance of the fetal gastrointestinal microbiota. Crit Rev Microbiol 43:352–369. https://doi.org/10.1080/1040841X.2016.1211088

Stoutenbeek CP, van Saene HK, Miranda DR, Zandstra DF (1984) The effect of selective decontamination of the digestive tract on colonisation and infection rate in multiple trauma patients. Intensive Care Med 10:185–192. https://doi.org/10.1007/BF00259435

Surewaard BGJ, Thanabalasuriar A, Zeng Z et al (2018) α-Toxin induces platelet aggregation and liver injury during Staphylococcus aureus sepsis. Cell Host Microbe 24:271–284.e3. https://doi.org/10.1016/j.chom.2018.06.017

Szajewska H, Mrukowicz JZ (2001) Probiotics in the treatment and prevention of acute infectious diarrhea in infants and children: a systematic review of published randomized, double-blind, placebo-controlled trials. J Pediatr Gastroenterol Nutr 33 Suppl 2:S17–S25. https://doi.org/10.1097/00005176-200110002-00004

Tacconelli E, Mazzaferri F, de Smet AM et al (2019) ESCMID-EUCIC clinical guidelines on decolonization of multidrug-resistant gram-negative bacteria carriers. Clin Microbiol Infect 25:807–817. https://doi.org/10.1016/j.cmi.2019.01.005

Tan J, Patil PC, Luzzio FA, Demuth DR (2018) In vitro and in vivo activity of peptidomimetic compounds that target the periodontal pathogen Porphyromonas gingivalis. Antimicrob Agents Chemother 62:e00400–e00418. https://doi.org/10.1128/AAC.00400-18

Tan GSE, Tay HL, Tan SH et al (2020) Gut microbiota modulation: implications for infection control and antimicrobial stewardship. Adv Ther 37:4054–4067. https://doi.org/10.1007/s12325-020-01458-z

Tang K, Su Y, Brackman G et al (2015) MomL, a novel marine-derived N-acyl homoserine lactonase from Muricauda olearia. Appl Environ Microbiol 81:774–782. https://doi.org/10.1128/AEM.02805-14

Tang F, Li L, Meng X-M et al (2019) Inhibition of alpha-hemolysin expression by resveratrol attenuates Staphylococcus aureus virulence. Microb Pathog 127:85–90. https://doi.org/10.1016/j.micpath.2018.11.027

Tannock GW, Tiong IS, Priest P et al (2011) Testing probiotic strain Escherichia coli Nissle 1917 (Mutaflor) for its ability to reduce carriage of multidrug-resistant E. coli by elderly residents in long-term care facilities. J Med Microbiol 60:366–370. https://doi.org/10.1099/jmm.0.025874-0

Teng M, Yao Y, Nair V, Luo J (2021) Latest advances of virology research using CRISPR/Cas9-based gene-editing technology and its application to vaccine development. Viruses 13:779. https://doi.org/10.3390/v13050779
Teo J, Kwa ALH, Loh J et al (2012) The effect of a whole-system approach in an antimicrobial stewardship programme at the Singapore General Hospital. Eur J Clin Microbiol Infect Dis 31:947–955. https://doi.org/10.1007/s10096-011-1391-y
Thomann A, de Mello Martins AGG, Brengel C et al (2016) Application of dual inhibition concept within looped autoregulatory systems toward antivirulence agents against Pseudomonas aeruginosa infections. ACS Chem Biol 11:1279–1286. https://doi.org/10.1021/acschembio.6b00117
Thorakkattu P, Khanashyam AC, Shah K et al (2022) Postbiotics: current trends in food and pharmaceutical industry. Foods 11:3094. https://doi.org/10.3390/foods11193094
Trentini DB, Suskiewicz MJ, Heuck A et al (2016) Arginine phosphorylation marks proteins for degradation by a Clp protease. Nature 539:48–53. https://doi.org/10.1038/nature20122
Turk BE (2007) Manipulation of host signalling pathways by anthrax toxins. Biochem J 402:405–417. https://doi.org/10.1042/BJ20061891
Utari PD, Setroikromo R, Melgert BN, Quax WJ (2018) PvdQ quorum quenching acylase attenuates Pseudomonas aeruginosa virulence in a mouse model of pulmonary infection. Front Cell Infect Microbiol 8:119. https://doi.org/10.3389/fcimb.2018.00119
Vad-Nielsen J, Lin L, Bolund L et al (2016) Golden Gate assembly of CRISPR gRNA expression array for simultaneously targeting multiple genes. Cell Mol Life Sci 73:4315–4325. https://doi.org/10.1007/s00018-016-2271-5
van Nood E, Vrieze A, Nieuwdorp M et al (2013) Duodenal infusion of donor feces for recurrent Clostridium difficile. N Engl J Med 368:407–415. https://doi.org/10.1056/NEJMoa1205037
Veenendaal AKJ, Sundin C, Blocker AJ (2009) Small-molecule type III secretion system inhibitors block assembly of the Shigella type III secreton. J Bacteriol 191:563–570. https://doi.org/10.1128/JB.01004-08
Versporten A, Zarb P, Caniaux I et al (2018) Antimicrobial consumption and resistance in adult hospital inpatients in 53 countries: results of an internet-based global point prevalence survey. Lancet Glob Health 6:e619–e629. https://doi.org/10.1016/S2214-109X(18)30186-4
Wade DS, Calfee MW, Rocha ER et al (2005) Regulation of Pseudomonas quinolone signal synthesis in Pseudomonas aeruginosa. J Bacteriol 187:4372–4380. https://doi.org/10.1128/JB.187.13.4372-4380.2005
Wan P, Cui S, Ma Z et al (2020) Reversal of mcr-1-mediated colistin resistance in Escherichia coli by CRISPR-Cas9 system. Infect Drug Resist 13:1171–1178. https://doi.org/10.2147/IDR.S244885
Wan F, Draz MS, Gu M et al (2021) Novel strategy to combat antibiotic resistance: a sight into the combination of CRISPR/Cas9 and nanoparticles. Pharmaceutics 13:352. https://doi.org/10.3390/pharmaceutics13030352
Wang H-X, Li M, Lee CM et al (2017a) CRISPR/Cas9-based genome editing for disease modeling and therapy: challenges and opportunities for nonviral delivery. Chem Rev 117:9874–9906. https://doi.org/10.1021/acs.chemrev.6b00799
Wang R, Li J, Chen W et al (2017b) A biomimetic mussel-inspired ε-poly-l-lysine hydrogel with robust tissue-anchor and anti-infection capacity. Adv Funct Mater 27:1604894. https://doi.org/10.1002/adfm.201604894
Wang X, Thompson CD, Weidenmaier C, Lee JC (2018) Release of Staphylococcus aureus extracellular vesicles and their application as a vaccine platform. Nat Commun 9:1379. https://doi.org/10.1038/s41467-018-03847-z
Wang S, Gao J, Wang Z (2019) Outer membrane vesicles for vaccination and targeted drug delivery. Wiley Interdiscip Rev Nanomed Nanobiotechnol 11:e1523. https://doi.org/10.1002/wnan.1523
Watanabe K (2016) Bacterial membrane vesicles (MVs): novel tools as nature- and nano-carriers for immunogenic antigen, enzyme support, and drug delivery. Appl Microbiol Biotechnol 100:9837–9843. https://doi.org/10.1007/s00253-016-7916-7
Wei X, Gao J, Wang F et al (2017) In situ capture of bacterial toxins for antivirulence vaccination. Adv Mater 29. https://doi.org/10.1002/adma.201701644

Weiss DS, Brotcke A, Henry T et al (2007) In vivo negative selection screen identifies genes required for Francisella virulence. Proc Natl Acad Sci U S A 104:6037–6042. https://doi.org/10.1073/pnas.0609675104

Whelan K (2013) Mechanisms and effectiveness of prebiotics in modifying the gastrointestinal microbiota for the management of digestive disorders. Proc Nutr Soc 72:288–298. https://doi.org/10.1017/S0029665113001262

Wilson R, Sykes DA, Watson D et al (1988) Measurement of Pseudomonas aeruginosa phenazine pigments in sputum and assessment of their contribution to sputum sol toxicity for respiratory epithelium. Infect Immun 56:2515–2517. https://doi.org/10.1128/iai.56.9.2515-2517.1988

Wittekamp BH, Plantinga NL, Cooper BS et al (2018) Decontamination strategies and bloodstream infections with antibiotic-resistant microorganisms in ventilated patients: a randomized clinical trial. JAMA 320:2087–2098. https://doi.org/10.1001/jama.2018.13765

Biometric nanosponges as a broad-spectrum countermeasure to biological threats—HDIAC WO2017103225A1.pdf (2021)

Wong TH, Morton CA, Collier N et al (2019) British Association of Dermatologists and British Photodermatology group guidelines for topical photodynamic therapy 2018. Br J Dermatol 180:730–739. https://doi.org/10.1111/bjd.17309

Wood S, Metcalf D, Devine D, Robinson C (2006) Erythrosine is a potential photosensitizer for the photodynamic therapy of oral plaque biofilms. J Antimicrob Chemother 57:680–684. https://doi.org/10.1093/jac/dkl021

World Health Organization (2022) Global Antimicrobial Resistance and Use Surveillance System (GLASS) report 2022. World Health Organization, pp 71

Wright KJ, Hultgren SJ (2006) Sticky fibers and uropathogenesis: bacterial adhesins in the urinary tract. Future Microbiol 1:75–87. https://doi.org/10.2217/17460913.1.1.75

Wu K, Yang Y, Zhang Y et al (2015) Antimicrobial activity and cytocompatibility of silver nanoparticles coated catheters via a biomimetic surface functionalization strategy. Int J Nanomedicine 10:7241–7252. https://doi.org/10.2147/IJN.S92307

Wu J, Wang K, Wang X et al (2021a) The role of the gut microbiome and its metabolites in metabolic diseases. Protein Cell 12:360–373. https://doi.org/10.1007/s13238-020-00814-7

Wu Y, Battalapalli D, Hakeem MJ et al (2021b) Engineered CRISPR-Cas systems for the detection and control of antibiotic-resistant infections. J Nanobiotechnology 19:401. https://doi.org/10.1186/s12951-021-01132-8

Xavier GRS, Carmona-Ribeiro AM (2017) Cationic biomimetic particles of polystyrene/cationic bilayer/gramicidin for optimal bactericidal activity. Nanomaterials (Basel) 7:422. https://doi.org/10.3390/nano7120422

Xi Y, Ge J, Guo Y et al (2018) Biomimetic elastomeric polypeptide-based nanofibrous matrix for overcoming multidrug-resistant bacteria and enhancing full-thickness wound healing/skin regeneration. ACS Nano 12:10772–10784. https://doi.org/10.1021/acsnano.8b01152

Xiong M, Lee MW, Mansbach RA et al (2015) Helical antimicrobial polypeptides with radial amphiphilicity. Proc Natl Acad Sci U S A 112:13155–13160. https://doi.org/10.1073/pnas.1507893112

Xu C, Zhu H, Qiu P (2019) Aging progression of human gut microbiota. BMC Microbiol 19:236. https://doi.org/10.1186/s12866-019-1616-2

Yang G, Chen S, Zhang J (2019a) Bioinspired and biomimetic nanotherapies for the treatment of infectious diseases. Front Pharmacol 10:751. https://doi.org/10.3389/fphar.2019.00751

Yang S, Li X, Yang F et al (2019b) Gut microbiota-dependent marker TMAO in promoting cardiovascular disease: inflammation mechanism, clinical prognostic, and potential as a therapeutic target. Front Pharmacol 10:1360. https://doi.org/10.3389/fphar.2019.01360

Yao X, Lyu P, Yoo K et al (2021) Engineered extracellular vesicles as versatile ribonucleoprotein delivery vehicles for efficient and safe CRISPR genome editing. J Extracell Vesicles 10:e12076. https://doi.org/10.1002/jev2.12076

Yeh YC, Huang TH, Yang SC et al (2020) Nano-based drug delivery or targeting to eradicate bacteria for infection mitigation: a review of recent advances. Front Chem 8:1–22. https://doi.org/10.3389/fchem.2020.00286

Yosef I, Manor M, Kiro R, Qimron U (2015) Temperate and lytic bacteriophages programmed to sensitize and kill antibiotic-resistant bacteria. Proceedings of the National Academy of Sciences of the United States of America 112, 7267–7272. https://doi.org/10.1073/pnas.1500107112

Yu Y-J, Wang X-H, Fan G-C (2018) Versatile effects of bacterium-released membrane vesicles on mammalian cells and infectious/inflammatory diseases. Acta Pharmacol Sin 39:514–533. https://doi.org/10.1038/aps.2017.82

Zhang B, Teng Z, Li X et al (2017a) Chalcone attenuates Staphylococcus aureus virulence by targeting sortase A and alpha-hemolysin. Front Microbiol 8:1715. https://doi.org/10.3389/fmicb.2017.01715

Zhang Q-Z, Zhao K-Q, Wu Y et al (2017b) 5-aminolevulinic acid-mediated photodynamic therapy and its strain-dependent combined effect with antibiotics on Staphylococcus aureus biofilm. PLoS One 12:e0174627. https://doi.org/10.1371/journal.pone.0174627

Zhang Y, Sass A, Van Acker H et al (2018) Coumarin reduces virulence and biofilm formation in Pseudomonas aeruginosa by affecting quorum sensing, type III secretion and C-di-GMP levels. Front Microbiol 9:1952. https://doi.org/10.3389/fmicb.2018.01952

Zhu Y, Wu S, Sun Y et al (2022) Bacteria-targeting photodynamic nanoassemblies for efficient treatment of multidrug-resistant biofilm infected keratitis. Adv Funct Mater 32:1–11. https://doi.org/10.1002/adfm.202111066

Zivkovic A, Sharif O, Stich K et al (2011) TLR 2 and CD14 mediate innate immunity and lung inflammation to staphylococcal Panton-Valentine leukocidin in vivo. J Immunol 186:1608–1617. https://doi.org/10.4049/jimmunol.1001665

Zorba A, Nguyen C, Xu Y et al (2018) Delineating the role of cooperativity in the design of potent PROTACs for BTK. Proc Natl Acad Sci U S A 115:E7285–E7292. https://doi.org/10.1073/pnas.1803662115

Anti-Quorum-Sensing Techniques: A Potential Strategy to Manage Antibiotic Resistance

6

Srishti Singh, Alok Kumar Singh, Mayank Gangwar, and Gopal Nath

Abstract

Science's most potent invention for combating bacterial illnesses is antibiotics. However, careless use of antibiotics at higher doses has increased the resistance of hostile pathogens to antibiotics. Bacterial intra- and intercellular communication mechanisms known as quorum sensing (QS) can detect the development of microbial pathogenic features, such as the generation of virulence factors and biofilm formation. Quorum sensing (QS) regulates the virulence factors of bacteria using cell-to-cell signaling networks with the help of autoinducers (AIs). During the stationary phase, tiny signaling molecules called AIs are created. These molecules serve as mirrors that reflect the inoculum's density and control the expression of bound genes once bacterial cultures reach a particular growth stage. Most bacteria can use both types to modify the target gene's expression. However, Gram-positive bacteria use the peptide derivatives of these signaling molecules, whereas Gram-negative bacteria use the fatty acid derivatives. Because of this, we must prevent quorum sensing, which is accomplished through various types of quorum-sensing inhibitors (QSIs). Natural and synthetic QSIs

S. Singh · A. K. Singh (✉)
Department of Botany, CMP Degree College, University of Allahabad, Prayagraj, Uttar Pradesh, India

M. Gangwar
Faculty of Medicine, Viral Research and Diagnostic Laboratory, Department of Microbiology, Institute of Medical Sciences, Banaras Hindu University, Varanasi, Uttar Pradesh, India

Clinical Studies and Trials Unit, Division of Development Research, Indian Council of Medical Research, Department of Health Research, Ministry of Health and Family Welfare, Govt. of India, V. Ramalingaswami Bhawan, New Delhi, India

G. Nath
Department of Microbiology, Institute of Medical Sciences, Banaras Hindu University, Varanasi, Uttar Pradesh, India

M. Gangwar, G. Nath (eds.), *Emerging Paradigms for Antibiotic-Resistant Infections: Beyond the Pill*, https://doi.org/10.1007/978-981-97-5272-0_6

are essential for inhibiting QS. This chapter provides a general introduction, mechanisms, and a few examples of QSIs, including synthetic and natural sources. It further concentrates on the utilization of QSIs in various domains and the suppression of QS processes. They also highlight the recently authorized patent on discovering new bioactive agents, analogues, and approaches interfering with QS signaling.

Keywords

Anti-quorum · Antibiotic resistance · Inhibitors · Autoinducers · Signaling

6.1 General Introduction to Quorum Sensing

The quorum-sensing (QS) process, in which bacteria send chemical signals and express virulence genes depending on cell density controls the behavior of bacteria within biofilms (Bhargava et al. 2010). When their population density reaches a certain point, bacteria can sense this behavior change and behave like "multicellular" organisms instead of single-celled ones. By communicating through tiny signaling molecules, they can express genes for many phenotypes, particularly those in charge of their pathogenic behavior. This behavior is known as the QS system of bacteria. Since bacteria use many chemical signals or autoinducers, their communication can be multilingual. Depending on the kind of autoinducer, quorum-sensing systems can be divided into five major categories (Turan et al. 2017): *N*-acyl homoserine lactones (AHLs), autoinducer-2 (AI-2), autoinducer-3 (AI-3), *Pseudomonas* quinolone signal (PQS), and autoinducing peptides (AIPs).

6.1.1 *N*-Acyl Homoserine Lactones (AHLs)

More than 70 different types of Gram-negative bacteria create AHLs, which permeate the cell membrane and attach to the regulatory proteins inside the cell (Galloway et al. 2011). *N*-Acyl homoserine lactone (AHL) autoinducer signal molecules are used by Gram-negative bacteria for quorum sensing. *N*-Acyl homoserine lactone (AHL)-dependent quorum sensing controls several virulence factors in Gram-negative bacteria, including their adhesion, biofilm formation, exozyme release, and pigment creation (Rutherford and Bassler 2012). All AHL-based quorum-sensing systems consist of three essential parts:

- The synthase molecule of the LuxI type
- The molecular signaling agent AHL
- The receptor protein of the LuxR type

6.1.2 Autoinducing Peptides (AIPs)

AIPs are created as pro-AIPs inside of the bacterial cell, and, depending on the organism, these can be digested and changed either inside or outside of the cell (Okada et al. 2005). Because AIPs are impermeable to cell membranes, unlike AHL signaling molecules, they require specific transport proteins for inward and outward conveyance. Gram-positive bacteria have a peptide-based QS system mediated by membrane-bound receptor histidine kinases. Quorum-sensing systems control competence mechanisms such as *Bacillus subtilis* sporulation and the formation of pathogenicity in *Staphylococcus aureus*, *Listeria monocytogenes*, *Clostridium perfringens*, and *Enterococcus faecalis* (Bhatt 2018). There are two ways that Gram-positive bacteria sense quorum:

- A signal transduction system consisting of two components
- The internalization phase

6.1.3 Autoinducer-2 (AI-2)

Numerous Gram-positive and Gram-negative bacteria recognize autoinducer-2, the third class of signaling molecules in quorum-sensing systems. Furanosyl borate diester, one of the few biomolecules with boron integrated into it, is made by the LuxS protein. The synthesis of the virulent traits and pathogenicity within the host systems of different infectious agents, such as *Vibrio cholerae*, *Haemophilus influenza*, *Streptococcus pneumonia*, and enterohemorrhagic *Escherichia coli*, is regulated by AI-2. In mixed-population organisms, like those seen in biofilms, the AI-2 language enables a given species to react to variations not only in its cell count but also in the cell counts of other species coexisting in the same community. Due to its identification in more than 60 species and different bacterial groupings, AI-2 is regarded as necessary in interspecies communication. In *Salmonella typhimurium* and *Vibrio harveyi*, AI-2 controls the ABC transporter and bioluminescence, respectively. The production of AI-2 in *V. harveyi* depends on the synthase LuxS, encoded by the *LuxS* gene. Nevertheless, the precise function of AI-2 in other bacterial species is still unknown (Pereira et al. 2013).

6.1.4 Autoinducer-3 (AI-3)

Specific enteric pathogens and bacteria in the human intestinal microbiota create an aromatic signaling molecule called autoinducer-3 (AI-3). The genes that code for the synthesis of AI-3 and its molecular structure are unknown. The intestinal pathogens *Escherichia coli* and *Salmonella typhimurium* detect AI-3 through the sensor kinase QseC, which phosphorylates the response regulator QseB to initiate target gene transcription. The structure of AI-3 is believed to mimic the two hormones because the same QseC/B two-component system is also employed to detect the presence of epinephrine and norepinephrine produced by the host (Kendall and Sperandio 2007; Moreira et al. 2010).

6.1.5 PQS

In quorum-signaling studies, *P. aeruginosa* has emerged as a model organism. The four components of *P. aeruginosa* are PQS (*Pseudomonas* quinolone signal), IQS (integrated quorum-sensing signal), RhlI/RhlR, and LasI/LasR. They utilize different signaling molecules, such as IQS, PQS, OdDHL (*N*-oxododecanoyl-L-homoserine lactone), and BHL (*N*-butanoyl-L-homoserine lactone). *Pseudomonas aeruginosa* QS signals have an impact on virulence factors such as LasA protease (which is controlled by OdDHL that disrupts the epithelial layer), LasB elastase (which degrades down matrix protein structures like BHL and OdDHL), rhamnolipids (controlled by BHL, causing immune cell necrosis), protease with an alkaline pH (which degrades proteins involved in the host's defense controlled by OdDHL), LecA lectin (which is controlled by PQS, enhancing colonization), and pyocyanin (which is engaged in defensive evading controlled by BHL, PQS, and OdDHL). In various hosts, such as zebra fish, fruit flies, mice, and nematodes (models of injuries from burns, persistent lung infections, and pneumonia), *P. aeruginosa*'s quorum-sensing system is necessary for complete pathogenicity. The quinolone-based QS system acts mainly via 2-heptyl-3-hydroxy-4-quinolone, called the Pseudomonas quinolone signal (PQS), which connects the LysR-type transcriptional regulator PqsR to stimulate some virulence genes (Table 6.1) (Lee and Zhang 2015).

Table 6.1 Distinct QS routes in bacteria

Route	Signals	Bacteria	References
CAI-1	Hydroxy ketones	Gram-negative *Vibrio cholerae*	Higgins et al. (2007)
Autoinducer-3 pathway	Autoinducer-3	Gram-negative	Kendall et al. (2007)
4Q pathway	HHL, HHQ, and PQS	Gram-negative	Diggle et al. (2006)
N-acyl homoserine lactone (AI-1) pathway	*N*-acyl homoserine lactone (AHL)	Gram-negative	Ravn et al. (2001)
Autoinducing peptides	AIP signals	Gram-negative	MDowell et al. (2001)
Autoinducer-2 (AI-2) pathway	AI-2 (two types)	Both Gram-positive and Gram negative	Chen et al. (2002)

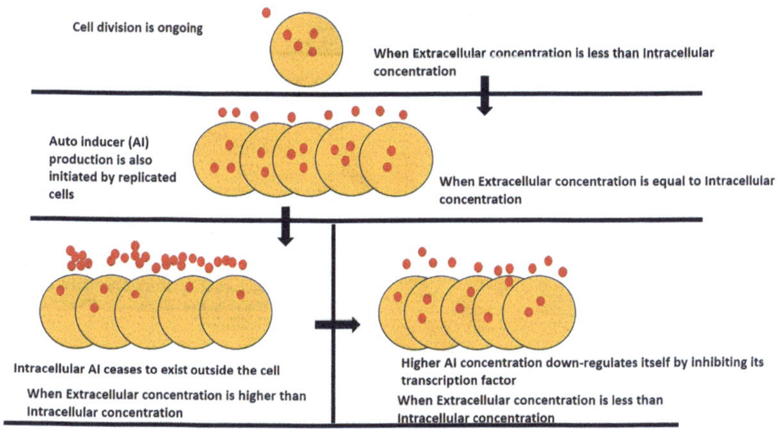

Fig. 6.1 Mechanism of quorum sensing

6.2 Mechanisms of Quorum Sensing

Individual bacteria produce autoinducers during their reproductive cycle. Once generated, autoinducers exit particular cells. Because bacteria proliferate, more individual cells create autoinducers, and their extracellular concentration rises until it reaches a "critical mass." This threshold increases intracellular autoinducers' intracellular concentration by making it energetically unfavorable for them to continue leaving the cell. When autoinducers reach a higher concentration inside cells, they attach to their receptors and set off signaling cascades that change transcription factor activity and, in turn, gene expression. Reduced autoinducer production in a negative feedback loop is one of the numerous features of bacteria for which changes in gene expression occur (Fig. 6.1).

6.2.1 Role of Quorum Sensing in Controlling Virulence Factors in Bacteria

Numerous virulent elements are regulated by the QS communication network, which comprises cell-to-cell signaling pathways. It has been demonstrated that QS controls the factors responsible for virulence generation in a variety of aquatic diseases (Priya et al. 2023). Quorum sensing plays a critical regulatory role in developing microbial biofilms, regulating drug efflux pumps, the production of toxic substances, and the expression of infectious genes, and in creating extracellular polysaccharides (Fig. 6.2; Table 6.2) (Bäuerle et al. 2018).

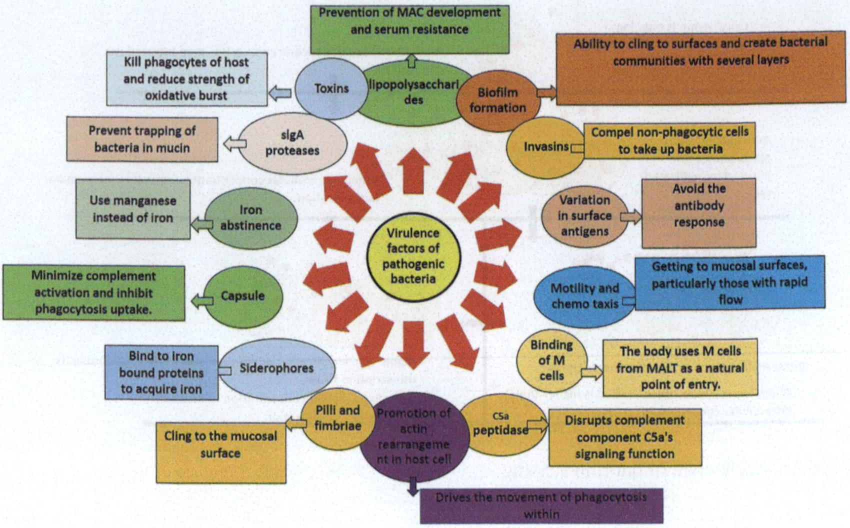

Fig. 6.2 Various virulence characteristics of harmful bacteria regulated by quorum sensing

6.3 Silencing of Bacteria: An Anti-Quorum Strategy

Quorum quenching (QQ) is the process by which bacteria are rendered "silent" by interfering with the quorum-sensing system. QS inhibition and the control of pathogenic symptoms' onset primarily depend on the reduction of cell communication and the obstruction of their formation. A mechanism consisting of signal generation, signal accumulation, and signal detection is shared by all QS systems. At least one of these three processes is targeted by every QS inhibitor (QSI) that has been mechanistically identified.

6.3.1 AI (Autoinducer) Synthesis Blockage

It is possible to stop the development of AI by creating structural analogues. At lower dosages, some antibiotics can suppress the production of AI synthesis. For example, molecules like L/D-*S*-adenosyl homocysteine, *S*-adenosyl-L-cysteine, and sinefungin suppress the production of AHLs.

6.3.2 Inactivation of AIs

AHLs in Gram-negative bacteria can be explicitly inactivated by acylase, lactonase, and oxidoreductase enzymes. As a result, AHL accumulation in the extracellular environment is prevented and QS-regulated genes are not expressed.

Table 6.2 Examples of virulence factors in plant, animal, and human bacterial pathogens controlled by quorum sensing

Pathogens	Signal molecule	Phenotypes	References
Pathogens affecting plants			
Agrobacterium tumefaciens	3OC8HSL	Ti plasmid conjugation, pathogenicity	Von Bodman et al. (2003)
Bacillus thuringiensis	PapR	Exoenzymes	Rocha-Estrada et al. (2010)
Erwinia carotovora	3OC6HSL	Virulence, carbapenem, and exoenzymes	Barnard and Salmond (2007)
Pseudomonas syringae	OHL	Extracellular polysaccharides, pathogenicity, and motility	Helman and Chernin (2015)
Pectobacterium carotovorum	HHL, OHHL, OOHL	Enzymes that break down extracellular cell walls	Lee and Zhang (2015)
Xanthomonas campestris	DSF	Extracellular polysaccharides, development of biofilms, and pathogenicity	Zhou et al. (2017)
Pathogens affecting animals			
Aeromonas hydrophila	BHL	Protease, virulence	Natrah et al. (2012)
Cortinarius violaceus	C6HSL	Antibiotics, violacein, and exoenzymes	McClean et al. (1997)
Photorhabdus luminescens	Alkylresorcinols	Virulence	Brameyer et al. (2015)
Vibrio harveyi	OH-BHL, AI-2, CAI-1	Siderophore, type III secretion, virulence, and protease	Defoirdt et al. (2008)
	Indole	Production of biofilms, motility, and pathogenicity	Yang et al. (2017)
Pathogens affecting humans			
E. faecalis	GBAP	Protease and gelatinase	Nakayama et al. (2001)
Escherichia coli O157:H7	Indole	Development of biofilms and motility	Bansal et al. (2007)
Helicobacter pylori	AI-2	Motility	Rader et al. (2007)
Pseudomonas aeruginosa	OdDHL, BHL, PQS	Hemolysin, rhamnolipids, protease, and pathogenicity	Lee and Zhang (2015)
S. aureus	AIP	Exotoxins, virulence, and biofilm dispersal	Thoendel et al. (2011)
Streptococcus pneumoniae	CSP	Autolysis, virulence, and competence	Rocha-Estrada et al. (2010)
Serratia liquefaciens	C4HSL	Formation of biofilms and swarming motility	Labbate et al. (2004)

6.3.3 Inhibition of AI Signal Reception

It is possible to stop quorum sensing by preventing the AI molecule from attaching to its receptor. Molecules that bind to the receptor more favorably than the AI molecule may be the source of competitive inhibition. Various plants, algae, and other species create chemicals that obstruct quorum sensing (Table 6.3).

6.4 Types of Agents that Interfere with the QS System

Agents must either promote or inhibit quorum sensing-regulated gene expression, based on the type of regulation (pathogenicity-induced or repressed). Agents that interfere with quorum sensing can be synthetic or natural materials that function as agonists or inhibitors of signal transduction, detection, or biosynthesis; they can also be an enzyme that renders the signal molecules inactive or antibodies that sequester the signal molecules and trigger an immune reaction (Table 6.4).

6.5 Applications of Quorum-Sensing Inhibitors in Biotechnology

6.5.1 Biotransformation

The global regulator (Hha13D6) and the cyclic diguanylate-binding protein (BdcAE50Q), two biofilm-dispersing proteins controlled by the *P. aeruginosa* QS signal 3OC12HSL, were engineered to achieve biofilm dispersal. The disperser cells produced 14 times more of the cognate QS signal molecule (3OC12HSL) than planktonic cells, although growing at a rate 14% slower than the initial colonizers. By adding 2-mM isopropyl-β-D-thiogalactopyranoside (IPTG), the disperser cells could induce hha13D6, thus allowing them to disperse biofilms completely. Removal of the disperser cells was thus made possible by this chemical switch. This intriguing approach can be used for self-regulation for various industrial applications, including biosensors, biocorrosion, biofouling, and bioremediation (Hong et al. 2012).

6.5.2 Wastewater Treatment

Antifouling treatments may include impairing QQ enzymes, adding QSIs to antifouling coatings, or using marine animals that have been genetically modified to release QSIs (Choudhary and Schmidt-Dannert 2010). It was discovered that a porcine kidney's acylase I might decrease biofilm development by 20–24% (Paul et al. 2009). Yeon et al. 2009 created a membrane based on the porcine kidney acylase I (QQ) enzyme to regulate biofouling.

Table 6.3 QS system inhibition strategies

Quenching way	Function method	Biological effect	References
Blockage of AI (autoinducer) synthesis			
FabI derivatives	Impede the activity of an enzyme	Stop the synthesis of autoinducers	Kalia et al. (2019)
(2-Nitrophenyl) methanol derivatives	Inhibit the production of the signal molecule's enzyme	Stop the synthesis of autoinducers	Storz et al. (2014)
TNRHNPHHLHHV (peptide)	Limits the activity of the LuxS enzyme	Prevents the formation of AI-2	Han and Lu (2009)
MT-DADMe-ImmA	An inhibitor at the picomolar level	Prevents the formation of AI-2	Schramm et al. (2008)
Degradation of AI			
The *aiiA* gene in *Bacillus* sp. 240B1	The *aiiA* gene encodes enzymes that degrade AHLs	Degrades AHL signaling molecules	Dong et al. (2001)
Thermostable AHL lactonase (AidB)	Helps in breaking down the HSL ring's ester bond by hydrolysis	Degrades AHLs	Zhang et al. (2019)
Externally added addition of ATP and LsrK	Phosphorylation and degradation of AI-2	Reduced QS reaction	Roy et al. (2010)
A recombinant strain named BbMomL	MomL causes loss of AHLs	Degrades AHLs	Zhang et al. (2019)
Crude extracts from Lb. custom ZHG 2–1	Crude extracts can degrade AHLs	Degrades AHLs by 37.1–87.6% at three sub-MICs	Zhang et al. (2019)
Imidazole	Degrades AI-2	Inhibits AI-2 function	Cui et al. (2020)
Inhibition of AI signal reception			
N-(3-oxododecanoyl) homoserine lactone derivatives	Block the QS molecule's binding site	Impede the growth of biofilms and raise the susceptibility to antibiotics	Bortolotti et al. (2019)
Medicinal herb extracts (MHEs)	MHEs are agents of competition	Stop QS system	Wei et al. (2020)
Flavonoids compounds	Decrease the intensity of the QS signal	Stop the QS signal	Truchado et al. (2015)
A small peptide 5906	Stops homodimer synthesis	Stops LuxS activity	Sun and Zhang (2016)
D-Galactose	Inhibits the action of AI-2	Stops AI-2 activity	Ryu et al. (2016)
Haloquinone analogues	Impede Wnt-driven transcription endogenously	Stop Wnt/β-catenin signaling	Halbedl et al. (2013)
Alkyl-quinoxalin-2(1*H*)-one derivatives	Exhibit QS inhibition	Limit or prevent the growth of biofilms	Blöcher et al. (2018)
2*H*-pyran-3(6*H*)-one derivatives	As an inhibitor	Inhibit signals	Takayama et al. (2013)

Table 6.4 Agents interfering with QS: several examples

Factors	Bacteria of interest	Molecular target	Activity	References
Natural substances				
Ajoene	*P. aeruginosa*, *Staphylococcus aureus*	Small regulatory RNAs	Inhibitory effect	Jakobsen et al. (2013)
Allicin	*P. aeruginosa*	Biofilm formation	Inhibitory effect	Bjarnsholt et al. (2005)
Ascorbic acids	*Clostridium perfringens*	Autoinducer-2	Inhibitory effect	Ali et al. (2009)
Allyl isothiocyanate	*E. coli*, *S. aureus*, *L. monocytogenes*, and *P. aeruginosa*	Biofilm formation, adhesion, and motility	Inhibitory effect	Borges et al. (2014)
Baicalein, hamamelitannin	*S. aureus*	Biofilm formation	Inhibitory effect	Brackman et al. (2016)
Citrus limonoids	*Vibrio harveyi*	Cell signaling	Inhibit signaling	Vikram et al. (2011)
Curcumin	*P. aeruginosa*	Bacterial pathogenicity is diminished	Inhibitory effect	Rudrappa and Bais (2008)
Cinnamon oil, ferula oil, Dorema oil	*P. aeruginosa*	QS-related phenotypes; production of pyocyanin, alginate, and rhamnolipid	Interfering	Kalia et al. (2019)
Clove oil	*C. violaceus* *P. aeruginosa*	Violacein production swarming ability	Interfering and reducing effect	Khan et al. (2009)
Coriander oil	*Campylobacter coli* *Campylobacter jejuni*	Biofilm formation and lipid peroxidation	Inhibitory effect	Duarte et al. (2016)
Cinnamon bark oil	*E. coli*	Permeability of the outer membrane	Modifying and inhibitory effect	Yap et al. (2015)
Epigallocatechin gallate Catechin	*E. coli* *Pseudomonas putida* *Staphylococcus* species	Biofilm formation Extracellular polymeric substances	Inhibitory effect	Xu et al. (2012)
Ellagic acid	*Eikenella corrodens* *Erwinia carotovora*	Biofilm production AHL production	Inhibitory effect	Truchado et al. (2015)
Through high-throughput screening, five compounds were identified as QSIs	*Pseudomonas aeruginosa*	AHL receptor	Inhibitory effect	Tan et al. (2013)

(continued)

Table 6.4 (continued)

Factors	Bacteria of interest	Molecular target	Activity	References
Flavonoids	*Pseudomonas aeruginosa*	AHL receptor	Receptor inhibition	Paczkowski et al. (2017)
Glycyrrhiza glabra flavonoids	*Acinetobacter baumannii*	Motility and biofilm formation	Interfering and reducing effect	Bhargava et al. (2015)
Giganteone A	*E. coli*	QS-related activity Biosensors	Reducing effect	Sivasothy et al. (2016)
Gingerone	*P. aeruginosa* PAO1	Swarming and biofilm-forming capacity	Reducing effect	Kumar et al. (2015)
Gallic acids	*S. mutans*	Biofilm formation	Inhibitory effect	Kang et al. (2008)
Malabaricone C	*P. aeruginosa*	Pyocyanin production and biofilm formation	Reducing effect	Chong et al. (2011)
Naringenin	*Pseudomonas aeruginosa*	AHL generation and identification	Inhibitory effect	Vandeputte et al. (2011)
Oregano oil	*C. violaceus*	Violacein production	Inhibitory effect	Rodriguez-Garcia et al. (2016)
Proanthocyanidins	*P. aeruginosa*	Production of QS-regulated virulence determinants	Reducing effect	Maisuria et al. (2016)
Pyrazine-2-carboxylic acid	*V. cholerae*	Biofilm formation	Inhibitory effect	Hema et al. (2016)
Rosamarinic acid	*P. aeruginosa*	Protease and elastase production, biofilm formation, and virulence factors	Influencing effect	Marwat et al. (2011)
Rose oil, geranium oil, lavender oil, rosemary oil	*C. violaceus* *E. coli*	AHL production Violacein pigmentation	Reducing effect	Szabó et al. (2010)
Resveratrol, piceatannol, oxyresveratrol	*Chromobacterium violaceum* CV026 *P. aeruginosa* PAO1	Violacein and pyocyanin production	Reducing effect	Sheng et al. (2015)
Salicylic acids	*P. aeruginosa*	AHL production, twitching, and swimming motility	Interfering and reducing effect	Bandara et al. (2006)

(continued)

Table 6.4 (continued)

Factors	Bacteria of interest	Molecular target	Activity	References
Sulforaphane, erucin	*P. aeruginosa*	Transcriptional activator of LasR biofilm formation	Antagonistic and inhibitory effect	Ganin et al. (2013)
Thyme oil	*C. violaceum* *P. fluorescens* KM121	Flagella gene expression, biofilm formation	Reducing effect	Myszka et al. (2016)
Tea polyphenols (*Camellia sinensis* L.)	*P. aeruginosa*	Proteolytic activity, elastase, swarming motility, and biofilm formation	Reducing effect	Yin et al. (2015)
Linalool	*A. baumannii*	Biofilm formation, adhesion	Alternation and inhibitory effect	Alves et al. (2016)
Iberin	*P. aeruginosa*	Rhamnolipid production and gene expression of lasB and rhlA	An interfering agent	Jakobsen et al. (2013)
Indole-3-acetic acid	*Vibrio harveyi*	Indole pathway	With both affinity and intrinsic efficacy	Yang et al. (2017)
Synthetic compounds				
Brominated thiophenes	*Vibrio harveyi*	Cell signaling	Inhibit the signaling pathway	Yang et al. (2015)
3-Acylpyrroles	*Vibrio cholerae*	CAI-1 receptor	Agonistic	Perez et al. (2014)
Five compounds were identified to act as QSIs	*Burkholderia mallei*, *Yersinia pestis*	AHL biosynthesis	Inhibitory effect	Christensen et al. (2013)
Thiazolidinediones and dioxazaborocanes	*Vibrio harveyi*	AI-2 receptor	Inhibitory effect	Brackman et al. (2013)
Antibodies				
AP4–24 H11 antibody	*S. aureus*	Acts as a sequester of AIP-4	Reduces the pathogenicity of QS both in vivo and in vitro	Naga et al. (2023)
HSL-4/MAb HSL-2	*Pseudomonas aeruginosa*	*N*-acyl homoserine lactone	Activates immune responses	Palliyil et al. (2014)

(continued)

Table 6.4 (continued)

Factors	Bacteria of interest	Molecular target	Activity	References
Monoclonal antibodies	*S. aureus* agr pathogen and *E. faecalis* agr-like system	Interfere with QS mechanisms and neutralize AIPs	Disrupt signaling	Singh et al. (2016)
XYD-11G2	*P. aeruginosa*	Stop the synthesis of pyocyanin and suppress the 3-oxo-C12-HSL signal	Inhibitory effect	De Lamo Marin et al. (2007)
Enzymes				
AHL lactonase aiiA	*Bacillus* sp. 240B1	AHL signal molecules	Degradation	Dong and Zhang (2005)
AHL lactonase homologues aiiA	*Bacillus thuringiensis*	AHL signal molecules	Degradation	Dong and Zhang (2005)
AHL lactonase homologues aiiA	*Bacillus cereus*	AHL signal molecules	Degradation	Dong and Zhang (2005)
AHL lactonase homologues aiiA	*Bacillus mycoides*	AHL signal molecules	Degradation	Dong and Zhang (2005)
AHL lactonase homologues aiiA	*Bacillus anthracis*	AHL signal molecules	Degradation	Dong and Zhang (2005)
AHL lactonase attM, aiiB	*Agrobacterium tumefaciens*	AHL signal molecules	Degradation	Dong and Zhang (2005)
AHL acylase aiiD	*Ralstonia* strain XJ12B	AHL signal molecules	Degradation	Dong and Zhang (2005)
AHL lactonase ahlD	*Arthrobacter* sp. IBN110	AHL signal molecules	Degradation	Dong and Zhang (2005)
AHL lactonase ahlK	*Klebsiella pneumoniae*	AHL signal molecules	Degradation	Dong and Zhang (2005)
Acylase I	Porcine (kidney)	–	–	Dong and Zhang (2005)
AHL acylase ND	*Variovorax paradoxus* VAI-C	AHL signal molecules	Degradation	Dong and Zhang (2005)
Achromobacter xylosoxidans strain Q19	*Pseudomonas aeruginosa*	PQS signal molecules	Degradation	Soh et al. (2015)

(continued)

Table 6.4 (continued)

Factors	Bacteria of interest	Molecular target	Activity	References
Bacillus sp. strain NFMI-C	*Vibrio harveyi*	AHL signal molecules	Degradation	Pande et al. (2015)
Four bacterial strains from plants	*Xanthomonas citri*	DSF signal molecules	Degradation	Caicedo et al. (2016)
Five bacterial strains from plants	*Xanthomonas campestris*	DSF signal molecules	Degradation	Newman et al. (2008)
Lactonase from *Bacillus* sp. strain QSI-1	*Aeromonas hydrophila*	AHL signal molecules	Degradation	Chu et al. (2014)
Modified acylase PvdQ	*Burkholderia cenocepacia*	AHL signal molecules	Degradation	Koch et al. (2014)
PONs lactonase	Human (airway epithelia)	–	–	Dong and Zhang (2005)

6.5.3 Aquaculture

Disease outbreaks can significantly impact the well-established aquaculture food industry (Defoirdt et al. 2011). The addition of enrichment cultures, primarily composed of *Bacillus* spp. bacteria that degrade AHLs, was later demonstrated to enhance the survival of various aquaculture species, including gilthead sea beam *Sparus aurata* larvae (Tinh et al. 2008), gigantic prawns from freshwater *Macrobrachium rosenbergii* (Nhan et al. 2010), and turbot larvae *Scophthalmus maxima* (Avella et al. 2010). The application of QQ in bacteria–eukaryote interactions in the marine contexts was broadened by the fish pathogen *Tenacibaculum maritimum* NCIMB2154T's degradation of long-chain AHLs via an acylase-type QQ enzyme (Romero et al. 2010). QSIs are produced by certain micro- and macroalgae, such as *Chlorella saccharophila* CCAP211, which affects the growth but not bioluminescence in *V. harveyi* or violacein production in *C. violaceum* CV026 (Natrah et al. 2011). Moreover, some QS-mediated behaviors, such as luminescence, were suppressed by *Chlamydomonas reinhardtii* microalga (Teplitski et al. 2004). Halogenated furanone synthesized by the red macroalgae *Delisea pulchra* has also been shown to block phenotypes regulated by QS (Manefield et al. 2000). *Aphrocallistes bocagei*, *Haliclona megastoma*, and *Clathria atrasanguinea* are among the sponges whose effectiveness against the synthesis of violacein pigment in *C. violaceum* CV026 and ATCC 12,472 has been approved. They also prevented *Serratia marcescens* from producing hemolysin, protease, and biofilms (Annapoorani et al. 2012).

6.5.4 Plant Cultivation

Potato plants may be protected from bacterial illnesses such as *P. carotovorum*, which causes soft rot disease, by the naturally occurring AHL-degrading bacterium

Microbacterium testaceum on their leaf surfaces (Wang et al. 2010). Plants such as *Medicago tranculata* and *Pisum sativum* can emit QS mimics when infected with bacteria. These AHL signal mimics that alter virulence protect plants from pathogenic microorganisms (Table 6.2) (Teplitski et al. 2004).

6.5.5 QSI Compounds in Animal Infection Models

QSI compounds can regulate bacterial pathogenicity in vivo, as demonstrated by a study on the marine luminous bacterium *V. harveyi*. This bacterium uses QS to control the production of T1, a 100-kDa toxin, and is a severe disease-causing pathogen in a range of commercially farmed invertebrates. Both bioluminescence and the manufacturing of the toxin T1 were reduced in the presence of a furanone molecule (Manefield et al. 2000). In the *Caenorhabditis elegans* infection model, *P. aeruginosa* quickly kills every worm; however, the mortality decreases to 5% or 40%, respectively, if 4-nitro-pyridine-N-oxide (4-NPO) or garlic extract is added to the growth medium (Bjarnsholt et al. 2005). In an additional trial, the mortality rate for mice administered a placebo was 88%, whereas the incidence of death for mice given furanone treatment was just 55% (Rasmussen et al. 2005).

6.5.6 Recombinant Gene Expression

Biological engineering is witnessing the emergence of quorum-sensing techniques for protein synthesis in both bacterial and eukaryotic models. Bulter et al. (2004) developed an artificial gene expression system in *E. coli* regulated by quorum sensing. Neddermann et al. (2003) generated a eukaryotic transactivator by fusing the ligand- and DNA-binding domains of the bacterial quorum sensor TraR.

6.5.7 Human Health

The expansion of QS system vehicles in cancer therapy has its roots in identifying the *P. aeruginosa* QS signal 3OC12HSL, which both induced and suppressed apoptosis in human breast cancer cell lines (Li et al. 2004). Applying genetically engineered bacteria with QS signals is one effective strategy for focusing on cancer cells (Choudhary and Schmidt-Dannert 2010). Using polyurethanes and plastics coated with QSIs on medical implants and catheters is one practical method for medical applications. For example, dispersion B may stop *Staphylococcus epidermidis* from forming a biofilm. When dispersion B and specific antibiotics like triclosan were combined, they showed a synergistic impact and prevented the growth of QS and *S. aureus*, *S. epidermidis*, and *E. coli* biofilms (Table 6.2) (Arciola et al. 2011).

6.5.8 Antibiotics and the Way They Work in Tandem with QSIs

Antibodies have also been reported to inhibit QS signaling molecules (Chen et al. 2002). Certain organisms possess receptors that impede QS molecules, and, when exposed to antigens, adaptive mammalian immune systems generate antibodies. In *Staphylococcus* pathogenesis, QSIs and antibiotics work together, with the therapeutically prescribed medicines being improved by RNA III-inhibiting peptides (Balaban et al. 2003). The antibiotics baicalein and ampicillin demonstrated this synergism against *P. aeruginosa* (Zeng et al. 2008). Furanone C30, penicillin acid, and patulin are QSIs that make *P. aeruginosa* susceptible to tobramycin (Hentzer and Givskov 2003). Combining baicalin hydrate (a QSI) with the antibiotic tobramycin lowers the number of microorganisms in the lung tissue (Brackman et al. 2011).

6.6 Recently Approved Patents on Novel Bioactive Substances, Their Analogues, and Methods for Disrupting QS Signaling (Lu et al. 2022)

In recent years, notable advancements in the field of bioactive substance discovery and quorum-sensing (QS) disruption have been granted patent approval. In 2019, a breakthrough was achieved with the invention of a novel brominated furanone derivative, demonstrating potent suppressive effects on biofilm development and QS. This innovation held promising therapeutic implications for inflammation-related conditions and dental disorders, particularly periodontal disease. Another significant development in the same year involved the use of 3,4,5-methyl trihydroxybenzoate to inhibit bacterial QS system activity, leading to a substantial reduction in the expression of pathogenic factors in *C. violaceum* and *P. aeruginosa*. The application of this methodology extended to the management of drug resistance in addition to the treatment of oral bacterial diseases through the inhibition of QS with D glucose. Furthermore, diverse techniques and materials were identified to impede biofilm growth by utilizing a functional ligand with specific binding regions for QS peptides and proteases. In 2020, patents were granted for substances capable of modulating AgrC receptors in *S. aureus* and related *Staphylococcus* species, offering potential applications in treating bacterial infections when combined with antibiotics. These recent innovations have significantly contributed to the ongoing efforts in developing strategies for combating bacterial infections and enhancing therapeutic interventions.

6.7 Conclusions

Addressing one of the foremost challenges in our society involves the ongoing pursuit of alternative therapeutic approaches. A particularly promising avenue in this quest is the exploration of anti-virulence therapy, which revolves around the utilization of agents capable of disrupting quorum sensing in bacterial pathogens. This strategy has garnered substantial attention within the scientific community and is actively being researched.

The escalating concern of antibiotic resistance poses a substantial threat to both human and veterinary medicine. Anti-virulence therapy offers a potential solution by targeting the communication mechanisms employed by bacteria. Quorum sensing, a pivotal system governing bacterial behavior, has been identified as a key focus. Multiple substances that interfere with quorum sensing have been identified, demonstrating efficacy in reducing illness in various plant and animal models.

Significantly, the quorum-sensing (QS) system plays a crucial role in controlling the production of biofilms in multidrug-resistant bacteria and regulating bacterial efflux pumps. As these mechanisms contribute to antibiotic resistance, disrupting quorum sensing through anti-virulence therapy presents a novel strategy. Beyond merely limiting bacterial invasiveness and infectiousness, it has been observed that quorum-quenching (QQ) methods can modulate the expression of efflux pump genes, further enhancing their potential against microbial resistance.

Looking ahead, the coming decades are anticipated to witness a substantial surge in the discovery of quorum-quenching compounds. These molecules hold promise for diverse applications, including their use as coating agents in medical devices. Such innovative approaches may gradually replace traditional bacteriostatic and bactericidal antibiotics in vivo, ushering in a new era in the battle against microbial resistance. The continued exploration of anti-virulence therapies and quorum-quenching compounds is crucial for the development of effective strategies to combat the evolving landscape of infectious diseases.

References

Ali B, Sabri AN, Ljung K, Hasnain S (2009) Auxin production by plant associated bacteria: impact on endogenous IAA content and growth of *Triticum aestivum* L. Lett Appl Microbiol 48(5):542–547. https://doi.org/10.1111/j.1472-765X.2009.02565.x

Alves S, Duarte A, Sousa S, Domingues FC (2016) Study of the major essential oil compounds of *Coriandrum sativum* against *Acinetobacter baumannii* and the effect of linalool on adhesion, biofilms and quorum sensing. Biofouling 32(2):155–165. https://doi.org/10.1080/08927014.2015.1133810

Annapoorani A, Parameswari R, Pandian SK, Ravi AV (2012) Methods to determine antipathogenic potential of phenolic and flavonoid compounds against urinary pathogen *Serratia marcescens*. J Microbiol Methods 91(1):208–211. https://doi.org/10.1016/j.mimet.2012.06.007

Arciola CR, Montanaro L, Costerton JW (2011) New trends in diagnosis and control strategies for implant infections. Int J Artif Organs 34(9):727–736. https://doi.org/10.5301/IJAO.2011.8784

Avella MA, Gioacchini G, Decamp O, Makridis P, Bracciatelli C, Carnevali O (2010) Application of multi-species of bacillus in sea bream larviculture. Aquaculture 305(1–4):12–19. https://doi.org/10.1016/j.aquaculture.2010.03.029

Balaban N, Giacometti A, Cirioni O, Gov Y, Ghiselli R, Mocchegiani F, Viticchi C, Del Prete MS, Saba V, Scalise G, Dell'Acqua G (2003) Use of the quorum-sensing inhibitor RNAIII-inhibiting peptide to prevent biofilm formation in vivo by drug-resistant *Staphylococcus epidermidis*. J Infect Dis 187(4):625–630. https://doi.org/10.1086/345879

Bandara MBK, Zhu H, Sankaridurg PR, Willcox MDP (2006) Salicylic acid reduces the production of several potential virulence factors of *Pseudomonas aeruginosa* associated with microbial keratitis. Investig Opthalmol Vis Sci 47(10):4453. https://doi.org/10.1167/iovs.06-0288

Bansal T, Englert D, Lee J, Hegde M, Wood TK, Jayaraman A (2007) Differential effects of epinephrine, norepinephrine, and indole on *Escherichia coli* O157:H7 chemotaxis, colonization, and gene expression. Infect Immun 75(9):4597–4607. https://doi.org/10.1128/IAI.00630-07

Barnard AML, Salmond GPC (2007) Quorum sensing in Erwinia species. Anal Bioanal Chem 387(2):415–423. https://doi.org/10.1007/s00216-006-0701-1

Bäuerle T, Fischer A, Speck T, Bechinger C (2018) Self-organization of active particles by quorum sensing rules. Nat Commun 9(1):3232. https://doi.org/10.1038/s41467-018-05675-7

Bhargava N, Sharma P, Capalash N (2010) Quorum sensing in Acinetobacter: an emerging pathogen. Crit Rev Microbiol 36(4):349–360. https://doi.org/10.3109/1040841X.2010.512269

Bhargava N, Singh SP, Sharma A, Sharma P, Capalash N (2015) Attenuation of quorum sensing-mediated virulence of *Acinetobacter baumannii* by *Glycyrrhiza glabra* flavonoids. Future Microbiol 10(12):1953–1968. https://doi.org/10.2217/fmb.15.107

Bhatt VS (2018) Quorum sensing mechanisms in Gram positive bacteria. In: Bramhachari PV (ed) Implication of quorum sensing system in biofilm formation and virulence. Springer, Singapore, pp 297–311. https://doi.org/10.1007/978-981-13-2429-1_20

Bjarnsholt T, Jensen PØ, Rasmussen TB, Christophersen L, Calum H, Hentzer M, Hougen H-P, Rygaard J, Moser C, Eberl L, Høiby N, Givskov M (2005) Garlic blocks quorum sensing and promotes rapid clearing of pulmonary *Pseudomonas aeruginosa* infections. Microbiology 151(12):3873–3880. https://doi.org/10.1099/mic.0.27955-0

Blöcher R, Rodarte Ramírez A, Castro-Escarpulli G, Curiel-Quesada E, Reyes-Arellano A (2018) Design, synthesis, and evaluation of alkyl-quinoxalin-2(1H)-one derivatives as anti-quorum sensing molecules, inhibiting biofilm formation in Aeromonas caviae Sch3. Molecules 23(12):3075. https://doi.org/10.3390/molecules23123075

Borges A, Serra S, Cristina Abreu A, Saavedra MJ, Salgado A, Simões M (2014) Evaluation of the effects of selected phytochemicals on quorum sensing inhibition and in vitro cytotoxicity. Biofouling 30(2):183–195. https://doi.org/10.1080/08927014.2013.852542

Bortolotti D, Trapella C, Bragonzi A, Marchetti P, Zanirato V, Alogna A, Gentili V, Cervellati C, Valacchi G, Sicolo M, Turrin G, Fantinati A, Di Luca D, Rizzo R (2019) Conjugation of LasR quorum-sensing inhibitors with ciprofloxacin decreases the antibiotic tolerance of *P. aeruginosa* clinical strains. J Chem 2019:1–13. https://doi.org/10.1155/2019/8143739

Brackman G, Cos P, Maes L, Nelis HJ, Coenye T (2011) Quorum sensing inhibitors increase the susceptibility of bacterial biofilms to antibiotics in vitro and in vivo. Antimicrob Agents Chemother 55(6):2655–2661. https://doi.org/10.1128/AAC.00045-11

Brackman G, Al Quntar AAA, Enk CD, Karalic I, Nelis HJ, Van Calenbergh S, Srebnik M, Coenye T (2013) Synthesis and evaluation of thiazolidinedione and dioxazaborocane analogues as inhibitors of AI-2 quorum sensing in *Vibrio harveyi*. Bioorg Med Chem 21(3):660–667. https://doi.org/10.1016/j.bmc.2012.11.055

Brackman G, Breyne K, De Rycke R, Vermote A, Van Nieuwerburgh F, Meyer E, Van Calenbergh S, Coenye T (2016) The quorum sensing inhibitor hamamelitannin increases antibiotic susceptibility of Staphylococcus aureus biofilms by affecting peptidoglycan biosynthesis and eDNA release. Sci Rep 6:20321. https://doi.org/10.1038/srep20321

Brameyer S, Kresovic D, Bode HB, Heermann R (2015) Dialkylresorcinols as bacterial signaling molecules. Proc Natl Acad Sci 112(2):572–577. https://doi.org/10.1073/pnas.1417685112

Bulter T, Lee S-G, Wong WW, Fung E, Connor MR, Liao JC (2004) Design of artificial cell–cell communication using gene and metabolic networks. Proc Natl Acad Sci 101(8):2299–2304. https://doi.org/10.1073/pnas.0306484101

Caicedo JC, Villamizar S, Ferro MIT, Kupper KC, Ferro JA (2016) Bacteria from the citrus phylloplane can disrupt cell–cell signalling in *Xanthomonas citri* and reduce citrus canker disease severity. Plant Pathol 65(5):782–791. https://doi.org/10.1111/ppa.12466

Chen X, Schauder S, Potier N, Van Dorsselaer A, Pelczer I, Bassler BL, Hughson FM (2002) Structural identification of a bacterial quorum-sensing signal containing boron. Nature 415(6871):545–549. https://doi.org/10.1038/415545a

Chong YM, Yin WF, Ho CY, Mustafa MR, Hadi AHA, Awang K, Narrima P, Koh C-L, Appleton DR, Chan K-G (2011) Malabaricone C from *Myristica cinnamomea* exhibits anti-quorum sensing activity. J Nat Prod 74(10):2261–2264. https://doi.org/10.1021/np100872k

Choudhary S, Schmidt-Dannert C (2010) Applications of quorum sensing in biotechnology. Appl Microbiol Biotechnol 86(5):1267–1279. https://doi.org/10.1007/s00253-010-2521-7

Christensen QH, Grove TL, Booker SJ, Greenberg EP (2013) A high-throughput screen for quorum-sensing inhibitors that target acyl-homoserine lactone synthases. Proc Natl Acad Sci 110(34):13815–13820. https://doi.org/10.1073/pnas.1313098110

Chu W, Zhou S, Zhu W, Zhuang X (2014) Quorum quenching bacteria *Bacillus* sp. QSI-1 protect zebrafish (*Danio rerio*) from *Aeromonas hydrophila* infection. Sci Rep 4(1):5446. https://doi.org/10.1038/srep05446

Cui T, Bai F, Sun M, Lv X, Li X, Zhang D, Du H (2020) *Lactobacillus crustorum* ZHG 2-1 as novel quorum-quenching bacteria reducing virulence factors and biofilms formation of *Pseudomonas aeruginosa*. LWT 117:108696. https://doi.org/10.1016/j.lwt.2019.108696

De Lamo Marin S, Xu Y, Meijler MM, Janda KD (2007) Antibody catalyzed hydrolysis of a quorum sensing signal found in Gram-negative bacteria. Bioorg Med Chem Lett 17(6):1549–1552. https://doi.org/10.1016/j.bmcl.2006.12.118

Defoirdt T, Boon N, Sorgeloos P, Verstraete W, Bossier P (2008) Quorum sensing and quorum quenching in Vibrio harveyi: lessons learned from in vivo work. ISME J 2(1):19-26. https://doi.org/10.1038/ismej.2007.92

Defoirdt T, Sorgeloos P, Bossier P (2011) Alternatives to antibiotics for the control of bacterial disease in aquaculture. Curr Opin Microbiol 14(3):251–258. https://doi.org/10.1016/j.mib.2011.03.004

Diggle SP, Cornelis P, Williams P, Cámara M (2006) 4-Quinolone signalling in *Pseudomonas aeruginosa*: old molecules, new perspectives. Int J Med Microbiol 296(2–3):83–91. https://doi.org/10.1016/j.ijmm.2006.01.038

Dong YH, Zhang LH (2005) Quorum sensing and quorum-quenching enzymes. J Microbiol 43(1):101–109

Dong Y-H, Wang L-H, Xu J-L, Zhang H-B, Zhang X-F, Zhang L-H (2001) Quenching quorum-sensing-dependent bacterial infection by an *N*-acyl homoserine lactonase. Nature 411(6839):813–817. https://doi.org/10.1038/35081101

Duarte A, Luís Â, Oleastro M, Domingues FC (2016) Antioxidant properties of coriander essential oil and linalool and their potential to control campylobacter spp. Food Control 61:115–122. https://doi.org/10.1016/j.foodcont.2015.09.033

Galloway WRJD, Hodgkinson JT, Bowden SD, Welch M, Spring DR (2011) Quorum sensing in Gram-negative bacteria: small-molecule modulation of AHL and AI-2 quorum sensing pathways. Chem Rev 111(1):28–67. https://doi.org/10.1021/cr100109t

Ganin H, Rayo J, Amara N, Levy N, Krief P, Meijler MM (2013) Sulforaphane and erucin, natural isothiocyanates from broccoli, inhibit bacterial quorum sensing. MedChemComm 4(1):175–179. https://doi.org/10.1039/C2MD20196H

Halbedl S, Kratzer M-C, Rahm K, Crosta N, Masters K-S, Zippert J, Bräse S, Gradl D (2013) Synthesis of novel inhibitors blocking Wnt signaling downstream of β-catenin. FEBS Lett 587(5):522–527. https://doi.org/10.1016/j.febslet.2013.01.034

Han XG, Lu CP (2009) Detection of autoinducer-2 and analysis of the profile of luxS and pfs transcription in Streptococcus suis serotype 2. Curr Microbiol 58(2):146–152. https://doi.org/10.1007/s00284-008-9291-9

Helman Y, Chernin L (2015) Silencing the mob: disrupting quorum sensing as a means to fight plant disease. Mol Plant Pathol 16(3):316–329. https://doi.org/10.1111/mpp.12180

Hema M, Princy SA, Sridharan V, Vinoth P, Balamurugan P, Sumana MN (2016) Synergistic activity of quorum sensing inhibitor, pyrizine-2-carboxylic acid and antibiotics against multi-drug resistant *V. cholerae*. RSC Adv 6(51):45938–45946. https://doi.org/10.1039/C6RA04705J

Hentzer M, Givskov M (2003) Pharmacological inhibition of quorum sensing for the treatment of chronic bacterial infections. J Clin Invest 112(9):1300–1307. https://doi.org/10.1172/JCI20074

Higgins DA, Pomianek ME, Kraml CM, Taylor RK, Semmelhack MF, Bassler BL (2007) The major *Vibrio cholerae* autoinducer and its role in virulence factor production. Nature 450(7171):883–886. https://doi.org/10.1038/nature06284

Hong SH, Hegde M, Kim J, Wang X, Jayaraman A, Wood TK (2012) Synthetic quorum-sensing circuit to control consortial biofilm formation and dispersal in a microfluidic device. Nat Commun 3(1):613. https://doi.org/10.1038/ncomms1616

Jakobsen TH, Bjarnsholt T, Jensen PØ, Givskov M, Høiby N (2013) Targeting quorum sensing in *Pseudomonas aeruginosa* biofilms: current and emerging inhibitors. Future Microbiol 8(7):901–921. https://doi.org/10.2217/fmb.13.57

Kalia M, Yadav VK, Singh PK, Dohare S, Sharma D, Narvi SS, Agarwal V (2019) Designing quorum sensing inhibitors of *Pseudomonas aeruginosa* utilizing FabI: an enzymic drug target from fatty acid synthesis pathway. 3 Biotech 9(2):40. https://doi.org/10.1007/s13205-019-1567-1

Kang M-S, Oh J-S, Kang I-C, Hong S-J, Choi C-H (2008) Inhibitory effect of methyl gallate and gallic acid on oral bacteria. J Microbiol 46(6):744–750. https://doi.org/10.1007/s12275-008-0235-7

Kendall MM, Sperandio V (2007) Quorum sensing by enteric pathogens. Curr Opin Gastroenterol 23(1):10–15. https://doi.org/10.1097/MOG.0b013e3280118289

Kendall MM, Rasko DA, Sperandio V (2007) Global effects of the cell-to-cell signaling molecules Autoinducer-2, Autoinducer-3, and epinephrine in a luxS mutant of Enterohemorrhagic *Escherichia coli*. Infect Immun 75(10):4875–4884. https://doi.org/10.1128/IAI.00550-07

Khan MSA, Zahin M, Hasan S, Husain FM, Ahmad I (2009) Inhibition of quorum sensing regulated bacterial functions by plant essential oils with special reference to clove oil. Lett Appl Microbiol 49(3):354–360. https://doi.org/10.1111/j.1472-765X.2009.02666.x

Koch G, Nadal-Jimenez P, Reis CR, Muntendam R, Bokhove M, Melillo E, Dijkstra BW, Cool RH, Quax WJ (2014) Reducing virulence of the human pathogen Burkholderia by altering the substrate specificity of the quorum-quenching acylase PvdQ. Proc Natl Acad Sci 111(4):1568–1573. https://doi.org/10.1073/pnas.1311263111

Kumar L, Chhibber S, Kumar R, Kumar M, Harjai K (2015) Zingerone silences quorum sensing and attenuates virulence of *Pseudomonas aeruginosa*. Fitoterapia 102:84–95. https://doi.org/10.1016/j.fitote.2015.02.002

Labbate M, Queck SY, Koh KS, Rice SA, Givskov M, Kjelleberg S (2004) Quorum sensing-controlled biofilm development in *Serratia liquefaciens* MG1. J Bacteriol 186(3):692–698. https://doi.org/10.1128/JB.186.3.692-698.2004

Lee J, Zhang L (2015) The hierarchy quorum sensing network in *Pseudomonas aeruginosa*. Protein Cell 6(1):26–41. https://doi.org/10.1007/s13238-014-0100-x

Li L, Hooi D, Chhabra SR, Pritchard D, Shaw PE (2004) Bacterial *N*-acylhomoserine lactone-induced apoptosis in breast carcinoma cells correlated with down-modulation of STAT3. Oncogene 23(28):4894–4902. https://doi.org/10.1038/sj.onc.1207612

Lu L, Li M, Yi G, Liao L, Cheng Q, Zhu J, Zhang B, Wang Y, Chen Y, Zeng M (2022) Screening strategies for quorum sensing inhibitors in combating bacterial infections. J Pharm Anal 12(1):1–14. https://doi.org/10.1016/j.jpha.2021.03.009

Maisuria VB, Los Santos YL, Tufenkji N, Déziel E (2016) Cranberry-derived proanthocyanidins impair virulence and inhibit quorum sensing of *Pseudomonas aeruginosa*. Sci Rep 6(1):30169. https://doi.org/10.1038/srep30169

Manefield M, Harris L, Rice SA, De Nys R, Kjelleberg S (2000) Inhibition of luminescence and virulence in the black Tiger prawn (*Penaeus monodon*) pathogen *Vibrio harveyi* by intercellular signal antagonists. Appl Environ Microbiol 66(5):2079–2084. https://doi.org/10.1128/AEM.66.5.2079-2084.2000

Marwat SK, Khan MS, Ghulam S, Anwar N, Mustafa G, Usman K (2011) Phytochemical constituents and pharmacological activities of sweet Basil-Ocimum basilicum L. (Lamiaceae). Asian J Chem 23(9):3773

McClean KH, Winson MK, Fish L, Taylor A, Chhabra SR, Camara M, Daykin M, Lamb JH, Swift S, Bycroft BW, Stewart GSAB, Williams P (1997) Quorum sensing and *Chromobacterium viola-*

ceum: exploitation of violacein production and inhibition for the detection of *N*-acylhomoserine lactones. Microbiology 143(12):3703–3711. https://doi.org/10.1099/00221287-143-12-3703

MDowell P, Affas Z, Reynolds C, Holden MTG, Wood SJ, Saint S, Cockayne A, Hill PJ, Dodd CER, Bycroft BW, Chan WC, Williams P (2001) Structure, activity and evolution of the group I thiolactone peptide quorum-sensing system of *Staphylococcus aureus*. Mol Microbiol 41(2):503–512. https://doi.org/10.1046/j.1365-2958.2001.02539.x

Moreira CG, Weinshenker D, Sperandio V (2010) QseC mediates *Salmonella enterica* serovar typhimurium virulence in vitro and in vivo. Infect Immun 78(3):914–926. https://doi.org/10.1128/IAI.01038-09

Myszka K, Schmidt MT, Majcher M, Juzwa W, Olkowicz M, Czaczyk K (2016) Inhibition of quorum sensing-related biofilm of *Pseudomonas fluorescens* KM121 by *Thymus vulgare* essential oil and its major bioactive compounds. Int Biodeterior Biodegrad 114:252–259. https://doi.org/10.1016/j.ibiod.2016.07.006

Naga NG, El-Badan DE, Ghanem KM, Shaaban MI (2023) It is the time for quorum sensing inhibition as alternative strategy of antimicrobial therapy. Cell Commun Signal 21(1):133. https://doi.org/10.1186/s12964-023-01154-9

Nakayama J, Cao Y, Horii T, Sakuda S, Akkermans ADL, De Vos WM, Nagasawa H (2001) Gelatinase biosynthesis-activating pheromone: a peptide lactone that mediates a quorum sensing in *Enterococcus faecalis*. Mol Microbiol 41(1):145–154. https://doi.org/10.1046/j.1365-2958.2001.02486.x

Natrah FMI, Defoirdt T, Sorgeloos P, Bossier P (2011) Disruption of bacterial cell-to-cell communication by marine organisms and its relevance to aquaculture. Mar Biotechnol 13(2):109–126. https://doi.org/10.1007/s10126-010-9346-3

Natrah FMI, Alam MI, Pawar S, Harzevili AS, Nevejan N, Boon N, Sorgeloos P, Bossier P, Defoirdt T (2012) The impact of quorum sensing on the virulence of *Aeromonas hydrophila* and *Aeromonas salmonicida* towards burbot (*Lota lota* L.) larvae. Vet Microbiol 159(1–2):77–82. https://doi.org/10.1016/j.vetmic.2012.03.014

Neddermann P, Gargioli C, Muraglia E, Sambucini S, Bonelli F, De Francesco R, Cortese R (2003) A novel, inducible, eukaryotic gene expression system based on the quorum-sensing transcription factor TraR. EMBO Rep 4(2):159–165. https://doi.org/10.1038/sj.embor.embor734

Newman KL, Chatterjee S, Ho KA, Lindow SE (2008) Virulence of plant pathogenic bacteria attenuated by degradation of fatty acid cell-to-cell signaling factors. Mol Plant Microbe Interact 21(3):326–334. https://doi.org/10.1094/MPMI-21-3-0326

Nhan DT, Cam DTV, Wille M, Defoirdt T, Bossier P, Sorgeloos P (2010) Quorum quenching bacteria protect *Macrobrachium rosenbergii* larvae from *Vibrio harveyi* infection: quorum quenching bacteria protect prawn. J Appl Microbiol 109(3):1007–1016. https://doi.org/10.1111/j.1365-2672.2010.04728.x

Okada M, Sato I, Cho SJ, Iwata H, Nishio T, Dubnau D, Sakagami Y (2005) Structure of the *Bacillus subtilis* quorum-sensing peptide pheromone ComX. Nat Chem Biol 1(1):23–24. https://doi.org/10.1038/nchembio709

Paczkowski JE, Mukherjee S, McCready AR, Cong J-P, Aquino CJ, Kim H, Henke BR, Smith CD, Bassler BL (2017) Flavonoids suppress *Pseudomonas aeruginosa* virulence through allosteric inhibition of quorum-sensing receptors. J Biol Chem 292(10):4064–4076. https://doi.org/10.1074/jbc.M116.770552

Palliyil S, Downham C, Broadbent I, Charlton K, Porter AJ (2014) High-sensitivity monoclonal antibodies specific for homoserine lactones protect mice from lethal *Pseudomonas aeruginosa* infections. Appl Environ Microbiol 80(2):462–469. https://doi.org/10.1128/AEM.02912-13

Pande GSJ, Natrah FMI, Flandez AVB, Kumar U, Niu Y, Bossier P, Defoirdt T (2015) Isolation of AHL-degrading bacteria from micro-algal cultures and their impact on algal growth and on virulence of *Vibrio campbellii* to prawn larvae. Appl Microbiol Biotechnol 99(24):10805–10813. https://doi.org/10.1007/s00253-015-6918-1

Pereira CS, Thompson JA, Xavier KB (2013) AI-2-mediated signalling in bacteria. FEMS Microbiol Rev 37(2):156–181. https://doi.org/10.1111/j.1574-6976.2012.00345.x

Perez LJ, Karagounis TK, Hurley A, Bassler BL, Semmelhack MF (2014) Highly potent, chemically stable quorum sensing agonists for *Vibrio cholerae*. Chem Sci 5(1):151–155. https://doi.org/10.1039/C3SC52572D

Priya PS, Boopathi S, Murugan R, Haridevamuthu B, Arshad A, Arockiaraj J (2023) Quorum sensing signals: aquaculture risk factor. Rev Aquac 15(4):1294–1310. https://doi.org/10.1111/raq.12774

Paul D, Kim Y, Ponnusamy K, Kweon JH (2009) Application of quorum quenching to inhibit biofilm formation. Environ Eng Sci 26:1319–1324

Rader BA, Campagna SR, Semmelhack MF, Bassler BL, Guillemin K (2007) The quorum-sensing molecule autoinducer 2 regulates motility and flagellar morphogenesis in *Helicobacter pylori*. J Bacteriol 189(17):6109–6117. https://doi.org/10.1128/JB.00246-07

Rasmussen TB, Skindersoe ME, Bjarnsholt T, Phipps RK, Christensen KB, Jensen PO, Andersen JB, Koch B, Larsen TO, Hentzer M, Eberl L, Hoiby N, Givskov M (2005) Identity and effects of quorum-sensing inhibitors produced by Penicillium species. Microbiology 151(5):1325–1340. https://doi.org/10.1099/mic.0.27715-0

Ravn L, Christensen AB, Molin S, Givskov M, Gram L (2001) Methods for detecting acylated homoserine lactones produced by Gram-negative bacteria and their application in studies of AHL-production kinetics. J Microbiol Methods 44(3):239–251. https://doi.org/10.1016/S0167-7012(01)00217-2

Rocha-Estrada J, Aceves-Diez AE, Guarneros G, De La Torre M (2010) The RNPP family of quorum-sensing proteins in Gram-positive bacteria. Appl Microbiol Biotechnol 87(3):913–923. https://doi.org/10.1007/s00253-010-2651-y

Rodriguez-Garcia I, Cruz-Valenzuela MR, Silva-Espinoza BA, Gonzalez-Aguilar GA, Moctezuma E, Gutierrez-Pacheco MM, Tapia-Rodriguez MR, Ortega-Ramirez LA, Ayala-Zavala JF (2016) Oregano (*Lippia graveolens*) essential oil added within pectin edible coatings prevents fungal decay and increases the antioxidant capacity of treated tomatoes. J Sci Food Agric 96(11):3772–3778. https://doi.org/10.1002/jsfa.7568

Romero M, Avendaño-Herrera R, Magariños B, Cámara M, Otero A (2010) Acylhomoserine lactone production and degradation by the fish pathogen *Tenacibaculum maritimum*, a member of the Cytophaga–Flavobacterium–Bacteroides (CFB) group. FEMS Microbiol Lett 304(2):131–139. https://doi.org/10.1111/j.1574-6968.2009.01889.x

Roy V, Fernandes R, Tsao C-Y, Bentley WE (2010) Cross species quorum quenching using a native AI-2 processing enzyme. ACS Chem Biol 5(2):223–232. https://doi.org/10.1021/cb9002738

Rudrappa T, Bais HP (2008) Curcumin, a known phenolic from *Curcuma longa*, attenuates the virulence of *Pseudomonas aeruginosa* PAO1 in whole plant and animal pathogenicity models. J Agric Food Chem 56(6):1955–1962. https://doi.org/10.1021/jf072591j

Rutherford ST, Bassler BL (2012) Bacterial quorum sensing: its role in virulence and possibilities for its control. Cold Spring Harb Perspect Med 2(11):a012427. https://doi.org/10.1101/cshperspect.a012427

Ryu E-J, Sim J, Sim J, Lee J, Choi B-K (2016) D-Galactose as an autoinducer 2 inhibitor to control the biofilm formation of periodontopathogens. J Microbiol 54(9):632–637. https://doi.org/10.1007/s12275-016-6345-8

Schramm VL, Gutierrez JA, Cordovano G, Basu I, Guha C, Belbin TJ, Evans GB, Tyler PC, Furneaux RH (2008) Transition state analogues in quorum sensing and SAM recycling. Nucleic Acids Symp Ser 52(1):75–76. https://doi.org/10.1093/nass/nrn038

Sheng J-Y, Chen T-T, Tan X-J, Chen T, Jia A-Q (2015) The quorum-sensing inhibiting effects of stilbenoids and their potential structure–activity relationship. Bioorg Med Chem Lett 25(22):5217–5220. https://doi.org/10.1016/j.bmcl.2015.09.064

Singh RP, Desouky SE, Nakayama J (2016) Quorum quenching strategy targeting Gram-positive pathogenic bacteria. In: Donelli G (ed) Advances in microbiology, infectious diseases and public health, vol 901. Springer International Publishing, pp 109–130. https://doi.org/10.1007/5584_2016_1

Sivasothy Y, Krishnan T, Chan K-G, Abdul Wahab S, Othman M, Litaudon M, Awang K (2016) Quorum sensing inhibitory activity of Giganteone A from *Myristica cinnamomea* King against *Escherichia coli* biosensors. Molecules 21(3):391. https://doi.org/10.3390/molecules21030391

Soh EY, Chhabra SR, Halliday N, Heeb S, Müller C, Birmes FS, Fetzner S, Cámara M, Chan K, Williams P (2015) Biotic inactivation of the *Pseudomonas aeruginosa* quinolone signal molecule. Environ Microbiol 17(11):4352–4365. https://doi.org/10.1111/1462-2920.12857

Storz MP, Allegretta G, Kirsch B, Empting M, Hartmann RW (2014) From in vitro to in cellulo: structure–activity relationship of (2-nitrophenyl)methanol derivatives as inhibitors of PqsD in *Pseudomonas aeruginosa*. Org Biomol Chem 12(32):6094–6104. https://doi.org/10.1039/C4OB00707G

Sun B, Zhang M (2016) Analysis of the antibacterial effect of an Edwardsiella tarda LuxS inhibitor. Springerplus 5:92. https://doi.org/10.1186/s40064-016-1733-4

Szabó MÁ, Varga GZ, Hohmann J, Schelz Z, Szegedi E, Amaral L, Molnár J (2010) Inhibition of quorum-sensing signals by essential oils. Phytother Res 24(5):782–786. https://doi.org/10.1002/ptr.3010

Takayama H, Jia Z, Kremer L, Bauer JO, Strohmann C, Ziegler S, Antonchick AP, Waldmann H (2013) Discovery of inhibitors of the Wnt and hedgehog signaling pathways through the catalytic enantioselective synthesis of an Iridoid-inspired compound collection. Angew Chem Int Ed 52(47):12404–12408. https://doi.org/10.1002/anie.201306948

Tan SY-Y, Chua S-L, Chen Y, Rice SA, Kjelleberg S, Nielsen TE, Yang L, Givskov M (2013) Identification of five structurally unrelated quorum-sensing inhibitors of *Pseudomonas aeruginosa* from a natural-derivative database. Antimicrob Agents Chemother 57(11):5629–5641. https://doi.org/10.1128/AAC.00955-13

Teplitski M, Chen H, Rajamani S, Gao M, Merighi M, Sayre RT, Robinson JB, Rolfe BG, Bauer WD (2004) *Chlamydomonas reinhardtii* secretes compounds that mimic bacterial signals and interfere with quorum sensing regulation in bacteria. Plant Physiol 134(1):137–146. https://doi.org/10.1104/pp.103.029918

Thoendel M, Kavanaugh JS, Flack CE, Horswill AR (2011) Peptide signaling in the staphylococci. Chem Rev 111(1):117–151. https://doi.org/10.1021/cr100370n

Tinh NTN, Yen VHN, Dierckens K, Sorgeloos P, Bossier P (2008) An acyl homoserine lactone-degrading microbial community improves the survival of first-feeding turbot larvae (*Scophthalmus maximus* L.). Aquaculture 285(1–4):56–62. https://doi.org/10.1016/j.aquaculture.2008.08.018

Truchado P, Larrosa M, Castro-Ibáñez I, Allende A (2015) Plant food extracts and phytochemicals: their role as quorum sensing inhibitors. Trends Food Sci Technol 43(2):189–204. https://doi.org/10.1016/j.tifs.2015.02.009

Turan NB, Chormey DS, Büyükpınar Ç, Engin GO, Bakirdere S (2017) Quorum sensing: little talks for an effective bacterial coordination. TrAC Trends Anal Chem 91:1–11. https://doi.org/10.1016/j.trac.2017.03.007

Vandeputte OM, Kiendrebeogo M, Rasamiravaka T, Stévigny C, Duez P, Rajaonson S, Diallo B, Mol A, Baucher M, El Jaziri M (2011) The flavanone naringenin reduces the production of quorum sensing-controlled virulence factors in *Pseudomonas aeruginosa* PAO1. Microbiology 157(7):2120–2132. https://doi.org/10.1099/mic.0.049338-0

Vikram A, Jesudhasan PR, Jayaprakasha GK, Pillai SD, Patil BS (2011) Citrus limonoids interfere with *Vibrio harveyi* cell–cell signalling and biofilm formation by modulating the response regulator LuxO. Microbiology 157(1):99–110. https://doi.org/10.1099/mic.0.041228-0

Von Bodman SB, Bauer WD, Coplin DL (2003) Quorum sensing in plant-pathogenic bacteria. Annu Rev Phytopathol 41(1):455–482. https://doi.org/10.1146/annurev.phyto.41.052002.095652

Wang W-Z, Morohoshi T, Ikenoya M, Someya N, Ikeda T (2010) AiiM, a novel class of *N*-acylhomoserine lactonase from the leaf-associated bacterium *Microbacterium testaceum*. Appl Environ Microbiol 76(8):2524–2530. https://doi.org/10.1128/AEM.02738-09

Wei Q, Bhasme P, Wang Z, Wang L, Wang S, Zeng Y, Wang Y, Ma LZ, Li Y (2020) Chinese medicinal herb extract inhibits PQS-mediated quorum sensing system in *Pseudomonas aeruginosa*. J Ethnopharmacol 248:112272. https://doi.org/10.1016/j.jep.2019.112272

Xu X, Zhou XD, Wu CD (2012) Tea catechin epigallocatechin gallate inhibits *Streptococcus mutans* biofilm formation by suppressing gtf genes. Arch Oral Biol 57(6):678–683. https://doi.org/10.1016/j.archoralbio.2011.10.021

Yang Q, Aamdal Scheie A, Benneche T, Defoirdt T (2015) Specific quorum sensing-disrupting activity (AQSI) of thiophenones and their therapeutic potential. Sci Rep 5(1):18033. https://doi.org/10.1038/srep18033

Yang Q, Pande GSJ, Wang Z, Lin B, Rubin RA, Vora GJ, Defoirdt T (2017) Indole signalling and (micro)algal auxins decrease the virulence of *Vibrio campbellii*, a major pathogen of aquatic organisms. Environ Microbiol 19(5):1987–2004. https://doi.org/10.1111/1462-2920.13714

Yap PSX, Krishnan T, Chan K-G, Lim SHE (2015) Antibacterial mode of action of *Cinnamomum verum* bark essential oil, alone and in combination with piperacillin, against a multi-drug-resistant *Escherichia coli* strain. J Microbiol Biotechnol 25(8):1299–1306. https://doi.org/10.4014/jmb.1407.07054

Yeon KM, Cheong WS, Oh HS, Lee WN, Hwang BK, Lee CH, Beyenal H, Lewandowski Z (2009) Quorum sensing: a new biofouling control paradigm in a membrane bioreactor for advanced wastewater treatment. Environ Sci Technol 43(2):380–385. https://doi.org/10.1021/es8019275

Yin H, Deng Y, Wang H, Liu W, Zhuang X, Chu W (2015) Tea polyphenols as an antivirulence compound disrupt quorum-sensing regulated pathogenicity of *Pseudomonas aeruginosa*. Sci Rep 5(1):16158. https://doi.org/10.1038/srep16158

Zeng Z, Qian L, Cao L, Tan H, Huang Y, Xue X, Shen Y, Zhou S (2008) Virtual screening for novel quorum sensing inhibitors to eradicate biofilm formation of *Pseudomonas aeruginosa*. Appl Microbiol Biotechnol 79(1):119–126. https://doi.org/10.1007/s00253-008-1406-5

Zhang J-W, Xuan C-G, Lu C-H, Guo S, Yu J-F, Asif M, Jiang W-J, Zhou Z-G, Luo Z-Q, Zhang L-Q (2019) AidB, a novel thermostable *N*-acylhomoserine lactonase from the bacterium Bosea sp. Appl Environ Microbiol 85(24):e02065–e02019. https://doi.org/10.1128/AEM.02065-19

Zhou L, Zhang L-H, Cámara M, He Y-W (2017) The DSF family of quorum sensing signals: diversity, biosynthesis, and turnover. Trends Microbiol 25(4):293–303. https://doi.org/10.1016/j.tim.2016.11.013

Revolutionizing Healthcare: Unleashing the Potential of Immunotherapies to Strengthen the Body's Immune Defense Against Antibiotic-Resistant Pathogen

7

Anasuya Patil, Vaishali P. Nagulwar, Sandeep Kumar Yadav, Fauzia Tabassum, and Singh Shekhar Gautam

Abstract

This book chapter explores the transformative potential of immunotherapy in revolutionizing healthcare, particularly in the context of combating antibiotic-resistant pathogens. As the efficacy of traditional antibiotics wanes in the face of evolving microbial threats, there is an urgent need for innovative approaches to bolster the body's immune defenses. Immunotherapies offer a promising alternative by harnessing the power of the immune system to target and eliminate pathogens. This chapter examines the mechanisms underlying immunotherapies, their applications in combating antibiotic resistance, and the challenges and opportunities they present in reshaping the landscape of infectious disease treatment. Through a comprehensive analysis of current research and clinical advancements, this chapter highlights the pivotal role of immunotherapies in advancing healthcare and safeguarding global public health against the threat of antibiotic resistance.

A. Patil (✉)
KLE College of Pharmacy, Bengaluru, Karnataka, India

V. P. Nagulwar
Government College of Pharmacy, Chhatrapati Sambhajinagar, Chhatrapati Sambhajinagar, India

S. K. Yadav
Department of Pharmacology, SGT College of Pharmacy, Shree Guru Gobind Singh Tricentenary (SGT) University, Gurugram, Haryana, India

F. Tabassum
Department of Pharmacology, Vision College, Riyadh, Kingdom of Saudi Arabia

S. S. Gautam
Educational Institute of Pharmacy, Prayagraj, Uttar Pradesh, India

M. Gangwar, G. Nath (eds.), *Emerging Paradigms for Antibiotic-Resistant Infections: Beyond the Pill*, https://doi.org/10.1007/978-981-97-5272-0_7

Keywords
Immunotherapy · Antibiotic resistance · Immune defense · Pathogen · Healthcare revolution

7.1 Introduction

The rising mortality rate from serious infectious illnesses caused by drug-resistant bacteria emphasizes the critical need for novel therapeutic techniques. The increasing use of standard antimicrobial medications and associated circumstances has resulted in the growth of drug-resistant bacteria, demanding innovative remedies. Immunotherapy seems to be a potential technique for addressing issues in infectious illness treatment. According to studies, immunotherapy uses innate and adaptive immune responses to destroy sick cells. It comprises both passive and active immunotherapies. Passive immunotherapy provides ex vivo-generated immunological components, such as antibodies, without rousing the host's immune system. On the other hand, active immunotherapy stimulates the patient's immune response by promoting the production of certain immunological effectors through techniques such as vaccination treatment (Qadri et al. 2023).

The idea behind immunotherapy is to strengthen the body's defense mechanisms so they can more effectively target and fight illnesses. Immunotherapy, in contrast to conventional medicine, gives the immune system the ability to identify and eliminate aberrant cells, offering a targeted and long-lasting remedy for several illnesses, such as cancer and autoimmune diseases. To combat the danger posed by antibiotic-resistant bacteria, immunotherapy is essential. Changing to identify and fight a variety of infections provide a focused remedy by lowering the selection pressure that leads to the formation of resistance strains. Immunotherapy increases the overall efficacy of treatment when paired with conventional antibiotics, especially in cases of persistent infections when conventional regimens may not be sufficient. It may also be used prophylactically to avoid infections in high-risk persons or environments.

Immunotherapy is significant because of its position as a multimodal strategy for efficient infection management, as well as its tailored flexibility and decreased selection pressure. Continuous investigation and advancement in immunotherapy lead to creative approaches to tackle the problems caused by antibiotic resistance, therefore making it an essential part of the changing face of infectious disease control (Fig. 7.1) (Naran et al. 2018).

7.2 Monoclonal Antibodies-Based Immunotherapies

Monoclonal antibodies (mAbs)-based immunotherapy is a novel method to medical treatment that takes use of the specificity and adaptability of monoclonal antibodies to target and attack a wide range of illnesses. These antibodies are precisely recognized and attach to particular molecules on the surface of cells, pathogens, or

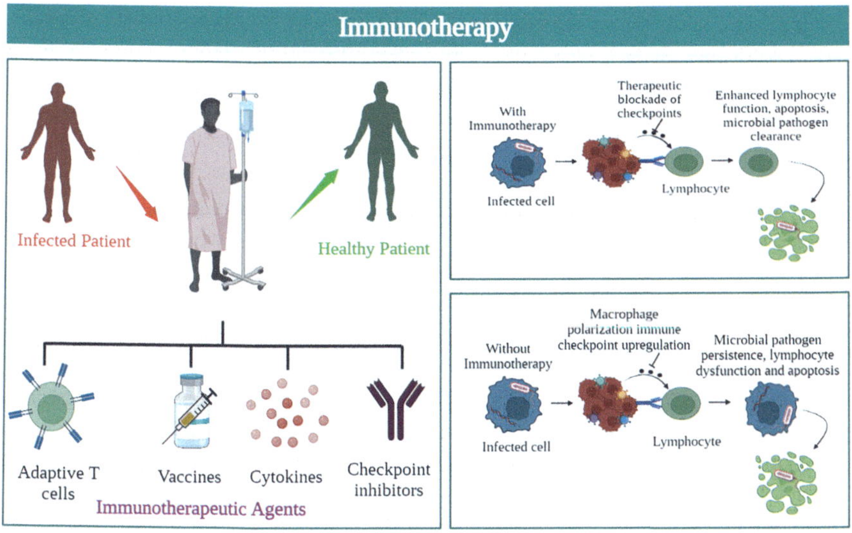

Fig. 7.1 Diagrammatic illustration of the immunotherapy as a potential treatment in the case of infected persons

aberrant proteins because they are clones of a single kind of immune cell created in a lab. Monoclonal antibodies are created when two cells fuse together to form a hybrid cell, or hybridoma, that has both immortality and the ability to manufacture a single, targeted antibody. Dr. Milstein's significant work has earned him the title of father of modern immunology. In addition to treating a wide range of other illnesses, monoclonal antibodies are a kind of active immunity that are used to treat cancer, including colorectal, breast, and lymphoma. Additionally, they may be used with medicinal substances to provide cytotoxic effects on cancer cells (Varadé et al. 2021).

7.3 Application of Monoclonal Antibodies

There are several potential applications for mAbs in the treatment of antibiotic-resistant illnesses. Targeting the resistant microorganisms directly is one way to get beyond the special defenses they have developed against conventional antibiotics. Using mAbs to alter the host immune response is a second tactic that helps the body identify and eliminate antibiotic-resistant bacteria more effectively. With mAbs, we can circumvent the resistance mechanisms that bacteria acquire against traditional antibiotics, providing a respectable substitute for therapeutic intervention. The ability of mAbs to lessen the selection pressure on bacteria to develop resistance is likely their most well-known advantage; this is good news for manufacturers as well as practitioners. While it is true that most of the selective forces acting on conventional antibacterial compounds are thought to be strong, in the case of mAbs, which are used as immunotherapy, the selective forces may work to strengthen the immune system, reducing the possibility of promoting additional resistance.

7.3.1 Against Bacteria

A vital starting point for the creation of successful vaccines is provided by monoclonal antibodies, which generally have excellent specificity whereas the majority of antibiotics have a wide and nonselective spectrum of action. Research into monoclonal antibodies that target bacteria like *Escherichia coli, Bacillus anthracis, Pseudomonas aeruginosa, Staphylococcus aureus, and Clostridium difficile* is still ongoing. Human IgG is the most common kind of these antibodies, and the complementarity determining regions (CDR) of the antibody determine which pathogenic bacterial antigens or strains it will bind to (Polonelli et al. 2008).

Finding a trustworthy antigen structure in the intended pathogenic microbe, however, is necessary for effective administration. The human monoclonal antibody 514G3, which is intended to precisely target the cell wall protein *S. aureus*-specific staphylococcal protein A (SpA), is an example of such advancement. Phase I and II trials have tested this antibody, demonstrating a targeted strategy for treating *Staphylococcus aureus* infections.

Monoclonal antibodies have shown potential use in eliminating MRSA colonization of the nasopharynx in animal models. Additionally, there is evidence that *S. aureus* biofilm-associated infections may be successfully treated with monoclonal antibodies (Raafat et al. 2019). The creation of monoclonal and polyclonal antibodies, as well as the continuous creation of vaccines against the bacteria *C. difficile*, represent yet another triumph in the search for treatments and preventative measures to be used against germs resistant to antibiotics. When whey from *C. difficile*-infected cows was first used to treat *C. difficile*-associated diarrhea (CDI) in people, it demonstrated both high tolerance and effectiveness in preventing infection recurrences. During the investigation, it was determined that toxins A and B were pertinent immune target structures. Bezlotoxumab, a monoclonal antibody that targets toxin B in CDI, is effective and may lower the disease's recurrence rate by around 10% (Ebbers et al. 2021).

Aside from that, the FDA has authorized the use of Raxibacumab (commercial name ABthrax) and Obiltoxaximab (commercial name Anthim) against *Bacillus anthracis* (Anthrax toxin) for the prevention and treatment of inhalation of *Bacillus anthracis* (Varadé et al. 2021). Monoclonal antibodies (MAbs) are gaining more attention as a potential treatment for recently discovered bacterial illnesses. Antibody profiles during latent tuberculosis (TB) infection highlight the defensive role of such antibodies and promote macrophage killing of intracellular bacteria, demonstrating improved Fc-mediated immune effector action, and highlighting the important role that antibodies play in immunomodulation during TB infection (Lu et al. 2016). Thus far, MAbs have not proven effective. However, a number of specifically designed monoclonal antibodies have progressed to clinical trials against *Pseudomonas aeruginosa* and *Staphylococcus aureus*. MEDI3902 (AstraZeneca

PLC) is a bispecific IgG1 antibody that targets *P. aeruginosa's* PcrV protein (host cell cytotoxicity) and *Pslexopolysaccharide* (tissue adhesion and colonization) that is being developed for the treatment of pneumonia in high-risk patients. Furthermore, the targets may allow for widespread coverage since they are common to all *P. aeruginosa* strains globally. In cases of alpha toxin-regulated host cell damage, AR-301 (Aridis Pharmaceuticals), an MAb with the ability to neutralize virulence factors, conferred protection when administered as a supplemental treatment to patients suffering from methicillin-tolerant *S. aureus* (MRSA) pneumonia (François et al. 2018). A phase II clinical study is underway for MEDI4893 (AstraZeneca PLC), a novel long-acting monoclonal antibody that targets alpha-toxin. It provides effective immunoprophylaxis in cases of *S. aureus* infection and sustains serum concentrations in healthy individuals after intravenous infusion (Ramamurthy et al. 2021).

7.3.2 Against Fungi

MAbs may influence the fungal organism in question by modulating its biological processes, including creating extracellular virulence determinants from vesicles. The wide antigenic differences that exist between human beings and fungus species support the use of MAbs in immunotherapy to treat a variety of fungal infections. Creating MAbs against intracellular targets effectively strengthens the host defense system (Qadri et al. 2023). The capsule of *C. neoformans* is a crucial part of its pathogenicity, which may compromise the host's defenses. A possible explanation is that MAbs that target the polysaccharide of cryptococcal bacteria attach to the polysaccharide of *Cryptococcus neoformans* (*C. neoformans*) and promote the extraction of the polysaccharide antigen from human and animal blood hence improving microbial opsonization in vitro and in vivo (Lee et al. 1995). Further MAbs that target capsular polysaccharides have also been shown to extend the survival of infected mice and reduce the fungal burden in the tissue. Furthermore, these antibodies have been shown to enhance the efficacy of fluconazole and amphotericin-B antifungals against *C. neoformans* (Dromer and Charreire 1991). In a clinical trial, the murine IgG1 MAb against *C. neoformans* polysaccharide, or 18B7, was administered to HIV patients who had successfully recovered from cryptococcal meningitis. The goal of this study was to determine the maximum tolerated dose and safety of MAb 18B7 in humans. The results showed that at high dosages, 18B7 was both safe and efficient for reducing serum glucuronoxylomannan (Larsen et al. 2005).

7.3.3 Against Parasitic Disease

Strong mAbs, such as mAbs 2530, 2544, 2586, and 2587 against *P. falciparum* transmission-blocking vaccine candidate Pfs25, have been identified as a result of antibodies defined by the humoral immune response of volunteers participating in clinical studies (Scally et al. 2017). Similarly, *P. falciparum* circumsporozoite protein binding by Fab of mAbs 311 and 317, which were obtained from human donors involved in an RTS, S/AS01B clinical study, prevents the growth of parasites in vivo in C57BL/6 mice. Promising in vitro suppression of *P. falciparum* has been demonstrated by mAbs such as humABAMA1 and humAB10.1–10.3, which are generated from semi-immune donors and target the merozoite antigens Apical Membrane Antigen 1 (AMA1) and Merozoite Surface Protein 10 (MSP10), respectively. Recently, an MSP1 subunit antigen was used to produce an opsonization-driven neutrophilic respiratory burst response to merozoites in vitro, highlighting the importance of opsonizing antibodies as a study focus (Jäschke et al. 2017).

7.3.4 Against Viral Diseases

Therapeutic monoclonal antibodies for H1N1, SARS, MERS, and SARS-CoV-2 have been produced or are in the development. VIS-410, motavizumab, and palivizumab for RSV, and MEDI-8852, MHAA4549, CR8020, CT-P27, and diridavumab for influenza are a few examples of human or humanisedmAbs that were created as antivirals and have progressed past Phase 1 research. Eli Lilly's mAb LY-CoV555 (in collaboration with AbCellera) was the subject of the first clinical trial for treating COVID-19, which was started on May 28, 2020 (NCT04411628). It is a human IgG1 that targets the spike protein and is made from recovered patients' human B cells. It was well tolerated at all tested doses, and there have been no significant drug-related severe adverse events (SAEs) to date. Many Phase 3 trials for LY-CoV555 are presently being conducted for COVID-19 inpatients (NCT04501978), outpatients (NCT04518410), staff members of nursing homes (NCT04497987), and infected individuals. A Phase 3 study (NCT04427501) combining LY-CoV555 and LY-CoV016 is also being conducted for individuals with mild to moderate COVID-19 illness (Cruz-Teran et al. 2021).

7.4 Current Challenges and Future Prospects

All of these attempts to create mAb as a treatment for ARIs have, however, had less than enthusiastic outcomes. Regretfully, despite being given at extremely high dosages (e.g., 3000 mg per patient for MEDI-8852 and 8400 mg per patient for MHAA4549), they failed to demonstrate any discernible therapeutic benefit. Palivizumab, also marketed as Synagis®, is the only one that has been approved for use as a prophylactic; nevertheless, due to its high cost and low effectiveness, its clinical usage is restricted to the sickest preterm newborns (Cruz-Teran et al. 2021).

Although mAb therapies have been shown to have positive health effects, debates over their economic feasibility will persist due to their high cost. However, expensive costs are typically linked to novel, early therapies, and Mabs are not the only medications with exorbitant costs. The fact that many Mabs are still covered by patents contributes to their high cost. The vast arsenal of various anti-infective medications, including vaccines and antibiotics, can be partly blamed for the sluggish advancement of Mab therapies for infectious disorders. Mab medicines cover a smaller range of patients than standard treatments, which makes them less appealing commercially because they are specific to a particular infection. Furthermore, unlike other anti-infectives that may be taken orally, Mabs must be supplied by intravenous or subcutaneous injection, making them inappropriate for use by patients in impoverished nations with limited access to healthcare. Additionally, compared with vaccinations, mabs offer only temporary prophylaxis and are more successful in avoiding infections than in curing ones that have already occurred. Lastly, Mabs' low success rate in gaining permission to treat infectious illnesses and their costly development and manufacturing costs reduce their commercial attractiveness (Reichert and Dewitz 2006).

Its application for treating mixed infections may be hampered by Mab therapies' pathogen-specificity. Using a combination of Mabs to target the wide variety of antigens that viruses carry might be one way to address this issue. An approach like this would closely resemble the immune system's natural reaction, in which the body naturally produces several antibodies in reaction to the antigens a virus presents, each of which binds to a distinct antigen. The immune system uses this variety of antibodies to its advantage while battling the invader. Rabies therapy with a combination of Mabs is already under investigation. New prospects for serum treatment have been made possible by recent developments in Mab engineering. Significantly, Mabs provide a way to make standardized agents, which can result in a solution that is more accurate and effective than conventional serum treatment when mixed into a cocktail. Little commercial investment has been made in the development of anti-infective Mab products thus far. This is partially due to the short market share of infectious illnesses (Quinteros et al. 2017).

7.5 Vaccine-Based Immunotherapy

Vaccines are very effective in the battle against antimicrobial resistance (AMR) because of a number of characteristics. First of all, vaccination usually has little effect on the growth of microorganisms other than the strains that are targeted. This is because vaccinations work by enabling the immune system to recognize highly precise antigens for the illnesses they are intended to prevent (Jit et al. 2020). It is maintained that vaccine resistance poses less of a threat to human and animal health and welfare than medicine resistance since it is less likely to develop and, when it does, does less harm. Second, because vaccinations are unique, it is possible to produce vaccines that specifically target disease strains that are more dangerous or likely to develop resistance. This was the situation with the *S. pneumoniae* vaccinations, which were created based on virulence characteristics most likely to cause severe sickness. Thirdly, Rynkiewicz et al. suggested that vaccinations and antimicrobial drugs may work effectively (Rynkiewicz et al. 2011). Vaccines have the potential to slow down the pace of infection, which prolongs the time it takes for a disease to become resistant to treatment.

Ultimately, by preventing illness, immunizations can be administered just a few times and have a lasting impact on the population as a whole. This is achieved in part by the near-lifelong effects of vaccinations and in part by the development of herd immunity, which stops the spread of infectious agents (Alghamdi 2021).

7.6 Exploring the Development and Mechanism of Therapeutic Vaccines

A revolutionary approach to healthcare, vaccine-based immunotherapy harnesses the immune system's capacity to both prevent and cure illness. Conventional vaccinations against infectious illnesses continue to be essential, and new developments in viral vector and mRNA technologies have made it easier to react quickly to new dangers. The efficiency of vaccines is further optimized by adjuvants and immune-modulating substances. The promise of vaccine-based immunotherapy in complete disease care is highlighted by its combination with other therapies, such as checkpoint inhibitors or chemotherapy. Personalized vaccinations that are based on a person's genetic makeup and illness profile have been made possible by advancements in precision medicine. Vaccine-based immunotherapy is at the vanguard of medical innovation, offering a variety of innovative and promising approaches to harness the immune system for better health outcomes, whether the goal is to prevent infections or cure preexisting disorders (Micoli et al. 2021).

There are still many AMR infections that need to be addressed, despite the development and use of efficacious vaccinations against select AMR germs such as *N. meningitis, S. pneumoniae*, and *H. influenzae*.

As of right now, vaccines against *C. difficules*, extraintestinal *E. coli* (apart from the rather controversial whole-cell-based Solco-Urac and Uro-Vaxom), *S. aureus, N. gonorrheae, P. auregenosa, K. pneumoniae*, and *Salmonella* spp. are not available (Brodie et al. 2020).

There are vaccine candidates on exhibit that are at different stages of development. Many vaccine technologies, including protein vaccines, glycoconjugates, synthetic conjugates, bioconjugates, outer membrane vesicles (OMVs), and liposomal vaccines (LAVs), are being employed to find and create these vaccines (Fig. 7.2) (Micoli et al. 2021).

It is possible to extract a patient's immune cells and utilize them to make a vaccine, which is subsequently given back into the body to boost the immune system and perhaps stop cancer cells from growing, decrease tumors, and prevent recurrence. These vaccinations can frequently be used in conjunction with other medicinal substances to strengthen the immune system.

The most popular therapeutic use of antibiotics acts on many different bacteria to treat existing illnesses, which raises the likelihood that resistant clones may develop. A single target, such as the bacterial cell wall or the translation machinery, is often the sole mechanism of action of an antibiotic. The resistance mechanisms of bacteria include blocking access to antibiotic targets, drug efflux, altering the drug targets, and changing or inactivating the antibiotic itself. Bacteria can also acquire resistance through acquisition or development. So, for instance, a single mutation might make the antibiotic useless by changing the therapeutic target. Furthermore,

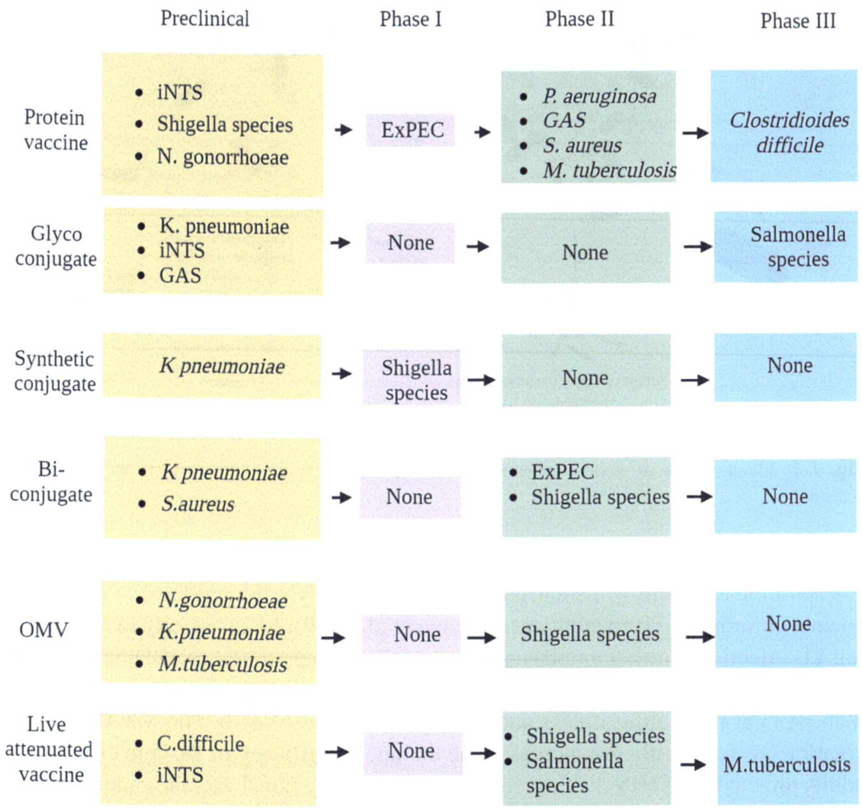

Fig. 7.2 Highlights the several vaccinations for these infections that are being developed at the moment

the generation of resistant clones is favored by the selection pressure that antibiotics apply. Vaccines reduce the likelihood that resistant clones will be chosen by serving as a preventative measure. Since vaccines frequently target several antigens and/or numerous epitopes of the same antigen (polyclonal antibodies), multiple mutations affecting distinct epitopes would be necessary for the formation of vaccine escape variants. Resistant clones might, nonetheless, develop as a result of serotype substitution or mutations (Fig. 7.3) (Romanò et al. 2015).

7.7 Notable Examples of Vaccines and Their Impact

7.7.1 Against Bacteria

Mycobacterium tuberculosis (Mtb) is the primary cause of infectious disease-associated mortality and the causative organism of TB. Although the only approved

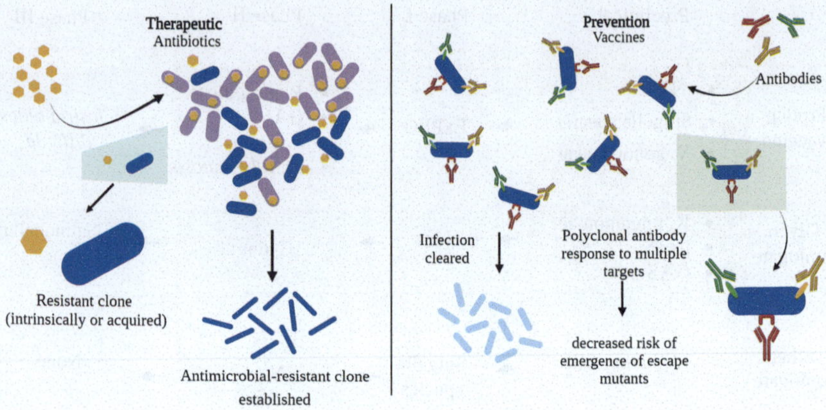

Fig. 7.3 Mechanisms of action of vaccination and antibiotic activity, as well as the rise of resistance

TB vaccine, Bacillus Calmette-Guérin (BCG) often protects against the most severe extrapulmonary forms of pediatric TB, it provides only a limited level of immunity against pulmonary TB in adult cases (Sable et al. 2019). Its incapacity to end ongoing TB infections further emphasises the need for novel strategies, despite its widespread usage as a TB vaccine. The ideal TB vaccination should be more effective than BCG at preventing illness and halting the spread of Mtb. The MVA85A vaccination has reportedly not increased the therapeutic efficacy of BCG in children or adults infected with HIV-1 (Tameris et al. 2013). Some novel vaccine candidates are either now undergoing or have just completed clinical trials, with differing degrees of success. The M72:AS01E subunit vaccine, which combines the adjuvant AS01E from GlaxoSmithKline with the immunogenic fusion protein (M72) derived from two *M. tuberculosis* antigens, showed apparent promise as it exhibited a 49.7% efficiency in inducing immunity in cases of TB disease in HIV-negative individuals harboring latent TB infection.

It is important to consider combining various vaccine candidates since *P. aeruginosa* exhibits a range of virulence mechanisms and adapts to its host environment (e.g., by producing biofilms). A novel reverse vaccinology method identified many antigens that, when added together, significantly decreased *P. aeruginosa* infection in a mouse model of acute pneumonia. Numerous vaccine targets against *K. pneumoniae* have been found within the last few decades. Bioconjugate vaccines comprising CPS from two *K. pneumoniae* serotypes have been shown in recent studies to be immunogenic and efficacious in protecting mice against lethal infection. It was recently shown that a vaccine candidate derived from *K. pneumoniae* outer membrane vesicles might offer protection in a preclinical animal model; the mechanism involved was dependent on both humoral and cellular immunity. This discovery emphasizes the potential of GMMA (generalized modules for membrane antigens) technology in the development of vaccines (Qadri et al. 2023).

There are currently no *C. difficile* vaccinations on the market. Clinical trials are underway to produce vaccines that specifically target the main pathogenic agents, toxins A and B (TcdA and TcdB, respectively). Sanofi initiated a phase III clinical research to evaluate the effectiveness of preventing initial symptomatic episodes in patients over 50 years of age who are susceptible to *C. difficile* infection. Nevertheless, an independent data monitoring committee that reviewed the interim data for the experiment decided that there was little chance the research would have achieved its effectiveness goal, leading to the trial's termination. In a phase III clinical study, Pfizer is presently evaluating a genetically engineered full-length TcdA and TcdB toxoid vaccine. A phase II experiment utilizing a vaccine (VLA84) that combines a fusion protein comprising shortened versions of TcdA and TcdB were successfully concluded by Valneva. Both vaccinations produced antibodies that neutralized the poisons and were very immunogenic. Recently, a phase I study evaluating the immunogenicity and safety of an experimental vaccine based on the F2 antigen was initiated by GlaxoSmithKline (GSK). Adjuvant AS01B may or may not be used in the formulation of the vaccine, which is given to healthy people between the ages of 18 and 45 and 50 and 70 (US National Library of Medicine 2019).

There are currently no extraintestinal pathogenic *Escherichia coli* (ExPEC) vaccines on the market, and only a small number of candidates are being developed. This leaves *E. coli* exposed to subpar whole-cell-based vaccines, such as the less effective Solco-Urovac (an inactivated polymicrobial vaccine) and Uro-Vaxom (made up of membrane proteins of 18 strains). The bacterial adhesin protein FimH and an adjuvant, which make up the vaccine being developed by Sequoia Sciences, were examined in a phase I clinical study including 67 women, 30 of whom had a history of recurrent UTIs that dated back 2 years. According to preliminary findings, the vaccination may lessen the incidence of UTIs (Starks et al. 2021).

A vaccine known as ExPEC4V is being developed by Janssen Vaccines and LimmaTech Biologics. It is based on O antigens, which correlate to four common serotypes. A phase Ib multicenter study involving healthy women with a history of recurrent UTIs examined the vaccine, which was created by the bioconjugation technique. ExPEC4V produced potent, long-lasting, and functional immune responses in this experiment while also being safe and well tolerated. The incidence of UTIs caused by *E. coli* of any serotype decreased, despite the study's lack of power to identify a substantial drop in the incidence of UTIs caused by the vaccine-specific serotypes (Huttner et al. 2017).

The effectiveness of three potential vaccinations against *S. aureus* has been assessed in human clinical studies. Nabi Biopharmaceuticals produced the conjugate vaccine StaphVAX, which targets capsular polysaccharides type 5 (CP5) and CP8; however, it did not demonstrate effectiveness in reducing *S. aureus* bacteremia in patients receiving haemodialysis who had end-stage renal illness. In cardiothoracic surgery patients, Merck's V710 vaccine, which targets the iron-scavenging protein IsdB, was tested in phase IIb and phase III studies to determine whether the vaccine was effective in lowering the percentage of patients who had deep sternal wound infections or postoperative *S. aureus* bacteremia. The experiment was terminated due to safety concerns and an interim analysis that indicated a low likelihood of vaccination effectiveness (Fowler et al. 2013).

Pfizer proceeded with a phase IIb study for their four-component vaccine candidate, SA4ag, which includes CP5, CP8, and the two surface protein antigens, MntC, and ClfA. The study assessed the effectiveness of the vaccination in preventing deep incisional or organ/space surgical site infections, as well as postoperative *S. aureus* bloodstream infections, in patients having elective open posterior multilevel spinal surgery. A scheduled interim observation analysis revealed a low statistical chance that the research would fulfill the predetermined primary efficacy end goals, leading to the trial's termination. Preclinical research has found new and promising vaccine candidates, including the virulence factor SpA, the pore-forming toxins leucocidins, and innovative adjuvants that boost cell-mediated immunity and boost vaccination effectiveness (Pozzi et al. 2017).

In endemic nations like South Asia, South-East Asia, and sub-Saharan Africa, as well as for tourists, a vaccination against *Salmonella enterica* subsp. enterica serovar *Typhi* will be especially helpful for newborns and premature children. Several more vaccines are being developed, and there are currently 20 available ones that protect against *Salmonella Typhi*. A limited length of antibody response, multiple dosage requirements, lack of immunological memory, affinity maturation, and live attenuated Ty21a oral vaccines and Vi polysaccharide vaccines are among the inherent limitations of the marketed vaccines. In order to get around these restrictions, new tactics have been put forth. For example, conjugating Vi polysaccharide to the right carrier protein allows the antigen to go from being T cell-independent to T cell-dependent. A phase III research on a combination vaccine against *Salmonella Typhi* Vi demonstrated over 90% effectiveness in children aged 2–5 years. However, the lack of a clear financial motivation to produce vaccines against *Salmonella Typhi* hindered the release of these vaccines into the market (Lin et al. 2001).

Vi glycoconjugate vaccinations have recently received licenses in China and India, thanks to the growing network of vaccine producers in emerging markets. Global health vaccine institutes like the International Vaccine Institute in Seoul, South Korea, and the GSK Vaccines Institute for Global Health (GVGH) in Siena, Italy, along with important academic institutions like the U.S. National Institutes of Health and the Centre for Vaccine Development and Global Health (CVD) at the University of Maryland, USA, have also provided incentives for the development of *Salmonella Typhi* vaccines. The WHO recently prequalified Typbar TCV, a typhoid conjugate vaccine that comprises Salmonella Typhi Vi polysaccharide conjugated to tetanus toxoid carrier protein, and it is presently undergoing efficacy studies in several countries (Jin et al. 2017).

7.7.2 Against Fungi

The growing number of fungal pathogenic species from *Aspergillus, Candida, Cryptococcus,* and other related genera—which are the leading causes of mycotic deaths—is the primary focus of vaccine research efforts. The development of vaccines against fungus is hampered by a number of factors, including host risk factors and unique fungal pathogenic processes. Given that different fungal pathogenic species use different antigens or epitopes, targeted vaccines are necessary to combat the spread of fungal infections (Segal et al. 2006). An effective strategy to prevent invasive fungal infections, particularly invasive candidiasis (IC), is to elicit an immunological response that mimics the innate immune responses against candida. This strategy may also help prevent other invasive infections that are comparable. The PEV7 and NDV3 vaccines have passed phase I testing for cytotoxicity and immunogenicity after being assessed using this technology. The efficacy of the PEV7 vaccination to prevent chronic vulvovaginal candidiasis (VVC) is being studied. Pevion BiotechAG is now testing shortened recombinant-secreted aspartic protease 2 (Sap2) in people. An effective strategy to prevent invasive fungal infections, particularly invasive candidiasis (IC), is to elicit an immunological response that mimics the innate immune responses against *candida*. This strategy may also help prevent other invasive infections that are comparable. The PEV7 and NDV3 vaccines have passed phase I testing for cytotoxicity and immunogenicity after being assessed using this technology. The efficacy of the PEV7 vaccination to prevent chronic vulvovaginal candidiasis (VVC) is being studied. Pevion BiotechAG is now testing shortened recombinant secreted aspartic protease 2 (Sap2) in people (Ibrahim et al. 2013).

7.7.3 Against Viral Disease

Clinical studies are now being conducted to create mRNA vaccines that express the pre-membrane (prM) and envelope (E) proteins of the *Zika virus (ZIKV);* however, there are presently no licensed RNA vaccines for human use (NCT04064905; NCT03014089). An mRNA vaccine expressing the stabilized perfusion SARS-CoV-2 spike protein began clinical development (NCT04283461) shortly after the SARS-CoV-2 genome was discovered. Every participant experienced anti-SARS-CoV-2 immune responses during phases I and II, and none of them experienced adverse effects that would have ended the experiment. Since then, the vaccine has advanced to phase III (NCT04470427), where its effectiveness against COVID-19 is presently being studied. The main component of DNA vaccines is pDNA backbones with an inserted eukaryotic expression cassette that codes for viral antigen. After the vaccine was absorbed into living cells, it stimulated the production of high-capacity target genes, which started an immune response specific to the antigen. Numerous clinical trials are still in progress even though there aren't any approved human DNA vaccines at the moment (Ramamurthy et al. 2021).

Recombinant viral-based vectors, either in live attenuated or non-replicative forms, are used in viral vaccines to express a target antigen. Many vaccines based on viral vectors are undergoing clinical trials; they include two SARS-CoV-2 viral vaccines in phase III and an adenovirus-based ZIKV vaccine (NCT03356561). The former is based on a type 5 adenovirus that expresses the SARS-CoV-2 spike (S) protein (NCT04341389), whereas the latter is based on an adenoviral vector derived from chimpanzees (ISRCTN89951424) and expresses the same protein (Trovato et al. 2020).

7.7.4 Against Parasitic Disease

The first human vaccination against a parasite illness is RTS, S/AS01 (Mosquirix, GlaxoSmithKline), which demonstrated low effectiveness (<50%) against *Plasmodium falciparum* in children. A thorough pipeline of noteworthy antimalarial vaccines under development is covered in a study by Philips et al. and several subunit and irradiation sporozoite vaccines against malaria are presently under development (Phillips et al. 2017).

Given that multiple approved dog vaccines have demonstrated excellent efficiency against Canine leishmaniasis, vaccinations can be a valuable weapon in the fight against zoonotic transmission. The combination vaccine therapy approaches are working well to improve therapeutic effectiveness at a lower dosage in the treatment of parasite illness. A combination of recombinant Tc24 C4 antigen and benznidazole, one of the two chemotherapeutic agents used to treat Chagas disease, worked synergistically to abrogate the parasite load in mice by inducing a balanced Th1/Th2 response and releasing IFN-γ and IL-4 specifically in response to the antigen. There are now several number of mRNA vaccines being researched to prevent toxoplasmosis, leishmaniasis, and malaria. In a mouse infection investigation, it was demonstrated that the treatment of *Trypanosoma cruzi* with GM-CSF-encoding plasmid cruzipain (Cz) induced a Th1 response through IgG2a production in mice and improved survival (Ramamurthy et al. 2021).

7.8 Potential Applications and Ongoing Research

Although vaccine-based immunotherapy is quite successful in avoiding infectious illnesses, there are still a number of ongoing difficulties. Long-lasting vaccine development is hampered by the fast antigenic changes seen in some diseases, such as HIV and influenza. As evidenced by the changing strains of SARS-CoV-2, vaccination efficacy necessitates ongoing modification in response to the development of new variations. Finding all-purpose vaccinations that offer comprehensive defense against various strains is still a difficult task. Effective vaccine-based immunotherapies depend on modifying vaccinations for immunocompromised people, organizing international responses, and comprehending the dynamics of long-term immunity. In spite of these difficulties, the area is still being advanced by continued research, international cooperation, and public health campaigns with the goal of removing barriers and improving the influence of vaccinations on world health (Kennedy et al. 2020).

The process of creating vaccines is incredibly difficult and takes longer than we would want, despite the tremendous progress that immunization has made in combating infectious illnesses. It is always necessary for researchers to create vaccinations that can combat a variety of virulent serotypes since some bacteria have multiple virulent serotypes. For instance, the quadrivalent vaccination has to include all six serogroups (A, B, C, Y, W-135, and X) of *N. meningitidis* since these serogroups account for the majority of meningococcal illness cases globally (Rouphael and Stephens 2012).

Still, the existing approach to developing vaccines is being revolutionized by the introduction of new R&D technologies including bioconjugation, structural vaccinology, reverse vaccinology 2.0, and nanoparticles. It is also impossible to overestimate the importance of molecular biology to the development of vaccines and conceptual understanding. Our understanding of virulence and host immune response mechanisms has improved due to molecular techniques like cloning and recombinant DNA technology. These techniques have also shortened the development time of vaccines against SARS-Cov2, such as the Covid-19 vaccines developed by Pfizer, Moderna, Astrazeneca, and others, to less than a year. Vaccines against AMR pathogens are also being developed using these strategies (Lurie et al. 2020).

7.9 Checkpoint Inhibitors in Immunotherapy

Immune checkpoint molecules are essential immune system regulators that control immunological responses by producing inhibitory receptors on immune cells. These molecules are essential for self-tolerance maintenance, immune response regulation (intensity and duration to minimize collateral tissue damage), and prevention of autoimmune reactions. One important effect of immunological checkpoint molecules signaling, particularly on T cells, is the induction of a state referred to as "exhaustion." Reduced effector function, prolonged expression of immunological checkpoint molecules such programmed cell death protein 1 (PD1), weakened recall responses, and a different transcriptional state from functioning effector or memory T cells are the hallmarks of T cell fatigue.

A strategy used by many infections and malignancies to elude immune regulation is the exploitation of inhibitory connections mediated by immunological checkpoint proteins. In order to combat this, the development of cancer therapies has concentrated on blocking immunological checkpoint molecules using particular antibodies, resulting in the licensing of a variety of monoclonal antibodies (MAbs). PD1 is targeted by pembrolizumab and nivolumab, cytotoxic T lymphocyte antigen 4 (CTLA4) is targeted by ipilimumab, and programmed cell death 1 ligand 1 (PDL1) is targeted by atezolizumab, avelumab, and durvalumab (Shiravand et al. 2022).

Although the use of immunotherapies to treat infectious diseases has not received as much attention, viruses' common use of inhibitory pathways for immune evasion raises the possibility of their application in both prevention and treatment of infectious diseases. In order to reverse T cell fatigue seen in chronic infectious illnesses, checkpoint blockage is the subject of current research. Additionally, there appears to be hope for using this strategy to create long-term immunity in acute infections. The development of vaccinations against infectious illnesses including HIV, hepatitis B virus (HBV), and malaria may benefit from the use of immune checkpoint blocking (Qadri et al. 2023).

Research utilizing both animal and ex vivo infection models demonstrates that immune checkpoint molecules regulate host immunological responses by reducing immune-mediated tissue damage while also facilitating pathogen clearance. Checkpoint inhibition has complex impacts; experimental infection models have shown both beneficial and harmful results. Limited, uncontrolled trials indicate that checkpoint molecule blockage may enhance patient outcomes for certain illnesses.

Reversing immunosuppressive receptor–ligand interactions using MAbs effectively lifts the inhibitory mechanisms that stop the immune system from attacking normal cells, which has led to the recent popularity of checkpoint inhibition treatment. With the help of this therapeutic approach, the immune system can resume normal function and mount greater immune responses against stronger ligands by blocking disease-related altered immune checkpoint activation. A number of important proteins involved in immune control are specifically targeted by checkpoint inhibitors, including CTLA4, TIM3, and PD-L1 (Wykes and Lewin 2018).

In summary, research on immune checkpoint blockade's promise in infectious illnesses is only beginning, even though it has transformed the treatment of cancer. Comprehensive research is necessary to fully use immunotherapies in the treatment of infectious diseases and the production of vaccines, given the delicate balance between triggering immune responses and averting immunopathology. These strategies' flexibility and specificity provide viable ways to handle the changing problems that infectious pathogens present.

7.10 Success Stories in Using Checkpoint Inhibitors for Infections

7.10.1 Against Bacteria

There have been varying assessments of the function of checkpoint molecules in the pathophysiology of various bacterial illnesses. PD-L1 on gastric epithelial cells, CTLA-4 on mononuclear cells, and PD-1 on Th cells are all found in significant concentrations in the digestive biopsies of individuals with Helicobacter pylori infection. Anti-CTLA-4 mAb treatment decreased stomach inflammation without changing the bacterial load in mice with *H. pylori* infection.

PD-1-deficient animals enhanced survival and eliminated the germs in a deadly *Listeria monocytogenes* infection scenario. The effector capabilities of various cell types (macrophages, NK cells, and CTLs) that provide protection against intracellular pathogens were found to be decreased in a different research following the injection of anti-PD-L1 mAb prior to Listeria infection.

T cell expression of PD-1 and CTLA-4 as well as antigen-presenting cell (APC) expression of PD-L1 are all increased in active TB infection (Abers et al. 2019).

As evidenced by PD-1 suppression in mutant mice, cellular investigations, and epidemiological research, immune checkpoint expression in TB may be understood as a physiological reaction to the persistent Mtb pathogen. Its inhibition may also exacerbate infection and pathogenesis. Therefore, the choice to use immune checkpoint inhibition for TB treatment will probably rely on a number of parameters, including the host (immunocompetence and HIV status) and particular mycobacterial factors (drug resistance and Mtb strain) (Langan et al. 2020).

7.10.2 Against Parasitic Disease

The research of immunological checkpoint proteins in malaria infections is relatively recent, as vaccinations have been the main focus of efforts to manage the disease. Field research in malaria-endemic Kenya and Mali revealed that *P. falciparum*-infected people expressed PD1 on CD4+ and CD8+ T cells, suggesting that this molecule is involved in immune evasion. In mice, it has been demonstrated that mAbs blocking the PD1/PD-L1 interaction have therapeutic potential for treating leishmaniasis and malaria. The investigations demonstrate the therapeutic potential of such antibodies and the mAbs utilised against PD-L1 in dendritic cells and against CD4+ lymphocytes that are targeted by the PD1 receptor in both illnesses. Conversely, PD-L2 mediates Th1 immunity and inhibits the PD1/PD-L1 connection, providing protection against malarial infections. In mice afflicted with cerebral or fatal malaria, multimeric PDL2 fused with the Fc region of immunoglobulin (PDL2–Fc) has been demonstrated to decrease infections (Ramamurthy et al. 2021).

7.10.3 Against Viral Disease

Immune checkpoint proteins have been the subject of much research during HIV infection; initially, this research focused on the virus's natural history and T cell activity; more recently, it has expanded to include consequences. T cell fatigue is a common symptom of HIV and other persistent viral infections. Many immune checkpoint proteins, such as PD1, CTLA4, TIM3, and LAG3, are expressed more often on CD4+ and CD8+ T cells in HIV infection that is not being treated. Last but not least, ex vivo blocking with anti-PD1 or anti-PDL1 antibodies enhanced the activity of CD8+ T cells specific to HIV and eliminated target cells that were infected.

The delivery of anti-PD1 antibody to rhesus macaques infected with SIV led to a fast increase of virus-specific CD8+ T cells with enhanced functional quality, as evidenced by a delayed time to macaque death and decreased levels of SIV RNA in plasma.

After giving the lowest dosage of anti-PDL1 antibody treatment to six HIV-positive people on antiretroviral therapy (ART), a phase II dose-escalation research was recently stopped by Bristol-Myers Squibb. Interestingly, in two out of the six subjects, there was a noticeable increase in CD4+ and CD8+ T cells specific to the HIV virus, despite no changes in HIV RNA or DNA levels. After receiving anti-PDL1 antibody treatment for many months, one of the six subjects experienced hypophysitis (Gay et al. 2017).

In conclusion, immune checkpoint inhibitors may be able to help find a cure or enable people to safely discontinue antiretroviral therapy (ART). First, taking these medications may improve the ability of HIV-specific T cells to eradicate HIV-positive cells. Second, as with ipilimumab and other modified antibodies against PD1, they may result in the direct removal of infected cells that exhibit the pertinent immune checkpoint signature, especially when employing a depleting antibody that activates Fc receptors. Third, immune checkpoint inhibitors could be able to make HIV latency disappear. Lastly, immune checkpoint inhibitors have the ability to boost vaccination responsiveness, which means they may be used in conjunction with other therapeutic vaccines that are currently being developed (Zhou et al. 2013).

It has been shown in several ex vivo experiments that utilizing blood drawn from patients with chronic HBV infection that suppression of PD1, CTLA4, 2B4, and TIM3 increases the activity of HBV-specific CD8+ T cells. Using checkpoint blockade alone or in combination has the potential to boost the generation of CD8+ T cells specific to HBV and even antibodies against HBsAg. A recent open-label trial of 20 individuals with chronically suppressed HBV infection indicated that nivolumab (anti-PD1 antibody; 0.3 mg/kg) was safe and well tolerated with or without an HBV vaccination. Additionally, one person had HBsAg seroconversion.

Clinical outcomes and the clearance of the hepatitis B virus (HBV) are linked to polymorphisms in checkpoint molecules. Checkpoint molecules are increased on HBV-specific CTLs during acute infection, and a subsequent decrease in PD-1 expression is linked to viral clearance (Zhang et al. 2008). During the course of a dose-escalation, noncomparative trial (NCT01658878)ii, no HBV-infected patients had seroconversion in response to nivolumab as a therapy for hepatocellular carcinoma (HCC). Several organizations have used the woodchuck hepatitis virus (WHV) to study the function of the PD-1/PD-L1 pathway in an HBV infection model in woodchucks. PD-1 expression increased by almost threefold after a prolonged WHV infection, and this was linked with WHV VL. Combination treatment with entecavir, therapeutic vaccination, and anti-PD-1 mAb raised the number of WHV-specific CTLs in the liver by almost threefold in one trial.

In a single-arm trial (NCT01008358)iii of patients with HCC who were infected with HCV, anti-CTLA-4 treatment reduced VL by three logarithmic points. In 45% of the patients, there were transient abnormalities in the liver function tests. 15% of participants in a randomized controlled trial (RCT) of anti-PD-1 treatment (NCT00703469)iv were able to achieve viral control with a single dosage of the medication. After solid organ transplantation (SOT) and allogeneic hematopoietic stem cell transplantation (HSCT), *cytomegalovirus (CMV)* infections are often seen. Managing CMV in immunocompromised individuals presents substantial hurdles due to antiviral medication toxicity and resistance. While the effectiveness of various immunotherapies in CMV patients has not been fully investigated, early research indicates that CPIs may improve T cell immunity against CMV.

Only people with impaired immune systems can develop progressive multifocal leukoencephalopathy (PML), a fatal central nervous system (CNS) infection caused by the JC virus (JCV). The severity of the condition greatly influences its prognosis. When a patient has PML, their macrophages express more PD-L1 and their T cells express more PD-1. Pembrolizumab's therapeutic impact was recently assessed in eight patients with PML by an uncontrolled prospective research (NCT01730131). A fivefold decrease in the percentage of CSF T cells expressing PD-1 was observed, and pembrolizumab entirely removed PD-1+ T cells from the blood in every patient. In CSF compared with blood, PD-1 expression recovered more quickly after pembrolizumab (Abers et al. 2019).

7.10.4 Against Fungal Disease

The management of invasive fungal infections (IFIs) with conventional antifungal therapy is still difficult because there are few effective medications, many toxicities, drug–drug interactions, and the spread of resistant pathogens like azole-resistant Aspergillus fumigatus and echinocandin-resistant *Candida* species throughout the world. The investigation of currently available oncological immune therapies, such as immune checkpoint inhibitors (ICIs), as adjuvant therapy for infectious illnesses is prompted by mounting evidence that the immunological etiology of infections and malignancies is comparable (Wykes and Lewin 2018). Very few clinical studies have been conducted on antifungal ICI treatment; a few of these are included below:

The first case of A patient with an invasive fungal illness who was not responding to antifungal medication was seen in 2017 and received supplementary ICI treatment as salvage therapy. Combination immunotherapy was started with five doses of IFN-γ, and one dose of the PD-1 inhibitor nivolumab is due to the patient's poor prognosis, severe immune paralysis, high expression of PD-1 on Th cells, and signs of poor monocytic maturation. After receiving nivolumab, PD-1 expression returned to levels seen in healthy controls in 4 days. Over 2 weeks, there was also an improvement in lymphocytopenia and monocyte maturation. After a few weeks, the patient is fully recovered. Another recent case report had a 38-year-old female patient who had previously been in good condition being treated for an extensive polymicrobial soft tissue infection that resulted from a minor thoracic trauma using combination immunotherapy with nivolumab and IFN-γ. Enterococci, several gram-negative bacteria, Aspergillus species, Mucor species, *L. ramosa*, and Rhizopusarrhizus were among the pathogens discovered by culture and genetic approaches. Comparably, a 56-year-old diabetic patient with COVID-19 who had received dexamethasone and tocilizumab (anti-IL-6) was described to have undergone combinatorial immunotherapy with nivolumab and IFN-γ. One week after being admitted to the intensive care unit (ICU), the patient developed invasive pulmonary aspergillosis, which was subsequently complicated by two Mucoralean brain abscesses and ethmoidal sinusitis (Wurster et al. 2022).

The cause of histoplasmosis in patients with advanced AIDS and SOT, *Histoplasma capsulatum*, increases the production of PD-L1 macrophages, which suppresses T cell activation and proliferation. Mice lacking PD-1 exhibited 100% survival in a mouse model of histoplasmosis wherein 100% of wild-type mice perish within 20 days of infection. Anti-PD-1 medication was also given to wild-type mice, which resulted in an extended life period following *H. capsulatum* challenge. This treatment dramatically boosted fungal clearance and decreased airspace edema and necrosis (Lázár-Molnár et al. 2008).

Anti-CTLA-4, anti-PD-1, and anti-PD-L1 mAbs greatly extend the life span of mice with candidemia; these effects are linked to APCs' elevation of MHC-II expression and T cells' enhanced production of IFN-γ. Similarly, mice infected with Cryptococcus neoformans survive much longer when treated with anti-PD-1 and anti-CTLA-4 mAbs; these effects are linked to better fungal clearance and a skewing towards Th1 immunity (Abers et al. 2019).

7.11 Addressing Concerns and Limitations

Among the most important immune system regulators are immunological checkpoints. Immune checkpoints are now comparable with immune system inhibitor regulators in the majority of mentioned circumstances. There are more immune checkpoint molecules in addition to those we have already covered, including BTLA, KIR, A2AR, B7-H4, NOX2, HO-1, and SIGLEC7. In addition, immune treatment appears to be a good target for the stimulatory immunological checkpoints, including OX40, GITR, CD40, CD122, and CD137. T cells are able to identify and eliminate any threats by using neoantigens that are produced on tumour cells. Tumour cells demand inhibitory substances to bind and mute immune cells in order to evade host immunity. Meanwhile, developing new drugs is never an easy ride. Numerous clinical studies aimed at creating novel drugs that target immunological checkpoints have been cancelled because of lethal adverse events and poor reactivity. We face an insurmountable mountain in the form of ICI-induced IrAEs, with death being the direct outcome. Furthermore, further information is still needed about the ICI processes (Cai et al. 2021).

Research on the interaction between immune checkpoint proteins and immunological activation and suppression has demonstrated the significance of these proteins in the development of infectious illnesses and cancer. Numerous research studies have indicated that immune checkpoint blockade may also be highly relevant for treating a number of infectious diseases, such as TB, HIV, HBV, and malaria; in these diseases, drug resistance is still a challenge, effective vaccine development has not been achieved, or lifelong medication treatment is required. It is important to acknowledge that immune checkpoint blockage may potentially result in immunological-related side effects since CTLA4, PD1, LAG3, and TIM3 are also involved in the control of peripheral tolerance, which helps to avoid autoimmunity. It is also unclear whether different immune checkpoint blockade responses will occur and how infectious disease clinical outcomes will pan out. All the same, immune checkpoint blockage might be a significant new approach to treating persistent illnesses for which there are now no reliable treatments or vaccines (Wykes and Lewin 2018).

7.12 Future Directions and Challenges of Immunotherapy

7.12.1 Future Challenges in Immunotherapy

The use of immunotherapy has grown significantly in recent years. This encompasses the introduction of soluble components like cytokines and monoclonal antibodies in addition to the development of cellular immunotherapy. Researchers are tackling a variety of issues in the vaccination field to increase both adaptability and efficacy. Vaccine combinations, injection reductions, and improved stability and durability as a result of lyophilization and nanotechnology techniques are a few of them. The development of universal vaccinations that give protection against all serotypes of certain pathogens—especially highly variable viruses like influenza—is a primary priority. Alternative modes of administration, such as oral, inhalation, intranasal, cutaneous, rectal, and vaginal, are also being investigated. This is to get around common problems with oral administration, such as immunological tolerance. Early childhood vaccination is another topic under investigation in an effort to save infants from common illnesses. This is an important area of research because of their immune system's immaturity, which makes them susceptible to infections. The requirement for the development of fresh, more potent vaccinations against common, widespread infections is one of the other topics. The ongoing requirement to combat newly developing pandemics, like the most recent SARS-CoV-2 (CoViD-19) epidemic, requires the swift creation of efficacious vaccinations (Ferris and Hood 2018). Therefore, it is important to recognize some of the crucial concerns, such the difficulties in formulating vaccines, the necessity to detect immune-related elements, the use of novel, creative techniques in their quick production, and the adjustments required for quick vaccination during outbreaks. Genetic vaccines have previously undergone extensive research and are undergoing clinical testing. It is obvious that these vaccinations have several benefits, including the capacity for quicker manufacture and simpler transportation. Nonetheless, a significant amount of research is needed to significantly enhance the immunogenicity and safety of these vaccinations. Meanwhile, novel adjuvants including ligands, which trigger the innate immune response, are being developed. These adjuvants have a host of potential benefits, including increased immunogenicity and safety. Lastly, research has been done on a novel idea known as taught immunity. This phenomenon is mostly indicative of the memory of the innate immune system. Numerous options exist for improving vaccine design as a result of these immunoepigenetic alterations of innate immunity, such as via knowledge of and control over the immunological memory-like characteristics of macrophages and natural killer cells (Wykes and Lewin 2018).

There is great interest in applying comparable tactics to treat various pathogenic microbial infections, including bacterial and fungal infections in people, due to immunotherapy's capacity to dramatically change cancer therapeutics. Little study has been done on how well these tactics translate to human illness in a clinical setting, despite first findings in preclinical models being constructive. Out of over 1700 immunotherapy trials, there are currently very few clinical trials for infectious disorders (Hotchkiss and Opal 2020). Considering the looming problem of drug resistance, immunotherapy might fill a vital gap in the market for adjuvant or substitute therapies for antibiotics. Future studies should focus on how preclinical discoveries may be used to treat infectious illnesses in humans without running the risk of making the situation worse. Research on the relationship between immunotherapies and traditional antimicrobial treatments is necessary since immunotherapy has been shown to either interfere with or even enhance medication tolerance. Checkpoint and cytokine inhibitors appear particularly plausible due to their ease of manufacturing and administration. These inhibitors will become more and more useful when their patents expire and the cost of a therapy drops dramatically. There are still a lot of unanswered questions regarding the use of immunotherapy to treat various microbial illnesses. These will need to be addressed in the near future by doing more preclinical research and appropriate clinical trials.

Immuno-based approaches hold particular promise in their quest to overcome the shortcomings of conventional chemotherapeutics, such as efficacy, cytotoxicity, and the growing issue of drug tolerance. It is evident that, like most therapies for any ailment, fighting the myriad infectious illnesses requires a multifaceted approach. The next pre-clinical and clinical studies have to incorporate the optimal combination of methodologies that will yield the best results for patients. In addition to standard alternative therapies, a combination of several immunotherapeutic techniques is necessary to guarantee the eradication of diseases like malaria or TB (Ramos-Espinosa et al. 2018). The advantage of a strategy that prioritizes optimal clinical outcomes is that it contributes to the development of medicines with high levels of selectivity and specificity and ushers in the era of precision medicine.

Numerous innovative technology approaches are being created to counteract the rising number of infectious illnesses. Antibody–antibiotic conjugates preserve the antibiotics' bactericidal properties while assisting them in getting where they're required. In a phase I clinical trial, the anti-*S. aureus* antibody–antibiotic combination DSTA4637S (Genentech) demonstrated favorable protection and pharmacokinetic characteristics. The antibiotic agent contained a MAb targeted against the cell wall teichoic acids of *S. aureus*. A recent confirmatory research study combined antivirulence immunotherapy with antimicrobial sonodynamic treatment in the form of a nanocapturer. These are made of nanovesicles with a neutralizing antibody coated on their surface. When ultrasonic waves are applied to the nanovesicles, the antibody generates reactive oxygen species (ROS), which destroys the pathogen and accelerates the elimination of virulence to eradicate MRSA in mice (Qadri et al. 2023).

The use of phage displays libraries for large-scale antibody fragment screening, improvements in cell culture conditions, identification of appropriate expression hosts, molecular technique advancements to fine-tune and enhance antibody durability and effectiveness (thus lowering dosage), and other factors have all contributed to the decreased cost of MAb-based therapies, which has increased accessibility to these treatments. As a result, many of the obstacles that previously hindered the broad use of therapeutic antibodies have been removed, allowing for their increased use and utilization. Adoption of novel vaccination strategies, such as those based on viral vector, mRNA, and DNA vaccines, provides options that may lead to faster and less expensive pipelines for vaccine creation, resolving the shortcomings of earlier peptide-based vaccines. Concurrently, targeted delivery and vehicle technologies have advanced, demonstrating the potential to alleviate concerns about pharmaceutical safety arising from systemic immunotherapy treatment and perhaps improving options for experimental therapies. This means that advances in immunotherapy are increasingly attractive treatments for a variety of infectious illnesses. Consequently, it appears that these diverse immune therapies provide interesting new avenues for the prevention and treatment of human pathogenic microbiological illnesses. Two increasingly popular therapy options are vaccinations and Mammalian antigen-binding antibodies (MABs). MABs have been utilized in the treatment of several medical illnesses, including cancer and autoimmune disorders. Owing to their distinct and targeted characteristics, these immunotherapies are especially versatile. In the years to come, they will benefit the general population and advance medical treatment.

References

Abers MS, Lionakis MS, Kontoyiannis DP (2019) Checkpoint inhibition and infectious diseases: a good thing? Trends Mol Med 25(12):1080–1093

Alghamdi S (2021) The role of vaccines in combating antimicrobial resistance (AMR) bacteria. Saudi J Biol Sci 28(12):7505–7510

Brodie A, El-Taji O, Jour I, Foley C, Hanbury D (2020) A retrospective study of immunotherapy treatment with uro-vaxom (OM-89®) for prophylaxis of recurrent urinary tract infections. Curr Urol 14(3):130–134

Cai X, Zhan H, Ye Y, Yang J, Zhang M, Li J, Zhuang Y (2021) Current progress and future perspectives of immune checkpoint in cancer and infectious diseases. Front Genet 12:785153

Cruz-Teran C, Tiruthani K, McSweeney M, Ma A, Pickles R, Lai SK (2021) Challenges and opportunities for antiviral monoclonal antibodies as COVID-19 therapy. Adv Drug Deliv Rev 169:100–117

Dromer F, Charreire J (1991) Improved amphotericin B activity by a monoclonal anti-*Cryptococcus neoformans* antibody: study during murine cryptococcosis and mechanisms of action. J Infect Dis 163(5):1114–1120

Ebbers M, Hemmer CJ, Müller-Hilke B, Reisinger EC (2021) Immunotherapy and vaccination against infectious diseases. Wiener klinischeWochenschrift 133:714–720

Ferris MT, Hood DW (2018) Host genetic regulation of immune-based and infectious diseases: introduction to mammalian genome special issue: genetics of infectious disease. Mamm Genome 29:365–366

Fowler VG, Allen KB, Moreira ED, Moustafa M, Isgro F, Boucher HW, Sobanjo-terMeulen A (2013) Effect of an investigational vaccine for preventing *Staphylococcus aureus* infections after cardiothoracic surgery: a randomized trial. JAMA 309(13):1368–1378

François B, Mercier E, Gonzalez C, Asehnoune K, Nseir S, Fiancette M, Laterre PF (2018) Safety and tolerability of a single administration of AR-301, a human monoclonal antibody, in ICU patients with severe pneumonia caused by *Staphylococcus aureus*: first-in-human trial. Intensive Care Med 44:1787–1796

Gay CL, Bosch RJ, Ritz J, Hataye JM, Aga E, Tressler RL, AIDS Clinical Trials 5326 Study Team (2017) Clinical trial of the anti-PD-L1 antibody BMS-936559 in HIV-1 infected participants on suppressive antiretroviral therapy. J Infect Dis 215(11):1725–1733

Hotchkiss RS, Opal SM (2020) Activating immunity to fight a foe—a new path. N Engl J Med 382(13):1270–1272

Huttner A, Hatz C, van den Dobbelsteen G, Abbanat D, Hornacek A, Frölich R, Fonck VG (2017) Safety, immunogenicity, and preliminary clinical efficacy of a vaccine against extraintestinal pathogenic *Escherichia coli* in women with a history of recurrent urinary tract infection: a randomised, single-blind, placebo-controlled phase 1b trial. Lancet Infect Dis 17(5):528–537

Ibrahim AS, Luo G, Gebremariam T, Lee H, Schmidt CS, Hennessey JP Jr, Edwards JE Jr (2013) NDV-3 protects mice from vulvovaginal candidiasis through T-and B-cell immune response. Vaccine 31(47):5549–5556

Jäschke A, Coulibaly B, Remarque EJ, Bujard H, Epp C (2017) Merozoite surface protein 1 from *Plasmodium falciparum* is a major target of opsonizing antibodies in individuals with acquired immunity against malaria. Clin Vaccine Immunol 24(11):e00155–e00117

Jin C, Gibani MM, Moore M, Juel HB, Jones E, Meiring J, Pollard AJ (2017) Efficacy and immunogenicity of a Vi-tetanus toxoid conjugate vaccine in the prevention of typhoid fever using a controlled human infection model of Salmonella Typhi: a randomised controlled, phase 2b trial. Lancet 390(10111):2472–2480

Jit M, Anderson M, Cooper B (2020) Quantifying the benefits of vaccines in combating antimicrobial resistance. Eur Secur 26(1):16–19

Kennedy RB, Ovsyannikova IG, Palese P, Poland GA (2020) Current challenges in vaccinology. Front Immunol 11:1181

Langan EA, Graetz V, Allerheiligen J, Zillikens D, Rupp J, Terheyden P (2020) Immune checkpoint inhibitors and tuberculosis: an old disease in a new context. Lancet Oncol 21(1):e55–e65

Larsen RA, Pappas PG, Perfect J, Aberg JA, Casadevall A, Cloud GA, Dismukes WE (2005) Phase I evaluation of the safety and pharmacokinetics of murine-derived anticryptococcal antibody 18B7 in subjects with treated cryptococcal meningitis. Antimicrob Agents Chemother 49(3):952–958

Lázár-Molnár E, Gácser A, Freeman GJ, Almo SC, Nathenson SG, Nosanchuk JD (2008) The PD-1/PD-L costimulatory pathway critically affects host resistance to the pathogenic fungus *Histoplasma capsulatum*. Proc Natl Acad Sci 105(7):2658–2663

Lee SC, Kress Y, Dickson DW, Casadevall A (1995) Human microglia mediate anti-*Cryptococcus neoformans* activity in the presence of specific antibody. J Neuroimmunol 62(1):43–52

Lin FYC, Ho VA, Khiem HB, Trach DD, Bay PV, Thanh TC, Kaufman JB (2001) The efficacy of a Salmonella typhi Vi conjugate vaccine in two-to-five-year-old children. N Engl J Med 344(17):1263–1269

Lu LL, Chung AW, Rosebrock TR, Ghebremichael M, Yu WH, Grace PS, Alter G (2016) A functional role for antibodies in tuberculosis. Cell 167(2):433–443

Lurie N, Saville M, Hatchett R, Halton J (2020) Developing Covid-19 vaccines at pandemic speed. N Engl J Med 382(21):1969–1973

Micoli F, Bagnoli F, Rappuoli R, Serruto D (2021) The role of vaccines in combatting antimicrobial resistance. Nat Rev Microbiol 19(5):287–302

Naran K, Nundalall T, Chetty S, Barth S (2018) Principles of immunotherapy: implications for treatment strategies in cancer and infectious diseases. Front Microbiol 9:3158

Phillips MA, Burrows JN, Manyando C (2017) Nature reviews disease primers. Malaria 3:17050

Polonelli L, Pontón J, Elguezabal N, Moragues MD, Casoli C, Pilotti E, Travassos LR (2008) Antibody complementarity-determining regions (CDRs) can display differential antimicrobial, antiviral and antitumor activities. PLoS One 3(6):e2371

Pozzi C, Olaniyi R, Liljeroos L, Galgani I, Rappuoli R, Bagnoli F (2017) Vaccines for *Staphylococcus aureus* and target populations. In: *Staphylococcus aureus*: microbiology, pathology, immunology, therapy and prophylaxis. Springer, pp 491–528

Qadri H, Shah AH, Alkhanani M, Almilaibary A, Mir MA (2023) Immunotherapies against human bacterial and fungal infectious diseases: a review. Front Med 10:1135541

Quinteros DA, Bermúdez JM, Ravetti S, Cid A, Allemandi DA, Palma SD (2017) Therapeutic use of monoclonal antibodies: general aspects and challenges for drug delivery. In: Nanostructures for drug delivery. Elsevier, pp 807–833

Raafat D, Otto M, Reppschläger K, Iqbal J, Holtfreter S (2019) Fighting *Staphylococcus aureus* biofilms with monoclonal antibodies. Trends Microbiol 27(4):303–322

Ramamurthy D, Nundalall T, Cingo S, Mungra N, Karaan M, Naran K, Barth S (2021) Recent advances in immunotherapies against infectious diseases. Immunother Adv 1(1):ltaa007

Ramos-Espinosa O, Islas-Weinstein L, Peralta-Álvarez MP, López-Torres MO, Hernández-Pando R (2018) The use of immunotherapy for the treatment of tuberculosis. Expert Rev Respir Med 12(5):427–440

Reichert JM, Dewitz MC (2006) Anti-infective monoclonal antibodies: perils and promise of development. Nat Rev Drug Discov 5(3):191–195

Romanò L, Paladini S, Galli C, Raimondo G, Pollicino T, Zanetti AR (2015) Hepatitis B vaccination: are escape mutant viruses a matter of concern? Hum Vaccines Immunother 11(1):53–57

Rouphael NG, Stephens DS (2012) *Neisseria meningitidis*: biology, microbiology, and epidemiology. In: *Neisseria meningitidis*: advanced methods and protocols. Springer, pp 1–20

Rynkiewicz D, Rathkopf M, Sim I, Waytes AT, Hopkins RJ, Giri L, Nielsen CJ (2011) Marked enhancement of the immune response to BioThrax®(Anthrax Vaccine Adsorbed) by the TLR9 agonist CPG 7909 in healthy volunteers. Vaccine 29(37):6313–6320

Sable SB, Posey JE, Scriba TJ (2019) Tuberculosis vaccine development: progress in clinical evaluation. Clin Microbiol Rev 33(1):10–1128

Scally SW, McLeod B, Bosch A, Miura K, Liang Q, Carroll S, Julien JP (2017) Molecular definition of multiple sites of antibody inhibition of malaria transmission-blocking vaccine antigen Pfs25. Nat Commun 8(1):1568

Segal BH, Kwon-Chung J, Walsh TJ, Klein BS, Battiwalla M, Almyroudis NG, Romani L (2006) Immunotherapy for fungal infections. Clin Infect Dis 42(4):507–515

Shiravand Y, Khodadadi F, Kashani SMA, Hosseini-Fard SR, Hosseini S, Sadeghirad H, Kulasinghe A (2022) Immune checkpoint inhibitors in cancer therapy. Curr Oncol 29(5):3044–3060

Starks CM, Miller MM, Broglie PM, Cubbison J, Martin SM, Eldridge GR (2021) Optimization and qualification of an assay that demonstrates that a FimH vaccine induces functional antibody responses in women with histories of urinary tract infections. Hum Vaccines Immunother 17(1):283–292

Tameris MD, Hatherill M, Landry BS, Scriba TJ, Snowden MA, Lockhart S, McShane H (2013) Safety and efficacy of MVA85A, a new tuberculosis vaccine, in infants previously vaccinated with BCG: a randomised, placebo-controlled phase 2b trial. Lancet 381(9871):1021–1028

Trovato M, Sartorius R, D'Apice L, Manco R, De Berardinis P (2020) Viral emerging diseases: challenges in developing vaccination strategies. Front Immunol 11:2130

US National Library of Medicine (2019) ClinicalTrialsgov. https://clinicaltrials.gov/ct2/show/NCT04026009

Varadé J, Magadán S, González-Fernández Á (2021) Human immunology and immunotherapy: main achievements and challenges. Cell Mol Immunol 18(4):805–828

Wurster S, Watowich SS, Kontoyiannis DP (2022) Checkpoint inhibitors as immunotherapy for fungal infections: promises, challenges, and unanswered questions. Front Immunol 13:1018202

Wykes MN, Lewin SR (2018) Immune checkpoint blockade in infectious diseases. Nat Rev Immunol 18(2):91–104

Zhang Z, Zhang JY, Wherry EJ, Jin B, Xu B, Zou ZS, Wang FS (2008) Dynamic programmed death 1 expression by virus-specific CD8 T cells correlates with the outcome of acute hepatitis B. Gastroenterology 134(7):1938–1949

Zhou J, Cheung AK, Tan Z, Wang H, Yu W, Du Y, Chen Z (2013) PD1-based DNA vaccine amplifies HIV-1 GAG-specific CD8+ T cells in mice. J Clin Invest 123(6):2629–2642

Beyond Antibiotics: Pioneering Strategies in Infection Control to Counter Antibiotic Resistance's Rising Tide

8

Neha Dhami, Mayank Gangwar, Dileep Kumar, Avinash Kumar Rao, and Sandeep Kumar

Abstract

Antibiotic resistance poses a formidable global health crisis and necessitates innovative strategies to preserve the efficacy of infection control. This chapter explores the transformative potential of advanced nonantibiotic therapies, such as phage therapy, probiotics, immunotherapy, and gene editing, as alternatives to traditional antibiotics. The primary objective of this study was to critically examine the role of nonantibiotic therapies in addressing antibiotic resistance. We aimed to elucidate the mechanisms and applications of each therapy, emphasizing its precision, targeted action, and potential synergy with conventional antibiotics. By outlining the objectives, we aim to contribute to a comprehensive

N. Dhami (✉)
Department of Social Work, Uttarakhand Open University, Haldwani, India

Department of Education, Kumaun University, Nainital, Uttarakhand, India

M. Gangwar
Faculty of Medicine, Viral Research and Diagnostic Laboratory, Department of Microbiology, Institute of Medical Sciences, Banaras Hindu University, Varanasi, Uttar Pradesh, India

Clinical Studies and Trials Unit, Division of Development Research, Indian Council of Medical Research, Department of Health Research, Ministry of Health and Family Welfare, Govt. of India, V. Ramalingaswami Bhawan, New Delhi, India

D. Kumar
Poona College of Pharmacy, Bharati Vidyapeeth (Deemed to be) University, Pune, Maharashtra, India

A. K. Rao
Patel College of Pharmacy, Madhyanchal Professional University (MPU), Bhopal, Madhya Pradesh, India

S. Kumar
MM College of Pharmacy, Maharishi Markandeshwar (Deemed to be) University, Ambala, Haryana, India

M. Gangwar, G. Nath (eds.), *Emerging Paradigms for Antibiotic-Resistant Infections: Beyond the Pill*, https://doi.org/10.1007/978-981-97-5272-0_8

understanding of the evolving landscape of infection control. The methodology involved a systematic review of the current literature, clinical trials, and emerging research in the field of advanced nonantibiotic therapies. Each therapy was meticulously analyzed, considering the mechanisms of action, applications, and challenges. We explored case studies and advancements to provide an up-to-date and nuanced perspective on the state of nonantibiotic therapies. Our findings underscore the potential impact of nonantibiotic therapies on combating antibiotic-resistant infections. Phage therapy exhibits remarkable specificity; probiotics restore microbial equilibrium; and advanced therapies, such as immunotherapy and gene editing, offer precision in modulating host responses. These results emphasize the potential synergy between these therapies and traditional antibiotics, presenting a multifaceted approach to infection control. In conclusion, advanced antibacterial therapies represent a paradigm shift in infection control. Precision-reduced antibiotic overuse and the potential for personalized interventions offer hope in the battle against antibiotic resistance. Although challenges exist, our exploration reveals the transformative potential of these therapies in shaping the future of infectious disease management. By embracing nonantibiotic alternatives, we pave the way for more resilient and sustainable approaches to combating antibiotic resistance ultimately redefining the landscape of infection control.

Keywords
Nonantibiotic therapies · Precision medicine · Phage therapy · Probiotics · Immunotherapy

8.1 Introduction

Antibiotic resistance stands as an imminent threat to global health, presenting a pressing challenge that demands innovative solutions. The overuse and misuse of antibiotics have led to the emergence of multidrug-resistant pathogens, rendering traditional treatments increasingly ineffective. As a result, the need for alternative strategies in infection control has become paramount (Walesch et al. 2022). In recent years, researchers and medical professionals have been exploring nonantibiotic advanced therapies as potential alternatives. These therapies include, but are not limited to, phage therapy, probiotics, immunotherapy, and gene editing. In this era of heightened urgency, the exploration of nonantibiotic advanced therapies marks a pivotal paradigm shift in infection control strategies. Nonantibiotic advanced therapies encompass a diverse range of innovative approaches that transcend the limitations of traditional antibiotic treatments. From phage therapy and probiotics to advanced therapies like immunotherapy and gene editing, these novel interventions hold the promise of revolutionizing the way we combat infectious diseases. By harnessing the power of phages, which are viruses that specifically target and destroy bacteria, we can directly eliminate multidrug-resistant pathogens. Probiotics, on the other hand, aim to restore the balance of beneficial bacteria in the body, creating an environment that is unfavorable for antibiotic-resistant strains to thrive.

Immunotherapy utilizes the body's own immune system to fight against infections by boosting its ability to identify and eliminate pathogens (MacNair et al. 2023). Additionally, gene editing techniques like CRISPR-Cas offer the potential to directly modify the genetic material of bacteria, rendering them unable to cause harm. In summary, nonantibiotic advanced therapies present a promising and innovative approach to combating the surge of antibiotic resistance. These nonantibiotic advanced therapies present a multifaceted approach to infection control, addressing the limitations of traditional antibiotics, and offering new avenues for treatment and prevention. Nonantibiotic advanced therapies have the potential to revolutionize infection control by providing alternative approaches to treating infectious diseases that are not reliant on traditional antibiotics alone. They offer a more sustainable and resilient solution to combat the surge of antibiotic resistance by targeting specific mechanisms in bacteria, disrupting their virulence, and enhancing the immune response. These innovative therapies aim to redefine the landscape of infection control by providing effective and targeted interventions that can overcome the challenges posed by antibiotic resistance. Within the broader context of the book, **"Emerging Paradigms for Antibiotic-Resistant Infections: Beyond the Pill,"** our chapter seeks to articulate a comprehensive understanding of the role and potential of nonantibiotic advanced therapies in addressing the global challenge of antibiotic resistance (Theuretzbacher and Piddock 2019). By examining the precision of these therapies, their impact on microbiome equilibrium, and the cutting-edge advancements in immunotherapy and gene editing, we aim to contribute valuable insights that extend beyond the confines of traditional antibiotic-centric approaches. The goal of this chapter is to give readers a thorough understanding of the various nonantibiotic advanced therapies available for infection control (Murugaiyan et al. 2022). This includes examining their precision, impact on microbiome equilibrium, and advancements in immunotherapy and gene editing. These therapies have the potential to redefine the paradigm of infection control and address the global challenge of antibiotic resistance. The integration of nonantibiotic advanced therapies in infection control represents a paradigm shift in the field of medicine. It offers new avenues for combating antibiotic resistance, overcoming the limitations of traditional antibiotic treatments, and promoting sustainable healthcare practices. In conclusion, nonantibiotic advanced therapies have the potential to revolutionize infection control by providing targeted interventions that can effectively combat antibiotic resistance (Mantravadi et al. 2019) and promote the long-term sustainability of healthcare (Aslam et al. 2018).

8.2 Background

8.2.1 Historical Context of Antibiotic Use and Emergence of Resistance

The introduction of antibiotics in the mid-twentieth century marked a groundbreaking era in medicine, providing unprecedented efficacy in treating bacterial infections. The advent of penicillin, followed by a myriad of other antibiotics, revolutionized healthcare and significantly reduced mortality rates associated with infectious diseases. However, the success of antibiotics was short-lived as bacteria quickly developed resistance mechanisms, rendering these once highly effective drugs ineffective against certain pathogens. This led to the emergence and spread of multidrug-resistant organisms, creating a global public health crisis. To address this crisis, there is a need to explore alternative strategies for infection control that go beyond traditional antibiotic treatments. One such strategy is the development and implementation of nonantibiotic advanced therapies, which offer a sustainable and resilient solution to combat the surge of antibiotic resistance. In recent years, there has been a growing recognition of the limitations of traditional antibiotic treatments and the need for innovative approaches to infection control. Nonantibiotic advanced therapies encompass a wide range of strategies that target bacterial infections through mechanisms other than antibiotics. These therapies include bacteriophages, phage therapy, antimicrobial peptides, nanotechnology-based approaches, and immunotherapy. The success of modern medicine in areas such as organ transplantation, cancer therapy, and major surgeries has relied heavily on effective antibiotic treatment to control bacterial infections. However, the rise of antibiotic resistance has put these advancements at risk. If effective global action plans are not adopted soon, we might encounter terrible complications for social, medical, and economic prospects. Nonantibiotic advanced therapies have the potential to revolutionize infection control by providing targeted interventions that can effectively combat antibiotic resistance and promote the long-term sustainability of modern medicine.

8.2.2 Overview of Contemporary Challenges in Infection Control

Contemporary infection control faces multifaceted challenges in the wake of antibiotic resistance. The rise of multidrug-resistant pathogens has exacerbated the complexity of treating infections, leading to prolonged illnesses, higher healthcare costs, and increased mortality rates. Healthcare-associated infections further strain the efficacy of existing treatments, creating a pressing need for innovative approaches to break the cycle of infection and resistance (Dancer 2013). In this context, nonantibiotic advanced therapies offer new avenues for infection control. These therapies have diverse mechanisms of action, including targeting specific bacterial strains with bacteriophages, disrupting biofilm with nanotechnology-based approaches, and enhancing the immune system's ability to combat infections through immunotherapy (Makabenta et al. 2020). By utilizing these alternative strategies, nonantibiotic advanced therapies have the potential to overcome the limitations of traditional antibiotics and pave the way for a new era of infection control (Ioannou et al. 2024). These therapies have shown promise in preclinical and clinical studies, demonstrating their ability to effectively target antibiotic-resistant bacteria and reduce infection rates. Furthermore, nonantibiotic advanced therapies can contribute to the preservation of existing antibiotics by reducing their overuse and misuse. By harnessing the power of genomics, nanotechnology, and other emerging fields, non-antibiotic advanced therapies can provide targeted and efficient treatments that address the root causes of antibiotic resistance and promote long-term sustainability in infection control. To summarize, it is critical that we prioritize the development and implementation of nonantibiotic advanced therapies for infection control (Hebert and Weber 2011). Revolutionizing infection control requires a comprehensive approach that includes the development of new antimicrobial agents, the implementation of comprehensive strategies, and harnessing innovative technologies such as genomics and nanotechnology. By taking urgent action and intensifying research and innovation, we can hope to stem the tide of antibiotic resistance and safeguard the future of modern medicine.

8.2.3 Rationale for Exploring Alternative Strategies to Address Antibiotic Resistance

The urgency to explore alternative strategies stems from the realization that traditional antibiotics alone are insufficient to combat the growing menace of resistance. The overreliance on these drugs has led to a diminishing arsenal of effective treatments, underscoring the necessity for diversified and innovative solutions. Paul Farmer, a physician-anthropologist who has labored for decades to bring the world's attention to the phenomenon of multidrug-resistant tuberculosis, has also played a key role in capturing global political attention (Control 2005; Farmer 1999). Antibiotics and hormonal mimics are not just analogous but completely intertwined at a material level, from antibacterial soaps with endocrine-disrupting properties to the novel micro-environments housing bacteria in fragments of plastic floating in the ocean to the interactions of xenobiotic hormone-like molecules and bacteria in the animal microbiome (Feehan and Garcia-Diaz 2020). These interconnected factors highlight the urgent need to explore alternative strategies beyond antibiotics to address the complex issue of antibiotic resistance. Consequently, a multidisciplinary approach is necessary to combat antibiotic resistance, including reducing the misuse and overuse of antibiotics in various sectors, such as human medicine, animal husbandry, and agriculture (Hutchison et al. 2018). By addressing the root causes of antibiotic resistance and adopting a comprehensive approach, we can hope to revolutionize infection control and ensure long-term sustainability. One potential nonantibiotic advanced therapy that has shown promise in infection control is the use of phage therapy. Phage therapy involves the use of bacteriophages, which are viruses that can specifically target and kill bacteria (Pires et al. 2020). The use of bacteriophages as an alternative to antibiotics could potentially revolutionize infection control and combat the surge of antibiotic resistance.

8.3 Nonantibiotic Advanced Therapy

Nonantibiotic advanced therapies, such as phage therapy, probiotics, and immunotherapy, are revolutionizing infection control and combating the surge of antibiotic resistance. These innovative treatments offer targeted and precision-based approaches that minimize collateral damage to the microbiome and address the root causes of resistance.

8.3.1 Phage Therapy

8.3.1.1 Mechanisms of Action

Phage therapy involves the use of bacteriophages, which are viruses that target and lyse specific bacteria. These phages attach to the bacterial cell surface and inject their genetic material, which then hijacks the bacterial machinery to produce more phages. The newly replicated phages cause the bacterial cell to lyse, effectively killing the bacteria (Burrowes et al. 2011). Phage therapy has shown great promise in clinical trials and case reports, demonstrating its ability to effectively treat infections caused by antibiotic-resistant bacteria. Phage therapy has several advantages over traditional antibiotics. First, phages specifically target and kill bacteria, leaving the rest of the microbiome untouched (Pires et al. 2020). This helps to minimize the disruption of beneficial bacteria and reduce the risk of dysbiosis. Second, phages have the ability to evolve and adapt to changes in bacterial populations, making them less susceptible to resistance development (Burrowes et al. 2011). These characteristics make phage therapy a promising alternative to antibiotics in the fight against antibiotic resistance. Phages can target bacteria in biofilm, which are notoriously difficult to eradicate with antibiotics. Overall, phage therapy offers a promising alternative to antibiotics in the battle against antibiotic resistance and has the potential to revolutionize infection control strategies. Furthermore, susceptible bacterial hosts inject their genetic material, redirect bacterial machinery for their replication, and release newly formed phages, continuing the cycle. This targeted approach contrasts with broad-spectrum antibiotics, offering a tailored solution to combat bacterial infections. Overall, phage therapy offers a promising alternative to antibiotics for the treatment of various bacterial infections, including multidrug-resistant strains.

8.3.1.2 Applications and Success Stories

Phage therapy has shown successful applications in the treatment of various bacterial infections, including those caused by antibiotic-resistant strains (Rohde et al. 2018). For example, a case study in Belgium reported the successful use of phage therapy to treat a patient with a severe *Pseudomonas aeruginosa* infection that was unresponsive to antibiotics (Lebeaux et al. 2021). The patient experienced significant improvement after receiving phage treatment, indicating the effectiveness of phage therapy against difficult-to-treat infections (Haddad et al. 2018). Additionally, a case report from the United States documented the successful use of phage therapy to treat a patient with a life-threatening *Acinetobacter baumannii* infection. The infection was resolved, and the patient recovered fully without the need for antibiotics. These success stories highlight the potential of phage therapy as a viable treatment option for antibiotic-resistant infections. Phage therapy offers several advantages over antibiotics in the fight against antibiotic resistance. These include: (1) Selective targeting of specific bacterial strains: Phages have the ability to specifically target and infect only certain bacterial strains, minimizing harm to beneficial bacteria and reducing the risk of disrupting the body's natural microbiota. (2) Rapid replication and evolution: Phages have a short generation time and high mutation rates, allowing them to rapidly adapt to changes in bacterial populations and overcome bacterial resistance mechanisms. (3) Ability to penetrate biofilm: Biofilms are communities of bacteria that form protective layers, making them highly resistant to antibiotics. However, phages have been shown to effectively penetrate biofilms and target bacteria within, making them a potential solution for infections associated with biofilm formation. (4) Limited impact on host cells: Phages have a narrow host range and generally do not infect or harm eukaryotic human cells, reducing the risk of adverse effects on the host during phage therapy. These successful case reports demonstrate the potential of phage therapy as an effective and safe treatment for antibiotic-resistant infections. Phage therapy has also shown promise in the agricultural industry setting, where it is used to control plant pathogens and bacterial pathogens in livestock and crop plants and reduce the need for chemical pesticides. Some success stories in the field of phage therapy include the treatment of chronic and persistent infections caused by bacteria such as *Pseudomonas aeruginosa* and *Xanthomonas citric* in citrus plants. Phage therapy has emerged as a promising alternative to antibiotics for the treatment of bacterial infections, particularly those caused by multidrug-resistant strains. It offers advantages such as specificity, limited impact on the patient's microbiome, and potential avoidance of overgrowth of secondary pathogens. Overall, phage therapy has the potential to be a game-changer in the fight against antibiotic resistance by providing a safe and effective alternative treatment for bacterial infections. Beyond clinical applications, phage therapy exhibits promise in agricultural settings to combat plant pathogens, reducing reliance on chemical pesticides. These applications underscore the versatility and potential impact of phage therapy across diverse environments. Figure 8.1 shows challenges and application of phage theory.

8.3.1.3 Exploration of the Challenges and Considerations

While phage therapy holds immense potential, it is not without challenges. One key consideration is the necessity for a precise match between the phage and the target bacterial strain. Phage specificity can limit the broad applicability of this therapy, requiring a thorough understanding of the infecting bacteria's genomic characteristics. Another challenge is the potential for the development of phage resistance in bacteria, similar to antibiotic resistance (Burrowes et al. 2011). Therefore, ongoing research and development are essential to identify and address these challenges. Additionally, regulatory frameworks and guidelines need to be established to ensure the safe and effective use of phage therapy in clinical settings. Recombinant phages

CHALLENGES AND APPLICATIONS OF PHAGE THEORY

Applications	Challenges
Treatment of Antibiotic-resistant Infections.	Limited Understanding of Phage-bacteria Dynamics.
Biotechnology and Gene Engineering	Ethical Legal Considerations.
Diagnostic tools.	Phage Stability and Storage.
Targeted Therapy for Specific Bacterial Strains.	Host Range Specifity.
Biofilm Disruption	Regulatory Hardles.

Fig. 8.1 Challenges and application of phage theory

are developed to deliver antimicrobial proteins in target bacteria. The therapy can be used topically on open wounds, inhaled for respiratory infections, or administered orally for gastrointestinal infections. However, there are some serious concerns associated with phage therapy. The major one is their fine specificity toward host bacterial species. This specificity means that a specific phage will only infect and kill certain types of bacteria, limiting its effectiveness against a wide range of bacterial infections (Burrowes et al. 2011). Therefore, extensive research and understanding of the target bacteria's genomic characteristics are necessary to ensure a precise match between the phage and the bacterial strain. This precludes their application as empiric therapy for acute infections. Additionally, the production and purification of phages on a large scale can be challenging and costly. To overcome these challenges, ongoing research is focused on developing phage cocktails that contain multiple phages with different target specificities to increase the range of bacteria that can be effectively targeted. By combining different phages in a cocktail, it increases the likelihood of targeting a broader spectrum of bacterial pathogens (Chan and Abedon 2012). One potential avenue for further exploration in phage therapy is the development of recombinant phages that can deliver antimicrobial proteins to target bacteria. These antimicrobial proteins could enhance the efficacy of phage therapy and overcome some of the limitations associated with phage specificity (Hatfull et al. 2022). They could potentially be used in conjunction with antibiotics to increase their effectiveness and overcome antibiotic resistance (Haddad et al. 2018). Regulatory frameworks and standardization present challenges in the widespread implementation of phage therapy. The need for well-defined protocols, rigorous testing, and clear guidelines for phage preparation and administration is essential to ensuring safety and efficacy. In navigating these challenges, phage therapy emerges as a promising frontier in nonantibiotic advanced therapies. Its precision, coupled with ongoing research addressing challenges, positions it as a valuable tool in the evolving landscape of infection control.

8.3.2 Probiotics

8.3.2.1 Restoration of Microbiome Equilibrium

Probiotics have emerged as a potential strategy for restoring microbiome equilibrium and combating bacterial infections. These living microorganisms, typically bacteria or yeasts, are believed to confer health benefits when consumed in adequate amounts. They can help promote a healthy balance of gut bacteria, which plays a crucial role in immune function and protection against infections. Probiotics have shown promise in preventing and treating various bacterial infections, including antibiotic-associated diarrhea, urinary tract infections, and respiratory tract infections. In addition to their direct antimicrobial effects, probiotics may also modulate the immune response and enhance the efficacy of antibiotics. Overall, the development and use of phage therapy and probiotics offer innovative approaches to combating antibiotic resistance and controlling bacterial infections. Probiotics can help restore the microbiome equilibrium (Federici et al. 2020) and strengthen immune defenses, making them a promising strategy in infection control. These live microorganisms have been shown to have beneficial effects, ranging from improving gut health to boosting the immune system. Furthermore, studies have demonstrated that certain strains of probiotics can inhibit the growth and adhesion of pathogenic bacteria, reducing the risk of infection (Wittebole et al. 2013). In addition, probiotics have been shown to enhance the efficacy of antibiotic treatment by reducing the incidence of antibiotic-associated diarrhea and preventing the overgrowth of opportunistic pathogens such as *Clostridium difficile*. In summary, probiotics offer a multifaceted approach to infection control by restoring microbiome equilibrium, strengthening immune defenses, inhibiting the growth of pathogenic bacteria, and enhancing the efficacy of antibiotic treatment. By restoring and maintaining microbiome equilibrium, probiotics offer a multifaceted approach to infection control. Overall, the use of probiotics provides a nonantibiotic approach to infection control by promoting a healthy microbiome and enhancing the body's natural defenses against pathogens (Burrowes et al. 2011). Overall infection control, the use of probiotics, and phage therapy offer innovative approaches that target bacterial infections and combat antibiotic resistance. Revolutionizing infection control involves shifting toward nonantibiotic therapies such as probiotics and phage therapy.

8.3.2.2 Strain-Specific Benefits and Challenges

The efficacy of probiotics is inherently strain-specific, with each strain conferring distinct benefits. Certain strains exhibit antimicrobial properties, directly inhibiting the growth of pathogenic bacteria. Others enhance mucosal barrier function or modulate immune responses. Understanding the unique attributes of each strain is crucial for tailoring probiotic interventions to specific infections. For example, some strains of *Lactobacillus* have been shown to be effective against urinary tract infections, while certain strains of bifid bacterium have been found to be effective against gastrointestinal infections (Imane and Doumandji 2018). Additionally, it is important to consider the potential challenges associated with probiotic use. These challenges include the risk of carrying antibiotic-resistant genes and concerns about the safety of live bacteria in probiotic foods. Nonantibiotic advanced therapies, such as probiotics, offer a promising approach to combating the surge of antibiotic resistance. They provide a targeted and strain-specific intervention that can enhance the immune response, inhibit the growth of pathogens, and restore microbiome equilibrium (Gueimonde et al. 2013).

They also offer the potential to minimize the adverse effects associated with antibiotics and reduce the risk of antibiotic resistance development. Overall, the use of probiotics as nonantibiotic advanced therapies offers a promising solution to combat antibiotic resistance and prevent infections while also promoting overall host well-being. The adoption of nonantibiotic advanced therapies, such as probiotics, in infection control has the potential to revolutionize the field by providing targeted and effective interventions that can combat antibiotic resistance and prevent infections. These therapies work by modulating the immune response, inhibiting the growth of pathogens, and restoring microbiome equilibrium (Moal and Servin 2014). Additionally, ongoing research is focused on identifying and optimizing the most effective mechanisms of action for probiotics. These mechanisms may include producing antimicrobial agents like bacteriocins, enhancing mucosal barrier function, and modulating immune responses. In the context of myocarditis, the use of nonantibiotic advanced therapies like fecal microbiota transplantation and live bio therapeutic agents is an area that warrants further investigation. Overall, nonantibiotic advanced therapies, including probiotics, offer a promising solution to combat antibiotic resistance and prevent infections. Challenges associated with probiotics include strain variability in terms of survival and persistence in the host gastrointestinal tract. Environmental factors, such as pH and bile salts, can impact probiotic viability, necessitating careful selection and formulation. Moreover, the potential for transitory colonization poses challenges in maintaining a sustained therapeutic effect. Figure 8.2 shows the detailed of probiotics and its types, challenges, and applications.

8.3.2.3 Examination of Synergistic Approaches with Traditional Antibiotics

Synergistic approaches involving probiotics and traditional antibiotics present a promising strategy for infection control. Probiotics can enhance the efficacy of antibiotics by promoting a balanced microbiome, preventing overgrowth of opportunistic pathogens, and mitigating antibiotic-associated side effects, such as diarrhea. This approach can potentially reduce the reliance on high doses of antibiotics and minimize the development of antibiotic resistance (Murugaiyan et al. 2022). Additionally, probiotics can also interact with traditional antibiotics to enhance their antimicrobial activity. For example, certain probiotic strains have been shown

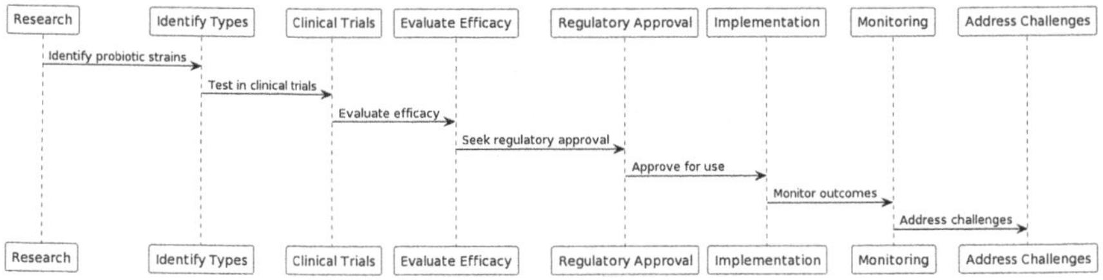

Fig. 8.2 Probiotics: types, challenges and applications

to produce antimicrobial substances that can enhance the effectiveness of antibiotics against resistant bacteria. Studies suggest that certain probiotic strains may exert direct antimicrobial effects, complementing the action of antibiotics. Moreover, probiotics can modulate the host immune response, potentially reducing inflammation associated with infections. These synergistic approaches between probiotics and traditional antibiotics have the potential to not only enhance treatment outcomes but also reduce the emergence of antibiotic resistance. Overall, the integration of nonantibiotic advanced therapies, such as probiotics, in infection control strategies, provides a promising avenue for combating antibiotic resistance and revolutionizing the field of infection control. By harnessing the power of probiotics to enhance host immune responses, promote a balanced microbiome, and synergistically interact with traditional antibiotics, we can develop more effective and sustainable approaches to combat infectious diseases (Nii-Trebi 2017). However, caution is required in navigating these synergistic approaches, as the interactions between probiotics and antibiotics are complex and context-dependent. Consideration of antibiotic type, dosage, and timing is essential to maximize benefits while minimizing potential antagonistic effects. Furthermore, additional research is needed to identify the most effective probiotic strains and combinations for specific infections, as well as optimal dosing regimens. Overall, the use of probiotics as nonantibiotic advanced therapies in infection control shows great potential for combating the surge of antibiotic resistance. Probiotics offer a promising alternative to antibiotics in the field of infection control (Ghafur et al. 2012). In conclusion, probiotics emerge as valuable contributors to infection control, offering a holistic approach to restoring and maintaining microbiome equilibrium. Understanding the strain-specific benefits and challenges, coupled with exploring synergistic approaches with traditional antibiotics, positions probiotics as integral components in the evolving landscape of nonantibiotic advanced therapies (Murugaiyan et al. 2022; Dutta 2020). In summary, the integration of probiotics as nonantibiotic advanced therapies in infection control strategies has the potential to revolutionize the field by enhancing treatment outcomes, reducing the emergence of antibiotic resistance, and promoting more effective and sustainable approaches to combat infectious diseases.

8.3.3 Advanced Therapies (Immunotherapy and Gene Editing)

8.3.3.1 Precision Medicine in Infection Control

The integration of precision medicine into infection control represents a revolutionary advancement in treatment strategies. Precision medicine tailor's interventions to individual patients are based on their genetic, environmental, and lifestyle factors. In the context of infection control, this approach allows for a more targeted and effective response to infectious agents. By utilizing advanced technologies such as immunotherapy and gene editing, precision medicine holds the potential to revolutionize the field of infection control (Simundza 2018; Friedmann 1989; Mahon and Hafner 2017). By understanding the unique characteristics of the host and the infecting pathogen, precision medicine facilitates the selection of treatments that are most likely to succeed. This individualized approach minimizes the risk of adverse reactions and enhances overall treatment outcomes. The utilization of advanced diagnostics, such as genomic sequencing, enables clinicians to identify specific vulnerabilities in pathogens, guiding the selection of the most appropriate therapeutic interventions. In summary, the integration of precision medicine into infection control has the potential to revolutionize treatment strategies by tailoring interventions to individual patients based on their unique genetic and environmental factors. This personalized approach improves treatment outcomes, reduces the emergence of antibiotic resistance, and allows for the development of advanced therapies such as immunotherapy and gene editing (Newman and Ausubel 2016; Naran et al. 2018). The development of multidrug-resistant organisms is a significant challenge in infection control, and it is largely fueled by the misuse and overuse of antibiotics in various sectors, including human medicine, animal husbandry, and agriculture. To combat the surge of antibiotic resistance, urgent action is necessary in the form of intensified research and innovation. The scientific community must unite to explore novel antimicrobial agents, harnessing innovative technologies, and interdisciplinary approaches (Bothe et al. 2023). Genomics, nanotechnology, and other emerging fields offer promising avenues for developing targeted and efficient antimicrobial treatments. By prioritizing the development of new antimicrobial agents and implementing comprehensive strategies, we can hope to stem the tide of AMR and safeguard the future of modern medicine (Fig. 8.3).

8.3.3.2 Leveraging the Body's Immune System

Precision medicine also holds great potential for leveraging the body's immune system to combat infections. By identifying specific characteristics of a patient's immune response, such as genetic markers or immune cell profiles, precision medicine can guide the development of immunotherapies that enhance the immune response against pathogens. These personalized immunotherapies can be designed to target specific vulnerabilities in the pathogen or boost the body's natural defense mechanisms. Immunotherapy, a key component of advanced therapies, harnesses the body's immune system to combat infections. Unlike traditional treatments that directly target pathogens, immunotherapy focuses on enhancing the host's immune response. This can involve the administration of immune-modulating agents, such

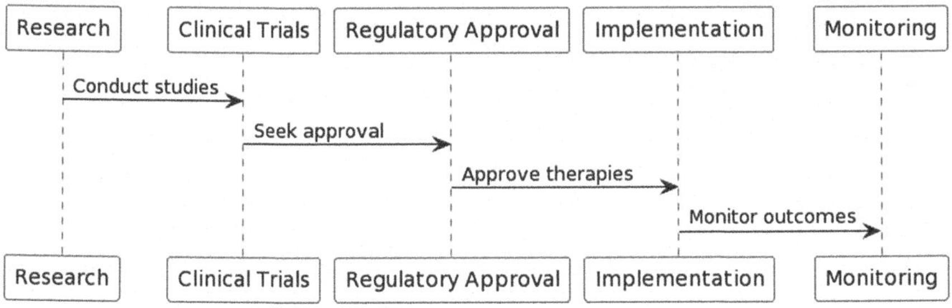

Fig. 8.3 Applications and challenges of advanced therapies

as monoclonal antibodies, or the stimulation of the immune system through vaccines. By targeting the immune system, immunotherapy has the potential to provide long-term protection against infections and reduce the reliance on antibiotics. In addition to immunotherapy, gene editing technologies have also shown promise for revolutionizing infection control. Gene editing technologies, such as CRISPR-Cas9, allow for precise modifications of the genetic material of microorganisms. This enables researchers to target specific genes related to pathogen city or antibiotic resistance, potentially rendering the microorganisms less harmful or susceptible to treatment. By leveraging the body's immune system and utilizing gene editing technologies, advanced therapies in infection control have the potential to revolutionize the way we combat infections (Javed et al. 2023). These therapies offer a promising alternative to traditional antibiotics, especially in the face of multidrug-resistant organisms. In addition, it is important to address the inadequate surveillance and data collection systems that currently exist in the battle against AMR (Murugaiyan et al. 2022). Without accurate and up-to-date information on the prevalence of resistant pathogens, it becomes challenging to devise effective management strategies and monitor the impact of interventions. This emphasizes the need for improved surveillance systems and comprehensive data collection, which can inform treatment decisions and identify emerging trends in AMR. Through these approaches, we can revolutionize infection control by tackling the surge of antibiotic resistance and ensuring that effective treatment options remain available for future generations. By implementing these strategies, we can combat the spread of antimicrobial resistance, reduce the burden on healthcare systems, and safeguard the future of modern medicine. To effectively combat the surge of antibiotic resistance and safeguard the future of healthcare, a multifaceted approach is necessary (Murugaiyan et al. 2022). This approach should involve intensifying research and innovation in the development of novel antimicrobial agents, harnessing innovative technologies such as genomics and nanotechnology, and improving surveillance and data collection systems. In conclusion, to effectively combat the surge of antibiotic resistance and safeguard the future of healthcare, a multifaceted approach is necessary.

8.3.3.3 Exploration of the Potential of Gene Editing to Enhance Resistance

Gene editing technologies such as CRISPR/Cas9 offer great potential for enhancing resistance against antibiotics. By targeting specific genes related to pathogen city or antibiotic resistance, these technologies can potentially render microorganisms less harmful or less susceptible to treatment. Gene editing can also be utilized to modify the genomes of commensal bacteria or bacteriophages, allowing them to target and eliminate antibiotic-resistant bacteria. This approach should involve intensified research and innovation to develop novel antimicrobial agents, harnessing innovative technologies like genomics and nanotechnology. Furthermore, implementing international collaboration and cooperation is essential to addressing the global nature of antibiotic resistance. The economic implications of the rise of antibiotic resistance are severe, leading to prolonged illnesses, increased hospital stays, and the need for more expensive treatments. The potential for widespread infections and pandemics also looms large, as antibiotic resistance can facilitate the rapid dissemination of resistant microbes on a global scale (Bothe et al. 2023). To effectively address these challenges, a shift in the understanding and approach to antibiotic resistance is necessary (Hutchison et al. 2018). CRISPR-Cas9 has demonstrated its potential in enhancing resistance against antibiotics by enabling precise modification of antibiotic resistance genes in bacteria. By harnessing the power of CRISPR-Cas systems, we have the ability to revolutionize infection control and overcome the challenges posed by antibiotic resistance. This approach could potentially revolutionize infection control and mitigate the challenges posed by antibiotic resistance, providing a nonantibiotic alternative to combat multidrug-resistant organisms and reducing the reliance on traditional antibiotics (Javed et al. 2023). Moreover, the use of CRISPR-Cas systems in infection control can also address the issue of microbial dysbiosis. By specifically targeting pathogenic bacteria and leaving commensal bacteria unharmed. CRISPR-Cas systems have the potential to revolutionize infection control by specifically targeting antibiotic-resistant bacteria and reducing the reliance on traditional antibiotics. Despite the progress and potential of gene editing technologies like CRISPR-Cas9 in enhancing resistance against antibiotics, there are still challenges to overcome, such as off-target effects and the risk of creating unintended consequences in the bacterial genome. The potential of gene editing lies in its ability to precisely modify the genetic material of both pathogens and host cells. By targeting and editing antibiotic resistance genes in bacteria, gene editing technologies like CRISPR-Cas9 have the potential to enhance resistance against antibiotics. CRISPR-Cas9, a type II CRISPR-Cas system, has emerged as a powerful tool in the fight against antibiotic resistance. A comprehensive approach encompassing improved antimicrobial stewardship, infection prevention and control measures, the development of novel antimicrobial agents, enhanced international collaboration, and responsible use of antimicrobials is crucial to combat the surge of antibiotic resistance and protect public health (Bothe et al. 2023). In summary, advanced therapies, encompassing precision medicine, immunotherapy, and gene editing, represent a paradigm shift in infection control. The focus on precision, adaptability, and targeted interventions positions these approaches as invaluable tools in the ongoing battle against antibiotic resistance, offering hope for a future where infections are managed with unprecedented efficacy and minimal collateral damage. However, it is crucial to address ethical, safety, and delivery challenges in order to fully harness the potential of gene editing technologies like CRISPR-Cas9 in combating antibiotic resistance and revolutionizing infection control.

8.4 Challenges and Considerations

Implementing nonantibiotic advanced therapies, such as phage therapy, probiotics, and gene editing, necessitates the establishment of robust standardization and regulatory frameworks. Standardization ensures consistency in therapy administration, manufacturing processes, and quality control. Achieving consensus on protocols, dosages, and treatment regimens is critical for the widespread adoption and efficacy of these therapies. Antibiotic therapies have transformed treatments for bacterial and fungal infections. However, widespread infections are posing threats to human and animal health, with several pathogens developing resistance to available antibiotics. The normal role of antibiotics in the social context of microbial communities is not clear. Additionally, ethical considerations arise when using gene editing technologies like CRISPR-Cas9 for infection control. Questions surrounding consent, privacy, and the potential for unintended consequences demand careful deliberation and ethical oversight. To fully address the challenge of emerging antimicrobial-resistant pathogens, a comprehensive approach is required. This approach should encompass not only the development of novel therapies but also improvements in surveillance, infection prevention and control measures, and the responsible use of antibiotics. Moreover, the financial implications of implementing advanced therapies must be considered. The cost effectiveness of nonantibiotic advanced therapies needs to be evaluated, along with strategies for their affordable access and widespread availability. In conclusion, nonantibiotic advanced therapies such as gene editing technologies like CRISPR-Cas9 hold immense promise for revolutionizing infection control and combating antibiotic resistance. However, there are several challenges and considerations that need to be addressed in order to fully harness their potential. These challenges include standardization and regulation, ethical considerations, comprehensive approaches to addressing antimicrobial resistance, and financial implications. In order to revolutionize infection control and combat the surge of antibiotic resistance, nonantibiotic advanced therapies such as gene editing technologies like CRISPR-Cas9 require careful consideration and collaboration from various stakeholders. They should involve healthcare providers, researchers, policymakers, ethicists, and the general public to ensure that these therapies are safe, effective, and ethically sound. Clear guidelines and oversight are essential to address safety concerns, monitor treatment outcomes, and facilitate the responsible use of these innovative interventions. In order to fully harness the potential of nonantibiotic advanced therapies like gene editing technologies such as CRISPR-Cas9, it is important to invest in research and development, strengthen regulatory frameworks, promote interdisciplinary collaborations, and prioritize global cooperation. Furthermore, public awareness and education about the potential of nonantibiotic advanced therapies are crucial. It is important to empower individuals with knowledge and understanding so that they can make informed decisions about their healthcare and contribute to the responsible use of these therapies. In today's rapidly changing world, the significance of accurate surveillance, infection prevention and control measures, and the responsible use of antibiotics cannot be overstated. These strategies are vital in combating the rise of antibiotic resistance and ensuring the effectiveness of current and future antimicrobial treatments. One of the challenges associated with nonantibiotic advanced therapies is the limited understanding of interactions within complex microbial ecosystems. The microbiome, comprising diverse microbial communities, interacts dynamically with therapeutic agents, impacting treatment outcomes. Understanding the nuanced interplay between phages, probiotics, and the host microbiota is crucial for optimizing therapeutic interventions (Theuretzbacher and Piddock 2019). In conclusion, addressing the challenges and considerations associated with nonantibiotic advanced therapies requires collaborative efforts from researchers, regulatory bodies, and ethical committees. By navigating these complexities, we can unlock the full potential of these innovative approaches and pave the way for a future where infection control is characterized by precision, safety, and ethical responsibility.

8.5 Impact and Future Directions

8.5.1 Discussion on the Reduction of Antibiotic Overuse and Misuse

The adoption of nonantibiotic advanced therapies holds the potential to significantly reduce the overuse and misuse of traditional antibiotics. By providing targeted and specific treatments, these alternatives minimize the selective pressure driving antibiotic resistance. Phage therapy, probiotics, and advanced therapies like immunotherapy and gene editing offer tailored solutions that address infections without resorting to broad-spectrum antibiotics thereby contributing to the preservation of antibiotic efficacy. Furthermore, the integration of precision medicine approaches in patient stratification and treatment selection can further optimize outcomes and minimize side effects. In the future, advancements in genomics, bioinformatics, and systems biology will continue to enhance our understanding of pathogens and the immune response, leading to the development of even more effective and personalized nonantibiotic therapies. Overall, revolutionizing infection control through nonantibiotic advanced therapies requires a multidisciplinary approach and collaboration between various sectors (Murugaiyan et al. 2022). The reduction of antibiotic overuse and misuse not only curtails the emergence of resistant strains but also mitigates the associated side effects and complications. It is essential to implement comprehensive strategies that educate healthcare professionals, veterinarians, and the agricultural industry on appropriate antibiotic prescribing and use. Editing offers promising solutions to combat antibiotic resistance. In addition to targeting specific pathogens, it is critical to understand and preserve the microbiome as well as promote positive immune responses. This will help avoid unintended consequences, such as the emergence of new resistant strains or disruptions to the natural balance of microbial communities. By integrating advanced technologies such as CRISPR/Cas systems, nanotechnology, and phage therapy, we can specifically target pathogens while minimizing harm to beneficial bacteria and host cells (Khan et al. 2018). A shift toward nonantibiotic therapies aligns with a sustainable approach to infection control, ensuring that antibiotics remain effective for treating severe infections while minimizing the risk of resistance development. Furthermore, nonantibiotic therapies have the potential to provide targeted and personalized treatments, considering individual differences in microbial communities and immune responses (Pires et al. 2020).

8.5.2 Exploration of Personalized Medicine and Targeted Interventions

Nonantibiotic, advanced therapies pave the way for personalized medicine and targeted interventions. Understanding the unique microbial profiles, genetic variations, and immune responses of individual patients allows for tailored treatment approaches. By utilizing advanced technologies and comprehensive data collection, healthcare professionals can better identify the most appropriate nonantibiotic therapies for each patient, optimizing treatment outcomes and minimizing the risk of resistance. Moreover, nonantibiotic therapies offer the potential to address infections caused by multidrug-resistant organisms, which have become a significant global health concern (Murugaiyan et al. 2022). Precision medicine, facilitated by advanced diagnostics, enables healthcare providers to select therapies based on specific characteristics, optimizing efficacy, and minimizing adverse effects. Additionally, targeted interventions can be developed to disrupt the mechanisms of antibiotic resistance in pathogens. This approach includes techniques that target enzymes or proteins responsible for antimicrobial resistance, as well as strategies that prevent the transfer of resistance genes between bacteria. The exploration of personalized medicine extends beyond pathogen-specific interventions to include patient-centric approaches. These patient-centric approaches consider not only the microbial profile but also individual genetic variations, immune responses, and lifestyle factors. By integrating this comprehensive understanding of the patient, healthcare providers can develop personalized treatment plans that optimize efficacy and minimize the risk of adverse effects. Inadequate surveillance and data collection hinder the fight against AMR (Murugaiyan et al. 2022). By considering individual factors, such as host genetics and microbiome composition, nonantibiotic therapies present an opportunity to redefine treatment paradigms. This shift toward personalized interventions contributes to improved patient outcomes and enhances the overall effectiveness of infection control strategies.

8.5.3 Insight into the Potential Synergy with Traditional Antibiotics

Nonantibiotic advanced therapies offer the potential for synergy with traditional antibiotics, presenting a holistic approach to infection control. Combining these approaches may enhance treatment efficacy while reducing the risk of resistance development. This combination therapy approach can target both the pathogen directly and boost the host immune response, creating a comprehensive defense against infections. Furthermore, the use of nonantibiotic therapies can help preserve the effectiveness of traditional antibiotics by reducing their overuse and minimizing the development of resistance. These alternative therapies, such as quorum-sensing inhibitors, phage therapy, and monoclonal antibodies, provide promising avenues for combating drug-resistant pathogens and reducing the reliance on antibiotics as the primary treatment option (Morris and Cerceo 2020; Lv et al. 2021). For instance, probiotics can mitigate antibiotic-associated dysbiosis, and phage therapy can target specific antibiotic-resistant strains, thereby augmenting the overall impact of infection control strategies. Moreover, advances in precision medicine and microbiome research offer new possibilities for personalized interventions. These approaches leverage individual-specific traits, such as microbiome composition and genetic variations, to tailor treatment plans. This personalized approach significantly improves the effectiveness of treatment, reduces the risk of adverse effects, and helps overcome the limitations of antibiotic therapy. Furthermore, nonantibiotic advanced therapies have the potential to target a broader range of pathogens, including pathobionts and commensals. This broad spectrum of targeting, along with the potential for combination therapy with traditional antibiotics, revolutionizes infection control and paves the way for more effective strategies to combat the surge of antibiotic resistance (Murugaiyan et al. 2022). One of the key advantages of nonantibiotic advanced therapies is their potential to enhance patient outcomes. By targeting the underlying mechanisms of infection and boosting the immune response, these therapies can improve treatment efficacy and reduce the likelihood of recurrent or persistent infections (Maxson and Mitchell 2016). Synergistic approaches enable the strategic use of both nonantibiotic and traditional antibiotic treatments (Rex et al. 2019), fostering a balance between precision and broad applicability. This integrative approach not only maximizes the therapeutic benefits but also minimizes the potential for resistance development (Murugaiyan et al. 2022). The complementary nature of these therapies holds promise for overcoming challenges associated with individual approaches and maximizing the effectiveness of infection control in diverse clinical scenarios.

8.5.4 Reflection on the Long-Term Impact on Public Health

The long-term impact of embracing nonantibiotic advanced therapies on public health is profound. A reduction in antibiotic resistance translates into improved treatment outcomes, reduced healthcare costs, and a lower burden of infectious diseases. The integration of personalized medicine and targeted interventions contributes to a paradigm shift in healthcare, emphasizing individualized care and preventive strategies. Public health initiatives can focus on promoting awareness about the risks of antibiotic resistance and the potential benefits of nonantibiotic advanced therapies (Rubens et al. 2014). By fostering synergy with traditional antibiotics and addressing the root causes of resistance, nonantibiotic therapies contribute to the sustainability of infection control practices (Infection Control 2022). This paradigm shifts not only benefits individual patients but also strengthens global public health resilience against emerging infectious threats (Murugaiyan et al. 2022; Lushniak 2014). In conclusion, the impact and future directions of nonantibiotic advanced therapies hold immense promise for reshaping infection control strategies. The reduction of antibiotic overuse, personalized medicine, synergistic approaches, and long-term public health benefits collectively contribute to a transformative vision of infection control that is effective, sustainable, and tailored to the needs of both individuals and populations. Figure 8.4 shows side-by-side comparison of antibiotic and nonantibiotic therapies, including their treatment approach, main mode of action, resistance development, applications, and challenges. This provides a clear visual representation of the differences between the two types of therapies. Figure 8.5 shows the alternatives to antibiotics.

8.6 Conclusion

In recapitulating our exploration of nonantibiotic advanced therapies, including phage therapy, probiotics, immunotherapy, and gene editing, it becomes evident that these innovative approaches mark a turning point in the battle against antibiotic resistance. The key finding underscores the precision, targeted efficacy, and potential synergy with traditional antibiotics that these therapies offer. Furthermore, the potential for personalized medicine and individualized approaches to infection control holds promise for improving treatment outcomes and reducing the burden of infectious diseases (Pires et al. 2020). The significance of nonantibiotic therapies in reshaping infection control cannot be overstated. They provide a nuanced and individualized approach to treatment, addressing the limitations and challenges posed by antibiotic resistance (Ioannou et al. 2024). Phage therapy's precision (Hatfull et al. 2022) Probiotic microbiome restoration (Moal and Servin 2014), immunotherapy's leverage of the body's immune system (Mahon and Hafner 2017), and gene editing's potential to confer genetic resistance collectively redefine the landscape of infectious disease management (Bobate et al. 2023). As we stand at the cusp of this transformative era, a call-to-action echoes. Continued research is imperative to unravel the complexities of these therapies, understand their long-term effects, and refine their applications. Investment in research, development, and clinical trials will accelerate the translation of promising concepts into tangible treatments (Foster 2019; Moghadam et al. 2020). The seamless integration of nonantibiotic therapies into mainstream medicine practices demand collaborative efforts from researchers, healthcare (Rubens et al. 2014), providers, regulators, and

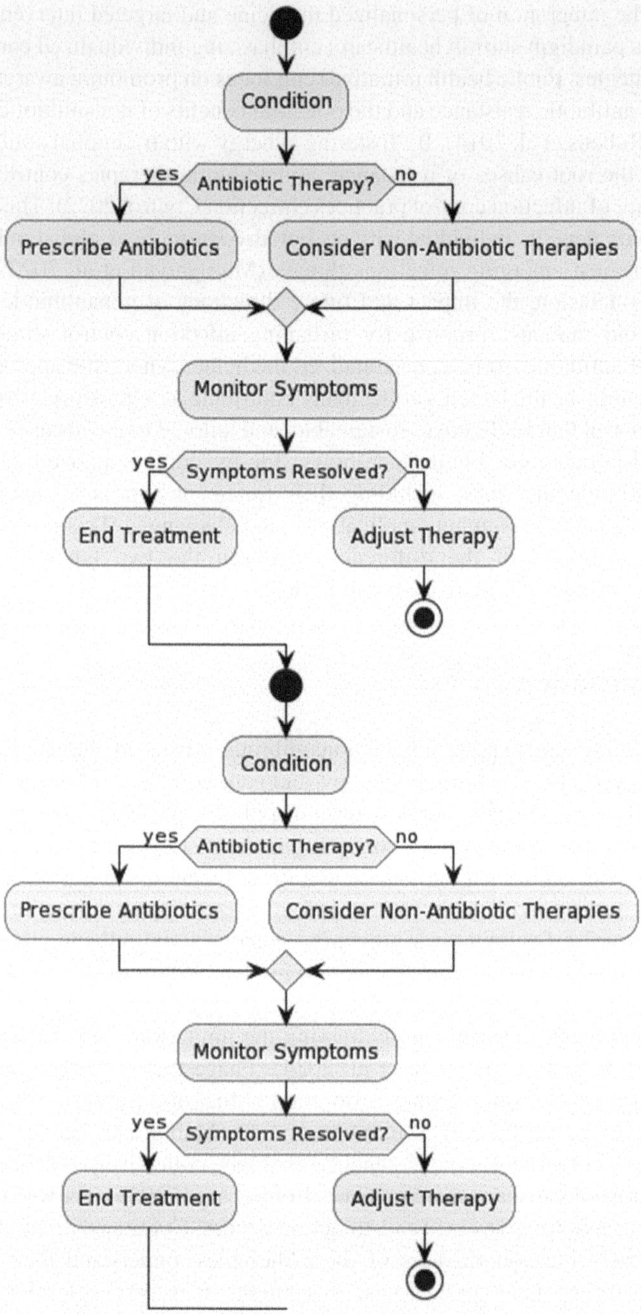

Fig. 8.4 Comparison of antibiotic and nonantibiotic therapies

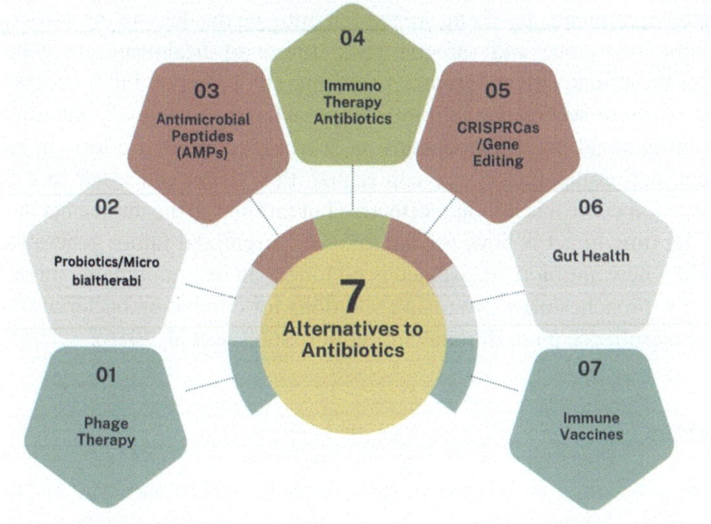

Fig. 8.5 Alternatives to antibiotics illustrates various alternatives to antibiotics, including phage therapy, probiotics, immunotherapy, and gene editing. Each alternative offers unique mechanisms and applications for combating bacterial infections while addressing challenges associated with antibiotic resistance

ethical committees. Together, we can seize the opportunity to revolutionize infection control and combat the surge of antibiotic resistance (Lushniak 2014). The urgency of our call to action lies in the potential of these therapies to not only combat antibiotic resistance but also to revolutionize the entire paradigm of infection control. As we navigate the challenges and opportunities presented by nonantibiotic advanced therapies, we pave the way for a future where infections are managed with unprecedented precision, efficacy, and sustainability. The integration of nonantibiotic therapies represents a crucial step forward in combating the global crisis of antibiotic resistance. It is essential to invest in the research, development, and clinical trials of these therapies to fully understand their potential and ensure their safe and effective use (Pires et al. 2020). These innovative approaches offer hope for a future where we can effectively treat bacterial infections without relying solely on traditional antibiotics. By leveraging the power of phages, immunotherapy, microbiome-modifying therapies, and other non-traditional treatments, we have the opportunity to overcome the limitations of antibiotics and usher in a new era of infection control. This shift towards nonantibiotic therapies is not without its challenges (Mandal et al. 2014). In terms of phage therapy, further research is needed to address concerns such as phage resistance and the development of regulatory frameworks for their clinical use (Pires et al. 2020). Moreover, the ethical implications and potential risks of these therapies must also be carefully considered and addressed. To ensure the successful integration of nonantibiotic therapies, collaboration among researchers, healthcare providers, regulators, and ethical committees is crucial. It is imperative that we work together to accelerate the translation of promising concepts into tangible treatments. By doing so, we can mitigate the devastating consequences of antibiotic resistance and safeguard the future of healthcare. In conclusion, embracing these innovative therapies is not merely an option but a necessity. The vision of a future where nonantibiotic approaches is seamlessly integrated into mainstream medical practices beckons us. Through concerted efforts in research, investment, and collaboration, we can realize this vision and usher in a new era where infection control is not only effective but tailored to the unique needs of each patient. The time to act is now, for the sake of current and future generations, and for a world where antibiotic resistance is no longer an insurmountable threat. These innovative approaches offer promising solutions for combating bacterial infections and reducing reliance on traditional antibiotics (Hauser et al. 2016).

References

Aslam B, Wang W, Arshad M, Khurshid M, Muzammil S, Rasool MH, Nisar MA, Alvi RF, Aslam M, Qamar MU, Salamat MKF, Baloch Z (2018) Antibiotic resistance: a rundown of a global crisis. Front Public Health 2:145. https://doi.org/10.2147/idr.s173867

Bobate S, Mahalle S, Dafale NA, Bajaj A (2023) Emergence of environmental antibiotic resistance: mechanisms, monitoring, and management. Elsevier. https://doi.org/10.1016/j.envadv.2023.100409

Bothe H, Chhajed SS, Bothe S (2023) The next pandemic? Antimicrobial resistance and its societal implications: a comprehensive review. https://doi.org/10.20944/preprints202309.0652.v2

Burrowes B, Harper DR, Anderson JS, McConville M, Enright MC (2011) Bacteriophage therapy: potential uses in the control of antibiotic-resistant pathogens. Expert Rev Anti Infect Ther 9(9):775–785. https://doi.org/10.1586/eri.11.90

Chan BK, Abedon ST (2012) Phage therapy, pharmacology, and phage cocktails. Adv Appl Microbiol 78:1–23. https://www.sciencedirect.com/science/article/pii/B9780123948052000014

Control A (2005) Paul Farmer. https://en.wikipedia.org/wiki/Paul_Farmer

Dancer S (2013) Infection control in the post-antibiotic era. Healthc Infect 18(2):51–60. https://doi.org/10.1071/hi12042

Dutta TK (2020) Alternative therapeutic approaches in the era of antimicrobial resistance: an overview. Indian J Anim Health 59(1):23–23. https://doi.org/10.36062/ijah.59.1.2020.23-28

Farmer P (1999) Pathologies of power: rethinking health and human rights. Am J Public Health 89(10):1486–1496. https://doi.org/10.2105/ajph.89.10.1486

Federici S, Suez J, Elinav E (2020) Our microbiome: on the challenges, promises, and hype. Springer, pp 539–557. https://doi.org/10.1007/978-3-030-51849-3_20

Feehan AK, Garcia-Diaz J (2020) Bacterial, gut microbiome-modifying therapies to defend against multidrug-resistant organisms. Microorganisms 8(2):166. https://doi.org/10.3390/microorganisms8020166

Foster JA (2019) Decoding microbiome research for clinical psychiatry. Can J Psychiatry 65(1):19–20. https://doi.org/10.1177/0706743719890725

Friedmann T (1989) Progress toward human gene therapy. Science 244(4910):1275–1281. https://doi.org/10.1126/science.2660259

Ghafur A, Nagvekar V, Thilakavathy S, Chandra K, Gopalakrishnan R, Vidyalakshmi P (2012) "Save antibiotics, save lives": an Indian success story of infection control through persuasive diplomacy. Antimicrob Resist Infect Control 1(1):29–29. https://doi.org/10.1186/2047-2994-1-29

Gueimonde M, Sánchez B, Reyes-Gavilán CGDL, Margollés A (2013) Antibiotic resistance in probiotic bacteria. Front Microbiol 4:202. https://doi.org/10.3389/fmicb.2013.00202

Haddad LE, Harb CP, Gebara MA, Stibich M, Chemaly RF (2018) A systematic and critical review of bacteriophage therapy against multidrug-resistant ESKAPE organisms in humans. Clin Infect Dis 69(1):167–178. https://doi.org/10.1093/cid/ciy947

Hatfull GF, Dedrick RM, Schooley RT (2022) Phage therapy for antibiotic-resistant bacterial infections. Annu Rev Med 73(1):197–211. https://doi.org/10.1146/annurev-med-080219-122208

Hauser AR, Mecsas J, Moir DT (2016) Beyond antibiotics: new therapeutic approaches for bacterial infections. Clin Infect Dis 63:89. https://doi.org/10.1093/cid/ciw200

Hebert C, Weber SG (2011) Common approaches to the control of multidrug-resistant organisms other than methicillin-resistant *Staphylococcus aureus* (MRSA). Infect Dis Clin 25(1):181–200. https://www.sciencedirect.com/science/article/pii/S0891552010000930

Hutchison C, Knight GM, Stabler R, Chandler C (2018) The modern era must end. Antiiotic resistance helps us rethink medicine and farming. BMJ https://researchonline.lshtm.ac.uk/id/eprint/4648945/

Imane HA, Doumandji A (2018) Characterization and screening of the potential probiotic lactic acid bacteria and Bifidobacterium strains isolated from different biotopes. Mediterr J Nutr Metab 11(2):145–173. https://doi.org/10.3233/mnm-17191

Infection Control (2022). https://pubmed.ncbi.nlm.nih.gov/30085559/

Ioannou P, Baliou S, Samonis G (2024) Nanotechnology in the diagnosis and treatment of antibiotic-resistant infections. Antibiotics (Basel) 13:121. https://doi.org/10.3390/antibiotics13020121

Javed MU, Hayat MT, Mukhtar H, Imre K (2023) The CRISPR-Cas9 system: a prospective pathway toward combating antibiotic resistance. Antibiotics 12(6):1075. https://doi.org/10.3390/antibiotics12061075

Khan S, Mahmood MS, Rahman SU, Zafar H, Habibullah S, Khan Z, Ahmad A (2018) CRISPR/Cas9: the Jedi against the dark empire of diseases. J Biomed Sci 25:29. https://doi.org/10.1186/s12929-018-0425-5

Lebeaux D, Merabishvili M, Caudron É, Lannoy D, Simaey LV, Duyvejonck H, Guillemain R, Thumerelle C, Podglajen I, Compain F, Kassis N, Mainardi J, Wittmann J, Rohde C, Pirnay J, Dufour N, Vermeulen S, Gansemans Y, Nieuwerburgh FV, Vaneechoutte M (2021) A case of phage therapy against Pandrug-resistant *Achromobacter xylosoxidans* in a 12-year-old lung-transplanted cystic fibrosis patient. Viruses 13(1):60. https://doi.org/10.3390/v13010060

Lushniak BD (2014) Antibiotic resistance: a public health crisis. Public Health Rep 129(4):314–316. https://doi.org/10.1177/003335491412900402

Lv Y, Tong Z, Shi J, Jia Y, Deng T, Wang Z (2021) Reversion of antibiotic resistance in multidrug-resistant pathogens using the nonantibiotic pharmaceutical benzydamine. Commun Biol 4:1328. https://doi.org/10.1038/s42003-021-02854-z

MacNair CR, Rutherford ST, Tan M (2023) Alternative therapeutic strategies to treat antibiotic-resistant pathogens. Nat Rev Microbiol 22:262. https://doi.org/10.1038/s41579-023-00993-0

Mahon RN, Hafner R (2017) Applying precision medicine and immunotherapy advances from oncology to host-directed therapies for infectious diseases. Front Immunol 8:688. https://doi.org/10.3389/fimmu.2017.00688

Makabenta JM, Nabawy A, Li CH, Schmidt-Malan SM, Patel R, Rotello VM (2020) Nanomaterial-based therapeutics for antibiotic-resistant bacterial infections. Nat Rev Microbiol 19:23. https://doi.org/10.1038/s41579-020-0420-1

Mandal SM, Roy A, Ghosh AK, Hazra TK, Basak A, Franco OL (2014) Challenges and future prospects of antibiotic therapy: from peptides to phage utilization. Front Pharmacol 5:105. https://doi.org/10.3389/fphar.2014.00105

Mantravadi PK, Kalesh KA, Dobson RC, Hudson AO, Parthasarathy A (2019) The quest for novel antimicrobial compounds: emerging trends in research, development, and technologies. Antibiotics 8(1):8. https://doi.org/10.3390/antibiotics8010008

Maxson T, Mitchell DA (2016) Targeted treatment for bacterial infections: prospects for pathogen-specific antibiotics coupled with rapid diagnostics. Tetrahedron 72(25):3609–3624. https://doi.org/10.1016/j.tet.2015.09.069

Moal VL, Servin AL (2014) Anti-infective activities of lactobacillus strains in the human intestinal microbiota: from probiotics to gastrointestinal anti-infectious biotherapeutic agents. Clin Microbiol Rev 27(2):167–199. https://doi.org/10.1128/cmr.00080-13

Moghadam MT, Amirmozafari N, Shariati A, Hallajzadeh M, Mirkalantari S, Khoshbayan A, Jazi FM (2020) How phages overcome the challenges of drug-resistant bacteria in clinical infections. Infect Drug Resist 13:45–61. https://doi.org/10.2147/idr.s234353

Morris S, Cerceo E (2020) Trends, epidemiology, and management of multi-drug-resistant Gram-negative bacterial infections in the hospitalized setting. Antibiotics 9(4):196. https://doi.org/10.3390/antibiotics9040196

Murugaiyan J, Kumar P, Rao GS, Iskandar K, Hawser S, Hays JP, Mohsen Y, Adukkadukkam S, Awuah WA, Alvarez MJR, Sylvia N, Nansubuga EP, Tilocca B, Roncada P, Roson-Calero N, Moreno-Morales J, Amin R, Kumar BK, Kumar A, Dongen MV (2022) Progress in alternative strategies to combat antimicrobial resistance: focus on antibiotics. Antibiotics 11(2):200. https://doi.org/10.3390/antibiotics11020200

Naran K, Nundalall T, Chetty S, Barth S (2018) Principles of immunotherapy: implications for treatment strategies in cancer and infectious diseases. Front Microbiol 9:3158. https://doi.org/10.3389/fmicb.2018.03158

Newman MA, Ausubel FM (2016) Introduction to gene editing and manipulation using CRISPR/Cas9 technology. Curr Protoc Mol Biol 115(1):14. https://doi.org/10.1002/cpmb.14

Nii-Trebi NI (2017) Emerging and neglected infectious diseases: insights, advances, and challenges. Biomed Res Int 2017:1–15. https://doi.org/10.1155/2017/5245021

Pires DP, Costa AR, Pinto G, Meneses L, Azeredo J (2020) Current challenges and future opportunities of phage therapy. FEMS Microbiol Rev 44(6):684–700. https://doi.org/10.1093/femsre/fuaa017

Rex JH, Lynch HF, Cohen IG, Darrow JJ, Outterson K (2019) Designing development programs for non-traditional antibacterial agents. Nat Commun 10(1):9. https://doi.org/10.1038/s41467-019-11303-9

Rohde C, Wittmann J, Kutter E (2018) Bacteriophages: a therapy concept against multi-drug-resistant bacteria. Surg Infect 19(8):737–744. https://doi.org/10.1089/sur.2018.184

Rubens M, Ramamoorthy V, Shehadeh N (2014) Public health in the twenty-first century: the role of advanced technologies. Front Public Health 2:224. https://doi.org/10.3389/fpubh.2014.00224

Simundza J (2018) Infection and immunity: insights and therapeutic strategies through genomic analysis of the host, pathogen, and host-pathogen interaction. Genome Med 10(1):72. https://doi.org/10.1186/s13073-018-0583-9

Theuretzbacher U, Piddock LJV (2019) Non-traditional antibacterial therapeutic options and challenges. Cell Host Microbe 26(1):61–72. https://doi.org/10.1016/j.chom.2019.06.004

Walesch S, Birkelbach J, Jézéquel G, Haeckl FPJ, Hegemann JD, Hesterkamp T, Hirsch AKH, Hammann P, Müller R (2022) Fighting antibiotic resistance—strategies and (pre)clinical developments to find new antibacterials. EMBO Rep 24(1):e56033. https://doi.org/10.15252/embr.202256033

Wittebole X, Roock SD, Opal SM (2013) A historical overview of bacteriophage therapy as an alternative to antibiotics for the treatment of bacterial pathogens. Virulence 5(1):226–235. https://doi.org/10.4161/viru.25991

Revolutionizing Infection Control: Harnessing the Power of Stem Cells for Precision Antimicrobial Therapy

9

Arup Dey, Pankaj Kumar Sharma, Ashutosh Pathak, Sanjay Kumar Yadav, Jashanjit Singh, and Sourav Ghosh

Abstract

As antibiotic resistance spreads, infectious illnesses continue to represent a major danger to human health, emphasizing the urgent need for novel and targeted treatment methods. Stem cells have emerged as a promising source for antimicrobial therapy research due to their regenerative and immunomodulatory properties. This chapter explores stem cells' potential as a focused approach to treating infections, emphasizing their capacity to differentiate into particular cell types that can specifically target and eradicate microorganisms, as well as their capacity to alter the immune system and strengthen host defense. The article examines the limitations and prospects for clinical use of stem cell-derived antibacterial treatments. Among these challenges include the necessity for standardized techniques for cell isolation, growth, and distribution, as well as the importance of thorough safety and effectiveness testing. Despite this, the use of stem cells as a targeted antibacterial method has tremendous potential for the development of effective and long-lasting therapeutic therapies against infectious illnesses. This detailed investigation navigates the changing environment of stem cell-based antimicrobial therapy, promoting further research into effective and focused treatment techniques in the fight against antibiotic resistance.

A. Dey · S. Ghosh (✉)
Department of Pharmaceutical Technology, School of Health and Medical Science, Adamas University, Kolkata, West Bengal, India

P. K. Sharma
Apex Institute of Pharmaceutical Sciences, Jaipur, Rajasthan, India

A. Pathak · S. K. Yadav
Dr. Shakuntala Mishra National Rehabilitation University, Lucknow, Uttar Pradesh, India

J. Singh
Gurukul Institute of Pharmaceutical Sciences, Fatehgarh Sahib, Punjab, India

M. Gangwar, G. Nath (eds.), *Emerging Paradigms for Antibiotic-Resistant Infections: Beyond the Pill*, https://doi.org/10.1007/978-981-97-5272-0_9

Keywords

Stem cell · Mesenchymal stem cells (MSCs) · Antimicrobial resistance · Antimicrobial therapy

Abbreviations

AMR	Antimicrobial resistance
LMICs	Low- and middle-income countries
MSCs	Mesenchymal stem cells/Mesenchymal stromal cells
RPE	Retinal pigment epithelium
NSCs	Neural stem cells
CAR-T	Chimeric antigen receptor T cells
CD4	Clusters of differentiation 4
CD8	Clusters of differentiation 8
IL-10	Interleukin 10
IL-35	Interleukin 35
TGF-β	Transforming growth factor-beta
DNA	Deoxyribonucleic acid
RNA	Ribonucleic acid
EVs	Extracellular vehicles
UTI	Urinary tract infection
CRISPR	Clustered regularly interspaced short palindromic repeats
NMs	Nanomaterials
NPs	Nanoparticles
ESCs	Embryonic stem cell

9.1 Introduction

A variety of factors cause antimicrobial resistance (AMR). But antibiotic misuse is the most significant factor. Antibiotic usage grew by 65% between 2000 and 2015 globally. There was a significant increase in low- and middle-income countries (LMICs) (Klein et al. 2018). AMR represents a challenge to the efficient fight against and cure of a growing number of illnesses. Viruses, bacteria, parasites, and fungi cause these illnesses. AMR occurs when microbes, like parasites, fungi, and viruses, evolve. They no longer react to medications. This makes infections harder to cure. It also raises the risk of disease spreading, severe illness, and death. As a result, the treatments become ineffective. Illnesses remain in the body, boosting the danger of spreading to others (World Health Organization 2019). AMR is a major worldwide public health and development problem. Bacterial AMR caused 1.27 million deaths globally in 2019. It also contributed to 4.95 million fatalities, as expected (Fernandez 2022). Bacteriophages are viruses that preferential infect and destroy bacteria. Researchers have used phage treatment to treat individuals with severe multidrug-resistant illnesses. Microbiome-based techniques, including live microbiome-based medicinal solutions, can treat antibacterial-resistant infections. In clinical studies, optimizing existing antibiotic treatments have been more successful at curing infections and inhibiting resistance (Akshay et al. 2023). Mesenchymal stem cells (MSC)-based therapy is a novel approach to antimicrobial treatment. MSCs form the basis (Alcayaga-Miranda et al. 2017).

Stem cells can specialize into particular cell types. A stem cell has two distinguishing characteristics. It renews itself. It has high potential for future therapeutic use in tissue regeneration and repair (Biehl and Russell 2009). We must develop different strategies. We need substitutes for or improvements to current conventional antimicrobials. Mesenchymal stromal cells (MSCs) contain antibacterial characteristics. This might aid with this issue (Russell et al. 2020). Researchers have long studied mesenchymal stromal cells (MSCs) as a raw material for tissue engineering. They also studied MSCs as immunomodulators to treat inflammatory illnesses (Devireddy et al. 2017). MSCs have recently emerged as a promising therapy for AMR. Researchers are still discovering the mechanisms of MSCs' antibacterial characteristics. Scientists have discovered that stem cells release antimicrobial compounds that interact with pathogens. They also release other factors that enhance the antimicrobial activity of host immune cells. In proof-of-concept experiments, MSCs have shown high synergy with conventional antibiotic therapies to penetrate biofilm infections. Additionally, they can act as antifungal, antiviral, and antiparasitic drugs. In this review, we will explore the principles that underpin stem cell-based antimicrobial treatment. We'll also cover the current advances and problems associated with stem cell-based therapy and consider upcoming viewpoints and potential applications (Alcayaga-Miranda et al. 2017).

9.2 Mechanism of Action of Stem Cells Combat Infections

MSCs can also affect innate and adaptive immune responses. This provides therapeutic advantages in the host body through paracrine actions and by stimulating host cells (Weiss and Dahlke 2019). MSCs can change the tissue environment besides acting as tissue progenitor cells. MSCs may trigger an immunosuppressive response in various immune cells. This includes monocytes, macrophages, dendritic cells, T cells, B cells, and natural killer cells. As a result, MSCs can alter the environment directly by creating cytokines, chemokines, and growth factors or indirectly by releasing extracellular vesicles (Song et al. 2020). These immunomodulatory actions are already helping to design treatments. For example, donors provide MSCs. These MSCs can activate the patient's T cells. They are suitable for developing CAR-T cell treatments. Another example is the use of MSC-derived extracellular vesicles in clinical studies. They treat various disorders including severe and critical illnesses in COVID-19 patients. These immunomodulatory effects hold promise for developing more effective cell and gene treatments. They may treat inflammatory, autoimmune, and cardiovascular illnesses, as well as carcinoma (Javed et al. 2022).

MSC activation by inflammatory cytokines is also known as inflammatory cytokine licensing. The immune system and MSCs work close together. Immunosuppressive factors are released, outer molecule expression is changed, and growth factor production is affected by the inflammatory cytokines produced by T cells and antigen-presenting cells during an immune response. These substances are necessary for the immunoregulation and tissue repair of MSCs (Fig. 9.1) (Shi et al. 2012).

There are several parts of the immune system. These parts consist of immune cells and immunological chemicals. They shield the body from internal cancer and external infections. The host could not be protected by an under-stimulated immune system. However, overstimulation would damage the host's healthy tissues and cells. It would cause damage to organs or tissues. As a result, several mechanisms must be used to tightly modulate the immune response. It is essential to uncover the intricate workings of this regulatory network. It aids in our comprehension of the pathophysiology of illnesses linked to immune malfunction. It is also essential for creating novel therapeutic approaches. Numerous cell types have the ability to avoid immune system overstimulation, including M2 macrophages, suppressive dendritic cells, Breg (B regulatory cells), CD8+ Treg, natural and induced CD4+ Treg (T regulatory cells), and Breg (B regulatory cells) (Li et al. 2014). The immunological

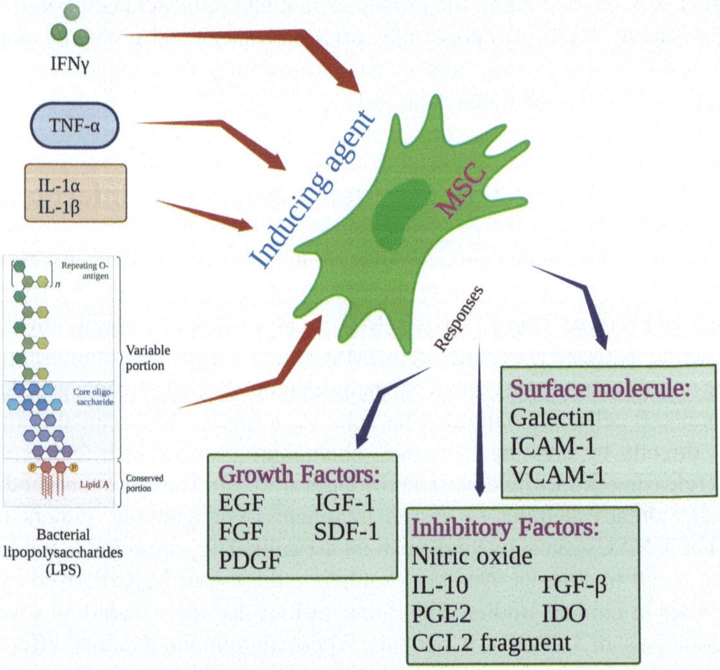

Fig. 9.1 Activation factors of MSC and the responses

response is modulated by these cells. They achieve this by secreting cytokines that restrict growth, including IL-10, TGF-β, and IL-35. Additionally, they employ inhibitory ligands and receptors, such as PD-1 and PD-L1. Additionally, these cells control the survival, development, and differentiation of immune cells (García-González et al. 2016).

For MSCs to have their immune-suppressive effects, direct cell interaction is necessary. However, research has indicated that MSC-expressed immune modulators are more important. Indoleamine 2,3-dioxygenase (IDO), prostaglandin E2 (PGE2), CD39, CD73, tumor necrosis factor-stimulated gene 6 (TSG6), transforming growth factor beta (TGF-β), hepatocyte growth factor (HGF), histocompatibility locus antigen-G (HLA-G), interleukin-10 (IL-10), interleukin-1 receptor antagonist (IL1RA), galectins, C–C motif chemokine ligand 2 (CCL2), programmed cell death ligands 1 and 2 (PD-L1 and PD-L2), heme oxygenase 1 (HO-1), and complement system-related factors are among them (Jiang and Xu 2020).

MSCs have become recognized as an adult stem cell that actively participates in tissue healing. When tissue injury occurs, MSCs in the near area or those produced from bone marrow are thought to move into the affected tissue. Details on their movement, differentiation, and survival strategies at the damage sites, however, still need to be discovered, owing to a lack of good tracking markers. Nonetheless, once they arrive at the site of injury, MSCs must interact intimately with numerous stromal and inflammatory cells to engage in tissue healing. The processes underlying MSC-mediated tissue healing are complicated, although MSC-derived trophic factors play a crucial role (Shi et al. 2012).

Beyond immunomodulation and microbiological elimination, stem cells have the ability to restore damaged tissue and aid in the healing process. By secreting neurotrophic factors that facilitate tissue repair and regeneration through the fine-tuning of mitogenic, angiogenic, anti-apoptotic, and reduction activities, stem cells can protect against photoreceptor degeneration (Lamba et al. 2006). Even if common antimicrobial treatments can completely remove germs from the eyes, the patient's extended vision will be disrupted since the ocular tissue that was damaged during the infection cannot be restored. Given its potential for repairing damaged ocular layers, stem cell therapy needs to be given top priority as the most effective treatment approach going forward. Parallel to this, stem cells' potent immunosuppressive qualities enable both autologous and allogeneic transplantation without the need for medication immunosuppression or genetic modification due to a decrease in the generation of pro-inflammatory cytokines (Chen et al. 2011). When eyes are wounded and infected with microbes, vision can recover. Replacing photoreceptor cells and retinal pigment epithelium (RPE) will restore the natural visual-processing pathways for proper visual perception, making this an attractive and novel therapy approach. In ocular tissue damaged by microbial diseases, stem cells are the best source of cells because they can proliferate and differentiate into functional photoreceptor cells and RPE (Teh et al. 2018).

Stem cells can display direct antibacterial characteristics due to their ability to differentiate into diverse cell lineages capable of identifying and eliminating microbial infections. MSCs can differentiate into macrophages, which are immune cells that phagocytose and eliminate invading pathogens (Li and Hua 2017). Stem cells are capable of producing and releasing AMPs and proteins, which can directly inhibit or kill bacteria. Peptides and proteins have the ability to disrupt microorganism cell membranes, inhibit microbial DNA replication, hinder protein synthesis, and interfere with overall microorganism physiology. MSCs can release cathelicidin, an AMP that kills microorganisms by interfering with their cell membranes (Esfandiyari et al. 2019). The production of cytokines and chemokines may be regulated by stem cells, which promotes an anti-inflammatory response and reduces tissue damage brought on by inflammation. Research has demonstrated that stem cells can inhibit immune cells that cause tissue inflammation and injury. MSCs have the ability to restrict the activity of T-cells, an essential subset of immune cells that significantly increases tissue damage in inflammatory and autoimmune diseases (Burr et al. 2013). In order to preserve immunological homeostasis and prevent unwarranted immune activation, stem cells have the ability to induce the production of regulatory T-cells. Regulatory T-cells have the dual ability to promote tissue repair and inhibit the activities of other immune cells that cause inflammation and tissue damage. MSCs have been shown to increase the production of regulatory T-cells and reduce inflammation in animal models of bacterial pneumonia, resulting in improved bacterial clearance and pulmonary function (Wang et al. 2021).

The paracrine role of MSCs, which causes a horizontal transfer of proteins and nucleic acids between injured cells and MSCs, has at least partially contributed to the current research's intense interest in EVs produced by MSCs (Bruno et al. 2009). Furthermore, evidence of bacterial or inflammatory activity is shown. When LL-37 is released, it has a bactericidal effect on *Escherichia coli* and *S. aureus*, although β-defensin-2 solely affects *E. coli*. Hepcidin, which was discovered to inhibit the growth of a polybacterial mix taken from the microbiome of mice, was released by MSCs produced from menstrual secretions. MSCs can also release extracellular vesicles, which may contain active molecules with an antibacterial impact. This will require further research in the future (Fig. 9.2) (Alcayaga-Miranda et al. 2017).

9.3 Current Advances and Challenges

A wide range of pathogenic microorganisms causes ocular diseases. These include fungi, viruses, parasites, and bacteria. It has become a huge public health concern during the last two decades. Typical treatments for visual impairment and blindness caused by microbial infections include antimicrobial therapy. It kills the bacteria. In extreme cases, treatment may include surgical procedures. The affected region may be replaced with a retinal prosthesis (Teh et al. 2018). In the recent decade, the area of retinal research has achieved considerable progress in bone marrow-derived MSCs. This is especially true in the treatment of blindness caused by retinal degenerative disorders (Sun et al. 2007). Stem cell therapy shows promise in rebuilding damaged ocular layers. In the coming years, it should focus on it as the best therapeutic approach (Chen et al. 2011). Microbial-infected damaged eyes can recover vision. This appealing treatment is new. It can restore photoreceptor cells and

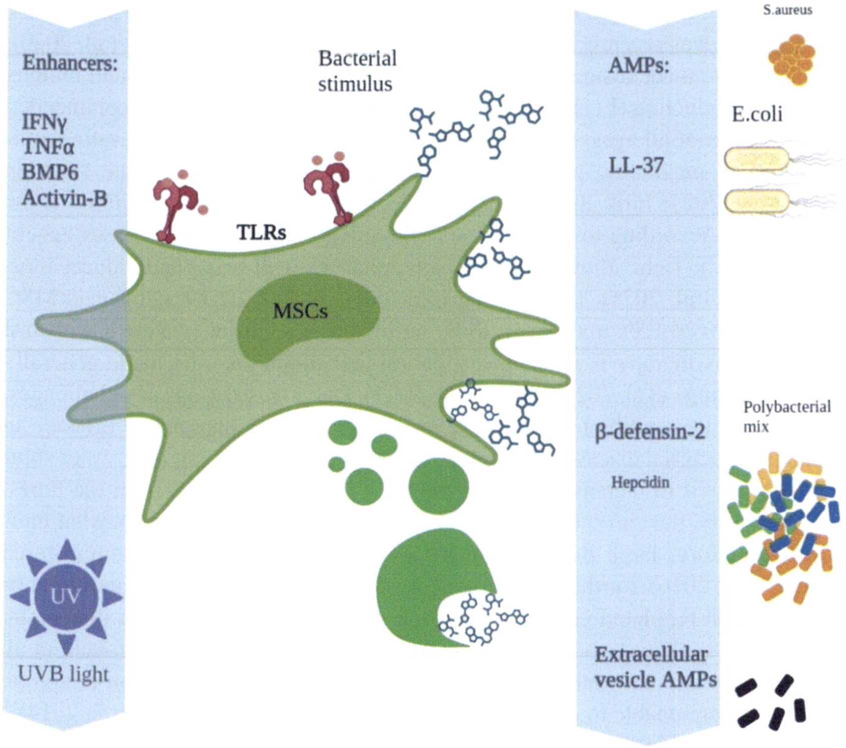

Fig. 9.2 Mesenchymal stem cells (MSCs) secretion of different AMPs and their expression

RPE. This restoration can restore the original visual processing pathways for proper visual perception. There is also great promise for improved microbial-damaged ocular tissue healing. Finally, doctors might use stem cells to treat severe infection-related ocular damage. They have antibacterial action and regenerative characteristics. They might also restore eyesight in microbial-infected individuals (Teh et al. 2018). Conventional antimicrobial therapy has problems. Injured tissue cannot be restored, even with intensive therapy, because of them. This usually results in visual impairment (Garg 2012). When microbial infections occur in the eyes and the causative bacteria cannot be identified, the supplied antimicrobial medicine is determined empirically (Johnson et al. 1997). Treatment varies based on the patient's age, the pathogen species, and the time between injury and treatment. It also depends on the level of ocular tissue damage and the delay in administering effective treatment (Foster et al. 1996).

The terrible development of a new virus, SARS-CoV-2, which causes the illness COVID-19, has spread over the world, producing a pandemic. It typically affects the respiratory system and lungs, in rare cases causing serious organ damage and pneumonia as a result of excessive immunological reactions (Esquivel et al. 2021). Conventional therapies, such as hydroxychloroquine, are typically ineffective and are not utilized in therapy. However, this medication is given by certain physicians. RamedSavier was recently introduced as an antiviral medication (Dai et al. 2020). Remdesivir is an adenosine analog that targets viral RNA polymerase and inhibits viral RNA production (Li et al. 2004). Similarly, intravenous IVIG is recommended for individuals of all ages. By reducing inflammatory cytokines and elevating anti-inflammatory mediators, prescribed thymosin α can enhance immune function (Wong et al. 2021; Idris 2023). One novel method of treating this condition is the use of MSC. According to clinical research, patients with serious illnesses benefit greatly from the right adjuvant care, which enhances well-being and reduces mortality (Fadai et al. 2021). By injecting intravenously COVID-19, allogeneic MSC and MSC-derived EVs may prove to be an effective adjuvant treatment with fast diffusion. This therapy is under investigation and might soon be made generally accessible (Bange et al. 2021; Saichi et al. 2021). Due to restrictions and challenges, current inhalation antibiotics do not demonstrate optimal therapeutic efficacy in eliminating bacterial disease (Ho et al. 2019). However, significant drug water solubility is required to increase the bioavailability of drugs breathed from the lungs. Simultaneously, the current generation of inhaled anti-infectives is somewhat ineffective; therefore, large dosages (up to several 100 mg) must be administered (Hastedt et al. 2016). Furthermore, frequent administration of large dosages of antibiotics without regulated release and without particular targeting to the illness site may result in toxicity to healthy lung cells (d'Angelo et al. 2014; Hoffmann et al. 2002). Recent clinical trials have demonstrated that existing inhalation antibiotic formulations are unable to completely eradicate the illness (Tolker-Nielsen 2014).

Urinary tract infection (UTI) is a major healthcare issue globally, affecting outpatient clinics, emergency department visits, and hospitalized patients (AbouHeidar et al. 2019). Antibiotics and immunizations are typically used to treat UTIs. Cranberry juice may also be used as a home remedy for vaginal yeast infections and burning sensations induced by a UTI in the vaginal area. Because pathogens are multidrug resistant, identifying a viable pharmacological source or system is critical (Sharma et al. 2020). Advances in stem cell therapy provide fresh insights into treating lower urinary tract problems. Adipose tissue is one of the most often utilized stem cell sources due to its ease of availability, abundance, minimal risk of serious consequences, and absence of ethical concerns (Liu et al. 2022). Most research did not look into the ideal therapy dosage of ASCs, and the association between damage severity and number of transplants has yet to be fully understood. Currently, no ideal strategy for ASC administration has been found, which most likely determines whether stem cells reach the afflicted area and how well they are retained (Wang et al. 2020; Siddiqui et al. 2017).

9.4 Future Prospective

The use of Cas9 and clustered regularly interspaced short palindromic repeats (CRISPR) has revolutionized genome engineering recently. By using this cutting-edge technique in stem cell research, illness models may be created to look into potential new treatments. Treating degenerative disorders is an especially appealing use of novel molecular knowledge systems in clinical research (Valenti et al. 2019). Later research revealed that CRISPR loci are transcribed (Tang et al. 2002) and that archaeal cells with spacers that correspond to their own genomes are immune to infection by bacteriophages (Mojica et al. 2005). The earliest proof that the CRISPR system functions as a microbial molecular immunological memory and viral defense mechanism came from researchers seeking to extend the lifespan of bacterial cultures used to make yoghurt and ice cream. These systems include a group of CRISPR-associated (Cas) genes, noncoding RNAs, and a unique set of repetitive elements (Barrangou et al. 2007). A major difficulty in stem cell treatments is identifying an effective strategy to control their proliferation and differentiation. RNA, DNA, proteins, and peptides are some of the macromolecules that can regulate the division of SCs. These macromolecules need to be well distributed throughout the SCs in order to function. Traditional methods of delivering macromolecular compounds are seldom successful and have a lot of drawbacks (Zhu et al. 2014). Due to the special properties of nanomaterials, nanotechnology can enhance cell-based therapy and augment SC treatment (Corradetti and Ferrari 2016). Moreover, the combination of these technologies will give rise to a brand-new, interdisciplinary discipline that emphasizes in-depth study (Farzamfar et al. 2019). According to Santos et al. endogenous and exogenous neural stem cells (NSCs) can be stimulated to proliferate, self-renew, and differentiate through the interaction of nanomaterials (NMs) and nanoparticles (NPs) with proneurogenic elements inside the stem cell niche (Santos et al. 2012). Moreover, Zhang et al. reported that superparamagnetic NPs labeled with peptides may be injected intravenously into the affected area and monitored by magnetic resonance imaging techniques (Zhang et al. 2018).

The understanding of MSCs' immune modulatory pathways is thus crucial for developing therapy procedures that are personalized to each individual patient. The variety of possible responses to MSCs or any other stem cells precludes the development of a single treatment regimen for a specific illness (Rameshwar 2009). The patient profile can be further segmented into parameters such as age, gender, prior medical history, and general health. Each of these characteristics can influence how a person responds to stem cell therapy. Both genetic and environmental influences might have unanticipated consequences, necessitating the investigation of personalized stem cell treatment (Buchholz and Ganser 2009). The present conventional therapies do not consider the patients' particular molecular and other characteristics. Personalized stem cell therapy will provide customized treatment with maximum efficacy and minimal side effects. Modern personalized stem cell therapy's effectiveness will be determined by present efforts to employ these cells in the framework of pharmacogenomics principles, among other signs. With strong scientific studies on the different genetic and cellular profiles of patients, the possibilities for customized treatment using stem cells are encouraging (Patel et al. 2010).

Stem cell-based therapeutics have several challenges that must be addressed immediately. The greatest recurring problem is the ethical conflict surrounding the use of stem cells. They are presently embroiled in their own ethical debate, which concerns their endless ability for differentiation and the possibility that these cells will be used in human cloning in the future. The difficulty is now being addressed by creating optimized methods to ensure their safety, as well as worldwide development (Cyranoski 2019). MSCs have been widely deemed safe, but future research should focus on continuous monitoring and long-term follow-up to reduce the likelihood of cancer growth following treatments (Volarevic et al. 2018). The most apparent hurdle to personalized embryonic stem cell treatment (ESCs) is immunogenicity, which can result in cell rejection after implanting. Thousands or millions of ESC lines must be employed to accommodate genetically different patient groups. Generating such a huge number of ESC lines would be problematic due to ethical and scientific constraints on collecting these cells (Patel et al. 2010).

9.5 Conclusion

Overestimated immune response and inflammation with infections produce tissue damage, which is one of the leading causes of infectious disease-related death. However, MSC therapy has been shown to give therapeutic effects in a variety of disease models by lowering inflammation, pathogen burden, and tissue damage. MSCs resist tissue damage at the area of infection by speeding up pathogen clearance by AMP production and direct phagocytosis, as well as decreasing inflammation through the release of various anti-inflammatory cytokines. MSCs play a crucial role in tissue regeneration by secreting mitogens and transforming them into cells from the target tissue. MSC-derived EVs include multiple therapeutically helpful cytokines and miRNAs, and therapy with MSC-EVs yielded encouraging outcomes. However, caution should be used when using MSCs for treatment since they can harbor hazardous infections and may produce unfavorable results in individuals with preexisting diseases.

References

AbouHeidar NF, Degheili JA, Yacoubian AA, Khauli RB (2019) Management of urinary tract infection in women: a practical approach for everyday practice. Urol Ann 11(4):339

Akshay SD, Deekshit VK, Mohan Raj J, Maiti B (2023) Outer membrane proteins and efflux pumps mediated multi-drug resistance in salmonella: rising threat to antimicrobial therapy. ACS Infect Dis 9(11):2072–2092

Alcayaga-Miranda F, Cuenca J, Khoury M (2017) Antimicrobial activity of mesenchymal stem cells: current status and new perspectives of antimicrobial peptide-based therapies. Front Immunol 8:339

Bange EM, Han NA, Wileyto P, Kim JY, Gouma S, Robinson J, Huang AC (2021) CD8 T cells compensate for impaired humoral immunity in COVID-19 patients with hematologic cancer. Res Sq

Barrangou R, Fremaux C, Deveau H, Richards M, Boyaval P, Moineau S, Horvath P (2007) CRISPR provides acquired resistance against viruses in prokaryotes. Science 315(5819):1709–1712

Biehl JK, Russell B (2009) Introduction to stem cell therapy. J Cardiovasc Nurs 24(2):98

Bruno S, Grange C, Deregibus MC, Calogero RA, Saviozzi S, Collino F, Camussi G (2009) Mesenchymal stem cell-derived microvesicles protect against acute tubular injury. J Am Soc Nephrol 20(5):1053–1067

Buchholz S, Ganser A (2009) Hematopoietic stem cell transplantation. Indications, foundations and perspective: Indikationen, Grundlagen und Ausblick. Internist 50:572–580

Burr SP, Dazzi F, Garden OA (2013) Mesenchymal stromal cells and regulatory T cells: the Yin and Yang of peripheral tolerance? Immunol Cell Biol 91(1):12–18

Chen PM, Yen ML, Liu KJ, Sytwu HK, Yen BL (2011) Immunomodulatory properties of human adult and fetal multipotent mesenchymal stem cells. J Biomed Sci 18(1):49

Corradetti B, Ferrari M (2016) Nanotechnology for mesenchymal stem cell therapies. J Control Release 240:242–250

Cyranoski D (2019) The potent effects of Japan's stem-cell policies. Nature 573(7775):482–486

Dai L, Zheng T, Xu K, Han Y, Xu L, Huang E, Gao GF (2020) A universal design of betacoronavirus vaccines against COVID-19, MERS, and SARS. Cell 182(3):722–733

d'Angelo I, Conte C, La Rotonda MI, Miro A, Quaglia F, Ungaro F (2014) Improving the efficacy of inhaled drugs in cystic fibrosis: challenges and emerging drug delivery strategies. Adv Drug Deliv Rev 75:92–111

Devireddy LR, Boxer L, Myers MJ, Skasko M, Screven R (2017) Questions and challenges in the development of mesenchymal stromal/stem cell-based therapies in veterinary medicine. Tissue Eng Part B Rev 23(5):462–470

Esfandiyari R, Halabian R, Behzadi E, Sedighian H, Jafari R, Fooladi AAI (2019) Performance evaluation of antimicrobial peptide II-37 and hepcidin and β-defensin-2 secreted by mesenchymal stem cells. Heliyon 5(10):e02652

Esquivel D, Mishra R, Soni P, Seetharaman R, Mahmood A, Srivastava A (2021) Stem cells therapy as a possible therapeutic option in treating COVID-19 patients. Stem Cell Rev Rep 17:144–152

Fadai NT, Sachak-Patwa R, Byrne HM, Maini PK, Bafadhel M, Nicolau DV Jr (2021) Infection, inflammation and intervention: mechanistic modelling of epithelial cells in COVID-19. J R Soc Interface 18(175):20200950

Farzamfar S, Nazeri N, Salehi M, Valizadeh A, Marashi SM, SavariKouzehkonan G, Ghanbari H (2019) Will nanotechnology bring new hope for stem cell therapy? Cells Tissues Organs 206(4–5):229–241

Fernandez G (2022) Turning the juggernaut. Lancet Planet Health 6(2):E75–E75

Foster RE, Martinez JA, Murray TG, Rubsamen PE, Flynn HW Jr, Forster RK (1996) Useful visual outcomes after treatment of *Bacillus cereus* endophthalmitis. Ophthalmology 103(3):390–397

García-González P, Ubilla-Olguín G, Catalán D, Schinnerling K, Aguillón JC (2016) Tolerogenic dendritic cells for reprogramming of lymphocyte responses in autoimmune diseases. Autoimmun Rev 15(11):1071–1080

Garg P (2012) Fungal, mycobacterial, and Nocardia infections and the eye: an update. Eye 26(2):245–251

Hastedt JE, Bäckman P, Clark AR, Doub W, Hickey A, Hochhaus G, Weers JG (2016) Scope and relevance of a pulmonary biopharmaceutical classification system AAPS/FDA/USP Workshop March 16–17th, 2015 in Baltimore, MD, vol 2. Springer

Ho DK, Nichols BL, Edgar KJ, Murgia X, Loretz B, Lehr CM (2019) Challenges and strategies in drug delivery systems for treatment of pulmonary infections. Eur J Pharm Biopharm 144:110–124

Hoffmann IM, Rubin BK, Iskandar SS, Schechter MS, Nagaraj SK, Bitzan MM (2002) Acute renal failure in cystic fibrosis: association with inhaled tobramycin therapy. Pediatr Pulmonol 34(5):375–377

Idris I (2023) ACE and ARB does not increase the risk of dying from COVID-19. Diabetes Obes Metab Now 1(1):e00016

Javed A, Karki S, Sami Z, Khan Z, Shree A, Sah BK, Saxena S (2022) Association between mesenchymal stem cells and COVID-19 therapy: systematic review and current trends. BioMed Res Int 2022:9346939

Jiang W, Xu J (2020) Immune modulation by mesenchymal stem cells. Cell Prolif 53(1):e12712

Johnson MW, Doft BH, Kelsey SF, Barza M, Wilson LA, Barr CC, Volkert DC (1997) The endophthalmitis vitrectomy study: relationship between clinical presentation and microbioloaic spectrum. Ophthalmology 104(2):261–272

Klein EY, Van Boeckel TP, Martinez EM, Pant S, Gandra S, Levin SA, Laxminarayan R (2018) Global increase and geographic convergence in antibiotic consumption between 2000 and 2015. Proc Natl Acad Sci 115(15):E3463–E3470

Lamba DA, Karl MO, Ware CB, Reh TA (2006) Efficient generation of retinal progenitor cells from human embryonic stem cells. Proc Natl Acad Sci 103(34):12769–12774

Li N, Hua J (2017) Interactions between mesenchymal stem cells and the immune system. Cell Mol Life Sci 74:2345–2360

Li T, Qiu Z, Zhang L, Han Y, He W, Liu Z, Wang A (2004) Significant changes of peripheral T lymphocyte subsets in patients with severe acute respiratory syndrome. J Infect Dis 189(4):648–651

Li S, Xie Q, Zeng Y, Zou C, Liu X, Wu S, Dai Z (2014) A naturally occurring CD8+ CD122+ T-cell subset as a memory-like Treg family. Cell Mol Immunol 11(4):326–331

Liu M, Chen J, Cao N, Zhao W, Gao G, Wang Y, Fu Q (2022) Therapies based on adipose-derived stem cells for lower urinary tract dysfunction: a narrative review. Pharmaceutics 14(10):2229

Mojica FJ, Díez-Villaseñor CS, García-Martínez J, Soria E (2005) Intervening sequences of regularly spaced prokaryotic repeats derive from foreign genetic elements. J Mol Evol 60:174–182

Patel SA, King CC, Lim PK, Habiba U, Dave M, Porecha R, Rameshwar P (2010) Personalizing stem cell research and therapy: the arduous road ahead or missed opportunity? Curr Pharmacogenomics Person Med 8(1):25–36

Rameshwar P (2009) Casting doubt on the safety of "off-the-shelf" mesenchymal stem cells for cell therapy. Mol Ther 17(2):216–218

Russell KA, Garbin LC, Wong JM, Koch TG (2020) Mesenchymal stromal cells as potential antimicrobial for veterinary use—a comprehensive review. Front Microbiol 11:606404

Saichi M, Ladjemi MZ, Korniotis S, Rousseau C, AitHamou Z, Massenet-Regad L, Soumelis V (2021) Single-cell RNA sequencing of blood antigen-presenting cells in severe COVID-19 reveals multi-process defects in antiviral immunity. Nat Cell Biol 23(5):538–551

Santos T, Maia J, Agasse F, Xapelli S, Ferreira L, Bernardino L (2012) Nanomedicine boosts neurogenesis: new strategies for brain repair. Integr Biol 4(9):973–981

Sharma P, Netam AK, Singh R (2020) Prevalence and in vitro antibiotic susceptibility pattern of bacterial strains isolated from tribal women suffering from urinary tract infections in District Anuppur, Madhya Pradesh, India. Biomed Res Ther 7(8):3944–3953

Shi Y, Su J, Roberts AI, Shou P, Rabson AB, Ren G (2012) How mesenchymal stem cells interact with tissue immune responses. Trends Immunol 33(3):136–143

Siddiqui ZA, Abboudi H, Crawford R, Shah S (2017) Intraurethral bulking agents for the management of female stress urinary incontinence: a systematic review. Int Urogynecol J 28:1275–1284

Song N, Scholtemeijer M, Shah K (2020) Mesenchymal stem cell immunomodulation: mechanisms and therapeutic potential. Trends Pharmacol Sci 41(9):653–664

Sun X, Jiang H, Yang H (2007) In vitro culture of bone marrow mesenchymal stem cells in rats and differentiation into retinal neural-like cells. J Huazhong Univ Sci Technol 27(5):598–600

Tang TH, Bachellerie JP, Rozhdestvensky T, Bortolin ML, Huber H, Drungowski M, Hüttenhofer A (2002) Identification of 86 candidates for small non-messenger RNAs from the archaeon *Archaeoglobus fulgidus*. Proc Natl Acad Sci 99(11):7536–7541

Teh SW, Mok PL, Abd Rashid M, Bastion MLC, Ibrahim N, Higuchi A, Subbiah SK (2018) Recent updates on treatment of ocular microbial infections by stem cell therapy: a review. Int J Mol Sci 19(2):558

Tolker-Nielsen T (2014) Pseudomonas aeruginosa biofilm infections: from molecular biofilm biology to new treatment possibilities. APMIS 122:1–51

Valenti MT, Serena M, DalleCarbonare L, Zipeto D (2019) CRISPR/Cas system: an emerging technology in stem cell research. World J Stem Cells 11(11):937

Volarevic V, Markovic BS, Gazdic M, Volarevic A, Jovicic N, Arsenijevic N, Stojkovic M (2018) Ethical and safety issues of stem cell-based therapy. Int J Med Sci 15(1):36

Wang Y, Zhou S, Yang R, Rahman M, Sequeira RC, Cao N, Fu Q (2020) Magnetic targeting of super-paramagnetic iron oxide nanoparticle labeled myogenic-induced adipose-derived stem cells in a rat model of stress urinary incontinence. Nanomedicine 30:102281

Wang MY, Zhou TY, Zhang ZD, Liu HY, Zheng ZY, Xie HQ (2021) Current therapeutic strategies for respiratory diseases using mesenchymal stem cells. MedComm 2(3):351–380

Weiss ARR, Dahlke MH (2019) Immunomodulation by mesenchymal stem cells (MSCs): mechanisms of action of living, apoptotic, and dead MSCs. Front Immunol 10:1191

Wong KK, Lee SWH, Kua KP (2021) *N*-Acetylcysteine as adjuvant therapy for COVID-19—a perspective on the current state of the evidence. J Inflamm Res 14:2993

World Health Organization (2019) WHO. Antimicrobial resistance [Internet]. https://www.who.int/health-topics/antimicrobial-resistance

Zhang G, Khan AA, Wu H, Chen L, Gu Y, Gu N (2018) The application of nanomaterials in stem cell therapy for some neurological diseases. Curr Drug Targets 19(3):279–298

Zhu K, Li J, Lai H, Yang C, Guo C, Wang C (2014) Reprogramming fibroblasts to pluripotency using arginine-terminated polyamidoamine nanoparticles based non-viral gene delivery system. Int J Nanomed 9:5837–5847

10 Combatting Antibiotic Resistance: Leveraging Fecal Microbial Transplantation for Gut Health

Prashasti Upreti, Lalchand D. Devhare, Lara Hashim Abdulmageed, Yamjala Ganesh Kumar, Rajesh Kumar, and G. Dharmamoorthy

Abstract

This chapter provides a comprehensive exploration of the relationship between gut microbes and human health. It explores the pivotal roles gut microbiota play in maintaining homeostasis, influencing metabolism, modulating the immune system, and even impacting neurological functions. The disruptive phenomenon of dysbiosis, characterized by an imbalance in the gut microbial community, is thoroughly examined, highlighting its association with various diseases, including inflammatory bowel diseases and antibiotic-resistant infections. Fecal microbiota transplantation emerges as a revolutionary therapy that seeks to restore healthy microbial balance in the gut along with addressing various gastrointestinal disorders and MDRO, presenting key insights from existing studies and case

P. Upreti (✉)
Department of Microbiology, Graphic Era Deemed to be University, Dehradun, Uttarakhand, India

L. D. Devhare
Manwatkar College of Pharmacy, Chandrapur, Maharashtra, India

L. H. Abdulmageed
University of Anbar, Ramadi, Anbar, Iraq

Y. G. Kumar
Department of Pharmacognosy and Phytochemistry, Sree Dattha Institute of Pharmacy, Hyderabad, Telangana, India

R. Kumar
Maharaja Agrasen School of Pharmacy, Maharaja Agrasen University, Solan, Himachal Pradesh, India

G. Dharmamoorthy
Department of Pharmaceutical Analysis, MB School of Pharmaceutical Sciences (Erstwhile SreeVidyanikethan College of Pharmacy), Mohan Babu University, Tirupati, Andhra Pradesh, India

M. Gangwar, G. Nath (eds.), *Emerging Paradigms for Antibiotic-Resistant Infections: Beyond the Pill*, https://doi.org/10.1007/978-981-97-5272-0_10

reports. As a promising avenue in combating antibiotic-resistant infections, FMT not only showcases its efficacy in MDRO decolonization but also holds future prospects in refining its applications and protocols.

Keywords
FMT · Dysbiosis · CDI · IBS · UC · MDRO

Abbreviations

ABR	Antibiotic resistant
BBB	Blood–brain barrier
CDI	*Clostridium difficile* infection
CNS	Central nervous system
CPE	Carbapenemase-producing Enterobacteriacea
CR	Carbapenem resistant
ESBL	Extended spectrum beta-lactamase
FDA	Food and Drug Administration
FMT	Fecal microbiota transplant
GALT	Gut-associated lymphoid tissues
GI	Gastrointestinal
HFD	High fat diet
HIV	Human immunodeficiency virus
HMO	Human milk oligosaccharides
IBS	Irritable Bowel syndrome
MDR	Multiple drug resistant
MDRO	Multidrug resistant organism
RDA	Recommended Daily Allowance
SCFA	Short chain fatty acids
UC	Ulcerative colitis
UTI	Urinary tract infections
VRE	Vancomycin Resistant Enterococcus

10.1 Introduction

Fecal microbiota transplantation (FMT) is the process of directly altering the recipient's microbial composition and conferring a health benefit by administering a donor's fecal matter solution into the recipient's digestive tract. Ge Hong, a Chinese physician from the fourth century, is credited with describing the use of excrement as a medicine for treating a number of ailments including diarrhea (Gupta et al. 2016). While FMT has been used to treat recurrent CDI (bacteria causing diarrhea and colitis), there is growing evidence from several clinical trials that suggests it may also have therapeutic potential for other conditions such as liver, gastrointestinal diseases, cancer, autoimmune infections, brain disorders, obesity, and metabolic syndrome (Biazzo and Deidda 2022).

There is an advancing implementation of FMT specifically after its proven effectiveness against antibiotics for recurrent CDI. Moreover, its application in idiopathic conditions other than CDI is evident with the inclusion of this fecal therapy in the treatment repository of clinics offering reliable medical assistance to the respective patients. FMT is becoming the consenter of research with its novel insights into the gut microbiota–a microbial human organ playing a crucial role in the immunity and energy metabolism (Borody and Khoruts 2012; Baunwell et al. 2021). *Escherichia coli, Acinetobacter,* and other taxa were synergistically declining, while a notable high level of beneficial bacteria was present in the gut microbiota of the FMT group. The results indicate a positive host–microbiome association for the management of pathogen-associated illnesses and gut infections after FMT (Wang et al. 2023).

Recent years have seen a surge in interest in FMT, with numerous placebo-controlled trials examining the procedure yielding encouraging findings about symptom alleviation and gut microbiota modulation in IBS. However, before FMT inclusion as a tailored treatment for IBS in clinical practice, further information about long-term consequences is required (Dale and Lied 2020). Particularly, an array of research investigations implies that bowel donors' microbial constitution and richness affect the efficiency of FMT, which gives rise to the theory that there are FMT super donors. Upcoming designed bacteriotherapy strategies are expected to be made possible by the detection and consequent assessment of super donor gut microbiomes, which will undoubtedly increase our knowledge regarding the microbiological aspect of chronic medical conditions (Wilson et al. 2019). The emergence of more virulent, MDRO like VRE, extended-spectrum β lactamase-producing bacteria, carbapenem-resistant *Enterobacteriacea*, and others, along with treatment-related complications are likely the main reasons for the growing major applications of FMT. In the setting of intensive medical treatment, several of these bacteria create reservoirs within the gut when antibiotic pressure is regularly administered (Khoruts and Sadowsky 2016).

10.2 Understanding the Gut Microbiota

The microorganisms residing the digestive systems of animals including plethora of bacteria, archaea, fungi, and some viruses are referred to as gut microbiome. Along with its impact on intestinal epithelium, the gut flora has significant influence on other bodily functions comprising colonization, immune responses, food and drug metabolism, and behavior via gut–brain axis (Wikipedia 2024; Neuroscience 2024). The GI tract comprises of a series of organs equipped with various microscopic habitants comprising around 10^{14} including more than 500 different species. With stomach being the most acidic, the GI tract becomes more alkaline and anaerobic as it descends from stomach to colon passing through duodenum, jejunum ilium, and cecum. The differential range of housing provided by the gut exerts selective pressure on the microbial ecosystem (Sasso et al. 2023). This diverse microbial colonization makes an individual unique as it is linked with the gut environment one holds and maintain governing the aspects of immunologic, digestive, and pH factors.

The two most common phyla in an adult gut mucosa are *Firmicutes* and *Bacteroidetes,* while *Actinobacteria, Proteobacteria,* and *Verrucomicrobia* are less common (Zhang et al. 2022). There is no denial in the unwavering fact about microbial omnipresence in human gut. Along with millions of commensally bacteria residing the body, recent investigations uncover the complex heterogeneity of human microbial ecosystem.

Apart from the macro environment, the rapidly developing domain of microbiome research is unveiling the numerous manners in which our microenvironment comprising gut microbiota has the potential to influence our body's physiology, encompassing our immune system, growth, metabolism, digestion, and even our conduct. Indeed, it is estimated that the quantity of microbial cells in our bodies is approximately equivalent to that of human cells (Johnson 2020).

The ability to sequence entire genomes has changed the focus of research on the human gut microbiome from identifying individual microorganisms to using metagenomics techniques that show not only the species of microbes but also the relationships between their metabolism and human health. Sequence-based identification of human-associated microbiota by Tang (2022). An individual's genome is accountable for molding a distinct microbial association within the GI tract at the molecular level. As for externally, microorganisms are first introduced to a neonate through the birth canal and mother's breast milk. Human milk oligosaccharides, or HMO, are complex carbohydrates found in breast milk enriched with bacteria like Bacteroides and *Bifid bacterium* providing a competitive edge to the new born. Hence a mother's germ treasure is the primary source of microbial colostrum's for an infant who later on gets deviated on account of environmental exposures and various other factors including diet. Striking the right balance is crucial for imbibing the beneficial aspects of healthy gut protecting an individual from various diseases by diminishing their susceptibility to ailments (Ursell et al. 2012; Vatanen et al. 2022).

10.2.1 Importance of Balanced Microbiota for Health

Physical activity is critical when it comes to maintaining the energy levels of the body. Any deterioration in the physical conditioning accompanied by rise in calories intake is a gateway to metabolic syndrome that is characterized by obesity, elevated levels of cholesterol, hypertension, and irregulated glycemic levels. Short chain fatty acids (SCFA)-byproducts of microbial fermentation have been documented in alleviating the effects of various disorders like metabolic syndrome, gastrointestinal diseases, and certain cancers (Den Besten et al. 2013). The microbiota is a specialized organ capable of instructing the immune system to elevate its endurance to the determinants governing numerous immune responses. It has the ability to catabolize edibles that are otherwise indigestible, synthesize specific vitamins like vitamin B (Xu and Gordon 2003).

A study reported that over 85% of recommended daily allowance (RDA) of vitamin B6 along with significant concentrations of other forms of B family vitamins could be made available by the assistance of gut biome. Moreover, it decomposes hazardous food compounds like oxalate and acts as a blockade against pathogenic bacteria (Sasso et al. 2023). Studies have revealed several significant impacts of the microbiota on the host immune system. For example, germ-free mice have significant defects in the development of the gut-associated lymphoid tissues (GALT). Evidences from immunology, epidemiology, and clinical settings suggests that alterations in the gut microbiota may play a major role in the occurrence of a variety of inflammatory diseases (Round and Mazmanian 2009). There have been incidences where the absence of certain bacterial species is linked to the neural impairment that can be compensated by furnishing distinct microbial assemblage and can be a reliable clinical alternative. One such evidence rationalizing the microbial impact on brain function is depicted in a study involving maternally deprived rats. These rats were kept away from maternal association from early life stages when being administered with *Bifidobacterium infantis*; a beneficial gut inhabitant rebounded the immune responses while retracting the behavioral shortfall. Furthermore, *B. longum* is correlated with alleviating stress-related responses (Rogers et al. 2016).

10.3 Dysbiosis and Its Implications

Dysbiosis refers to an imbalance in the microbial communities that inhabit the human body, particularly the gut microbiota. The imbalance can result in an overgrowth of harmful bacteria, fungi, or other microorganisms, leading to disruptions in the normal functioning of the digestive system and potential negative impacts on overall health (Marchesi et al. 2016). Dysbiosis is a significant cause leading to numerous gastrointestinal disorders along with harming other distal organs and systems. Not long ago, it was found that gut bacteria can affect the CNS functions (Sasso et al. 2023). The gut often designated to be the second brain is not only a home to distinct association of microbiota but also acts as a portal connection to the brain via gut microbiome–brain axis. This axis acts as a conjunction of various systems of the body comprising immune system, neural connection via vagus nerve, hormonal signaling network, and number of microbial metabolites (Bravo et al. 2011). These pathways when disturbed during dysbiosis might cause alterations in the permeability of the blood–brain barrier (BBB), becoming a potent cause of neurodevelopment and neurodegenerative disorders (Rutsch et al. 2020). Worldwide observational studies depicted that Alzheimer's patients were found to inhabit lower microbiota heterogeneity as compared with the normal controls (Cammann et al. 2023). There are various factors contributing to such microbial imbalance (Fig. 10.1).

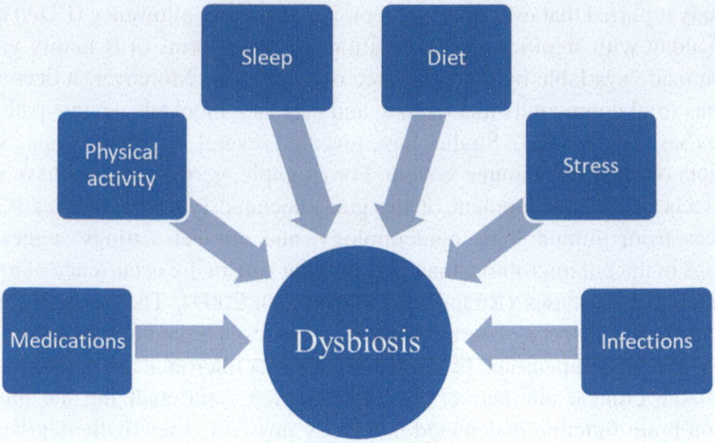

Fig. 10.1 Factors contributing to dysbiosis

10.3.1 Diet

Nutritional index of food plays a significant role in propelling the microbial dynamics toward normal gut health. Research concludes the influence of high fat diet (HFD) on disturbed microbiota ecosystem leaving an individual susceptible to numerous health complications spiking anxiety levels with diminished synaptic plasticity. HFD hamper the composition of gut assemblage such as decreased amount of abundant gut bacteria *Ruminococcaceae* while elevating the levels of Rickenellaceae leading to microbial skewness. Excess sucrose in diet negatively impacts the neurophysiology of memory-based cognitive responses. Research studies uncover the correlation of depleted microbial diversity with exhilarated incidences of insulin resistance, atherosclerosis, inflammation, and adiposity (Rogers et al. 2016; Daniel et al. 2014).

10.3.2 Medications

Along with the preexisting side effects of antibiotic exploitation, studies confirmed another limitation of antibiotic consumption linked with gut microbiome. Certain drugs belonging to the class of broad-spectrum antibiotics lead to microbial perturbation by reducing bacterial diversity specifically those linked with a health gut, allowing the growth of certain microbial colonies responsible for leaving the individual prone to infections (Wen and Duffy 2017). Temporary downfall in microbial conglomeration of some advantageous bacterial groups such as *Bifid bacteria* and *Lactobacilli* has been reported along with transient spike in *Clostridium* culture imparting an inimical influence on the gut well-being. A study discovered some damaging outcomes on the abundance of bacteroides, with a subsequent increase in resistant strains, after administering a drug whose impact lasted for 2 years (Daisy Coyle 2017).

With the deviating consequence on the plentitude of microbial composition of gut ecosystem, there has been associated rise in the inflammatory reactions particularly related to pulmonary disorders, via catastrophic alteration of certain immune cells occupying the colon (Russell et al. 2012).

10.3.3 Stress and Circadian Rhythm

Numerous studies have demonstrated a correlation between stress and alterations in gut microbial composition. Stress, whether psychological or physical, has been associated with a decrease in microbial diversity and shifts in the abundance of specific taxa (Foster et al. 2017). The intricate communication between the gut and the brain, known as the gut–brain axis, is believed to mediate these changes, influencing the microbial community through neural and endocrine pathways (Kelly et al. 2015). The circadian rhythm, an intrinsic biological clock regulating physiological processes over a 24-h cycle, extends its influence to the gut microbiota. Disruptions in the host's circadian rhythm have been linked to alterations in microbial clocks and community structure (Thaiss et al. 2014). The expression of clock genes in the gut epithelium and immune cells establishes a connection between the host's circadian system and the temporal dynamics of the microbiota (Voigt et al. 2014).

10.3.4 Other Factors

Diseases caused by certain pathogens including Rotavirus or other bacteria can cause dysbiosis (microbial imbalance within gut), which includes severe diarrhea disorders that can even turn fatal. Furthermore, GI infections increase the likelihood of IBS (Moré and Swidsinski 2015). Studies have consistently shown that physical activity influences the abundance of specific microbial taxa in the gut. For instance, increased exercise has been associated with a rise in the affluence of beneficial bacteria, including but not limited to members of the *Fermicutes* and *Bacteroidetes phyla* (Monda et al. 2017).

10.4 FMT Procedure

FMT, also known as fecal transplant, is a therapeutic intervention that involves the transfer of fecal material from a healthy donor to a recipient, aiming to restore a balanced and diverse gut microbiota. The rationale behind FMT lies in the critical role of the gut microbiota in maintaining health and its association with various diseases when dysregulated.

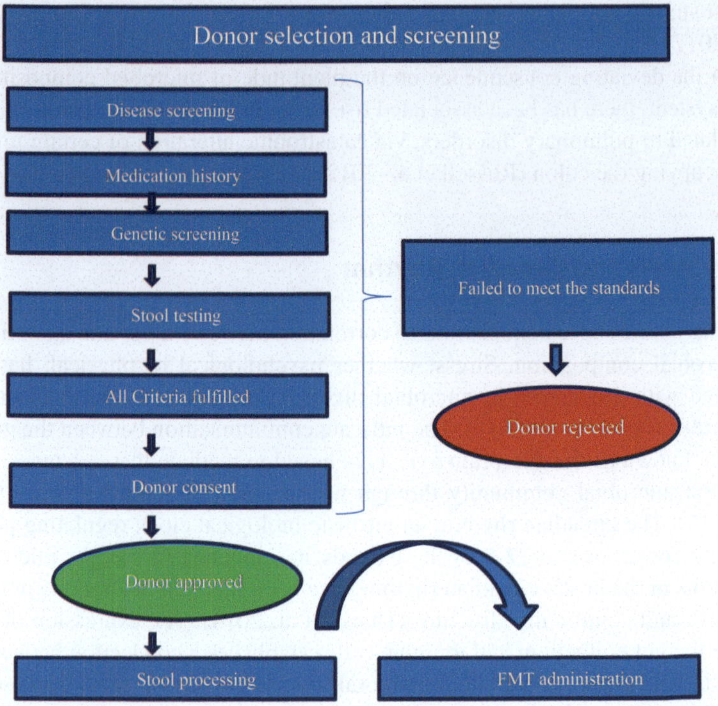

Fig. 10.2 Outline showing the chronological guidelines to be followed before FMT administration

10.4.1 FMT Donor and Selection

The donor selection process typically follows a set of meticulous criteria to minimize potential risks and optimize the chances of successful outcome. Several key considerations in the donor and selection process are shown in Fig. 10.2.

10.4.1.1 Screening for Infectious Diseases

The selection of healthy donors is crucial to ensure the safety and efficacy of FMT. Donors should undergo rigorous screening including but not limited to HIV, hepatitis, and various bacterial, viral, and parasitic infections. This is essential to prevent the transmission of potentially harmful pathogens during FMT (Brandt 2013). Donors may also be asked for urinalysis and oral medical history. If the volunteer's examination indicates recent antibiotic treatment, controlled substance use, and autoimmune conditions or has a family background in cancer, they may not be considered a viable fecal donor (Cleveland Clinic Fecal Transplant 2023). There has been raising attention on strengthening the healthcare system by improving the routes of FMT delivery encouraging stool donation and guaranteeing donor screening and management (Hocking et al. 2023). A thorough assessment of their gastrointestinal health is essential to ensure the absence of any gastrointestinal disorders or conditions that may impact the composition of their microbiota (Bakken et al. 2011).

10.4.1.2 Evaluation of Medication History and Lifestyle

Donors should provide a detailed medication history, and those with a background of antibiotic use known to alter the gut should be carefully considered, as these factors can influence the microbial composition (Paramsothy et al. 2019). Lifestyle factors, such as diet and exercise, can significantly impact the gut microbiota. Donors with a balanced and diverse diet, rich in fibers, and varied nutrients are preferred (David et al. 2014).

10.4.1.3 Genetic Screening for Microbial Dysbiosis

Some studies suggest that the genetic makeup of donors may play a distinct role in the success of FMT (Goodrich et al. 2014). Genetic screening enables a comprehensive analysis of microbial DNA, identifying key taxa, and their genetic variations associated with dysbiosis (Belizário and Napolitano 2015). Moreover, this approach facilitates the prediction of FMT outcomes by assessing the compatibility between donor and recipient microbiota. Understanding the genetic makeup of both parties enhances the precision of FMT, minimizing the risk of adverse events (Bajaj et al. 2017).

10.4.1.4 Donor Stool Testing

Fecal testing in FMT is a meticulous and critical process that profoundly influences the success and safety of the procedure. Excreta profile allows for the comprehensive analysis of microbial diversity and composition. Studies emphasize that a diverse and balanced microbial community in the donor sample positively correlates with the positive outcomes of FMT (Li et al. 2016). Tailoring donor stool selection to match the specific needs of the recipient is a growing area of interest, reducing the risk of adverse events of the technique (Ott et al. 2017). Advance sequencing techniques, such as 16S rRNA gene sequencing, can provide detailed insights in delivering the profile of microbial composition with consequent selection of well-balanced and diverse microbiota (Kang et al. 2017).

10.4.1.5 Consent and Donor Education

Obtaining informed consent is crucial to ensure that the recipient fully understands the procedure, associated risks, and potential benefits. The consent process should include a detailed explanation of the FMT procedure, potential side effects, and the experimental nature of the treatment (Bakken et al. 2011; DeFilipp et al. 2019). Probable risks include transfer of unknown pathogens leading to negative transformation of the recipient's immune response, and the theoretical risk of long-term complications. Providing recipients with a detailed overview of potential risks allow making well-informed decisions about participating in FMT (Cammarota et al. 2019). Reference to a standardized consent form, as suggested by the FDA, helps maintain consistency and clarity in conveying information to the recipient (Fig. 10.2).

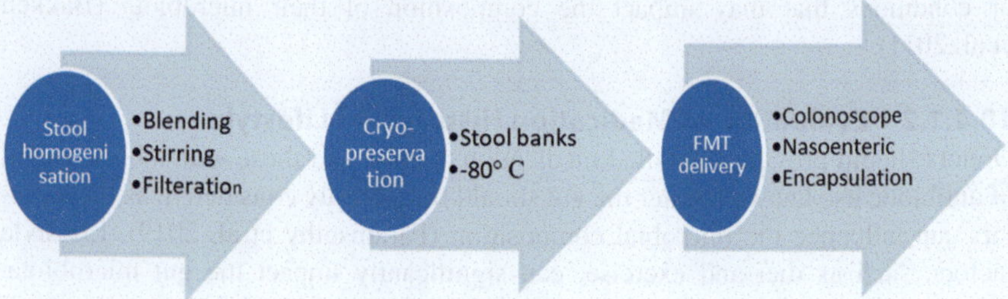

Fig. 10.3 Process of FMT delivery from stool processing to route of administration

10.4.2 Stool Processing and Administration of FMT

Once qualified, donors provide stool samples for FMT. Stool collection methods may involve a combination of self-collection or supervised collection in a clinical setting. Donors are instructed to collect a representative sample that reflects their typical microbial composition (Fig. 10.3) (Cammarota et al. 2019; Kelly et al. 2016).

10.4.2.1 Homogenization of Stool Material

After collection, stool material (minimum quantity of 50 g) undergoes homogenization to create a consistent and well-mixed suspension. Various techniques, including blending, stirring, or mechanical homogenization, are employed to ensure an even distribution of microbial content throughout the sample. The homogenized stool material may undergo dilution to achieve a standardized concentration suitable for administration. Filtration methods, such as using mesh screens or centrifugation, are often employed to remove solid particles, ensuring a smoother and more uniform fecal suspension (Cammarota et al. 2015; Costello et al. 2019; Kim et al. 2019).

Rigorous quality control measures are guaranteed to assess the viability and safety of the microbial community. Microbiota analysis, through techniques like 16S rRNA sequencing, metagenomics, or shotgun sequencing, provides insights into composition and diversity of the donor stool (Li et al. 2016; Cammarota et al. 2019).

10.4.2.2 Cryopreservation and Storage

After the preparation of the fecal suspension, cryopreservation is employed to freeze and store the microbial content effectively. Emphasis on the importance of considering storage duration is crucial highlighting the need for standardized protocols for optimal long-term preservation. Studies suggest that frozen at ultra-low temperatures, typically below −80 °C, is effective in maintaining microbial diversity over an extended period (Hamilton et al. 2012; Youngster et al. 2014). The establishment of stool banks has become a significant development in FMT, allowing for standardized procedures and improved accessibility to screen and well-characterized fecal material. Stool banks follow rigorous protocols for donor screening, sample processing, and storage to ensure consistency and safety across FMT procedures (Alang and Kelly 2015).

10.4.2.3 Different Administration Process

A commonly used method involves delivering fecal material directly into the colon using a colonoscopy allowing targeted placement of the microbial aliquot in the lower GI tract, often leading to successful microbial engraftment (Kao et al. 2017; Cammarota et al. 2017). Second approach comprises of encapsulation of fecal material into oral capsules offering a noninvasive alternative to the recipient. Due to the simplicity and patient acceptance, capsule administration has gained a lot of popularity recently indicating comparable efficacy to colonoscopy delivery, particularly in recurrent *Clostridium difficile* infection cases (Hvas et al. 2019; Youngster et al. 2014). Another method is the nasogastric or nasoenteric tube delivery approach in which a tube is inserted through the patient's nose and advanced into the stomach or further into small intestine, depending on the target location for microbial delivery and the prepared fecal suspension is then infused through the tube (Zhang et al. 2018; Tariq et al. 2019; Feuerstadt et al. 2022). Fecal aliquot can also be infused into the upper GI permitting precise placement of the microbial concoction in the upper digestive system. This procedure is typically performed under endoscopic guidance, often utilizing a gastroscopy aiming to restore a balanced gut microbiota (Fig. 10.3) (Ianiro et al. 2016; Allegretti et al. 2020).

In some cases, pre-antibiotic treatment regimen is followed before FMT delivery. This approach aims to reduce the abundance of potentially interfering bacteria in the recipient's gut; however, the course of medication has to be tapered just prior to FMT, creating a more favourable environment for engraftment of the donor. The results of an investigation indicate that one variable risk factor for a subsequent failure of FMT is not receiving a tapering period of oral vancomycin prior to FMT. Patients were 89% prone to fail if they did not undergo this taper before FMT (Patron et al. 2017).

10.4.3 Applications of FMT

FMT has shown promise in various clinical applications, primarily related to the restoration of a healthy gut microbiota. Some notable implementations of faecal transplant include:

10.4.3.1 *Clostridium difficile* Infection

Data suggest that *Clostridium difficile* infection (inflammation of the colon) is known to affect about half a million populations in United States with every one in six individual suffering from this infection has shown to be endured with its recurrence within 2–8 weeks (Borji et al. 2023). CDI has turned out to be a global setback not only limited to the hospitalized patients but also be troublesome for the society grabbing variable degree of attention amongst different income countries (Monaghan et al. 2022). United States with an incident rate of 101.3 cases for every 100,000 people as of 2020 spent around 6.3 billion USD on healthcare assistance in diagnosis and prevention of CDI (Liu et al. 2023). CDI is a gastrointestinal illness caused by the bacterium *Clostridium difficile*, a gram-positive spore-forming bacterium known to produce toxins that can damage the lining of the colon, leading to a range of symptoms from mild diarrhea to severe colitis. CDI often occurs following the disruption of the normal gut microbiota, typically due to the use of antibiotics. Antibiotics can alter the balance of beneficial bacteria in the colon, providing an opportunity for *C. difficile* to proliferate. The bacterium produces spores that are resistant to harsh environmental conditions, enabling them persists in the environment and spreads easily (Leffler and Lamont 2015; Gerding et al. 2014; Kelly 2012; Khoruts and Sadowsky 2016).

FMT has emerged as a highly effective treatment for recurrent and refractory cases of CDI. By introducing a healthy donor's fecal material into the patient's GI tract, FMT aims to restore the balance of gut microbiota, providing competition against *C. difficile* and suppressing its growth. A landmark randomized controlled trial by van Nood and colleagues demonstrated the superiority of FMT over vancomycin in treating recurrent CDI. The study reported an impressive success rate of over 90% in the FMT group, emphasizing FMT as a highly effective intervention (van Nood et al. 2013).

Cammarota and colleagues conducted a prospective study evaluating the use of FMT in patients with recurrent CDI. The study reported a success rate of 94% after a single FMT procedure, highlighting the sustained efficacy of FMT in preventing CDI recurrence (Cammarota et al. 2017). Hence, it can be concluded that data support the efficacy of FMT as a treatment for recurrent CDI with distinguishable results depending upon the method of delivery and administration frequency producing best outcomes with lower GI endoscopy.

The world's first regulatory approval was recently granted to Biome Bank by the Therapeutic Goods Administration of Australia for their product Biomictra (biologic microbiome) intended to treat CDI. Additionally, Ferring Pharmaceuticals has produced a live human derived fecal drug named RBX2600 (Reboyta), which has recently been legalized by the U.S. FDA (Sasso et al. 2023).

Certain data suggest that patients with recurrent CDI have shown to accumulate various MDRO. However, FMT has proven to be an excellent treatment in ameliorating the burden of ABR genes along with increasing the load of beneficial bacteria (Millan et al. 2016).

10.4.3.2 Ulcerative Colitis

Ulcerative colitis (UC) is a chronic inflammatory bowel disease (IBD) characterized by inflammation and ulceration of the colon and rectum accompanied by abdominal pain, diarrhea, rectal bleeding, and weight loss. The infection is found to be linked with damaged cells lining the colon, defective mucus, or epithelial barrier. Furthermore, it is believed to involve a complex interplay of genetic, environmental, and immune factors (Feng et al. 2023). While UC is not caused by specific pathogens, it is associated with an imbalance in the gut microbiota and dysregulation of the immune response. The microbiome in individuals with UC often exhibits alteration in the abundance and diversity of certain bacteria (Guo et al. 2020; Matsuoka 2021).

FMT with its powerful applicability for microbial restoration and Immune modulation has proven to be a reliable therapeutic alternative in case of UC. Prominent studies have exhibited noteworthy enhancements in terms of clinical manifestations, level of illness, and life satisfaction. A trial demonstrated that patients with UC showed improved clinical and endoscopic outcomes after FMT in a randomized controlled setting demonstrating the promise of the strategy (Rossen et al. 2015). Significant rates of remission in patients with inactive UC have been observed in another groundbreaking study presenting the idea of multicolor intensive FMT (Costello et al. 2019; Paramsothy et al. 2017).

10.4.3.3 Irritable Bowel Syndrome

The gastrointestinal ailment known as irritable bowel syndrome (IBS) is a global problem affecting about 11.2% of adult population. IBS is characterized by recurrent abdominal pain, discomfort, and altered bowel habits without any structural or biochemical abnormalities (El-Salhy and Mazzawi 2018). It is a complex condition with contributing factors, including altered gut motility, visceral hypersensitivity, changes in the gut microbiota. Disruptions in the balance of beneficial and harmful bacteria, along with microbial dysbiosis, may contribute to the symptoms observed in individuals with IBS (Singh et al. 2022; Mazzawi et al. 2021; Halkjær et al. 2023). There is a profound skewness associated with IBS patients accountable for the decreased level of few bacterial groups including *Bacteroidetes, Bifidobacterium,* and *Faecalibacterium* along with subsequent spike in the member of Firmicutes (Rajilić-Stojanović et al. 2011). Restoring normal intestinal homeostasis by FMT may lead to symptomatic relief since the changed intestinal microbiota in IBS patients has been linked to immunological dysfunction and altered neurological function, including lower pain threshold and increased anxiety (Shaikh et al. 2023).

Enterococcus are normal inhabitants of the human GI tract, but certain strains have become problematic due to their ability to acquire resistance to multiple antibiotics, including vancomycin referred to as vancomycin-resistant enterococcus (VRE). VRE is often attributed to acquired genetic elements, specifically the presence of Van genes. Endocarditis (inflammation of inner lining of heart), bacteremia (Bacteria in blood), UTI (urinary tract infections), and intra-abdominal infections are some frequently occurred infections caused due to the interference of VRE. Certain strains of enterococcus particularly *Enterococcus faecalis* and *Enterococcus faecium* are the commonly found culprits in various enterococcal infections (Levitus et al. 2023; Kim et al. 2018).

Study conducted by Stripling et al. (2015) was the first to examine the efficacy of FMT in lowering the burden of VRE in colon. It has demonstrated a positive correlation between fecal VRE dominance and the risk of VRE bacteremia to a multiple of nine times (Stripling et al. 2015). The initial findings of the study supported the FMT's apparent safety, proposing that it may have a long-term effect on VRE colonization that aids in outbreak containment with a success rate of 87.5% (Davido et al. 2019). A study by Hyun et al. (2022) demonstrated the several advantages of

FMT, including reduction in the abundance of antibiotic resistance gene and expression in the gut microbiome. Furthermore, it is found to be effective in MDRO decolonization and composition restoration (Hyun et al. 2022). In a rat model, the synthesis of Lantibiotic (peptide antibiotic) by a four-strain consortium of commensal bacteria, including *Blautia producta*, was shown to lower the susceptibility to VRE (Kim et al. 2019).

10.4.3.4 MDRO Decolonization

Genes responsible for antibiotic resistance are omnipresent in the current scenario of antibiotic exploitation by humans. This decade old practice has resulted in significant selection pressure and became a proven aid in the proliferation of resistant microbial strains in the gut microenvironment. Furthermore, loss of competitive exclusion paved the way to amiable growth and sustenance of disease-causing *Salmonella* despite antibiotic treatment (Modi et al. 2014). Additional studies on mice have depicted the effect of antibiotics in elevating the levels of sialic acid, a monosaccharide, specifically in the gut region that can act as a potent growth stimulant for some opportunistic pathogens like *Salmonella typhimurium* and *Clostridium difficile* (Ng et al. 2013).

Multiple drug resistance (MDR) refers to the ability of microorganisms, such as bacteria, to withstand the effects of multiple antimicrobial drugs. ESKAPE (*Enterococcus faecium, Staphylococcus aureus, Klebsiella pneumoniae, Acinetobacter baumannii, Pseudomonas aeruginosa, and Enterobacter species*) is the acronym used for the list of resistant bacteria flagged for the stringent requirement of developing novel therapies. This phenomenon poses a significant challenge in healthcare settings as it limits the effectiveness of traditional antibiotic treatments. MDR strains can persist in the human body, leading to prolonged infections and increased difficulty in managing and treating diseases (Ghani et al. 2022). Apart from being a possible home for MDROs like VRE, and Carbapenemase-producing enterobacteriacea (CPE), the gut microbiome also functions as a repository for antibiotic resistance genes (Hyun et al. 2022).

10.5 Clinical Evidences

Numerous randomized clinical trials have demonstrated the effectiveness and safety of FMT in recurrent CDI, and recommendations consider FMT as a second-line treatment. The treatment of the recurrent CDI has been shown to have a success rate of around 92%. FMT is being assessed as a preliminary treatment for a number of additional conditions including recurrent CDI and IBS, such as nonalcoholic steatohepatitis, hepatic encephalopathy, obesity, and neurological disorders (Kim et al. 2019). Additionally, it has been observed that faecal donation is safe and beneficial for immunocompromised individuals, including transplant recipients (Table 10.1) (Stripling et al. 2015).

The first case study including a 14-year-old girl with a condition of hemophagocytic lymphohistiocytosis accompanied by resistant *Klebsiella pneumoniae* after

Table 10.1 List of case studies/reports for MDRO decolonization by administration of FMT (Manges et al. 2016; Bilinski et al. 2017; Bahl et al. 2020; Wang et al. 2021)

Case studies	MDRO	Route of administration	Pre-treatment	Post FMT outcomes
1. 60 years old with kidney transplant	ESBL producing *E. coli*	Nasoduodenal tube	None	ESBL *E. coli* free for 3 months
2. 13-year-old girl with otomastoiditis and multiple conditions	Carbapenemase producing *K. pneumoniae*	Nasoduodenal tube	None	Carbapenemase producing *K. pneumoniae free for 8 months*
3. Five patients with conditions like pancreatic cancer, Crohn's disease	MRSA enterocolitis	Nasojejunal tube	Vancomycin	No MRSA strains for 3 months
4. Ventilator-ridden 66-year-old quadriplegic	Carbapenemase producing *P. aeruginosa* and VRE	Colonoscope	Vancomycin	Reduction in MDR infections
5. 33-year-old female with kidney and cardiac transplant	VRE, recurrent CDI	Nasogastric tube	Vancomycin	Recurrent CDI and VRE free for a year
6. 11 patients VRE	VRE	RBX2660 via enema	–	73% success rate, VRE free for 6 months to a year
7. 82-year-old woman	OXA-48 carbapenemase producing *K. pneumoniae*	Nasojejunal tube	Oral gentamicin and colomycin	Free from carbapenemase producing *K. pneumoniae*
8. 20 participants with some affected by neutropenia	Antibiotic-resistant bacteria (ARB)	Intraduodenal	None	75% complete ARB decolonization
9. 69-year-old female with multiple organ failure and recurrent CDI	Carbapenemase producing *K. pneumoniae*	Colonoscope	Fidaxomicin	Negative MDRO for a week with follow-up duration over 2 years
10. 37-year-old woman with renal transplant	Carbapenemase producing *K. pneumoniae*	Nasogastric tube	–	Negative for carbapenemase producing *K. pneumoniae*

being administered by FMT demonstrated a noteworthy decolonization of *Klebsiella pneumoniae*, depicting the efficacy of the therapy in treating recurring clinical infections (Neuroscience 2024). Case study of 60-year-old male who underwent renal allo-transplantation twice, later suffered with recurrent episodes of transplant pyelonephritis caused by an Extended Spectrum β-Lactamase (ESBL) producing *Escherichia coli* when given FMT at week two the rectal culture became negative for ESBL and the subsequent rectal cultures remained negative hereafter during 12 weeks of follow-up. During this same follow-up period, the patient did not develop any symptoms of urinary tract infection (Singh et al. 2014). Another case study of a 66-year-old male accompanied with a load of VRE, MDR *Enterobacteriaceae*, and carbapenem-resistant pseudomonas witnessed a prominent improvement in residing symptoms of sepsis and infections along with notable reduction in MDRO colonization (Crum-Cianflone et al. 2015). A paper published by Manges and associates summarizes many case studies demonstrating the efficacy of FMT in eliminating MDRO, examining the application of FMT for MDR bacterial decontamination in intestine with no adverse effects. Following is a list of case studies along with others (Manges et al. 2016).

10.6 Conclusions

In conclusion, the intricate world of gut microbes plays a pivotal role in maintains our health and influencing various physiological processes. The balance and diversity of these microbes contribute to the proper functioning of the immune system, digestion, and overall well-being. However, when this delicate balance is disrupted, giving rise to dysbiosis, a spectrum of health issues can emerge. FMT has emerged as a groundbreaking therapeutic approach, leveraging the potential of the gut microbiome in treating a spectrum of conditions, ranging from recurrent CDI to antibiotic-resistant strains. The success of FMT in reshaping the gut microbiota and achieving positive clinical outcomes underscores its transformative impact on healthcare. As research continues to unravel the complexities of the gut microbiome and refine FMT protocols, the future prospects of this technique are promising. FMT not only holds potential for addressing a diverse array of gastrointestinal disorders but also offers opportunities to explore its applications in other systemic conditions, opening new avenues for personalized and microbiome-based medicine. As we delve deeper into the dynamic interplay between health and the microbial communities within us, FMT stands at the forefront as a revolutionary tool with the potential to redefine treatment paradigms and improve patient outcomes in the years to come.

References

Alang N, Kelly CR (2015) Weight gain after fecal microbiota transplantation. Open Forum Infect Dis 2(1):ofv004. https://doi.org/10.1093/ofid/ofv004

Allegretti JR, Kassam Z, Mullish BH, Chiang A, Carrellas M, Hurtado J, Marchesi JR, McDonald JAK, Pechlivanis A, Barker GF, Miguéns Blanco J, Garcia-Perez I, Wong WF, Gerardin Y, Silverstein M, Kennedy K, Thompson C (2020) Effects of fecal microbiota transplantation with oral capsules in obese patients. Clin Gastroenterol Hepatol 18(4):855–863.e2. https://doi.org/10.1016/j.cgh.2019.07.006

Bahl MI, Jørgensen SMD, Skriver AH (2020) Faecal microbiota transplantation for eradication of co-infection with *Clostridioides difficile* and extensively drug-resistant KPC-producing *Klebsiella pneumoniae*. Scand J Gastroenterol 55:626–630

Bajaj JS, Kassam Z, Fagan A, Gavis EA, Liu E, Cox IJ, Kheradman R, Heuman D, Wang J, Gurry T, Williams R, Sikaroodi M, Fuchs M, Alm E, John B, Thacker LR, Riva A, Smith M, Taylor-Robinson SD, Gillevet PM (2017) Fecal microbiota transplant from a rational stool donor improves hepatic encephalopathy: a randomized clinical trial. Hepatology (Baltimore, Md.) 66(6):1727–1738. https://doi.org/10.1002/hep.29306

Bakken JS, Borody T, Brandt LJ, Brill JV, Demarco DC, Franzos MA, Kelly C, Khoruts A, Louie T, Martinelli LP, Moore TA, Russell G, Surawicz C, Fecal Microbiota Transplantation Workgroup (2011) Treating *Clostridium difficile* infection with fecal microbiota transplantation. Clin Gastroenterol Hepatol 9(12):1044–1049. https://doi.org/10.1016/j.cgh.2011.08.014

Baunwell SMD, Terveer EM, Dahlerup JF, Erikstrup C, Arkkila P, Vehreschid MJ (2021) The use of Faecal Microbiota Transplantation (FMT) in Europe: a Europe-wide survey. Lancet Reg Health Eur 9:100181. https://doi.org/10.1016/j.lanepe.2021.100181

Belizário JE, Napolitano M (2015) Human microbiomes and their roles in dysbiosis, common diseases, and novel therapeutic approaches. Front Microbiol 6:1050. https://doi.org/10.3389/fmicb.2015.01050

Biazzo M, Deidda G (2022) Fecal microbiota transplantation as new therapeutic avenue for human diseases. J Clin Med 11(14):4119. https://doi.org/10.3390/jcm11144119. PMID: 35887883; PMCID: PMC9320118

Bilinski J, Grzesiowski P, Sorensen N, Madry K, Muszynski J, Robak K, Basak GW (2017) Fecal microbiota transplantation in patients with blood disorders inhibits gut colonization with antibiotic-resistant bacteria: results of a prospective, single-center study. Clin Infect Dis 65(3):364–370. https://doi.org/10.1093/cid/cix252

Borji S, Kadivarian S, Dashtbin S et al (2023) Global prevalence of *Clostridioides difficile* in 17,148 food samples from 2009 to 2019: a systematic review and meta-analysis. J Health Popul Nutr 42:36. https://doi.org/10.1186/s41043-023-00369-3

Borody T, Khoruts A (2012) Fecal microbiota transplantation and emerging applications. Nat Rev Gastroenterol Hepatol 9:88–96. https://doi.org/10.1038/nrgastro.2011.244

Brandt LJ (2013) American Journal of Gastroenterology Lecture: intestinal microbiota and the role of fecal microbiota transplant (FMT) in treatment of *C. difficile* infection. Am J Gastroenterol 108(2):177–185. https://doi.org/10.1038/ajg.2012.450

Bravo JA, Forsythe P, Chew MV, Escaravage E, Savignac HM, Dinan TG, Bienenstock J, Cryan JF (2011) Ingestion of lactobacillus strain regulates emotional behavior and central GABA receptor expression in a mouse via the vagus nerve. Proc Natl Acad Sci USA 108(38):16050–16055. https://doi.org/10.1073/pnas.1102999108

Cammann D, Lu Y, Cummings MJ et al (2023) Genetic correlations between Alzheimer's disease and gut microbiome genera. Sci Rep 13:5258. https://doi.org/10.1038/s41598-023-31730-5

Cammarota G, Masucci L, Ianiro G, Bibbò S, Dinoi G, Costamagna G, Sanguinetti M, Gasbarrini A (2015) Randomised clinical trial: faecal microbiota transplantation by colonoscopy vs. vancomycin for the treatment of recurrent *Clostridium difficile* infection. Aliment Pharmacol Ther 41(9):835–843. https://doi.org/10.1111/apt.13144

Cammarota G, Ianiro G, Tilg H, Rajilić-Stojanović M, Kump P, Satokari R, Sokol H, Arkkila P, Pintus C, Hart A, Segal J, Aloi M, Masucci L, Molinaro A, Scaldaferri F, Gasbarrini G, Lopez-Sanroman A, Link A, de Groot P, de Vos WM, European FMT Working Group (2017) European consensus conference on faecal microbiota transplantation in clinical practice. Gut 66(4):569–580. https://doi.org/10.1136/gutjnl-2016-313017

Cammarota G, Ianiro G, Kelly CR et al (2019) International consensus conference on stool banking for faecal microbiota transplantation in clinical practice. Gut 68(12):2111–2121. https://doi.org/10.1136/gutjnl-2019-319548. PMID: 31563878; PMCID: PMC6872442

Cleveland Clinic Fecal Transplant (2023). https://my.clevelandclinic.org/health/treatments/25202-fecal-transplant

Costello SP, Hughes PA, Waters O, Bryant RV, Vincent AD, Blatchford P, Katsikeros R, Makanyanga J, Campaniello MA, Mavrangelos C, Rosewarne CP, Bickley C, Peters C, Schoeman MN, Conlon MA, Roberts-Thomson IC, Andrews JM (2019) Effect of fecal microbiota transplantation on 8-week remission in patients with ulcerative colitis: a randomized clinical trial. JAMA 321(2):156–164. https://doi.org/10.1001/jama.2018.20046

Crum-Cianflone NF, Sullivan E, Ballon-Landa G (2015) Fecal microbiota transplantation and successful resolution of multidrug-resistant-organism colonization. J Clin Microbiol 53:1986. https://doi.org/10.1128/jcm.00820-15

Daisy Coyle APD (2017). https://www.healthline.com/nutrition/8-things-that-harm-gut-bacteria

Dale HF, Lied GA (2020) Gut microbiota and therapeutic approaches for dysbiosis in irritable bowel syndrome: recent developments and future perspectives. Turk J Med Sci 50(SI-2):1632–1641. https://doi.org/10.3906/sag-2002-57

Daniel H, Gholami AM, Berry D, Desmarchelier C, Hahne H, Loh G, Mondot S, Lepage P, Rothballer M, Walker A, Böhm C, Wenning M, Wagner M, Blaut M, Schmitt-Kopplin P, Kuster B, Haller D, Clavel T (2014) High-fat diet alters gut microbiota physiology in mice. ISME J 8(2):295–308. https://doi.org/10.1038/ismej.2013.155. Epub 2013 Sep 12. PMID: 24030595; PMCID: PMC3906816

David LA, Maurice CF, Carmody RN, Gootenberg DB, Button JE, Wolfe BE, Ling AV, Devlin AS, Varma Y, Fischbach MA, Biddinger SB, Dutton RJ, Turnbaugh PJ (2014) Diet rapidly and reproducibly alters the human gut microbiome. Nature 505(7484):559–563. https://doi.org/10.1038/nature12820

Davido B, Batista R, Fessi H, Michelon H, Escaut L, Lawrence C, Denis M, Perronne C, Salomon J, Dinh A (2019) Fecal microbiota transplantation to eradicate vancomycin-resistant enterococci colonization in case of an outbreak. Med Maladies Infectieuses 49(3):214–218. https://doi.org/10.1016/j.medmal.2018.11.002

DeFilipp Z, Bloom PP, Torres Soto M, Mansour MK, Sater MRA, Huntley MH, Turbett S, Chung RT, Chen YB, Hohmann EL (2019) Drug-resistant *E. coli* bacteremia transmitted by fecal microbiota transplant. N Engl J Med 381(21):2043–2050. https://doi.org/10.1056/NEJMoa1910437

den Besten G, van Eunen K, Groen AK, Venema K, Reijngoud DJ, Bakker BM (2013) The role of short-chain fatty acids in the interplay between diet, gut microbiota, and host energy metabolism. J Lipid Res 54(9):2325–2340. https://doi.org/10.1194/jlr.R036012. Epub 2013 Jul 2. PMID: 23821742; PMCID: PMC3735932

El-Salhy M, Mazzawi T (2018) Fecal microbiota transplantation for managing irritable bowel syndrome. Expert Rev Gastroenterol Hepatol 12(5):439–445. https://doi.org/10.1080/17474124.2018.1447380

Feng J, Chen Y, Liu Y et al (2023) Efficacy and safety of fecal microbiota transplantation in the treatment of ulcerative colitis: a systematic review and meta-analysis. Sci Rep 13:14494. https://doi.org/10.1038/s41598-023-41182-6

Feuerstadt P, Aroniadis OC, Svedlund FL, Garcia M, Stong L, Boules M, Khanna S (2022) Heterogeneity of randomized controlled trials of fecal microbiota transplantation in recurrent *Clostridioides difficile* infection. Dig Dis Sci 67(7):2763–2770. https://doi.org/10.1007/s10620-021-07141-9

Foster JA, Rinaman L, Cryan JF (2017) Stress and the gut–brain axis: regulation by the microbiome. Neurobiol Stress 7:124–136. https://doi.org/10.1016/j.ynstr.2017.03.001

Gerding DN, Johnson S, Rupnik M, Aktories K (2014) *Clostridium difficile* binary toxin CDT: mechanism, epidemiology, and potential clinical importance. Gut Microbes 5(1):15–27. https://doi.org/10.4161/gmic.26854

Ghani R, Mullish BH, Roberts LA, Davies FJ, Marchesi JR (2022) The potential utility of fecal (or intestinal) microbiota transplantation in controlling infectious diseases. Gut Microbes 14(1):2038856. https://doi.org/10.1080/19490976.2022.2038856

Goodrich JK, Waters JL, Poole AC, Sutter JL, Koren O, Blekhman R, Beaumont M, Van Treuren W, Knight R, Bell JT, Spector TD, Clark AG, Ley RE (2014) Human genetics shape the gut microbiome. Cell 159(4):789–799. https://doi.org/10.1016/j.cell.2014.09.053

Guo XY, Liu XJ, Hao JY (2020) Gut microbiota in ulcerative colitis: insights on pathogenesis and treatment. J Dig Dis 21(3):147–159. https://doi.org/10.1111/1751-2980.12849

Gupta S, Allen-Vercoe E, Petrof EO (2016) Fecal microbiota transplantation: in perspective. Ther Adv Gastroenterol 9(2):229–239. https://doi.org/10.1177/1756283X15607414

Halkjær SI, Lo B, Cold F, Højer Christensen A, Holster S, König J, Brummer RJ, Aroniadis OC, Lahtinen P, Holvoet T, Gluud LL, Petersen AM (2023) Fecal microbiota transplantation for the treatment of irritable bowel syndrome: a systematic review and meta-analysis. World J Gastroenterol 29(20):3185–3202. https://doi.org/10.3748/wjg.v29.i20.3185

Hamilton MJ, Weingarden AR, Sadowsky MJ, Khoruts A (2012) Standardized frozen preparation for transplantation of fecal microbiota for recurrent *Clostridium difficile* infection. Am J Gastroenterol 107(5):761–767. https://doi.org/10.1038/ajg.2011.482

Hocking L, Ianiro G, Leong RW, Iqbal T, Kao D, Cabling M, Stockwell S, Romanelli RJ, Marjanovic S (2023) Faecal microbiota transplantation for recurrent *C. difficile* infections: challenges and improvement opportunities for clinical practice and healthcare systems. Aliment Pharmacol Ther 57(5):549–564. https://doi.org/10.1111/apt.17309

Hvas CL et al (2019) Fecal microbiota transplantation is superior to fidaxomicin for treatment of recurrent *Clostridium difficile* infection. Gastroenterology 156(5):1324–1332

Hyun J, Lee SK, Cheon JH et al (2022) Faecal microbiota transplantation reduces amounts of antibiotic resistance genes in patients with multidrug-resistant organisms. Antimicrob Resist Infect Control 11:20. https://doi.org/10.1186/s13756-022-01064-4

Ianiro G, Tilg H, Gasbarrini A (2016) Antibiotics as deep modulators of gut microbiota: between good and evil. Gut 65(11):1906–1915. https://doi.org/10.1136/gutjnl-2016-312297

Johnson KV (2020) Gut microbiome composition and diversity are related to human personality traits. Hum Microbiome J 15:100069. https://doi.org/10.1016/j.humic.2019.100069

Kang DW, Adams JB, Gregory AC, Borody T, Chittick L, Fasano A, Khoruts A, Geis E, Maldonado J, McDonough-Means S, Pollard EL, Roux S, Sadowsky MJ, Lipson KS, Sullivan MB, Caporaso JG, Krajmalnik-Brown R (2017) Microbiota transfer therapy alters gut ecosystem and improves gastrointestinal and autism symptoms: an open-label study. Microbiome 5(1):10. https://doi.org/10.1186/s40168-016-0225-7

Kao D, Roach B, Silva M, Beck P, Rioux K, Kaplan GG, Chang HJ, Coward S, Goodman KJ, Xu H, Madsen K, Mason A, Wong GK, Jovel J, Patterson J, Louie T (2017) Effect of oral capsule- vs colonoscopy-delivered fecal microbiota transplantation on recurrent *Clostridium difficile* infection: a randomized clinical trial. JAMA 318(20):1985–1993. https://doi.org/10.1001/jama.2017.17077

Kelly CP (2012) Can we identify patients at high risk of recurrent *Clostridium difficile* infection? Clin Microbiol Infect 18(Suppl 6):21–27. https://doi.org/10.1111/1469-0691.12046

Kelly JR, Kennedy PJ, Cryan JF, Dinan TG, Clarke G, Hyland NP (2015) Breaking down the barriers: the gut microbiome, intestinal permeability and stress-related psychiatric disorders. Front Cell Neurosci 9:392. https://doi.org/10.3389/fncel.2015.00392

Kelly CR, Khoruts A, Staley C, Sadowsky MJ, Abd M, Alani M, Bakow B, Curran P, McKenney J, Tisch A, Reinert SE, Machan JT, Brandt LJ (2016) Effect of fecal microbiota transplantation on recurrence in multiply recurrent *Clostridium difficile* infection: a randomized trial. Ann Intern Med 165(9):609–616. https://doi.org/10.7326/M16-0271

Khoruts A, Sadowsky M (2016) Understanding the mechanisms of faecal microbiota transplantation. Nat Rev Gastroenterol Hepatol 13(8):508–516. https://doi.org/10.1038/nrgastro.2016.98

Kim HS, Kim DH, Yoon HJ, Lee WJ, Woo SH, Choi SP (2018) Factors associated with vancomycin-resistant enterococcus colonization in patients transferred to emergency departments in Korea. J Korean Med Sci 33(48):e295. https://doi.org/10.3346/jkms.2018.33.e295

Kim SG, Becattini S, Moody TU, Shliaha PV, Littmann ER, Seok R, Gjonbalaj M, Eaton V, Fontana E, Amoretti L, Wright R, Caballero S, Wang ZX, Jung HJ, Morjaria SM, Leiner IM, Qin W, Ramos RJJF, Cross JR, Narushima S, Honda K, Peled JU, Hendrickson RC, Taur Y, van den Brink MRM, Pamer EG (2019) Microbiota-derived lantibiotic restores resistance against vancomycin-resistant enterococcus. Nature 572(7771):665–669. https://doi.org/10.1038/s41586-019-1501-z

Leffler DA, Lamont JT (2015) *Clostridium difficile* infection. N Engl J Med 372(16):1539–1548. https://doi.org/10.1056/NEJMra1403772

Levitus M, Rewane A, Perera TB (2023) Vancomycin-resistant Enterococci. In: StatPearls [Internet]. StatPearls Publishing, Treasure Island (FL). https://www.ncbi.nlm.nih.gov/books/NBK513233/

Li SS, Zhu A, Benes V, Costea PI, Hercog R, Hildebrand F, Huerta-Cepas J, Nieuwdorp M, Salojärvi J, Voigt AY, Zeller G, Sunagawa S, de Vos WM, Bork P (2016) Durable coexistence of donor and recipient strains after fecal microbiota transplantation. Science 352(6285):586–589. https://doi.org/10.1126/science.aad8852

Liu C, Monaghan T, Yadegar A, Louie T, Kao D (2023) Insights into the evolving epidemiology of *Clostridioides difficile* infection and treatment: a global perspective. Antibiotics (Basel) 12(7):1141. https://doi.org/10.3390/antibiotics12071141

Manges AR, Steiner TS, Wright AJ (2016) Fecal microbiota transplantation for the intestinal decolonization of extensively antimicrobial-resistant opportunistic pathogens: a review. Infect Dis (Lond) 48(8):587–592. https://doi.org/10.1080/23744235.2016.1177199

Marchesi JR, Adams DH, Fava F, Hart A (2016) The gut microbiota and host health: a new clinical frontier. Gut 65(2):330–339. https://doi.org/10.1136/gutjnl-2015-309990

Matsuoka K (2021) Fecal microbiota transplantation for ulcerative colitis. Immunol Med 44(1):30–34. https://doi.org/10.1080/25785826.2020.1792040

Mazzawi T, El-Salhy M, Lied GA, Hausken T (2021) The effects of fecal microbiota transplantation on the symptoms and the duodenal Neurogenin 3, Musashi 1, and Enteroendocrine cells in patients with diarrhea-predominant irritable bowel syndrome. Front Cell Infect Microbiol 11:524851. https://doi.org/10.3389/fcimb.2021.524851

Millan B, Park H, Hotte N, Mathieu O, Burguiere P, Tompkins TA, Kao D, Madsen KL (2016) Fecal microbial transplants reduce antibiotic-resistant genes in patients with recurrent *Clostridium difficile* infection. Clin Infect Dis 62(12):1479–1486. https://doi.org/10.1093/cid/ciw185

Modi SR, Collins JJ, Relman DA (2014) Antibiotics and the gut microbiota. J Clin Invest 124(10):4212–4218. https://doi.org/10.1172/JCI72333

Monaghan TM, Biswas R, Satav A et al (2022) *Clostridioides difficile* epidemiology in India. Anaerobe 74:102517. https://doi.org/10.1016/j.anaerobe.2022.102517

Monda V, Villano I, Messina A, Valenzano A, Esposito T, Moscatelli F, Viggiano A, Cibelli G, Chieffi S, Monda M, Messina G (2017) Exercise modifies the gut microbiota with positive health effects. Oxidative Med Cell Longev 2017:3831972. https://doi.org/10.1155/2017/3831972

Moré MI, Swidsinski A (2015) *Saccharomyces boulardii* CNCM I-745 supports regeneration of the intestinal microbiota after diarrheic dysbiosis—a review. Clin Exp Gastroenterol 8:237–255. https://doi.org/10.2147/CEG.S85574

Neuroscience (2024) The gut microbiome and its impact on the brain. http://neuroscience.openetext.utoronto.ca/chapter/chapter-1-the-gut-microbiome-and-its-impact-on-the-brain/

Ng KM, Ferreyra JA, Higginbottom SK, Lynch JB, Kashyap PC, Gopinath S, Naidu N, Choudhury B, Weimer BC, Monack DM, Sonnenburg JL (2013) Microbiota-liberated host sugars facilitate post-antibiotic expansion of enteric pathogens. Nature 502(7469):96–99. https://doi.org/10.1038/nature12503

Ott SJ, Waetzig GH, Rehman A, Moltzau-Anderson J, Bharti R, Grasis JA, Cassidy L, Tholey A, Fickenscher H, Seegert D, Rosenstiel P, Schreiber S (2017) Efficacy of sterile fecal filtrate transfer for treating patients with *Clostridium difficile* infection. Gastroenterology 152(4):799–811.e7. https://doi.org/10.1053/j.gastro.2016.11.010

Paramsothy S, Kamm MA, Kaakoush NO, Walsh AJ, van den Bogaerde J, Samuel D, Leong RWL, Connor S, Ng W, Paramsothy R, Xuan W, Lin E, Mitchell HM, Borody TJ (2017) Multidonor intensive faecal microbiota transplantation for active ulcerative colitis: a randomised placebo-controlled trial. Lancet 389(10075):1218–1228. https://doi.org/10.1016/S0140-6736(17)30182-4

Paramsothy S, Nielsen S, Kamm MA, Deshpande NP, Faith JJ, Clemente JC, Paramsothy R, Walsh AJ, van den Bogaerde J, Samuel D, Leong RWL, Connor S, Ng W, Lin E, Borody TJ, Wilkins MR, Colombel JF, Mitchell HM, Kaakoush NO (2019) Specific bacteria and metabolites associated with response to fecal microbiota transplantation in patients with ulcerative colitis. Gastroenterology 156(5):1440–1454.e2. https://doi.org/10.1053/j.gastro.2018.12.001

Patron RL, Hartmann CA, Allen S, Orenstein R (2017) Vancomycin taper and risk of failure of fecal microbiota transplantation in patients with recurrent *Clostridium difficile* infection. Clin Infect Dis 65(7):1214–1217. https://doi.org/10.1093/cid/cix511

Rajilić-Stojanović M, Biagi E, Heilig HG, Kajander K, Kekkonen RA, Tims S, de Vos WM (2011) Global and deep molecular analysis of microbiota signatures in fecal samples from patients with irritable bowel syndrome. Gastroenterology 141(5):1792–1801. https://doi.org/10.1053/j.gastro.2011.07.043

Rogers GB, Keating DJ, Young RL, Wong ML, Licinio J, Wesselingh S (2016) From gut dysbiosis to altered brain function and mental illness: mechanisms and pathways. Mol Psychiatry 21(6):738–748. https://doi.org/10.1038/mp.2016.50. Epub 2016 Apr 19. PMID: 27090305; PMCID: PMC4879184

Rossen NG, Fuentes S, van der Spek MJ, Tijssen JG, Hartman JH, Duflou A, Löwenberg M, van den Brink GR, Mathus-Vliegen EM, de Vos WM, Zoetendal EG, D'Haens GR, Ponsioen CY (2015) Findings from a randomized controlled trial of fecal transplantation for patients with ulcerative colitis. Gastroenterology 149(1):110–118.e4. https://doi.org/10.1053/j.gastro.2015.03.045

Round JL, Mazmanian SK (2009) The gut microbiota shapes intestinal immune responses during health and disease. Nat Rev Immunol 9(5):313–323. https://doi.org/10.1038/nri2515

Russell SL, Gold MJ, Hartmann M, Willing BP, Thorson L, Wlodarska M, Gill N, Blanchet MR, Mohn WW, McNagny KM, Finlay BB (2012) Early life antibiotic-driven changes in microbiota enhance susceptibility to allergic asthma. EMBO Rep 13(5):440–447. https://doi.org/10.1038/embor.2012.32

Rutsch A, Kantsjö JB, Ronchi F (2020) The gut–brain axis: how microbiota and host inflammasome influence brain physiology and pathology. Front Immunol 11:604179. https://doi.org/10.3389/fimmu.2020.604179

Sasso JM, Ammar RM, Tenchov R, Lemmel S, Kelber O, Grieswelle M, Zhou QA (2023) Gut microbiome–brain alliance: a landscape view into mental and gastrointestinal health and disorders. ACS Chem Neurosci 14(10):1717–1763. https://doi.org/10.1021/acschemneuro.3c00127

Shaikh SD, Sun N, Canakis A, Park WY, Weber HC (2023) Irritable bowel syndrome and the gut microbiome: a comprehensive review. J Clin Med 12(7):2558. https://doi.org/10.3390/jcm12072558

Singh R, van Nood E, Nieuwdorp M, van Dam B, ten Berge IJ, Geerlings SE, Bemelman FJ (2014) Donor feces infusion for eradication of extended spectrum beta-lactamase producing *Escherichia coli* in a patient with end stage renal disease. Clin Microbiol Infect 20(11):O977–O978. https://doi.org/10.1111/1469-0691.12683

Singh P, Alm EJ, Kelley JM, Cheng V, Smith M, Kassam Z, Nee J, Iturrino J, Lembo A (2022) Effect of antibiotic pretreatment on bacterial engraftment after Fecal Microbiota Transplant (FMT) in IBS-D. Gut Microbes 14(1):2020067. https://doi.org/10.1080/19490976.2021.2020067

Stripling J, Kumar R, Baddley JW, Nellore A, Dixon P, Howard D, Ptacek T, Lefkowitz EJ, Tallaj JA, Benjamin WH, Morrow CD, Rodriguez JM (2015) Loss of vancomycin-resistant enterococcus fecal dominance in an organ transplant patient with *Clostridium difficile* colitis after fecal microbiota transplant. Open Forum Infect Dis 2(2):ofv078. https://doi.org/10.1093/ofid/ofv078

Tang L (2022) Sequence-based identification of human-associated microbiota. https://www.nature.com/articles/d42859-019-00011-5

Tariq R, Pardi DS, Bartlett MG, Khanna S (2019) Low cure rates in controlled trials of fecal microbiota transplantation for recurrent *Clostridium difficile* infection: a systematic review and meta-analysis. Clin Infect Dis 68(8):1351–1358. https://doi.org/10.1093/cid/ciy721

Thaiss CA, Zeevi D, Levy M, Zilberman-Schapira G, Suez J, Tengeler AC, Abramson L, Katz MN, Korem T, Zmora N, Kuperman Y, Biton I, Gilad S, Harmelin A, Shapiro H, Halpern Z, Segal E, Elinav E (2014) Transkingdom control of microbiota diurnal oscillations promotes metabolic homeostasis. Cell 159(3):514–529. https://doi.org/10.1016/j.cell.2014.09.048

Ursell LK, Metcalf JL, Parfrey LW, Knight R (2012) Defining the human microbiome. Nutr Rev 70(Suppl 1):S38–S44. https://doi.org/10.1111/j.1753-4887.2012.00493.x. PMID: 22861806; PMCID: PMC3426293

van Nood E, Vrieze A, Nieuwdorp M, Fuentes S, Zoetendal EG, de Vos WM, Visser CE, Kuijper EJ, Bartelsman JF, Tijssen JG, Speelman P, Dijkgraaf MG, Keller JJ (2013) Duodenal infusion of donor feces for recurrent *Clostridium difficile*. N Engl J Med 368(5):407–415. https://doi.org/10.1056/NEJMoa1205037

Vatanen T, Jabbar KS, Ruohtula T, Honkanen J, Avila-Pacheco J, Siljander H, Stražar M, Oikarinen S, Hyöty H, Ilonen J (2022) Mobile genetic elements from the maternal microbiome shape infant gut microbial assembly and metabolism. Cell 185:4921–4936. https://doi.org/10.1016/j.cell.2022.11.023

Voigt RM, Forsyth CB, Green SJ, Mutlu E, Engen P, Vitaterna MH, Turek FW, Keshavarzian A (2014) Circadian disorganization alters intestinal microbiota. PLoS One 9(5):e97500. https://doi.org/10.1371/journal.pone.0097500

Wang J, Li X, Wu X, Wang Z, Wu X, Wang S, Jing G, Yan T (2021) Fecal microbiota transplantation as an effective treatment for Carbapenem-resistant *Klebsiella pneumoniae* infection in a renal transplant patient. Infect Drug Resist 14:1805–1811. https://doi.org/10.2147/IDR.S308308

Wang Y, He Y, Liang Y, Liu H, Chen X, Kulyar MF, Shahzad A, Wei K, Li K (2023) Fecal microbiota transplantation attenuates *Escherichia coli* infected outgrowth by modulating the intestinal microbiome. Microb Cell Factories 22(1):30. https://doi.org/10.1186/s12934-023-02027-z

Wen L, Duffy A (2017) Factors influencing the gut microbiota, inflammation, and type 2 diabetes. J Nutr 147(7):1468S–1475S. https://doi.org/10.3945/jn.116.240754. Epub 2017 Jun 14. PMID: 28615382; PMCID: PMC5483960

Wikipedia (2024). https://en.wikipedia.org/wiki/Gut_microbiota#cite_note-10

Wilson BC, Vatanen T, Cutfield WS, O'Sullivan JM (2019) The super-donor phenomenon in fecal microbiota transplantation. Front Cell Infect Microbiol 9:2. https://doi.org/10.3389/fcimb.2019.00002

Xu J, Gordon JI (2003) Honor thy symbionts. Proc Natl Acad Sci U S A 100(18):10452–10459. https://doi.org/10.1073/pnas.1734063100. Epub 2003 Aug 15. PMID: 12923294; PMCID: PMC193582

Youngster I, Russell GH, Pindar C, Ziv-Baran T, Sauk J, Hohmann EL (2014) Oral, capsulized, frozen fecal microbiota transplantation for relapsing *Clostridium difficile* infection. JAMA 312(17):1772–1778. https://doi.org/10.1001/jama.2014.13875

Zhang F, Cui B, He X, Nie Y, Wu K, Fan D, FMT-Standardization Study Group (2018) Microbiota transplantation: concept, methodology and strategy for its modernization. Protein Cell 9(5):462–473. https://doi.org/10.1007/s13238-018-0541-8

Zhang H, Chen Y, Wang Z, Xie G, Liu M, Yuan B, Chai H, Wang W, Cheng P (2022) Implications of gut microbiota in neurodegenerative diseases. Front Immunol 13:785644. https://doi.org/10.3389/fimmu.2022.785644

11 A Journey Beyond Resistance: Discovering the Promise of Non-Antibiotic Therapies

Nitish Kumar Singh, Aprajita Singh, and Raghvendra Dubey

Abstract

For decades, numerous studies have demonstrated the emergence of antimicrobial resistance (AMR) and the limits of traditional antibiotics; in the realm of modern medicine, a quiet revolution is underway, reshaping the landscape of infection treatment: the rise of non-antibiotic advanced (NAA) therapies. These innovative approaches, driven by scientific curiosity, offer a glimpse of a future in which infections are fought with precision and effectiveness. The NAA therapies represent a variety of strategies for fighting microbial infections and investigating the mechanisms of AMR in eukaryotic organisms. To understand the intricacies of NAA therapies, it is important to consider their effectiveness, safety profile, cost effectiveness, and interactions with microbial communities. This chapter provides more information about using NAA therapies, including different modalities, AMR peptides and probiotics, phytochemicals and plant-derived compounds, and CRISPR-Cas system. These NAA therapies offer promising new ways to treat infections and reduce the selective pressure of AMR. Furthermore, this article defines the multidisciplinary nature of non-antibiotic therapies and their regulation, viz., the use of various methods of clinical practices along with the evaluation process. Therefore, with their potential to sidestep the looming specter of antimicrobial resistance, these therapies stand as sentinels of hope, ushering in a new era of infection management rooted in resilience and adaptability. Eventually, this will boost the future infectious disease management paradigm.

N. K. Singh (✉) · R. Dubey
Department of Life and Basic Sciences, Jaipur National University, Jaipur, India
e-mail: nitish.singh@jnujaipur.ac.in

A. Singh
Department of Microbiology, Institute of Medical Sciences, Banaras Hindu University, Varanasi, Uttar Pradesh, India

M. Gangwar, G. Nath (eds.), *Emerging Paradigms for Antibiotic-Resistant Infections: Beyond the Pill*, https://doi.org/10.1007/978-981-97-5272-0_11

Keywords

Antimicrobial resistance · Non-antibiotic advanced therapies · Scientific curiosity · Phytochemicals · CRISPR-Cas system

11.1 Introduction

Innovative methods of treating illnesses without the use of conventional antibiotics are referred to as non-antibiotic advanced (NAA) therapy. In order to battle microbial infections, minimize the danger of antimicrobial resistance (AMR), and lessen the negative consequences of antibiotic use, these therapies incorporate a variety of tactics and technology. Advances in science and medicine have had a significant role in the creation and uptake of novel treatment approaches, even though it is difficult to pinpoint a single person or organization as the originator of NAA therapies. Sustained investigation and cooperation are necessary to enhance and broaden the range of non-antibiotic treatments in order to tackle the intricate problems associated with antibiotic resistance. AMR impedes the advancement of global health goals, such as the Sustainable Development Goals (SDGs) of the United Nations, on a larger scale. Antimicrobial resistance (AMR) is a serious and growing hazard to public health worldwide, with far-reaching effects in the social, medical, and economic spheres. AMR fundamentally reduces the efficacy of antibiotics, making it harder or even impossible to treat illnesses that were previously curable. International agencies including the World Health Organization (WHO), the Centers for Disease Control and Prevention (CDC), and the World Bank acknowledge that antimicrobial resistance (AMR) poses a threat to the security of global health. Coordinated initiatives at the local, national, and international levels are necessary to address the global repercussions of AMR (Cassini et al. 2019). Promoting antimicrobial stewardship, enhancing infection prevention and control procedures, funding the discovery of novel antibiotics and alternative therapies, and fortifying healthcare systems to guarantee that all populations have access to efficient treatments are some of the tactics used to combat AMR. Researchers and pioneers in healthcare have ventured into therapeutic domains that have never been explored before in their pursuit of novel remedies. The emergence of NAA therapy signals a new chapter in the ongoing fight against AMR and a period of hope and resiliency in the treatment of infectious diseases (Ryan et al. 2012). These therapies provide a way ahead toward a future where diseases can be efficiently managed, antibiotics can be saved for future generations, and global health can be protected for the welfare of humanity via innovation, collaboration, and a dedication to sustainability. NAA therapies without antibiotics is a vibrant field where science and creativity collide, with precision medicine taking center stage. Non-antibiotic treatments, which take their cues from nature's arsenal and use state-of-the-art technology, are revolutionizing the field of infectious disease management and treatment (Di Somma et al. 2020). They represent a therapeutic alternative, providing customized approaches that precisely target infections while preserving the delicate balance of the human microbiome. The risk of bacterial resistance may be reduced by non-antibiotic therapy. They might not apply selective pressure to bacteria that are not their intended targets because they are only targeting particular types of bacteria, which could impede the emergence of resistance. With every new development, we get a little bit closer to a day when illnesses are not only managed but also outwitted, resistance develops into resilience, and powerful yet individualized treatment is the norm (Czaplewski et al. 2016). This paradigm change makes NAA therapy a viable therapeutic choice as well as a symbol of human resourcefulness, fortitude, and the unrelenting quest for a more sustainable, healthy planet. Some of the important NAA therapies are as follows:

- Antimicrobial Peptides
- Probiotics and Prebiotics
- Phytochemicals and Plant-derived compounds
- CRISPR-Cas system

11.2 Antimicrobial Peptides

Host defense peptides (HDPs), another name for antimicrobial peptides (AMPs), are short chains of amino acids that have strong antibacterial action against a variety of pathogens, such as bacteria, fungi, viruses, and parasites. AMPs typically consist of 10–50 amino acids. These peptides are essential to the innate immune system, serving as a frontline defense mechanism against invading microorganisms. Zeya and Spitznagel are credited with making the initial discovery of antimicrobial peptides in 1966 (Zasloff 2002). Through the isolation of an antibacterial material from rabbit leukocytes, they came up with the name "cationic proteins." Later on, it was discovered that these proteins belonged to a significant class of antimicrobial peptides called defensins. As a result, many AMPs with different structures and antibacterial activity were identified and characterized. They are present in tissues, organs, and a variety of body fluids, including tears, perspiration, saliva, and mucus. Antimicrobial peptides are produced by all types of life. The mechanisms of action of antimicrobial peptides were uncovered by scientists in the 1990s and early 2000s, thanks to developments in molecular biology and biophysical approaches. The therapeutic potential of antimicrobial peptides as substitute antimicrobial agents was investigated by researchers as antimicrobial resistance became an increasingly pressing topic. The human body uses a variety of cell types and tissues in a convoluted process to manufacture Host Defense Peptides (HDPs) (Yeung et al. 2011). The predominant source of these peptides is the epithelium that lines the outside of organs and tissues such as the skin, gastrointestinal tract, respiratory tract, and genitourinary tract that are exposed to the outside world. A number of variables, including as microbial exposure, environmental cues, and genetic predisposition, influence the synthesis of HDPs. Immune cells and epithelial cells are prompted to release signaling molecules known as chemokines and cytokines when they come into contact with microbial pathogens or inflammatory signals. The elevation of HDP gene expression results from the intracellular signaling pathways that these signaling molecules initiate. These peptides exhibit broad-spectrum antimicrobial activity, regulating the production of cytokines and attracting immune cells to modulate immunological responses, through cell migration, proliferation, and angiogenesis—all essential for regaining tissue integrity—they aid in wound healing and promote tissue repair (Hancock and Sahl 2006). Antimicrobial peptides are generally encoded by multigene families, that is distinct AMPs are encoded by various genes within the genome. These multigene families frequently comprise closely related genes with comparable structural characteristics and biological roles. The diversity in AMPs made possible by the presence of multigene families gives the host organism the ability to fight off a variety of microbial invaders. AMPs can be constitutively expressed or produced, contingent on a number of variables, including the presence of microbial pathogens, inflammatory signals, and tissue injury. Growing interest has been seen in the biotechnological uses of antimicrobial peptides in recent years. The goal has been to create new antimicrobial peptides that are more stable, effective, and selective. The study of antimicrobial peptides is still in its early stages, with goals including improving our knowledge of their structure–function correlations, maximizing their potential as therapeutic agents, and overcoming obstacles to their commercialization and clinical translation. Some examples of antimicrobial peptides (AMPs) are as follows (Wang et al. 2016a).

11.2.1 Defensins

Small cationic peptides with a positive charge at physiological pH are known as defensins. Their primary amino acid sequences contain an excess of positively charged amino acid residues, like arginine and lysine, which contribute to their cationic character. Defensins interact with microbial membranes and damage their integrity because of their compact tertiary structure, which is composed of beta-sheets and alpha-helices (Mahlapuu et al. 2016). They are a family of endogenous antimicrobial peptides that typically have a conserved cysteine-stabilized alpha-beta (CSαβ) structural motif, contributing to their stability. Defensins are generally short peptides with lengths varying from 29 to 45 amino acids. However, post-translational modifications, sequence differences, and other variables including evolutionary divergence and species-specific adaptations might cause changes in length. Defensins' molecular weight also fluctuates based on their length and makeup of amino acids. Defensins typically range in molecular weight from around 3 to 6 kilodaltons (kDa). They target both gram-positive and gram-negative bacteria, including common pathogens including *Bacillus subtilis*, *Pseudomonas aeruginosa*, *Escherichia coli*, and *Streptococcus pyogenes*. They could be used to treat infections that are resistant to many drugs since they are very effective against bacterial types that are resistant to antibiotics. Additionally, they show antifungal action against a variety of fungal pathogens that can cause both superficial and systemic fungal infections, such as *Aspergillus fumigatus*, *Cryptococcus neoformans*, and *Candida albicans*. Defensins have antiviral properties that enable them to combat a wide range of viruses, including respiratory syncytial virus (RSV) and enveloped viruses like influenza (Brogden 2005). With their exceptional capacity to distinguish between the membranes of microorganisms and host cells, defensins—like other antimicrobial peptides—are able to precisely target and eliminate infections while leaving host cells unharmed. Negatively charged molecules, such as lipopolysaccharides (LPS) in gram-negative bacteria and lipoteichoic acid (LTA) in gram-positive bacteria, are frequently abundant in microbial membranes. The cationic defensins are drawn to these negatively charged components, which makes it easier for them to bind to microbial membranes. Defensins' affinity for host cells is decreased by the generally greater neutral charge of host cell membranes. The increased ratio of cardiolipin to phosphatidylglycerol in microbial membranes contributes to their negative charge and makes them more prone to defensin binding. Defensin-induced permeabilization is less likely to occur in host cell membranes because they have higher levels of sphingolipids and cholesterol than microbial membranes (Singh et al. 2017a). Based on their expression patterns, localization, and structure, they are divided into multiple categories. The following are the different kinds of defensins in Table 11.1 (Gao et al. 2021).

11.2.2 Cathelicidins

A family of tiny cationic peptides known as cathelicidins antimicrobial peptides (CAMPs) is important for the innate immune system's defense against microbial infections. Vertebrates are the main host species of CAMPs. Their antimicrobial activity is broad-spectrum, exhibiting protection against a range of pathogens such as viruses, bacteria, fungi, and parasites (Gao et al. 2021). The CAMP gene is found on chromosome 3p21.3 in humans. The precursor protein that is used to make cathelicidin peptides is encoded by the CAMP gene. There are two primary domains in the precursor protein that the CAMP gene encodes. The cathelin-like domain, located at the N-terminus of the precursor protein, is structurally similar to the protease inhibitor cathelin, which is present in neutrophil granules. The antimicrobial peptide domain is located at the C-terminus and exhibits sequence variation among cathelicidin peptides. The active cathelicidin peptide is released via proteolytic cleavage of the precursor protein. The functional element of cathelicidins is the antimicrobial peptide domain, which is released upon cleavage and has antimicrobial action (Singh et al. 2017b). Different signaling pathways control the expression of the CAMP gene in response to microbial infections, inflammatory stimuli, and other immunological challenges. Numerous immune system cells, such as neutrophils, macrophages, epithelial cells, and keratinocytes, express the CAMP gene. Research into the CAMP gene and the cathelicidin peptides it encodes is still able to shed light on host defense processes and offer hope for new treatments to fight inflammatory and infectious diseases. The clinically significant pathogens *Staphylococcus aureus*, *Streptococcus pyogenes*, *Escherichia coli*, *Pseudomonas aeruginosa*, and *Mycobacterium tuberculosis* are among the gram-positive and gram-negative bacteria that cathelicidins can target. They could be used to treat infections that are resistant to many drugs since they are very effective against bacterial types that are resistant to antibiotics. Additionally, it demonstrates antifungal

Table 11.1 Different kinds of defensins

Alpha-defensins	Beta-defensins	Theta-defensins
Location and function		
Defensins of this class are mostly present in neutrophils and specific types of epithelial cells. Alpha-defensins are mostly expressed in the small intestine's Paneth cells and neutrophil granules in humans. They are essential in protecting the gastrointestinal system and other mucosal surfaces against microbiological infections	These are a different class of defensins that are present in the skin, respiratory system, and urogenital tract, among other tissues and organs. The epithelial cells that line mucosal surfaces in human express beta-defensins, which help the host fight against microbial infections	They are cyclic peptides in certain nonhuman primates, such as rhesus monkeys. Theta-defensins exhibit potent antimicrobial activity and have been studied for their potential therapeutic applications
Chromosome involved		
Human chromosome 8p23.1 has six genes that encode alpha-defensins	In humans, a gene cluster located on chromosome 8p23.1 encodes beta-defensins	Theta-defensins are not expressed in humans due to a genetic mutation that inactivates the theta-defensin gene
Structure		
Three intramolecular disulfide bonds and a six-cysteine residue motif are characteristic features of alpha-defensins. Alpha-defensins' tertiary structure is stabilized by these disulfide connections, creating a recognizable three-dimension	In contrast to alpha-defensins, beta-defensins feature a distinct pattern of cysteine residues. Their disulfide bond configuration varies, but they usually have a six-cysteine pattern	Theta-defensins are cyclic peptides, forming a closed-loop structure due to the N- and C-termini linkage. This structure is formed by the head-to-tail cyclization of two nonapeptides through a process known as proteolytic processing
Genes involved		
The DEFA gene cluster comprises the following genes: DEFA1, DEFA3, DEFA4, DEFA5, DEFA6, and DEFA7	The DEFB1, DEFB4, DEFB103A, DEFB104, DEFB105, and other genes are part of the beta-defensin gene cluster	"DEFT" stands for "defensin theta," the gene that codes for theta defensins. The DEFT gene is thought to be a pseudogene in humans
Molecular weight		
Human alpha-defensins, such as human neutrophil peptides (HNP) are typically around 29–30 amino acids in length and have molecular weights of approximately 3–4 kDa	Conversely, human beta-defensins (HBDs) are comparatively long peptides, usually measuring between 36 and 45 amino acids, and having molecular weights between 4 and 5 kDa	Theta defensins generally consist of around 18–20 amino acid residues. The molecular weight of theta defensins would be roughly 1.98–2.2 kDa
Example		

(continued)

Table 11.1 (continued)

Alpha-defensins	Beta-defensins	Theta-defensins
Example such as HNP-1, HNP-2, HNP-3, and HNP-4	Human examples of beta-defensins such as human Beta Defensin 1 (hBD-1) (hBD-2) (hBD-3)	An example of a theta defensin is rhesus theta-defensin 1 (RTD-1) isolated from the bone marrow and certain other tissues of rhesus macaques

action against a range of fungal pathogens that can cause both superficial and systemic fungal infections, such as *Aspergillus fumigatus* and *Candida albicans*. They exhibit antiviral activity against specific viruses, which include enveloped viruses like *human immunodeficiency virus* (HIV) and *herpes simplex virus* (HSV) (Brogden 2005). The only known human cathelicidin is LL-37, out of 30 other types of cathelicidins. LL-37 is a 37-amino acid peptide that is generated from the precursor protein hCAP18 (human cationic antimicrobial protein 18). Human cathelicidin LL-37 is well known for its quick destruction of microbiological membranes, especially bacterial ones (Ridyard and Overhage 2021). Cathelicidins have antimicrobial activity through multiple methods, one of which is the breakdown of bacteria' lipoprotein membranes. Although LL-37 is best known for its antibacterial properties, new research indicates that it also has significant effects on cell migration and proliferation, especially in the context of wound healing and tissue repair. LL-37 stimulates different stages of the repair process, which aids in the healing of wounds. It has the ability to accelerate the migration of keratinocytes, which are the main epidermal cells and are necessary for re-epithelialization, a vital stage in the closure of wounds. In order to supply nutrition and oxygen to the wound site, LL-37 also encourages angiogenesis, the formation of new blood vessels. By regulating immune responses and lowering excessive inflammation at the wound site, LL-37 demonstrates anti-inflammatory qualities. Because fibroblasts produce extracellular matrix components like collagen, which give tissues structural support, they are important participants in wound healing and tissue repair. LL-37 has the ability to increase fibroblast migration and proliferation, which will aid in the production and deposition of collagen at the wound site (Fig. 11.1) (Sancho-Vaello et al. 2020).

11.2.3 Dermcidin

It is a little antimicrobial peptide that is primarily present in perspiration from humans. The eccrine sweat glands secrete it as a component of the innate immune system's defense against microbiological invaders. The DCD gene, also called the dermcidin gene, is found on the short arm of human chromosome 12 (12p13.31), and it encodes the protein. Its transcription is regulated by a number of variables such as hormone signals and environmental stimuli. It is composed of several exons and intones (Alaoui Mdarhri et al. 2022). A precursor protein that is processed to

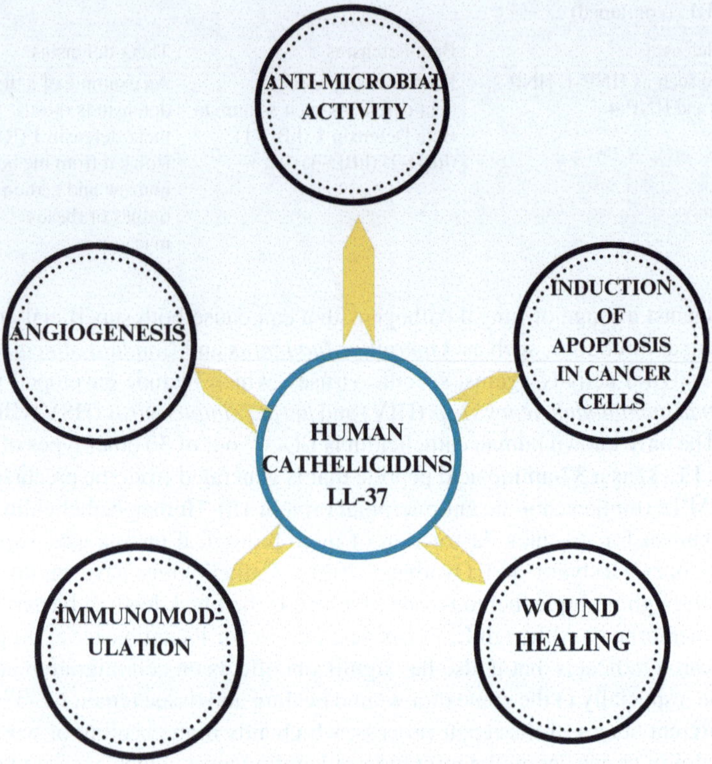

Fig. 11.1 Various functions associated with human Cathelicidin LL-37

produce mature dermcidin peptides is encoded by the DCD gene. The precursor protein has a dermcidin domain that is broken by proteases to release active dermcidin peptides and a signal sequence that controls its secretion from sweat glands. With 110 amino acids and a molecular weight of about 11 kDa, it is a cationic peptide (Tricarico et al. 2022). Dermcidin interacts with microbial membranes through the amphipathic structure, which combines hydrophobic and hydrophilic regions. This interaction compromises the integrity of the microbial membranes and enhances the skin's antibacterial qualities. The N- and C-terminal portions of dermcidin's structure are crucial to the peptide's ability to operate as an antibiotic. A signal sequence found in the N-terminal portion of dermcidin usually controls the secretion of the protein from eccrine sweat glands (Alaoui Mdarhri et al. 2022). This signal sequence aids in the appropriate processing and secretion of dermcidin into sweat, where it can act as an antibacterial agent on the skin's surface. The antimicrobial peptide domain of dermcidin is located in its C-terminal region. Microbial infections on the skin's surface are directly eliminated by this domain. It works to keep the skin clean and guard against infections by keeping microorganisms from growing too much on the skin's surface. Both gram-positive and gram-negative bacteria, including common skin pathogens like *Pseudomonas aeruginosa*,

Streptococcus pyogenes, *Escherichia coli*, and *Staphylococcus aureus*, can be targeted by dermcidin. Additionally, it exhibits effectiveness against fungi that cause infections of the skin and mucous membranes, such *Candida albicans* (Alaoui Mdarhri et al. 2022; Tricarico et al. 2022). Furthermore, microbial resistance mechanisms are unable to readily undermine dermcidin's antibacterial effects, rendering it an invaluable constituent of the host's defense against infections. According to research, dermcidin may have a role in controlling the operation of the sweat glands, the skin's moisture balance, and the healing of wounds. Its presence in sweat demonstrates the complex nature of eccrine gland secretions, which play a vital role in the skin's immune defense system in addition to assisting in temperature regulation.

11.3 Mode of Antimicrobial Mechanism of AMPs

By interacting with microbial membranes, AMPs disrupt the unification of microbial membranes and subsequently induce microbial death, which is the molecular mechanism of AMPs. There are multiple crucial phases involved in providing a comprehensive understanding of the molecular mechanisms that underlie the function of AMPs (Kumar et al. 2018).

Because AMPs contain basic amino acid residues like arginine and lysine, they are cationic peptides, which means they have a net positive charge. Microbial membranes, particularly those found in bacteria and fungi, usually possess a net negative charge because of the presence of negatively charged lipopolysaccharides and phospholipids. It is made easier for AMPs to initially bind to microbial membranes by this electrostatic attraction. AMPs have hydrophobic and hydrophilic areas within their structure, giving them amphipathic characteristics. They spread throughout the membrane surface after attaching to microbial membranes and integrating themselves into the lipid bilayer. The orderly arrangement of lipids is upset by this insertion, which leads to membrane instability and increased permeability. AMPs have the ability to aggregate or oligomerize on the membrane surface, creating temporary channels or pores. These holes cause disruptions to the membrane potential and ion gradients by allowing water, ions, and tiny molecules to enter and exit. Crucial cellular functions like waste removal and nutrition intake are hampered as a result. Microbial cell death is the result of cellular homeostasis disturbance and membrane integrity breakdown. The entry of water and ions into the cell might cause microorganisms to undergo osmotic lysis, or they can cause metabolic problems and apoptosis. AMPs can cause different types of cell death, and this can vary based on the microbial species and concentration. Apart from their immediate antimicrobial actions, AMPs have the ability to regulate immunological reactions by stimulating immune cells, encouraging chemotaxis, and augmenting the attraction of phagocytes to infection sites. Additionally, they have immunomodulatory qualities that control inflammation and aid in wound healing and tissue regeneration. Because of their complex mode of action, AMPs are powerful antimicrobial agents that may combat a variety of infections (Fig. 11.2) (Kumar et al. 2018).

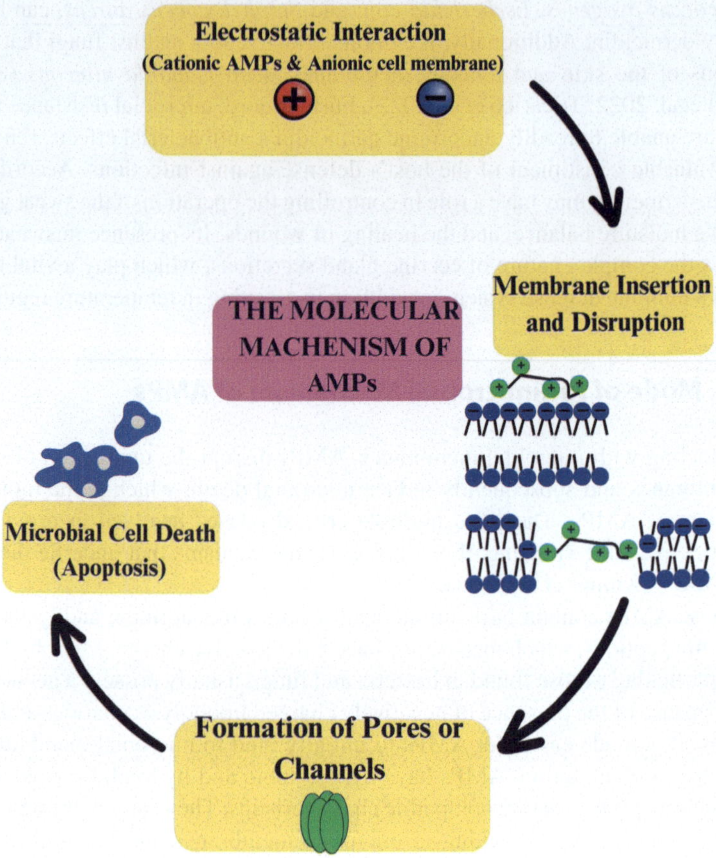

Fig. 11.2 Molecular mechanism of AMPs

11.4 Probiotics and Prebiotics

Probiotics are live bacteria that give the host health advantages when given in sufficient doses (WHO). Usually, they are yeast or bacteria that resemble the helpful microorganisms that are normally present in the human gastrointestinal tract. The proper balance of intestinal flora can be preserved with the use of probiotics. Probiotic microorganisms are usually consumed orally because in order to colonize and start providing benefits, they need to make it past the acidic environment of the stomach and the bile salts in the small intestine (Sanders et al. 2019). One crucial factor in determining the effectiveness of probiotics is their capacity to endure passage through the digestive system. Probiotics have been around since ancient times when people realized that foods that had undergone fermentation had health benefits. For thousands of years, people have utilized fermentation—the process of employing microbes like bacteria and yeast to turn carbohydrates into alcohol or acids—to preserve food and improve its flavor (Hill et al. 2014). Ancient people often ate fermented foods like yoghurt, kefir, sauerkraut, and miso, which contained healthy bacteria. These people included those in China, India, Egypt, and Greece. In the late nineteenth and early twentieth centuries, probiotics were first understood in their current form (Holscher 2017). The "Father of Probiotics," Élie Metchnikoff, was a Russian scientist who studied the advantages of fermented dairy products for health. Metchnikoff postulated in the early 1900s that eating fermented milk products containing lactic acid bacteria could change the gut microbiota and stop the formation of pathogenic bacteria, therefore extending life and improving health (Huys et al. 2013). Probiotics can help alleviate symptoms of irritable bowel syndrome (IBS), strengthen the immune system, prevent antibiotic-associated diarrhea, enhance digestion, and shorten the duration and severity of infectious diarrhea. Probiotics include a range of bacteria, including *Saccharomyces boulardii*, *Bifidobacterium bifidum*, *Bifidobacterium longum*, *Bifidobacterium breve*, *Lactobacillus acidophilus*, *Lactobacillus casei*, *Lactobacillus rhamnosus*, *Streptococcus thermophilus*, and *Lactobacillus plantarum* (Markowiak-Kopeć and Śliżewska 2020). Probiotics' efficacy is contingent upon several aspects, including strain specificity, viability, and dosage. Therefore, for maximum health advantages choose products that contain well-researched strains and adhere to suggested dosages. Prebiotics are indigestible carbohydrates or fibers that provide sustenance for probiotic bacteria and other advantageous microorganisms found in the gastrointestinal tract. They aid in preserving a balanced and healthy gut microbiota by encouraging the development and activity of these advantageous bacteria. They are made up of intricate carbohydrates that are difficult for the human digestive system to break down. Prebiotics are not broken down in the stomach or small intestine before entering the colon, where they are fermented by gut microorganisms. While preventing the growth of dangerous bacteria, they specifically promote the development and activity of advantageous bacteria, such as *Lactobacillus* and *Bifidobacterium* species. They thereby feed these advantageous bacteria, promoting their growth and proliferation.

Prebiotic substances that are frequently found in food include oligofructose, fructooligosaccharides (FOS), galactooligosaccharides (GOS), inulin, and certain dietary fibers. Synbiotics are a beneficial combination of probiotics and prebiotics that work synergistically (Holscher 2017). Probiotics and prebiotics work together in synbiotics to support gut health and general well-being. By giving probiotic microorganisms the right conditions in the form of prebiotics, synbiotics seek to increase their viability and activity. Probiotic–host interactions are intricate and multidimensional, incorporating metabolic, immunomodulatory, and neuroendocrine signaling pathways (Guarner and Malagelada 2003). To fully utilize probiotics' therapeutic potential for improving health and preventing disease, it is imperative to comprehend these interactions.

11.5 The Mechanism by Which Probiotics Provide Health Benefits

11.5.1 Improving Digestion

Lactase, an enzyme produced by probiotics, facilitates the breakdown of lactose, the sugar present in dairy products. This enzymatic action facilitates the better digestion of dairy by those who are lactose intolerant. Food fibers and other complex carbohydrates that the human digestive system would not be able to process are fermented by it. Short-chain fatty acids (SCFAs), such as butyrate, acetate, and propionate, are produced during this fermentation process and support gut health by feeding colonocytes (Belkaid and Hand 2014). Nondigestible carbohydrates are fermented in the colon by *Bifidobacterium breve*, which results in SCFAs. Probiotics can attach to intestinal epithelial cells thanks to surface proteins called adhesins, which helps in colonization and host interaction. In order to promote gut colonization and immunological regulation, *Lactobacillus rhamnosus* produces SpaCBA pili, which facilitates its adherence to intestinal epithelial cells (Sanders et al. 2014). By increasing the expression of tight junction proteins like ZO-1 and occludin, *Streptococcus thermophilus* reduces gut permeability and strengthens the intestinal barrier. Certain probiotic strains have the ability to bind to minerals and create chelates or complexes that the body can absorb more readily. These chelates can improve absorption in the intestines by shielding minerals from precipitation or binding with other materials in the gut lumen. Probiotic *Lactobacillus acidophilus* has been demonstrated to enhance intestinal calcium absorption, which may lower the risk of osteoporosis and bone fractures. Studies have looked into *Bifidobacterium bifidum's* capacity to improve the gut's absorption of magnesium, which is critical for the health of bones and muscles (Reid et al. 2003).

11.5.2 Boosting Immune Function

Probiotics have an interaction with immune cells in the mucosa-associated lymphoid tissue (MALT) and gut-associated lymphoid tissue (GALT), promoting the secretion of pro-inflammatory cytokines like tumor necrosis factor-alpha (TNF-α) while inhibiting the production of anti-inflammatory cytokines like interleukin-10 (IL-10) (Belkaid and Hand 2014). By increasing the expression of tight junction proteins, they fortify the intestinal barrier and support the integrity of the gut epithelium. By acting as a barrier, infections and poisons are kept from moving from the gut lumen into the circulation. When dendritic cells' Toll-like receptors (TLRs) are activated by *Bifidobacterium longum*, anti-inflammatory cytokines like interleukin-10 (IL-10) and tumor necrosis factor-alpha (TNF-α) are produced, which support immunological tolerance (Reid et al. 2003). It has been demonstrated that the *Lactobacillus casei* strain stimulates the intestinal regulatory T cell (Treg) development process. It has been demonstrated that *Lactobacillus rhamnosus* increases the activity of natural killer (NK) cells, which are essential for the innate immune response against tumor cells and viral infections.

11.5.3 Preventing Antibiotic-Associated Diarrhea (AAD)

Antibiotics kill both helpful and dangerous bacteria, upsetting the balance of the gut microbiota. By replenishing good microorganisms and inhibiting the expansion of opportunistic pathogens, probiotics aid in the restoration of this balance (Sanders et al. 2019). It generates antimicrobial substances such hydrogen peroxide, organic acids, and bacteriocins, which stop the formation of antibiotic-associated diarrhea and stop the growth of bacteria resistant to antibiotics. *Saccharomyces boulardii* is technically a yeast, but because of its advantageous effects on gut health, it is sometimes categorized as a probiotic. It has been demonstrated to lower the risk of AAD by modifying immunological responses in the gut and preventing the development of harmful bacteria.

11.5.4 Alleviating the Symptoms of Irritable Bowl Syndrome

Probiotics interact with the gastrointestinal tract's smooth muscle cells and enteric nervous system to affect gut motility. They can aid in controlling bowel motions and easing IBS-related constipation or diarrhea symptoms. By regulating immunological responses and inhibiting the release of pro-inflammatory mediators including interleukin-6 (IL-6) and interleukin-8 (IL-8), it has anti-inflammatory actions in the gut (Hill et al. 2014). For those with IBS, this decrease in intestinal inflammation can ease pain and discomfort in the abdomen. Investigation has explored the potential of *Lactobacillus acidophilus* to enhance gut health and mitigate symptoms of IBS, such as bloating, irregular bowel movements, and abdominal pain.

11.5.5 Reducing Severity and Duration of Infectious Diarrhea

Probiotics inhibit the colonization and growth of pathogens that cause infectious diarrhea by competing with harmful bacteria for resources and adhesion sites in the gastrointestinal tract. They generate antimicrobial substances like lactic acid, which make the surrounding environment acidic and prevent the growth of harmful bacteria. They also create tiny antimicrobial peptides called bacteriocins, which are specific to particular types of bacteria. Probiotics aid in the body's defense against infectious agents by boosting the synthesis of secretory IgA antibodies and antimicrobial peptides, which kill infections (Huys et al. 2013). Additionally, they enhance the immune response to invasive infections by modulating the function of immune cells including dendritic cells and macrophages. By absorbing available resources and attaching to intestinal epithelial cells, *Lactobacillus casei* outcompetes *Clostridium difficile* and lowers the risk of diarrhea associated with it.

11.5.6 Neurotransmitter and Hormone Regulation

Probiotics may interact with the host's neuroendocrine system, according to newly available data. Additionally, probiotics may change the gut's synthesis of hormones and neurotransmitters, which could have an impact on mood, hunger, and stress reactions. Neurotransmitters that control mood, stress reactions, and behavior, such as gamma-aminobutyric acid (GABA) and serotonin, can be produced by them (Markowiak-Kopeć and Śliżewska 2020). Probiotics can help control the body's reaction. *Lactobacillus rhamnosus* can affect the gut–brain axis by modifying neurotransmitter activity in the gut and the hypothalamic–pituitary–adrenal (HPA) axis.

11.5.7 Maintenance of Vaginal Health

Supporting the balance of bacteria in the vaginal microbiota is a key component of using particular probiotics to maintain vaginal health. Lactobacilli are the most common form of bacteria found in the vagina, when given orally or intravaginally, probiotic lactobacilli strains like *Lactobacillus rhamnosus GR-1* and *Lactobacillus reuteri RC-14* have been demonstrated to colonize the vaginal tract. By doing this, they support the maintenance and restoration of the vaginal population of helpful bacteria, which can be disturbed by a variety of causes like the use of antibiotics, changes in hormone levels, and sexual activity (Guarner and Malagelada 2003). Lactic acid and hydrogen peroxide are metabolic byproducts produced by lactobacilli. The pH range of these compounds is normally between 3.5 and 4.5, which results in an acidic environment in the vagina. The growth of pathogenic bacteria and yeast, particularly those linked to bacterial vaginosis (BV) and yeast infections, is inhibited by the acidic pH.

11.6 Phytochemicals and Plant-Derived Compounds

Phytochemicals, sometimes referred to as phytonutrients, are organic substances that are found naturally in plants and give fruits, vegetables, grains, legumes, herbs, and spices their vibrant hues, flavors, and scents. Plants use phytochemicals for a variety of purposes, such as defense against diseases, pests, UV rays, and environmental stresses. They support the defense systems of the plant and aid in ensuring its survival. Phytochemicals provide protective roles in plants, but when taken by people, they may also have a number of health benefits. Numerous phytochemicals with distinct chemical structures and biological activities exist. Phytochemicals present a viable substitute for traditional antibacterial agents and hold promise for becoming innovative therapies in the fight against infectious illnesses (Tanaka and Kashiwada 2021). It is crucial to remember that the antimicrobial activity of phytochemicals can change based on a number of variables, including formulation, concentration, and the particular microbe that is being targeted.

11.6.1 Different Types of Phytochemicals

11.6.1.1 Flavonoids

These are polyphenolic substances with a carbon skeleton that reads C6–C3–C6. They are made up of a heterocyclic pyran ring (C) connecting two benzene rings (A and B). Quercetin, kaempferol, and epigallocatechin gallate (EGCG) are a few examples. By blocking microbial enzymes, rupturing microbial cell membranes, and interfering with microbial nucleic acid production, they exhibit antimicrobial activity in multiple ways (Bagheri Farahani et al. 2017). Quercetin, for instance, has been demonstrated to have bacteriostatic effects against a variety of bacteria by inhibiting bacterial DNA gyrase and topoisomerase. It is thought that quercetin binds to DNA gyrase ATP-binding site and inhibits its ATPase function. This stops ATP from being hydrolyzed, which is required for gyrase energy-dependent DNA supercoiling function. Reactive oxygen species (ROS) can be produced in bacterial cells by quercetin, which also has antioxidant qualities. ROS have the ability to oxidatively damage DNA, which can result in strand breaks and other types of damage (Lau et al. 2019). Certain flavonoids prevent viruses from spreading by preventing them from adhering to host proteins or blocking viral enzymes, which prevents viruses from connecting to and entering host cells. Flavonoids such as hesperidin and naringenin have antiviral activity against the *Influenza virus*, *Herpes simplex virus* (HSV), *and Human immunodeficiency virus* (HIV) by inhibiting viral replication and reducing viral infectivity. Flavonoids such as quercetin, kaempferol, and catechins have been shown to damage the cell membranes of bacteria, including *Escherichia coli*, *Pseudomonas aeruginosa*, and *Staphylococcus aureus*.

11.6.1.2 Terpenoids

They have different structures and are formed from isoprene units. They consist of diterpenes, triterpenes, monoterpenes, and sesquiterpenes. Thymol, carvacrol, and menthol are a few examples. By rupturing microbial cell membranes, changing the permeability of membranes, and interfering with microbial enzyme function, terpenoids demonstrate their antimicrobial activity. Bacteria like *Staphylococcus aureus*, *Candida albicans*, and *Escherichia coli* have their cell membranes disrupted by certain terpenoids, notably thymol and carvacrol, which are present in thyme and oregano essential oils (Rusu et al. 2022). Terpenoids have the potential to disrupt the expression or function of virulence factors, including immune evasion proteins, adhesins, and toxins, which are generated by *Staphylococcus aureus*. Terpenoids can reduce *S. aureus* pathogenicity and improve the efficiency of the host immune response against the bacteria by focusing on virulence factors (Tallei et al. 2020). Bacteria such as *Pseudomonas aeruginosa* and *Vibrio harveyi* have quorum sensing pathways that are interfered with by thymol and linalool, which hinders their ability to produce biofilm and leads to infections. Both gram-positive and gram-negative bacteria can be effectively combated by them. It has been demonstrated that several terpenoids cause ROS or disrupt antioxidant defense systems to cause oxidative stress in microbial cells. This oxidative stress causes critical metabolic processes to be disrupted and destroys cellular components, which ultimately results in the death of microbial cells.

11.6.1.3 Alkaloids

These are variously shaped compounds that include nitrogen. These consist of isoquinoline alkaloids, tropane, indole, and pyrrolidine. Berberine, nicotine, and caffeine are a few examples. By rupturing microbial cell membranes, reducing microbial enzyme activity, and interfering with microbial nucleic acid production, alkaloids have antimicrobial effects. By attaching to bacterial DNA, for instance, berberine prevents RNA synthesis and bacterial DNA replication. By interfering with the production of microbial cell walls, primarily in bacteria and fungi, it also damages bacterial cell membranes (Leung and Covasa 2021). Through the inhibition of enzymes involved in the formation of cell walls, alkaloids impair the structural integrity of microbial cells, increasing their vulnerability to lysis and destruction. The cell signaling pathways involved in the development, survival, and virulence of microorganisms can be modulated by alkaloids. Alkaloids can disrupt microbial infections and strengthen host defense mechanisms by interfering with these pathways. Certain alkaloids cause microbial cells to undergo programmed cell death, or apoptosis. This method aids in the control of microbial infections and the removal of infected cells.

11.6.1.4 Phenolic Acids

These are aromatic compounds that have a phenyl ring linked to one or more hydroxyl groups. Gallic acid, ferulic acid, and caffeic acid are a few examples. By rupturing microbial cell membranes, reducing microbial enzyme activity, and interfering with the formation of microbial cell walls, phenolic acids exhibit antimicrobial action. A combination of hydrogen bonding, π–π stacking interactions, electrostatic interactions, and minor groove binding are involved in the binding of caffeic acid to microbial nucleic acids. Caffeic acid inhibits microbial development and causes cell death by interfering with vital cellular functions in microbes, including transcription, translation, and replication (Firdous 2014). This is achieved through its interaction with nucleic acids. Through direct binding to enzymes, chelation of metal cofactors, modification of enzyme regulation, loss of enzyme stability, and selectivity toward target enzymes, ferulic acid limits the activity of microbial enzymes. Studies have demonstrated that bacteria like *Escherichia coli*, *Staphylococcus aureus*, and *Pseudomonas aeruginosa* had their cell membranes disrupted by gallic acid and caffeic acid. Certain phenolic acids prevent the synthesis of virulence factors and interfere with cell-to-cell communication.

11.6.1.5 Saponins

These glycosides consist of one or more hydrophilic sugar chains joined to a hydrophobic aglycone (sapogenin). Often, the aglycone part has a steroid or triterpene structure. Their ability to generate frothy solutions in water is what makes them unique. Ginsenosides, glycyrrhizin, and digitonin are a few examples (Wink 2015).

By rupturing microbial cell membranes, creating complexes with membrane sterols, and reducing microbial enzyme activity, saponins have antimicrobial effects. Digitonin's molecular structure contains both hydrophilic (which loves water) and hydrophobic (which repels water) sections, giving it amphipathic characteristics. This enables it to interact with the lipid bilayer found in the membranes of microorganisms. Digitonin's hydrophilic part extends into the aqueous environment, while its hydrophobic portion penetrates the lipid bilayer (Kawacka et al. 2020). The integrity and organization of the cell membrane are compromised by this. Digitonin has a somewhat selective effect on the membranes of microorganisms. When contrast to mammalian cells, it usually targets microbial cells more specifically. The distinction between the structure and composition of mammalian and microbial cell membranes is thought to be the cause of this selectivity. The lipid makeup of microbial cell membranes differs from that of mammalian cell membranes and frequently contains larger amounts of sterols. Its selective antibacterial activity is facilitated by its interaction with these particular microbial membrane constituents. For instance, ginsenosides interact with the phospholipids and cholesterol in bacterial cell membranes to cause disruption, which results in cellular leakage and bacterial cell death. By boosting the activity of immune cells including macrophages, natural killer cells, and lymphocytes, saponins activate the immune system and support host defense and pathogen clearance (Leung and Covasa 2021; Kawacka et al. 2020). It is crucial to remember, nevertheless, that digitonin's toxicity and other adverse effects, especially at higher concentrations, may restrict its use in clinical settings.

11.6.1.6 Essential Oils

These are plant-extracted volatile chemicals. Tea tree oil, eucalyptus oil, and lavender oil are a few examples. Essential oils modify the permeability of microbial cell membranes, cause disruptions to them, and reduce the activity of microbial enzymes. For instance, tea tree oil exhibits broad-spectrum antibacterial action against viruses, bacteria, and fungus. The *influenza virus* is susceptible to the antiviral properties of peppermint oil (Menthapiperita) (Firdous 2014). They might alter the signaling pathways of microbial cells, such as those that control quorum sensing, the development of biofilms, and stress reactions. Essential oils interfere with microbial cell signaling, limiting their ability to multiply and cause illnesses, which disrupts microbial communication and coordination. Essential oils influence ion channels and transporters on the cell membrane, which upsets the pH and ion balance of microbiological cells. Ion gradient disruption impairs vital biological processes and metabolic activities, preventing microbial growth and reproduction (Suez et al. 2018). Numerous bioactive chemicals, mainly 1,8-cineole (eucalyptol) and α-terpineol, are present in eucalyptus oil. It has anti-inflammatory qualities that may aid in lowering inflammation brought on by microbial infections. Eucalyptus oil reduces the inflammatory response brought on by microbial invasion by preventing the release of cytokines and other pro-inflammatory mediators (Bagheri Farahani et al. 2017).

11.6.1.7 Isothiocynates (ITCs)

Are substances that include sulfur and are present in cruciferous vegetables. Sulforaphane, allyl isothiocyanate, and phenethyl isothiocyanate are a few examples. Isothiocyanates interfere with microbial cell metabolism, break down microbial cell membranes, and decrease microbial enzyme function. The electron transport cycle, which produces ATP during aerobic respiration through oxidative phosphorylation, can be disrupted by ITCs (Lacaille-Dubois and Wagner 1996). ITCs hinder ATP synthesis when the passage of electrons along the ETC is disrupted, which lowers energy production and causes metabolic dysfunction in microbial cells. Direct bactericidal effects are demonstrated by phenethyl isothiocyanate (PEITC), which breaks down the cell membrane integrity of *Candida albicans* and *Helicobacter pylori* (Tanaka and Kashiwada 2021). Because PEITC is lipophilic, it can pass through bacterial cell membranes, rupturing the lipid bilayer structure and causing an increase in membrane permeability. Cell death and contents leaking are the results of this disturbance. By producing ROS, PEITC can also cause oxidative stress in *Helicobacter pylori* cells (ROS). ROS buildup impairs lipids, proteins, and nucleic acids—all of which are essential for cellular viability and survival—thus making bacteria less able to survive. Bacteria such as *Salmonella* spp., *Listeria monocytogenes*, and *Escherichia coli* have their cell membranes disrupted by allyl isothiocyanate (AITC), which is present in mustard and wasabi (Rusu et al. 2022).

11.6.1.8 Tannins

These are plant tissues that contain polyphenolic chemicals. Tannic acid, ellagic acid, and catechins are a few examples. Iron and manganese, two metal ions necessary for microbial development and metabolism, can be chelated by tannins, depriving bacteria of these nutrients and preventing their growth (Lau et al. 2019). They precipitate microbial enzymes, bind to proteins and cell membranes of microorganisms, and stop their growth. Ellagic acid, for instance, prevents bacterial adherence and damages bacterial cell membranes. Ellagic acid inhibits the synthesis of extracellular matrix components and microbial adhesins, which are essential for adhesion and biofilm development, by modulating gene expression.

11.6.1.9 Lignans

They are phenolic chemicals that can be discovered in nuts, grains, and seeds. Matairesinol, pinoresinol, and secoisolariciresinol are a few examples. Microbial virulence factors are chemicals produced by pathogens to aid in colonization, invasion, and evasion of host immune defenses. Lignans have the ability to interfere with the generation or function of these molecules. Lignans interfere with the ability of microbial pathogens to adhere to host tissues, induce tissue damage, elude immune surveillance, and create chronic infections by targeting virulence factors such as adhesins, toxins, proteases, and biofilm formation proteins. Through their interactions with transcription factors and regulatory proteins, they can modify the expression of genes linked to virulence and growth in microorganisms (Rusu et al. 2022). Lignans assist in the eradication of microbial pathogens and the prevention of infections by encouraging the activation of immune cells, such as macrophages, dendritic cells, and T lymphocytes. Additionally, lignans have anti-inflammatory qualities that help reduce excessive inflammation and tissue damage brought on by microbial infections, aiding in the healing process of the host and the eventual cure of the infection.

11.6.1.10 Artemisinin

It is a strong antimalarial substance that is extracted from the *Artemisia annua* plant, sometimes known as sweet wormwood, and is used to treat malaria. Iron is necessary for artemisinin to start working against Plasmodium parasites. When iron, namely heme, which is a result of the parasite's consumption of hemoglobin, is present, artemisinin is activated by an endoperoxide bridge cleavage. Artemisinin causes the parasite cells to produce free radicals, especially ROS, after it is activated (Bagheri Farahani et al. 2017). These extremely reactive free radicals harm proteins, lipids, and nucleic acids, among other components of the organism. The membranes of the parasite's organelles and structures, including as the mitochondria and endoplasmic reticulum, are harmed by the buildup of ROS. The parasite dies as a result of the membrane integrity being disrupted, which obstructs vital cellular functions (Rusu et al. 2022). Due to their effectiveness against drug-resistant types of malaria parasites, artemisinin and its derivatives are essential parts of combination therapies used to treat malaria all over the world. It is noteworthy that the World Health Organization (WHO) recommends artemisinin-based combination therapies (ACTs) as initial treatments for uncomplicated malaria because of its high efficacy and capacity to lower the risk of drug resistance. A plentiful supply of antimicrobial compounds with a variety of modes of action and therapeutic potential against microbial infections can be found in phytochemicals (Carlet et al. 2012).

11.7 CRISPR-Cas System

The CRISPR-Cas system, first identified in the late 1980s as a mysterious component of bacterial immune systems, has quickly transformed from a perplexing genetic phenomenon to a revolutionary tool in contemporary molecular biology and biotechnology (de Jong et al. 2014). Short repeating DNA sequences called CRISPRs—clustered regularly interspaced short palindromic repeats—are present in the genomes of bacteria and archaea. Through groundbreaking research in the early 2000s, the adaptive immunological function of CRISPR-Cas in bacteria and archaea was clarified (Barrangou and Marraffini 2014). It was discovered that CRISPR-Cas accurately targets and cleaves foreign DNA sequences to protect against viral and plasmid infections. 2012 saw the important discovery by Emmanuelle Charpentier, Jennifer Doudna, and associated teams that the CRISPR-Cas system could be used to modify eukaryotic cells' genomes programmably (Bikard et al. 2014). The CRISPR-Cas system is made up of two basic parts: the Cas proteins and the CRISPR array. The repetitive and spacer sequences that make up the CRISPR array are a molecular memory of previous infections. Enzymes known as CRISPR-associated proteins, or Cas proteins, are vital to the immune response.

There are various kinds of Cas proteins, and they all have different purposes. To precisely modify the DNA sequence, the Cas9 endonuclease is guided by a synthetic single-guide RNA (sgRNA) (Citorik et al. 2014). The French microbiologist Emmanuelle Charpentier and the American biochemist Jennifer A. Doudna were awarded the Nobel Prize in Chemistry 2020 for their groundbreaking work in unraveling the CRISPR-Cas system and transforming CRISPR-Cas9 into a revolutionary genetic editing tool. Their achievements have had a significant impact on science and have great potential to help solve some of the most important problems facing humanity. With the use of CRISPR-Cas9 technology, scientists can precisely edit genes, modifying genetic sequences in living things with previously unheard-of accuracy and efficiency. Although CRISPR-Cas9 is most recognized for its functions in gene control and genome editing, it may also be used in antibacterial applications. It is crucial to understand that, in contrast to antimicrobial peptides or conventional medicines, CRISPR-Cas9 does not directly inhibit microbial growth. Its capacity to target and break particular DNA sequences, particularly those of harmful microbes, is what gives it its antimicrobial potential instead (Deveau et al. 2010). Targeting pathogenic genes and creating innovative tactics to fight infectious diseases are made possible by the accurate and adaptable CRISPR-Cas9 system. Its uses go beyond fundamental research to include diagnostics and possible therapeutic procedures that try to lessen the negative effects of pathogenic bacteria on human health (Fig. 11.3).

Targeting the virulence factor genes of harmful bacteria, such *Staphylococcus aureus*, a major cause of infections in humans, has proven to be a potential application of CRISPR-Cas9 technology. It has a variety of virulence characteristics that allow it to infect host tissues, get past immune system defenses, and spread illness (Gomaa et al. 2014). These virulence factors, which are essential to the pathophysiology of staphylococcal infections, include adhesins, toxins, and immune evasion

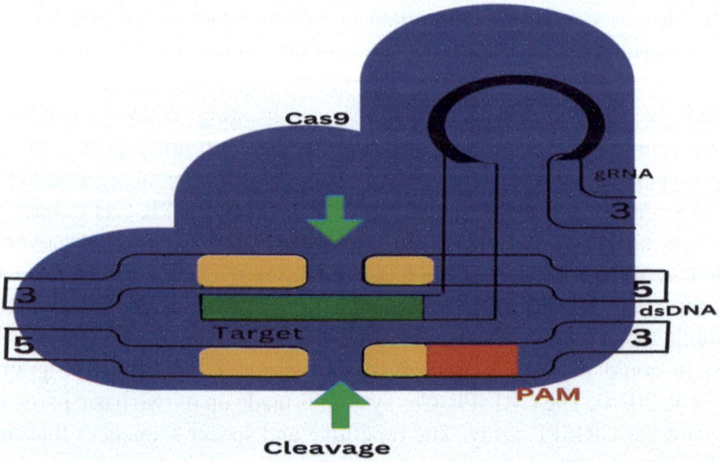

Fig. 11.3 CRISPR-Cas9 system

proteins. Creating guide RNAs (gRNAs) complementary to particular regions in the target genes is one method of targeting virulence factor genes in *Staphylococcus aureus* using CRISPR-Cas9. Double-strand breaks (DSBs) in the bacterial DNA are caused by the Cas9 nuclease, which is guided by these gRNAs to the appropriate chromosomal region. The target genes may undergo insertions, deletions, or mutations as a result of the bacterial cell's repair mechanism repairing these DSBs, so impairing their ability to function. Toxins are one type of virulence factor that CRISPR-Cas9 in *Staphylococcus aureus* targets. It generates a range of toxins that increase the severity of the disease, tissue damage, and immune evasion, such as hemolysins, leukocidins, and exotoxins (Hu et al. 2023). CRISPR-Cas9 can lessen the pathogenicity of the bacterium by lowering its capacity to damage tissue and elude human immune responses by specifically targeting the genes generating these toxins. Another class of virulence protein called adhesins helps *Staphylococcus aureus* adhere to host tissues and create biofilms, which are colonies of bacteria covered in an extracellular polymeric matrix. Adhesin-encoding genes can be disrupted by CRISPR-Cas9, which stops the bacteria from attaching to host cells and surfaces and stops biofilm development (Bosi et al. 2016). This interference hinders the bacterium's capacity to spread diseases and encourages the host immune system to eliminate it. Moreover, *Staphylococcus aureus* produces immune evasion proteins that CRISPR-Cas9 can target to prevent the bacterium from being recognized and eliminated by the host immune system. The host immune response is disrupted by these immune evasion proteins in a number of ways, including phagocytosis, complement activation, and cytokine generation. CRISPR-Cas has been utilized to target the genes that cause methicillin resistance in strains of *Staphylococcus aureus* (MRSA) (Høyland-Kroghsbo et al. 2017). The goal of study is to restore susceptibility to beta-lactam antibiotics and reverse antibiotic resistance by modifying genes that encode penicillin-binding proteins, such as mecA or mecC. CRISPR-Cas9 can lessen the severity and duration of an infection by improving the host's capacity to identify and eradicate *Staphylococcus aureus* infections by mutating the genes encoding these proteins (Jiang et al. 2013).

Targeted gene editing is the main method used to apply CRISPR-Cas systems against *Mycobacterium tuberculosis* (Mtb) in order to comprehend the bacterium's pathogenicity and drug resistance processes, as well as to perhaps develop novel therapeutic approaches. The ability of Mtb to become resistant to several antibiotics, such as isoniazid, rifampicin, and others, is well known (Jiang et al. 2013). Scientists create complementary gRNAs (gRNAs) that match particular sequences in the Mtb genome. Usually, these sequences aim to disrupt genes linked to virulence, resistance to antibiotics, or vital biological processes. Mtb cells are treated with the CRISPR components, which include the Cas protein and the intended gRNA (Jiang and Marraffini 2015). Numerous delivery techniques, including plasmid vectors, phage-mediated transduction, and electroporation, can do this. The Cas protein is guided to the target location within the bacterial genome by the gRNA once it has entered the Mtb cell. Until it comes upon a sequence that is complementary to the gRNA, the gRNA-Cas ribonucleoprotein (RNP) complex searches the DNA. The Cas protein causes a DSB at that spot in the Mtb genome when the gRNA-Cas combination identifies the target sequence. The Non-Homologous End Joining (NHEJ) pathway is frequently used in the repair process of Mtb, and it can lead to minor insertions or deletions (indels) at the break site. These mutations have the ability to modify or wipe out the targeted gene by interfering with its normal function (Karvelis et al. 2013). Researchers can examine the function of particular genes in virulence, pathogenesis, and treatment resistance in Mtb by altering those genes. For example, CRISPR-mediated deletion of antibiotic-resistant genes can be used to find new targets for drugs or comprehend the processes that underlie drug resistance in Mtb strains (Alaoui Mdarhri et al. 2022).

A prevalent bacteria in the human stomach, *Escherichia coli* can cause a variety of illnesses, such as bloodstream infections, gastroenteritis, and urinary tract infections. The acquisition of resistance genes encoding enzymes that inactivate antibiotics, change drug targets, or promote antibiotic efflux from the bacterial cell is a common cause of antibiotic resistance in *E. coli*. Researchers have used CRISPR-Cas systems to target and delete specific genes linked to resistance mechanisms in order to tackle antibiotic resistance in *E. coli*. For example, CRISPR-Cas technology has been used to alter genes generating extended-spectrum beta-lactamases (ESBLs), such as blaCTX-M and blaTEM, which give resistance to broad-spectrum beta-lactam antibiotic (Louwen et al. 2014). CRISPR-Cas systems have been used to disrupt genes eVncodingcarbapenemases, such as blaKPC and blaNDM, which make *E. coli* resistant to carbapenem antibiotics (Rupp and Fey 2003). The goal of this research is to modify these resistance genes so that *E. coli* strains are once again susceptible to carbapenems, which are essential antibiotics for treating serious infections (Sternberg et al. 2012). Additionally, genes involved in efflux pump systems—which expel antibiotics from bacterial cells, lowering their intracellular concentration and efficacy—have been targeted with CRISPR-Cas technology (Makarova et al. 2011). The efficiency of antibiotic therapy can be increased and *E. coli's* susceptibility to antibiotics can be restored by disrupting the genes encoding efflux pumps.

In order to improve treatment outcomes and lessen the burden of antibiotic resistance linked to biofilm-related illnesses, CRISPR-Cas technology presents a possible method of lowering the virulence and persistence of *Pseudomonas aeruginosa* infections. The adaptable and opportunistic pathogen *Pseudomonas aeruginosa* is well known for its capacity to create strong biofilms and control gene expression through quorum sensing systems. Encased in a self-produced matrix of extracellular polymeric substances (EPS), biofilms are intricate communities of bacteria that cling to surfaces and shield the bacteria from external stressors and antimicrobial treatment (Mojica et al. 2005). In clinical settings, *Pseudomonas aeruginosa* biofilms can cause persistent infections in wounds, people with cystic fibrosis, and medical devices. For these reasons, biofilms are particularly hazardous (Rath et al. 2015). A kind of cell-to-cell communication known as quorum sensing depends on the synthesis, release, and detection of signaling molecules known as autoinducers. A possible method to reduce the virulence and durability of *Pseudomonas aeruginosa* infections is CRISPR-Cas technology. One essential part of *Pseudomonas aeruginosa* biofilms' extracellular matrix is alginate. Alginate biosynthesis is synthesized and regulated by genes belonging to the alg operon, including algD, algA, and algC. Changes in these genes can affect the stability and structural integrity of the biofilm, making the bacterium more vulnerable to immune clearance and antimicrobial agents (Selle et al. 2015). This could improve the effectiveness of treatment for infections caused by *Pseudomonas aeruginosa* strains that form biofilms.

Gram-negative *Helicobacter pylori* is a common human infection linked to gastrointestinal conditions such as gastritis, peptic ulcers, and stomach cancer. Antibiotic resistance genes in the *Helicobacter pylori* genome can be precisely and specifically edited thanks to CRISPR-Cas9 technology, offering a possible solution to the problem of antimicrobial resistance in this bacterium (Singh et al. 2017a). *Helicobacter pylori's* resistance to clarithromycin is mostly caused by point mutations in the 23S rRNA gene's peptidyl transferase region. The efficiency of clarithromycin-based eradication regimens is decreased by these mutations, which include A2142G, A2142C, and A2143G (Wang et al. 2016b). These mutations interfere with clarithromycin ability to attach to the ribosome. These particular 23S rRNA gene alterations can be targeted and edited using CRISPR-Cas9 technology, potentially restoring susceptibility to clarithromycin. Mutations in the rdxA and frxA genes, which encode enzymes involved in the activation and detoxification of metronidazole, respectively, are frequently linked to metronidazole resistance in *Helicobacter pylori* (Wiedenheft et al. 2012). These genes' loss-of-function mutations or deletions hinder the bacterium's ability to activate metronidazole, making it resistant to the antibiotic. The potential to alter or disrupt the rdxA and frxA gene sequences using CRISPR-Cas9 editing allows for the restoration of metronidazole susceptibility in *H. pylori* strains that are resistant to the drug (Zetsche et al. 2015). Mutations in the quinolone resistance-determining region (QRDR) of the gyrA gene, which codes for the DNA gyrase enzyme, are frequently the cause of fluoroquinolone resistance in *Helicobacter pylori*. Fluoroquinolone resistance is conferred by mutations such as alterations at sites 87 and 91 of the gyrA gene, which change the drug's binding affinity to the DNA gyrase complex. Targeted changes in the gyrA gene's QRDR can be introduced using CRISPR-Cas9 technology, which may be able to reverse fluoroquinolone resistance in *Helicobacter pylori* strains that carry these mutations (Du et al. 2018).

11.8 Implication of NAA Therapy

NAA therapy holds significant promise and importance in the field of medicine and public health due to several key factors (Sweileh and Moh'd Mansour 2020; Prats-Ejarque 2021).

11.8.1 Addressing Antimicrobial Resistance (AMR)

The possibility of NAA therapy to overcome antimicrobial resistance is one of the most important factors supporting its significance. Numerous traditional antibiotics are no longer effective due to the advent of multidrug-resistant bacteria caused by the overuse and misuse of antibiotics. Alternative strategies that get around established antibiotic mechanisms are provided by non-antibiotic medicines, which lessen the selection pressure that leads to the development of resistance.

11.8.2 Broadening Treatment Options

The therapeutic options available to healthcare providers are increased by non-antibiotic therapy. Clinical professionals can increase the chance of treatment success by customizing treatments for particular patients and specific infections by using a variety of therapeutic modalities, such as probiotics, phytochemicals, and antimicrobial peptides.

11.8.3 Precision and Targeted Therapy

Advanced non-antibiotic therapies frequently offer accurate and focused infection treatments. By choosing phages particular to the bacterial strain that is infecting, for instance, bacteriophage therapy reduces the risk of harm to helpful microbes. This accuracy enhances patient outcomes while lowering the possibility of upsetting the body's microbiome.

11.8.4 Reducing Side-Effects and Toxicity

Numerous non-antibiotic medicines have favorable safety profiles, in contrast to certain antibiotics that may have harmful side effects and toxicity. Antimicrobial peptides that come from natural sources, for example, are less likely to cause negative patient reactions. As a result, patients stick to treatment plans better and the strain on healthcare systems is lessened.

11.8.5 Fostering Innovation and Research

Research and innovation in infectious diseases are driven by non-antibiotic medicines. Researchers and scientists are looking into innovative approaches to create cutting-edge therapeutic molecules, improve delivery methods, and boost the effectiveness of treatments. In addition to expanding scientific understanding, this work opens the door for more innovations in infection control in the future.

11.8.6 Promoting Sustainable Healthcare Practices

Advanced non-antibiotic medicines lessen the need for traditional antibiotics, which promotes sustainability in healthcare. Reducing the negative effects of antibiotic residues on the environment, maintaining antibiotic efficacy for upcoming generations, and encouraging conscientious antimicrobial stewardship in hospital settings are examples of sustainable approaches.

11.8.7 Global Healthcare Implications

The use of non-antibiotic medicines has a big impact on world health, especially in areas where access to effective antibiotics is few or antimicrobial resistance is common. These treatments provide scalable and affordable ways to fight infectious diseases, enhance patient outcomes, and lessen global healthcare inequities.

11.8.8 Synergistic and Combinatorial Strategies

Combinatorial and synergistic approaches can be created by integrating non-antibiotic medicines with currently available therapeutic modalities. Clinicians can improve therapy efficacy, overcome resistance mechanisms, and improve patient outcomes by utilizing the complimentary mechanisms of many medicines.

11.8.9 Potential for Long-Term Solution

Non-antibiotic medicines provide the possibility of long-term remedies, in contrast to conventional antibiotics, which might offer temporary relief but exacerbate long-term problems like AMR. These therapies address the underlying causes of infection and foster long-lasting resilience against microbial threats by focusing on fundamental features of microbial physiology and host-pathogen interactions.

11.9 Conclusions

We must consider the broader consequences of these interventions for public health, healthcare systems, and international efforts to control infectious diseases as we negotiate the difficult field of non-antibiotic therapy. Antibiotic resistance can be lessened, resistant bacteria can spread more slowly, and the efficacy of current antibiotics can be maintained for future generations through non-antibiotic therapy. We can fully utilize non-antibiotic therapies to address the complex issues surrounding infectious diseases in the twenty-first century by adopting the values of sustainability and creativity. To sum up, NAA therapy offers novel ways to address antibiotic resistance and enhance patient outcomes, marking a paradigm shift in the management of infectious illnesses. Non-antibiotic therapies hold the potential to revolutionize the fields of precision medicine, focused medicines, and interdisciplinary collaboration in the diagnosis, treatment, and prevention of microbial infections. The scientific literature on non-antibiotic therapies is growing, and it is becoming clear that these strategies have great potential to change the way that infectious disease management is thought about. These innovative approaches address the complex issues of multidrug-resistant (MDR) microorganisms and developing infectious illnesses while reducing the overuse of antibiotics.

References

Alaoui Mdarhri H, Benmessaoud R, Yacoubi H, Seffar L, Guennouni Assimi H, Hamam M, Boussettine R, Filali-Ansari N, Lahlou FA, Diawara I, Ennaji MM, Kettani-Halabi M (2022) Alternatives therapeutic approaches to conventional antibiotics: advantages, limitations and potential application in medicine. Antibiotics (Basel, Switzerland) 11(12):1826

Bagheri Farahani Z, Mirzaie A, Ashrafi F, RahimpourHesari M, Chitgar A, Noorbazargan H, Rahimi A (2017) Phytochemical composition and biological activities of *Artemisia quettensis* Podlechethanolic extract. Nat Prod Res 31(21):2554–2558

Barrangou R, Marraffini LA (2014) CRISPR-Cas systems: prokaryotes upgrade to adaptive immunity. Mol Cell 54(2):234–244

Belkaid Y, Hand TW (2014) Role of the microbiota in immunity and inflammation. Cell 157(1):121–141

Bikard D, Euler CW, Jiang W, Nussenzweig PM, Goldberg GW, Duportet X, Marraffini LA (2014) Exploiting CRISPR-Cas nucleases to produce sequence-specific antimicrobials. Nat Biotechnol 32(11):1146–1150

Bosi E, Monk JM, Aziz RK, Fondi M, Nizet V, Palsson BØ (2016) Comparative genome-scale modelling of *Staphylococcus aureus* strains identifies strain-specific metabolic capabilities linked to pathogenicity. Proc Natl Acad Sci 113(26):E3801–E3809

Brogden KA (2005) Antimicrobial peptides: pore formers or metabolic inhibitors in bacteria? Nat Rev Microbiol 3(3):238–250

Carlet J, Jarlier V, Harbarth S, Voss A, Goossens H, Pittet D, Participants of the 3rd World Healthcare-Associated Infections Forum (2012) Ready for a world without antibiotics? The Pensières antibiotic resistance call to action. Antimicrob Resist Infect Control 1:1–13

Cassini A, Högberg LD, Plachouras D, Quattrocchi A, Hoxha A, Simonsen GS, Hopkins S (2019) Attributable deaths and disability-adjusted life-years caused by infections with antibiotic-resistant bacteria in the EU and the European Economic Area in 2015: a population-level modelling analysis. Lancet Infect Dis 19(1):56–66

Citorik RJ, Mimee M, Lu TK (2014) Sequence-specific antimicrobials using efficiently delivered RNA-guided nucleases. Nat Biotechnol 32(11):1141–1145

Czaplewski L, Bax R, Clokie M, Dawson M, Fairhead H, Fischetti VA, Rex JH (2016) Alternatives to antibiotics—a pipeline portfolio review. Lancet Infect Dis 16(2):239–251

de Jong E, van der Oost J, Brouns SJ (2014) Molecular mechanisms of CRISPR-mediated microbial immunity. Cell Mol Life Sci 76(16):3049–3073

Deveau H, Garneau JE, Moineau S (2010) CRISPR/Cas system and its role in phage-bacteria interactions. Ann Rev Microbiol 64:475–493

Di Somma A, Moretta A, Canè C, Cirillo A, Duilio A (2020) Antimicrobial and antibiofilm peptides. Biomolecules 10(4):652

Du D, Wang-Kan X, Neuberger A, Van Veen HW, Pos KM, Piddock LJ, Luisi BF (2018) Multidrug efflux pumps: structure, function and regulation. Nat Rev Microbiol 16(9):523–539

Firdous SM (2014) Phytochemicals for treatment of diabetes. EXCLI J 13:451–453

Gao Y, Zhang X, Wei Q, Wen D, Zhang W (2021) Recent progress on antimicrobial peptides: varieties, mechanisms of action, and drug delivery systems. Eur J Med Chem 230:113853

Gomaa AA, Klumpe HE, Luo ML, Selle K, Barrangou R, Beisel CL (2014) Programmable removal of bacterial strains by use of genome-targeting CRISPR-Cas systems. MBio 5(1):10–1128

Guarner F, Malagelada JR (2003) Gut flora in health and disease. Lancet 361(9356):512–519

Hancock RE, Sahl HG (2006) Antimicrobial and host-defense peptides as new anti-infective therapeutic strategies. Nat Biotechnol 24(12):1551–1557

Hill C, Guarner F, Reid G, Gibson GR, Merenstein DJ, Pot B, Canani RB (2014) Expert consensus document: the international scientific Association for Probiotics and Prebiotics consensus statement on the scope and appropriate use of the term probiotic. Nat Rev Gastroenterol Hepatol 11(8):506–514

Holscher HD (2017) Dietary fibre and prebiotics and the gastrointestinal microbiota. Gut Microbes 8(2):172–184

Høyland-Kroghsbo NM, Paczkowski J, Mukherjee S, Broniewski J, Westra E, Bondy-Denomy J, Bassler BL (2017) Quorum sensing controls the *Pseudomonas aeruginosa* CRISPR-Cas adaptive immune system. Proc Natl Acad Sci 114(1):131–135

Hu M, Chen S, Ni Y, Wei W, Mao W, Ge M, Qian X (2023) CRISPR/Cas9-mediated genome editing in vancomycin-producing strain *Amycolatopsis keratiniphila*. Front Bioeng Biotechnol 11:1141176

Huys G, Botteldoorn N, Delvigne F, De Vuyst L, Heyndrickx M, Pot B, Dubois JJ, Daube G (2013) Microbial characterization of probiotics—advisory report of the Working Group "8651 Probiotics" of the Belgian Superior Health Council (SHC). Mol Nutr Food Res 57(8):1479–1504

Jiang W, Marraffini LA (2015) CRISPR-Cas: new tools for genetic manipulations from bacterial immunity systems. Ann Rev Microbiol 69:209–228

Jiang W, Bikard D, Cox D, Zhang F, Marraffini LA (2013) RNA-guided editing of bacterial genomes using CRISPR-Cas systems. Nat Biotechnol 31(3):233–239

Karvelis T, Gasiunas G, Miksys A, Barrangou R, Horvath P, Siksnys V (2013) crRNA and tracrRNA guide Cas9-mediated DNA interference in *Streptococcus thermophilus*. RNA Biol 10(5):841–851

Kawacka I, Olejnik-Schmidt A, Schmidt M, Sip A (2020) Natural plant-derived chemical compounds as *Listeria monocytogenes* inhibitors in vitro and in food model systems. Pathogens (Basel, Switzerland) 10(1):12

Kumar P, Kizhakkedathu JN, Straus SK (2018) Antimicrobial peptides: diversity, mechanism of action and strategies to improve the activity and biocompatibility in vivo. Biomol Ther 8(1):4

Lacaille-Dubois MA, Wagner H (1996) A review of the biological and pharmacological activities of saponins. Phytomed Int J Phytother Phytopharmacol 2(4):363–386

Lau KM, Yue GG, Chan YY et al (2019) A review on the immunomodulatory activity of *Acanthopanax senticosus* and its active components. Chin Med 14:25

Leung R, Covasa M (2021) Do gut microbes taste? Nutrients 13(8):2581

Louwen R, Staals RH, Endtz HP, van Baarlen P, van der Oost J (2014) The role of CRISPR-Cas systems in virulence of pathogenic bacteria. Microbiol Mol Biol Rev 78(1):74–88

Mahlapuu M, Håkansson J, Ringstad L, Björn C (2016) Antimicrobial peptides: an emerging category of therapeutic agents. Front Cell Infect Microbiol 6:194

Makarova KS, Haft DH, Barrangou R, Brouns SJ, Charpentier E, Horvath P, Koonin EV (2011) Evolution and classification of the CRISPR–Cas systems. Nat Rev Microbiol 9(6):467–477

Markowiak-Kopeć P, Śliżewska K (2020) The effect of probiotics on the production of short-chain fatty acids by human intestinal microbiome. Nutrients 12(4):1107

Mojica FJ, Díez-Villaseñor CS, García-Martínez J, Soria E (2005) Intervening sequences of regularly spaced prokaryotic repeats derive from foreign genetic elements. J Mol Evol 60:174–182

Prats-Ejarque G (2021) Exploring the pharmacological properties of human antimicrobial ribonucleases. UAB

Rath D, Amlinger L, Rath A, Lundgren M (2015) The CRISPR-Cas immune system: biology, mechanisms and applications. Biochimie 117:119–128

Reid G, Jass J, Sebulsky MT, McCormick JK (2003) Potential uses of probiotics in clinical practice. Clin Microbiol Rev 16(4):658–672

Ridyard KE, Overhage J (2021) The potential of human peptide LL-37 as an antimicrobial and anti-biofilm agent. Antibiotics (Basel, Switzerland) 10(6):650

Rupp ME, Fey PD (2003) Extended spectrum beta-lactamase (ESBL)-producing Enterobacteriaceae: considerations for diagnosis, prevention and drug treatment. Drugs 63(4):353–365

Rusu ME, Fizeșan I, Vlase L, Popa DS (2022) Antioxidants in age-related diseases and anti-aging strategies. Antioxidants (Basel, Switzerland) 11(10):1868

Ryan EM, Alkawareek MY, Donnelly RF, Gilmore BF (2012) Synergistic phage-antibiotic combinations for the control of *Escherichia coli* biofilms in vitro. FEMS Immunol Med Microbiol 78(3):214–218

Sancho-Vaello E, Gil-Carton D, François P et al (2020) The structure of the antimicrobial human cathelicidin LL-37 shows oligomerization and channel formation in the presence of membrane mimics. Sci Rep 10:17356

Sanders ME, Lenoir-Wijnkoop I, Salminen S, Merenstein DJ, Gibson GR, Petschow BW, Pot B (2014) Probiotics and prebiotics: prospects for public health and nutritional recommendations. Ann N Y Acad Sci 1309(1):19–29

Sanders ME, Merenstein DJ, Reid G, Gibson GR, Rastall RA (2019) Probiotics and prebiotics in intestinal health and disease: from biology to the clinic. Nat Rev Gastroenterol Hepatol 16(10):605–616

Selle K, Klaenhammer TR, Barrangou R (2015) CRISPR-based screening of genomic island excision events in bacteria. Proc Natl Acad Sci 112(26):8076–8081

Singh V, Braddick D, Dhar PK (2017a) Exploring the potential of genome editing CRISPR-Cas9 technology. Gene 599:1–18

Singh S, Singh SK, Chowdhury I, Singh R (2017b) Understanding the mechanism of bacterial biofilms resistance to antimicrobial agents. Open Microbiol J 11:53–62

Sternberg SH, Haurwitz RE, Doudna JA (2012) Mechanism of substrate selection by a highly specific CRISPR endoribonuclease. RNA 18(4):661–672

Suez J, Zmora N, Zilberman-Schapira G, Mor U, Dori-Bachash M, Bashiardes S, Korem T (2018) Post-antibiotic gut mucosal microbiome reconstitution is impaired by probiotics and improved by autologous FMT. Cell 174(6):1406–1423

Sweileh WM, Moh'd Mansour A (2020) Bibliometric analysis of global research output on antimicrobial resistance in the environment (2000–2019). Glob Health Res Policy 5(1):1–11

Tallei TE, Tumilaar SG, Niode NJ, Fatimawali F, Kepel BJ, Idroes R, Effendi Y, Sakib SA, Emran TB (2020) Potential of plant bioactive compounds as SARS-CoV-2 main protease (M^{pro}) and spike (S) glycoprotein inhibitors: a molecular docking study. Scientifica 2020:6307457

Tanaka N, Kashiwada Y (2021) Phytochemical studies on traditional herbal medicines based on the ethnopharmacological information obtained by field studies. J Nat Med 75(4):762–783

Tricarico PM, Gratton R, Santos-Silva CAD, Moura RRD, Ura B, Sommella E, Crovella S (2022) A rare loss-of-function genetic mutation suggest a role of dermcidin deficiency in hidradenitis suppurativa pathogenesis. Front Immunol 13:1060547

Wang H, La Russa M, Qi LS (2016a) CRISPR/Cas9 in genome editing and beyond. Annu Rev Biochem 85:227–264

Wang G, Li X, Wang Z (2016b) APD3: the antimicrobial peptide database as a tool for research and education. Nucleic Acids Res 49(D1):D255–D263

Wiedenheft B, Sternberg SH, Doudna JA (2012) RNA-guided genetic silencing systems in bacteria and archaea. Nature 482(7385):331–338

Wink M (2015) Modes of action of herbal medicines and plant secondary metabolites. Medicines 2(3):251–286

Yeung AT, Gellatly SL, Hancock RE (2011) Multifunctional cationic host defence peptides and their clinical applications. Cell Mol Life Sci 68:2161–2176

Zasloff M (2002) Antimicrobial peptides of multicellular organisms. Nature 415(6870):389–395

Zetsche B, Gootenberg JS, Abudayyeh OO, Slaymaker IM, Makarova KS, Essletzbichler P, Zhang F (2015) Cpf1 is a single RNA-guided endonuclease of a class 2 CRISPR-Cas system. Cell 163(3):759–771

The Rise of Antibiotic Resistance: A Global Threat, Origin, and Evolution of Antibiotic Resistance: Current Scenario and Future Prospective

12

Vishal Bhojyawal, Mansi Kesarwani, and Sweety Gupta

Abstract

The worldwide health issue of antibiotic resistance has become a serious threat to public health, having a substantial impact on medical care, healthcare systems, and socioeconomic stability. Antibacterial resistance is becoming more prevalent in both developed and developing countries and as a result, it is necessary to determine both the cause of the increase and its treatment. In antibiotic resistance, bacteria become resistant to the drugs intended to kill them through natural selection, where bacteria mutate their genetic sequence, enhance the use of existing medications, as well as for the design of new drugs less susceptible to resistance development, and novel strategies to combat resistance and develop ways to survive antibiotic treatment. Although the passage of genes and bacteria is restricted by several obstacles, pathogens constantly pick up new resistance factors from other species, limiting our ability to prevent and treat bacterial infections. If the issue of antibiotic resistance continues to grow, then infection will become more difficult to treat and raise the possibility of complications and death. Evolutionary events that lead to the generation of novel resistance factors in pathogens are uncommon and challenging to anticipate, but they may have massive repercussions. In this book, we explore recent developments in the origin and evolution of antibiotic resistance, highlighting the current scenario and prospects as well as mechanisms underlying the development of resistance in bacteria, including genetic mutations and horizontal gene transfer. Furthermore, we discuss the contributing factors such as inappropriate antibiotic use, overuse

V. Bhojyawal (✉) · S. Gupta
Department of Zoology, T. D. P. G. College, V. B. S. Purvanchal University, Jaunpur, Uttar Pradesh, India

M. Kesarwani
Department of Biochemistry, University of Allahabad, Prayagraj, Uttar Pradesh, India

M. Gangwar, G. Nath (eds.), *Emerging Paradigms for Antibiotic-Resistant Infections: Beyond the Pill*, https://doi.org/10.1007/978-981-97-5272-0_12

of antibiotics, how resistance genes contribute to the biology of the host, new structural details of relevant molecular events, and the identification of novel resistance gene families and the ways in which they interact. The misuse and overuse of antibiotics are the main causes of the development of drug-resistant. Finally, we explore how we could apply this knowledge to create next-generation medicine that could tackle antibiotic resistance problems. The escalating prevalence of multidrug-resistant pathogens poses challenges in the management of infectious diseases, leading to increased morbidity, mortality, and healthcare costs. Addressing antibiotic resistance requires a multifaceted approach enhanced surveillance, stewardship programs, novel drug discovery, and alternative treatment strategies. By understanding the complexities of antibiotic resistance and implementing concerted efforts globally, we can mitigate its impact and safeguard the effectiveness of antibiotics for future generations.

Keywords

Antibiotics resistance · Metal nanomaterials · Antimicrobial resistance · Mutation · Antibiotic stewardship program

12.1 Introduction

Just like cancer, diabetes, and high blood pressure, antibiotic resistance in medical healthcare is very common and spreads like the net. Before comprehending antibiotic resistance, we must first understand antibiotics and their impact on health as well as their significant role in medical science. Antibiotics are probably one of the most effective types of chemotherapy in medical history and start cleaning pathogenic microorganisms, saving millions of lives from disease (Aminov 2010). The term "antibiotic" originated from the word "antibiosis," which means "against life." When an ill individual consumes antibiotics, antibiotics kill the bacterial cell in the targeted location by either stopping it from reproducing or altering a vital biological process or function. However, this definition has been modified in modern times an antibiotic in low concentrations either kills off other microbes or prevents them from growing. The mode of antibiotic action is often classified into two types bactericidal, which kills other bacteria completely, and bacteriostatic, which inhibits bacterial development. The first antibiotic, Penicillin, was found in September 1928 by the late Sir Alexander Fleming, an English bacteriologist who accidentally obtained the antibiotic from a soil-inhabiting fungus. However, the discovery was first reported in 1929, and human clinical trials were carried out in 1940 (Etebu and Arikekpar 2016). The bacterial cell wall, the cell membrane, protein synthesis, DNA and RNA synthesis, and the metabolism of folic acid (vitamin B9) are the five main targets of antibiotics (Halawa et al. 2024). Nowadays, several distinct kinds of antibiotics are categorized according to their modes of action like aminoglycoside, chloramphenicol, tetracycline macrolide, and streptothricin, which limit protein synthesis, while rifampin and quinolones interact with DNA and RNA synthesis, other groups such as β-lactams and glycopeptides inhibit or destroy bacterial cell walls (Kapoor et al. 2017).

After studying the introduction of antibiotics, let us understand why antibiotic resistance is a top worldwide public health problem and why this topic has garnered the attention of several medical organizations. Various bacterial species developed the ability to tolerate antibiotics long before humans produced antibiotics to cure and prevent infectious illnesses. Antibiotic resistance can result from mutations in a bacterium's preexisting genome as well as the incorporation of foreign DNA (Larsson and Flach 2022). A report by the World Health Organization (WHO) in 2019 mentioned via antimicrobial resistance (AMR) that about 700,000 people were killed annually and by 2050, that number is expected to have increased to 20 million deaths, at a cost of more than $2.9 trillion (Uddin et al. 2021). Antibiotic-resistant bacterial infections may cause serious sickness, higher death rates, and a higher chance of complications and hospital admission. In the context of new antibiotics in the drug development pipeline, the WHO describes the future of a post-antibiotic world and warns that not only will this eliminate the improvements in healthcare made over the past 100 years but also that common infections could become unmanageable and life-threatening (Llor and Bjerrum 2014). To address this, several measures have been taken, including enhancing educational technology, organizing awareness camps, educating patients, the public, and schools by holding various seminars and events to educate students, and—most importantly—forming a team to make sure that no one purchases antibiotics without a prescription (Lee et al. 2015).

12.2 Historical Perspective

12.2.1 Emergence of Antibiotics

Infectious diseases have been a significant threat throughout human history, leading to numerous deaths and epidemics. Thankfully, advancements in medicine, sanitation, and public health measures have significantly reduced their impact in modern times. For much of history, understanding of infectious diseases was limited, and effective strategies for prevention and treatment were lacking. It wasn't until the development of modern medicine and the understanding of germ theory in the nineteenth century that significant progress was made in combating infectious diseases. It is a key milestone in the history of antibiotic research. The discovery of the healing effects of molds, such as penicillin by Alexander Fleming, marked a turning point in the fight against bacterial infections. Advancements in microscopy also allowed scientists to observe microorganisms, leading to the understanding of their role in causing diseases. Refutation of the abiogenesis theory helped clarify the origins of microorganisms, further advancing our understanding of infectious diseases. These breakthroughs paved the way for rapid progress in antibiotic research and development (Mohr 2016).

Over 2500 years ago, the ancient Chinese were the first known to utilize antibiotics. They recognized the healing properties of moldy soybeans and applied this substance to treat ailments like furuncles, carbuncles, and related infections. Many ancient civilizations, such as the Egyptians and Greeks, utilized molds and plants for treating infections, harnessing antibiotic properties from these organisms, although the compounds responsible for antibiotic action remained undiscovered during that era. The initial observations of microorganisms were made in 1665 by Robert Hooke, who, using a simple microscope, examined the structures of small organisms, plants, as well as microbes, and fungi (Kourkouta et al. 2018).

12.2.2 Early Signs of Resistance

After the widespread implementation of antibiotics in clinical settings, in the middle of the twentieth century, the first indications of antibiotic resistance started to appear. The emergence of penicillin resistance in *Staphylococcus aureus* in the late 1940s, a few years after penicillin became generally accessible, is among the first cases that have been described. As bacteria developed a variety of defense mechanisms to withstand the effects of antibiotics, this signaled the start of an ongoing arms competition between antibiotics and bacteria. Drug stewardship initiatives, new drug development, and ongoing surveillance are required to counteract the growing threat of antibiotic resistance. Over the ensuing decades, resistance spread to other bacteria and antibiotics.

Resistance has been observed in nearly all antibiotics that have been developed over time. Vancomycin was introduced for clinical use in 1972 to address methicillin resistance in both *S. aureus* and coagulase-negative *staphylococci*. While initially believed unlikely to occur in a clinical setting due to its resistance difficulty, cases of vancomycin resistance in coagulase-negative *staphylococci* were reported between 1979 and 1983. During the late 1960s to the early 1980s, the pharmaceutical industry introduced numerous new antibiotics to combat resistance. However, following this period, the antibiotic pipeline dwindled, resulting in fewer new drugs being introduced. Consequently, in 2015, many decades after the initial use of antibiotics, bacterial infections resurged as a significant threat.

12.3 Reasons Behind Antibiotic Resistance

On Earth, every organism extends its life by generating offspring. The primary goal of microbes is the same: live, replicate, and spread as quickly as possible. As a result, microorganisms adapt to their surroundings and evolve in ways that ensure their continued survival. If anything inhibits their ability to develop, then genetic and molecular modification may occur, rendering the bacteria resistant to the medicine and allowing them to survive. The evolution of drug resistance is the natural process of bacteria. Antibiotic misuse and abuse, inaccurate diagnosis and inappropriate antibiotic prescription, patient sensitivity loss and self-medication, poor healthcare settings, inadequate personal cleanliness, and extensive agricultural use are some additional factors that contribute to antibiotic resistance (Berendsen et al. 2010).

12.4 Mechanisms of Antibiotic Resistance

Among bacterial species, resistance levels might differ greatly within related bacterial groups. Susceptibility and resistance are often assessed in terms of the minimum inhibitory concentration (MIC), which inhibits bacterial growth. If a species' average MIC falls within the resistant range, the species is considered to have intrinsic resistance to that drug (Reygaert 2018). Additionally, bacteria can pick up resistance genes from other related organisms, the degree of resistance varies depending on the species and the genes received. Bacteria can develop intrinsic or acquired resistance; intrinsic resistance is a naturally existing trait that arises from the biology of the organism. It is always manifested in the species—for example, vancomycin resistance in *Escherichia coli*. Acquired resistance occurs when a bacterium susceptible to antibiotics becomes resistant, this can happen through mutation or the acquisition of new DNA.

In bacteria, resistance develops genetically, either through mutation or the acquisition of new resistance genes via genetic exchange processes. Antibiotics are known to activate pathogens' resistance mechanisms through chromosomal gene mutations, horizontal gene transfer of chromosomes, gene transfer, or plasmid-mediated antibiotic resistance (Munita and Arias 2016).

12.4.1 Mutational Resistance

A mutation is a spontaneous event that occurs whether an antibiotic is present. A portion of bacterial cells from a vulnerable population develop mutations in genes that impact the drug's effectiveness, preserving cell viability in the presence of an antibacterial compound. Generally speaking, the following pathways are used by mutations that lead to AMR to change the way antibiotics work: modifications to the antimicrobial target (lowering the drug's affinity) (a) the activation of efflux mechanism to expel the toxic chemical, (b) a reduction in drug absorption, (c) changes in important metabolic pathways through modulation of regulatory networks (Munita and Arias 2016). When a resistant mutant forms, the antibiotic destroys the susceptible population, allowing the resistant germs to thrive. For example, in gram-negative bacteria, fluoroquinolones are most commonly used against DNA gyrase so mutations in gyrase A are identified in isolates with low-level resistance, while higher MICs are associated with further mutations, particularly in parC but also in gyrase B and, less commonly, in parE. Resistance mediated by these mutations is increased by efflux pump expression. In gram-positive species, the main target in gram-positive species is typically parC for example sparfloxacin targets *gyrA*, while ciprofloxacin targets *parC* (Woodford and Ellington 2007).

12.4.2 Horizontal Gene Transfer Resistance

Horizontal gene transfer (HGT) is one of the main reasons for the acquisition of foreign DNA material and the dissemination of antibiotic resistance in vitro and in vivo environments (Sun et al. 2019). Mechanisms mediating the horizontal transfer of ARGs include transformation, conjugation transfer, transduction, membrane vesicles (MVs), and DNA packaged into virus-like particles, Bacterial mobile genetic elements (MGEs), such as Integrative and conjugative elements and conjugative plasmid, have been identified as the key vehicles for the spread of pathogens and antibiotic-resistant determinants (Tao et al. 2022).

Edward Tatum and Joshua Lederberg originally demonstrated conjugation in 1946. They demonstrated that bacteria could transfer genetic information unidirectionally across DNA, with a component known as the F (Fertility) mediating this process (Virolle et al. 2020). In other words, Conjugation is the contact-dependent process by which mobile genetic elements like plasmids and integrating and conjugation elements (ICEs) are transferred between bacteria that are close to one another through a pilus or pore (Partridge et al. 2018). The conjugation of two species or members of the same genus can transfer resistance genes. For example gram-negative bacteria can spread their plasmids containing carbapenemase resistance genes (blaKPC, blaNDM, and blaOXA-48) to other bacteria that are vulnerable (Hasan et al. 2021). Transformation is the most basic kind of horizontal gene transfer (HGT), in which recipient bacteria absorb extracellular DNA from lysed donor bacteria and incorporate it into their genomes to gain new features (Yu et al. 2022). Antibiotic resistance can be acquired by *Neisseria gonorrhoeae, Vibrio cholerae, and Streptococcus pneumoniae* by metamorphosis. Research has demonstrated that *E. coli* can be transformed by plasmid DNA, suggesting that the bacterium can take up DNA in the gut. Consequently, it is plausible that transformation plays a role in the spread of ARGs (Lichev et al. 2019). Plasmids and transposons are the most important MGEs, as they play critical roles in the establishment and spread of antibiotic resistance in therapeutically relevant species. Transduction employs mild bacteriophage as a carrier to transmit chromosomal and extrachromosomal DNA from the donor bacteria to the recipient bacteria and enable the recipient bacteria to acquire new features. *Staphylococcus aureus* is more likely to undergo resistance transduction (Foster 2017). Through phage-mediated transduction, mecA gene-carrying bacteria transfer resistance to methicillin-resistant *Staphylococcus aureus* (MRSA) (Craft et al. 2019). Finally, lysogenic conversion, transposition, protoplast fusion, and integrons represent one of the most effective mechanisms for the accumulation of AMR genes. Integrons are site-specific recombination systems that can recruit open-reading frames to form movable gene cassettes. Integrons provide an efficient and relatively easy technique for adding new genes to bacterial chromosomes, as well as the requisite apparatus to ensure their expression. They represent a robust genetic exchange approach and one of the primary drivers of bacterial evolution (Bhat et al. 2023).

12.5 Factors that Lead to Resistance to Antibiotics

Bacteria eventually develop a natural resistance to drugs. However, a few things can expedite the procedure, such as.

12.5.1 Antibiotic Misuse and Overuse

Antibiotic use when not necessary contributes to antibiotic resistance. Sir Alexander Fleming first highlighted the concern about antibiotic misuse in 1945 when he cautioned that the "**public will demand [the drug and] ... then will begin an era ... of abuses.**" Epidemiological studies conclude that there is a direct connection between the use of antibiotics and the development and spread of resistant bacterial strains, abuse and overuse result in complex, incurable, and protracted infections in people, raising medical expenses and occasionally resulting in death (Ventola 2015).

12.5.2 Agricultural Factors

Antibiotics are not limited to use in clinical settings; they can also be prescribed for livestock farming, where they can be used to treat animal diseases, and increase feed conversion efficiency, and prevent disease, and in concentrate animal feed at sub-therapeutic levels to promote animal growth. The level and rate of antibiotic utilization in the farming sector do not follow any precautions, and the level of antibiotics continues to increase and agricultural bacteria become resistant as well enhancing the level of resistance (Manyi-Loh et al. 2018). When antibiotics were first used in human medicine in the 1950s, they were investigated for their potential application in the treatment of plant diseases. Nearly 40 antibiotics were evaluated to combat plant diseases. Streptomycin has been the primary antibiotic used in crops (Plantomycin, Agrept, and Fructocin) to defend against fire blight since its introduction in 1955. The annual use of antibiotics in cash crops in the United States is predicted to be between 19,550 and 65,227 kilograms (Koch et al. 2021). When these animals release their waste into soil the soil absorbs antibiotics that directly enter the human body. For example, certain antibiotics, such as tylosin, chloramphenicol, and TCN (a powder mixture of oxytetracycline, chloramphenicol, and neomycin), have been linked to the development of aplastic anemia in humans, carcinogenicity, and the aggravation of renal disease (neomycin). It appears that there is a need to call for cooperation or team collaboration from farmers, individuals, veterinarians, consumers, and local pharmaceutical product vendors who must work together to ensure the prudent use of antibiotics in both clinical and agricultural settings throughout the world (Berendsen et al. 2010).

12.5.3 Environmental Factors

Antibiotic pollution is the increased concentration of antibiotics in the environment as a result of the careless and excessive use of antibiotics. Antibiotics can be released into the environment via streams of human waste, veterinary waste, and waste from raising livestock (Manyi-Loh et al. 2018). Vegetables and fruits can become contaminated by animal waste. Additionally, microorganisms found in vegetable plants may be capable of harming plants by developing antibiotic resistance. Furthermore, following spraying, a large percentage of antibiotics are left unused. Root vegetables that are grown in biosolids-containing soil might have higher levels of ARGs, resulting in adverse effects on the food chain and gene pool (Koch et al. 2021).

12.5.4 Not the Right Prescribing Patterns

Antibiotics that are given incorrectly have a major role in fostering AMR. **"Not the Right Prescribing Patterns"** describes the act of prescribing an antibiotic when it is not required, or when an antibiotic is chosen improperly or with the incorrect dosage and duration. According to a survey, 50% of patients received at least one antibiotic throughout their hospital stay without a valid reason. A 2017 report from the Centers for Disease Control and Prevention (CDC) stated that almost one-third of hospital patients had antibiotic prescriptions written without proper testing and for longer periods. Even worse conditions exist in nursing homes, where around 75% of antibiotic prescriptions are written improperly, with erroneous dosages and lengths of time.

12.5.5 Inadequate Personal Hygiene (Poor Hygiene)

Additional obstacles that affect the global development of AMR include the issues of hazardous waste, sewage, manure runoffs, water contamination, hygiene, and sanitation. In some nations, there is a discharge of waste from slaughterhouses and butcheries into the environment. Untreated animal waste is used as fertilizer, and surface water is a resource that both people and animals share.

Hazardous waste that is not properly managed can end up in drinking water, soils, food crops, groundwater, and sediments. The latter may result in harmful environmental impacts that affect animals and produce germs resistant to antibiotics, leading to a generalized outbreak of ailments in the population (EssackS. 2021).

12.6 Consequences of Antibiotic Resistance

Globally, antibiotic resistance significantly contributes to both economic burden and mortality. In areas where antibiotic consumption is higher, antibiotic resistance may be more common. Antibiotic use is widely unregulated and uncontrolled, and this issue needs to be addressed on a global scale. An investigation conducted in the United Kingdom found that 11.3% of participants did not complete their last antibiotic treatment as directed. When asked why they didn't finish the course, 65% of the participants said they felt better or didn't remember to take an antibiotic on time (Zaman et al. 2017). AMR has killed approximately 50,000 people annually in Europe and the USA alone, and thousands more people have died in other nations. Between 2015 and 2050, antibiotics resistance (ABR) might claim the lives of almost 2.4 million people in Europe, North America, and Australia if appropriate action is not taken. The majority of previous research indicates that patients with resistant infections require more expensive medication, require longer hospital stays, and more diagnostic testing, and have higher mortality rates than patients with nonresistant infections. In low-income nations, second- and third-line medications for drug-resistant infections are unavailable, increasing the chance of resistance to first-line antibiotics. The advancements in modern medicine are being jeopardized by antibiotic resistance. Organ transplants, chemotherapy, and surgeries become risky without the right antibiotics (Poudel et al. 2023).

12.7 Current Strategies for Combating Antibiotic Resistance

In the above, we covered every aspect that raises the likelihood of antibiotic resistance and its mechanism Now, we will talk about many ways to tackle AMR, focusing on antibiotic resistance.

12.7.1 International Measures

WHO recommended approaches such as

1. Enhanced cooperation amongst international agencies, professional associations, nongovernmental organizations, and governments.
2. Incentives for the investigation and creation of novel medications and vaccines,
3. International strategy for the management of fake antibiotics.
4. New networks that monitor the use of antibiotics and the AMR.

12.7.2 National Strategies

Like national committees with intersectoral coordination and regulatory actions, state Antibiotics should only be available with a prescription; producers should be legally required to publish statistics on antimicrobial distribution; and there should be incentives for sensible, second one National AMR Policy, India comprehend the origins, dissemination, and contributing variables of AMR, establish an antimicrobial program, justify the use of antimicrobials, and promote the development of novel, potent antimicrobials.

12.7.3 In a Medical Facility or Hospital

Forming a committee for infection prevention and control (IPC), effective infection diagnosis and management, monitoring the usage and resistance of antibiotics, and appropriate microbiological procedures (Uchil et al. 2014).

Antimicrobial Stewardship Programs (AMS) are designed to maximize antibiotic utilization by using particular guidelines, approaches, and plans to reduce antibiotic resistance and ensure that antimicrobial medications are appropriate and effective. In simple terms, it refers to a coordinated set of efforts to encourage the safe use of antimicrobials. AMS is one of the three pillars of a comprehensive strategy to strengthen the healthcare system. The other two are patient and medication safety as well as IPC. The AMS programs principles, which emphasize the prudent use of the agents, also apply to antimicrobials in the agriculture and livestock industries. The following objectives guide the AMS programs to collaborate with medical professionals to prescribe the **five "Ds" of** antimicrobial therapy: the appropriate **D**rug, recommended **D**ose, route of **D**ose, right **D**uration, and prompt **D**e-escalation to pathogen-directed therapy, to reduce resistance, stop the overuse, abuse, and misuse of antibiotics in community, inpatient, and outpatient settings—as well as in the agricultural sector, to lower the cost of healthcare-related expenses, and to lessen the negative effects of antibiotics, for instance, *Clostridioides difficile* (Shrestha et al. 2023).

Antibiotic combination therapy has also been shown to be an effective method for restoring bacteria susceptibility to antibiotics. Depending on the infectious microorganisms' susceptibility pattern, two or more medicines may be combined in a way that maximizes treatment effectiveness by producing a synergistic effect. When using antibiotics in combination, there are three potential modes of action (a) blockage of the same target using different mechanisms, (b) suppression of a different target within the same pathway (i.e., a combination of sulfamethoxazole and trimethoprim), and (c) blockage of the target in many pathways. In the current clinical settings, the first line of treatment for MDR infections will be combinational antibiotic therapy. For instance, colistin and polymyxin are used to treat infections brought on by multidrug-resistant gram-negative infections (Tängdén 2014).

Research on nontraditional antibiotic therapy options to fight bacterial resistance has exploded recently. Antimicrobial peptides (AMPs), inhibitors as adjuvant (β-lactamase inhibitors typically work in conjunction with β-lactam antibiotics by attaching themselves to the enzyme's active site and preventing it from doing its work. This keeps the antibiotic's lactam ring from hydrolyzing, preserving its structural integrity and having an antibacterial effect), metal nanomaterials (such as gold, silver, and zinc can be used directly for the detection and treatment of bacterial infections), photo-sensitive materials, and cationic polymers (chitosan is a natural polymer material with good biocompatibility, biodegradability, and antibacterial properties), liposomal drug delivery systems, carbon-based nano-delivery systems, biomimetic nano-delivery systems, polymer-based antibiotic delivery systems (PADSs) (Yao et al. 2023).

12.8 Future Perspectives

The challenge of antibiotic resistance is the predictable outcome of how we created and utilized antibiotics since their discovery. The future of the resistance issue is clear if we proceed with the development, use, and safeguarding of antibiotics for the next 80 years in the same manner as we did for the previous 80 years (Spellberg and Gilbert 2014). If we want to influence the future and have long-term access to effective antibiotic medication for illnesses, we must think disruptively and question long-held and sometimes beloved notions. New technologies that can more efficiently disinfect people, surfaces, and food—so much so that no antibiotics are needed for sterilization—must support conventional infection-prevention measures. Additionally, we require technology that allows for extensive medical treatment without requiring the implantation of extraneous elements like metal or plastic. We treat infections by addressing host targets rather than microbial targets because the infusion of monoclonal antibodies or white cells that attack bacteria offers promise for curing infections. New preclinical studies show that we can effectively implement treatments that either suppress the inflammatory reaction to an infection or restrict the growth of microorganisms by preventing them from accessing host resources, all without trying to kill the germs. *Acinetobacter baumannii,* for instance, was not killed by an antibiotic of a novel class (LpxC inhibitors), but it was able to stop the bacterium from causing illness in vivo by blocking the manufacture of gram-negative lipopolysaccharide (Spellberg et al. 2023).

Several approaches have been used to find new antimicrobials. Recently, the host defense peptide sequence was coupled to traditional antibiotics, improving their selectivity and efficacy against bacteria. Last but not least, AMPs have been nanoencapsulated using a variety of techniques, including hydrogels, dendritic polymers, liposomes, self-assembly, nanospheres, nanocapsules, carbon nanotubes, and DNA cages. These tactics provide increased antibacterial efficacy, decreased side effects, and unmistakable resistance against metabolic deterioration (Mandal et al. 2014).

Because there were more multidrug-resistant (MDR) strains, antimicrobial compounds and examination techniques such as lipopolysaccharide (LPS) inhibitors, efflux pump inhibitors, phytochemicals, phage therapy, and antimicrobial peptides (AMPs) were developed to control infections caused by MDR bacterial pathogens. To summarize, it is critical to create novel classes of antibiotics or antimicrobial medicines that target multidrug-resistant bacteria through distinct mechanisms of action. The treatment for bacterial infections often involves the use of combination drugs, yet even this pattern of combination dosage may cause pathogen resistance. It is critical to look for antimicrobial substances with a novel molecular approach to combat the problems caused by antibiotic resistance (Konwar et al. 2022).

12.9 Conclusions

Bacteria developing resistance is a global problem but not a novel one, which was cautioned by the antibiotic inventor Alexander Flamingo. He had said that if its use is not in an effective way it could lead to fatal circumstances and the same situation we are witnessing. As WHO in its reports cautioned about this emerging global problem, antibiotics are substances produced by many microorganisms that prevent the growth of other germs and eventually eradicate them. Microbes that are continuously exposed to these compounds develop resistance as a result of mutations that alter their gene sequence. Therefore, when these compounds are introduced to bacteria again, they do not affect them. Chromosomal gene mutations, horizontal gene transfer of chromosomes, gene transfer, or plasmid-mediated antibiotic resistance are some of the hotspot reasons that develop resistance in microbes toward antibiotics. In society, sometimes misuse, administering them unnecessarily, or not finishing the recommended dosage, due to less awareness, inadequate personal hygiene result in antibiotic resistance, which is currently regarded as a serious and prominent medical concern. The massive threat posed by multidrug resistance can be controlled by developing new medicines that specifically target bacterial pathogenicity. It is possible to modify the structural properties of antibiotics by creating powerful structures from preexisting antibiotics. The metallo antibiotics can be quite helpful in this situation. Phages might be very important in the treatment of bacterial infections in people. Phage therapy in conjunction with antibiotics is probably a future option. We should also focus on the regulation side of antibiotic distribution, prescription, and sale. For this, we need a proper mechanism to tackle ground-level supervision of small stores and to hospital level.

In summary, it is imperative to develop new classes of antibiotics or antimicrobial medications that specifically target bacteria that are resistant to several drugs by using unique mechanisms of action. Several techniques are being developed to overcome resistance such as immunity AMS Programs, combinational therapy, photo-sensitive materials, cationic polymers, phage therapy protons efflux pumps, and different drug delivery methods are introduced, but these ways are not sufficient. As a result, there is a strong need to create new tactics for controlling bacterial infections as well as their resistance to antibiotics.

References

Aminov RI (2010) A brief history of the antibiotic era: lessons learned and challenges for the future. Front Microbiol 1(134):134. https://doi.org/10.3389/fmicb.2010.00134

Berendsen B, Stolker L, de Jong J, Nielen M, Tserendorj E, Sodnomdarjaa R, Cannavan A, Elliott C (2010) Evidence of natural occurrence of the banned antibiotic chloramphenicol in herbs and grass. Anal Bioanal Chem 397(5):1955–1963. https://doi.org/10.1007/s00216-010-3724-6. Epub 2010 Apr 30. PMID: 20431869; PMCID: PMC2886120

Bhat BA, Mir RA, Qadri H, Dhiman R, Almilaibary A, Alkhanani M, Mir MA (2023) Integrons in the development of antimicrobial resistance: critical review and perspectives. Front Microbiol

31(14):1231938. https://doi.org/10.3389/fmicb.2023.1231938. PMID: 37720149; PMCID: PMC10500605

Craft KM, Nguyen JM, Berg LJ, Townsend SD (2019) Methicillin-resistant Staphylococcus aureus (MRSA): antibiotic-resistance and the biofilm phenotype. Medchemcomm 10(8):1231–1241. https://doi.org/10.1039/c9md00044e. PMID: 31534648; PMCID: PMC6748282

EssackS. (2021) Water, sanitation and hygiene in national action plans for antimicrobial resistance. Bull World Health Organ 99(8):606–608. https://doi.org/10.2471/BLT.20.284232. Epub 2021 Jun 1. PMID: 34354316; PMCID: PMC8319870

Etebu E, Arikekpar I (2016) Antibiotics: classification and mechanisms of action with emphasis on molecular perspectives. Int J Appl Microbiol Biotechnol Res 4(2016):90–101

Foster TJ (2017) Antibiotic resistance in Staphylococcus aureus. Current status and future prospects. FEMS Microbiol Rev 41(3):430–449. https://doi.org/10.1093/femsre/fux007. PMID: 28419231

Halawa EM, Fadel M, Al-Rabia MW, Behairy A, Nouh NA, Abdo M, Olga R, Fericean L, Atwa AM, El-Nablaway M, AbdeenA. (2024) Antibiotic action and resistance: updated review of mechanisms, spread, influencing factors, and alternative approaches for combating resistance. Front Pharmacol 12(14):1305294. https://doi.org/10.3389/fphar.2023.1305294. PMID: 38283841; PMCID: PMC10820715

Hasan CM, Dutta D, Nguyen ANT (2021) Revisiting antibiotic resistance: mechanistic foundations to evolutionary outlook. Antibiotics (Basel). 11(1):40. https://doi.org/10.3390/antibiotics11010040. PMID: 35052917; PMCID: PMC8773413

Kapoor G, Saigal S, Elongavan A (2017) Action and resistance mechanisms of antibiotics: a guide for clinicians. J Anaesthesiol Clin Pharmacol 33(3):300–305. https://doi.org/10.4103/joacp.JOACP_349_15. PMID: 29109626; PMCID: PMC5672523

Koch N, Islam NF, Sonowal S, Prasad R, SarmaH. (2021) Environmental antibiotics and resistance genes as emerging contaminants: methods of detection and bioremediation. Curr Res Microb Sci 14(2):100027. https://doi.org/10.1016/j.crmicr.2021.100027. PMID: 34841318; PMCID: PMC8610363

Konwar AN, Hazarika SN, Bharadwaj P, Thakur D (2022) Emerging non-traditional approaches to combat antibiotic resistance. Curr Microbiol 79(11):330. https://doi.org/10.1007/s00284-022-03029-7. PMID: 36155858; PMCID: PMC9510247

Kourkouta L, Tsaloglidou A, Koukourikos K, Iliadis C, Plati P, Dimitriadou A (2018) History of antibiotics. Sumerianz J Med Healthcare 1(2):51–54

Larsson DGJ, Flach CF (2022) Antibiotic resistance in the environment. Nat Rev Microbiol 20(5):257–269. https://doi.org/10.1038/s41579-021-00649-x. Epub 2021 Nov 4. PMID: 34737424; PMCID: PMC8567979

Lee CR, Lee JH, Kang LW, Jeong BC, Lee SH (2015) Educational effectiveness, target, and content for prudent antibiotic use. Biomed Res Int 2015:214021. https://doi.org/10.1155/2015/214021. Epub . PMID: 25945327; PMCID: PMC4402196

Lichev A, Angelov A, Cucurull I, Liebl W (2019) Amino acids as nutritional factors and (p)ppGpp as an alarmone of the stringent response regulate natural transformation in Micrococcus luteus. Sci Rep 9(1):11030. https://doi.org/10.1038/s41598-019-47423-x. PMID: 31363120; PMCID: PMC6667448

Llor C, Bjerrum L (2014) Antimicrobial resistance: risk associated with antibiotic overuse and initiatives to reduce the problem. Ther Adv Drug Saf 5(6):229–241. https://doi.org/10.1177/2042098614554919. PMID: 25436105; PMCID: PMC4232501

Mandal SM, Roy A, Ghosh AK, Hazra TK, Basak A, Franco OL (2014) Challenges and future prospects of antibiotic therapy: from peptides to phages utilization. Front Pharmacol 13(5):105. https://doi.org/10.3389/fphar.2014.00105. PMID: 24860506; PMCID: PMC4027024

Manyi-Loh C, Mamphweli S, Meyer E, Okoh A (2018) Antibiotic use in agriculture and its consequential resistance in environmental sources: potential public health implications. Molecules 23(4):795. https://doi.org/10.3390/molecules23040795. PMID: 29601469; PMCID: PMC6017557

Mohr KI (2016) History of antibiotics research. Curr Top Microbiol Immunol 398:237–272. https://doi.org/10.1007/82_2016_499. PMID: 27738915

Munita JM, Arias CA (2016) Mechanisms of antibiotic resistance. Microbiol Spectr 4(2). https://doi.org/10.1128/microbiolspec.VMBF-0016-2015. PMID: 27227291; PMCID: PMC4888801

Partridge SR, Kwong SM, Firth N, Jensen SO (2018) Mobile genetic elements associated with antimicrobial resistance. Clin Microbiol Rev 31(4):e00088–e00017. https://doi.org/10.1128/CMR.00088-17. PMID: 30068738; PMCID: PMC6148190

Poudel AN, Zhu S, Cooper N, Little P, Tarrant C, Hickman M, Yao G (2023) The economic burden of antibiotic resistance: a systematic review and meta-analysis. PLoS One 18(5):e0285170. https://doi.org/10.1371/journal.pone.0285170. PMID: 37155660; PMCID: PMC10166566

Reygaert WC (2018) An overview of the antimicrobial resistance mechanisms of bacteria. AIMS Microbiol 4(3):482–501. https://doi.org/10.3934/microbiol.2018.3.482. PMID: 31294229; PMCID: PMC6604941

Shrestha J, Zahra F, Cannady P Jr (2023) Antimicrobial stewardship. In: StatPearls [Internet]. StatPearls Publishing, Treasure Island, FL. 2024 Jan–. PMID: 34283434

Spellberg B, Gilbert DN (2014) The future of antibiotics and resistance: a tribute to a career of leadership by John Bartlett. Clin Infect Dis 59 Suppl 2(Suppl 2):S71–S75. https://doi.org/10.1093/cid/ciu392. PMID: 25151481; PMCID: PMC4176319

Spellberg B, Bartlett JG, Gilbert DN (2023) The future of antibiotics and resistance. N Engl J Med 368(4):299–302. https://doi.org/10.1056/NEJMp1215093. PMID: 23343059; PMCID: PMC3617123

Sun D, Jeannot K, Xiao Y, Knapp CW (2019) Editorial: horizontal gene transfer mediated bacterial antibiotic resistance. Front Microbiol 10:1933. https://doi.org/10.3389/fmicb.2019.01933. PMID: 31507555; PMCID: PMC6718914

Tängdén T (2014) Combination antibiotic therapy for multidrug-resistant gram-negative bacteria. Ups J Med Sci 119(2):149–153. https://doi.org/10.3109/03009734.2014.899279. Epub 2014 Mar 26. PMID: 24666223; PMCID: PMC4034552

Tao S, Chen H, Li N, Wang T, Liang W (2022) The spread of antibiotic resistance genes in vivo model. Can J Infect Dis Med Microbiol 18(2022):3348695. https://doi.org/10.1155/2022/3348695. PMID: 35898691; PMCID: PMC9314185

Uchil RR, Kohli GS, Katekhaye VM, Swami OC (2014) Strategies to combat antimicrobial resistance. J Clin Diagn Res 8(7):ME01–ME04. https://doi.org/10.7860/JCDR/2014/8925.4529. Epub 2014 Jul 20. PMID: 25177596; PMCID: PMC4149102

Uddin TM, Chakraborty AJ, Khusro A, Zidan BRM, Mitra S, Emran TB, Dhama K, Ripon MKH, Gajdács M, Sahibzada MUK, Hossain MJ, Koirala N (2021) Antibiotic resistance in microbes: history, mechanisms, therapeutic strategies and future prospects. J Infect Public Health 14(12):1750–1766. https://doi.org/10.1016/j.jiph.2021.10.020. Epub 2021 Oct 23. PMID: 34756812

Ventola CL (2015) The antibiotic resistance crisis: part 1: causes and threats. P T 40(4):277–283. PMID: 25859123; PMCID: PMC4378521

Virolle C, Goldlust K, Djermoun S, Bigot S, LesterlinC. (2020) Plasmid transfer by conjugation in gram-negative bacteria: from the cellular to the community level. Genes (Basel) 11(11):1239. https://doi.org/10.3390/genes11111239. PMID: 33105635; PMCID: PMC7690428

Woodford N, Ellington MJ (2007) The emergence of antibiotic resistance by mutation. Clin Microbiol Infect 13(1):5–18. https://doi.org/10.1111/j.1469-0691.2006.01492.x. PMID: 17184282

Yao J, Zou P, Cui Y, Quan L, Gao C, Li Z, Gong W, Yang M (2023) Recent advances in strategies to combat bacterial drug resistance: antimicrobial materials and drug delivery systems. Pharmaceutics 15(4):1188. https://doi.org/10.3390/pharmaceutics15041188. PMID: 37111673; PMCID: PMC10141387

Yu Z, Wang Y, Henderson IR, Guo J (2022) Artificial sweeteners stimulate horizontal transfer of extracellular antibiotic resistance genes through natural transformation. ISME J 16(2):543–554. https://doi.org/10.1038/s41396-021-01095-6. Epub 2021 Sep 1. PMID: 34465899; PMCID: PMC8776823

Zaman SB, Hussain MA, Nye R, Mehta V, Mamun KT, Hossain N (2017) A review on antibiotic resistance: alarm bells are ringing. Cureus 9(6):e1403. https://doi.org/10.7759/cureus.1403. PMID: 28852600; PMCID: PMC5573035

13 Functional Nanomaterials in Antimicrobial Therapy for Antibiotic-Resistant Bacterial Infections

Gudikandula Krishna, Dasari Thrimothi, and V. Srilekha

Abstract

Nanotechnology's clinical applications are gaining popularity, especially as a potential therapy method for infectious disorders. Multiple drug-resistant microbes (MDROs) are a worldwide concern for the occurrence of diseases leading to illness and death. The rise in illnesses brought on by resistance to antibiotics bacterial strains has increased the need for investigating and creating new bactericidal medications or unconventional approaches that can effectively address antimicrobial resistance. Nanomaterial-based therapies show great potential in addressing severe bacterial infections by circumventing resistance to antibiotic pathways. Strategies based on nanomaterials, particularly those that do not depend on tiny-molecule antibiotics, exhibit promise as they circumvent drug-resistant bacterial mechanisms. Nps, also known as Nanoparticles, possess a size that allows them to traverse the membranes of cells of harmful bacteria and disrupt crucial molecular pathways. In addition, they can specifically target biofilms and eradicate infections that have demonstrated resistance to conventional treatment methods.

Keywords

Antibiotics · Antibiotic resistance · Nanomaterial · Nanoparticles

G. Krishna (✉) · D. Thrimothi
Department of Microbiology, Kakatiya University, Warangal, Telangana, India

V. Srilekha
Department of Biotechnology, Chaitanya (Deemed to be University), Warangal, Telangana, India

M. Gangwar, G. Nath (eds.), *Emerging Paradigms for Antibiotic-Resistant Infections: Beyond the Pill*, https://doi.org/10.1007/978-981-97-5272-0_13

13.1 Introduction

In the past, infectious diseases have triggered pandemics and epidemics, leading to a significant loss of life. The Plague of Athens, which occurred during the Peloponnesian War in 430 BC, and the COVID-19 pandemic, which caused 14.9 million fatalities according to the WHO in 2022, both showed the significant influence of illnesses on the worldwide population. Antimicrobial drugs are developed to decrease infection mortality rates (WHO 2022). Santonin, cathelicidins, and amino glycosides are potent antibacterial substances obtained from microbes, plants, and animals (Swamy et al. 2017; de Barros et al. 2019).

Antibiotic resistance continues to be a significant concern for public health, showing a steady increase in resistance rates over recent decades. This condition affects many microorganisms, making many antibiotics less efficient or ineffective against infections. As a result, common diseases that were once easily curable have become more challenging, leading to more extended hospital stays, increased healthcare costs, and higher fatality rates. The slow rate of new antibiotic discovery is a primary concern linked to antibiotic resistance. The poor financial performance of antibiotics has forced several pharmaceutical companies to stop their research and development of these drugs. Developing a new antibiotic is a lengthy and expensive process, which has resulted in less interest in creating innovative solutions in this crucial medical field. Improper and inaccurate use of antibiotics is a significant factor in developing resistance. Antibiotics are often provided unnecessarily in healthcare facilities and for common community ailments. Moreover, using antibiotics in farming and animal breeding worsens the expansion of the resistance, mainly when antibiotics are used to boost growth in livestock, which might transfer people via the chain of food.

The occurrence of antibiotics in the environment is a result of using antibiotics excessively and inappropriately. Upon ingestion, antibiotics undergo enzymatic degradation in the body, converting the antibiotic into its active form. This active component then inhibits bacterial growth thereby treating the infection. There are higher amounts of antibiotics in nature as a result of unmetabolized antibiotics or antibiotic residues that are excreted through urine or feces and reach the environment through sewage systems. The existence of elevated antibiotic concentrations in the environment triggers a process of natural selection in bacteria, leading to the development of mutations and the acquisition of resistance to current antimicrobials (Kollef et al. 2017). Superbugs, bacteria resistant to antibiotics, have obtained genes that allow them to withstand antibiotics by employing specific processes. As an illustration, methicillin-resistant *S. aureus* (MRSA) acquired resistance to methicillin by genetic mutations in the structural gene that is resistant to methicillin (mecA gene or SCCmec gene). Over 100,000 infections were caused by MRSA in 2019, posing a serious risk to public health.

To tackle this difficulty, different long-term strategies have been employed to suggest practical solutions for combating antibiotic-resistant bacterial infections (Sun et al. 2021), including identifying new inhibitors (Allen et al. 2014), disrupting quorum sensing (Gunaratnam et al. 2021), and creating novel antibiotics (Miethke et al. 2021). Considerable and extensive work has been dedicated to short-term tactics, such as chemically altering current antibiotics to combat drug resistance. The development of novel antibiotics and the chemical alteration of current medications often lead to reduced target selectivity, eliminating many powerful treatments from clinical use (Yacoby et al. 2007). To overcome this limitation, a possible approach is the rational functionalization of antibacterial drugs through enhancing medication selectivity and targeting capability. This can be achieved by linking medicines that bind to specific targets that bind to bacterial cells (Bertrand et al. 2014). Targeting capability is achieved through particular ligands that may identify biomolecules or receptors in high amounts on the surface of bacterial cells or infected tissues, leading to increased drugs' interactions with bacterial cells. For instance, Vitamin B9, a small molecule-targeted ligand, has been extensively used in clinical treatment (Friedman et al. 2013). Using targeted ligands has numerous benefits, including robust binding affinity, precise specificity, inherent stability, and widespread commercial availability (Deusenbery et al. 2021). The specific systems demonstrate a notable increase in the local concentrations of antibacterial drugs in targeted areas compared to off-target regions. Recently, targeted ligands, including antibodies, aptamers, and peptides, have been extensively used and have shown strong inhibitory effects against bacteria. In the past 10 years, nanotechnology has garnered significant interest for its antibacterial properties, demonstrating the ability to improve the treatment of antibiotic-resistant microorganisms. Nanomaterials possess antibacterial properties that can enhance the effectiveness of current antibiotics or work independently to combat bacteria (Afrasiabi et al. 2020).

Nanomaterials offer numerous benefits for antibacterial purposes, particularly in addressing antibiotic-resistant microorganisms. Nanomaterials can serve as antibiotic transporters, aiding in enhancing drug accumulation, decreasing overall drug exposure, promoting bacterial absorption, and improving drug stability (Santos et al. 2018). These antibiotic carriers can penetrate the cell membrane barrier and transport antibiotics to eradicate antibiotic-resistant bacteria. In addition, nanoparticles can deliver various antibiotics and release them in a controlled manner to stop the development of microorganism's resistant to antibiotics (Makabenta et al. 2021). Their novel bactericidal processes, which differ from antibiotics, have been proven. New research shows that these nanomaterials' antibacterial properties come from their ability to disrupt bacterial cell membranes through non-oxidative bactericidal pathways, release metal ions, and make reactive oxygen species (ROS) (Wang et al. 2017a, b).

Nanomaterials' physical features, such as size, shape, and structure, can be readily customized and managed. Finally, nanomaterial surfaces can be customized, allowing for various antibacterial properties. It is crucial to comprehend the inherent properties and antibacterial mechanisms to create innovative and efficient antibacterial techniques utilizing the enhanced features of nanomaterials (Harish et al. 2023). This review examines nanomaterial-based targeted antibacterial techniques that enhance bacteria death by efficiently accumulating nanomaterials passively or actively at infection sites or around bacteria cells.

13.2 Global Antibiotic Consumption

In recent times, there has been a global surge in the utilization of antibiotics, driven mainly by improvements in the field of antibiotic discovery and the continuous escalation of infectious diseases. Klein et al. (2018) reported a significant 65% increase in global antibiotic usage from 2000 to 2015. The total number of defined daily doses (DDDs) rose from 2.11 billion to 34.8 billion DDDs during this period, mostly in low-income countries. The utilization of antibiotics in India increased significantly from 3.2 billion to 6.5 billion Defined Daily Doses (DDD). Compared with industrialized nations such as Japan, the rise in Defined Daily Dose (DDD) was just 6.96% between 2009 and 2013, corresponding to a range of 14,700 DDDs to 15,800 DDDs. Between 2009 and 2013, Japan experienced the most significant increase in the usage of antibacterial quinolones, with a substantial surge of 20.79% (Muraki et al. 2016).

13.3 Origin of Antibiotic Resistance

The ability of bacteria to resist the effects of antibiotics intended to eradicate or hinder their growth is known as antibiotic resistance. A bacterial strain's ability to survive through resistance mechanisms and the degree of resistance it expresses are two examples of factors that affect antibiotic resistance (Orfali et al. 2022). Microbial strains can exhibit natural resistance or react aggressively to the transfer of genetic material between bacteria by various mechanisms such as transgenes, genetic elements, plasmids and phages, or by changes in cell genes, leading to cross-resistance (Davies and Davies 2010). Resistant microbes can rapidly multiply if resistance genes are present on plasmids. Biochemical processes operate to defend the bacterial cell wall against different substances by causing changes in the target, enzymatic breakdown and altering the absorption of efflux pump proteins. Antibiotics of the first-generation encounter resistance to antibiotics in various therapeutic settings due to natural mechanisms (Bartlett et al. 2013).

Antibiotic resistance constitutes a worldwide threat to human health due to its association with elevated levels of mortality and morbidity. Gram-negative and gram-positive bacterial infections have become challenging to treat because of multidrug resistance, rendering standard antibiotics ineffective. Antibiotic resistance has significantly reduced the effectiveness of antibiotics in clinical settings both before antibiotics were available and after they were introduced (Velez and Sloand 2016). Antibiotic resistance was first recorded when sulfonamides were introduced in 1930, indicating the potential for environmental resistance in its inherent state before the antibiotic era but not confirming the presence of highly resistant pathogens. The development and utilization of antibiotics in various areas of life marked the start of the antibiotic era. High concentrations of antibiotics resulting from human activities led to a significant change in their effectiveness and the rapid proliferation of antibiotic-resistant bacteria.

Antibiotic resistance has existed since the early days of antibiotic use in clinical settings, and it has become widespread and resistant because pharmacological agents can deactivate the cell walls of bacteria. Studies indicate that chemical alterations can be made to antibiotics to prevent or reduce their breakdown by penicillinases (β-lactamases) (D'Costa et al. 2006). Detecting and deactivating bacterial penicillinase might improve the effectiveness of antibiotics by utilizing genes or components from other beneficial microbes, as indicated by many research studies. Antibiotic resistance in bacterial strains has been influenced by increased antibiotic use, genetic modification, and the presence of Antibiotic resistance genes before human intervention. Following their release from natural genetic sources, they swiftly proliferated among commensal and infectious bacteria belonging to several taxonomic groups.

The lack of effective preventive measures, challenges in managing bacterial infections and related diseases, and limited development of new antibacterial drugs highlight the need for innovative treatment strategies and wide-spectrum antibiotic regimens. Patients with bacteremia and other severe infections lack prompt identification of pathogenic microorganisms and their antibiotic susceptibility patterns. Researchers suggest that understanding infectious disease-induced biochemical processes and specific strategies for virulence can provide new ways to target and interact with important pathogenicity factors or characteristics of bacterial infectivity without inducing resistance through evolution (Fig. 13.1).

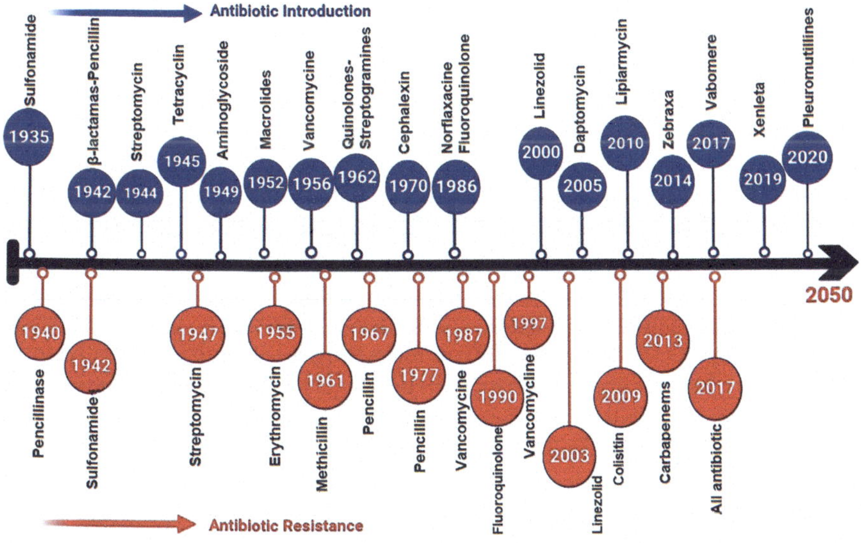

Fig. 13.1 A timeline that shows the development of antibiotics (Helmy et al. 2023)

13.4 The Necessity for Anti-Bacterial Nanomaterials and the Burden of Antibiotic Resistance

Microbes' resistance to antimicrobial assault demonstrates adaptation and represents the highest point of evolution. Research indicates that microbial infection causes over three million fatalities in developing countries each year. Approximately ten million deaths occur each year due to this issue, primarily in tropical regions, despite significant progress in diagnoses and treatments (Fenollar and Mediannikov 2018). Developed countries are also affected. Antimicrobial drug resistance is becoming an important worldwide public health concern in medicine. The economic impact linked to multidrug-resistant (MDR) microorganisms in the United States is estimated to be around $20 billion annually (Munita and Arias 2016). Clinicians often lack a dependable option to care for people with infections due to the challenge of resistance. The problem has been exacerbated by the proliferation of resistant to multiple drugs and super-resistant bacteria, which are resistant to virtually all antibiotics. This problem is closely linked to prescribing too many antibiotics, which then encourages the growth and spread of strains that are genetically and physically different (Wang et al. 2017a, b).

Microorganisms that are extensively and repeatedly resistant to drugs continue to be a significant public health concern since it is associated with increased rates of death, illness, and expensive treatment (Roca et al. 2015). According to Lebeaux et al. (2013), biofilms produced by various bacterial strains are associated with around 65–80% of human disorders. This exacerbates the situation further. Cells within biofilms exhibit a significantly lower susceptibility to antibiotics compared with planktonic cells, often ranging from 100 to 1000 times less vulnerable (Saginur et al. 2006). Biofilms become drug-resistant for several reasons, including less drug getting through the extracellular matrix, lower drug levels, slower bacterial metabolic rates, and the passing on of resistant genes (Hall and Mah 2017). Drug-resistant isolates of multiple drugs of *Staphylococcus aureus* and *Pseudomonas aeruginosa* primarily cause chronic infections associated with biofilms worldwide. These infections are characterized by the following: slow healing of wounds, malfunction of catheters, and longer stays in the hospital (Nathwani et al. 2014).

Antimicrobial drug resistance has been recognized by the World Health Organization (WHO) as one of the three major global public health issues in the twenty-first century. The ESKAPE group, consisting of *Staphylococcus, Enterococcus, Pseudomonas, Actinobacter, Klebsiella* and *Enterobacter* are the most critical and concerning pathogens. They are all linked to a high rate of death. According to WHO statistics, over 80% of multi-drug resistant or extensively drug-resistant bacteria result from the inappropriate use and excessive use of antibiotics, leading to infections with significant side effects. Currently, the options for treatment and prevention are restricted because of the proliferation of multidrug-resistant bacteria and other resistant infections. Alternative therapy approaches are required for microbial infections. Due to the ineffectiveness of many antibiotics, it is essential to explore improved therapeutic alternatives. Using medications that require the smallest possible dose to treat the infection is a suitable strategy (Morgan et al. 2011).

Nanotechnology shows (Hussain et al. 2018) promise as a therapeutic approach because of its high effectiveness and therapeutic index against germs. Nanoparticles provide a practical option for treating many bacterial illnesses, particularly those caused by multi-drug resistance pathogens. This can be employed alone or with antibiotics to achieve synergistic solid effects. Nanomaterials that may react to various internal and external stimuli to eliminate microorganisms and improve drug delivery and release are considered promising approaches (Qiu et al. 2018).

13.5 Antibacterial Nanoparticles' Mode of Action

Even though the mechanism of action of metal oxide and metal nanoparticles has yet to be completely comprehended, numerous different modes of action have been proposed for certain metals. NPs, regardless of their chemical makeup, can combat bacteria in multiple ways, making it challenging for bacteria to build resistance. Additionally, nanoparticles can transport antibiotics to bacteria as drug carriers, leading to increased therapeutic effectiveness and reduced total drug exposure. NPs combat several infections mainly through cell wall and cytoplasmic membrane rupture and the creation of ROS, causing oxidative stress. Additionally, they may inhibit enzymes, alter gene expression, and deactivate proteins to a lesser degree (Gao and Zhang 2021).

NPs aggregate on the surface, forming "pits" in the bacterial wall. Consequently, they can breach the cell wall, leading to alterations in the cell membrane, structural harm, and cell demise (Muthukrishnan et al. 2019). Positively charged ions generated by nanoparticles have a bactericidal effect that is strengthened by attaching to bacteria's negatively charged surface by biosorption involving phosphate and carboxyl groups. In addition, electrostatic attachment to the cell wall results in membrane potential changes, membrane depolarization, and integrity damage. This ultimately leads to the stoppage of energy transduction and the cell's death (Pelgrift and Friedman 2013). Due to the dense peptidoglycan coating of gram-positive bacteria, nanoparticles have difficulty in penetrating the bacterium and instead interact only with the bacterial surface (Sarwar et al. 2016).

Reactive oxygen species (ROS) induces oxidative stress due to its high positive redox potential. ROS can be triggered directly by respiratory chain disturbances or NPs (Natan and Banin 2017). Singlet oxygen (O_2), superoxide radical (O_{2-}), hydroxyl radical (OH), and hydrogen peroxide (H_2O_2) are examples of ROS. Various nanoparticles produce distinct combinations of ROS, leading to varied antibacterial effects. Ag and Cu nanoparticles have all ROS types.

In contrast, MgO nanoparticles only generate the ZnO nanoparticles, and superoxide radical exclusively produces a mix of hydroxyl extreme and hydrogen peroxide. ROS are caused due to flaws and vacancies in the crystal lattice during the fundamental process (Malka et al. 2013). The generation and elimination of ROS are typically in equilibrium. High levels of stress lead to increased generation of ROS, altering the permeability of the cell membrane and resulting in bacterial damage (Yuan et al. 2017). Multiple methods facilitate ROS generation. The

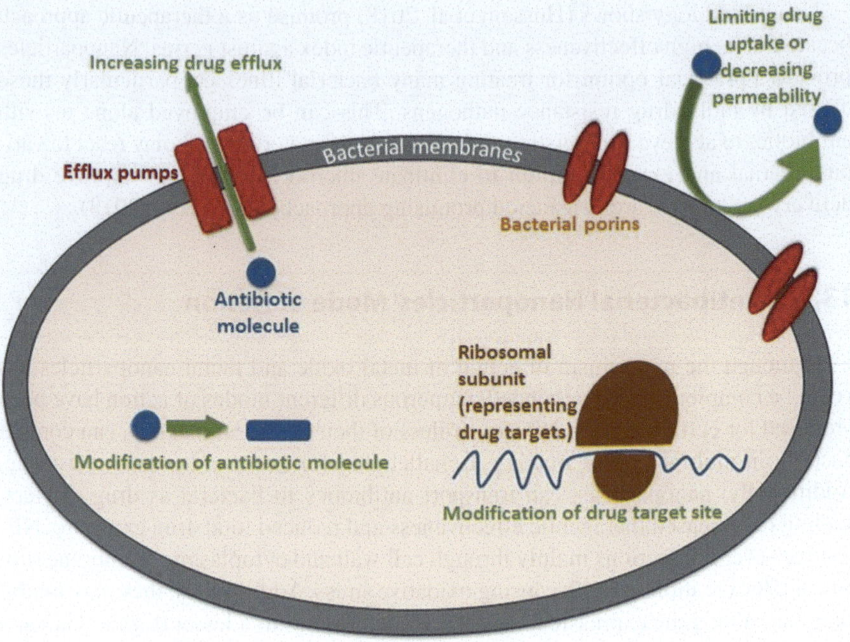

Fig. 13.2 Antimicrobial resistance mechanisms in bacterial cells (Helmy et al. 2023)

photocatalytic hypothesis suggests that when nanoparticles are exposed to energy higher than the band gap, conduction band is stimulated by valence band electrons, creating a hole in the valence band and generating reactive substances on and within the material. ROS inside and outside the cell can damage the cell membrane via oxidation of lipids, leading to the generation of free radicals (Ortiz-Benítez et al. 2019). The thickness and negative charge of the composition of gram-positive bacteria's cell walls make it harder for oxygen radicals like OH^- to permeate. Cellular components may be harmed by oxidative stress such as proteins, lipids, enzymes, nucleic acids, and ROS, leading to lipid peroxidation, protein modification, enzyme inhibition, and RNA or DNA damage. The formation of ROS from zinc oxide (ZnO), silver (Ag), iron oxide, and titanium dioxide (TiO_2) NPs has been observed to have notable antimicrobial effects (Fig. 13.2) (Quinteros et al. 2016; Tian et al. 2018; Kariminezhad et al. 2020; Priyadarshini et al. 2020; Arakha et al. 2015).

13.6 Promising Antibacterial Agents Are Metal Oxide and Metallic Nanoparticles

Silver nanoparticles (AgNps) are considered the future of antibiotics. This is due to their tremendous effectiveness in suppressing bacteria. At the moment, AgNps are the most prevalent nanoparticles in the realm of commercially available nanomaterials (Arya et al. 2019). Antimicrobial research focusing on nanoparticles has expanded gradually due to their reduced toxicity in comparison with alternative nanoparticles. The first phase of the cytotoxic mechanism of AgNps nanoparticles entails their adhesion and infiltration into the surface of the microbial membrane. The Ag^+ ions cause damage to the interior components, resulting in oxidative stress and ROS generation. The presence of ROS and stress impacts the function of the K^+/Na^+ ATPase pump and transduction of signaling pathways (Flores-López et al. 2019).

The presence of phosphorus-containing molecules in DNA allows their interaction with Ag^+ ions and AgNps. This contact leads to the protein inactivation and ultimately, the death of cells. The presence of thiols, chlorine, sulfur, and oxygen can considerably influence the release of Ag^+ ions (Maurer and Meyer 2016). The size of the Ag^+ ion has a considerable impact on its release rate (Abuayyash et al. 2018). Small AgNps have a high ability to enter the cell wall. They modify the design and structural stability of the membrane, leading to an increase in permeability and ultimately causing cell death. The bacterial species type also influences the action of AgNps. Cell wall composition, thickness, and layout changes are responsible for this phenomenon (Tamayo et al. 2014). It was explained by Hamouda et al. (2019) that AgNps kill bacteria by creating ROS, which damage the cell wall and plasma membrane. This is a result of membrane lipid peroxidation and protein inactivation. The membrane's structural integrity is altered by these processes, leading to the malfunction of the proteins involved in trafficking. Additionally, it leads to the release of potassium.

13.6.1 Gold NPs

Gold NPs often kill bacteria by stopping tRNA from attaching to the ribosome during DNA transcription or by sticking to the bacteria's cell wall and changing its abilities. This leads to a drop in adenosine triphosphate levels, cell contents leak, and ultimately cell death (Gupta et al. 2016). Gold NPs exhibit greater efficacy against gram-negative bacteria because they are more easily integrated into the bacterial cells (Shikha et al. 2020). Gold NPs may possess a mechanism that is not reliant on ROS, making them appear safer for mammalian cells (Lee and Lee 2018).

13.6.2 Titanium Dioxide NPs

Titanium dioxide NPs possess photocatalytic capabilities that enable them to eliminate bacteria by exposure to ultraviolet light. ROS, which are produced as a result of this process, leads to oxidative stress and damage proteins, lipids, and DNA (Ripolles-Avila et al. 2019). Nevertheless, TiO_2 NPs retain their antibacterial activities even without light. In this scenario, NPs adhere to the surface and directly interact with the cell wall, disrupting membrane integrity (Khater et al. 2020). Adding metal NPs through doping improves their antibacterial characteristics and aids in combating bacteria (Györgyey et al. 2016; Tallósy et al. 2014).

13.6.3 Zinc Oxide NPs

Zinc oxide NPs demonstrate that potent photocatalytic capabilities are equivalent to TiO_2 nanoparticles. UV irradiation or not, they generate ROS that may hinder DNA replication, denature proteins, or disrupt cell membranes, leading to antibacterial solid action (Karimzadeh et al. 2020). Another suggested way it works is by releasing zinc ions, leading to the buildup of ZnO NPs on the membrane of bacteria due to electrostatic forces. This process stops transfer of electrons across membrane or cell entry, which stops enzymes from working and damages DNA or mitochondria. This stops bacteria from growing and kills cells (Lianga et al. 2020). Kadiyala et al. (2018) proposed a method that involves altering energy metabolism in cells, leading to heightened synthesis of pyrimidines (particularly uridine monophosphate production), metabolism of carbohydrates, and reduced synthesis of amino acids. This mechanism also accounts for more antibacterial effectiveness against *S. aureus* than *E. coli*, as *E. coli* does not depend on uridine for anaerobic growth.

13.6.4 Iron Oxide NPs

Iron oxide NPs function by liberating metal ions that disturb bacterial cells through the membrane's permeation and disruption of electron transport. Alternatively, they can produce ROS, which damage DNA and proteins (Baptista et al. 2018).

13.6.5 Copper and Copper Oxide NPs

Copper NPs can alter shapes and interact with cell membranes, leading to a reduction in transmembrane electrochemical potential due to the ions they release. This leads to membrane damage and causes cell death (Roy et al. 2016). Additionally, they have the ability to release ROS, which may damage DNA, lipid peroxidation, and mitochondria. ROS, which are also produced by copper oxide, can damage DNA and break membranes, impairing vital enzymes and leading to cell death (Rajivgandhi et al. 2019; Bezza et al. 2020). When CuO NPs are present, they change the expression of crucial proteins. This has a major impact on the metabolism of bacteria, including active transport and electron transfer, as well as bacterial denitrification (Su et al. 2015).

13.6.6 Platinum NPs

Similar to other metal NPs, platinum NPs penetrate the cytoplasmic membrane and cell wall, resulting in DNA damage, ROS, cell accumulation during the cell cycle's S-phase, and ultimately, cell death. (Hashimoto et al. 2017).

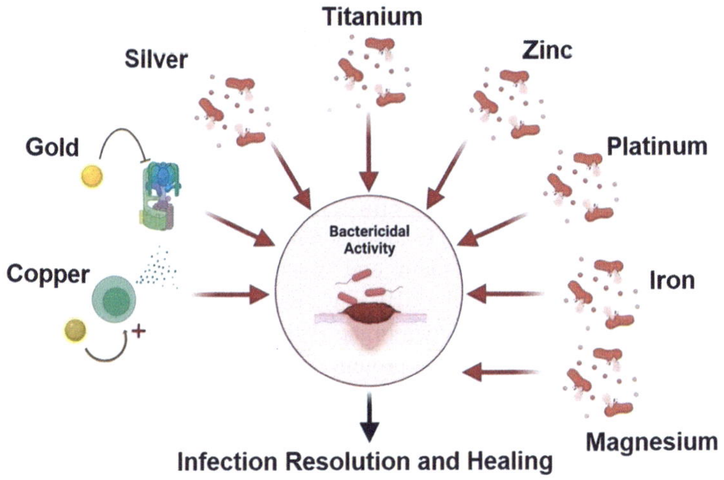

Fig. 13.3 Mechanisms of action of NPs

13.6.7 Magnesium Oxide NPs

Both the particle size and pH influence the antibacterial properties of magnesium oxide. Elevated pH levels have the capability to harm the membrane of the cell ultimately resulting in the demise of bacterial cells (Cai et al. 2018). ROS are formed on the surface of magnesium oxide NPs. These break down cell walls, releasing contents inside cells and eventually killing the cells (Fig. 13.3) (Bhattacharya et al. 2019).

13.7 NPs Have Synergistic Antibacterial Activity

The efficacy of nanoparticles can be significantly enhanced by conjugating or coating them with other materials. The use of NPs with antibiotics can substantially reduce bacterial resistance. Strain variations affect antibiotic resistance by altering their mode of action, while microbes are more susceptible to nanoparticles. If a strain resists one antimicrobial agent, another agent may cause death through a different mechanism. Antibiotics are transportable by NPs, which makes it easier for the drugs to enter cell walls of bacteria. The cell wall is weakened by the antibiotic, slowing the NPs' quick entry and become complicated.

In 2005, Li et al. conducted a study demonstrating a significant reduction in bacterial growth when amoxicillin was combined with AgNps. Amoxicillin was facilitated across the microbial membrane by using hydrophobic AgNPs, which interacted with the cell membrane. In their study, Duran et al. (2010) provided evidence that the combined impact of AgNps and amoxicillin is due to the formation of sulfur bridges between the two medicines. In a 2010 study, Fayaz et al. demonstrated that the AgNps-complex of ampicillin compound controls crosslinks from forming in the peptidoglycan layer, which kills the cells. This combination suppressed the process of DNA unwinding. In their study, Panáček et al. (2018) found that a small quantity of AgNps can augment the efficacy of antibiotics without inducing any cellular damage. Wan et al. (2016) found that removing *Acinetobacter baumannii* is better when polymyxin B, rifampicin, and silver nanoparticles (AgNps) are used together. A study by Banoee et al. (2010) says that the ZnONps-complex of ciprofloxacin is terrible for the outer membrane factor (OMF) proteins and the NorA endogenous efflux transporter. These proteins facilitate the cellular uptake of ciprofloxacin. Roy et al. (2010) conducted a study to examine the collective impact of TiO_2Nps and several antibiotics on MRSA. Significant bactericidal action was seen. However, the exact mechanism of how it functions in unison has yet to be completely elucidated. In 2010, Chamundeeswari et al., discovered that combining AgNps and ampicillin kills *S. aureus, E. coli*, and *K. mobilize* germs more effectively than amoxicillin alone.

The nanoparticle–antibiotic complex has a substantial synergistic effect against pathogenic bacteria. Combining two distinct nanoparticles can enhance their activity and decrease their toxicity. Encapsulating NPs with biomaterials can boost their effectiveness. Encapsulating AgNPs with titanium dioxide or metallic oxide substances enables their antibacterial properties. According to Hu et al. (2018), ZnO/Ag bi-metallic nanoflowers have increased antibacterial activity. Cobos et al. (2020) and Jaworski et al. (2018) discovered that graphene oxide nanocomposites coated with AgNps exhibit antibacterial solid characteristics. In their study, Rodriguez Nunez et al. (2019) proved that nanohybrids containing AgNps had great effectiveness against both gram-positive and gram-negative microbes.

13.8 Conclusion

Nanomaterials offer a promising solution for combating biofilm infections and bacteria resistant to drugs that are unresponsive to traditional antibiotics. NPs can be customized with specific surface characteristics to enhance therapeutic efficacy and minimize damage to the host. Nanomaterials possess many antimicrobial systems that can efficiently hinder or halt the emergence of medication resistance, providing a hopeful a resolution to the problems of the post-antibiotic era. Insufficient understanding of nanoparticle toxicity, clearance, and metabolism hinders the practical use of nanomaterials. To apply this knowledge in therapeutic contexts, one must have a solid understanding of the pharmacodynamics and pharmacokinetics of nanoparticles. NPs are designed to react to particular stimuli in various environments at infection sites that may offer a potential approach to battle drug-resistant bacteria.

In order to develop effective antimicrobial nanomaterials, chemists, engineers, microbiologists, and biomedical researchers must work together. Bringing antimicrobial NPs into therapeutic applications requires cooperation between fundamental, translational, and industry research institutes. Nanomaterial-based medicines provide a promising substitute for antibiotics in treating severe illnesses, and the advancement of antimicrobial nanoparticles could transform the medical industry. Although barriers remain, the potential advantages of utilizing nanomaterials to address resistance to antibiotic are significant. Further study in this field is crucial to surmount challenges and use of these treatments in clinical settings.Conflict of InterestAll the authors declare there is no conflict of interest.

References

Abuayyash A, Ziegler N, Gessmann J, Sengstock C, Schildhauer TA, Ludwig A, Köller M (2018) Antibacterial efficacy of Sacrifical anode thin films combining silver with platinum group elements within a bacteria-containing human plasma clot. Adv Eng Mater 20(2):1700493

Afrasiabi S, Pourhajibagher M, Raoofian R, Tabarzad M, Bahador A (2020) Therapeutic applications of nucleic acid aptamers in microbial infections. J Biomed Sci 27(1):6

Allen RC, Popat R, Diggle SP, Brown SP (2014) Targeting virulence: can we make evolution-proof drugs? Nat Rev Microbiol 12(4):300–308

Arakha M, Pal S, Samantarrai D, Panigrahi TK, Mallick BC, Pramanik K et al (2015) Antimicrobial activity of iron oxide nanoparticle upon modulation of nanoparticle-bacteria interface. Sci Rep 5(1):14813

Arya G, Sharma N, Mankamna R, Nimesh S (2019) Antimicrobial silver nanoparticles: future of nanomaterials. Microbial Nanobionics: volume 2, basic research and applications. pp 89–119

Banoee M, Seif S, Nazari ZE, Jafari-Fesharaki P, Shahverdi HR, Moballegh A et al (2010) ZnO nanoparticles enhanced antibacterial activity of ciprofloxacin against Staphylococcus aureus and Escherichia coli. J Biomed Mater Res B Appl Biomater 93(2):557–561

Baptista PV, McCusker MP, Carvalho A, Ferreira DA, Mohan NM, Martins M, Fernandes AR (2018) Nano-strategies to fight multidrug resistant bacteria—"A Battle of the Titans". Front Microbiol 9:1441

Bartlett JG, Gilbert DN, Spellberg B (2013) Seven ways to preserve the miracle of antibiotics. Clin Infect Dis 56(10):1445–1450

Bertrand N, Wu J, Xu X, Kamaly N, Farokhzad OC (2014) Cancer nanotechnology: the impact of passive and active targeting in the era of modern cancer biology. Adv Drug Deliv Rev 66:2–25

Bezza FA, Tichapondwa SM, Chirwa EM (2020) Fabrication of monodispersed copper oxide nanoparticles with potential application as antimicrobial agents. Sci Rep 10(1):16680

Bhattacharya P, Swain S, Giri L, Neogi S (2019) Fabrication of magnesium oxide nanoparticles by solvent alteration and their bactericidal applications. J Mater Chem B 7(26):4141–4152

Cai L, Chen J, Liu Z, Wang H, Yang H, Ding W (2018) Magnesium oxide nanoparticles: effective agricultural antibacterial agent against Ralstoniasolanacearum. Front Microbiol 9:790

Chamundeeswari M, Sobhana SL, Jacob JP, Kumar MG, Devi MP, Sastry TP, Mandal AB (2010) Preparation, characterization and evaluation of a biopolymeric gold nanocomposite with antimicrobial activity. Biotechnol Appl Biochem 55(1):29–35

Cobos M, De-La-Pinta I, Quindós G, Fernández MJ, Fernández MD (2020) Graphene oxide–silver nanoparticle nanohybrids: synthesis, characterization, and antimicrobial properties. Nano 10(2):376

D'Costa VM, McGrann KM, Hughes DW, Wright GD (2006) Sampling the antibiotic resistome. Science 311(5759):374–377

Davies J, Davies D (2010) Origins and evolution of antibiotic resistance. Microbiol Mol Biol Rev 74(3):417–433 de Barros E, Gonçalves RM, Cardoso MH, Santos NC, Franco OL, Cândido ES (2019) Snake venom cathelicidins as natural antimicrobial peptides. Front Pharmacol 10:1415. https://doi.org/10.3389/fphar.2019.01415

Deusenbery C, Wang Y, Shukla A (2021) Recent innovations in bacterial infection detection and treatment. ACS Infect Dis 7(4):695–720

Durán N, Marcato PD, Conti RD, Alves OL, Costa F, Brocchi M (2010) Potential use of silver nanoparticles on pathogenic bacteria, their toxicity and possible mechanisms of action. J Braz Chem Soc 21:949–959

Fayaz AM, Balaji K, Girilal M, Yadav R, Kalaichelvan PT, Venketesan R (2010) Biogenic synthesis of silver nanoparticles and their synergistic effect with antibiotics: a study against gram-positive and gram-negative bacteria. Nanomedicine 6(1):103–109

Fenollar F, Mediannikov O (2018) Emerging infectious diseases in Africa in the 21st century. New Microbes New Infect 26:S10–S18

Flores-López LZ, Espinoza-Gómez H, Somanathan R (2019) Silver nanoparticles: electron transfer, reactive oxygen species, oxidative stress, beneficial and toxicological effects. Mini review. J Appl Toxicol 39(1):16–26

Friedman AD, Claypool SE, Liu R (2013) The smart targeting of nanoparticles. Curr Pharm Des 19(35):6315–6329

Gao W, Zhang L (2021) Nanomaterials arising amid antibiotic resistance. Nat Rev Microbiol 19(1):5–6

Gunaratnam S, Millette M, McFarland LV, DuPont HL, Lacroix M (2021) Potential role of probiotics in reducing Clostridioides difficile virulence: interference with quorum sensing systems. Microb Pathog 153:104798

Gupta A, Landis RF, Rotello VM (2016) Nanoparticle-based antimicrobials: surface functionality is critical. F1000Research 5:F1000 Faculty Rev-364

Györgyey Á, Janovák L, Ádám A, Kopniczky J, Tóth KL, Deák Á et al (2016) Investigation of the in vitro photocatalytic antibacterial activity of nanocrystalline TiO2 and coupled TiO2/Ag containing copolymer on the surface of medical grade titanium. J Biomater Appl 31(1):55–67

Hall CW, Mah TF (2017) Molecular mechanisms of biofilm-based antibiotic resistance and tolerance in pathogenic bacteria. FEMS Microbiol Rev 41(3):276–301

Hamouda RA, Hussein MH, Abo-Elmagd RA, Bawazir SS (2019) Synthesis and biological characterization of silver nanoparticles derived from the cyanobacterium Oscillatorialimnetica. Sci Rep 9(1):13071

Harish V, Ansari MM, Tewari D, Yadav AB, Sharma N, Bawarig S et al (2023) Cutting-edge advances in tailoring size, shape, and functionality of nanoparticles and nanostructures: a review. J Taiwan Inst Chem Eng 149:105010

Hashimoto M, Yanagiuchi H, Kitagawa H, Honda Y (2017) Inhibitory effect of platinum nanoparticles on biofilm formation of oral bacteria. Nano Biomed 9(2):77–82

Helmy YA, Taha-Abdelaziz K, Hawwas HAEH, Ghosh S, AlKafaas SS, Moawad MM et al (2023) Antimicrobial resistance and recent alternatives to antibiotics for the control of bacterial pathogens with an emphasis on foodborne pathogens. Antibiotics 12(2):1–52

Hu M, Li C, Li X, Zhou M, Sun J, Sheng F et al (2018) Zinc oxide/silver bimetallic nanoencapsulated in PVP/PCL nanofibres for improved antibacterial activity. Artific Cells Nanomed Biotechnol 46(6):1248–1257

Hussain S, Joo J, Kang J, Kim B, Braun GB, She ZG et al (2018) Antibiotic-loaded nanoparticles targeted to the site of infection enhance antibacterial efficacy. Nat Biomed Eng 2(2):95–103

Jaworski S, Wierzbicki M, Sawosz E, Jung A, Gielerak G, Biernat J et al (2018) Graphene oxide-based nanocomposites decorated with silver nanoparticles as an antibacterial agent. Nanoscale Res Lett 13:1–17

Kadiyala U, Turali-Emre ES, Bahng JH, Kotov NA, Scott VJ (2018) Unexpected insights into antibacterial activity of zinc oxide nanoparticles against methicillin resistant: *Staphylococcus aureus* (MRSA). Nanoscale 10(10):4927–4939

Kariminezhad H, Mousapour M, Khorram S, Amani H (2020) Photodynamic inactivation of Staphylococcus epidermidis: application of PEGylated gold nanoparticles. Arab J Sci Eng 45:71–79

Karimzadeh MR, Soltanian S, Sheikhbahaei M, Mohamadi N (2020) Characterization and biological activities of synthesized zinc oxide nanoparticles using the extract of Acantholimonserotinum. Green Process Synthes 9(1):722–733

Khater MS, Kulkarni GR, Khater SS, Gholap H, Patil R (2020) Study to elucidate effect of titanium dioxide nanoparticles on bacterial membrane potential and membrane permeability. Mater Res Express 7(3):035005

Klein EY, Van Boeckel TP, Martinez EM, Pant S, Gandra S, Levin SA, Goossens H, Laxminarayan R (2018) Global increase and geographic convergence in antibiotic consumption between 2000 and 2015. Proc Natl Acad Sci 115:E3463–E3470. https://doi.org/10.1073/pnas.1717295115

Kollef MH, Bassetti M, Francois B, Burnham J, Dimopoulos G, Garnacho Montero J, Lipman J, Luyt C-E, Nicolau DP, Postma MJ, Torres A, Welte T, Wunderink RG (2017) The intensive care medicine research agenda on multidrug resistant bacteria, antibiotics, and stewardship. Intensive Care Med 43:1187–1197. https://doi.org/10.1007/s00134-017-4682-7

Lebeaux D, Chauhan A, Rendueles O, Beloin C (2013) From in vitro to in vivo models of bacterial biofilm-related infections. Pathogens 2(2):288–356

Lee H, Lee DG (2018) Gold nanoparticles induce a reactive oxygen species-independent apoptotic pathway in Escherichia coli. Colloids Surf B: Biointerfaces 167:1–7

Li P, Li J, Wu C, Wu Q, Li J (2005) Synergistic antibacterial effects of β-lactam antibiotic combined with silver nanoparticles. Nanotechnology 16(9):1912

Lianga SXT, Wong LS, Lim YM, Djearamanea S, Lee PF (2020) Effects of zinc oxide nanoparticles on streptococcus pyogenes. South Afr J Chem Eng 34(1):63–71

Makabenta JMV, Nabawy A, Li CH, Schmidt-Malan S, Patel R, Rotello VM (2021) Nanomaterial-based therapeutics for antibiotic-resistant bacterial infections. Nat Rev Microbiol 19(1):23–36

Malka E, Perelshtein I, Lipovsky A, Shalom Y, Naparstek L, Perkas N et al (2013) Eradication of multi-drug resistant bacteria by a novel Zn-doped CuO nanocomposite. Small 9(23):4069–4076

Maurer LL, Meyer JN (2016) A systematic review of evidence for silver nanoparticle-induced mitochondrial toxicity. Environ Sci Nano 3(2):311–322

Miethke M, Pieroni M, Weber T, Brönstrup M, Hammann P, Halby L et al (2021) Towards the sustainable discovery and development of new antibiotics. Nat Rev Chem 5(10):726–749

Morgan DJ, Okeke IN, Laxminarayan R, Perencevich EN, Weisenberg S (2011) Non-prescription antimicrobial use worldwide: a systematic review. Lancet Infect Dis 11(9):692–701

Munita JM, Arias CA (2016) Mechanisms of antibiotic resistance. Virulence mechanisms of bacterial pathogens.pp 481–511

Muraki Y, Yagi T, Tsuji Y, Nishimura N, Tanabe M, Niwa T, Watanabe T, Fujimoto S, Takayama K, Murakami N, Okuda M (2016) Japanese antimicrobial consumption surveillance: first report on oral and parenteral antimicrobial consumption in Japan (2009–2013). J Glob Antimicrob Resist 7:19–23. https://doi.org/10.1016/j.jgar.2016.07.002

Muthukrishnan L, Chellappa M, Nanda A (2019) Bio-engineering and cellular imaging of silver nanoparticles as weaponry against multidrug resistant human pathogens. J Photochem Photobiol B Biol 194:119–127

Natan M, Banin E (2017) From nano to micro: using nanotechnology to combat microorganisms and their multidrug resistance. FEMS Microbiol Rev 41:302–322

Nathwani D, Raman G, Sulham K, Gavaghan M, Menon V (2014) Clinical and economic consequences of hospital-acquired resistant and multidrug-resistant Pseudomonas aeruginosa infections: a systematic review and meta-analysis. Antimicrob Resist Infect Control 3(1):1–16

Orfali R, Perveen S, AlAjmI MF, Ghaffar S, Rehman MT, AlanzI AR et al (2022) Antimicrobial activity of Dihydroisocoumarin isolated from WadiLajab sediment-derived fungus *Penicilliumchrysogenum*: in vitro and in silico study. Molecules 27(11):3630

Ortiz-Benítez EA, Velázquez-Guadarrama N, Durán Figueroa NV, Quezada H, Olivares-Trejo JDJ (2019) Antibacterial mechanism of gold nanoparticles on Streptococcus pneumoniae. Metallomics 11(7):1265–1276

Panáček A, Kvítek L, Smékalová M, Večeřová R, Kolář M, Röderová M et al (2018) Bacterial resistance to silver nanoparticles and how to overcome it. Nat Nanotechnol 13(1):65–71
Pelgrift RY, Friedman AJ (2013) Nanotechnology as a therapeutic tool to combat microbial resistance. Adv Drug Deliv Rev 65(13–14):1803–1815
Priyadarshini S, Mainal A, Sonsudin F, Yahya R, Alyousef AA, Mohammed A (2020) Biosynthesis of TiO 2 nanoparticles and their superior antibacterial effect against human nosocomial bacterial pathogens. Res Chem Intermed 46:1077–1089
Qiu M, Wang D, Liang W, Liu L, Zhang Y, Chen X et al (2018) Novel concept of the smart NIR-light–controlled drug release of black phosphorus nanostructure for cancer therapy. Proc Natl Acad Sci 115(3):501–506
Quinteros MA, Aristizábal VC, Dalmasso PR, Paraje MG, Páez PL (2016) Oxidative stress generation of silver nanoparticles in three bacterial genera and its relationship with the antimicrobial activity. Toxicol In Vitro 36:216–223
Rajivgandhi G, Maruthupandy M, Muneeswaran T, Anand M, Quero F, Manoharan N, Li WJ (2019) Biosynthesized silver nanoparticles for inhibition of antibacterial resistance and biofilm formation of methicillin-resistant coagulase negative Staphylococci. Bioorg Chem 89:103008
Ripolles-Avila C, Martinez-Garcia M, Hascoët AS, Rodríguez-Jerez JJ (2019) Bactericidal efficacy of UV activated TiO2 nanoparticles against gram-positive and gram-negative bacteria on suspension. CyTA-J Food 17(1):408–418
Roca I, Akova M, Baquero F, Carlet J, Cavaleri M, Coenen S et al (2015) The global threat of antimicrobial resistance: science for intervention. New Microbes New Infect 6:22–29
Rodriguez Nunez YA, Castro RI, Arenas FA, López-Cabaña ZE, Carreño G, Carrasco-Sánchez V et al (2019) Preparation of hydrogel/silver nanohybrids mediated by tunable-size silver nanoparticles for potential antibacterial applications. Polymers 11(4):716
Roy AS, Parveen A, Koppalkar AR, Prasad MA (2010) Effect of nano-titanium dioxide with different antibiotics against methicillin-resistant Staphylococcus aureus. J Biomater Nanobiotechnol 1(1):37
Roy K, Sarkar CK, Ghosh CK (2016) Antibacterial mechanism of biogenic copper nanoparticles synthesized using Heliconiapsittacorum leaf extract. Nanotechnol Rev 5(6):529–536
Saginur R, StDenis M, Ferris W, Aaron SD, Chan F, Lee C, Ramotar K (2006) Multiple combination bactericidal testing of staphylococcal biofilms from implant-associated infections. Antimicrob Agents Chemother 50(1):55–61
Santos RS, Figueiredo C, Azevedo NF, Braeckmans K, De Smedt SC (2018) Nanomaterials and molecular transporters to overcome the bacterial envelope barrier: towards advanced delivery of antibiotics. Adv Drug Deliv Rev 136:28–48
Sarwar S, Chakraborti S, Bera S, Sheikh IA, Hoque KM, Chakrabarti P (2016) The antimicrobial activity of ZnO nanoparticles against Vibrio cholerae: variation in response depends on biotype. Nanomedicine 12(6):1499–1509
Shikha S, Chaudhuri SR, Bhattacharyya MS (2020) Facile one pot greener synthesis of sophorolipid capped gold nanoparticles and its antimicrobial activity having special efficacy against gram negative Vibrio cholerae. Sci Rep 10(1):1463
Su Y, Zheng X, Chen Y, Li M, Liu K (2015) Alteration of intracellular protein expressions as a key mechanism of the deterioration of bacterial denitrification caused by copper oxide nanoparticles. Sci Rep 5(1):15824
Sun Y, Meng L, Zhang Y, Zhao D, Lin Y (2021) The application of nucleic acids and nucleic acid materials in antimicrobial research. Curr Stem Cell Res Ther 16(1):66–73
Swamy MK, Arumugam G, Kaur R, Ghasemzadeh A, Yusoff MM, Sinniah UR (2017) GC-MS based metabolite profiling, antioxidant and antimicrobial properties of different solvent extracts of malaysianplectranthusamboinicus leaves. Evid Based Complement Alternat Med 2017:e1517683. https://doi.org/10.1155/2017/1517683
Tallósy SP, Janovák L, Ménesi J, Nagy E, Juhász Á, Balázs L et al (2014) Investigation of the antibacterial effects of silver-modified TiO 2 and ZnO plasmonic photocatalysts embedded in polymer thin films. Environ Sci Pollut Res 21:11155–11167

Tamayo LA, Zapata PA, Vejar ND, Azócar MI, Gulppi MA, Zhou X et al (2014) Release of silver and copper nanoparticles from polyethylene nanocomposites and their penetration into Listeria monocytogenes. Mater Sci Eng C 40:24–31

Tian Y, Li G, Zhang H, Xu L, Jiao A, Chen F, Chen M (2018) Construction of optimized Au@ Ag core-shell nanorods for ultralow SERS detection of antibiotic levofloxacin molecules. Opt Express 26(18):23347–23358

Velez R, Sloand E (2016) Combating antibiotic resistance, mitigating future threats and ongoing initiatives. J Clin Nurs 25(13–14):1886–1889

Wan G, Ruan L, Yin Y, Yang T, Ge M, Cheng X (2016) Effects of silver nanoparticles in combination with antibiotics on the resistant bacteria Acinetobacterbaumannii. Int J Nanomedicine 11:3789–3800

Wang L, Hu C, Shao L (2017a) The antimicrobial activity of nanoparticles: present situation and prospects for the future. Int J Nanomedicine 12:1227–1249

Wang Y, Harrington OD, Wang Y, Murray CK, Hamblin MR, Dai T (2017b) In vivo investigation of antimicrobial blue light therapy for multidrugresistantAcinetobacterbaumannii burn infections using bioluminescence imaging. J Vis Exp 122:e54997

World Health Organization (WHO) (2022) 14.9 million excess deaths associated with the COVID-19 pandemic in 2020 and 2021. URL https://www.who.int/news/item/05-05-2022-14.9-million-excess-deaths-were-associated-with-the-covid-19-pandemic-in 2020-and-2021

Yacoby I, Bar H, Benhar I (2007) Targeted drug-carrying bacteriophages as antibacterial nanomedicines. Antimicrob Agents Chemother 51(6):2156–2163

Yuan YG, Peng QL, Gurunathan S (2017) Effects of silver nanoparticles on multiple drug-resistant strains of Staphylococcus aureus and Pseudomonas aeruginosa from mastitis-infected goats: an alternative approach for antimicrobial therapy. Int J Mol Sci 18(3):569

14 Association and Contribution of Dentistry in Creating and Sharing the Burden of Antibiotic-Resistant Microbial World

Alka Shukla, Sukesh Sharma, Mayank Gangwar, Gaurav Sharma, and B. Apurva Rao

Abstract

The whole world is aware of antibiotic revolution and how it saved the entire human race during World War II. The discovery of penicillin and other synthetic bacteriostatic/bactericidal antibiotics lead to a golden phase of human health where bacterial infections were conveniently confronted. However, in due course microbes evolved to develop resistance against these "magic bullets" and there is continuous decline in antibiotic susceptibility. The contributing factors include wide and extensive usage of antibiotic in every part of biosphere like medical science, animal husbandry, aquaculture, agriculture, etc. There is no untouched-domain where antibiotic has not been utilized in direct or indirect manner. One of human healthcare sector that requires use of antibiotics is dentistry. Many oral infections are caused due to bacteria and thus role of antibiotics is critical to

A. Shukla (✉)
Faculty of Medicine, Viral Research and Diagnostic Laboratory, Department of Microbiology, Institute of Medical Sciences, Banaras Hindu University, Varanasi, Uttar Pradesh, India

S. Sharma
Department of Paediatrics, Motilal Nehru Medical College, Prayagraj, Uttar Pradesh, India

M. Gangwar
Faculty of Medicine, Viral Research and Diagnostic Laboratory, Department of Microbiology, Institute of Medical Sciences, Banaras Hindu University, Varanasi, Uttar Pradesh, India

Clinical Studies and Trials Unit, Division of Development Research, Indian Council of Medical Research, Department of Health Research, Ministry of Health and Family Welfare, Govt. of India, V. Ramalingaswami Bhawan, New Delhi, India

G. Sharma
Department of Public Health Dentistry, SCB Dental College and Hospital, Cuttack, Odisha, India

B. A. Rao
Department of Paediatric Dentistry, SDM College, Dharwad, Karnataka, India

M. Gangwar, G. Nath (eds.), *Emerging Paradigms for Antibiotic-Resistant Infections: Beyond the Pill*, https://doi.org/10.1007/978-981-97-5272-0_14

combat them. Since there is surge in antibiotic resistance among pathogenic microbes, we are bound to search for alternatives agents and strategies. Lately, many potential substitutes have gained the lime light which include bacteriophages, ultraviolet radiations, photodynamic therapy, surgical approaches, plant extracts, etc. Emphasis is to minimize the application and use of antibiotics in any form (topical/local/systemic) for any objective (curative/preventive). This chapter briefly brushes through antimicrobial-resistant world and mechanism involved behind it. It thoroughly explains and highlights the role and contribution of dentistry in reducing the piled-up burden of antibiotic resistance and its consequences faced by the world.

Keywords
Antibiotics · Antimicrobial resistance · Dentistry · Alternative to antibiotics · Bacteriophages

14.1 Introduction

In 1928, the discovery of penicillin led to an outbreak of the antibiotic era, and its beneficial effects were reflected in the form of a decrease in mortality and an increased life expectancy of the human race (Ezzariai et al. 2018). Nevertheless, every single antibiotic, irrespective of its source of origin (natural or synthetic), is susceptible to resistance and thus has a restricted duration of usefulness (Blair et al. 2015). Over the years, the application of antibiotics expanded from the medical field to various other territories, including animal science, plant science, etc., and such extensive usage triggered an unwanted chain of events in the environment. The past few decades have witnessed the emergence of antibiotic/antimicrobial resistance as one of the uppermost unnerving threats across the globe which can result in compromised health status, increased morbidity and mortality rates, and decreased quality of life (Xu et al. 2024). There is an ambit of antibiotics, including aminoglycoside, tetracycline, macrolide, chloramphenicol, glycopeptide, sulfonamide, trimethoprim, beta-lactams, fluoroquinolone, quinolones, and so on, against which microbes have developed resistance (Ezzariai et al. 2018; Singh et al., 2019). Based on a study, it was revealed that in 2019, approximately 1.27 million people died owing to antimicrobial resistance (Salehi et al. 2022). Moreover, in 2016, the U.K. Government released a report predicting a ten million count of people may be affected each year because of AMR by 2050 (O'Neill 2016). Therefore, it is a matter of prime concern to act upon and limit the irreversible damage.

14.2 Development of Resistance: Mechansim

Antibiotic resistance comes into existence when microbes adapt to grow in the presence of antibiotics. Over the period of time, microbes have evolved to shield themselves and tarnished the success of antibiotics. The rigorous use and over-use of antibiotics have led many resistant components to pass on vertically or horizontally to other sensitive microbes and have disseminated globally. Moreover, studies have revealed that the development of resistance to multiple antibiotics is not just a quick phenomenon, it seems to transfer from the resistant strains to other strains (Davies 2007). The development of antimicrobial resistance among microbes is accredited to antibiotic-resistant genes (ARGs) that are to project resistance against activity of any antibiotics (Finley et al. 2013; Larsson et al. 2022).

There are many mechanisms through which a microbe tries to cope with the stress caused by an antibiotic and can develop resistance against it, but six of them are widely accepted and briefly described as follows (Xu et al. 2024).

(a) Microbes can alter the available concentration of antibiotic within cell by accelerating the action of efflux pumps.
(b) Microbes can follow chemical approach to inactivate the antibiotics by causing hydrolysis and redox reactions or by modifying them by acyltransfer, phosphorylation, glycosylation, nucleotidylation, ribosylation, and thiol transfer (Wright 2005).
(c) Bacteria has the potential to mutate genes that can lead to alternation in the target site of antibiotics (Lambert 2005). One of the simple examples of such mutational resistance is where a point mutation in *rpoB* gene affects the RNA polymerase.
(d) Bacteria can also evolve to cause replacement of the antibiotic target site. This approach results in the evolution of novel targets that are not affected by antibiotics but mimic the same original biochemical function of the cell (Munita et al. 2016).
(e) By protecting the antibiotic target site that prohibits the antibiotic from arriving at its binding site. Examples of drugs that face such resistance include tetracycline and fluoroquinolones (Munita et al. 2016).
(f) Cellular protection with poor permeability of antibiotics. This is mainly seen with gram-negative bacteria. For certain drugs, it is mandatory to cross the outer membrane and reach the intracellular territory of the cell to bring about the desired antimicrobial action. Therefore, bacteria decrease their permeability to prevent intake of such drug molecules. Figure 14.1 summarizes the aforementioned routes of development of resistance mechanism by microbe.

14.3 Contributions of Dentistry in Development of Antimicrobial-Resistant World

Dentists are an integral part of health system and have immense contribution toward antibiotic resistance world. Oral biogeography is heavily contributed by diverse indigenous microbes and harbors around six billion bacteria (Edlund et al. 2015). The majority of it lives in a symbiotic environment and seldom causes any harm to the host (Welch et al. 2020). But when there is a bit of turmoil in the balanced microbial world of oral cavity, many diseases crop up. Oral biofilm is attributed as the main culprit behind ambit of such diseases which include dental caries,

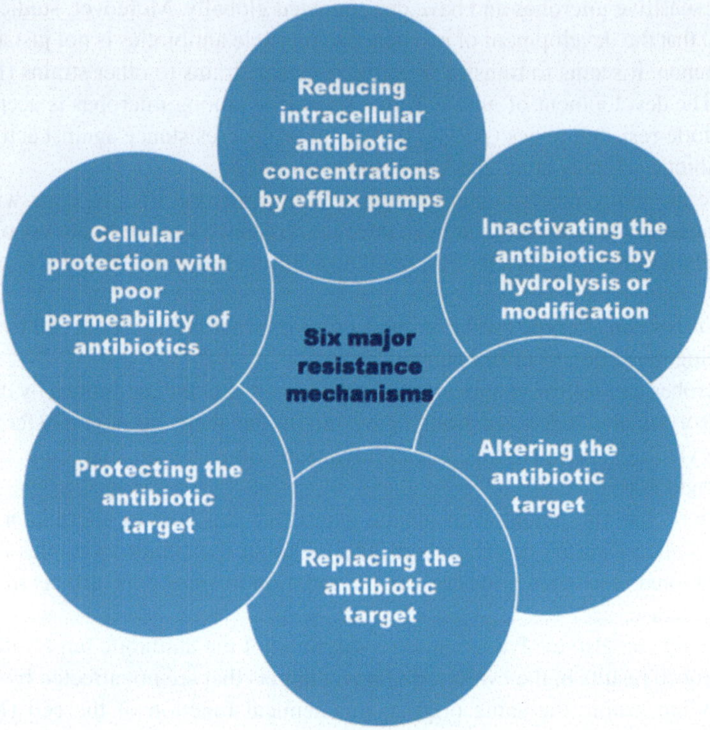

Fig. 14.1 Mechanism of action through which microbes develop resistance against any antibiotic

periodontal diseases, peri-implantitis, and oropharyngeal candidiasis (Yin et al. 2019). Oral disease accounts for high rate of morbidity for instance, in China dental caries morbidity rate is approximately 90% which is an alarming figure, whereas it has low outpatient rate (Xiang et al. 2024). In such cases since the cause is bacteria, antibiotics become the desirable candidates to combat them. Antibiotics do not just pose as potential option for the treatment, but it is also involved as key elements in many prophylactic regimens (Haque et al. 2019). Reports have been published stating that irrational and over-prescription of antibiotics to cure dental infections is one of the major reasons for the development of antibiotic-resistant microbial world (Haque et al. 2019). Data suggests 66% of antibiotics prescriptions in dentistry are not clinically required (Patrick et al. 2018). A report published in the year 2015 disclosed that across the globe, around 40–50% of prescribed antibiotics were unjustified out of which 3–11% were dental prescriptions (Carlet et al. 2015; Sukumar et al. 2020). Moreover, studies have also revealed that only upto 12% of dental surgeons prescribe antibiotics appropriately (Ahmadi et al. 2021). One of the prominent ARGs that is *tet*(M) is frequently found in the oral microbes (Sukumar et al. 2020). Due to the rise in antibiotic resistance and escalation in superbug population, along with judicious use or minimal use of antibiotics, hunt and application of all possible alternatives is the need of the hour.

14.4 How Dentist Can Help?

Dentists can play a vital role in fighting against an exponential rise in antibiotic resistance by judicious prescription of antibiotics thus minimizing the over-usage of it and employing alternative agents.

The following strategies can be proved beneficial in achieving aforementioned goal:

(a) Prevention: Being a key element in healthcare system, dentists have opportunities to directly educate the population regarding preventive and prophylactic measures that will assure the prevention of any infection and thus eliminate the scope to use antibiotics to treat them. Good hygiene practices and preventive measures are the topmost things that can be incorporated into one's lifestyle to combat infectious elements. Also, the use of proper sterilization techniques and sterilized instruments in dental clinic can eliminate odds of acquiring microbial infections from the clinic. A study demonstrated that the microbial contamination of a dental clinic's waiting area touches the mark of 732 CFU/m^3, which is beyond the acceptable range (Baudet et al. 2021). Proper disinfection and sterilization measures ensure the higher chances of prevention of transmission of microbes along with antibiotic-resistant bacteria which again is a desirable outcome.

(b) Proper diagnosis of disease and adequate prescriptions of antibiotics: If the dentist could make correct diagnosis, it will help in identifying the exact etiological agent, and appropriate antibiotic (narrow spectrum specific to the etiological microbes) could be prescribed without exposing the patient to unnecessary broad-spectrum antibiotics. A study conducted in Newcastle demonstrated that 47% decline was registered in vancomycin-resistant Enterococcus and 42% in Methicillin-resistant Staphylococcus aureus (MRSA) cases just by restoring broad-spectrum drug with narrow spectrum drug (Sukumar et al. 2020). In France, a study observed that around 12% of antibiotic prescription were dental prescription (Dormoy et al. 2021), thus dentist have to be careful and sure about the worthiness of what they are prescribing and how will it impact the patient's overall health and environment.

(c) Considering alternate treatment agents: Oral microbial infections can be cured and prevented using alternative approaches that completely eliminate antibiotics from the picture (fig. 14.2). Marine-derived biomaterials like chitosan (Cicciù et al. 2019), bacteriophage therapy, lasers, and antimicrobial photodynamic therapy (Ahmadi et al. 2021.). Nanomaterials like silver nanoparticles (Ahmed 2022), natural plant extracts like *Camellia sinensis, Melaleuca alternifolia* oil, *Azadirachtaindica* (dos Santos Cardoso et al. 2021.) are a few of the recently emerged substitutes that have given promising results in the field of dentistry. These new tools have been described in detail in the following headings.

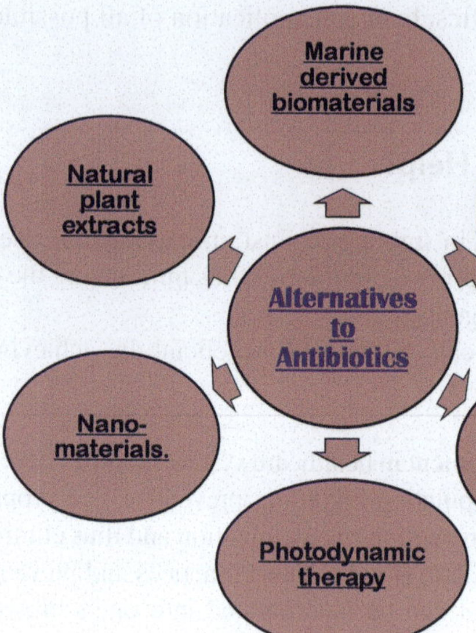

Fig. 14.2 Various alternatives that can replace traditional antibiotics

14.5 Antimicrobial Photodynamic Therapy

Photodynamic therapy is yet another untapped area that lately has been proven sincerely beneficial in treating infectious wounds (Moldalievich et al. 2024). It is based on a simple principle of photo-excitation wherein a phototoxic response against microbes is generated by an exposure to a beam of light of a specific wavelength with the help of a photosensitizer, and the efficiency of treatment relies on class and specifications of the light being used (Gonçalves et al. 2023; Feuerstein 2012). Several in-vitro and clinical trials have been reported in the literature where application of PDT to target biofilm formed in the inner surfaces of periodontal pockets and on dental hard tissues have been employed (Mocuta et al. 2021; Alsaif et al. 2021; Dalvi et al. 2021; Koochaki et al. 2021; Alvarenga et al. 2019). PDT can also be used to decontaminate oral cavity prior to any dental procedure to avert any cross-infections (Besegato et al. 2022). Standardly, methylene blue, curcumin, erythrosine, and porphyrin dyes are being used as photosensitizers in PDT procedure along with blue or red light (Gonçalves et al. 2023). Bacteria that have been targeted with PDT in dental setup include *Streptococcus mutans*, *Lactobacillus casei,* and *Candida albicans* (Romão et al. 2018; Gong et al. 2019; Gonçalves et al. 2021). It can also be used along with the other therapeutic agents like bacteriophage, hydrogen peroxide, etc. to get synergistic effects (Moldalievich et al. 2024). An enhanced antimicrobial activity, where reactive oxygen species genesis is involved, has been noted with the integrated use of blue light and hydrogen peroxide (Feuerstein 2012). A case report by Moldalievich et al. established that purulent wounds of oral cavity showed promising results when treated with a light of wavelength of 450 nm along

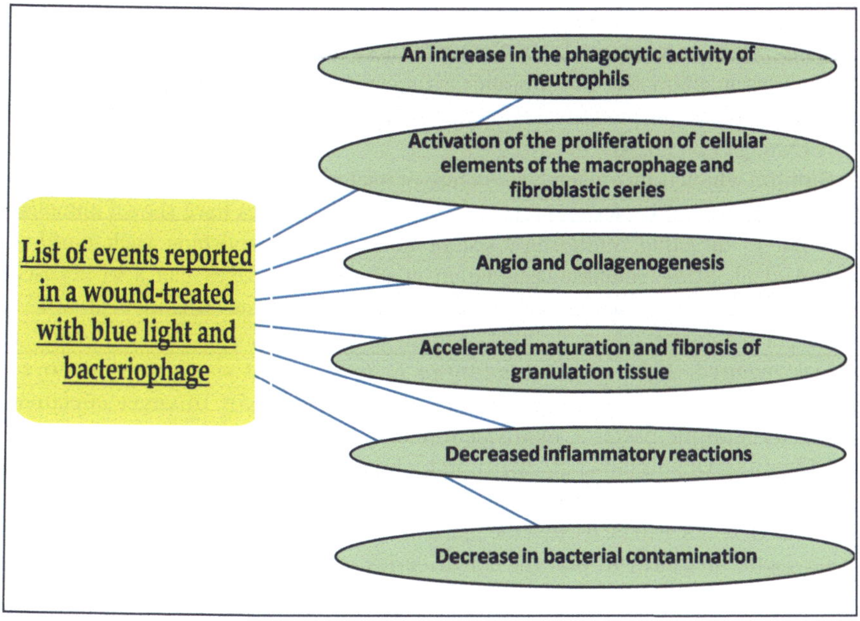

Fig. 14.3 Series of events noticed in a wound postexposure to the combination of blue light and bacteriophage

with bacteriophage (Moldalievich et al. 2024). In this case, the characteristic that could be noted in the wound area, including improved healing consists of increased rate of maturation of fibroblasts present in granulation tissue, decreased inflammatory reactions, dynamic genesis of new collagen fibres, etc. listed in the Fig. 14.3 (Moldalievich et al. 2024).

Apart from devoid of possibility to develop antimicrobial resistance, PDT has several other merits such as prompt commencement of action, use of photosensitizer agents limits the detrimental action to the involved sites (Gonçalves et al. 2023), accessibility to bacteria in biofilm state which on other hand is diminished in case of antibiotics because of decreased diffusion rate (Feuerstein 2012).

To make it easily available to the patients, light emitting toothbrushes that are intended to damage dental plaque and reduce gingival inflammation have been tested, and the results showed up to 97.5% decline in microbial flora (Genina et al. 2015). Therefore, it can be considered as an effective alternative to other antibiotics regimens to cure or prevent gingival infections.

14.6 Bacteriophage

These are the natural entities present in abundance in biosphere which infect and affect bacteria to utilize their machinery for its genomic multiplication (Hendrix 2002). Based on the life cycle and the extent of activity against bacteria, bacteriophages (phages) are classified into two categories: (a) lytic phages and (b) lysogenic phages (Xiang et al. 2024). Out of these two, lytic phages have earned the spotlight as an alternative to antibiotics owing to its significant host-specific nature, self-maintenance, adaptability, and high efficiency in dealing with microbial infections (Xiang et al. 2024).

Biofilm, which is an organized structure of bacteria, poses a significant challenge during treatment of bacterial infection. Nevertheless, phages have shown immense potential in targeting multidrug-resistant bacteria and abolishing such biofilms (Xiang et al. 2024; Shanmugasundaram et al. 2024). Phages can also be used sequentially along with other agents like sodium hypochlorite (Khalifa et al. 2015), blue light (Moldalievich et al. 2024), etc. to obtain superior results. In a study, a 2-week matured, 100 μm thick biofilm of *E. faecalis* was successfully removed using phage and sodium hypochlorite (Khalifa et al. 2015). To target infections caused by multiple bacteria, multivalent phage cocktails can also be brought into use (Gordillo Altamirano et al. 2019). In high bacterial load infections, antibiotic–bacteriophage synergy can also be employed to combat the infections (Bhargava et al. 2021). It can also be used in the place of mouthwash containing antibiotics, for prophylactic measures to prevent dental caries and periodontal infections (Zhao et al. 2023).

After a phage identifies and attacks the vulnerable bacterium, it takes up either of the two replication cycles: lytic or lysogenic. The mechanism of action of bacteriophage on bacteria can be understood with the help of Fig. 14.4 (Kasman et al. 2022).

Since it is very specific to the target and can penetrate biofilm, bacteriophage has immense potential to manage microbial infection without hampering the commensal flora of the environment and thus it definitely qualifies to be on the list of traditional antibiotic alternatives.

14.7 Marine-Derived Product: Chitosan

The quest for more efficacious antimicrobial agents has led to revelation of chitosan, which is a compound derived from chitin—a polymer found in many marine life forms like crustaceans, mollusks, etc. (Pellis et al. 2022.). For the past two decades, chitosan has been studied thoroughly, and these studies have demonstrated its antimicrobial property against good number of bacteria, fungi, and yeast (Kong, et al. 2010; Yan et al. 2021). Moreover, chitosan has a broad-spectrum bactericidal activity against microbes, whereas low harmful activity against mammalian tissue (Kong, et al. 2010).

The proposed mechanism of action of chitosan on bacteria is attributed to chelating property of chitosan/chitosan derivatives and is depicted in Fig. 14.5 (Kong, et al. 2010). Certain factors including pH of environment, physical state of chitosan, availability of metal ions (if any), hydrophilic characteristic of chitosan compound, chelation potential (Rabea et al. 2003), molecular weight of chitosan, etc., influence the degree of antimicrobial efficacy of chitosan (Kong, et al. 2010).

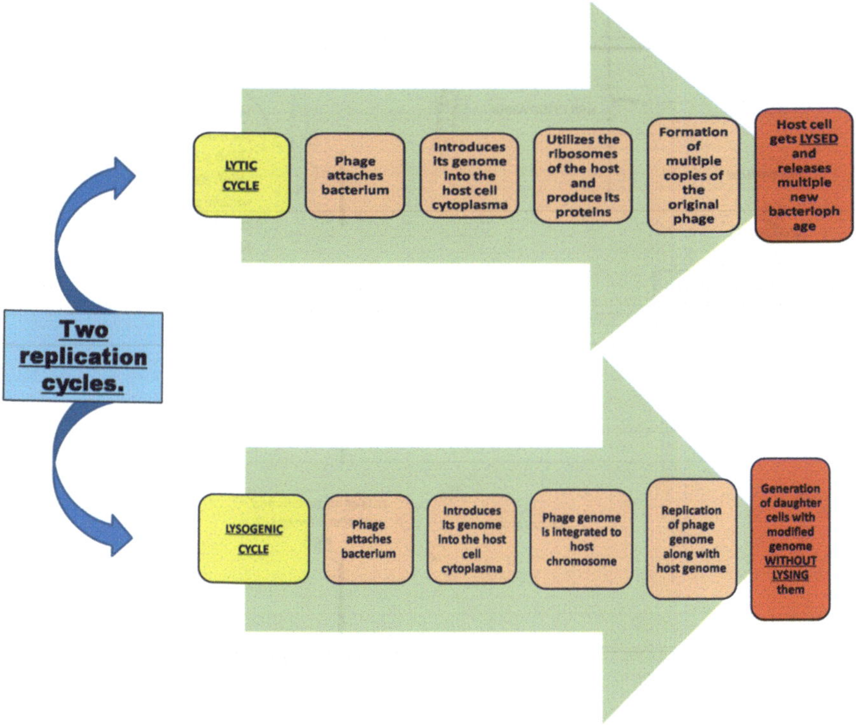

Fig. 14.4 Two pathways which a bacteriophage takes up to replicate its genomic material after it attaches to its target bacteria

In dentistry, chitosan has shown significant contribution toward combating microbial infections. It can be combined with other dental materials and can easily be applied. Along with its antimicrobial effect, it has also projected wound healing property which again is a desirable element in treating infectious wounds (Cicciù et al. 2019). Thus, it is a potential candidate to look forward to in reducing dependency on traditional antibiotics.

14.8 Lasers

It is an acronym that stands for "light amplification by stimulated emission of radiation" (Laser). It is a part of advanced technology and has been extensively researched by researches for its application in different sectors.

In dentistry, it was introduced by Weichman and Johnson in 1970s (Weichman et al. 1971) and since then it has continued to serve in dentistry for a range of purposes. The mechanism of action involves one of the three kinds of laser-cell interactions, explained in the Fig. 14.6 (Elavarasu et al. 2012).

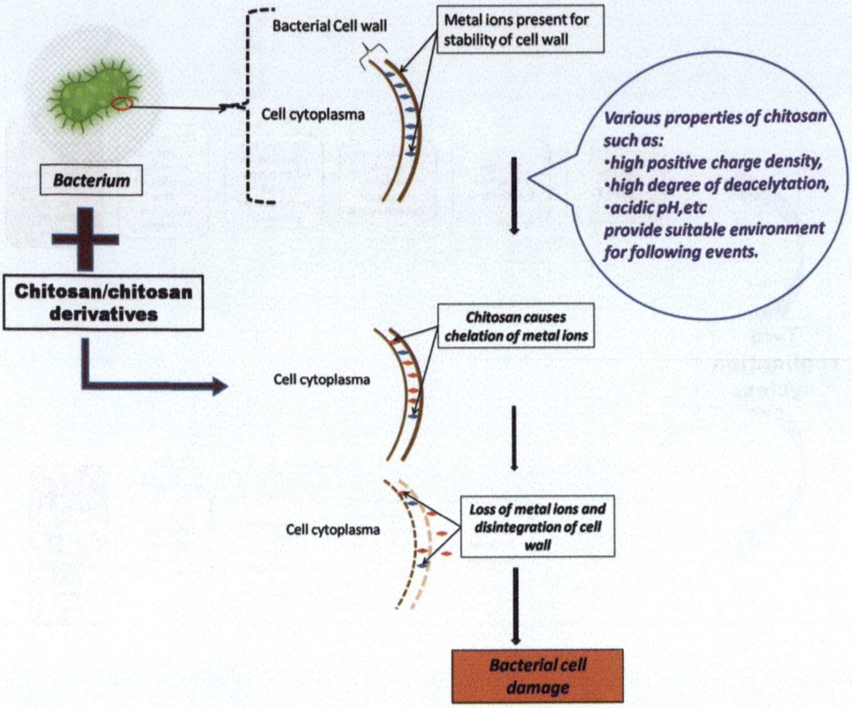

Fig. 14.5 Mechanism of action of chitosan on a bacterium

In the field of dentistry, lasers have occupied a significant place as it assists in treating many microbial infections such as periodontitis, root canal infections, dental plaque removal, etc. as well as for surgical and esthetic treatments such as gingivectomy, gingival depigmentation, etc. (Karoussis et al. 2017; Mossaad et al. 2021). Various laser cells that are in practice involve Neodymium Yttrium Aluminium Garnet (Nd:YAG) laser, Erbium Yttrium Aluminium Garnet (Er:YAG) laser, Carbon dioxide laser, and Argon lasers (Hedge et al. 2018).

Additionally, laser can also assist to increases the efficiency of traditional antibiotic drugs such as Moxifloxacin by increasing its concentration in serum through Intravenous Blood Laser Irradiation (IBLI) technique (Zahra et al. 2021).

IBLI also prompts a series of reactions in blood serum such as it triggers bactericidal property of the serum, reduces available amount of c-reactive protein & plasma toxicity, and elevates amount of antibodies in the serum; all of these events result in reduction of inflammation cause by microbial infection (Zahra et al. 2021). Therein, such beneficial properties of laser along with no scope for microbes to the development of tolerance/resistance against it demands its more and more application in medical world to tame the microbial infections.

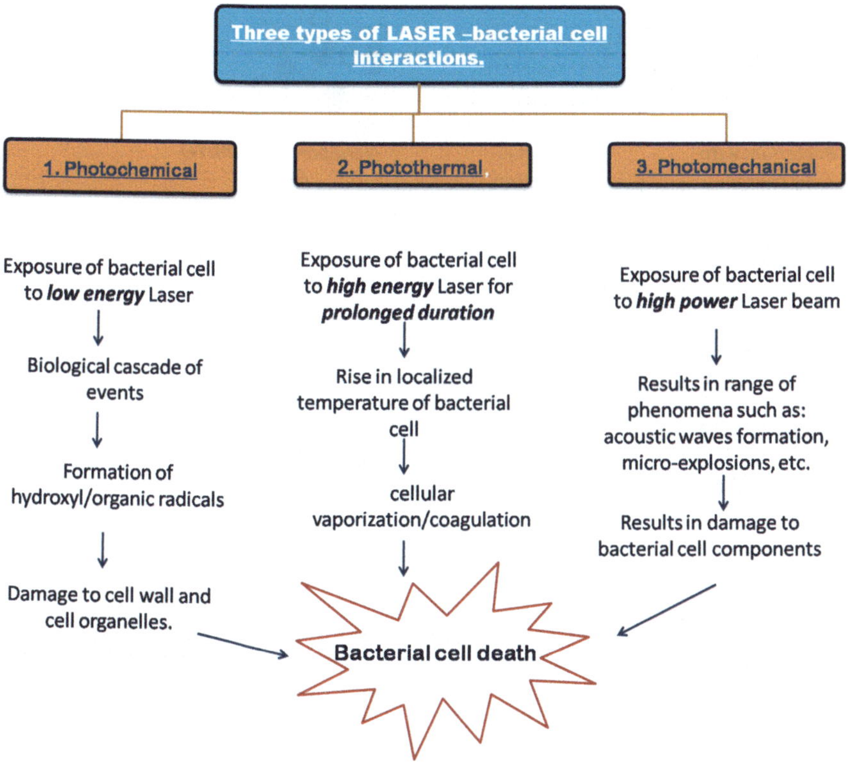

Fig. 14.6 Three types of laser-cell interactions which lead to the damage of the bacterial cell

14.9 Metal Nanoparticles

Nanoparticles (NPs) are inorganic particles derived from metal and are result of unceasing evolution in science. Literature mentions that the antibiotic property possessed by these nano entities is not just limited to the microbes, but it also impacts of biomolecules which favor the microbes to build up resistance. They are believed to have a nonspecific mechanism of action against bacteria which widens its antibiotic spectrum, including a good number of gram-positive, gram–negative bacteria, and multi-resistant bacteria and makes them less susceptible to the development of resistance against them (Sánchez-López et al. 2020). However, their efficacy is governed by certain elements like size of the particle, roughness, shape, and surface energy (Wang et al. 2017). The mechanism of action of nanoparticles on bacteria is based on electrostatic interactions and is explained with the help of Fig. 14.7. Bacteria have negative charge on their surface which attracts positively charged metallic nanoparticles leading to the formation of strong bonds between metal ions and cell membrane/wall. This results in damage of the cell wall and sequential increases in cell wall/membrane permeability. The metallic particles also release metal ions that enter the bacterial cell and stimulate the production of reactive

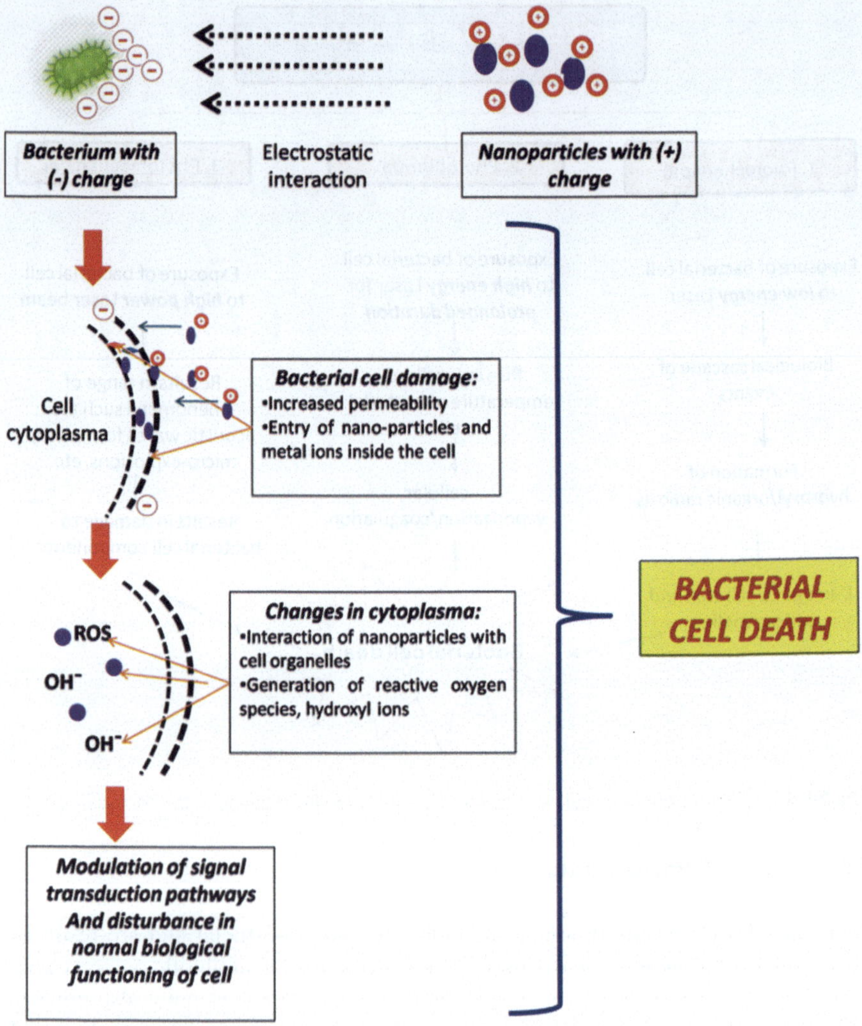

Fig. 14.7 Mechanism of action of nanoparticles on a bacterial cell

oxygen species (ROS) and free radicals like hydrogen peroxide, superoxide anion, hydroxyl radical, which in turn hampers biological events of the cell and ultimately causes cell death (Stensberg et al. 2011; Sánchez-López et al. 2020; Dakal et al. 2016).

Various nanoparticles that have been studied and used in the medical world so far include silver nanoparticles (AgNPs) and hydroxyl apatite NPs (HANPs).

In dentistry, nanoparticles are being used not just to cure the existing problem but also as a prophylactic tool to prevent the formation of biofilms. A study by Kulshrestha et al. in 2014 demonstrated successful attempt to prevent the formation of *Streptococcus mutans* biofilm on the implant surface by application of graphene/

zinc oxide nanocomposite (GZNC) (Kulshrestha et al. 2014). The nanotechnology has been incorporated in many toothpastes and mouthwashes replacing traditional antibiotic solutions to form a protective layer on tooth enamel and prevent plaque formation followed by the establishment of dental caries (Pepla et al. 2014).

Nanotechnology can also be amalgamated with other components to enhance the beneficial effects and minimize the unfavorable outcomes from it. Plant extract-silver nanoparticle has been studied, and it was concluded that it can be safely applied as an antimicrobial agent in handling of oral microbial infections (Ahmed et al. 2022). The advancement in this technology puts application of traditional antibiotics on the back foot, and thus this area needs more attention and exploration.

14.10 Plant Extracts/Herbal Products

Ayurvedic medicine including its materials is not a new zone to look upon, it has been a backbone of the ancient medical healthcare system and still promises to serve the mankind with touch of advanced technology (Amanpour et al. 2023). Along with its near-to-none adverse effect, as it seen with many synthetic compounds, a continuous increase in antibiotic resistance has rekindled the researcher's interest in herbal and plant medicine. Also, it is more economic and easily available (Jain et al. 2019).

To defend themselves from microbes and other predators, plants produce certain compounds that can be extracted and utilized for preventive and curative purpose (Cowan et al. 1999). Not just plant extract, but essential oils derived from plants also acquire antibacterial and antifungal properties (Kalembaet al. 2003). The mechanism of action of essential oil relies on its hydrophilic or lipophilic property.

Certain components also interfere with proton translocation which is followed by a disturbance in energy metabolism pathway of the bacteria, whereas a few components like phenolic alcohols disturb the membrane-linked enzyme proteins and eventually stop their activity (Knobloch et al. 1989). The antimicrobial property of the essential oil is dependent on its chemical constituent which in turn is influenced by the place of origin of the plant, section of the plant, plant-development growth stage, distillation conditions, and storage conditions (Kalembaet al. 2003).

In the war against microbes where traditional antibiotics pose a threat of AMR and other adverse effects, natural compounds/elements obtained from plants and herbs prove to be a suitable alternative. Out of 500,000 plants, only 1% has been phytochemically researched thus posing a great scope for discovery of newer potential natural components (Palombo et al. 2011).

14.11 Conclusion

In the past, post discovery of penicillin, undoubtedly the traditional antibiotics have proved their magical effect in curing and controlling microbial infections. But unfortunately, microbes have natural evolutionary mechanisms wherein they modify themselves and guard themselves against harmful effects. Thus, every antibiotic has a limited span of life, which makes them completely "unproductive pill." The emerging paradigms for antibiotic-resistant infections call for newer preventive and therapeutic therapies and tools, which can withstand the load of a fast-expanding resistant microbial world. Science and technology have presented us with many alternates that have enlightened the lamp of hope in the darkness.

Although many alternatives have emerged, prevention is always a better remedy than anything else. Hence, it demands the increasing attention of clinicians and the public in all sectors that are directly or indirectly contributing to or dealing with microbes and infections. The preventive strategies and judicious use of our resources will definitely lead to a wonderful symbiotic world, where all the species can live without interfering with or damaging each other's life cycles.

References

Ahmadi H, Ebrahimi A, Ahmadi F (2021) Antibiotic therapy in dentistry. Int J Dent 2021(1):6667624

Ahmed O, Sibuyi NRS, Fadaka AO, Madiehe MA, Maboza E, Meyer M, Geerts G (2022) Plant extract-synthesized silver nanoparticles for application in dental therapy. Pharmaceutics 14(2):380

Alsaif A, Tahmassebi JF, Wood SR (2021) Treatment of dental plaque biofilms using photodynamic therapy: a randomised controlled study. Eur Arch Paediatr Dent 22(5):791–800

Alvarenga LH, Gomes AC, Carribeiro P, Godoy-Miranda B, Noschese G, Ribeiro MS et al (2019) Parameters for antimicrobial photodynamic therapy on periodontal pocket—randomized clinical trial. Photodiagn Photodyn Ther 27:132–136

Amanpour S, Javar MA, Sarhadinejad Z, Doustmohammadi M, Moghadari M, Sarhadynejad Z (2023) A systematic review of medicinal plants and herbal products' effectiveness in oral health and dental cure with health promotion approach. J Educ Health Promot 1:306

Baudet A, Guillaso M, Grimmer L, Regad M, Florentin A, Mediqai Study Group (2021) Microbiological contamination of the office environment in dental and medical practice. Antibiotics (Basel, Switzerland) 10(11):1375

Besegato JF, de Melo PBG, Tamae PE, Alves APAR, Rondón LF, Leanse LG et al (2022) How can biophotonics help dentistry to avoid or minimize cross infection by SARS-CoV-2? Photodiagn Photodyn Ther 37:102682

Bhargava K, Nath G, Bhargava A, Aseri GK, Jain N (2021) Phage therapeutics: from promises to practices and prospectives. Appl Microbiol Biotechnol 105:9047–9067

Blair JM, Webber MA, Baylay AJ, Ogbolu DO, Piddock LJ (2015) Molecular mechanisms of antibiotic resistance. Nat Rev Microbiol 13(1):42–51

Carlet J (2015) The world alliance against antibiotic resistance: consensus for a declaration. Clin Infect Dis 60(12):1837–1841

Cicciù M, Fiorillo L, Cervino G (2019) Chitosan use in dentistry: a systematic review of recent clinical studies. Mar Drugs 17(7):417

Cowan MM (1999) Plant products as antimicrobial agents. Clin Microbiol Rev 12(4):564–582

Dakal TC, Kumar A (2016) Mechanistic basis of antimicrobial actions of silver nanoparticles. Front Microbiol 7:231711

Dalvi S, Benedicenti S, Sălăgean T, Bordea IR, Hanna R (2021) Effectiveness of antimicrobial photodynamic therapy in the treatment of periodontitis: a systematic review and meta-analysis of in vivo human randomized controlled clinical trials. Pharmaceutics 13(6):836

Davies J (2007) Microbes have the last word: a drastic re-evaluation of antimicrobial treatment is needed to overcome the threat of antibiotic-resistant bacteria. EMBO Rep 8(7):616–621

Dormoy J, Vuillemin MO, Rossi S, Boivin JM, Guillet J (2021) Perceptions of antibiotic use and resistance: are antibiotics the dentists' anxiolytics? Antibiotics 10(6):735
dos Santos Cardoso VF, Roppa RHA, Antunes C, Moraes ANS, Santi L, Konrath EL (2021) Efficacy of medicinal plant extracts as dental and periodontal antibiofilm agents: a systematic review of randomized clinical trials. J Ethnopharmacol 281:114541
Edlund A, Santiago-Rodriguez TM, Boehm TK, Pride DT (2015) Bacteriophage and their potential roles in the human oral cavity. J Oral Microbiol 7(1):27423
Elavarasu S, Naveen D, Thangavelu A (2012) Lasers in periodontics. J Pharm Bioallied Sci 4(Suppl 2):S260–S263
Ezzariai A, Hafidi M, Khadra A, Aemig Q, El Fels L, Barret M et al (2018) Human and veterinary antibiotics during composting of sludge or manure: global perspectives on persistence, degradation, and resistance genes. J Hazard Mater 359:465–481
Feuerstein O (2012) Light therapy: complementary antibacterial treatment of oral biofilm. Adv Dent Res 24(2):103–107
Finley RL, Collignon P, Larsson DJ, McEwen SA, Li XZ, Gaze WH et al (2013) The scourge of antibiotic resistance: the important role of the environment. Clin Infect Dis 57(5):704–710
Genina EA, Titorenko VA, Belikov AV, Bashkatov AN, Tuchin VV (2015) Adjunctive dental therapy via tooth plaque reduction and gingivitis treatment by blue light-emitting diodes tooth brushing. J Biomed Opt 20(12):128004–128004
Gonçalves MLL, Santos EM, Renno ACM, Horliana ACRT, de Almeida Cruz M, Parisi JR et al (2021) Erythrosine as a photosensitizer for antimicrobial photodynamic therapy with blue light-emitting diodes–An in vitro study. Photodiagn Photodyn Ther 35:102445
Gonçalves MLL, Sobral APT, Gallo JMAS, Gimenez T, Ferri EP, Ianello S et al (2023) Antimicrobial photodynamic therapy with erythrosine and blue light on dental biofilm bacteria: study protocol for randomised clinical trial. BMJ Open 13(9):e075084
Gong J, Park H, Lee J, Seo H, Lee S (2019) Effect of photodynamic therapy on multispecies biofilms, including Streptococcus mutans, Lactobacillus casei, and Candida albicans. Photobiomodul Photomed Laser Surg 37(5):282–287
Gordillo Altamirano FL, Barr JJ (2019) Phage therapy in the postantibiotic era. Clin Microbiol Rev 32(2):10–1128
Haque M, Sartelli M, Haque SZ (2019) Dental infection and resistance—global health consequences. Dent J 7(1):22
Hedge MN, Garg P, Hedge ND (2018) Lasers in dentistry: an unceasing evolution. J Otolaryngol Ent Res 10(6):422–448
Hendrix RW (2002) Bacteriophages: evolution of the majority. Theor Popul Biol 61(4):471–480
Jain V (2019) Role of Quercusinfectoria in health and oral health â€A review. Int J Green Pharm 13(3)
Kalemba DAAK, Kunicka A (2003) Antibacterial and antifungal properties of essential oils. Curr Med Chem 10(10):813–829
Karoussis IK, Kyriakidou K, Psarros C, Lang NP, Vrotsos IA (2017) Nd:YAG laser radiation (1.064 nm) accelerates differentiation of osteoblasts to osteocytes on smooth and rough titanium surfaces in vitro. Clin Oral Implants Res 28(7):785–790
Kasman LM, Porter LD (2022) Bacteriophages. In: StatPearls [Internet]. StatPearls Publishing, Treasure Island, FL
Khalifa L, Brosh Y, Gelman D, Coppenhagen-Glazer S, Beyth S, Poradosu-Cohen R et al (2015) Targeting Enterococcus faecalis biofilms with phage therapy. Appl Environ Microbiol 81(8):2696–2705
Knobloch K, Pauli A, Iberl B, Weigand H, Weis N (1989) Antibacterial and antifungal properties of essential oil components. J Essent Oil Res 1(3):119–128
Kong M, Chen XG, Xing K, Park HJ (2010) Antimicrobial properties of chitosan and mode of action: a state of the art review. Int J Food Microbiol 144(1):51–63
Koochaki M, Hendi A, Ghasemi M, Seyedjafari E, Hamidain M, Chiniforush N (2021) Comparative evaluation of the effects of antimicrobial photodynamic therapy with an LED and a laser on

the proliferation of human gingival fibroblasts on the root surface: an in vitro study. J Lasers Med Sci 12:e47
Kulshrestha S, Khan S, Meena R, Singh BR, Khan AU (2014) A graphene/zinc oxide nanocomposite film protects dental implant surfaces against cariogenic Streptococcus mutans. Biofouling 30(10):1281–1294
Lambert PA (2005) Bacterial resistance to antibiotics: modified target sites. Adv Drug Deliv Rev 57(10):1471–1485
Larsson DG, Flach CF (2022) Antibiotic resistance in the environment. Nat Rev Microbiol 20(5):257–269
Mocuta DE, Miron MI, Hogea E, Muntean C, Todea DC (2021) Dental-plaque decontamination around dental brackets using antimicrobial photodynamic therapy: an in vitro study. Int J Environ Res Public Health 18(23):12847
Moldalievich YA, Mukhtaralievich AN, Tumakovich MZ (2024) Modern approaches to the treatment of purulent wounds of the bottom of the oral cavity and neck with the use of blue light and bacteriophages. World Sci 1(83)
Mossaad AM, Abdelrahman MA, Kotb AM, Alolayan AB, Elsayed SAH (2021) Gummy smile management using diode laser gingivectomy versus botulinum toxin injection-A prospective study. Ann Maxillofac Surg 11(1):70
Munita JM, Arias CA (2016) Mechanisms of antibiotic resistance. Microbiol Spectr 4(2). https://doi.org/10.1128/microbiolspec.VMBF-0016-2015
O'Neill, J. (2016). Tackling drug-resistant infections globally: final report and recommendations
Palombo EA (2011) Traditional medicinal plant extracts and natural products with activity against oral bacteria: potential application in the prevention and treatment of oral diseases. Evid Based Complement Alternat Med 2011(1):680354
Patrick A, Kandiah T (2018) Resistance to change: how much longer will our antibiotics work? Fac Dent J 9(3):103–111
Pellis A, Guebitz GM, Nyanhongo GS (2022) Chitosan: sources, processing and modification techniques. Gels (Basel, Switzerland) 8(7):393
Pepla E, Besharat LK, Palaia G, Tenore G, Migliau G (2014) Nano-hydroxyapatite and its applications in preventive, restorative and regenerative dentistry: a review of literature. Ann Stomatol 5(3):108
Rabea EI, Badawy MET, Stevens CV, Smagghe G, Steurbaut W (2003) Chitosan as antimicrobial agent: applications and mode of action. Biomacromolecules 4(6):1457–1465
Romão IQ, Cavalcante SIA, Leite HLA, Goncalves LM, Branco-de-Almeida LS, Paschoal MAB (2018) Effect of combining erythrosine with a high-power dental curing light appliance on the viability of a planktonic culture of Streptococcus mutans. Photomed Laser Surg 36(12):676–679
Salehi B, Abu-Darwish MS, Tarawneh AH, Cabral C, Gadetskaya AV, Salgueiro L et al (2022) Antimicrobial resistance collaborators global burden of bacterial antimicrobial resistance in 2019: a systematic analysis. Lancet 399:629–655
Sánchez-López E, Gomes D, Esteruelas G, Bonilla L, Lopez-Machado AL, Galindo R, Souto EB (2020) Metal-based nanoparticles as antimicrobial agents: an overview. Nano 10(2):292
Shanmugasundaram S, Nayak N, Puzhankara L, Kedlaya MN, Rajagopal A, Karmakar S (2024) Bacteriophages: the dawn of a new era in periodontal microbiology? Crit Rev Microbiol 50(2):212–223
Singh R, Singh AP, Kumar S, Giri BS, Kim KH (2019) Antibiotic resistance in major rivers in the world: a systematic review on occurrence, emergence, and management strategies. J Clean Prod 234:1484–1505
Stensberg MC, Wei Q, McLamore ES, Porterfield DM, Wei A, Sepúlveda MS (2011) Toxicological studies on silver nanoparticles: challenges and opportunities in assessment, monitoring and imaging. Nanomedicine 6(5):879–898
Sukumar S, Martin FE, Hughes TE, Adler CJ (2020) Think before you prescribe: how dentistry contributes to antibiotic resistance. Aust Dent J 65(1):21–29

Wang L, Hu C, Shao L (2017) The antimicrobial activity of nanoparticles: present situation and prospects for the future. Int J Nanomedicine 12:1227–1249
Weichman JA, Johnson FM (1971) Laser use in endodontics: a preliminary investigation. Oral Surg Oral Med Oral Pathol 31(3):416–420
Welch JLM, Ramírez-Puebla ST, Borisy GG (2020) Oral microbiome geography: micron-scale habitat and niche. Cell Host Microbe 28(2):160–168
Wright GD (2005) Bacterial resistance to antibiotics: enzymatic degradation and modification. Adv Drug Deliv Rev 57(10):1451–1470
Xiang Y, Song F, Li H, Ji X (2024) Phage therapy: a renewed approach against oral diseases caused by Enterococcus faecalis infections. Microb Pathog 189:106574
Xu W, Pan Z, Wu Y, An XL, Wang W, Adamovich B et al (2024) A database on the abundance of environmental antibiotic resistance genes. Sci Data 11(1):250
Yan D, Li Y, Liu Y, Li N, Zhang X, Yan C (2021) Antimicrobial properties of chitosan and chitosan derivatives in the treatment of enteric infections. Molecules 26(23):7136
Yin W, Wang Y, Liu L, He J (2019) Biofilms: the microbial "protective clothing" in extreme environments. Int J Mol Sci 20(14):3423
Zahra AT (2021) Improvement of antibiotics absorption and regulation of tissue oxygenation through blood laser irradiation. Heliyon 7(4):e06863
Zhao X, Chen F, Wei H, Li Y (2023) Preparation and evaluation of bacteriophage LysP53 mouthwash. J Oral Sci Res 39(6):553

Part III

Phage Therapy: Viral Warrior Against Bacterial Invaders

Bacteriophages: The Natural Combatants to Fight AMR

15

Aiswarya Mohan, Daman Saluja, and Urmi Bajpai

Abstract

Viruses infecting eukaryotes are often harmful to their host. The extensive damage done to humankind by the SARS-CoV-2 virus shall remain etched in our collective memory for a long time. However, viruses of bacteria called bacteriophages are hugely beneficial for us and our ecosystem. As with any other of their kind, this group of viruses also requires a living host to propagate, and in the case of bacteriophages, it is a bacterium. Bacteriophages (phages) are also called "bacteria-eaters." They are the most abundant and ubiquitous biological entity on the planet. By keeping a check on the bacterial population, they play a significant role in maintaining the microbial equilibrium in any ecosystem. The therapeutic purpose of bacteriophages against infectious bacterial pathogens can be traced back to the pre-antibiotic period. However, their exploration as anti-bacterial

---

Supplementary Information The online version contains supplementary material available at https://doi.org/10.1007/978-981-97-5272-0_15.

A. Mohan
Department of Biomedical Science, Acharya Narendra Dev College, University of Delhi, Delhi, India

Department of Epidemiology and Public Health, Central University of Tamil Nadu, Nilakkudi, Tamil Nadu, India

D. Saluja
Dr B R Ambedkar Centre of Biomedical Research, University of Delhi, Delhi, India

U. Bajpai (✉)
Department of Biomedical Science, Acharya Narendra Dev College, University of Delhi, Delhi, India
e-mail: urmibajpai@andc.du.ac.in

M. Gangwar, G. Nath (eds.), *Emerging Paradigms for Antibiotic-Resistant Infections: Beyond the Pill*, https://doi.org/10.1007/978-981-97-5272-0_15

agents was thwarted after the discovery of antibiotics, except in some East European countries. Now, with the acute rise in antimicrobial resistance (AMR), where most bacterial pathogens have gained resistance to available antibiotics, a rebirth of phage therapy is being observed, and phage-immune system synergies are being investigated. Whole phages (natural and engineered), phage-derived enzymes, and phage–antibiotic combinations are emerging as potential alternatives/complements to antibiotics and, hence, deserve attention not only from researchers, clinicians, and regulatory bodies but also from the common people. The strength of virulent phages as therapeutics lies in their high target-specificity, low or no adverse effects, and abundant presence in nature, which assures a continuous discovery of novel phages and a populated phage pipeline. The spectrum of antibacterial activities of phages is beyond their applications in humans; it is extended to food crops, animals, aquaculture, and poultry farms. Besides, bacteriophages are also effective as biocontrol agents and disinfectants, prophylactics, in microbiome-editing, cancer treatment, etc. Their deployment in technologies such as phage display and drug-delivery systems make them genuinely versatile and promising agents. Hence, advancing our understanding of phage biology and its applications shall reap rich dividends.

Keywords

Bacteriophage · Antimicrobial resistance (AMR) · Phage therapy

15.1 Introduction

Viruses are bad and to be feared. All of them? Nah! There are good, friendly viruses, too, which could save us from the deadly scourge of antibiotic-resistant bacterial infections. These viruses are called bacteriophages.

We live in a world of ubiquitous, multifarious viruses showing diversity in structure, genetic makeup, and host cell preferences. Viruses are present in almost all ecosystems, infecting and predating on their hosts, including humans, animals, plants, bacteria, and so on. Ever since the viruses were discovered, defining or answering the seemingly simple yet complex question of whether they are living has not been straightforward. Viruses were not included in the Tree of Life, though they help shape it in various ways. A simple definition of a virus can be given as small infectious particles that have one type of nucleic acid (DNA or RNA) protected by a protein coat (capsid) (MH Adams 1959). They are inert outside a host and can withstand harsh conditions for a long time. Infectious virions contact the host cell through random collisions and bind through their receptor-binding proteins. They initiate the infection by injecting their genetic material, followed by hijacking the host cell machinery to shut down the host gene expression and synthesize their own progeny (Rakhuba et al. 2010), which are released at the end of the infection cycle. In higher organisms, the infection continues in the surrounding cells to damage tissue or organs, leading to severe effects.

Viruses that infect bacterial cells are classified as bacteriophages (or simply "phages") as they show specific lysis patterns in the bacterial culture. Phages are the most abundant life form on earth, estimated to be 10^{31} in number. As in any case of prey and predator, their abundance in an ecosystem is highly correlated (Chibani-Chennoufi et al. 2004). Hence, studying phage biology and its influence on the evolution and pathogenicity of the host bacteria can provide deeper insights into bacterial life cycles and their infection mechanisms. Given the rising antimicrobial resistance crisis, an improved understanding of the antibacterial potential of phages and applying them as therapeutics (phage therapy) is the need of the hour.

15.2 History of Phage Therapy

Bacteriophages were extensively used as therapeutics before the discovery of antibiotics. The period between 1920 and 1940 was considered the golden period for phage therapy when the therapy was being successfully used to treat several ailments associated with bacterial infections. However, due to less stringent safety practices in phage preparations, lack of suitable controls, and a limited understanding of the pharmacokinetics and pharmacodynamics of the administered phages, their efficacy as reliable antibacterial agents couldn't get the medical community's confidence. Also, the advent and popularity of antibiotics caused a decline in phage therapy in the Western world. However, it continued to be practiced in some of Eastern European countries such as Russia, Georgia, and Poland (Summers 2005).

15.2.1 Discovery of Bacteriophages and Pioneers of Phage Therapy

While monitoring cholera outbreaks in India, Ernest Hankin, a British physician, noticed a lack of decay in the Ganga and Jamuna rivers despite extensive waste disposal and several other human and animal activities. He expected but did not find much bacterial contamination in the rivers and suspected the water to possess antibacterial properties. He attributed the antibacterial properties to some substances, which were heat labile, could be filtered through porcelain, and demonstrated bactericidal activity against *Vibrio cholerae* (Hankin 1896). In 1898, Gamaleya, a Russian bacteriologist, observed a similar phenomenon when working with *Bacillus subtilis* (Samsygina and Boni 1984). Various other researchers also had related observations to make. About two decades later, Frederick William Twort, a British physician, observed clear zones or transparent areas in a *Staphylococcus aureus* culture, which he proposed to be caused by ultra-microscopic viruses. He published his findings in the Lancet (Twort 1915, 1936), where he explained that the cleared zones were seen more in younger bacterial cultures than the older ones, and also, heat-killed and dead bacterial cultures didn't produce any clear spots. However, he didn't pursue further to ascertain his hypothesis.

Around the same time, a French-Canadian scientist, Felix d'Herelle, working in a dysentery outbreak-controlling team at the Pasteur Institute in Paris, also observed clear circular zones, now known as plaques, on his bacterial culture plates. He referred to the clear zones as *taches vierges* (Summers 1999) and coined the term "bacteriophage," meaning "bacteria eater" in Greek. He continued with phage research, used his phage preparation to cure dysentery and similar infections, and published his early findings in 1917 (Publications Service 2007). The first bacteriophage treatment was given to a 12-year-old boy suffering from severe dysentery. To establish the procedure's safety to others, D'Herelle and other accompanying officials self-administered the phage suspension (Summers 1999) before administering it to the patient. He found patient's symptoms to disappear, leading to full recovery. This pioneering use of phages by D'Herelle, demonstrating their efficacy as antibacterial agents, led to phage therapy against various bacterial diseases such as avian typhosis, septicemia, and bubonic plague. The practice was termed "phagotherapy". This approach of using phage was depicted in a book named "Arrowsmith" written by Sinclair Lewis in 1925. Later, the invention of the electron microscope in the 1940s enabled visualization of phages, revealing their distinct morphological features, such as the shape and size of their head and tail, which helped in understanding the initial steps of phage interactions with the host (Luria et al. 1943; Ruska 1940). Electron microscopy considerably strengthened phage research and aided in the classification of bacteriophages.

The first institute for phage research was started in Tbilisi, Georgia, by the renowned bacteriologist Giorgi Eliava in 1923 in collaboration with Felix d'Herelle (Chanishvili 2012). The institute was mainly centered on researching and producing bacteriophages and continued to become one of the most extensive facilities for bacteriophage research. The Hirszfeld Institute of Immunology and Experimental Therapy in Poland, established in 1952, was also a prominent facility for bacteriophage research.

15.3 Structure, Life Cycle, and Classification of Phages

The antecedents of phage classification were Sir McFarlane Burnet and H Ruska, who independently classified phages based on the differences in their size, physiological preferences (Burnet 1933), and morphology (Ruska 1943), respectively. Lwoff, Horne, and Tournier, 1962 proposed another classification based on the nucleic acid and the virion properties. The International Committee on Taxonomy of Viruses (ICTV) (https://talk.ictvonline.org/taxonomy/w/ictv-taxonomy, previously known as the Provisional Committee on Nomenclature of Viruses (PCNV), was founded in 1965, which is the only international body now concerned with virus taxonomy. ICTV classifies phages considering all possible criteria, including morphology, physiological characteristics, and the nature of nucleic acid as the genetic material (Table 15.1).

Table 15.1 Classification of bacteriophages. Source: https://talk.ictvonline.org/

Order	Family	Characteristics	Example
Caudovirales	Ackermannviridae	Non-enveloped, head–tail structure, linear dsDNA, average genome size of 155 kb	*Serratia* phage MAM1
	Autographiviridae	Non-enveloped, head–tail structure, linear dsDNA, average genome size of 40 kb	*Vibrio* phage Vc1
	Myoviridae	Non-enveloped, head–tail structure, contractile tail, linear dsDNA, average genome size is 200 kb	*Erwinia* phage M7
	Podoviridae	Non-enveloped, head–tail structure, noncontractile tail, linear dsDNA, average genome size is 40 kb	Streptococcus phage Cp1
	Siphoviridae	Non-enveloped, head–tail structure, long noncontractile tail, linear dsDNA, average genome size is 50 kb	*Mycobacterium* phage Brujita
Norzivirale	Fiersviridae	Non-enveloped, spherical, monopartite, linear, ssRNA, average genome size is 4 kb	Escherichia phage MS2
Tubulavirales	Inoviridae	Non-enveloped, rod of filaments, helical capsid, circular ssDNA, average genome size is 8 kb	*Xanthomonas* phage Cf1c
Vinavirales	Corticoviridae	The icosahedral capsid, circular supercoiled dsDNA, average genome size is 10 kb	Pseudoalteromonas phage PM2
Kalamavirales	Tectiviridae	Non-enveloped, icosahedral capsid, linear dsDNA, average genome size is 15 kb	*Pseudomonas* phage PRD1
Petitvirales	Microviridae	Non-enveloped, icosahedral capsid, circular ssDNA, average genome size is 5 kb	*Escherichia* phage alpha3
Mindivirales	Cystoviridae	Enveloped, spherical double-capsid structure, segmented, linear dsRNA, total genome size 13 kb	*Pseudomonas* phage Phi6
	Plasmaviridae	Enveloped, spherical or pleomorphic, circular dsDNA, average genome size is 12 kb	Acholeplasma phage L2

15.3.1 Structural Classification

A bacteriophage contains genetic material, most commonly double-stranded DNA (dsDNA) and, in some cases, single-stranded DNA or RNA or double-stranded RNA, protected inside a protein coat called capsid. The capsid is made of protein subunits called capsomeres; in some cases, lipids are also present (Ackermann

2006). Some phages have an extra lipid layer outside the capsid called the envelope, which is usually the plasma membrane from the bacterial host and gets incorporated during the membrane budding process (Poranen and Bamford 2012). The size of a phage particle depends upon its genome size, indicating that phages with large capsids have large and complex genomes (Savva 2020).

Bacteriophages were earlier classified to the order *Caudovirales,* and the major families included were myoviridae, siphoviridae, and podoviridae; classified according to their contractile, noncontractile, and short tails, respectively. Although tailed phages constitute a monophyletic evolutionary group, they differ in genetic and structural aspects. Their genome sizes vary between 17 kb and 500 kb, and the tail length and contraction efficiency also vary in different families of phages (Abedon 2005). Other orders and families of phages exhibit different structures, such as filamentous, isometric, pleomorphic, etc. and are classified into various groups according to their nucleic acid content and genetic makeup. M13 and PM2 are two representative phages from the filamentous and isometric phages, respectively (Fig. 15.1). However, recently, a major overhaul in the phage classification system has been made by the Bacterial Viruses Subcommittee of the International Committee on Taxonomy of Viruses (ICTV) and a binomial system of nomenclature, consistent with other taxonomies in biology has been established (Zerbini et al. 2022). The morphology-based families *Myoviridae*, *Podoviridae*, and *Siphoviridae* have been abolished, and the order *Caudovirales* has been replaced by the class *Caudoviricetes.* This is done to group all tailed bacterial and archaeal viruses with icosahedral capsids and a double-stranded DNA genome Fig. 15.1 (Turner et al. 2023).

15.3.2 The Life Cycle of Phages

Based on their life cycle, phages are broadly classified as lytic (virulent) and temperate phages. A lytic phage infects the bacterial host cell, hijacks its cellular machinery, and replicates inside to make phage progeny, which are released from the host, resulting in cell lysis. In contrast, a temperate phage incorporates its genetic material into the host genome after infection and replicates along with it or stays extrachromosomal (Inal 2003; Guttman et al. 2005) (Fig. 15.1). Integrase, a phage-encoded enzyme, facilitates the integration of the phage genome into the host genome. The infected bacterium is termed a lysogen, and the phage genome integrated into the host genome is called prophage. When the host encounters a stressful event, the temperate phage can switch to the lytic cycle through induction. Recently, pseudo-lysogeny and chronic phage infection have been discovered as other forms of phage life cycle. Pseudo-lysogeny occurs when the phage DNA enters the host cell but stays independently in the cytoplasm, neither integrated into the host genome nor as a plasmid. Depending on environmental conditions, they undergo a lytic or lysogenic cycle (Abedon 2009). Chronic infection is similar to a lytic cycle where the phage successfully infects and amplifies inside the host cell but doesn't

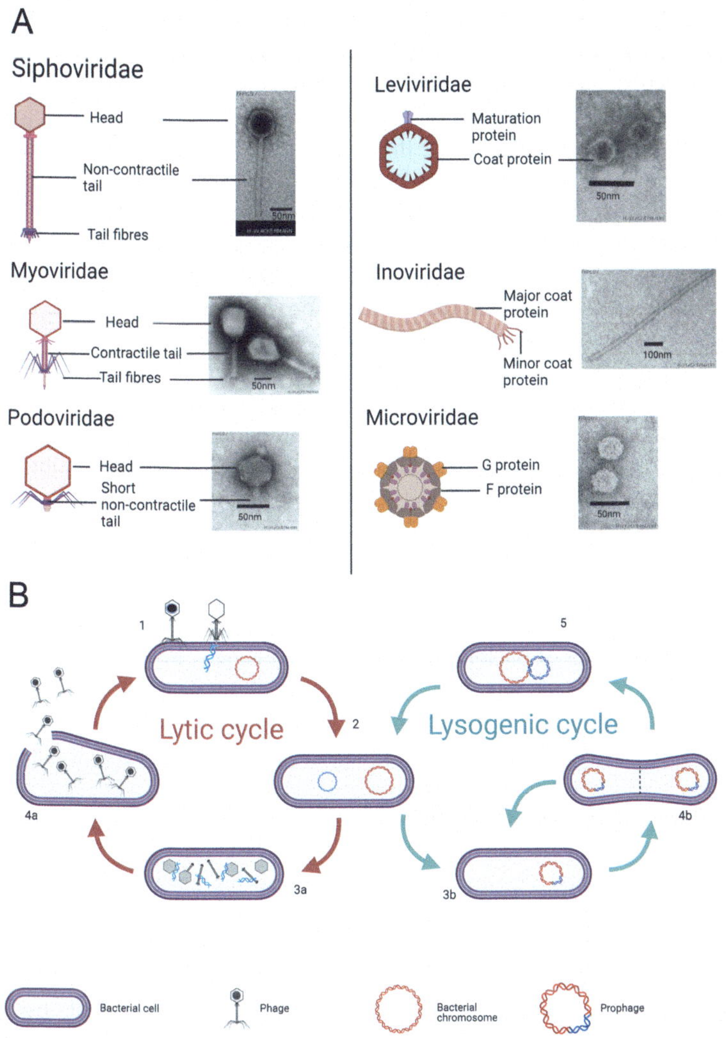

Fig. 15.1 Structural and functional classification (old system) of bacteriophages. (**a**) Classification based on phage morphology. A schematic representation and electron micrograph of each morphology from the respective family is shown. (Electron micrographs of phages are taken from Félix d'Hérelle Reference Center for Bacterial Viruses of the Université Laval, in Memorium Prof. Dr. Hans-Wolfgang Ackermann). (**b**) **Replicative cycle of bacteriophages.** Bacteriophages can be classified according to the replicative cycle they follow —Virulent and Temperate phages. (**1**) **Infection** Attachment of phage to the host cell receptors, followed by injection of phage genetic material into the host cell. (**2**) **Replication** Phage DNA replicates inside the cell to produce copies of itself. (**3a**) **Biosynthesis** Utilizing the host cell resources, phage genes are transcribed and translated to produce the structural proteins and enzymes. (**4a**) **Assembly and release of phage particles** Structural components and phage DNA assemble to form mature phage particles, which are released by lysing the host cell using the lytic enzymes synthesized during the late stage. (**3b** and **4b**) **Lysogeny** Phage DNA gets integrated into the host DNA and replicates along with it for several generations without causing any harm to the host. (**5**) **Induction** is the transition process from lysogeny to lytic cycle triggered by any chemical or physical stress resulting in the excision of phage DNA from the host DNA. *According to the recent ICTV report, Leviviridae has been renamed Fiersviridae

result in the lysis of the host cell; instead, the progenies are released using specialized export mechanisms (Olszak et al. 2017).

The molecular mechanisms underlying the infection process of phage are partially understood. The interaction between the receptor-binding phage proteins and the host cell receptors determines the host range of a phage (Weinbauer and Rassoulzadegan 2004). Double-stranded DNA viruses synthesize an enzyme called viral-associated peptidoglycan hydrolase (VAPGH), which facilitates the injection of genetic material into the host cell by attaching and disrupting the host cell membrane structures (Rodríguez-Rubio et al. 2013). Endolysins are the enzymes the nonfilamentous double-stranded DNA viruses produce, which hydrolyze the peptidoglycan to release viral progeny. These endolysins are a part of the Holin-endolysin system, where the enzymes are synthesized at the late stage of viral replication, and Holin permeabilizes the bacterial cell membrane to translocate endolysins into the periplasm (Wang et al. 2000). However, not all infections result in successful virion production. To counter predation, bacteria evolved to have several antiphage systems such as restriction endonucleases (Kelly and Smith 1970; Smith and Welcox 1970), CRISPR/Cas (Barrangou et al. 2007; Pourcel et al. 2005), Defense Island System Associated with Restriction Modification (DISARM) (Ofir et al. 2018), Bacteriophage Exclusion (BREX) (Goldfarb et al. 2015), etc., which degrade the injected phage DNA or inhibit their replication (Azam and Tanji 2019). A comprehensive understanding of bacterial defense mechanisms is important for designing successful phage therapy and is required to be reviewed well (Egido et al. 2022; Oromí-Bosch et al. 2023; Teklemariam et al. 2023).

15.4 Application of Phages

Phages are called good viruses due to their numerous beneficial applications in clinical and nonclinical settings. The therapeutic use of phages to treat bacterial infection in humans, animals, poultry, and plants is among the most promising applications of phages where they emerge as credible alternatives/complements to antibiotics. Other than phage therapy, phages and their encoded proteins are also used in biocontrol, diagnostics, as biosensors, delivery vehicles, and phage display libraries (Botsaris et al. n.d.; Choi et al. 2011; Kovacs et al. 2007).

Phages are constituents of the gut, skin, and other microbiomes in human and animal systems, where several microbial communities interact and maintain equilibrium (De Paepe et al. 2014). Minor fluctuations in the abundance of any species in a microbiome can cause alterations in the dynamicity of the equilibrium, which could manifest as a disease (Hannigan et al. 2018). The phage community plays a prominent role in maintaining a healthy equilibrium among the microbial population and also keeps a check by controlling the number of microorganisms in the system. The predation by phage is a factor controlled by the abundance of the host because phage encounters the host through random collisions (Rakhuba et al. 2010). When one of the bacterial species becomes copious, the possibility of infection by the phage increases, leading to the number of host cells returning to normal. This is explained as the "kill the winner" fashion of phage–bacteria interaction. "Community shuffling" is another model in which the phage undergoing a lysogenic cycle activates the lytic cycle due to some stress conditions. This results in the lysis of the host, leading to an imbalance in microbiota, which eventually ends up as a disease in the host animal (De Paepe et al. 2014) (Fig. 15.3).

15.4.1 Human Phage Therapy

The discovery of penicillin by Alexander Fleming in 1928, which was introduced as a drug in the 1940s and the development of sulfonamides in the 1930s were marked as the most outstanding achievements in the history of medicine. The period after that, spanning decades, was considered the golden age of antibiotics (Davies and Davies 2010). Over 40 antibiotics were discovered (Aminov 2010), saving millions of lives and increasing healthcare standards worldwide. Many diseases, once incurable, could be treated, causing the mortality rate due to microbial infections to drop drastically (Levy 1992). However, bacteria soon began to gain resistance to the available antibiotics. Though antibiotic resistance is a natural phenomenon, the extended misuse and overuse of antibiotics in humans and poultry, aquaculture, agriculture, etc., have led to widespread antimicrobial resistance, becoming a significant healthcare scare. Given a steep rise in mortality and morbidity due to resistant microbial infections every year, which demands new therapeutic interventions, the World Economic Forum Global Risk Report (2019) listed antibiotic resistance as one of the greatest threats to human health.

Phage therapy is the use of lytic bacteriophages for treating bacterial infections. Bacteriophages show extraordinary specificity toward their bacterial hosts. They have a benign presence in the environment and are an excellent and relatively inexpensive resource as antibacterial agents (Loc-Carrillo and Abedon 2011). Phages are also gaining traction for their roles as anti-inflammatory agents, immunomodulators, and key components in the human gut microbiome (Dalmasso et al. 2014). The phage properties that make them suitable as antibacterial therapeutics include their specificity, abundant and benign availability in nature, and ability to kill antibiotic-resistant strains of bacterial pathogens (Loc-Carrillo and Abedon 2011).

The narrow host range of phages makes them highly specific and offers significant advantages over broad-spectrum antibiotics. However, in the instances where bacteria develop resistance to phages by alterations of their cell surface receptors or co-receptors, which phages use to infect the cells (Abedon et al. 2011), a mixture of phages (phage cocktail) is used. In the cocktail, phages bind to the target bacterium through different receptors on its cell surface. This approach can considerably reduce resistance to phages (Fig. 15.2). In the case of personalized therapy, phages specific to the disease-causing clinical strains are identified and characterized. A single phage or a cocktail of these phages has been proven to be a successful intervention in treating infections that do not respond to premade phage preparations (Mattila et al. 2015). The challenge, however, could be the isolation/identification of phages within the time frame required in the treatment. Hence, building libraries of

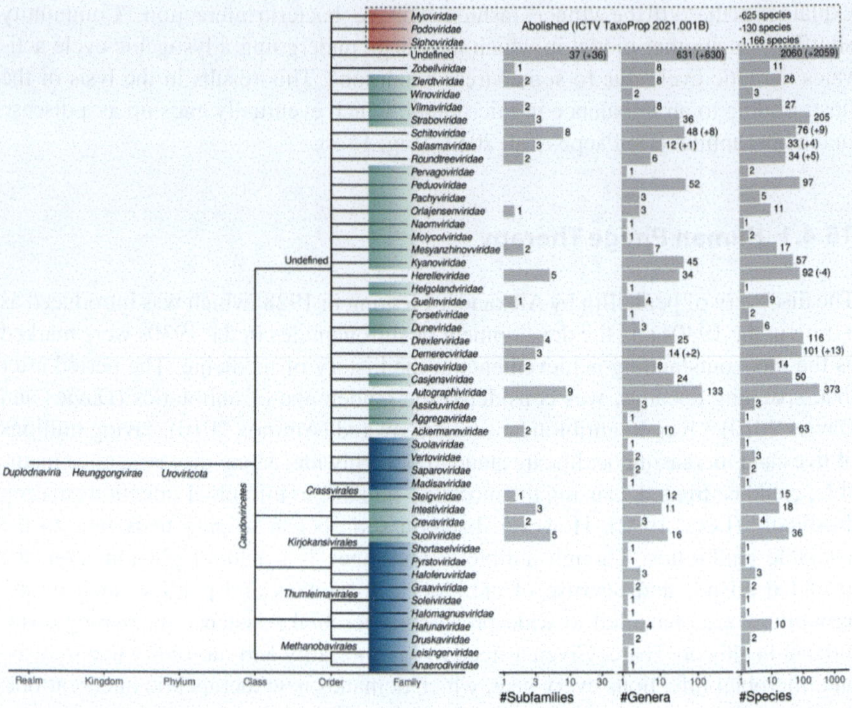

Fig. 15.2 New bacterial and archaeal virus family classification system, created in 2021. "Cladogram depicts the taxonomic structure of the order *Caudoviricetes*". New bacterial and archaeal virus families are highlighted in green or blue, respectively, and the abolished Myoviridae, Podoviridae, and Siphoviridae are in red. Numbers of subfamilies, genera, and species are depicted as histograms, adjacent to each family. Numbers in parentheses indicate changes recorded from ratified taxonomic proposals". *The image is reproduced from* Turner, D., Shkoporov, A.N., Lood, C. et al. (2023). Abolishment of morphology-based taxa and change to binomial species names: 2022 taxonomy update of the ICTV bacterial viruses subcommittee. *Arch Virol* **168**, 74. https://doi.org/10.1007/s00705-022-05694-2

well-characterized phages against bacterial pathogens is crucial as they can be screened to find the most effective phage both for mono-phage therapy and for preparing a phage cocktail (Chan et al. 2013). While phages against bacterial species can be readily isolated from the environment, clinical specimen or sewage, it can be daunting for some bacterial species, such as *S. aureus* or *E. faecalis* phage (Mattila et al. 2015). Therefore, besides hunting for phages from natural environments, engineered phages can be created using synthetic biology approaches (Khalifa et al. 2018; Yehl et al. 2019), where genome engineering can diversify their receptor-binding proteins hence expanding their host range (Fig. 15.5).

Phage-encoded antibacterial enzymes such as depolymerase, endolysins, and VAPGH are also used as antibacterial agents (Fig. 15.2). Phages synthesize these enzymes at different stages of their infectious cycle and act from within the cell (endolysins) or without (VAPGH). Lysins are among the most promising

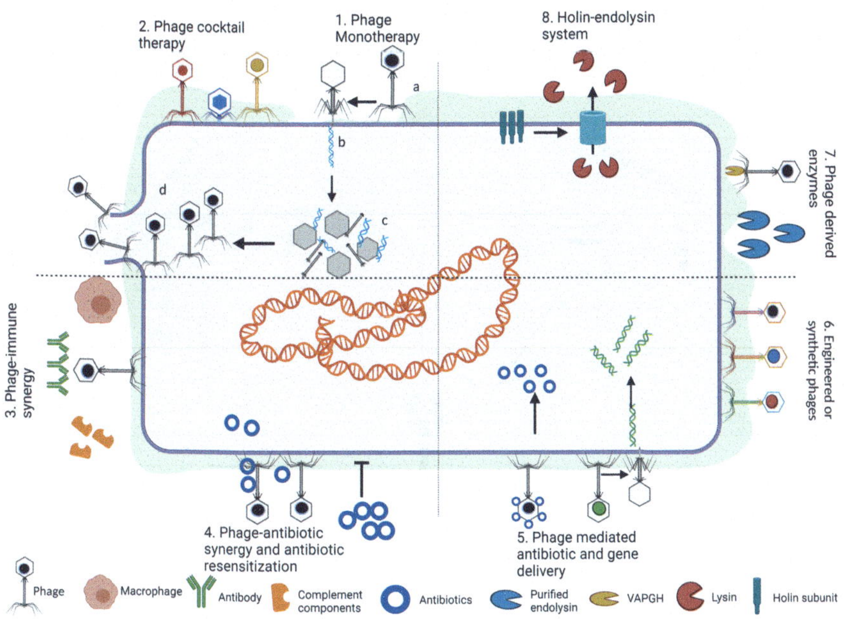

Fig. 15.3 Phages and phage-derived products used as therapeutics: A schematic representation. (1) Mono-phage therapy with the lytic cycle of the phage includes adsorption, infection, biosynthesis, assembly, and release of phage virions. (2) Cocktail therapy is where multiple phages are used against the same bacterium, increasing efficiency and reducing the occurrence of phage resistance. (3) Phage-immune synergy depicts activation of phagocytic cells and opsonization mediated by antibodies and complement activation. (4) Phage–antibiotic synergy in eliminating the pathogen. Phage infection also sometimes causes antibiotic resensitization. (5) Genetically engineered phages to facilitate targeted antibiotic delivery and gene delivery. (6) Engineered or synthetic phages to increase host range and enhanced infectivity. (7) Phage-derived lysins and hydrolases used as enzybiotics. (8) The holin-endolysin system is part of the lytic cycle, where the enzymes are synthesized in the late stage, enabling the lysis of cell wall components for the release of the phage particles. *This figure is adapted from* (Gordillo Altamirano and Barr 2019)

phage-derived proteins, put forward as an alternative and viable options under the "non-antibiotic" approach for treating bacterial infectious diseases. Compared with standard-of-care antibiotics, which are broad-spectrum, lysins are narrow-spectrum and do not harm the normal commensal microflora. In addition, lysins can kill growing and nongrowing bacterial cells (Fenton et al. 2010) and antibiotic-resistant bacteria. Also, since their target is peptidoglycan, one of the essential components of the bacterial cell wall, the development of bacterial resistance against these enzymes is considred less likely or very slow to occur since the probability of bacteria modifying their peptidoglycan is remote. The activity of lysins in gram-positive and gram-negative bacteria varies because of the structural differences in the cell wall composition. Gram-positive bacteria are more susceptible to lysins because of the absence of an outer membrane. In the case of gram-negative bacteria, lysin enzymes are required to go across the outer membrane to reach the peptidoglycan

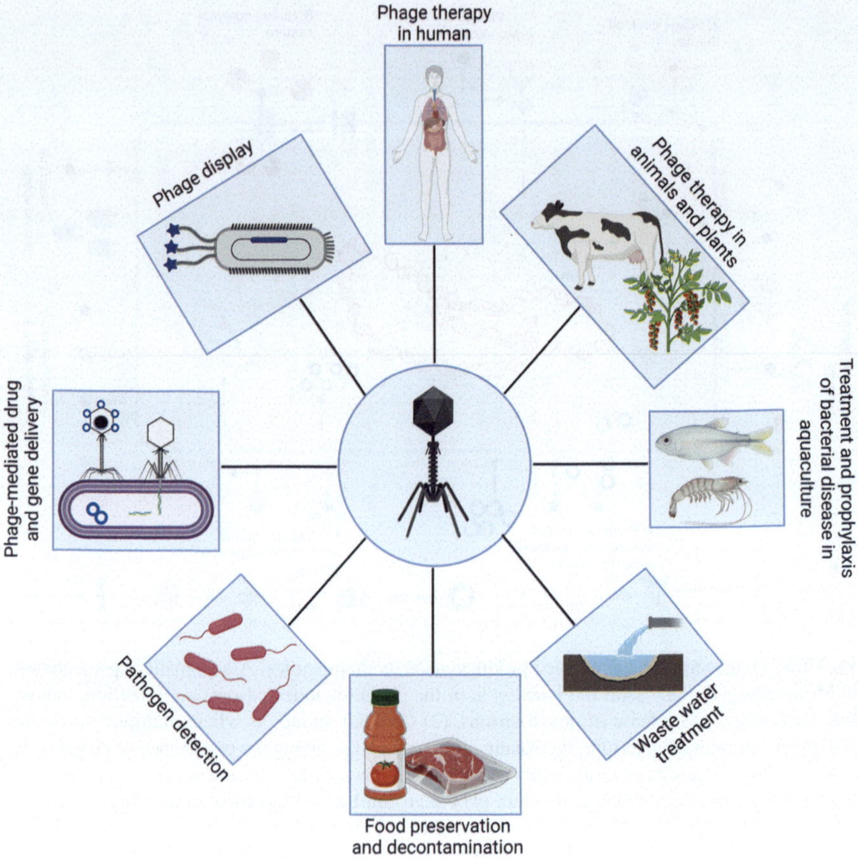

Fig. 15.4 Application of phages. From top to right: phages and phage-derived enzymes used in treating bacterial infections in humans; Phage-based preparations used in animals and agriculture for treatment of infections and sanitization; Phage-based feeds used in aquaculture for treatment and prophylaxis, Usage of phages in wastewater treatment to eliminate bacterial load, Phages and phage-based products for decontamination and preservation of food products, Phage-based assays for pathogen detection, Phages as carriers for drug and gene delivery, phage as an expression vector in phage display technique

layer. For this, outer membrane permeabilizing agents such as EDTA, chloroform, polycationic agents, and peptides can be used (Briers et al. 2007, 2011, 2014; Yang et al. 2015). Fischetti and his colleagues applied these lysins against gram-positive pathogens and termed them as "enzybiotics" (Nelson et al. 2001). There are several examples of commercial applications of lysins, too. Micreos company (Wageningen, The Netherlands) developed an endolysins-based aseptic liquid solution, Staphefekt™, which explicitly targets MRSA infection of the skin (Totté et al. 2017). ContraFect Corporation (NY, USA) developed CF-301 (Exebacase), a bacteriophage-derived lysin (used in combination with standard-of-care antibiotic) in the treatment of patients with *Staphylococcus aureus* infection in the bloodstream and

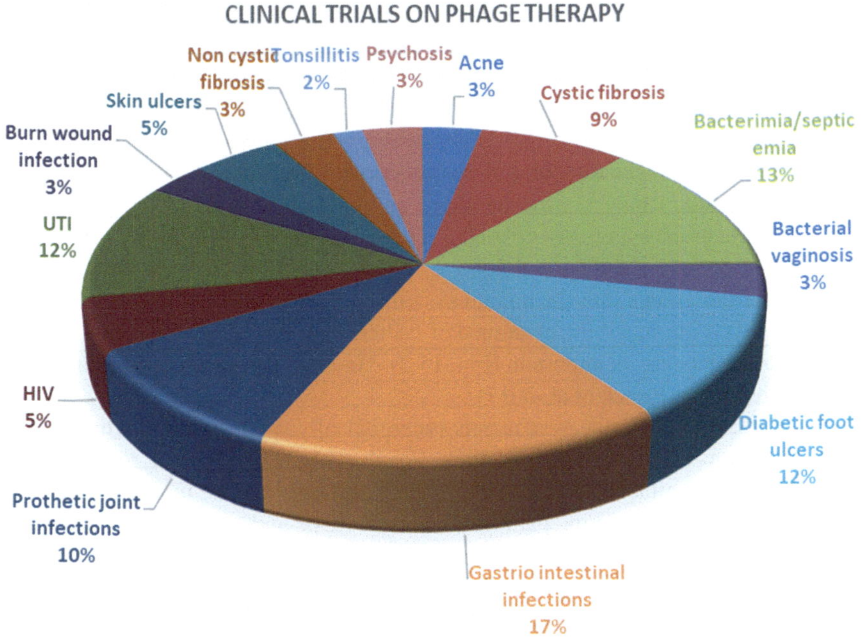

Fig. 15.5 Number of clinical trials on phage therapy targeting different bacterial infections. Clinical trials of phage therapy on various disease conditions and their interventions were retrieved from the NLM website—clinicaltrials.gov.in. Search was done by using the terms "Phage", "Phage therapy", and "Phage applications". Obtained data were exported and curated in Microsoft Excel. Duplicates and incomplete data were removed. The resulting dataset is given in Supplementary Table S15.1 and the summary is depicted in the pie chart. From the analysis, the major share of clinical trials on phage therapy emerge to focus on gastrointestinal infections (17%) followed by septicemia (13%) and diabetic foot ulcers

having infective endocarditis (IE) (Fowler et al. 2020; Schuch et al. 2017). It is the first and only lysin to complete a Phase 2 clinical trial. Several reviews (Abdelrahman et al. 2021; Bai et al. 2016; Chang 2020; De Maesschalck et al. 2020; Murray et al. 2021; Rodríguez-Rubio et al. 2015; Schmelcher and Loessner 2016; Shah et al. 2023; Sharma and Paul 2017) have described the recent advances in the use of phage-derived enzymes beyond clinical applications like diagnosis, food safety, decontamination of environmental pathogens, etc.

15.5 How Does Human Phage Therapy Work?

The U.S. FDA grants Emergency Use Authorization (EUA) permission for bacteriophage therapy. This allows the compassionate use of phage therapy for a patient with no other treatment options under the physician's discretion in a controlled clinical environment (US FDA 2021). During the COVID-19 pandemic, patients with multidrug-resistant infections and secondary infections received phage therapy and recovered successfully (Nicholls and Aslam 2022; Wu et al. 2022).

To gain an insight into how phage therapy works in humans, a deep understanding of the combined interaction of the pathogen, phages, and the human immune system and microbiome is required. Phages are the natural commensals in the human microbiome and are found in various parts of the human body like skin, gut, vagina, etc. Studies have shown the phage and bacterial populations to be related both antagonistically and synergistically (Van Belleghem et al. 2018). Similar to antibiotics, their availability, activity, etc., change when the phages are delivered through various routes into the human system. For phage therapy to be successful, immune intervention, for the clearance of bacterial debris, phagocytosis of phage–bacteria complex, etc. is important (Fig. 15.2). "Immunophage synergy" is used for this collaboration (Roach et al. 2017).

Sometimes, the immune system acts synergistically with phages to remove bacterial pathogens (Abd El-Aziz et al. 2019; Roach et al. 2017). The macrophages can recognize phages attached to the bacterial cells through specific motifs on the capsids, ultimately leading to the pathogen's phagocytosis and clearance. The immune response to phage therapy is highly dependent on the mode of phage administration. Humoral immune response to the oral administration of phage is usually low. However, intraperitoneal administration has been observed to produce immunoglobulins against the phage capsid proteins (Hodyra-Stefaniak et al. 2015; Majewska et al. 2015). Other administration routes include intravenous, topical, inhalation, intranasal, intratracheal…etc.

While considering oral administration of phage preparations for therapy, studying their efficiency and stability in the gastrointestinal tract is important. Phage stability at low pH is often challenging; hence, acidity neutralizers are usually prescribed along with oral administration of phage preparations. However, some studies have shown no significant role of acidity neutralizers in gut transit (Majewska et al. 2015; Yongsheng et al. 2008). Neutral and near-neutral pH (6.0–8.0) can be considered acceptable for the less stable phages, while stable phages could withstand a 2.0–11.0 pH (Jończyk et al. 2011). Encapsulation of phages for therapeutic usage is also an effective method to protect the phage particles from acidic environments and enhance anti-bacterial activity (Colom et al. 2017). In Bangladesh, clinical trials were carried out for treating diarrhea in children aged 4–60 months using bacteriophages against *E. coli* (Sarker et al. 2017). Though the study, which was called "Antibacterial Treatment Against Diarrhea in Oral Rehydration Solution" (August 2009–January 2013), did not improve diarrhea outcome, it showed a safe gut transit of the phages without disturbing the gut microbiome of the healthy individuals and a small number of phages were recovered from the feces. Also, no immune reactions were observed to be initiated against the phages in the healthy subjects. In another case, a phase I randomized, double-blind, placebo-controlled crossover trial was conducted in 32 adults with moderate gastrointestinal disorders. The subjects were given capsules containing four different bacteriophages targeting gastrointestinal pathogens. Compared with the placebo, the treatment was safe and effective in reducing colon and pain in the small intestine, gastrointestinal dysfunction, etc. (Gindin et al. 2019).

Based on the early reports on phage therapy in treating systemic infections such as typhoid and bacteremia, phages were likely to be effective and safe via the intravenous (IV) route (Speck and Smithyman 2016). However, the intravenous delivery of phages for therapy has less commonly been practiced as it was feared to elicit toxin shock or serum sickness in the case of administered phage preparation being contaminated. Notwithstanding these apprehensions, the last decade has witnessed an increasing number of case studies using the IV route for phage therapy (Aslam et al. 2020; Le et al. 2023; Pirnay et al. 2023). In 2017, the famous case of successful intravenous and intracavitary phage–antibiotic synergy therapy given on compassionate grounds to a 68-year-old diabetic patient with systemic MDR *A. baumanii* infection played a pivotal role in highlighting the safety of IV phage therapy (Fig. 15.5) (Schooley et al. 2017).

Bioengineered phages as therapeutics are under various stages of research and clinical trials. Engineered and synthetic phages are used to counter bacterial resistance and increase the efficiency of phage killing. SASPject is a product that integrates small acid-soluble spore proteins (SASPs) from *Bacillus subtilis* into the phage genome. In-vitro studies using the product showed remarkable bactericidal activity against various *Staphylococcus aureus* isolates, including drug-resistant ones (Cass et al. 2021). Phage-immune synergy and phage–antibiotic synergy are the major focus areas in clinical studies. In-vitro studies on phage-antibiotic synergy have also shed light on the possibility of "phage adjuration," a phenomenon where the sensitivity against antibiotics is restored when used with phages in antibiotic-resistant bacteria (Liu et al. 2020). A recent study published in Nature Communications reported the case of a 30-year-old fracture-related infection of pan-drug-resistant *Klebsiella pneumoniae*. The infectious agent showed pan-antibiotic resistance and thus was treated with preadapted bacteriophages on compassionate grounds. Phage therapy accompanied by antibiotics, Tigecyclin, and Avibactam showed improvement in the patient's symptoms within 2 days. Three months after surgical intervention and continued phage–antibiotic combinational treatment resulted in a complete recovery from infection. The study also shows the effective phage–antibiotic synergy in vitro suspensions and biofilms (Eskenazi et al. 2022). Several other case reports also showed efficient phage–antibiotic synergy in patients suffering from various multidrug-resistant infections of *A. baumanii, K. pneumoniae,* and *E. coli* (Aslam et al. 2020; Bao et al. 2020; Kuipers et al. 2019; Nir-Paz et al. 2019; Qin et al. 2021; Rubalskii et al. 2020).

Secondary bacterial infections are prevalent in viral infection cases, immunocompromised people, and patients on immunosuppressive drugs (Das et al. 2022; Manohar et al. 2020). In 2019, a 15-year-old with cystic fibrosis and other comorbidities who developed a disseminated infection of *M. abscessus* after lung transplantation was given a cocktail of natural and engineered lytic phages. This case marked the first-ever report of successful phage therapy against a mycobacterium infection using engineered phages. In another case, a patient with disseminated cutaneous *Mycobacterium chelonae* infection and necrotizing granulomas and lesions received phage treatment after several months of irresponsive antibiotic treatments. Surgical interventions along with antibiotic and phage therapy showed significant improvement in patient symptoms without any adverse effects of using phages (Little et al. 2022). Secondary infections in COVID-19 patients have also been successfully treated using phage cocktails. Wu et al. (2021) describe a case where four patients in the COVID ICU with carbapenem-resistant *Acinetobacter baumanii* infection received a cocktail of two phages, ΦAb121 and ΦAb124. While two patients completely recovered, the other two patients though were cleared of *A. baumanii* infections but died of respiratory failure. The combination of phages was formulated to avoid phage resistance in bacteria and was administered via inhalation and topically. As per the registered clinical trials done around the world, the majority focuses on gastrointestinal infections followed by septicemia and diabetic foot ulcers (Fig. 15.4).

In addition to using phages for treating antibiotic-resistant, recalcitrant infections, phage-mediated microbiome editing for treating non-infectious diseases is also gaining importance, which has been successfully demonstrated in animal models in several reports (Dahlman et al. 2021; Khan Mirzaei and Deng 2021). Phage-guided nanoparticles have also been shown to reduce the side effects of chemotherapy drugs and promote the growth of commensal *Clostridium butyricum* in mouse models (Zheng et al. 2019).

15.5.1 Phage Applications in Treating Bacterial Infections in Animals and Plants

Bacteriophages also find applications in detecting bacterial pathogens, drug delivery, wastewater treatment food safety, aquaculture, and agriculture. Poultry and meat are chief sources of food-borne pathogens and infections. Bacteriophages against these pathogens have been used in poultry to reduce the incidence of food-borne diseases caused by *Clostridium perfringens* (Miller et al. 2010), *Salmonella enterica* (Nabil et al. 2018), *Campylobacterjejuni* (Richards et al. 2019) among others. Commercial preparations like BioTector by Korean CheilJedang Corporation were introduced in 2010 to reduce the *Salmonella* load in poultry. Phages-based food additives and food decontaminants have also been developed and used widely. The commercial product Listex P-100 by the Dutch company EBI based on Listeria phages is approved by the U.S. FDA and listed under Generally Recognized As Safe (GRAS) as a food additive in storing foods (Soni et al. 2010).

Bacteriophages and phage-based products have also been used widely in agriculture and aquaculture. Phage-treated feeds and pellets have effectively treated *Pseudomonas* and *Flavobacterium psychrophilum* infections in fishes (Donati et al. 2021; Park and Nakai 2003). Phages alone and in combination with other crop protection technologies were effective in treating bacterial infections in plants like bacterial wilt and spots caused by *Ralstonia solanacearum* and *Xanthomonas campestris*, respectively (Papaianni et al. 2020; Wang et al. 2019). This led to the development of a commercial biocontrol product named AgriPhage by the American company Omnilytics. The product has shown efficacy in treating most plant bacterial infections and is compatible with organic food production (Ingram and Lu 2009). Spraying the phage suspension at various stages of the plant growth has successfully reduced the incidence of *Pseudomonas syringae* infection in kiwi plants, as demonstrated by Fiorillo et al. (2023).

15.6 Future Perspective

Ten years ago, while reviewing the future impact of antimicrobial resistance based on the available data, a British economist, Jim O'Neill and his team assessed a considerable human and economic cost and predicted ten million deaths attributed to AMR annually by 2050 and a loss of about 100 trillion dollars if the antimicrobial resistance was left unchecked (2014). Despite the glaring statistics, the world hasn't yet imagined a world without antimicrobials. Antimicrobial resistance is an evolutionary inevitability compounded by the rampant misuse and overuse of antibiotics. To mitigate the impact of this crisis of a global scale, conserving existing antibiotics and finding new treatment options quickly are paramount. The interconnected, interwoven world we live in and the growing acknowledgment of the One Health concept highlight the drivers of the spread of resistance and urge exploring alternatives without delay.

Bacteriophages have significantly contributed to our understanding of molecular genetics in the past and form the basis of several ground-breaking technologies, such as phage display CRISPR/Cas, among others. Being the natural predators of bacteria, phages and phage-derived products are fast emerging as promising antibacterial agents. Amidst the present crisis of antibiotic resistance and increased occurrence of infectious diseases, a renaissance of phage therapy in its second century is being witnessed globally. A greater emphasis on phage biology and its clinical applications is important for phage therapy to become mainstream or even be used on compassionate grounds as personalized therapy.

One of the challenges in phage therapy is identifying virulent phages with a high potential to infect bacterial strains successfully. Another is predicting the likelihood of phages developing resistance to their bacterial host. A rise in the pace of natural phage discovery doesn't guarantee a proportional increase in the number of phages suitable for phage therapy. The future use of artificial intelligence (AI) and machine learning lies in screening large datasets of phage–host interactions and developing algorithms to select effective phages that best match the clinical strains and monitor the efficacy of phage therapy in real time. This would help clinicians make informed choices and optimize treatment options. Also, by predicting phage resistance, the technologies can help design phage cocktails with an enhanced host range. By analyzing clinical data, AI can also improve clinical trial success rates by optimizing key steps in phage therapy. Hence, AI is likely to be leveraged extensively to design the next generation of therapeutic phages and improve treatment outcomes of phage therapy. Another aspect deserving greater attention is determining the pharmacokinetics and pharmacodynamics of phages as drugs, their interactions with the bacterial hosts and the immune responses of the therapy recipient with limited data. Randomized, controlled clinical trials are required to generate data that can establish the efficacy and safety of the therapy.

Bacteriophages are among the most promising arsenals we have to avert or delay the antibiotic apocalypse. Creating awareness among scientists, clinicians, biotech industries and start-ups, the regulatory authorities, and the public is essential to address the challenges, dispel the associated myths, draw funds, and highlight the significant promise phages hold in medicine.

Acknowledgments We thank the Indian Council of Medical Research (ICMR), New Delhi, India, for supporting the project (5/8/5/38/2019-ECD-I). AM received the ICMR project fellowship. We thank Acharya Narendra Dev College (ANDC) and Dr. B.R. Ambedkar Centre for Biomedical Research, University of Delhi for the infrastructural support.

Figures were created using BioRender.com (https://biorender.com).

Availability of Data and Materials

All data analyzed during this study and its supplementary information (Table S15.1) file are included in the article.

References

Abd El-Aziz AM, Elgaml A, Ali YM (2019) Bacteriophage therapy increases complement-mediated lysis of bacteria and enhances bacterial clearance after acute lung infection with multidrug-resistant Pseudomonas aeruginosa. J Infect Dis 219(9):1439–1447. https://doi.org/10.1093/infdis/jiy678

Abdelrahman F, Easwaran M, Daramola OI, Ragab S, Lynch S, Oduselu TJ, Khan FM, Ayobami A, Adnan F, Torrents E, Sanmukh S, El-Shibiny A (2021) Phage-encoded endolysins. Antibiotics 10(2):124. https://doi.org/10.3390/antibiotics10020124

Abedon S (2005) The bacteriophages. https://books.google.com/books?hl=en&lr=&id=6J9vQgBfUK0C&oi=fnd&pg=PR5&dq=3.%09Abedon,+S.T.,+2005.+The+bacteriophages.+Oxford+University+Press.&ots=lLrzGLbmo0&sig=SFkZIrTbwZr-worVzzYeRwS9SrI

ST Abedon (2009) Disambiguating bacteriophage pseudolysogeny: an historical analysis of lysogeny, pseudolysogeny, and the phage carrier state. Academia. EduST Abedon Contemporary Trends in Bacteriophage Research, 2009•academia.Edu. Retrieved March 7, 2024, from https://www.academia.edu/download/53529769/036_2009g__Abedon__Contemp._Trends_in_Bacteriophage_Res._scanned__Disambiguating_Pseudolyosgeny.pdf

Abedon ST, Kuhl SJ, Blasdel BG, Kutter EM (2011) Phage treatment of human infections. Bacteriophage 1(2):66–85. https://doi.org/10.4161/bact.1.2.15845

Ackermann HW (2006) The bacteriophages, vol 2. Oxford University Press, p 746

MH Adams (1959) Bacteriophages. Cabdirect.Org. Retrieved March 3, 2024, from https://www.cabdirect.org/cabdirect/abstract/19602204111

Aminov RI (2010) A brief history of the antibiotic era: lessons learned and challenges for the future. Front Microbiol 1:134. https://doi.org/10.3389/fmicb.2010.00134

Aslam S, Lampley E, Wooten D, Karris M, Benson C, Strathdee S, Schooley RT (2020) Lessons learned from the first 10 consecutive cases of intravenous bacteriophage therapy to treat multidrug-resistant bacterial infections at a single center in the United States. Open Forum Infect Dis 7(9):ofaa389. https://doi.org/10.1093/ofid/ofaa389

Azam AH, Tanji Y (2019) Bacteriophage-host arm race: an update on the mechanism of phage resistance in bacteria and revenge of the phage with the perspective for phage therapy. Appl Microbiol Biotechnol 103(5):2121–2131. https://doi.org/10.1007/s00253-019-09629-x

Bai J, Kim Y-T, Ryu S, Lee J-H (2016) Biocontrol and rapid detection of food-borne pathogens using bacteriophages and endolysins. Front Microbiol 7. https://doi.org/10.3389/fmicb.2016.00474

Bao J, Wu N, Zeng Y, Chen L, Li L, Yang L, Zhang Y, Guo M, Li L, Li J, Tan D, Cheng M, Gu J, Qin J, Liu J, Li S, Pan G, Jin X, Yao B et al (2020) Non-active antibiotic and bacteriophage synergism to successfully treat recurrent urinary tract infection caused by extensively drug-resistant Klebsiella pneumoniae. Emerg Microbes Infect 9(1):771–774. https://doi.org/10.1080/22221751.2020.1747950

Barrangou R, Fremaux C, Deveau H, Richards M, Boyaval P, Moineau S, Romero DA, Horvath P (2007) CRISPR provides acquired resistance against viruses in prokaryotes. Science (New York, N.Y.) 315(5819):1709–1712. https://doi.org/10.1126/SCIENCE.1138140

Botsaris G, Liapi M, Kakogiannis C, et al. (n.d.) Application of a rapid and sensitive combined phage-PCR method for the detection of Mycobacterium avium subspecies paratuberculosis in raw milk. Cabdirect.Org. Proceedings of the 10th International Colloquium on Paratuberculosis.. Retrieved March 3, 2024, from https://www.cabdirect.org/cabdirect/abstract/20123141253

Briers Y, Volckaert G, Cornelissen A, Lagaert S, Michiels CW, Hertveldt K, Lavigne R (2007) Muralytic activity and modular structure of the endolysins of Pseudomonas aeruginosa bacteriophages phiKZ and EL. Mol Microbiol 65(5):1334–1344. https://doi.org/10.1111/J.1365-2958.2007.05870.X

Briers Y, Walmagh M, Lavigne R (2011) Use of bacteriophage endolysin EL188 and outer membrane permeabilizers against Pseudomonas aeruginosa. J Appl Microbiol 110(3):778–785. https://doi.org/10.1111/J.1365-2672.2010.04931.X

Briers Y, Walmagh M, Van Puyenbroeck V, Cornelissen A, Cenens W, Aertsen A, Oliveira H, Azeredo J, Verween G, Pirnay JP, Miller S, Volckaert G, Lavigne R (2014) Engineered endolysin-based "Artilysins" to combat multidrug-resistant gram-negative pathogens. MBio 5(4):e01379–e01314. https://doi.org/10.1128/MBIO.01379-14

Burnet FM (1933) The classification of dysentery-coli bacteriophages. III. A correlation of the serological classification with certain biochemical tests. J Pathol Bacteriol. Retrieved March 3, 2024, from https://www.cabdirect.org/cabdirect/abstract/19342700140

Cass J, Barnard A, Fairhead H (2021) Engineered bacteriophage as a delivery vehicle for antibacterial protein, SASP. Pharmaceuticals (Basel, Switzerland) 14(10):1038. https://doi.org/10.3390/PH14101038

Chan BK, Abedon ST, Loc-Carrillo C (2013) Phage cocktails and the future of phage therapy. Future Microbiol 8(6):769–783. https://doi.org/10.2217/FMB.13.47

Chang Y (2020) Bacteriophage-derived endolysins applied as potent biocontrol agents to enhance food safety. Microorganisms 8(5). https://doi.org/10.3390/MICROORGANISMS8050724

Chanishvili N (2012) Phage therapy-history from Twort and d'Herelle through Soviet experience to current approaches. Adv Virus Res 83:3–40. https://doi.org/10.1016/B978-0-12-394438-2.00001-3

Chibani-Chennoufi S, Bruttin A, Dillmann ML, Brüssow H (2004) Phage-host interaction: an ecological perspective. J Bacteriol 186(12):3677–3686. https://doi.org/10.1128/JB.186.12.3677-3686.2004

Choi J, Kotay S, Goel R (2011) Bacteriophage-based biocontrol of biological sludge bulking in wastewater. Bioeng Bugs 2(4):214–217. https://doi.org/10.4161/bbug.2.4.16211

Colom J, Cano-Sarabia M, Otero J, Aríñez-Soriano J, Cortés P, Maspoch D, Llagostera M (2017) Microencapsulation with alginate/CaCO3: a strategy for improved phage therapy. Sci Rep 7. https://doi.org/10.1038/SREP41441

Dahlman S, Avellaneda-Franco L, Barr JJ (2021) Phages to shape the gut microbiota? Curr Opin Biotechnol 68:89–95. https://doi.org/10.1016/J.COPBIO.2020.09.016

Dalmasso M, Hill C, Ross RP (2014) Exploiting gut bacteriophages for human health. Trends Microbiol 22(7):399–405. https://doi.org/10.1016/J.TIM.2014.02.010

Das R, Kotra K, Singh P, Loh B, Leptihn S, Bajpai U (2022) Alternative treatment strategies for secondary bacterial and fungal infections associated with COVID-19. Infect Dis Ther 11(1):53–78. https://doi.org/10.1007/S40121-021-00559-8

Davies J, Davies D (2010) Origins and evolution of antibiotic resistance. Microbiol Mol Biol Rev 74(3):9–16. https://doi.org/10.1128/MMBR.00016-10

De Maesschalck V, Gutiérrez D, Paeshuyse J, Lavigne R, Briers Y (2020) Advanced engineering of third-generation lysins and formulation strategies for clinical applications. Crit Rev Microbiol 46(5):548–564. https://doi.org/10.1080/1040841X.2020.1809346

De Paepe M, Leclerc M, Tinsley CR, Petit MA (2014) Bacteriophages: an underestimated role in human and animal health? Front Cell Infect Microbiol 5. https://doi.org/10.3389/FCIMB.2014.00039/FULL

Donati VL, Dalsgaard I, Sundell K, Castillo D, Er-Rafik M, Clark J, Wiklund T, Middelboe M, Madsen L (2021) Phage-mediated control of Flavobacterium psychrophilum in aquaculture: in vivo experiments to compare delivery methods. Front Microbiol 12. https://doi.org/10.3389/FMICB.2021.628309

Egido JE, Costa AR, Aparicio-Maldonado C, Haas PJ, Brouns SJJ (2022) Mechanisms and clinical importance of bacteriophage resistance. FEMS Microbiol Rev 46(1):fuab048. https://doi.org/10.1093/FEMSRE/FUAB048

Eskenazi A, Lood C, Wubbolts J, Hites M, Balarjishvili N, Leshkasheli L, Askilashvili L, Kvachadze L, van Noort V, Wagemans J, Jayankura M, Chanishvili N, de Boer M, Nibbering P, Kutateladze M, Lavigne R, Merabishvili M, Pirnay JP (2022) Combination of pre-adapted bacteriophage therapy and antibiotics for treatment of fracture-related infection due to pandrug-resistant Klebsiella pneumoniae. Nat Commun 13(1). https://doi.org/10.1038/S41467-021-27656-Z

Fenton M, Ross P, Mcauliffe O, O'Mahony J, Coffey A (2010) Recombinant bacteriophage lysins as antibacterials. Bioeng Bugs 1(1):9–16. https://doi.org/10.4161/BBUG.1.1.9818

Fiorillo A, Frezza D, Di Lallo G, Visconti S (2023) A phage therapy model for the prevention of pseudomonas syringae pv. actinidiae infection of kiwifruit plants. Plant Dis 107(2):267–271. https://doi.org/10.1094/PDIS-02-22-0348-SC/ASSET/IMAGES/LARGE/PDIS-02-22-0348-SCF3.JPEG

Fowler VG, Das AF, Lipka-Diamond J, Schuch R, Pomerantz R, Jáuregui-Peredo L, Bressler A, Evans D, Moran GJ, Rupp ME, Wise R, Ralph Corey G, Zervos M, Douglas PS, Cassino C (2020) Exebacase for patients with Staphylococcus aureus bloodstream infection and endocarditis. J Clin Invest 130(7):3750–3760. https://doi.org/10.1172/JCI136577

Gindin M, Febvre HP, Rao S, Wallace TC, Weir TL (2019) Bacteriophage for gastrointestinal health (PHAGE) study: evaluating the safety and tolerability of supplemental bacteriophage consumption. J Am Coll Nutr 38(1):68–75. https://doi.org/10.1080/07315724.2018.1483783

Goldfarb T, Sberro H, Weinstock E, Cohen O, Doron S, Charpak-Amikam Y, Afik S, Ofir G, Sorek R (2015) BREX is a novel phage resistance system widespread in microbial genomes. EMBO J 34(2):169–183. https://doi.org/10.15252/EMBJ.201489455

Gordillo Altamirano FL, Barr JJ (2019) Phage therapy in the postantibiotic era. Clin Microbiol Rev 32(2). https://doi.org/10.1128/CMR.00066-18

Guttman B, Raya R, Kutter E (2005) Basic phage biology. Bacteriophages: Biology and Applications Retrieved March 3, 2024, from https://books.google.com/books?hl=en&lr=&id=6sTBd2EJpmYC&oi=fnd&pg=PA29&dq=45.%09Guttman,+B.,+Raya,+R.+and+Kutter,+E.,+2005.+Basic+phage+biology.+Bacteriophages:+Biology+and+applications,+4.&ots=j26Y-hi4ko&sig=hBK0TAGLDB5he8srhIvtC6SSGg4

Hankin E (1896) L'action bactericide des eaux de la Jumna et du Gange sur le vibrion du cholera. Cir Nii Ac Jp. Retrieved March 3, 2024, from https://cir.nii.ac.jp/crid/1370567185920754061

Hannigan GD, Duhaime MB, Koutra D, Schloss PD (2018) Biogeography and environmental conditions shape bacteriophage-bacteria networks across the human microbiome. PLoS Comput Biol 14(4):e1006099. https://doi.org/10.1371/JOURNAL.PCBI.1006099

Hodyra-Stefaniak K, Miernikiewicz P, Drapała J, Drab M, Jonczyk-Matysiak E, Lecion D, Kazmierczak Z, Beta W, Majewska J, Harhala M, Bubak B, Kłopot A, Górski A, Dabrowska

K (2015) Mammalian Host-Versus-Phage immune response determines phage fate in vivo. Sci Rep 5. https://doi.org/10.1038/SREP14802

JM Inal (2003) Archivum immunologiae et therapiae. Phage therapy: a reappraisal of bacteriophages as antibiotics. Europepmc.Org. Retrieved March 3, 2024, from https://europepmc.org/article/med/12956433

Ingram D, Lu S-E (2009) Evaluation of foliar sprays of bacteriophages for the management of bacterial canker in greenhouse tomatoes. Plant Health Progress 10(1):–19. https://doi.org/10.1094/PHP-2009-0512-01-RS

Jończyk E, Kłak M, Międzybrodzki R, Górski A (2011) The influence of external factors on bacteriophages--review. Folia Microbiol 56(3):191–200. https://doi.org/10.1007/S12223-011-0039-8

Kelly TJ, Smith HO (1970) A restriction enzyme from Hemophilus influenzae. II. J Mol Biol 51(2):393–409. https://doi.org/10.1016/0022-2836(70)90150-6

Khalifa L, Gelman D, Shlezinger M, Dessal AL, Coppenhagen-Glazer S, Beyth N, Hazan R (2018) Defeating antibiotic- and phage-resistant enterococcus faecalis using a phage cocktail in vitro and in a clot model. Front Microbiol 9:326. https://doi.org/10.3389/FMICB.2018.00326

Khan Mirzaei M, Deng L (2021) Sustainable microbiome: a symphony orchestrated by synthetic phages. Microb Biotechnol 14(1):45–50. https://doi.org/10.1111/1751-7915.13697

Kovacs EW, Hooker JM, Romanini DW, Holder PG, Berry KE, Francis MB (2007) Dual-surface-modified bacteriophage MS2 as an ideal scaffold for a viral capsid-based drug delivery system. Bioconjug Chem 18(4):1140–1147. https://doi.org/10.1021/BC070006E

Kuipers S, Ruth MM, Mientjes M, de Sévaux RGL, van Ingen J (2019) A Dutch case report of successful treatment of chronic relapsing urinary tract infection with bacteriophages in a renal transplant patient. Antimicrob Agents Chemother 64(1):e01281–e01219. https://doi.org/10.1128/AAC.01281-19

Le T, Nang SC, Zhao J, Yu HH, Li J, Gill JJ, Liu M, Aslam S (2023) Therapeutic potential of intravenous phage as standalone therapy for recurrent drug-resistant urinary tract infections. Antimicrob Agents Chemother 67(4):e0003723. https://doi.org/10.1128/AAC.00037-23

SB Levy (1992) From tragedy the antibiotic age is born. The antibiotic paradox: how miracle drugs are destroying the miracle. Springer. Retrieved March 3, 2024, from https://link.springer.com/chapter/10.1007/978-1-4899-6042-9_1

Little JS, Dedrick RM, Freeman KG, Cristinziano M, Smith BE, Benson CA, Jhaveri TA, Baden LR, Solomon DA, Hatfull GF (2022) Bacteriophage treatment of disseminated cutaneous mycobacterium chelonae infection. Nat Commun 13(1). https://doi.org/10.1038/S41467-022-29689-4

Liu CG, Green SI, Min L, Clark JR, Salazar KC, Terwilliger AL, Kaplan HB, Trautner BW, Ramig RF, Maresso AW (2020) Phage-antibiotic synergy is driven by a unique combination of antibacterial mechanism of action and stoichiometry. MBio 11(4):1–19. https://doi.org/10.1128/MBIO.01462-20

Loc-Carrillo C, Abedon ST (2011) Pros and cons of phage therapy. Bacteriophage 1(2):111–114. https://doi.org/10.4161/BACT.1.2.14590

Luria SE, Delbrück M, Anderson TF (1943) Electron microscope studies of bacterial viruses. J Bacteriol 46(1):57–77. https://doi.org/10.1128/JB.46.1.57-77.1943

Lwoff A, Horne RW, Tournier P (1962) [A virus system]. Comptes Rendus Hebdomadaires Des Seances de l'Academie Des Sciences 254:4225–4227. https://pubmed.ncbi.nlm.nih.gov/14467544/

Majewska J, Beta W, Lecion D, Hodyra-Stefaniak K, Kłopot A, Kazmierczak Z, Miernikiewicz P, Piotrowicz A, Ciekot J, Owczarek B, Kopciuch A, Wojtyna K, Harhala M, Mąkosa M, Dąbrowska K (2015) Oral application of T4 phage induces weak antibody production in the gut and in the blood. Viruses 7(8):4783–4799. https://doi.org/10.3390/V7082845

Manohar P, Loh B, Nachimuthu R, Hua X, Welburn SC, Leptihn S (2020) Secondary bacterial infections in patients with viral pneumonia. Front Med 7. https://doi.org/10.3389/FMED.2020.00420

Mattila S, Ruotsalainen P, Jalasvuori M (2015) On-demand isolation of bacteriophages against drug-resistant bacteria for personalized phage therapy. Front Microbiol 6:1271. https://doi.org/10.3389/FMICB.2015.01271

Miller RW, Skinner EJ, Sulakvelidze A, Mathis GF, Hofacre CL (2010) Bacteriophage therapy for control of necrotic enteritis of broiler chickens experimentally infected with Clostridium perfringens. Avian Dis 54(1):33–40. https://doi.org/10.1637/8953-060509-REG.1

Murray E, Draper LA, Ross RP, Hill C (2021) The advantages and challenges of using endolysins in a clinical setting. Viruses 13(4):680. https://doi.org/10.3390/V13040680

Nabil NM, Tawakol MM, Hassan HM (2018) Assessing the impact of bacteriophages in the treatment of Salmonella in broiler chickens. Infect Ecol Epidemiol 8(1):1539056. https://doi.org/10.1080/20008686.2018.1539056

Nelson D, Loomis L, Fischetti VA (2001) Prevention and elimination of upper respiratory colonization of mice by group a streptococci by using a bacteriophage lytic enzyme. Proc Natl Acad Sci USA 98(7):4107–4112. https://doi.org/10.1073/PNAS.061038398

Nicholls P, Aslam S (2022) Role of bacteriophage therapy for resistant infections in transplant recipients. Curr Opin Organ Transplant 27(6):546–553. https://doi.org/10.1097/MOT.0000000000001029

Nir-Paz R, Gelman D, Khouri A, Sisson BM, Fackler J, Alkalay-Oren S, Khalifa L, Rimon A, Yerushalmy O, Bader R, Amit S, Coppenhagen-Glazer S, Henry M, Quinones J, Malagon F, Biswas B, Moses AE, Merril G, Schooley RT et al (2019) Successful treatment of antibiotic-resistant, poly-microbial bone infection with bacteriophages and antibiotics combination. Clin Infect Dis 69(11):2015–2018. https://doi.org/10.1093/CID/CIZ222

Ofir G, Melamed S, Sberro H et al (2018) DISARM is a widespread bacterial defence system with broad anti-phage activities. Nat Microbiol 3(1):90–98. Retrieved March 3, 2024, from https://www.nature.com/articles/s41564-017-0051-0

Olszak T, Latka A, Roszniowski B, Valvano MA, Drulis-Kawa Z (2017) Phage life cycles behind bacterial biodiversity. Curr Med Chem 24(36):3987–4001. https://doi.org/10.2174/0929867324666170413100136

Oromí-Bosch A, Antani JD, Turner PE (2023) Developing phage therapy that overcomes the evolution of bacterial resistance. Ann Rev Virol 10(1):503–524. https://doi.org/10.1146/ANNUREV-VIROLOGY-012423-110530

Papaianni M, Paris D, Woo SL, Fulgione A, Rigano MM, Parrilli E, Tutino ML, Marra R, Manganiello G, Casillo A, Limone A, Zoina A, Motta A, Lorito M, Capparelli R (2020) Plant dynamic metabolic response to bacteriophage treatment after Xanthomonas campestris pv. campestris infection. Front Microbiol 11. https://doi.org/10.3389/FMICB.2020.00732

Park SC, Nakai T (2003) Bacteriophage control of pseudomonas plecoglossicida infection in ayu Plecoglossus altivelis. Dis Aquat Org 53(1):33–39. https://doi.org/10.3354/DAO053033

Pirnay J-P, Djebara S, Steurs G, Griselain J, Cochez C, De Soir S, Glonti T, Spiessens A, Vanden Berghe E, Green S, Wagemans J, Lood C, Schrevens E, Chanishvili N, Kutateladze M, de Jode M, Ceyssens P-J, Draye J-P, Verbeken G et al (2023) Retrospective, observational analysis of the first one hundred consecutive cases of personalized bacteriophage therapy of difficult-to-treat infections facilitated by a Belgian consortium. MedRxiv 2023(08):28.23294728. https://doi.org/10.1101/2023.08.28.23294728

Poranen MM, Bamford DH (2012) Assembly of large icosahedral double-stranded RNA viruses. Adv Exp Med Biol 726:379–402. https://doi.org/10.1007/978-1-4614-0980-9_17

Pourcel C, Salvignol G, Vergnaud G (2005) CRISPR elements in Yersinia pestis acquire new repeats by preferential uptake of bacteriophage DNA, and provide additional tools for evolutionary studies. Microbiology 151(Pt 3):653–663. https://doi.org/10.1099/MIC.0.27437-0

Publications Service (2007) On an invisible microbe antagonistic toward dysenteric bacilli: brief note by Mr. F. D'Herelle, presented by Mr. Roux. Res Microbiol 158(7):553–554. https://doi.org/10.1016/J.RESMIC.2007.07.005

Qin J, Wu N, Bao J, Shi X, Ou H, Ye S, Zhao W, Wei Z, Cai J, Li L, Guo M, Weng J, Lu H, Tan D, Zhang J, Huang Q, Zhu Z, Shi Y, Hu C et al (2021) Heterogeneous Klebsiella pneumoniae co-infections complicate personalized bacteriophage therapy. Front Cell Infect Microbiol 10. https://doi.org/10.3389/FCIMB.2020.608402

Rakhuba D, Kolomiets E et al (2010) Bacteriophage receptors, mechanisms of phage adsorption and penetration into host cell. Pol J Microbiol 59(3):145–155. https://doi.org/10.33073/pjm-2010-025

Richards PJ, Connerton PL, Connerton IF (2019) Phage biocontrol of campylobacter jejuni in chickens does not produce collateral effects on the gut microbiota. Front Microbiol 10:476. https://doi.org/10.3389/FMICB.2019.00476

Roach DR, Leung CY, Henry M, Morello E, Singh D, Di Santo JP, Weitz JS, Debarbieux L (2017) Synergy between the host immune system and bacteriophage is essential for successful phage therapy against an acute respiratory pathogen. Cell Host Microbe 22(1):38–47.e4. https://doi.org/10.1016/J.CHOM.2017.06.018

Rodríguez-Rubio L, Martínez B, Donovan DM, Rodríguez A, García P (2013) Bacteriophage virion-associated peptidoglycan hydrolases: potential new enzybiotics. Crit Rev Microbiol 39(4):427–434. https://doi.org/10.3109/1040841X.2012.723675

Rodríguez-Rubio L, García P, Rodríguez A, Billington C, Hudson JA, Martínez B (2015) Listeriaphages and coagulin C23 act synergistically to kill listeria monocytogenes in milk under refrigeration conditions. Int J Food Microbiol 205:68–72. https://doi.org/10.1016/J.IJFOODMICRO.2015.04.007

Rubalskii E, Ruemke S, Salmoukas C, Boyle EC, Warnecke G, Tudorache I, Shrestha M, Schmitto JD, Martens A, Rojas SV, Ziesing S, Bochkareva S, Kuehn C, Haverich A (2020) Bacteriophage therapy for critical infections related to cardiothoracic surgery. Antibiotics (Basel, Switzerland) 9(5):232. https://doi.org/10.3390/ANTIBIOTICS9050232

Ruska H (1940) Visualizing bacteriophage lysis in the supermicroscope. Nat Sci 28:45–46. https://doi.org/10.1007/BF01486931

Ruska H (1943) Attempt to organize the types of viruses. Arch Virol 2:480–498. https://doi.org/10.1007/BF01244584

Samsygina GA, Boni EG (1984) Bacteriophages and phage therapy in pediatric practice. Pediatriia 4:67–70

Sarker SA, Berger B, Deng Y, Kieser S, Foata F, Moine D, Descombes P, Sultana S, Huq S, Bardhan PK, Vuillet V, Praplan F, Brüssow H (2017) Oral application of Escherichia coli bacteriophage: safety tests in healthy and diarrheal children from Bangladesh. Environ Microbiol 19(1):237–250. https://doi.org/10.1111/1462-2920.13574

Savva R (2020) Bacteriophages - Perspectives and Future. IntechOpen. https://doi.org/10.5772/intechopen.73439

Schmelcher M, Loessner MJ (2016) Bacteriophage endolysins: applications for food safety. Curr Opin Biotechnol 37:76–87. https://doi.org/10.1016/J.COPBIO.2015.10.005

Schooley RT, Biswas B, Gill JJ, Hernandez-Morales A, Lancaster J, Lessor L, Barr JJ, Reed SL, Rohwer F, Benler S, Segall AM, Taplitz R, Smith DM, Kerr K, Kumaraswamy M, Nizet V, Lin L, McCauley MD, Strathdee SA et al (2017) Development and use of personalized bacteriophage-based therapeutic cocktails to treat a patient with a disseminated resistant Acinetobacter baumannii infection. Antimicrob Agents Chemother 61(10). https://doi.org/10.1128/AAC.00954-17

Schuch R, Khan BK, Raz A, Rotolo JA, Wittekind M (2017) Bacteriophage Lysin CF-301, a potent Antistaphylococcal biofilm agent. Antimicrob Agents Chemother 61(7). https://doi.org/10.1128/AAC.02666-16

Shah S, Das R, Chavan B, Bajpai U, Hanif S, Ahmed S (2023) Beyond antibiotics: phage-encoded lysins against gram-negative pathogens. Front Microbiol 14. https://doi.org/10.3389/FMICB.2023.1170418

Sharma U, Paul VD (2017) Bacteriophage lysins as antibacterials. Crit Care (London, England) 21(1):99. https://doi.org/10.1186/S13054-017-1681-6

Smith HO, Welcox KW (1970) A restriction enzyme from Hemophilus influenzae. I. Purification and general properties. J Mol Biol 51(2):379–391. https://doi.org/10.1016/0022-2836(70)90149-X

Soni KA, Nannapaneni R, Hagens S (2010) Reduction of listeria monocytogenes on the surface of fresh channel catfish fillets by bacteriophage Listex P100. Foodborne Pathog Dis 7(4):427–434. https://doi.org/10.1089/FPD.2009.0432

Speck P, Smithyman A (2016) Safety and efficacy of phage therapy via the intravenous route. FEMS Microbiol Lett 363(3). https://doi.org/10.1093/FEMSLE/FNV242

Summers W (1999) Felix dHerelle and the origins of molecular biology. https://books.google.com/books?hl=en&lr=&id=o3uxoUfbBnQC&oi=fnd&pg=PR7&dq=105.%09Summers,+W.C.,+1999.+Felix+dHerelle+and+the+origins+of+molecular+biology.+Yale+University+Press.&ots=iC9w8ncWiK&sig=GXH7G41bI3UH6dTHnE8PCfckpfA

WC Summers (2005) Bacteriophage research: early history. Bacteriophages: biology and applications. Books.Google.Com. Retrieved March 3, 2024, from https://books.google.com/books?hl=en&lr=&id=6sTBd2EJpmYC&oi=fnd&pg=PA5&dq=106.%09Summers,+W.C.,+2005.+Bacteriophage+research:+early+history.+Bacteriophages:+Biology+and+applications,+pp.5-27.&ots=j26Y-gq4gv&sig=9fDQY8txYbc58NLJV1s6VhyvdRk

Teklemariam AD, Al-Hindi RR, Qadri I, Alharbi MG, Ramadan WS, Ayubu J, Al-Hejin AM, Hakim RF, Hakim FF, Hakim RF, Alseraihi LI, Alamri T, Harakeh S (2023) The battle between bacteria and bacteriophages: a conundrum to their immune system. Antibiotics (Basel, Switzerland) 12(2):381. https://doi.org/10.3390/ANTIBIOTICS12020381

Totté JEE, van Doorn MB, Pasmans SGMA (2017) Successful treatment of chronic Staphylococcus aureus-related dermatoses with the topical Endolysin Staphefekt SA.100: a report of 3 cases. Case Rep Dermatol 9(2):19–25. https://doi.org/10.1159/000473872

Turner D, Shkoporov AN, Lood C, Millard AD, Dutilh BE, Alfenas-Zerbini P, van Zyl LJ, Aziz RK, Oksanen HM, Poranen MM, Kropinski AM, Barylski J, Brister JR, Chanisvili N, Edwards RA, Enault F, Gillis A, Knezevic P, Krupovic M et al (2023) Abolishment of morphology-based taxa and change to binomial species names: 2022 taxonomy update of the ICTV bacterial viruses subcommittee. Arch Virol 168(2):74. https://doi.org/10.1007/S00705-022-05694-2

Twort FW (1915) An investigation on the nature of ultra-microscopic viruses. Lancet 186(4814):1241–1243. https://doi.org/10.1016/S0140-6736(01)20383-3

Twort FW (1936) Further investigations on the nature of ultra-microscopic viruses and their cultivation. J Hyg 36(2):204–235. https://doi.org/10.1017/S0022172400043606

U.S. Food and Drug Administration (2021) National Institute of Allergy and Infectious Diseases Center for Biologics Evaluation and Research (CBER) Science and Regulation of Bacteriophage Therapy Workshop. U.S. Food and Drug Administration, Silver Spring, MD, USA

Van Belleghem JD, Dąbrowska K, Vaneechoutte M, Barr JJ, Bollyky PL (2018) Interactions between bacteriophage, bacteria, and the mammalian immune system. Viruses 11(1):10. https://doi.org/10.3390/V11010010

Wang IN, Smith DL, Young R (2000) Holins: the protein clocks of bacteriophage infections. Ann Rev Microbiol 54:799–825. https://doi.org/10.1146/ANNUREV.MICRO.54.1.799

Wang X, Wei Z, Yang K, Wang J, Jousset A, Xu Y, Shen Q, Friman VP (2019) Phage combination therapies for bacterial wilt disease in tomato. Nat Biotechnol 37(12):1513–1520. https://doi.org/10.1038/S41587-019-0328-3

Weinbauer MG, Rassoulzadegan F (2004) Are viruses driving microbial diversification and diversity? Environ Microbiol 6(1):1–11. https://doi.org/10.1046/J.1462-2920.2003.00539.X

Wu N, Dai J, Guo M, Li J, Zhou X, Li F, Gao Y, Qu H, Lu H, Jin J, Li T, Shi L, Wu Q, Tan R, Zhu M, Yang L, Ling Y, Xing S, Zhang J et al (2021) Pre-optimized phage therapy on secondary Acinetobacter baumannii infection in four critical COVID-19 patients. Emerg Microbes Infect 10(1):612–618. https://doi.org/10.1080/22221751.2021.1902754

Wu N, Chen LK, Zhu T (2022) Phage therapy for secondary bacterial infections with COVID-19. Curr Opin Virol 52:9–14. https://doi.org/10.1016/J.COVIRO.2021.11.001

Yang H, Wang M, Yu J, Wei H (2015) Antibacterial activity of a novel peptide-modified Lysin against Acinetobacter baumannii and Pseudomonas aeruginosa. Front Microbiol 6:1471. https://doi.org/10.3389/FMICB.2015.01471

Yehl K, Lemire S, Yang AC, Ando H, Mimee M, Torres MDT, de la Fuente-Nunez C, Lu TK (2019) Engineering phage host-range and suppressing bacterial resistance through phage tail fiber mutagenesis. Cell 179(2):459–469.e9. https://doi.org/10.1016/J.CELL.2019.09.015

Yongsheng M, Pacan JC, Wang Q, Xu Y, Huang X, Korenevsky A, Sabour PM (2008) Microencapsulation of bacteriophage Felix O1 into chitosan-alginate microspheres for oral delivery. Appl Environ Microbiol 74(15):4799–4805. https://doi.org/10.1128/AEM.00246-08

Zerbini FM, Siddell SG, Mushegian AR, Walker PJ, Lefkowitz EJ, Adriaenssens EM, Alfenas-Zerbini P, Dutilh BE, García ML, Junglen S, Krupovic M, Kuhn JH, Lambert AJ, Łobocka M, Oksanen HM, Robertson DL, Rubino L, Sabanadzovic S, Simmonds P et al (2022) Differentiating between viruses and virus species by writing their names correctly. Arch Virol 167(4):1231–1234. https://doi.org/10.1007/S00705-021-05323-4

Zheng DW, Dong X, Pan P, Chen KW, Fan JX, Cheng SX, Zhang XZ (2019) Phage-guided modulation of the gut microbiota of mouse models of colorectal cancer augments their responses to chemotherapy. Nat Biomed Eng 3(9):717–728. https://doi.org/10.1038/S41551-019-0423-2

Reviving Phage Therapy: A Novel Approach to Treat Respiratory-Associated Infections

16

Ramesh Nachimuthu, Ramya Juliet, S. Bershiyal, and Oishi Mitra

Abstract

Bacteriophages or phages in general are bacterial viruses that can kill bacteria and play a fundamental role in regulating bacterial populations. One notable application of phages is in the treatment of bacterial infections, where phages' lytic capabilities enable them to effectively eliminate bacterial populations. In the face of rising antibiotic resistance, especially multi-, extensive-, and pan-drug-resistant bacteria, there is an urgent need for alternative therapies, like phage therapy. This chapter highlights the clinical applications of phage therapy against drug-resistant respiratory-associated infections and also covers ongoing research on phages and challenges. Isolating bacteriophages against specific pathogens, characterizing them therapeutically for their lytic activity, host range, and life cycle, genome analysis, and purification are essential before considering the phages for preclinical and clinical studies. Further, this review also addresses the challenges that have been faced in translating phage therapy from benchwork to clinical settings, providing an in-depth understanding of the optimization of the mode of administration, pharmacokinetics, and pharmacodynamics of phages and its various formulations against pulmonary infections. This review also points to the compassionate use of phage therapy in respiratory-associated infections. The emergence of superbugs has left clinicians with minimal or no drug of choice. Phage therapy could serve as a promising alternative or be used in combination with antibiotics to address this crisis. This chapter provides an update on the compassionate use of phages in therapy against respiratory infections and the associated challenges with it.

R. Nachimuthu (✉) · R. Juliet · S. Bershiyal · O. Mitra
School of Biosciences and Technology, Vellore Institute of Technology, Vellore, Tamil Nadu, India
e-mail: ramesh.n@vit.ac.in

M. Gangwar, G. Nath (eds.), *Emerging Paradigms for Antibiotic-Resistant Infections: Beyond the Pill*, https://doi.org/10.1007/978-981-97-5272-0_16

Keywords
Phage therapy · Respiratory infections · Multi-drug resistance · Challenges · Clinical applications

16.1 Introduction

Antibiotics, "a compound made by microbes to kill other microbes," is one of the breakthrough discoveries in modern medicine, nevertheless, its rapid and improper use has led to antibiotic resistance (Waksman et al. 2010). Following the rapid discovery of antibiotics during the "Golden Age of Antibiotics" between the 1940s and 1960s, there was a dry pipeline in the discovery of new antibiotics from the 1970s, after which the antibiotics discovered hardly reached clinical trials (Katz and Baltz 2016). Antimicrobial resistance (AMR) is a global challenge severely affecting healthcare settings and the national economy, leading to extended hospital stays, and high mortality rates (Dadgostar 2019; Sharma et al. 2023). The emergence of superbugs with Multidrug Resistance (MDR), Extensive Drug Resistance (XDR), and Pan Drug Resistance (PDR) (Basak et al. 2016) is worrisome (Fig. 16.1). The acronym ESKAPE pathogens coined by the World Health Organization (WHO) as priority pathogens causes major nosocomial infections (Manohar et al. 2018a, b; De Oliveira et al. 2020). Misuse/overuse of antibiotics in agriculture, animal husbandry, poultry, and healthcare settings can be a reason for this rapid AMR increase. Other factors include mutation, evolution, and horizontal gene transfer in bacteria (Michael et al. 2014). At present, the state of treatment care in clinical settings is almost equal to the pre-antibiotic era since antibiotics are no longer effective.

Patients with underlying health conditions like cystic fibrosis, pneumonia, tuberculosis, lung transplantation, lung cancers, lung implants, or other pulmonary functional impairments or respiratory disorders are affected by secondary bacterial infections that are drug-resistant (Congedi et al. 2023; Mohanty et al. 2023; Pai and Memish 2016; Perikleous et al. 2023). Antibiotic resistance in both gram-positive and gram-negative respiratory pathogens has accelerated treatment failure in respiratory infections in the medical sector (Zumla et al. 2014). The scientific community is on the hunt for therapeutic strategies to defeat antibiotic resistance caused in respiratory diseases. This opens the gateway for possible options like the discovery of new antibiotics, antimicrobial peptides, nanoparticles, fecal microbiota transplant, bacteriocins, quorum sensing inhibitors, probiotics, plant-based products, phage therapy, and phage-derived enzymes like endolysins (Łojewska and Sakowicz 2021; Mdarhri et al. 2022).

Phage therapy is the usage of obligately lytic phages, which infect and lyse bacteria (Lin et al. 2017). Phages have advantages over antibiotics, such as they are species/strain-specific, do not affect normal microbiota, can lyse antibiotic-resistant bacteria, auto-dosing, naturally occurring, and can penetrate biofilms, with no adverse effects (Manohar et al. 2018a, b). Major limitations of phage therapy include phage resistance, narrow host range, pharmacokinetics and pharmacodynamic properties, and production (Loc-Carrillo and Abedon 2011). The limitations

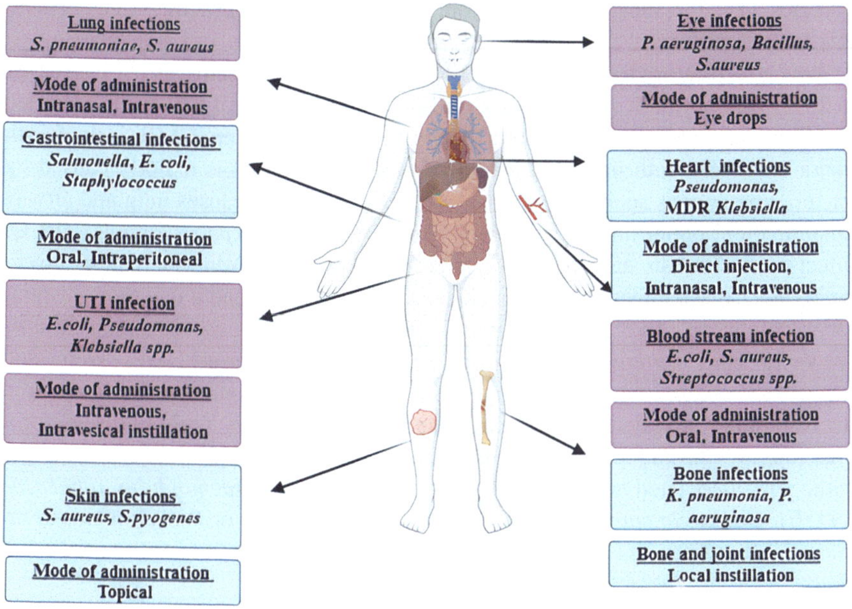

Fig. 16.1 Common bacterial infections at various body sites and their preferred mode of phage delivery system

can also be circumvented by phage–antibiotic synergy and engineered phages. This chapter particularly deals with phage therapy as an alternative or adjunctive therapy along with antibiotics to combat multi-drug resistance in pulmonary infections.

16.2 Role of Bacteria in Respiratory Tract Infections

Bacteria are ubiquitous and only a small portion of them cause infections. In the past, bacterial infections were easier to treat than viral infections. But, recently, treating bacterial infections is harder than viral infections because of their resistance to antibiotics (Doron and Gorbach 2008). Initially, lungs were considered sterile and devoid of bacteria, later with the development of technology and advancement in sequencing, it has been revealed that lungs contain diverse microorganisms and form lung microbiota (Dickson et al. 2016). Opportunistic pathogens can infect and colonize the respiratory tract of healthy individuals and cause infections. The availability of large epithelial surfaces helps in the easy colonization of bacteria (Niederman 1994). Viruses mostly cause upper respiratory tract infections (Robson-Ansley et al. 2012) and lower are by bacteria (Klekotka et al. 2015; Robson-Ansley et al. 2012). The colonizing bacteria leads to increased mucus secretion, inflammation, blockage of airways, and other lung function impairments (Dasaraju and Liu 1996). When antibiotics are misused or the antibiotic treatment fails to eradicate the bacterial infections usually caused as secondary infections, during viral diseases, allergies, or colds, they mature as chronic lung infections (Miravitlles and Anzueto 2017; Nord 1995).

Lower respiratory tract infections (LRIs) cause massive deaths, as per the survey in 2019, 488.9 million cases, and 2.4 million deaths because of LRIs were recorded globally (Safiri et al. 2022). Chronic lower respiratory infection (CLRI) is the fourth major cause of death in global health (Heron 2019). CLRI is majorly caused by bacterial pathogens such as *Pseudomonas aeruginosa* that causes lung and airway inflammation, opportunistic pathogens like *Burkholderia cepacia* complex (BCC) infect cystic fibrosis and immuno-compromised patients, *Non-tuberculous mycobacterium* causes infections in bronchiectasis (Blasi 2018) and a few more respiratory pathogens like *Haemophilus influenzae, Streptococcus pneumoniae, Staphylococcus aureus, and Moraxella catarrhalis* (Heron 2019; Welp and Bomberger 2020) colonize the lung during infections. Colonization of bacteria in the respiratory airways and viral existence in the lungs plays a role in disease progression of Chronic Obstructive Pulmonary Disease (COPD). Major bacterial pathogens associated with COPD involve non-typeable *Haemophilus influenzae* (NTHi), *Moraxella catarrhalis*, *Streptococcus pneumoniae*, or *Pseudomonas aeruginosa* (Miravitlles and Anzueto 2017).

16.3 Respiratory Pathogens and Their Resistance to Antibiotics

H. influenzae causes invasive and lung infections. The emergence of MDR *H. influenzae* can be seen globally. In 2007, extensively drug-resistant *H.influenzae* was reported in Taiwan (Su et al. 2020). Currently, the prevalence of extended-spectrum beta-lactamases and the incidence of MDR *H.influenzae* are at 34.9% and 23.1%, respectively (Abavisani et al. 2024; Su et al. 2020). The occurrence of *Streptococcus pneumoniae,* which causes community-acquired respiratory tract infections (Yahiaoui et al. 2016), is increasing and its resistance to fluoroquinolones, macrolides, and the rise of MDRs is alarming (Chen et al. 2009; Mykhalko et al. 2017). *Moraxella catarrhalis* recurrent infections in the upper respiratory tract, its beta-lactamase production, and resistance toward macrolides and tetracyclines have been reported in various countries (Hsu et al. 2012; Król-Turmińska et al. 2020; Murphy and Parameswaran 2009). *Methicillin-resistant Staphylococcus aureus* apart from skin and soft tissue infections can also result in severe community-acquired pneumonia (SCAP) (Defres et al. 2009; He and Wunderink 2020). MRSA resistance to erythromycin (Zarfel et al. 2013), fluoroquinolones (Alseqely et al. 2021), oxacillin (Hosaka et al. 2022), rifampicin (Alseqely et al. 2021; Murugan et al. 2015), clindamycin (Kot et al. 2020), and multidrug resistance in *MRSA* leads to chronic illnesses and mortality.

Expression of multidrug resistance phenotype and carbapenem resistance in *Pseudomonas aeruginosa* is threatening (Teo et al. 2021); it can easily evolve to pan-drug resistance because of their ability to produce beta-lactamase enzymes, aminoglycoside-modifying enzymes, up-regulation of efflux pumps (Bonomo and

Szabo 2006), and phenotypic macrolide resistance (Chen et al. 2020). The antibiotic-resistant mechanisms though studied briefly haven't been fully explored in *Burkholderia cepacia*, their intrinsic resistance toward polymyxins because of their modifications in lipopolysaccharides and fewer porins in the outer membrane provide less access for the antibiotics to enter, and mutations make them multidrug-resistant (Rhodes and Schweizer 2016). It mainly affects cystic fibrosis patients and treatment is difficult (Horsley et al. 2016). *Non-tuberculous mycobacterium* has a complex intrinsic multifaceted drug-resistant system that helps them withstand pulmonary infections amidst antibiotics (Luthra et al. 2018; Pfaeffle et al. 2021). Patients who are in prolonged ventilator support are more likely to die because of MDR infections, this is because of extended hospital stay and exposure to antibiotics for long periods (Michels-Zetsche et al. 2024). This drug resistance in respiratory pathogens and its drastic rise, the emergence of superbugs, and the lack of new/effective antibiotics push the scientific community to find nonantibiotic approaches against respiratory infections.

16.4 Phages as an Alternative to Treat Respiratory Bacterial Infections

Antimicrobial resistance (AMR) in respiratory tract infections is considered to be one of the major human health hazards thus leading to the need for better alternatives to the already available antibiotics. The enormous demand for new classes of antibiotics that could combat AMR is also stagnating due to economic and commercial viabilities (Hatfull et al. 2022). Bacteriophages, viruses that can infect and replicate within a bacterial cell, have a different bactericidal mechanism when compared with antibiotics and serve as a potential alternative to AMR. Phage therapy involves the therapeutic application of bacteriophages to cure harmful pathogenic infections (Kortright et al. 2019). Lytic or virulent phages are favored for phage therapy over lysogenic phages owing to their ability to replicate inside the bacterial cell without incorporating its DNA into the bacterial genome thus causing lysis of the cell with the help of phage-encoded lysins. Phages demonstrate exceptional host-specificity, being able to infect a single species of bacterium or a specific strain within a species (Golkar et al. 2014). Phages bind to their bacterial target in the first stage of the lytic cycle by a particular interaction between the phage's tail fibers and receptors on the bacterial cell surface. This binding event makes it easier for phage-encoded enzymes to penetrate the bacterial cell wall and introduce phage genetic information into the bacterium. Once within, the phage captures control of the bacteria's cellular mechanism, directing the production of structural proteins for the phage's head, sheath, and tail fibers. Phage-produced lysins damage the host cell after the replication, eventually leading to cell lysis and the release of a large number of new phage particles (Schmalstig et al. 2021). A distinctive benefit of phage therapy is the rapid rise in the phage population and auto-dosing, which results from the lytic cycle. In contrast to traditional medications, phage infection increases the phage concentration at the location of the infection, enabling a precise and effective strategy against bacterial infections (Rehman et al. 2019). Phages also have the potential to coevolve with the pathogen thus adapting to the changes and combating resistance.

16.5 Pharmacokinetics and Pharmacodynamics of Phages in Respiratory Infections

Bacteriophages/phage therapy was in practice even before the discovery of antibiotics but eventually lost its identity because of a lack of knowledge of their pharmacokinetics and pharmacodynamic properties, which posed a major challenge in translating in vitro work to clinical settings (Nang et al. 2023). Despite the advances in the world of scientific research, the mystery remains the same for bacteriophages. However, there have been studies trying to explore the pharmacokinetics of phages in in vitro models (Dhungana et al. 2021). The infectivity of phages depends on the mode of administration. For pulmonary infections, nebulized phages, or phages that are given through the nasal route, effectively eliminate bacteria from the respiratory tract (Schmalstig et al. 2021). Intravenous administration of phages might undergo elimination or inactivation during metabolism, whereas orally administered phages cannot withstand the acid secreted by the gastrointestinal tract (Abedon and Thomas-Abedon 2010), these discrepancies are majorly avoided during the treatment of lung infections, by direct application of phages into the site of infection (lungs) through inhalation or spraying.

In a study conducted by Semler et al in mice, it was seen that the phages administered intravenously for respiratory infections cannot reach the lungs effectively, whereas aerosol treatment reduced the bacterial load in the pulmonary *Burkholderia cepacia* infection (Semler et al. 2014). Like antibiotics, the metabolism of phages occurs in the liver and spleen, phages accumulated will either get inactivated or neutralized or spread throughout the body and finally be excreted in the urine or phagocytosed by macrophages. Phages can penetrate all the bodily tissues and even cross the blood–brain barrier, sometimes triggering the immune response and causing hypersensitivity reactions (Dąbrowska and Abedon 2019). The purity of the phage lysate, devoid of endotoxins, is crucial in phage therapy (Van Belleghem et al. 2017). The limitation is that there should be at least a threshold level of bacterial concentration in the infection, for phages to get multiplied thus performing the phenomenon of auto-dosing and eliminating the bacteria (Ochieng' Oduor et al. 2016).

16.6 Application of Phage Therapy in Lung Transplantations and Implants

Lung transplantation failures are by rejections and post-transplantation infections (Burguete et al. 2013). Recipients of pulmonary transplantations develop bacterial infections post-surgery, leading to severe morbidity, and mortality, despite taking all the antibiotic precautions (Witt et al. 2012). In particular, in cystic fibrosis patients, superinfections and co-infections of pre-transplant pathogens are recurring and cause chronic lung dysfunction. The state of immunocompromised to reduce rejections makes the patient even more vulnerable to infections (McCort et al. 2021). These types of infections can be because of several opportunistic pathogens mostly gram-negatives such as MDR *Pseudomonas aeruginosa* (Tebano et al. 2016), *Burkholderia* spp., (Dorgan and Hadjiliadis 2014), and also a few gram-positives such as vancomycin-resistant *Enterococcus*, *Clostridium difficile* (Cusini et al. 2018), *Corynebacterium* spp., (Los-Arcos et al. 2018), and *Mycoplasma hominis* (Dixit et al. 2017). The infections can also be because of the implants like stents and arterial catheters that are kept during transplantation.

Phage therapy usage after organ transplantation has been reported in numerous studies with successful clinical outcomes, and there are no side effects (Rubalskii et al. 2020). There is a global increase in lung transplants for people with terminal pulmonary illnesses around 34,000 per year (Chambers et al. 2021). Phages have been in use for several respiratory-related MDR infections on a compassionate basis, especially in lung transplants (Aslam 2023). Phage therapy has been administered to lung transplant recipients on compassionate grounds who either had predisposed CF or post-transplant pneumonia or bacteremia. The phages were given either intravenously or in a nebulized form, sometimes both. Phages were prescribed as phage cocktails (more than one phage) or single phage in multiple dosages. From 2019 to 2023, nine compassionate phage therapies in lung transplantation cases have taken place (published and unpublished) with substantial clinical improvements (Aslam 2023).

16.7 Phage Therapy During the Pandemic Era

COVID-19, an uninvited guest, has killed more than 6.87 million people worldwide, though the engrossment has decreased, there are cases of COVID-19 or its variants being reported, occasionally. Relative to the actual COVID-19 infection, deaths because of secondary infections during the pandemic were an indirect reason for the increase in mortality rate (Russell et al. 2021). Inflammation and infection played major roles in SARS-CoV-2 infected patients. Inflammation and accumulation of fluid provide bacteria with a nutrition-rich environment thus causing secondary infections. Phage therapy can be an effective way of controlling secondary infection, particularly in corona virus-infected patients as it has anti-inflammatory properties (Międzybrodzki et al. 2009) and controls inflammation in the lungs (Pabary et al. 2015), by modulating the immune system (Van Belleghem et al. 2017) and lytic activity to control bacterial infection. Two phage therapy studies where phages were used as an emergency Investigational New Drug (eIND) approved by the U.S Food and Drug Administration (FDA) in COVID-19 patients with secondary bacterial infection of carbapenem-resistant *Acinetobacter baumannii* have been reported (Wu et al. 2021). These case studies showcase the various applications of phage therapy in treating bacterial infections in severely acute respiratory syndrome cases.

16.8 Phage–Antibiotic Synergy (PAS) to Tackle Phage Resistance

The combined effect of antibiotics and bacteriophages to treat pathogenic infections is referred to as phage–antibiotic synergy (PAS) (Manohar et al. 2022). This integrated strategy becomes more vital with an increase in superbugs and phage-resistant mutants during phage therapy. Phages minimize damage to human cells by targeting specific pathogens, whereas antibiotics offer broad-spectrum protection. When amalgamated, they have a synergistic effect that may help overcome the process of bacterial resistance. This approach could lead to more precise, effective, personalized therapies. The phage–antibiotic synergistic effect is designed to overcome the drawbacks of using antibiotics or phages solely, providing an effective alternative to address the rising threat of antibiotic-resistant infections worldwide and enhance the general efficacy of therapeutic interventions (Fig. 16.2). Lin et al. have proven that an inhalable powder formulation of phage PEV20 and ciprofloxacin enhanced the bactericidal effects on multidrug-resistant *Pseudomonas aeruginosa* causing respiratory tract infections (Lin et al. 2019). In laboratory conditions of a study conducted by Oechslin et al., both in vitro, fibrin clots and rats with aortic endocarditis were subjected to treatment using an anti-*Pseudomonas* phage cocktail alone or in synergy with ciprofloxacin. Conjugation of phage and ciprofloxacin demonstrated a high level of synergy, effectively eliminating over 6 log CFUs/g of vegetations within 6 h and successfully treating 64% of the rats (7 out of 11) thereby confirming the efficacy of the PAS (Oechslin et al. 2017). A great advantage of

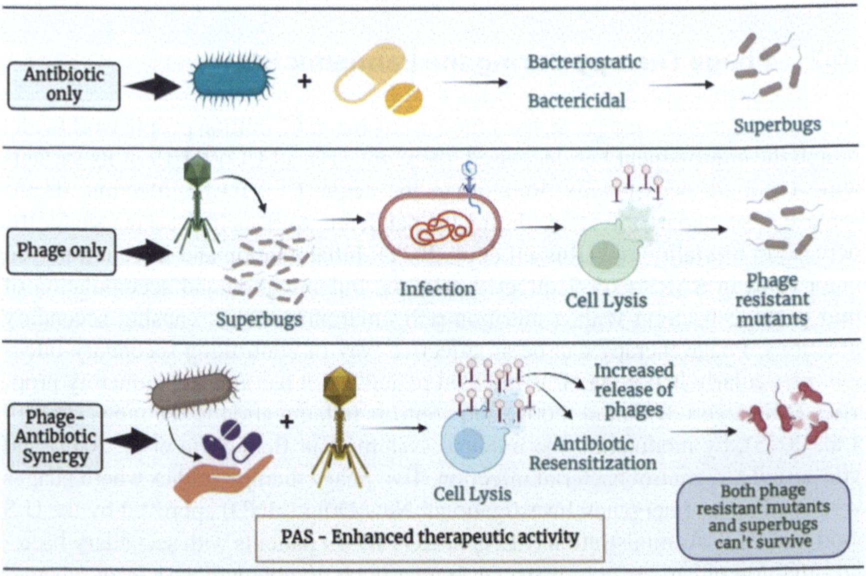

Fig. 16.2 Diagrammatic representation of phage antibiotic synergy against superbugs. Resulting in minimal phage resistance, increased phage progeny release, and antibiotic resensitization

using the synergistic effect of both phages and antibiotics is that it prevents the availability of bacterial mutants and superbugs with double resistance, as it is evident that a bacterium can't develop resistance to both phages and antibiotics simultaneously (Segall et al. 2019).

The analysis of the dosage of each for the combinational effect is important as deciding the point of administering the phage–antibiotic combination to have an optimal effect over the resistant pathogens (Manohar et al. 2020). When dealing with PAS, it is crucial to know the correct order of administering the combination. Szelachowska et al. suggested that administration of phages before a sublethal dosage of antibiotics would be beneficial in rupturing the biofilms formed, and the later dosage of antibiotic will also help in preventing the further formation of biofilm by the pathogens thus facilitating the reduction in bacterial load (Łusiak-Szelachowska et al. 2020). Torres-Barceló et al. in their experiment suggested that the administration of antibiotics before phages can alter the virulence of the phage thus reducing the bactericidal potential of the phages (Torres-Barceló et al. 2018). This brings in another sector of combinational therapy which is phage–antibiotic antagonism, where one individual therapy is beneficial over the conjugated effect. There is a requirement of a minimum level of bacterial concentration for the phages to replicate and increase their concentrations, in case the bacterial concentration is lower than required due to the prior effect of the antibiotics, the phage can't cause further reduction in the bacterial concentration thus suggesting the application of one individual therapy for this kind of case (Payne and Jansen 2002). In a study conducted by Valente et al., the combined impact of systemic daptomycin and nebulized bacteriophages for the treatment of pneumonia caused by methicillin-resistant *Staphylococcus aureus* (MRSA) was assessed. Utilizing a rat model for ventilator-associated pneumonia by MRSA, administering intravenous daptomycin along with nebulized phages did not demonstrate superiority over aerophage therapy individually in terms of enhancing animal survival (55% compared to 50%), nor did it show a significant reduction in bacterial loads in the lungs or spleen. Consequently, this combination did not appear to offer any advantages for managing MRSA pneumonia in patients (Valente et al. 2021). The appropriate determination of the doses and mode of administration of these combinations with sufficient efficacy testing will be fruitful for enhanced recovery from respiratory tract and other pathogenic infections.

16.9 Challenges in Administering Phage Therapy for Respiratory Infections

Phage administrations are determined by the site of infection, whether it can be administered orally, topically, or intravenously (Fig. 16.2). Mainly for infections associated with the respiratory tract, phages can be administered through dry powder inhalation, nebulization, soft mist inhaler, and endotracheal tube administration. In inhalation, phages can be delivered directly to the respiratory tract by allowing for targeted therapy. Studies dating as early as 1936 where phages are given for pulmonary pathogens through the inhalation method. But when it comes to dry powder inhalation, the aerosol particles larger than 5 μm remain deposited in the throat and fail to reach the lungs (Morello et al. 2011). To prepare the dry powder of phages, the phages should undergo extreme temperature conditions that may lead to a drop-in phage titer (Chang et al. 2018). In nebulization, phages are nebulized into small droplets for inhalation. This method can be used to deliver phages deep into the lungs. Unfortunately, air jet nebulizers and mesh nebulizers cause tail detachment that leads to broken phages that are inefficient in killing the pathogens (Astudillo et al. 2018). A study conducted by Carrigy et al. shows that vibrating nebulizers are a good choice when compared with jet nebulizers and soft mist inhalers due to their reduction in the titers. Soft mist inhalers can be efficiently used for self-administration (Carrigy et al. 2017). VAP-associated infections play a major role in nosocomial infections due to their biofilm-forming capability. To eradicate the formation of biofilms, phages should contain enzymes responsible for eradicating the biofilms, which is another challenge, and mostly these biofilms are polymicrobial, containing multiple organisms (Rodrigues et al. 2017). Phage cocktails (Manohar et al. 2019) are a wise choice to eradicate such complications but when phages are given in the cocktails, their proportions, phage stability, mode of administration, endotoxin removals, and phage titers should be checked thoroughly. Thus, it is important to analyze the phage efficacy in different routes of administration and to understand their pharmacodynamics before using phages for administration.

16.10 Success Stories and Insights Gained from Clinical Trials of Phage Therapy in Respiratory Infections

The treatment for lung-associated diseases such as cystic fibrosis (CF) is complicated because of their persistent respiratory infections, additionally rise in multidrug-resistant infections in CF patients (Sriramulu 2013) prompts medical practitioners to look for alternatives. Eliava Phage Therapy Centre (EPTC) in Georgia has successfully treated CF patients with respiratory bacterial infections using phage therapy, obtaining good clinical improvement. In a compassionate use case study, a CF patient with complex respiratory infections of *S. aureus, P. aeruginosa*, and MDR *Achromobacter xylosoxidans* was treated with phage therapy. The patient was treated with customized phage cocktails for *Achromobacter* (more than one phage) and antibiotics, administered orally and by inhalation. There was no phage resistance, and significant improvement in lung function and symptoms were observed (Hoyle et al. 2018). Phage cocktails and administration of phages together with antibiotics helped in the betterment of treatment outcomes.

Another case study from EPTC reported a patient with *P. aeruginosa*-associated lower respiratory tract infection was treated with Pyo and Intesti phage lysates. Both orally and through a nebulizer, when phage resistance showed up, the treatment was switched to customized phage therapy. Initial susceptibility, the emergence of resistance to even customized phages and resensitization to Pyo and Intesti phages were observed throughout the treatment plan. During the treatment course, the bacterial load decreased, but there was not a time frame where there was no *P.aeruginosa*. Pulse field gel electrophoresis analysis revealed the coexistence of both susceptible and resistant strains of *P. aeruginosa* in the patient. A similar study from EPTC reported that a patient with a lung infection and CF did not respond to Pyo and Intesti bacteriophages from the first culture; therefore, customized phages were used from the start orally, and after a long phage-only treatment, finally, the patient was culture negative for *P. aeruginosa* (Zaldastanishvili et al. 2021). The occurrence of phage resistance, co-infections with different strains of the same bacterial species with varied phenotypic and genotypic properties, and the inability of the patient to adapt to the mode of administration of phages make even the customized phage therapy trials challenging (Zaldastanishvili et al. 2021).

A 61-year-old patient suffered from MDR *P. aeruginosa* infection along with several post-lung transplant complications. AB-PA01 phage cocktail consisting of four phages targeting MDR *P. aeruginosa* along with antibiotics such as colistin and piperacillin-tazobactam was used for the treatment. After 29 days of treatment, the patient recovered but had a second episode of mucoid MDR *P. aeruginosa* infection causing respiratory failure and septic shock. Antibiotic treatment along with AB-PA01-m1 (prefixed combination AB-PA01 + 1 new specific phage) and Navy phage cocktail 1 (personalized phage cocktail) was started and followed for 3 months, the patient recovered fully with any sign of MDR *P. aeruginosa* or adverse effects. In a different case, a patient with post-lung transplant MDR *P. aeruginosa* and *Mycobacterium abscessus* infection was treated with AB-PA01 phage and colistin antibiotic, resulting in successful recovery. A 28-year-old patient who had bilateral lung transplantation and was affected with MDR *Burkholderia dolosa* was susceptible only to minocycline. Phage therapy was considered, but the appropriate phage was not available by the time her health got worse. Later a single lytic phage BdPF16phi4281 was identified, showed initial improvement in health but later died because of other complications, and had no adverse effects from phage therapy (Aslam et al. 2019). Though phages are ubiquitous, finding an appropriate phage is still a problem because of a lack of standard isolation protocols.

AmpliPhi Biosciences manufactured phage cocktail, AB-PA01 (two Myoviridae and two Podoviridae) was administered both intravenously and through a nebulizer as adjunctive therapy to a pulmonary *Pseudomonal* infection patient, despite the complete recovery of the patient, the duration of the phage treatment and mode of administration to treat airways and chest wall remained confusing for the medical practitioners (Maddocks et al. 2019). In a compassionate phage therapy study conducted by Wu et al., phages were administered against carbapenem-resistant *Acinetobacter baumannii* as a secondary infection in a COVID-19 patient. When phage resistance reemerged, a tailored phage from the phage library was promptly administered (Wu et al. 2021). This study reveals the importance of establishing a phage library or phage bank containing a broad host range of lytic phages characterized according to therapeutic standards for immediate phage intervention.

In 20 CF patients with non-tuberculosis *Mycobacterium abscessus,* compassionate ground phage treatment was provided, as phage cocktails and engineered phages. Phage neutralization, immune responses, and phage resistance were observed during the study (Dedrick et al. 2023). Apart from clinical improvement and recovery from 11 patients, inconclusive results and no treatment responses from others were the major challenges. A lot of other successful interventions of phage therapy in treating pulmonary-resistant infections have been reported (Nick et al. 2022; Rao et al. 2022) and still, there is a long way to go in the future in terms of phage therapy to translate from bench to clinical settings (Fig. 16.3).

16.11 Conclusion

The growing burden of AMR all over the world has worsened since the COVID-19 pandemic leading to various respiratory infections. The usage of phage therapy on compassionate levels for patients with lung infections gives us a beacon of hope to fight against multidrug-resistant pulmonary infections. Phage therapy has paved its way amidst antibiotics as a standard therapeutic option in the foreseeable future for treating chronic respiratory illnesses. Bacteria with its innate ability to develop resistance when exposed to phage treatment for long gives rise to a possibility of bacteriophage resistance. A strategic synergy between phages and antibiotics plays

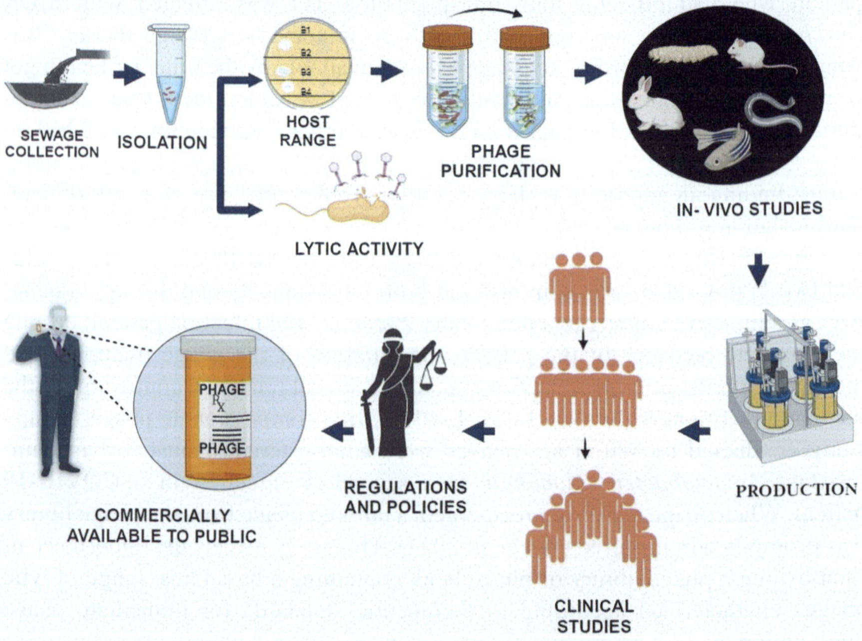

Fig. 16.3 Decoding phage therapy from benchwork to clinical settings

a major role in combating the same, enhancing the efficiency of compassionate therapy. Despite the renewed interest in phage therapy and groundbreaking outcomes in compassionate usage, there is a lot more to dig into phage therapy against pulmonary infections in terms of stability and efficacy of phages during delivery and mode of administration, duration, standard administration protocols, phage resistance, immunogenic response, availability of well-characterized phages, and regulatory guidelines before phage lysates are commercially available to the public. Irrespective of the above-mentioned challenges, ongoing research and clinical trials raise the hopes on phages for combating antibiotic resistance in a better way.

Authorship Contribution RN, RJ, BS, OM wrote and finalized the manuscript.

Conflicts of Interest The authors declare no competing financial or non-financial interests.

Ethical Approval This article does not contain any studies with human participants or animals by any of the authors.

Data Availability Not applicable.

References

Abavisani M, Keikha M, Karbalaei M (2024) First global report about the prevalence of multidrug resistant Haemophilusinfluenzae: a systematic review and meta-analysis. BMC Infect Dis 24(1):1–8

Abedon ST, Thomas-Abedon C (2010) Phage therapy pharmacology. Curr Pharm Biotechnol 11(1):28–47

Alseqely M, Newton-Foot M, Khalil A, El-Nakeeb M, Whitelaw A, Abouelfetouh A (2021) Association between fluoroquinolone resistance and MRSA genotype in Alexandria, Egypt. Sci Rep 11(1):4253. https://doi.org/10.1038/s41598-021-83578-2

Aslam S (2023) Phage therapy in lung transplantation: current status and future possibilities. Clin Infect Dis 77(Supplement_5):S416–S422

Aslam S, Courtwright AM, Koval C, Lehman SM, Morales S, Furr CL, Rosas F, Brownstein MJ, Fackler JR, Sisson BM, Biswas B, Henry M, Luu T, Bivens BN, Hamilton T, Duplessis C, Logan C, Law N, Yung G et al (2019) Early clinical experience of bacteriophage therapy in 3 lung transplant recipients. Am J Transpltant 19(9):2631–2639. https://doi.org/10.1111/ajt.15503

Astudillo A, Leung SSY, Kutter E, Morales S, Chan H-K (2018) Nebulization effects on structural stability of bacteriophage PEV 44. Eur J Pharm Biopharm 125:124–130

Basak S, Singh P, Rajurkar M (2016) Multidrug resistant and extensively drug resistant bacteria: a study. J Pathog 2016:4065603. https://doi.org/10.1155/2016/4065603

Blasi F (2018) Lung diseases: chronic respiratory infections. Int J Mol Sci 19(10):3051. https://doi.org/10.3390/ijms19103051

Bonomo RA, Szabo D (2006) Mechanisms of multidrug resistance in Acinetobacter species and Pseudomonas aeruginosa. Clin Infect Dis 43(Suppl 2):S49–S56. https://doi.org/10.1086/504477

Burguete SR, Maselli DJ, Fernandez JF, Levine SM (2013) Lung transplant infection. Respirology 18(1):22–38. https://doi.org/10.1111/j.1440-1843.2012.02196.x

Carrigy NB, Chang RY, Leung SSY, Harrison M, Petrova Z, Pope WH, Hatfull GF, Britton WJ, Chan H-K, Sauvageau D, Finlay WH, Vehring R (2017) Anti-tuberculosis bacteriophage

D29 delivery with a vibrating mesh nebulizer, jet nebulizer, and soft mist inhaler. Pharm Res 34(10):2084–2096
Chambers DC, Perch M, Zuckermann A, Cherikh WS, Harhay MO, Hayes D Jr, Hsich E, Khush KK, Potena L, Sadavarte A, Lindblad K, Singh TP, Stehlik J, International Society for Heart and Lung Transplantation (2021) The international thoracic organ transplant registry of the International Society for Heart and Lung Transplantation: thirty-eighth adult lung transplantation report–2021; focus on recipient characteristics. J Heart Lung Transplant 40(10):1060–1072
Chang RYK, Chen K, Wang J, Wallin M, Britton W, Morales S, Kutter E, Li J, Chan H-K (2018) Proof-of-principle study in a murine lung infection model of antipseudomonal activity of phage PEV20 in a dry-powder formulation. Antimicrob Agents Chemother 62(2):10–128. https://doi.org/10.1128/AAC.01714-17
Chen R, Chen Y, Black S, Hao C-L, Ding Y-F, Zhang T, Zhao G-M (2009) Antibiotic resistance patterns and serotype distribution in Streptococcus pneumoniae from hospitalized pediatric patients with respiratory infections in Suzhou, China. J Trop Pediatr 56(3):204–205
Chen Q, Lu W, Zhou D, Zheng G, Liu H, Qian C, Zhou W, Lu J, Ni L, Bao Q, Li A, Xu T, Xu H (2020) Characterization of two macrolide resistance-related genes in multidrug-resistant Pseudomonas aeruginosa isolates. Pol J Microbiol 69(3):349–356. https://doi.org/10.33073/pjm-2020-038
Congedi S, Navalesi P, Boscolo A (2023) Multidrug-resistant organisms in lung transplant: a narrative review. Curr Opin Organ Transplant 28(3):174–179
Cusini A, Béguelin C, Stampf S, Boggian K, Garzoni C, Koller M, Manuel O, Meylan P, Mueller NJ, Hirsch HH, Weisser M, Berger C, van Delden C (2018) Clostridium difficile infection is associated with graft loss in solid organ transplant recipients. Am J Transpl 18(7):1745–1754. https://doi.org/10.1111/ajt.14640
Dąbrowska K, Abedon ST (2019) Pharmacologically aware phage therapy: pharmacodynamic and pharmacokinetic obstacles to phage antibacterial action in animal and human bodies. Microbiol Mol Biol Rev 83(4):10–128. https://doi.org/10.1128/MMBR.00012-19
Dadgostar P (2019) Antimicrobial resistance: implications and costs. Infect Drug Resist 12:3903
Dasaraju PV, Liu C (1996) Infections of the respiratory system. https://pubmed.ncbi.nlm.nih.gov/21413304/
De Oliveira DMP, Forde BM, Kidd TJ, Harris PNA, Schembri MA, Beatson SA, Paterson DL, Walker MJ (2020) Antimicrobial resistance in ESKAPE pathogens. Clin Microbiol Rev 33(3):10–128. https://doi.org/10.1128/cmr.00181-19
Dedrick RM, Smith BE, Cristinziano M, Freeman KG, Jacobs-Sera D, Belessis Y, Whitney BA, Cohen KA, Davidson RM, van Duin D, Gainey A, Garcia CB, Cr RG, Haidar G, Ip W, Iredell J, Khatami A, Little JS, Malmivaara K et al (2023) Phage therapy of mycobacterium infections: compassionate use of phages in 20 patients with drug-resistant mycobacterial disease. Clin Infect Dis 76(1):103–112. https://doi.org/10.1093/cid/ciac453
Defres S, Marwick C, Nathwani D (2009) MRSA as a cause of lung infection including airway infection, community-acquired pneumonia and hospital-acquired pneumonia. Eur Respir J 34(6):1470–1476
Dhungana G, Nepal R, Regmi M, Malla R (2021) Pharmacokinetics and pharmacodynamics of a novel virulent Klebsiella phage Kp_Pokalde_002 in a mouse model. Front Cell Infect Microbiol 11:684704
Dickson RP, Erb-Downward JR, Martinez FJ, Huffnagle GB (2016) The microbiome and the respiratory tract. Annu Rev Physiol 78:481–504
Dixit A, Alexandrescu S, Boyer D, Graf EH, Vargas SO, Silverman M (2017) Mycoplasma hominis empyema in an 18-year-old stem cell and lung transplant recipient: case report and review of the literature. J Pediatr Infect Dis Soc 6(4):e173–e176. https://doi.org/10.1093/jpids/pix049
Dorgan DJ, Hadjiliadis D (2014) Lung transplantation in patients with cystic fibrosis: special focus to infection and comorbidities. Expert Rev Respir Med 8(3):315–326. https://doi.org/10.1586/17476348.2014.899906
Doron S, Gorbach SL (2008) Bacterial infections: overview. International Encyclopedia of Public Health, p 273

Golkar Z, Bagasra O, Pace DG (2014) Bacteriophage therapy: a potential solution for the antibiotic resistance crisis. J Infect Dev Ctries 8(02):129–136

Hatfull GF, Dedrick RM, Schooley RT (2022) Phage therapy for antibiotic-resistant bacterial infections. Annu Rev Med 73:197–211

He H, Wunderink RG (2020) Staphylococcus aureus pneumonia in the community. Semin Respir Crit Care Med 41(4):470–479. https://doi.org/10.1055/s-0040-1709992

Heron M (2019) Deaths: leading causes for 2017. Ntl Vital Stat Rep 68(6):1–77

Horsley A, Jones AM, Lord R (2016) Antibiotic treatment for Burkholderiacepacia complex in people with cystic fibrosis experiencing a pulmonary exacerbation. Cochrane Database Syst Rev 2016(1):CD009529. https://doi.org/10.1002/14651858.CD009529.pub3

Hosaka Y, Yahara K, Clark A, Kitagawa H, Hisatsune J, Sugai M, Shibayama K, Stelling J (2022) Surveillance of multi-drug resistance phenotypes in Staphylococcus aureus in Japan and correlation with whole-genome sequence findings. J Hosp Infect 123:34–42. https://doi.org/10.1016/j.jhin.2022.02.011

Hoyle N, Zhvaniya P, Balarjishvili N, Bolkvadze D, Nadareishvili L, Nizharadze D, Wittmann J, Rohde C, Kutateladze M (2018) Phage therapy against Achromobacterxylosoxidans lung infection in a patient with cystic fibrosis: a case report. Res Microbiol 169(9):540–542. https://doi.org/10.1016/j.resmic.2018.05.001

Hsu SF, Lin YT, Chen TL, Siu LK, Hsueh PR, Huang ST, Fung CP (2012) Antimicrobial resistance of Moraxella catarrhalis isolates in Taiwan. J Microbiol Immunol Infect 45(2):134–140. https://doi.org/10.1016/j.jmii.2011.09.004

Katz L, Baltz RH (2016) Natural product discovery: past, present, and future. J Ind Microbiol Biotechnol 43(2–3):155–176

Klekotka RB, Mizgała E, Król W (2015) The etiology of lower respiratory tract infections in people with diabetes. Pneumonol Alergol Pol 83(5):401–408

Kortright KE, Chan BK, Koff JL, Turner PE (2019) Phage therapy: a renewed approach to combat antibiotic-resistant bacteria. Cell Host Microbe 25(2):219–232

Kot B, Wierzchowska K, Piechota M, Grużewska A (2020) Antimicrobial resistance patterns in methicillin-resistant Staphylococcus aureus from patients hospitalized during 2015-2017 in hospitals in Poland. Med Princ Pract 29(1):61–68. https://doi.org/10.1159/000501788

Król-Turmińska K, Olender A, Bogut A (2020) Tetracycline resistance in Moraxella Catarrhalis clinical strains isolated in Poland. New Microbiol 43(3):103–106. https://pubmed.ncbi.nlm.nih.gov/32596738/

Lin DM, Koskella B, Lin HC (2017) Phage therapy: an alternative to antibiotics in the age of multi-drug resistance. World J Gastrointest Pharmacol Ther 8(3):162

Lin Y, Chang RYK, Britton WJ, Morales S, Kutter E, Li J, Chan HK (2019) Inhalable combination powder formulations of phage and ciprofloxacin for P. aeruginosa respiratory infections. Eur J Pharm Biopharm 142:543–552

Loc-Carrillo C, Abedon ST (2011) Pros and cons of phage therapy. Bacteriophage 1(2):111–114

Łojewska E, Sakowicz T (2021) An alternative to antibiotics: selected methods to combat zoonotic foodborne bacterial infections. Curr Microbiol 78(12):4037

Los-Arcos I, Len O, Martín-Gómez MT, Baroja A, Berastegui C, Deu M, Sacanell J, Román A, Gavaldà J (2018) Clinical characteristics and outcome of lung transplant recipients with respiratory isolation of Corynebacterium spp. J Clin Microbiol 56(8):e00142–e00118. https://doi.org/10.1128/JCM.00142-18

Łusiak-Szelachowska M, Weber-Dąbrowska B, Górski A (2020) Bacteriophages and lysins in biofilm control. Virol Sin 35(2):125–133

Luthra S, Rominski A, Sander P (2018) The role of antibiotic-target-modifying and antibiotic-modifying enzymes in Mycobacterium abscessus drug resistance. Front Microbiol 9:2179. https://doi.org/10.3389/fmicb.2018.02179

Maddocks S, Fabijan AP, Ho J, Lin RCY, Nl BZ, Dugan C, Kliman I, Branston S, Morales S, Iredell JR (2019) Bacteriophage therapy of ventilator-associated pneumonia and empyema caused by Pseudomonas aeruginosa. Am J Respir Crit Care Med 200(9):1179–1181. https://doi.org/10.1164/rccm.201904-0839LE

Manohar P, Tamhankar AJ, Lundborg CS, Ramesh N (2018a) Isolation, characterization and in vivo efficacy of Escherichia phage myPSH1131. PLoS One 13(10):e0206278

Manohar P, Nachimuthu R, Lopes BS (2018b) The therapeutic potential of bacteriophages targeting gram-negative bacteria using Galleria mellonella infection model. BMC Microbiol 18:1–11

Manohar P, Tamhankar AJ, Lundborg CS, Nachimuthu R (2019) Therapeutic characterization and efficacy of bacteriophage cocktails infecting Escherichia coli, Klebsiella pneumoniae, and Enterobacter species. Front Microbiol 10:574

Manohar P, Loh B, Athira S, Nachimuthu R, Hua X, Welburn SC, Leptihn S (2020) Secondary bacterial infections during pulmonary viral disease: phage therapeutics as alternatives to antibiotics? Front Microbiol 11:1434

Manohar P, MadurantakamRoyam M, Loh B, Bozdogan B, Nachimuthu R, Leptihn S (2022) Synergistic effects of phage–antibiotic combinations against Citrobacteramalonaticus. ACS Infect Dis 8(1):59–65

McCort M, MacKenzie E, Pursell K, Pitrak D (2021) Bacterial infections in lung transplantation. J Thorac Dis 13(11):6654–6672

Mdarhri HA, Benmessaoud R, Yacoubi H, Seffar L, Assimi HG, Hamam M, Boussettine R, Filali-Ansari N, Lahlou FA, Diawara I, Ennaji MM, Kettani-Halabi M (2022) Alternatives therapeutic approaches to conventional antibiotics: advantages, limitations and potential application in medicine. Antibiotics 11(12):1826. https://doi.org/10.3390/antibiotics11121826

Michael CA, Dominey-Howes D, Labbate M (2014) The antimicrobial resistance crisis: causes, consequences, and management. Front Public Health 2:145. https://doi.org/10.3389/fpubh.2014.00145

Michels-Zetsche JD, Gassmann V, Jasuja JK, Neetz B, Höger P, Meis J, Britsch S, Sommerwerck U, Fähndrich S, Bornitz F, Müller MM, Herth FJF, Trudzinski FC (2024) Role of multidrug-resistant bacteria in weaning from invasive mechanical ventilation. Respir Res 25(1):69

Międzybrodzki R, Fortuna W, Weber-Dąbrowska B, Górski A (2009) A retrospective analysis of changes in inflammatory markers in patients treated with bacterial viruses. Clinical and experimental medicine 9:303–12

Miravitlles M, Anzueto A (2017) Chronic respiratory infection in patients with chronic obstructive pulmonary disease: what is the role of antibiotics? Int J Mol Sci 18(7):1344

Mohanty S, Feemster K, Yu KC, Watts JA, Gupta V (2023) Trends in antimicrobial resistance in US children: a multicenter evaluation. Open Forum Infect Dis 10(3):ofad098

Morello E, Saussereau E, Maura D, Huerre M, Touqui L, Debarbieux L (2011) Pulmonary bacteriophage therapy on Pseudomonas aeruginosa cystic fibrosis strains: first steps towards treatment and prevention. PLoS One 6(2):e16963. https://doi.org/10.1371/journal.pone.0016963

Murphy TF, Parameswaran GI (2009) Moraxella catarrhalis, a human respiratory tract pathogen. Clin Infect Dis 49(1):124–131. https://doi.org/10.1086/599375

Murugan K, Kavitha K, Al-Sohaibani S (2015) Rifampicin resistance among multi-resistant MRSA clinical isolates from Chennai, India, and their molecular characterization. Genet Mol Res 14(1):2716–2725. https://doi.org/10.4238/2015.March.31.1

Mykhalko YO, Duhovych TV, Kish PP (2017) Susceptibility of streptococcus pneumoniae to fluoroquinolones and macrolides in upper respiratory tract infections. Wiad Lek 70(2):224–226

Nang SC, Lin Y-W, PetrovicFabijan A, Chang RYK, Rao GG, Iredell J, Chan H-K, Li J (2023) Pharmacokinetics/pharmacodynamics of phage therapy: a major hurdle to clinical translation. Clin Microbiol Infect 29(6):702–709

Nick JA, Dedrick RM, Gray AL, Vladar EK, Smith BE, Freeman KG, Malcolm KC, Epperson LE, Hasan NA, Hendrix J, Callahan K, Walton K, Vestal B, Wheeler E, Rysavy NM, Poch K, Caceres S, Lovell VK, Hisert KB et al (2022) Host and pathogen response to bacteriophage engineered against Mycobacterium abscessus lung infection. Cell 185(11):1860–1874.e12. https://doi.org/10.1016/j.cell.2022.04.024

Niederman MS (1994) The pathogenesis of airway colonization: lessons learned from the study of bacterial adherence [review of the pathogenesis of airway colonization: lessons learned from the study of bacterial adherence]. Eur Respir J 7(10):1737–1740

Nord CE (1995) The role of anaerobic bacteria in recurrent episodes of sinusitis and tonsillitis. Clin Infect Dis 20(6):1512–1524

Ochieng' Oduor JM, Onkoba N, Maloba F, Nyachieo A (2016) Experimental phage therapy against haematogenous multi-drug resistant Staphylococcus aureus pneumonia in mice. S Afr J Lab Clin Med 5(1):435. https://doi.org/10.4102/ajlm.v5i1.435

Oechslin F, Piccardi P, Mancini S, Gabard J, Moreillon P, Entenza JM et al (2017) Synergistic interaction between phage therapy and antibiotics clears Pseudomonas aeruginosa infection in endocarditis and reduces virulence. J Infect Dis 215(5):703–712

Pabary R, Singh C, Morales S, Bush A, Alshafi K, Bilton D, Alton EW, Smithyman A, Davies JC (2015) Antipseudomonal bacteriophage reduces infective burden and inflammatory response in murine lung. Antimicrob Agents Chemother 60(2):744–751. https://doi.org/10.1128/AAC.01426-15

Pai M, Memish ZA (2016) Antimicrobial resistance and the growing threat of drug-resistant tuberculosis. J Epidemiol Glob Health 6(2):45–47

Payne RJ, Jansen VA (2002) Evidence for a phage proliferation threshold? J Virol 76(24):13123–13124

Perikleous EP, Gkentzi D, Bertzouanis A, Paraskakis E, Sovtic A, Fouzas S (2023) Antibiotic resistance in patients with cystic fibrosis: past, present, and future. Antibiotics 12(2):217. https://doi.org/10.3390/antibiotics12020217

Pfaeffle HOI, Alameer RM, Marshall MH, Houpt ER, Albon DP, Heysell SK (2021) Clofazimine for treatment of multidrug-resistant non-tuberculous mycobacteria. Pulm Pharmacol Ther 70:102058. https://doi.org/10.1016/j.pupt.2021.102058

Rao S, Betancourt-Garcia M, Kare-Opaneye YO, Swierczewski BE, Bennett JW, Horne B, Fackler J, Lp SH, Brownstein MJ (2022) Critically ill patient with multidrug-resistant acinetobacterbaumannii respiratory infection successfully treated with intravenous and nebulized bacteriophage therapy. Antimicrob Agents Chemother 66(1):e0082421. https://doi.org/10.1128/AAC.00824-21

Rehman S, Ali Z, Khan M, Bostan N, Naseem S (2019) The dawn of phage therapy. Rev Med Virol 29(4):e2041

Rhodes KA, Schweizer HP (2016) Antibiotic resistance in Burkholderia species. Drug Resist Updates 28:82–90. https://doi.org/10.1016/j.drup.2016.07.003

Robson-Ansley P, Howatson G, Tallent J, Mitcheson K, Walshe I, Toms C, Toit GDU, Smith M, Ansley L (2012) Prevalence of allergy and upper respiratory tract symptoms in runners of the London marathon. Med Sci Sports Exerc 44(6):999–1004

Rodrigues ME, Lopes SP, Pereira CR, Azevedo NF, Lourenço A, Henriques M, Pereira MO (2017) Polymicrobial ventilator-associated pneumonia: fighting in vitro Candida albicans-Pseudomonas aeruginosa biofilms with antifungal-antibacterial combination therapy. PLoS One 12(1):e0170433. https://doi.org/10.1371/journal.pone.0170433

Rubalskii E, Ruemke S, Salmoukas C, Boyle EC, Warnecke G, Tudorache I, Shrestha M, Schmitto JD, Martens A, Rojas SV, Ziesing S, Bochkareva S, Kuehn C, Haverich A (2020) Bacteriophage therapy for critical infections related to cardiothoracic surgery. Antibiotics (Basel, Switzerland) 9(5):232. https://doi.org/10.3390/antibiotics9050232

Russell CD, Fairfield CJ, Drake TM, Turtle L, Seaton RA, Wootton DG, Sigfrid L, Harrison EM, Docherty AB, de Silva TI, Egan C (2021) Co-infections, secondary infections, and antimicrobial use in patients hospitalised with COVID-19 during the first pandemic wave from the ISARIC WHO CCP-UK study: a multicentre, prospective cohort study. The Lancet Microbe 2(8):e354–65

Safiri S, Mahmoodpoor A, Kolahi A-A, Nejadghaderi SA, Sullman MJM, Mansournia MA, Ansarin K, Collins GS, Kaufman JS, Abdollahi M (2022) Global burden of lower respiratory infections during the last three decades. Front Public Health 10:1028525

Schmalstig AA, Freidy S, Hanafin PO, Braunstein M, Rao GG (2021) Reapproaching old treatments: considerations for PK/PD studies on phage therapy for bacterial respiratory infections. Clin Pharmacol Therap 109(6):1443–1456

Segall AM, Roach DR, Strathdee SA (2019) Stronger together? Perspectives on phage-antibiotic synergy in clinical applications of phage therapy. Curr Opin Microbiol 51:46–50. https://doi.org/10.1016/j.mib.2019.03.005

Semler DD, Goudie AD, Finlay WH, Dennis JJ (2014) Aerosol phage therapy efficacy in Burkholderiacepacia complex respiratory infections. Antimicrob Agents Chemother 58(7):4005–4013. https://doi.org/10.1128/AAC.02388-13

Sharma A, Thakur N, Thakur A, Chauhan A, Babrah H (2023) The challenge of antimicrobial resistance in the Indian healthcare system. Cureus 15(7):e42231. https://doi.org/10.7759/cureus.42231

Sriramulu D (2013) Evolution and impact of bacterial drug resistance in the context of cystic fibrosis disease and nosocomial settings. Microbiol Insights 6:29

Su P-Y, Huang A-H, Lai C-H, Lin H-F, Lin T-M, Ho C-H (2020) Extensively drug-resistant Haemophilusinfluenzae—emergence, epidemiology, risk factors, and regimen. BMC Microbiol 20(1):102

Tebano G, Geneve C, Tanaka S, Grall N, Atchade E, Augustin P, Thabut G, Castier Y, Montravers P, Desmard M (2016) Epidemiology and risk factors of multidrug-resistant bacteria in respiratory samples after lung transplantation. Transpl Infect Dis 18(1):22–30. https://doi.org/10.1111/tid.12471

Teo JQ, Lim JC, Tang CY, Lee SJ, Tan SH, Sim JH, Ong RT, Kwa AL (2021) Ceftolozane/tazobactam resistance and mechanisms in carbapenem-nonsusceptible Pseudomonas aeruginosa. mSphere 6(1):10–128. https://doi.org/10.1128/mSphere.01026-20

Torres-Barceló C, Gurney J, Gougat-Barberá C, Vasse M, Hochberg ME (2018) Transient negative effects of antibiotics on phages do not jeopardise the advantages of combination therapies. FEMS Microbiol Ecol 94(8):fiy107

Valente LG, Federer L, Iten M, Grandgirard D, Leib SL, Jakob SM et al (2021) Searching for synergy: combining systemic daptomycin treatment with localised phage therapy for the treatment of experimental pneumonia due to MRSA. BMC Res Notes 14(1):1–4

Van Belleghem JD, Clement F, Merabishvili M, Lavigne R, Vaneechoutte M (2017) Pro- and anti-inflammatory responses of peripheral blood mononuclear cells induced by Staphylococcus aureus and Pseudomonas aeruginosa phages. Sci Rep 7(1):8004. https://doi.org/10.1038/s41598-017-08336-9

Waksman SA, Schatz A, Reynolds DM (2010) Production of antibiotic substances by actinomycetes. Ann N Y Acad Sci 1213:112–124

Welp AL, Bomberger JM (2020) Bacterial community interactions during chronic respiratory disease. Front Cell Infect Microbiol 10:536183

Witt CA, Meyers BF, Hachem RR (2012) Pulmonary infections following lung transplantation. Thorac Surg Clin 22(3):403–412. https://doi.org/10.1016/j.thorsurg.2012.04.006

Wu N, Dai J, Guo M, Li J, Zhou X, Li F, Gao Y, Qu H, Lu H, Jin J, Li T, Shi L, Wu Q, Tan R, Zhu M, Yang L, Ling Y, Xing S, Zhang J et al (2021) Pre-optimized phage therapy on secondary Acinetobacterbaumannii infection in four critical COVID-19 patients. Emerg Microb Infect 10(1):612–618. https://doi.org/10.1080/22221751.2021.1902754

Yahiaoui RY, den Heijer C, van Bijnen EM, Paget WJ, Pringle M, Goossens H, Bruggeman CA, Schellevis FG, Stobberingh EE (2016) Prevalence and antibiotic resistance of commensal Streptococcus pneumoniae in nine European countries. Future Microbiol 11:737–744. https://doi.org/10.2217/fmb-2015-0011

Zaldastanishvili E, Leshkasheli L, Dadiani M, Nadareishvili L, Askilashvili L, Kvatadze N, Goderdzishvili M, Kutateladze M, Balarjishvili N (2021) Phage therapy experience at the eliava phage therapy center: three cases of bacterial persistence. Viruses 13(10):1901. https://doi.org/10.3390/v13101901

Zarfel G, Krziwanek K, Johler S, Hoenigl M, Leitner E, Kittinger C, Masoud L, Feierl G, Grisold AJ (2013) Virulence and antimicrobial resistance genes in human MRSA ST398 isolates in Austria. Epidemiol Infect 141(4):888–892. https://doi.org/10.1017/S0950268812001343
Zumla A, Memish ZA, Maeurer M, Bates M, Mwaba P, Al-Tawfiq JA, Denning DW, Hayden FG, Hui DS (2014) Emerging novel and antimicrobial-resistant respiratory tract infections: new drug development and therapeutic options. Lancet Infect Dis 14(11):1136–1149

The Art and Science of Defeating Superbugs: Present and Future of Phage Therapy

17

Sunil Kumar Raman, Chunna Yadav, Himanshu Bansode, Shilpi Mishra, Unnati Sharma, and Niharika Mishra

Abstract

Addressing the escalating challenges of antibiotic resistance, stemming from the excessive usage and production of antibiotics, prompts a shift towards alternative non-antibiotic treatment. In the ongoing battle against microbial infections, the identification of chemical antibiotics emerged as a formidable tool between the 1920s and 1940s. Excessive use and misuse of antibiotics have spurred muta-

The original version of the chapter has been revised. A correction to this chapter can be found at https://doi.org/10.1007/978-981-97-5272-0_39

Significance and Contribution: In-depth analysis of advancements in phage therapy focusing on enhancing safety, efficacy, and delivery mechanisms is pivotal for the successful translation of this approach in countering antimicrobial resistances (AMRs). The discourse aims to drive innovative non-antibiotic therapeutic strategies against AMRs. An integral aspect involves disseminating knowledge among emerging researchers, clinicians, and epidemiologists, fostering awareness, and stimulating the formulation of policies. This initiative is essential to scale up phage therapy technology, transforming it into a versatile tool with broad-spectrum clinical applications and significance of refining phage therapy to combat the evolving challenges posed by AMRs.

S. K. Raman (✉) · C. Yadav · H. Bansode
CSIR-Central Drug Research Institute, Jankipuram Extension, Lucknow, Uttar Pradesh, India

S. Mishra
Institute of Pharmaceutical Sciences, Department of Pharmacology, University of Lucknow, Lucknow, India

U. Sharma
School of Pharmaceutical Sciences, Chhatrapati Shahuji Maharaj University, Kanpur, Uttar Pradesh, India

N. Mishra
Krishna Institute of Pharmacy and Sciences, Kanpur, Uttar Pradesh, India

M. Gangwar, G. Nath (eds.), *Emerging Paradigms for Antibiotic-Resistant Infections: Beyond the Pill*, https://doi.org/10.1007/978-981-97-5272-0_17

tions, rapidly reduced antibiotic susceptibility and ultimately contributing to antibiotic resistance. However, the abrupt acceleration of antibiotic resistance, a natural occurrence, has raised concerns among healthcare professionals, microbiologists, clinicians, and pharmaceutical researchers globally. The continuous microbial mutations and inevitable horizontal gene transfer underscore the need to reassess strategies against drug-resistant infections. The potential reduction in effectiveness of traditional antibiotic therapies prompts exploration of alternative approaches, such as non-antibiotic therapies, to address challenges posed by drug-resistant microbes. Phage therapy (PT) has resurfaced as a promising domain for delivering patient care against both pathogenic and drug-resistant microbial infections with enhanced reliability. Researchers are diligently committed to contributing to the development and advancement of this antibiotic alternative. It is suitable for both primary treatment and as an adjuvant therapy alongside antimicrobial agents, providing standard care to patients and offering undoubted potential in combating microbial infections effectively.The primary findings emphasize the detrimental impact of antibiotic misuse, demonstrating a swift escalation of antibiotic resistance. Furthermore, the resurgence of phage therapy is highlighted as an evident alternate of antibiotics, with ongoing global initiatives aimed at establishing its clinical safety and efficacy against opportunistic pathogens and antibiotic-resistant microbial infections, positioning phage therapy as a potential avenue for personalized therapeutic interventions.

Keywords

Phage therapy · Antibiotic-resistant infections · Adjuvant therapy · Clinical efficacy · Phage-host interaction

17.1 Introduction

17.1.1 Antimicrobial Resistances

Antimicrobial resistances (AMRs) is the nightmare to the not just for epidemiologists, clinicians, and drug discovery professionals but for all population irrespective of land boundaries. Public health policy maker and all stakeholders are awakened to curb the spread of AMRs in both developing countries and developed ones. The spread of infectious diseases has been never down to baseline since the cross-nation transportation advancements has been achieved. The biological evolution of microorganisms is inevitable and same has evolved to grow resistant towards antimicrobials. Due to over exposure to the antimicrobials like Antibiotics and antifungals, these therapies now fail to contain the infection by those microbes either in MIC which were effective earlier or complete resistance towards these drugs is observed in the various strains of microbes. The bacterial Infections namely "*Enterococcus faecium, Staphylococcus aureus, Klebsiellapneumoniae, Acinetobacterbaumannii, Pseudomonas aeruginosa, Enterobacter* species," now infamous as "ESKAPE pathogens," the pathogens are the flag bearers of AMRs. These pathogens are also termed as "superbugs" (Fig. 17.1) (Prestinaci et al. 2015).

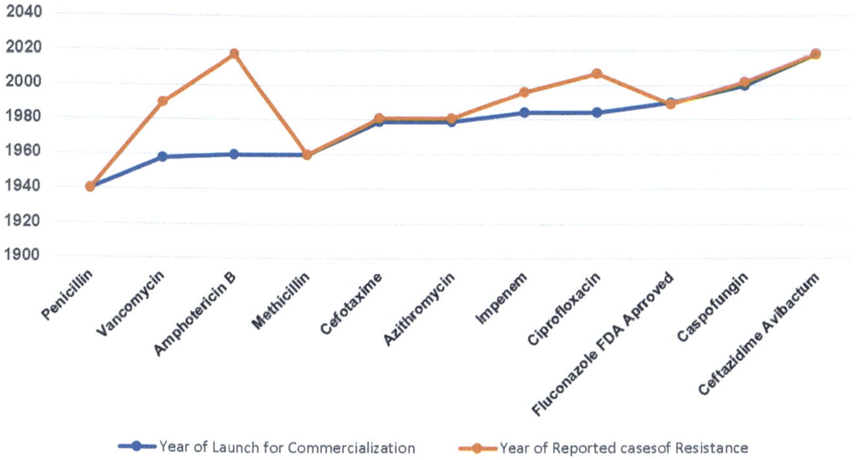

Fig. 17.1 Emergence of resistance

17.1.1.1 Why Are Bacterial Infection Most Dreaded?

Though the world has witness reoccurant invasion of zoonotic infection prevailed by viruses, still bacterial infection is equally or even greater topic of health hazard for the human, animal, and plants too unspared. The genetic diversity in bacteria is greater than virus since their genome is more diverse and larger than the later. Making genetic variation an obvious due to various natural phenomenons like, Horizontal gene transfer, genetic shift, mobile genetic elements, influence of environments. These causes enhancement in survival of bacteria in different topographies and extreme weathers (Rahman et al. 2020).

The estimates of impact of AMRs gives estimate of five million people directly or indirectly getting affected by drug resistance, another estimate gives a phantomizing report of ten million deaths due to AMRs by 2050 which is not too far. We can keep discussing the vulnerability due to the superbug on environment, economy, public health, and it gives no happy hint unless we put collective effort to counter and contain this ever-growing drug resistance. On our end we can target reduction of misuse as well as overuse of antibiotics and have dedicated collaborative research to discover reliable short-term and long-term approaches to save ourselves and the population whom we have already put into vulnerable zone. The graph below shows the rapidity of development of drug resistance in pathogens against antibiotics launched in the timeline. We can observe we now almost get AMRs as soon as the antibiotics as launched commercially (Dixit et al. 2019).

17.1.1.2 Rationale for Exploring Alternative Therapies Like Phage Therapy

In pursuit of alternate therapeutics against antibiotics resistant pathogens, microbiologists, and clinical researchers have initiated or better say resumed research on Phage Therapy. A rediscovered arena of therapeutics holding evidential proof of successful activity of killing the pathogens. We shall discuss the Phage Therapy in this chapter.

17.2 Bacteriophages

17.2.1 Definition and Characteristics of Bacteriophages

The word "Bacteriophage" is derived from two Greek words "bacterium" and "phagein" denoting "bacterium" and "eat," respectively. According to the morphology and nucleic acid, Phages are classified into different families viz. Myoviridae, Siphoviridae, Podoviridae, Lipothrixviridae, Rudiviridae, Microviridae, Rudiviridae, Inoviridae, Corticoviridae, and Finnlakeviridae. Bacteriophages are the viruses that infects the bacteria by attaching to a susceptible host bacterium's cell wall and injects its genetic material (RNA or DNA) into the host cell cytoplasm, disrupt bacterial metabolism and cause the bacterium to lyse. The biotic part of ecosystem, bacteriophages or simply phages are naturally occurring viruses, having an outer membrane of glycoproteins termed as capsid. Capsid encapsulates genetic material DNA or RNA in it. These viruses have special ability to infect bacterium specific to genus or species of bacteria and use host mechanism to multiply itself (Abedon et al. 2011; Dy et al. 2014).

The process of evolution has best depicted the law of conservation of energy, mass, with this statement we want to highlight this fact that the ecological population of bacteria has more than ten bacteriophages against each species of bacteria. Based on it mechanism of multiplication it can be categorized broadly into two different types of phages namely lytic and lysogenic (temperate) phages.

17.2.2 Historical Background

17.2.2.1 Early Discoveries of Bacteriophages

Though the evolution took millions of years to result in what and how we are today, our co evolving partner bacteriophages followed same processed but was very late to come into books of biology, until in the year 1896, when Ernest Hanbury Hankin observed in water of ganga river that they did not spread cholera ganga river water could destroy cholera bacteria.

Further it too nearly two more decades after initial observations made by Hankin in 1896. In 1917 an innovative microbiologist Felix d'herelle published on Bacteriophages as "Bacteria eating viruses." It was Felix who correlated possible application of Bacteriophages against bacteria causing infections. He would isolate phage and was relatively successful in using it clinically for treating dysentery in 1917 and afterwards (Fig. 17.2) (Stone et al. 2019).

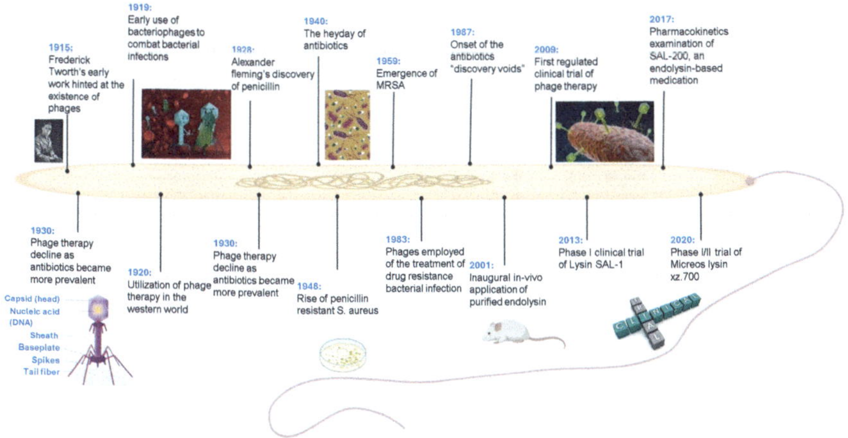

Fig. 17.2 Evolution of bacteriophages journey from discovery to the present

17.2.2.2 Initial Applications and Challenges

The clinical application of isolated bacteriophages ready to launch but the thriller was yet to be announced. In parallel to phage utility as therapeutics another lab observation made by Alexander Flemming in his lab led to identification of the revolutionary antibiotic renowned as Penicillin. This Penicillin further created history in global healthcare infrastructure that we all are aware of. In mean time rise of antibiotics at industrial scale shadowed the development of bacteriophage as drug product what we call as Phage therapy today.

17.2.3 Phage–Host Interactions at the Molecular Level

The displacement of bacteriophage are brought about by diffusion, flow, Brownian motions biosynthesis, ages when are in proximity fire most they initiate the mechanism to recognize the cell wall of bacterium species, once recognized and in case the specific bacterium is detected by the sticky proteinious limb like projection called tail fiber of phage they form a reversible bond on the surface of the bacteria and once attachment is taken place, the injection of nucleic acid into the bacterial cell is done via its spikes, followed by the hijack of the intracellular biochemical machinery of the bacterium host by penetration, transcription, biosynthesis, and lysis of bacteria after maturing (Letarov and Letarova 2023).

17.2.4 Fate of Phage–Host Interaction

17.2.4.1 How Bacteriophages Target and Infect Bacteria?

The phage host interaction studies have established the mechanisms of bacteriophages infecting the selective specie of bacteria. The bacteriophages attack the

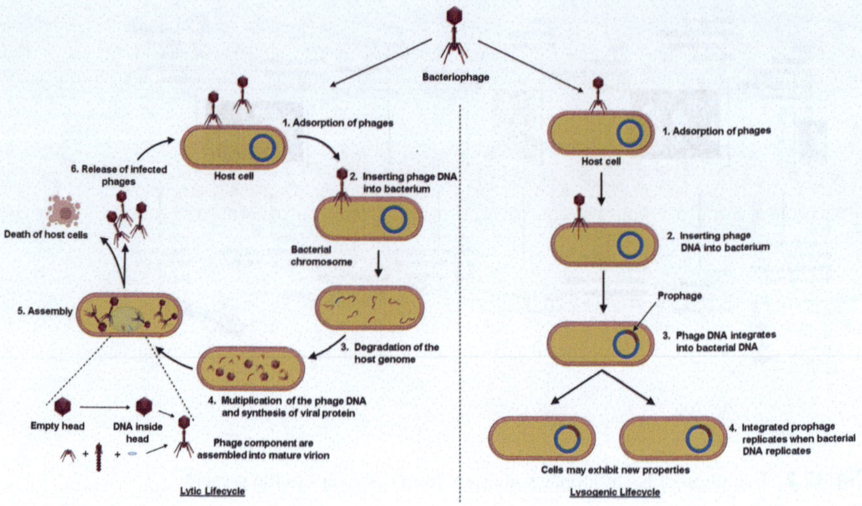

Fig. 17.3 Life cycle of bacteriophage

bacterium by namely two pathways Lytic pathway and lysogenic pathway as mentioned above prior to which steps of phage attachment must have taken place as Interaction cannot take place without attachment at first place.

In the lytic pathway the bacteriophages degrade the cell wall with its enzymes endolysins and the high osmotic pressure inside the bacterial cell wall bursts, this led to cellular death of the bacterium where intra cellular components are no more protected by integrated cell wall. There were claims by some researchers that this bursting of bacterium's intracellular components can cause endotoxin related immunological response in the human/animals in which those bacteria resides, but this claim has been ruled out since no any such report are found elsewhere where phage therapy is administered for therapeutic application either in human or animal. Another activity running parallelly with lytic path is the release of new phages, ready to infect other bacterium in the available milleu. In the lysogenic pathway bacteriophages incorporates its viral DNA into bacterium genome and forms prophage which later causes lysis of bacterium on triggering responses, e.g., temperature, pH, osmotic pressure, and nutrients available. To differentiate lytic pathway firm lysogenic pathway from lytic pathway, to confirm it we need to perform electron microscopy, DNA extraction, or propagation on sensitive strain (Fig. 17.3).

17.2.4.2 Bactericidal Mechanism

- Disruption in cell wall with high osmotic pressure internally bursts the bacteria.
- Depolymerase acts on biofilms (if any).
- Mycolic acid layer degradation.
- Release of intracellularly multiplied new viral copies from tens to hundreds.

17.2.5 Advantages of Phage Therapy

The use of bacteriophage for therapeutics intend is phage therapy, the bacterial infection poses a great deal of thereat to global population and the growing failure rate of antibiotics must be overcome and the phage therapy is a promising area of research for improving the therapeutics and the greater advantage of PT over antibiotics are both biological affinity due to specificity for target bacteria particular species, strain and the cost of development and production of phage therapy is also very less then developing novel antibiotics to launch in market.

Reduced impact on commensal microbiota is another great advantage of phage therapy over antibiotics, which do not spare the other commensally or benefit bacteria.

Potential to overcome antibiotic resistance is the recently flourishing arena which has built the research base in a century old discovered therapy against gut infection back then which has a diverse utility against pathogens, drug resistant pathogens specially.

The hypothesis of stating another advantage of phage therapy is we have already seen the consequences of using antibiotics, the synthesized molecules against a live bacterium, which has greatly trained and has helped the bacteria's to adapt to express rapid mutations and the result of which Is AMRs. The biology researchers have put ahead the concept of evolutionary trade off, by which we can restore antibiotic sensitivity in bacteria using phage therapy which shall be discussed in phage host interaction (Table 17.1) (Loc-Carrillo and Abedon 2011).

Table 17.1 Highlights of phage therapy

Insights	Advantages
Overcoming cross resistance	The greatest advantage is it has applicability against drug resistant bacteria
Specificity	Targets specific pathogenic bacteria, commensal bacteria are unharmed
Active against biofilms	Depolymerase action on polysaccharides complex
Single dose potential	Can be administered as single dose
Low dosage use	Low dose has in site amplification capability lysing the host bacteria
Immune compatibility	High purity phage therapy attracts negligible neutralization immune response
Null side effects	Since phages doesn't bind to mammalian cells, no side effects are reported
Safety	Safe for administration via various routes without therapeutic incompatibilies
Environmental impact	No detrimental affect like antibiotics, safe for environment being natural biotic organism unless genetically engineered, doesn't take part in biological magnification, a greener approach

17.2.6 Challenges and Limitations

- Phages could be deactivated by neutralizing immune reaction.
- Extra dose of phages could be eliminated by immune cells which can reduce the number of active phages at target infected tissue sites.
- Purification part of phages are essential to eliminate the risk of therapeutic anaphylaxis due to presence of endotoxins before deploying Phage therapy.

17.2.7 Phage Isolation and Characterization

17.2.7.1 Isolation, Amplification, and Purification

Phages are isolated from various site location, i.e., soil, sewage, canals, freshwater lakes etc. by using specific phage buffer viz. Tris-phage buffer, SM buffer, MP buffer etc. Buffers are selected based on their ability to maintain the stability and activity of phages throughout the isolation process. Some common types of buffers used in the isolation of phages are shown in Table 17.2. After isolation the phages are amplified followed by purification through various methods, i.e., CsCl gradient method, Filtration method. With the help of spot test, we can check the activity of phages against different pathogens.

17.2.7.2 Phage Titer

The phage titer refers to the concentration of viable bacteriophages in a stock suspension. Number of bacteriophage capable of infecting a cell effectively are measured as plaque-forming unit, i.e., PFU.

Table 17.2 Composition of buffers

S. no.	Components	For 500 mL
Tris-phage buffer		
1.	1 M Tris (pH = 7.5)	5 mL
2.	1 $MgSO_4$	5 mL
3.	NaCl	2gm
4.	0.1 $CaCl_2$	5 mL
Mycobacteriophage (MP) buffer		
1.	1 M Tris (pH = 7.5)	25 mL
2.	5 M NaCl	15 mL
3.	1 M $MgCl_2$	5 mL
4.	1 M $CaCl_2$	1 mL
Saline-magnesium (SM) buffer		
1.	1 M Tris (pH = 7.5)	25 mL
2.	8 mM $MgSO_4.7H_2O$	1 g
3.	100 mM NaCl	2.9 g

17.2.7.3 PFU Formula

$$PFU / mL = \frac{\text{No.of Plaques counted}}{\text{Volume}(mL) \times \text{Dilution factor}}$$

Characterization

Phages are characterized by performing the Next Generation Whole Genome Sequencing (NG-WGS) and Transmission Electron Microscopy methods. The sequencing data are assembled, annotated, and compared to the reference mycobacteriophage.

17.2.8 Selection of Phage for Therapy

Phages are selected by considering their effectiveness and safety for therapeutic use. Some of the factors to be considered for phage selection are host range, virulence, stability, resistance development etc. Hence, these are determined for specific pathogens by performing in-vitro experiments on suitable base agar plate supplemented with growth supplement followed by amplification.

One of the major aspects of Phage selection also includes selection of phages based on specific characteristics required for each route as cited in Table 17.3.

Table 17.3 Selection of phages based on route-specific characteristics

S. no.	Route of administration	Formulation	Selection
1.	Oral	Lyophilized powder or liquid suspension	Should be able to withstand acidic environment of the stomach and shall have broad activity against GIT target pathogens
2.	Inhalation	Aerosols or nebulized solutions	Should be safe for pulmonary administration and shall be able to reach the lower respiratory tract
3.	Topical	Ointment, gel and solution	Should have the ability to penetrate the affected tissues to reach the infection site
4.	Intravenous	Sterile injectable solutions	Should be free from endotoxins which may cause adverse reactions
5.	Other routes (intramuscular, intraperitoneal, Intravesical)	Depending on the specific requirements of the route	Should be compatible as per specific formulation

17.2.9 Selection of Phages for Clinical Isolates

Firstly, the clinical isolates taken from patients are determine which bacterial infection the patient has. After that will isolate the phages against that bacterial species as described in the Sect. 17.1. The phage formulation preparation can be given to the patient keeping in mind the route of administration, patient compliance, infection site, survival of phages at various physiological condition at different pH dependent upon the severity of the diseased (Elahi et al. 2021; Hyman 2019; Shende et al. 2017).

17.2.10 Phage Formulations

17.2.10.1 Strategies for Phage Formulation

Depending upon the target tissue infection site, the formulation strategies are decided to execute upon, whether the target tissue site is topical, visceral etc. to avoid the washing off the phages from systemic circulation in case of intravenous administration we have to use validated methods and identify add use critical quality attributes (CQA), meet quality target product profile (QTPP) for each dosage forms, e.g., Liquid, solid, semisolids which are delivered through various routes, i.e., oral, topical, IV, and inhalation (Rosner and Clark 2021).

17.2.10.2 Phage Cocktails

One of major hurdle in the due process of identification of pathogen from clinical samples may encounter phages being resisted by bacteria or may be case of poly microbial vectors, in that case we apply phage cocktail or bacteriophage cocktail. Researcher are exploring the use of bacteriophage cocktails, which are mixtures of different bacteriophages targeting the same bacterial pathogen. This approach aims to increase the therapeutic efficacy and reduce the risk of bacterial resistance against PT. The bacteriophages are being widely investigated to be used for treating various infectious diseases from tubercular infections, to *Escherichia coli*, Pseudomonas, and other deadly infections which has evolved to be antibiotic resistant now. Some case shows the bacterial resistance against the bacteriophages most likely when administered in relatively lower dose. This can also be understood as the bacterial resistance against phages and its lytic enzymes. This phenomenon is usually natural and unwantedly can result in therapy failure in context of phage therapy. This could seem to be dreadful news for the researcher's fraternity and to overcome this resistance against phages we deploy the phage cocktail. This strategy not only rules out the resistance developing in bacteria against previously administered phages in single phage composition but also ensure the lysis of the bacteria earlier exposed to the phages. We apply two or more known and tested phages against the species of the pathogen to ensure the successful therapeutic outcome in the patient.

17.2.11 Genetically Engineered Phages

Researcher are investigating the possibility of genetically engineered (GE) phages by modifying phages to enhance their specificity, virulence, or other therapeutic properties. This includes the use of synthetic biology techniques (CRISPR-CAS) to design phages with customized features to develop phage therapy as personalized medicine. Genetic and chemical engineering is a reliable method to increase the potency of PT in greater magnitude to meet the objectives like reduction in possible endotoxins and endotoxins triggered immune responses, increasing antibiotic sensitivity in pathogen, removing biofilms so that phages or antimicrobials can reach to the pathogen and show desired pharmacological action, or to be employed as an adjuvant therapy with antibiotics. GE phages also opens avenue into deeper research like identifying genes which causes lysogenicity and down regulate it and increase intra genus specificity for enhanced lytic activity rather the inter genus targeting like in event of broad spectrum antibiotics. GE phage also can be studied for amplifying in vitro without phage host bacteria (Guo et al. 2021).

17.2.12 Phage-Encoded Enzymes

Some studies focus on using specific enzymes produced by phages rather than whole phage particles. These enzymes, such as endolysins, Viron-associated lysin (VAL), Holins, Spanins, and Bacterial polysaccharide depolymerase can have antibacterial effects and may be used independently or in conjunction with other treatments (Guo et al. 2021; Oliveira et al. 2018).

17.2.13 Combination Therapies

Investigating the synergistic effects of combining phage therapy with traditional antibiotics or other antimicrobial agents is gaining attention. This approach aims to enhance treatment outcomes and overcome challenges like bacterial resistance. Evolutionary trade-off is a situation where natural selection is improved adopting one attribute and losing the other. This case is correlated in phage therapy and phage resistance in such a way that, at present phage therapy is showing bactericidal action and less occurrence of phage resistance in bacteria is reported, in times when bacteria would increase phage resistance, it could lose the antibiotic insensitivity and in that way, the degree of polymorphic variation at binding site for antibiotics, efflux pump that eliminates antibiotics out of bacteria' cell would reduce its ability. in both way Phage therapy would still be helpful for adjuvant therapy in combination with antibiotics (Diallo and Dublanchet 2022; Kim et al. 2018).

17.2.14 Clinical Applications

17.2.14.1 Numerous Clinical Trials Have Been Conducted in the Field of Phage Therapy

The phage therapy is in clinical practice for over a century in west European countries like Poland, Georgia, and Russia. Since than they have gained their expertise in isolation, amplification and has been applying in patient care and patients visit there for medical tourism. In recent, shift on phage therapy research has bloomed exponentially from Americas, whole of Europe, China, and Indian subcontinent in awake of rising cases of various pathological conditions including AMRs mentioned in Table 17.4 (Furfaro et al. 2018 #17; Hitchcock et al. 2023 #18).

17.2.15 Phage Pharmacokinetics

Pharmacokinetics is the dynamic process governing the movement of administered substances through and out of the body, encompassing absorption, distribution, metabolism, and excretion. In the context of phage therapy, the success of eradicating pathogenic bacteria hinges on achieving optimal phage concentrations at the target site. This necessitates a thorough comprehension of phage pharmacokinetics (PK), which differs markedly from traditional antimicrobials due to the unique tripartite dynamic relationship between phages, their host bacteria, and the human immune system(Wahida et al. 2021). Phages exhibit intricate PK profiles, replicating in the presence of susceptible bacteria, adding complexity to their behavior compared to conventional antibiotics. Their half-life in animal models ranges from 2.2 to 4.5 h, persisting for 2–3 days even in the absence of a bacterial host. Two primary means of achieving sufficient phage concentrations involve in situ replication (active treatment) or conventional dosing (passive treatment), both requiring robust strategies to counteract phage reduction mechanisms.

Injecting phages intramuscularly or through inhalation generally results in higher concentrations, while subcutaneous administration leads to lower blood concentrations (Heo et al. 2009). The highest concentrations of phages were observed in uninfected mice injected intraperitoneally (Chhibber et al. 2008). Peak concentrations in the blood occur at 6 h after intraperitoneal injection (Jun et al. 2014), and at 12 h after oral administration (Chhibber and Kumari 2012; Yen et al. 2017). No phages were detected after 48 h on both routes. These findings support the two-compartment pharmacokinetic model, which suggests that phages are initially distributed to organs and then slowly eliminated from the body (Uchiyama et al. 2014). The complexity of the interactions between phages and the tissues of individual organ systems is also likely to contribute to these dynamics. For example, sub diffusive movement of phages in mucous secretions, increases the interactions between phages and bacteria (Barr et al. 2015).

In contrast to intraperitoneal therapy, the lungs exhibit higher concentrations of phages following inhalational therapy, Subcutaneous administration, on the other hand, results in lower and more delayed blood concentrations when compared to intraperitoneal injection (Carmody et al. 2010; Pouillot et al. 2012). Intravenous and

Table 17.4 The registered clinical trials of PT for various ailments

Clinical trials in phage therapy	• **Phage therapy for chronic rhinosinusitis (CRS):** – ClinicalTrials.gov identifier: NCT04697673 – Recent status: Ongoing (enrolment ongoing; interim analysis pending) – Purpose: Assessing the safety and effectiveness of intranasal bacteriophage therapy in patients with (CRS) who have not responded to conventional treatments • **Utilizing phage therapy to treat infections in burn wounds:** – ClinicalTrials.gov identifier: NCT04577317 – Recent status: Ongoing (recruitment ongoing; data collection in progress) – Purpose: Researching the safety and effectiveness of applying topical phage therapy to treat burn wound infections caused by antibiotic-resistant bacteria • **Implementing phage therapy as a treatment for urinary tract infections (UTIs):** – ClinicalTrials.gov identifier: NCT04687508 – Recent status: Recruiting (actively recruiting participants; data collection expected to start soon) – Purpose: Evaluating the effectiveness of oral bacteriophage therapy in patients experiencing recurrent urinary tract infections (UTIs) caused by multidrug-resistant bacteria • **Phage therapy aimed at treating lung infections in individuals with cystic fibrosis:** – ClinicalTrials.gov identifier: NCT04470025 – Recent status: Active, not recruiting (data collection completed; analysis and reporting underway) – Purpose: Assessing the safety and effectiveness of inhaled phage therapy in patients diagnosed with cystic fibrosis and suffering from chronic lung infections triggered by Pseudomonas aeruginosa bacteria • **Phage therapy for *Clostridioides difficile* infection (CDI):** – ClinicalTrials.gov identifier: NCT04690994 – Recent status: Completed (trial completed; results published or pending publication – Purpose: Investigating the use of oral bacteriophage therapy as an adjunct to standard antibiotic treatment in patients with recurrent CDI • **Phage therapy for diabetic foot infections:** – ClinicalTrials.gov identifier: NCT04390337 – Recent status: Recruiting (actively recruiting participants; early results expected in the coming months) – Purpose: Evaluating the safety and effectiveness of applying topical phage therapy in diabetic patients with foot ulcers that are infected by antibiotic-resistant bacteria • **Utilizing phage therapy to treat acute bacterial skin and skin structure infections (ABSSSI):** – ClinicalTrials.gov identifier: NCT04834132 – Recent status: Ongoing (enrolment ongoing; preliminary data collection underway) – Purpose: Assessing the safety and effectiveness of intravenous bacteriophage therapy in patients diagnosed with acute bacterial skin and skin structure infections (ABSSSI) caused by Staphylococcus aureus or streptococcus pyogenes • **Phage therapy in tonsillitis acute tonsillitis:** – ClinicalTrials.gov identifier: NCT04682964 – Status: Active, not recruiting – Purpose: The objective is to assess the safety and effectiveness of bacteriophage therapy in the treatment of acute tonsillitis caused by a range of bacterial pathogens, such as staphylococcus spp., enterococcus spp., streptococcus spp., and Enteropathogenic *Escherichia coli* (*E. coli*) • **Using phage therapy as a treatment approach for initial cases of chronic prosthetic joint infections (ACTIVE1)** – ClinicalTrials.gov identifier: NCT05269121 – Sponsor: Adaptive phage therapeutics, Inc. – Purpose: The study is likely aimed at assessing the safety and effectiveness of bacteriophage therapy as a treatment for chronic prosthetic joint infections (PJI) for the first time. Chronic PJI, a severe complication of joint replacement surgery, is frequently attributed to bacterial biofilms • **Clinical investigation focusing on the efficacy and safety of phage therapy for chronic constipation** – ClinicalTrials.gov ID NCT05973721 – Status: Recruiting – Purpose: The study aims to assess PiB-specific phage's efficacy in treating refractory constipation through clinical trials, focusing on target exposure, occupation, and functional effects post-action • **Phage therapy for Avis-PHAGE in LYON clinic** – ClinicalTrials.gov ID NCT05883995 – Status: Recruiting – Purpose: The Avis-PHAGE in LYON clinic study, conducted by Hospices Civils de Lyon, is a single-site non-interventional retrospective and prospective study. Its objective is to quantify the requests for phage therapy received at the clinic from 2015 to 2028. The study will involve 1500 patients referred to Hospices Civils de Lyon seeking medical advice on phage therapy • **Phage treatment for *Staphylococcus aureus*-induced prosthetic joint infections managed with debridement, antibiotics, and implant retention (PhagoDAIRI)** – ClinicalTrials.gov ID NCT00663091 – Purpose: A prospective, randomized, double-blind controlled study is assessing the safety and effectiveness of WPP-201 for the treatment of venous leg ulcers • **Bacteriophage therapy for challenging or difficult-to-treat infections (BT100)** – ClinicalTrials.gov ID NCT05498363 – Purpose: A retrospective observational analysis investigates the first one hundred consecutive cases of bacteriophage therapy for difficult infections, backed by a Belgian consortium

(continued)

Table 17.4 (continued)

Phage therapy in veterinary medicine	• Phage therapy holds promise as a valuable tool in the veterinary arsenal for combating bacterial infections in animals • Phage therapy in veterinary medicine utilizes diverse administration routes based on infection type and location, such as topical, oral, inhalation, or injection. While historically employed, recent interest in clinical trials and case studies seeks to validate its efficacy and safety across animal species, advancing its adoption in veterinary practice • "Acinetobacterbaumanii" infections in mice • "Pseudomonas aeruginosa" infections in mice • "Staphylococcus aureus infections" in mice and cows • "Clostridium difficile infections" in hamsters • **Phage therapy for *salmonella* infections in poultry:** – Phages effective against salmonella typhimurium are commonly sourced from bird feces in poultry farms and dovecotes – The Salmonella typhimurium phage (type B1 Siphoviridae) exhibits potent lytic activity against salmonella enterica – Experimental trials on salmonella-infected mice and rabbits showed 90–100% survival rates with phage treatment, compared to 0–10% in the control group (Lenev 2013) • **Phage therapy for *managing Colibacillosis* infection in poultry:** – Colibacillosis, caused by avian pathogenic *Escherichia coli* (APEC), is a significant cause of mortality in poultry – Simultaneous administration of phages reduced mortality by 25% at the lowest concentration and by 5% at the highest concentration in treated chickens – Combining two phages, DAF6 and SPR02, administered intramuscularly, resulted in over 80% chicken survival • ***Escherichia coli* infections in mice, poultry, calves, piglets, and lambs**: – Bacteriophages are effective against Enterobacteriaceae, including *E. coli* – An experiment on escherichiosis in piglets used T4 phage from the Myoviridae family, providing up to 100% protection at all dosages – In another experiment, calves orally infected with 10^10 K1+ *E. coli* showed promising results with phage treatment, as three out of four infected calves remained healthy after 8 h, and the fourth showed only mild symptoms • **Phage therapy in bovine species:** – Mastitis, a common condition in cattle caused by staphylococci, streptococci, and enterobacteria, leads to significant economic losses in cattle farms – CS1 and DW2 bacteriophages were found effective in lysing *S. aureus*, a common mastitis-causing bacteria – A phage cocktail comprising CS1, DW2, and phage K was administered intramammarily, showing promising therapeutic efficacy in reducing *S. aureus* infections – The phage preparation could potentially be used in teat-washes for prophylaxis against *S. aureus* mastitis • **Phage therapy in swine species:** – Swine species commonly suffer from various bacterial infections, including respiratory, gastrointestinal, and skin infections – Phage therapy has shown promise in treating bacterial infections in pigs, particularly those caused by antibiotic-resistant bacteria • **Phage therapy in companion animals:** – Companion animals may suffer from bacterial infections similar to those seen in humans, including skin infections, urinary tract infections, and respiratory infections – Phage therapy offers a potential alternative or adjunct to traditional antibiotics in treating bacterial infections in companion animals, especially those caused by multidrug-resistant bacteria • ***Vibrio cholerae* in rabbits:** – In an experiment using an in vivo rabbit model, 10 *V. cholerae* phages (B1—B10) at concentrations ranging from 10^10 to 10^11 phages/mL were tested. However, no prophylactic or therapeutic effect was observed, unlike the successful phage treatment seen in human cholera cases

intraperitoneal therapies share similarities, but the latter results in higher concentrations of phages within the peritoneal cavity. The distribution of phages in mice after intravenous administration has been studied, with lung, kidney, spleen, and liver showing low, but significant phage titers, around 100 times lower than in the blood. Following intraperitoneal administration, phages can be detected in these organs at concentrations ranging from 10 to 10^6 PFU/g, as reported by Matsuzaki. Oral administration of phages to mice and human volunteers without an antacid resulted in the detection of phages in the feces, suggesting that phages can cross the stomach barrier, as noted in two studies that reported the presence of bacteriophage against *A. baumannii* in the serum of a human patient following single intravenous injections for 50 and 120 m, respectively (Taati Moghadam et al. 2020).

The ongoing presence of phages within infected organisms highlights the importance of examining interactions between phages and their bacterial hosts. Factors such as the route of administration, phage concentration, and susceptible bacterial levels can impact phage concentrations in specific organs. For example, phages may persist in the spleen of infected fish for up to 5 days after injection, whereas in non-infected fish, they only persist for up to 24 h, potentially owing to in situ phage multiplication (Nakai et al. 1999). Phages have been seen in the blood for at least 96 h in the case of systemic MRSA infection (Ochieng'Oduor et al. 2016). Oral *P. aeruginosa* was used to elicit gut-derived sepsis, which was then treated with oral phages (Watanabe et al. 2007). High-phage concentrations persisted in the blood and liver after 8 days, even in the presence of high bacterial concentrations, suggesting that phages could continue to reproduce as long as their bacterial host was there. A study in which lytic phages were administered intravenously in a mouse model of *E. coli* sepsis. Phages were detected in the spleens of surviving mice up to 2 weeks after infection (Schneider et al. 2018). Another study compared phage concentrations in mice with *S. aureus* sepsis and in those administered phages alone. The results showed that phage concentrations in the blood, lungs, and liver were significantly higher in infected mice (Takemura-Uchiyama et al. 2014). In a *P. aeruginosa* pneumonia model phages were administered to infected mice and showed higher concentrations of phages in the lungs compared to mice administered phages only (Debarbieux et al. 2010). In the separated study, evaluated the effects of the route of administration and phage concentration on the dissemination of three lytic *E. coli* phages in the organs of chickens (lungs, liver, duodenum, and spleen). Higher phage titers were detected in all organs 3 h after phage administration, and no phage was detected in these organs 10 h after challenge (Oliveira et al. 2016).

17.2.16 Role of Phage Therapy in Various Field

The results of phage therapy are so robust against the bacterial infections. Tuberculosis infection is one of most dreaded infection with high fatality rate, long-term medication regime and has higher chances of converting into multi drug resistant infections since the oral dose of DOTs are relatively high and the peripheral organ toxicities are reported often. Below Table 17.5 comprises of the PT employed for treating tuberculosis (Zalewska-Piątek 2023).

Table 17.5 The case study of PT deployed for treating mycobacterium tuberculosis infection

Role of bacteriophages	Challenges and limitation in TB treatment	Reference
Phage proteins, resulting in target identification and a "bacterial target-driven screening" to find phage-like bacteria-encoded inhibitors of recognized presumptive bacterial pathogens targets	Identification of epitope targets of *Mtb* host pathogenic factor	Wan et al. (2021)
Treatment with mycobacteriophage led to altered pathology, characterized by reduced granuloma formation	Limited curative efficacy of isoniazid monotherapy	Zemskova and Dorozhkova (1991)
Phages may combine anti-bacterial activity with anti-inflammatory and immunoregulatory properties, implying that they may have therapeutic implications	The non resolving feature of TB inflammation is driven by tubercle bacilli's advanced intracellular survival mechanisms	Gorski et al. (2018b)
Phages have the capacity to bind endotoxins, limit excessive reactive oxygen species generation, and reduce inflammatory infiltration	*M. tuberculosis*, delivers its strong exotoxin via the ESX-4 type VII secretion system. To transfer proteins, including poisons, across cell membranes, bacteria employ molecular machinery	Miedzybrodzki et al. (2012)
Patient's initial immune status upregulation of phagocytosis has been associated with a good prognosis of PT	TB inhibits phagolysosome acidification. MTB slows phagocytosis maturation by decreasing phagosome acidification and then remains alive in a significantly lower acidic environment	Gorski et al. (2018b)
The phage may promote the creation of IL-10, which is known to play a key role in homeostasis. This impact may be comparable to that of probiotics, which increase Il-10 release by phagocytes	IL-10 is an immunomodulatory cytokine with effects on both innate and adaptive immune responses. It has been associated with TB reactivation in humans and the development of TB disease in animals	Kakinuma et al. (2017)
Phage adapts the *M. tb* cell membrane to the many settings it meets during infection [e.g., interaction with ALF, within the phagosome, extracellular, within granulomas or cavities, or when being transferred]	Drug delivery-access to intracellular TB	Allue-Guardia et al. (2021)

(continued)

Table 17.5 (continued)

Role of bacteriophages	Challenges and limitation in TB treatment	Reference
Intestinal epithelial cells (IEC) continuously produce an anti-inflammatory cytokine known as IL-1 receptor antagonist (IL1RN). Phages have the capacity to enhance the synthesis of IL1RN; suggesting that they could potentially counteract the elevated production of IL-1 in cases of intestinal inflammatory disorders	IL-1 R1 knockout mice exhibited notably larger granulomatous lesions in their lungs, accompanied by neutrophil infiltration	Gorski et al. (2018a), Sugawara et al. (2001)
Utilizing bacteriophages to decrease the generation of ROS could offer advantages in the management of inflammatory bowel diseases (IBD), as oxidative stress is thought to play a role in their development and progression	The host enzymes play a crucial role in controlling *mycobacterium tuberculosis* by generating ROS and RNS	Miedzybrodzki et al. (2008); Tyagi et al. (2015)
Innovative phage delivery methods are in demand, and a potential approach has been recently identified. This approach involves transporting bacteriophages into macrophages through nonvirulent "carrier" bacteria	Eradicating intracellular pathogens like *M. tuberculosis* is a formidable task because the phage must navigate through the mammalian cell membrane to infect the target bacteria and endure harsh intracellular conditions, including acidic pH levels	Broxmeyer et al. (2002)

17.2.17 Phage Therapy for NTM Infection

Throughout their life cycle, these endolysins eliminate the majority of Mtb. *Mycobacterium bovis*, NTM A lytic phage may be able to remove Mtb in infected areas. Inhibiting lipid metabolism and the normal generation of long-chain fatty acids. Increasing the permeability of the cell wall and the efficacy of several antibiotics. Increasing Mtb susceptibility to heat shock, H2O2, SDS, and low PH. Several shortened forms of Lysin A were created, and their actions were investigated using zymography and expression in both Escherichia coli and Mycobacterium smegmatis. Grover N, Paskaleva Lysin A has two catalytically active domains, both of which exhibit E.coli cell lysis when expressed just in the periplasmic area However, only one of these domains and the full-length Lysin A triggered cell lysis in M. smegmatis. In E. coli periplasm, full-length protein remained inactive. According to our findings, the inactivity is caused by a C-terminal domain that interacts with the N-terminal domain. Surface plasmon resonance confirmed this connection. There are a few critical prerequisites for an efficient TB phage cocktail. The candidate phage must infect both M. tuberculosis and M. smegmatis in order for substantial numbers required for clinical usage to be generated safely and effectively on the fast-growing, non-pathogenic M. smegmatis strain outside of strict biosafety conditions (Hashemi Shahraki and Mirsaeidi 2021). Lytic phages are favored because they typically leave few bacterial cells alive after infection, and they naturally decline when no more bacterial hosts are present. In contrast, temperate phages have the capability to create lysogens, which survive and propagate but may transfer genes to bacteria, potentially enhancing host pathogenicity or antibiotic resistance. Temperate phages can undergo genetic modifications to prevent the shutdown of lytic genes or chromosomal integration, making them suitable candidates. Additionally, it's crucial to create phage cocktails capable of targeting a broad spectrum of clinical isolates while also ensuring diversity to reduce the likelihood of emerging phage-resistant strains. Sequence research reveals that LysA is a probable endolysin responsible for peptidoglycan cleavage in mycobacteria cell walls, whereas LysB likely works at the mycolylarabinogalactan bond to liberate free mycolic acid. Either LysA or LysB alone could produce a significant change in the cell shape of *M. smegmatis* as well as cell death. LysA-treated *M. smegmatis* had crinkled and uneven spheroidal shape, according to scanning electron microscopy (SEM) results. The peptidoglycan in mycobacteria's cell wall was largely destroyed by the action of LysA, and the membrane tension drove the cell to adopt a distinctive spherical shape. However, how the exogenous LysA penetrates the mycolic acid-rich environment remains unknown (Zalewska-Piątek 2023).

17.2.18 Efficacy of Mycobacteriophage Therapy in Animal Models

Few studies have been published on the use of animal models to evaluate the potential of phage therapy for mycobacterial infections. Although these studies cannot replace human trials, they provide valuable information on phages, dosing strategies, endpoints, pharmacokinetics, immunogenicity, and in vivo safety. Additionally, these studies provide data on the efficacy of bacteriophages in treating mycobacterial infections, which can guide future human trials. Below, I summarize these studies. It is important to note that most of these treatments involve a single phage treatment as opposed to cocktail of mycobacteriophages recommended to overcome the development of phage resistance during treatment period (Table 17.6).

17.2.19 Safety and Regulatory Considerations

For the universal acceptance of phage therapy not just limited to only in vivo experiments, there requires numerous regulatory approvals, translational studies and evidence that the formulation of PT are safe and efficacious. Unless safety and efficacy is not proven with considerable statistical and scientific proofs and clinically relevant outcomes, bringing the formulation at premarket level still remains uncertain. Though the magisterial use of phage therapy for human use is permitted in USA,

Table 17.6 In-vivo assessment of mycobacteriophage therapy in animal models: exploring pathogen specificity, treatment routes, and results summary (Loponte et al. 2021)

Animal model	Pathogen	Mycobacteriophage treatment	Route of treatment	Results summary	References
Guinea pig	*M. tuberculosis*	Polonus	Subcutaneous injection in the inguinal region every alternate day on six occasions	>Survival	Macandergolcz (1962)
		DS-6A, GR-21/T and My-327	Twice a week subcutaneously for 10 weeks	Spleen index	Sula et al. (1981)
		DS-6A	Not available	<Tuberculous inflammation	Zemskova and Dorozhkova (1991)
C57 BL/6 mice	*M. avium*	*M. smegmatis* with TM4 phage	7 days later, treated either once or twice with TM4 phage (7.9 × 10(10) PFU/mL), *M. smegmatis* (4 × 10(8) cfu/mL), or *M. smegmatis* with TM4 phage delivered intravenously (i.v.)	<*M. avium* count in the spleen	Lia Danelishvili et al. (2006)
Guinea pig	*M. tuberculosis*	D29	10^6 PFU/0.1 mL intranasally with an interval of 3 times per week for 4 weeks	<Gross organ lesions and bacteria	Peng Li and Meyer (2009)
Mice (C57BL/6)	*M. tuberculosis*	D29	7.4 ± 0.1 $\log_{10}$(PFU/mouse) delivered to the lungs by nose only inhalation device (vibrating mesh nebulizer)	Prophylactic protection against inhaled *M. tuberculosis*	Carrigy et al. (2017)
Mice (BALB/c)	*M. tuberculosis*	D29	50 μL D29 phage suspension with a concentration of 10^{10} pfu/mL by endotracheal route using Penn-century aerosolizer; Collison 6-jet and spinning top aerosol nebulizers (STAG)	Deposition and distribution of aerosolized phage D29 particles in naive BALB/c mice	Liu et al. (2016)
Mice	*M. ulcerans*	D29	A single dose of mycobacteriophage D29 containing 8 $\log_{10}$ PFU by subcutaneous injection administered 33 days after bacterial challenge	Decreased pathology and prevent ulceration. >Cytokine levels (including IFN-γ), both in footpads and draining lymph nodes	Trigo et al. (2013)

(continued)

Table 17.6 (continued)

Animal model	Pathogen	Mycobacteriophage treatment	Route of treatment	Results summary	References
Humanized NSG-SGM3 mice	*M. tuberculosis*	DS6A	From day 3 of Mtb infection, every other day with ten doses of DS6A (10^{11}phage particles/dose) via intravenous administration	<Gross lung pathology with less inflammation and other pathological changes <*M. tuberculosis* count in lung and spleen Bacilli completely eradicated in the spleens in six of nine mice. Lung bacillary load three times less in phage-treated mice	Yang et al. (2023)
SCID mice	D29-LysB	*M. abscessus*	Treatment started 2 days post infection using 50 μL of aerosolized endotoxin-free D29-LysB at 40 mM for 6 days	<*M. abscessus* count in lung	Hurst-Hess et al. (2023)
BALB/c mice	D29-LysB	*M. ulcerans*	Treatment was initiated when footpads of mice were swollen to 2.7 mm and was performed by two s.c. injections in the infected footpad with 50 μg of LysB in PBS at 10 and 13-days post-infection	Mice were sacrificed after the emergence of ulceration	Fraga et al. (2019)

Table 17.7 Current regulatory statuses in various countries on PT Policies (Yang et al. 2023)

Safety and regulatory considerations	• Assessing risks like phage toxicity, immune responses, and horizontal gene transfer is vital for phage therapy safety. Safety and regulatory considerations are crucial to ensure effectiveness, minimize risks, and integrate phage therapy into clinical practice. Collaboration among researchers, regulators, healthcare providers, and patients is essential for navigating these complexities and advancing phage therapy • **Regulation in India:** Phage therapy in India is provided under compassionate phage therapy (cPT) on a compassionate basis. Regulations for phage therapy in India are overseen by the declaration of Helsinki and coordinate with central drugs standard control organization (CDSCO) enabling the implementation of cPT in India • **Regulation in United States:** It falls under the oversight of the united state Food and Drug Administration (USFDA). The USFDA regulates phage therapy as a biological product, ensuring its safety, efficacy, and quality through rigorous evaluation processes and adherence to regulatory standards • **Regulation in Australia:** The TGA oversees phage therapy products as biologicals, ensuring safety, quality, and efficacy. In Australia, patients can access phage therapy through the SAS, CTN, or CTX pathways, following a referral from a doctor to an infectious disease specialist associated with phage Australia • **Regulation in UK, France and Belgium**: – In the UK, there are no licensed phage therapy products. The MHRA oversees phage therapy, classifying natural phages as biological medicines and regulating compassionate use – France regulates phage therapy through ANSM, aiming to establish production and legal frameworks by 2019 – Belgium's unique regulatory framework for phage therapy is managed by the Belgian Federal Agency for Medicines • **Regulation in Poland, Georgia and Russia:** Regulations for phage therapy in Poland, Georgia, and Russia are overseen by respective health regulatory authorities: URPLWMiPB (Office for Registration of medicinal products, medical devices, and biocidal products) in Poland, NARP (National Agency for regulation of pharmacists) in Georgia, and Roszdravnadzor (Federal Service for Surveillance in healthcare) in Russia. Collaboration among regulators, healthcare providers, and researchers is crucial to navigate and advance phage therapy in these countries

UK and elsewhere, now a days the awareness for phage therapy is also spreading and the need of harmonized GMP method for phage productions are into policy formulation stages. The supportive statuses with various regulatory bodies are tabulated below (Table 17.7).

17.3 Future Perspectives

17.3.1 Emerging Technologies and Strategies

Genetic engineering, chemical engineering, mechanobiology can help to understand and increase reliability upon the phage therapy. But in case the phage resistance becomes prevalently irreversible alike antibiotics. Hopefully we have sufficient time to research in these aspects considering the fate of the phages to still be used as natural product up to some more decades. The genetically mass produced endolysins, lysins, holins, and other enzymes may have fate like antibiotics, therefore we can expect other researchers to let natural evolution of phages yields the active enzymes responsible for lysis (Azimi et al. 2019).

17.3.2 Ethical Considerations

Balancing benefits and risks. The evolution can take place in dosage form/formulation since they are live biological agents unlike any other protein, attenuated vaccines, and other biologics, hence, to harmonize the uniformity of selection of phages against the specific pathogens the formulator should consider all aspects of preclinical, translational factors which can improve the formulation in terms of stability. Here stability in term of phage therapy means viability of the phages on intermediate and long-term storage.

17.3.3 Ghost Phage

In case of the phage formulation will fail to kill the mentioned strains, then we propose to take out the genetic material (single standard DNA) present inside the isolated and purified bacteriophage and replace it with the drugs. The hypothesis is that the phages with drugs will act as a carrier.

17.4 Conclusion

Phage therapy is a rising non-antibiotic approach against drug-resistant bacterial infections. After reviewing discussions and recent developments on research and regulatory advancements in phage therapy globally, we can conclude that phage therapy is an emerging field of research promising enhanced therapeutic efficacy and safety. Phage formulations can be scaled up to meet a larger number of patients suffering from drug-resistant bacterial infections and to improve quality of life.

References

Abedon ST, Thomas-Abedon C, Thomas A, Mazure H (2011) Bacteriophage prehistory: is or is not Hankin, 1896, a phage reference? Bacteriophage 1(3):174–178

Allue-Guardia A, Saranathan R, Chan J, Torrelles JB (2021) Mycobacteriophages as potential therapeutic agents against drug-resistant tuberculosis. Int J Mol Sci 22(2). https://doi.org/10.3390/ijms22020735

Azimi T, Mosadegh M, Nasiri MJ, Sabour S, Karimaei S, Nasser A (2019) Phage therapy as a renewed therapeutic approach to mycobacterial infections: a comprehensive review. Infect Drug Resist 12:2943–2959

Barr JJ, Auro R, Sam-Soon N, Kassegne S, Peters G, Bonilla N, Hatay M, Mourtada S, Bailey B, Youle M (2015) Subdiffusive motion of bacteriophage in mucosal surfaces increases the frequency of bacterial encounters. Proc Natl Acad Sci 112(44):13675–13680

Broxmeyer L, Sosnowska D, Miltner E, Chacon O, Wagner D, McGarvey J, Barletta RG, Bermudez LE (2002) Killing of Mycobacterium avium and Mycobacterium tuberculosis by a mycobacteriophage delivered by a nonvirulent mycobacterium: a model for phage therapy of intracellular bacterial pathogens. J Infect Dis 186(8):1155–1160. https://doi.org/10.1086/343812

Carmody LA, Gill JJ, Summer EJ, Sajjan US, Gonzalez CF, Young RF, LiPuma JJ (2010) Efficacy of bacteriophage therapy in a model of Burkholderia cenocepacia pulmonary infection. J Infect Dis 201(2):264–271

Carrigy AA (2017) Nutrient additions alter semi-arid grassland response to fire

Chhibber S, Kumari S (2012) Application of therapeutic phages in medicine. Bacteriophages:139–158

Chhibber S, Kaur S, Kumari S (2008) Therapeutic potential of bacteriophage in treating Klebsiella pneumoniae B5055-mediated lobar pneumonia in mice. J Med Microbiol 57(12):1508–1513

Danelishvili L, Young LS, Bermudez LE (2006) In vivo efficacy of phage therapy for Mycobacterium avium infection as delivered by a nonvirulent mycobacterium. Microbial drug resistance 12(1):1–6

Debarbieux L, Leduc D, Maura D, Morello E, Criscuolo A, Grossi O, Balloy V, Touqui L (2010) Bacteriophages can treat and prevent Pseudomonas aeruginosa lung infections. J Infect Dis 201(7):1096–1104

Diallo K, Dublanchet A (2022) Benefits of combined phage-antibiotic therapy for the control of antibiotic-resistant bacteria: a literature review. Antibiotics (Basel, Switzerland) 11(7):839–839

Dixit A, Kumar N, Kumar S, Trigun V (2019) Antimicrobial resistance: progress in the decade since emergence of New Delhi metallo-β-lactamase in India. Indian J Community Med 44(1):4

Dy RL, Richter C, Salmond GP, Fineran PC (2014) Remarkable mechanisms in microbes to resist phage infections. Ann Rev Virol 1:307–331

Elahi Y, Nowroozi J, Fard RMN (2021) Isolation and characterization of bacteriophages from wastewater sources on enterococcus spp. isolated from clinical samples. Iran J Microbiol 13(5):671

Fraga AG et al (2019) Antimicrobial activity of Mycobacteriophage D29 Lysin B during Mycobacterium ulcerans infection. PLoS neglected tropical diseases 13(8):e0007113

Furfaro LL, Payne MS, Chang BJ (2018) Bacteriophage therapy: clinical trials and regulatory hurdles. Front Cell Infect Microbiol 8:376

Gorski A, Jonczyk-Matysiak E, Lusiak-Szelachowska M, Miedzybrodzki R, Weber-Dabrowska B, Borysowski J (2018a) Bacteriophages targeting intestinal epithelial cells: a potential novel form of immunotherapy. Cell Mol Life Sci 75(4):589–595. https://doi.org/10.1007/s00018-017-2715-6

Gorski A, Jonczyk-Matysiak E, Miedzybrodzki R, Weber-Dabrowska B, Lusiak-Szelachowska M, Baginska N, Borysowski J, Lobocka MB, Wegrzyn A, Wegrzyn G (2018b) Phage therapy: beyond antibacterial action. Front Med (Lausanne) 5:146. https://doi.org/10.3389/fmed.2018.00146

Guo D, Chen J, Zhao X, Luo Y, Jin M, Fan F, Park C, Yang X, Sun C, Yan J (2021) Genetic and chemical engineering of phages for controlling multidrug-resistant bacteria. Antibiotics 10(2):202

Hashemi Shahraki A, Mirsaeidi M (2021) Phage therapy for mycobacterium abscessus and strategies to improve outcomes. Microorganisms 9(3):596

Heo Y-J, Lee Y-R, Jung H-H, Lee J, Ko G, Cho Y-H (2009) Antibacterial efficacy of phages against Pseudomonas aeruginosa infections in mice and Drosophila melanogaster. Antimicrob Agents Chemother 53(6):2469–2474

Hitchcock NM, Devequi Gomes Nunes D, Shiach J, Valeria Saraiva Hodel K, Dantas Viana Barbosa J, Alencar Pereira Rodrigues L, Coler BS, Botelho Pereira Soares M, Badaró R (2023) Current clinical landscape and global potential of bacteriophage therapy. Viruses 15(4):1020

Hurst-Hess K et al (2023) Intrapulmonary treatment with mycobacteriophage LysB rapidly reduces Mycobacterium abscessus burden. Antimicrobial Agents and Chemotherapy 67(6):e00162–e00123

Hyman P (2019) Phages for phage therapy: isolation, characterization, and host range breadth. Pharmaceuticals 12(1):35

Jun JW, Shin TH, Kim JH, Shin SP, Han JE, Heo GJ, De Zoysa M, Shin GW, Chai JY, Park SC (2014) Bacteriophage therapy of a Vibrio parahaemolyticus infection caused by a multiple-antibiotic–resistant O3: K6 pandemic clinical strain. J Infect Dis 210(1):72–78

Kakinuma Y, Kimura T, Watanabe Y (2017) Possible involvement of liver resident macrophages (Kupffer cells) in the pathogenesis of both intrahepatic and extrahepatic inflammation. Can J Gastroenterol Hepatol 2017:2896809. https://doi.org/10.1155/2017/2896809

Kim M, Jo Y, Hwang YJ, Hong HW, Hong SS, Park K, Myung H (2018) Phage-antibiotic synergy via delayed lysis. Appl Environ Microbiol 84(22):e02085–e02018

Lenev S (2013) Bacteriophages in chicken salmonellosis.. Bacteriophages in human and veterinary medicine

Letarov AV, Letarova MA (2023) The burden of survivors: how can phage infection impact non-infected bacteria? Int J Mol Sci 24(3):2733

Li P-Y, Meyer KE (2009) Contextualizing experience effects in international business: A study of ownership strategies. Journal of World Business 44(4):370–382

Liu K et al (2016) Inhalation study of mycobacteriophage D29 aerosol for mice by endotracheal route and nose-only exposure. Journal of aerosol medicine and pulmonary drug delivery 29(5):393–405

Loc-Carrillo C, Abedon ST (2011) Pros and cons of phage therapy. Bacteriophage 1(2):111–114

Loponte R, Pagnini U, Iovane G, Pisanelli G (2021) Phage therapy in veterinary medicine. Antibiotics 10(4):421

Macander-golcz V (1962) Trials of Treatment of Guineapigs inoculated with Mycobacterium tuberculosis by means of a Mycobacteriophage "Polonus"

Miedzybrodzki R, Switala-Jelen K, Fortuna W, Weber-Dabrowska B, Przerwa A, Lusiak-Szelachowska M, Dabrowska K, Kurzepa A, Boratynski J, Syper D, Pozniak G, Lugowski C, Gorski A (2008) Bacteriophage preparation inhibition of reactive oxygen species generation by endotoxin-stimulated polymorphonuclear leukocytes. Virus Res 131(2):233–242. https://doi.org/10.1016/j.virusres.2007.09.013

Miedzybrodzki R, Borysowski J, Weber-Dabrowska B, Fortuna W, Letkiewicz S, Szufnarowski K, Pawelczyk Z, Rogoz P, Klak M, Wojtasik E, Gorski A (2012) Clinical aspects of phage therapy. Adv Virus Res 83:73–121. https://doi.org/10.1016/B978-0-12-394438-2.00003-7

Nakai T, Sugimoto R, Park K-H, Matsuoka S, Mori K-I, Nishioka T, Maruyama K (1999) Protective effects of bacteriophage on experimental Lactococcus garvieae infection in yellowtail. Dis Aquat Org 37(1):33–41

Ochieng'Oduor JM, Onkoba N, Nyachieo A, Maloba F (2016) Experimental phage therapy against haematogenous multi-drug resistant Staphylococcus aureus pneumonia in mice. Afr J Lab Med 5(1):1–7

Oliveira H, Vilas Boas D, Mesnage S, Kluskens LD, Lavigne R, Sillankorva S, Secundo F, Azeredo J (2016) Structural and enzymatic characterization of ABgp46, a novel phage endolysin with broad anti-gram-negative bacterial activity. Front Microbiol 7:208

Oliveira H, São-José C, Azeredo J (2018) Phage-derived peptidoglycan degrading enzymes: challenges and future prospects for in vivo therapy. Viruses 10(6):292

Pouillot F, Chomton M, Blois H, Courroux C, Noelig J, Bidet P, Bingen E, Bonacorsi S (2012) Efficacy of bacteriophage therapy in experimental sepsis and meningitis caused by a clone O25b: H4-ST131 Escherichia coli strain producing CTX-M-15. Antimicrob Agents Chemother 56(7):3568–3575

Prestinaci F, Pezzotti P, Pantosti A (2015) Antimicrobial resistance: a global multifaceted phenomenon. Pathog Glob Health 109(7):309–318

Rahman MT, Sobur MA, Islam MS, Ievy S, Hossain MJ, El Zowalaty ME, Rahman AT, Ashour HM (2020) Zoonotic diseases: etiology, impact, and control. Microorganisms 8(9):1405

Rosner D, Clark J (2021) Formulations for bacteriophage therapy and the potential uses of immobilization. Pharmaceuticals 14(4):359

Schneider G, Szentes N, Horváth M, Dorn Á, Cox A, Nagy G, Doffkay Z, Maróti G, Rákhely G, Kovács T (2018) Kinetics of targeted phage rescue in a mouse model of systemic Escherichia coli K1. BioMed Res Int 2018:7569645

Shende R, Hirpurkar S, Sannat C, Rawat N, Pandey V (2017) Isolation and characterization of bacteriophages with lytic activity against common bacterial pathogens. Vet World 10(8):973

Stone E, Campbell K, Grant I, McAuliffe O (2019) Understanding and exploiting phage–host interactions. Viruses 11(6):567

Sugawara I, Yamada H, Hua S, Mizuno S (2001) Role of interleukin (IL)-1 type 1 receptor in mycobacterial infection. Microbiol Immunol 45(11):743–750. https://doi.org/10.1111/j.1348-0421.2001.tb01310.x

Sula L, Sulova J, Stolcpartova M (1981) Therapy of experimental tuberculosis in guinea pigs with mycobacterial phages DS-6A, GR-21 T, My-327. Czechoslovak Medicine 4(4):209–214

Taati Moghadam M, Amirmozafari N, Shariati A, Hallajzadeh M, Mirkalantari S, Khoshbayan A, Masjedian Jazi F (2020) How phages overcome the challenges of drug resistant bacteria in clinical infections. Infect Drug Resist 13:45–61

Takemura-Uchiyama I, Uchiyama J, Osanai M, Morimoto N, Asagiri T, Ujihara T, Daibata M, Sugiura T, Matsuzaki S (2014) Experimental phage therapy against lethal lung-derived septicemia caused by Staphylococcus aureus in mice. Microbes Infect 16(6):512–517

Trigo G et al (2013) Phage therapy is effective against infection by Mycobacterium ulcerans in a murine footpad model. PLoS neglected tropical diseases 7(4):e2183

Tyagi P, Dharmaraja AT, Bhaskar A, Chakrapani H, Singh A (2015) Mycobacterium tuberculosis has diminished capacity to counteract redox stress induced by elevated levels of endogenous superoxide. Free Radic Biol Med 84:344–354. https://doi.org/10.1016/j.freeradbiomed.2015.03.008

Uchiyama J, Takemura-Uchiyama I, Sakaguchi Y, Gamoh K, Kato S-I, Daibata M, Ujihara T, Misawa N, Matsuzaki S (2014) Intragenus generalized transduction in staphylococcus spp. by a novel giant phage. ISME J 8(9):1949–1952

Wahida A, Tang F, Barr JJ (2021) Rethinking phage-bacteria-eukaryotic relationships and their influence on human health. Cell Host Microbe 29(5):681–688

Wan X, Hendrix H, Skurnik M, Lavigne R (2021) Phage-based target discovery and its exploitation towards novel antibacterial molecules. Curr Opin Biotechnol 68:1–7. https://doi.org/10.1016/j.copbio.2020.08.015

Watanabe R, Matsumoto T, Sano G, Ishii Y, Tateda K, Sumiyama Y, Uchiyama J, Sakurai S, Matsuzaki S, Imai S (2007) Efficacy of bacteriophage therapy against gut-derived sepsis caused by Pseudomonas aeruginosa in mice. Antimicrob Agents Chemother 51(2):446–452

Yang Q, Le S, Zhu T, Wu N (2023) Regulations of phage therapy across the world. Front Microbiol 14:1250848

Yen M, Cairns LS, Camilli A (2017) A cocktail of three virulent bacteriophages prevents vibrio cholerae infection in animal models. Nat Commun 8(1):14187

Zalewska-Piątek B (2023) Phage therapy—challenges, opportunities and future prospects. Pharmaceuticals 16(12):1638

Zemskova ZS, Dorozhkova IR (1991) Pathomorphological assessment of the therapeutic effect of mycobacteriophages in tuberculosis. PRO Tuberk 11:63–66. (Patomorfologicheskaia otsenka lechebnogo deĭstviia mikobakteriofagov pri tuberkuleze)

Phage Therapy: Challenges in Its Development and Clinical Implementation

18

Vijaya Lakshmi Nag, Yashik Bansal, and Mahesh Jansari

Abstract

The use of bacteriophages for therapeutic purposes, termed phage therapy was met with inconsistent reports of treatment outcome. Antibiotics were discovered around the same time and were regarded as a magic bullet for the treatment of infectious diseases. This led to the loss of interest in phage therapy. Almost a century later, the non-judicious use of antibiotics has contributed to the rise of drug-resistant bacteria that show increasing resistance to the last line of antibiotics. Consequently, interest has been renewed in phage therapy to treat multidrug-resistant infections. The future of phage therapy looks more promising, as the trials have extended globally. The wider use of phage therapy, however, has to navigate numerous obstacles that are a hurdle to its successful implementation. These include the issues related to the product (phage selection, preparation, extreme specificity for host, emergence of resistance to phages, development of immune response against the phages, pharmacokinetics, and pharmacodynamics), patient (lesser awareness and hesitancy in accepting a live virus based treatment option for combating an infection) and regulatory issues (related to clinical trials, drug/phage manufacturing, licensing, and patents). This chapter discusses various challenges to the use of phage therapy.

V. L. Nag (✉)
Department of Microbiology, All India Institute of Medical Sciences, Jodhpur, Rajasthan, India

Y. Bansal
Department of Microbiology, Employees State Insurance Corporation Medical College Hospital, Alwar, Rajasthan, India

M. Jansari
Department of Pulmonary Medicine, All India Institute of Medical Sciences, Kalyani, West Bengal, India

M. Gangwar, G. Nath (eds.), *Emerging Paradigms for Antibiotic-Resistant Infections: Beyond the Pill*, https://doi.org/10.1007/978-981-97-5272-0_18

Keywords

Phage therapy · Bacteriophage therapy · Clinical trial · Antimicrobial resistance · Multidrug resistance

18.1 Introduction

Bacteriophage (or phage) discovery is credited to Felix d'Hercule and Frederick Twort along with d' Herelle who brought wider attention to their existence with his paper in 1917. Ever since their discovery at the beginning of the twentieth century, phages have been known to possess antibacterial therapeutic properties which clinicians and researchers have tried to utilize for the treatment of infectious diseases (Dublanchet and Bourne 2007; d'Herelle 1918), a decade from the discovery of Penicillin. However, its use declined with the invention of antibiotics. The limited use of phage therapy mainly in Georgia, Poland, and Russia (Myelnikov 2018) resulted in publications in the non-English language (Diallo and Dublanchet 2023) further keeping data away from the Western world.

The classical clinical approach during the early years of discovery was to utilize phage therapy for infections notably dysentery (Almeida and Sundberg 2020) and cholera (Summers 1993), skin infections (Crutchfield and Stout 1930), septicemia (MacNeal and Frisbee 1932) and gas gangrene (Chanishvili 2012) besides other conditions such as peritonitis, otitis externa, and surgical site infections (Wittebole et al. 2014). These studies were mainly non-randomized trials due to the evolving nature of biomedical research principles in those times, where a potential treatment could not be denied to a patient leading to issues with the quality of data generated (Suh et al. 2022).

Broadly, there are two types of phages (Campbell 2003) based on their biological activity, (i) *lytic phages* that infect a bacteria by attaching to its receptor, multiply and release numerous bacteriophages after lysis of the cell. (ii) *temperate phages* undergo a lysogenic cycle where the phage DNA gets integrated into the host bacteria's chromosome as a prophage (Fig. 18.1) (Lin et al. 2022). The phage DNA keeps on replicating along with the bacteria and the daughter cells carry this prophage until the lytic phase gets activated resulting in the release of newly assembled virions (Fortier and Sekulovic 2013). The phages of clinical interest are morphologically tailed viruses, have a double-stranded DNA as the nucleic acid material and are lytic in nature (Lin et al. 2017). Temperate phages, on the other hand, can be modified by genetic engineering to become lytic or therapeutically more useful (Monteiro et al. 2019).

It is estimated that the number of phages in nature ranges up to 10^{31}–10^{32} (Balcazar 2014) outnumbering bacteria by almost 10 times (Batinovic et al. 2019), making them the most abundant biological entity on our planet. So, theoretically, it should be possible to isolate phages against most bacteria causing infections in humans. An accurate classification of the phages is important along with building a data bank on the genome of these phages for their effective utilization in phage

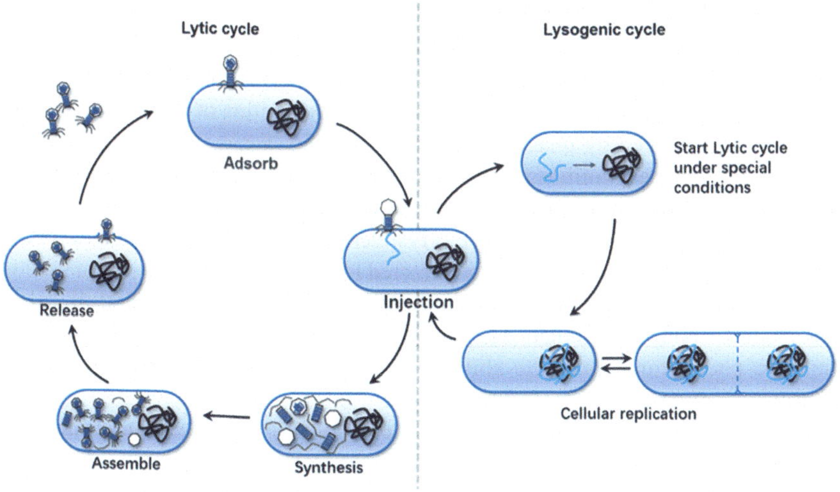

Fig. 18.1 Comparison of lytic and lysogenic life cycle of phages. From Lin et al. 2022, used under Creative Commons CC-BY license

therapy. This has been achieved (in the last decade) by using genome sequencing methods and by updating the phage taxonomy. Traditionally, the taxonomy of phages was based on their morphology and at least 10 types of phages were identified on these grounds (Zhu et al. 2022; Petrovic Fabijan et al. 2023). These include the tailed, DNA phages Caudovirales which constitute more than 95% of the known phages (Ackermann 2011). The families include myoviridae, siphoviridae, podoviridae, microviridae, corticoviridae, corticoviridae, leviviridae, cystoviridae, inoviridae, and plasmaviridae (Ackermann 2011). The International Committee on Taxonomy of Viruses has recently updated the phage taxonomy by eliminating some families and orders based on phage morphology and adding their members to class Caudoviricetes, which now comprises of majority of the phages used in therapy (Hitchcock et al. 2023; Turner et al. 2021).

Rising interest in phage therapy has been witnessed due to alarming bacterial drug resistance (Karn et al. 2023). It is estimated that 700,000 people die annually as a consequence of antimicrobial resistance (AMR). This death toll is projected to reach ten million by 2050, and the United Nations has therefore deemed AMR as the greatest and most urgent global risk (Laxminarayan et al. 2013). In the last 3 years, many clinical trials have been registered to ascertain their clinical utility and safety in the treatment of life-threatening multidrug-resistant (MDR) infections. In contrast to chemical therapeutic agents, the pharmacokinetics of a biological, live product is different. The phages are amenable to amplification in the human body. It can interact with antibiotics in a favorable or an unfavorable manner. It could carry virulence factors between different organisms via transduction or can be suppressed by the immune system of the humans. This chapter discusses these challenges that would ultimately seal the fate of this promising therapy in the coming times.

18.2 Mechanism of Action

The antibacterial properties of phages are exerted by two predominant mechanisms: (i) breakdown of the bacterial cells directly by the lytic action of the phages and (ii) the enzymes produced by phages that possess activity against bacteria as well as their products such as biofilms (Wang et al. 2024). The lytic phages are capable of instant killing of the organisms whereas the lysogenic phages exhibit delayed killing depending upon the duration of the lysogeny. Moreover, the lysogenic phase imparts superinfection immunity to the infected bacteria that protects it from getting infected by another phage (Bondy-Denomy et al. 2016; Mavrich and Hatfull 2019). Thus, even in phage combinations, known as phage cocktails, the presence of a temperate phage can impair the ability of other phages in instant killing of the bacteria.

Phages need to establish contact with a bacterial cell for attachment in order to inject its nucleic acid inside the cells (Abedon 2023a). This happens via the receptors located on the bacteria, In many scenarios, bacteria secrete an extracellular matrix around them that organizes to form a biofilm (Di Martino 2018; Karygianni et al. 2020). This biofilm protects the bacteria from antibacterial agents, including antimicrobials and phages (Visnapuu et al. 2022). Phages secrete a variety of enzymes to counteract them, such as polysaccharide depolymerases and peptidoglycan hydrolases. Polysaccharide depolymerases are capable of breaking down the bacterial capsule (or slime layer), lipopolysaccharide and biofilms (Hughes et al. 1998; Liu et al. 2022) and peptidoglycan hydrolases damage the bacterial cell wall (Oliveira et al. 2018).

The depolymerases produced by the phages have multiple clinical utility. It can be used as adjuvant therapy to augment innate immune responses, as co-therapy with antibiotics, to prevent or reduce the anti-phage resistance and as anti-biofilm therapeutic agents (Wang et al. 2024). Their safety and efficacy are well documented in pre-clinical studies. Clinical trials under research may provide its clinical utility in the coming future.

18.3 Present Status

With the emergence of MDROs and pan-drug resistant organisms causing infections in humans, the renewed interest in phage therapy has led to numerous studies and trials beyond Eastern Europe to the US, Western Europe, Canada, Australia, and Asia.

18.3.1 Pre-Clinical Development

Pre-clinical trials have been conducted in different animal models such as mice, rat, dog, pig, mink, sheep, rabbit, and *Drosophila melanogaster*. The commonest animal models used in these trials are the BALB/c mice and Wistar rats (Melo et al.

2020; Mehmood Khan et al. 2023). The pre-clinical trials aimed at studying the utility of phage therapy and its efficacy and safety in numerous clinical scenarios involving infections of the skin and soft tissue (wounds, diabetic foot ulcers, burns and postoperative wounds), gastrointestinal tract, respiratory system (both upper as well as lower respiratory infections), eye (keratitis), ear (otitis externa, otitis media), dental infections of the root canal, urinary tract infections and bacteremia (Melo et al. 2020).

A majority of the studies have been conducted against *Pseudomonas aeruginosa*, *Staphylococcus aureus*, *Klebsiella pneumoniae*, *Acinetobacter baumannii,* and *Escherichia coli*. Other organisms targeted during the pre-clinical trials include *Enterococcus faecalis*, *Vibrio cholera*, *Vibrio parahemolyticus*, *Salmonella enterica*, *Salmonella enteritidis*, *Clostridioides difficile*, and *Mycobacterium ulcerans* (Melo et al. 2020; Abdelkader et al. 2019).

Different approaches were studied in the pre-clinical phase that included monotherapy with a single phage preparation, a cocktail of multiple phages, and combination therapy of phages and antibiotics. In the majority of the studies, a phage cocktail was more effective in combating infection as compared to preparations containing single phages (Nale and Clokie 2021). Phage therapy in combination with antibiotics also showed excellent results (Gordillo Altamirano et al. 2022; Li et al. 2021; Diallo and Dublanchet 2022) except for a study on the efficacy of a combination of phage therapy with clindamycin for the treatment of *Staphylococcus aureus* pneumonia in mice (Melo et al. 2020; Oduor et al. 2016). The combination therapy did not perform better than the phage therapy which was attributed to the antagonistic effect of clindamycin exerted due to its bacteriostatic activity owing to inhibition of protein synthesis, that interfered with the action of phages.

The routes and dosing of the phages were also studied and it has been consistently observed that phage therapy performed better when the phages were delivered through the intraperitoneal route compared to other routes such as intramuscular, intravenous, or topical route (Malik et al. 2017). An additional advantage of the intraperitoneal route is that the phages get secreted into the urine and therefore, this approach works for the treatment of urinary tract infections as well (Melo et al. 2020; Malik et al. 2017). In contrast, respiratory tract infections were best treated through the intranasal route or aerosol dissemination (Semler et al. 2014) whereas others report better outcomes with intraperitoneal route (Malik et al. 2017). Phage dose is calculated as multiplicity of infection (MOI) which is defined as the ratio of infecting agent (phage in the present context) to target cell (bacterial cells in the present context) (Nang et al. 2023). A dose of MOI of 10 or more (up to 100–200) was reported to be associated with better outcomes as compared to doses with a MOI of 1 or less (Melo et al. 2020).

18.3.2 Clinical Trials

At present, much of the literature on phage therapy comprises qualitative data in the form of case reports and case series. The renewed interest in phage therapy, especially in the Western counties, saw the beginning of clinical trials in the last two decades that picked up pace in the last 4–5 years. A search for registered phage therapy trials was conducted on two sites from US and India. At the National Center for Biotechnology Information (NCBI) managed website ClinicalTrials.gov, a search for "bacteriophage therapy OR phage therapy" yielded a total of 38 results out of which two results were excluded as they were studying the utility of phages in non-infectious conditions. Table 18.1 shows the summary of the 36 studies registered from 2005 to 2024 so far (ClinicalTrials.gov 2024a, b, c, d, e).

Twenty two studies have been registered since 2022 indicating increasing interest towards phage therapy. Out of 36 studies, 28 are interventional studies while the rest are either observational or expanded access studies. Among interventional studies, three are in phase I, thirteen in phase I/II, six in phase II, three in phase III trial, whereas no information is available on the remaining three studies. Thirteen studies out of total 36 are in recruitment phase, three are active but not recruiting participants/subjects, three are not yet recruiting participants/subjects, one study is enrolling by invitation, five studies have been completed, the status of four studies is unknown, three are no longer available, two studies were terminated whereas two have been withdrawn. The reason for study termination includes no benefit of oral phages compared to the standard of care in diarrhea patients (study NCT00937274) and changes in the development strategy (study NCT04287478). Study NCT04787250 was withdrawn due to changes in study protocol while study NCT05269134 was withdrawn because of the sponsor decision. A similar search on the Clinical Trials Registry of India (CTRI) portal (CTRI 2024; Patel et al. 2021) showed that four phage therapy trials are registered from 2018 so far as summarized in Table 18.2.

18.3.3 Therapeutic Use in Clinical Practice

Even though the phage therapy was largely abandoned by the western countries after the success of antimicrobials for the treatment of infectious diseases, phage therapy was continued in the erstwhile Soviet Union and its successor states (Myelnikov 2018). To this date, phage therapy has especially been practiced in Poland and Georgia, the latter also helping international patients seek phage therapy (McCallin et al. 2019).

The clinical use of this therapy has largely been based on the delivery of a tailored/personalized phage therapy using customized phages based on the species and/or strain of the causative bacteria. Such use is approved as a compassionate use in serious or life-threatening infections with no other treatment option (Abedon et al. 2017). It is pertinent to note that during such emergency use, a faster administrative system is required to save precious time. The principles of ethics laid down in the Declaration of Helsinki guide the protocols of delivering treatment to the patient being treated on compassionate grounds (Pirnay and Verbeken 2023; Yang et al. 2023). Of note, the potential of phage therapy in the treatment of drug-resistant tuberculosis (TB) deserves a special mention here since TB has been one of the

Table 18.1 Clinical trials of phage therapy registered on ClinicalTrials.gov website (ClinicalTrials.gov 2024a, b, c, d, e)

Year	NCT number	Concise study title (acronym)	Phage name	Phase	Sample size	Status	Location
2005	NCT00945087	Experimental phage therapy of bacterial infections		NA		UNK	Poland
2006	NCT00663091	Safety and efficacy of WPP-201 treatment of venous leg ulcers	WPP-201	I	64	C	USA
2008	NCT05498363	Bacteriophage therapy of difficult-to-treat infections (BT100 study)[a]			100	C	Belgium
2009	NCT00937274	Antibacterial treatment against diarrhea in ORS	T4 phage	NA	120	T	Bangladesh
2015	NCT02116010	Phage therapy for the treatment of *Escherichia coli* and *Pseudomonas aeruginosa* wound infections in burned patients (PHAGOBURN)		I/II	–	UNK	Belgium, France, Switzerland
2018	NCT03395743	Individual patient expanded access for AB-PA01, an investigational anti-*Pseudomonas aeruginosa* bacteriophage therapeutic[a]	AB-PA01	NA	–	NLA	–
2018	NCT03395769	Individual patient expanded access for AB-SA01, an investigational anti-*Staphylococcus aureus* bacteriophage therapeutic[a]	AB-SA01	NA	–	NLA	–
2020	NCT04287478	Bacteriophage therapy in patients with UTI		I/II	1	T	USA
2020	NCT04636554	Personalized phage treatment in Covid-19 patients with bacterial co-infections (pneumonia or bacteremia/septicemia)[a]		NA	–	NLA	–

(continued)

Table 18.1 (continued)

Year	NCT number	Concise study title (acronym)	Phage name	Phase	Sample size	Status	Location
2020	NCT04682964	Bacteriophage therapy in tonsillitis		III	128	ANR	Uzbekistan
2020	NCT04596319	Safety and tolerability of inhaled AP-PA02 in subjects with chronic *P. aeruginosa* lung infections and cystic fibrosis (SWARM-Pa)[b]	AP-PA02	I/II	29	C	USA
2021	NCT05177107	Bacteriophage therapy in patients with diabetic foot osteomyelitis (DANCE study)		II	126	R	USA
2021	NCT04684641	CYstic fibrosis bacterioPHage study at yale (CYPHY)[b]	YPT-01	I/II	8	C	USA
2021	NCT04803708	Bacteriophage therapy TP-102 in diabetic foot ulcers (REVERSE)	TP-102	I/II	20	C	Israel
2022	NCT05269121	Bacteriophage therapy in first time chronic PJI (ACTIVE1 study)		I/II	20	NR	USA
2022	NCT02664740	Phage therapy versus placebo for diabetic foot ulcers infected by *S. aureus* (PhagoPied study)		I/II	60	NR	France
2022	NCT05369104	Phage therapy in PJI due to *Staphylococcus aureus* treated with DAIR (PhagoDAIRI study)		II	64	R	France
2022	NCT04787250	Bacteriophage therapy in patients with PJI		I/II	0	W	USA
2022	NCT04323475	Phage therapy for the prevention and treatment of wound infections in burned patients		I	12	UNK	–
2022	NCT04650607	Phage safety cohort study (PHA-SA-CO)[a]		NA	100	R	France

2022	NCT05010577	Nebulized bacteriophage therapy in cystic fibrosis patients with chronic *Pseudomonas aeruginosa* pulmonary infection	BX004-A	I/II	32	ANR	USA, Prague Czechia, Israel, Netherlands, Spain
2022	NCT04815798	Phage therapy for the prevention and treatment of pressure ulcers		I/II	69	UNK	USA
2022	NCT05453578	Safety and microbiological activity of bacteriophage therapy in cystic fibrosis subjects colonized with *Pseudomonas aeruginosa*	WRAIR-PAM-CF1	I/II	72	R	USA
2022	NCT05184764	Safety, tolerability, and efficacy of intravenous AP-SA02 in subjects with *S. aureus* bacteremia (diSArm)	AP-SA02	I/II	50	R	USA, Australia
2022	NCT05488340	LBP-EC01 in the treatment of acute uncomplicated UTI caused by drug resistant *E. coli* (ELIMINATE Trial)	LBP-EC01	II	318	R	USA
2023	NCT05537519	Phage therapy for the treatment of UTI		I/II	1	ANR	Canada
2023	NCT05973721	Efficacy and safety of phage therapy for chronic constipation	Pib specific Phage	NA	50	R	China
2023	NCT05269134	Bacteriophage therapy in patients with PJI (ACTIVE2)		II	0	W	USA
2023	NCT05883995	Avis-PHAGEinLYON clinic[a]		NA	1500	R	France
2023	NCT05967130	Treatment chronic UTI post kidney transplant		III	20	R	Iran
2023	NCT05948592	Bacteriophage therapy TP-102 in patients with diabetic foot infection (REVERSE2)	TP-102	II	80	R	USA

(continued)

Table 18.1 (continued)

Year	NCT number	Concise study title (acronym)	Phage name	Phase	Sample size	Status	Location
2023	NCT05616221	Safety, phage kinetics, and efficacy of inhaled AP-PA02 in subjects with non-cystic fibrosis bronchiectasis and chronic pulmonary *Pseudomonas aeruginosa* infection (Tailwind)	AP-PA02	II	60	R	USA
2023	NCT06185920	PHAGEinLYON Clinic Cohort study: a descriptive study of severe infections treated with phage therapy at the HCL[a]		NA	250	R	France
2023	NCT05272579	PrePhage—faecal bacteriophage transfer for enhanced gastrointestinal tract maturation in preterm infants	Fecal filtrate transfer	I (Early)	20	R	Denmark
2024	NCT06262282	Mycobacteriophage treatment of non-tuberculosis mycobacteria (POSTSTAMP)[a]		NA	10	EI	USA
2024	NCT05590195	Effect of PreforProÂ on urinary and vaginal health	Preforpro	III	50	NR	Canada

ORS oral rehydration solution, *UTI* urinary tract infection, *PJI* prosthetic joint infection, *NA* not available, *USA* United States of America, *R* recruiting, *EI* enrolment by invitation, *T* terminated, *NR* not yet recruiting, *C* completed, *ANR* active, not recruiting, *NLA* no longer available, *UNK* unknown status, *W* withdrawn

[a] Indicates observational or expanded access studies (studies other than interventional studies)

[b] Studies with results

Table 18.2 Clinical trials of phage therapy registered on Clinical Trials Registry of India (CTRI 2024)

Year	CTRI number	Concise study title (acronym)	Phage	Phase	Sample size	Status
2018	CTRI/2018/07/015130	Cure of non-healing ulcer in Bacteriophage therapy[a]	Monophage from environment	NA	48	C
2019	CTRI/2019/04/018476	Removal of harmful bacteria from nose[b]	–	NA	30	NR
2021	CTRI/2021/12/038527	Chronic non-healing ulcer and bacteriophage therapy	Customized phages	II	46	NR
2023	CTRI/2023/11/059899	Study to assess the effectiveness and safety of Bacteriophage in the treatment of diabetic foot infection	TP-102	II	96	NR

NR not yet recruiting, *C* completed, *CTRI* Clinical Trials Registry of India
[a] Studies with results. See reference Patel et al. 2021
[b] Indicates observational or expanded access studies (studies other than interventional studies)

leading causes of death globally. At least 23 cases of treatment of Mycobacterium abscessus in cystic fibrosis and bronchiectasis patients have been described by Ouyang et al. (2023) in which 13 improved clinically, whereas 5 failed to show improvement. The remaining five had an inconclusive outcome.

The phages used in such type of tailored treatment are known as magistral phages, so these phage preparations need to follow the regulatory guidelines for therapeutic medications prepared in the pharmacies and clinics, rather than Good Manufacturing Practices (GMP) applicable on manufacture of drugs at commercial scale, except European countries such as Germany, Switzerland (Pirnay and Verbeken 2023) and also Georgia in the near future shall consider implementing GMP requirements on phage therapy (Yang et al. 2023). In the United States, such emergency use expanded access is provided by the Food and Drug Administration, through a special access scheme by the Therapeutic Goods Administration in Australia, Agence Nationale de Sécurité du Médicaments et les Produits de Santé (L'ANSM) in France, through the relevant health ministry in Israel and Georgia and through a National framework or act in Poland and magistral phages monograph in Belgium (Pirnay and Verbeken 2023).

18.4 Challenges in Development and Clinical Implementation

Randomized clinical trials will solve the mystery of efficacy and safety related to phage therapy. It is not easy to have robust research within a short time to address multiple facets of phage therapy as many challenges exist. These challenges are related to the product (phages) itself, the human host or the administrative/regulatory bodies.

18.4.1 Product Related

18.4.1.1 Limited Spectrum of Activity

For its success, phage therapy needs to cater to a variety of infectious diseases caused by bacteria. Considering the fact that phages demonstrate specificity to particular bacteria, it has always been acknowledged that a large collection of phage library needs to be built to be an effective treatment of choice in clinical practice. However, data has shown extreme selection for bacterial strains, with many phages infecting as little as 1–10% of the strains within a given species (Chen et al. 2019; Hyman 2019; Leptihn and Loh 2022). Leptihn and Loh (2022) have estimated that assuming a phage action on 1% isolates of a bacterial species along with a chance of bacterial resistance to the phages, a phage therapy center would require at least 300–600 phages per species. The notorious drug-resistant organisms, collectively labeled as ESKAPE pathogens would thus require more than 3500 phages.

The challenges posed by the limited spectrum of phages have two important clinical implications: (i) phage therapy cannot be easily used as an empiric therapy, (ii) the identification of the causative organisms in a diagnostic microbiology laboratory would need to be precise in order to offer a tailored phage therapy to the patient. To overcome these issues, a phage cocktail can prove beneficial where multiple phages are utilized for the treatment of an infection. The development of such cocktails can be aimed at a specific disease type (McCallin et al. 2013; Abedon et al. 2011) rather than individual organism type or prioritized towards few MDR pathogens that are causing majority of the multidrug resistant infections in the community.

In order to develop a catalog of phages and select those with desirable properties, i.e., lytic action and known spectrum of activity, genomic sequencing is being done to build the bacteriophage genome data in NCBI's GenBank with more than 44,000 genomes being available till date (Hitchcock et al. 2023). Analysis of these genomes shows that majority of these phages are aimed at common pathogens such as *Staphylococcus*, *Streptococcus*, *Escherichia*, *Pseudomonas*, *Salmonella*, *Mycobacterium*, *Lactococcus* and *Gordonia* (Hitchcock et al. 2023). A fairly large library shall be able to overcome the issue of limited spectrum of phage action. Another method of overcoming this challenge is to employ genetic engineering to modify the host range of phages (Jia et al. 2023). As part of evolutionary changes in the phages, this host range is liable to changes in the nature (Yehl et al. 2019) as well as in the laboratory through phage training (Borges 2021).

18.4.1.2 Emergence of Anti-Phage Resistance in Bacteria

The stress exerted from the environment induces genetic modifications in every entity to impart survival benefits. Bacteria undergo such changes that manifest in the form of resistance to therapy. This resistance, though expressed after treatment with a new drug was started, could have been present before the exposure to the drug or could have been acquired after the said exposure. Similar to antimicrobial resistance, the resistance-related genes could be transferred horizontally between the bacteria as well as vertically during multiplication by binary fission.

Resistance to phages in the bacteria is generally mediated by one or more of the following three mechanisms (Egido et al. 2022):

(i) alterations in the receptors for phage binding,
(ii) active mechanisms directed against the infecting phages,
(iii) superinfection immunity: infection by a temperate phage imparting resistance to other phages.

The alteration in the receptors is a continuous process in nature and it includes either the loss of the receptor, its masking or a structural change in the receptor leading to non-recognition by the phage (Wittebole et al. 2014). These structural changes are brought upon by mutations in the receptor genes. Masking of the receptors can be achieved through the production of an extracellular matrix over the cells, by producing proteins or binding the receptors with lipoproteins during cell lysis to stop further lysis of the bacteria (Decker et al. 1994). By altering the gene expression, the downregulation of the receptors is another mechanism of reducing the adsorption of the phages onto the bacterial cell (Gencay et al. 2018).

Active mechanisms against the attacking phages include the enzyme restriction endonucleases (R-M system) that help in cleavage of the foreign DNA by the endonucleases but protect the bacterial DNA by performing its methylation (Birkholz et al. 2022). A more elaborate mechanism involves the CRISPR-Cas system where the clustered regularly interspaced short palindromic repeats and associated proteins (CRISPR) array integrates the phage DNA in the form of a spacer, producing its complementary RNA (cRNA) sequences that combine with the cas proteins and then scan the integrated spacer in the CRISPR array to identify the complementary sequences, which are then cleaved to ultimately destroy the phage DNA (Dupuis et al. 2013). Other mechanisms include aborting the infection by the phages by undergoing critical changes such as bacterial cell death, arrest of the protein maturation by the bacteria, depletion of cellular enzymes and nucleotides required for the replication of DNA by the phages, or production of chemicals that act as DNA intercalating agents and prevent circularization of the phage DNA, thereby interfering with the phage DNA replication process (Egido et al. 2022; Kronheim et al. 2018).

Phages also contribute to the resistance against other phages by imparting superinfection immunity to the bacteria, achieved through production of masking proteins that mask the receptors from other phages (Pedruzzi et al. 1998), formation of membrane-associated proteins to block channels allowing entry of phage DNA through the receptors (Labrie et al. 2010), or attacking the DNA of the other phages by using a R-M system described above or using repressor proteins that binds to the incoming new phage DNA and arrests its replication process (Pope et al. 2011).

It has been observed that mutations conferring resistance against phage infection described above, are fairly common in vitro, occurring at a rate of approximately frequency of around 10^{-5} to 10^{-7} but are relatively less common in vivo (Wittebole et al. 2014). It has also been reported that these mutations are often associated with other disadvantages for the bacteria such as loss or decrease in virulence, a decrease in their rate of growth or increase/return of sensitivity towards antimicrobials (Abedon 2023b). Resistance can also be countered by the use of cocktails with more than one phage type against the bacterial isolate, that possess ability to infect the bacterial cells through different receptors (Oechslin 2018).

18.4.1.3 Negative Interactions with Antibiotics

Phages have often been co-administered along with antibiotics. This can have a better treatment outcome by the synergistic action of the antibiotic along with phages (Li et al. 2021). This synergism is best observed at low doses of the antibiotics that are otherwise sub-inhibitory. This phenomenon is known as phage antibiotic synergy (Gu Liu et al. 2020). Various mechanisms of this synergy are postulated that include alteration in the cell size by beta-lactam antibiotics that damage the cell wall responsible for imparting rigidity and shape to the cell, decrease in the mutants imparting property of resistance against either antibiotics or the phages when used alone, increase in susceptibility of the bacteria in the presence of combination therapy, the reduced minimum inhibitory concentration of the antibiotic in the presence of phages and/or the release of depolymerases by the phages resulting in increased diffusion of the antibiotic inside the bacterial cell (Łusiak-Szelachowska et al. 2022).

The simultaneous exposure of antibiotics can also interfere with the phage action due to reduction in the number of bacterial cells available for amplification of the phages (Abedon 2022). Proliferation threshold is the minimum required density of bacterial cells to support the replication and of the phages resulting in amplification (Payne and Jansen 2002; Kasman et al. 2002). If the antibiotic kills and reduces the number of bacterial cells below this threshold before phages could be amplified, the result of this interaction will be detrimental to the aim of treating the infection (Łusiak-Szelachowska et al. 2022). The antagonistic effect is also seen with antibiotics that interfere with the DNA replication and protein synthesis such as vancomycin or gentamicin combination with phage for the treatment of *S aureus*, tobramycin and phage combination for the treatment of *P aeruginosa*, nalidixic acid, novobiocin, and rifampicin (Łusiak-Szelachowska et al. 2022; Chaudhry et al. 2017).

18.4.1.4 Phages as Vectors for Transmission of Bacterial Genes

Phages can mediate the movement of non-phage DNA from one bacteria to another by a process known as transduction (Doub 2021). Such type of DNA transfer can occur, in theory, (i) between the bacteria used as vectors during phage manufacturing, (ii) from targeted bacteria to other non-targeted bacteria in the humans and (iii) from non-targeted bacteria inside human body to targeted bacteria. In real conditions, such movement is restricted when phages with a very limited host range or a very narrow spectrum are used (PetrovicFabijan et al. 2023). Nevertheless, it is an important concern owing to the fact that such movement of non-phage DNA can help in transfer of virulence factors and resistance related genes/plasmids from one bacteria to another (Penadés et al. 2015; Colavecchio et al. 2017).

This is considered to be a lesser issue as transduction is not an in vivo phenomenon seen during the therapeutic application of phages but is a general process occurring commonly due to the abundance of temperate phages in the nature (PetrovicFabijan et al. 2023). So, therapeutic application is not likely to be an important contributor towards transfer of such genes containing virulence factors and resistance genes. Also, phages used in therapy are generally lytic phages and temperate or transducing phages are altogether avoided in clinical use (Hibstu et al. 2022).

18.4.2 Human Host Related

18.4.2.1 Immune Response Against Phages

Phages, just like any other micro-organism are recognized by the immune system as foreign antigens and a humoral immune response is mounted against the phages by the body (Van Belleghem et al. 2018). Immune response is generally a trivial concern when phage therapy is given for a short duration using a topical application for skin infections or when the patient is immunosuppressed. However, phages administered through other routes such as the intravenous route are more likely to experience an immune response from the human body (Hatfull et al. 2022).

The immune response generally seen in patients comprises of humoral immune response where antibodies that could be neutralizing or non-neutralizing (Górski et al. 2012) are formed in varying proportions and concentrations, depending upon the type of phage, the purity/quality of the preparation, and dose. Treatment failures due to neutralizing antibody response against phages have been observed and reported in the literature (Dedrick et al. 2021). Other studies have reported favorable outcomes even in the presence of antibody formation against the phages (Żaczek et al. 2016). Phages being present in the gastrointestinal tract, are tolerated well immunologically when administered through the oral route. Lastly, the bacterial cell lysis by the phages in case of treatment of infections with Gram negative bacteria can release endotoxins (Liu et al. 2021) which are capable of generating immune response through an inflammatory response.

18.4.2.2 Pharmacokinetics

Successful treatment outcomes with phage therapy are dependent upon similar pharmacokinetic principles as those seen with antibiotics, such as maximum phage concentration achieved at the target site, time for which such therapeutic concentrations are maintained, area under the curve, minimum numbers of phages required for successful bacterial lysis (akin to the minimum inhibitory concentration) and the dosage of phages in the preparation. Much of the factors related to these pharmacokinetic aspects are not adequately understood and are either under investigation or need more consideration (Petrovic Fabijan et al. 2023).

Indeed, phage therapy trials have failed in the past owing to these pharmacokinetics-related challenges such as low titers of phages used in the trial, insufficient coverage of phage spectrum or low numbers of bacteria at the target site that resulted in poor amplification of the phages during the trial (Górski et al. 2020). As far as the dosing is concerned, phage therapy has been associated with unfavorable outcomes when phages were used in a dose of 10^2–10^5 PFU/mL (PFU = plaque forming unit). A phage preparation with a titer as high as 10^9 to 10^{10} has been used without safety issue that was associated with a favorable outcome whereas Aslam et al. have reported side effects such as wheezing, fever, and dyspnea in a patient under phage therapy against *P aeruginosa* infection when a phage cocktail at a titer of 10^{11} PFU/mL was administered (Aslam et al. 2020).

Phage concentrations are also likely to decrease over time and this rate of elimination of the phage could vary depending upon the route of administration of the phages (Dąbrowska 2019). Therefore, it is important to standardize and document the phage dosing and frequency based on the route of administration, the type of bacteria and infection being treated, the phage concentrations achieved in the serum and phage auto dosing due to in situ amplification (PetrovicFabijan et al. 2023). This is very challenging due to vast number of phages that are available for therapeutic purposes.

18.4.3 Other Challenges

18.4.3.1 Regulatory Framework

The regulations around phage therapy are complicated, and not yet refined. These range from allowing commercial phage cocktails to be bought over the counter to limiting their use as compassionate care in emergency conditions using personalized phages. Countries like Russia and Georgia allow the purchase of certain phage cocktails over the counter without even the requirement of a medical prescription (Yang et al. 2023). But, in contrast to Georgia and Poland, Russia does not allow the use of personalized phage therapy and only licensed cocktails are allowed to be used. These countries identify phage therapy as a pharmaceutical/medicinal product and their respective Ministries govern the licensing of phages for clinical use.

The regulatory framework at present is designed as per three scenarios where phages are used: (i) personalized phage therapy given to an individual patient, (ii) licensed phage cocktail preparations, including phage products used over the counter and (iii) clinical trials to assess the safety and efficacy of phages in clinical care. The pioneering work on personalized phage therapy regulation has been done by Belgium that has a regulation on the use of magistral phages. These personalized phages are allowed to be prepared under the responsibility of the clinician and pharmacist. Most Western European countries recognize phages as medicinal products and their use is governed by Good Manufacturing Practices (GMP), clinical trials and licensing for authorization for use on commercial scale (Bretaudeau et al. 2020). United Kingdom, however, allows the use of phages produced with GMP certificate to be imported for unlicensed use (Yang et al. 2023).

The regulations surrounding the use of phage therapy are constantly evolving as new evidence is generated in clinical trials and are expected to become more stringent with the advancement of clinical trials. Bretaudeau et al. (2020) recommend a slightly relaxed and less stringent policy for exotic phages meant for a limited number of patients and that the stringent GMP-related regulations are more appropriate for large-scale phage production.

18.4.3.2 Funding and Investment

In contrast to antibiotics, in which the development and patenting of drugs have been relatively straightforward, the development of phage therapy could face more challenges. Phages exist everywhere and a bespoke approach of phage therapy coupled with the ubiquitous availability of phages in the nature could discourage large investments in the research and development of phage therapy from businesses, for whom the incentives could be less lucrative, putting greater onus of funding acquisition on the public sector/governments. Yang et al. (2023) report a total of nine investigator-initiated trials from 2005 to 2021 that set up the foundation of reports of adequate safety of the therapy. As of today, the early developments are promising as many biotech and pharmaceutical companies have shown promising participation in the development of commercial phage therapy.

18.4.3.3 Acceptance of Phages (Viruses) as Treatment of Bacterial Infections

Phages are essentially viruses and due to large variations in the social economic profiles of the communities around the world, the level of education and awareness among the general public could impact the acceptance of a therapeutic option based on delivering single or multiple live viruses (phage cocktails) inside the body to fight a bacteria, especially in times when viruses have been the causative agents of many outbreaks and pandemics in the recent years (Garcia-Blanco et al. 2022). Also, the phage therapy has been around for more than a century with limited application till date. This could spark skepticism against the treatment option as well (Petrovic Fabijan et al. 2023; Summers 2012). Educational and awareness initiatives are appropriate tools for countering these challenges given the fact that humans are routinely exposed to phages on a daily basis and they are one of the commonest viruses in the human microbiome (Łusiak-Szelachowska et al. 2020).

18.5 Chapter Summary

Phage therapy has regained interest after a century of abandonment due to rising antimicrobial resistance and the preliminary data on safety and tolerance from continued use in some countries is encouraging to set up randomized, double-blind controlled clinical trials. Application of GMP can improve the quality of phage preparations so that clinical trials do not suffer from lower quality materials. Phage therapy offers numerous benefits over conventional antibiotics, the biggest being their non-action on the commensal flora/microbiome in the body. The highly specific attack on a particular strain of a bacterial species warrants the necessity of personalized phage therapy in clinical practice and their efficacy can be augmented by combining phages into a cocktail or administering them as a co-therapy with low dose antibiotics for synergy or phage enzymes (depolymerases).

The development of Phage therapy is riddled with unanswered potential roadblocks such as immune response against phage therapy by human body, development of anti-phage resistance, transduction, evolving phage regulations, the varying pharmacokinetics, and the acceptance of phage therapy by the patients. Many clinical trials have been registered in the last 2–3 years, with a few trials moving into phase III indicating a likelihood that phage therapy could be a reality in the next 4–5 years.Conflict of InterestThe authors declare that they have no conflicts of interest.

References

Abdelkader K, Gerstmans H, Saafan A, Dishisha T, Briers Y (2019) The preclinical and clinical progress of bacteriophages and their lytic enzymes: the parts are easier than the whole. Viruses 11:96

Abedon ST (2022) Further considerations on how to improve phage therapy experimentation, practice, and reporting: pharmacodynamics perspectives. Phage (New Rochelle) 3:98–111

Abedon ST (2023a) Bacteriophage adsorption: likelihood of virion encounter with bacteria and other factors affecting rates. Antibiotics (Basel) 12:723

Abedon ST (2023b) Ecology and evolutionary biology of hindering phage therapy: the phage tolerance vs. phage resistance of bacterial biofilms. Antibiotics (Basel) 12:245

Abedon ST, Kuhl SJ, Blasdel BG, Kutter EM (2011) Phage treatment of human infections. Bacteriophage 1:66–85

Abedon ST, García P, Mullany P, Aminov R (2017) Editorial: phage therapy: past, present and future. Front Microbiol 8:981

Ackermann HW (2011) Bacteriophage taxonomy. Microbiol Aust 32:90–94

Almeida GMF, Sundberg LR (2020) The forgotten tale of Brazilian phage therapy. Lancet Infect Dis 20:e90–e101

Aslam S, Lampley E, Wooten D, Karris M, Benson C, Strathdee S, Schooley RT (2020) Lessons learned from the first 10 consecutive cases of intravenous bacteriophage therapy to treat multidrug-resistant bacterial infections at a single center in the United States. Open Forum Infect Dis 7:ofaa389

Balcazar JL (2014) Bacteriophages as vehicles for antibiotic resistance genes in the environment. PLoS Pathog 10:e1004219

Batinovic S, Wassef F, Knowler SA et al (2019) Bacteriophages in natural and artificial environments. Pathogens 8:100

Birkholz N, Jackson SA, Fagerlund RD, Fineran PC (2022) A mobile restriction-modification system provides phage defence and resolves an epigenetic conflict with an antagonistic endonuclease. Nucleic Acids Res 50:3348–3361

Bondy-Denomy J, Qian J, Westra ER, Buckling A, Guttman DS, Davidson AR, Maxwell KL (2016) Prophages mediate defense against phage infection through diverse mechanisms. ISME J 10:2854–2866

Borges AL (2021) How to train your bacteriophage. Proc Natl Acad Sci U S A 118:e2109434118

Bretaudeau L, Tremblais K, Aubrit F, Meichenin M, Arnaud I (2020) Good manufacturing practice (GMP) compliance for phage therapy medicinal products. Front Microbiol 11:1161

Campbell A (2003) The future of bacteriophage biology. Nat Rev Genet 4:471–477

Chanishvili N (2012) Phage therapy-history from Twort and d'Herelle through Soviet experience to current approaches. Adv Virus Res 83:3–40

Chaudhry WN, Concepción-Acevedo J, Park T, Andleeb S, Bull JJ, Levin BR (2017) Synergy and order effects of antibiotics and phages in killing Pseudomonas aeruginosa biofilms. PLoS One 12:e0168615

Chen Y, Batra H, Dong J, Chen C, Rao VB, Tao P (2019) Genetic engineering of bacteriophages against infectious diseases. Front Microbiol 10:954

Clinical Trials database (2024a) National Library of Medicine, Bethesda. https://clinicaltrials.gov/search?intr=bacteriophage%20therapy%20OR%20phage%20therapy. Accessed 26 Feb 2024

Clinical Trials database (2024b) National Library of Medicine, Bethesda. https://clinicaltrials.gov/study/NCT00937274?intr=bacteriophage%20therapy%20OR%20phage%20therapy&rank=23. Accessed 26 Feb 2024

Clinical Trials database (2024c) National Library of Medicine, Bethesda. https://clinicaltrials.gov/study/NCT04287478?intr=bacteriophage%20therapy%20OR%20phage%20therapy&rank=1. Accessed 26 Feb 2024

Clinical Trials database (2024d) National Library of Medicine, Bethesda. https://clinicaltrials.gov/study/NCT04787250?intr=bacteriophage%20therapy%20OR%20phage%20therapy&rank=17. Accessed 26 Feb 2024

Clinical Trials database (2024e) National Library of Medicine, Bethesda. https://clinicaltrials.gov/study/NCT05269134?intr=bacteriophage%20therapy%20OR%20phage%20therapy&rank=10. Accessed 26 Feb 2024

Clinical Trials Registry of India (2024) Indian Council of Medical Research—National Institute of Medical Statistics, New Delhi. https://ctri.nic.in/Clinicaltrials/login.php. Accessed 26 Feb 2024

Colavecchio A, Cadieux B, Lo A, Goodridge LD (2017) Bacteriophages contribute to the spread of antibiotic resistance genes among foodborne pathogens of the enterobacteriaceae family - a review. Front Microbiol 8:1108

Crutchfield ED, Stout BF (1930) Treatment of Staphylococcic infections of the skin by the bacteriophages. Arch Dermatol 22:1010–1021

d'Herelle F (1918) Sur le rôle du microbe filtrantbactériophagedans la dysentériebacillaire. Comptes Rendus Acad Sci 167:970–972

Dąbrowska K (2019) Phage therapy: what factors shape phage pharmacokinetics and bioavailability? Systematic and critical review. Med Res Rev 39:2000–2025

Decker K, Krauel V, Meesmann A, Heller KJ (1994) Lytic conversion of Escherichia coli by bacteriophage T5: blocking of the FhuA receptor protein by a lipoprotein expressed early during infection. Mol Microbiol 12:321–332

Dedrick RM, Freeman KG, Nguyen JA et al (2021) Potent antibody-mediated neutralization limits bacteriophage treatment of a pulmonary Mycobacterium abscessus infection. Nat Med 27:1357–1361

Di Martino P (2018) Extracellular polymeric substances, a key element in understanding biofilm phenotype. AIMS Microbiol 4:274–288

Diallo K, Dublanchet A (2022) Benefits of combined phage-antibiotic therapy for the control of antibiotic-resistant bacteria: a literature review. Antibiotics (Basel) 11:839

Diallo K, Dublanchet A (2023) A century of clinical use of phages: a literature review. Antibiotics (Basel) 12:751

Doub JB (2021) Risk of bacteriophage therapeutics to transfer genetic material and contain contaminants beyond endotoxins with clinically relevant mitigation strategies. Infect Drug Resist 14:5629–5637

Dublanchet A, Bourne S (2007) The epic of phage therapy. Can J Infect Dis Med Microbiol 18:15–18

Dupuis MÈ, Villion M, Magadán AH, Moineau S (2013) CRISPR-Cas and restriction-modification systems are compatible and increase phage resistance. Nat Commun 4:2087

Egido JE, Costa AR, Aparicio-Maldonado C, Haas PJ, Brouns SJJ (2022) Mechanisms and clinical importance of bacteriophage resistance. FEMS Microbiol Rev 46:fuab048

Fortier LC, Sekulovic O (2013) Importance of prophages to evolution and virulence of bacterial pathogens. Virulence 4:354–365

Garcia-Blanco MA, Ooi EE, Sessions OM (2022) RNA viruses, pandemics and anticipatory preparedness. Viruses 14:2176

Gencay YE, Sørensen MCH, Wenzel CQ, Szymanski CM, Brøndsted L (2018) Phase variable expression of a single phage receptor in Campylobacter jejuni NCTC12662 influences sensitivity toward several diverse CPS-dependent phages. Front Microbiol 9:82

Gordillo Altamirano FL, Kostoulias X, Subedi D, Korneev D, Peleg AY, Barr JJ (2022) Phage-antibiotic combination is a superior treatment against Acinetobacterbaumannii in a preclinical study. EBioMedicine 80:104045

Górski A, Międzybrodzki R, Borysowski J et al (2012) Phage as a modulator of immune responses: practical implications for phage therapy. Adv Virus Res 83:41–71

Górski A, Borysowski J, Międzybrodzki R (2020) Phage therapy: towards a successful clinical trial. Antibiotics (Basel) 9:827

Gu Liu C, Green SI, Min L, Clark JR et al (2020) Phage-antibiotic synergy is driven by a unique combination of antibacterial mechanism of action and stoichiometry. MBio 11:e01462–e01420

Hatfull GF, Dedrick RM, Schooley RT (2022) Phage therapy for antibiotic-resistant bacterial infections. Annu Rev Med 73:197–211

Hibstu Z, Belew H, Akelew Y, Mengist HM (2022) Phage therapy: a different approach to fight bacterial infections. Biologics 16:173–186

Hitchcock NM, Devequi Gomes Nunes D, Shiach J et al (2023) Current clinical landscape and global potential of bacteriophage therapy. Viruses 15:1020

Hughes KA, Sutherland IW, Jones MV (1998) Biofilm susceptibility to bacteriophage attack: the role of phage-borne polysaccharide depolymerase. Microbiology (Reading) 144:3039–3047

Hyman P (2019) Phages for phage therapy: isolation, characterization, and host range breadth. Pharmaceuticals (Basel) 12:35

Jia HJ, Jia PP, Yin S, Bu LK, Yang G, Pei DS (2023) Engineering bacteriophages for enhanced host range and efficacy: insights from bacteriophage-bacteria interactions. Front Microbiol 14:1172635

Karn SL, Gangwar M, Kumar R, Bhartiya SK, Nath G (2023) Phage therapy: a revolutionary shift in the management of bacterial infections, pioneering new horizons in clinical practice, and reimagining the arsenal against microbial pathogens. Front Med (Lausanne) 10:1209782

Karygianni L, Ren Z, Koo H, Thurnheer T (2020) Biofilm matrixome: extracellular components in structured microbial communities. Trends Microbiol 28:668–681

Kasman LM, Kasman A, Westwater C, Dolan J, Schmidt MG, Norris JS (2002) Overcoming the phage replication threshold: a mathematical model with implications for phage therapy. J Virol 76:5557–5564

Kronheim S, Daniel-Ivad M, Duan Z et al (2018) A chemical defence against phage infection. Nature 564:283–286

Labrie SJ, Samson JE, Moineau S (2010) Bacteriophage resistance mechanisms. Nat Rev Microbiol 8:317–327

Laxminarayan R, Duse A, Wattal C et al (2013) Antibiotic resistance-the need for global solutions. Lancet Infect Dis 13:1057–1098

Leptihn S, Loh B (2022) Complexity, challenges and costs of implementing phage therapy. Future Microbiol 17:643–646

Li X, He Y, Wang Z et al (2021) A combination therapy of phages and antibiotics: two is better than one. Int J Biol Sci 17:3573–3582

Lin DM, Koskella B, Lin HC (2017) Phage therapy: an alternative to antibiotics in the age of multi-drug resistance. World J Gastrointest Pharmacol Ther 8:162–173

Lin J, Du F, Long M, Li P (2022) Limitations of phage therapy and corresponding optimization strategies: a review. Molecules 27:1857

Liu D, Van Belleghem JD, de Vries CR et al (2021) The safety and toxicity of phage therapy: a review of animal and clinical studies. Viruses 13:1268

Liu S, Lu H, Zhang S, Shi Y, Chen Q (2022) Phages against pathogenic bacterial biofilms and biofilm-based infections: a review. Pharmaceutics 14:427

Łusiak-Szelachowska M, Weber-Dąbrowska B, Żaczek M, Borysowski J, Górski A (2020) The presence of bacteriophages in the human body: good, bad or neutral? Microorganisms 8:2012

Łusiak-Szelachowska M, Międzybrodzki R, Drulis-Kawa Z et al (2022) Bacteriophages and antibiotic interactions in clinical practice: what we have learned so far. J Biomed Sci 29:23

MacNeal WJ, Frisbee FC (1932) Bacteriophage as therapeutic agent in Staphylococcus bacteriemia. JAMA 99:1150–1155

Malik DJ, Sokolov IJ, Vinner GK et al (2017) Formulation, stabilisation and encapsulation of bacteriophage for phage therapy. Adv Colloid Interf Sci 249:100–133

Mavrich TN, Hatfull GF (2019) Evolution of superinfection immunity in cluster a mycobacteriophages. MBio 10:e00971–e00919

McCallin S, AlamSarker S, Barretto C et al (2013) Safety analysis of a Russian phage cocktail: from metagenomic analysis to oral application in healthy human subjects. Virology 443:187–196

McCallin S, Sacher JC, Zheng J, Chan BK (2019) Current state of compassionate phage therapy. Viruses 11:343

Mehmood Khan F, Manohar P, Singh Gondil V et al (2023) The applications of animal models in phage therapy: an update. Hum Vaccin Immunother 19:2175519

Melo LDR, Oliveira H, Pires DP, Dabrowska K, Azeredo J (2020) Phage therapy efficacy: a review of the last 10 years of preclinical studies. Crit Rev Microbiol 46:78–99

Monteiro R, Pires DP, Costa AR, Azeredo J (2019) Phage therapy: going temperate? Trends Microbiol 27:368–378

Myelnikov D (2018) An alternative cure: the adoption and survival of bacteriophage therapy in the USSR, 1922-1955. J Hist Med Allied Sci 73:385–411

Nale JY, Clokie MR (2021) Preclinical data and safety assessment of phage therapy in humans. Curr Opin Biotechnol 68:310–317

Nang SC, Lin YW, Petrovic Fabijan A et al (2023) Pharmacokinetics/pharmacodynamics of phage therapy: a major hurdle to clinical translation. Clin Microbiol Infect 29:702–709

Oduor JMO, Onkoba N, Maloba F, Nyachieo A (2016) Experimental phage therapy against haematogenous multi-drug resistant Staphylococcus aureus pneumonia in mice. Afr J Lab Med 5:435

Oechslin F (2018) Resistance development to bacteriophages occurring during bacteriophage therapy. Viruses 10:351

Oliveira H, São-José C, Azeredo J (2018) Phage-derived peptidoglycan degrading enzymes: challenges and future prospects for in vivo therapy. Viruses 10:292

Ouyang X, Li X, Song J et al (2023) Mycobacteriophages in diagnosis and alternative treatment of mycobacterial infections. Front Microbiol 14:1277178

Patel DR, Bhartiya SK, Kumar R, Shukla VK, Nath G (2021) Use of customized bacteriophages in the treatment of chronic nonhealing wounds: a prospective study. Int J Low Extrem Wounds 20:37–46

Payne RJ, Jansen VA (2002) Evidence for a phage proliferation threshold? J Virol 76:13123; author reply 13123–4

Pedruzzi I, Rosenbusch JP, Locher KP (1998) Inactivation in vitro of the Escherichia coli outer membrane protein FhuA by a phage T5-encoded lipoprotein. FEMS Microbiol Lett 168:119–125

Penadés JR, Chen J, Quiles-Puchalt N, Carpena N, Novick RP (2015) Bacteriophage-mediated spread of bacterial virulence genes. Curr Opin Microbiol 23:171–178

Petrovic Fabijan A, Iredell J, Danis-Wlodarczyk K, Kebriaei R, Abedon ST (2023) Translating phage therapy into the clinic: recent accomplishments but continuing challenges. PLoS Biol 21:e3002119

Pirnay JP, Verbeken G (2023) Magistral phage preparations: is this the model for everyone? Clin Infect Dis 77(Supple 5):S360–S369

Pope WH, Jacobs-Sera D, Russell DA et al (2011) Expanding the diversity of mycobacteriophages: insights into genome architecture and evolution. PLoS One 6:e16329
Semler DD, Goudie AD, Finlay WH, Dennis JJ (2014) Aerosol phage therapy efficacy in Burkholderiacepacia complex respiratory infections. Antimicrob Agents Chemother 58:4005–4013
Suh GA, Lodise TP, Tamma PD et al (2022) Considerations for the use of phage therapy in clinical practice. Antimicrob Agents Chemother 66:e0207121
Summers WC (1993) Cholera and plague in India: the bacteriophage inquiry of 1927-1936. J Hist Med Allied Sci 48:275–301
Summers WC (2012) The strange history of phage therapy. Bacteriophage 2:130–133
Turner D, Kropinski AM, Adriaenssens EM (2021) A roadmap for genome-based phage taxonomy. Viruses 13:506
Van Belleghem JD, Dąbrowska K, Vaneechoutte M, Barr JJ, Bollyky PL (2018) Interactions between bacteriophage, bacteria, and the mammalian immune system. Viruses 11:10
Visnapuu A, Van der Gucht M, Wagemans J, Lavigne R (2022) Deconstructing the phage-bacterial biofilm interaction as a basis to establish new antibiofilm strategies. Viruses 14:1057
Wang H, Liu Y, Bai C, Leung SSY (2024) Translating bacteriophage-derived depolymerases into antibacterial therapeutics: Challenges and prospects. Acta Pharm Sin B 14:155–169
Wittebole X, De Roock S, Opal SM (2014) A historical overview of bacteriophage therapy as an alternative to antibiotics for the treatment of bacterial pathogens. Virulence 5:226–235
Yang Q, Le S, Zhu T, Wu N (2023) Regulations of phage therapy across the world. Front Microbiol 14:1250848
Yehl K, Lemire S, Yang AC et al (2019) Engineering phage host-range and suppressing bacterial resistance through phage tail fiber mutagenesis. Cell 179:459–469
Żaczek M, Łusiak-Szelachowska M, Jończyk-Matysiak E et al (2016) Antibody production in response to Staphylococcal MS-1 phage cocktail in patients undergoing phage therapy. Front Microbiol 7:1681
Zhu Y, Shang J, Peng C, Sun Y (2022) Phage family classification under Caudoviricetes: a review of current tools using the latest ICTV classification framework. Front Microbiol 13:1032186

Breakthroughs in the Non-Antibiotic Therapies for Infectious Disease Treatment from Phages to Peptides and Beyond

19

Lokesh and Sruti Patel

Abstract

The development of antibiotics has changed the course of bacterial disease treatment and probably saved billions of human lives, making them an essential achievement in medicine. However, with the surge in the human population, extensive use and misuse of antibiotics have led microorganisms to develop resistance against these antibiotics, which is known as antimicrobial resistance. This threatens human health, leading to catastrophic consequences in the healthcare system. The proposed chapter prescribes answers to the need for innovative approaches that offer insights into fighting microbial resistance other than the traditional antibiotics. The range of non-antibiotic therapies currently under study includes phage therapy, antimicrobial peptides, CRISPR-Cas against antimicrobial-resistant pathogens, immunotherapies and vaccines, probiotics, postbiotics and symbiotic, and bacteriophage-derived enzymes. We have also discussed the challenges for the development and implementation of these therapies, such as regulatory hurdles, safety concerns, and budget issues, simultaneously highlighting their benefits relative to conventional antibiotics with different modes of action. We provide a brief analysis of the current state of affairs and the ethical issues pertaining to non-antibiotic therapies for antimicrobial resistance and suggestions for future research.

Lokesh (✉)
Department of Biotechnology, Shri Alpesh N. Patel Post Graduate Institute of Science and Research, Affiliated with Sardar Patel University, Anand, Gujarat, India

S. Patel
Dr. Vikram Sarabhai Institute of Cell and Molecular Biology, Faculty of Science, The Maharaja Sayajirao University of Baroda, Vadodara, Gujarat, India

M. Gangwar, G. Nath (eds.), *Emerging Paradigms for Antibiotic-Resistant Infections: Beyond the Pill*, https://doi.org/10.1007/978-981-97-5272-0_19

Keywords

Antibiotics · Non-antibiotic therapies · Phage therapy · Phage dynamics · Antimicrobial peptides · Drug development · Regulatory hurdles · Ethical issues

19.1 Introduction

Majority of bacteria play a vital role in preserving the ecosystem in which we live. Only a small fraction of bacteria is known to cause infection and diseases, significantly impacting human health (Doron and Gorbach 2008). These infections are mainly transmitted through air droplets, direct or indirect contact with unsanitary surfaces, vectors, and contaminated food or water. Once infected, these bacteria cause inflammation and systemic disturbances in the human body. In the late 1800s, German scientist Paul Enrlich discovered a chemical agent specifically designed to kill bacteria. This discovery was fruitful in 1907 when an arsenic-containing organic compound named *Salvarsan*. The antimicrobial had a severe side effect on humans because of arsenic. In 1928, Alexander Fleming discovered Penicillin, the first modern antibiotic. Antibiotics are chosen as preventative prognosis to fight against the infection. Their principle mechanism of action includes inhibition of nucleic acid synthesis, due to which protein synthesis, ATP synthesis, cell wall synthesis, and all other metabolic pathways are affected. This results in the decrease in the population of bacteria which helps our immune system to clear up the infection. However, few bacteria might escape from these antibiotics and tend to develop resistance. These antibiotic-resistance bacteria grow and take over. Moreover, antibiotic resistance in bacteria can transfer their genomic mutations vertically to other bacterial cells; those microorganisms can also horizontally acquire mobile genetic elements that encode resistance genes (Zhang et al. 2022).

The WHO's 2022 global antimicrobial resistance report highlights the alarming resistance rates among prevalent bacterial pathogens. Antimicrobial resistance (AMR) is a phenomenon that evolves through natural selection, but anthropogenic activities can accelerate its evolution. A common cause for the alarming increase of resistant microorganisms is the use and misuse of antibiotics (Ajulo and Awosile 2024).

To tackle antimicrobial resistance, non-antibiotic techniques such as phage therapy and antimicrobial peptides (AMPs) are used nowadays. Bacteriophages are used to treat bacterial infections in such cases. AMPs are small peptides that act on multiple drug targets on the plasma membrane and intracellular targets of pathogenic bacteria and have an advantage over drug-resistant pathogens (Table 19.1).

19.2 Phage Therapies

Phages are small viruses ranging from 20 to 200 nm in size, which solely target and kill specific bacterial strains, without affecting other microbiota in our bodies. Phage therapy is the use of these bacteriophage viruses to treat bacterial infections

Table 19.1 Comparative analysis of bacteriophages and antibiotics

Bacteriophages	Antibiotics	Ref.
Highly specific; therefore, chances of dysbiosis and non-targeted bacterial (gut flora) infection are avoided	Non-specific; therefore, can lead to complications by disrupting the patient's natural flora	Semler et al. (2012)
Multiply at the infection site, so they are always present where needed	Being metabolized by the body, antibiotics are cleared before they can be concentrated at the site of infection	
No adverse effects have been associated.	Various unwanted reactions, such as gut problems, allergies, and fungal infections were observed	
Bacteria that resist one phage can still be killed by other phages with the same host range	Bacteria that resist one antibiotic can often resist other antibiotics as well	
New phages can be quickly isolated and tested against resistant bacteria	New antibiotics take a long time to develop and test against resistant bacteria	
Less evidence to support use	Lots of clinical experience	

that have been around for over a hundred years (The Promise of Phages 2023). Phage was first discovered and characterized in 1915 by Frederick William Twort and by Felix d'Hèrelle in 1917. Although this therapy gained some popularity in the 1920s, it was quickly surpassed by the discovery of broad-spectrum antibiotics (Murray et al. 2022). There were several reasons for the decline of Phage therapy, including the storage of phages using preservatives like phenols, which resulted in its inactivation. Furthermore, there was little knowledge about the specificity of the phages and the clinical trials of the phage therapies hadn't been done to show that it's efficacious (Cook and Wright 2022). However, in recent years, the emergence of multidrug-resistant bacteria has left the world scrambling for alternative treatments. This global issue has the potential to revitalize phage therapy.

Phages are obligate parasites that essentially take over the cellular machinery of the host cell to divide and proliferate. Phages begin to attach itself to the host bacterium by interacting with receptors, followed by tight adsorption onto the cell and consequently injecting its genome. Thereafter, the phage can undergo two different life cycles, lytic and lysogenic. The lytic cycle includes hijacking the host replication mechanism to produce its viral progeny. The phages produce a protein called endolysin that degrades the bacterial peptidoglycan layer and are released into the immediate environment upon lysis of the host. On the other hand, during the lysogenic cycle, the bacteriophage integrates its genome into that of the bacteria's as a prophage. Under stress conditions, the prophage might exit lysogeny and begin to replicate and produce new viral phages that are released via lysis of the bacterium (Murray et al. 2022).

The secretion of endolysin enzyme by phages makes for a potential alternative to the classical antibiotics. This enzyme is responsible for the elimination of most bacterial infections in plants, mammals, and aquatic life (Fig. 19.1) (Jamal et al. 2018a, b).

The left side of the cycle shows the lytic cycle, and right side shows the lysogenic cycle. In the lytic cycle, phages will attach to the bacterial surface with the help of a base plate and hook. After they inject the genetic material into the bacterial, the phage genome will take control of all bacterial mechanisms and produce the phage protein and other essential things. After assembly and packing, phages will be released in the surrounding environment with the help of the endoysin enzyme, infecting other bacteria. Another cycle is lysogeny, where, after injecting its genetic material, it will be integrated into the host genome via homologous recombination and replicate with it.

19.3 Different Approaches for Phage Therapy

The five different kinds of approaches for phage therapy includes: (1) Phage Cocktail, (2) Phage-derived enzyme, (3) Phage and Phage Enzyme combined with Antibiotic, (4) Phage engineering, (5) Phage with Crisper.

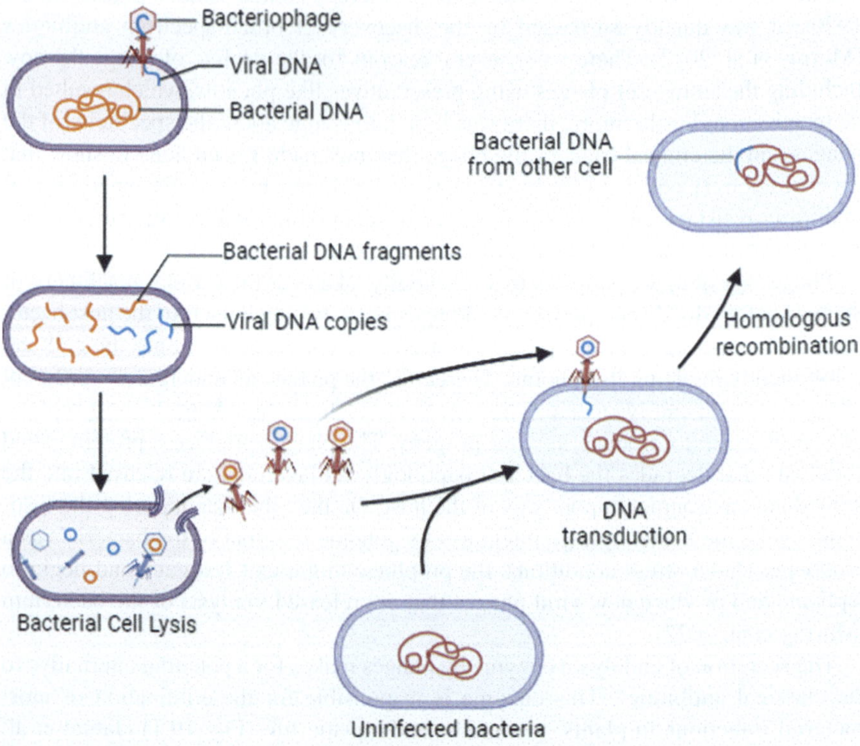

Fig. 19.1 Represents the life cycle of the bacteriophage during infection

19.3.1 Phage Cocktail

Phage cocktails combine various virus types to eliminate one specific or several bacterial infections. This also prevents bacteria from becoming resistant to phages and removes the limitations brought forth by the limited phage lysis spectrum (Jasim et al. 2018). Since the study showed significant variations in phage, metagenomic research may be applied to treat skin and infection of wounds. In comparison with monophage therapy, phage cocktail therapy significantly lowers bacterial resistance to phages. Methicillin-resistant *Staphylococcus aureus* and *Clostridiumdifficile* can be effectively combated using phage combinations (Yang et al. 2020).

19.3.2 Phage-Derived Enzyme

There are various enzymes encoded in the phage that could penetrate the bacterial cells, which are virion-associated peptidoglycan hydrolases (VAPGH), endolysins, and polysaccharide depolymerases. Although not much clinical research has been done, VAPGH, discovered in the basal plate region of the phage, is injected when the phage attacks the bacteria (Comeau et al. 2007). Jacob et al. in 1958 identified yet another enzyme that can lyse bacteria. It encodes a family of proteins called endolysin (lysin). Phage lysins are remarkably species-specific; they do not harm biological tissues or disturb the normal flora of bacteria. Secondly, lysins exhibit a broader spectrum of lysis activity in comparison to phages, and they are less likely to induce the emergence of disease resistance against them. These enzymes can be commercially manufactured easily and is going to promote the use of such compounds for future purposes (Wei et al. 2020).

19.3.3 Phage and Phage Enzyme Combined with Antibiotic

Recent studies have shown that phage combinations with antibiotics have greater therapeutic effect than monophage treatments. Combination, also called phage-antibiotic synergy or PAS, originated in 2007 (Kronheim et al. 2018). Bacteria forming biofilms prevent the action of antibiotics leading to resistance; however, this synergic combination of phage and conventional antibiotics helps to degrade the biofilms substantially (Abedon 2019). Some antibiotics affect the phages and this in turn affect the synergy effect on patients, hence more clinical studies are required to show their effectiveness (Segall et al. 2019).

19.3.4 Phage Engineering

In phage therapy, genetic engineering is the most important technique used to overcome the resistance and minimize the lysis spectrum of phages. Genetical engineering enhances the phage-based therapy by changing the delivery system and the lysis pattern. Modifications of genes that encode receptor-binding proteins (RBPs) in phage tail fibers or spikes are a significant aspect of genetically engineered phages (Koonin et al. 2017).

19.3.5 Phage with CRISPR

The novel combination of both phages and the CRISPR-Cas system has shown promising results in approaching specific targets and eliminate pathogen bacteria, including resistant bacteria. An effective way to fight the bacteria is by delivering exogenous CRISPR nucleases for pathogens that do not have CRISPR-CAS system. According to reports, phagemids with an exogenous CRISPR-Cas9 system target bacterium quickly and precisely. The phage paired with the CRISPR-Cas system offers a promising future in the prevention and treatment of drug-resistant bacteria (Liu et al. 2019).

19.4 Phage Therapy in Human

Phage therapy was majorly used in humans because of its selectivity against pathogenic bacteria. The primary determinant of a phage's suitability for therapy in either humans or animals is its exclusive lytic (virulent) or temperate character. As phages naturally occur in the environment, humans are constantly interacting with them without suffering any harm. If they can reach their intended bacterial host species after administration, then their population grows.

While some bacteria are harmful to human beings, others are beneficial as well, such as the body's normal flora (Jikia et al. 2005; Summers 1999). The pathogenic bacteria like *Pneumococci, Staphylococci, and Streptococci,* present in the environment are also known to have been found in the human body (Weber-Dabrowska et al. 2001). Certain phage lytic enzymes are used for the decrease of nasopharyngeal (nose and pharynx) carriage of pathogens. Other phages are also used to treat neonatal suffering from meningitis of brain and spinal cord. *S. aureus* is generally known to be a multidrug—methicillin, penicillin, and ciprofloxacin-resistant bacteria, whose infection is removed using phage therapy (Jikia et al. 2005). Phage lysin enzymes are also used for treating anthrax caused by *Bacillus anthracis*. Similarly, lung infections caused by *Staphylococcus* or cystic fibrosis caused by *P. aeruginosa*, both of which can be successfully cured using phages. Therefore, pertaining to the diverse bacterial infections, phages are shown to be the most effective by Eastern Europe researchers.

19.5 Phage Therapy in Agriculture

The word "bacteriophage biocontrol" is more frequently applied to plants. Many research findings on phage biocontrol of various significant bacterial plant diseases have been published over the past few years. They only tend to remain in large numbers in any setting as long as the host is present. Phages, unlike copper-based insecticides, do not build up in the soil (Buttimer et al. 2017). Usually restricted to stains belonging to a certain species of bacterium, phages have a limited host spectrum.

Implication of phage solutions by some commercial corporations helps in getting rid of bacterial infections in crops like potatoes, apples, peppers and tomatoes among others. Some countries, such as the USA, Scotland, or Hungary, declared phage therapy as the official treatment of their antibiotic-resistant MRSA (Methicillin-resistant *Staphylococcus aureus*) infections. In order to distinguish the Phage drink from other drinks, its formula will be a variable one. While the mutually recognized authorization of medicinal products would save time and money on re-registration in the EU-wide market, it is still quite a complicated issue during product licensing (Buttimer et al. 2017).

The multiple attributes of phages can enhance phage controls. Being organic farming-friendly makes them a good option. Phage cocktails with specific host ranges can be developed using these attributes. Phages have the ability to evolve to combat resistance and new bacterial strains. They can be used along with other control agents. Sensitivity to UV light and specific soil conditions may limit their effectiveness they can be used. They are sensitive to UV light and specific soil conditions, which could be a drawback to their utilization (Buttimer et al. 2017).

19.6 Phage Formulation

Bacteriophages can be incorporated into therapeutic formulations involving stabilizing substances such as powder, semi-solids, and nanofibers. This provides greater possibilities for effective delivery at the site of infection, increases safety, and improves patient outcomes. Various techniques of freeze drying (lyophilization), spray drying, emulsions, polymeric microparticles, nanoparticles, and liposomes are used for phage encapsulation (Rosner and Clark 2021).

The problems relating to the stability of phages had already been reported by earlier scientists, which consist of the sensitivity of phages to organic solvents, pH, temperature, and salinity. Since then, several attempts have been made to overcome these issues. Phages that commonly exist at 4 °C with a limited drop in titer are observed in most cases. However, to increase their long-term stability, the phages can be stored by freezing at −80 °C. In some cases, where the phages undergo rapid degradation, the addition of gelatin, magnesium ions, and glycerol aids in their stability (Malik et al. 2017).

Bacteriophage formulations pose several additional challenges for therapeutic delivery, apart from the storage of free phage lysates. Factors that are considered in developing stable phage formulations mainly focus on the methods of phage delivery to target bacteria, establishing the stability of a given formulation under a range of conditions and improving phage survival during the production process. Encapsulation by emulsions has been shown to improve phage stability as well as keeping intact their bioactivity. These formulations depend on droplet size, changing of the emollient or emulsifier, and other additives into the mixture (Malik et al. 2017).

19.7 Phage-Bacterium Population Dynamics

Phage pharmacokinetics is the body's response to phages, whereas pharmacodynamics is the phage's impact on our body. Unlike antibiotics, PK/PD of phage therapy is not very well understood, mainly because of the self-replicating ability of phages, leading to a complex phage PK (Levison and Levison 2009). Elimination of bacteria by phages vastly depends on the absorption of its particles by the specific bacterium (Cairns et al. 2009). The binding kinetics of the phage and bacteria is considered being a first-order reaction, depending on the concentrations of the bacteria and phage Levin and Bull 2004). Bacteriophage diversity plays an important role in the formulation process, requiring extensive comparative studies and better understanding of phage diversity (Levin and Bull 2004).

The potential variability of pathogenic bacterial strains should also be considered while designing phage cocktails to counteract any variations with maximum coverage, resulting in a cost-effective and more stable product. A mathematical model was proposed by Payne and Jansen, highlighting the phage infection of bacteria during the exponential growth phase of bacteria (Payne and Jansen 2001). This model was further modified by including the effect of bacterial mutation which resulted in phage resistant population, in turn effecting our immune response to fight bacterial infection (Levin and Bull 2004).

Mathematical models help determine the effects of parameters like the dosage of phage, the rate at which it is delivered or bacterial growth and clearance rate, on phage therapy outcomes. It is also useful for interpreting the results of in vivo studies and design future experiments (Malik et al. 2017).

Phage growth is regulated by a number of factors: the phage-bacterium absorption rate, burst size (the number of phage released after each bacterial infection), the bacterial growth rate, latent period (that is the time required by phage to complete its replication cycle in a bacteria), phage and bacterial elimination rates due to host absorption and the initial concentrations of both. Considering all the factors, encapsulation of phages and their slow release enables phages to maintain a concentration that is therapeutically effective in vivo. It has been shown that in vivo phage concentration at the infection site is an important aspect in governing the efficacy of phage therapy. Phage cocktails require precise formulation conditions including the individual formulation of each phage to ensure viability and stability during storage and administration (Leung and Weitz 2016).

Additionally, these formulations should be considered to be easy while scaling up the production. Formulations which can be stored for a long time at a suitable temperature will be favored (Malik et al. 2017).

19.8 Phage Pharmacokinetics

According to Alisky et al. (1998) and Weber-Dabrowska et al. (2001), the phages used in phage therapy have traditionally been administered orally for intestinal and systemic infections, Trans nasally for lung infections, transrectally for prostatitis, Trans urinary for bladder and kidney, intravenously for septicemia, and topically for wound infection, otitis, and eye infection. Phage therapies can specifically target and eradicate microbes. However, the phages are meant to treat an intestinal infection should be administering directly at the site of infection because generally phages that are injected orally or transrectally need to get past the layers of epithelial cells in the gut and then reach the blood and lymph (Table 19.2).

Table 19.2 Comparison of phage therapy cases across countries, diseases, and their target

Country	Administration route	Target bacteria	Disease	Year	References
USA	Topically (dripped, covered with gauze)	*Aureus*	Toe ulcers	2016	Fish et al. (2016)
	Percutaneous drainage and intravenously	*Baumannii*	Systemic infection	2017	Zhvania et al. (2017)
	Topically (around the ulcer once)	*Aureus*	Digital osteomyelitis	2018	Schooley et al. (2017)
	Intravenously	*A. baumannii*	Craniectomy site infection	2018	Jennes et al. (2017)
Georgia	Topically (spray cream) and orally	*Aureus*	Skin disorder	2017	Fish et al. (2018)
	Orally and via inhalation	*Ach. xylosoxidas*	Cystic fibrosis	2018	Hoyle et al. (2018)
Belgium	Intravenously	*P. aeruginosa*	Systemic infection	2017	Jault et al. (2019)
Belgium, France, Switzerland	Topically (an alginate template	*P. aeruginosa*	Burn infection	2018	
Poland	Intrarectally	*E. faecalis*	Prostatitis	2018	

19.9 Current Progress and Challenges in Clinical Phage Therapy Research

The lack of validated and sufficiently controlled clinical trials poses a challenge in the journey of phage therapy reaching the clinic. Trial design includes several pharmacological considerations that are unique to the phages. Their dosage regulation is extremely important as they are self-replicating and their dose would most likely increase exponentially upon reaching the site of infection. Additionally, the phages need to be administered with high precision as they require direct contact with the bacterium. It should also be considered that combination therapy such as phage cocktails increase the risk of an inflammatory response, phage resistance development for all phages in the cocktail (Furfaro et al. 2018).

To ensure the safety, sterility, and purity of phages, the formulations should include good manufacturing practices or equivalent quality assurance standards. Secondly, the possibility of the onset of toxic shocks as a result of the bactericidal effect of phages must be considered (Furfaro et al. 2018).

19.9.1 Industry-Sponsored Trials

Across global different companies focusing on phage therapy with USA as the leading country in the world. Majority of companies are in the third stage of clinical trials, whereas in the US, the FDA has approved to enforce of phage therapy banks following the Investigation New Drug (IND) steps (Yang et al. 2023). Table 19.3 shows a list of recent clinical trials which are completed and some are currently recruiting (Data sourced from https://clinicaltrials.gov/).

19.9.2 Investigator-Initiated Trials

Discovered in 1922, phage therapy was used in different fields to fight bacterial infections but recently it is very demanding due to the emergence of antibiotic-resistant bacteria and it showed no or few side effects when compared with the control group. Considering that randomized controlled trials are limited but also one of the most important parameters, the safety and effectiveness of treatment need to be examined thoroughly. During the course of phage therapy tests, the lack of concentrated effect in the clinical trial are now been looked at with studies including the comparative safety of phage therapy and its efficiency, against standard treatment (Yang et al. 2023). Various studies of increasing the dosage rate are also being done as suggested by the clinical data in Table 19.4.

Table 19.3 Few lists of Industry based clinical trials adapted from https://clinicaltrials.gov/

Study title	Study status	Study results	Interventions	Study type	NCT. Id
Bacteriophage therapy of difficult-to-treat infections	Completed	No	Bacteriophage therapy	Observational	NCT05498363
A prospective, randomized, double-blind controlled study of Wpp-201 for the safety and efficacy of treatment of venous leg ulcers	Completed	No	Drug: Bacteriophage|Drug: Wpp-201 Bacteriophage|Drug: Bacteriophages	Interventional	NCT00663091
Bacteriophage therapy Tp-102 in diabetic foot ulcers	Completed	No	Biological: Tp-102	Interventional	NCT04803708
Ph 1/2 study evaluating safety and tolerability of inhaled Ap-Pa02 in subjects with chronic pseudomonas aeruginosa lung infections and cystic Fibrosi	Completed	Yes	Biological: Ap-Pa02|Other: Placebo	Interventional	NCT04596319
Cystic fibrosis bacteriophage study at Yale (Cyphy)	Completed	Yes	Drug: Standard Dose Ypt-01|Other: Placebo	Interventional	NCT04684641
Interleukin-2 plus anti-Hiv Therapy InHiv-infected children with weakened immune	Completed	No	Biological: Diphtheria and tetanus toxoids and acellular pertussis vaccine adsorbed|biological: Diphtheria and tetanus toxoids adsorbed|biological: Tetanus	Interventional	NCT00006066

(continued)

Table 19.3 (continued)

Study title	Study status	Study results	Interventions	Study type	NCT. Id
Bacteriophage therapy of difficult-to-treat infections	Completed	No	Bacteriophage therapy	Observational	NCT05498363
Bacteriophage therapy in tonsillitis	Active-not recruiting	No	Drug: Nebulizer inhalation irrigation of the mucous membranes of the tonsils with a bacteriophage	Interventional	NCT04682964
A study of LBP-EC01 in the treatment of acute uncomplicated UTI caused by drug resistant E. Coli (ELIMINATE trial)	Recruiting	No	DRUG: LBP-EC01 0.1 × IV dose|DRUG: LBP-EC01 0.01 × IV dose|DRUG: LBP-EC01 IV infusion dose|DRUG: Placebo|DRUG: LBP-EC01	Interventional	NCT05488340
Study to evaluate the safety, phage kinetics, and efficacy of inhaled AP-PA02 in subjects with non-cystic fibrosis bronchiectasis	Recruiting	No	DRUG: TMP/SMX|non-cystic fibrosis bronchiectasis *Pseudomonas aeruginosa* Lung infection	Interventional	NCT05616221

19.10 Regulations of Phage Therapy Across the World

Although there is no widely accepted approach for giving human medicine, Georgia's healthcare workers have frequently employed phage therapy to treat patients in the Georgian Eliava phage Therapy center. It develops different healthcare items among which its pioneering phage formulation is being sold the most (Kutateladze 2015). Some doctors in Poland tailored phage products to particular individuals for personalized therapy (Kutter et al. 2010).

As per the suggestion of Górski et al. (2018), the country has standardized treatments for patients who have no hope of recovery and their options have been exhausted and hence specific medicine is their last option. This may be different from region to region. This approach gained a slight edge over the conventional protocols, but still guidelines needed to be formulated for the phages use as

Table 19.4 Various clinical trails conducted by the investigator-initiated trials

Name	Phase	Administration route	Target	Outcome	References
Microgen coil protein	I	Oral	*E. coli*	Safety demonstrated	AlamSarker et al. (2016)
WPP-201	I	Topical	*P. aerogenosa, E. coli,* and *S. aureus*	Safety demonstrated	Rhoads et al. (2009)
PP1131	I/II	Topical	*P. aeruginosa*	Safety demonstrated; the control group reduced bacterial burden at a faster pace	Jault et al. (2019)
NS	I/II	Oral	NS	Safety demonstrated	Bruttin and Brussow (2005)
PYOPHAGE	II/III	IU	Multiple common bacteria	Safety profile favored and inferior to standard of care antibiotic treatment	Leitner et al. (2020)
Microgen coil protein	I	Oral	*E. coli*	Safety demonstrated	AlamSarker et al. (2016)

treatment. In Europe, an evaluation of problems and the making of guidelines for phage therapy is done.

Russia, Georgia, and Poland—who are the front runners in the development of phage therapy—have formed guidelines for its application. As a member of the EU since 2004, Poland has been bound by the EU's rules and regulations (Międzybrodzki et al. 2018). Phages have been used in a more methodical and strictly controlled manner, in recent times. In Poland, clinical trials such as clinical trial directives 2001/20/EC and 2005/28/EC regulate good clinical practice (Żaczek et al. 2022). Nevertheless, the year 2023 has witnessed a new body that is the European Medicines Agency highlighting it's research on phage therapy, originally founded in 2021. Even though, it is classified as a pharmaceutical product in the Republic of Georgia and Russia, it is not widely accepted in the West therefore there is no possibility of conducting double-binding clinical trials (Międzybrodzki et al. 2018).

Phage studies began in India in 1896. The Indian government established a facility in 1940 to study biological products, which included phages (Yang et al. 2023). Before the founding of Vitalis Phage Therapy in 2017, phage was not often utilized. This firm now provides phage treatment in collaboration with a center in Georgia. Phage treatment is now given as compassionate therapy in India and is governed by the Central Drugs Standard Control Organization and the Declaration of Helsinki. Phage treatment is supported in India because of its potential. Researchers are gathering at the Indian Council of Medical Research to talk about regulating phage treatment and setting up research facilities. Thus, in the future, it can be anticipated that there will be more specialized laws and phage treatment research facilities in India (Antimicrobial Resistance and Phage Therapy in India 2023).

In China, phage therapy was first reported in 1958, and it was first customized and given to patients in 2019. Thereafter, many hospitals have utilized this phage therapy, but the government has restricted private firms from carrying out independent clinical trials. There are two ways to commercialize phage treatment, one being fixed-ingredient phage products, and the other being customized phage treatment (Liang et al. 2023).

In the USA, the FDA regulates the norms that include viruses used for medical treatment; they are classified as biological products (Furfaro et al. 2018). But it should have to follow certain rules and regulations in manufacturing, preclinical research, and clinical trials. However, not all phage therapy trails are authorized by the FDA until now in the United States, therefore many are still in the developping stages. The FDA has also created a phage bank, which is a positive step in thee same direction (Adaptive Phage Therapeutics, 2021). IND applications for phage therapy are similarly subjected to FDA review. Clinical trials are thereby approved if FDA does not hold on the application after 30 days (Yang et al. 2023). FDA also allows extended access to INDs under exceptional conditions for patients who have not enrolled in clinical trials. The Institutional Review board and FDA should be made aware of the prolonged usage of phage therapy (Yang et al. 2023).

19.11 Antimicrobial Peptides (AMPs)

AMPs play a significant role in fighting the battle against antibiotic resistance. These are small molecular peptides (<100 amino acids) that play a crucial role in fighting against microbes like bacteria, fungi, parasites and viruses (Wang et al. 2018). They also play a role in several biological activities like immune regulation, angiogenesis, wound healing, and antitumor activity. They are shown to be of amphiphatic nature, containing both hydrophobic and hydrophilic regions (Talapko et al. 2022). Certain other AMPs make use of metal ions, forming cationic salt bridges with negatively charged particles on the microbial membranes, allowing for their penetration. Upon reaching the cytoplasm, they interact with the ribosome, inhibiting its ribonuclease activity (Wang et al. 2018).

It was first discovered in 1939 by microbiologist Rene Dubos while working on an antimicrobial agent named gramicidin that protected mice from pneumococcal infection (Moretta et al. 2021). Several AMPs were discovered thereafter including tyrocidine, produced by *Bacillus brevis* which show antimicrobial activity against bacterial species, purothionin, from the plant*Triticumaestivu*, against fungi and bacteria (Kang et al. 2019). According to AMP database there are more than 3000 AMPs that have been isolated and examined till now.

The three different pathways for AMP biosynthesis include the Traditional ribosomal synthesis, non-ribosomal synthesis, and proteolytic digestion of proteins. Ribosomal translation machinery is used for the synthesis of AMPs such as histatins and human β-defensins. Non-ribosomal peptide synthesis mechanisms are commonly found in filamentous fungi and bacteria. Lastly, cryptic peptides are the ones produced by proteolytic cleavage of bigger protein molecules. AMPs are generally expressed throughout the entire lifespan, but they are stored in granules and only released at the specific site of infection or inflammation, under necessary conditions (Kang et al. 2019).

19.12 Drug Development

The chemical synthesis of AMPs as APIs (Active Pharmaceutical ingredients) is highly effective but complex and expensive. Genetic engineering is considered as one of the best suitable method for designing high quality AMPs. To increase the yield of AMPs, certain recombinant technologies have been implemented by making use of competent bacteria, yeast or transgenic plants and animals. As the manufacturing cost of AMPs are significantly high, biotechnology engineering or fermentation should be considered as cheaper alternatives. Initially, in vitro screening is done to analyze the physio-chemical properties and antimicrobial activity to check the efficiency of the AMPs (Zhang et al. 2021).

According to Dosler and Mataraci (2013), the results of in vitro experimental studies of AMPs individually and in combination with antibiotics suggested that AMPs improve the pharmacokinetic efficacy of antibiotics. Additionally, the bioavailability of AMPs decreased due to the unsuitable pharmacodynamics of AMPs, including their instability resulting from proteolytic degradation, neutralization by negatively charged proteins, and rapid elimination by the kidney and liver (Moretta et al. 2021).

PK/PD modeling and simulation help identify the multifaceted correlation between the specific AMP, it's target bacterium, and the host organism's response it it's administration. This is done to optimize the dosage regime and establish novel antimicrobial peptides (Moretta et al. 2021).

19.13 AMP Formulation and Delivery

The development of AMP-based therapy faces several challenges with respect to the stability, efficacy, and toxicity of AMPs. Nanotechnology is used to increase the stability, resulting in nano-structure that improves the pharmacodynamics of AMPs by inhibiting renal clearance and enhancing permeability though bacterial membranes. Nano-delivery systems including metal nanoparticles, carbon nanotubes, and polymers-based nanostructures are known to efficient and they improve bioactivity and antimicrobial effectiveness of AMPs (Zhang et al. 2021).

AMPs are not stable in the GIT track due to high pH concentration, so this will reduce the activity of peptides. Various chemical strategies like metabolic stability and lipophilicity, is carried out to enhance permeability (Greber and Dawgul 2016).

There are various common techniques, including cyclization, to enhance metabolic stability and maintain the original bioactive conformation. Another approach is by replacing genetically encoded amino acids (noncanonical residues), which results in peptidomimetics with improved stability and permeability, and also by modifying N-/C-terminals to enhance permeability and prevent proteolytic degradation. Altering backbone like replacing peptide bonds with enzyme-stable bioisosteres can also be used (Matsuzaki 2019).

Encapsulation with porous materials was employed to AMPs in order to protect them from proteolytic degradation and offer increased stability. As for AMP release, silica nanoparticles with smaller pore size is found to release it faster than those with larger ones. Hence, pore size matters while designing encapsulates. Another way of delivery of AMP is via attachment on a surface. The various hydrophilic, cationic, and hydrophobic groups on AMPs are used to form physical interactions with different materials. Surface attachment is done using polymer micelles, Polymer nanofibers, liposomes, lipid nanoparticles, and lipid crystals (Wang et al. 2021).

AMPs can also be covalently modified via conjugation without compromising the antimicrobial activities. This technique has a certain advantage over physical interactions because it provides greater chemical modifiability of the peptides. Nanoparticle peptide conjugates require conjugation with chitosan, gold, and silver nanoparticles that are shown to boost its functionality. Other conjugation techniques include biomacromolecule peptide conjugates with proteins, DNA, and polysaccharides. This technique reduces the level of toxicity of the conjugate. AMPs can also be conjugated with antibiotics, peptide nucleic acid to mitigate the risk of adverse side effects. Conjugation with lipid molecules exhibits a broad-spectrum activity against pathogens and a lower possibility of drug resistance (Wang et al. 2021).

19.14 Drug Combinations

A potential therapeutic approach is designed in combination with antibiotics to overcome antibiotic resistance, in turn enhancing the antibacterial activity and simultaneous reduction in the toxicity of AMPs. AMP (e.g.: Nisin) combined with antibiotics like penicillin, chloramphenicol, ciproflaxin, indolicidin, or azithromycin, tetracycline antibiotics showed promising results in fighting multidrug-resistant bacteria. A combination of two or more types AMPs can also result in higher efficacy. Hence the antibacterial activity can be further increased by different combination therapies involving synthetic antimicrobial polymers (Zhang et al. 2021).

19.15 Current Clinical Trials and Challenges with AMPs

AMPs, due to their low toxicity levels, wide range of targets, low accumulation in tissue, and inability to any immune responses, can potentiate therapy to treat skin infection and soft tissue. They can also fight multi-microbial infections. AMPs are already approved by the FDA and the EMA (Mahlapuu et al. 2016).

Clinical trials are most important because they aim to try a new drug on a volunteer human before giving it to a large population to check its efficacy, safety, and dosage requirements. A drug takes time to enter the market in terms of cost and time. All this depends on drug regulatory agencies around the world, like the FDA and the European Medicines Agency in the European Union, which are the most powerful agencies. After their approval, drugs are available in the market, and patents are granted to novel drugs. Clinical trials have four major phases that always abide by the standards of Good Clinical Practice (GCP) (Costa et al. 2019).

In the first stage of clinical trials, companies check the drug for its side effects, dosage requirements, and safety. There are two important factors: PK (Pharmacokinetics) and another PD (Pharmacodynamics). The major focus is on PK (absorption, distribution, metabolism, elimination, and toxicity—ADMET). During research, pharmaceutical pharmacodynamics (PD) is examined in a small group of healthy people (50–100) In Phase II, up to 600 human volunteers are included in the trial to study the effectiveness of the drug but not focusing on safety. This study has been done to establish preliminary data on how the drug treats people with such diseases compared to standard treatment and placebo. In Phase III, more people are considered to evaluate the effect of dosage range, safety, effectiveness, and combination with other drugs on the market. This stage is crucial because if the FDA, US, or EMA authorizes it, it can come on the market. The final stage is Phase IV, where the drug is monitored after being placed in the market and this involves a diverse population (Costa et al. 2019).

In preclinical studies, various AMPs give positive results, while in clinical trials, they are not able to translate because, in preclinical studies, there is an absence of real physiological conditions like, high ionic strength, and presence of protease enzymes. This condition varies due to different in vivo and in vitro data, and vice versa (Mahlapuu et al. 2016).

In Table 19.5. Various Clinical trials data from Phase I to III of AMPs against Bacterial infection and all adapted from (DrugDataBase 2022; Usmani et al. 2017).

19.16 Regulations of Peptide Therapy Across the World

The development of AMPs that started decades ago, the optimism that exists today is resisted when facing difficulties in translating R & D to create novel products (Fox 2013). Due to the complex regulatory guidelines, biotech firms and research are caught up and it became difficult for AMPs to enter into clinical trials. After overcoming this hurdle, another problem in clinical trials was the pathological state, target population, and various delivery methods of AMPs (Matsuzaki 2019).

In order for a particular AMP to be successfull, it must undergo a land and expensive iterative process. This is why it is more difficult for biotech startups even if they make partnerships with Big Pharma companies. The post-antibiotic erarequired a strong solution in the medical field so the regulators have updated guidelines to give more attention to promote the clinical application of new therapies such as AMPs. Pexiganan, an AMP that did not get approved by the US FDA for treatment of

Table 19.5 Antimicrobial peptide clinical trails

Peptide name	Target	Administration	Phase	Mechanism	References
Nisin	Gram-positive bacteria	Oral	I/II	Depolarization of cell membrane	Prince et al. (2016)
Gramicidin	Infected wounds and ulcers	Topical	III	Membrane disruption/ immunomodulation	David and Rajasekaran (2015)
Polymyxin B	Gram-negative bacteria	Topical	III	Membrane disruption/ immunomodulation	Morrison and Jacobs (1976)
Polymyxin E (Colistin)	*A. baumannii*/pneumonia	Intravenous	III	Membrane disruption/ immunomodulation	Yu et al. (2015)
Daptomycin	Skin infection/bacteremia	Intravenous	III	Membrane disruption/ immunomodulation	Taylor and Palmer (2016)
Melittin	Inflammation	Intradermal	I/II	Membrane disruption/ immunomodulation	Lee and Bae (2016)
Friulimicin	MRSA/pneumonia	Intravenous	I	Membrane disruption	Schneider et al. (2009)
Murepavadin (POL7080)	*P. aeruginosa, K. pneumoniae*	Intravenous	II	Binding to LptD	Srinivas et al. (2010)
Neuprex® (rBPI21)	Pediatric meningococcemia	Intravenous	III	Membrane disruption	Schultz and Weiss (2007)
Iseganan (IB-367)	Pneumonia/oral mucositis	Topical	III	Membrane disruption	Orlov et al. (1805)
Pexiganan (MSI-78)	Diabetic foot ulcers	Topical	III	Membrane disruption/ immunomodulation	Gottler and Ramamoorthy (2009)
Surotomycin (CB-315)	*C. difficile*	Oral	III	Membrane disruption	Alam et al. (2015)
XOMA-629 (XMP-629)	Impetigo/acne rosacea	Topical	III	Immunomodulation	Easton et al. (2009)
Omiganan (MBI-226)	Antisepsis/catheter infection	Topical	Completed III	Membrane disruption/ immunomodulation	Rubinchik et al. (2009)
NVB-302	*C. difficile*	Oral	I	Inhibition of cell wall synthesis	Crowther et al. (2013)
OP-145	Chronic middle ear infection	Ear drops	I/II	Membrane disruption/ immunomodulation	Malanovic et al. (2015)

Peptide name	Target	Administration	Phase	Mechanism	References
LTX-109	MRSA/impetigo	Topical	I/II	Membrane disruption	Isaksson et al. (2011), Sivertsen et al. (2014)
EA-230	Sepsis	Intravenous	II	Immunomodulation	van Groenendael et al. (2018)
SGX942 (Dusquetide)	Oral mucositis	Oral rinse	III	Immunomodulation	Kudrimoti et al. (2016)
hLF1–11	Bacterial infection	Intravenous	I/II	Membrane disruption/ immunomodulation	Welling et al. (2018)
C16G2	*Streptococcus mutans*	Mouth wash	II	Membrane disruption	Guo et al. (2015)
Ramoplanin (NTI-851)	*C. difficile*, VRE	Oral	III	Inhibition of cell wall synthesis	Fulco and Wenzel (2006)
p2TA (AB103)	Necrotic tissue infection	Intravenous	III	Immunomodulation	Bulger et al. (2014)
D2A21	Burn wound infections	Topical	III	Membrane disruption	Muchintala et al. (2020)
LFF571	*C. difficile*	Oral	II	Inhibition of protein synthesis	Leeds et al. (2012)
GSK1322322 (Lanopepden)	Bacterial skin infection	Oral	II	Peptide deformylase inhibitor	Peyrusson et al. (2015)
PMX-30063 (Brilacidin)	Acute bacterial skin infection	Intravenous	II	Membrane disruption/ immunomodulation	Mensa et al. (2014)
XF-73 (Exeporfinium chloride)	*Staphylococcal* infection	Topical	II	Membrane disruption	Ooi et al. (2009)
Ghrelin	Chronic respiratory infection	Intravenous	II	Immunomodulation	Gualillo et al. (2003)
Wap-8294A2 (Lotilibcin)	Gram-positive bacteria	Topical	I/II	Membrane disruption	Itoh et al. (2017)
PL-5	Skin infections	Topical	I	Membrane disruption	Miyake et al. (2004)

infected diabetic ulcers because it had no advantage over oral ofloxacin would have a different outcome in the present scenario (Matsuzaki 2019). Few AMPs are conjugated with antibiotics and work synergistically making them multidrug approaches to speed up the elimination of bacterial infection to prevent pathogen resistance. Overall, AMPs have a promising future as a novel and effective class of antibiotics (Matsuzaki 2019).

19.17 Conclusions

Bacteriophages have several potential therapeutic properties such as their high specific and effective lysing property, they are safe, rapidly modifiable to combat new bacterial infections. Phage therapy is mostly used for the treatment of antibiotic-resistant bacterial infections and for its advantage over classical antibiotics. While using phage therapy as a treatment method, the pharmacodynamics of the phages should be taken into consideration. As phages are live viruses that are being injected into the humans, proper consent should be taken and the patients should be made aware of the side effects with respect to the therapy. Not only that, proper evaluation of the clinical trial data should be done with a thorough screening of the results to release the product into the market. Pertaining to the antibiotic-resistant bacteria known as the superbugs, phage therapy has gained popularity in its treatment in the recent years. Antimicrobial peptides are another such novel therapeutic treatment option that is mostly used in conjugation with antibiotics. These peptides themselves penetrate through the bacterial wall and inactivate their replicating machinery. However, when administered with antibiotics, they are shown to increase its efficacy for better bacterial killing. Similar to phages, AMPs require vigorous rounds of clinical trials before it can be approved and released in the market. Even though the government and the regulatory agencies are approving this phage and peptide therapy slowly, more efforts and attention should be put into its research and development by the biotech or the pharmaceutical companies to make these treatments readily available and effective to the common people.

References

Abedon ST (2019) Phage-antibiotic combination treatments: antagonistic impacts of antibiotics on the pharmacodynamics of phage therapy? Antibiotics 8(4):182. https://doi.org/10.3390/antibiotics8040182

Adaptive Phage Therapeutics (2021, February 10) Announces FDA Clearance of IND Application for PhageBankTM for the Treatment of Prosthetic Joint Infections. BioSpace. https://www.biospace.com/adaptive-phage-therapeutics-announces-fda-clearance-of-ind-application-for-phagebankfor-the-treatment-of-prosthetic-joint-infections

Ajulo S, Awosile B (2024) Global antimicrobial resistance and use surveillance system (GLASS 2022): investigating the relationship between antimicrobial resistance and antimicrobial consumption data across the participating countries. PLoS One 19(2):e0297921–e0297921. https://doi.org/10.1371/journal.pone.0297921

Alam MZ, Wu X, Mascio C, Chesnel L, Hurdle JG (2015) Mode of action and bactericidal properties of surotomycin against growing and nongrowing Clostridium difficile. Antimicrob Agents Chemother 59(9):5165–5170. https://doi.org/10.1128/aac.01087-15

AlamSarker S, Berger B, Deng Y, Kieser S, Foata F, Moine D, Descombes P, Sultana S, Huq S, Bardhan PK, ValérieVuillet FP, Brüssow H (2016) Oral application of *Escherichia coli* bacteriophage: safety tests in healthy and diarrheal children from Bangladesh. Environ Microbiol 19(1):237–250. https://doi.org/10.1111/1462-2920.13574

Alisky J, Iczkowski K, Rapoport A, Troitsky N (1998). Bacteriophages show promise as antimicrobial agents. Journal of Infection/ the Journal of Infection, 36(1):5–15. https://doi.org/10.1016/s0163-4453(98)92874-2

Antimicrobial Resistance and Phage Therapy in India (2023) The microbiologist. https://www.themicrobiologist.com/features/antimicrobial-resistance-and-phage-therapy-in-india/1386.article

Bruttin A, Brussow H (2005) Human volunteers receiving Escherichia coli phage T4 orally: a safety test of phage therapy. Antimicrob Agents Chemother 49(7):2874–2878. https://doi.org/10.1128/aac.49.7.2874-2878.2005

Bulger EM, Maier RV, Sperry J, Joshi M, Henry S, Moore FA, Moldawer LL, Demetriades D, Talving P, Schreiber M, Ham B, Cohen M, Opal S, Segalovich I, Maislin G, Kaempfer R, Shirvan A (2014) A novel drug for treatment of necrotizing soft-tissue infections. JAMA Surg 149(6):528. https://doi.org/10.1001/jamasurg.2013.4841

Buttimer C, McAuliffe O, Ross RP, Hill C, O'Mahony J, Coffey A (2017) Bacteriophages and bacterial plant diseases. Front Microbiol 8:34. https://doi.org/10.3389/fmicb.2017.00034

Cairns BJ, Timms AR, Jansen VAA, Connerton IF, Payne RJH (2009) Quantitative models of in vitro bacteriophage–host dynamics and their application to phage therapy. PLoS Pathog 5(1):e1000253. https://doi.org/10.1371/journal.ppat.1000253

Comeau AM, Tétart F, Trojet SN, Prère M-F, Krisch HM (2007) Phage-antibiotic synergy (PAS): β-lactam and quinolone antibiotics stimulate virulent phage growth. PLoS One 2(8):e799. https://doi.org/10.1371/journal.pone.0000799

Cook MA, Wright GD (2022) The past, present, and future of antibiotics. Sci Transl Med 14(657):eabo7793. https://doi.org/10.1126/scitranslmed.abo7793

Costa F, Teixeira C, Gomes P, Martins MCL (2019) Clinical Application of AMPs. Advances in Experimental Medicine and Biology 281–298. https://doi.org/10.1007/978-981-13-3588-4_15

Crowther GS, Baines SD, Todhunter SL, Freeman J, Chilton CH, Wilcox MH (2013) Evaluation of NVB302 versus vancomycin activity in an in vitro human gut model of Clostridium difficile infection. J Antimicrob Chemother 68(1):168–176. https://doi.org/10.1093/jac/dks359

David JM, Rajasekaran AK (2015) Gramicidin A: a new mission for an old antibiotic. J Kidney Cancer VHL 2(1):15–24. https://doi.org/10.15586/jkcvhl.2015.21

Doron S, Gorbach SL (2008) Bacterial infections: overview. Elsevier EBooks, pp 273–282. https://doi.org/10.1016/b978-012373960-5.00596-7

Dosler S, Mataraci E (2013) In vitro pharmacokinetics of antimicrobial cationic peptides alone and in combination with antibiotics against methicillin resistant Staphylococcus aureus biofilms. Peptides 49:53–58. https://doi.org/10.1016/j.peptides.2013.08.008

DrugBank Online | Database for Drug and Drug Target Info (2022). Drugbank.com; DrugBank. https://go.drugbank.com/

Easton DM, Nijnik A, Mayer ML, Hancock REW (2009) Potential of immunomodulatory host defense peptides as novel anti-infectives. Trends Biotechnol 27(10):582–590. https://doi.org/10.1016/j.tibtech.2009.07.004

Fish R, Kutter E, Wheat G, Blasdel B, Kutateladze M, Kuhl S (2016) Bacteriophage treatment of intransigent diabetic toe ulcers: a case series. J Wound Care 25(Sup7):S27–S33. https://doi.org/10.12968/jowc.2016.25.sup7.s27

Fish R, Kutter E, Bryan D, Wheat G, Kuhl S (2018) Resolving digital staphylococcal osteomyelitis using bacteriophage—A case report. Antibiotics 7(4):87–87. https://doi.org/10.3390/antibiotics7040087

Fox JL (2013) Antimicrobial peptides stage a comeback. Nat Biotechnol 31(5):379–382. https://doi.org/10.1038/nbt.2572

Fulco P, Wenzel RP (2006) Ramoplanin: a topical lipoglycodepsipeptide antibacterial agent. Expert Rev Anti-Infect Ther 4(6):939–945. https://doi.org/10.1586/14787210.4.6.939

Furfaro LL, Payne MS, Chang BJ (2018) Bacteriophage therapy: clinical trials and regulatory hurdles. Front Cell Infect Microbiol 8:376. https://doi.org/10.3389/fcimb.2018.00376

Górski A, Międzybrodzki R, Łobocka M, Głowacka-Rutkowska A, Bednarek A, Borysowski J, Jończyk-Matysiak E, MarzannaŁusiak-Szelachowska, Weber-Dąbrowska B, Bagińska N, Letkiewicz S, Dąbrowska K, Scheres J (2018) Phage therapy: what have we learned? Viruses 10(6):288–288. https://doi.org/10.3390/v10060288

Gottler LM, Ramamoorthy A (2009) Structure, membrane orientation, mechanism, and function of pexiganan—A highly potent antimicrobial peptide designed from magainin. Biochim Biophysic Acta 1788(8):1680–1686. https://doi.org/10.1016/j.bbamem.2008.10.009

Greber KE, Dawgul M (2016) Antimicrobial peptides under clinical trials. Curr Top Med Chem 17(5):620–628. https://doi.org/10.2174/1568026616666160713143331

Gualillo O, Lago F, Gómez-Reino J, Casanueva FF, Dieguez C (2003) Ghrelin, a widespread hormone: insights into molecular and cellular regulation of its expression and mechanism of action. FEBS Lett 552(2–3):105–109. https://doi.org/10.1016/s0014-5793(03)00965-7

Guo L, McLean JS, Yang Y, Eckert R, Kaplan CW, Kyme P, Sheikh O, Varnum B, Lux R, Shi W, He X (2015) Precision-guided antimicrobial peptide as a targeted modulator of human microbial ecology. Proc Natl Acad Sci 112(24):7569–7574. https://doi.org/10.1073/pnas.1506207112

Hoyle N, Zhvaniya P, Balarjishvili N, Bolkvadze D, Nadareishvili L, Nizharadze D, Wittmann J, Rohde C, Kutateladze M (2018) Phage therapy against Achromobacterxylosoxidans lung infection in a patient with cystic fibrosis: a case report. Res Microbiol 169(9):540–542. https://doi.org/10.1016/j.resmic.2018.05.001

Isaksson J, Brandsdal BO, Engqvist M, Flaten GE, Svendsen JS, Stensen W (2011) A synthetic antimicrobial peptidomimetic (LTX 109): stereochemical impact on membrane disruption. J Med Chem 54(16):5786–5795. https://doi.org/10.1021/jm200450h

Itoh H, Tokumoto K, Kaji T, Paudel A, Panthee S, Hamamoto H, Sekimizu K, Inoue M (2017) Total synthesis and biological mode of action of WAP-8294A2: a menaquinone-targeting antibiotic. J Org Chem 83(13):6924–6935. https://doi.org/10.1021/acs.joc.7b02318

Jamal M, Ahmad W, Andleeb S, Jalil F, Imran M, Nawaz MA, Hussain T, Ali M, Rafiq M, Kamil MA (2018a) Bacterial biofilm and associated infections. J Chin Med Assoc 81(1):7–11. https://doi.org/10.1016/j.jcma.2017.07.012

Jamal M, Bukhari SMAUS, Andleeb S, Ali M, Raza S, Nawaz MA, Hussain T, Rahman SU, Shah SSA (2018b) Bacteriophages: an overview of the control strategies against multiple bacterial infections in different fields. J Basic Microbiol 59(2):123–133. https://doi.org/10.1002/jobm.201800412

Jasim HN, Hafidh RR, Abdulamir AS (2018) Formation of therapeutic phage cocktail and endolysin to highly multi-drug resistant Acinetobacterbaumannii: in vitro and in vivo study. DOA 21(11):1100–1108. https://doi.org/10.22038/ijbms.2018.27307.6665

Jault P, Leclerc T, Jennes S, Pirnay JP, Que YA, Resch G, Rousseau AF, Ravat F, Carsin H, Le Floch R, Schaal JV (2019) Efficacy and tolerability of a cocktail of bacteriophages to treat burn wounds infected by Pseudomonas aeruginosa (PhagoBurn): a randomised, controlled, double-blind phase 1/2 trial. Lancet Infect Dis 19(1):35–45. https://doi.org/10.1016/s1473-3099(18)30482-1

Jennes S, Merabishvili M, Soentjens P, Pang KW, Rose T, Keersebilck E, Soete O, François PM, Teodorescu S, Verween G, Verbeken G, De Vos D, Pirnay J-P (2017) Use of bacteriophages in the treatment of colistin-only-sensitive Pseudomonas aeruginosa septicaemia in a patient with acute kidney injury—a case report. Crit Care 21(1):129. https://doi.org/10.1186/s13054-017-1709-y

Jikia D, Chkhaidze N, Imedashvili E, Mgaloblishvili I, Tsitlanadze G, Katsarava R, Glenn Morris J, Sulakvelidze A (2005) The use of a novel biodegradable preparation capable of the sustained release of bacteriophages and ciprofloxacin, in the complex treatment of multidrug-resistant Staphylococcus aureus-infected local radiation injuries caused by exposure to Sr90. Clin Exp Dermatol 30(1):23–26. https://doi.org/10.1111/j.1365-2230.2004.01600.x

Kang X, Dong F, Shi C, Zheng H (2019) DRAMP 2.0, an updated data repository of antimicrobial peptides. ResearchGate; Springer Nature. https://www.researchgate.net/publication/335144811_DRAMP_20_an_updated_data_repository_of_antimicrobial_peptides

Koonin EV, Makarova KS, Zhang F (2017) Diversity, classification and evolution of CRISPR-Cas systems. Curr Opin Microbiol 37:67–78. https://doi.org/10.1016/j.mib.2017.05.008

Kronheim S, Daniel-Ivad M, Duan Z, Hwang S, Wong AI, Mantel I, Nodwell JR, Maxwell KL (2018) A chemical defence against phage infection. Nature 564(7735):283–286. https://doi.org/10.1038/s41586-018-0767-x

Kudrimoti M, Curtis A, Azawi S, Worden FP, Katz S, Adkins D, Bonomi M, Elder J, Sonis ST, Straube R, Donini O (2016) Dusquetide: A novel innate defense regulator demonstrating a significant and consistent reduction in the duration of oral mucositis in preclinical data and a randomized, placebo-controlled phase 2a clinical study. J Biotechnol 239:115–125. https://doi.org/10.1016/j.jbiotec.2016.10.010

Kutateladze M (2015) Experience of the eliava institute in bacteriophage therapy. Virol Sin 30(1):80–81. https://doi.org/10.1007/s12250-014-3557-0

Kutter E, De Vos D, Gvasalia G, Alavidze Z, Gogokhia L, Kuhl S, Abedon S (2010) Phage therapy in clinical practice: treatment of human infections. Curr Pharm Biotechnol 11(1):69–86. https://doi.org/10.2174/138920110790725401

Lee G, Bae H (2016) Anti-inflammatory applications of melittin, a major component of bee venom: detailed mechanism of action and adverse effects. Molecules 21(5):616. https://doi.org/10.3390/molecules21050616

Leeds JA, Sachdeva M, Mullin S, Dzink-Fox J, LaMarche MJ (2012) Mechanism of action of and mechanism of reduced susceptibility to the novel anti-Clostridium difficile compound LFF571. Antimicrob Agents Chemother 56(8):4463–4465. https://doi.org/10.1128/AAC.06354-11

Leitner L, Ujmajuridze A, Chanishvili N, Goderdzishvili M, Chkonia I, Rigvava S, Chkhotua A, Changashvili G, McCallin S, Schneider MP, Liechti MD, Mehnert U, Bachmann LM, Sybesma W, Kessler TM (2020) Intravesical bacteriophages for treating urinary tract infections in patients undergoing transurethral resection of the prostate: a randomised, placebo-controlled, double-blind clinical trial. Lancet Infect Dis 21(3):427–436. https://doi.org/10.1016/S1473-3099(20)30330-3

Leung CY, Weitz J S (2016) Synergistic elimination of bacteria by phage and the innate immune system. https://doi.org/10.1101/057927

Levin BR, Bull JJ (2004) Population and evolutionary dynamics of phage therapy. Nat Rev Microbiol 2(2):166–173. https://doi.org/10.1038/nrmicro822

Levison ME, Levison JH (2009) Pharmacokinetics and pharmacodynamics of antibacterial agents. Infect Dis Clin N Am 23(4):791–815. https://doi.org/10.1016/j.idc.2009.06.008

Liang S, Qi Y, Yu H, Sun W, Raza SHA, Alkhorayef N, Alkhalil SS, Salama EEA, Zhang L (2023) Bacteriophage Therapy as an Application for Bacterial Infection in China. Antibiotics 12(2):417. https://doi.org/10.3390/antibiotics12020417

Liu Y, Mi Z, Mi L, Huang Y, Li P, Liu H, Yuan X, Niu W, Jiang N, Bai C, Gao Z (2019) Identification and characterization of capsule depolymerase Dpo48 from *Acinetobacterbaumannii* phage IME200. PeerJ 7:e6173–e6173. https://doi.org/10.7717/peerj.6173

Mahlapuu M, Håkansson J, Ringstad L, Björn C (2016) Antimicrobial peptides: an emerging category of therapeutic agents. Front Cell Infect Microbiol 6:194. https://doi.org/10.3389/fcimb.2016.00194

Malanovic N, Leber R, Schmuck M, Kriechbaum M, Cordfunke RA, Drijfhout JW, de Breij A, Nibbering PH, Kolb D, Lohner K (2015) Phospholipid-driven differences determine the action of the synthetic antimicrobial peptide OP-145 on gram-positive bacterial and mammalian membrane model systems. Biochim Biophys Acta 1848(10):2437–2447. https://doi.org/10.1016/j.bbamem.2015.07.010

Malik DJ, Sokolov IJ, Vinner GK, Mancuso F, Cinquerrui S, Vladisavljevic GT, Clokie MRJ, Garton NJ, Stapley AGF, Kirpichnikova A (2017) Formulation, stabilisation and encapsulation of bacteriophage for phage therapy. Adv Colloid Interf Sci 249:100–133. https://doi.org/10.1016/j.cis.2017.05.014

Matsuzaki K (2019) Antimicrobial peptides. Springer

Mensa B, Howell GL, Scott RW, DeGrado WF (2014) Comparative mechanistic studies of brilacidin, daptomycin, and the antimicrobial peptide LL16. Antimicrob Agents Chemother 58(9):5136–5145. https://doi.org/10.1128/aac.02955-14

Międzybrodzki R, Hoyle N, Zhvaniya F, Łusiak-Szelachowska M, Weber-Dąbrowska B, Łobocka M, Borysowski J, Alavidze Z, Kutter E, Górski A, Gogokhia L (2018) Current updates from the long-standing phage research centers in Georgia, Poland, and Russia. Bacteriophages 1–31. https://doi.org/10.1007/978-3-319-40598-8_31-1

Miyake O, Ochiai A, Hashimoto W, Murata K (2004) Origin and diversity of alginate lyases of families PL-5 and -7 in *Sphingomonas* sp. strain A1. J Bacteriol 186(9):2891–2896. https://doi.org/10.1128/jb.186.9.2891-2896.2004

Moretta A, Scieuzo C, Petrone AM, Salvia R, Manniello MD, Franco A, Lucchetti D, Vassallo A, Vogel H, Sgambato A, Falabella P (2021) Antimicrobial peptides: a new hope in biomedical and pharmaceutical fields. Front Cell Infect Microbiol 11:668632. https://doi.org/10.3389/fcimb.2021.668632

Morrison DC, Jacobs DM (1976) Binding of polymyxin B to the lipid A portion of bacterial lipopolysaccharides. Immunochemistry 13(10):813–818. https://doi.org/10.1016/0019-2791(76)90181-6

Muchintala D, Suresh V, Raju D, Sashidhar RB (2020) Synthesis and characterization of cecropin peptide-based silver nanocomposites: its antibacterial activity and mode of action. Mater Sci Eng C 110:110712. https://doi.org/10.1016/j.msec.2020.110712

Murray CJL, Ikuta KS, Sharara F, Swetschinski L, Aguilar GR, Gray A, Han C, Bisignano C, Rao P, Wool E, Johnson SC, Browne AJ, Chipeta MG, Fell F, Hackett S, Haines-Woodhouse G, Bahar, Kumaran EAP, McManigal B, Achalapong S (2022) Global burden of bacterial antimicrobial resistance in 2019: a systematic analysis. Lancet 399(10325):629–655. https://doi.org/10.1016/s0140-6736(21)02724-0

Ooi N, Miller K, Hobbs J, Rhys-Williams W, Love W, Chopra I (2009) XF-73, a novel antistaphylococcal membrane-active agent with rapid bactericidal activity. J Antimicrob Chemother 64(4):735–740. https://doi.org/10.1093/jac/dkp299

Orlov D, Hong T, Menzel LP, Azimov R, Falla TJ, Waring AJ, et al. (1805) Bactericidal mechanism of iseganan (IB-367), a rapidly acting antimicrobial protegrin peptide, in Proceedings of the 41st Annual Interscience Conference on Antimicrobial Agents and Chemotherapy, Chicago

Payne RJH, Jansen VAA (2001) Understanding bacteriophage therapy as a density-dependent kinetic process. J Theor Biol 208(1):37–48. https://doi.org/10.1006/jtbi.2000.2198

Peyrusson F, Butler D, Tulkens PM, Van Bambeke F (2015) Cellular pharmacokinetics and intracellular activity of the novel peptide deformylase inhibitor GSK1322322 against Staphylococcus aureus laboratory and clinical strains with various resistance phenotypes: studies with human THP-1 monocytes and J774 murine macrophages. Antimicrob Agents Chemother 59(9):5747–5760. https://doi.org/10.1128/aac.00827-15

Prince A, Sandhu P, Ror P, Dash E, Sharma S, Arakha M, Jha S, Akhter Y, Saleem M (2016) Lipid-II independent antimicrobial mechanism of Nisin depends on its crowding and degree of oligomerization. Sci Rep 6(1):37908. https://doi.org/10.1038/srep37908

Rhoads DD, Wolcott RD, Kuskowski MA, Wolcott BM, Ward LS, Sulakvelidze A (2020) Bacteriophage therapy of venous leg ulcers in humans: results of a phase I safety trial. J Wound Care 18(6):237–243. https://doi.org/10.12968/jowc.2009.18.6.42801

Rosner D, Clark J (2021) Formulations for bacteriophage therapy and the potential uses of immobilization. Pharmaceuticals 14(4):359–359. https://doi.org/10.3390/ph14040359

Rubinchik E, Dugourd D, Algara T, Pasetka C, Friedland HD (2009) Antimicrobial and antifungal activities of a novel cationic antimicrobial peptide, omiganan, in experimental skin colonisation models. Int J Antimicrob Agents 34(5):457–461. https://doi.org/10.1016/j.ijantimicag.2009.05.003

Schneider T, Gries K, Josten M, Wiedemann I, Pelzer S, Labischinski H, Sahl H-G (2009) The lipopeptide antibiotic friulimicin B inhibits cell wall biosynthesis through complex formation with bactoprenol phosphate. Antimicrob Agents Chemother 53(4):1610–1618. https://doi.org/10.1128/aac.01040-08

Schooley RT, Biswas B, Gill JJ, Hernandez-Morales A, Lancaster J, Lessor L, Barr JJ, Reed SL, Rohwer F, Benler S, Segall AM, Taplitz R, Smith DM, Kerr K, Kumaraswamy M, Nizet V, Lin L, McCauley MD, Strathdee SA, Benson CA (2017) Development and use of personalized bacteriophage-based therapeutic cocktails to treat a patient with a disseminated resistant acinetobacterbaumannii infection. Antimicrob Agents Chemother 61(10):10–128. https://doi.org/10.1128/aac.00954-17

Schultz H, Weiss JP (2007) The bactericidal/permeability-increasing protein (BPI) in infection and inflammatory disease. Clin Chim Acta 384(1–2):12–23. https://doi.org/10.1016/j.cca.2007.07.005

Segall AM, Roach DR, Strathdee SA (2019) Stronger together? Perspectives on phage-antibiotic synergy in clinical applications of phage therapy. Curr Opin Microbiol 51:46–50. https://doi.org/10.1016/j.mib.2019.03.005

Semler DD, Lynch KH, Dennis JJ (2012) The promise of bacteriophage therapy for burkholderiacepacia complex respiratory infections. Front Cell Infect Microbiol 1:27. https://doi.org/10.3389/fcimb.2011.00027

Sivertsen A, Isaksson J, Leiros HKS, Svenson J, Svendsen JS, Brandsdal BO (2014) Synthetic cationic antimicrobial peptides bind with their hydrophobic parts to drug site II of human serum albumin. BMC Structural Biology 14(1). https://doi.org/10.1186/1472-6807-14-4

Srinivas N, Jetter P, Ueberbacher BJ, Werneburg M, Zerbe K, Steinmann J, Van der Meijden B, Bernardini F, Lederer A, Dias RLA, Misson PE, Henze H, Zumbrunn J, Gombert FO, Obrecht D, Hunziker P, Schauer S, Ziegler U, Käch A, Eberl L (2010) Peptidomimetic antibiotics target outer-membrane biogenesis in *Pseudomonas aeruginosa*. Science 327(5968):1010–1013. https://doi.org/10.1126/science.1182749

Summers WC (1999) Félix d'Herelle and the origins of molecular biology. Yale University Press

Talapko J, Meštrović T, Juzbašić M, Tomas M, Erić S, Aleksijević LH, Bekić S, Schwarz D, Matić S, Neuberg M, Škrlec I (2022) Antimicrobial peptides—mechanisms of action, antimicrobial effects and clinical applications. Antibiotics 11(10):1417–1417. https://doi.org/10.3390/antibiotics11101417

Taylor SD, Palmer M (2016) The action mechanism of daptomycin. Bioorg Med Chem 24(24):6253–6268. https://doi.org/10.1016/j.bmc.2016.05.052

The Promise of Phages (2023) Nat Biotechnol 41(5):583–583. https://doi.org/10.1038/s41587-023-01807-7

Usmani SS, Bedi G, Samuel JS, Singh S, Kalra S, Kumar P, Ahuja AA, Sharma M, Gautam A, Raghava GPS (2017) THPdb: Database of FDA-approved peptide and protein therapeutics. PLoS ONE 12(7):e0181748. https://doi.org/10.1371/journal.pone.0181748

van Groenendael R, Kox M, van Eijk LT, Pickkers P (2018) Immunomodulatory and kidney-protective effects of the human chorionic gonadotropin derivate EA-230. Nephron 140(2):148–151. https://doi.org/10.1159/000490772

Wang J, Dou X, Song J, Lyu Y, Zhu X, Xu L, Li W, Shan A (2018) Antimicrobial peptides: promising alternatives in the post feeding antibiotic era. Med Res Rev 39(3):831–859. https://doi.org/10.1002/med.21542

Wang C, Hong T, Cui P, Wang J, Xia J (2021) Antimicrobial peptides towards clinical application: delivery and formulation. Adv Drug Deliv Rev 175:113818–113818. https://doi.org/10.1016/j.addr.2021.05.028

Weber-Dabrowska B, Mulczyk M, Górski A (2001) Bacteriophage therapy for infections in cancer patients. Clin Appl Immunol Rev 1(3–4):131–134. https://doi.org/10.1016/s1529-1049(01)00015-0

Wei J, Peng N, Liang Y, Li K, Li Y (2020) Phage Therapy: Consider the Past, Embrace the Future. Applied Sciences 10(21):7654. https://doi.org/10.3390/app10217654

Welling M, Brouwer C, Roscini L, Cardinali G, Corte L, Casagrande Pierantoni, D, et al. (2018) Structure-activity relationship study of synthetic variants derived from the highly potent human antimicrobial peptide hLF(1-11). Cohesive J. Microbiol. Infect. Dis. https://doi.org/10.31031/CJMI.2018.01.000512

Yang Y, Shen W, Zhong Q, Chen Q, He X, Baker JL, Xiong K, Jin X, Wang J, Hu F, Le S (2020) Development of a bacteriophage cocktail to constrain the emergence of phage-resistant Pseudomonas aeruginosa. Front Microbiol 11:327. https://doi.org/10.3389/fmicb.2020.00327

Yang Q, Le S, Zhu T, Wu N (2023) Regulations of phage therapy across the world. Front Microbiol 14:1250848. https://doi.org/10.3389/fmicb.2023.1250848

Yu Z, Qin W, Lin J, Fang S, Qiu J (2015) Antibacterial mechanisms of polymyxin and bacterial resistance. Biomed Res Int 2015:1–11. https://doi.org/10.1155/2015/679109

Żaczek M, Górski A, Weber-Dąbrowska B, Letkiewicz S, Fortuna W, Rogóż P, Pasternak E, Międzybrodzki R (2022) A thorough synthesis of phage therapy unit activity in Poland—its history, milestones and international recognition. Viruses 14(6):1170. https://doi.org/10.3390/v14061170

Zhang Q-Y, Yan Z-B, Meng Y-M, Hong X-Y, Shao G, Ma J-J, Cheng X-R, Liu J, Kang J, Fu C-Y (2021) Antimicrobial peptides: mechanism of action, activity and clinical potential. Mil Med Res 8(48):48. https://doi.org/10.1186/s40779-021-00343-2

Zhang Z, Zhang Q, Wang T, Xu N, Lu T, Hong W, Penuelas J, Gillings M, Wang M, Gao W, Qian H (2022) Assessment of global health risk of antibiotic resistance genes. Nat Commun 13(1). https://doi.org/10.1038/s41467-022-29283-8

Zhvania P, Hoyle NS, Nadareishvili L, Nizharadze D, Kutateladze M (2017) Phage therapy in a 16-year-old boy with Netherton syndrome. Front Med 4:94. https://doi.org/10.3389/fmed.2017.00094

Synergistic Effects of Phage and Antibiotic Combinations Against Pathogenic Bacteria

20

Minakshi Sahu, Ranjeet Kumar Vishwakarma, Deepak Kumar, Mayank Gangwar, Krupa Parmar, and Gopal Nath

Abstract

Due to the increasing threat posed by superbugs, renewed curiosity in phage as a highly promising treatment for challenging infections. Scientists and clinical communities are currently evaluating bacteriophage as a viable and practical approach to treating bacterial illnesses that are challenging to eliminate. Apart from the emergence of resistance to antibacterial agents in pathogenic bacteria, biofilms present a noteworthy obstacle in managing bacterial diseases. Due to the potential limitations associated with phages, a more appropriate inquiry would pertain to the comparative efficacy of employing a phage with antibiotics to manage bacterial infections. Combination therapy offers significant advantages, including enhanced elimination of bacterial cells and lower development of resistance to the phage or antibiotic. Phage-antibiotic synergy is well documented, with its manifestation

M. Sahu · D. Kumar · G. Nath (✉)
Department of Microbiology Institute of Medical Sciences, Banaras Hindu University, Varanasi, Uttar Pradesh, India

R. K. Vishwakarma
Department of Physiology, Institute of Medical Sciences, Banaras Hindu University, Varanasi, Uttar Pradesh, India

M. Gangwar
Faculty of Medicine, Viral Research and Diagnostic Laboratory, Department of Microbiology, Institute of Medical Sciences, Banaras Hindu University, Varanasi, Uttar Pradesh, India

Clinical Studies and Trials Unit, Division of Development Research, Indian Council of Medical Research, Department of Health Research, Ministry of Health and Family Welfare, Govt. of India, V. Ramalingaswami Bhawan, New Delhi, India

K. Parmar
Division of Clinical Microbiology, Department of Laboratory Medicine and Pathology, Mayo Clinic, Rochester, MN, USA

M. Gangwar, G. Nath (eds.), *Emerging Paradigms for Antibiotic-Resistant Infections: Beyond the Pill*, https://doi.org/10.1007/978-981-97-5272-0_20

being dependent on the particular type of phage and antibiotic involved. Our chapter is focused on PAS, its mechanism, and the antagonistic effect.

Keywords

Bacteriophage · Phage-antibiotic synergy · Antagonistic · Combination therapy

20.1 Introduction

The extensive utilization and excessive application of antibiotics during the last nine decades have resulted in the inevitable emergence of bacterial resistance, leading to an increasing number of instances where medical treatments are ineffective, as reported globally (WHO 2022).

Before the beginning of twenty-first century, the issue was exacerbated by the notion that the commercialization of novel antibiotics might solve bacterial resistance. However, during the past 20 years, the antibiotic market has been dysfunctional as a result of the exorbitant expenses associated with the development of novel compounds (Diallo and Dublanchet 2022). With the diminishing prospect of replicating the therapeutic advancements of the twentieth century, there is an immediate necessity to create innovative methods for managing antibiotic-resistant microorganisms (Safir et al. 2020). The World Health Organization has issued a cautionary statement about an upcoming phase known as the "postantibiotic era," during which ten million individuals could perish each year due to illnesses that are resistant to antimicrobial treatments (O'Neill 2014). According to O'Neill (2014), it is projected that drug-resistant pathogens will result in the mortality of ten million individuals annually by 2050 (Pirnay et al. 2022). Phage therapy has been employed in certain regions of Eastern Europe meanwhile Felix d'Herelle initially administered it to pediatric patients with diarrhea in 1919. Western countries have predominantly disregarded phage therapy due to the advancement of antibiotics. However, the escalating worldwide problem of AMR has sparked renewed attention to phage therapy as a potential treatment for multi-drug resistant (MDR) bacterial illnesses (Hill et al. 2018). Phage-antibiotic combinations have demonstrated potential in laboratory settings against multidrug-resistant bacteria (Morrisette et al. 2022). Phage-antibiotic treatment is particularly crucial in situations when the pathogenic bacteria exhibit drug resistance or when the spread of antibiotics to the diseased area is limited. The distinctive mechanism by which phages have ability to reduce the occurrence of AMR and decrease antibiotic exposure (Morrisette et al. 2020).

Phage antibiotic synergism's presence can diminish the quantity of antibiotics employed in treatment and ultimately mitigate the occurrence of resistance to antibiotics in pathogenic bacteria (Jo et al. 2016; Coulter et al. 2014). Researchers hypothesized that combining phages with antibiotics may be more efficient in reducing bacteria than using either treatment alone, due to the differing mechanisms of bacterial killing between the two (Torres-Barceló and Hochberg 2016; Chaudhry et al. 2017). Li et al. observed that the concept of the PAS as a mechanism that combines phages and antibiotics to minimize the dosage of drugs and inhibit the emergence of AMR (Li et al. 2021a). PAS stands for the phenomenon of enhanced phage production following the administration of bactericidal drugs at levels that do not cause complete bacterial death. Furthermore, the use of the *S. aureus* phage in conjunction with drugs leads to a decrease in antibiotic resistance (Łusiak-Szelachowska et al. 2022).

20.2 Phage Antibiotic Synergism

PAS was introduced somewhat more than 10 years ago to explain the phenomenon in which low doses of antibiotics might enhance the replication of phages by bacterial cells (Morrisette et al. 2020). Comeau et al. initially used the term "phage-antibiotic synergy" (PAS) in 2007. In the initial study, the scientists noted that exposure to low levels of β-Lactam and quinolone antibiotics led to a notable increase in both the size and quantity of plaques. This suggests that the process of plaque formation involves a faster rate of attachment, a shorter latent period, and a more significant release of viral particles (Comeau et al. 2007; Kim et al. 2018). In present times, there seems to be an increasing curiosity about employing phages as adjunctive therapy in conjunction with primary antibiotic therapies. Phage effectiveness is progressively being supported by evidence in the literature, particularly in combination with standard-of-care antibiotics (Pirnay et al. 2022). The word PAS was initially introduced in research that explained how sublethal doses of antibiotics might trigger an increase in the replication of virulent phages by microbial cells (Morrisette et al. 2020). Although the effect of phage-antibiotic synergy has been previously reported in laboratory settings and animal experiments (Regeimbal et al. 2016; Oechslin et al. 2016), its potential for treating chronic infections caused by multi-drug resistant bacteria in people are receiving more interest (Hill et al. 2018). Lin et al. examined the connection between antibiotics and nebulized phages in *Pseudomonas aeruginosa,* and the study revealed that the synergy between the two was affected by the specific strain of bacteria and the type of antibiotic used (Lin et al. 2018). The potential advantage of employing a dual strategy could be the enhanced inhibition of pathogenic bacteria and the diminished ability of pathogenic bacteria to acquire resistance to antibiotics and phage (Torres-Barceló and Hochberg 2016). Multiple research examining the synergistic advantages of the phages and conventional antibiotics have yielded promising outcomes. Research has demonstrated that when antibiotics are present in low doses that do not prevent bacterial growth, they can enhance the production of the phages leading to a decrease in bacterial population (Tagliaferri et al. 2019).

The positive impact of combining phages and antibiotics has been documented in certain cases (Ryan et al. 2012; Kamal and Dennis 2015; Uchiyama et al. 2018), whereas it has not been observed in other instances (Torres-Barceló et al. 2018; Gelman et al. 2018). The observed synergistic effect is believed to be due to improved bacterial eradication, enhanced infiltration into biofilms, and reduced development of phage/antibiotic-resistant strains. Recently, the phenomenon of PAS has been seen even when antibiotics are administered at doses below the level that would normally be considered medically dangerous. This suggests the presence of unknown synergistic processes (Pirnay et al. 2022). In addition, the efficacy of phages may be contingent upon factors such as their rate of adsorption, burst size, latent time, and inoculum. However, the precise significance of these variables has not yet been completely comprehended (Morrisette et al. 2020). The effectiveness of PAS seems to rely on the order in which the two components are given, with superior results found when phage administration is offered before antibiotic delivery. Additionally, there appears to be a certain time period after the phage infection when the administration of antibiotics is necessary to provide the best possible post-antibiotic effect (Pirnay et al. 2022). Simon et al. demonstrated the synergistic effects of phage Sb-1 and oxacillin antibiotic on various *S. aureus* isolates. The oxacillin drug was tested at doses ranging from 5 to 100 µg/mL along with two different multiplicity of infection (MOI) levels. It was found that the combination of phage and antibiotic have a greater bactericidal activity as compared to antibiotic alone (Simon et al. 2021). Resensitization of microorganisms to antibiotics is a targeted approach that employs temperate phages. Another set of processes pertains to the increased susceptibility of lysogens to antibiotics as a result of their impact on the lytic cycle. Al-Anany et al. studied the synergistic interaction between temperate phages and antibiotics. This study demonstrates that the temperate phage *E. coli* HK97 exhibits a synergistic effect with ciprofloxacin, leading to the complete eradication of *E. coli* in a controlled laboratory setting. Notably, the way temperate phages and antibiotics work together is different from the way lytic phages and antibiotics work together. During synergistic effect, antibiotic enhance the phage production as well as affect the Rec A protein which play important role in bacterial SOS response. The observed behavior is caused by a reduction in the number of lysogens (Al-Anany et al. 2021). Fluoroquinolones, a type of antibiotics, have the ability to induce the lytic cycle in lysogenic bacteria, which is quite fascinating (Table 20.1) (Łusiak-Szelachowska et al. 2022).

20.2.1 Possible Mechanism of PAS

Multiple explanations can be suggested to elucidate the PAS effect (Fig. 20.1).

20.2.1.1 Decrease in the Frequency of Phage or Antibiotic-Resistant Mutations

Several studies have done that have assessed the effectiveness of phage-antibiotic combination against *E. faecium*. A study showed that a reduction in phage resistance was observed when phage and daptomycin antibiotics used in time kill analysis. (Łusiak-Szelachowska et al. 2022). In another study, daptomycin and phage cocktail were used to demonstrate the bactericidal activity. Addition of phage to the DAP reduced the development of phage resistance (Morrisette et al. 2022).

20.2.1.2 Antibiotics Promote an Increase in Plaque Size, Enhance Phage Proliferation, and Burst Size

In, PAS plaque size, phage replication and burst size were enhanced by the presence of antibiotic (Ryan et al. 2012). Low doses of tetracycline, linezolid and ketolide drugs can lead to a threefold increase plaque size caused by the *S. aureus* phage.

Table 20.1 In vitro effects of combining phage with antibiotic

Pathogens	Pages	Antibiotics	Synergistic effect	Reference
Escherichia coli	T4	Cefotaxime	Short latent period, large burst size, and plaque size	Ryan et al. (2012)
Escherichia coli	ΦMFP	Aztreonam, Monobactam, Cefixime and cephalosporin	Stimulate phage growth	Comeau et al. (2007)
Acinetobacter baumannii	IME-AB2	Colistin	Reduction in bacterial count and mutants	Mukhopadhyay et al. (2023)
Pseudomonas aeruginosa	σ-1	Ceftriaxone	Elongation and filamentation of cells	Knezevic et al. (2013)
Klebsiella pneumoniae	M1-Kp1	Ciprofloxacin and trimethoprim Sulfamethoxazole	Enhanced the activity of bacteriophage M1-Kp1, with the time	Posada Suárez (2019)
Escherichia coli	ΦEcSw	Ampicillin	Induce phage numbers	Easwaran et al. (2020)
Pseudomonas aeruginosa	Phage s-1	Ceftriaxone	Reduced bacterial growth, cell elongation	Knezevic et al. (2013)
Pseudomonas aeruginosa	KPP22	Piperacillin and ceftazidime	Reduction in bacterial count and increase lytic activity	Uchiyama et al. (2018)

However, quinolone and beta-lactam antibiotics did not have such effects (Kaur et al. 2012). It was observed that the combination of a *S. aureus* phage and some antibiotics (ciprofloxacin, cefotaxime, linezolid, tetracycline, and clarithromycin) resulted in an enhanced production of new bacteriophages (Li et al. 2021b). The subsequent process of PAS entails a decrease in the bacterial mutants that possess resistance to phages or antibiotics. A study found that even at low doses, meropenem and ciprofloxacin can totally inhibit the development of mutants of phage (Oechslin et al. 2016).

20.2.1.3 Antibiotics Induce Cell Elongation and Filamentation

Changes in the growth of *E. coli* were observed when exposed to a sub-inhibitory dose of b-lactams and fluoroquinolones. Instead of completely halting cell division, the lesser exposures resulted in a cell that divided poorly and had a more fragile cell wall. The distorted physiological state caused by cellular filamentation, where cells elongate but do not divide, was believed to contribute to the phenomenon known as PAS. This is based on the idea that the modified state of filamenting cells could result in a boost in the biosynthetic ability of the phages. In other words, there would be more precursor molecules available to support and improve the replication and maturation of the phages. This hypothesis supported the fact that β-lactams modify the cell wall, perhaps leading to increased effects of phage lysis genes (Morrisette et al. 2020).

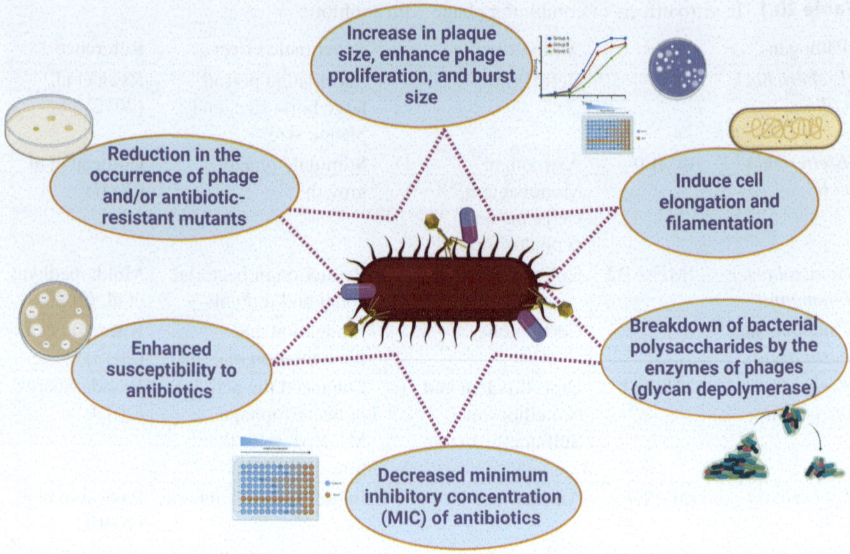

Fig. 20.1 Possible mechanisms of phage and antibiotic combination

20.2.1.4 Enhanced Susceptibility to Antibiotics as a Result of the Presence of the Phage

A combined method can help recover antibiotic susceptibility, especially when the phage interacts with the microbial drug efflux systems (Chan et al. 2016).

20.2.1.5 Decreased MIC of Antibiotics Following the Addition of Phages to the Antibiotic

Chang et al. (2019) demonstrated a synergistic effect of a phage and CIP (Ciprofloxacin) on many strains of *Pseudomonas aeruginosa* in a traditional biofilm paradigm. Synergistic effects was observed when the half of the CIP was mixed with 10^8 PFU/mL of a phage (Chang et al. 2019).

20.2.1.6 Breakdown of Bacterial Polysaccharides, Leading to Increased Diffusion and Penetration of Antibiotics into Cells

Phages encode polysaccharide depolymerase that may selectively break down bacterial structural carbohydrate polymers found in bacterial biofilms.

20.3 PAS in the Eradication of Biofilm

Shlezinger et al. demonstrated the effectiveness of PAS in biofilm models of *E. faecalis* (vancomycin resistance). However, treatment with the phages alone did not yield satisfactory results. Their use of confocal imaging and fluorescence-activated cell sorting allowed them to identify an alteration in the peptidoglycan of the cell wall, which may be connected to the reported outcomes (Shlezinger et al. 2019).

Nevertheless, the specific mechanism responsible for the proposed vancomycin resensitization remains unidentified Phages have the ability to synthesize enzymes such as depolymerases and endolysins. These enzymes help in breaking down the extracellular matrix of biofilms. As a result, the bacteria that were previously hidden within the biofilm become more exposed to antibiotics. In addition, the enzymes released by phages eliminate the microbes found on the outer layer of the biofilm. As a result, the microorganisms inside the biofilm become more active in their metabolic processes, as they receive more oxygen and nutrients. Consequently, they become more susceptible to the effects of phages and antibiotics (Abedon 2015).

The efficacy of a phage-antibiotic combination was observed to control biofilm formation in a laboratory setting. This study aims to assess the effects of different phage T4 titers combined with different doses of cefotaxime on *E. coli* biofilms, in comparison to using cefotaxime alone. The biofilms were cultured for 24 h and then exposed to cefotaxime for a further 24 h. These studies were then replicated using different phage titers. Minimum concentration required to completely eliminate the biofilm formation was 256 mg/mL cefotaxime alone. While this concentration was lowered to 128 mg/mL and 32 mg/mL when phages were added (Ryan et al. 2012). In another study using a biofilm model, it was found that combining tobramycin (2 mg/mL) with phage resulted in the enhanced killing of biofilm cells. This study also found that using this combination led to a reduction in phage-resistant cells (39%) and antibiotic-resistant cells (99.99%) when compared with the phage and antibiotic alone. A significant 99% decrease in phage-resistant cells and a 60% decrease in tobramycin-resistant cells in *P. aeruginosa* biofilms when using the combination compared to using phage and antibiotic separately (Coulter et al. 2014). The biofilm model demonstrated consistent findings, indicating that higher phage titers resulted in decreased minimum biofilm elimination levels (Ryan et al. 2012). The other in vitro experiments that were assessed employed varying phage inoculums. The most effective "dose" of phage is expected to vary depending on the specific phage, its corresponding bacterial host, and the combination of antibiotics used.

A different aspect of PAS is associated with the activity of phage depolymerases. Depolymerase in bacteriophage can be classified into the following types:

Hyaluronate lyases, peptidoglycan hydrolases, endorhamnosidases, alginate lyases, and endosialidases(Yan et al. 2014). Depolymerases in phages can exist in two distinct forms: as a constituent of a virion particle, typically as tail spikes or fiber proteins connected to the base plate, although they may also be found in other locations, and as a soluble protein produced during host lysis following phage maturation (Drulis-Kawa et al. 2015).

Exopolysaccharides have crucial functions in preserving the structural integrity of bacteria and enhancing their ability to cause disease. Depolymerases that break down extracellular polymeric substances (EPS) help phages to enter and infect bacteria that live in biofilms (Hughes et al. 1998). While depolymerases are typically specific, they may exhibit a wider range of activity compared to their parent phages.

However, it is recommended to use numerous depolymerases in order to successfully eliminate mixed biofilms (Harper et al. 2014). Depolymerases can also play a role in the PAS phenomenon by enhancing antibiotic dispersion and promoting cell penetration. For instance, the alginate lyase obtained from *Pseudomonas* phages can enhance the diffusion of aminoglycosides, which in turn inhibits the growth of *P. aeruginosa* (Hatch and Schiller 1998) or successfully eliminates *P. aeruginosa* biofilms (Table 20.2) (Alkawash et al. 2006).

20.4 Phage Antibiotic Antagonism

Antagonism between phages and antibiotics has been reported, primarily with rifampin. Rifampin antagonizes phage activity by inhibiting bacterial RNA polymerase, hence restricting the transcription of phage's genes and viral production. This theoretical process is corroborated by the observation that phages possessing their RNA polymerase are typically resistant to such antibiotics when tested in a controlled laboratory setting (Abedon 2019). Matos et al. discovered that *E. faecalis* V583 prophages play a role in pathogenicity. Matos et al. proposed that the connection between the use of antibiotics and the success of *E. faecalis* as a nosocomial pathogen is linked to the production of prophages caused by fluoroquinolones. This release promotes the spread of genes among different isolates (Matos et al. 2013).

Table 20.2 Eradication of biofilm formation through phage and antibiotic combination

Pathogens	Pages	Antibiotics	Synergistic effect	Ref.
Escherichia coli	vB_ Eco4M-7	Ciprofloxacin and rifampicin	Reduction in bacterial count within the biofilm and inhibit biofilm formation	Necel et al. (2022)
Escherichia coli	T4	Cefotaxime	Enhanced the eradication of bacterial biofilms	Ryan et al. (2012)
Enterococcus faecium	ATCC 113, NV-497, 503-01,503-02	Daptomycin, ampicillin	Eradicate of bacterial biofilms	Kunz Coyne et al. (2022)
Enterococcus faecalis	EFLK1	Vancomycin	87% reduction in biofilm biomass	Shlezinger et al. (2019)
Acinetobacter baumannii	pB3074	Cefotaxime or Meropenem	Eradication of mature biofilm	Luo et al. (2023)
Acinetobacter baumannii	φAB182	Colistin, Polymixin B, ceftazidime and cefotaxime	Eradication of biofilms	Vashisth et al. (2022)
Pseudomonas aeruginosa	(EPA1)	Gentamicin or ciprofloxacin	Reduction in bacterial count within the biofilm	Akturk et al. (2019)
Staphylococcus aureus	SAvB14	Gentamicin and Ceftriazone	Reduction in bacterial count in the biofilm	Horiuk et al. (2021)

The different finding suggest that the interaction between phages and antibiotics varied depending on what kind of antibiotics or phages are used (Table 20.3) (Verma et al. 2009).

20.5 Future Prospective

We draw conclusions that generate hypotheses for future experimental and clinical research, particularly in combination therapy. It is essential to assess the potential for MDR evolution through transduction process and explore different time gaps and titter of phages and antibiotics rather than solely focusing on their simultaneous administration. Future research should investigate the relationship between immune system and phages, as elevated phage levels have triggered a robust immunological reaction, potentially diminishing phage effectiveness against bacteria. It is crucial to examine certain combinations of phages and antibiotics to discover if they work together positively or negatively. Optimizing the phage and antibiotic delivery sequence is necessary for phage-antibiotic combos to become common treatment choices. The potential future applications for combined phage-antibiotic therapy comprise gene editing, using endolysins, and creating microencapsulation systems.

Table 20.3 The antagonistic effects of combining phage with antibiotic

Pathogens	Phages	Antibiotics	Antagonistic effect	Ref.
Ralstonia solanacearum	FRSF1,FRSL1, FRSL2, FRSB1	Rifampin	Phage replication blocked	Matsui et al. (2017)
Pseudomonas aeruginosa	LUZ19	Rifampin	Complete inhibition of phage	Ceyssens et al. (2014)
Pseudomonas aeruginosa	PAK_P3 or PAK_P4	Rifampin	Phage production blocked.	Blasdel et al. (2017)
Ralstonia solanacearum	RSL2 and RSF1	Rifampin	Did not produce progeny phages	Matsui et al. (2017)
Pseudomonas aeruginosa	UJB10	Amikacin, ciprofloxacin	Reducing the average burst size	Nicholls et al. (2023)
E. coli	ΦHP3	Ciprofloxacin	Inhibits both the bacterial and the phage topoisomerases	Gu Liu et al. (2020)

20.6 Conclusion

There needs to be more understanding of the specific mechanisms that influence the interactions between phages and antibiotics. Thus, it is challenging to forecast the most effective combinations for a particular bacterial disease. However, the promising results achieved thus far indicate that further exploration of phage/antibiotic combinations is a worthwhile pursuit that is likely to result in a powerful and effective solution against multi-drug resistant bacteria in the future. Phages and antibiotics have been widely employed together to improve the elimination of drug-resistant bacteria and reduce global antibiotic resistance. Various experimental approaches, such as plaque size analysis, biofilm tests, and animal trials, were employed to assess the combined impact of phages and antibiotics effectively. Multiple research has investigated the mechanism that causes synergy between phages and antibiotics. Phage exposure exerts a selective pressure on microorganisms hence enhancing the probability of genetic alterations. When subjected to this specific type of pressure, several essential elements of the microorganism that are connected to bacterial destruction, susceptibility to drugs, and factors that promote growth are either lost or suppressed. Phages and antibiotics can work together effectively against pathogens that are initially resistant, offering promise for treating pan-drug-resistant "superbugs."

References

Abedon S (2015) Ecology of anti-biofilm agents I: antibiotics versus bacteriophages. Pharmaceuticals 8:525–558. https://doi.org/10.3390/ph8030525

Abedon ST (2019) Phage-antibiotic combination treatments: antagonistic impacts of antibiotics on the pharmacodynamics of phage therapy? Antibiotics 8:182. https://doi.org/10.3390/antibiotics8040182

Akturk E, Oliveira H, Santos SB, Costa S, Kuyumcu S, Melo LDR, Azeredo J (2019) Synergistic action of phage and antibiotics: parameters to enhance the killing efficacy against mono and dual-species biofilms. Antibiotics 8:103. https://doi.org/10.3390/antibiotics8030103

Al-Anany AM, Fatima R, Hynes AP (2021) Temperate phage-antibiotic synergy eradicates bacteria through depletion of lysogens. Cell Rep 35:109172. https://doi.org/10.1016/j.celrep.2021.109172

Alkawash MA, Soothill JS, Schiller NL (2006) Alginate lyase enhances antibiotic killing of mucoid *Pseudomonas aeruginosa* in biofilms. APMIS 114:131–138. https://doi.org/10.1111/j.1600-0463.2006.apm_356.x

Blasdel BG, Chevallereau A, Monot M, Lavigne R, Debarbieux L (2017) Comparative transcriptomics analyses reveal the conservation of an ancestral infectious strategy in two bacteriophage genera. ISME J 11:1988–1996. https://doi.org/10.1038/ismej.2017.63

Ceyssens P-J, Minakhin L, Van Den Bossche A, Yakunina M, Klimuk E, Blasdel B, De Smet J, Noben J-P, Bläsi U, Severinov K, Lavigne R (2014) Development of giant bacteriophage ϕKZ is independent of the host transcription apparatus. J Virol 88:10501–10510. https://doi.org/10.1128/JVI.01347-14

Chan BK, Sistrom M, Wertz JE, Kortright KE, Narayan D, Turner PE (2016) Phage selection restores antibiotic sensitivity in MDR Pseudomonas aeruginosa. Sci Rep 6:26717. https://doi.org/10.1038/srep26717

Chang RYK, Das T, Manos J, Kutter E, Morales S, Chan H-K (2019) Bacteriophage PEV20 and ciprofloxacin combination treatment enhances removal of Pseudomonas aeruginosa biofilm isolated from cystic fibrosis and wound patients. AAPS J 21:49. https://doi.org/10.1208/s12248-019-0315-0

Chaudhry WN, Concepción-Acevedo J, Park T, Andleeb S, Bull JJ, Levin BR (2017) Synergy and order effects of antibiotics and phages in killing Pseudomonas aeruginosa biofilms. PLoS One 12:e0168615. https://doi.org/10.1371/journal.pone.0168615

Comeau AM, Tétart F, Trojet SN, Prère M-F, Krisch HM (2007) Phage-antibiotic synergy (PAS): β-lactam and quinolone antibiotics stimulate virulent phage growth. PLoS One 2:e799. https://doi.org/10.1371/journal.pone.0000799

Coulter L, McLean R, Rohde R, Aron G (2014) Effect of bacteriophage infection in combination with tobramycin on the emergence of resistance in Escherichia coli and Pseudomonas aeruginosa biofilms. Viruses 6:3778–3786. https://doi.org/10.3390/v6103778

Diallo K, Dublanchet A (2022) Benefits of combined phage–antibiotic therapy for the control of antibiotic-resistant bacteria: a literature review. Antibiotics 11:839. https://doi.org/10.3390/antibiotics11070839

Drulis-Kawa Z, Majkowska-Skrobek G, Maciejewska B (2015) Bacteriophages and phage-derived proteins—application approaches. Curr Med Chem 22:1757–1773. https://doi.org/10.2174/0929867322666150209152851

Easwaran M, De Zoysa M, Shin H (2020) Application of phage therapy: synergistic effect of phage EcSw (ΦEcSw) and antibiotic combination towards antibiotic-resistant *Escherichia coli*. Transbound Emerg Dis 67:2809–2817. https://doi.org/10.1111/tbed.13646

Gelman D, Beyth S, Lerer V, Adler K, Poradosu-Cohen R, Coppenhagen-Glazer S, Hazan R (2018) Combined bacteriophages and antibiotics as an efficient therapy against VRE enterococcus faecalis in a mouse model. Res Microbiol 169:531–539. https://doi.org/10.1016/j.resmic.2018.04.008

Gu Liu C, Green SI, Min L, Clark JR, Salazar KC, Terwilliger AL, Kaplan HB, Trautner BW, Ramig RF, Maresso AW (2020) Phage-antibiotic synergy is driven by a unique combination of antibacterial mechanism of action and stoichiometry. mBio 11:e01462–e01420. https://doi.org/10.1128/mBio.01462-20

Harper D, Parracho H, Walker J, Sharp R, Hughes G, Werthén M, Lehman S, Morales S (2014) Bacteriophages and biofilms. Antibiotics 3:270–284. https://doi.org/10.3390/antibiotics3030270

Hatch RA, Schiller NL (1998) Alginate lyase promotes diffusion of aminoglycosides through the extracellular polysaccharide of mucoid *Pseudomonas aeruginosa*. Antimicrob Agents Chemother 42:974–977. https://doi.org/10.1128/AAC.42.4.974

Hill C, Mills S, Ross RP (2018) Phages & antibiotic resistance: are the most abundant entities on earth ready for a comeback? Future Microbiol 13:711–726. https://doi.org/10.2217/fmb-2017-0261

Horiuk YV, Kukhtyn MD, Horiuk VV, Sytnik VA, Dashkovskyy OO (2021) Effect of phage SAvB14 combined with antibiotics on Staphylococcus aureus variant bovis. Regul Mech Biosyst 12:531–536. https://doi.org/10.15421/022173

Hughes KA, Sutherland IW, Jones MV (1998) Biofilm susceptibility to bacteriophage attack: the role of phage-borne polysaccharide depolymerase. Microbiology 144:3039–3047. https://doi.org/10.1099/00221287-144-11-3039

Jo A, Kim J, Ding T, Ahn J (2016) Role of phage-antibiotic combination in reducing antibiotic resistance in Staphylococcus aureus. Food Sci Biotechnol 25:1211–1215. https://doi.org/10.1007/s10068-016-0192-6

Kamal F, Dennis JJ (2015) Burkholderia cepacia complex phage-antibiotic synergy (PAS): antibiotics stimulate lytic phage activity. Appl Environ Microbiol 81:1132–1138. https://doi.org/10.1128/AEM.02850-14

Kaur S, Harjai K, Chhibber S (2012) Methicillin-resistant Staphylococcus aureus phage plaque size enhancement using sublethal concentrations of antibiotics. Appl Environ Microbiol 78:8227–8233. https://doi.org/10.1128/AEM.02371-12

Kim M, Jo Y, Hwang YJ, Hong HW, Hong SS, Park K, Myung H (2018) Phage-antibiotic synergy via delayed lysis. Appl Environ Microbiol 84:e02085–e02018. https://doi.org/10.1128/AEM.02085-18

Knezevic P, Curcin S, Aleksic V, Petrusic M, Vlaski L (2013) Phage-antibiotic synergism: a possible approach to combatting Pseudomonas aeruginosa. Res Microbiol 164:55–60. https://doi.org/10.1016/j.resmic.2012.08.008

Kunz Coyne A, Stamper K, Kebriaei R, Holger DJ, El Ghali A, Morrisette T, Biswas B, Wilson M, Deschenes MV, Canfield GS, Duerkop BA, Arias CA, Rybak MJ (2022) Phage cocktails with daptomycin and ampicillin eradicates biofilm-embedded multidrug-resistant enterococcus faecium with preserved phage susceptibility. Antibiotics 11:1175. https://doi.org/10.3390/antibiotics11091175

Li X, He Y, Wang Z, Wei J, Hu T, Si J, Tao G, Zhang L, Xie L, Abdalla AE, Wang G, Li Y, Teng T (2021a) A combination therapy of phages and antibiotics: two is better than one. Int J Biol Sci 17:3573–3582. https://doi.org/10.7150/ijbs.60551

Li X, Hu T, Wei J, He Y, Abdalla AE, Wang G, Li Y, Teng T (2021b) Characterization of a novel bacteriophage Henu2 and evaluation of the synergistic antibacterial activity of phage-antibiotics. Antibiotics 10:174. https://doi.org/10.3390/antibiotics10020174

Lin Y, Chang RYK, Britton WJ, Morales S, Kutter E, Chan H-K (2018) Synergy of nebulized phage PEV20 and ciprofloxacin combination against Pseudomonas aeruginosa. Int J Pharm 551:158–165. https://doi.org/10.1016/j.ijpharm.2018.09.024

Luo J, Xie L, Yang M, Liu M, Li Q, Wang P, Fan J, Jin J, Luo C (2023) Synergistic antibacterial effect of phage pB3074 in combination with antibiotics targeting cell wall against multidrug-resistant Acinetobacter baumannii in vitro and ex vivo. Microbiol Spectr 11:e00341–e00323. https://doi.org/10.1128/spectrum.00341-23

Łusiak-Szelachowska M, Międzybrodzki R, Drulis-Kawa Z, Cater K, Knežević P, Winogradow C, Amaro K, Jończyk-Matysiak E, Weber-Dąbrowska B, Rękas J, Górski A (2022) Bacteriophages and antibiotic interactions in clinical practice: what we have learned so far. J Biomed Sci 29:23. https://doi.org/10.1186/s12929-022-00806-1

Matos RC, Lapaque N, Rigottier-Gois L, Debarbieux L, Meylheuc T, Gonzalez-Zorn B, Repoila F, Lopes MDF, Serror P (2013) Enterococcus faecalis prophage dynamics and contributions to pathogenic traits. PLoS Genet 9:e1003539. https://doi.org/10.1371/journal.pgen.1003539

Matsui T, Yoshikawa G, Mihara T, Chatchawankanphanich O, Kawasaki T, Nakano M, Fujie M, Ogata H, Yamada T (2017) Replications of two closely related groups of jumbo phages show different level of dependence on host-encoded RNA polymerase. Front Microbiol 8:1010. https://doi.org/10.3389/fmicb.2017.01010

Morrisette T, Kebriaei R, Lev KL, Morales S, Rybak MJ (2020) Bacteriophage therapeutics: a primer for clinicians on phage-antibiotic combinations. Pharmacother J Hum Pharmacol Drug Ther 40:153–168. https://doi.org/10.1002/phar.2358

Morrisette T, Lev KL, Canfield GS, Duerkop BA, Kebriaei R, Stamper KC, Holger D, Lehman SM, Willcox S, Arias CA, Rybak MJ (2022) Evaluation of bacteriophage cocktails alone and in combination with daptomycin against daptomycin-nonsusceptible Enterococcus faecium. Antimicrob Agents Chemother 66:e01623–e01621. https://doi.org/10.1128/AAC.01623-21

Mukhopadhyay S, Zhang P, To KKW, Liu Y, Bai C, Leung SSY (2023) Sequential treatment effects on phage–antibiotic synergistic application against multi-drug-resistant Acinetobacter baumannii. Int J Antimicrob Agents 62:106951. https://doi.org/10.1016/j.ijantimicag.2023.106951

Necel A, Bloch S, Topka-Bielecka G, Janiszewska A, Łukasiak A, Nejman-Faleńczyk B, Węgrzyn G (2022) Synergistic effects of bacteriophage vB_Eco4-M7 and selected antibiotics on the biofilm formed by Shiga toxin-producing Escherichia coli. Antibiotics 11:712. https://doi.org/10.3390/antibiotics11060712

Nicholls P, Clark JR, Gu Liu C, Terwilliger A, Maresso AW (2023) Class-driven synergy and antagonism between a Pseudomonas phage and antibiotics. Infect Immun 91:e00065–e00023. https://doi.org/10.1128/iai.00065-23

O'Neill J (2014) Antimicrobial resistance: tackling a crisis for the health and wealth of nations

Oechslin F, Piccardi P, Mancini S, Gabard J, Moreillon P, Entenza JM, Resch G, Que Y-A (2016) Synergistic interaction between phage therapy and antibiotics clears *Pseudomonas aeruginosa* infection in endocarditis and reduces virulence. J Infect Dis 215:703–712. https://doi.org/10.1093/infdis/jiw632

Pirnay J-P, Ferry T, Resch G (2022) Recent progress toward the implementation of phage therapy in Western medicine. FEMS Microbiol Rev 46:fuab040. https://doi.org/10.1093/femsre/fuab040

Posada Suárez L (2019) Phage-antibiotic synergy with bacteriophage M1-Kp1 in clinical strain of Klebsiella pneumoniae. Universidad de los Andes Facultad de Ciencias Departamento de Ciencias Biológicas

Regeimbal JM, Jacobs AC, Corey BW, Henry MS, Thompson MG, Pavlicek RL, Quinones J, Hannah RM, Ghebremedhin M, Crane NJ, Zurawski DV, Teneza-Mora NC, Biswas B, Hall ER (2016) Personalized therapeutic cocktail of wild environmental phages rescues mice from Acinetobacter baumannii wound infections. Antimicrob Agents Chemother 60:5806–5816. https://doi.org/10.1128/AAC.02877-15

Ryan EM, Alkawareek MY, Donnelly RF, Gilmore BF (2012) Synergistic phage-antibiotic combinations for the control of *Escherichia coli* biofilms *in vitro*. FEMS Immunol Med Microbiol 65:395–398. https://doi.org/10.1111/j.1574-695X.2012.00977.x

Safir MC, Bhavnani SM, Slover CM, Ambrose PG, Rubino CM (2020) Antibacterial drug development: a new approach is needed for the field to survive and thrive. Antibiotics 9:412. https://doi.org/10.3390/antibiotics9070412

Shlezinger M, Coppenhagen-Glazer S, Gelman D, Beyth N, Hazan R (2019) Eradication of vancomycin-resistant enterococci by combining phage and vancomycin. Viruses 11:954. https://doi.org/10.3390/v11100954

Simon K, Pier W, Krüttgen A, Horz H-P (2021) Synergy between phage Sb-1 and oxacillin against methicillin-resistant Staphylococcus aureus. Antibiotics 10:849. https://doi.org/10.3390/antibiotics10070849

Tagliaferri TL, Jansen M, Horz H-P (2019) Fighting pathogenic bacteria on two fronts: phages and antibiotics as combined strategy. Front Cell Infect Microbiol 9:22. https://doi.org/10.3389/fcimb.2019.00022

Torres-Barceló C, Hochberg ME (2016) Evolutionary rationale for phages as complements of antibiotics. Trends Microbiol 24:249–256. https://doi.org/10.1016/j.tim.2015.12.011

Torres-Barceló C, Gurney J, Gougat-Barberá C, Vasse M, Hochberg ME (2018) Transient negative effects of antibiotics on phages do not jeopardise the advantages of combination therapies. FEMS Microbiol Ecol 94. https://doi.org/10.1093/femsec/fiy107

Uchiyama J, Shigehisa R, Nasukawa T, Mizukami K, Takemura-Uchiyama I, Ujihara T, Murakami H, Imanishi I, Nishifuji K, Sakaguchi M, Matsuzaki S (2018) Piperacillin and ceftazidime produce the strongest synergistic phage–antibiotic effect in Pseudomonas aeruginosa. Arch Virol 163:1941–1948. https://doi.org/10.1007/s00705-018-3811-0

Vashisth M, Yashveer S, Anand T, Virmani N, Bera BC, Vaid RK (2022) Antibiotics targeting bacterial protein synthesis reduce the lytic activity of bacteriophages. Virus Res 321:198909. https://doi.org/10.1016/j.virusres.2022.198909

Verma V, Harjai K, Chhibber S (2009) Restricting ciprofloxacin-induced resistant variant formation in the biofilm of Klebsiella pneumoniae B5055 by complementary bacteriophage treatment. J Antimicrob Chemother 64:1212–1218. https://doi.org/10.1093/jac/dkp360

WHO Regional Office for Europe/European Centre for Disease Prevention and Control (2022) Antimicrobial resistance surveillance in Europe 2022–2020 data. WHO, Copenhagen

Yan J, Mao J, Xie J (2014) Bacteriophage polysaccharide depolymerases and biomedical applications. BioDrugs 28:265–274. https://doi.org/10.1007/s40259-013-0081-y

Phage Therapy Reimagined: Viral Solutions for Antibiotic Resistant Challenges

21

Babita, Seema Sangwan, and Nishu Sehrawat

Abstract

The rise of multi-drug resistant bacterial infections is a major public health concern and presents a significant challenge for the management of the deadliest diseases due to the low effectiveness of commercially available antibiotics which demands the development of novel therapeutic methods. Phage therapy, utilizing Bacteriophages (viruses that infect bacteria) that selectively target and eradicate multi-drug resistant bacteria without harming eukaryotic cells becomes an effective alternative or adjunct to traditional antibiotics. In contrast to antibiotics, Bacteriophages are host specific and they have no effect on beneficial bacteria. Furthermore, it is difficult for the bacterium to acquire resistance to Bacteriophages because they can evolve alongside the host. The worldwide increase in multi-drug resistance (MDR) has rekindled phage therapy as a potential treatment for diseases that conventional antibiotics are unable to treat. Phages are versatile and natural particles that can be engineered or utilized for their lytic enzymes in industrial and medical settings. Phage therapy acts as a targeted approach wherein Bacteriophages infect and eliminate particular pathogenic bacteria with the ability to spread and treat illnesses throughout the body. Importantly, phages multiply in the presence of bacterial targets and have no adverse effects on non-target bacteria or tissues in the body. They also demonstrate rapid bacterial death, often within minutes after infection. This novel technique offers a potentially effective way to face the critical global challenge of fighting infections caused by multidrug-resistant bacteria.

Keywords

Phage therapy · Bacteriophages · Antibiotic resistance · Multi-drug resistance

Babita (✉) · S. Sangwan · N. Sehrawat
Department of Microbiology, CCS Haryana Agricultural University, Hisar, Haryana, India

M. Gangwar, G. Nath (eds.), *Emerging Paradigms for Antibiotic-Resistant Infections: Beyond the Pill*, https://doi.org/10.1007/978-981-97-5272-0_21

21.1 Introduction

Global problem of antibiotic resistance creates a serious concern for public health and effectiveness of the present-day medicines. Treatment of common diseases is severely restricted by the evolution and spread of multidrug-resistant bacteria, which raises morbidity, death, and healthcare costs. Therefore, the development of innovative approaches for treating bacterial illnesses resistant to traditional antibiotics is of utmost importance. Phage therapy is one such possible approach for treating bacterial diseases that are resistant to traditional antibiotics. It employs Bacteriophages, also known as phages, which are viruses that target and destroy bacteria as therapeutic agents. Phages are a diverse group of the most abundant living entities on Earth, which have co-evolved with their bacterial hosts for billions of years. As phages are highly selective, they only affect the targeted bacterial strains and have no effect on the beneficial microbiota or the human host (Bhargava et al. 2021). They can also evolve to bacterial resistance through mutation or exchange of their genetic material with other phages. Furthermore, phages can disrupt or penetrate biofilms which represent complex bacterial communities linked to chronic illnesses and frequently resistant to antibiotics. Since its discovery and use in therapeutic settings at the beginning of the twentieth century, phage therapy is longstanding and has been a subject of debate. However, the development of antibiotics and the lack of scientific rigorousness in phage research contributed to the withdrawal of phage therapy in Western countries, but it remained in use within the former Soviet Union and certain Eastern European countries. Phage therapy has been the topic of renewed interest due to the increase in antibiotic resistance. Numerous clinical trials and case reports have proved its safety and effectiveness in treating a range of bacterial infections, including sepsis, diabetic foot ulcers, urinary tract infections, and cystic fibrosis. Though, phage therapy has been found effective in a few cases where antibiotics have failed, it also has certain challenges, including identifying the appropriate phage for each infection, combating phage resistance, and resolving ethical and regulatory concerns. In collation with standardized and static strategy of antibiotics, phage therapy necessitates a distinctive and dynamic approach. It also needs more clinical trials, scientific proof, and public awareness to become a widely used treatment for bacterial infections. This chapter, include the current status of phage therapy, its possible advantages and disadvantages as well as future research and development directions. We contend that phage therapy can be renewed as a viral remedy for antibiotic-resistant bacterial infections, but it needs additional innovation, cooperation, and education to reach its full potential.

21.2 Bacteriophages: The Precise Weapons Found in Nature

Bacteriophages are widespread viruses that exist in all types of ecosystems. They can be found in a wide range of habitats, from hot springs to arctic inland waters, and are remarkably abundant and diverse. Phages are especially abundant in seawater, where on surface waters concentrations can reach up to ten million viruses per milliliter. They are also found in agricultural soils and forest floors, with counts varying from 10^8 to 10^9 per gram of soil. Their prevalence highlights their significance in ecosystems across the globe. Since their discovery in the early 1900s (Herelle et al. 1930; Twort 1961), phage therapy has been studied as a promising approach to tackle bacterial superbugs. As biotechnological tools for the treatment of pathogenic bacteria, phage particles have gained increasing popularity in various fields. As viruses, they affect host populations and ecosystem functions by infecting bacteria, archaea, and micro-eukaryotes hence; they play a vital part in the global bio-geochemical cycles.

21.3 Mechanism of Bacteriophage Action on Target Bacteria

Phages chosen for therapeutic purposes are typically from the *Caudovirales* order, enclosing ten families (Adriaenssens et al. 2020). These phages possess virions comprised a capsid containing linearly arranged double-stranded DNA (dsDNA) and a tail. The tail end is equipped with proteins facilitating phage attachment to the designated host enabling infiltration of the bacterial cellular barrier to introduce phage DNA (Ackermann 1998). As they are bacterial parasites, they must bind to particular surface receptors of the host bacterium in order to survive and infect their host. Phages replicate within host bacteria through lytic or lysogenic modes (Fig. 21.1). While targeting a particular species or subset of bacteria, a phage

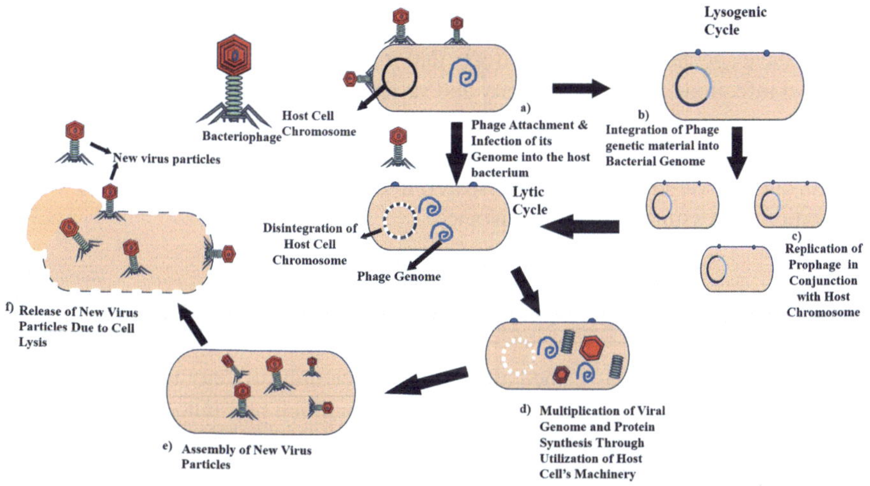

Fig. 21.1 Illustration of underlying principle of Bacteriophage therapy. Steps of Bacteriophage action on target bacteria: (**a**) Phage attachment and infection of its genome into the host bacterium (**b**) Integration of phage genetic material into bacterial genome (**c**) Replication of prophase in conjunction with host chromosome (**d**) Multiplication of viral genome and protein synthesis through utilization of host cell's machinery (**e**) Assembly of newly formed viral particles (**f**) Release of new virus particles due to cell lysis

particle penetrates and replicates inside the host cell by using the host cell's machinery, which includes energy production and protein synthesis. In lytic mode of propagation, phages bind to particular receptors located on the surface of bacteria, inject their genetic material into the bacterial cell, and ultimately destroy the host bacterium to release new virus particles. In contrast, during the lysogenic cycle, the genetic material of the phage fused with the genome of the host cell, forming a prophase and ultimately replicates in conjunction with it, thus, potentially changing the host's properties. Lysogenic phages can also act as vehicles for horizontal gene transfer and the dissemination of antibiotic resistance genes, making them less effective for acute infections. Therefore, for Bacteriophage therapy, lytic or virulent phages are preferred as they effectively kill bacterial cells. This is particularly significant in the context of an increase in the spread of multidrug-resistant bacterial strains. The widespread use of antibiotics, although initially popular due to their broad-spectrum activity, has led to concerns about their impact on the human microbiome (Fair and Tor 2014; Cho and Blaser 2012). Unlike antibiotics, Bacteriophages are specific to their target bacteria, offering a more targeted approach which minimize disruption to commensal microbiota and makes them safer for therapy against bacterial infections (Divya Ganeshan and Hosseinidoust 2019). They typically exhibit variable host specificity and are restricted to bacteria that carry their receptors. Also, bacterial defense mechanisms against phages have evolved over time and they prevent phage infection through receptor modification and CRISPR/Cas, (Labrie et al. 2010) but phages can co-evolve with their hosts and respond with recognition of receptors and anti-CRISPR genes (Koskella and Brockhurst 2014). Thus, increasing their ability to infect them. Several bacterial genera, such as mycoplasmas, archaea, and cyanobacteria, are susceptible to phage infection. Overall, phages have special antibacterial effects that make them effective weapons against bacterial infections. Their multiplicity and variety demonstrate how important a role they play in controlling microbial populations and influencing ecosystem dynamics. Phage therapy and its biotechnological applications are expected to become more significant in a variety of fields as research into them continues, particularly in the context of addressing antibiotic resistance.

21.4 Bacteriophage Therapy: Successful Case Studies

Bacteriophage therapy has gained attention as a potential strategy to combat bacterial diseases amidst the increasing threat of antimicrobial resistance. While historical evidence and various experimental studies have highlighted the synergistic effects of combining phage therapy with antibiotics, as well as the efficiency of Bacteriophages alone across different models, the field still lacks validated and well-controlled clinical trials to support its widespread adoption (Schooley et al. 2017; Furfaro et al. 2018; Aslam et al. 2019). The advancement of phage therapy is hindered by challenges such as the lack of adequately designed medical trials and the requirement of careful consideration of therapeutic variables, including dosage and the composition of phage cocktails (Aslam et al. 2019; Payne and Jansen 2003).

Despite these challenges, there are notable successes in employing Bacteriophage therapy as adjunctive or standalone in treating illnesses caused by multidrug-resistant organisms (Table 21.1); (Kortright et al. 2019; Bhargava et al. 2021). Some case studies involving successful trials of Bacteriophage therapy are listed below:

21.4.1 Bone and Joint Infections (BJIs)

Bone and joint infections (BJIs) present considerable challenges in management and can lead to severe morbidity. Neglecting BJIs can result in acute sepsis, bone and joint degradation, intense pain, sinus formation, and lasting disability. Effective management typically requires thorough microbiological assessment to tailor antibiotic therapy accurately (Colston and Atkins 2018). Recent research has proposed a successful approach for managing multidrug-resistant (MDR) *Pseudomonas aeruginosa* infections in prosthetic joints (PJs) and osteomyelitis, involving local phage administration, antibiotic spacers, and systemic antibiotics. (Kuipers et al. 2019). Ferry et al. (2018) demonstrated recurrent *Staphylococcus aureus* prosthetic joints (PJs) condition cured using a combination of phage therapy and antibiotics applied locally. Additionally, a case report demonstrated successful treatment of left tibial infection caused by multidrug-resistant *Klebsiella pneumoniae* and extensively drug-resistant *Acinetobacter baumannii* using a mixture of antibiotics, Bacteriophages, leading limb preservation and tissue recovery (Nir-Paz et al. 2019). These discoveries underline the potential of phage therapy as an adjunct to or standalone treatment for drug-resistant bacterial infections in orthopedic surgery. Given the significant impact of BJIs placing strain on healthcare resources because of urbanization and an aging population, there is a need for well-coordinated multidisciplinary collaboration to achieve successful therapy.

21.4.2 Infections of the Urinary Tract (UTIs)

Urinary tract ailments are prevalent and contribute to substantial loss of productivity, financial burden, and strain on healthcare systems. UTI complications, particularly those involving antimicrobial-resistant (AMR) pathogens, can lead to sepsis and mortality. Several studies have demonstrated the phage potential in managing UTIs. Valerio et al. (2017) explored the efficacy of Bacteriophages either alone or in conjunction with antibiotics for curing *E. coli*-induced urinary tract infections (UTIs), as well as the sensitivity of bacteria to selected antibacterial agents when Bacteriophages are present. In a case of repeated urinary tract infection (UTI) by *Klebsiella pneumoniae* which is sensitive to broad-spectrum beta-lactamase in a patient with a ureteral stent and indwelling urostomy, Kuipers et al. (2019) utilized phage therapy in combination with antibiotics. These cases highlight the potency of phage therapy as a viable therapeutic approach for UTIs caused by AMR pathogens.

Table 21.1 The clinical studies involving bacteriophages for treatment of multi-drug resistant bacterial infections

Disease	Bacterial pathogen	Therapeutic phage	Mode of administration	References
Osteomyelitis (left tibia infection)	Multidrug-resistant *Klebsiella pneumoniae* and extensively drug-resistant *Acinetobacter baumannii*	MK278861, AbKT21phi3 and MK278859	In combination-phages MK278861, AbKT21phi3 and MK278859 with colistin and meropenem antibiotics which are administered intravenously	Nir-Paz et al. (2019)
Prosthetic joint infection	*Staphylococcus aureus*	Staphylococcal phages 1815 and 1493	Oral administration of amoxicillin, clindamycin and daptomycin and injection of 1815 and 1493 phages in joints	Ferry et al. (2018)
Urinary tract infections	*Escherichia coli*	ECA2	Phage ECA2 and direct inoculation of chloramphenicol, ampicillin, tetracycline, ciprofloxacin, piperacillin and kanamycin antibiotics	Valerio et al. (2017)
Gastrointestinal bacteremia	*E. coli*	HP3	Intraperitoneal administration of phage HP3 alone	Green et al. (2017)
Biofilm-induced frontal sinusitis	*P. aeruginosa*	Pa223 & 222, Pa204, Pa193	Intranasal administration of bacteriophage cocktail CT-PA	Fong et al. (2019)
Biofilm infection	*P. aeruginosa* and *E. coli*	PB1-ATCC15692-B3 and T4-ATCC11303-B4	Synergistic combination of phages PB1-ATCC15692-B3 and T4-ATCC11303-B4 with direct inoculation of tobramycin	Coulter et al. (2014)
Aortic graft infection	Multidrug-resistant *Pseudomonas aeruginosa*	OMKO1	Co-application of phage OMKO1 and ceftazidime antibiotics (direct application of ceftazidime antibiotic in mediastinal fistula	Chan et al. (2018)

(continued)

Table 21.1 (continued)

Disease	Bacterial pathogen	Therapeutic phage	Mode of administration	References
Lung infections	Multidrug-resistant *Acinetobacter baumannii*	PBAB93, PBAB80, PBAB08, PBAB68 and PBAB25	Intranasal administration of a cocktail of phages containing PBAB93, PBAB80, PBAB08, PBAB68 and PBAB25	Cha et al. (2018)
Bacteremia	*P. aeruginosa*	Cocktail of two phages provided by the U.S. Navy	Combination of antibiotics polymyxin B, meropenem, colistin, tobramycin and aztreonam and intravenous administration of a cocktail of two phages provided by the U.S. Navy	Duplessis et al. (2018)

21.4.3 Gastrointestinal Infections (GIs)

Perturbations in the gut microbiota can have adverse effects on the health of an individual, resulting in a variety of gastrointestinal infections (GIs). Global concerns regarding bacterial resistance in GIs have prompted the exploration of innovative therapeutic interventions. Phage therapy has emerged as a promising approach to combat gastrointestinal bacterial infections while preserving the commensal flora. Galtier et al. (2017) noted that administering phages as a preventive measure 4 days post-challenge with adherent-invasive *E. coli* reduced bacterial colonization and prevented colitis progression in dextran sodium sulfate-treated mice. Green et al. (2017) proposed the efficacy of phages taken from the environment in treating severe *E. coli* infections in immune-compromised patients, mitigating the risk of translocation-related complications. Chaturvedi and Nath (2018) observed the elimination of MDR *Klebsiella pneumoniae* in the gut of albino mice upon oral administration of MDR-specific phages, thereby preventing potential nosocomial infections upon translocation.

21.4.4 Biofilm Infections

Biofilm infections pose a significant challenge in healthcare as they are highly resistant to conventional treatments like antibiotics and disinfectants. These persistent infections, affecting millions worldwide, often lead to chronic conditions that are difficult to control. Harper et al. (2014) highlighted the formidable nature of biofilm-associated infections. However, Fong et al. (2019) conducted research using a phage cocktail to combat *Pseudomonas aeruginosa* biofilm-induced frontal sinusitis in animal models, demonstrating its efficacy and safety. The study elucidated that Bacteriophages replicate inside the target bacterium and secrete enzymes that dismantle the extracellular matrix substance of biofilm thereby eliminating the pathogen. Additionally, Coulter et al. (2014) showcased the synergistic effect of combining Bacteriophages with antibiotics, resulting in complete eradication of biofilms and reducing bacterial resistance to both agents. This approach holds promise in effectively managing biofilm infections by utilizing the unique mechanisms of Bacteriophages alongside conventional antibiotic therapy.

21.4.5 Heart/Pulmonary Infections

Heart and pulmonary infections pose significant challenges, often leading to septicemia and affecting other organs if left untreated. In light of the evolving resistance of pathogenic microorganisms to conventional treatments, there is an urgent need for innovative therapeutic approaches. Chan et al. (2018) successfully isolated a lytic phage that targets multi-drug resistant strains of *Pseudomonas aeruginosa*, specifically those possessing the multi-drug systemic porin M on their outer membrane. This Bacteriophage effectively disrupts the efflux pumping system of the bacteria thereby, sensitizing the pathogen to a wide range of antibacterial drugs. In view of limitations of antibiotics in combating resistant infections, this phage-based approach presents a promising avenue for addressing the challenges posed by antimicrobial resistance. They are specific to their hosts, leading to increased antibiotic sensitivity when phage resistance develops (Chan et al. 2018). In vivo studies in mice by Cha et al. (2018) demonstrated that administration of a recently discovered combination of phages notably increased the survival rates of individuals who suffered from multi-drug resistant *Acinetobacter baumannii* lung infections when compared to control groups that did not get treatment.

21.4.6 Bloodstream Infections

21.4.6.1 *Pseudomonas* Sepsis Case Study

A pediatric patient diagnosed with DiGeorge syndrome and congenital heart disease experienced *Pseudomonas aeruginosa* bacteremia subsequent to multiple surgeries, including pacemaker insertion (Duplessis et al. 2018). Initially, conventional antipseudomonal antibiotics effectively managed the infection, but later proved ineffective. Complications arose due to adverse reactions to fluoroquinolones and cephalosporins, limiting antibiotic options. Consequently, a cocktail of two phages, identified through screening provided by the U.S. Navy, was administered intravenously. Following phage therapy initiation, blood cultures fluctuated between positive for *P. aeruginosa* and falling below the detection limit over several days. A temporary halt in phage therapy on day 11, coinciding with a brief cessation due to progressive heart failure saw blood cultures negative for *P. aeruginosa* for 4 days.

Although phage therapy appeared to mitigate bloodstream infection, it failed to control the infection source, as therapy progressed, blood cultures reverted to a positive state.

21.4.7 Phage Interactions with Antibiotics

Antibiotic resistance remains a significant concern, prompting exploration of Bacteriophage-based therapy as a feasible substitute or adjunct to antibiotics. Current recommendations advocate for therapeutic use of Bacteriophage in combination with antibiotics or alone, their combination with antibiotics is supposed to improve effectiveness (Gkartziou et al. 2021). The interaction between antibiotics and phage can result in synergistic, additive, and antagonistic effects. Antibiotic selection typically depends on the patient's health and susceptibility data to achieve a "phage adjuvant" effect when coupled with phages (Gu Liu et al. 2020). Optimal antibiotic concentrations are crucial to prevent mutations in bacterial cells at semi-lethal levels, resulting in strain-specific phages ineffective (Saha and Mukherjee 2019). However, some antibiotics may hinder phage therapy by eliminating the bacterial host and impeding phage reproduction, necessitating rigorous preclinical evaluations before in vivo administration.

21.5 Phage Engineering for Enhancing Its Therapeutic Efficiency

Phage engineering has emerged as a potential remedy to address the increasing danger posed by antibiotic-resistant bacterial pathogens, which contribute to approximately 700,000 fatalities globally each year due to incurable bacterial infections (Arias and Murray 2009; Ventola 2015; Ruckert et al. 2020). Bacteriophages are considered promising alternatives to antibiotics however no individual phage exhibits the necessary attributes of an ideal therapeutic agent. Furthermore, bacteria can quickly develop resistance to phages, rendering approved therapies effectiveness and perpetuating the investigation of more effective eco-friendly phages (Zalewska-Piątek 2023). An alternative approach to rapidly acquire phages with desired properties while minimizing costs and time spent on acquisition, characterization, and approval for therapy involves targeted genome modifications of known phage isolates. A recent example of this approach's success is the treatment of a severe *Mycobacterium abscessus* infection in a teenage patient using an engineered phage (Łobocka et al. 2021). Due to their relatively small genomes and ease of manipulation in laboratory settings, phages are ideal candidates for genetic modification studies aimed at improving their efficacy in treating human and veterinary bacterial diseases. Phage genetic modification usually involves gene substitution/replacement, gene integration, and mutation through molecular methodologies. These modifications aim to broaden the phage host range, enhance antibacterial effects, increase persistence within the host, modulate interactions with the host

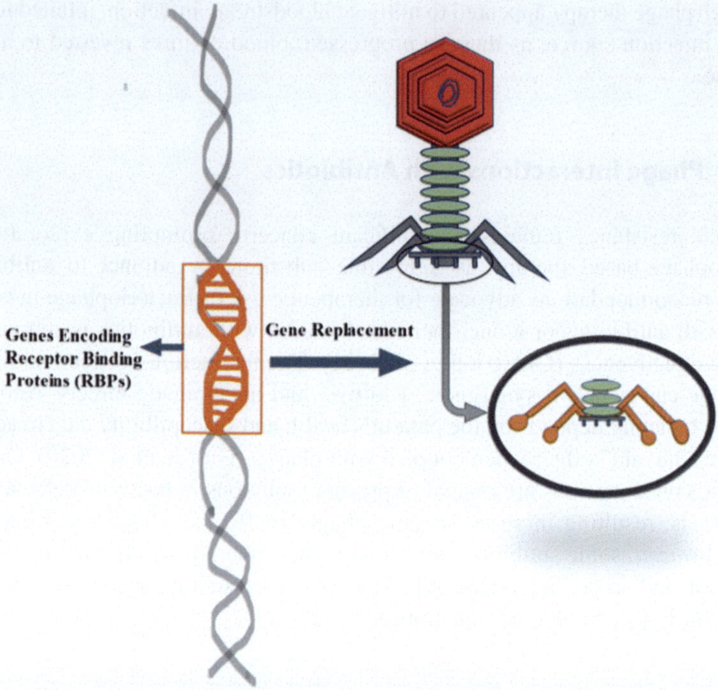

Fig. 21.2 Illustration of whole gene replacement technique. Alteration of genes encoding phage receptor binding proteins to expand the host specificity

immune system, and minimize host-toxicity effects upon bacterial population lysis. Tail fiber protein genes (receptor binding protein-encoding genes) are often targeted for mutation or replacement to broaden the phage host range (Fig. 21.2), while foreign gene integration into non-coding regions of its genome can be harmful to its host. This genetic modification approach is specifically applied to address concerns regarding various aspects of phage biology relevant to disease treatment.

21.6 Innovations in Developing Formulation Strategies for Bacteriophage Therapy

Phage therapy typically employs Bacteriophages in a liquid state, usually freshly prepared. When considering ideal drug characteristics, pharmacists prioritize specificity, high affinity, solubility, stability, and safety (Tomlinson 2004). Bacteriophages meet four of these criteria, excelling in specificity by targeting only infecting pathogens while sparing commensal microbiota and demonstrating high affinity for rapid host adsorption in aqueous environments. Although Bacteriophages exist in the form of colloidal dispersion, their stability remains a challenge, akin to other protein structures (Ackermann et al. 2004; Gill and Hyman 2010; Jończyk et al. 2011;

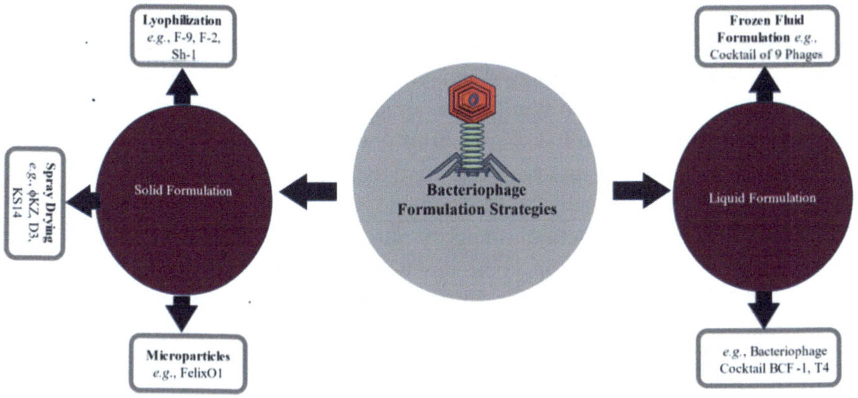

Fig. 21.3 Various strategies for the development of bacteriophage formulation for phage therapy

Tovkach et al. 2012). Unlike other criteria, stability is a significant hurdle as Bacteriophagesexhibit only partial stability when dissolved in a solution, require cooling for storage over an extended period. This limitation impacts the application, transportation, and shelf life of phage-based therapeutics. Fortunately, research has focused on stabilizing Bacteriophages, exploring methods such as the introduction of substances that improve stability. Transforming them into stable forms through processes such as spray drying, Lyophilization or integration into different matrices like biodegradable polymers, ointments, or micro-particles (Fig. 21.3). These strategies aim to extend the shelf life and enhance the practicality of Bacteriophage therapy by mitigating the structural instability of phages in aqueous suspensions. Efforts in this area continue to evolve, offering promising avenues to address the stability challenge and optimize the efficacy of phage-based therapeutics (Vandenheuvel et al. 2015).

21.6.1 Recent Advances in Enhancing Stability of Phage-Based Therapeutics

Recent advancements in phage-based therapeutics necessitate a focus on enhancing their stability, a critical aspect for the development of effective protein-based pharmaceuticals. Typically, proteins exhibit metastability in solution, (Vagenende et al. 2009) emphasizing the importance of stabilizing their native structure and biological activity. To address this, established stabilization approaches utilized in the development of protein-based drugs are being adapted for logical aspects of phage therapeutics. Phages, akin to large protein complexes encapsulating genetic material, often exhibit limited stability in solution, posing challenges in meeting pharmaceutical regulations due to unacceptable titer drops. Various excipients are utilized in pharmaceutical formulations for proteins in both liquid and dry forms to bolster stability. Strategies include adding cosolutes such as salts, amino acids, polyols, saccharides, and surfactants to stabilize proteins in solution, or converting solutions into dry formulations via Lyophilization (freeze drying) or spray drying (Ameri and Maa 2006; Wang 2000). Spray-dried and lyophilized formulations require the inclusion of additives that serve to stabilize the structural integrity of proteins and simultaneously enhance the amount of resulting powder. Trehalose, a disaccharide, is recognized for its efficacy as a lyo-, cryo, and thermoprotectant, (Crowe et al. 1996) forming an amorphous matrix post-drying to shield embedded proteins through direct interactions or structural integrity through glass formation is essential. However, while Lyophilization continues to be the preferred drying method for heat-sensitive pharmaceuticals, it is time-consuming, energy-intensive, and impractical for large-scale processing. (Ameri and Maa 2006). Additionally, it lacks precise control often, involves manipulation of particle characteristics and frequently necessitates micronization to decrease particle size. In contrast, spray drying is emerging as a viable alternative, particularly for drying thermally sensitive biomolecules, providing enhanced regulation of particle characteristics. These advancements underscore the ongoing efforts to optimize stability in phage-based therapeutics through tailored formulation techniques.

21.6.1.1 Strategies for Formulating Liquid Phages

Recent advancements have been made in improving the stability of phages, particularly in liquid formulations (Merabishvili et al. 2013a, b) which are commonly used in various applications comprising food additives, injectables, wound washings, and for pulmonary delivery (Borie et al. 2009; Golshahi et al. 2008; Hoe et al. 2013; Tsonos et al. 2014; Merabishvili et al. 2013a, b; Semler et al. 2012).

Traditionally, liquid phage suspensions have had limited stability, typically lasting around 1 year when stored at 4 °C. In tackling this problem, scientists have investigated diverse approaches, including the incorporation of salts or bovine serum albumin, gelatin to improve stability. However, the utilization of animal-derived proteins in pharmaceuticals presents obstacles. Despite stability being crucial for clinical purposes, there remains a dearth of comprehensive data on stabilizing phages in liquid suspensions (Fortier and Moineau 2009; Tovkach et al. 2012). A study aimed at enhancing the shelf-life of unstable *Erwinia* phages 49 and ZF40 by elevating the magnesium ion concentration (up to 10 mM concentration of magnesium chloride or sulphate, Tovkach et al. 2012). This new buffer composition significantly enhanced the stability of Bacteriophages, maintaining physical and genetic integrity at 4 °C for storage periods lasting 1–2 years compared to significant titer loss observed previously after 1 year. Alternative research focused on investigating basic phage suspensions, as utilized in preclinical investigations and clinical trials. For instance, phage T4 was diluted in mineral water and stored at refrigeration temperature demonstrating sustained phage titer for 1 month or more (Bruttin and Brüssow 2005). Similarly, a phage cocktail (BFC-1) used in a patient trial for burn wounds in Belgium, consisting of a Bacteriophage lysate that has undergone filter-sterilized and is devoid of endotoxins, diluted in saline solution (0.9% NaCl) maintained its titer for 1 year at 4 °C (Merabishvili et al. 2009). According to Cooper et al. (2014), a different phage cocktail meant for nebulization that was suspended in PBS buffer maintained its titer for 6 months at temperatures between 4 and 22 °C. A minimum stability of 18 months is advised, though, in order to create a phage-based pharmaceutical that is both stable and commercially viable. It is crucial to limit titer reductions during this shelf life so that they don't go over 1 log unit, maintaining the formulation's therapeutic efficacy.

21.6.1.2 Strategies for Formulating Solid Phages

In addition to liquid formulations, efforts have also been directed towards solid phage formulations. Phages are large protein capsids that contain genetic material thus, drying techniques commonly used in protein processing have been explored to create dry phage formulations. These approaches aim to enhance stability and extend shelf life by reducing moisture content and preventing degradation.

Lyophilization

Lyophilization, commonly known as cryo- or freeze drying, is a well-known method in the pharmaceutical sector for drying proteins, resulting in enhanced stability compared to proteins in solution (Wang 2000). Researchers have explored the application of Lyophilization to phage-containing powders and tablets to mitigate moisture-related damage, a process crucial for maintaining phage titer stability over time (Schade and Caroline 1944a, b). Studies have shown that phage titer in lyophilized powders and tablets remained stable for up to 12 months, provided that humidity levels during storage were kept low. Early investigations into Lyophilization effects on phages included the examination of stabilizers and cryoprotectants (Schade and Caroline 1944a, b). These researchers discovered that the addition of various substances such as egg white, lecithin, yeast extract, and certain organ extracts could protect phages from Lyophilization-induced damage. However, many of these substances are not suitable for pharmaceutical formulations. Lyophilized powders containing these additives remained stable at 37 °C temperature for 12 months when stored together with a desiccant. Recent research has focused on identifying pharmaceutically acceptable excipients to enhance phage durability during Lyophilization. A Study found that with the inclusion of trehalose or sucrose, titer reductions were confined to one order of magnitude after Lyophilization, with an additional order of magnitude decreased after more than 3 years of storage at temperature of 4 °C using a 0.5 M sucrose concentration. (Merabishvili et al. 2013a, b). However, elevated levels of additives could lead to faster phage inactivation, highlighting the delicate balance required in formulation (Puapermpoonsiri et al. 2010). Further investigations by Golshahi et al. (2011) revealed that lyophilized phages remained stable in combination with a lactose/lactoferrin mixture (60:40 weight-to-weight). Although there was a fall of 1–2 order or log unit in titer at initial stage, after Lyophilization, the titer stayed stable for 3 months at either 4 °C or 22 °C. Thus, Lyophilization offers a promising method for enhancing the stability of phage preparations, with studies focusing on identifying suitable stabilizers and excipients to maintain phage viability throughout the dehydration or drying procedure and subsequent storage.

Spray Drying

Spray drying is a technique widely utilized in the pharmaceutical industry to transform liquid mixtures into dry powders while preserving control over particle properties (Ameri and Maa 2006). These spray-dried powders have versatile applications, including reconstitution for injection or wound washing and can be further processed to create a variety of dosage forms, including pills, bandages, tablets, and powder for inhalation. (Ohtake et al. 2010)). Recent advancements have seen the application of spray drying in creating stable Bacteriophage formulations (Matinkhoo et al. 2011; Vandenheuvel et al. 2013). Matinkhoo et al. (2011) utilized spray drying to process different phages from the phage family *Myoviridae*, incorporating trehalose and leucine as stabilizing agents. They examined the impact of chemical surfactants such as Pluronic F68, casein, and Tyloxapol on phage durability. When used in powdered form, the resultant inhalable fine powder showed excellent aerosol, activity with more than 50% of the dose entering the lungs and smallest possible loss in titer (<1 Log unit) following the spray drying technique. Stability tests performed after preserving in refrigerator for more than 3 months revealed negligible loss in titer (<0.15 order or log units), despite the restricted size of the produced powder samples. Another study (Vandenheuvel et al. 2013) examined the impacts of spray drying *Myo-* and *Podoviruses* with various saccharide excipients present. Trehalose demonstrated superior stabilizing properties compared to dextran and lactose because of its capacity to establish hydrogen bonds with the Bacteriophage particles and engage in direct interaction with it (Carpenter and Crowe 1989). Additionally, trehalose can create an amorphous trehalose glass that maintains its form stability while exhibiting a high transition temperature. (Franks 1991). This amorphous matrix mitigated harmful chemical processes by lowering molecular mobility within the virion capsid proteins. Additionally, the resulting hollow sphere powder particles displayed favorable aerosolization characteristics. Subsequent long-term stability assessments revealed that high humidity conditions (54%) led to the crystallization of amorphous trehalose, resulting in decreased phage titer (Vandenheuvel et al. 2014). Optimal stability was observed during refrigeration at zero humidity. These initial findings regarding drying of phages by this method hold promise for the development of respirable Bacteriophage formulations aimed at combating pulmonary diseases. The versatility and stability of spray-dried phage powders suggest their potential as effective therapeutic agents for pulmonary delivery.

Microspheres and Microcrystals

Alvarez-Gonzalez et al. (2012) demonstrated an alternative to Lyophilization and spray drying for producing phage-containing dry powder by rapidly drying *Siphoviridae* into microcrystals. Stabilization of the phages in organic solvents required the addition of trehalose, glycine, or BSA (bovine serum albumin). However, these microcrystals exhibited a fall of 3–4 order of magnitude/log units in phage titer over 1 month at ambient conditions. Another method involves encapsulating phages in microspheres of biodegradable polymers, as evidenced by-Ma et al. 2008; Puapermpoonsiri et al. 2009. FelixO1 phage was effectively embedded in a biodegradable, miscible polymer (Ma et al. 2008). Wet microspheres maintained a consistent titer for more than 6 weeks when stored in refrigerator, but the addition of trehalose was necessary to prevent phage inactivation upon drying. Despite some loss in titer after storage, it was less pronounced compared to phages stored at room temperature.

21.7 Challenges and Future Prospects

Bacteriophage therapy, functioning as a dynamic therapeutic agent, arises as a compelling tool in contemporary medicine, providing an alternative strategy to combat antibiotic-resistant pathogens. A great deal of scientific and clinical research is necessary before it can go from an experimental to a standard worldwide treatment. It is critical to address and disprove misconceptions that phases could be the source of human disorders (Ali et al. 2023; Hitchcock et al. 2023; Suh et al. 2022). Implementation of phage therapy (PT) involves identifying and cultivating the clinical isolate from a patient, attributed to a particular illness, which needs to be tested for its phage susceptibility. Physicians require accessibility to a variety of Bacteriophages that target a particular bacterium ideally sourced from phage repositories. Extensive phage libraries facilitate high-throughput screening for suitable phages or phage cocktails, a key aspect of personalized medicine. A substantial phage bank is necessary to prepare Bacteriophages selected for a specific patient in the form of a medicinal product (Furfaro et al. 2018; Luong et al. 2020; Pirnay et al. 2018). Escalating bacterial resistance to antibiotics, the continuous exploration for novel phages and their customization as agents for therapy remains imperative. The commercialization of phage formulations is uncommon because of significant financial expenses that could be invested elsewhere. Challenges related to securing patents for products based on phages and the absence of managing endorsement, including from the FDA, pose significant obstacles (Lin et al. 2021; Suh et al. 2022; Sulakvelidze et al. 2001). Overcoming such regulatory hurdles and establishing standardized phage-based therapeutic guidelines is crucial for widespread acceptance. Future technological advancements may resolve issues related to patenting phage preparations, potentially increasing specificity and availability of PT. Genome editing techniques, including homologous recombination, sequencing and phage engineering using CRISPR/Cas may enable the modification of phages to selectively target antibiotic-resistant bacteria. Customized phages or combinations of phages could potentially be patented and brought to market more readily. The development of phage archives by international research groups and phage therapy centers could help meet the current demand for readily available phage products. Development of ideal or safe Bacteriophage drugs, either for customized or commercial use, needs stringent quality control measures. Factors including, sterility, stability, and the absence of harmful impurities must be considered. (Chen et al. 2019; Mahler et al. 2023; Pirnay et al. 2015; Skurnik 2022; Zhang et al. 2022). Despite potential technological progress, experts assert that phage-based treatment will not replace antibiotics but could complement them or act as a final barricade to patients battling infections resistant to other therapeutic methods. In light of rising frequency of life-threatening drug-resistant infections, revisiting the function of PT and other alternatives versus antibiotics is essential. Phage-based therapy merits renewed consideration in Western medicine (Ali et al. 2023; Skurnik 2022).

21.8 Conclusion

The increase in antibiotic-resistant (AR) bacterial illnesses has spurred interest in Bacteriophages as a promising solution. Bacteriophages, being highly specific to their host bacteria, demonstrate efficacy with minimal serious side effects. Clinical case studies indicate that a combination of Bacteriophages and antibiotics, or phages alone, holds potential as a therapeutic option against recalcitrant bacterial infections. This approach may offer advantages such as enhanced bacterial clearance, improved biofilm adsorption, and a reduced likelihood of developing phage resistance. While Bacteriophage therapy shows promise in combating bacterial infections hostile to multiple drugs, there must be a necessity for a thorough comprehension of interactions between phages, microbiome, and infected human host before widespread implementation. Recent findings highlight discrepancies in immunemodulatory effects, host range, and the potential for horizontal gene transfer. To advance Bacteriophage therapy and assure regulatory approval, further research is essential. Specifically, investigations on interactions between phages and their respective hosts in the scenario of useful microflora are crucial. This will contribute to a better understanding of Bacteriophage therapy, facilitating its integration into modern medicine while addressing concerns related to immunomodulation and gene transfer. A comprehensive grasp of these aspects will accelerate the regulatory approval and utilization of the Bacteriophage treatment as a feasible substitute or complement to antibiotic in the perspective of antibiotic resistance.

References

Ackermann HW (1998) Tailed bacteriophages: the order *Caudovirales*. Adv Virus Res 51:135–201

Ackermann HW, Tremblay D, Moineau S (2004) Long-term bacteriophage preservation. WFCC Newsl 38:35–40

Adriaenssens EM, Sullivan MB, Knezevic P, van Zyl LJ, Sarkar BL, Dutilh BE et al (2020) Taxonomy of prokaryotic viruses: 2018-2019 update from the ICTV bacterial and archaeal viruses subcommittee. Arch Virol 165(5):1253–1260

Ali Y, Inusa I, Sanghvi G, Mandaliya V, Bishoyi AK (2023) The current status of phage therapy and its advancement towards establishing standard antimicrobials for combating multi drug-resistant bacterial pathogens. Microb Pathog 181:106199

Alvarez-Gonzalez E, Alfadhel M, Mane P, Ford SJ, Moore BD, van der Walle CF (2012) Bioprocessing of bacteriophages via rapid drying onto microcrystals. Biotechnol Prog 28(2):540–548

Ameri M, Maa YF (2006) Spray drying of biopharmaceuticals: stability and process considerations. Dry Technol 24(6):763–768

Arias CA, Murray BE (2009) Antibiotic-resistant bugs in the 21st century—a clinical super-challenge. N Engl J Med 360(5):439–443

Aslam S, Courtwright AM, Koval C, Lehman SM, Morales S, Furr CLL, Schooley RT (2019) Early clinical experience of bacteriophage therapy in 3 lung transplant recipients. Am J Transplant 19(9):2631–2639

Bhargava K, Nath G, Bhargava A, Aseri GK, Jain N (2021) Phage therapeutics: from promises to practices and prospectives. Appl Microbiol Biotechnol 105:9047–9067

Borie C, Sánchez ML, Navarro C, Ramírez S, Morales MA, Retamales J, Robeson J (2009) Aerosol spray treatment with bacteriophages and competitive exclusion reduces *Salmonella enteritidis* infection in chickens. Avian Dis 53(2):250–254

Bruttin A, Brüssow H (2005) Human volunteers receiving *Escherichia coli* phage T4 orally: a safety test of phage therapy. Antimicrob Agents Chemother 49(7):2874–2878

Carpenter JF, Crowe JH (1989) An infrared spectroscopic study of the interactions of carbohydrates with dried proteins. Biochemistry 28(9):3916–3922

Cha K, Oh HK, Jang JY, Jo Y, Kim WK, Ha GU et al (2018) Characterization of two novel bacteriophages infecting multidrug-resistant (MDR) *Acinetobacter baumannii* and evaluation of their therapeutic efficacy *in vivo*. Front Microbiol 9:696

Chan BK, Turner PE, Kim S, Mojibian HR, Elefteriades JA, Narayan D (2018) Phage treatment of an aortic graft infected with *Pseudomonas aeruginosa*. Evol Med Public Health 2018(1):60–66

Chaturvedi A, Nath G (2018) Oral administration of *Klebsiella pneumoniae*-specific bacteriophage eradicates the bacteria in albino mice. Indian J Med Microbiol 36(2):293–294

Chen Y, Batra H, Dong J, Chen C, Rao VB, Tao P (2019) Genetic engineering of bacteriophages against infectious diseases. Front Microbiol 10:954

Cho I, Blaser MJ (2012) The human microbiome: at the interface of health and disease. Nat Rev Genet 13(4):260–270

Colston J, Atkins B (2018) Bone and joint infection. Clin Med 18(2):150

Cooper CJ, Denyer SP, Maillard JY (2014) Stability and purity of a bacteriophage cocktail preparation for nebulizer delivery. Lett Appl Microbiol 58(2):118–122

Coulter LB, McLean RJ, Rohde RE, Aron GM (2014) Effect of bacteriophage infection in combination with tobramycin on the emergence of resistance in *Escherichia coli* and *Pseudomonas aeruginosa* biofilms. Viruses 6(10):3778–3786

Crowe LM, Reid DS, Crowe JH (1996) Is trehalose special for preserving dry biomaterials? Biophys J 71(4):2087–2093

Divya Ganeshan S, Hosseinidoust Z (2019) Phage therapy with a focus on the human microbiota. Antibiotics 8(3):131

Duplessis C, Biswas B, Hanisch B, Perkins M, Henry M, Quinones J et al (2018) Refractory *Pseudomonas* bacteremia in a 2-year-old sterilized by bacteriophage therapy. J Pediatric Infect Dis Soc 7(3):253–256

Fair RJ, Tor Y (2014) Antibiotics and bacterial resistance in the 21st century. Perspect Med Chem 6:PMC-S14459

Ferry, T., Leboucher, G., Fevre, C., Herry, Y., Conrad, A., Josse, J., & Laurent, F. (2018). Salvage debridement, antibiotics and implant retention ("DAIR") with local injection of a selected cocktail of bacteriophages: is it an option for an elderly patient with relapsing *Staphylococcus aureus* prosthetic-joint infection? In Open forum infectious diseases (5, 11, ofy269). Oxford University Press

Fong SA, Drilling AJ, Ooi ML, Paramasivan S, Finnie JW, Morales S et al (2019) Safety and efficacy of a bacteriophage cocktail in an *in vivo* model of *Pseudomonas aeruginosa* sinusitis. Transl Res 206:41–56

Fortier LC, Moineau S (2009) Phage production and maintenance of stocks, including expected stock lifetimes. In: Bacteriophages: methods and protocols, Isolation, characterization, and interactions, vol 1, pp 203–219

Franks F (1991) Material science and the production of shelf stable biologicals. Bio Pharm 4:38–42

Furfaro LL, Payne MS, Chang BJ (2018) Bacteriophage therapy: clinical trials and regulatory hurdles. Front Cell Infect Microbiol 8:376

Galtier M, Sordi LD, Sivignon A, De Vallée A, Maura D, Neut C et al (2017) Bacteriophages targeting adherent invasive *Escherichia coli* strains as a promising new treatment for Crohn's disease. J Crohn's Colitis 11(7):840–847

Gill JJ, Hyman P (2010) Phage choice, isolation, and preparation for phage therapy. Curr Pharm Biotechnol 11(1):2–14

Gkartziou F, Giormezis N, Spiliopoulou I, Antimisiaris SG (2021) Nanobiosystems for antimicrobial drug-resistant infections. Nanomaterials 11(5):1075

Golshahi L, Seed KD, Dennis JJ, Finlay WH (2008) Toward modern inhalational bacteriophage therapy: nebulization of bacteriophages of *Burkholderia cepacia* complex. J Aerosol Med Pulm Drug Deliv 21(4):351–360

Golshahi L, Lynch KH, Dennis JJ, Finlay WH (2011) *In vitro* lung delivery of bacteriophages KS4-M and ΦKZ using dry powder inhalers for treatment of *Burkholderia cepacia* complex and *Pseudomonas aeruginosa* infections in cystic fibrosis. J Appl Microbiol 110(1):106–117

Green SI, Kaelber JT, Ma L, Trautner BW, Ramig RF, Maresso AW (2017) Bacteriophages from ExPEC reservoirs kill pandemic multidrug-resistant strains of clonal group ST131 in animal models of bacteremia. Sci Rep 7(1):46151

Gu Liu C, Green SI, Min L, Clark JR, Salazar KC, Terwilliger AL et al (2020) Phage-antibiotic synergy is driven by a unique combination of antibacterial mechanism of action and stoichiometry. MBio 11(4):10–1128

Harper DR, Parracho HM, Walker J, Sharp R, Hughes G, Werthén M et al (2014) Bacteriophages and biofilms. Antibiotics 3(3):270–284

Herelle F, Malone RH, Lahiri MN (1930) Studies on Asiatic cholera. Studies on Asiatic cholera., (Memoir No. 14)

Hitchcock NM, Devequi Gomes Nunes D, Shiach J, Valeria Saraiva Hodel K, Dantas Viana Barbosa J, Alencar Pereira Rodrigues L et al (2023) Current clinical landscape and global potential of bacteriophage therapy. Viruses 15(4):1020

Hoe S, Semler DD, Goudie AD, Lynch KH, Matinkhoo S, Finlay WH et al (2013) Respirable bacteriophages for the treatment of bacterial lung infections. J Aerosol Med Pulm Drug Deliv 26(6):317–335

Jończyk E, Kłak M, Międzybrodzki R, Górski A (2011) The influence of external factors on bacteriophages. Folia Microbiol 56:191–200

Kortright KE, Chan BK, Koff JL, Turner PE (2019) Phage therapy: a renewed approach to combat antibiotic-resistant bacteria. Cell Host Microbe 25(2):219–232

Koskella B, Brockhurst MA (2014) Bacteria–phage coevolution as a driver of ecological and evolutionary processes in microbial communities. FEMS Microbiol Rev 38(5):916–931

Kuipers S, Ruth MM, Mientjes M, de Sévaux RG, van Ingen J (2019) A Dutch case report of successful treatment of chronic relapsing urinary tract infection with bacteriophages in a renal transplant patient. Antimicrob Agents Chemother 64(1):10–1128

Labrie SJ, Samson JE, Moineau S (2010) Bacteriophage resistance mechanisms. Nat Rev Microbiol 8(5):317–327

Lin RC, Sacher JC, Ceyssens PJ, Zheng J, Khalid A, Iredell JR, Network TAPB (2021) Phage biobank: present challenges and future perspectives. Curr Opin Biotechnol 68:221–230

Łobocka M, Dąbrowska K, Górski A (2021) Engineered bacteriophage therapeutics: rationale, challenges and future. BioDrugs 35(3):255–280

Luong T, Salabarria AC, Roach DR (2020) Phage therapy in the resistance era: where do we stand and where are we going? Clin Ther 42(9):1659–1680

Ma Y, Pacan JC, Wang Q, Xu Y, Huang X, Korenevsky A, Sabour PM (2008) Microencapsulation of bacteriophage Felix O1 into chitosan-alginate microspheres for oral delivery. Appl Environ Microbiol 74(15):4799–4805

Mahler M, Costa AR, van Beljouw SP, Fineran PC, Brouns SJ (2023) Approaches for bacteriophage genome engineering. Trends Biotechnol 41(5):669–685

Matinkhoo S, Lynch KH, Dennis JJ, Finlay WH, Vehring R (2011) Spray-dried respirable powders containing bacteriophages for the treatment of pulmonary infections. J Pharm Sci 100(12):5197–5205

Merabishvili M, Pirnay JP, Verbeken G, Chanishvili N, Tediashvili M, Lashkhi N, Vaneechoutte M (2009) Quality-controlled small-scale production of a well-defined bacteriophage cocktail for use in human clinical trials PloS one, 4(3), e4944

Merabishvili M, Pirnay JP, Verbeken G, Chanishvili N, Tediashvili M, Lashkhi N et al (2013a) Quality-controlled small-scale production of a well-defined bacteriophage cocktail for use in human clinical trials. PLoS One 4(3):e4944

Merabishvili M, Vervaet C, Pirnay JP, De Vos D, Verbeken G, Mast J et al (2013b) Stability of *Staphylococcus aureus* phage ISP after freeze-drying (lyophilization). PLoS One 8(7):e68797

Nir-Paz R, Gelman D, Khouri A, Sisson BM, Fackler J, Alkalay-Oren S, Hazan R (2019) Successful treatment of antibiotic-resistant, poly-microbial bone infection with bacteriophages and antibiotics combination. Clin Infect Dis 69(11):2015–2018

Ohtake S, Martin RA, Yee L, Chen D, Kristensen DD, Lechuga-Ballesteros D, Truong-Le V (2010) Heat-stable measles vaccine produced by spray drying. Vaccine 28(5):1275–1284

Payne RJ, Jansen VA (2003) Pharmacokinetic principles of bacteriophage therapy. Clin Pharmacokinet 42:315–325

Pirnay JP, Blasdel BG, Bretaudeau L, Buckling A, Chanishvili N, Clark JR et al (2015) Quality and safety requirements for sustainable phage therapy products. Pharm Res 32:2173–2179

Pirnay JP, Verbeken G, Ceyssens PJ, Huys I, De Vos D, Ameloot C, Fauconnier A (2018) The magistral phage. Viruses 10(2):64

Puapermpoonsiri U, Spencer J, van der Walle CF (2009) A freeze-dried formulation of bacteriophage encapsulated in biodegradable microspheres. Eur J Pharm Biopharm 72(1):26–33

Puapermpoonsiri U, Ford SJ, Van der Walle CF (2010) Stabilization of bacteriophage during freeze drying. Int J Pharm 389(1–2):168–175

Ruckert A, Fafard P, Hindmarch S, Morris A, Packer C, Patrick D et al (2020) Governing antimicrobial resistance: a narrative review of global governance mechanisms. J Public Health Policy 41:515–528

Saha D, Mukherjee R (2019) Ameliorating the antimicrobial resistance crisis: phage therapy. IUBMB Life 71(7):781–790

Schade AL, Caroline L (1944a) The preparation of a polyvalent dysentery bacteriophage in a dry and stable form: III. Stability of the dried bacteriophage towards heat, humidity, age and acidity. J Bacteriol 48(2):243–251

Schade AL, Caroline L (1944b) The preparation of a polyvalent dysentery bacteriophage in a dry and stable form: II. Factors affecting the stabilization of dysentery bacteriophage during lyophilization. J Bacteriol 48(2):179–190

Schooley RT, Biswas B, Gill JJ, Hernandez-Morales A, Lancaster J, Lessor L, Hamilton T (2017) Development and use of personalized bacteriophage-based therapeutic cocktails to treat a patient with a disseminated resistant *Acinetobacter baumannii* infection. Antimicrob Agents Chemother 61(10):10–1128

Semler DD, Lynch KH, Dennis JJ (2012) The promise of bacteriophage therapy for *Burkholderia cepacia* complex respiratory infections. Front Cell Infect Microbiol 1:27

Skurnik M (2022) Can bacteriophages replace antibiotics? Antibiotics 11(5):575

Suh GA, Lodise TP, Tamma PD, Knisely JM, Alexander J, Aslam S et al (2022) Considerations for the use of phage therapy in clinical practice. Antimicrob Agents Chemother 66(3):e0207121

Sulakvelidze A, Alavidze Z, Morris JG Jr (2001) Bacteriophage therapy. Antimicrob Agents Chemother 45(3):649–659

Tomlinson IM (2004) Next-generation protein drugs. Nat Biotechnol 22(5):521–522

Tovkach FI, Zhuminska GI, Kushkina AI (2012) Long-term preservation of unstable bacteriophages of enterobacteria. Mikrobiol Z 74(2):60–66

Tsonos J, Oosterik LH, Tuntufye HN, Klumpp J, Butaye P, De Greve H et al (2014) A cocktail of *in vitro* efficient phages is not a guarantee for *in vivo* therapeutic results against avian colibacillosis. Vet Microbiol 171(3–4):470–479

Twort FW (1961) An investigation on the nature of ultra-microscopic viruses. Acta Kravsi

Vagenende V, Yap MG, Trout BL (2009) Mechanisms of protein stabilization and prevention of protein aggregation by glycerol. Biochemistry 48(46):11084–11096

Valerio N, Oliveira C, Jesus V, Branco T, Pereira C, Moreirinha C, Almeida A (2017) Effects of single and combined use of bacteriophages and antibiotics to inactivate *Escherichia coli*. Virus Res 240:8–17

Vandenheuvel D, Singh A, Vandersteegen K, Klumpp J, Lavigne R, Van den Mooter G (2013) Feasibility of spray drying bacteriophages into respirable powders to combat pulmonary bacterial infections. Eur J Pharm Biopharm 84(3):578–582

Vandenheuvel D, Meeus J, Lavigne R, Van den Mooter G (2014) Instability of bacteriophages in spray-dried trehalose powders is caused by crystallization of the matrix. Int J Pharm 472(1–2):202–205

Vandenheuvel D, Lavigne R, Brüssow H (2015) Bacteriophage therapy: advances in formulation strategies and human clinical trials. Ann Rev Virol 2:599–618

Ventola CL (2015) The antibiotic resistance crisis: part 1: causes and threats. P T 40(4):277

Wang W (2000) Lyophilization and development of solid protein pharmaceuticals. Int J Pharm 203(1–2):1–60

Zalewska-Piątek B (2023) Phage therapy—challenges, opportunities and future prospects. Pharmaceuticals 16(12):1638

Zhang X, Zhang C, Liang C, Li B, Meng F, Ai Y (2022) CRISPR–Cas9 based bacteriophage genome editing. Microbiol Spect 10(4):e00820–e00822

Part IV

Immunotherapies: Boosting the Body's Defense

22 Diverse Properties of *Tinospora cordifolia* (Giloy, Heart Leaved Moonseed) World Wild Use for Immunotherapies; Boosting the Body's Defense and Immune Support

Preeti Semwal, Biresh Kumar Sarkar, Kumari Sonu, Garima Chand, Lalchand D. Devhare, and Tanuj Hooda

Abstract

Tinospora cordifolia is a climbing shrub that grows on other trees, from the botanical family Menispermaceae. The plant is native to India but is also found in tropical areas of many countries. Although people have been using Giloy, since ancient times, generally at that time it was used for animals but in a global pandemic like COVID-19, people used it in large quantities and only after that the process of research on it started. The infection of COVID-19 severity is variable and affect all ages people and especially elderly people whose immune system is very weak. This Herbs have various potential compounds which used

P. Semwal (✉)
Shri Guru Ram Rai University, Dehradun, Uttarakhand, India

Department of Pharmaceutical Science, Hemvati Nandan Bahuguna Garhwal Central University, Srinagar Garhwal, Uttarakhand, India

B. K. Sarkar
Central Ayurveda Research Institute, CCRAS, Ministry of AYUSH, New Delhi, India

K. Sonu
IIMT college of Pharmacy, Knowledge Park III, Greater Noida, Uttar Pradesh, India

G. Chand
College of Pharmacy, Graphic Era Hill University, Bhimtal, Uttarakhand, India

L. D. Devhare
Manwatkar College of Pharmacy, Ghodpeth, Maharashtra, India

T. Hooda
MM College of Pharmacy, Maharishi Markandeshwar (deemed to be) University, Mullana, Ambala, Haryana, India

M. Gangwar, G. Nath (eds.), *Emerging Paradigms for Antibiotic-Resistant Infections: Beyond the Pill*, https://doi.org/10.1007/978-981-97-5272-0_22

influences immunity and maintenance of the haemostasis of inflammation, anti-inflammatory and scientists are still working to understand its useful properties. Giloy is generally used for fever urinary problems, asthma dysentery, diarrhea, skin infections, Hansen's disease (formerly called leprosy), diabetes, jaundice, anorexia, eye conditions and it act as a natural disease reliever for the human body. The wide range of uses and potential health benefits from giloy come from numerous beneficial plant compounds found throughout the plant. *Tinospora cordifolia* has secondary metabolite compounds like terpenoids, alkaloids, lignans, steroids. The compounds show they have antimicrobial, antiviral, anticancer, and antidiabetic properties also they are known to have therapeutic benefits high, law blood pressure (BP), pain relief, cancer, malaria, intestinal spasms, ovarycystic infection, ovarian cancer. In view of its antioxidant effect, the Giloy is widely used for immunotherapies and other immune system benefits viz., allergies, bacteria, and pathogens. It's able to prevent the growth of *Klebsiella pneumoniae*, *Escherichia coli*, *Pseudomonas* spp., *Proteus* spp. This led us to believe that Giloy has a multitalented potential, immunity booster activity is to be explored, and further studies are needed to ensure this activity. In this chapter, we have discussed the diverse properties of Giloyin in greater detail.

Keywords
Immunotherapy · *Tinospora cordifolia* · Medicinal herb · Tropical diseases

22.1 Introduction

The Ayurveda knowledge system of India's a long past and traditional medicinal heritage from ancient era. History revels that *Tinospora cardifolia* (giloy) plants have been used since 4000–5000 B.C. for therapeutic purposes. *Tinospora cardifolia* a semi evergreens deciduous climbing shrub commonly known as "Guduchi" in Sanskrit and belonging to Menispermaceae family of flowering plants (Singh et al. 2003). It is a genetically diverse, large with greenish-yellow typical flower, found in higher altitudes of different areas along with countries. Based on the information provided, this appears to be describing the plant Giloy (*Tinospora cordifolia*), which is a climbing vine found in various parts of India as well as other regions like North Africa, South Africa, Bangladesh, Malaysia, Myanmar, and Vietnam. Giloy is found up to 1200 m elevation in parts of India like Kumaon, Garhwal, Assam, West Bengal, Bihar, Deccan, Karnataka, and Kerala. It is a large climber that spreads over tall trees in forests, with aerial roots growing from its branches. The dried aerial roots are light brown or whitish, odorless, and have a bitter taste. Various Ayurvedic institutions recommend consuming Giloy stems/roots to boost immunity, including against COVID-19. An Ayurvedic organization called Patanjali Ayurveda created a medicine kit called "Coronil", which contains Giloy, Ashwagandha, and Tulsi as main ingredients, promoted for COVID-19 (Mohanraj et al. 2018; Sharma et al. 2019). It is a large climber that spreads over tall trees in forests, with aerial roots growing from its branches (Sharma et al. 2019). In Ayurveda, Giloy is used either alone or with other herbs to treat a wide range of conditions from fever and digestion issues to urinary tract disorders, arthritis, anemia, asthma, and more.

The potential of this healing plant as a therapeutic agent against SARS-COV-2 virus has also been studied. Recently the plant is of great interest to researchers across the globe because of its diverse pharmacological properties to fight against various types of diseases like; antidiabetic, antiperiodic, antispasmodic, anti-inflammatory, antiarthritic, antioxidant, antiallergic, antistress, anticleprotic, anti-malarial, hepatoprotective, antineoplastic, immunomodulatory active role. These medicinal plants are rich in secondary metabolites and have essentials therapeutic importance. All the parts of plants of the plant are enormously helpful due to the presence of different compounds of pharmaceutical importance belonging to various groups as alkaloids, steroids, diterpenoids lactones, glycosides, sesquiterpenoid, and phenolics (Verma et al. 2021). Since times antediluvian, medicinal plants have been acceptable virtually in social culture for the treatment of diverse disorders and ailments. The widespread use of herbal remedies and health care preparations as those described in old age texts such as the Vedas and the bible and commonly used traditional medicinal plants have been uncovering the amount of natural products with medicinal properties (Jayswal 2021). In the recent past, research has been mainly focused on various plants that serve for utilization of health care management. In this respect, medicinal plant *Tinospora cordifolia* (giloy) has caught the devotion of researchers recently, as one of the versatile rejuvenating shrubs possessing various therapeutic application viz. immune modulatory, antitumor, antihypertensive, antioxidant, hepatoprotection etc. almost all parts of the plants are reported to be useful in ethnobotanical surveys conducted by ethnobotanists (Badjatya et al. 2020).

The aim of this book chapter is to provide in-depth and comprehensive information regarding, *Tinospora cardifolia* (TC) and its immune support & health-related properties, elucidating the physiological role in the human body and focusing on its diverse role in immunotherapies.

22.2 Taxonomic Classifications of *Tinospora cordifolia*

The kingdom Plantae includes all multicellular, eukaryotic photosynthetic organisms. As a member of the division Magnoliophyta, this organism is a flowering plant. Its class Magnoliopsida indicates it is a dicot, and the order Ranunculales suggests it is related to buttercups and other Ranunculaceae (Sharma et al. 2021). Being in the family Menispermaceae means this plant is likely a woody vine. The genus Tinospora includes several climbing shrubs found in tropical areas. And the specific epithet cordifolia indicates this particular species has heart-shaped leaves. So, in summary, this classification describes *Tinospora cordifolia*, a woody climbing vine with heart-shaped leaves, which is likely native to tropical regions. According to Botanical Description of *Tinospora cordifolia* (Menispermaceae) is a deciduous climbing shrub found in tropical and subtropical regions of Asia (Khan et al. 2017). It has the following botanical characteristics. Growth habit, Perennial

Table 22.1 Regional names of *Tinospora cordifolia* (Giloy)

S.N.	Language	Regional name
1.	English	Heart-Leaved Moonseed
2.	Hindi	Giloy
3.	Marathi	Gulvel, Giroli
4.	Kannad	Amrutaballi
5.	Tamil	Amritavali
6.	Telgu	Tippateega
7.	Malyalam	Chittamruthu
8.	Bengali	Gurch
9.	Sanskrit	Amrita
10.	Gujrati	Galac, Garo
11.	Punjabi	Gilow
12.	Urdu	Gilo
13.	Persian	Gulbel
14.	Oriya	Guluchi

climber with long, fleshy stems that can reach lengths of 6 m. The stems are succulent, papery, and light grey, with lenticels. Leaves have an Alternate, simple, cordate with a heart-shaped base. The lamina is membranous and can be entire or lobed, with 3–7 nerves radiating from the base. Leaves are up to 9 cm long. Inflorescence Axillary cymes, small and yellowish green in color. Plants are dioecious with male and female flowers on separate plants. Flowers: Small, yellowish green, with 6 sepals and no petals. Male flowers are arranged in fascicles. Female flowers are usually solitary. Fruit has Globose drupe, 5–8 mm in diameter, fleshy, glorious, ripening red to orange red. Contains 1–2 seeds. Seeds have Ovoid, with a hard seed coat. Giloy Flowering during summer and Fruits mature in winter. Propagation has Stem cuttings. Ecology Found in tropical forests, thickets, and wastelands. Climbs up trees and spreads vegetatively (Tiwari et al. 2018). The regional names of Giloy are depicted in Table 22.1.

22.3 The Antibacterial Effect of Giloy (*Tinospora cordifolia*) and Its Role in Opposing Antimicrobial Resistance

Giloy is an herb used in Ayurvedic medicine it has been shown to have antibacterial properties against several types of bacteria. The studies show that various active compounds in Giloy exhibit antibacterial activity, including tinosporaside, cordifolioside, syringin, berbelineine, giloin, gilosterol, and tinosporol.

Giloy extracts have been found to be effective against common disease-causing bacteria like Staphylococcus aureus, Bacillus subtilis, Escherichia coli, and *Klebsiella pneumoniae*. The antibacterial effects are believed to be due to inhibition of cell membrane synthesis, alteration of bacterial cell permeability, binding to bacterial proteins, and interference with bacterial enzyme systems. Giloy may help combat antimicrobial resistance by providing an alternative antibacterial treatment option and reducing the need for conventional antibiotic use which drives resistance. Using Giloy along with antibiotics could potentially have a synergistic effect by attacking bacteria through multiple mechanisms making it more difficult for resistance to develop (Sharma et al. 2012). Further research is still needed to fully characterize the antimicrobial mechanisms of giloy, identify the most effective extracts/compounds, and conduct clinical trials to validate its efficacy and safety as an antibacterial therapy in humans. Giloy shows promise as a natural antibacterial agent that could be useful for fighting infections and addressing the growing public health threat of antimicrobial resistance. More studies are warranted to harness the antibacterial potential of this traditional Ayurvedic herb.

Tinospora cordifolia is a plant that shows potential antibacterial activity against bacteria that cause urinary tract infections. Its alkaloids, carbohydrates, and flavonoids make it effective against clinical isolates of drug-resistant bacteria. Plant-based therapies like Giloy could be useful alternatives to tackle emerging drug resistance in healthcare globally. There is limited exploratory research from India on the biological and therapeutic utility of Giloy and Using the bioactive compounds of *Tinospora cordifolia* to develop new antimicrobials could help combat multidrug-resistant (MDR) organisms. Phytochemicals from plants like *Tinospora cordifolia* may help reverse antibiotic resistance. More plant-based research could accelerate clinical approval and use of these phytochemicals against MDR pathogens. The present study aims to assess and explore the antibacterial properties of *Tinospora cordifolia* to combat antimicrobial resistance by MDR pathogens isolated from patients.

The key goals are to evaluate *Tinospora cordifolia*'s antibacterial activity against drug-resistant bacteria, and explore its potential to reverse antibiotic resistance, leading to new antimicrobial therapies. The research could accelerate the development of plant-derived treatments for multidrug-resistant infections (Sharma et al. 2012).

The various studies found antibacterial properties of *Tinospora cordifolia* and the urgent problem of multidrug-resistant (MDR) bacteria. Antimicrobial resistance (AMR) is a rising concern about human health and 700,000 people surrender each year to diseases caused due to AMR. Antibiotic enters widespread clinical use resistance often emerges rapidly. Bacteria have a remarkable ability to adapt to antibiotics through various genetic mutations and acquisition of resistance genes. The ESKAPE pathogens you mentioned are prime examples of highly drug-resistant organisms causing difficult-to-treat infections in hospitals and communities globally. They exemplify the ability of bacteria to develop resistance to multiple antibiotic classes. Continued overuse and misuse of existing antibiotics in healthcare and agriculture are key drivers of rising AMR. But the antibiotic pipeline has slowed, with not enough new drugs in development. Fighting AMR will require a multifaceted global response, including more judicious antibiotic use, improved infection control, development of new antibiotics and alternative therapies, and efforts to minimize environmental antibiotic pollution. Overall, AMR is a pressing challenge requiring urgent action across many sectors to avoid a post-antibiotic era where common infections could once again become untreatable and deadly. Ongoing surveillance, research, innovation, and policy efforts will be crucial to get ahead of this threat.

Pseudomonas aeruginosa is the fifth most common nosocomial (hospital-acquired) pathogen. It is multidrug-resistant (MDR) and found frequently in medical devices like ventilators (Emori and Gaynes 1993). It causes infections in immune-compromised patients with conditions like COPD, cystic fibrosis, cancer, trauma, burns, sepsis, and ventilator-associated pneumonia including those caused by COVID-19 (Blanco et al. 2021; Kreutmair et al. 2021). The Studies from the Indian Council of Medical Research (ICMR) show Pseudomonas species are resistant to common antibiotics like penicillin's, cephalosporins, fluoroquinolones, tetracyclines, and macrolides (Javiya et al. 2008; Ramana and Chaudhury 2012). Due to antibiotic resistance, there is a need to look into alternative sources like medicinal plants to find potential new drug leads against *Pseudomonas aeruginosa* infections. *Pseudomonas aeruginosa* is a challenging multidrug-resistant pathogen causing many hospitals-acquired infections, and new treatment options are needed to combat its antibiotic resistance. Exploring medicinal plants is a suggested approach to find new drug leads based on the information provided.

Functionalized nanoparticles (NPs) have potential for usage in industries like pharmaceuticals and biomedicine due to their unique size, chemical, and physical properties. Common functionalized nanoparticles (NPs) include gold (Au), silver (Ag), silica (Si), PEG, PLGA, and PCL, with applications in catalysis, chemical sensing, biolabeling, photonics, and drug/biomolecule delivery.

Gold and silver nanoparticles (NPs) are particularly useful. Silver nanoparticles (NPs) are often used as antibacterial agents. Hybrid materials of Ag nanoparticles (NPs) and macromolecules have antimicrobial properties. Gold nanoparticles are extensively studied and used as antibacterial agents due to their stability, low toxicity, large surface area, and ease of functionalization. The key references cited: Subbiah et al. (2010): Introduced functionalized nanoparticles (NPs) and their potential applications. Mahalingam et al. (2004), Sinha et al. (2009), Mukherjee et al. (2010): Discussed applications of various nanoparticles (NPs). Astete and Sabliov (2006): Focused on PLGA NPs. Connor et al. (2005), Dizaj et al. (2014): Highlighted antimicrobial properties of Au NPs. Computer-based tools have enabled advanced drug discovery techniques like virtual screening of medicinal plant compounds. This allows for more efficient identification of bioactive chemicals. The study focused on *Tinospora cordifolia*, assessing its potential as an antibacterial agent and identifying its most active phytochemical constituent. The researcher's nano-functionalized silver and gold nanoparticles with *T. cordifolia* plant extracts, as nano-functionalization can enhance medicinal plant bioactivity (Sliwoski et al. 2014). They isolated and purified a secondary metabolite from *T. cordifolia* using FCS (fluorescence correlation spectroscopy) (Maurya et al. 2012). They tested the antimicrobial activity of this isolated compound with the silver and gold nanoparticles. More information is still needed on the specific bioactive compounds in *Tinospora cordifolia*. This study aimed to identify a molecular target for the most reactive *T. cordifolia* (Giloy) compound using computational methods like Density Functional Theory (DFT) and molecular docking. The study evaluated *Tinospora cordifolia*'s antibacterial potential, nano-functionalized its extracts, isolated an active secondary metabolite, and computationally analysed its molecular interactions identifying this medicinal plant as a source for new antibiotic drug leads.

The study evaluated the biological importance and antimicrobial properties of *Tinospora cordifolia* (Giloy), a medicinal plant. *T. cordifolia* stem nanoparticles showed excellent antibacterial activity against drug-resistant *P. aeruginosa* strains, even at minimal concentrations. This makes them a promising source of antibacterial agents. The nanoparticles enhanced the therapeutic benefits and effectiveness of *T. cordifolia.* Out of the five Cordifolisides identified, Cordifoliside C was found to be the most reactive. In silico studies showed Cordifoliside C had good binding affinity with the TolB protein, which is important for *P. aeruginosa's* resistance and growth. Docking studies also showed good binding interactions between Cordifoliside C and TolB. The Research results demonstrate the potential of using compounds from *T. cordifolia* for antimicrobial drug design, especially against drug-resistant *P. aeruginosa. T. cordifolia* nanoparticles and compounds like Cordifolioside C show promise as novel antimicrobial agents, especially against resistant strains, deserving further drug development research.

22.4 Active Constituents of Giloy (*Tinospora cordifolia*)

The stems contain a variety of active compounds including tinosporine, tinosporide, tinosporaside, cordifolide, cordifol, heptacosanol, clerodane Furano diterpene, diterpenoid furanolactone tinosporidine, columbin, and beta-sitosterol. The stems are often used for their immune boosting, anti-inflammatory, antioxidant, antipyretic and hepatoprotective properties. The leaves contain compounds like syringin, cordifolioside A, magnoflorine, tinocordiside, and tinocordifolioside. The leaves exhibit antidiabetic, antioxidant, anti-inflammatory, and immunomodulatory activities. The roots are rich in compounds like syringin, giloin, gilenin, tinosporide, and tinosporidine) (Saha and Ghosh 2012), the key active constituents found in giloy (*Tinospora cordifolia*): is one of Alkaloids. The main alkaloids in giloy are berberine, giloin, gilonin, and tinosporine. Berberine has antioxidant, anti-inflammatory, and immune-modulating properties. Second constituents of Giloy is Glycosides, these contains glucokinins, cordioside, and syringin. These glycosides are thought to contribute to giloy's anti-inflammatory, antipyretic, and immunomodulatory effects. Third constituents of Giloy are Steroids. Steroids like tinosporol, tinosporidine, cordifol, heptacosanol are present. They account for the antistress, antiallergic, and anti-inflammatory actions. Fourth constituents of Giloy are Polysaccharides. Polysaccharides in giloy have immunostimulant properties. Tinosporide, cordifol, cordifolide, syringin, cleroindicin, furanolactone, and giloin are other key phytochemicals found in giloy. The wide range of alkaloids, glycosides, steroids, and other active compounds in giloy contribute to its broad therapeutic effects and medicinal uses in Ayurveda.

The roots demonstrate anti-inflammatory, antiarthritic, antiallergic, antistress, and antimalarial effects. The fruits contain diterpenoid lactones like tinosporidine, tinosporide, cordifolide, tinosporic acid, and syringin. The fruits possess antioxidant, anti-diabetic, and gastroprotective properties. The stems and leaves are most widely used and studied for their therapeutic benefits, followed by the roots and then the fruits. The stems contain maximum active constituents like tinosporine, tinosporide, cordifolide, etc. responsible for giloy's pharmacologic activities. Table 22.2 represents the therapeutic constituents and uses of Giloy plant parts (Chi et al. 2016).

22.5 Physical Properties and Chemical Properties of Giloy

In physical properties, growth habit of Giloy has Continuous climber it can grow up to 20 m in length. Giloy Stem is woody, Barrel-shaped, and hollow. The leaves of Giloy plants have alternate, ovate, and palmately shaped with pointed tips. The leaf tips have 8–10 cm long and 5–6 cm broad. Flower of Giloy has Small, greenish yellow, borne in clusters. Giloy Fruit has Oval shaped, greenish-yellow, 2–3 cm long (Mittal et al. 2014) it is a long climbing vine with hollow woody stems, palmate leaves, small yellowish flowers in clusters, and oval yellowish fruits of 2–3 cm length. The detailed morphological and anatomical properties indicate it is likely aspecies of cucurbit plant.

On chemical properties basis the plant contains active compounds like flavonoids, terpenoids, glycosides, and phenolic compounds that have pharmacological effects. Flavonoids like quercetin and kaempferol have anti-inflammatory and antioxidant effects that can modulate the immune system. Terpenoids like menthol and eugenol have antibacterial and antiviral properties that can fight infections. Phenolic compounds like rosmarinic acid and caffeic acid are antioxidants that can neutralize free radicals and reduce oxidative stress. Glycosides like ginsenosides and anthraquinones have diverse pharmacological activities including immunomodulation, antiinflammation, and antimicrobial effects. The combination of these bioactive compounds gives the plant adaptogenic, immune-protective, and chemo-preventive

Table 22.2 Therapeutic constituents and uses of *Tinospora cordifolia* plant parts

Plant part	Active component	Key compounds	Therapeutic benefits
Leaves	Alkaloids	Axillarin, axillarine	Antibacterial, antifungal, Parkinson's disease treatment
Aerial root	Steroids	Beta-sitosterol	Anti-inflammatory, osteoporosis treatment, antiviral, anticancer, antidiabetic, analgesic
Stem	Glycosides	Tinocordiside	Neuroprotective
Entire plant	Terpenoids	Jateorine, Columbin	Antimicrobial, reaxing vessels, antihypertensive, vasorelaxant, germicide
Roots	Lignans	Giloin, Cordifol	Anxiolytic, analgesic, anxi-olytic, HIV protease inhibitor
Flower	Essential oil	Cadinenes, geraniol	Diuretic, anti-spasmodic, bone health, anti-inflammatory

properties to enhance overall health and resilience against diseases. However, the exact mechanisms of action and chemical structure–activity relationships of many active constituents are still being researched for full elucidation (Patel et al. 2009) the medicinal properties of this plant can be attributed to its unique phytochemical profile containing pharmacologically active compounds that have been validated by scientific research for their therapeutic effects and benefits.

Tinospora cordifolia (common name Giloy, heart-leaved moonseed) is a large climbing shrub found in tropical regions of India and some parts of China. It has been used in traditional Ayurvedic medicine in India to treat various conditions like general weakness, indigestion, urinary infections, ovary cystic infection, etc. The pharmacologically active compounds are present in various parts of the plant like stems, roots, leaves, etc. Over 200 phytochemicals from different classes have been identified, including diterpenoids, sesquiterpenoids, glycosides, steroids, phenolic compounds, etc. Some key phytoconstituents are tinosporine, tinosporide, tinosporaside, cordifolide, cordifol, heptacosanol, clerodane furano diterpene, tinosporidine, columbin, beta-sitosterol. Specific compounds found in the stalks are berberine, palmatine, tembertarine, magniflorine, choline, tinosporin. The wide range of pharmacological activities is attributed to the diverse phytochemical composition with different classes of active compounds present in the plant.

22.6 Therapeutic Uses of *Tinospora cordifolia* to Treatment of Tropical Diseases

Anti-allergic: Reduces symptoms of allergic rhinitis like sneezing, runny nose, itchy eyes, etc., due to its antihistamine properties. Giloy helps inhibit the release of histamine which is responsible for allergic reactions. It Helps treat skin allergies and conditions like eczema, urticaria, and dermatitis. Giloy has anti-inflammatory and antioxidant properties that help reduce skin inflammation, rashes and itching. Giloy Provides relief in respiratory allergies like dust allergy and hay fever. It reduces respiratory congestion and breathing difficulties caused by an allergic response. It's also Reduces asthma symptoms triggered by allergens. Giloy has a broncho dilating action that helps open up constricted airways and improves airflow. it helps counter allergies to foods like nuts, seafood, etc. It inhibits the release of mast cell chemicals that cause food allergy symptoms. It Provides relief from insect bite allergies. Giloy extract when applied topically can reduce rashes and itching due to insect bites. It Reduces the risk of anaphylaxis or severe allergic reaction by decreasing histamine levels overall. Giloy is an effective ayurvedic herb to treat a variety of allergies through its antihistaminic, anti-inflammatory, and antioxidant properties. It can provide symptomatic relief in most types of allergic conditions (Aher and Wahi 2011).

Immune booster, anti-inflammatory Giloy has anti-inflammatory properties that can help reduce inflammation, fever, and aid in autoimmune conditions. The compounds in giloy like Gilledin are responsible for its anti-inflammatory effects. Antioxidant Giloy acts as a natural antioxidant by removing harmful free radicals from the body. This helps boost immunity and prevent chronic diseases caused by oxidative stress. Fever relief Giloy is often used in Ayurveda to bring down mild to moderate fevers. It helps improve the body's natural response to infection. Giloy improves digestion improving gut health and promoting regular bowel movements, giloy helps boost overall immunity. Its antispasmodic action soothes digestive issues like IBS. Fights respiratory problems giloy juice/decoction helps manage respiratory disorders like cough, cold, and tonsillitis. It helps get rid of phlegm and mucus accumulation in the lungs. It Manages diabetic symptoms Giloy regulates blood sugar, reduces high blood pressure, and prevents complications of diabetes like nephropathy. This protects the body from diabetes-related immune weakness. It Reduces stress and fatigue as an adaptogenic herb, giloy reduces mental stress, anxiety, improves stamina, and helps fight fatigue. This reenergizes the immune system. Dosage most experts recommend 500 mg to 1 g of giloy powder or extract daily. It can be taken along with honey, warm water, or milk. Giloy contains antioxidants and anti-inflammatory compounds that help fight free radicals, strengthen cells, and combat illnesses. It helps eliminate toxins and improve blood quality.

Giloy has antimicrobial properties that aid in quick recovery. Giloy boosts macrophage activity (More and Pai 2012). Macrophages are immune cells that fight microbes and pathogens. It has Compounds like cordifolioside, syringin, and tinocordioside in giloy have immunomodulatory and anticancer effects. Giloy inhibits immune hemolysis and complements the activity of IgG antibodies. This demonstrates its ability to modulate the immune system. Giloy stem extract can stimulate lymphocyte cells and increase levels of protective enzymes like catalase. Giloy extract enhances macrophage function and neutrophil activity, further boosting the immune response. Giloy has scientifically demonstrated immunoprotected abilities due to its antioxidant, anti-inflammatory, and immunomodulatory properties (Patel et al. 2009). It supports overall immune function through multiple mechanisms. Giloy works holistically to improve immunity by fighting infection, reducing inflammation, managing chronic diseases like diabetes, and lowering mental stress.

Hyperthermia or high fever: Giloy has antipyretic properties which means it can help reduce high body temperatures associated with fever. The aqueous extracts of giloy stem have been found to reduce fever in rats. Giloy acts as a mild analgesic or pain reliever. This can help provide relief from body aches associated with fevers. Giloy has anti-inflammatory properties. This helps reduce inflammation that often accompanies fever and associated pain. Giloy helps modulate the immune system. During fever, it can help balance the immune response and prevent the overreaction of the inflammatory cytokines. Giloy contains antioxidants that can help counter the oxidative stress caused by fever on the body's cells and tissues. Giloy has liver protecting effects. This helps support the liver which is working hard to metabolize the pyrogens causing the fever. Giloy can help reduce high fevers through multiple mechanisms including antipyretic, analgesic, anti-inflammatory, immune modulating, and tissue-protective effects (Patel et al. 2009; Gupta and Sharma 2011). It is considered a safe and effective traditional remedy for fever and associated symptoms. Drink giloy juice or decoction daily. Giloy is known to improve digestion and boost metabolism. Consume it first thing in the morning on an empty stomach. Make a giloy powder and take it with lukewarm water. Mix 1 teaspoon giloy powder in a glass of water. Drink this before meals to fire up digestion. Add giloy to your meals. You can add giloy powder or extract to soups, curries, smoothies, etc. This will help improve absorption and digestion of nutrients from food. Take giloy with honey. Mix giloy powder with raw organic honey. Having this combo on an empty stomach can give metabolism a good push. Try giloy and amla juice. Amla is a rich source of vitamin C and antioxidants. Combining it with giloy can boost metabolism efficiently. Makegiloy tea. Steep giloy powder or leaves in hot water for 5–10 min. Sip this tea through the day to keep metabolic function optimal. Use giloy in kadha. Boil giloy powder with herbs like tulsi, cinnamon, black pepper to make an immunity-boosting kadha. Sipping this can also perk up sluggish digestion. Take giloy supplements after consulting an Ayurvedic practitioner about the correct dosage for your needs. The Key is to consume giloy regularly to keep metabolic function at its peak. Along with giloy, also ensure a balanced diet, regular exercise and satisfactory sleep.

Help lower blood sugar levels: Giloy is an Ayurvedic herb that has been shown to help lower blood sugar levels. The active compounds in giloy help improve insulin secretion and sensitivity. To help treat hyperglycaemia, take 1–2 g of giloy powder twice daily. You can mix the powder into warm water or add it to teas. Taking giloy capsule supplements can also be effective. Drink giloy juice on an empty stomach in the morning. Combine giloy powder with amla and neem powder, then add honey and drink. This can help lower fasting blood sugar having giloy decoction made by boiling giloy stems in water can also be beneficial. Drink this regularly to help manage hyperglycaemia apply a paste made from giloy leaves externally on the skin. This is said to help control blood sugar levels (Jagetia and Rao 2006).

Always monitor your blood glucose carefully when using giloy or other herbs. Be sure to consult your doctor before using giloy if you take diabetes medications, as the dosage may need adjustment. In addition to giloy, follow a healthy diabetes diet, exercise regularly, stay hydrated, and take prescribed diabetes medications. Controlling carbohydrate intake is key for managing hyperglycaemia.

Treating respiratory conditions: Giloy acts as an immunomodulator and has anti-inflammatory properties which help improve overall respiratory health. It helps reduce respiratory tract infections. Giloy has broncho dilating abilities which help improve airflow to the lungs and provide relief in asthma. It also helps reduce histamine levels which are responsible for asthma symptoms. Giloy helps counter allergic responses in the respiratory system. Its antihistamine properties help treats allergic rhinitis and hay fever. Giloy has laxative abilities. It helps remove toxins and waste from the lungs and improves overall respiratory health. Giloy acts as a rasayan (rejuvenator). It helps enhance respiratory stamina. This makes it useful for athletes and people exposed to dust and pollution giloy helps manage respiratory conditions through its immunomodulatory, anti-inflammatory, antiviral, bronchodilating and detoxifying properties. It is effective in treating chronic respiratory problems (Patel and Mishra 2011).

The claim that Giloy can tremendously improve vision or reduce eye inflammation lacks strong scientific evidence. More research is needed to substantiate these vision-related benefits. Some preliminary studies suggest giloy may have antioxidant and antiaging properties that could potentially benefit skin health, but the evidence is limited (Ly et al. 2007). The giloy is remarkably effective in HIV treatment is not well-supported by current scientific research. While it may have some immunomodulatory effects, there is no convincing evidence that it can treat or cure HIV infection. Giloy reducing HIV's drug resistance, increasing treatment results, and the described effects on immune cells are unsubstantiated and require further rigorous clinical study. In general, while Giloy has been used traditionally for various ailments, many of the specific claims made in the passage about its benefits for vision, skin, and HIV treatment are either exaggerated or lack solid scientific backing from well-designed clinical trials (Fig. 22.1).

More research is needed to validate these therapeutic applications before making definitive recommendations (Karpova et al. 1991).

Giloy has demonstrated anticancer properties in animal models using root extracts. Giloy has been shown to have a radioprotective effect, increasing the body weight, spleen weight, and cellularity significantly. The active components of Giloyare believed to enhance the host's immune system by increasing the leukocyte count and stimulating the proliferation and activation of stem cells. Giloy is widely used in Ayurvedic medicine and is known for its numerous medicinal benefits. It is believed to be effective in treating diseases by boosting immunity and is also used as a blood purifier. The potential anticancer and immunomodulatory properties of Giloy, have been observed in animal studies. These properties are attributed to its ability to enhance the immune system, particularly by increasing leukocyte count

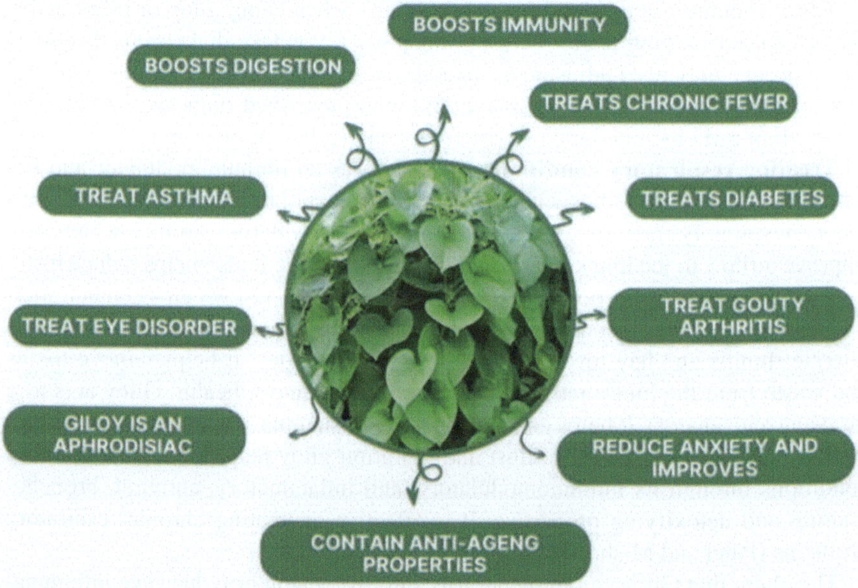

Fig. 22.1 Schematic diagram representing the multiple benefits of Giloy (*Tinospora cardifolia*) for the human body

and activating stem cells. Giloy is a highly regarded herb in Ayurvedic medicine for its various therapeutic benefits, including its use in cancer treatment and as a blood purifier (Kapil and Sharma 1997).

Giloy has anti-inflammatory, antioxidant, and immunomodulatory properties that can help reduce inflammation and speed healing. It's often used for joint pain, arthritis, and musculoskeletal injuries. It can be taken internally as a powder, tablet, or liquid extract. For injuries, it's usually mixed into a paste and applied topically after boiling the stem in water. Giloy is considered safe for short-term use for most people, but it can cause side effects like diarrhea, headache, or nausea in some cases. Those with diabetes or autoimmune conditions should use it cautiously. In addition to giloy, managing pain, resting the injury, icing, elevating, and bandaging may be part of a comprehensive treatment plan. Consulting an Ayurvedic practitioner or doctor is recommended (Chen et al. 2000). Proper diagnosis should be made before using giloy or any herbal medicine to treat an injury. Mild injuries can often heal on their own, while serious injuries may require medical treatment. Immunomodulatory properties: Giloy is believed to have the ability to modulate or regulate the immune system's activity. Anti-inflammatory properties: It is thought to possess anti-inflammatory effects, which can help reduce inflammation. Antioxidant properties: Giloy is considered to have antioxidant effects, which can help neutralize free radicals and protect cells from oxidative stress. The herb is particularly useful in wound healing due to its ability to promote collagen production and the regeneration of epithelial cells (cells that line the surfaces of the body). Giloy is believed to have antibacterial and antifungal properties, which can help prevent infections in wounds. Giloy has been used in Indian Ayurvedic medicine for a long time, highlighting its traditional importance in various therapeutic applications. Giloy is considered a valuable herb in Ayurvedic medicine, particularly for its potential immunomodulatory, anti-inflammatory, antioxidant, wound healing, and antimicrobial properties. Its traditional use and purported benefits make it a subject of interest for further research and potential applications in modern medicine.

22.7 Functional Analysis and Genome Sequencing Is of a Multitalented Medicinal Herb *Tinospora cordifolia* (Giloy)

In recent research scientists phylogenetically explain the characteristics of Giloy. *Tinospora cordifolia*, commonly known as giloy, is an important medicinal plant used in Ayurvedic medicine. It has been used traditionally for its immunomodulatory, anti-inflammatory, antipyretic, antiarthritic, antiallergic, antistress, antileprotic, antimalarial, hepatoprotective, and antioxidant properties. Modern scientific research has validated many of giloy's traditional uses through identifying its active phytochemical constituents and pharmacological actions. Some of its major bioactive compounds include polysaccharides, alkaloids, diterpenoid lactones, steroids, sesquiterpenoid, phenolics, aliphatic compounds and polysaccharides. Functional analysis of giloy using genomic and proteomic tools can help identify genes involved in biosynthesis of its medicinally important phytochemicals. This can help in targeted metabolic engineering to improve desired compound yields. Genome sequencing of giloy will reveal its genetic blueprint and allow for identification of functional genes related to biosynthesis, regulatory mechanisms, growth, and development, etc. This information can be used to improve cultivation practices and isolate useful genes. Comparative genomics across different Tinospora species can provide insights intoevolution of medicinal properties and inter/intra-species genetic diversity. Integrating functional genomics data from transcriptomics, proteomics, metabolomics along with genome sequence can give a -level understanding of giloy's medicinal properties. This can help in identifying gene targets for synthetic biology applications (Rana et al. 2012; Mahajan et al. 2024).

Its genome was sequenced using a combination of 10× Genomics linked read and Nanopore long read technologies. The draft genome assembly was 1.01 Gbp, making it the first sequenced genome from the Menispermaceae plant family. Genome size estimation found the genome to be 1.13 Gbp (Mahajan et al. 2024). The transcriptome was also sequenced from leaf tissue. Phylogenetic analysis positioned *T. cordifolia* as a basal eudicot. Comparative evolutionary analysis revealed gene family expansions/contractions and evidence of adaptive evolution. Adaptive evolution was found in genes involved in key medicinal pathways like benzylisoquinoline alkaloid, terpenoid, lignin, and flavonoid biosynthesis. Adaptively evolved genes were also found in pathways related to endosymbiosis, adventitious root formation, peroxisome biogenesis, phytohormone biosynthesis, and stress tolerance. Overall, genome and transcriptome sequencing provided insights into the genomic basis for the diverse medicinal properties and evolutionary adaptations of *Tinospora cordifolia* applying modern functional genomics and genome sequencing approaches to explore the medicinal herb giloy can reveal new insights into its therapeutic properties and allow for targeted improvement of its pharmacological qualities.

The study reports the current genome of *T. cordifolia* and provides important insights into the genomic basis of its medicinal properties due to the presence of diverse secondary metabolites. However, the presence of these metabolites also posed problems during the DNA and RNA extraction steps, and also severely reduced the data throughput from the ONT nanopore sequencing technologies. Thus, it is required to develop improved nucleic acid extraction protocols to deal with such medicinal plants. Further, the presence of high heterozygosity and repetitive content in *T. cordifolia* genome posed challenges during the assembly process and limited the assembly statistics such as the N50 score. The availability of more data from both second and third-generation sequencing technologies is much needed to achieve better coverage and a more contiguous assembly of this species. Further, due to the absence of genomes from the Tinospora genus and Menispermaceae family, species belonging to the same order was used for evolutionary analysis to eliminate the effect of large genetic distances. The availability of more genomes from closely related plant species would offer a deeper understanding of the species-specific or family-specific characteristics of *T. cordifolia*. Additionally, conducting experimental validation on the role of adaptively evolved genes in secondary metabolites biosynthesis will provide further support to the results of this analysis and will help in developing better approaches to benefit from the medicinal properties of this plant (Nath et al. 2023).

22.8 Conclusion

Giloy is a climber plant that is native to the Indian subcontinent and parts of Southeast Asia. It has been used extensively in Ayurvedic and Unani systems of traditional medicine. The plant is considered to have immunomodulatory, anti-inflammatory, antioxidant, antidiabetic, and antimicrobial properties due to the presence of various bioactive compounds like alkaloids, steroids, diterpenoids, etc. Different parts of the giloy plant like the stem, roots, leaves, and whole plant extracts are used to prepare various formulations like decoctions, tablets, powders, juices, etc. In Ayurveda, giloy is recommended for treating fevers, jaundice, diabetes, gout, rheumatoid arthritis, viral hepatitis, respiratory illnesses like asthma, and as an immunomodulator. Modern research has explored giloy's potential therapeutic applications for conditions like cancer, HIV/AIDS, allergies, osteoporosis, and neurodegenerative diseases, though more clinical studies are still needed. Giloy is generally considered safe for consumption, but excessive doses or prolonged use may cause side effects. Proper medical guidance is recommended before using it, especially by pregnant women or those with preexisting conditions. So, giloy is a multipurpose medicinal plant with anti-inflammatory, immunomodulatory and other beneficial properties as per traditional medicine systems and emerging research. Giloy has been used for centuries in Ayurvedic and Unani medicine systems, primarily in India and China. It is known for its immune boosting properties and is used to treat a wide range of conditions including fever, diabetes, arthritis, diarrhea, headaches, etc. Giloy is a versatile plant all parts of it can be used to make powders, pastes, juices and other products. It has diverse medicinal properties and antimicrobial properties that are beneficial for overall health and wellbeing. Daily consumption of giloy in some form is recommended in Ayurveda and Unani medicine for its immunity boosting and therapeutic effects. It is easy to grow, making it accessible and cost effective as a medicinal plant. In summary, giloy is an important medicinal plant in traditional Indian and Chinese medicine that has a variety of health benefits and medicinal use especially for boosting immunity. Its easy availability makes it a valuable therapeutic agent for improving overall health and preventing diseases.

References

Aher V, Wahi A (2011) Biotechnical approach to evaluate immunomodulatory activity of ethanolic extract of *Tinospora cordifolia* stem (mango plant climber). Iran J Pharm Res 3:863–872

Astete CE, Sabliov CM (2006) Synthesis and characterization of PLGA nanoparticles. J Biomater Sci Polym Ed 17(3):247–289

Badjatya JK, Chaudhari PV, Bheda A (2020) Prevention, treatment and cure for covid-19. Int J Drug Regul Affairs 8(3):36–55

Bal NC, Maurya SK, Sopariwala DH, Sahoo SK, Gupta SC, Shaikh SA, Pant M, Rowland LA, Bombardier E, Goonasekera SA, Tupling AR (2012) Sarcolipin is a newly identified regulator of muscle-based thermogenesis in mammals. Nat Med 18(10):1575–1579

Blanco J-R, Cobos-Ceballos M-J, Navarro F, Sanjoaquin I, de Las Revillas FA, Bernal E, Buzon-Martin L, Viribay M, Romero L, Espejo-Perez S et al (2021) Pulmonary long-term consequences of covid-19 infections after hospital discharge. Clin Microbiol Infect 27(6):892–896

Chen S, Wu K, Knox R (2000) Structure-function studies of dtdiaphorase (nqo1) and nrh: quinone oxidoreductase (nqo2). Free Radic Biol Med 29(3–4):276–284

Chi S, She G, Han D, Wang W, Liu Z, Liu B, et al (2016) Genus tinospora: ethnopharmacology, phytochemistry, and pharmacology. Evidence-based complementary and alternative medicine

Connor EE, Mwamuka J, Gole A, Murphy CJ, Wyatt MD (2005) Gold nanoparticles are taken up by human cells but do not cause acute cytotoxicity. Small 1(3):325–327

Dizaj SM, Lotfipour F, Barzegar-Jalali M, Zarrintan MH, Adibkia K (2014) Antimicrobial activity of the metals and metal oxide nanoparticles. Mater Sci Eng C 44:278–284

Emori TG, Gaynes RP (1993) An overview of nosocomial infections, including the role of the microbiology laboratory. Clin Microbiol Rev 6(4):428–442

Gupta R, Sharma V (2011) Ameliorative effects of *Tinospora cordifolia* root extract on histopathological and biochemical changes induced by aflatoxin-b1 in mice kidney. Toxicol Int 18(2):94

Jagetia GC, Rao SK (2006) Evaluation of the antineoplastic activity of guduchi (*Tinospora cordifolia*) in ehrlich ascites carcinoma bearing mice. Biol Pharm Bull 29(3):460–466

Javiya VA, Ghatak SB, Patel KR, Patel JA (2008) Antibiotic susceptibility patterns of pseudomonas aeruginosa at a tertiary care hospital in Gujarat, India. Indian J Pharm 40(5):230

Jayswal MG (2021) A comprehensive review on *Tinospora cordifolia* (giloy): the medicinalplant. Int J Multidiscipl Res Dev 8(6):80–85

Kapil A, Sharma S (1997) Immunopotentiating compounds from *Tinospora cordifolia*. J Ethnopharmacol 58(2):89–95

Karpova E, YaV V, Dudukina T, Tsvetkova I (1991) 4-trifluoromethylumbelliferyl glycosides as new substrates for revealing diseases connected with hereditary deficiency of lysosome glycosidases. Biochem Int 24(6):1135–1144

Khan MM, dul Haque MS, Chowdhury MSI (2017) Medicinal use of the unique plant *Tinospora cordifolia*: evidence from the traditional medicine and recent research. Asian J Med Biol Res 2(4):508–512

Kreutmair S, Unger S, Núñez NG, Ingelfinger F, Alberti C, De Feo D, Krishnarajah S, Kauffmann M, Friebel E, Babaei S et al (2021) Distinct immunological signatures discriminate severe covid-19 from non-sars-cov-2-driven critical pneumonia. Immunity 54(7):1578–1593

Ly P, Singh S, Shaw C (2007) Novel environmental toxins: steryl glycosides as a potential etiological factor for age-related neurodegenerative diseases. J Neurosci Res 85(2):231–237

Mahajan S, Chakraborty A, Bisht MS, Sil T, Sharma VK (2024) Genome sequencing and functional analysis of a multipurpose medicinal herb *Tinospora cordifolia* (giloy). Sci Rep 14(1):2799

Mahalingam K (2004) Involution codes: with application to DNA strand design. University of South Florida

Mittal J, Sharma MM, Batra A (2014) *Tinospora cordifolia*: a multipurpose medicinal plant-a. J Med Plants 2(2)

Mohanraj K, Karthikeyan BS, Vivek-Ananth R, Chand RB, Aparna S, Mangalapandi P, Samal A (2018) Imppat: a curated database of i ndian m edicinal p lants, p hytochemistry a nd t herapeutics. Sci Rep 8(1):4329

Mukherjee R, De UK, Ram GC (2010) Evaluation of mammary gland immunity and therapeutic potential of Tinospora cordifolia against bovine subclinical mastitis. Trop Anim Health Prod 42:645–651

More P, Pai K (2012) In vitro nadh-oxidase, nadph-oxidase and myeloperoxidase activity of macrophages after *Tinospora cordifolia* (guduchi) treatment. Immunopharmacol Immunotoxicol 34(3):368–372

Nath H, Khataniar A, Bania KK, Mukerjee N, Al-Hussain SA, Zaki ME, Rajkhowa S (2023) Nano-functionalization and evaluation of antimicrobial activity of *Tinospora cordifolia* against the tolb protein of pseudomonas aeruginosa–an antibacterial andcomputational study. Front Microbiol 14:1138106

Patel MB, Mishra S (2011) Hypoglycemic activity of alkaloidal fraction of *Tinospora cordifolia*. Phytomedicine 18(12):1045–1052

Patel, S. S., Shah, R. S., and Goyal, R. K. (2009). Antihyperglycemic, antihyperlipidemic and antioxidant effects of dihar, a polyherbal ayurvedic formulation in streptozotocin induced diabetic rats

Ramana B, Chaudhury A (2012) Prevalence of uropathogens in diabetic patients and their resistance pattern at a tertiary care Centre in South India. Int J Biol Med Res 3(1):1433–1435

Rana V, Thakur K, Sood R, Sharma V, Sharma T (2012) Genetic diversity analysis of *Tinospora cordifolia* germplasm collected from northwestern himalayan region of India. J Genet 91(1):99

Saha S, Ghosh S (2012) *Tinospora cordifolia*: one plant, many roles. Anc Sci Life 31(4):151

Sharma U, Bala M, Kumar N, Singh B, Munshi RK, Bhalerao S (2012) Immunomodulatory active compounds from *Tinospora cordifolia*. J Ethnopharmacol 141(3):918–926

Sharma P, Dwivedee BP, Bisht D, Dash AK, Kumar D (2019) The chemical constituents and diverse pharmacological importance of *Tinospora cordifolia*. He-liyon 5(9):e02437

Sharma H, Rao PS, Singh AK (2021) Fifty years of research on *Tinospora cordifolia*: from botanical plant to functional ingredient in foods. Trends Food Sci Technol 118:189–206

Singh S, Pandey S, Srivastava S, Gupta V, Patro B, Ghosh A (2003) Chemistry and medicinal properties of *Tinospora cordifolia* (guduchi). Indian J Pharm 35(2):83–91

Sinha S, Pan I, Chanda P, Sen SK (2009) Nanoparticles fabrication using ambient biological resources. J Appl Biosci 19:1113–1130

Sliwoski G, Kothiwale S, Meiler J, Lowe EW (2014) Computational methods in drug discovery. Pharmacol Rev 66(1):334–395

Subbiah R, Veerapandian M, S Yun K (2010) Nanoparticles: functionalization and multifunctional applications in biomedical sciences. Curr Med Chem 17(36):4559–4577

Tiwari P, Nayak P, Prusty SK, Sahu PK (2018) Phytochemistry and pharmacology of *Tinospora cordifolia*: a review. Syst Rev Pharm 9(1):70–78

Verma DK, Kumar P, El-Shazly M et al (2021) Unmasking the many faces of giloy (*Tinospora cordifolia* l.): a fresh look on its phytochemical and medicinal properties. Curr Pharm Des 27(22):2571–2581

Part V

Natural Antimicrobial Solutions: Nature's Weapon

Antimicrobial Efficacy of Rutin: Bridging the Gap Between Traditional Medicine and Modern Science

23

Divya Gupta, Raghvendra Raman Mishra, Pragya Mishra, Shubham Singh, Sadhana Shrivastava, Sangeeta Shukla, and Subodh Kumar

Abstract

Rutin is a flavonoid compound commonly found in various plants and other traditionally used medicines such as propolis. It is commonly used for its purported therapeutic benefits such as hepatoprotective, cardioprotective, nephroprotective, and so on. The antimicrobial efficacy of rutin has gained significant attention in modern scientific research due to its antimicrobial properties and no reported side effects. This exertion aims to bridge the gap between traditional medicinal knowledge and contemporary understanding by concluding the current scientific evidence on rutin antimicrobial activity, through a comprehensive

D. Gupta · S. Shrivastava · S. Shukla
Reproductive Biology and Toxicology Lab, School of Studies in Zoology, Jiwaji University, Gwalior, Madhya Pradesh, India

R. R. Mishra
Medical Laboratory Technology, DDU KAUSHAL KENDRA, RGSC, Banaras Hindu University,
Varanasi, Uttar Pradesh, India

P. Mishra
Food Processing and Management, DDU KAUSHAL KENDRA, RGSC, Banaras Hindu University,
Varanasi, Uttar Pradesh, India

S. Singh
Department of Zoology, Guru Ghasidas Vishwavidyalaya, Bilaspur, Chhattisgarh, India

S. Kumar (✉)
Department of Medical Laboratory Technology, School of Allied Health Sciences and Management, Delhi Pharmaceutical Sciences and Research University, New Delhi, India

Centre for Precision Medicine and Pharmacy, Delhi Pharmaceutical Sciences and Research University, New Delhi, India

M. Gangwar, G. Nath (eds.), *Emerging Paradigms for Antibiotic-Resistant Infections: Beyond the Pill*, https://doi.org/10.1007/978-981-97-5272-0_23

analysis of relevant literature, including in vitro and in vivo studies. This manifestation elucidates rutin's potential mechanisms of action against a spectrum of pathogenic microorganisms and also highlights the challenges and opportunities in translating its antimicrobial properties into clinical applications emphasizing the importance of further research to fully harness its therapeutic potential. Overall, this display underscores rutin as a promising natural compound with significant antimicrobial activity, suggesting avenues for future investigation and development in both traditional medicine and modern healthcare practices.

Keywords

Rutin · Antimicrobial efficacy · Natural compound · Traditional medicine · Healthcare practices

23.1 Introduction

In the realm of healthcare, the convergence of traditional wisdom and modern scientific inquiry often yields remarkable discoveries. One such intriguing intersection lies in the exploration of natural compounds for their antimicrobial properties (Cheng et al. 2023). Rutin, a flavonoid abundantly found in various plants, has garnered significant attention due to its promising therapeutic potential (Valdes et al. 2023). The objective of the present task is to delve into the antimicrobial efficacy of rutin, shedding light on its capacity to combat microbial infections and potentially bridge the longstanding gap between traditional medicine and modern scientific approaches.

Traditional curative systems mainly including Ayurveda, Siddha, and Unani across several cultures have long relied on plants and their derivatives for treating various ailments, including microbial infections (Cock et al. 2023). Rutin, known for its presence in fruits, vegetables, and medicinal herbs, has been historically used in folk medicine for its purported healing properties (Singh et al. 2008). However, with the advent of modern scientific methods, there arises an opportunity to explore and validate the efficacy of rutin in a more systematic and evidence-based manner.

In the current scientific era, advancements in microbiology and biochemistry have enabled researchers to unravel the mechanisms underlying the antimicrobial activity of rutin. Several studies have suggested the potential antimicrobial activity of rutin to inhibit the growth of a wide range of pathogenic microorganisms, including bacteria, fungi, and viruses (Garro et al. 2015; Mokhtar et al. 2017; Alnour et al. 2022; Reddy et al. 2023). Moreover, rutin's multifaceted pharmacological properties, including antioxidant and anti-inflammatory effects, further contribute to its therapeutic appeal in combating microbial infections. Despite the promising findings, there remains a gap between traditional knowledge and contemporary scientific validation regarding the antimicrobial efficacy of rutin. Bridging this gap entails rigorous scientific inquiry, encompassing preclinical and clinical studies to elucidate the molecular mechanisms and assess the therapeutic potential of rutin-based interventions (Farha et al. 2022).

This text document endeavor to consolidate existing evidence, providing a comprehensive overview of the antimicrobial properties of rutin and its potential applications in modern healthcare. By integrating traditional wisdom with scientific rigor, we aspire to not only validate the efficacy of rutin but also foster greater synergy between traditional medicine and modern scientific approaches in combating infectious diseases. Through this interdisciplinary exploration, we contribute to the advancement of therapeutic strategies that harness the inherent healing potential of natural compounds like rutin.

23.2 Occurrence, Structure, and Pharmacology of Rutin

The name rutin comes from the plant *Rutagraveolens*, whose aerial parts contain rutin. Rutin is a flavonoid glycoside, also known by other names like purple quercitrin or vitamin P. It can be found in several plant species and natural products derived from plants such as propolis, oranges, plums, onions, grapes, cherries, apricots, and buckwheat seeds (Nouri et al. 2020). Rutin is also called rutinoside or quercetin-3-rutinoside or sophorin. Chemically, rutin is a glycoside that combines the flavonol quercetin with the disaccharide rutinose (rhamnose and glucose). Although rutin is a common dietary flavonoid, slightly or rutin is very poorly absorbed intact since gut microflora metabolizes into a variety of compounds (Pandey et al. 2021). These compounds may be absorbed by the small intestine wall, such as quercetin and phenol derivatives. Studies have shown 50 mg of either rutin or quercetin were used in the diet of healthy humans, plasma level of quercetin was found to be only 3.5 μmol/L. Moreover, up to 50% of an ingested dose of 75 mg/kg rutin was found as microbial metabolites from urine of humans (Semwal et al. 2021). Rutin metabolites, such as 3,4-dihydroxyphenylacetic acid and 3,4-dihydroxytoluene, which are formed in the gut from dietary rutin used up by a variety of fruits and vegetables consumed, are potent inhibitors of AGEs (Fig. 23.1). Therefore, it is recommended that diets with plenty of fruits and vegetables are potent inhibitors of advanced glycosylation end products (AGEs). The name rutin comes from the plant *Rutagraveolens*, whose aerial parts contain rutin. Its and vegetables contribute to the prevention of processes resulting in aging and diabetes complications, and rutin consumption has been correlated with good health (Riva et al. 2020).

Although rutin is a common dietary flavonoid. Slight or no dietary rutin is absorbed intact since gut microflora metabolizes it into a variety of compounds (Pandey et al. 2021). These compounds may be absorbed by the small intestine wall, such as quercetin and phenol derivatives. Studies have shown that when doses of 50 mg of either rutin or quercetin were used in the diet of healthy humans, plasma levels of quercetin were found to be only 3.5 μmol/L. Moreover, up to 50% of an ingested dose of 75 mg rutin was recovered as microbial metabolites from urine in humans (Semwal et al. 2021). Rutin metabolites, such as 3,4-dihydroxyphenylacetic acid and 3,4-dihydroxytoluene, which are formed in the gut from dietary rutin

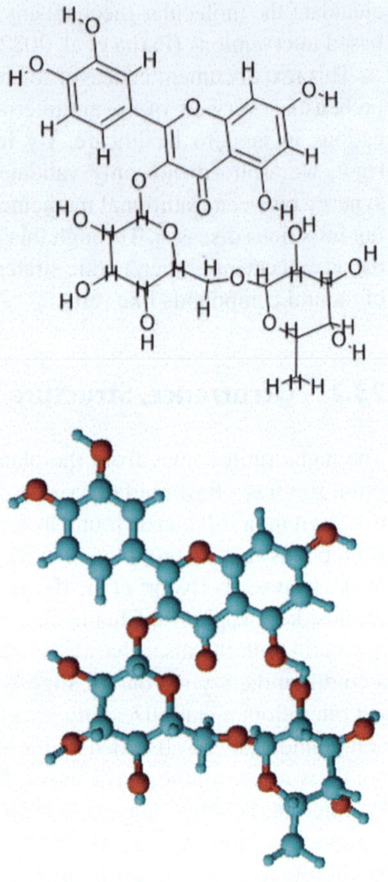

Fig. 23.1 Chemical structure of rutin. (Chemical Drawing Software. ChemSketch. ACD/Labs (acdlabs.com). Advanced Chemistry Development Inc., Canada, UK)

consumed in the form of a variety of fruits and vegetables, are potent inhibitors of advanced glycation end products (AGEs) (Table 23.1). Thus, it has been suggested that diets rich in fruits and vegetables contribute to the prevention of processes resulting in aging and diabetes complications, and rutin consumption has been correlated with good health (Riva et al. 2020).

23.3 Extraction of Rutin

The extraction step plays a vital role in regaining bioactive compounds from natural sources. There are certain requirements that the extraction step must meet, such as adaptability, ease of use, efficacy, cost-effectiveness, and the ability to extract and preserve the major fraction of the natural bioactive substances present in the plant resources. Various conventional extraction techniques exist, such as Soxhlet extraction, which has been extensively used for the extraction of bioactive compounds. However, this method has significant drawbacks, including thermal degradation of

Table 23.1 Physiochemical properties of rutin (https://pubchem.ncbi.nlm.nih.gov/compound/Rutin)

S. No.	Parameter	Depiction
1.	Molecular formula	$C_{27}H_{30}O_{16}$
2.	Molecular weight	610.52 g/mol
3.	Form and color	Powder yellow to green
4.	Solubility	125 mg/L
5.	Melting point	125 °C
6.	Hydrogen bond Donar count	10
7.	Hydrogen bond acceptor count	16
8.	Site of absorption	Small intestine
9.	Excretion	Urine (about 10%), rest unchanged in feces

target compounds and potential harm to human well-being and the environment associated with the use of harmful organic solvents (Chua 2013; Singh et al. 2015). Various modern techniques have been developed in recent years for extracting high-value compounds like rutin from plants. These methods include ultrasonics, microwave, infrared irradiation, pressurized liquid, enzymes, mechanochemical methods, and supercritical fluids, which are considered green extraction techniques. To efficiently extract rutin from plant matrices, solvents such as ethanol, methanol, acetone, diethyl ether, isopropanol, ethyl acetate, and their mixtures with water are commonly used. The use of aqueous solutions containing organic solvents has been found to improve the efficiency of extraction of bioactive compounds (Gullon et al. 2017).

Ultrasound Assisted Extraction (UAE) is a technique that has recently gained popularity for extracting bioactive compounds. It offers several benefits over conventional extraction methods, such as simplicity, quickness, low solvent volume requirements, low cost, and eco-friendliness. The mechanism of UAE is attributed to acoustic cavitation, which causes the sample matrix to break down, facilitating solvent diffusion into the cellular material, and thus enhancing extraction efficiency (Ameer et al. 2017).

Another useful method is Microwave-Assisted Extraction (MAE), which uses microwave energy to trigger molecular motion by ionic transmission and dipole rotation. This method increases temperature and pressure, changing the cell structure and refining solvent penetration across the sample matrix. The main advantage of this technique is the reduced extraction time and solvent volume compared to traditional methods such as Soxhlet extraction and maceration (Angiolillo et al. 2015).

Infrared-Assisted Extraction (IRAE) is considered an alternative to conventional techniques for extracting natural flavonoids from various bioresources. This technique uses infrared energy to efficiently heat the solvent and partition analytes from the cell wall into the surrounding medium (Huang et al. 2014).

Pressurized Liquid Extraction (PLE), also known as accelerated solvent extraction, is another widely used method for extracting natural products. PLE uses solvents at high pressure and temperature, increasing the solubility of the target analytes. Ionic Liquid-based Pressurized Liquid Extraction (IL-PLE) has been reported to be effective in the extraction of rutin and quercetin from Chinese medicinal plants (Carro et al. 2013).

Enzyme-Assisted Extraction (EAE) is another promising and eco-friendly technique for extracting bioactive compounds from plant sources. This method involves the use of enzymes to catalyze the deprivation and rupturing of cell walls, thereby improving the production of target compounds. Enzymes such as pectinases, cellulases, and hemicellulases have been effectively used to extract flavonoids (Xu et al. 2013).

23.4 Biological Activity of Rutin

Rutin is a flavonoid that is commonly found in our diet. It is a naturally derived compound that is usually non-toxic. Rutin has many convenient activities like anticancer, antioxidant, anti-diabetic, anti-inflammatory, antiseptic, fungicidal, hepatoprotective, renoprotective, anti-arthritis, and parasiticidal. (Ma et al. 2023; Ganeshpurkar and Saluja 2017). In the following section, we will briefly describe the main biological properties of rutin.

23.4.1 Anticancer Property

Cancer is caused by due to various types of carcinogens including bacteria, viruses, and many types of mutations/genetic aberrations (Maurya et al. 2012; Mishra et al. 2011; Gupta et al. 2023a, b; Tewari et al. 2013). Rutin, a flavonoid, has been studied for its potential to prevent cancer, sensitize chemotherapy, and act as an anticancer agent for different types of cancer (Peuget et al. 2024). Researchers found that injecting mice with rutin at 120 mg/kg every 4 days significantly reduced development of human leukaemia tumors (Lin et al. 2012a, b). Alonso-Castro et al. (2013) demonstrated that rutin was cytotoxic against human colon adenocarcinoma cells in vitro and had anti-tumor and anti-angiogenic properties in vivo without any toxic effects on nude mice with tumors. Combining rutin with cisplatin increased cytotoxicity and apoptosis in malignant murine ascites Dalton's lymphoma (DL) cells, leading to higher host survivability and decreased glutathione in cancerous cell lines (Prasad and Prasad 2021). Rutin attained after Schinusmolle also demonstrated antitumor movement against the human leukemic monocyte's lymphoma (U-937) in a study conducted by Calzada et al. (2018). Rutin extracted from the methanolic extract of straw had chemopreventive and antitumor properties against DMBA and croton oil-induced membrane carcinogenesis in mice, as revealed by Dixit (2014). Rutin has demonstrated its potential as a chemosensitizer, amplifying the anticancer efficacy of various drugs on chemosensitized cells. In a study by Iriti et al. (2017), rutin sensitized human breast cancer cells to cyclophosphamide and effectively halted cell cycle progression. Furthermore, numerous animal models have revealed rutin'schemopreventive properties, such as its ability to inhibit azoxymethane-induced colonic tumorigenesis in mice and rats (Mahmoud 2012).

23.4.2 Antioxidant Property

Antioxidants are compounds that can stop or slow down cell injury triggered by free radicals. Free radicals are unbalanced particles produced by the body during several metabolic processes. Antioxidants work by inhibiting other molecules' oxidation and balancing the cells' oxidative state (Gupte et al. 2022). They are also known as "free-radical scavengers." Shrivastava et al. 2023 demonstrate that flavonoids are excellent free radical scavengers. Rutin is an antioxidant that has a wide variety of pharmacologic activities and is applied as a protecting mediator in various ailments and toxic manifestations. Several reports have shown that administration of rutin can improve the antioxidant security arrangement against hepatic oxidative stress induced by iron overload. This protection results in the liver of iron because of together antioxidant and its role in metal chelation. Rutin has also been found to exhibit a protective effect on the liver, kidney, and brain against acrylamide-induced oxidative challenge (Shrivastava et al. 2023). Additionally, rutin has been reported to decrease reactive oxygen species (ROS) formation by enhancing the movement of numerous antioxidants within the cells. According to a study by Patil et al. (2017), rutin and quercetin have a protective effect in contrast to radiation resulted DNA injury in humanoid lymphocytes. This effect may be credited to the antioxidant activity of rutin, which involves the hunting of radiation-mediated free radical formation and reserve of radiation resulted in oxidative injury.

23.4.3 Anti-inflammatory Property

Inflammation is not synonymous with infection. Inflammation is a protective reaction involving immune cells, blood vessels, and molecular mediators that occurs when bodily tissues are exposed to damaging stimuli, such as pathogens, damaged cells, or irritants. Rutin has been shown to improve ulcerative colitis healing through radical scavenging and anti-inflammatory properties (Hussein et al. 2014). Rutin has been shown in vitro and in vivo to be a potential therapeutic agent for the treatment of various severe vascular inflammatory diseases by inhibiting the high mobility group box 1 (HMGB1) protein signaling pathway, which acts as a late mediator of severe vascular inflammatory conditions (Yoo et al. 2014). The study of the anti-inflammatory effects of flavonoid-rich common buckwheat sprout (CBS) and Tartary buckwheat sprout (TBS) revealed that TBS extracts contained roughly 3.6 times more rutin than CBS extracts. Encapsulated rutin's anti-inflammatory effect has been shown to be enhanced by inhibiting IL-6 and NF-kB (Jantrawut et al. 2017).

23.4.4 Antimicrobial Activity

Pathogenic microorganisms posed a threat to humans since the beginning of time, causing a significant amount of human illness and death. Prior to the discovery of antibiotics like penicillin in 1928 and sulfone drugs in the 1930s, the only options for fighting infectious diseases were plant extracts and toxic arsenic (Gupta et al. 2023a, b). Antibiotics have played a vital role in treating bacterial and fungal infections over the past 60 years, but the emergence of dangerous and antibiotic-resistant bacteria has become increasingly common in recent decades (Kundapur et al. 2023). Drug resistance can occur through various mechanisms, making it a challenging problem to overcome (Kumar et al. 2023). Flavonoids are known for their various beneficial properties, including antioxidant, anti-inflammatory, anti-allergic, anticancer, antiviral, and antifungal properties. As plants produce flavonoids in response to microbial infections, there is a growing interest in their antibacterial properties and their potential application in treating human diseases (Table 23.2).

23.4.5 Antifungal Effect

Flavonoids have been suggested as a possible treatment for fungal pathogens in humans, as they can prevent spore germination in plant pathogens (Harborne and Williams 2000). There are various fungal infections such as candidiasis, aspergillosis, ringworm, blastomycosis, etc., which can be treated with antifungal agents. However, the development of antifungal agents has been slower compared to antibacterial agents. A study has investigated the antifungal agent of combining amphotericin B, a antifungal material, with rutin or quercetin, and the results have shown that this mixture improves the fungicidal action of amphotericin B in contrast to *Cryptococcus neoformans*. Rutin and auraptene, which are obtained from *Polygala paniculata* commonly found in Brazil, have also demonstrated fungicidal action in contrast to *Cryptococcus gattii* (Oliveira et al. 2016). Rutin has been found to have anti-arthritic and antifungal effects on septic arthritis caused by *Candida albicans* (Han 2009). In one more experiment, rutin isolated from tobacco leaves has been shown to be an effective antifungal and antibacterial agent against *Candida albicans*, *Staphylococcus aureus*, *Bacillus subtilis*, *Escherichia coli*, *Klebsiella*, and *oxytoca* (Araruna et al. 2012). The examination of the whole leaf aqueous extract has shown high fungicidal and antibacterial activity against *Candida albicans*, *Cryptococcus neoformans*, *B. cereus*, *B. subtilis*, *S. aureus*, *E. coli*, *P. aeruginosa*, *K. pneumoniae* bacteria (Oliveira et al. 2014). The extract contains seven phenolic compounds, including rutin. The use of olive leaves as nutraceuticals has been proposed to reduce the risk of microbial infections.

Table 23.2 Antimicrobial effect of rutin on various microbes

S. No.	Microorganism	References
1	*Staphylococcus aureus*, *Staphylococcus glurance* and *Eschericia coli*	Soni et al. (2013)
2	*Pseudomonas aeruginosa* and *Klebsiella pneumoniae*	Singh et al. (2008)
3	*Salmonella enteritidis*	Arima et al. (2002)
4	*Herpes simplex virus (HSV)*, *Respiratory syncytial virus*, *Poliovirus*, and *Sindbis virus*	Middleton and Kandaswami (1994)
5	*Staphylococcus aureus*	Stojković et al. (2013)
6	*E. coli*, *P. auruginosa*, *S. aureus*, *K. oxytoca*, *B. subtilis*, *C. albicans*; *A. galli*, *P. phostuma* and *S. aegypti*	Dubey et al. (2013)
7	*Pseudomonas aeruginosa*, *Acinetobacterbaumannii*, *Staphylococcus aureus*, and *Candida krusei*	Orhan et al. (2010)
8.	*Cryptococcus gattii*, *Candida tropicalis*, *Candida parapsilosis*	Arima et al. (2002)
9.	Escherichia coli, Klebsiella pneumoniae, Staphylococcus aureus, and *methicillin-resistant* S. aureus	Saleemi et al. (2022)
10.	Lactobacillus casei, Enterococcus faecalis, Escherichia coli, Streptococcus sang, Streptococcus oralis, Agreggatibacteractinomyc etemomitans,Staphylococcusaures	Gutiérrez-Venegas et al. (2019)
11	*Staphylococcus aureus*, *Sarcinalutea*, *Bacillus cereus*, *Bacillus subtilis*, *Escherichia coli*, *Candida albicans*	Lupaşcus et al. (2010)
12	*Methicillin-resistant*, *Staphylococcus aureus*	Amin et al. (2015)
13	*Helicobacter* pylori	Martini et al. (2009)
14	*Escherichia coli*, *Staphylococcus aureus*, *Enterococcus faecium*, *Proteus vulgaris*, *Pseudomonas aeruginosa*, *Klebsiella pneumonia*, *Candida albicans*, and *Saccharomyces cerevisiae*	Samsonowicz et al. (2015)
15	*Staphylococcus aureus* and *Escherichia coli*; *Candida albicans*, *C. krusei* and *C. tropicalis*	Araruna et al. (2012)
16	*Epidermophytonfloccosum (Hartz) Langeron e Milochevitch*, *Trichophytonrubrum (Castellani) Saboraud*, *TrichophytontonsuransMalmsten*, *Trichophytonviolaceum*, *Trichophytonmentagrophytes (Robin) Blanchard*, *Microsporumgypseum (Bodin) Guiart e Grigorakis*, *NannizziacajetaniAjello*, *MicrosporumcanisBodin*, *Nannizziagypsea*	Baldisserotto et al. (2015)
17	*Escherichia coli*, *Proteus mirabilis*, *Klebsiella pneumoniae*	Alnour et al. (2022)
18	*Streptococcus mutans*, *Staphylococcus aureus*, *Escherichia coli*, *Candida albicans*	Reddy et al. (2023)
20	*P. aeruginosa*	Mokhtar et al. (2017)
21	*Helicobacter pylori*	Garro et al. (2015)

23.4.6 Anti-bacterial Property

Bacterial infections have caused a wide range of illnesses and deaths in humans, including pneumonia, tetanus, cholera, TB, bloodstream infections (sepsis), and sexually transmitted infections such as gonorrhoea. Antibiotics are commonly referred to as antibacterial agents since they are used to kill or inhibit the growth of germs. Singh et al. (2008) investigated the antibacterial efficacy of methanolic extracts of *Pterisvittata*, also known as "Brake fern," against eight intestinal pathogens. They discovered that a 70% aqueous methanolic extract exhibited high action against *Pseudomonas aeruginosa* and *Klebsiella pneumonia*, and the extract contained the flavonoid rutin. Another study (Basile et al. 2000) discovered that *Castanea sativa* extract, when combined with an aqueous solution of sulfuric acid, had substantial antibacterial activity against a variety of Gram-positive and Gram-negative bacteria. It has been observed that mixing several flavonoid molecules can increase antibacterial activity. In a study on the effect of rutin in combination with other flavonoids and antibiotics in methicillin-resistant *Staphylococcus aureus* (MRSA), it was found that quercetin with morin and rutin with amoxicillin, ampicillin, cephradine, ceftriaxone, imipenem, and methicillin showed synergism, while an additive relationship was indicated between morin and rutin and amoxicillin, cephradine, ceftriaxone, imipenem, and methicillin. *Streptococcus suis* develops biofilms and causes serious illnesses in pigs and people such as meningitis and streptococcal toxic shock syndrome. Rutin was discovered to inhibit CPS synthesis and reduce *S. suis* biofilm formation without compromising growth. As a result, in vitro rutin could be employed as a novel natural biofilm inhibitor and to prevent biofilm-related illnesses (Turgeon et al. 1994; Wang et al. 2017). Rutin exhibits strong antibacterial properties against *Streptococcus pyogenes*, *Enterococcus faecalis*, *Bacillus cereus*, *Pseudomonas aeruginosa*, *Klebsiella pneumoniae*, and *Escherichia coli* (Danciu et al. 2023). The honey produced by the stingless bee, Meliponacompressipesmanaosensis, had antibacterial action against Gram-positive and Gram-negative bacteria, and rutin was discovered in these honey samples (Table 23.2) (Pimentel et al. 2013).

23.4.7 Antiviral Effect

Flavonoids such as rutin and quercetin have been found to exhibit antiviral effects in addition to their antibacterial and antifungal effects. Middleton and Kandaswami (1994) and Selway (1986) have reported the antiviral effect of rutin against various virus strains, such as respiratory syncytial virus, herpes simplex virus, Inbis Virus, and polio virus. Rutin's antiviral effect may be due to its inhibition of viral polymerase and binding of viral capsid protein and viral nucleic acid (Selway 1986). Rutin's activity was evaluated against NS3 serine protease in vitro (Zuo et al. 2005), and it was found to have a strong inhibitory effect on hepatitis C virus (HCV) by inhibiting the early entry stage of the HCV life cycle when isolated from plum (Bose et al. 2017). It shows an IC50 of 110 μM against Enterovirus-A71 strain

CMUH01 (B5) and inhibits its replication stage (Lin et al. 2012a, b). Rutin was also found to reduce the infectivity of the C4 subgenotype EV-A71 with an IC50 of 200 μM (Wang et al. 2015). Sulphatedrutin has a potent antiviral activity against HIV-1x 4 virus IIIB and HIV-1 R5 isolated Ada-M and Ba-L strains, possibly by blocking viral entry and viral cell fusion through interacting with HIV-1 envelope glycoprotein (Tao et al. 2007). Rutin has also been tested against Dengue virus type 2 and influenza virus. Antiviral testing for selected compounds against murine norovirus infection revealed that rutin has anti-murine norovirus activity in cell-based assays (Zandi et al. 2011). Rutin has also shown effects on hepatic monooxygenase activities in experimental influenza virus infection (EIVI) (Savov et al. 2006). Molecular docking studies have suggested that rutin can act as a novel hepatitis B virus inhibitor (Parvez et al. 2019). Another study conducted using rutin and other flavonoids present in the extract of ficusbenjamina found that rutin has antiviral activity against HSV-1 and HSV2 (Yarmolinsky et al. 2012). The mode of action of rutin is by inhibiting HIV glycoprotein-mediated cell–cell fusion.

23.5 Antimicrobial Techniques

23.5.1 Agar Disk Diffusion Technique

This method was developed in 1940 (Heatley 1944) and is regularly used in clinical microbiology laboratories to test antimicrobial susceptibility. Not all types of bacteria can be tested using this method, but certain pathogenic bacteria such as streptococci, Haemophilus influenza, Haemophilusparainfluenzae, Neisseria gonorrhoeae, and Neisseria meningitidis can be tested using specific culture media, various incubation conditions and interpretive criteria for inhibition. To perform this well-known procedure, agar plates are inoculated with a standardized inoculum of the test microorganism. Filter paper discs containing the test compound at a desired concentration, which are approximately 6 mm in diameter, are then placed on the agar surface. The Petri dishes are then incubated under suitable conditions. The antimicrobial agent generally diffuses into the agar, inhibiting germination and growth of the test microorganism. The diameters of the inhibition growth zones are then measured. Minimum Inhibitory concentration cannot be calculated by this method (Nijs et al. 2003).

23.5.2 Antimicrobial Gradient Method

The antimicrobial gradient method is a combination of dilution and diffusion techniques to determine the MIC value. This method works by creating a concentration gradient of the antimicrobial agent in the agar medium. The commercial version of this technique is known as Etests by BioMérieux. To perform this method, we place a strip with an increasing concentration gradient of the antimicrobial agent from one end to the other on the agar surface that has been previously inoculated with the microorganism being tested. The antimicrobial gradient method is commonly used for determining the MIC value of antibiotics, antifungals, and antimycobacterial (Jorgensen et al. 2010). The MIC value is determined at the intersection of the strip and the growth inhibition ellipse. This technique is simple to implement and is frequently used to meet clinicians' demands. However, it can be expensive when testing more drugs, as each strip costs around 2–4 dollars (Hausdorfer et al. 1998).

23.5.3 Agar Well Diffusion Method

This method, which is similar to the disk-diffusion method, is frequently employed to evaluate how effectively antimicrobial drugs work against microbes. This technique involves covering the surface of an agar plate with a microbial inoculum. Next, a sterile cork borer or point is used to make a 6–8 mm diameter hole in the agar. The antimicrobial agent or extract solution is then added to the well at a volume of 20–100 mL. After that, the agar plates are incubated in an environment that is appropriate for the test microorganism. The microbial strain under investigation is prevented from growing as the antimicrobial drug diffuses throughout the agar media (Balouiri et al. 2016).

23.5.4 Agar Plug Diffusion Method

The agar plug diffusion method is frequently used to illustrate how microorganisms are antagonistic to one another (Elleuch et al. 2010). This procedure is similar to the disk-diffusion approach. Initially, the microorganism of interest is grown on an agar medium by creating narrow streaks on the plate's surface. Molecules secreted by the growing microbial cells diffuse across the agar medium. Following incubation, an agar plot or cylinder is aseptically sliced with a sterile cork borer and placed on the agar surface of a different plate that has already been infected with the test microorganism. From the plug, the molecules spread out across the agar media. Lastly, the development of the inhibition zone surrounding the agar plug indicates the antibacterial activity of the chemicals released by the microorganism.

23.5.5 Poisoned Food Method

The poisoned food method is a widely used technique to evaluate how well antifungal medications work against Mold (Kumar et al. 2014). To obtain the appropriate concentration, the antifungal drug or extract is first added to melted agar and thoroughly mixed. After that, the mixture is put into Petri dishes. A mycelia disc, measuring between 2 and 5 mm, is positioned in the middle of the plate following an overnight pre-incubation. After that, the plates are further incubated in settings appropriate for the fungal strain under investigation. Both the control and sample plates' fungal growth diameters are measured, and the antifungal impact is calculated using the following formula: ((Dc − Ds)/Dc) × 100 = antifungal activity%, where Dc is the diameter of growth in the control plate, and Ds is the diameter of growth in the plate containing the tested antifungal agent. Sporulation can also be compared to the control. If standardization of the method used fails, the researcher must include a positive control with a known antimicrobial molecule to compare the results and ensure the experimental approach is correct.

23.6 Mechanism of Antimicrobial Action of Rutin: A Bioflavonoid

The antimicrobial mechanism of action of rutin involves several pathways, which collectively contribute to its efficacy against various microorganisms.

Rutin can achieve this by causing membrane disruption, inhibiting nucleic acid synthesis (including Dihydrofolate reductase (DHFR), Helicase, and Gyrase), inhibiting bacterial virulence (toxin production), inhibiting bacterial communication (Quorum sensing), inhibiting cell envelope synthesis (Fatty Acid Synthase (FAS)), inhibiting peptidoglycan synthesis, inhibiting efflux pumps (which can reverse antimicrobial resistance), inhibiting respiratory chain complex, and inhibiting ATP synthesis (Fig. 23.2).

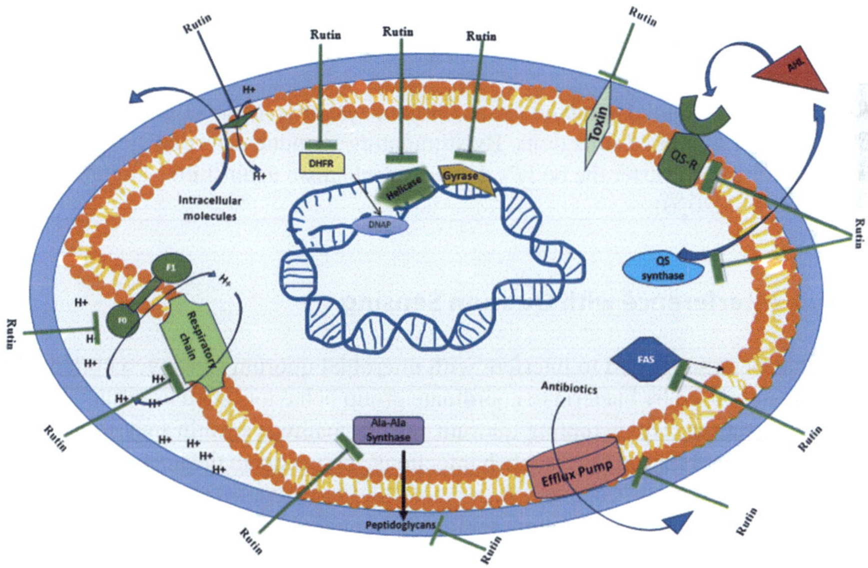

Fig. 23.2 A diagrammatic representation of how rutin works to inhibit bacterial cells

23.6.1 Disruption of Cell Membrane Integrity

Rutin has been shown to disrupt the integrity of microbial cell membranes. This disruption may lead to leakage of essential cellular components, resulting in cell lysis and eventual death of the microorganisms (Gullon et al. 2017).

23.6.2 Inhibition of Enzymatic Activity

Rutin can inhibit the activity of microbial enzymes crucial for their survival and proliferation. By targeting enzymes involved in processes such as cell wall synthesis, nucleic acid biosynthesis, and replication, energy metabolism, rutin interferes with microbial growth and viability (Farha et al. 2022).

23.6.3 Generation of Reactive Oxygen Species

Rutin possesses excellent antioxidant properties and may generate reactive oxygen species within microbial cells. The accumulation of reactive oxygen species may cause oxidative stress, leading to damage to cellular components such as proteins, lipids, nucleic acids, ultimately impairing microbial survival (Sharma et al. 2013).

23.6.4 Modulation of Immune Response

Rutin has been reported to modulate the immune response, enhancing the hosts ability to combat microbial infections. By stimulating immune cells and cytokine production, rutin can bolster the body's defence mechanism against invading microbes (da Silva et al. 2020).

23.6.5 Interference with Quorum Sensing

Rutin has been suggested to interfere with microbial quorum sensing, a communication system used by bacteria to coordinate group behaviors such as virulence and biofilm formation. By disrupting quorum sensing pathways, rutin may inhibit the ability of microorganisms to establish infections and persist in the host environment (Ivanov et al. 2022).

Overall, the multifaceted antimicrobial mechanisms of rutin make it a promising candidate for the development of novel antimicrobial agents and highlight its potential in combating microbial infections. Further research is warranted to elucidate the precise molecular mechanisms underlying rutin's antimicrobial activity and to explore its therapeutic applications in clinical settings.

23.7 Conclusion

In conclusion, the exploration of rutin antimicrobial activity serves as a compelling bridge between the realms of traditional medicine and modern scientific understandings. Through a thorough investigation of rutin properties, mechanism of action, and empirical evidence, this chapter highlights the significant potential of this natural compound in combating microbial infections. By integrating traditional knowledge with rigorous scientific inquiry, we gain valuable insights that can inform the development of novel antimicrobial agents without any side effects. Rutin exemplifies the synergy between ancient wisdom and contemporary research, offering promising avenues for addressing the pressing challenges of antibiotic resistance and microbial infectious diseases. As we continue to unravel the mysteries of rutin antimicrobial properties, we pave the way for innovative therapeutic strategies that harmonize the wisdom of the past with the advancements of the present.

References

Alnour TM, Ahmed-Abakur EH, Elssaig EH, Abuduhier FM, Ullah MF (2022) Antimicrobial synergistic effects of dietary flavonoids rutin and quercetin in combination with antibiotics gentamicin and ceftriaxone against E. coli (MDR) and P. mirabilis (XDR) strains isolated from human infections: implications for food–medicine interactions. Italian J Food Sci 34(2):34–42

Alonso-Castro AJ, Domínguez F, García-Carrancá A (2013) Rutin exerts antitumor effects on nude mice bearing SW480 tumor. Arch Med Res 44(5):346–351

Ameer K, Shahbaz HM, Kwon JH (2017) Green extraction methods for polyphenols from plant matrices and their byproducts: a review. Compr Rev Food Sci Food Saf 16(2):295–315

Amin MU, Khurram M, Khattak B, Khan J (2015) Antibiotic additive and synergistic action of rutin, morin and quercetin against methicillin resistant Staphylococcus aureus. BMC Complement Altern Med 15:1–12

Angiolillo L, Del Nobile MA, Conte A (2015) The extraction of bioactive compounds from food residues using microwaves. Curr Opin Food Sci 5:93–98

Araruna MK, Brito SA, Morais-Braga MF, Santos KK, Souza TM, Leite TR, Costa JG, Coutinho HD (2012) Evaluation of antibiotic & antibiotic modifying activity of pilocarpine & rutin. Indian J Med Res 135(2):252

Arima H, Ashida H, Danno GI (2002) Rutin-enhanced antibacterial activities of flavonoids against Bacillus cereus and Salmonella enteritidis. Biosci Biotechnol Biochem 66:1009–1014

Baldisserotto A, Vertuani S, Bino A, De Lucia D, Lampronti I, Milani R, Gambari R, Manfredini S (2015) Design, synthesis and biological activity of a novel Rutin analogue with improved lipid soluble properties. Bioorg Med Chem 23(1):264–271

Balouiri M, Sadiki M, Ibnsouda SK (2016) Methods for in vitro evaluating antimicrobial activity: A review. Journal of pharmaceutical analysis 6(2):71–79

Basile A, Sorbo S, Giordano S, Ricciardi L, Ferrara S, Montesano D, Cobianchi RC, Vuotto ML, Ferrara L (2000) Antibacterial and allelopathic activity of extract from Castanea sativa leaves. Fitoterapia 71:S110–S116

Bose M, Kamra M, Mullick R, Bhattacharya S, Das S, Karande AA (2017) Identification of a flavonoid isolated from plum (Prunusdomestica) as a potent inhibitor of hepatitis C virus entry. Sci Rep 7(1):3965

Calzada F, Solares-Pascasio JI, Valdes M, Garcia-Hernandez N, Velázquez C, Ordoñez-Razo RM, Barbosa E (2018) Antilymphoma potential of the ethanol extract and rutin obtained of the leaves from schinusmollelinn. Pharm Res 10(2):119

Carro AM, González P, Lorenzo RA (2013) Applications of derivatization reactions to trace organic compounds during sample preparation based on pressurized liquid extraction. J Chromatogr A 1296:214–225

Cheng Y, Wan S, Yao L, Lin D, Wu T, Chen Y, Zhang A, Lu C (2023) Bamboo leaf: a review of traditional medicinal property, phytochemistry, pharmacology, and purification technology. J Ethnopharmacol 306:116166

Chua LS (2013) A review on plant-based rutin extraction methods and its pharmacological activities. J Ethnopharmacol 150(3):805–817

Cock IE, Ndlovu NA, Van Vuuren SF (2023) The traditional use of southern African medicinal plants to treat oral pathogens, and studies into their relevant antimicrobial properties. S Afr J Bot 153:258–279

da Silva AB, Coelho PLC, das Neves Oliveira M, Oliveira JL, Amparo JAO, da Silva KC, Soares JRP, Pitanga BPS, dos Santos Souza C, de Faria Lopes GP, da Silva VDA (2020) The flavonoid rutin and its aglycone quercetin modulate the microglia inflammatory profile improving antiglioma activity. Brain Behav Immun 85:170–185

Danciu BM, Oprescu D, Popa CI, Balanescu L, Stanciu GD, Simionescu AA (2023) Can obstetricians prevent the occurrence of necrotizing enterocolitis in newborns? In: Proceedings of national congress of fetal and neonatal medicine, p 18

Dixit S (2014) Anticancer effect of rutin isolated from the methanolic extract of Triticumaestivum straw in mice. Med Sci 2(4):153–160

Dubey S, Ganeshpurkar A, Bansal D, Dubey N (2013) Experimental studies on bioactive potential of rutin. Chron Young Sci 4:153

Elleuch L, Shaaban M, Smaoui S, Mellouli L, Karray-Rebai I, Fourati-Ben Fguira L, Shaaban KA, Laatsch H (2010) Bioactive secondary metabolites from a new terrestrial Streptomyces sp. TN262. Appl Biochem Biotechnol 162:579–593

Farha AK, Gan RY, Li HB, Wu DT, Atanasov AG, Gul K, Zhang JR, Yang QQ, Corke H (2022) The anticancer potential of the dietary polyphenol rutin: current status, challenges, and perspectives. Crit Rev Food Sci Nutr 62(3):832–859

Ganeshpurkar A, Saluja AK (2017) The pharmacological potential of rutin. Saudi Pharm J 25(2):149–164

Garro MF, Ibáñez AGS, Vega AE, Sosa ACA, Pelzer L, Saad JR, Maria AO (2015) Gastroprotective effects and antimicrobial activity of Lithraeamolleoides and isolated compounds against Helicobacter pylori. J Ethnopharmacol 176:469–474

Gullon B, Lú-Chau TA, Moreira MT, Lema JM, Eibes G (2017) Rutin: a review on extraction, identification and purification methods, biological activities and approaches to enhance its bioavailability. Trends Food Sci Technol 67:220–235

Gupta D, Singh S, Shukla S, Shrivastava S (2023a) Oncofertility: treatment options from bench to bedside. Cancer Pathog Therapy 1:E021–E021

Gupta D, Singh S, Soni R, Gupte SS, Rathour A, Shrivastava S, Shukla S (2023b) Possible metabolic effect of acrylamide on biological system. Food Safety Health 1(2):126–138

Gupte SS, Rathour A, Gupta D, Soni R, Shrivastava S, Bhaduria M, Nirala SK, Singh S, Sharma A, Yadav D, Rakshit S (2022) Tephrosiapurpurea ameliorates oxidative and histological alterations induced by aflatoxin B1 in rats. Int J Pharm Investig 12(3):340

Gutiérrez-Venegas G, Gómez-Mora JA, Meraz-Rodríguez MA, Flores-Sánchez MA, Ortiz-Miranda LF (2019) Effect of flavonoids on antimicrobial activity of microorganisms present in dental plaque. Heliyon 5(12):e03013

Han Y (2009) Rutin has therapeutic effect on septic arthritis caused by Candida albicans. Int Immunopharmacol 9(2):207–211

Harborne JB, Williams CA (2000) Advances in flavonoid research since 1992. Phytochemistry 55(6):481–504

Hausdorfer J, Sompek E, Allerberger F, Dierich MP, Rüsch-Gerdes S (1998) E-test for susceptibility testing of mycobacterium tuberculosis. Int J Tuberc Lung Dis 2(9):751–755

Heatley NG (1944) A method for the assay of penicillin. Biochem J 38(1):61

Huang T, Chen N, Wang D, Lai Y (2014) Infrared-assisted extraction coupled with high performance liquid chromatography (HPLC) for determination of liquiritin and glycyrrhizic acid in licorice root. Anal Methods 6(15):5986–5991

Hussein SA, Abou Zaid OA, Abdel-Maksoud HA, Khadija AA (2014) Anti-inflammatory and antioxidant effect of rutin on 2,4,6-trinitrobenzene sulfonic acid induced ulcerative colitis in rats. BVM J 27:208–220

Iriti M, Kubina R, Cochis A, Sorrentino R, Varoni EM, Kabała-Dzik A, Azzimonti B, Dziedzic A, Rimondini L, Wojtyczka RD (2017) Rutin, a quercetin glycoside, restores chemosensitivity in human breast cancer cells. Phytother Res 31(10):1529–1538

Ivanov M, Novović K, Malešević M, Dinić M, Stojković D, Jovčić B, Soković M (2022) Polyphenols as inhibitors of antibiotic resistant bacteria—mechanisms underlying rutin interference with bacterial virulence. Pharmaceuticals 15(3):385

Jantrawut P, Phongpradist R, Muller M, Viernstein H (2017) Enhancement of anti-inflammatory activity of polyphenolic flavonoid rutin by encapsulation. Pak J Pharm Sci 30(5):1521

Jorgensen JH, McElmeel ML, Fulcher LC, Zimmer BL (2010) Detection of CTX-M-type extendedspectrum beta-lactamase (ESBLs) by testing with MicroScan overnight and ESBL confirmation panels. Journal of clinical microbiology 48(1):120–123

Kumar SN, Nambisan B, Sundaresan A, Mohandas C, Anto RJ (2014) Isolation and identification of antimicrobial secondary metabolites from Bacillus cereus associated with a rhabditid entomopathogenic nematode. Ann Microbiol 64(1):209–218

Kumar D, Singh S, Kundapur RR, Gupta D, Shukla S (2023) Introduction and history of insect biotechnology. In: Introduction to insect biotechnology. Springer International Publishing, Cham, pp 1–15

Kundapur R, Kumar D, Singh S, Shukla S, Gong C (2023) Insect metagenomics. In: Introduction to insect biotechnology. Springer International Publishing, Cham, pp 81–91

Lin YJ, Chang YC, Hsiao NW, Hsieh JL, Wang CY, Kung SH, Tsai FJ, Lan YC, Lin CW (2012a) Fisetin and rutin as 3C protease inhibitors of enterovirus A71. J Virol Methods 182(1–2):93–98

Lin JP, Yang JS, Lin JJ, Lai KC, Lu HF, Ma CY, Sai-Chuen Wu R, Wu KC, Chueh FS, Gibson Wood W, Chung JG (2012b) Rutin inhibits human leukemia tumor growth in a murine xenograft model *in vivo*. Environ Toxicol 27(8):480–484

Lupaşcus D, Tuchiluş CRISTINA, Profire LENUŢA (2010) Physico-chemical and antimicrobial properties of novel rutin derivatives with 6-aminopenicillanic acid. Farmacia 56(5):501–506

Ma X, Ren X, Zhang X, Griffin N, Liu H, Wang L (2023) Rutin ameliorates perfluorooctanoic acid-induced testicular injury in mice by reducing oxidative stress and improving lipid metabolism. Drug Chem Toxicol 46(6):1223–1234

Mahmoud AM (2012) Influence of rutin on biochemical alterations in hyperammonemia in rats. Exp Toxicol Pathol 64(7–8):783–789

Martini S, D'Addario C, Braconi D, Bernardini G, Salvini L, Bonechi C, Figura N, Santucci A, Rossi C (2009) Antibacterial Activity of Grape Extracts on cag A-Positive and-Negative Helicobacter pylori Clinical Isolates. Journal of Chemotherapy 21(5):507–513

Maurya SK, Tewari M, Mishra RR, Shukla HS (2012) Genetic aberrations in gallbladder cancer. Surg Oncol. 21(1):37–43

Middleton JE, Kandaswami C (1994) The impact of plant flavonoids on mammalian biology: implications for immunity, inflammation and cancer. In: Harborne JB (ed) The flavonoids: advances in research since 1986. Chapman and Hall, London, UK, pp 619–652

Mishra RR, Tewari M, Shukla HS (2011) Helicobacter pylori and pathogenesis of gallbladder cancer. J Gastroenterol Hepatol 26(2):260–266

Mokhtar M, Ginestra G, Youcefi F, Filocamo A, Bisignano C, Riazi A (2017) Antimicrobial activity of selected polyphenols and capsaicinoids identified in pepper (Capsicum annuum L.) and their possible mode of interaction. Curr Microbiol 74:1253–1260

Nijs A, Cartuyvels R, Mewis A, Peeters V, Rummens JL, Magerman K (2003) Comparison and evaluation of Osiris and Sirscan 2000 antimicrobial susceptibility systems in the clinical microbiology laboratory. J Clin Microbiol 41(8):3627–3630
Nouri Z, Fakhri S, Nouri K, Wallace CE, Farzaei MH, Bishayee A (2020) Targeting multiple signaling pathways in cancer: the rutin therapeutic approach. Cancers 12(8):2276
Oliveira LFSD, Fuentefria AM, Klein FDS, Machado MM (2014) Antifungal activity against Cryptococcus neoformans strains and genotoxicity assessment in human leukocyte cells of Euphorbia tirucalli L. Braz J Microbiol 45:1349–1355
Oliveira VM, Carraro E, Auler ME, Khalil NM (2016) Quercetin and rutin as potential agents antifungal against Cryptococcus spp. Braz J Biol 76:1029–1034
Orhan DD, Özçelik B, Özgen S, Ergun F (2010) Antibacterial, antifungal, and antiviral activities of some flavonoids. Microbiol Res 165:496–504
Pandey P, Khan F, Qari HA, Oves M (2021) Rutin (bioflavonoid) as cell signaling pathway modulator: prospects in treatment and chemoprevention. Pharmaceuticals 14(11):1069
Parvez MK, Rehman MT, Alam P, Al-Dosari MS, Alqasoumi SI, Alajmi MF (2019) Plant-derived antiviral drugs as novel hepatitis B virus inhibitors: cell culture and molecular docking study. Saudi Pharm J 27(3):389–400
Patil SL, Swaroop K, Kakde N, Somashekarappa HM (2017) *In vitro* protective effect of rutin and quercetin against radiation-induced genetic damage in human lymphocytes. Indian J Nucl Med 32(4):289
Peuget S, Zhou X, Selivanova G (2024) Translating p53-based therapies for cancer into the clinic. Nat Rev Cancer 24:1–24
Pimentel RBDQ, da Costa CA, Albuquerque PM, Junior SD (2013) Antimicrobial activity and rutin identification of honey produced by the stingless bee Meliponacompressipesmanaosensis and commercial honey. BMC Complement Altern Med 13(1):1–14
Prasad R, Prasad SB (2021) Modulatory effect of rutin on the antitumor activity and genotoxicity of cisplatin in tumor-bearing mice. Adv Pharm Bull 11(4):746
Reddy JRSM, Kannan KP, Sankaran K, Rengasamy G, Priya VV, Sathishkumar P (2023) Eradication of dental pathogens using flavonoid rutin mediated silver-selenium nanoparticles. Inorg Chem Commun 157:111391
Riva A, Kolimár D, Spittler A, Wisgrill L, Herbold CW, Abrankó L, Berry D (2020) Conversion of rutin, a prevalent dietary flavonol, by the human gut microbiota. Front Microbiol 11:585428
Saleemi MA, Alallam B, Yong YK, Lim V (2022) Synthesis of zinc oxide nanoparticles with bioflavonoid rutin: characterisation, antioxidant and antimicrobial activities and *in vivo* cytotoxic effects on artemia nauplii. Antioxidants 11(10):1853
Samsonowicz M, Kamińska I, Kalinowska M, Lewandowski WJSAPAM (2015) Alkali metal salts of rutin–synthesis, spectroscopic (FT-IR, FT-Raman, UV–VIS), antioxidant and antimicrobial studies. Spectrochim Acta A Mol Biomol Spectrosc 151:926–938
Savov VM, Galabov AS, Tantcheva LP, Mileva MM, Pavlova EL, Stoeva ES, Braykova AA (2006) Effects of rutin and quercetin on monooxygenase activities in experimental influenza virus infection. Exp Toxicol Pathol 58(1):59–64
Semwal R, Joshi SK, Semwal RB, Semwal DK (2021) Health benefits and limitations of rutin—a natural flavonoid with high nutraceutical value. Phytochem Lett 46:119–128
Selway JT (1986) Antiviral activity of flavones and flavans. Progress in clinical and biological research 213:521–536
Sharma S, Ali A, Ali J, Sahni JK, Baboota S (2013) Rutin: therapeutic potential and recent advances in drug delivery. Expert Opin Investig Drugs 22(8):1063–1079
Shrivastava S, Gupta D, Gupte SS, Uthra C, Sharma V, Shukla S (2023) Curative efficacy of polyphenols against acrylamide induced toxicity and analysis of acrylamide in Indian food by LCMS method. Food Chem Adv 3:100522
Singh M, Govindarajan R, Rawat AKS, Khare PB (2008) Antimicrobial flavonoid rutin from Pterisvittata L. against pathogenic gastrointestinal microflora. Am Fern J 98(2):98–103
Singh A, Ahmad S, Ahmad A (2015) Green extraction methods and environmental applications of carotenoids-a review. RSC Adv 5(77):62358–62393

Soni H, Malik J, Singhai AK, Sharma S (2013) Antimicrobial and antiinflammatory 1214 activity of the hydrogels containing rutin delivery. Asian J Chem 25:8371–8373

Stojković D, Petrović J, Soković M, Glamočlija J, Kukić-Marković J, Petrović S (2013) In situ antioxidant and antimicrobial activities of naturally occurring caffeic acid, p-coumaric acid and rutin, using food systems. J Sci Food Agric 93:3205–3208

Tao J, Hu Q, Yang J, Li R, Li X, Lu C, Chen C, Wang L, Shattock R, Ben K (2007) *In vitro* anti-HIV and-HSV activity and safety of sodium rutin sulfate as a microbicide candidate. Antivir Res 75(3):227–233

Tewari M, Agarwal A, Mishra RR, Meena RN, Shukla HS (2013) Epigenetic changes in carcinogenesis of gallbladder. Indian J Surg Oncol 4(4):356–361

Turgeon PL, Higgins R, Gottschalk M, Beaudoin M (1994) Antimicrobial susceptibility of Streptococcus suis isolates. Br Vet J 150(3):263–269

Valdes M, Calzada F, Martínez-Solís J, Martínez-Rodríguez J (2023) Antihyperglycemic effects of Annona cherimola miller and the flavonoid Rutin in combination with Oral antidiabetic drugs on Streptozocin-induced diabetic mice. Pharmaceuticals 16(1):112

Wang C, Wang P, Chen X, Wang W, Jin Y (2015) Saururuschinensis (Lour.) Baill blocks enterovirus 71 infection by hijacking MEK1–ERK signaling pathway. Antivir Res 119:47–56

Wang X, Peng D, Wang W, Xu Y, Zhou X, Hesketh T (2017) Massive misuse of antibiotics by university students in all regions of China: implications for national policy. Int J Antimicrob Agents 50(3):441–446

Xu MS, Chen S, Wang WQ, Liu SQ (2013) Employing bifunctional enzymes for enhanced extraction of bioactives from plants: flavonoids as an example. J Agric Food Chem 61(33):7941–7948

Yarmolinsky L, Huleihel M, Zaccai M, Ben-Shabat S (2012) Potent antiviral flavone glycosides from Ficusbenjamina leaves. Fitoterapia 83(2):362–367

Yoo H, Ku SK, Baek YD, Bae JS (2014) Anti-inflammatory effects of rutin on HMGB1-induced inflammatory responses *in vitro* and *in vivo*. Inflamm Res 63:197–206

Zandi K, Teoh BT, Sam SS, Wong PF, Mustafa MR, AbuBakar S (2011) *In vitro* antiviral activity of fisetin, rutin and naringenin against dengue virus type-2. J Med Plants Res 5(23):5534–5539

Zuo GY, Li ZQ, Chen LR, Xu XJ (2005) *In vitro* anti-HCV activities of Saxifragamelanocentra and its related polyphenolic compounds. Antivir Chem Chemother 16(6):393–398

Beyond Pharmaceuticals: Tapping into Nature's Weapons Against Pathogenic Resilience

24

Ritu Raj Patel, Aradhana Mishra, Sudhir Kumar Singh, and Meenakshi Singh

Abstract

Infectious diseases caused by pathogenic microorganisms' impact a vast population of humans every year and spread significantly with alarmingly high fatality rates. The recent drug abuse during the COVID-19 outbreak appears to have led to an increase in multidrug-resistant-associated pathogenic microbial species, which has further exacerbated the already heightened drug resistance. The majority of antimicrobials now in use have limited efficacy, which restricts their use in clinical settings. The escalating threat of antibiotic-resistant infections necessitates alternative solutions beyond traditional pharmaceuticals. Nature has endowed every species, from prokaryotic organisms to enormous eukaryotes, with remarkable systems that contain naturally occurring antimicrobial compounds. The antimicrobial properties of the majority of naturally occurring bioactive compounds obtained from the rich reservoir of plants, bacteria, fungus, animals, marine life, and other sources such as honey, propolis are now being investigated. These compounds have demonstrated the potential to eliminate microbial infections associated with the emergence of drug resistance and tolerance. Apart from these, viruses such as bacteriophages have the ability to invade bacterial cells and cause the bacterium to lyse by disrupting its metabolism. The chapter explores the dominion of natural antimicrobials and dig into the formidable arsenal of antimicrobial properties residing in various natural compounds and provides a comprehensive overview of nature's potential in combating resil-

R. R. Patel · A. Mishra · M. Singh (✉)
Department of Medicinal Chemistry, Faculty of Ayurveda, Institute of Medical Sciences, Banaras Hindu University, Varanasi, Uttar Pradesh, India
e-mail: meenakshisingh@bhu.ac.in

S. K. Singh
Department of Microbiology, Institute of Medical Sciences, Banaras Hindu University, Varanasi, Uttar Pradesh, India

M. Gangwar, G. Nath (eds.), *Emerging Paradigms for Antibiotic-Resistant Infections: Beyond the Pill*, https://doi.org/10.1007/978-981-97-5272-0_24

ient pathogens. It explores the inhibitory actions of natural antimicrobial compounds and their potential mechanisms against several pathogenic fungal and bacterial species. The chapter refines the antimicrobial drugs currently in use while establishing new therapeutic approaches based on natural compounds to tackle the rising multidrug-resistant-associated infections.

Keywords

Antimicrobial resistance · Animal-derived compounds · Plant products · Regulatory action

24.1 Introduction

Antimicrobial resistance (AMR) is considered the most critical health concern worldwide. Antibiotics, long heralded as miracle medications, revolutionized medicine and played a significant role in the improvement of life expectancy. The twentieth century antibiotic era significantly decreased the risk of infectious diseases and undoubtedly saved billions of lives. Yet, the microbiological susceptibility to currently available drugs which are the cause of critical point drug resistance in communities and hospitals has decreased over time. The indiscriminate and extensive use of these medications have unintentionally fostered the arrival of antibiotic-resistant bacteria (Savoia 2012). Drug resistance, commonly referred to as AMR occurs as a result of microorganisms including bacteria, parasites, viruses, fungi changing in a way that makes the drugs used to treat the disease they cause ineffective.

The emergence of resistance in a majority of clinical isolates of bacteria, scarcity of innovative antibiotics, and recurring infections brought on by persistent bacteria make it difficult to treat infections successfully. Moreover, the rise of multidrug-resistant infections, once confined to hospital settings, has now become a global threat, challenging our ability to treat common infections and undermining the success of medical procedures that rely on effective antimicrobial agents. Even the newest groups of antimicrobial medicines have limited lifespan because of the widespread occurrence of resistance. Single-target drugs can have a variety of adverse effects and frequently result in drug resistance. One of two approaches is either multidrug therapy or a single drug that acts upon multi-targets (Upadhyay 2021). Microorganisms that develop resistance to the majority of antimicrobials are commonly known as "superbugs." According to estimates, bacterial AMR caused fatalities of 4.95 million people globally in 2019 and it led to the deaths of 1.27 million individuals among those deaths. The six most common pathogens associated with resistance-related deaths were *Acinetobacter baumannii*, *Pseudomonas aeruginosa*, *Staphylococcus aureus*, *Escherichia coli*, *Streptococcus pneumoniae*. In 2019, these six pathogens had caused 929,000 deaths attributed to the AMR and around 3.57 million fatalities associated with it (Murray et al. 2022). The United Kingdom government commissioned an assessment on AMR in 2014 that was published by Jim O'Neill and his colleagues. According to the analysis, by 2050, AMR might result in ten million annual fatalities (O'Neill 2014). This estimate has been used frequently by specialists, lay media, and public health organizations, and it has become a common refrain. This alarming prediction, often cited without context, underscores the significant clinical and public health challenges posed by resistance in microorganisms (de Kraker et al. 2016). Acknowledging the huge growing burden with the need for immediate action, emphasizes the urgency of addressing this global health crisis. The World AMR Awareness Week (WAAW) is an international initiative to promote best practices and increase awareness as well as understanding of AMR to reduce the development of drug-resistant diseases. In fact, "Preventing antimicrobial resistance together" was the theme for WAAW 2023.

The historical account of antimicrobial agents' development combines the achievements of medical advancement with the difficulties presented by newly emerging pathogens as shown in Fig. 24.1. Early in the twentieth century, Paul Ehrlich's groundbreaking research established a foundation for the development and discovery of antibiotics, fostering the well-accepted idea of "magic bullets" and leading to the identification of salvarsan, the first successful syphilis therapy. The decades that followed witnessed the introduction of synthetic drugs such as sulfa

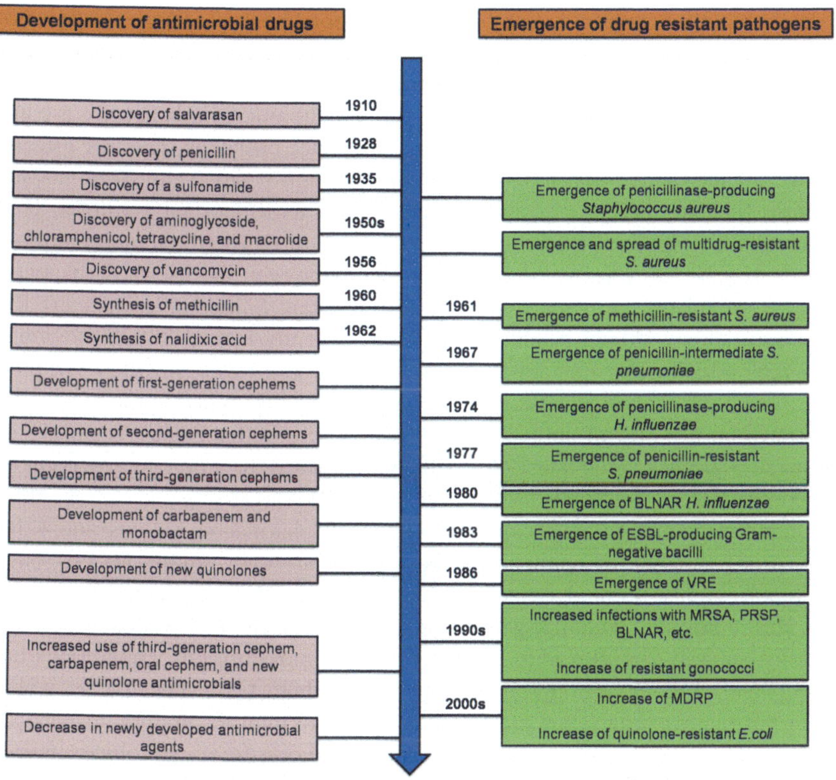

Fig. 24.1 Antimicrobial drug development and emergence of resistant pathogens

medicines, a major breakthrough being Prontosil in the 1930s. The breakthrough moment, however, came in 1928 when Alexander Fleming discovered penicillin, ushering in the age of antibiotics and completely changing the field of medicine. The widespread production of penicillin during World War II transformed medicine by treating previously fatal bacterial infections. Penicillin's were originally effective against Gram-positive species like *S. aureus*. The discovery of several antibiotic classes, including tetracycline, chloramphenicol, macrolide, glycopeptide, and streptomycin, in the 1940s and 1950s brought in the Golden Age of Antibiotics. However, soon after the arrival of penicillin-resistant pathogens in the mid-1960s, methicillin was subsequently developed in order to combat particularly *S. aureus*, bacteria that produced the enzyme penicillinase and hydrolyzed penicillin. A significant instance of resistance surfaced with methicillin-resistant *Staphylococcus aureus* (MRSA) and penicillin-resistant *Streptococcus pneumoniae* (PRSP) were isolated in the year 1961 in the UK highlighted the adaptability of bacteria in the face of antibiotics and became emblematic of this challenge (Aminov 2010).

Developed in the 1960s, cephems are widely used and can be divided into multiple generations according to their antibacterial spectra. Cefazolin is one of the first-generation cephems that exhibit potent activity against Gram-positive organisms and *E. coli*. Second-generation cephems, such cefotiam, extend their range by covering more *Enterobacteriaceae* and other Gram-negative pathogens. Ceftazidime and cefotaxime, two third-generation cephems, show increased efficacy against Gram-negative pathogens. Similar to penicillins and cephalosporins, carbapenems constitute a class of wide-spectrum beta-lactam antibiotics. Carbapenems are antibiotics that have been developed to combat resistance problems. They are effective against a wide range of bacteria including both Gram-positive and Gram-negative bacteria. Ipenem, meropenem, and ertapenem are a few important examples. These antibiotics are often reserved for severe infections or cases where other antibiotic classes may be less effective due to resistance. However, the rise of carbapenem-resistant organisms, particularly carbapenem-resistant *Enterobacteriaceae* (CRE), become a major worldwide health concern. Subsequently, in the 1980s, *Haemophilus influenzae* which was β-Lactamase-Negative Ampicillin-Resistant (BLNAR) posed a challenge by evolving resistance mechanisms beyond β-lactamase production. Vancomycin, a powerful antibiotic, started to lose its effectiveness against some strains of *S. aureus* in the late 1990s, which has caused the worrying emergence of Vancomycin-Resistant *Staphylococcus aureus* (VRSA) (Lessing and Raftery 1998).

Quinolone antimicrobials exhibited enhanced pharmacodynamics and safety. Quinolones are a type of synthetic antibiotic that leads to the inhibition of the bacterial topoisomerase IV and DNA gyrase, preventing DNA repair and replication. This class comprises both first-generation quinolones (nalidixic acid) and later generations, such as ciprofloxacin, levofloxacin, and moxifloxacin. Quinolones have huge potential as an antimicrobial agent against both Gram-negative as well as Gram-positive bacteria. They are frequently used to treat respiratory, urinary, and gastrointestinal infections. However the misuse and overuse of quinolones have led to the emergence of resistance, and certain regulatory authorities have restricted their usage due to potential side effects such as tendonitis and tendon rupture.

Additionally, the decline in the discovery of new antibiotics in twenty-first century and the emergence of pathogens resistant to multiple drugs such as Multidrug-Resistant *Pseudomonas aeruginosa* (MDRP), Pan Drug-Resistant (PDR) pathogens and Extensively Drug-Resistant (XDR) remains a major concern (Saga and Yamaguchi 2008).

Therefore, in the face of rising AMR, there is a vital need for novel approaches that go beyond the usual dependence on pharmaceuticals. The objective of this chapter is to look into developing paradigms for combating antibiotic-resistant diseases, with a particular emphasis on the idea of reverting to a pre-antibiotic era and using nature's resources as an acceptable solution. As the conventional antibiotic pipeline struggles to keep up with the rising issue of resistance, investigating alternate sources and mechanisms becomes vital.

24.2 Nature's Arsenal

24.2.1 Biodiversity as a Source of Novel Therapeutics

Biodiversity stands as an invaluable reservoir of novel therapeutics as shown in Fig. 24.2, offering a vast and diverse array of organisms, ecosystems, and genetic material that has proven instrumental in the discovery and development of pharmaceutical compounds. This abundant supply of biological diversity, which

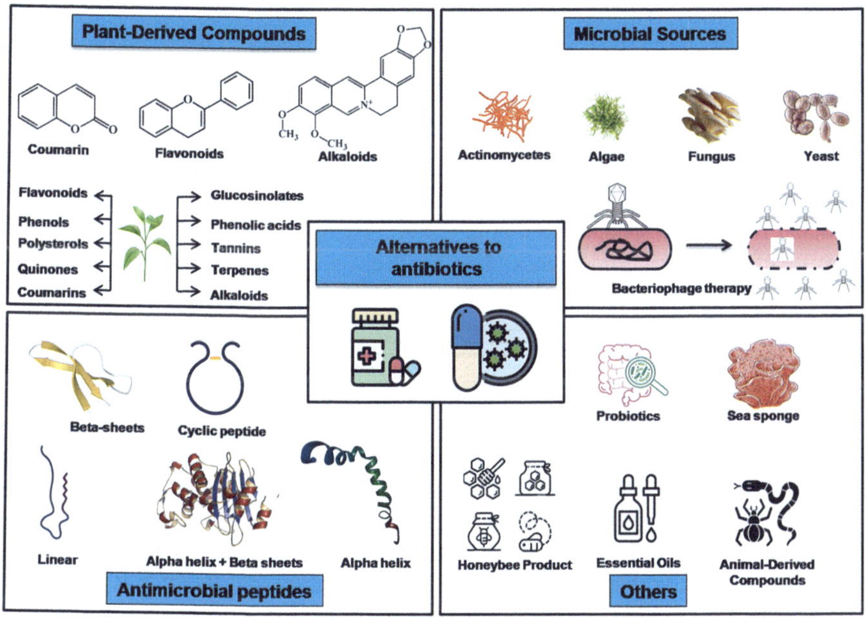

Fig. 24.2 Biodiversity as a source of novel therapeutics as an alternative to antibiotic against antimicrobial resistance

includes microbes as well as plants and animals, offers a remarkable reservoir of bioactive substances with possible therapeutic uses. Bioprospecting, the systematic search for valuable compounds in natural environments, has been a key strategy in tapping into this diversity. Indigenous tribes have a deep understanding of the local flora and fauna, which helps them, identify medicinal plants and traditional cures. These identifications, when paired with contemporary scientific methods, reveal novel bioactive compounds which could be an alternative against antimicrobial resistance.

24.2.2 Plant-Derived Compounds

In the past few decades, the scientific and pharmaceutical communities have been paying more attention to research on the compounds derived from plants and their possible antimicrobial activity because of the rising prevalence of MDR pathogens. These bioactive plant-derived compounds (PDC) represent a largely unexplored source of antimicrobial chemotypes that are employed in various traditional medical systems around the world. Many different secondary metabolites found in plants can be used as alternate strategies to control disease caused by MDR pathogens of the estimated 250,000–500,000 known plant species on Earth, only a limited subset have been thoroughly investigated for possible antibacterial compounds. It is noteworthy that somewhat more than 10% of these plant species is having great potential in the pharmaceuticals with utilization extending to 50,000 species and more (Chen et al. 2016). In India, 7500 of the 17,000 higher plant species are known to have therapeutic properties. The huge medicinal importance of the plants is known due to their therapeutic effects in any country on the planet (Kala et al. 2006). In the Indian subcontinent, Ayurveda which is the oldest medical system has identified almost 2000 different kinds of medicinal plants. Presently, over 25% of medicines is derived from medicinal plants and a huge number of those are synthetic equivalents made from prototype molecules found in modern pharmacopoeia that are extracted from plant species (Rao et al. 2004). Therefore, it is essential to employ innovative scientific approaches to discover and validate novel botanical therapeutics.

24.2.2.1 Flavonoids

Flavonoids are the promising class of phototherapeutic metabolites against drug-resistant pathogens. The presence of aromatic rings i.e. A, B, and C make up the basic structure of flavonoids, two of which are made of six carbon atoms each that are connected to the three-carbon heterocyclic ring (C) as shown in Fig. 24.3.

Flavonoids are classified into six sub-classes: flavanols, flavanones, isoflavones, flavonols, anthocyanins, and flavones, depending on the heterocycle type that varies (Sopjani et al. 2024). The majority of studies discovered that flavonoids are effective against drug-resistant, biofilm-forming bacteria and very efficient when used in concert with conventional antibiotics as mentioned in Table 24.1. The flavonoids that were studied the most frequently were luteolin, kaempferol, apigenin, and

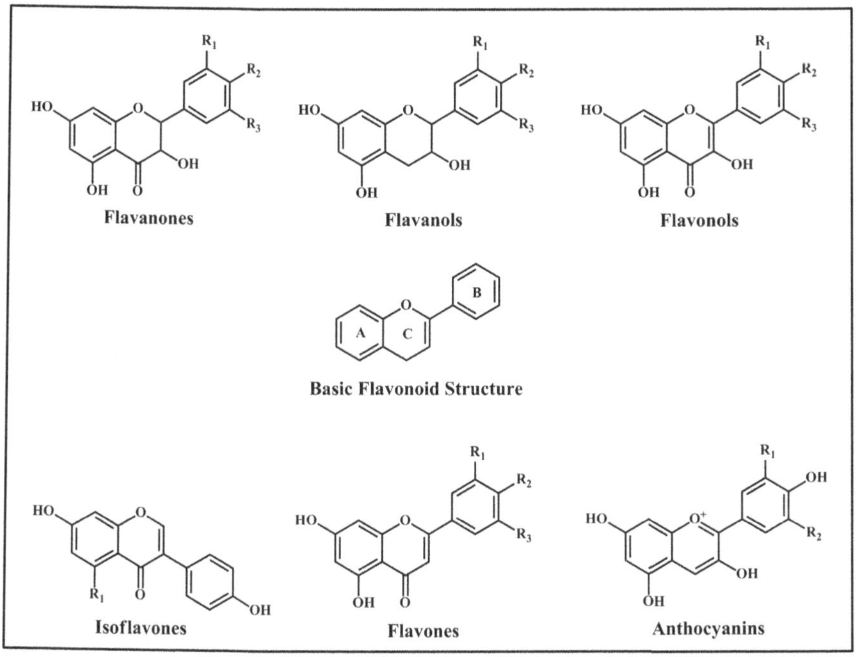

Fig. 24.3 Structure of basic flavonoid and its subclasses

quercetin (Pathak and Mazumder 2023). Flavonoids have been shown to interfere with microbial cell membrane integrity, inhibit microbial enzymes, alter microbial gene expression, block the synthesis of nucleic acids, energy metabolism inhibition, inhibit cell attachment, and biofilm formation (Wu et al. 2024). Several research investigations have also confirmed the low toxicity and safety of flavonoids. Further research is required to evaluate the therapeutic importance of flavonoids as adjuvants to conventional antimicrobial medications and develop flavonoid-based therapy for MDR diseases.

24.2.2.2 Coumarin

A broad-spectrum anti-infective strategy that targets quorum sensing (QS) and biofilm inhibition is acknowledged as shown in Fig. 24.4, to treat infections with drug-resistant pathogenic bacteria. Coumarin is considered a phytobased compound as it is isolated from the natural species and known as the quorum sensing inhibitor (QSi) for its antibacterial efficacy.

In a recent coumarin derivative assessment against *P. aeruginosa*, compound 4t emerged as an important biofilm inhibitor. This chemical not only inhibits quorum sensing but it also functions as an iron chelator, competing with pyoverdine. Consequently, it induces the deficiency of iron in *P. aeruginosa* (Liu et al. 2023). Further, grandivittin, aegelinol benzoate, and agasyllin, and are the three notable coumarins that were derived from the roots of *Ferulagocampestris*. These

Table 24.1 Plant-derived compounds with their structures and inhibitory concentrations

Plant-derived compounds	Plant	Structure	Inhibitory concentration	References
A. Flavonoids				
1. *5-hydroxy-6,7,8,3′,4′-pentamethoxyflavone*	*Citrus grandis* Linn	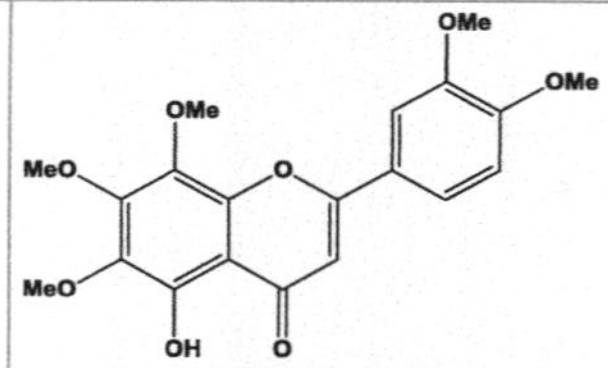	128–256 μg/mL	Phetkul et al. (2023)
2. Luteolin	*Euphorbia humifusa*	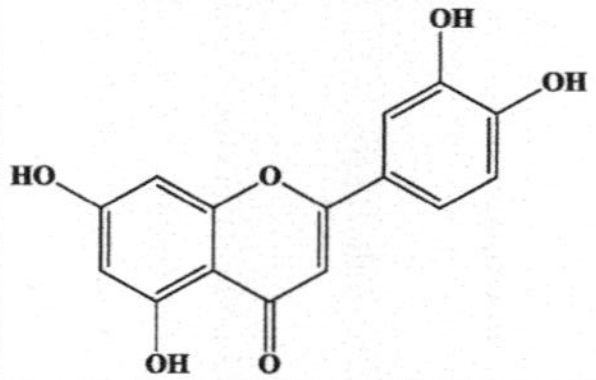	32–64 μg/mL	Wu et al. (2023)
3. *5,6,7,8,4′-pentamethoxyflavone*	*Citrus grandis* Linn	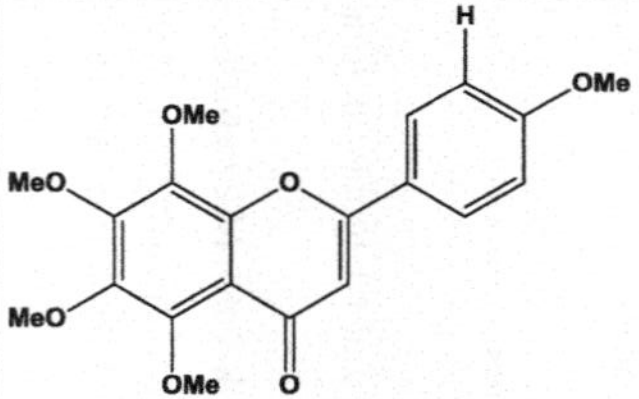	256 μg/mL	Phetkul et al. (2023)

4. Naringin	*Citrus grandis* Linn	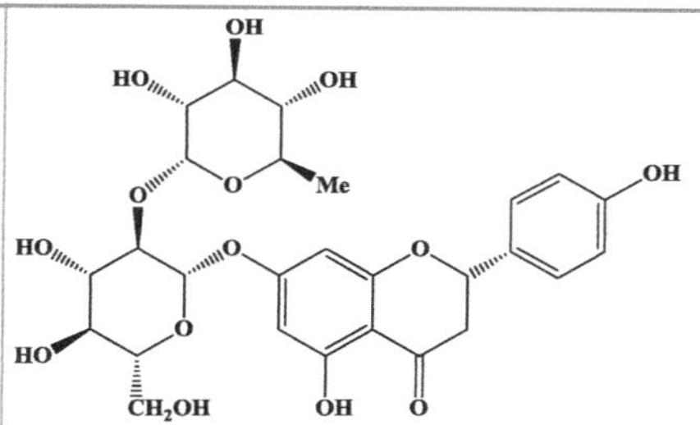	>2048 μg/mL	Phetkul et al. (2023)
5. Brazilin	*Caesalpinia sappan*	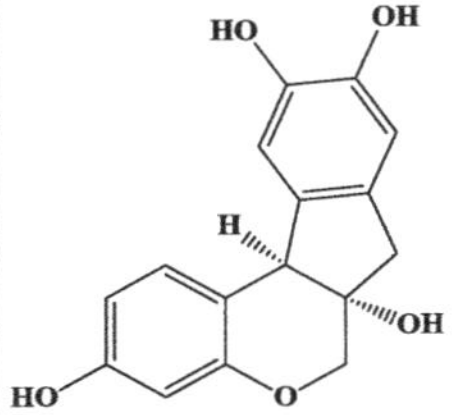	32 μg/mL	Peng et al. (2018)
6. Kaempferol	*Euphorbia humifusa*	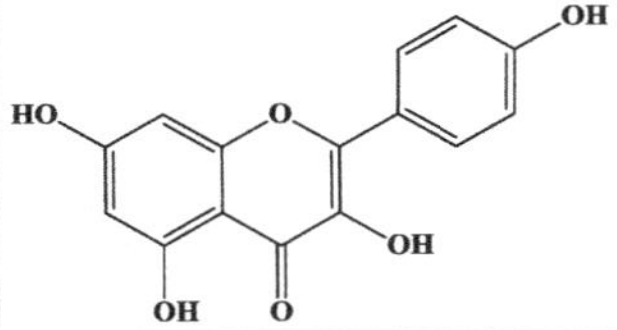	512 μg/mL	Wu et al. (2023)

(continued)

Table 24.1 (continued)

Plant-derived compounds	Plant	Structure	Inhibitory concentration	References
7. Biochanin A	*Dalbergia oliveri*	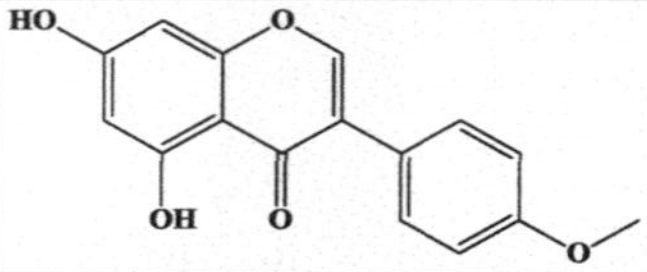	64–256 μg/mL	Bai et al. (2023)
8. Quercetin	*Euphorbia humifusa*	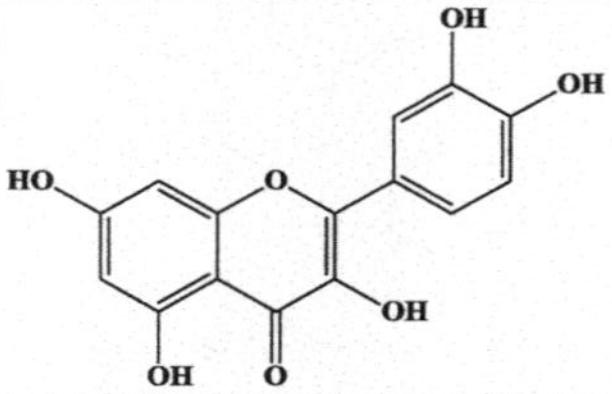	256 μg/mL	Wu et al. (2023)
B. Coumarin				
1. *8-(2,3-dihydroxy-3-methylbutyl)-7-methoxy2H-1-benzopyran-2-one*	*Citrus grandis* Linn	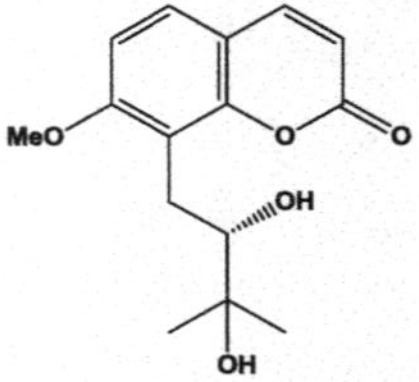	>2048 μg/mL	Phetkul et al. (2023)
2. *N-hydroxy-7-((6-methyl-2-oxo-2H-chromen-4-yl)oxy) heptanamide*	Coumarin derivative	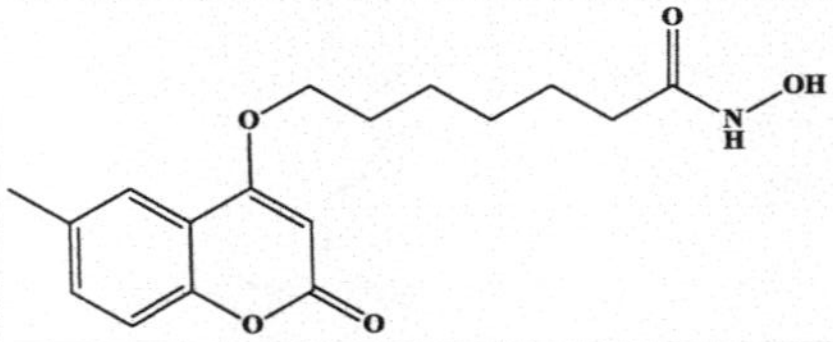	3.6 μM	Liu et al. (2023)

3. *8-Acetyl-7-hydroxy-4--methylcoumarin*	Coumarin derivative	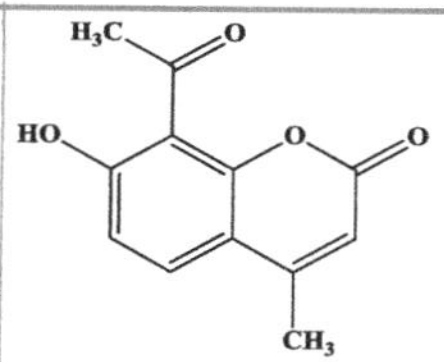	≤256 µg/mL	Martin et al. (2023)
4. *6′,7′-Dihydroxybergamottin*	*Peucedanumluxurians*	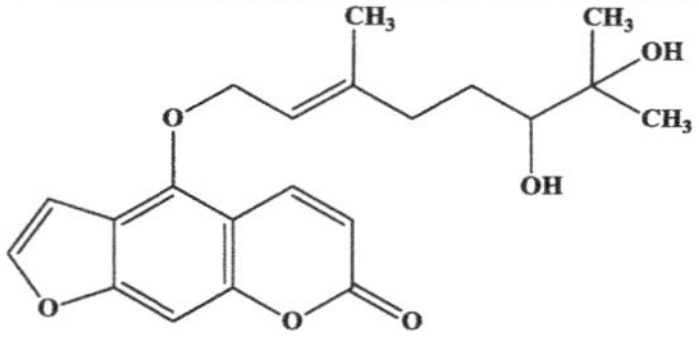	1.20–2.10 mg/mL	Widelski et al. (2018)
5. Peucedanin	*Peucedanumluxurians*	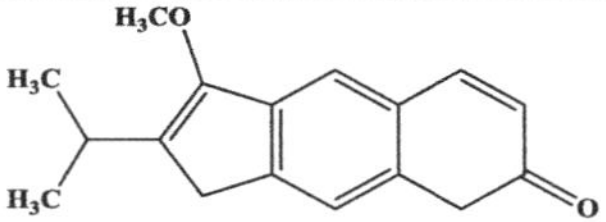	1.40–2.75 mg/mL	Widelski et al. (2018)
6. Officinalinisobutyrate	*Peucedanumluxurians*	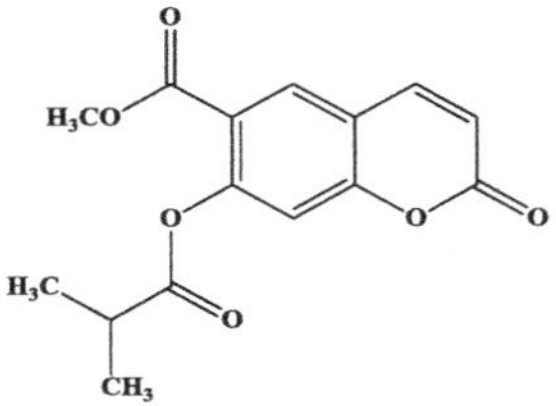	2.25–4.80 mg/mL	Widelski et al. (2018)
7. Scopoletin	*Mitracarpusfrigidus*	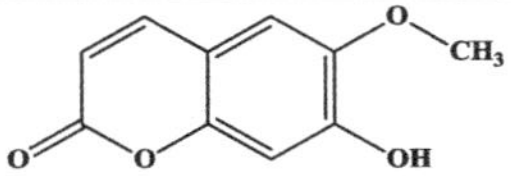	50 µg/mL	Lemos et al. (2020)

(continued)

Table 24.1 (continued)

Plant-derived compounds	Plant	Structure	Inhibitory concentration	References
8. Aegelinol	*Ferulagocampestris*	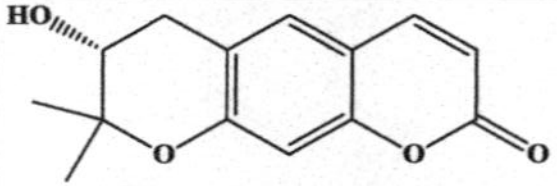	16 μg/mL	Basile et al. (2009)
C. Alkaloids				
1. Berberine	–	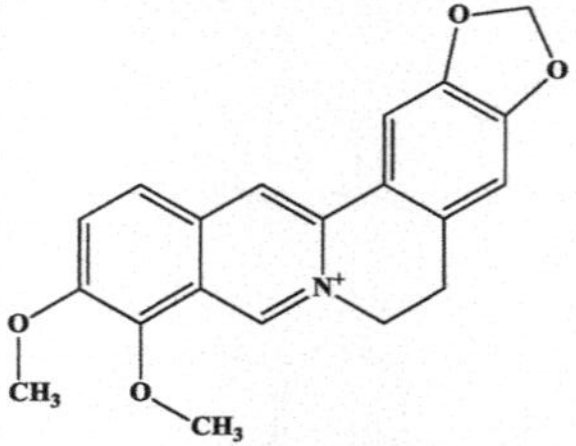	125 μg/mL	Mohtar et al. (2009)
2. Tetrandrine	*Stephaniatetrandra*	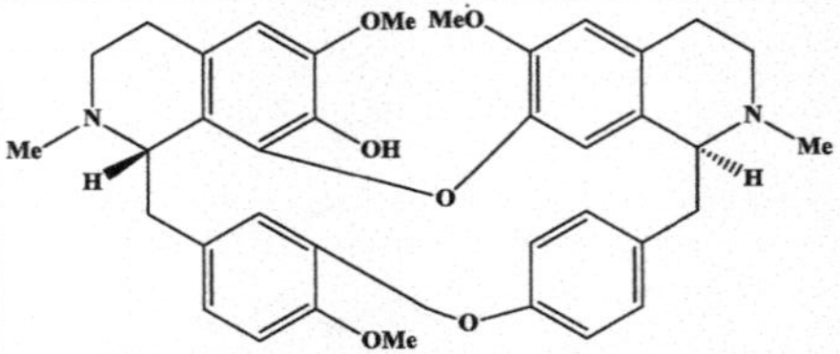	64–128 μg/mL	Zuo et al. (2011)

3. Demethyltetrandrine	*Stephaniatetrandra*	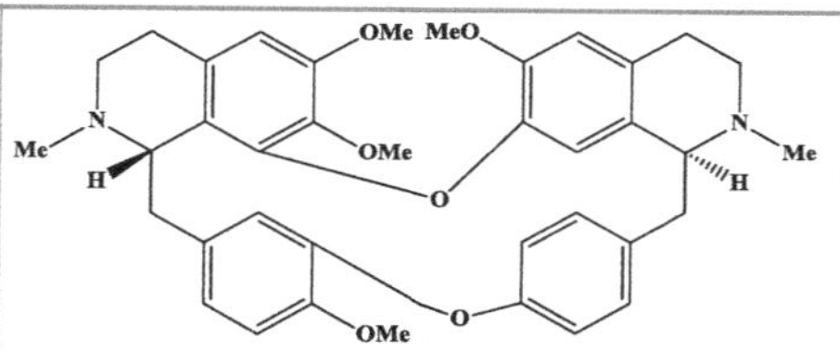	64–128 μg/mL	Zuo et al. (2011)
4. Palmatine	*Tinospora capillipes*	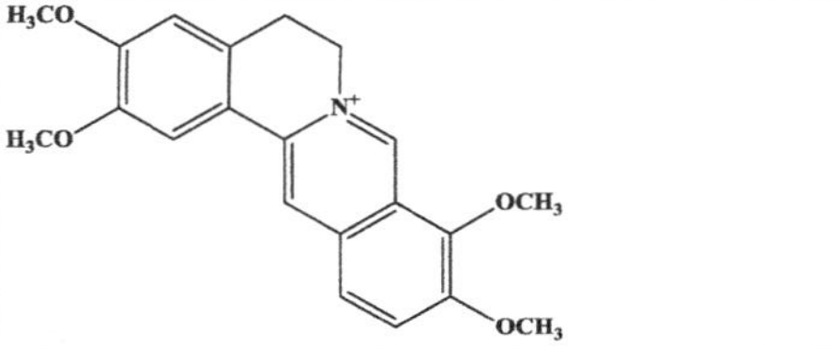	0.1–0.8 mg/mL	Deng et al. (2012)
5. Jatrorrhizine	*Tinospora capillipes*	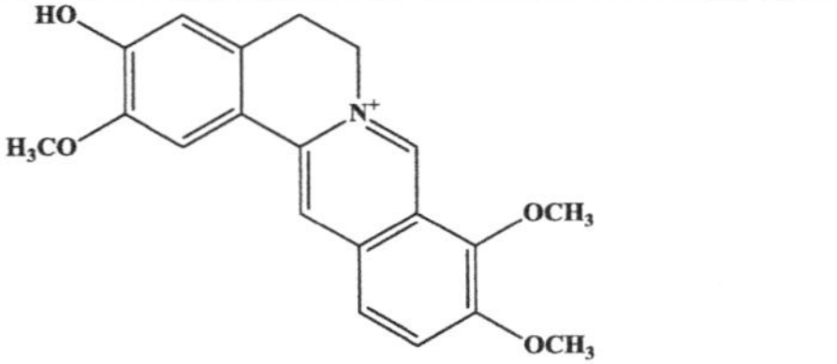	0.1–0.6 mg/mL	Deng et al. (2012)
6. 8-acetylnorchelerythrine	*Toddaliaasiatica*	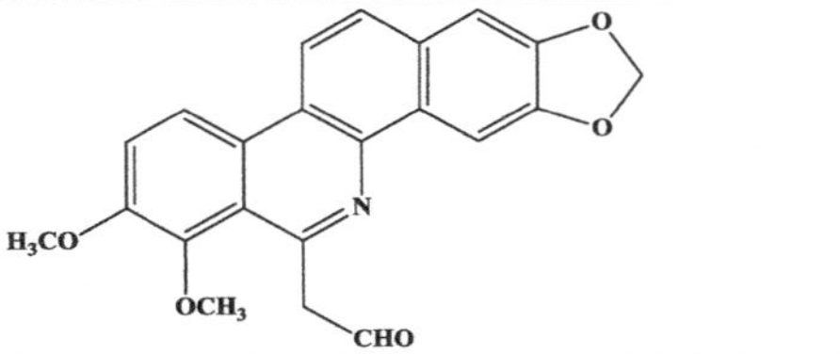	0.06–0.15 mg/ mL	Hu et al. (2014)

(continued)

Table 24.1 (continued)

Plant-derived compounds	Plant	Structure	Inhibitory concentration	References
7. *O*-methylmoschatoline	*Guatteriahispida*	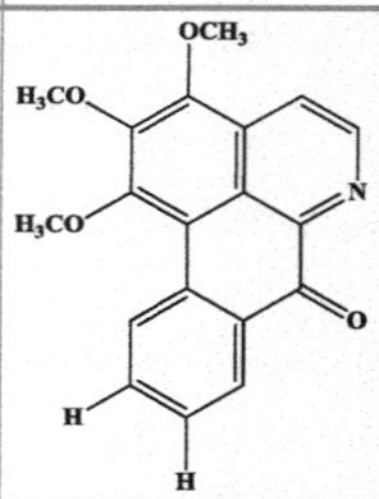	12.5–25 μg/mL	Costa et al. (2010)
8. Lysicamine	*Guatteriahispida*	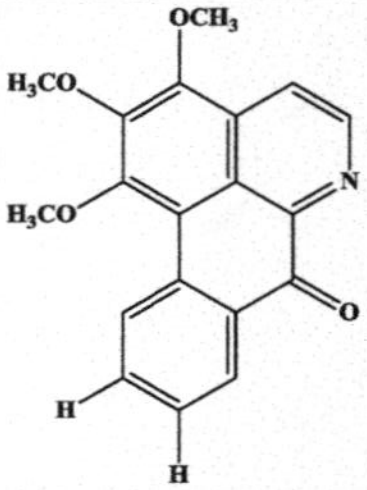	50–100 μg/mL	Costa et al. (2010)

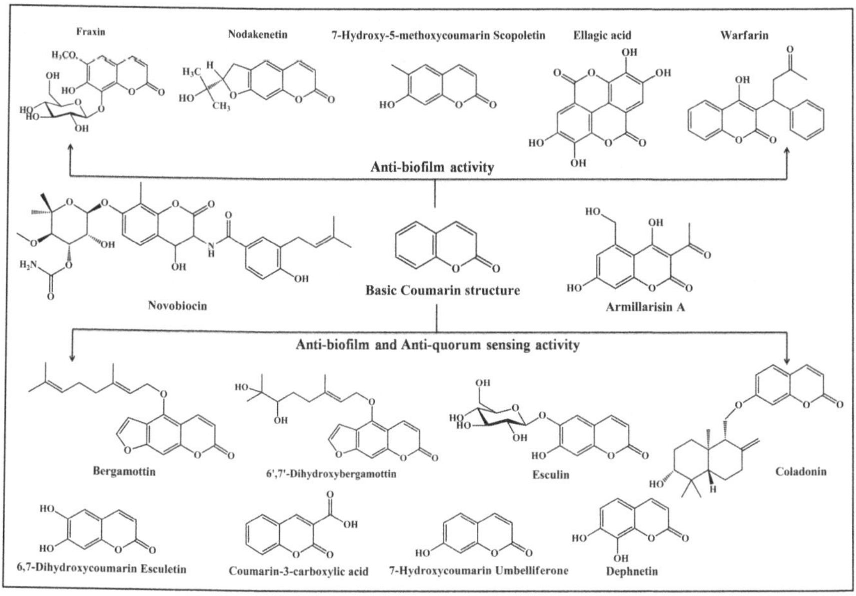

Fig. 24.4 Various coumarin structures with its anti-quorum sensing and anti-biofilm properties

compounds were found to possess antibacterial properties. Aegelinol and agasyllin demonstrated antibacterial efficacy against *Enterobacter cloacae*, *Salmonella typhi*, *S. aureus*, and *Enterobacterearogenes* (MIC = 16 and 32 μg/mL, respectively). Remarkably, these substances additionally exhibited antimicrobial properties against *Helicobacter pylori*, exhibiting dose-dependent suppression from 5 to 25 μg/mL (Basile et al. 2009). In a separate study, the anti-biofilm activity of coumarin was investigated. The results of the investigation showed that at concentrations higher than 400 μM, coumarin hampered *Porphyromonas gingivalis* growth and altered cell shape. Sub-MIC coumarin concentrations had no effect on early biofilm development, adhesion, aggregation, or characteristics of the cell surface. Nonetheless, at the late stage and dispersal of pre-formed biofilms, they showed anti-biofilm effects. Following coumarin treatment, biofilms became sporadic and gene expression associated with biofilms was downregulated. Based on a molecular docking study, it was revealed that coumarin interacts with HmuY protein and inhibits AI-2 activity. The study found that coumarin reduced the production of *P. gingivalis* biofilms through a quorum-sensing mechanism. Two furocoumarins that are derived from the grapefruit juice, namely dihydroxybergamottin and bergamottin, demonstrated anti-biofilm and anti-QS activity by successfully suppressing the biofilm production in *E. coli* O157:H7 (Girennavar et al. 2008).

24.2.2.3 Alkaloids

Plants produce secondary metabolites in response to both abiotic and biotic environmental changes. Alkaloids are one type of secondary metabolite that helps plants fight against predators and pathogens. The protective characteristics of alkaloids serve as a powerful weapon against antibiotic-resistant microorganisms (Mittal and Jaitak 2019). Alkaloids formed the foundation for the development of various antibiotics with a wide range of activity. A study was carried out to test alkaloid compounds that work as possible antagonistic agents against three MRSA isolates which showed the difference in their susceptibility to vancomycin and demonstrated active efflux activity. Berberine showed a modest inhibitory effect against two MRSA isolates, demonstrating a MIC value of 125 µg/mL (Mohtar et al. 2009). The antimicrobial effectiveness of the two phyto-derived bisbenzylisoquinoline alkaloids, i.e., demethyltetrandrine (d-Tet) and tetrandrine (Tet), were determined individually as well as combining with antibiotics such as cefazolin (CFZ), azithromycin (AZM), levofloxacin (LEV), and ampicillin (AMP) against ten clinically isolated MRSA strains. Tet and d-Tet exhibited MIC ranging 64–128 µg/mL. A noteworthy synergy was observed with the Tet/CFZ combination, reducing MICs by 75–94% against 90% of the isolates (Zuo et al. 2011). Furthermore, a comprehensive phytochemical analysis of *Guatteriahispida* bark was conducted. The results revealed three oxoaporphine alkaloids, *O*-methylmoschatoline, lysicamine, and liriodenine, demonstrated efficacy against *C. dubliniensis* and *S. epidermidis*, exhibiting MIC values ranging from 12.5 to 100 µg/mL.*O*-methylmoschatoline demonstrated higher activity against *S. epidermidis* in comparison with chloramphenicol (positive control) and exhibited comparable activity against *C. dubliniensis* in comparison with ketoconazole (positive control), while lysicamine, and liriodenine had the same action to chloramphenicol against *S. epidermidis* (Costa et al. 2010). Table 24.1 presents different PDC alongside their structures and inhibitory concentrations.

24.2.3 Microbial Sources

Microorganisms such as actinomycetes, archaea, fungi, yeast, and microalgae provide valuable bioactive chemical compounds. As the global threat of AMR grows, the study of microbial origins has become increasingly important. Researchers are looking at various microbial ecosystems, such as soils, oceans, and extreme environments, to find novel compounds that can combat drug-resistant bacterial strains. Beyond typical antibiotics, microorganisms help to develop novel therapies such as bacteriophages, which target specific bacterial species. Bacteria, notably the *Streptomyces* genus, have been prolific producers of antibiotics, which serve as an important defense mechanism in their natural habitats. For decades, microbial-derived antibiotics such as penicillin and streptomycin have played an important role in human medicine. It has been reported that novobiocin, an aminocoumarin molecule produced by *Streptomyces niveus*, significantly known to inhibit the response to DNA damage in Gram-positive were also discovered to inhibit the expression of error-prone DNA polymerases in an MDR *A. baumannii*. This dual action inhibits the pathogen's ability to acquire AMR via DNA damage-induced mutagenesis (Jara et al. 2016). In another work, 181 strains of Actinomycetes were isolated from the samples in Vietnam and were screened for their antimycobacterial and antimicrobial activity. Following a bioassay-guided approach, *Streptomyces alboniger* strain A121 was chosen for the further extraction of the bioactive metabolites. The identified compounds, obscurolide $B_{2\beta}$ (1) and chartreusin (2) were tested for their antimycobacterial efficacy against *M. smegmatis*. Notably, chartreusin showed the considerable antimycobacterial and antibacterial activity against several bacteria (Gram-positive) such as *Lactobacillus fermentum*, *S. aureus*, and *Bacillus subtilis* (Huynh et al. 2024).

In the hunt for novel antibiotic compounds derived from marine fungi in tropical seas, secondary metabolites derived from *Aspergillus candidus* KUFA 0062 cultures were investigated. The compound (2S, 3S, 5R)-2-benzyl-1-methyl-5--nonylpyrrolidin-3-ol (5a), isolated from *A. candidus* inhibited *S. aureus* ATCC 29213, *E. faecalis* ATCC 29212, and both MRSA and vancomycin-resistant *Enterococci* (VRE) strains (Buttachon et al. 2018). Hence, using microbial sources for producing new antimicrobial compounds not only broadens our defense methods against AMR but also aligns with the ideas of sustainable and environmentally friendly solutions.

The extensive use of bacteriophages, or phages, against the AMR has emerged as a potential method known as phage therapy. Bacteriophages are viruses that only target and infect bacteria, providing a highly accurate and targeted strategy for combating bacterial diseases. In the context of AMR, when standard antibiotics are losing effectiveness, phage therapy offers a feasible alternative. Phages function by adhering to and entering bacterial cells, resulting in their breakdown. Unlike broad-spectrum antibiotics, phages are very selective, targeting just infected bacterial strains while leaving the beneficial microbiome alone. This specialization helps reduce resistance development, as phages can evolve to keep up with bacterial changes. Phage therapy has been considered as an effective alternative in treating infections against the MDR bacteria, especially to those which are resistant to conventional antibiotics. The rise of MDR *Klebsiella pneumoniae* leads to serious health concern, necessitating novel treatment strategies. A study investigated the efficacy of bacteriophages is considered as an effective treatment for wound infections caused by MDR and biofilm-producing *K. pneumoniae*, comparing them to the conventional antibiotic gentamicin. A lytic phage that targets drug resistant *K. pneumoniae* was discovered and identified as belonging to the *Drexlerviridae* family. A phage similar to gentamicin successfully cleared pathogens from wounds in a mouse model. Notably, mice that received phage therapy showed improvement physically. These findings highlight the efficacy of phage therapy in treating *K. pneumoniae*-infected wounds, indicating that topical phage therapy is a safe and effective alternative for wound infection treatment (Kelishomi et al. 2024). In an innovative approach by using phage therapy with nanotechnology, Elsayed et al. isolated novel bacteriophages against MRSA and MDR *S. aureus* strains and encapsulated it with chitosan nanoparticles. MRSA phages, both non-encapsulated and encapsulated that were characterized through transmission electron microscopy (TEM). It was found that the MRSA phage that was encapsulated with CS-NPs has increased lytic activity as well as resilience to unfavorable circumstances such as pH and temperature (Elsayed et al. 2024).

24.2.4 Marine Organisms

Marine organisms have emerged as a rich, yet mostly unexplored resource. The oceans, which are rich in biodiversity, include a vast variety of microbes, sponges, algae, and other marine animals that produce unique bioactive substances capable of combating drug-resistant infections. Researchers are investigating deep water, coral reefs, and marine sediments to isolate and characterize bioactive compounds with antibacterial characteristics. A high quantity of cnidocysts was extracted from *Para zoanthusaxinellae* (Mediterranean zoanthid coral), encouraging further investigation into their antibacterial properties. The extract obtained from these cnidocysts showed notable antibacterial effectiveness against most of the human pathogens that are known to have resistance against conventional drugs, such as *Coccus* species and *Streptococcus agalactiae*. Furthermore, the extract also showed antibacterial efficacy against a variety of *Vibrio* spp. (Stabili et al. 2024). The antimicrobial property of crude extracts from four different soft coral species: *Dendronephthya hemprichi*, *Sarcophyton ehrenbergi*, *Sinularia leptoclados* and *Nephthyamolle*, was determined. The results showed that *D. hemprichi* extract had much higher antibacterial activity than other extracts with a MIC value of 0.1 mg/mL against *A. hydrophila* ATCC 13037 and *E. faecalis* ATCC 29212, and 0.5 mg/mL against *S. aureus* ATCC 6538 and *P. aeruginosa* ATCC 9027 (Mohammed El-Sawi et al. 2024). The antibacterial efficacy of *Bacillus cereus* MH997647 IA5 extract (symbiont to marine sponge *Agelas* species) was determined against the *K. pneumoniae* extended-spectrum β-lactamases (ESBL). The screening for antibacterial activity revealed that the symbiont derived from the sponge exhibited inhibitory effects demonstrating the zone of inhibition of 7.72 mm against *K. pneumoniae* ESBL (Gultom et al. 2024).

24.2.5 Animal-Derived Compounds

Animal-derived compounds have emerged as a focus in the hunt for innovative AMR strategies. The vast and diverse biochemistry of distinct animal species provides a wealth of bioactive substances capable of combating drug-resistant diseases. Venoms from snakes, spiders, and marine creatures were particularly explored to identify antimicrobial activity present in them. The antimicrobial efficacy of L-amino acid oxidases i.e. BmooLAAO-I extracted from the venom of *Bothropsmoojeni* and BjussuLAAO-II extracted from the venom of *Bothrops jararacussu* snake were examined. Both BmooLAAO-I as well as Bjussu LAAO-II displayed notable bacteriostatic properties against *S. aureus* (MIC = 0.12–0.24 μg/mL) and *E. coli* (MIC = 4.88–62.5 μg/mL). Furthermore, they reduced biofilm formation in *S. aureus* (88–89%) and *E. coli* (53–89%). This demonstrates the potential of L-amino acid oxidases derived from snake venomin limiting bacterial growth as well as reducing biofilm formation, providing opportunities for new antibacterial treatments (Alves de Melo Fernandes et al. 2024). A novel peptide with antimicrobial properties, Opisin derived from the venom of *Opistophthalmus glabrifrons* has been found. Opisin being a α-helical molecule, cationic and amphipathic in nature are made up of 19 amino acid residues. Opisin effectively inhibits Gram-positive bacteria (MIC = 4.0–10.0 μM) as well as vancomycin- and methicillin-resistant bacteria significantly (MIC = 2.0–6.0 μM) (Bao et al. 2015).

24.2.6 Essential Oils

In the scientific community, there has been a noticeable increase in the investigation of essential oils as potential antibiotic alternatives. These oils, which are extracted from a range of aromatic plants, are currently being thoroughly studied. They have unique fragrances and medicinal properties because they are made up of complex mixtures of volatile substances including terpenes, phenols, and aldehydes. Numerous researchers have reported the antibacterial efficacy of the essential oils against a variety of fungi, bacteria, and even drug-resistant species. Tea tree oil, oregano oil, and thyme oil have been shown to effectively prevent the growth of MDR bacteria (MRSA and VRE). The antibacterial efficacy and chemical composition of *Ptychotisverticillata* essential oil (PVEO) was investigated along with its potential synergies with established antibiotics such as ampicillin, erythromycin and amoxicillin. Analysis conducted through Gas Chromatography-Mass Spectrometry (GC-MS) identified carvacrol (6.99%), thymol (18.07%), p-cymene (19.38%), and γ-terpinene (38.96%) as the predominant bioactive compounds. PVEO had considerable antibacterial activity (MIC = 0.2–2 mg/mL) against *Micrococcus luteus* and *E coli*. The Fractional Inhibitory Concentration Index (FICI) was utilized for the detection of synergistic effects, and PVEO and amoxicillin showed significant synergy against *M. luteus* (FICI = 0.43) (Taibi et al. 2024). Additionally, GC-MS analysis was utilized to determine the chemical composition of *Citrus limon* essential oil (CLEO), uncovering predominant constituents like γ-terpinene (10.3%), β-pinene (12.6%), and limonene (60.7%). This finding encouraged additional research into CLEO's antibacterial and antibiofilm characteristics. CLEO had good antibacterial effectiveness against *E. coli* with MIC_{90} = 22.80 mg/mL and MIC_{50} = 22.61 mg/mL. The crystal violet method confirmed CLEO's antibiofilm action against *S. enterica*, with MIC_{50} and MIC_{90} values of 1.37 and 1.47 mg/mL, respectively (Kačániová et al. 2024).

24.2.7 Honeybee Products

Honeybee products, namely honey and propolis, have emerged as promising alternatives in the fight against AMR. Honey, well-known for its ancient therapeutic use, has broad-spectrum antibacterial capabilities due to its low pH, osmotic effect, and the presence of different bioactive components. Researchers have investigated the potency of honey against a wide range of bacteria that also includes strains resistant to conventional antibiotics. The emergence of extensively drug-resistant (XRD) *Salmonella typhi* strains is increasingly problematic in clinical environments, especially for individuals with septicemia. These highly virulent variants exhibit resistance to first- and second-line medications, such as ceftriaxone, with the exception of azithromycin. An investigation was conducted to determine the potential of conA-unbound protein fraction (F1) derived from *Melipona beecheii* honey in inhibiting the biofilm formation and it was found that it has a strong inhibitory effect on *P. aeruginosa* biofilm growth. The F1 fraction as well as crude protein extract (CPE) showed strong inhibition, exceeding 80% at doses of 1.3 μg/mL and 4 μg/mL, respectively. Notably, these proteins not only prevent biofilm formation but also affect biofilm structure, regulating *fleQ* and *fleR* genes expression mainly involved in *P. aeruginosa* biofilm regulation. This study highlights the effectiveness of *M. beecheii* honey-derived F1 fraction proteins in preventing and modifying the production of the biofilm by *P. aeruginosa* (Pool-Yam et al. 2024). Propolis, a resinous material gathered by honeybees from plants, enhances honey's antibacterial qualities. Propolis includes a wide variety of bioactive components that include flavonoids, phenolic acids, terpenes contributing to its anti-inflammatory, antioxidant, and antibacterial potency. In another study, Cunha et al. reported that the biological characteristics of cinnamoyloxy-mammeisin (CNM), a notable 4-phenyl coumarin that was extracted from the geopropolis of stingless bees. Research was done on strains of *S. aureus* that were reported to have resistance against methicillin as well as those that were susceptible to it, despite the fact that CNM's antibacterial efficacy against this bacterium had not been previously investigated. CNM showed bacteriostatic activity with a MIC of 11.3 μM. Scanning electron microscopy also revealed that CNM, at 5.7 μM, inhibited the formation of biofilms and decreased bacterial adhesion to human keratinocytes, hence influencing cell survival and architecture (da Cunha et al. 2020).

24.2.8 Antimicrobial Peptides

In the search for novel antimicrobial drugs, LC-AMP-F1 (a cationic antimicrobial peptide) was derived from the cDNA library from *Lycosa coelestis* venom gland received a lot of interest. This work investigated the secondary structure, physicochemical properties as well as the sequence of LC-AMP-F1 for assessing its antibacterial property, cytotoxicity, stability, and antibiofilm capabilities. Compared to the well-studied melittin, LC-AMP-effectively impeded the proliferation of numerous bacteria, encompassing MDR strains. Furthermore, it significantly reduced biofilm formation while disrupting mature biofilms. Notably, LC-AMP-F1 combination with the conventional antimicrobial drugs produced synergistic therapeutic effects. These findings highlight LC-AMP-F1 as a promising antimicrobial peptide for combating drug-resistant infections (Song et al. 2024).

24.3 Mechanism of Action of Plant Based Drug

24.3.1 Diversity of Bioactive Compounds in Plants

The bioactive compound is mainly the primary and secondary metabolite derived from the plants. The primary metabolites play a crucial role in plant growth regulation and development while of secondary metabolites play a vital role in plant defense (Teoh 2016). So now days, the most promising compounds which are having huge potential in therapeutic uses are known to be secondary metabolites. The classification of the secondary metabolite is shown in the Fig. 24.5.

Nowadays, antibiotic-resistant properties have become a challenging and emerging topic which leads to a serious challenge in pharmaceutical company to develop the new drugs hence the natural product-based phyto secondary metabolite generally the terpenes exhibit the antimicrobial property against both antibiotic-sensitive and resistant bacteria. Terpenes rupture the cell and inhibit the transcription and the translational machinery (Álvarez-Martínez et al. 2021). Aromatic medicinal plants generally produce terpenes which play an important role in disease resistance. However, the terpenes have been reported due to its antibacterial action against *S. aureus*. Phenolics exert antimicrobial activity by inhibiting the gyrase activity, efflux pump activity and by disrupting the cell membrane integrity (Wang et al. 2020).

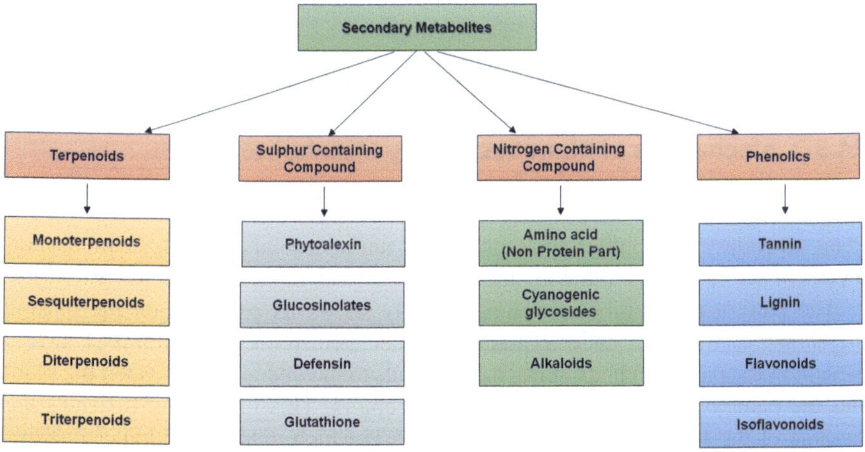

Fig. 24.5 Classification of secondary metabolites

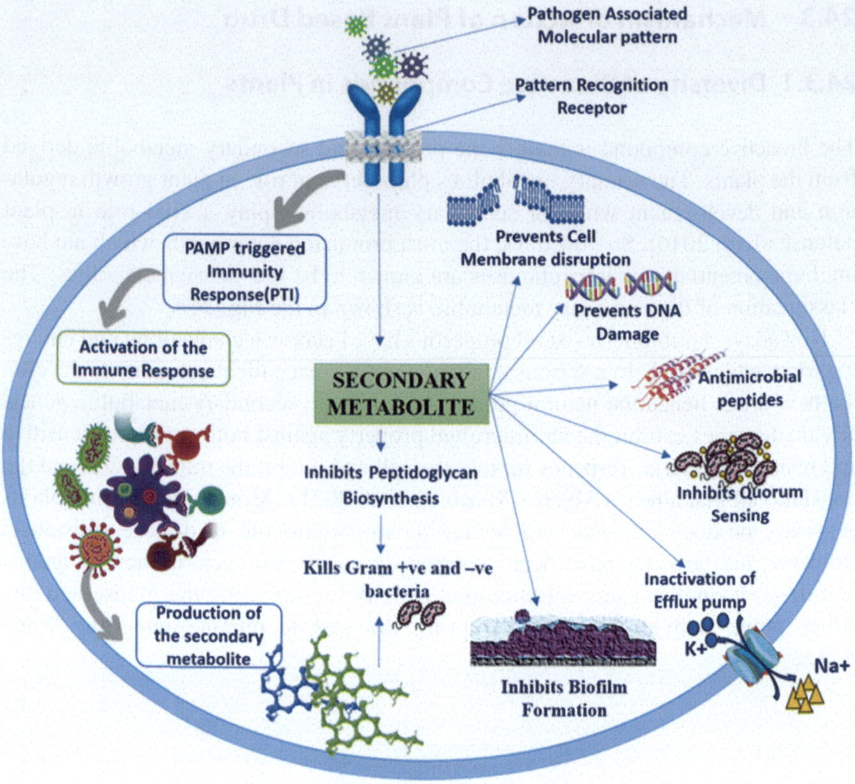

Fig. 24.6 Mechanism of action of phytocompounds as an antimicrobial agent

24.3.2 Mechanism of Plant Secondary Metabolite as an Antimicrobial Activity

Plant secondary metabolites show promising antimicrobial properties with several modes of action via the following mechanisms shown in the figure below Fig. 24.6.

The pattern recognition receptor generally shows its presence on the plasma membrane of the plant and the pathogen-associated molecular patterns (PAMPs) present in the pathogen body surface. This leads to the interaction between the plant and the pathogen's body surface via the receptors. This interaction activates the cascade of the PAMP-triggered immunity (PTI) which induces the hypersensitive response via effector-triggered immunity. This cascade of signaling leads to the production of the antimicrobial secondary metabolite which kills the infecting pathogen (Ali et al. 2021).

24.3.2.1 Berberine

Rhizoma coptidis contains the isoquinolone called berberine which works against the MRSA by inhibiting the formation of the biofilm. The study on berberine have been reported as it shows the synergistic effect when combined with the other antibiotics (Zuo et al. 2014). The combination of azithromycin and berberine leads to a reduction in the minimum inhibitory concentration by 59–96.9% and also it reduces the surface attachment (adhesion) and intracellular pathogenicity of the MRSA (Chu et al. 2016).

24.3.2.2 Allicin

Allicin belongs to the sulphur-containing bioactive compound which is obtained from the *Allium sativum*. The modification of protein in bacteria i.e. S-allylmercapto modification of the thiol done by the Allicin which leads to the reduction in the glutathione level, inactivation of the beneficial enzymes, induces the aggregation of the protein (Nakamoto et al. 2019). Allicin exhibits broad spectrum antimicrobial properties as it works against both the Gram-positive as well as Gram-negative bacteria i.e. MRSA, *Salmonella entericaserovar Typhimurium* (Müller et al. 2016).

24.3.2.3 Piperine

The alkaloid piperidine extracted from the *Piper nigrum*, *Piper longum* which shows antimicrobial efficacy against the Gram-positive and Gram-negative bacteria it acts as an inhibitor of *S. aureus* efflux pump when combined with the antibiotic ciprofloxacin (Umar et al. 2013).

24.3.2.4 Ajoene

The Ajoene is sulfur-containing bioactive compound as it exhibits antibacterial activity and works in a similar way like Allicin. Ajoene exhibits antimicrobial properties mainly it works against Gram-positive and Gram-negative bacteria like *Helicobacter pylori* and *Mycobacterium* species (Wong and Chow 2024).

24.3.2.5 Resveratrol

It is a naturally occurring polyphenolic compound that might be isolated from bananas, beans, pomeogranates, soyabeans. Resveratrol shows its potential by inhibiting the bacterial gene expression level, motility, biofilm formation, and it also interferes with the mechanism of quorum sensing over a variety of the, bacterial, viral, and fungal species (Vestergaard and Ingmer 2019).

24.3.2.6 Eugenol

It is the naturally occurring bioactive compound which might be extracted from the *Lauraceae*, *Lamiaceae*, and *Myristicaceae* families as it inhibits the biofilm formation, disruption of the cell membrane (*S. typhi*), reduces the toxin at gene expression level (*S. aureus*) (Yadav et al. 2015).

24.4 Ethical and Regulatory Considerations

The ethnobotany tells about the dynamic relation between plants and humans which has shown the emergence in some implications for the conservation of biodiversity, sustainable development. In order to overcome these major issues, intellectual property rights have become prevalent. Intellectual property rights by means of the ethnobotanical perspective provide the legal right to protect the traditional knowledge related to the phyto-based resources. The approaches based on community by means of ethnobotany and its traditional knowledge lead to the promising strategy for the protection of the rights and interests of the local communities. Some databases and digital platforms play a key role in the knowledge preservation and provide a secure and accessible repository for the sharing of knowledge related to ethnobotanical approaches. Several case studies with concern of their ethical regulation lead to the many complexities and challenges in traditional knowledge protection and their mutual benefits. The encouragement between the researchers and the local communities facilitates mutual learning, knowledge exchange and mutual benefits in the field of ethnobotanical research. For developing new drug products, the clinical research has to be placed under the ethical standards. The ethical issues that arise generally by the drug or phyto-based nanomedicines include the size of the particle, its impact on human health, ecofriendly nature. Hence there are several approaches to overcome the size of the particle i.e. Nanocurcumin with particle size of 11.3 nm produced by the alginate-curcumin nanocomposite method, PLGA nanoparticles method (Famna Roohi et al. 2020). The concept of ethics in the field of nanosciences is constantly being explored. The use of synthetic medicine has been proven it exhibit limited clinical efficiency which gives motivational importance to the plant as herbal medicine for therapeutic uses.

24.5 Conclusion and Future Prospects

Infectious diseases caused by pathogenic microorganisms pose a significant threat to global health, with alarmingly high fatality rates and the emergence of multidrug-resistant strains further exacerbating the situation. Limited efficacy of current antimicrobials highlights the need for alternative solutions. Nature provides a rich source of antimicrobial compounds across various organisms, offering the potential for combating drug-resistant infections. The chapter explores the diverse range of natural antimicrobials and their mechanisms against pathogens. By refining our understanding and developing novel therapeutic approaches, we can address the challenge of multidrug-resistant infections. Continued exploration of natural antimicrobials and interdisciplinary collaborations are crucial. Ethical considerations regarding sustainable sourcing and equitable access must be prioritized. Integrating traditional wisdom with scientific innovation holds promise for combating resilient pathogens and ensuring a healthier future.

References

Ali SA, Singh G, Datusalia AK (2021) Potential therapeutic applications of phytoconstituents as immunomodulators: pre-clinical and clinical evidences. Phytother Res 35(7):3702–3731. https://doi.org/10.1002/ptr.7068

Álvarez-Martínez FJ, Barrajón-Catalán E, Herranz-López M, Micol V (2021) Antibacterial plant compounds, extracts and essential oils: an updated review on their effects and putative mechanisms of action. Phytomedicine 90:153626. https://doi.org/10.1016/j.phymed.2021.153626

Alves de Melo Fernandes T, Rafaella Costa T, de Paula Menezes R, Arantes de Souza M, Gomes Martins CH, Junior NN, Gobbi Amorim F, Quinton L, Polloni L, Teixeira SC, Ferro AV, E., Soares, A. M., & de Melo Rodrigues Ávila, V. (2024) Bothrops snake venom L-amino acid oxidases impair biofilm formation of clinically relevant bacteria. Toxicon 238:107569. https://doi.org/10.1016/j.toxicon.2023.107569

Aminov RI (2010) A brief history of the antibiotic era: lessons learned and challenges for the future. Front Microbiol 1. https://doi.org/10.3389/fmicb.2010.00134

Bai X, Shen Y, Zhang T, Meng R, Zhang Y, Deng Y, Guo N (2023) Anti-biofilm activity of biochanin A against Staphylococcus aureus. Appl Microbiol Biotechnol 107(2–3):867–879. https://doi.org/10.1007/s00253-022-12350-x

Bao A, Zhong J, Zeng X-C, Nie Y, Zhang L, Peng ZF (2015) A novel cysteine-free venom peptide with strong antimicrobial activity against antibiotics-resistant pathogens from the scorpion *Opistophthalmus glabrifrons*: A NOVEL CYSTEINE-FREE VENOM PEPTIDE FROM THE SCORPION. J Pept Sci 21(10):758–764. https://doi.org/10.1002/psc.2801

Basile A, Sorbo S, Spadaro V, Bruno M, Maggio A, Faraone N, Rosselli S (2009) Antimicrobial and antioxidant activities of coumarins from the roots of Ferulago campestris (Apiaceae). Molecules 14(3):939–952. https://doi.org/10.3390/molecules14030939

Buttachon S, Ramos A, Inácio Â, Dethoup T, Gales L, Lee M, Costa P, Silva A, Sekeroglu N, Rocha E, Pinto M, Pereira J, Kijjoa A (2018) Bis-Indolyl Benzenoids, Hydroxypyrrolidine derivatives and other constituents from cultures of the marine sponge-associated fungus aspergillus candidus KUFA 0062. Mar Drugs 16(4):119. https://doi.org/10.3390/md16040119

Chen S-L, Yu H, Luo H-M, Wu Q, Li C-F, Steinmetz A (2016) Conservation and sustainable use of medicinal plants: problems, progress, and prospects. Chin Med 11(1):37. https://doi.org/10.1186/s13020-016-0108-7

Chu M, Zhang M, Liu Y, Kang J, Chu Z, Yin K, Ding L, Ding R, Xiao R, Yin Y, Liu X, Wang Y (2016) Role of Berberine in the treatment of methicillin-resistant Staphylococcus aureus infections. Sci Rep 6(1):24748. https://doi.org/10.1038/srep24748

Costa EV, Pinheiro MLB, Barison A, Campos FR, Salvador MJ, Maia BHLNS, Cabral EC, Eberlin MN (2010) Alkaloids from the bark of *Guatteria hispida* and their evaluation as antioxidant and antimicrobial agents. J Nat Prod 73(6):1180–1183. https://doi.org/10.1021/np100013r

da Cunha MG, de Cássia Orlandi Sardi J, Freires IA, Franchin M, Rosalen PL (2020) Antimicrobial, anti-adherence and antibiofilm activity against Staphylococcus aureus of a 4-phenyl coumarin derivative isolated from Brazilian geopropolis. Microb Pathog 139:103855. https://doi.org/10.1016/j.micpath.2019.103855

de Kraker MEA, Stewardson AJ, Harbarth S (2016) Will 10 million people die a year due to antimicrobial resistance by 2050? PLoS Med 13(11):e1002184. https://doi.org/10.1371/journal.pmed.1002184

Deng Y, Zhang M, Luo H (2012) Identification and antimicrobial activity of two alkaloids from traditional Chinese medicinal plant Tinospora capillipes. Ind Crop Prod 37(1):298–302. https://doi.org/10.1016/j.indcrop.2011.12.006

Elsayed MM, Elkenany RM, El-Khateeb AY, Nabil NM, Tawakol MM, Hassan HM (2024) Isolation and encapsulation of bacteriophage with chitosan nanoparticles for biocontrol of multidrug-resistant methicillin-resistant Staphylococcus aureus isolated from broiler poultry farms. Sci Rep 14(1):4702. https://doi.org/10.1038/s41598-024-55114-5

Famna Roohi NK, Bhaskaran M, Babu S, Gowda DV (2020) Mitochondrial drug delivery for the liver diseases: comprehensive review. Int J Res Pharm Sci 11(1):1172–1181. https://doi.org/10.26452/ijrps.v11i1.1955

Girennavar B, Cepeda ML, Soni KA, Vikram A, Jesudhasan P, Jayaprakasha GK, Pillai SD, Patil BS (2008) Grapefruit juice and its furocoumarins inhibits autoinducer signaling and biofilm formation in bacteria. Int J Food Microbiol 125(2):204–208. https://doi.org/10.1016/j.ijfoodmicro.2008.03.028

Gultom ES, Harahap U, Suryanto D, Sipahutar H, Restuati M (2024) Antibacterial activity of ethanol extract of the marine sponge (Agelas sp) symbiont Bacillus cereus MH997647 IA5 against Klebsiella pneumoniae ESBL. Tropical journal of. Nat Prod Res 8(1). https://doi.org/10.26538/tjnpr/v8i1.37

Hu J, Shi X, Chen J, Mao X, Zhu L, Yu L, Shi J (2014) Alkaloids from Toddalia asiatica and their cytotoxic, antimicrobial and antifungal activities. Food Chem 148:437–444. https://doi.org/10.1016/j.foodchem.2012.12.058

Huynh NTN, Pham NT, Tran CV, Nguyen DT, Dinh MTN, Nguyen VT, Nguyen MH, Ngo HV, Ho AN, Cheng J, Suh J-W, Tran SV, Nguyen TKN, Tran TTP (2024) Screening for Antimycobacterial activity of Actinomycetes collected in Vietnam—isolation and activity of metabolites from *Streptomyces alboniger* (A121). Nat Prod Commun 19(1):1934578X231224994. https://doi.org/10.1177/1934578X231224994

Jara LM, Pérez-Varela M, Corral J, Arch M, Cortés P, Bou G, Aranda J, Barbé J (2016) Novobiocin inhibits the antimicrobial resistance acquired through DNA damage-induced mutagenesis in Acinetobacter baumannii. Antimicrob Agents Chemother 60(1):637–639. https://doi.org/10.1128/AAC.01810-15

Kačániová M, Čmiková N, Vukovic NL, Verešová A, Bianchi A, Garzoli S, Ben Saad R, Ben Hsouna A, Ban Z, Vukic MD (2024) Citrus Limon essential oil: chemical composition and selected biological properties focusing on the antimicrobial (i*n vitro*, *in situ*), antibiofilm, insecticidal activity and preservative effect against salmonella enterica inoculated in carrot. Plan Theory 13(4):524. https://doi.org/10.3390/plants13040524

Kala CP, Dhyani PP, Sajwan BS (2006) Developing the medicinal plants sector in northern India: challenges and opportunities. J Ethnobiol Ethnomed 2(1):32. https://doi.org/10.1186/1746-4269-2-32

Kelishomi FZ, Nikkhahi F, Amereh S, Ghayyaz F, Marashi SMA, Javadi A, Shahbazi G, Khakpour M (2024) Evaluation of the therapeutic effect of a novel bacteriophage in the healing process of infected wounds with Klebsiella pneumoniae in mice. J Glob Antimicrob Resist 36:371–378. https://doi.org/10.1016/j.jgar.2024.01.018

Lemos ASO, Florêncio JR, Pinto NCC, Campos LM, Silva TP, Grazul RM, Pinto PF, Tavares GD, Scio E, Apolônio ACM, Melo RCN, Fabri RL (2020) Antifungal activity of the natural coumarin scopoletin against planktonic cells and biofilms from a multidrug-resistant Candida tropicalis strain. Front Microbiol 11:1525. https://doi.org/10.3389/fmicb.2020.01525

Lessing M, Raftery M (1998) Vancomycin-resistant Staphylococcus aureus. Lancet 351(9102):601–602. https://doi.org/10.1016/S0140-6736(05)78596-2

Liu J, Zhao S-Y, Hu J-Y, Chen Q-X, Jiao S-M, Xiao H-C, Zhang Q, Xu J, Zhao J-F, Zhou H-B, Zheng J-X, Sun P-H (2023) Novel coumarin derivatives inhibit the quorum sensing system and iron homeostasis as antibacterial synergists against *Pseudomonas aeruginosa*. J Med Chem 66(21):14735–14754. https://doi.org/10.1021/acs.jmedchem.3c01268

Martin ALAR, De Menezes IRA, Sousa AK, Farias PAM, dos Santos FAV, Freitas TS, Figueredo FG, Ribeiro-Filho J, Carvalho DT, Coutinho HDM, Fonteles MMF (2023) *In vitro* and in silico antibacterial evaluation of coumarin derivatives against MDR strains of Staphylococcus aureus and Escherichia coli. Microb Pathog 177:106058. https://doi.org/10.1016/j.micpath.2023.106058

Mittal RP, Jaitak V (2019) Plant-derived natural alkaloids as new antimicrobial and adjuvant agents in existing antimicrobial therapy. Curr Drug Targets 20(14):1409–1433. https://doi.org/10.2174/1389450120666190618124224

Mohammed El-Sawi N, Mohamed Hamed M, Hefny Gad M, Hefny AA, Y. (2024) Comparative assessment of the antibacterial potential of four soft corals collected from the Red Sea (Egypt) against selected pathogenic bacteria. Sohag J Sci 9(2):121–126. https://doi.org/10.21608/sjsci.2023.240439.1134

Mohtar M, Johari SA, Li AR, Isa MM, Mustafa S, Ali AM, Basri DF (2009) Inhibitory and resistance-modifying potential of plant-based alkaloids against methicillin-resistant Staphylococcus aureus (MRSA). Curr Microbiol 59(2):181–186. https://doi.org/10.1007/s00284-009-9416-9

Müller A, Eller J, Albrecht F, Prochnow P, Kuhlmann K, Bandow JE, Slusarenko AJ, Leichert LIO (2016) Allicin induces thiol stress in bacteria through S-Allylmercapto modification of protein cysteines. J Biol Chem 291(22):11477–11490. https://doi.org/10.1074/jbc.M115.702308

Murray CJL, Ikuta KS, Sharara F, Swetschinski L, Robles Aguilar G, Gray A, Han C, Bisignano C, Rao P, Wool E, Johnson SC, Browne AJ, Chipeta MG, Fell F, Hackett S, Haines-Woodhouse G, Kashef Hamadani BH, Kumaran EAP, McManigal B et al (2022) Global burden of bacterial antimicrobial resistance in 2019: a systematic analysis. Lancet 399(10325):629–655. https://doi.org/10.1016/S0140-6736(21)02724-0

Nakamoto M, Kunimura K, Suzuki J, Kodera Y (2019) Antimicrobial properties of hydrophobic compounds in garlic: allicin, vinyldithiin, ajoene and diallyl polysulfides (review). Exp Ther Med 19:1550. https://doi.org/10.3892/etm.2019.8388

O'Neill J (2014) Antimicrobial resistance: tackling a crisis for the health and wealth of nations. Rev Antimicrob Resist

Pathak D, Mazumder A (2023) Potential of flavonoids as promising Phytotherapeutic agents to combat multidrug-resistant infections. Curr Pharm Biotechnol 25. https://doi.org/10.2174/0113892010271172231108190233

Peng D, Chen A, Shi B, Min X, Zhang T, Dong Z, Yang H, Chen X, Tian Y, Chen Z (2018) Preliminary study on the effect of brazilin on biofilms of Staphylococcus aureus. Exp Ther Med 16:2108. https://doi.org/10.3892/etm.2018.6403

Phetkul U, Vongkul A, Chaichan K, Paosen S, Voravuthikunchai SP, Daus M, Maungchanburi S (2023) Isolation and structural elucidation of Coumarin and flavonoids from Citrus grandis Linn. (Tubtim Siam Pomelo) and their biological activities. Trends Sci 21(2):7129. https://doi.org/10.48048/tis.2024.7129

Pool-Yam L, Ramón-Sierra J, Oliva AI, Zamora-Bustillos R, Ortiz-Vázquez E (2024) Effect of conA-unbound proteins from Melipona beecheii honey on the formation of Pseudomonas aeruginosa ATCC 27853 biofilm. Arch Microbiol 206(1):54. https://doi.org/10.1007/s00203-023-03783-7

Rao MR, Palada MC, Becker BN (2004) Medicinal and aromatic plants in agroforestry systems. Agrofor Syst 61–62(1–3):107–122. https://doi.org/10.1023/B:AGFO.0000028993.83007.4b

Saga T, Yamaguchi K (2008) History of antimicrobial agents and resistant bacteria. J Jpn Med Assoc 137(3):513–517

Savoia D (2012) Plant-derived antimicrobial compounds: alternatives to antibiotics. Future Microbiol 7(8):979–990. https://doi.org/10.2217/fmb.12.68

Song Y, Wang J, Liu X, Yu S, Tang X, Tan H (2024) LC-AMP-F1 derived from the venom of the wolf spider Lycosa coelestis, exhibits antimicrobial and antibiofilm activities. Pharmaceutics 16(1):129. https://doi.org/10.3390/pharmaceutics16010129

Sopjani M, Falco F, Impellitteri F, Guarrasi V, Nguyen Thi X, Dërmaku-Sopjani M, Faggio C (2024) Flavonoids derived from medicinal plants as a COVID-19 treatment. Phytother Res 38(3):1589–1609. https://doi.org/10.1002/ptr.8123

Stabili L, Piraino S, Rizzo L (2024) The Mediterranean Zoanthid Parazoanthusaxinellae as a novel source of antimicrobial compounds. J Mar Sci Eng 12(2):354. https://doi.org/10.3390/jmse12020354

Taibi M, Elbouzidi A, Haddou M, Loukili EH, Bellaouchi R, Asehraou A, Douzi Y, Addi M, Salamatullah AM, Nafidi H-A, Bourhia M, Dauelbait M, Guerrouj BE, Chaabane K (2024) Chemical profiling, antibacterial efficacy, and synergistic actions of *Ptychotis verticillata* Duby essential oil in combination with conventional antibiotics. Nat Product Commun 19(1):1934578X231222785. https://doi.org/10.1177/1934578X231222785

Teoh ES (2016) Secondary metabolites of plants. In: Teoh IES (ed) Medicinal orchids of Asia. Springer International Publishing, pp 59–73. https://doi.org/10.1007/978-3-319-24274-3_5
Umar S, Golam Sarwar AHM, Umar K, Ahmad N, Sajad M, Ahmad S, Katiyar CK, Khan HA (2013) Piperine ameliorates oxidative stress, inflammation and histological outcome in collagen induced arthritis. Cell Immunol 284(1–2):51–59. https://doi.org/10.1016/j.cellimm.2013.07.004
Upadhyay HC (2021) Coumarin-1,2,3-triazole hybrid molecules: an emerging scaffold for combating drug resistance. Curr Top Med Chem 21(8):737–752. https://doi.org/10.2174/1568026621666210303145759
Vestergaard M, Ingmer H (2019) Antibacterial and antifungal properties of resveratrol. Int J Antimicrob Agents 53(6):716–723. https://doi.org/10.1016/j.ijantimicag.2019.02.015
Wang Z, Li S, Ge S, Lin S (2020) Review of distribution, extraction methods, and health benefits of bound phenolics in food plants. J Agric Food Chem 68(11):3330–3343. https://doi.org/10.1021/acs.jafc.9b06574
Widelski J, Luca S, Skiba A, Chinou I, Marcourt L, Wolfender J-L, Skalicka-Wozniak K (2018) Isolation and antimicrobial activity of coumarin derivatives from fruits of Peucedanum luxurians Tamamsch. Molecules 23(5):1222. https://doi.org/10.3390/molecules23051222
Wong YY, Chow Y (2024) Exploring the potential of spice-derived phytochemicals as alternative antimicrobial agents. eFood 5(1):e126. https://doi.org/10.1002/efd2.126
Wu X, Ma G-L, Chen H-W, Zhao Z-Y, Zhu Z-P, Xiong J, Yang G-X, Hu J-F (2023) Antibacterial and antibiofilm efficacy of the preferred fractions and compounds from Euphorbia humifusa (herba euphorbiae humifusae) against Staphylococcus aureus. J Ethnopharmacol 306:116177. https://doi.org/10.1016/j.jep.2023.116177
Wu X, Wang H, Xiong J, Yang G-X, Hu J-F, Zhu Q, Chen Z (2024) Staphylococcus aureus biofilm: formulation, regulatory, and emerging natural products-derived therapeutics. Biofilms 7:100175. https://doi.org/10.1016/j.bioflm.2023.100175
Yadav MK, Chae S-W, Im GJ, Chung J-W, Song J-J (2015) Eugenol: a phyto-compound effective against methicillin-resistant and methicillin-sensitive Staphylococcus aureus clinical strain biofilms. PLoS One 10(3):e0119564. https://doi.org/10.1371/journal.pone.0119564
Zuo G-Y, Li Y, Wang T, Han J, Wang G-C, Zhang Y-L, Pan W-D (2011) Synergistic antibacterial and antibiotic effects of Bisbenzylisoquinoline alkaloids on clinical isolates of methicillin-resistant staphylococcus aureus (MRSA). Molecules 16(12):9819–9826. https://doi.org/10.3390/molecules16129819
Zuo G-Y, Li Y, Wang G-C, Li Z-S, Han J (2014) Synergistic effects of berberines with antibiotics on clinical multi-drug resistant isolates of methicillin-resistant Staphylococcus aureus (MRSA). Med Chem Res 23(5):2439–2444. https://doi.org/10.1007/s00044-013-0844-5

Exploring the Potential of Phytochemicals: A Herbal Boon Against the MDR Bacteria Associated with the Chronic Wounds

25

Poonam Pal, M. Sree Vidhya, E. Karthikeyan, N. S. Disha, Niravsinh M. Rathod, and Sanjeev Kumar

Abstract

The Current study aimed at potential of phytochemical, applications in herbs, used to treat and prevent many MDR Bacteria which is associated with chronic wound. Multidrug-resistant (MDR) pathogens are a global threat and have created problems in providing adequate treatment of many infectious diseases. Although the conventional antimicrobial agents are quite effective against several pathogens, yet there is a need for more effective antimicrobial agents against MDR pathogens. Herbal drugs and phytochemicals have been used for their effective antimicrobial activity since ancient times and there is an increasing trend for the development of plant-based natural drugs for the preventing and cures of chronic wounds. The phytochemical investigations of traditionally used medicinal plants have shown the diversified useful compound classes like alka-

P. Pal · S. Kumar (✉)
Department of Dravyaguna, Faculty of Ayurveda, Institute of Medical Sciences, Banaras Hindu University, Varanasi, Uttar Pradesh, India
e-mail: kumarsanjeevdg@bhu.ac.in

M. S. Vidhya
Department of Pharmacology, Saveetha Institute of Medical and Technical Sciences, Chennai, Tamil Nadu, India

E. Karthikeyan
Department of Pharmaceutical Chemistry, Saveetha Institute of Medical and Technical Sciences, Chennai, Tamil Nadu, India

N. S. Disha
Department of Analysis, Saveetha Institute of Medical and Technical Sciences, Chennai, Tamil Nadu, India

N. M. Rathod
Parul Institute of Pharmacy, Faculty of Pharmacy, Parul University, Waghodia, Vadodara, Gujarat, India

M. Gangwar, G. Nath (eds.), *Emerging Paradigms for Antibiotic-Resistant Infections: Beyond the Pill*, https://doi.org/10.1007/978-981-97-5272-0_25

loid, phenolic, tannin, terpene, steroid, flavonoid, glycoside, and fatty acid. These phytochemical components play an important role in scavenging free radicals, fighting infection, and promoting a faster chronic wound-healing process. Phytochemicals can serve as a cost-effective alternative therapeutic agent for human diseases.

Keywords
Herbs · Phytochemicals · MDR bacteria · Antibacterial activity · Chronic wound

25.1 Introduction

Nowadays, Chronic wound is a major global health issue, millions of patients worldwide and cause an economic major problem for whole healthcare system. In the current time herbs or medicinal plants and their derivatives, such as isolated compounds, essential oils, have been considered by researchers and clinicians to be sources for the production of new agents which effective in chronic wounds (Salas-Oropeza et al. 2020). Around 80% of the global population, according to estimates from the World Health Organization (WHO), obtains their primary healthcare from traditional methods, with 85% of this population using plants. Given the difficulties these people encounter in accessing conventional methods of care, the World Health Organization recommends their widespread use as a means of sustaining health and advocates for the enactment of laws incorporating them into the official health systems of some nations that are members. WHO estimates indicate that improper wound healing may be the cause of almost five million deaths annually. Herbal-based extracts and isolates promote tissue regrowth in many ways, and this typically combines to improve the overall healing steps. Phytochemicals have shown great potential for avoiding the development and treatment of infections of microbes on wounds (Ayaz et al. 2015). Herbs present the highest amount of antioxidant, phytochemical compounds these are encourage clotting of blood, fight infection, and promote the healing of wounds through preventing lipid peroxidation, which increases collagen fibril viability and prevents cell injury (Getie et al. 2002). Medicinal plants rich in polyphenols such as flavonoid, phenols, etc. are reported which is promoted mainly ulcer healing activity due to their astringent agents, antimicrobial compounds, and free radical scavenging properties (Zohra et al. 2019; Deshmukh et al. 2009). The use of herbal products as potentially helpful agents in the wound-healing process has been the target of increasing number of study investigations. Primary benefits of these herbal treatments are their affordable cost, widespread availability, and not much adverse effects. A wide range of bioactive phytochemicals are found in plants, primarily in the polyphenols such as alkaloids, carotenoids, flavonoids, tannins, terpenoids, saponins, and phenolic compound categories (Edeoga et al. 2005). Chronic wounds are caused by a variety of multidrug-resistant bacteria and fungi. The most common bacterial pathogens found in the wound such as Staphylococcus spp., Corynebacterium spp., Pseudomonas spp., and anaerobes such as *Anaerococcus* spp. and *Finegoldia* spp. (Kalan et al. 2019). The development of chronic injuries can be influenced by many factors, including diversity, and the presence of particular pathogenic species. Certain streptococci of the group A and *S. aureus* strains have been demonstrated to induce critical infections and to set off effective inflammation reactions, which can hinder the healing process (Daeschlein 2013). Along with producing many virulence factors, such as the production of toxins and enzymes that promote additional tissue injury, they create biofilms with antiphagocytic properties. Seventy-eight percent of surgical patients die from nosocomial conditions, of which one-third are chronic wound infections. The majority of nosocomial infections are caused by endogenous bacterial flora, though many critically ill patients ultimately acquire resistant bacterial strains in their colonization (Guo and DiPietro 2010). Furthermore, large numbers of *S. aureus* and *P. aeruginosa* can impact wound healing since these microbes are commonly resistant to antibiotics. They form biofilms with antiphagocytic activity and wounds can be classified as vascular ulcers (such as venous and pressure ulcers (PUs). Excessive inflammation, chronic infections, the development of drug-resistant microbial biofilms, and the inability of dermal and/or epidermal cells to respond to repairing stimuli are all characteristic features shared by all these wounds (Nunan et al. 2014; Woo et al. 2007). Largest amount of collection samples from venous leg ulcers, Pressure ulcer and evaluating bacterial diversity using 16s rDNA sequencing, commonly found bacteria in these sites mention in Table 25.1 (Versey et al. 2021). Venous Leg Ulcers, when the legs' veins are not able to return blood to the heart as effectively as they should. Blood thus builds up through the blood vessels, increasing pressure. A buildup stress then prevents nourishment and oxygen from reaching the tissues, which results in harm to tissues with cell death, which eventually creates a wound. Most venous leg ulcers arise above the ankle and heal poorly without care (Ontario Health (Quality) 2021). The most accurate description of a pressure ulcer is probably "distortion or deformation that occurred damage". Pressure ulcers are a form of injury that happens beneath the skin when an area of skin is subjected to constant stress for a long period of time, resulting in tissue ischemia, a lack of nourishment and oxygen delivery to the tissues, and, eventually, tissue necrosis (Gebhardt 2002). Mainly causes of pressure ulcers are two way one is the direct cause and other one is the indirect causes. Direct causes such as pressure, shear, immobility, friction, and loss of sensation. And in indirect causes are health conditions like diabetes, renal failure, and heart failure, etc., poor nutrition, aging skin, and mobility problems. Approximately 25% of pressure ulcers are seen in lower extremities (Cannon and Cannon 2004).

The most commonly occurring bacterial species across all types of wounds, according to their findings, were *Finegoldia magna*, *S. aureus*, *S. epidermidis*, *E. faecalis*, *P. aeruginosa*, and *S. maltophilia*. These anaerobic bacteria, which include *Finegoldia* spp. and *Anaerococcus* spp., are highly prevalent in wound samples. 24% and 25%, respectively, as well as a diversity of commensal bacteria such as *Staphylococci* and *Corynebacteria* across wound types (Wolcott et al. 2016). Many antibiotics are used in patients that time in has critical infections, most antibiotic are resistant to bacteria alternative use of herbs in ancient times for the treatment of bacterial infections and other diseases also they found good effect with less

Table 25.1 Most common bacterial species associated with chronic wounds in different site

(i) Venous leg ulcers

Aerobes		Anaerobes	
Gram-positive	Gram-negative	Gram-positive	Gram-negative
Actinotignum schaalii	*Achromobacter xylosoxidans*	*Anaerococcus vaginalis*	*Bacteroides tectus*
Alcaligenes faecalis	*Acinetobacter baumannii*	*Finegoldia magna*	*Flavobacterium succinicans*
Brevibacterium casei	*Citrobacter spp.*	*Peptoniphilus harei*	*Fusobacterium gonidiaformans*
Corynebacterium	*Delftia acidovorans*	*Peptostreptococcus assacharolyticus*	
Amycolatum	*Enterobacter cloacae*	*Cutibacterium acnes*	
C. jeikeium	*E. hormaechei*		
C. simulans	*Klebsiella oxytoca*		
C. striatum	*Proteus spp.*		
C. tuberculostearicum	*Pseudomonas aeruginosa*		
Enterococcus faecalis	*P. fluorescens*		
Staphylococcus aureus	*Serratia nematodiphila*		
S. epidermidis	*Stenotrophomonas maltophilia*		
S. haemolyticus			
S. lugdunensis			
S. pettenkoferi			
Streptococcus agalactiae			

(ii) Pressure ulcers

Aerobes		Anaerobes	
Gram-positive	Gram-negative	Gram-positive	Gram-negative
Corynebacterium jeikeium	*Acinetobacter baumannii*	*Allobaculum* spp.	*B. fragilis*
C. striatum	*Delftia acidovorans*	*Anaerococcus vaginalis*	*Dialister* spp.
C. tuberculostearicum	*Enterobacter hormaechei*	*Eubacterium dolichum*	*Fusobacterium nucleatum*
Enterococcus faecalis	*Escherichia* spp.	*Finegoldia magna.*	*Prevotella bivia*
Staphylococcus aureus	*P. mirabilis*	*Peptococcus* spp.	
S. epidermidis	*P. aeruginosa*	*Peptoniphilus ivorii*	
S. haemolyticus	*Serratia* spp.		
S. lugdunensis			
S. agalactiae			
S. dysgalactiae			

side effect also. Such instances, the pathogen's resistance profile and the effectiveness of the treatment are directly correlated (Hernandez 2006). The first-line treatment for cases of resistant drug frequently involves an antibiotic like vancomycin, which is followed by second-line medications like quinupristin-dalfopristin, linezolid, and daptomycin. Macrolides, β-lactams (such as cephalosporin, amoxicillin), penicillinase-resistant penicillin, trimethoprim-sulfamethoxazole, fluoroquinolone, tigecycline, and clindamycin are additional antibiotics that are used to treat chronic wounds (Lipinski et al. 2012). The diverse microbial flora on wounds complicates patient therapy, particularly when a substantial number of resistant bacteria are identified on chronic wound biofilms. The synthesis of biofilm matrix, horizontal gene transfer, routine antibiotic usage, and microbial mutation are all contributing factors to the rise in resistance. The manufacturing of biofilm matrix, horizontal gene transfer, routine antibiotic usage, and microbial mutation are all contributing factors to the rise in resistance. One major issue in the management of chronic wounds, and resistance of biofilm-forming bacteria to antibiotics (Stewart and Costerton 2001).

25.2 Abnormality in Chronic Wound

Chronic wounds are wounds that are unable to heal or heal in long time (Mustoe et al. 2006). Chronic wounds are more common in all cases of wounds. A chronic wound is mostly caused by direct microbiological contact with the wound, as well as various abnormalities: hypoxia, scarring and fibrosis, infection/biofilm, hyperproliferative epidermis, persistent prolonged hypoxia, Scarring and fibrosis, infection/biofilm, hyperproliferative epidermis, persistent inflammation, impaired angiogenesis, fibrin cuffs, elevated MMPs, malnutrition, platelet-derived growth factor, basic fibroblast growth factor, epidermal growth factor, and transforming growth factor β, there is a drop in tissue growth factors and senescent cells excess (Falanga 2004; Singer and Clark 1999).

25.3 Chronic Wound Healing with Herbs

Since the beginning of time, several types of wounds can be treated and managed with the help of herbs and herbal-based ingredients. Throughout, the history of mankind, medications, and herbal treatments have been crucial in the treatment of disease (Sharma et al. 2021). Herbal remedies can be used to disinfect wounds, debride wounds, and provide a moist environment that encourages the formation of a suitable natural healing climate. Many plants are utilized in traditional cultures as treatments for burns, wounds, and cuts. These are some of the main phytochemicals that may be found in plants and which are known to successfully reduce inflammatory and promote healing (Fig. 25.1) (Yuan et al. 2016).

Fig. 25.1 Wound healing herbs are (A) *Cassia obtusifolia*, (B) *Achilleamillefolium* L., (C) *Cymbopogon Proximus*, (D) *Ormocarpum cochinchinense*, (E) *Casearia sylvestris*, (F) *Bletillastriata*, (G) *Berberis lyceum*, (H) *Azadirachta Indica*, (I) *Chamomilla Recutita*, (J) *Malvasylvestris*

25.4 Commonly Used Herbal Plants for MDR Bacteria

25.4.1 *Achilleamillefolium* L.

Achillea millefolium L. or Yarrow, is a member of the Asteraceae family that has been used traditionally in India and other nations to cure liver disorders, wound healing, and other medical issues. Many research (Saeidnia et al. 2011; Tadić et al. 2017; Benedek et al. 2007) have clearly shown the benefits of *A. millefolium* essential oil, including the inhibition of proteases, treatment of skin irritation, and prevention of sunburns. Many active components, including borneol, α-terpineol, β-pinene, camphor, and eucalyptol, were found in *Achilleamillefolium* Essential Oil (AMEO) by GC-Mass analysis (Guo et al. 2021). These components benefit in the healing of wounds. The isolated essential oil has both antioxidant and antibacterial properties. Which is linked to MDR bacteria that cause chronic wounds, including *pseudomonas aeruginosa*, *Acinetobacterlwoffii*, *Mycobacterium smegmatis*, *Staphylococcus aureus*, and *Streptococcus pneumoniae.* The most widespread bacteria that cause wounds are Staphylococcus aureus and *Pseudomonas aeruginosa* (Candan et al. 2003). According to Mazandarani et al. (2009), *S. epidermidis* and *S. aureus* had the strongest antibacterial activity. In *P. aeruginosa* had a larger inhibitory zone than other bacteria when tested against different extracts (hexane, petroleum ether, and methanol) from the aerial section of *A. millefolium* (Stojanović et al. 2005). Additionally, a study was conducted against *Salmonella typhimurium*, *Bacillus cereus*, *Pseudomonas aeruginosa*, *Escherichia coli*, *Enterococcus faecalis*, *Staphylococcus aureus*, and *Staphylococcus epidermidis*. The antibacterial activity of an ethanol extract made from the aerial portions of *A. millefolium* was tested against *B. cereus*, *C. albicans*, *P. aeruginosa*, *S. enteritidis, and E. coli*. The bacteria that showed the greatest MIC value were *S. enteritidis and B. cereus*. The other three examined strains, however, showed no signs of activity (Kokoska et al. 2002). According to these investigations, *A. millefolium* is one of the most potentially effective alternatives for treating microbial infections because it has been shown to have ethnopharmacological uses.

25.4.2 *Bletillastriata*

This plant, whose belongs to the *Orchidaceae* family, has traditional Chinese medicine uses. *Bletillastriata* is used to treat wounds, burns, bruises, bleeding, ulcers, and parched skin in addition to promoting wound healing (He et al. 2016). It has been used for thousands of years in Chinese pharmacopeias to treat trauma-related bleeding, ulcers, edema, and chapped skin. Its astringency has a significant impact on hemostasis and analgesia as well. Medical practitioners in Korea and Japan have used *Bletillastriata* to treat parasitic illnesses, whooping cough, stomach/duodenal ulcers, and tuberculosis. *B. striata* contains polysaccharides, bibenzyls, phenanthrenes, triterpenoids and its saponins, and steroids (He et al. 2016; Rhee et al. 1982).

25.4.3 *Casearia sylvestris*

Casearia sylvestris is belonging to the Family of Salicaceae, it is a popular herbal medicine. Leaves of *C. sylvestris*, contain many chemical constituents, it is used for the treatment of wound healing in different areas of Brazil (Ferreira et al. 2011). Which is predominantly composed of volatile oils, tannins, and triterpene content (Bueno et al. 2016), and are responsible for a wide range of biological activities, including antioxidant and cytotoxic effects (Albano et al. 2013). Different extracts (ethanolic, hydroethanolic, and aqueous extracts) and bioactive chemicals from *C. sylvestris* have been proven in pharmacology studies to have antitumoral, antiulcerogenic, anti-inflammatory, and wound-healing activities. Ellagic acid and other casearin-like clerodanediterpenes found in *C. sylvestris* leaves are assumed to be responsible for their anti-inflammatory and wound-healing abilities. Both in vitro and in vivo experiments on *C. sylvestris* extracts and extracted diterpenes showed low toxicity levels (Maistro et al. 2004; Basile et al. 1990).

Researchers utilized a biofilm coated with *C. sylvestris* extract to treat a rat burn wound model. Notable fibroblast proliferation, epithelialization, neovascularization, and healing area extension were seen. Based on these findings, *C. sylvestris* might be helpful in treating second-degree scald burn injuries (Albano et al. 2013). When *Caseariasylvestris* leaf extract was tested on wound healing in beef cattle, it was reported to improve the macroscopic appearance of cutaneous lesions in cattle during the first two treatment days (Lipinski et al. 2012). To create wound-healing membranes, an alcohol extract of the *Caseariasylvestris* leaf (diterpene concentrated fraction and casearin J) with anti-inflammatory and wound-healing properties were combined with an angiogenic latex generated from natural plant sources. Permeability and retention studies revealed cutaneous penetration of phenolic compounds from the membrane with ethanolic extract, as well as casearin-like clerodanediterpenes from all membranes (Carvalho et al. 2018).

25.4.4 *Malvasylvestris*

Malvasylvestris (MS), sometimes known as Mallow is a plant that is a member of the Malvaceae family. It grows in Europe, North Africa, and Asia, and folk and traditional medicine use it worldwide (Contardi et al. 2022). Major phytochemicals such as polysaccharides, carotenoids, polyphenols, fatty acids, ascorbic acid, and tocopherol, as well as essential oil, coumarins, and terpenoids, have all been extracted in varying concentrations from the edible part of the *M. sylvestris* plant (Cutillo et al. 2006; Pirbalouti et al. 2012). According to Pirbalouti et al. (2010), plants have antibacterial, anti-inflammatory, and antioxidant qualities. Benefits because of a number of active compounds found in its flowers (anthocyanins and anthocyanidins), leaves have terpenoids, tocopherols, and coumarins, which are derivatives of hydroxycinnamic acid), sterols, and fatty acids. Due to their demonstrated anti-inflammatory and antioxidant properties, these substances may prove beneficial in the management of chronic wounds, which are primarily marked by a higher level of free radicals and alterations in the inflammatory response (Bueno et al. 2012).

25.4.5 *Ormocarpum cochinchinense*

One of the significant and commonly valued medicinal plant in the Fabaceae family is *Ormocarpum cochinchinense*. There have been reports of this plant in the Eastern Ghats (Pullaiah and Sri Ramamurthy 2001). Amongst these phytochemicals such as flavonoids, alkaloids, phenolic groups, tannins, etc. are the most significant role. Phytochemical screening was done on leaf extracts from *Ormocarpum cochinchinense* using various solvents. Bioactive compounds including terpenoids, alkaloids, glycosides, steroids, phenols, tannins, flavonoids, and saponins were detected in leaf extracts using conventional techniques. The most phytochemicals were found in the aqueous extract, which was followed by the methanolic extract among the several extracts examined. Each of the following extracts included alkaloids: methanol, ethyl acetate, ethanol, and aqueous. Alkaloids in the leaf sample were determined and the results showed. Betacyanin and Cardiac glycosides are reported in ethanol, water, methanol and petroleum ether. Yet only two extracts—methanol and aqueous—contained flavonoids. Methanol, ethanol, chloroform, and aqueous extracts were all found to contain saponins. All of the solvent extracts contained no anthocyanins or glycosides. Nearly all six solvent extracts included phenols, and the plant's phenolic components suggest that it may have potential applications as an anti-microbial agent as well as bone fracture also (Pazhanisamy and Ebenezer 2013). MeOH solvent reaction exhibited the strongest antibacterial activity against all human infections, according to research (Thirumal 2019). In a combination of hydrogel and natural plant leaf extract *Ormocarpum cochinchinense* (OC), synthetic polymers polyvinyl alcohol (PVA) and polyvinyl pyrrolidone (PVP) were used as a biodegradability and biocompatibility agent in wound dressing applications (Narayanan et al. 2023). Using a solvent-casting technique, a composite film (CHPCHO) comprising carboxymethyl cellulose (CMC), hydroxypropyl methylcellulose (HPMC), cephalexin, honey, and *Ormocarpum cochinchinense* leaf extract was created using polyvinyl alcohol (PVA) was produced using the solvent-casting technique for wound dressing with success. Antibacterial efficacy against Gram-positive bacteria, specifically *S. aureus*. Potential applications for the CHPCHO composite film include wound dressing and antimicrobial properties also (Narayanan et al. 2024).

25.4.6 *Azadirachta Indica*

Azadirachta Indica is a frequently utilized medicinal herb with a wide range of natural activity. *It is a popular home cure in Ayurveda for several ailments and active ingredients, such as nimbidin, nimbin, and nimbidol, have antibacterial, anti-inflammatory, antifungal, and antiviral properties that help to speed up the healing process. According to* Subapriya and Nagini *(*2005*), high concentration of ingredients that are essential for wound healing during the proliferation phase.* Susceptible to *Staphylococcus aureus* and *Escherichia coli* (Mandal et al. 2012; Anbarasan and Bourdon 2019). *A. indica* extract, silk fibroin, and guar gum were added together to make Neem hydrogels (N-HG) with and without silk fibroin (N-SFB-HG). The components are entrapped by the gel-forming process. This suggests that in order to guarantee compliance for patients and the medical sector, the N-SFB-HG may be employed in the long run to develop innovative wound-healing treatments (Nasrine et al. 2023). *Azadirachta Indica* leaf extract is used to treat wound infections. Several substances that are helpful in treating resistant bacteria and mono- and poly-microbial diseases may be found in neem leaves (Hannan et al. 2014). Different extracts of *Azadirachta Indica* stem bark (water, ethanol-water, and ethanol) are utilized to evaluate the healing of wounds potential, as well as different factors including contraction of the wound, hydroxyproline, DNA content, total protein, nitric oxide content, along with strength are estimated in both excision and incision models. An excellent dressing agent for wounds is water extract from the stem bark of *Azadirachta Indica* (Maan et al. 2017). *Azadirachta Indica* extract combined with carbopol hydrogel is used to treat chronic burn wounds. The hydrogel preparation process was optimized using response surface methodology, in which the concentrations of carbopol, bacterial cellulose, and crosslinking agent were varied and various responses were recorded, including swelling percentage, entrapment efficacy, and water retention capacity (Choudary et al. 2023).

25.4.7 *Chamomilla Recutita*

Most widely used medicinal herbs is *Chamomilla Recutita*, member of the Asteraceae family. The medicinal properties of chamomile are attributed to many categories of potent compounds, such as spiroethers, flavonoids, alpha-bisabolol, and bisabolol oxides. These substances have anti-inflammatory, antibacterial, and antifungal qualities and also proved non-cytotoxic and able to improve the healing of experimental wounds on rats (Gupta et al. 2010). The aqueous extract used increased wound contraction, as well as wound-breaking strength and hydroxyproline content. In rats, chamomile extract applied as rubbing oil showed promise in hastening the healing of burn injuries. In a rat linear incisional wound model, topically applied aqueous extract has the ability to promote wound healing. According to Martins et al. (2009), given in animals treatment showed quicker wound healing. Acute otitis externa can be treated with the essential oil and extract which exhibited activity against three strains of *S. aureus* and the strains of *Candida* (Nogueira et al. 2008).

25.4.8 *Cassia obtusifolia*

Anthraquinones, terpenoids, flavonoids, and lipids are among the groups of biologically active metabolites that have been found in *C. obtusifolia* L (Anbarasan and Bourdon 2019). Medicinal herbs have been examined for their potential antibacterial properties, as many bacterial and fungal strains are resistant to a wide range of medications. *Bifidobacterium adolescentis*, *Bifidum*, *Bifidumlongum*, *Bifidobacterium breve*, *Clostridium perfringens*, *Escherichia coli*, and *Lactobacillus casei* were among the microbes that were effectively treated. According to Sung et al. (2004), isolated 1,2-dihydroxyanthraquinone strongly inhibited the growth of *C. perfringens* and *E. coli* while boosting the growth of *B. bifidum*. While ethanol extracts were sensitive to *A. fumigatus*, *C. obtrusifolia* L. leaf extract in petroleum ether and chloroform was susceptible to *E. faecalis*. Likewise, stem extracts of *C. obtusifolia* L. in petroleum ether, ethanol extracts, and chloroform form were all sensitive to *E. faecalis* (Dave and Ledwani 2012). The entire plant extract of *C. obtusifolia* strongly suppressed the development of *Staphylococcus aureus* MRSA8 (*E. coli* AG100, *Pseudomonas aeruginosa* PA01, *Enterobacter aerogenes* EA289) and *Klebsiella pneumoniae* KP55 (Voukeng et al. 2016). Strong antibacterial activity was demonstrated by the *C. obtusifolia* L. root extract and its components. Emodin, 2,5-dimethoxybenzoquione, questin, isotoralactone, and toralactone, all displayed strong antibacterial activity against *S. aureus* 209P and *E. coli* NIHJ, respectively (Sharon et al. 1992).

25.4.9 *Cymbopogon Proximus*

It is belonging to the Gramineae family of herbs, which are well-known around the world for their abundant essential oil content. Alkaloids, flavonoids, tannins, anthraquinones, saponins, and volatile oils were among the constituents of these plants that were tested phytochemically for (*Staphylococcus aureus* and *Bacillus subtilis*) and Gram-negative bacteria (*Escherichia coli* and *Pseudomonas aeruginosa*) (Kamali-EL and EL-Amir 2010). *Cymbopogonproximus* has a dependable hypoglycemic effect when used as a therapy, which enables this plant to be used for both hyperglycemia and its treatment. The effect of *Cymbopogonproximus* on the regeneration of an infected wound in situations of hyperglycemia produced by dexamethasone was comparable to that of antibacterial ointment (Inchina et al. 2020). The antimicrobial test demonstrated a considerable inhibitory effect on the development of multidrug-resistant bacteria investigations, with action shown at all tested doses. *Staphylococcus aureus*, *Pseudomonas aeruginosa*, *Proteus mirabilis*, *Escherichia coli*, and *Klebsiella pneumonia* were all successfully prevented by the Cymbopogonproximus (Moglad et al. 2020).

25.4.10 *Berberi's lyceum*

The Berberidaceae family with the exception of Australia, *Berberislycium* grows throughout the world's temperate and subtropical zones. Native to India, Nepal, and Pakistan, *Berberislycium* is found throughout the world. Berberine, berbamine, chinabine, karakoramine, palmatine, balauchistanamine, gilgitine, jhelumine, punjabine, sindamine, chinabine, acetic acid, maleic acid, and ascorbic acid are the chemical components of *Berberislycium* (Khare 2004). According to some researchers, berberis lyceum is among the greatest examples of a traditional remedy for bone fractures and wound healing. The primary alkaloid found in bone fracture plants is berberine, an is quinoline alkaloid, together with umbellitine. Bone healing is a physical process that our body uses to repair wounds. This is often extracted from the root or root bark of *Berberislycium* and other Berberis species that can be found in abundance in local woodlands Berberis lyceum water extract contains alkaloids, cardiac glycosides, saponins, and hydrolysable tannins. Fruits contain tannins, citric, tartaric, and malic acids. The root extract of *Berberislycium* was tested for wound healing capacity in Swiss Wistar rats. The roots were examined in methanolic and aqueous extracts using Swiss Wistar rats as test subjects.

The excision, incision, and deceased wound space models of wound repair were employed in the study. Both extracts showed improvements in breaking potency as well as in the area of epithelialization. According to the results, the methanolic extract was more effective than the aqueous extract (Asif et al. 2007).

25.5 Conclusions

The antimicrobial, anti-inflammatory, and anti-oxidative features make up these very similar properties in the herbs. Herbal medicines have fewer side effects and better bioavailability for healing wounds, they may be the source of future medications. Herbal products have been used as medicinal medications, nutraceuticals, and therapy choices for many years and generations without significant adverse effects, which has influenced researchers all around the world. Furthermore, because they are naturally occurring, these herbal therapeutic remedies are safer than synthetic medications. Understanding molecular pathways makes studying pharmacokinetics and pharmacodynamics extremely challenging due to the availability of many elements, complexity, and broad-spectrum bioactivity. This most recent study has added to a growing of knowledge suggesting that flavonoids, alkaloids, saponins, phenolic compounds, and polysaccharides are the active ingredients in pharmaceuticals that aid in the healing of wounds. All the mentioned herbs have potent phytochemical which is helpful for chronic wounds as well as antibacterial activity and anti-inflammatory activity.

References

Albano MN, da Silveira MR, Danielski LG, Florentino D, Petronilho F, Piovezan AP (2013) Anti-inflammatory and antioxidant properties of hydroalcoholic crude extract from *Caseariasylvestris* Sw. (Salicaceae). J Ethnopharmacol 147:612–617

Anbarasan T, Bourdon J-C (2019) The emerging landscape of p53 isoforms in physiology, cancer and degenerative diseases. Int J Mol Sci 20(24):6257

Asif A, Kakub G, Mehmood S, Khunum R, Gulfraz M (2007) Wound healing activity of root extracts of Berberis lycium royle in rats. Phytother Res 21:589–591

Ayaz M, Subhan F, Ahmed J, Khan A-U, Ullah F, Ullah I, Ali G, Syed N-I-H, Hussain S (2015) Sertraline enhances the activity of antimicrobial agents against pathogens of clinical relevance. J Biol Res 22:4

Basile AC, Sertié JAA, Panizza S, Oshiro TT, Azzolini CA (1990) Pharmacological assay of *Caseariasylvestris*. I: Preventive anti-ulcer activity and toxicity of the leaf crude extract. J Ethnopharmacol 30:185–197

Benedek B, Kopp B, Melzig MF (2007) *Achilleamillefolium* L. s.l.—is the anti-inflammatory activity mediated by protease inhibition? J Ethnopharmacol 113(2):312–317

Bueno JM, Sáez-Plaza P, Ramos-Escudero F, Jiménez AM, Fett R, Asuero AG (2012) Analysis and antioxidant capacity of anthocyanin pigments. Part II: chemical structure, color, and intake of anthocyanins. Crit Rev Anal Chem 42:126–151

Bueno PCP, Passareli F, Anhesine NB, Torres RB, Cavalheiro AJ (2016) Flavonoids from *Caseariasylvestris* Swartz Variety Lingua (Salicaceae) Biochem. Syst Ecol 68:23–26

Candan F, Unlu M, Tepe B, Daferera D, Polissiou M, Sökmen A et al (2003) Antioxidant and antimicrobial activity of the essential oil and methanol extracts of *Achilleamillefolium* subsp. millefoliumAfan. (Asteraceae). J Ethnopharmacol 87(2–3):215–220

Cannon BC, Cannon JP (2004) Management of pressure ulcers. Am J Health Syst Pharm 61:1895–1905

Carvalho FA, Uchina HS, Borges FA, Oyafuso MH, Herculano RD, Gremião MPD, Santos AG (2018) Natural membranes of *Hevea brasiliensis* latex as delivery system for *Casearias ylvestris* leaf components. Rev Bras 28:102–110

Choudary R et al (2023) Sustained release of *Azadirachta Indica* extract from stimuli responsive bacterial cellulose–carbapol hydrogel for chronic burn wound management. Biomed Mater Devices:1–30

Contardi M, Ayyoub AMM, Summa M, Kossyvaki D, Fadda M, Liessi N, Armirotti A, Fragouli D, Bertorelli R, Athanassiou A (2022) Self-adhesive and antioxidant poly(vinylpyrrolidone)/alginate-based bilayer films loaded with Malvasylvestris extracts as potential skin dressings. ACS Appl Bio Mater 5(6):2880–2893

Cutillo F, D'Abrosca B, Dellagreca M et al (2006) Terpenoids and phenol derivatives from *Malva* silvestris. Phytochemistry 67:481–485

Daeschlein G (2013) Antimicrobial and antiseptic strategies in wound management. Int Wound J 10:9–14

Dave H, Ledwani L (2012) A review on anthraquinones isolated from Cassia species and their applications. Indian J Nat Prod Resour 3:291–319

Deshmukh PT, Fernandes J, Atul A, Toppo E (2009) Wound healing activity of Calotropis gigantea root bark in rats. J Ethnopharmacol 125:178–181

Edeoga HO, Okwu DE, Mbaebie BO (2005) Phytochemical constituents of some Nigerian medicinal plants. Afr J Biotechnol 4:685–688

Falanga V (2004) The chronic wound: impaired healing and solutions in the context of wound bed preparation. Blood Cell Mol Dis 32(1):88–94

Ferreira PMP, Costa-Lotufo LV, Moraes MO, Barros FWA, Martins AMA, Cavalheiro AJ, Bolzani VS, Santos AG, Pessoa C (2011) Folk uses and pharmacological properties of *Caseariasylvestris*: a medicinal review. An Acad Bras Ciênc 83:1373–1384

Gebhardt KS (2002) Part 1. Causes of pressure ulcers. Nurs Times 98:4

Getie M, Gebre-Mariam T, Rietz R, Neubert RHH (2002) Evaluation of the release profiles of flavonoids from topical formulations of the crude extract of the leaves of Dodonea Viscosa (Sapindaceae). Pharmazie 57:320–322

Guo S, DiPietro LA (2010) Factors affecting wound healing. J Dental Res 89:219–229

Guo LP, Yang J, Zhou L, Wang S, Kang CZ, Huck CW (2021) Simultaneous quantification of 14 compounds in *Achillea millefolium* by GC-MS analysis and near-infrared spectroscopy combined with multivariate techniques. J Anal Methods Chem 2021:5566612

Gupta V et al (2010) Pharmacological potential of matricariarecutita—a review. Int J Pharm Sci Drug Res 2(1):12–16

Hannan A et al (2014) Antibacterial activity of neem (*Azadirachta Indica*) leaf extracts against wound causing bacteria. HealthMED 8(6):767–773

He XR, Wang XX, Fang JC, Zhao ZF, Huang LH, Hao G et al (2016) *Bletillastriata*: medicinal uses, phytochemistry and pharmacological activities. J Ethnopharmacol 195:20–38

Hernandez R (2006) The use of systemic antibiotics in the treatment of chronicwounds. Dermatol Ther 19:326–337

Inchina V et al (2020) Regeneration effect of an aqueous extract of Cymbopogon Proximus on infected wounds in a rodent model in steroid hyperglycemia. Archiv Eur Med 10(2):35–37

Kalan LR, Meisel JS, Loeschc MA, Horwinski J, Soaita I, Chen X et al (2019) Strain- and species-level variation in the microbiome of diabetic wounds is associated with clinical outcomes and therapeutic efficacy. Cell Host Microbe 25:641–655

Kamali-EL HH, EL-Amir MY (2010) Antibacterial activity and phytochemical screening of ethanolic extracts obtained from selected Sudanese medicinal plants. Curr Res J Biol Sci 2(2):143–146

Khare CP (2004) Indian herbal remedies. Springer, New York, pp 98–100

Kokoska L et al (2002) Screening of some Siberian medicinal plants for antimicrobial activity. J Ethnopharmacol 82(1):51–53

Lipinski LC, de Wouk AFP, da Silva NL, Perotto D, Ollhoff RD (2012) Effects of 3 topical plant extracts on wound healing in beef cattle. Afr J Tradit Complement Altern Med AJTCAM 9:542–547

Maan P, Yadav KS, Yadav NP (2017) Wound healing activity of *Azadirachta Indica* A. Juss Stem Bark in Mice Pharmacogn Mag 13(Suppl 2):S316–S320

Maistro EL, Carvalho JCT, Mantovani MS (2004) Evaluation of the genotoxic potential of the *Caseariasylvestris* extract on HTC and V79 cells by the comet assay. Toxicol In Vitro 18:337–342

Mandal J et al (2012) Antibiotic resistance pattern among common bacterial uropathogens with a special reference to ciprofloxacin resistant Escherichia coli. Indian J Med Res 136(5):842

Martins MD, Marques MM, Bussadori SK, Martins MA, Pavesi VC, Mesquita-Ferrari RA, Fernandes KP (2009) Comparative analysis between *Chamomilla Recutita* and corticosteroids on wound healing. An *in vitro* and *in vivo* study. Phytother Res 23(2):274–278

Mazandarani M, Ghaemi EA, Ghaffari F (2009) Antibacterial survey of different extracts of Peganumharmala L. different parts in north east of Golestan province (IncheBorun). J Plant Environ Physiol 15

Moglad EHO, Boon SKA, Ali HTO (2020) Various medicinal plants: a promising treatment for multidrug-resistant bacteria isolated from wound infection. Int J Pharm Sci Res 11(2):839–843

Mustoe TA, Shaughnessy KO, Kloeters O (2006) Chronic wound pathogenesis and current treatment strategies: a unifying hypothesis. Plast Reconstr Surg 117:35–41

Narayanan A et al (2023) Studies on the antimicrobial activity of *Ormocarpumcochinchinense* leaf extract/PVA-PVP blended polymer. Chem Select 8(10):e202203512

Narayanan A et al (2024) *In vitro* cytotoxicity and wound healing activities of quaternary blended carboxymethyl cellulose-hydroxypropyl methylcellulose composite film. Cellulose 31:1–14

Nasrine A et al (2023) Neem (*Azadirachta Indica*) and silk fibroin associated hydrogel: boon for wound healing treatment regimen. Saudi Pharm J 31(10):101749

Nogueira JC, DinizMde F, Lima EO (2008) *In vitro* antimicrobial activity of plants in acute otitis externa. Braz J Otorhinolaryngol 74(1):118–124

Nunan R, Harding KG, Martin P (2014) Clinical challenges of chronic wounds: searching for an optimal animal model to recapitulate their complexity. Dis Model Mech 7:1205–1213

Ontario Health (Quality) (2021) Skin substitutes for adults with diabetic foot ulcers and venous leg ulcers: a health technology assessment. Ont Health Technol Assess Ser 21(7):1–165

Pazhanisamy M, Ebenezer GAI (2013) Phytochemical screening of *Ormocarpumcochinchinense* leaf extracts. J Acad Indust Res 2(5):275

Pirbalouti AG, Azizi S, Koohpayeh A et al (2010) Wound healing activity of *Malvasylvestris* and Punicagranatum in alloxan-induced diabetic rats. Acta Pol Pharm 67:511–516

Pirbalouti AG, Azizi S, Koohpayeh A (2012) Healing potential of Iranian traditional medicinal plants on burn wounds in alloxan-induced diabetic rats. Rev Bras 22:397–403

Pullaiah TS, Sri Ramamurthy K (2001) The flora of eastern ghats hill ranges of south East India. Regency publication, New Delhi, p 990

Rhee JK, Kim PG, Baek BK, Lee SB, Ahn BZ (1982) Isolation of anthelmintic substance on clonorchissinensis from tuber of Bletillastriata. Korean J Parasitol 20:142

Saeidnia S et al (2011) A review on phytochemistry and medicinal properties of the genus Achillea. J Faculty Pharm 19(3):173

Salas-Oropeza J, Jimenez-Estrada M, Perez-Torres A, Castell-Rodriguez AE, Becerril-Millan R, Rodriguez-Monroy MA et al (2020) Wound healing activity of the essential oil of Burseramorelensis, in mice. Molecules 25(8):1795

Sharma A et al (2021) Medicinal plants and their components for wound healing applications. Future J Pharm Sci 7(1):1–13

Sharon A, Ghirlando R, Gressel J (1992) Isolation, purification, and identification of 2-(p-hydroxyphenoxy)-5,7-dihydroxychromone: a fungal-induced phytoalexin from Cassia obtusifolia. Plant Physiol 98:303–308

Singer AJ, Clark RA (1999) Cutaneous wound healing. N Engl J Med 341(10):738–746

Stewart PS, Costerton JW (2001) Antibiotic resistance of bacteria in biofilms. Lancet 358:135–138

Stojanović G et al (2005) *In vitro* antimicrobial activity of extracts of four Achillea species: the composition of Achilleaclavennae L. (Asteraceae) extract. J Ethnopharmacol 101(1–3):185–190

Subapriya R, Nagini S (2005) Medicinal properties of neem leaves: a review. Curr Med Chem Anticancer Agents 5(2):149–156

Sung BK, Kim MK, Lee WH, Lee DH, Lee HS (2004) Growth responses of Cassia obtusifolia toward human intestinal bac-teria. Fitoterapia 75:505–509

Tadić V, Arsić I, Zvezdanović J, Cvetković D, Pavkov S, Zugić A (2017) The estimation of the traditionally used yarrow (Achilleamillefolium L. Asteraceae) oil extracts with anti-inflamatory potential in topical application. J Ethnopharmacol 199:138–148

Thirumal S, Duraikannu G (2019) Fourier-transform infrared analysis and *in vitro* antibacterial activity of *Ormocarpumcochinchinense* (Elumbotti). Int J Pharm Biol Arch

Versey Z, da Cruz Nizer WS, Russell E, Zigic S, DeZeeuw KG, Marek JE, Overhage J, Cassol E (2021) Biofilm-innate immune Interface: contribution to chronic wound formation. Front Immunol 12:648554

Voukeng IK, Beng VP, Kuete V (2016) Antibacterial activity of six medicinal Cameroonian plants against gram-positive and gram-negative multidrug resistant phenotypes. BMC Complement Altern Med 16:388

Wolcott RDRA, Hanson JD, Rees EJ, Koenig LD, Phillips CD, Wolcott RDRA et al (2016) Analysis of the chronic wound microbiota of 2,963 patients by 16S rDNA pyrosequencing. Wound Repair Regen 24:163–174

Woo K, Ayello EA, Sibbald RG (2007) The edge effect: current therapeutic options to advance the wound edge. Adv Skin Wound Care 20:99–117

Yuan H, Ma Q, Ye L, Piao G (2016) The traditional medicine and modern medicine from natural products. Molecules 21(5):559

Zohra T, Ovais M, Khalil AT, Qasim M, Ayaz M, Shinwari ZK (2019) Extraction optimization, total phenolic, flavonoid contents, HPLC-DAD analysis and diverse pharmacological evaluations of Dysphania ambrosioides (L.) Mosyakin & Clemants. Nat Prod Res 33:136–142

26 Harnessing Nature's Pharmacy: Medicinal Plants Combatting Bacterial Infections in the Aquatic Environment

Ruksar, Tejpal Dahiya, Tannu Goyal, Abhimanyu, and Priyanka Priyanka Joshi

Abstract

Aquatic environment hosts diverse microbial communities crucial for ecosystem balance, yet some harbor pathogens dangerous to human health and the ecological balance of aquatic ecosystems. Antibiotic-resistant bacteria proliferation in water bodies challenges conventional treatments and aquaculture sustainability. To counteract this, renewed focus turns to natural antimicrobial sources including beneficial bacterial consortia and medicinal plants as potential solutions. This abstract elucidates medicinal plants' mechanisms, effectiveness, and implications in combating bacterial infections within aquatic ecosystems, underlining their promise for sustainable aquaculture and environmental preservation. Medicinal plants provide a rich source of bioactive chemicals with a variety of modes of action against aquatic bacterial infections and can be used in conjunction with modern scientific methods ethnopharmacology, and traditional knowledge. Notable examples include neem, garlic, turmeric, basil, aloe vera, and echinacea showing significant antibacterial properties. Integrating plant-based therapies into aquaculture demands rigorous evaluation encompassing efficacy, safety and sustainable utilization including toxicity assessment and conservation efforts. Medicinal plants have great potential to combat the global problem of antibiotic resistance in the aquatic environment and to maintain the resilience and health of aquatic ecosystems for future generations by appropriately utilizing natural sources.

Ruksar (✉) · T. Dahiya · Abhimanyu · P. P. Joshi
Department of Zoology, College of Basic Sciences and Humanities, Chaudhary Charan Singh Haryana Agricultural University, Hisar, Haryana, India

T. Goyal
Department of Zoology, Jan Nayak Chaudhary Devi Lal Memorial College, Chaudhary Devi Lal University, Sirsa, Haryana, India

M. Gangwar, G. Nath (eds.), *Emerging Paradigms for Antibiotic-Resistant Infections: Beyond the Pill*, https://doi.org/10.1007/978-981-97-5272-0_26

Keywords

Aquaculture · Bacterial infections · Medicinal plants · Antibiotic-resistant · Aquaculture sustainability

26.1 Introduction

Numerous microorganisms, including bacteria that can infect aquatic life and endanger human health, are abundant in the complex ecosystem that is the aquatic environment. Alternative methods of treating these illnesses must be investigated immediately due to the increase in antibiotic resistance in aquatic bacteria. The biochemical variety of medicinal plants makes them a possible source of new antibacterial drugs (Nascimento et al. 2000). This chapter focuses on their modes of action, effectiveness, and consequences for environmentally friendly conservation and sustainable aquaculture.

Scientists are looking more and more to nature's pharmacy for creative answers in an era where antibiotic resistance poses a danger to world health. For thousands of years, people have valued medicinal plants for their ability to treat many ailments. This is because they are a rich source of bioactive chemicals that have a wide range of chemical effects (Newman and Cragg 2016). To fight the growing issue of antibiotic-resistant bacteria in aquatic systems, scientists have recently started investigating the potential of these medicinal plants. The objective of the study is to create innovative, eco-friendly remedies by utilizing the antibacterial characteristics of plant extract. This multidisciplinary approach relies upon the knowledge domain of pharmacognosy, microbiology, ecology, and environmental science. Researchers aim to optimize the use of medicinal plants in the fight against diseases by investigating the relationships that arise between plants, bacterial pathogens, and the aquatic environment. First off, using plant-based remedies instead of synthetic ones that worsen environmental degradation and the development of antibiotic resistance is a sustainable way to combat antibiotic resistance. Furthermore, a great deal of medicinal plants is widely accessible and hold cultural significance in areas where conventional healthcare may be scarce, rendering them vital assets for marginalized groups. Furthermore, by maintaining the natural order of aquatic ecosystems, these plant-based therapies may have significant effects on ecosystem services and public health. In addition to providing vital resources and supporting biodiversity, healthy aquatic ecosystems are also important for controlling global biogeochemical cycles and reducing the effects of climate change (Costanza et al. 2014). Researchers are laying the groundwork for long-term solutions that preserve the integrity of the environment and human health by fusing traditional knowledge with cutting-edge scientific methods.

26.2 Bacterial Infections in the Aquatic Environment

Aquatics are susceptible to bacterial infections including freshwater bodies like rivers, lakes, and ponds, as well as marine ecosystems like seas and estuaries. The intricate relationships between bacteria, host organisms, environmental variables, and human influences must be considered to comprehend the dynamics of bacterial infections in these settings (Hatosy and Martiny 2015). Numerous pathways exist for bacterial pathogens to enter aquatic environments: sewage treatment plant discharge, runoff from agricultural areas carrying fertilizers and livestock waste, urban runoff containing pollutants and pathogens from impermeable surfaces, and wildlife faces (Prüss 1998). These pathogens can survive and spread across the ecosystem after they are placed in water and interact with aquatic animals and plants. Aquatic environments can be significantly impacted ecologically by bacterial infections. Fish population infections, for instance, can result in smaller populations, changed dynamics within communities, and disturbances ecosystem's function (Daskalov et al. 2007). Additionally, bacterial infections have the potential to harm sensitive ecosystems and keystone species, which could have a domino effect on the ecosystem (Harvell et al. 2002). The polluted or contaminated water poses dangers to human health through ingestion of contaminated seafood or direct contact. Particularly in susceptible groups like children, the elderly, and people with compromised immune systems, consuming or being exposed to contaminated water can result in gastrointestinal disorders, skin infections, lung infections, and other health issues (Bartram and Cairncross 2010). In an aquatic environment, *Salmonella* spp., *Vibrio cholerae*, *Escherichia coli*, and several *Pseudomonas* and *Aeromonas* species are found to be pathogenic. In addition, they cause sickening to humans who come into touch with contaminated water and are responsible for diseases in aquatic species like fish, amphibians, and invertebrates. Concern over the introduction and dissemination of microorganism's resistant to antibiotics in aquatic environments is growing (Pruden et al. 2006). Through wastewater discharge, agricultural runoff, and aquaculture techniques, antibiotics used in human and veterinary medicine can enter aquatic habitats and develop resistant bacterial strains. This can make antibiotic treatments less effective for treating infections in humans and animals, which presents serious problems for managing aquatic health and public health (Udikovic-Kolic et al. 2014).

26.2.1 The Rise of Antibiotic Resistance in Aquatic Bacteria

Microbial diversity is highest in aquatic habitats (Louca et al. 2016). However, the introduction and dissemination of antibiotic-resistant bacteria in these environments is a result of human activities including pollution, aquaculture, and the widespread use of antibiotics (Bengtsson-Palme et al. 2018; Martinez 2009). The bioaccumulation of antibiotic-resistant bacteria in aquatic creatures takes place through water absorption, sediment, food sources, and antibiotics (Zhang et al. 2015). These pollutants can bio-magnify once they are absorbed, which raises their concentrations in higher tropics levels (Berendonk et al. 2015). The transmission of antibiotic resistance throughout the food chain may be facilitated by this bioaccumulation and bio-magnification, which can also be dangerous for aquatic life (Knapp et al. 2010). Long after an antibiotic is first released into the water; antibiotic residues might linger there and continue to selectively pressure microorganisms to become resistant to the antibiotic (Kristiansson et al. 2011). Antibiotics have the ability, even at low concentrations, to maintain resistance genes within microbial populations and to encourage the growth of antibiotic-resistant bacteria (D'Costa et al. 2011). Additionally, bacteria can experience stress reactions from sublethal antibiotic concentrations, which might result in the activation of resistance pathways. The microbial communities in aquatic environments may change in terms of diversity and composition as a result of antibiotic resistance (Jutkina et al. 2016). The population structure of a community may shift because antibiotic-resistant bacteria outcompete susceptible strains. Ecological relationships, the cycling of nutrients, and ecosystem function can all be affected by these changes (Allen et al. 2010).

Intensive fish farming has promoted the growth of diseases, which led to an increase in the use of antimicrobials (Defoirdt et al. 2007, 2011). Fish do not effectively metabolize antibiotics and are largely unused back into the environment in faces. It is estimated that 75% of the antibiotics fed to fish are excreted into the water (Burridge et al. 2010). Their use as chemotherapeutic agents for the treatment of bacterial infectious diseases.

26.2.2 Antibiotics Consumption in the World

The global antimicrobial use in 2017 was estimated to be 10,259 tons, and it is projected to increase up to 13,600 tons in 2030 based on the OECD/FAO Agricultural Outlook and FAO Fish Stat production quantities (OECD/FAO Agricultural Outlook 2018). China alone accounted for 57.9% of the world's antibiotic usage in 2017, making the Asia-Pacific region the highest consumer of antibiotics globally (93.8%).

Cabello (2006) reported that "in Chile, statistics indicate that approximately 100–110 metric tons of quinolone antibiotics are used as veterinary medicine per year, most of them in aquaculture". Ciprofloxacin present in pond water samples is completely degraded after 3 months, whereas only 20% of oxolinic acid present in these samples is degraded after 5 months. Most antibiotics are usually below detectable limits in groundwater samples, although they are more stable upon adsorption to sediments. By frequency of use, the most widely used classes of antimicrobials worldwide were sulfonamides (14%), amphenicols (18%), quinolones (27%), and tetracyclines (20%). First-generation cephalosporins (85 mg/kg), lincosamides (33 mg/kg), and tetracyclines (23 mg/kg) have the highest levels. In Kerala, furazolidone, ciprofloxacin, and chloramphenicol are the three antibiotics most frequently used in aquaculture (Table 26.1).

Table 26.1 Antibiotics authorized for use in aquaculture

Antibiotic name	Treatment of
Oxytetracycline (1)	Furunculosis in salmonids (salmon or trout) caused by *Aeromonas Salmonicida* USP, 2000 g Gafkemia in lobsters
Florfenicol (2) Premix	Indicated in the treatment of furnculosis caused by susceptible strains of *Aeromonas salmonicida*
Sarafloxacin	Indicated in the treatment of furunculosis, vibriosis and enteric redmouth in Salmonidae
Erythromycin	In the treatment of bacterial kidney disease (*Renibbacterium salmoninarum*) and *streptococcosis* in yellowtail in Japan
Sulphonamides	Against furunculosis, enteric redmouth disease and vibriosis

26.2.3 Antibiotic Resistance in Different Bacteria Reported in India

Subashkumar et al. (2007) demonstrated similar results among *A. hydrophila* with 77.5% of their isolates resistant to amoxyclav. Cefalexin, being a third-generation cephalosporin, is used quite frequently against human infections signifying an up-rise in antibiotic resistance (Park 2014). Sreedharan et al. (2012), reported complete resistance to amoxyclav by motile aeromonads from ornamental fish culture system in Kerala. Suhet et al. (2011) also reported 100% resistance to amoxyclav by *A. caviae*, *A. hydrophila*, *A. jandaei*, *A. veronii*, and *A. sobria* isolated from fish. 100% of antibiotic resistance was shown in *Aeromonas* spp. of Kolkata, advising the role of sewage in the transfer and persistence of antibiotic resistance (Karkman et al. 2018).

26.2.4 Selection Pressure from Antibiotics

Aquatic bacteria frequently develop antibiotic resistance because of the selection pressure placed on them by the extensive usage and discharge of antibiotics into aquatic environments (Davies and Davies 2010; Martínez 2008). Antibiotics can infiltrate aquatic habitats through several different channels, such as sewage treatment plant output, runoff from farms where antibiotics are used for livestock production, aquaculture activities, and incorrect pharmaceutical disposal by end users (Knapp et al. 2010; Zhang et al. 2015). By preferring resistant strains and inhibiting vulnerable ones, medications once in the water can directly aid in the survival and growth of antibiotic-resistant bacteria (Kristiansson et al. 2011). Bacteria can transmit genetic material, including genes resistant to antibiotics, with one another through a process known as horizontal gene transfer (Ochman et al. 2000), which is facilitated by aquatic conditions. Through processes including conjugation, transformation, and transduction (Thomas and Nielsen 2005), bacteria can exchange genetic material with different bacterial species or strains to acquire resistant qualities. Horizontal gene transfer promotes the fast spread of antibiotic resistance in aquatic bacterial populations as well as the possibility of resistance genes spreading to human diseases (Wright 2010; Bhullar et al. 2012).

In aquatic conditions, bacteria frequently create biofilms, which are intricate microbial communities covered in a matrix of extracellular polymeric materials (Flemming and Wingender 2010). Bacteria are better sheltered by biofilms, which increases their resistance to antibiotics and other antimicrobial agents (Mah and O'Toole 2001). Because bacteria can exchange genetic material more effectively in biofilms, multidrug-resistant strains of the bacterium and the dissemination of genes resistant to antibiotics are encouraged (Donlan 2002). When contaminated water, contaminated seafood, or aerosols harboring resistant pathogens are in direct touch with humans, the health of such people is greatly endangered by antibiotic-resistant bacteria found in aquatic habitats (Ashbolt 2015). Humans can contract a variety of infections from resistant bacteria, such as respiratory, skin, and gastrointestinal disorders as well as urinary tract infections (Martínez 2008). In addition, the transfer of antibiotic resistance from environmental bacteria to human pathogens may make antibiotic treatments less effective, which could result in higher rates of morbidity, death, and medical expenses (Allen et al. 2010).

26.2.5 Loss of Biodiversity

Aquatic habitats may experience a decline in biodiversity due to the rise of antibiotic resistance in aquatic bacteria (Zhu et al. 2013). The total variety and abundance of microbial communities may decrease if susceptible species are supplanted by strains resistant to antibiotics (Forsberg et al. 2012). The ability of ecosystems to withstand changes in their surroundings and provide ecosystem services may be hampered by this loss of biodiversity (Cardinale et al. 2012). By interfering with microbial interactions and nutrient cycling processes, antibiotic resistance in aquatic bacteria might indirectly encourage the establishment of hazardous algal blooms (HABs). By contaminating seafood and exposing people to toxin-containing aerosols, HABs can produce poisons that damage aquatic life, deteriorate water quality, and endanger human health (Hallegraeff 2010; Anderson et al. 2012).

26.3 Medicinal Plants as a Source of Antimicrobial Agents

Throughout history, traditional medical systems have utilized medicinal plants to treat a wide range of illnesses, including bacterial infections. World Health Organization (WHO) has estimated that approximately 65% of the world's population relies mainly on plant-derived traditional medicines for their primary health care (Farnsworth et al. 1985). The use of medicinal plants as antimicrobial agents offers several advantages, including their widespread availability, cultural acceptance, and potential for sustainable use (Fabricant and Farnsworth 2001; Akinyemi et al. 2005). When applied to aquatic environments, the bioactive compounds of these plants can inhibit the growth, proliferation, and survival of pathogenic microorganisms, thereby reducing the prevalence and severity of infectious diseases in aquatic organisms.

26.3.1 Antimicrobial Activity of Medicinal Plants Against Aquatic Pathogens

Medicinal plants include bioactive chemicals with a wide range of intricate modes of action. These mechanisms involve interactions with different molecular targets and pathways within bacterial cells (Newman and Cragg 2012; Cushnie and Lamb 2011). These substances can suppress bacterial development, interfere with biological functions, and treat bacterial illnesses due to their antibacterial qualities.

26.3.2 Antibacterial Properties

Medicinal plants have been shown in numerous investigations to possess antibacterial activities against a broad spectrum of bacterial pathogens, encompassing both Gram-positive and Gram-negative bacteria (Cowan 1999; Cushnie and Lamb 2011). Several widely recognized instances include neem (*Azadirachta indica*), garlic (*Allium sativum*), ginger (*Zingiber officinale*), turmeric (*Curcuma longa*), tea tree (*Melaleuca alternifolia*), and oregano (*Origanum vulgare*) (Rios and Recio 2005; Cowan 1999). Bioactive substances found in several plants, including carvacrol, neem oil, gingerol, curcumin, and allicin, have been demonstrated to have strong antibacterial properties (Nascimento et al. 2000; Cushnie and Lamb 2011). Many of the bioactive substances found in medicinal plants, such as alkaloids, flavonoids, phenolics, terpenoids, saponins, glycosides, and essential oils, have antibacterial properties against bacteria (Cowan 1999; Cushnie and Lamb 2011). These substances may work singly or in combination to alter immune responses, interfere with cellular functions, damage bacterial cell membranes, and prevent bacterial growth (Rios and Recio 2005; Cushnie and Lamb 2011). A wealth of possible antibacterial agents with a variety of modes of action can be found in the diversity of bioactive chemicals found in medicinal plants. Combining several bioactive ingredients from several medicinal plants can have synergistic effects that increase the antibacterial activity of the ingredients and lower the possibility of resistance developing (Cushnie and Lamb 2011; Nascimento et al. 2000). To hinder the development of bacterial resistance, synergistic combinations may target several bacterial routes or mechanisms. Furthermore, in contrast to single-target synthetic antibiotics, the diverse chemical makeup of medicinal plants may make it more difficult for bacteria to build resistance (Cushnie and Lamb 2011).

26.3.3 Traditional and Indigenous Knowledge of Medicinal Plants as Antimicrobial Agents

For a very long time, indigenous civilizations around the world have used plant-based medicines to treat bacterial illnesses, and this practice has led to the widespread use of medicinal plants as antimicrobial agents (Alves et al. 2007; Ferreira et al. 2009). The generation-to-generation transmission of traditional medicinal plant knowledge offers important insights into the choice, processing, and application of plant-based remedies for a range of illnesses, including bacterial infections (Saslis-Lagoudakis et al. 2011). Considering that they can be grown, gathered, and processed sustainably, medicinal plants present a viable and environmentally acceptable substitute for synthetic antibiotics. Medicinal plants have no negative impact on ecosystems and may be grown organically, in contrast to manufactured antibiotics, which can lead to environmental contamination and the emergence of antibiotic resistance. Many medical plant preparations, such as decoctions, infusions, tinctures, poultices, salves, powders, and extracts, have been developed by indigenous communities. Depending on the desired therapeutic outcome, these preparations can be used topically, internally, or through inhalation.

By combining pharmacological research, biological science, and ethnobotanical knowledge in ethnopharmacology for anemia, studies on the therapeutic potential of medicinal plants in the prevention and treatment of bacterial infections are conducted. The efficacy, safety, and mechanisms of action of medicinal plants are validated by scientific approaches. At the same time, traditional and indigenous knowledge gathered by various cultures worldwide is also a part of this strategy.

26.3.4 Bioactive Compounds in Medicinal Plants

These are chemical compounds that have physiological effects on humans and fish alike. The therapeutic qualities of medicinal plants are attributed to these chemicals, which are also essential to their capacity to treat bacterial infections and other ailments (Nascimento et al. 2000; Cowan 1999). To clarify their pharmacological effects and create efficient treatments, it is crucial to comprehend the bioactive chemicals found in medicinal plants as well as their modes of action.

26.4 Mechanisms of Action

Research in ethnopharmacology clarifies the modes of action that underlie the antibacterial effectiveness of therapeutic herbs against bacterial illnesses (Saslis-Lagoudakis et al. 2011). The interactions between plant-derived chemicals and bacterial targets—such as cell membranes, enzymes, DNA replication, and protein synthesis pathways—are examined in pharmacological investigations (Cushnie and Lamb 2011). The molecular underpinnings of plant-bacteria interactions and the

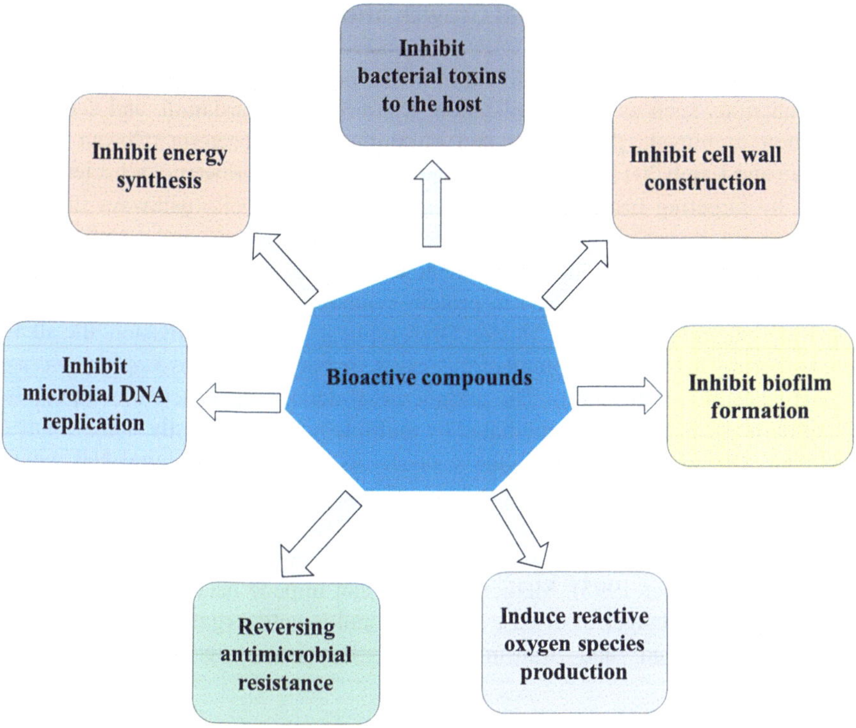

Fig. 26.1 Bioactive compounds and their antimicrobial mechanism of action

emergence of antibiotic resistance are studied using molecular biology approaches such as transcriptomics, proteomics, and genomics (Fig. 26.1).

Medicinal plants' bioactive components work in a variety of ways to produce their antimicrobial properties.

26.4.1 Broad-Spectrum Antimicrobial Activity

A vast variety of bioactive substances having broad-spectrum antibacterial activity against a broad spectrum of aquatic pathogens, such as bacteria, fungi, viruses, and protozoa, are found in medicinal plants. These substances are effective against a variety of infections and lower the likelihood of resistance development since they can target many microbial sites and pathways. Bioactive substances that possess broad-spectrum antibacterial activity comprise flavonoids, alkaloids, terpenoids, phenolic compounds, and essential oils (Cowan 1999; Cushnie and Lamb 2011).

26.4.2 Inhibition of Bacterial Growth and Proliferation

Numerous bioactive substances found in medicinal plants interfere with vital cellular functions such as DNA replication, transcription, translation, and cell wall formation to impede the growth and multiplication of bacteria (Cowan 1999; Cushnie and Lamb 2011). These substances may have bacteriostatic or bactericidal effects by targeting bacterial proteins, enzymes, or metabolic pathways that are important for bacterial metabolism and replication (Cushnie and Lamb 2011). Certain substances hinder the development and proliferation of bacteria by inhibiting important enzymes involved in protein synthesis, DNA replication, cell wall construction, and metabolism. To stop DNA replication and cell division, the alkaloid berberine, for example, inhibits DNA gyrase in bacteria such as *Staphylococcus aureus* (Stermitz et al. 2000). The antimicrobial actions of bioactive compounds found in medicinal plants are mediated by multiple pathways. Firstly, some chemicals bind with proteins, phospholipids, or sterols to damage bacterial cell membranes, resulting in cell death and leakage of contents (Cushnie and Lamb 2011). For instance, the outer membrane of Gram-negative bacteria such as *Escherichia coli* is disrupted by the polyphenol epigallocatechin gallate (EGCG) found in green tea (Hamilton-Miller 1995). Medicinal plants can impede bacterial development, which lowers the danger of disease outbreaks and aquatic organism death by preventing the spread and transmission of bacterial infections in aquatic environments.

26.4.3 Disruption of Microbial Biofilms

To defend against antimicrobial drugs and host immunological reactions, complex populations of bacteria called microbial biofilms are encased in a matrix of extracellular polymeric materials (Donlan and Costerton 2002; Flemming et al. 2016). Medicinal herbs contain bioactive substances that interfere with quorum sensing molecules, cell-cell communication pathways and extracellular matrix components to disrupt biofilm development, destabilize preexisting biofilms, and encourage biofilm dispersion (Borges et al. 2015; Roy et al. 2018). The ability of aquatic pathogens to cause infections and remain in aquatic habitats can be decreased by medicinal plants that disturb biofilms, making them more susceptible to host immune responses and antimicrobial treatments (De la Fuente-Núñez et al. 2014; Gupta et al. 2016).

26.4.4 Induction of Oxidative Stress

Reactive oxygen species (ROS) are produced within bacterial cells by some bioactive chemicals found in medicinal plants, which cause oxidative stress and have antimicrobial properties. These reactive oxygen species can harm lipids, proteins, and DNA in cells, which can result in oxidative damage, breakage in DNA strands and cell death (Kohen and Nyska 2002). It is well recognized that phenolic chemicals, such as flavonoids and tannins, have antioxidant qualities and can cause oxidative stress in bacterial cells, which prevents bacteria from growing and surviving (Cowan 1999). Medicinal herbs can effectively inhibit the growth of aquatic infections and lessen their virulence in aquatic environments by producing oxidative stress.

26.4.5 Modulation of Immune Responses

The identification, neutralization, and removal of bacterial pathogens in aquatic habitats can be improved by the body through the use of specific bioactive chemicals found in medicinal plants, which have the power to modify the human immunological response to bacterial infections. These substances might activate natural killer cells, neutrophils and macrophages—innate immune cells that aid in the synthesis of cytokines, phagocytosis and inflammatory reactions to bacterial invaders. Medicinal herbs can increase the host's defense systems against bacterial infections and promote overall health and resilience in aquatic creatures by increasing immune responses.

26.4.6 Synergistic Interactions and Combination Therapies

Complementary and synergistic effects against aquatic infections may be exhibited by the diverse combinations of bioactive chemicals found in medicinal plants (Rios and Recio 2005). Comparing synergistic combinations of bioactive compounds to single-compound treatments can increase the antibacterial effectiveness, expand the spectrum of activity and lower the chance of developing resistance (Cushnie and Lamb 2011). More efficient and long-lasting methods of treating bacterial infections in aquatic environments could be provided by combination therapies that combine several bioactive substances from several medicinal plants (Nostro et al. 2004).

26.5 Identification of Bioactive Compounds

The purpose of ethno-pharmacological research is to pinpoint the bioactive substances that provide medicinal plants with their antibacterial capabilities against bacterial infections (Heinrich et al. 1998). The chemical components of medicinal plants, such as phenolics, terpenoids, alkaloids and essential oils, are isolated, characterized and quantified using phytochemical analysis techniques like mass spectrometry, spectroscopy, and chromatography (Newman and Cragg 2020). Plant extracts and purified compounds are tested for antimicrobial activity against bacterial strains using bioassays and in vitro screening techniques, which aid in the identification of lead compounds for additional study (Cowan 1999).

26.6 Screening and Identification of Bioactive Compounds from Medicinal Plants

Identifying possible antimicrobial medicines to treat bacterial infections is a critical phase in the drug discovery and development process (Newman and Cragg 2016). There are several procedures involved in this process to identify, assess, and isolate the bioactive chemicals found in medicinal plants (Cragg and Newman 2005), such as:

26.6.1 Selection of Medicinal Plants

Medicinal plants are first screened by choosing them based on phytochemical diversity, ethnobotanical knowledge, and traditional applications (Fabricant and Farnsworth 2001). The plants that have been traditionally used in medicine to treat bacterial infections or that have antibacterial qualities are given priority for more research (Cowan 1999). Suggested plants for screening can be identified with the aid of ethnobotanical surveys, literature research, and consultations with traditional healers (Heinrich et al. 1998).

26.6.2 Collection and Extraction of Plant Material

Using morally and environmentally responsible harvesting techniques, plant material-including leaves, stems, roots, bark, flowers and fruits are gathered from farmed or wild sources. To enable the extraction of bioactive chemicals, the plant material is cleaned, dried, and crushed into a fine powder. Bioactive chemicals are extracted from plant material using a variety of extraction procedures; solvents with varying polarity are employed in maceration, Soxhlet extraction, ultrasound-assisted extraction, and supercritical fluid extraction, among others.

26.6.3 Isolation and Fractionation of Extracts

Fractionation techniques are applied to crude extracts derived from plant material to separate either specific bioactive chemicals or clusters of molecules sharing similar chemical properties (Newman and Cragg 2016). Solid-phase extraction, column chromatography, preparative thin-layer chromatography, and liquid–liquid extraction are examples of fractionation techniques. To determine which fractions, contain bioactive substances, each fraction is gathered, examined, and evaluated for biological activity (Fabricant and Farnsworth 2001).

26.6.4 Bioassays for Biological Activity

Bioassays are used to evaluate the antibacterial activities of extracts and fractions against bacterial pathogens to screen them for biological activity. Bioassays are utilized to assess the inhibitory effects of plant extracts or fractions on bacterial growth, viability, and biofilm formation. These assays may include agar diffusion, broth microdilution, time-kill kinetics, and biofilm inhibition tests (Cushnie and Lamb 2011). Antimicrobial substances with bioactive properties are present when bioassays yield positive findings.

26.6.5 Chemical Analysis and Structural Elucidation

Bioactive compounds isolated from plant extracts are subjected to chemical analysis and structural elucidation to identify their chemical composition and structural characteristics. Analytical techniques such as nuclear magnetic resonance (NMR), spectroscopy, mass spectrometry (MS), infrared (IR) spectroscopy, and high-performance liquid chromatography (HPLC) are used to analyze the chemical composition and determine the molecular structure of bioactive compounds (Wolfender et al. 2015).

26.6.6 Evaluation of Mechanisms of Action

Identified bioactive compounds are evaluated for their mechanisms of action against bacterial pathogens using molecular and pharmacological assays (Cragg and Newman 2013). Mechanistic studies may involve assessing the interaction of bioactive compounds with bacterial cell membranes, enzymes, DNA, or protein targets using in vitro and in vivo models (Tsuchiya 2015). Molecular biology techniques such as genomics, proteomics, and transcriptomics can provide insights into the molecular pathways and cellular processes targeted by bioactive compounds (Ghosh et al. 2012).

26.6.7 Structure–Activity Relationship (SAR) Studies

Structure–activity relationship (SAR) studies explore the relationship between the chemical structure of bioactive compounds and their biological activity against bacterial pathogens (Huigens III et al. 2013). SAR studies involve modifying the chemical structure of bioactive compounds through chemical synthesis or semi-synthesis to optimize their antimicrobial potency, selectivity, and pharmacokinetic properties (Lipinski et al. 2012). SAR analysis helps identify key structural features responsible for the antimicrobial activity of bioactive compounds and guides the design of novel analogs with improved therapeutic profiles.

26.7 Types of Bioactive Compounds (Phytocompounds)

Bioactive chemicals found in medicinal plants are diverse and can include, but are not limited to:

26.7.1 Flavonoids

According to Kumar and Pandey (2013), flavonoids are a broad class of polyphenolic chemicals with a benzo-γ-pyrone structure that are widely found in plants and are produced via the phenylpropanoid pathway. According to a recent analysis, flavonoids have been found in over 9000 different plant kingdom species to date, and at least several hundreds of those are edible components (Tuli 2019). Flavonoids (2-phenyl-benzo-γ-pyrone derivatives) are classified into flavanones, flavonols, flavones, isoflavones, and anthocyanins based on their molecular structure. One study states that the structure of flavonoids consists of a heterocyclic ring of pyron or pyran (ring C) connecting two benzene rings (rings A and B) to form a flavan backbone (Panche et al. 2016). The presence of a carbonyl group at the C ring's fourth carbon atom, a double bond between the ring's second and third carbon atoms, and several hydroxyl or other groups classify substances as flavonoids. Three hydroxyl groups are present in all naturally occurring flavonoids: two in ring A (positions 5 and 7) and one in ring B (position 3) (Pietta 2000).

According to Yamagata (2019), there are six major subclasses of flavonoids based on their underlying structure: flavonols, flavones, flavanones, isoflavones, flavan-3-ols, and anthocyanins. Fruits, vegetables, and herbs (*Camellia sinensis*) include the antioxidant quercetin, which is found in onions (*Allium cepa*), and the anti-inflammatory and antibacterial polyphenol catechins.

26.7.2 Alkaloids

The naturally occurring organic nitrogen-containing compounds known as alkaloids, which include morphine in opium poppies (*Papaver somniferum*) and caffeine in coffee (*Coffea arabica*), have a range of pharmacological effects. Alkaloids affect humans and other animals physiologically in a variety of significant ways. Although they have been identified in animals, insects, marine invertebrates, and certain microorganisms, they are isolated from plants (Dewick 2002). Alkaloids include substances including nicotine, morphine, sparteine, quinine, ephedrine, and strychine. According to Buckingham et al. (2010) over 40,000 alkaloid compounds have been found in nature (Fig. 26.2).

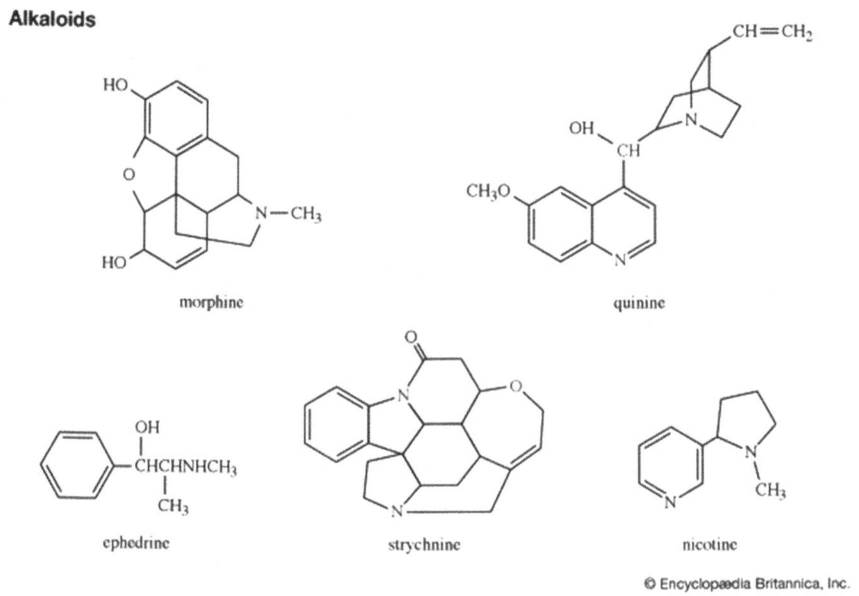

Fig. 26.2 Major alkaloids from phytochemicals

26.7.3 Phenolic Acids

One of the main types of phenolic compounds found in plants is phenolic acids, which are defined as phenolic compounds with one carboxylic acid group. Free phenolic acids, such as benzoic, phenylacetic, and cinnamic acids, have good water solubility and bioavailability in contrast to flavonoids. Fruits, herbs, and spices include aromatic chemicals like resveratrol in grapes (*Vitis vinifera*) and curcumin in turmeric (*Curcuma longa*). They possess antioxidant, antimicrobial, and anti-inflammatory qualities. A particular family of polyphenols known as phenolic acids is typically engaged in defense mechanisms against both biotic and abiotic stressors (Dicko et al. 2005). When included as functional components in emulsion model systems, phenolic acids have been demonstrated to reduce oxidative degradation due to their significant lipid and water solubility. Contrary to them, flavonoids cannot be absorbed in the stomach and are only weakly passively passed through the intestinal wall and into the bloodstream (Chen et al. 2020). Gallic acid, caffeic acid, rosmarinic acid, and carnosic acid are a few types of phenolic acids.

26.7.4 Terpenoids

Terpenoids, which are a broad family of naturally occurring chemicals present in almost all living things, are also referred to as isoprenoids or terpenes. One of the greatest classes of natural goods known is terpenoid structure, of which there are around 60,000 recognized. Compared to other living things, plants contain a comparatively high concentration of terpenoids. Bioavailable hydrocarbons derived from isoprene units include menthol, which is found in peppermint (*Mentha piperita*), and artemisinin, which is found in sweet wormwood (*Artemisia annua*). According to a survey, there may be hundreds or even millions of different terpenoid molecules in plants—terpenes and compounds with terpene moieties connected to other moieties originating from other pathways are all included in this encompassing phrase.

According to a recent study, terpenoids found in many plants have been utilized extensively in traditional medicine due to their potent anti-inflammatory and analgesic effects (Agatonovic-Kustrin and Morton 2018). It has long been assumed that the lineage-specific terpenoids that have developed during the development of green plants are involved in how plants interact ecologically with both biotic and abiotic elements of their surroundings. These include included signaling and rewarding beneficial creatures like pollinators and mycorrhiza, as well as providing defense against herbivores and diseases.

26.7.5 Saponins

One group of bioorganic substances is saponins. In more detail, these are naturally occurring glycosides that are characterized by their soap-like foaming ability, which causes them to froth when agitated in aqueous solutions.

A lipophilic triterpene molecule is joined with one or more hydrophilic glycoside sugar moieties to form saponins. According to scientific studies, saponins have therapeutic qualities like antibacterial and anti-inflammatory actions and are involved in the cytotoxic activity against malignant cells. Thus, saponins may offer a foundation for the creation of medications derived from natural sources (Biswas and Dwivedi 2019).

26.8 Case Studies Involving Medicinal Plants with Antibacterial Activity

26.8.1 Garlic (*Allium sativum*)

Garlic is a widely cultivated herb with a long history of medicinal use for various health conditions, including bacterial infections. It has been extensively studied for its antibacterial properties. Allicin, a sulfur-containing compound found in garlic, is primarily responsible for its antimicrobial activity (Bayan et al. 2014). Several studies have demonstrated garlic's efficacy against various bacterial pathogens, including *E. coli*, *S. aureus*, *P. aeruginosa*, *A. hydrophila*, *E. tarda*and *V. species*. The garlic extract inhibited the growth of multidrug-resistant strains of *S. aureus* and

E. coli isolated from clinical samples have been studied. Additionally, it has been found to disrupt bacterial cell membranes and biofilm formation, inhibit DNA replication, inhibit virulence factor expression in aquatic pathogens and modulate immune responses, contributing to its antibacterial effects in aquaculture systems (Table 26.2) (Hayat et al. 2022).

26.8.2 Tea Tree (*Melaleuca alternifolia*)

The Melaleuca is a tree of Australian origins. Tea tree oil is derived from the leaves of *Melaleuca alternifolia* by distillation method and it possesses strong antibacterial properties due to its high content of terpenes. The main bioactive component of tea tree is terpinen-4-ol ($C_{10}H_{18}O$), which exhibits potent antibacterial activity against a variety of bacterial pathogens. Numerous studies have demonstrated its oil efficacy against a wide range of bacterial pathogens, including *S. aureus*, *S. pyogenes*, *E. coli*, *V. species*, and *A. hydrophila*. Various research has shown that tea tree oil can inhibit microbial growth by disrupting bacterial cell membranes and enzyme functioning. It enhances the immune responses and leads to biofilm formation in aquatic pathogens leading to bactericidal effects. It has been observed clinically that tea tree oil has properties to treat bacterial infection, acne, and wounds (Table 26.2) (Hammer et al. 2012).

26.8.3 Turmeric (*Curcuma longa*)

Turmeric is a spice derived from the rhizomes of the Curcuma longa plant. It has religious views with numerous pharmaceutical properties. It contains the primary bioactive curcuminoid compounds (curcumin, demethoxy curcumin, and bisdemethoxy curcumin) which exhibit potent antibacterial activity against a wide range of bacterial pathogens. Several researches have shown that curcumin has a very effective role against the growth of Gram-positive bacteria such as *S. aureus* and *S. mutans* as well as Gram-negative bacteria such as *E. coli*, *P. aeruginosa*, and *H. pylori*. The bioactive compounds of *C. longa* are used as an antioxidant, antitumor, insecticide, repellent anti-inflammatory, antimicrobial, healing, and gastroprotective properties. Curcumin's antibacterial effects are attributed to its ability to disrupt the outer membrane of pathogens, inhibit bacterial enzymes and enhance the immune responses of the host against the invader pathogens. Clinical trials have also demonstrated that treating bacterial infections of the skin with turmeric extracts has been shown to inhibit bacterial growth and speedy wound healing (Table 26.2) (Mahato and Mahto 2023).

Table 26.2 Use of medicinal plants as an antibacterial

Medicinal plant name	Dose	Compound/extract(s)	Affected species	Effect	References
Astragalus membranaceus	100 mg/kg	Extract	*C. auratus*	Enhance resistance against illness	Wu (2020)
Adiantum capillus-veneris	2%	Leaves	*C. carpio*	Boost mucosal and serum bactericidal activity against several pathogenic bacteria	Hoseinifar et al. (2020)
Assam tea	2 g/kg	Extract	*O. niloticus*	Provide defense against infection caused by *S. agalactiae*	Van Doan et al. (2019)
Allium mongolicum Regel	40 mg/kg	Flavonoids	*Channaargus*	Enhance resistance to illness against *A. hydrophila*	Li et al. (2019)
Allium sativum	1.0 g	Extract	*O. niloticus*	Stop the spread of *S. iniae*	Foysal et al. (2019)
Allium sativum	200 mg/kg	Ethanolic extract	*O. mykiss*	Stop the growth of fungi in the culturing facility	Denji et al. (2020)
Berberine	1 g/kg	Hydroalcoholic extract	*O. niloticus*	Boost resistance to disease	Van Doan et al. (2020)
Cotinus coggygria	24 mg/32.34 g body weight	Methanolic extract	*O. mykiss*	Against infection caused by *A. hydrophila*	Bilen and Elbeshti (2019)
Coriandrum sativum	2%	Extract	*O. mykiss*	Boost defenses against infection by *Y. ruckeri*	Farsani et al. (2019)
Cucurbita mixta	4 and 6 g/kg	Powder	*O. mossambicus*	Boost resistance to *A. hydrophila* illness	Musthafa et al. (2017)
Eichhorniacrassipes	2.5% and 5%	Leaves	*Channapunctata*	Boost immunity against infections caused by *V. harveyi*	Verma et al. (2021)
Eleutherinebulbosa	12.5 g/kg	Powder	*L. vannamei*	Strengthen defenses against infections caused by *V. parahaemolyticus*	Munaeni et al. (2020)

Geniposide	100 mg/kg	Extract	*Crucian carp*	Increase resistance to illness from an infection with *A. hydrophila*	He et al. (2020)
Grape seed	20 and 30 g/kg	Extract	*C. carpio*	Boost resistance to *A. hydrophila*disease	Mehrinakhi et al. (2020)
Ginkgo biloba	10 g/kg	Leaf extract	*C. carpio*	Increase the common carp's survival rate following an *A. hydrophila* infection	Bao et al. (2019)
Ganoderma lucidum polysaccharides	2 g/kg	Extract	*M. rosenbergii*	Increase *A. hydrophila* resistance to illness	Mohan et al. (2019)
Jojoba	0.5–2 g/kg	Powder	*O. niloticus*	Against infection caused by *A. hydrophila*	Sarhan et al. (2019)
Juniperusexcelsa	4 and 8 mg/kg	Extract	*O. mykiss*	Offers defense against *Y. ruckeri*	Bilen et al. (2021)
Leucas aspera	8 g/kg	Extract	*O. niloticus*	Boost defenses against infection with *S. agalactiae*	Kurian et al. (2020)
Leucaena leucocephala	33%	Extract	Fish	Strengthen defenses against *P. aeruginosa* and *V. harveyi* infection	Verma et al. (2018)
Myrica esculenta	40 mg/L (immerse)	Ethanolic leaf extract	*O. mykiss*	Boost non-specific immunological markers and haematological indicators	Bhat et al. (2021)
Marjoram	200 mg/kg	Extract	*C. carpio*	Reduce the death rate of fish during *A. hydrophila*	Yousefi et al. (2021)
M. oleifera	40%	Leaves	*C. auratus gibelio*	Boost disease resistance through the TLR2 pathway against *A. hydrophila*	Zhang et al. (2020)

(continued)

Table 26.2 (continued)

Medicinal plant name	Dose	Compound/extract(s)	Affected species	Effect	References
Mentha longifolia	0.2%	Hydroalcoholic extract	*Rainbow trout*	Boost defenses against bacterial infections	Heydari et al. (2020)
Mucunapruriens	4–6 g/kg	Powder	*O. mossambicus*	Against infection caused by *A. hydrophila*	Musthafa et al. (2018)
Mentha piperita	3%	Extract	*O. mykiss*	Enhance defenses against *Y. Ruckeri*	Adel et al. (2016)
Melia azedarach	1.5 g/kg	Extract	*Catlacatla*	Against infection caused by *A. hydrophila*	Rajeshwari et al. (2016)
Mentha piperita	0.25%	Essential oil	*Nile tilapia*	Prevent against *Streptococcus agalactiae* infection	de Souza Silva et al. (2019)
Nigella sativa	7500 ppm	Extract	*L. vannamei*	Maintain vibriosis control	Nur et al. (2020)
Ocimum sanctum	100–200 mg/kg	Methanolic extract	*Epinephelustauvina*	Against infection with *V. harveyi*	Sivaram et al. (2004)
Ocimum gratissimum	12 g/kg	Leaf extract	*C. gariepinus*	Increase fish resistance to *Listeria monocytogenes*	Abdel-Tawwab et al. (2018)
Phyllanthus amarus	20 g/kg	Extract	*L. vannamei*	Increased resistance to *V. alginolyticus*	Huang et al. (2020)
Portulaca oleracea L.	1–3%	Leaf powder	*O. niloticus*	Enhance resistance to infections caused by *A. hydrophila*	Abdel-Razek et al. (2019)
Psidium guajava	1–10 mg/g	Leaf extracts	*O. mossambicus*	Boost defenses against *A. hydrophila*	Gobi et al. (2016)
Plant extracts	*Boesenbergia pandurata* 600 mg/L, *Solanum ferox* 900 mg/L and *Zingiber zerumbet* 200 mg/L	Mix *Boesenbergia pandurata*, *Solanum ferox* and *Zingiber zerumbet* extract	*O. niloticus*	Increase the resistance against *Pseudomonas fluorescens* and *A. hydrophila*	Hardi et al. (2019)

Rheum officinale	0.1, 1.0 g/kg	Extract	*E. coioides*	Against *Vibrio parahaemolyticus, V. vulnificus, V. alginolyticus, V. carchariae, A. hydrophila* and *E. tarda*	Kuo et al. (2020)
Rosemary	10 g/kg	Leaf powder	*O. niloticus*	Boost *O. niloticus*' ability to withstand illness	Naiel et al. (2020)
Radix Rehmanniae Preparata	0.2%	Extract	*Luciobarbuscapito*	Boost *L. capito*'s resilience to illness	Wu et al. (2019)
Rheum officinale	0.1%	Anthraquinone	*M. amblycephala*	Boost defenses against harmful pathogens	Liu et al. (2012)
Sargassum cristaefolium	10 and 20 g/kg	Leaves	*L. vannamei*	Prevent the growth of Vibrio spp. in the gut of *L.vannamei*	Jahromi et al. (2021)
San-Huang-San	500 mg/kg	Leaf powder	*L. vannamei*	Increased resistance against *V. parahaemolyticus*	Zhai and Li (2019)
Sonneratia caseolaris	3.17 mg/mL	Extract	*C. gariepinus*	Guard against the infection of *Edwardsiella tarda*	Aznan et al. (2018)
Tamarind	15 g/kg	Extract	*O. niloticus*L.	Guard against infection with *A. hydrophila*	Adeniyi et al. (2022)
Thymus vulgaris, Salvia officinalis, Eucalyptus globulus, Mentha piperita	0.77–6.18 mg/mL	Combination of essential oils	*S. aureus*	Potentially a fresh source of antimicrobial compounds for use in aquaculture in the sea	Mousavi et al. (2011)
Usnea barbata	4 mg/17.41 ± 0.3 g body weight/day	Extract	*O. mykiss*	Guard against infection from *Lactococcus garvieae*	Bilen et al. (2019)
Yucca	0.1%	Extract	*Nile tilapia*	Increase resistance to disease	El-Keredy and Naena (2020)

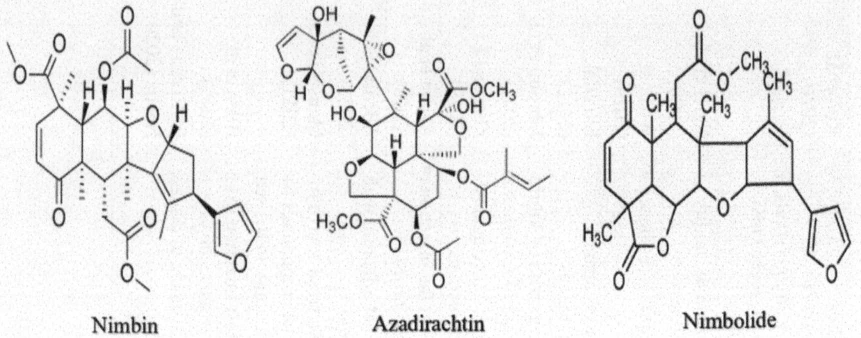

Fig. 26.3 Major phytochemicals of *Azadirachta indica*

26.8.4 Neem (*Azadirachta indica*)

Neem is a well-known medicinal plant, also known as Indian lilac, have been traditionally used in Ayurvedic medicine for its medicinal properties (Fig. 26.3). Neem extracts contain bioactive compounds such as Nimbin, NimbidinGedunin, Azadirachtin, Mahmoodin, and Cyclic trisulphide, which exhibit strong antimicrobial activity against various bacterial pathogens and antipyretic, anti-gastric ulcer, antifungal, and immunomodulatory properties (Eid et al. 2017). Its extract can inhibit the growth of Gram-positive bacteria such as *S. aureus* and *S. mutans*. Its bioactive compound disrupts the bacterial cell membranes by altering the chemistry of bacterial enzymes and metabolic compounds (Table 26.2).

Pharmaceutically it is also proved that neem-based products help in treating bacterial skin infections, dental caries, and periodontal diseases. Several studies also demonstrated that common pathogens in aquaculture viz. *Vibrio* species, *A. hydrophila* and *P. aeruginosa* are also inhibited by neem products. Neem extracts have been used in aquaculture settings to control bacterial infections in fish and shrimp (Sanjoy et al. 2013), demonstrating their potential as natural alternatives to synthetic antibiotics.

26.8.5 Oregano (*Origanum vulgare*)

Oregano essential oil, derived from the leaves of *Origanum vulgare*. It contains bioactive compounds such as carvacrol and thymol which exhibit potent antibacterial activity. Several studies have demonstrated oregano oil's efficacy against various bacterial pathogens including *Escherichia coli*, *Salmonella enterica*, and *Listeria monocytogenes*. Oregano feed supplements are used in aquaculture as immunomodulators, antioxidants, antiviral, antifungal, and antiparasitic. Carvacrol and thymol also have anti-stress, antipathogen, appetite-stimulating, and growth-promoting activities. It also improves the hematological parameters and immune response so these findings make the Oregano a natural antimicrobial agent for bacterial infections (Table 26.2) (Alagawany et al. 2020).

26.8.6 Aloe Vera (*Aloe barbadensis*)

Aloe vera extracts containing bioactive compounds such as polysaccharides, phenolics, and anthraquinones have been investigated for their antibacterial activity against aquatic bacterial pathogens. Aloe vera extract having aloin and emodin bioactive phenolic compounds has analgesic, antibiotic, and antiviral properties that inhibit the growth of *Aeromonas hydrophila*, *Vibrio species*, and *Pseudomonas aeruginosa*, which are the most common pathogens of aquaculture. There are several enzymes and vitamins found in its extract that help in proper digestion and health. Several studies demonstrated that it has growth-promoting and immunomodulatory properties that provide resistance against pathogens (Table 26.2) (Mahdavi et al. 2013).

26.8.7 Cranberry (*Vaccinium macrocarpon*)

Cranberry extracts containing bioactive compounds such as proanthocyanidins have been studied for their antibacterial activity against aquatic bacterial pathogens (Ofek et al. 1991). Studies have shown that cranberry extracts can inhibit the growth of bacteria such as *Escherichia coli*, *Salmonella species*, and *Pseudomonas aeruginosa* which are common contaminants in aquatic environments (Howell et al. 1998). Cranberry extracts have been shown to disrupt bacterial biofilms inhibit and quorum-sensing signaling (Bauer et al. 2005).

26.8.8 Basil (*Ocimumbasilicum*)

Basil extracts containing bioactive compounds such as eugenol and linalool have been investigated for their antibacterial activity against aquatic bacterial pathogens. Eugenol and linalool are phenolic compounds with potent antimicrobial properties. Studies have shown that basil extracts can inhibit the growth of bacteria such as *A. hydrophila*, *Vibrio species*, and *P. aeruginosa* which are common pathogens in aquatic environments. Basil extracts have been shown to disrupt bacterial biofilms, inhibit quorum sensing signaling, and modulate host immune responses (Table 26.2) (El-Ekiaby 2019).

26.9 Challenges and Future Perspective

Although there has been a surge in research on the identification and characterization of novel chemicals derived from plants, there has been limited success in their on-site application in agriculture, and their efficaciousness has been uneven (Lee et al. 2021). To understand the pharmacodynamics and underlying mechanism of phytocompounds, as well as the appropriate dosage, in vivo investigations are necessary. Aquaculture presents a challenge when it comes to translating in vitro studies into in vivo applications. Certain phytocompounds are unsafe to use by laypeople since overdoses are common with these compounds due to effective concentration (EC50) and acute toxicity (LC50) values (Yao et al. 2011).

Aquaculture capitalizes on the elements of feasibility and economic viability. Due to their high practicability and lower cost, aquatic animals can receive part or all the medicinal plants orally as feed additives instead of a refined phytocompound (Chong et al. 2020). This approach could have unfavorable effects; for instance, fish growth is stunted by a high dietary fiber content (Zhuo et al. 2021). Before the medicinal plant is added to fish feed, further procedures like fermentation have been suggested to lower its fiber content (Zhuo et al. 2021; Hassaan et al. 2018). For any medicinal plant of interest, in-depth research on a variety of topics is required, such as the preservation of bioactive phytocompounds and fermentation conditions. Additionally, these phytocompounds have demonstrated their effectiveness against fish diseases such as bacteria, viruses, and parasites by functioning as antimicrobial agents. Because the amount of product compounds can vary, it is also crucial to apply different extraction techniques that are appropriate for different compounds. Most notably, there is a limited production of phytocompounds isolated from plants, making them exclusively useful for research. Solving this issue is essential since more study is required to create the chemicals needed for the aquaculture sector.

There is a chance that phytocompounds will replace antibacterial agents in aquaculture. Advanced research and emerging technologies have demonstrated a broad spectrum of potential actions in phytocompounds, with medium to high levels of antibacterial activity against both marine and freshwater diseases (Vaseeharan and Thaya 2014). The use of these phytocompounds in aquaculture would undoubtedly be a step in the right direction for the use of herbs and medicinal plants. However, further study on natural products will be required in the future to fully utilize medicinal plants as well as to take advantage of the safety benefits that phytocompounds provide, which may significantly outweigh medicines derived from synthetic or semi-synthetic sources (Ghosh et al. 2012). To avoid or reduce the spread of infectious diseases in farmed fish, it is essential to use modern technologies for disease diagnosis in conjunction with efficient treatment, prevention, and control measures.

26.9.1 Potential Strategies for Maximizing the Benefits of Medicinal Plants in Combating Bacterial Infections in Aquatic Ecosystem

There are advantages and disadvantages to using medicinal plant's ability to treat bacterial infections in aquatic environments. These natural remedies offer promising alternatives to synthetic antibiotics and their effective integration into aquaculture practices. By overcoming barriers related to standardization, efficacy assessment, conservation, regulation, and knowledge sharing, we can unlock the full potential of medicinal plants as sustainable and effective remedies for promoting the health and resilience of aquatic ecosystems. Through concerted efforts and innovation, we can pave the way toward a future where nature's pharmacy plays a central role in safeguarding the health and sustainability of aquatic environments for generations to come.

26.9.1.1 Standardization and Quality Control

One of the primary challenges in utilizing medicinal plants for combating bacterial infections in the aquatic environment is ensuring standardization and quality control of plant-based products. Variability in plant composition, growing conditions, and extraction methods can lead to inconsistent efficacy and safety profiles. Establishing standardized protocols for plant cultivation, harvesting, extraction, and formulation is essential to ensure the reproducibility and reliability of medicinal plant-based therapies.

26.9.1.2 Efficacy and Safety Assessment

Rigorous evaluation of the efficacy and safety of medicinal plants is crucial for their successful integration into aquaculture practices. Comprehensive studies are needed to assess the antibacterial activity, pharmacokinetics, pharmacodynamics, and toxicity of plant extracts and compounds in aquatic organisms. In vitro and in vivo studies, as well as clinical trials in aquaculture settings can provide valuable insights into the therapeutic potential and safety profiles of medicinal plants.

26.9.1.3 Ecological Impact and Conservation

The sustainable utilization of medicinal plants requires careful consideration of their ecological impact and conservation status. Overharvesting, habitat destruction, and unsustainable harvesting practices can lead to depletion of plant populations and loss of biodiversity. Implementing sustainable harvesting methods, promoting the cultivation of medicinal plants in agro-forestry systems, and supporting community-based conservation initiatives are essential for preserving plant biodiversity and ensuring the long-term availability of medicinal plants.

26.9.1.4 Regulatory Considerations

Regulatory frameworks governing the use of medicinal plants in aquaculture vary between countries and regions, posing challenges for their widespread adoption. Clear guidelines and regulations are needed to ensure the safety, efficacy, and quality of plant-based products for aquatic use. Collaboration between regulatory agencies, researchers, industry stakeholders, and traditional healers can facilitate the development of evidence-based regulatory frameworks that support the responsible utilization of medicinal plants in aquaculture.

26.9.1.5 Knowledge Gaps and Research Priorities

Despite growing interest in medicinal plants for combating bacterial infections in the aquatic environment, there are still significant knowledge gaps and research priorities that need to be addressed. Further studies are needed to elucidate the mechanisms of action, optimize extraction methods, identify synergistic interactions between plant compounds, and explore novel plant species with antimicrobial properties. Interdisciplinary research collaborations and funding support for plant-based aquaculture research can help accelerate progress in this field.

26.9.1.6 Community Engagement and Knowledge Sharing

Engaging with local communities, traditional healers, and indigenous knowledge holders is essential for integrating traditional wisdom into modern scientific approaches to medicinal plant-based therapies. Knowledge sharing, capacity building, and participatory research approaches can foster mutual learning, empower local communities, and promote culturally sensitive and context-specific solutions for combating bacterial infections in aquatic ecosystems.

26.10 Conclusions

Various communities have been cultivating traditional medicinal plants for ages in order to treat a wide range of infectious ailments. Crude extracts and bioactive chemicals derived from plants may strengthen the effects of conventional antimicrobials, which may be more affordable, have less side effects, and increase the quality of therapy, according to numerous studies on the pharmacogenetics and kinetics of medicinal plants. Numerous studies have demonstrated the potential for plant extracts and their active compounds' antimicrobial activity to promote cell wall disruption and lysis, induce the production of reactive oxygen species, inhibit the formation of biofilms, inhibit the construction of cell walls, inhibit the replication of microbial DNA, inhibit the synthesis of energy, and inhibit the toxicity of bacteria to the host. Furthermore, these substances have the potential to inhibit antibacterial resistance and work in concert with antibiotics to eradicate harmful organisms. The production of traditional plant extracts and bioactive substances presents a possible treatment for disease-causing pathogenic microbial pathogens in aquaculture, based on their comprehensive antimicrobial processes.

The critical role of medicinal plants in addressing bacterial infections within aquatic ecosystems offers an ideal hope amidst the challenges posed by antibiotic-resistant pathogens and the sustainability of aquaculture. The diversity of microbial communities in aquatic environments is essential for ecosystem balance, yet the presence of pathogenic bacteria threatens both human health and ecological equilibrium. The proliferation of antibiotic-resistant bacteria further intensifies these concerns, necessitating innovative solutions beyond conventional treatments. In response to these challenges, there is a renewed emphasis on natural antimicrobial sources, with medicinal plants emerging as promising candidates. Through their rich repertoire of bioactive chemicals and diverse modes of action, medicinal plants exhibit significant efficacy against aquatic bacterial infections. Notable examples such as neem, garlic, turmeric, basil, aloe vera, and echinacea demonstrate potent antibacterial properties, offering a sustainable alternative to synthetic antibiotics. The integration of medicinal plants into aquaculture practices requires a comprehensive approach that encompasses scientific rigor, ethnopharmacological insights, and respect for traditional knowledge. Rigorous evaluation of efficacy, safety, and sustainability is paramount, encompassing toxicity assessment, conservation efforts, and the development of standardized protocols for plant-based therapies. By harnessing the potential of medicinal plants, we can address the global challenge of antibiotic resistance in aquatic environments while safeguarding the resilience and health of aquatic ecosystems for future generations. Through responsible utilization of natural sources and interdisciplinary collaboration, we can pave the way toward sustainable aquaculture practices and environmental preservation. Indeed, the utilization of medicinal plants represents a harmonious synergy between human innovation and nature's pharmacy, offering hope for a healthier, more resilient aquatic world.

References

Abdel-Razek N, Awad SM, Abdel-Tawwab M (2019) Effect of dietary purslane (*Portulaca oleracea* L.) leaves powder on growth, immunostimulation, and protection of Nile tilapia, *Oreochromis niloticus* against *Aeromonas hydrophila*infection. Fish Physiol Biochem 45(6):1907–1917

Abdel-Tawwab M, Adeshina I, Jenyo-Oni A, Ajani EK, Emikpe BO (2018) Growth, physiological, antioxidants, and immune response of African catfish, *Clarias gariepinus* (B.), to dietary clove basil, *Ocimum gratissimum*, leaf extract and its susceptibility to Listeria monocytogenes infection. Fish Shellfish Immunol 78:346–354

Adel M, Pourgholam R, Zorriehzahra J, Ghiasi M (2016) Hemato–Immunological and biochemical parameters, skin antibacterial activity, and survival in rainbow trout (*Oncorhynchus mykiss*) following the diet supplemented with *Mentha piperita* against *Yersinia ruckeri*. Fish Shellfish Immunol 55:267–273

Adeniyi OV, Olaifa FE, Emikpe BO (2022) Effects of dietary tamarind pulp extract on growth performance, nutrient digestibility, intestinal morphology, and resistance to *Aeromonas hydrophila* infection in Nile tilapia (*Oreochromis niloticus* L.). J Appl Aquac 34(1):43–63

Agatonovic-Kustrin S, Morton DW (2018) The cosmeceutical properties of compounds derived from marine algae. Mar Macro-Microalgae Overv 198

Akinyemi KO, Oladapo O, Okwara CE, Ibe CC, Fasure KA (2005) Screening of crude extracts of six medicinal plants used in south-west Nigerian unorthodox medicine for anti-methicillin resistant Staphylococcus aureus activity. BMC Complement Altern Med 5:1–7

Alagawany M, Farag MR, Salah AS, Mahmoud MA (2020) The role of oregano herb and its derivatives as immunomodulators in fish. Rev Aquac 12(4):2481–2492

Allen HK, Donato J, Wang HH, Cloud-Hansen KA, Davies J, Handelsman J (2010) Call of the wild: antibiotic resistance genes in natural environments. Nat Rev Microbiol 8(4):251–259

Alves RR, Rosa IL, Santana GG (2007) The role of animal-derived remedies as complementary medicine in Brazil. Bioscience 57(11):949–955

Anderson DM, Cembella AD, Hallegraeff GM (2012) Progress in understanding harmful algal blooms: paradigm shifts and new technologies for research, monitoring, and management. Annu Rev Mar Sci 4:143–176

Ashbolt NJ (2015) Environmental (saprozoic) pathogens of engineered water systems: understanding their ecology for risk assessment and management. Pathogens 4(2):390–405

Aznan AS, Lee KL, Low CF, Iberahim NA, Ibrahim WNW, Musa N et al (2018) Protective effect of apple mangrove *Sonneratia caseolaris* extract in Edwardsiellatarda-infected African catfish, *Clarias gariepinus*. Fish Shellfish Immunol 78:338–345

Bao L, Chen Y, Li H, Zhang J, Wu P, Ye KE et al (2019) Dietary Ginkgo biloba leaf extract alters immune-related gene expression and disease resistance to *Aeromonas hydrophila* in common carp *Cyprinus carpio*. Fish Shellfish Immunol 94:810–818

Bartram J, Cairncross S (2010) Hygiene, sanitation, and water: forgotten foundations of health. PLoS Med 7(11):e1000367

Bauer HW, Alloussi S, Egger G, Blümlein HM, Cozma G, Schulman CC, Multicenter UTI Study Group (2005) A long-term, multicenter, double-blind study of an *Escherichia coli* extract (OM-89) in female patients with recurrent urinary tract infections. Eur Urol 47(4):542–548

Bayan L, Koulivand PH, Gorji A (2014) Garlic: a review of potential therapeutic effects. Avicenna J Phytomed 4(1):1

Bengtsson-Palme J, Kristiansson E, Larsson DJ (2018) Environmental factors influencing the development and spread of antibiotic resistance. FEMS Microbiol Rev 42(1):fux053

Berendonk TU, Manaia CM, Merlin C, Fatta-Kassinos D, Cytryn E, Walsh F et al (2015) Tackling antibiotic resistance: the environmental framework. Nat Rev Microbiol 13(5):310–317

Bhat RAH, Rehman S, Tandel RS, Dash P, Bhandari A, Ganie PA et al (2021) Immunomodulatory and antimicrobial potential of ethanolic extract of Himalayan *Myricaesculanta* in *Oncorhynchus mykiss*: molecular modelling with *Aeromonas hydrophila* functional proteins. Aquaculture 533:736213

Bhullar K, Waglechner N, Pawlowski A, Koteva K, Banks ED, Johnston MD et al (2012) Antibiotic resistance is prevalent in an isolated cave microbiome. PLoS One 7(4):e34953

Bilen S, Elbeshti HTAG (2019) A new potential therapeutic remedy against *Aeromonas hydrophila* infection in rainbow trout (*Oncorhynchus mykiss*) using tetra, Cotinus coggygria. J Fish Dis 42(10):1369–1381

Bilen S, Sirtiyah AMA, Terzi E (2019) Therapeutic effects of beard lichen, *Usnea barbata* extract against *Lactococcus garvieae* infection in rainbow trout (*Oncorhynchus mykiss*). Fish Shellfish Immunol 87:401–409

Bilen S, Ispir S, Kenanoglu ON, Taştan Y, Güney K, Terzi E (2021) Effects of Greek juniper (*Juniperusexcelsa*) extract on immune responses and disease resistance against *Yersinia ruckeri* in rainbow trout (*Oncorhynchus mykiss*). J Fish Dis 44(6):729–738

Biswas T, Dwivedi UN (2019) Plant triterpenoid saponins: biosynthesis, in vitro production, and pharmacological relevance. Protoplasma 256:1463–1486

Borges A, Saavedra J, M., & Simoes, M. (2015) Insights on antimicrobial resistance, biofilms and the use of phytochemicals as new antimicrobial agents. Curr Med Chem 22(21):2590–2614

Buckingham J, Baggaley KH, Roberts AD, Szabo LF (eds) (2010) Dictionary of alkaloids with CD-ROM. CRC Press

Burridge L, Weis JS, Cabello F, Pizarro J, Bostick K (2010) Chemical use in salmon aquaculture: a review of current practices and possible environmental effects. Aquaculture 306(1–4):7–23

Cabello FC (2006) Heavy use of prophylactic antibiotics in aquaculture: a growing problem for human and animal health and for the environment. Environ Microbiol 8(7):1137–1144. https://doi.org/10.1111/j.1462-2920.2006.01054.x

Cardinale BJ, Duffy JE, Gonzalez A, Hooper DU, Perrings C, Venail P et al (2012) Biodiversity loss and its impact on humanity. Nature 486(7401):59–67

Chen J, Yang J, Ma L, Li J, Shahzad N, Kim CK (2020) Structure-antioxidant activity relationship of methoxy, phenolic hydroxyl, and carboxylic acid groups of phenolic acids. Sci Rep 10(1):2611

Chong CM, Murthy AG, Choy CY, Lai KS (2020) Phytotherapy in aquaculture: integration of endogenous application with science. J Environ Biol 41:1204–1214

Costanza R, De Groot R, Sutton P, Van der Ploeg S, Anderson SJ, Kubiszewski I et al (2014) Changes in the global value of ecosystem services. Glob Environ Chang 26:152–158

Cowan MM (1999) Plant products as antimicrobial agents. Clin Microbiol Rev 12(4):564–582

Cragg GM, Newman DJ (2005) Biodiversity: a continuing source of novel drug leads. Pure Appl Chem 77(1):7–24

Cragg GM, Newman DJ (2013) Natural products: a continuing source of novel drug leads. Biochim Biophys Acta 1830(6):3670–3695

Cushnie TT, Lamb AJ (2011) Recent advances in understanding the antibacterial properties of flavonoids. Int J Antimicrob Agents 38(2):99–107

D'Costa VM, King CE, Kalan L, Morar M, Sung WW, Schwarz C et al (2011) Antibiotic resistance is ancient. Nature 477(7365):457–461

Daskalov GM, Grishin AN, Rodionov S, Mihneva V (2007) Trophic cascades triggered by overfishing reveal possible mechanisms of ecosystem regime shifts. Proc Natl Acad Sci 104(25):10518–10523

Davies J, Davies D (2010) Origins and evolution of antibiotic resistance. Microbiol Mol Biol Rev 74(3):417–433

De la Fuente-Núñez C, Reffuveille F, Haney EF, Straus SK, Hancock RE (2014) Broad-spectrum anti-biofilm peptide that targets a cellular stress response. PLoS Pathog 10(5):e1004152

de Souza Silva LT, de Pádua Pereira U, de Oliveira HM, Brasil EM, Pereira SA, Chagas EC et al (2019) Hemato-immunological and zootechnical parameters of Nile tilapia fed essential oil of Mentha piperita after challenge with *Streptococcus agalactiae*. Aquaculture 506:205–211

Defoirdt T, Boon N, Sorgeloos P, Verstraete W, Bossier P (2007) Alternatives to antibiotics to control bacterial infections: luminescent vibriosis in aquaculture as an example. Trends Biotechnol 25(10):472–479

Defoirdt T, Sorgeloos P, Bossier P (2011) Alternatives to antibiotics for the control of bacterial disease in aquaculture. Curr Opin Microbiol 14(3):251–258

Denji KA, Soltani M, Islami HR, Kamali A (2020) The antifungal effects of Allium sativum and Artemisia sieberi extracts on hatching rate and survival of rainbow trout *Oncorhynchus mykiss* (Walbaum, 1792) larvae

Dewick PM (2002) Medicinal natural products: a biosynthetic approach. Wiley

Dicko MH, Gruppen H, Traoré AS, van Berkel WJ, Voragen AG (2005) Evaluation of the effect of germination on phenolic compounds and antioxidant activities in sorghum varieties. J Agric Food Chem 53(7):2581–2588

Donlan RM (2002) Biofilms: microbial life on surfaces. Emerg Infect Dis 8(9):881

Donlan RM, Costerton JW (2002) Biofilms: survival mechanisms of clinically relevant microorganisms. Clin Microbiol Rev 15(2):167–193

Eid A, Jaradat N, Elmarzugi N (2017) A review of chemical constituents and traditional usage of neem plant (*Azadirachta Indica*). Palest Med Pharm J 2(2):3

El-Ekiaby WT (2019) Basil oil nanoemulsion formulation and its antimicrobial activity against fish pathogen and enhance disease resistance against *Aeromonas hydrophila* in cultured Nile tilapia. Egypt J Aquat 9(4):13–33

El-Keredy MA, Naena NA (2020) Yucca plant as treatment for *Pseudomonas aeruginosa* infection in Nile tilapia farms with emphasis on its effect on growth performance

Fabricant DS, Farnsworth NR (2001) The value of plants used in traditional medicine for drug discovery. Environ Health Perspect 109(suppl 1):69–75

Farnsworth NR, Akerele RO, Bingel AS, Soejarto DD, Guo Z (1985) Medicinal plants in therapy. Bull World Health Organ 63:965–981

Farsani MN, Hoseinifar SH, Rashidian G, Farsani HG, Ashouri G, Van Doan H (2019) Dietary effects of *Coriandrum sativum* extract on growth performance, physiological and innate immune responses and resistance of rainbow trout (*Oncorhynchus mykiss*) against *Yersinia ruckeri*. Fish Shellfish Immunol 91:233–240

Ferreira FS, Brito SV, Ribeiro SC, Saraiva AA, Almeida WO, Alves RR (2009) Animal-based folk remedies sold in public markets in Crato and Juazeiro do Norte, Ceará, Brazil. BMC Complement Altern Med 9:1–8

Flemming HC, Wingender J (2010) The biofilm matrix. Nat Rev Microbiol 8(9):623–633

Flemming HC, Wingender J, Szewzyk U, Steinberg P, Rice SA, Kjelleberg S (2016) Biofilms: an emergent form of bacterial life. Nat Rev Microbiol 14(9):563–575

Forsberg KJ, Reyes A, Wang B, Selleck EM, Sommer MO, Dantas G (2012) The shared antibiotic resistome of soil bacteria and human pathogens. Science 337(6098):1107–1111

Foysal MJ, Alam M, Momtaz F, Chaklader MR, Siddik MA, Cole A et al (2019) Dietary supplementation of garlic (*Allium sativum*) modulates gut microbiota and health status of tilapia (*Oreochromis niloticus*) against Streptococcus iniae infection. Aquac Res 50(8):2107–2116

Ghosh S, Chisti Y, Banerjee UC (2012) Production of shikimic acid. Biotechnol Adv 30(6):1425–1431

Gobi N, Ramya C, Vaseeharan B, Malaikozhundan B, Vijayakumar S, Murugan K, Benelli G (2016) *Oreochromis mossambicus* diet supplementation with *Psidium guajava* leaf extracts enhance growth, immune, antioxidant response and resistance to *Aeromonas hydrophila*. Fish Shellfish Immunol 58:572–583

Gupta P, Sarkar S, Das B, Bhattacharjee S, Tribedi P (2016) Biofilm, pathogenesis and prevention—a journey to break the wall: a review. Arch Microbiol 198:1–15

Hallegraeff GM (2010) Ocean climate change, phytoplankton community responses, and harmful algal blooms: a formidable predictive challenge. J Phycol 46(2):220–235

Hamilton-Miller JM (1995) Antimicrobial properties of tea (*Camellia sinensis* L.). Antimicrob Agents Chemother 39(11):2375–2377

Hammer KA, Carson CF, Riley TV (2012) Effects of *Melaleuca alternifolia* (tea tree) essential oil and the major monoterpene component terpinen-4-ol on the development of single-and multistep antibiotic resistance and antimicrobial susceptibility. Antimicrob Agents Chemother 56(2):909–915

Hardi EH, Nugroho RA, Kusuma I, Apriza A (2019) Immunomodulatory effect and disease resistance from concoction three of Borneo plant extracts in tilapia, *Oreochromis niloticus*. J Aquac Indones 20:41–47

Harvell CD, Mitchell CE, Ward JR, Altizer S, Dobson AP, Ostfeld RS, Samuel MD (2002) Climate warming and disease risks for terrestrial and marine biota. Science 296(5576):2158–2162

Hassaan MS, Soltan MA, Mohammady EY, Elashry MA, El-Haroun ER, Davies SJ (2018) Growth and physiological responses of Nile tilapia, *Oreochromis niloticus* fed dietary fermented sunflower meal inoculated with *Saccharomyces cerevisiae* and *Bacillus subtilis*. Aquaculture 495:592–601

Hatosy SM, Martiny AC (2015) The ocean as a global reservoir of antibiotic resistance genes. Appl Environ Microbiol 81(21):7593–7599

Hayat S, Ahmad A, Ahmad H, Hayat K, Khan MA, Runan T (2022) Garlic, from medicinal herb to possible plant bioprotectant: a review. Sci Hortic 304:111296

He M, Liu G, Liu Y, Yang K, Qi X, Huang A et al (2020) Effects of geniposide as immunostimulant on the innate immune response and disease resistance in *Crucian carp*. Aquaculture 529:735713

Heinrich M, Ankli A, Frei B, Weimann C, Sticher O (1998) Medicinal plants in Mexico healers' consensus and cultural importance. Soc Sci Med 47(11):1859–1871

Heydari M, Firouzbakhsh F, Paknejad H (2020) Effects of Mentha longifolia extract on some blood and immune parameters, and disease resistance against yersiniosis in rainbow trout. Aquaculture 515:734586

Hoseinifar SH, Jahazi MA, Mohseni R, Raeisi M, Bayani M, Mazandarani M et al (2020) Effects of dietary fern (*Adiantum capillus-veneris*) leaves powder on serum and mucus antioxidant defence, immunological responses, antimicrobial activity and growth performance of common carp (*Cyprinus carpio*) juveniles. Fish Shellfish Immunol 106:959–966

Howell AB, Vorsa N, Marderosian AD, Foo LY (1998) Inhibition of the adherence of P-fimbriated *Escherichia coli* to uroepithelial-cell surfaces by proanthocyanidin extracts from cranberries. N Engl J Med 339(15):1085–1086

Huang HT, Lee PT, Liao ZH, Chen HY, Nan FH (2020) Effects of Phyllanthus amarus extract on nonspecific immune responses, growth, and resistance to *Vibrio alginolyticus* in white shrimp *Litopenaeusvannamei*. Fish Shellfish Immunol 107:1–8

Huigens RW III, Morrison KC, Hicklin RW, Flood TA Jr, Richter MF, Hergenrother PJ (2013) A ring-distortion strategy to construct stereochemically complex and structurally diverse compounds from natural products. Nat Chem 5(3):195–202

Jahromi ST, Pourmozaffar S, Jahanbakhshi A, Rameshi H, Gozari M, Khodadadi M et al (2021) Effect of different levels of dietary *Sargassum cristaefolium* on growth performance, hematological parameters, histological structure of hepatopancreas and intestinal microbiota of Litopenaeusvannamei. Aquaculture 533:736130

Jutkina J, Rutgersson C, Flach CF, Larsson DJ (2016) An assay for determining minimal concentrations of antibiotics that drive horizontal transfer of resistance. Sci Total Environ 548:131–138

Karkman A, Do TT, Walsh F, Virta MP (2018) Antibiotic-resistance genes in waste water. Trends Microbiol 26(3):220–228

Knapp CW, Dolfing J, Ehlert PA, Graham DW (2010) Evidence of increasing antibiotic resistance gene abundances in archived soils since 1940. Environ Sci Technol 44(2):580–587

Kohen R, Nyska A (2002) Invited review: oxidation of biological systems: oxidative stress phenomena, antioxidants, redox reactions, and methods for their quantification. Toxicol Pathol 30(6):620–650

Kristiansson E, Fick J, Janzon A, Grabic R, Rutgersson C, Weijdegård B et al (2011) Pyrosequencing of antibiotic-contaminated river sediments reveals high levels of resistance and gene transfer elements. PLoS One 6(2):e17038

Kumar S, Pandey AK (2013) Chemistry and biological activities of flavonoids: an overview. Sci World J 2013:1

Kuo IP, Lee PT, Nan FH (2020) Rheum officinale extract promotes the innate immunity of orange-spotted grouper (*Epinepheluscoioides*) and exerts strong bactericidal activity against six aquatic pathogens. Fish Shellfish Immunol 102:117–124

Kurian A, Van Doan H, Tapingkae W, Elumalai P (2020) Modulation of mucosal parameters, innate immunity, growth and resistance against Streptococcus agalactiae by enrichment of Nile tilapia (*Oreochromis niloticus*) diet with *Leucas aspera*. Fish Shellfish Immunol 97:165–172

Lee PT, Yamamoto FY, Low CF, Loh JY, Chong CM (2021) Gut immune system and the implications of oral-administered immunoprophylaxis in finfish aquaculture. Front Immunol 12:773193

Li M, Zhu X, Tian J, Liu M, Wang G (2019) Dietary flavonoids from *Allium mongolicum* regel promotes growth, improves immune, antioxidant status, immune-related signaling molecules and disease resistance in juvenile northern snakehead fish (*Channaargus*). Aquaculture 501:473–481

Lipinski CA, Lombardo F, Dominy BW, Feeney PJ (2012) Experimental and computational approaches to estimate solubility and permeability in drug discovery and development settings. Adv Drug Deliv Rev 64:4–17

Liu B, Xie J, Ge X, Xu P, Miao L, Zhou Q et al (2012) Comparison study of the effects of anthraquinone extract and emodin from *Rheum officinale* bail on the physiological response, disease resistance of *Megalobramaamblycephala* under high temperature stress. Turk J Fish Aquat Sci 12(4)

Louca S, Parfrey LW, Doebeli M (2016) Decoupling function and taxonomy in the global ocean microbiome. Science 353(6305):1272–1277

Mah TFC, O'Toole GA (2001) Mechanisms of biofilm resistance to antimicrobial agents. Trends Microbiol 9(1):34–39

Mahato D, Mahto H (2023) A review on the golden plant turmeric and its bioactive compound curcumin. In: Biosynthesis of bioactive compounds in medicinal and aromatic plants: manipulation by conventional and biotechnological approaches. Springer, pp 351–367

Mahdavi M, Hajimoradloo A, Ghorbani R (2013) Effect of Aloe vera extract on growth parameters of common carp (*Cyprinus carpio*). World J Med Sci 9(1):55–60
Martínez JL (2008) Antibiotics and antibiotic resistance genes in natural environments. Science 321(5887):365–367
Martinez JL (2009) Environmental pollution by antibiotics and by antibiotic resistance determinants. Environ Pollut 157(11):2893–2902
Mehrinakhi Z, Ahmadifar E, Sheikhzadeh N, Moghadam MS, Dawood MA (2020) Extract of grape seed enhances the growth performance, humoral and mucosal immunity, and resistance of common carp against. Ann Anim Sci 21(1):217–232
Mohan K, Muralisankar T, Uthayakumar V, Chandirasekar R, Rajan DK (2019) Dietary *Ganoderma lucidum* polysaccharides to enhance the growth, immune response and disease resistance of freshwater prawn *Macrobrachiumrosenbergii*. Aquac Rep 14:100203
Mousavi SM, Wilson G, Raftos D, Mirzargar SS, Omidbaigi R (2011) Antibacterial activities of a new combination of essential oils against marine bacteria. Aquac Int 19:205–214
Munaeni W, Yuhana M, Setiawati M, Wahyudi AT (2020) Effect in white shrimp *Litopenaeusvannamei* of *Eleutherinebulbosa* (Mill.) Urb. Powder on immune genes expression and resistance against *Vibrio parahaemolyticus* infection. Fish Shellfish Immunol 102:218–227
Musthafa MS, Ali ARJ, Kumar MSA, Paray BA, Al-Sadoon MK, Balasundaram C, Harikrishnan R (2017) Effect of *Cucurbita mixta* (L.) seed meal enrichment diet on growth, immune response and disease resistance in *Oreochromis mossambicus*. Fish Shellfish Immunol 68:509–515
Musthafa MS, Asgari SM, Kurian A, Elumalai P, Ali ARJ, Paray BA, Al-Sadoon MK (2018) Protective efficacy of *Mucunapruriens* (L.) seed meal enriched diet on growth performance, innate immunity, and disease resistance in *Oreochromis mossambicus* against *Aeromonas hydrophila*. Fish Shellfish Immunol 75:374–380
Naiel MA, Ismael NE, Negm SS, Ayyat MS, Al-Sagheer AA (2020) Rosemary leaf powder–supplemented diet enhances performance, antioxidant properties, immune status, and resistance against bacterial diseases in Nile tilapia (*Oreochromis niloticus*). Aquaculture 526:735370
Nascimento GG, Locatelli J, Freitas PC, Silva GL (2000) Antibacterial activity of plant extracts and phytochemicals on antibiotic-resistant bacteria. Braz J Microbiol 31:247–256
Newman DJ, Cragg GM (2012) Natural products as sources of new drugs over the 30 years from 1981 to 2010. J Nat Prod 75(3):311–335
Newman DJ, Cragg GM (2016) Natural products as sources of new drugs from 1981 to 2014. J Nat Prod 79(3):629–661
Newman DJ, Cragg GM (2020) Natural products as sources of new drugs over the nearly four decades from 01/1981 to 09/2019. J Nat Prod 83(3):770–803
Nostro A, Cannatelli MA, Crisafi G, Musolino AD, Procopio F, Alonzo V (2004) Modifications of hydrophobicity, in vitro adherence and cellular aggregation of *Streptococcus mutans* by *Helichrysumitalicum* extract. Lett Appl Microbiol 38(5):423–427
Nur I, Munaeni W, Abidin LOB (2020) Assessment of antibacterial and immunostimulating activity of black cumin (*Nigella sativa*) extract against vibriosis in white shrimp (*Litopenaeusvannamei*). Thai J Vet Med 50(4):549–557
Ochman H, Lawrence JG, Groisman EA (2000) Lateral gene transfer and the nature of bacterial innovation. Nature 405(6784):299–304
OECD/FAO Agricultural Outlook (2018–2027). http://stats.oecd.org. Accessed 30 July 2019
Ofek I, Goldhar J, Sharon N (1991) Anti-*Escherichia coli* adhesin activity of cranberry and blueberry juices. In: Toward anti-adhesion therapy for microbial diseases, vol 179
Panche AN, Diwan AD, Chandra SR (2016) Flavonoids: an overview. J Nutr Sci 5:e47
Park SH (2014) Third-generation cephalosporin resistance in gram-negative bacteria in the community: a growing public health concern. Korean J Intern Med 29(1):27
Pietta PG (2000) Flavonoids as antioxidants. J Nat Prod 63(7):1035–1042
Pruden A, Pei R, Storteboom H, Carlson KH (2006) Antibiotic resistance genes as emerging contaminants: studies in northern Colorado. Environ Sci Technol 40(23):7445–7450
Prüss A (1998) Review of epidemiological studies on health effects from exposure to recreational water. Int J Epidemiol 27(1):1–9

Rajeshwari S, Rajan MK, Pavaraj M, Sevarkodiyone SP (2016) Effect of *Melia azedarach* extract on some selected physiological parameters of (Catlacatla). Int J Aquac Fish Sci 2(1):27–30

Rios JL, Recio MC (2005) Medicinal plants and antimicrobial activity. J Ethnopharmacol 100(1–2):80–84

Roy R, Tiwari M, Donelli G, Tiwari V (2018) Strategies for combatting bacterial biofilms: a focus on anti-biofilm agents and their mechanisms of action. Virulence 9(1):522–554

Sanjoy B, Kim L, Mohamed S, Khatoon H, Yusoff FM (2013) Antibacterial activity of neem (*Azadirachtaindica*) leaves on Vibrio spp. isolated from cultured shrimp. Asian J Anim Vet Adv 8(2):355–361

Sarhan IA, Abdel-Aziz SA, Said AA, Abdel-Aleim AAF, Awad SM (2019) Effect of dietary supplementation of extracted jojoba meal on hematology, biochemical parameters and disease resistance in Nile tilapia (*Oreochromis niloticus*) infected by *Aeromonas hydrophila*. Egypt J Aquac 9(3):13–31

Saslis-Lagoudakis CH, Williamson EM, Savolainen V, Hawkins JA (2011) Cross-cultural comparison of three medicinal floras and implications for bioprospecting strategies. J Ethnopharmacol 135(2):476–487

Sivaram V, Babu MM, Immanuel G, Murugadass S, Citarasu T, Marian MP (2004) Growth and immune response of juvenile greasy groupers (*Epinephelustauvina*) fed with herbal antibacterial active principle supplemented diets against Vibrio harveyi infections. Aquaculture 237(1–4):9–20

Sreedharan K, Philip R, Singh ISB (2012) Virulence potential and antibiotic susceptibility pattern of motile aeromonads associated with freshwater ornamental fish culture systems: a possible threat to public health. Braz J Microbiol 43:754–765

Stermitz FR, Lorenz P, Tawara JN, Zenewicz LA, Lewis K (2000) Synergy in a medicinal plant: antimicrobial action of berberine potentiated by 5′-methoxyhydnocarpin, a multidrug pump inhibitor. Proc Natl Acad Sci 97(4):1433–1437

Subashkumar R, Vivekanandhan G, Raja SS, Natarajaseenivasan K, Thayumanavan T, Lakshmanaperumalsamy P (2007) Typing of *Aeromonas hydrophila* of fish and human diarrhoeal origin by outer membrane proteins and lipopolysaccharides. Indian J Microbiol 47:46–50

Suhet MI, Schocken-Iturrino RP, Amaral LA (2011) Hemolytic activity and resistance to antimicrobials by *Aeromonas* species isolated from intensive rearing of Nile tilapia (*Oreochromis niloticus*)/Atividadehemolíticaresistênciaaantimicrobianos por *Aeromonas* spp. isoladas de criação... Ars. Veterinaria 27(1):036–044

Thomas CM, Nielsen KM (2005) Mechanisms of, and barriers to, horizontal gene transfer between bacteria. Nat Rev Microbiol 3(9):711–721

Tsuchiya H (2015) Membrane interactions of phytochemicals as their molecular mechanism applicable to the discovery of drug leads from plants. Molecules 20(10):18923–18966

Tuli HS (ed) (2019) Current aspects of flavonoids: their role in cancer treatment. Springer Singapore

Udikovic-Kolic N, Wichmann F, Broderick NA, Handelsman J (2014) Bloom of resident antibiotic-resistant bacteria in soil following manure fertilization. Proc Natl Acad Sci 111(42):15202–15207

Van Doan H, Hoseinifar SH, Sringarm K, Jaturasitha S, Yuangsoi B, Dawood MA et al (2019) Effects of Assam tea extract on growth, skin mucus, serum immunity and disease resistance of Nile tilapia (*Oreochromis niloticus*) against *Streptococcus agalactiae*. Fish Shellfish Immunol 93:428–435

Van Doan H, Hoseinifar SH, Jaturasitha S, Dawood MA, Harikrishnan R (2020) The effects of berberine powder supplementation on growth performance, skin mucus immune response, serum immunity, and disease resistance of Nile tilapia (*Oreochromis niloticus*) fingerlings. Aquaculture 520:734927

Vaseeharan B, Thaya R (2014) Medicinal plant derivatives as immunostimulants: an alternative to chemotherapeutics and antibiotics in aquaculture. Aquac Int 22:1079–1091

Verma VK, Rani KV, Kumar SR, Prakash O (2018) *Leucaena leucocephala* pod seed protein as an alternate to animal protein in fish feed and evaluation of its role to fight against infection caused by *Vibrio harveyi* and *Pseudomonas aeruginosa*. Fish Shellfish Immunol 76:324–332

Verma VK, Prakash O, Kumar RSR, Rani KV, Sehgal N (2021) Water hyacinth (*Eichhorniacrassipes*) leaves enhances disease resistance in *Channapunctata* from *Vibrio harveyi* infection. J Basic Appl Zool 82:1–11

Wolfender JL, Marti G, Thomas A, Bertrand S (2015) Current approaches and challenges for the metabolite profiling of complex natural extracts. J Chromatogr A 1382:136–164

Wright GD (2010) Antibiotic resistance in the environment: a link to the clinic? Curr Opin Microbiol 13(5):589–594

Wu S (2020) Dietary Astragalus membranaceus polysaccharide ameliorates the growth performance and innate immunity of juvenile crucian carp (*Carassius auratus*). Int J Biol Macromol 149:877–881

Wu C, Shan J, Feng J, Wang J, Qin C, Nie G, Ding C (2019) Effects of dietary Radix Rehmanniae Preparata polysaccharides on the growth performance, immune response and disease resistance of *Luciobarbus capito*. Fish Shellfish Immunol 89:641–646

Yamagata K (2019) Metabolic syndrome: preventive effects of dietary flavonoids. Stud Nat Prod Chem 60:1–28

Yao JY, Zhou ZM, Li XL, Yin WL, Ru HS, Pan XY et al (2011) Antiparasitic efficacy of dihydrosanguinarine and dihydrochelerythrine from *Macleaya microcarpa* against *Ichthyophthirius multifiliis* in richadsin (*Squaliobarbuscurriculus*). Vet Parasitol 183(1–2):8–13

Yousefi M, Ghafarifarsani H, Hoseinifar SH, Rashidian G, Van Doan H (2021) Effects of dietary marjoram, Origanummajorana extract on growth performance, hematological, antioxidant, humoral and mucosal immune responses, and resistance of common carp, *Cyprinus carpio* against *Aeromonas hydrophila*. Fish Shellfish Immunol 108:127–133

Zhai Q, Li J (2019) Effectiveness of traditional Chinese herbal medicine, San-Huang-San, in combination with enrofloxacin to treat AHPND-causing strain of *Vibrio parahaemolyticus* infection in Litopenaeusvannamei. Fish Shellfish Immunol 87:360–370

Zhang QQ, Ying GG, Pan CG, Liu YS, Zhao JL (2015) Comprehensive evaluation of antibiotics emission and fate in the river basins of China: source analysis, multimedia modeling, and linkage to bacterial resistance. Environ Sci Technol 49(11):6772–6782

Zhang H, Ge X, Liu B, Teng T, Zhou Q, Sun C, Song C (2020) Comparative transcriptomic and proteomic analysis of the antibacterial activity of emodin on *Aeromonas hydrophila*. Aquaculture 529:735589

Zhu YG, Johnson TA, Su JQ, Qiao M, Guo GX, Stedtfeld RD et al (2013) Diverse and abundant antibiotic resistance genes in Chinese swine farms. Proc Natl Acad Sci 110(9):3435–3440

Zhuo LC, Yong ASK, Shapawi R, Lin YH (2021) Effects of fermented lemon peel supplementation in diet on growth, immune responses, and intestinal morphology of Asian Sea bass *Latescalcarifer*. Aquac Rep 21:100801

Essential Oil- Natural Weapons to Overcome Antimicrobial Resistance

27

Surabhi Shakya, Harikumar Pallathadka,
Nangthombam Grihalaxmi Devi, and Amit Kumar Verma

Abstract

Essential oils are long known to history and are used as an alternative approach for the treatment of different ailments. The market area of essential oil is broad and its uses are enormous. The volatile constituents present in the different types of oils like Clove oil, jojoba oil, palmarosa oil, rose oil, citronella oil, lemongrass oil, lavender oil, and many more are responsible for different pharmacological activities. Many of the oils have the same active constituents and can provide synergistic effects. The researchers of different areas concluded from their study that they are potent agents to combat the problem of antibiotic resistance. Crude oil and also its vapors are effective in many therapeutic activities. The synergistic effect of essential oils with conventional antibiotics is observed in many studies even in the cases of antibiotic-resistant bacteria. The Novel delivery system like nano-emulsion and micro-emulsion for the delivery of essential oil has the potential to enhance the therapeutic effect as compare to conventional form.

Keywords

Essential oil · Antibiotic resistance · Antimicrobial · Nanoemulsion · Microemulsion

S. Shakya
Department of Pharmacy, MJP Rohilkhand University, Bareilly, Uttar Pradesh, India

H. Pallathadka · N. G. Devi
Manipur International University, Imphal, Manipur, India

A. K. Verma (✉)
Department of Pharmacy, MJP Rohilkhand University, Bareilly, Uttar Pradesh, India

Manipur International University, Imphal, Manipur, India

M. Gangwar, G. Nath (eds.), *Emerging Paradigms for Antibiotic-Resistant Infections: Beyond the Pill*, https://doi.org/10.1007/978-981-97-5272-0_27

27.1 Introduction

Essential oils are the key components of aromatherapy and have established data in the management of many different ailments and disorders when used in different ways like inhalation, massage, and less likely to be used internally. The history of essential oil is as old as about 6000 years, and many civilizations like China, India, and Egypt used this approach in a holistic way (Ali et al. 2015). The use of essential oil since the ancient period is well-known in many cultural, spiritual, and medicinal activities. Being volatile in nature, they are obtained from plant sources and considered as "safe" agents for many health problems. Essential oils are the plant secondary metabolites synthesized in different parts of the plants. The synthesis of these aromatic compounds occurs in the oil sacs and the oil glands which are present in the depth of different parts (Herman et al. 2019). In the classical way of aromatherapy, the inhalation route is the most popular and easy to use in the case of essential oils. Persons used to inhale these essential oils through the nose for the systemic effect as inhalation of oil with water vapors or the fumes of burnt oil material reaches the lungs through the nasal passage and ultimately provides systemic effects after reaching the bloodstream or through the nasal passage it directly stimulates the brain. The undeniable fact of this therapy is that on targeting the lungs, respiratory tract, oral mucosa, or brain activity, the gastrointestinal tract remains unaffected (Michalak 2018). The fragrant smell of these essential oils is responsible for relieving stress and anxiety. The phytochemical of aromatic plants due to their volatile nature affect the nervous system and are also responsible for regulating hormonal secretions. The inhaled essential oils reach the olfactory system and stimulate it. The message further passes to the brain and stimulates the secretion of 5-HT, an important hormone for mood regulation. An enzyme, tryptophan hydroxylase catalyzed the synthesis of 5-HT from tryptophan. This 5-HT is stored in vesicles in the pre-synaptic neurons. Inhalation of aromatic oils triggers the release of 5-HT from the pre-synaptic neurons to the synaptic cleft and binds to the receptors present in post-synaptic neurons. This binding phosphorylates cyclic-AMP-response-element-binding protein (CREB) for the regulation of gene-related expression (Lv et al. 2013). 5-HT (5-hydroxy tryptamine) in the pre and post-synaptic neurons also affects serotonin secretion. The serotonin transporter gene encodes by the 5-HT transporter gene and is involved in the reuptake of serotonin which is a target factor for anti-depressant drugs especially SSRIs (Selective Serotonin reuptake inhibitors) (Caspi et al. 2003). The inhaled essential oil also reaches the lungs through the respiratory passage. The volatile components vaporize and reach into the nasal passage. The essential oil is absorbed systemically through the lower respiratory tract because there is 50% absorption in the pulmonary area and is also capable of exerting the local effect. The antimicrobial and antibacterial action of the essential oils is due to their local action on the oral mucosa, and they are effective in respiratory tract infections. The most odoriferous molecule stimulates the trigeminal system which is a part of both nasal cavity and olfactory system (Lucena de Vasconcelos et al. 2021).

Essential oils are a complex mixture containing compounds of different classes. The main components which are present in major amounts are mono- and sesquiterpenes. The volatile portion comprises terpenes, aldehydes, volatile acids, ketones, phenols, etc. These volatile compounds synthesize in different parts of plants which are used by them either for attracting pollinators or for protecting themselves from pests and predators. The essential oil contains different classes of compounds so they have the capability to target different aspects, one volatile oil can have many modes of action because of the presence of different classes of compounds. The highest phenol and aldehyde content in the essential oil showed good anti-bacterial activity as compared to the volatile oils rich in ketones. The essential oils are known to show more inhibition towards gram-positive bacteria than gram-negative bacteria (Dhifi et al. 2016).

27.1.1 Antimicrobial Resistance

The problem of antimicrobial resistance has been becoming a global concern and the emergence of resistant microorganism has been affecting the diagnosis and treatment of many serious ailments. Including this, the multi-drug resistance organisms are also a serious concern (WHO 2018). Estimation was done by the US Centers for Disease Control and Prevention about the serious issue of developing antimicrobial resistance in the population. In their report, they concluded that about two million patients each year develop the serious problem of antibiotic resistance (CDC 2013).

Some bacterial species in the population are found to be innately resistant toward antimicrobial agents. The main concern is about the acquired resistance which the bacteria develop during the course of time as the species which were initially susceptible will develop resistance towards the antimicrobial agents. There are several mechanisms through which the bacteria develop resistance towards antimicrobial agents, these are

- B-lactamase, gene encoding enzymes in the bacteria has the capability to degrade the antimicrobial agents.
- Presence of an efflux pump in the bacterial cell, reflux of antibacterial agent out of the cell before exerting its action.
- Several genes in the bacterial cell, that causes alteration in the bacterial cell wall. Due to alteration the cell wall does not contain the binding site for the respective antimicrobial agents.
- Mutation and selection may be one of the reasons for developing antimicrobial resistance.
- Resistant bacterial genes may get transferred to another bacterial cell though the process of conjugation, transformation or transduction and make the population resistant (Tenover 2006).

The common mechanism involved in the antifungal drug resistance

- Changes in the lipid composition of fungal cells and due to this the drug import into the cell is affected.
- Drug efflux protein gets over-expressed so the protrusion of the drug increased.
- The drug-targeting site gets changed due to changes in the genes.
- Glucan synthase, an enzyme important for the cell wall synthesis, Enchinocandins block the activity of this enzyme and affecting the cell integrity (Prasad et al. 2016).

27.1.2 Approaches of Essential Oil in Antimicrobial Resistance

The world is suffering from the problem of antibiotic resistance; the drugs available in the market no longer remain to be effective against resistant bacteria, viruses, and fungal strains. So, the approach towards essential oil is one of the major solutions to this problem. Many of the volatile components present in essential oil in which terpenes and their derivatives are significant. Essential oil rich in phenolic compounds has the ability to show antimicrobial activity by different mode of actions (Table 27.1). Different essential oil have their different effect on bacteria and fungi. The Gram-negative bacteria and Gram-positive bacteria both categories are affected by the essential oil but Gram-positive are better inhibited by the Gram-negative bacteria due to their cell wall structure (Sezen et al. 2019) (Fig. 27.1).

Table 27.1 Mechanism of action of essential oil on bacteria and fungi (Fig. 27.1) (Sezen et al. 2019; Nazzaro et al. 2017)

S. No.	Mechanism of action of essential oil on bacteria	Mechanism of action of essential oil on fungi
1.	Disruption of bacterial cell membrane structure	Disrupt the fungal cell membrane
2.	Act on bacterial protein and inhibit cell division	β-glucans activity was blocked by essential oil and ultimately inhibited the formation of the cell wall
3.	Inhibited the electron Transport System (ETS)	EO's have the capacity to inhibit both the function as mitochondrial function as well as proton pump in fungal cell
4.	Eo 's have the capability to interact with bacterial DNA and RNA	Efflux pump inhibited by essential oils
5.	Inhibition of enzymes by essential oil which are necessary for life	Inhibit the bacterial and fungal biofilm formation
6.	Some terpenes inhibit the ATPase activity in membrane channels	

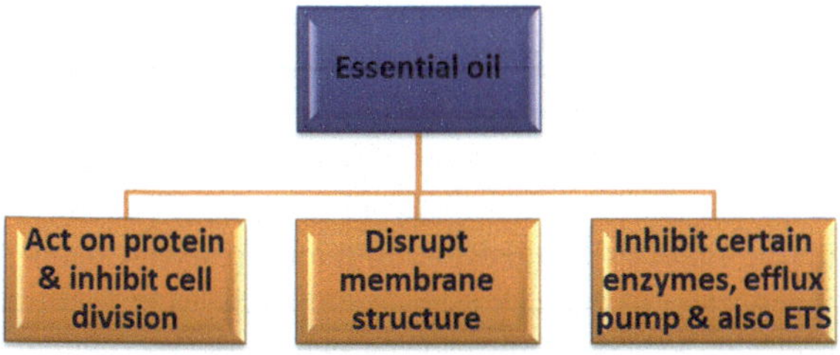

Fig. 27.1 Basic mechanism of essential oil on bacteria and fungi

27.1.2.1 Evidences: Replacement of Existing Antimicrobial Drugs or Antibiotics by Essential Oil

Brochot et al. (2017) studied the two different blends AB1 and AB2 of essential oil on different bacterial and fungal strains. The blends of four different essential oil tested against the bacteria *Streptococcus pyogenes* CIP 5641T, *Listeria monocytogenes* CIP 82110T, *Streptococcus pneumoniae* CIP 104471, and Staphylococcus aureus MRSA ATCC 33591; these four strains are gram-positive. *Escherichia coli* UTI 89, *Klebsiella pneumonia* CIP 8291T, *Pseudomonas aeruginosa* CIP 103467, *Proteus mirabilis* CIP 103181T, *Escherichia coli* ESBL, *Salmonella typhimurium* CIP 6062T, *Bacteriodes fragilis* ATCC 25285, *Yersinia enterocolitica* CIP 8027T, *Haemophilus influenza* IP 102514, and *Branhamella catarrhalis* CIP 7321T. *Pseudomonas aeruginosa* CIP 103467, *Escherichia coli* ESBL, and *Staphylococcus aureus* MRSA ATCC 3359 these ten strains of bacteria are gram-negative. *Candida albicans* DSM 1386, *Candida tropicalis* IP 2148.93, *Candida glabatra* DSM 11226, *Candida albicans* F26, *Candida albicans* F78, and *Candida albicans* F35 are six different fungal strains used in the study. Both the blend of essential oil showed bactericidal and bacteriostatic action against both types of bacteria i.e., gram-positive and gram-negative. Blend AB1 showed more effective results against viruses. AB1 blend contains the essential oil of *Daucuscarota, Cinnamomum zeylanicum, Eucalyptus globulus*, and *Rosmarinus officinalis*. AB2 blend contains the same composition but the essential oil of *Cinnamon zeylanicum* provided by a different supplier (Brochot et al. 2017).

A randomized clinical trial was done by Aydin and Yilmaz, 56 patients were studied who had undergone for surgery and had the problem of postoperative nausea and vomiting. The experimental group inhaled the peppermint oil for about 6–12, 12–24, and 24–48 h. The control group receives the same treatment but with less extent of nausea. The inhalation of peppermint oil was found to be effective in postoperative nausea (Aydın and Yılmaz 2018). This study suggested the avoidance of unnecessary use of drugs in case where natural components are effective.

Patent US7150888 by Vail WB and Vail M L related to the effect of strong vapors of two essential oils one is Eucalyptus oil and the other is Tea tree oil. The inhaled vapors of these two oils help to protect the respiratory diseases like cold, pneumonia, flu, tuberculosis, and other related infections. In their patent, they clearly defined that inhalation of aromatic vapors formed the protective layer of the respiratory system. Eucalyptus oil and Tea tree oil both are active against the pathogenic bacteria which are responsible for causing respiratory infections like *Staphylococcus aureus*, *Mycobacterium tuberculosis*, *Pseudomonas aeruginosa*, and *Burkholderia cepacia*. The authors of the patent exactly showed the composition of Tea tree oil from *Melaleuca alternifolia* and Eucalyptus oil from *Eucalyptus globules* (Vail and Vail 2006).

Iseppi et al. (2021) studied the four different essential oils alone and in combination with oils and antibiotics against the antibiotic-resistant pathogens and their biofilms respectively. The pathogens used in the study were Methicillin Resistant *Staphylococcus aureus*, Vancomycin-resistant enterococci, extended-spectrum lactamase (ESBL)—producing *Escherichia coli*. The essential oils tested against these pathogens were essential of *Citrus aurantium*, *Melaleuca alternifolia*, *Citrus x lemon*, *Cupressus sempervirens*, and *Eucalyptus globules*. These essential oils showed effective result against these pathogens and in combinations with essential oil and reference antibiotics, the synergistic actions displayed and out of all the essential oil Tea tree oil was found to be most effective.

Puškárová et al. (2017) reported the antifungal and antibacterial activity of six different essential oils. 10 mg of each essential oil like Clary sage oil, lavender oil, oregano essential oil, thyme essential oil, clove essential oil, and arborvitae essential oil was found to be effective antibacterial agent against all types of bacterial strains and these essential oils showed both fungistatics and fungicidal activity in both the forms either in direct application and in vapor forms. The strains used in study for the antimicrobial properties of essential oil were *Alternaria alternata*, *Cladosporium cladosporoides*, *Aspergillus fumigatus*, *Bacillus cereus*, *Escherichia coli*, *Chaetomiumglobosum*, *Enterococcus faecalis*, *Listeria monocytogenes*, *Penicillium chrysogenum*, *Pseudomonas fragi*, *Arthrobacter protophormiae*, *Salmonella typhimurium*, *Yersinia enterocolitica*, and *Staphylococcus aureus*.

Haba et al. (2014) prepared the emulsion based on essential oil and for the emulsification of the formulation rhamnolipid were used. Rhamnolipid with great emulsifying potential placed in the class of microbial surfactant. Essential oils of *Cinnamomum zeylanicum*, *Origanum compactum*, *Melaleuca alternifolia*, *Lavandula angustifolia* were used in the preparation of emulsion with rhamnolipid by ternary phase diagrams. The prepared emulsion evaluated for antimicrobial activity against *Candida albicans* and Methicillin Resistant *Staphylococcus aureus*. Emulsification increased the antimicrobial activity. Origanum essential oil was found to be more effective against *Candida albicans* and in case of methicillin-resistant *Staphyloccus aureus* cinnamomum essential oil was effective in comparison to others (Haba et al. 2014) (Table 27.2).

27.1.3 The Concept of Nanotechnology to Increase the Antimicrobial Potential of Essential Oil as Antimicrobial Agents

Essential oil proves to be effective against microbes but the approach of nanotechnology increases its potency and efficacy as antimicrobial agents (Table 27.3). The nanoemulsions (droplet size less than 100 nm) of essential oils consist of two

Table 27.2 Different essential oils and their associated pharmacological properties as antimicrobial

S. No.	Essential oil	Main component	Part used	Biological name	Pharmacological activity	References
1.	Palmarosa oil	Geraniol (major) Geranyl acetate	Leaves	*Cymbopogon martini*	Antifungal Antibacterial	Soni and Soni (2014)
2.	Clove oil	Eugenol (major) Chavicol Caryophyllene Eugenyl acetate	Flower bud	*Syzygiumaromaticum*	Antifungal Antimicrobial	Tambe and Gotmare (2020)
3.	Lavender oil	Linalool (major) Linalyl acetate (major) Terpinene-4-ol	Flowers	*Lavandula angustifolia*	Relief pain Antidepressant Antifungal	Prusinowska and Smigielski (2014)
4.	Tea tree oil	Terpinen-4-ol (major) ɣ-terpinene α-terpinene 1,8-cineole	Leaves	*Melaleuca alternifolia*	Wound healing Antimicrobial	Elmi et al. (2019)
5.	Sweet basil oil	Linalool (major) Geraniol 1,8-cineole a-bergamotene	Seeds	*Ocimumbasilicum* L.	Improved memory Relief from Asthmatic symptoms Reduced acne	Tripathi et al. (2014)
6.	Mint oil	Menthol (major) Menthone Isomenthone Neomenthyl acetate	Leaves and stem	Mentha arvensis	Herbicidal Antifungal Antioxidant Insecticidal Antimicrobial	Makkar et al. (2018)

(continued)

Table 27.2 (continued)

S. No.	Essential oil	Main component	Part used	Biological name	Pharmacological activity	References
7.	Lemongrass oil	Geranial (major) Neral β-myrcene	Leaves	Cymbopogon citratus Stapf	Antiseptic Antifever Analgesic Antidyspeptic Spasmolytic Anti-inflammatory	Hamad et al. (2017)
8.	Rosemary oil	1,8-cineole (major) Camphor α-pinene α-terpineole Borneol Camphene	Ariel parts	Rosmarinus officinalis L.	Antimicrobial Antioxidant Cytotoxic Anti-mutagenic	Stojiljkovic et al. (2018)
9.	Ylang-ylang oil	Linalool β Caryophyllene (major) α-humulene Geranyl acetate	Flower	*Cananga odorata* hook. F. and Thomson	Antioxidant Antifungal Antiprotozoal Anti-inflammatory	Tambe (2020); Tan et al. (2015)
10.	Peppermint	Menthol (major) Menthone Limonene Cineole	Ariel parts	*Menthapiperita* L.	Insect repellant Carminative Antimicrobial	Shrivastava (2009)
11.	Myrrh oil	2-Cyclohexen-1-one, 4-ethynyl-4-hydroxy-3,5,5-trimethyl- (major) γ-elemene Elemol Copaene	Stem bark	*Commiphora myrrh*	Bactericidal Fungicidal Anti-viral	Chen et al. (2013); Batiha et al. (2023)

12.	Frankincense oil	n-Octyl acetate (major) Nerolidolisobutyrate, 3,7,11-trimethyl-1,6,10-dodecatrien-3-ylester-formic acid, δ-elemene n-octanol	Stem bark	*Boswellia carterii*	Antifungal Anti-bacterial Anti-oxidant	Chen et al. (2013)
13.	Eucalyptus oil	1,8-cineole (major) α-pinene Linalool	Leaves	*Eucalyptus globulus*	Antimalarial Spasmodic Larvicidal Antifungal Stimulant Antiseptic	Surbhi et al. (2021)
14.	Patchouli oil	Patchouli alcohol (major), β-Patchoulene, Patchoulene epoxide, Pogostone	Ariel parts	*Pogostemon cablin* Benth	Anti-tumor Anti-oxidant Anti-fungal Anti-apoptotic	Junren et al. (2021)
15.	Thyme oil	Thymol (major) p-cymene γ-terpinene Aromadenrrene Ledol	Ariel parts	*Thymus vulgaris* L.	Antiseptic Anti-fungal Anti-bacterial	Abozid and Asker (2013)
16.	Cedarwood oil	Himachalenes (major) Atlantones Himachalene oxide Himachalol	Wood	*Cedrus atlantica*	Anti-inflammatory Anti-viral Anti-microbial Anti-fungal Molluscicidal	Chaudhary et al. 2011

(continued)

Table 27.2 (continued)

S. No.	Essential oil	Main component	Part used	Biological name	Pharmacological activity	References
17.	Geranium oil	Citronellol (major) Geraniol Citronellylformate 6,9-Guaiadiene	Ariel parts (leaves, flower, stalks)	*Pelargonium graveolens* Ait.	Antibacterial	Džamić et al. (2014)
18.	Ginger oil	β-zingiberene (major) Geraniol Neral β-bisabolene	Dried rhizome	*Zingiber officinale* Rosc	Antimicrobial Bronchodilator Anti-oxidant Anti-inflammatory Analgesic	Mahboubi (2019)
19.	Turmeric oil	Ar-turmerone (major) Ar-curcumene Curlone (β-turmerone) α-phellandrene	Rhizomes	*Curcuma longa*	Bactericidal Fungicidal	Singh et al. (2011)
20.	Citronella oil	Citronellal (major) Geraniol Citronellol	Leaves	*Cymbopogon nardus* Rendle	Mosquito repellant	Wany et al. (2013)
21.	Rose oil	B-citronellol (major) Nonadecane Nerol Geraniol	Petals	*Rosa Damascena* Mill	Mental relaxation Antibacterial Bacteriostatic	Mohebitabar et al. (2017)
22.	Camphor oil	D camphor (major) 1, 8-cineole 3-methyl-2-butanoic acid	Stem bark	*Cinnamomum camphora*	Anthelmintic activity Antimicrobial Antifungal Analgesic activity Antitussive activity	Sharma et al. (2021)

23.	Helichrysum oil or immortelle essential oil	Neryl acetate (major) α-pinene γ-curcumenenerylpropanoate	Ariel green parts	*Helichrysum italicum* (Roth) G. Don	Antioxidant Antimicrobial	Mollova et al. (2020)
24.	Juniper oil	α-pinene (major) myrcenesabinene limonene β-pinene	Berries	*Juniperus communis* L.	Antioxidant	Höferl et al. (2014)
25.	Artemisia oil	(E)-b-farnesene (major) germacra-4(15),5,10(14)-trien-1-ol 1,8-cineol Artemisia alcohol 3-methyl butanoate	Ariel parts	*Artemisia annua* L.	Antibacterial Antiradical	Zhigzhitzhapova et al. (2019)

Table 27.3 Nanoemulsions loaded with essential oil as an antimicrobial

S. no.	Nanoemulsions of essential oil	Purpose	Targeted species	References
1.	Thyme oil	Food	*Listeria monocytogenes* *Escherichia coli*	Ziani et al. (2011)
2.	Tea tree oil	Pharmaceuticals	*Staphylococcus aureus* *Escherichia coli*	Najafi et al. (2017)
3.	Eucalyptus oil	Food	*Staphyloccus aureus* *Bacillus cereus* *Escherichia coli*	Sugumar et al. (2012)
4.	Alhagimaurorum oil	Pharmaceuticals	*Escherichia coli* *Pseudomonas aeruginosa* *Acinetobacter baumannii* *Bacillus cereus* *Klebsiella pneumonia* *Staphylococcus aureus*	Hassanshahian et al. (2020)
5.	Clove oil	Pharmaceutical	*Candida albicans* *Aspergillus brasiliensis* *Cryptococcus neoformans* *Aspergillus fumigates* *Aspergillus flavus*	Shehabeldine et al. (2023)
6.	Origanum oil	Food	*Aspergillus niger* *Candida albicans* *Staphylococcus aureus*	Enayatifard et al. (2021)

phases, oil and water with a surfactant. The nanoemulsions provide the highest stability and they referred to be biocompatible and safe for use. The nanoemulsion of essential oil exhibits the small droplet size with more surfaces to volume ratio. This approach provides the delivery of oil in the controlled manner (Maurya et al. 2021).

27.1.4 Microemulsion of Essential Oil as Antimicrobial

Microemulsion served as the other important novel nano-drug delivery system recognized as the two types of biphasic, triphasic, or menopause forms. The small size of the dispersed particle and the ratio of surfactant and co-surfactant make the formulation more stable. The two phases designed to dissolve the water-soluble and the water-insoluble components to form w/o and o/w type emulsion (Fig. 27.2). The better protection to the thermo-labile substances provided to them with enhanced therapeutic efficacy and more bioavailability (Table 27.4) (Thakur et al. 2021).

Fig. 27.2 Micro and nano-formulations increase the targeted activity

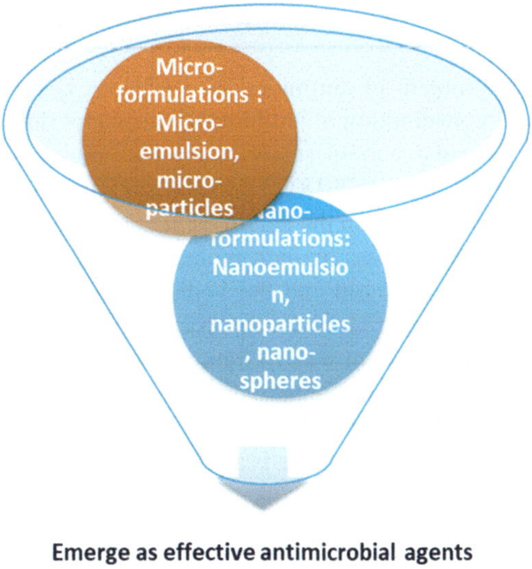

Table 27.4 Microemulsion loaded with essential oil as antimicrobial agents

S. No.	Essential oil	Purpose	Targeted species	References
1.	Lemon oil Litsea oil	Textiles	*Trichophyton rubrum* *Staphylococcus epidermidis* *Staphylococcus aureus* *Escherichia coli* *Pseudomonas aeruginosa*	Enayatifard et al. (2021)
2.	Thyme oil Eugenol Cinnamon bark oil	Pharmaceutical	*Salmonella enterica* *Listeria monocytogenes* *Salmonella enterica*	Ma et al. (2016)
3.	Cinnamon oil	Pharmaceutical	*Pseudomonas aeruginosa* *Aspergillus brasiliensis* *Candida albicans* *Staphylococcus aureus* *Escherichia coli*	Wang et al. (2023)
4.	Orange peel oil	Food	Total microbial count observed	Radi et al. (2017)
5.	Thyme oil	Food	Mold Yeast Lactic acid bacteria Staphylococcus aureus Coliforms	Almasi et al. (2021)

27.1.5 Conclusion and Future Prospects

The problem of antimicrobial resistance has continued to be increasing with the over consumption of antibiotics. Despite of doing lots of efforts this problem is not in control due to resistance of antimicrobial drugs and misuse and abuse of antibiotics (Klein et al. 2018). The antibacterial and antifungal resistance is creating the undue pressure on the health professionals as it directly and indirectly affecting the treatment process by increasing the human morbidity as well as creating financial problems. The old and the Vedic literature and the traditional knowledge to deal with this rising issue are a possible way of dealing. In this case, the use of essential oil is a prominent way to fight with this situation. Essential oils are excellent antibacterial and antifungal agents, their mechanism of action signifies that they can reduce the consumption of antibiotics and also gave synergistic effects if used in combinations. But there is a great need for future studies for exploring this area in a wider way as a single oil can provide multiple therapeutic actions.

Acknowledgement I wish to express my deep gratitude and sincere thanks to Prof. K.P. Singh, Vice Chancellor of MJP Rohilkhand University, Bareilly, U.P. for his encouragement and for all the facilities that he provided us for this work. I sincerely appreciate this magnanimity by taking us into this fold for which we shall remain indebted to him. Thank you for the generosity and support.

References

Abozid MM, Asker MMS (2013) Chemical composition, antioxidant and antimicrobial activity of essential oil of thyme and rosemary. Int J Acad Res 5(3):186–195. https://doi.org/10.7813/2075-4124.2013/5-3/A.26

Ali B, Ali Al-Wabel N, Shams S, Ahamad A, Khan SA, Anwar F (2015) Essential oils used in aromatherapy: a systemic review. Asian Pac J Trop Biomed 5(8):601–611

Almasi L, Radi M, Amiri S, McClements DJ (2021) Fabrication and characterization of antimicrobial biopolymer films containing essential oil-loaded microemulsions or nanoemulsions. Food Hydrocoll 117:106733

Aydın N, Yılmaz ÜD (2018) Effect of peppermint oil inhalation on postoperative nausea and vomiting. Cyprus j Med Sci 3:68–74

Batiha GES, Wasef L, Teibo JO, Shaheen HM, Zakariya AM, Akinfe OA, Teibo TKA, Al-kuraishy HM, Al-Garbee AI, Alexiou A, Papadakis M (2023) Commiphora myrrh: a phytochemical and pharmacological update. Naunyn Schmiedeberg's Arch Pharmacol 396:405–420

Brochot A, Guilbot A, Haddioui L, Roques C (2017) Antibacterial, antifungal, and antiviral effects of three essential oil blends. Microbiology 6(4):e00459. https://doi.org/10.1002/mbo3.459

Caspi A, Sugden K, Moffitt TE (2003) Influence of life stress on depression: moderation by a polymorphism in the 5-HTT gene. Science 301:386–389

CDC (2013) Antibiotic resistance threats in the United States. CDC

Chaudhary A, Sharma P, Nadda G, Tewary DK, Singh B (2011) Chemical composition and larvicidal activities of the Himalayan cedar, Cedrusdeodara essential oil and its fractions against the diamondback moth, Plutellaxylostella. J Insect Sci 11:157

Chen Y, Zhou C, Ge Z, Liu Y, Liu Y, Feng W, Li S, Chen G, Wei T (2013) Composition and potential anticancer activities of essential oils obtained from myrrh and frankincense. Oncol Lett 6:1140–1146

Dhifi W, Bellili S, Jazi S, Bahloul N, Mnif W (2016) Essential oils' chemical characterization and investigation of some biological activities: a critical review. Molecules 3(25):1–16

Džamić AM, Soković MD, Ristić MS, Grujić SM, Mileski KS, Marin PD (2014) Chemical composition, antifungal and antioxidant activity of Pelargonium graveolens essential oil. J Appl Pharm Sci 4(3):001–005

Elmi A, Ventrella D, Barone F, Carnevali G, Filippini G, Pisi A, Benvenuti S, Scozzoli M, Bacci ML (2019) *In vitro* effects of tea tree oil (melaleuca alternifolia essential oil) and its principal component Terpinen-4-ol on swine spermatozoa. Molecules 24(1071):1–15

Enayatifard R, Akbari J, Babaei A, Rostamkalaei SS, Hashemi SMH, Habibi E (2021) Antimicrobial potential of nano-emulsion form of essential oil obtained from aerial parts of Origanum Vulgare L. as food additive. Adv Pharm Bull 11(2):327–334

Haba E, Bouhdid S, Torrego-Solana N, Marques AM, Espuny MJ, Gracia-Celma MJ, Manresa A (2014) Rhamnolipids as emulsifying agents for essential oil formulations: antimicrobial effect against *Candida albicans* and methicillin-resistant *Staphylococcus aureus*. Int J Pharm 476:134–141. https://doi.org/10.1016/j.ijpharm.2014.09.039

Hamad A, Nuritasari A, Hartanti D (2017) Chemical composition and antimicrobial study of essential oil of lemongrass (Cymbopogoncitratus). Pharm Lett 9(5):109–116

Hassanshahian M, Saadatfar A, Masoumipour F (2020) Formulation and characterization of nanoemulsion from Alhagimaurorum essential oil and study of its antimicrobial, antibiofilm, and plasmid curing activity against antibiotic-resistant pathogenic bacteria. J Environ Health Sci Eng 18:1015. https://doi.org/10.1007/s40201-020-00523-7

Herman RA, Ayepa E, Shittu S, Fometu SS, Wang J (2019) Essential oils and their applications—a mini review. Adv Nutr Food Sci 4(4):1–13

Höferl M, Stoilova I, Schmidt E, Wanner J, Jirovetz J, Trifonova D, Krastev L, Krastanov A (2014) Chemical composition and antioxidant properties of Juniper berry (Juniperuscommunis L.) essential oil action of the essential oil on the antioxidant protection of saccharomyces cerevisiae model organism. Antioxidants 3:81–98. https://doi.org/10.3390/antiox3010081

Iseppi R, Mariani M, Condò C, Sabia C, Messi P (2021) Essential oils: a natural weapon against antibiotic-resistant bacteria responsible for nosocomial infections. Antibiotics 10:417. https://doi.org/10.3390/antibiotics10040417

Junren C, Xiaofang X, Mengting L, Qiuyun X, Gangmin L, Huiqiong Z, Guanru C, Xin X, Yanpeng Y, Fu P, Cheng P (2021) Pharmacological activities and mechanisms of action of Pogostemon cablin Benth: a review. Chin Med 16(5):1–20

Klein EY, Van Boeckel TP, Martinez EM, Pant S, Gandra S, Levin SA et al (2018) Global increase and geographic convergence in antibiotic consumption between 2000 and 2015. Proc Natl Acad Sci USA 115:E3463–E3470

Lucena de Vasconcelos D, De Vasconcelos AL, Lucena de Vasconcelos A (2021) AROMATHERAPY: PROPHYLACTIC AND THERAPEUTIC APPROACH TO DISEASES OF THE RESPIRATORY SYSTEM. Int J Biosci Med 4(15):1–13

Lv XN, Liu ZJ, Zhang HJ, Tzeng CM (2013) Aromatherapy and the central nerve system (CNS): therapeutic mechanism and its associated genes. Curr Drug Targets 14:872–879

Ma Q, Davidson PM, Zhong Q (2016) Antimicrobial properties of microemulsions formulated with essential oils, soybean oil, and tween 80. Int J Food Microbiol 226:20–25. https://doi.org/10.1016/j.ijfoodmicro.2016.03.011

Mahboubi M (2019) Zingiber officinale Rosc. essential oil, a review on its composition and bioactivity. Clin Phytoscience 5(6):1–12

Makkar MK, Sharma S, Kaur S (2018) Evaluation of Mentha arvensis essential oil and its major constituents for fungitoxicity. J Food Sci Technol 55(9):3840–3844

Maurya A, Singh VK, Das S, Prasad J, Kedia A, Upadhyay N, Dubey NK, Dwivedy AK (2021) Essential oil nanoemulsion as eco-friendly and safe preservative: bioefficacy against microbial food deterioration and toxin secretion, mode of action, and future opportunities. Front Microbiol 12(1–17):751062. https://doi.org/10.3389/fmicb.2021.751062

Michalak M (2018) Aromatherapy and methods of applying essential oils. Arch Physiother Glob Res 22(2):25–31

Mohebitabar S, Shirazi M, Bioos S, Rahimi R, Malekshahi F, Nejatbakhsh F (2017) Therapeutic efficacy of rose oil: a comprehensive review of clinical evidence. Avicenna J Phytomed 7(3):206–213
Mollova S, Fidan H, Antonova D, Bozhilov D, Stanev S, Kostova I, Stoyanova A (2020) Chemical composition and antimicrobial and antioxidant activity of Helichrysumitalicum (Roth) G.Don subspecies essential oils. Turk J Agric For 44:371–378
Najafi TR, Ghaemi B, Kharazi S, Rasoulikoohi S, Amani A (2017) Promising antibacterial effects of silver nanoparticle-loaded tea tree oil Nanoemulsion: a synergistic combination against resistance threat. Am Assoc Pharm Sci 19:1133. https://doi.org/10.1208/s12249-017-0922-y
Nazzaro F, Fratianni F, Coppola R, De Feo V (2017) Essential oils and antifungal activity. Pharmaceuticals 10(86):1–20
Prasad R, Shah AH, Rawal MK (2016) Antifungals: mechanism of action and drug resistance. In: Ramos J, Sychrová H, Kschischo M (eds) Yeast membrane transport. Advances in experimental medicine and biology, vol 892. Springer, Cham. https://doi.org/10.1007/978-3-319-25304-6_14
Prusinowska R, Smigielski KB (2014) Composition, biological properties and therapeutic effects of lavender (Lavandula angustifolia L.). A review. Rerva Polonica 60(2):56–66
Puškárová A, Bucˇková M, Kraková L, Pangallo D, Kozics K (2017) The antibacterial and antifungal activity of six essential oils and their cyto/genotoxicity to human HEL 12469 cells. Sci Rep 7:1–11. https://doi.org/10.1038/s41598-017-08673-9
Radi M, Darabi SA, Akhavan HR, Amiri S (2017) The use of orange peel essential oil microemulsion and nanoemulsion in pectin-based coating to extend the shelf life of fresh-cut orange. J Food Process Preserv 42:e13441. https://doi.org/10.1111/jfpp.13441
Sezen S, Gulluce M, Kesmez Can F, Alaylar B (2019) Essential oils and antimicrobial effects. Int J Sci Eng Res 10(9):65–70
Sharma N, Yadav Y, Kumar V, Sharma KC (2021) Folkare and modern pharmacology of camphor (Cinnamomum camphora Nees & Eberm): review. Sch Int J Tradit Complement Med 4(7):128–135
Shehabeldine AM, Doghish AS, El-Dakroury WA, Hassanin MMH, Al-Askar AA, AbdElgawad H, Hashem AH (2023) Antimicrobial, antibiofilm, and anticancer activities of Syzygium aromaticum essential oil Nanoemulsion. Molecules 28:5812. https://doi.org/10.3390/molecules28155812
Shrivastava A (2009) A review on peppermint oil. Asian J Pharm Clin Res 2(2):27–33
Singh S, Sankar B, Rajesh S, Sahoo K, Subudhi E, Nayak S (2011) Chemical composition of turmeric oil (Curcuma longa L. cv. Roma) and its antimicrobial activity against eye infecting pathogens. J Essent Oil Res 23(6):11–18. https://doi.org/10.1080/10412905.2011.9712275
Soni S, Soni UN (2014) In-vitro antifungal and antibacterial activity of selected essential oil. Int J Pharm Pharm Sci 6(6):586–591
Stojiljkovic J, Trajchev M, Nakov D, Petrovska M (2018) Antibacterial activities of rosemary essential oils and their components against pathogenic bacteria. Adv Cytol Pathol 3(4):93–96
Sugumar S, Nirmala J, Ghosh V, Anjali H, Mukherjee A, Chandrasekaran N (2012) Bio-based nanoemulsion formulation, characterization and antibacterial activity against food-borne pathogens. J Basic Microbiol 52:1–10
Surbhi AK, Singh S, Kumari P, Rasane P (2021) Eucalyptus: phytochemical composition, extraction methods and food and medicinal applications. Adv Tradit Med 23:369. https://doi.org/10.1007/s13596-021-00582-7
Tambe E (2020) FTIR & GC-MS analysis of YlangYlang oil, comparative study of its chemical composition and concern regarding diethyl phthalate. Res J Sci Eng Technol 10(2):35–48
Tambe E, Gotmare S (2020) Qualitative estimation of chemical composition of five different clove oils (Syzygiumaromaticum) by GC-MS. international journal of multidisciplinary. Educ Res 6(5):116–245
Tan LTH, Lee LH, Yin WF, Chan CK, Kadir HA, Chan KG, Goh BH (2015) Traditional uses, Phytochemistry, and bioactivities of Canangaodorata (Ylang-Ylang). Evid Based Complement Alternat Med 2015:1. https://doi.org/10.1155/2015/896314

Tenover FC (2006) Mechanisms of antimicrobial resistance in bacteria. AJIC 34(5):S3–S10. https://doi.org/10.1016/j.ajic.2006.05.219

Thakur D, Kaur G, Puri A, Nanda R (2021) Therapeutic potential of essential oil-based microemulsions: reviewing state-of-the-art. Curr Drug Deliv 18:1–16

Tripathi S, Srivastava HC, Shukla P, Shanker B (2014) Anti-oxidant and antimicrobial activities of sweet basil oils. Int J Pharm Sci Res 5(1):279–285

Vail WB, Vail ML (2006) Methods and apparatus to prevent colds, influenza, tuberculosis and opportunistic infections of the human respiratory system. (U.S. Patent No. US20067150888). U.S. Patent and Trademark Office

Wang W, Chen YF, Wei ZF, Jiang JJ, Peng JQ, He QT, Xu WY, Liu HM (2023) Microemulsion of cinnamon essential oil formulated with tea polyphenols, Gallic acid, and tween 80: antimicrobial properties, stability and mechanism of action. Microorganisms 11:2. https://doi.org/10.3390/microorganisms11010002

Wany A, Jha S, Nigam VK, Pandey DM (2013) Chemical analysis and therapeutic uses of citronella oil from *Cymbopogon winterianus*—a short review. Int J Adv Res 1(6):504–521

WHO (2018) Antimicrobial resistance. http://www.who.int/news-room/fact-sheets/detail/antimicrobial-resistance

Zhigzhitzhapova SV, Dylenova EP, Gulyaev SM, Randalova TE, Taraskin VV, Tykheev ZA, Radnaeva LD (2019) Composition and antioxidant activity of the essential oil of Artemisiaannua L. Nat Prod Res 34:2668. https://doi.org/10.1080/14786419.2018.1548461

Ziani K, Chang Y, McLandsborough L, McClements DJ (2011) Influence of surfactant charge on antimicrobial efficacy of surfactant stabilized thyme oil nanoemulsions. J Agric Food Chem 59(11):6247–6255

28 Antimicrobial Activity of Plant and Cyanobacterial Secondary Metabolites for Therapeutic Advancements in Modern Medicine

Rupanshee Srivastava, Nidhi Singh, Tripti Kanda, Sadhana Yadav, Shivam Yadav, and Neelam Atri

Abstract

In the realm of modern medicine, researchers worldwide are captivated by the exploration of secondary metabolites found in plants and cyanobacteria. Microbes play a significant role in global health by producing valuable secondary metabolites and approximately 80% of the global population utilizes plants and their metabolites for therapeutic purposes, making them a promising avenue for medicinal molecule development. These metabolites exhibit noteworthy pharmacological actions, including neurotoxicity, cytotoxicity, and antiviral activity, making them intriguing therapeutic options. Their ability to inhibit protein and enzyme activity by binding with them is considered beneficial in combating infections from pathogenic microorganisms. Additionally, secondary metabolites have been reported to disrupt ATP synthesis by creating an electrochemical gradient across the cell wall and interfering with the electron transport system (ETS) by rupturing the mitochondrial membrane to counteract microbial infections. Brentuximab vedotin, one such medication derived from marine peptides, has gained FDA approval for treating Hodgkin lymphoma and anaplastic large-cell lymphoma. Numerous publications have effectively categorized metabolites based on their biological actions however; this book chapter endeavors to illuminate the modes of action of various metabolites against disease-causing pathogens, providing insights that may complement alternative strategies for infection control.

R. Srivastava · N. Singh · T. Kanda · S. Yadav · N. Atri (✉)
Department of Botany, M.M.V., Banaras Hindu University, Varanasi, Uttar Pradesh, India
e-mail: neelam1409@bhu.ac.in

S. Yadav (✉)
Department of Botany, University of Allahabad, Prayagraj, Uttar Pradesh, India

M. Gangwar, G. Nath (eds.), *Emerging Paradigms for Antibiotic-Resistant Infections: Beyond the Pill*, https://doi.org/10.1007/978-981-97-5272-0_28

Keyword

Cyanobacteria · Infectious disease · Metabolites · Pathogens · Plants and therapeutic

28.1 Introduction

Antimicrobial resistance (AMR) is an important issue in international public health (Prestinaci et al. 2015) as the extensive use of several antibiotics has led to the development of resistance and reduced the number of effective treatment options. The problem is further exacerbated by various factors such as overuse of antibiotics, ignorance, incorrect diagnostic techniques, overpopulation, cross-infections, and mediocre health infrastructure (Manesh and Varghese 2021). The burgeoning AMR epidemic can be attributed to a number of negative reasons, including the scarcity of antimicrobial medications for the treatment of multidrug-resistant (MDR) microbiological infections (Qadri et al. 2021). As a result, innovative solutions to the growing issue of medication resistance in different harmful bacteria and fungi are desperately needed. Establishing molecular, genetic, and immune mechanisms is necessary for the effective management of the growing number of human infections (Qadri et al. 2022). Although synthetic antimicrobial medications are in use, some scientists are investigating the potential of naturally occurring compounds originating from microorganisms, plants, and animals as these drugs have proven to be effective in avoiding infections linked to increased medication tolerance (Rossiter et al. 2017).

Plants have served as medicinal agents since ancient times, playing a crucial role in human well-being. Approximately 80% of the global population relies on plants or herbs with medicinal properties to address their healthcare needs (World Health Organization 2019). Quinine, an alkaloid extracted from the bark of the cinchona tree, serves as an example as it is widely used to treat a number of infectious disorders, including pneumonia, typhoid fever, and common nasopharyngeal infections, in addition to malaria (Achan et al. 2011). Cinnamon holds historical importance as well, having been used as a versatile cure in ancient Chinese and Indian medicine. Cinnamaldehyde, the main biologically active ingredient has long been known to be an antibacterial (Vasconcelos et al. 2018). In response to growing concerns about the adverse effects of chemical drugs and their limited efficacy, there is a heightened interest in natural alternatives or complementary therapies. These plant metabolites, often viewed as alternatives to conventional medications, are at times the sole effective option available. Despite the vast potential, only a limited number of plant species have been explored for their therapeutic properties. As a result, ongoing research is focused on unravelling the therapeutic potential of various plant species (Baell 2016).

Microbes play a significant role in contributing to global health by serving as an abundant supply of secondary metabolites that are bioactive. Evidence indicates that exposure to antibiotics dates back to ancient times, as seen in the discovery of tetracycline in human skeletal remains from ancient Sudanese Nabia (Bassett et al. 1980). Pioneering work by Paul Ehrlich and Alexander Fleming in 1929 marked the onset of a new era in medicine known as "The Golden Age of Antibiotics," highlighting the potential of microbial secondary metabolites as valuable sources of drug leads. Despite their immense potential, only 5% of natural compounds contributing to drugs are derived from microbial secondary metabolites, with a focus primarily on chemoheterotrophs (Nelson et al. 2010). While chemoheterotrophs have been extensively studied for microbial drug discovery, attention needs to be directed towards various other groups. Cyanobacteria are a diverse group of prokaryotic organisms that have emerged as a novel source of metabolites. Their amenability to controlled fermenter studies adds to their appeal as a promising resource for drug exploration (Srivastava et al. 2022). The cyanobacterial metabolites have therapeutic properties as antibiotic, immunosuppressant, bleomycin, and histone deacetylase inhibitors (Schwarzer et al. 2003; Anjum et al. 2017). Out of which some has successfully progressed to phase II and phase III clinical trials.

Although, many literatures are available that deals with plant and cyanobacterial metabolites and have classified them on the basis of their phytochemistry or bioactivity. However, in the current chapter, an effort is made to give invaluable information about the mechanism of such metabolites which will be helpful in understanding their mode of action in combating various human pathogens.

28.2 Antimicrobial Activities of Plant and Cyanobacterial Metabolites

Bioactive compounds are the chemical substances that are present in various food and food products which has the capacity to regulate the metabolic processes of living beings that use them. They possess antioxidant activity, enzyme inhibition activity; can cause receptor activity and gene expression inhibition. Exploration of natural products derived from biological diversity has consistently resulted in the discovery of new chemical compositions, which have found applications in the field of drug development.

28.2.1 Antibacterial

The development of new antimicrobial agents and antibiotic classes, together with efficient medication prescription practices and a steer clear of antibiotic abuse are all necessary to help tackle the threat of antibiotic resistance. Plants, microalgae and cyanobacteria are the producers of several antibiotics and may serve the purpose of tackling antibiotic resistance. The initial stage of the drug development process involves examining the mode of action of bioactive metabolites.

The antibacterial activity of secondary metabolites derived from medicinal plants is attributed to a variety of mechanisms, including disruption of bacterial membranes, reduction of cellular metabolism, regulation of biofilm formation, inhibition of bacterial capsule production, attenuation of bacterial virulence through quorum-sensing, and reduction of microbial toxin production (Fig. 28.1). *P. gingivalis*, *F. nucleatum*, *A. actinomycetemcomitans*, and *P. intermedia* were all susceptible to the antibacterial action of cardamom fruit and seed extracts. The breakdown of the bacterial membrane caused by the cardamom extract was observed in *P. gingivalis*. These findings are consistent with earlier research showing that bacteria treated with 1,8-cineole, a major ingredient in cardamom extracts, exhibited membrane damage and shrinkage of the cells (Vasconcelos et al. 2018). The physiological processes linked to cell activity were disrupted when *Salmonella typhimurium* and *S. aureus* were exposed to olive oil polyphenol extract which led to decrease in intracellular ATP concentrations, cell membrane depolarization, bacterial protein depreciation, and leakage of cytoplasm (Guo et al. 2020). The Gram-negative bacteria outer membrane permeability was increased as a result of antibacterial mechanisms of *A. carpa*, *C. superba*, and *C. ornusmas* leaf extracts towards pathogenic bacteria, with *C. superba* extract being the most effective. The bacteria exhibited morphological changes primarily as a result of being exposed to the leaf extracts. These alterations included the formation of aggregates, the synthesis of EPS, irregular shapes, wrinkled cell surfaces, gaps in the cell wall, and cell shriveling (Mickymaray 2019).

However, mechanism of action of antibacterial compounds from cyanobacteria are insufficient but recently this area of research has been gaining much attention

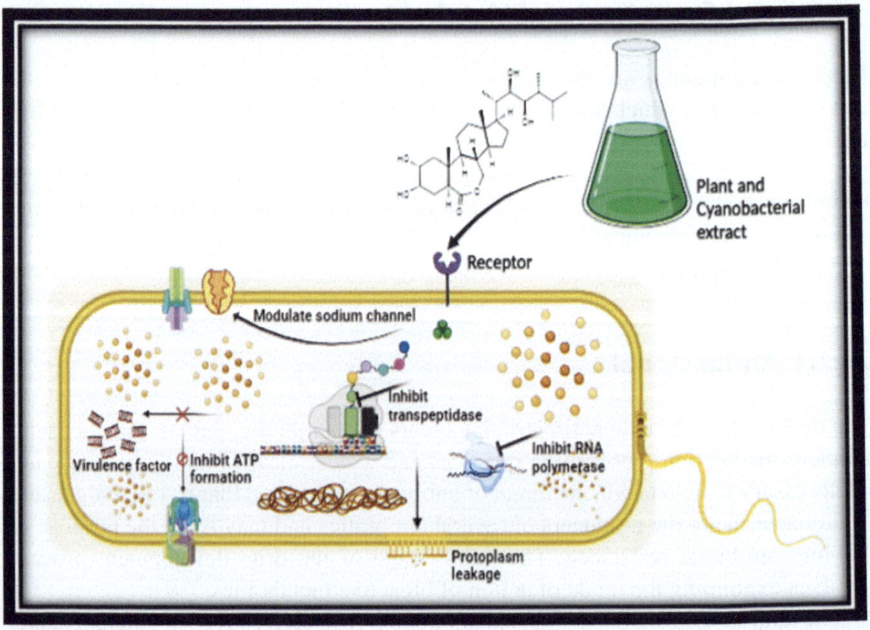

Fig. 28.1 A schematic diagram representing plant/cyanobacterial metabolites mode of action against bacteria

and it has been observed that like antibacterial from plants, they also target cell membrane and intracellular components of the bacteria (Benfield and Henriques 2020) (Fig. 28.1). It has been reported that malyngolide from *Lyngbya majuscule* hampers the quorum sensing of Gram-negative bacteria namely *Pseudomonas aeruginosa* and *Chromobacterium violaceum* CV017 which appears to be a promising approach to avoid resistant strain development (Dobretsov et al. 2010). *Scytonema hofmanni* produces scyptolin (act as serine protease) which inhibit the cell wall formation in various Gram-positive bacteria by restricting transpeptidase activity (Swain et al. 2017). Haplindole compounds which are alkaloid in nature possess antibacterial activity against Gram positive and negative cyanobacteria and have been found to inhibit bacterial RNA polymerase and modulator of sodium channel. Calothrixin A form *Calothrix* is one such example of bacterial RNA polymerase inhibitor (Thanh Doan et al. 2000). Some of them act by targeting the intracellular content of the bacterial cells, for instance, phosphatase B (mPTPB) from *M. tuberculosis* is required for its survival inside the macrophages and are the site for drug target and brunsvicamides B and C from *Tychonema* sp. exhibits inhibitory activity against mPTPB (Ruddraraju et al. 2020). Metabolites can also cause disruption of bacterial cell membranes however; there is very little evidence for the same.

28.2.2 Antifungal

Several researches have taken place to study the fungicide activity of natural products from plants and cyanobacteria to help develop medications to tackle infectious human pathogens. This is the need of the hour as the life-threatening fungal infection has enhanced mortality rate due to development of antifungal resistance and costly treatments (Hassan et al. 2022).

Plant extracts comprise bioactive compounds, like polyphenols and terpenoids, which exhibit promising antifungal activity, offering a potential alternative to synthetic pharmaceuticals with fewer side effects. *A. galanga* is a useful source of natural antifungal drugs for future pharmaceutical research, as evidenced by its suppressive effect on virulent hyphal development. The antifungal mechanism induced morphological changes in the size and shape of the cells produced enormous vacuoles in the yeast form, and prevented the transition to hyphae, but it is unlikely that this had an impact on the amount of chitin in *M. furfur* cell wall. The results show that the n-hexane extract of the rhizome of *A. galanga* has promising anti-Malassezia properties (Laokor and Juntachai 2021). *Thymus kotschyanus* (Lamiaceae family) is a medicinal plant that is utilized in traditional medicine. The study performed by Roudbary (Roudbary et al. 2023) for evaluation of the antifungal activity of *T. kotschyanus* extract and its impact on the CDR1 and CDR2 genes expression linked to drug resistance genes in *Candida albicans* were analysed, which showed reduced expression level of CDR1 and CDR2 genes.

According to Bui et al. (2014), the scytophycins, macrolides, and tolytoxfrom cyanobacteria are promising candidates for potent antifungal drugs. Lipopeptides such as laxaphicins B and C (obtained from *A. laxa*) possess fungicidal properties against *S. cerevisiae*, *P. notatum*, *A. oryzae*, *T. mentagrophytes*, and *C. albicans*. When extracted using hexane and methanol, some microalgae strains (green algae and cyanobacteria), such as *C. vulgaris* 012, *Chlamydomonas reinhardtii* 008, and *Chroococcus dispersus* 039, yield bioactive compounds with notable documented antifungal activity against *A. niger*, *A. fumigates*, *C. albicans*, and *C. kefyr*. Certain isolates of soil cyanobacteria have the ability to stop the growth of pathogenic fungus like *C. albicans* ATCC 14053 and *C. kefyr* ATCC 1140 (Soltani et al. 2005). These are some of the antifungal compounds reported from cyanobacteria however; a thorough literature survey revealed that the action mechanism of these bioactives has not been reported yet.

28.2.3 Antiprotozoal

Protozoans belong to the category of eukaryotic microorganisms, and certain individuals within this group exhibit parasitic behavior. The worldwide impact of protozoans' illnesses is worsened by the absence of approved vaccines, underscoring the importance of reliable and efficient medications for their prevention and cure. However, the efficacy of available drugs is progressively compromised by the emergence of drug-resistant parasites. This underscores the urgency for developing new medications, which fuels global research efforts in antiparasitic drug discovery. Such endeavors demand a variety of innovative approaches to ensure a continuous supply of promising compounds.

A potentially life-threatening infectious disease, malaria is brought on by protozoal parasites from the genus Plasmodium. The "Nobel medicine" *Artemisinin* and its derivatives have demonstrated potential use in antimalarial drug development initiatives and the two most widely accepted model regarding its mechanism are discussed. First, the reductive scission model in which ferrous or non-heme exogenous Fe^{2+} ions bind to artemisinin, leading to the cleavage of the peroxide bridge and the formation of oxygen-centre radicals, which then rearrange into carbon-centre radicals. This process can occur through various iron-peroxide interactions, resulting in the formation of primary carbon-centered radicals (via C3-C4 bond scission) or secondary carbon-centered (via 1,5 H-shifts). On the other hand, the open peroxide model suggests the involvement of a Fenton reaction between Fe^{2+} ions and the endoperoxide of artemisinin, leading to the generation of short-lived secondary carbon-centered radicals. These radicals are believed to be highly reactive and capable of damaging essential biomolecules within the malaria parasites (Pandey and Pandey-Rai 2016). While the specific mechanisms remain debated, extracts from *Cinchona* bark, *Azadirachta indica*, *Annona muricata*, *Allium sativum*, and *Carica papaya* possess anti-malarial qualities. These effects may stem from a variety of bioactive compounds, potentially disrupting the malaria parasite's life cycle, metabolism, or host immune response. The anti-leishmanial activity of *Emblica officinalis L.* fruit extract was investigated in a study performed by Rahat Ali and his co-researchers. The extract demonstrated inhibition of promastigotes and intra-macrophagic amastigotes in a dose-dependent manner. Additionally, morphological and ultrastructural changes were observed in parasites. It was also found to induce oxidative stress, mitochondrial dysfunction, DNA laddering, and apoptosis-like cell death in parasites (Ali et al. 2021).

Studies have reported antiprotozoal activities of various microalgae metabolites nevertheless, none of them have entered clinical testing for the same (Veas et al. 2020). Although cyanobacterial metabolites have been reported to produce metabolites against protozoa literature on the mechanism of action is scarce. Therefore, to provide insight into the mode of action of such metabolites against protozoa, other groups of algae were taken into consideration. For instance, Veas et al. (2020) (Veas et al. 2020) carried out a thorough investigation to show antiprotozoal properties of *T. cruzi* trypomastigotes using extracts from the microalgae species *C. reinhardtii*. Additionally, fluorescence microscopy revealed a decrease in the amount of parasite DNA in Vero cells, which was determined by morphological characteristics such nuclear size and the existence of a kinetoplast structure. Similarly, screening of trypanocidal components from various microalgal extract showed that the presence of secondary metabolites possessing trypanocidal activity as they have the ability to cause cellular structure modification and reduction in the expression of fundamental metabolites in *T.cruzi*. These metabolites can hamper the mitochondrial membrane potential, terpenes is one such metabolite which have the ability to cause the kinetoplast to dilate and vacuolate, which lowers the expression of metabolites such as succinate and acetate in *T. Cruzi* (Ramírez-Macías et al. 2012). Likewise, several studies have been conducted to study the efficacy of cyanobacterial metabolites against *Leishmania* and other protozoa. Fucoidan, a sulfated polysaccharide with polyanionic characteristics found in numerous brown algae, has been identified as an inhibitor of intracellular amastigote growth in *L. Donovani* (Kar et al. 2011). Methanolic extract of *Nannochloris* species SBL-1 (a microalgae isolated from Red Sea) reduced *L. Infantum* promastigotes viability to 62% (at 250 μg/mL concentration) (Dehghani et al. 2020). *Dunaliella* has shown to be a potent candidate for antileishmanial activity against *L. infantum* and *L. major* with IC_{50} 151–284 μg/mL. The result of the study indicated that *Dunaliella* could be a promising species for developing Leishmaniasis drugs (Gharbi et al. 2021). *Nostoc commune* and *Rivularia isolation* were tested against *L. donovani, T. cruzi, P. falciparum*, and *M. Tuberculosis* and were found to have antiprotozoal activity indicating the value of cyanotherapeutics (Broniatowska et al. 2011). Against certain drug-resistant strains, lactam, and dolastatins derived from *Oscillatoria* sp., and *Lyngbya majuscula*, respectively, demonstrated a protozocidal effect. Companeramides, also known as depsipeptide, have reportedly demonstrated antimalarial properties against strains of *Plasmodium falciparum* resistant to chloroquine (Wani et al. 2021).

28.2.4 Antitumor

Secondary metabolites are a rich source of cytotoxic bioactive compound possessing anticancer properties which can have role in chemotherapy. Herbal therapy offers a highly practical substitute for conventional cancer treatment. Medicinal herbs have a significant potential to treat cancer, according to a comprehensive evaluation of several plants. *Coptischinensis* extract is widely used to treat blood cancer, dysentery, cholera, diabetes, and lung cancer due to potent anti-bacterial/tumor action. Its active ingredients include berberine, an alkaloid which acts by suppressing telomerase and nucleophosmin/B23 activity; berberine alkaloids cause human leukemia HL-60 cells to undergo apoptosis. Curcumin's influence on COX-2 expression was investigated by treating human HT-29 colon cancer cells with varying doses of curcumin. Curcumin suppressed HT-29 cell proliferation. By suppressing the mRNA and protein expression of COX-2, curcumin had an effect; however, COX-1 was not similarly inhibited. Anticancerous and cell survival activity of curcumin on MCF-7 (human breast cancer cell lines) was done by examining lactate dehydrogenase and tetrazolium salt. The findings demonstrated the time- and concentration-dependent cytotoxicity and cell inhibition caused by curcumin (Khan et al. 2019). An in vitro investigation revealed that *G. oblongifolia* bark includes a number of significant secondary metabolites, such as oblongifolin A–G and oblongixanthones A–C. These metabolites have the highest levels of apoptotic activity in HeLa-C3 cell lines and the highest levels of cytotoxicity in cervical cancer cells. The HT-29 and HCT-116 cells (colon cancer cells), as well as IEC-6 and INT-407 (normal immortalized intestinal cells), exhibit favorable activity in response to *G. indicagarcinol*. All intestinal cells exhibit substantial inhibitory characteristics when treated for 72 h with garcinol at IC_{50} values ranging from 3.2 to 21.4 μM. Compared to normal immortalized cells, cancer cells exhibited greater anticancer capabilities (Feng et al. 2014). In a separate study conducted by Kikuchi et al., *Morus alba*'s root extract albanol A exhibited apoptosis-inducing and cytotoxic activity in the HL-60 cell line. Albanol A inhibited topoisomerase II with a reduction in pro-caspases 3, 8, and 9 levels (Kikuchi et al. 2010). Research is required to elucidate the anticancer activity mechanism of numerous previously studied and unstudied plants.

Cyanobacterial metabolites have also been reported to possess cytotoxic properties which can help in cancer treatments. These metabolites generally act by arresting cell cycle, disrupting the tubulin activity or by activating caspase. Symplostatin 1 obtained from marine cyanobacterium *Symploca hydnoides* acts against various cell lines like including NCI/ADR, A-10, HUVEC, SK-OV-3, and MDA-MB-435 by inhibiting mitosis and proliferation. They initiate cell death by Bcl-2 phosphorylation and activating the caspase-3 (Mooberry et al. 2003). Calothrixin A (*Calothrix* sp.) arrests the G2 phase in leukemia and also disrupts the replication during the S-phase (Khan et al. 2009) whereas Calothrixin B arrest the cell cycle in G1 phase (Khan et al. 2009). Calothrixin A also increases the reactive oxygen species within the cell which lead to DNA fragmentation in Jurkat human T cells (Chen et al. 2003). Similarly, Apratoxin A from *Lyngbya majuscule* also arrests the cycle in G1 phase in cervical carcinoma and modify the pathway of secretory proteins in osteosarcoma cell lines (Matthew et al. 2008). Curacin A from *Lyngbya majuscule* inhibits polymerization of tubulin and has a cytotoxic effect on L1210 leukemic cell lines (Burja et al. 2002). *L. majuscule* also produces lyngbyabellin A and B and they possess actin polymerization property alongwith cytotoxic effect against colorectal and breast cancer cell lines (Kang et al. 2018). Similarly, Belamide A (*Symploca* sp.) also polymerizes tubilin and causes arrest of cell cycle at G2 phase thereby demonstrating a significant activity against colon cancer cell (HCT-11) (Simmons et al. 2006). Another metabolite namely cryptophycin 1 has cytotoxic effect on acute leukemia by disrupting its microtubule assembly (Kerksiek et al. 1995). In addition, it also acts on cell cycle by arresting it in M-phase, thereby inhibiting proliferation of mammary adenocarcinoma, human lung cancer and ovarian cancer cells (Mooberry et al. 1997). Scytonemin, a pigment from *Stigonema* sp. which play role in combating acute T cell leukemia by inhibiting cell proliferation by arresting cell cycle at M-phase and inhibiting synthesis of new cells. *L. majuscule* also produces somocystinamide, which is a lipopeptide and promote cell death by activating caspase-8 (Wrasidlo et al. 2008). These are few of the many examples of metabolites reported from cyanobacteria possessing cytotoxic activity.

28.2.5 Antiviral

Viruses stand as a major contributor to human illness and mortality globally. The world has seen serious effects on human society throughout the past 20 years as a result of the spread of numerous deadly viral illnesses such as COVID-19, avian influenza, swine influenza, Ebola, and severe acute respiratory syndrome (SARS). Additionally, the rise of viral resistance and the negative impacts of antiviral medications have emerged as crucial challenges in the field of medicine. A multitude of studies have demonstrated that bioactive compounds possess the ability to inhibit viral activities.

In addition to synthetic combinations, plant-derived metabolites are also a focus of the search for new antiviral medicines. The mangrove plant *Acanthus ilicifolius L.*, a member of the Acanthaceae family, has strong antiviral activity in addition to its many other therapeutic uses. It suppressed cell cycle dysregulation-induced HIV, which causes HIV-1-infected individuals' CD4+ lymphocyte counts to rise. HSV-I viral envelope glycoproteins D and C inhibition has also been reported (Lin et al. 2008). It has been discovered that *Mangiferaindica L.* extract possesses antiviral properties against influenza viruses. In addition, it has a bioactive substance called mangiferin, which may be useful in preventing HSV-I from replicating and counteracting the cytotoxic effects of HIV (Amin 2019). A documented study highlights that extracts of *Allium sativum* contain numerous antiviral phytocompounds such as ajoene which exhibits protective effects against CD4+ cells destruction induced by HIV and boosts cellular immunity. It also hinders the attachment of the virus to host cells and inhibits the HIV-1 reverse transcriptase activity. Additionally, ajoene induces apoptosis in cells infected with Human Cytomegalovirus (HCMV). Allicin and allyl methyl thiosulfinate interfere with the entry of Herpes Simplex Virus types I and II, Human Rhinovirus 2 Vaccinia Virus, Parainfluenza Virus 3 and Vesicular Stomatitis Virus by disturbing viral envelopes and membranes. Allicin also impedes the *Reticuloendotheliosis* Virus replication by downregulating the ERK/MAPK pathway. Diallylsulfide, garlicin, and alliin, combat Dengue Virus by reducing inflammation via suppression of oxidative stress. Allitridin exhibits multifaceted properties against HCMV, by preventing DNA synthesis by disturbing viral immediate-early antigen expression, suppressing viral IEG gene transcription to inhibit viral replication, and enhancing the expansion of regulatory T cells and their anti-HCMV immunosuppressive activity (Wang et al. 2017; Rouf et al. 2020) (Fig. 28.2). Nonetheless, insufficient studies exist regarding the bioactive compounds structure relationship activity, which is essential for developing drugs effective against deadly viruses.

Due to their ease of cultivation and potential as a source of antiviral chemicals, cyanobacteria have drawn a lot of interest. The virus infects the host in three stages: adsorption and invasion into host cells, eclipse phase and maturity and release of viral progeny (Hassan et al. 2022) and various studies have reported that the cyanobacterial metabolites can inactivate the virus immediately after adsorption and invasion. Cyanovirin-N (obtained from *N. ellipsosporum*) causes inactivation of primary strains of HIV-1 in an irreversible manner and inhibits cellular transmission by preventing the fusion between HIV and the host cell by disrupting their glycoprotein (Burja et al. 2001). Cyanovirin-N is also found effective against Ebola, influenza virus and hepatitis C virus (Carpine and Sieber 2021). Micovirin (obtain from *M. aeruginosa*) is also found to inhibit a HIV-1 strains by inhibiting formation of syncytium between infected T cells and uninfected CD4$^+$ T cells however, it is less toxin than cyanovirin (Shahzad-ul-Hussan et al. 2011). Scytovirin blocks entry into

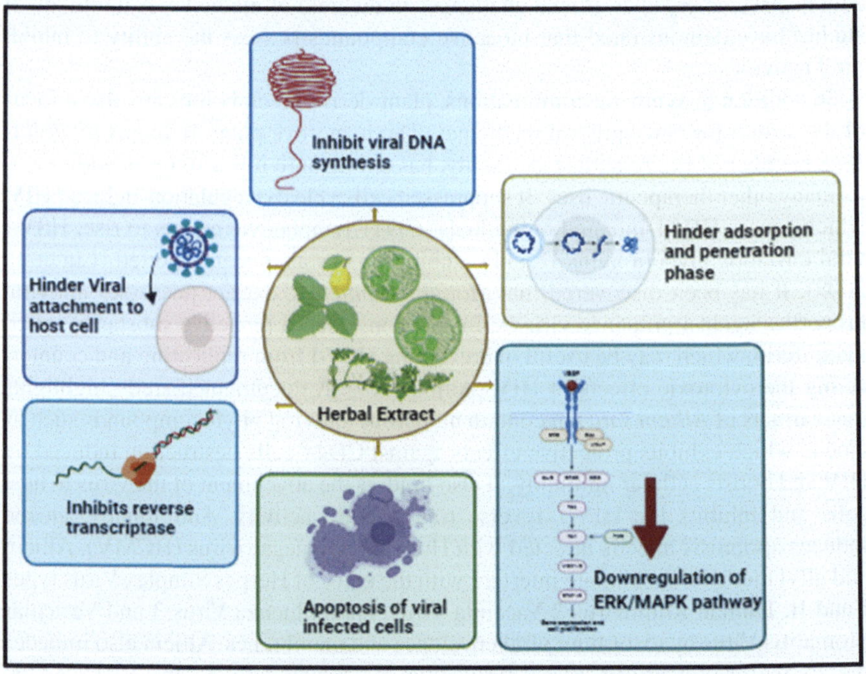

Fig. 28.2 Mode of action of plants metabolites against virus

target cells by binding with mannose-rich oligosaccharides present in the envelope glycoprotein and is isolated from *Scytonema varium* (Garrison et al. 2014). Allophycocyanin, a pigment from blue-green algae also acts as a source of antiviral metabolites. For instance, allophycocyanin obtained from *Spirulina platensis* delays viral nucleic acid synthesis and the apoptotic process. It neutralizes cytopathic effects in African green monkey kidney cells and human rhabdomyosarcoma cells which are induced by enterovirus 71 (Shih et al. 2003). Another class of metabolite which belongs to the functional category of carbohydrate are found to be effective antiviral agents on both enveloped and non-enveloped viruses like HIV-1, HSV-1, HCMV, influenza A, coxsackie, measles, polio, and mumps. They generally act by preventing the viral entry into the host cell as stated above (Ahmadi et al. 2015). For example, nostoflan from *Nostoc flagelliforme* act against HSV, HCMV and influenza A virus by the inhibiting its binding to the host cell (Kanekiyo et al. 2007). Similarly, calcium spirulina (from *S. platensis*) also interferes with viral adsorption and penetration phases. However, it does not inhibit non-enveloped virus like poliovirus and coxsackie virus suggesting its selective inhibition for enveloped virus (Carpine and Sieber 2021). Cyanobacterial bioactive compounds, like lectins and polysaccharides, are also being considered as potential medications against COVID-19 infection. Sulfated polysaccharides, in particular, have the ability to bind to the S-protein of COVID-19 which further prevents it from attaching to the host tissues heparin sulfate co-receptor, effectively arresting infection (Rajarshi et al. 2021).

28.3 Conclusion

Due to the growing resistance to current antibiotics and contemporary medicine, the worldwide AMR epidemic necessitates an immediate search for alternative and effective antimicrobials. It is now clear that developing more potent antibiotics based on antiquated theories and chemical classes is insufficient to combat antibiotic resistance; doing so will only have a limited, transient impact and increase resistance. The alarming rate at which new infections are continuing to arise raises severe concerns about public health as it threatens the current medical paradigm's ability to eradicate infectious illnesses.

In this context, natural bioactive compounds derived from plants and microbes offer a vast and sustainable source of antimicrobial agents. Plants and cyanobacteria, being readily available and easy to cultivate, produce metabolites with a broad spectrum of activity against pathogens. Additionally, these metabolites from plants exhibit immunomodulatory properties, benefiting overall health. Inspire of their strong bioactivities, only a small number of these metabolites have gone through clinical trials.

The hypothesis is that the pharmaceutical potential of these compounds will increase, particularly with the significant advancements in omics technology, notably metabolomics. This technology enables the identification of even trace amounts of active phytochemicals and the elucidation of specific molecular mechanisms underlying their effects on targeted pathogens. A comprehensive understanding of these molecular mechanisms is crucial for developing successful therapeutic approaches, and this is likely to be a focal point of future research.

To summaries, this chapter offers a thorough analysis of specific natural compounds and their antimicrobial mechanisms of action against a range of human diseases. The knowledge gathered from this review can help develop new approaches to combat the growing danger of AMR and improve the antimicrobial treatments that are already in use.

References

Achan J, Talisuna AO et al (2011) Quinine, an old anti-malarial drug in a modern world: role in the treatment of malaria. Malar J 10(1):1–12

Ahmadi A, Zorofchian Moghadamtousi S et al (2015) Antiviral potential of algae polysaccharides isolated from marine sources: a review. Biomed Res Int 2015:825203

Ali R, Islamuddin M et al (2021) *Embilica officinalis* L. inhibits the growth and proliferation of *Leishmania donovani* through the induction of ultrastructural changes, mitochondrial dysfunction, oxidative stress and apoptosis-like cell death. Biomed Pharmacother 143:112156

Amin A (2019) AL dulaimi and marwa AA AL rawi, antiviral activity of *Mangifera* extract on influenza virus cultivated in different cell cultures. J Pure Appl Microbiol 13(1):455–458

Anjum K, Abbas SQ et al (2017) Emerging biopharmaceuticals from bioactive peptides derived from marine organisms. Chem Biol Drug Des 90(1):12–30

Baell JB (2016) Feeling nature's PAINS: natural products, natural product drugs, and pan assay interference compounds (PAINS). J Nat Prod 79(3):616–628

Bassett EJ, Keith MS et al (1980) Tetracycline-labeled human bone from ancient Sudanese Nubia (AD 350). Science 209(4464):1532–1534

Benfield AH, Henriques ST (2020) Mode-of-action of antimicrobial peptides: membrane disruption vs. intracellular mechanisms. Front Med Technol 2:610997

Broniatowska B, Allmendinger A et al (2011) Antiprotozoal, antitubercular and cytotoxic potential of cyanobacterial (blue-green algal) extracts from Ireland. Nat Product Commun 6(5):1934578X1100600523

Bui T-H, Wray V et al (2014) Balticidins A–D, antifungal hassallidin-like lipopeptides from the Baltic Sea cyanobacterium *Anabaena cylindrica* Bio33. J Nat Prod 77(6):1287–1296

Burja AM, Banaigs B et al (2001) Marine cyanobacteria—a prolific source of natural products. Tetrahedron 57(46):9347–9377

Burja A, Abou-Mansour E et al (2002) Culture of the marine cyanobacterium, *Lyngbya majuscula* (Oscillatoriaceae), for bioprocess intensified production of cyclic and linear lipopeptides. J Microbiol Methods 48(2–3):207–219

Carpine R, Sieber S (2021) Antibacterial and antiviral metabolites from cyanobacteria: their application and their impact on human health. Curr Res Biotechnol 3:65–81

Chen X, Smith GD et al (2003) Human cancer cell (Jurkat) killing by the cyanobacterial metabolite calothrixin A. J Appl Phycol 15:269–277

Dehghani J, Adibkia K et al (2020) Towards a new avenue for producing therapeutic proteins: microalgae as a tempting green biofactory. Biotechnol Adv 40:107499

Dobretsov S, Teplitski M et al (2010) Malyngolide from the cyanobacterium *Lyngbya majuscula* interferes with quorum sensing circuitry. Environ Microbiol Rep 2(6):739–744

Feng C, Huang S-X et al (2014) Characterization of proapoptotic compounds from the bark of *Garcinia oblongifolia*. J Nat Prod 77(5):1111–1116

Garrison AR, Giomarelli BG et al (2014) The cyanobacterial lectin scytovirin displays potent *in vitro* and *in vivo* activity against Zaire Ebola virus. Antivir Res 112:1–7

Gharbi K, Fathalli A et al (2021) Tunisian inland water microflora as a source of phycobiliproteins and biological activity with beneficial effects on human health. Oceanol Hydrobiol Stud 50(4):385–397

Guo L, Gong S et al (2020) Antibacterial activity of olive oil polyphenol extract against *Salmonella typhimurium* and *Staphylococcus aureus*: possible mechanisms. Foodborne Pathog Dis 17(6):396–403

Hassan S, Meenatchi R et al (2022) Identification and characterization of the novel bioactive compounds from microalgae and cyanobacteria for pharmaceutical and nutraceutical applications. J Basic Microbiol 62(9):999–1029

Kanekiyo K, Hayashi K et al (2007) Anti-herpes simplex virus target of an acidic polysaccharide, nostoflan, from the edible blue-green alga *Nostoc flagelliforme*. Biol Pharm Bull 30(8):1573–1575

Kang HK, Choi M-C et al (2018) Therapeutic properties and biological benefits of marine-derived anticancer peptides. Int J Mol Sci 19(3):919

Kar S, Sharma G et al (2011) Fucoidan cures infection with both antimony-susceptible and-resistant strains of *Leishmania donovani* through Th1 response and macrophage-derived oxidants. J Antimicrob Chemother 66(3):618–625

Kerksiek K, Mejillano MR et al (1995) Interaction of cryptophycin 1 with tubulin and microtubules. FEBS Lett 377(1):59–61

Khan QA, Lu J et al (2009) Calothrixins, a new class of human DNA topoisomerase I poisons. J Nat Prod 72(3):438–442

Khan T, Ali M et al (2019) Anticancer plants: a review of the active phytochemicals, applications in animal models, and regulatory aspects. Biomol Ther 10(1):47

Kikuchi T, Nihei M et al (2010) Albanol A from the root bark of *Morus alba* L. induces apoptotic cell death in HL60 human leukemia cell line. Chem Pharm Bull 58(4):568–571

Laokor N, Juntachai W (2021) Exploring the antifungal activity and mechanism of action of Zingiberaceae rhizome extracts against *Malassezia furfur*. J Ethnopharmacol 279:114354

Lin T-P, Chen S-Y et al (2008) Inhibition of the Epstein–Barr virus lytic cycle by andrographolide. Biol Pharm Bull 31(11):2018–2023

Manesh A, Varghese GM (2021) Rising antimicrobial resistance: an evolving epidemic in a pandemic. Lancet Microbe 2(9):e419–e420

Matthew S, Schupp PJ et al (2008) Apratoxin E, a cytotoxic peptolide from a Guamanian collection of the marine cyanobacterium *Lyngbya bouillonii*. J Nat Prod 71(6):1113–1116

Mickymaray S (2019) Efficacy and mechanism of traditional medicinal plants and bioactive compounds against clinically important pathogens. Antibiotics 8(4):257

Mooberry SL, Busquets L et al (1997) Induction of apoptosis by cryptophycin 1, a new antimicrotubule agent. Int J Cancer 73(3):440–448

Mooberry SL, Leal RM et al (2003) The molecular pharmacology of symplostatin 1: a new antimitotic dolastatin 10 analog. Int J Cancer 104(4):512–521

Nelson ML, Dinardo A et al (2010) Brief communication: mass spectroscopic characterization of tetracycline in the skeletal remains of an ancient population from Sudanese Nubia 350–550 CE. Am J Phys Anthropol 143(1):151–154

Pandey N, Pandey-Rai S (2016) Updates on artemisinin: an insight to mode of actions and strategies for enhanced global production. Protoplasma 253:15–30

Prestinaci F, Pezzotti P et al (2015) Antimicrobial resistance: a global multifaceted phenomenon. Pathogens Global Health 109(7):309–318

Qadri H, Shah AH et al (2021) Novel strategies to combat the emerging drug resistance in human pathogenic microbes. Curr Drug Targets 22(12):1424–1436

Qadri H, Shah AH et al (2022) Natural products and their semi-synthetic derivatives against antimicrobial-resistant human pathogenic bacteria and fungi. Saudi J Biol Sci 29(9):103376

Rajarshi K, Khan R et al (2021) Essential functional molecules associated with SARS-CoV-2 infection: potential therapeutic targets for COVID-19. Gene 768:145313

Ramírez-Macías I, Marín C et al (2012) *In vitro* and *in vivo* studies of the trypanocidal activity of four terpenoid derivatives against *Trypanosoma cruzi*. Am J Trop Med Hygiene 87(3):481

Rossiter SE, Fletcher MH et al (2017) Natural products as platforms to overcome antibiotic resistance. Chem Rev 117(19):12415–12474

Roudbary M, Alimohammadi A et al (2023) Antifungal activity of Thymus kotschyanus extract: an *in vitro* study on the expression of CDR1 and CDR2 genes in clinical isolates of *Candida albicans*. J Herbal Med 38:100644

Rouf R, Uddin SJ et al (2020) Antiviral potential of garlic (*Allium sativum*) and its organosulfur compounds: a systematic update of pre-clinical and clinical data. Trends Food Sci Technol 104:219–234

Ruddraraju KV, Aggarwal D et al (2020) Highly potent and selective N-aryl oxamic acid-based inhibitors for Mycobacterium tuberculosis protein tyrosine phosphatase B. J Med Chem 63(17):9212–9227

Schwarzer D, Finking R et al (2003) Nonribosomal peptides: from genes to products. Nat Prod Rep 20(3):275–287

Shahzad-ul-Hussan S, Gustchina E et al (2011) Solution structure of the monovalent lectin microvirin in complex with Manα (1–2) man provides a basis for anti-HIV activity with low toxicity. J Biol Chem 286(23):20788–20796

Shih SR, Tsai KN et al (2003) Inhibition of enterovirus 71-induced apoptosis by allophycocyanin isolated from a blue-green alga *Spirulina platensis*. J Med Virol 70(1):119–125

Simmons TL, McPhail KL et al (2006) Belamide A, a new antimitotic tetrapeptide from a Panamanian marine cyanobacterium. Tetrahedron Lett 47(20):3387–3390

Soltani N, Khavari-Nejad R et al (2005) Screening of soil cyanobacteria for antifungal and antibacterial activity. Pharm Biol 43(5):455–459

Srivastava R, Prajapati R et al (2022) Phycochemistry and bioactivity of cyanobacterial secondary metabolites. Mol Biol Rep 49(11):11149–11167

Swain SS, Paidesetty SK et al (2017) Antibacterial, antifungal and antimycobacterial compounds from cyanobacteria. Biomed Pharmacother 90:760–776

Thanh Doan N, Rickards RW et al (2000) Allelopathic actions of the alkaloid 12-epi-hapalindole E isonitrile and calothrixin A from cyanobacteria of the genera *Fischerella* and *Calothrix*. J Appl Phycol 12:409–416

Vasconcelos N, Croda J et al (2018) Antibacterial mechanisms of cinnamon and its constituents: a review. Microb Pathog 120:198–203

Veas R, Rojas-Pirela M et al (2020) Microalgae extracts: potential anti-*Trypanosoma cruzi* agents? Biomed Pharmacother 127:110178

Wang L, Jiao H et al (2017) Allicin alleviates reticuloendotheliosis virus-induced immunosuppression via ERK/mitogen-activated protein kinase pathway in specific pathogen-free chickens. Front Immunol 8:1856

Wani AK, Akhtar N et al (2021) Cyanobacteria-derived small molecules: a new class of drugs. In: Volatiles and metabolites of microbes. Elsevier, pp 283–303

World Health Organization (2019) WHO global report on traditional and complementary medicine 2019. World Health Organization

Wrasidlo W, Mielgo A et al (2008) The marine lipopeptide somocystinamide A triggers apoptosis via caspase 8. Proc Natl Acad Sci 105(7):2313–2318

Antimicrobial Aspects of Brazilian Red Propolis and Its Compounds

29

Nagela Bernadelli Sousa Silva, Daniela Silva Gonçalves, Gabriel Guimarães Calefi, Diego Godina Prado, Luís Carlos Scalon Cunha, and Carlos Henrique Gomes Martins

Abstract

Antimicrobial resistance is the result of mutations suffered by bacteria, fungi, and viruses that render their respective antimicrobials ineffective in combating infections, what contributes to increased morbidity and mortality rates, mainly in hospitalized patients. The lack of therapeutic options makes this scenario alarming and the search for new antimicrobials extremely urgent. In this sense, natural products have been contributing significantly to the discovery of new medicines, particularly against infectious diseases. Among natural products, Brazilian Propolis stands out, especially from mangrove regions in northeastern Brazil, frequently referred to as Brazilian Red Propolis, this product is meticulously harvested by *Apismellifera* bees from the red resinous exudate found on the trunks of *Dalbergia Ecastophyllum* trees. Chemical classes present in Brazilian Red Propolis such as flavonoids, esters, quinones, alcohols, ketones, phenolic acids, and aldehydes are identified as those responsible for its antimicrobial activities. Brazilian Red Propolis has already been tested against Gram-negative and Gram-positive bacteria, particularly the multidrug-resistant *Staphylococcus aureus*, *Pseudomonas aeruginosa*, *Escherichia coli*, as well as yeasts such as *Candida* spp. and *Cryptococcus* spp., revealing promising Minimum Inhibitory Concentration (MIC) values. Its antiviral action, although less explored, has already been demonstrated against Chikungunya virus, Enterovirus surrogates,

N. B. S. Silva · D. S. Gonçalves · G. G. Calefi · C. H. G. Martins (✉)
Institute of Biomedical Sciences (ICBIM), Federal University of Uberlândia, Uberlândia, MG, Brazil
e-mail: carlos.martins2@ufu.br

D. G. Prado
Institute of Chemistry (IQ), Federal University of Uberlândia, Uberlândia, MG, Brazil

L. C. S. Cunha
Chemistry Department, Federal Institute of Triângulo Mineiro, Uberaba, MG, Brazil

M. Gangwar, G. Nath (eds.), *Emerging Paradigms for Antibiotic-Resistant Infections: Beyond the Pill*, https://doi.org/10.1007/978-981-97-5272-0_29

and human coronavirus. Based on that, one can conclude that Brazilian Red Propolis is a hopeful natural product against a panel of clinically relevant microorganisms. The study of its antimicrobial properties can contribute to the future use of Brazilian Red Propolis as a therapeutic option against various human infections. This chapter will address the antimicrobial properties of Brazilian Red Propolis and its constituents against various bacteria, fungi, and viruses of high clinical relevance, as well as chemical aspects and future perspectives of clinical trials.

Keywords
Brazilian Red Propolis · Antimicrobial resistance · MIC · Phytoconstituents

29.1 Introduction

The discovery of antibiotics and antifungals was a major milestone in healthcare, saving millions of lives since the twentieth century. Without these medications, simple surgeries or immunosuppressive therapies could pose a high risk to human health, increasing the number of deaths from microbial infections (Huemer et al. 2020). The approval of the first antiviral drug in 1963 also represented a breakthrough, as viral infections have caused millions of deaths across human civilization (Kausar et al. 2021). However, the indiscriminate use of antimicrobials resulted in the emergence of resistant microorganisms, which reduced the effectiveness of these medications, making microbial infections more persistent and difficult to treat (Morrison and Zembower 2020).

Antimicrobial resistance accounts for a setback for modern medicine, limiting therapeutic options and harming people's quality of life. According to a report published in 2019 by the United Nations Interagency Coordination Group, antimicrobial resistance is identified as responsible for 700,000 deaths every year and could cause up to ten million deaths by 2050 (WHO 2019). The costs associated with these infections to the United States healthcare system range from 28 to 45 billion dollars per year and more than 100 billion dollars worldwide, in addition to reducing the Gross Domestic Product (GDP) by 2–3.5% globally (Morrison and Zembower 2020; Nunes et al. 2022). Therefore, the creation of new strategies against microbial resistance becomes paramount, as available antimicrobials are not effective.

Several studies have developed in an attempt to find therapeutic options that provide an improvement in the quality of life of human beings. Nature has always been a source of great interest on the part of man, due to so many resources offered, many of which have not yet even been studied. Researchers are encouraged to look for new drugs with lower toxicity that actively combat/prevent infection by microorganisms, whether fungi, bacteria, or viruses. In this context, natural products can be a promising alternative. These products stand out compared to synthetic molecules due to their enormous diversity of secondary metabolites and structural complexity (Atanasov et al. 2021; Chopra and Dhingra 2021).

Since ancient times, natural products have been used as a therapeutic option for various diseases. Even today, according to Rawat et al. (2023), approximately 80% of the world's population consumes natural products as a tool against health disorders. In the era of multidrug resistance that the world is currently facing, natural products are promising in combating multidrug-resistant microorganisms, due to their ability to act on different targets as well as perform different mechanisms of action (Chopra and Dhingra 2021). According to Newman and Cragg (2020), from 1981 to 2019, 78 antibacterial drugs, one antiviral, and two antiparasitics derived from natural products have already been approved for use by the Food Drug Administration (FDA). However, no antifungal derived from natural products was approved for use until 2019, being in phase III testing, which shows the need for more studies of these products, especially against fungal infections.

Brazil has the greatest biodiversity on the planet, being the country where around 15% of all living species in the world are found. It has six different biomes, namely the Amazon Forest, "Caatinga," "Pantanal" "Pampa," "Cerrado" and Atlantic Forest and three maritime ecosystems (Valli et al. 2018). In several Brazilian states, several botanical sources of promising natural products can be found in the treatment of various diseases. In northeastern Brazil, a type of natural product is found that seems promising, due to its extensive variety of chemical compounds with pharmacological activity. Propolis is a product collected from the trunks of trees of different botanical origins with the function of protecting the hive from invaders such as bacteria and parasites, as well as maintaining the temperature of the colonies. Various types of propolis exist, characterized by distinct chemical compositions influenced by factors such as floral diversity, bee saliva enzymes, bee species, genetic diversity, and the geographic location of collection (Rodrigues et al. 2020; Zabaiou et al. 2017). In 2007, the 13th type of Brazilian propolis was described in the literature, being called red propolis due to its reddish color (De Freitas et al. 2017).

The Brazilian Red Propolis is a non-toxic natural product collected by bees from plant exudates. It is produced by bees of the species *Apismelifera* by collecting the red exudate found on the surface of the holes made by insects in the trunk of *Dalbergia Ecastophyllum*, popularly known as "rabo de bugio," considered to be of botanical origin (Aldana-Mejia et al. 2021a). Various chemical classes present in Brazilian Red Propolis have been identified, including flavonoids, isoflavonoids, neoflavonoids, pterocarpanes, chalcones, polyprenylated benzophenones hydroxynamic acids, terpenes, and tannins (Trusheva et al. 2006; Mayworm et al. 2014; Freires et al. 2016). Some components of a polar nature have also been identified, which makes Brazilian Red Propolis of great economic interest due to the easy solubilization for commercial purposes and the possibility of large-scale production (Miranda et al. 2019).

Currently, it ranks as the second most produced and commercialized type of Brazilian Propolis, found mainly in the coastal region of Brazil, particularly in the State of Alagoas (Reis et al. 2019). Several studies point to Brazilian Red Propolis as a promising natural product due to its antibacterial (Alencar et al. 2007; Botteon et al. 2021; Bueno-Silva et al. 2013a, 2013b; Dantas Silva et al. 2017;

Devequi-Nunes et al. 2018; Inui et al. 2014; Ripari et al. 2023) and antifungal (Oliveira et al. 2006; Sawaya et al. 2002; Siqueira et al. 2009; Das Neves et al. 2016; Martorano-Fernandes et al. 2020; Pippi et al. 2015) and antiviral potential (Silva-Beltrán et al. 2023; Silva and Dermody 2017; Silva et al. 2022).

Therefore, given the issue of microbial resistance and the urgency of discovering new antimicrobials, this chapter will address the antimicrobial properties of Brazilian Red Propolis and its isolated compounds have demonstrated efficacy against resistant bacteria, fungi, and viruses of high clinical relevance, as well as chemical aspects and future perspectives.

29.2 Chemical Aspects

Propolis consists of resins (50%), waxes (30%), essential oils (10%), pollen (5%), and other substances such as minerals and organic molecules (5%). The chemical composition of Propolis is directly influenced by geographic factors, climatic characteristics, type of soil, seasonality, the bee species, and the botanical sources available for its production (Regueira-Neto et al. 2017; Do Nascimento et al. 2019; dos Santos et al. 2019).

The most common compounds related to biological activities associated with natural products, including Propolis, are specialized metabolites (secondary metabolites) produced by living organisms from their biological functions (Oladipo et al. 2022). In Brazilian Red Propolis, the classes of metabolites such as flavonoids, isoflavonoids, pterocarpans, chalcones, polyprenylated benzophenones, and terpenes have already been related within its chemical composition (Trusheva et al. 2006; Mayworm et al. 2014; Do Nascimento et al. 2019). The major metabolites are phenolic compounds, mostly flavonoids (Rufatto et al. 2018).

Among the flavonoid classes, the most common are isoflavonoids and some derivatives, such as pterocarpanes. These compounds and some chalcones are the substances that differentiate the Brazilian Red Propolis from other Brazilian Propolis, and they are the most reported of this type of Propolis (Freires et al. 2016; Gomes Sá et al. 2023). The presence of compounds of these classes in Brazilian Red Propolis is assigned to *D. ecastophyllum*species, as shown in a work designed by Aldana-Mejía et al. (2021b). In this study the presence of some chemical markers was identified in Brazilian Red Propolis and the stems and resin of *D. ecastophyllum*, the following compounds were identified: medicarpin, neovestitol, vestitol, formononetin, isoliquiritigenin, calycosin, liquiritigenin, biochanin A, and 7-*O*-methylvestitol. Another work developed by Piccinelli et al. (2011), identified some of the cited metabolites, and the following compounds were identified: naringenin, 7,3′-dihydroxy-5′-methoxy-isoflavone, 3-hydroxy-8,9-dimethoxy-pterocarpan, and mucronulatol. Figure 29.1 shows the structures of the identified compounds and their respective classes.

Piccinelli et al. (2011), also identified two pigments that may be responsible for the red color of Propolis, both were isolated and elucidated from Brazilian Red

Flavanone

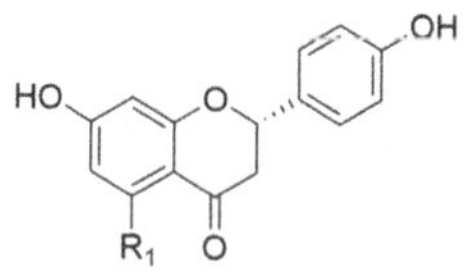

Liquiritigenin: $R_1 = H$
Naringenin: $R_1 = OH$

Isoflavan

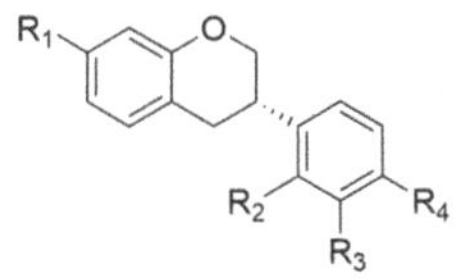

Vestitol: $R_1 = OH$, $R_2 = OH$, $R_3 = H$, $R_4 = OCH_3$
Neovestitol: $R_1 = OCH_3$, $R_2 = OH$, $R_3 = H$, $R_4 = OH$
7-*O*-methylvestitol :$R_1 = OCH_3$, $R_2 = OH$, $R_3 = H$, $R_4 = OCH_3$
Mucronulatol:$R_1 = OH$, $R_2 = OCH_3$, $R_3 = OH$, $R_4 = OCH_3$

Chalcone

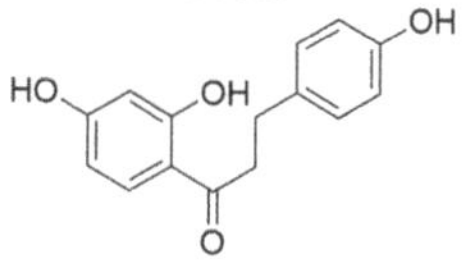

Isoliquiritigenin

Isoflavone

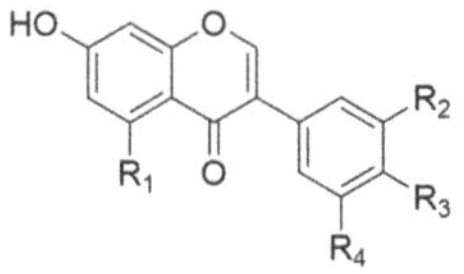

Calycosin: $R_1 = H$, $R_2 = OH$, $R_3 = OCH_3$. $R_4 = H$
Formononetin: $R_1 = H$, $R_2 = H$, $R_3 = OCH_3$, $R_4 = H$
Biochanin A: $R_1 = OH$, $R_2 = H$, $R_3 = OCH_3$, $R_4 = H$
7,3'-dihydroxy-5'-methoxy-isoflavone: $R_1 = H$, $R_2 = OCH_3$, $R_3 = H$, $R_4 = OH$

Pterocarpan

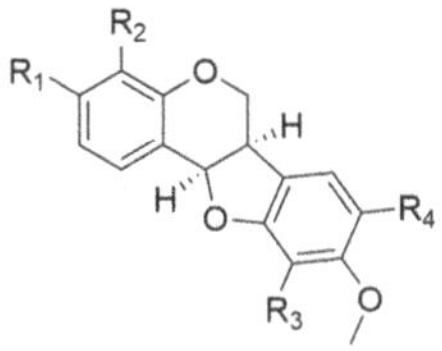

Medicarpin: $R_1 = H$, $R_2 = H$, $R_3 = H$, $R_4 = H$
3-hydroxy-8,9-dimethoxy-pterocarpan: $R_1 = OH$, $R_2 = OH$, $R_3 = H$, $R_4 = H$

Fig. 29.1 Major phenolic compounds reported in Brazilian Red Propolis

Propolis. These compounds were identified in the resin of *D. ecastophyllum* as isoflavonoid dimers: retusapurpurin A e retusapurpurin B (Fig. 29.2).

Besides these compounds, polyprenylated benzophenones are another class of specialized metabolites found only in Brazilian Red Propolis. Nevertheless, the study by Ccana-Ccapatinta et al. (2020) showed that compounds of this class were not related to *D. ecastophyllum*, these polyprenylated benzophenones were assigned to *Symphoniaglobulifera* species. This species is also found in mangrove swamps in the north of Brazil, where the Brazilian Red Propolis was produced, showing that other species may contribute to forming the Propolis. Figure 29.3 shows the structure of some of these benzophenones found in Brazilian Red Propolis (Piccinelli et al. 2011; Ccana-Ccapatinta et al. 2020).

Fig. 29.2 Pigments isolated from Brazilian Red Propolis

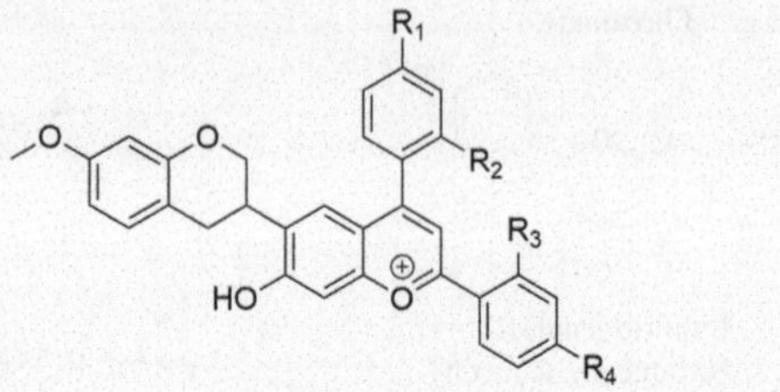

Retusapurpurin A: $R_1 = OH$, $R_2 = H$, $R_3 = OCH_3$, $R_4 = OH$
Retusapurpurin B: $R_1 = OCH_3$, $R_2 = OH$, $R_3 = H$, $R_4 = OH$

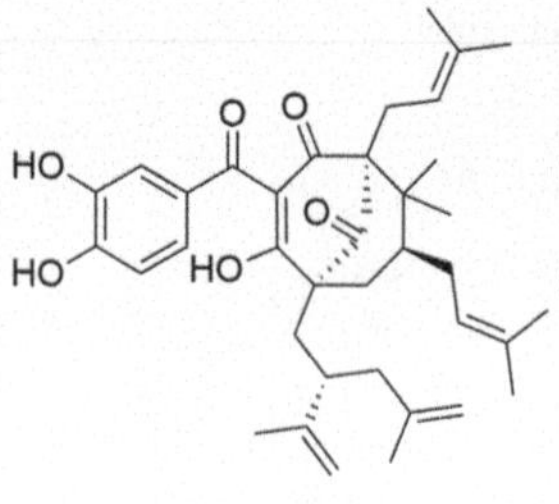

Xanthochymol

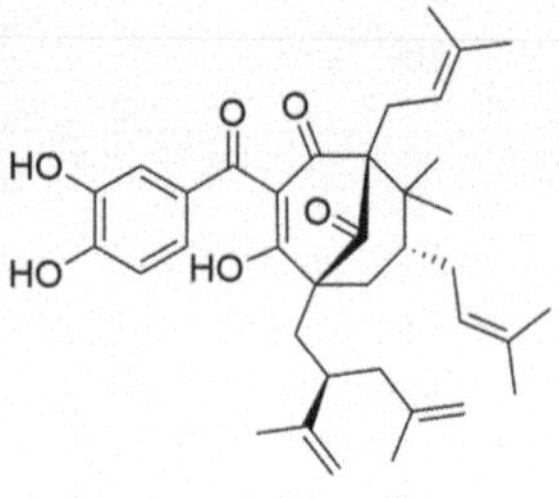

Guttiferone E

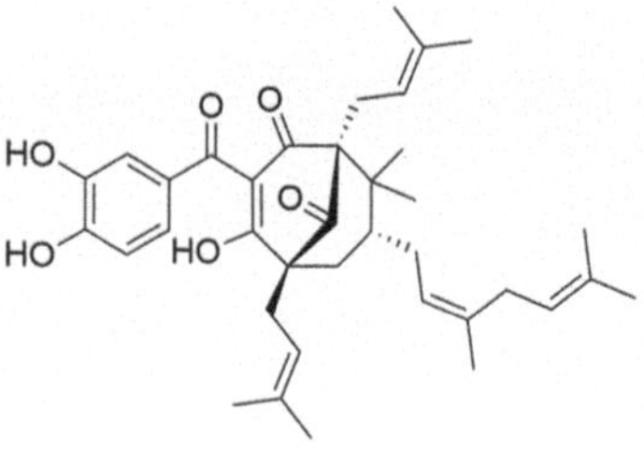

Oblongifolin A

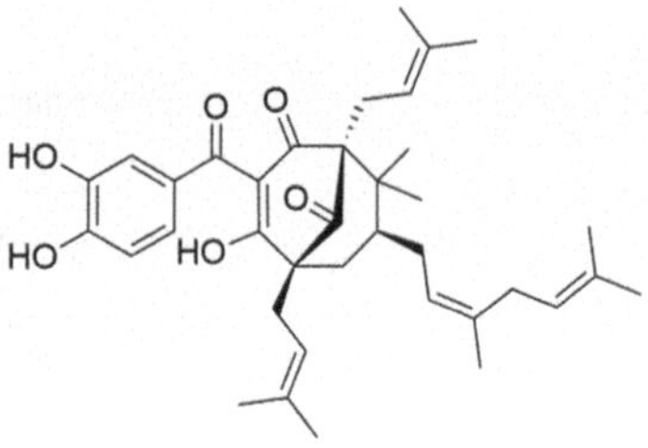

Oblongifolin B

Fig. 29.3 Polyprenylated benzophenones isolated from Brazilian Red Propolis

Recently, Brazilian Amazon Red Propolis has been identified as a novel variety of Brazilian Red Propolis, distinguished by a unique chemical marker distinct from that observed in red propolis from the Northeast. Calycosin (Fig. 29.1), an *O*-methylated isoflavone, was isolated as the major compound of this new variety of propolis. In addition to this marker, other compounds were noted from liquid chromatography coupled to mass spectrometry analyses, such as liquiritigenin, vestitol, 7,8,3′-trihydroxy-4′-methoxyisoflavone, formononetin, vesticarpan, calycosin, violanone, 3,8-dihydroxy-9-methoxy-pterocarpan, 3,4-dihydroxy-9-methoxy-pterocar pan, mucronulatol, 3-hydroxy-8,9-dimethoxy-pterocarpan, isoliquiritigenin, retusapurpurin B, biochanin A, and medicarpin. Polyprenylated benzophenones were absent in Brazilian Amazon Red Propolis. The researchers confirmed that it was a Brazilian Red Propolis due to its color and the presence of isoflavonoids (isoflavans, pterocarpans, and isoflavones). The absence of polyprenylated benzophenones was also an indication for this propolis to be considered a new variety (Dutra et al. 2023).

Besides the chemical markers that distinguish Brazilian Red Propolis from other Brazilian propolis, the literature has plenty of works that make the isolation or identification of many other metabolites found in Propolis. These studies bring the elucidation of isolated compounds for spectrometric techniques, such as Nuclear Magnetic Resonance (NMR). Moreover, there are works with identification proposals of the chemical composition of extracts obtained from Propolis, applying chromatographic techniques, such as High-Performance Liquid Chromatography (HPLC) and Gas Chromatography (GC) in tandem with Mass Spectrometry (MS), or comparing chemical properties, like UV/vis (ultraviolet/visible) absorbance, retention time, in relation to standards.

Among these works, a bioassay-guided fractionation study developed by Banzato et al. (2020), led to the isolation of eight isoflavonoid dimers (Propolones A-D, Propolonones A-C and Propolol A) from the hydroethanolic extract of Brazilian Red Propolis (Fig. 29.4).

In the study by Awale et al. (2008), 43 compounds were isolated from Brazilian Red Propolis. In addition to those previously mentioned in the studies conducted by Aldana-Mejía et al. (2021b) and Piccinelli et al. (2011), some flavonoids from other classes and other phenolic compounds were isolated and are shown in Fig. 29.5.

In another study, De Mendonça et al. (2015) identified some hydroxicinnamic acids, flavonols, and flavan-3-ols in the hydroethanolic extract by HPLC/MS (Fig. 29.5).

In the research conducted by Regueira-Neto et al. (2017), the compounds identified in Brazilian Red Propolis were apigenin, caffeic acid, chlorogenic acid, *p*-coumaric acid, ellagic acid, luteolin, quercetin, and rutin (Fig. 29.5). These works show the prevalence and diversity of phenolic compounds of various classes in the Brazilian Red Propolis.

Despite the predominance of phenolic compounds, there are other classes of specialized metabolites reported to Brazilian Red Propolis. One factor that can change the chemical composition of the extract is the kind of extraction, thus extracts or fractions obtained with non-polar solvents, show a chemical composition low in phenolic compounds. Trusheva et al. (2006) developed the fractionation of the ethanolic extract of Brazilian Red Propolis using petroleum ether, leading to the isolation of a triterpene and identification of five phenylpropene derivatives by gas chromatography coupled to mass spectrometry. Ccana-Ccapatinta et al. (2020) also obtained the isolation of two triterpenes from the hexane fraction. Figure 29.6 shows some low-polarity chemical structures found in Brazilian Red Propolis.

It is notable the diversity of compounds that can be found in Brazilian Red Propolis. Even so, it has a chemical composition with classes of metabolites and chemical markers that allow it to be distinguished from other types of propolis. A variety of phenolic compounds can be found in Brazilian Red Propolis. Among them, isoflavonoids, flavanones, flavones, isoflavans, pterocarpanes, and chalcones have been the most commonly identified or isolated compounds. Due to the fact that Brazilian Red Propolis is a rich source of special metabolites, its extracts and isolated compounds have shown important pharmacological properties, including antimicrobial.

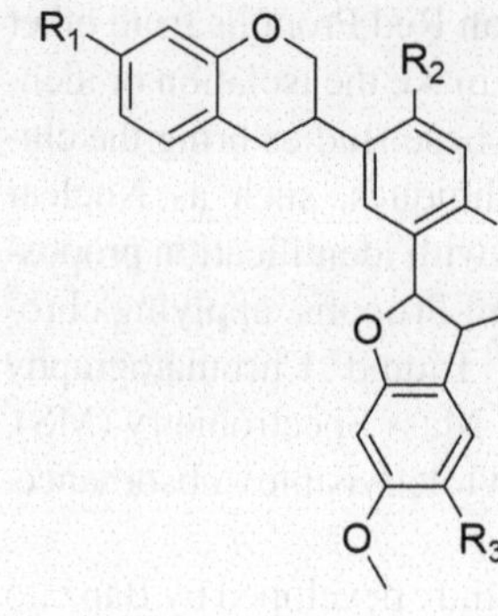

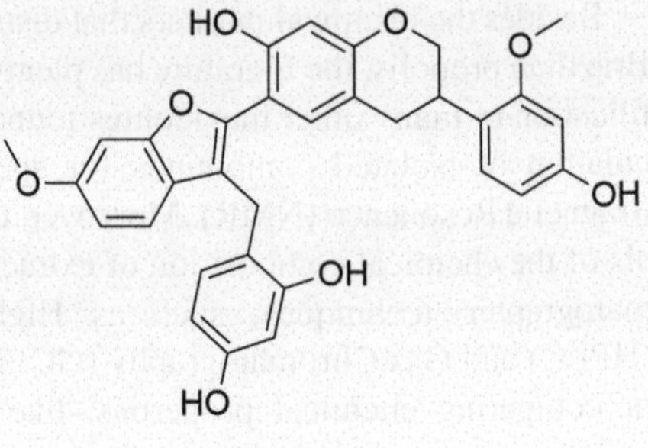

Propolol

Propolonone A = R_1 = OCH_3, R_2 = OH, R_3 =H
Propolonone B = R_1 = R_3= OCH_3, R_2 = OH
Propolonone C = R_1 = OH, R_2 = OCH_3, R_3 = H

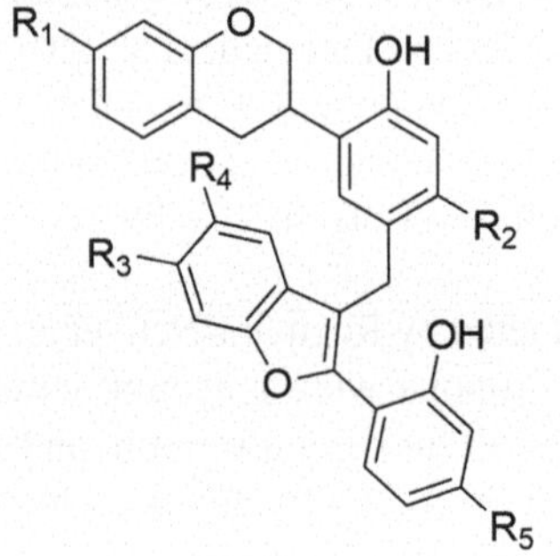

Propolone A = R_1 = R_3 = R_4 = OCH_3, R_2 = R_5 = OH
Propolone B = R_1 = R_3 = OCH_3, R_2 = R_5 = OH, R_4 = H
Propolone C = R_1 = R_5 = OCH_3, R_2 = R_3 = OH, R_4 = H
Propolone D = R_1 = R_5 = OH, R_2 = R_3 = OCH_3, R_4 = H

Fig. 29.4 Isoflavonoids dimers isolated from Brazilian Red Propolis

Various studies found in the literature suggest that qualitative and quantitative variations within the chemical composition of Brazilian Red Propolis may occur depending on its geographic origin, climatic conditions, type of soil, seasonal variations, botanical sources, types of extraction, and experimental conditions of analysis. Therefore, all these factors can affect the concentration of some compounds in the extracts and consequently promote different effects in biological assays.

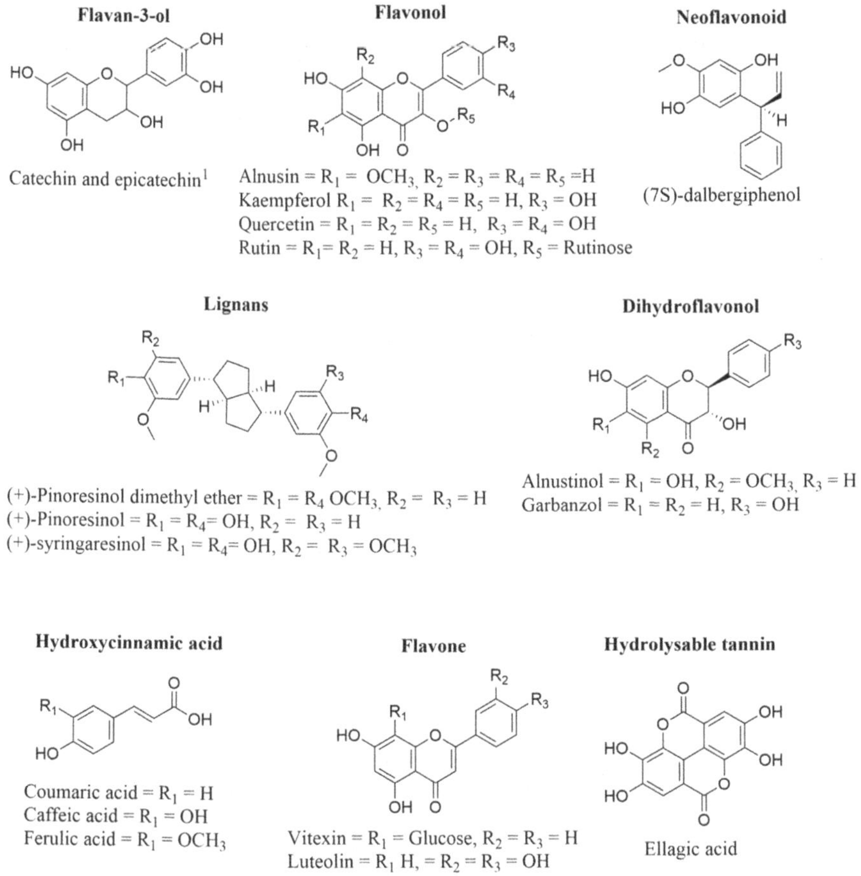

Fig. 29.5 Phenolic compounds isolated from Brazilian Red Propolis

29.3 Antimicrobial Properties

Brazilian Red Propolis products have been tested across the entire human-animal-environment interface, including antimicrobial activity against resistant microorganisms, of great concern and threat to global health.

De Souza et al. (2021) evaluated the hydroalcoholic extract of Red Propolis from the state of Bahia (Brazil) against *Enterococcus faecium* (ATCC 700221), resistant to teicoplanin and vancomycin, with Minimum Inhibitory Concentration (MIC)/Minimum Bactericidal Concentration (MBC) results of 62.5/250.0 μg/mL, hexane fractions with values of 31.25/31.25 μg/mL and dichloromethane of 62.5/125 μg/mL, guttiferone E, and xanthochymol compounds with concentrations of 7.81/15.6 μg/mL and oblongifolin B of 15.6/15.6 μg/mL. Likewise, for *Staphylococcus aureus* (ATCC 43300), resistant to methicillin and oxacillin, with results of 31.25/62.5 μg/mL, 15.6/62.5 μg/mL, 62.5/250 μg/mL, 7.81/15.6 μg/mL,

Phenylpropene derivatives

Trans-anetol = $R_1 = R_2 = H$
Trans-methyl isoeugenol = $R_1 = OCH_3$, $R_2 = H$
Trans-isoelemicin = $R_1 = R_2 = OCH_3$

Methyl eugenol = $R_1 = H$
Elemicin = $R_1 = OCH_3$

Triterpenes

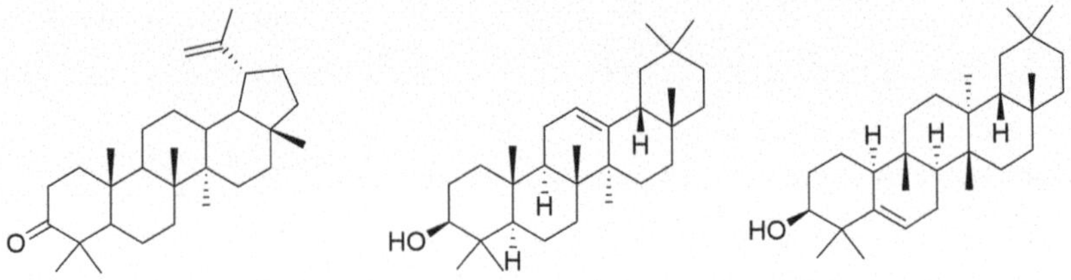

Lup-20(29)-en-3-one   *β*-amyrin   Glutinol

Fig. 29.6 Triterpenoids and phenylpropene derivatives identified in Brazilian Red Propolis

7.81/15.6 μg/mL, respectively. In continuity, guttiferone E and xanthochymol, as well as oblongifolin B, inhibited the formation of 50% of this biofilm at concentration of 3.12 μg/mL. Guttiferone E and xanthochymol were also capable of inhibiting coagulase activity at concentrations of 3.9 μg/mL (½MIC) and 1.95 μg/mL (¼MIC). In total, the compounds identified in this extract were neovestitol, formononetin, medicarpin, liquiritigenin, vestitol, 7-*O*-methylvestitol, a mixture of xanthochymol and guttiferone E, and oblongifolin B.

Ripari et al. (2023) explored the effects of the hydroalcoholic extract of Red Propolis (RPE) from the state of Bahia (Brazil) against six strains of methicillin-resistant *S. aureus* (MRSA): H44—clinical isolate; H55—clinical isolate; H97—clinical isolate; H119—clinical isolate; Community acquired (CA-MRSA)—USA300 and ATCC 33591. The values of MIC/MBC were: 0.0156–0.0625/2.0–>2.0 g/mL; 0.0312–0.0625/1.0 g/mL; 0.0156–0.0312/1.0–2.0 g/mL; 0.0625–0.125/1.0–2.0 g/mL; 0.0625–0.125/2.0 g/mL, and 0.062.5–0.125/0.50–1.0 g/mL, respectively. At concentrations of MIC 75%, RPE inhibited 37.78% of H44 biofilm; 61.46% of H55; 22.23% of H97; 40.67% of USA 300 and 38.65% of ATCC 33591 biofilm. Still regarding to inhibition of biofilm formation, MIC 25% RPE in combination with MIC 25% imipenem inhibited 56.50% of H44 biofilm; 19.80% of H55; 28.60% of H97; 46.90% of H119; 60.20% of USA300 and 64.40% of ATCC 33591 biofilm. MIC 25% RPE in combination with MIC 25% ertapenem inhibited 22% of H44 biofilm; 25.71% of H97; 31.7% of H199 and 57.50% USA300 biofilm. All biofilms were eradicated when RPE was combined with imipenem or ertapenem. The association of MIC 25% RPE with MIC 25% imipenem achieved 100% of membrane permeability of all strains, while RPE MBC alone achieved more than 80% of membrane permeability of H44. Furthermore, RPE presented a synergistic action with imipenem and an additive action with ertapenem against ATCC 33591. Last, bacterial growth was significantly inhibited after 24 h of exposure to REP with imipenem and was partially inhibited after 24 h of exposure to REP with ertapenem on the time-kill curve of ATCC 33591. The compounds identified in this extract were medicarpin, liquiritigenin, neovestitol, isoliquiritigenin, oblongifolin, guttiferone E, vestitol, and *O*-methylvestitol.

Regueira-Neto et al. (2019) used the hydroethanolic extract of Red Propolis (state of Alagoas/Brazil) against four strains of clinical isolates: *Escherichia coli* 06, resistant to cephalotin, cephalexin, cefadroxil, ceftriaxone; *Pseudomonas aeruginosa* 03, resistant to cefepime, ceftazidime, imipenem, ciprofloxacin, piperacillin, levofloxacin, meropenem, amikacin; *P. aeruginosa* 24, resistant to cefepime, ceftazidime, imipenem, ciprofloxacin, piperacillin, levofloxacin, meropenem, amikacin; *S. aureus* 10, resistant to cefadroxil, cephalexin, cephalotin, oxacillin, penicillin, ampicillin, amoxicillin, moxifloxacin, ciprofloxacin, levofloxacin, ampicillin plus sulbactam, amoxicillin plus clavulanic acid, eritromycin, clarithromycin, azithromycin, clindamycin with the following MIC results: ≥ 1,024 μg/mL; 512 μg/mL; 512 μg/mL; 53 μg/mL, respectively. The hydroethanolic extract presented synergistic action with imipenem at a concentration of 0.5 μg/mL for *P. aeruginosa* 03, with gentamicin at a concentration of 787 μg/mL for *P. aeruginosa* 03, 3.9 μg/mL for *P. aeruginosa* 24 and 2.5 μg/mL for *S. aureus* 10. The compounds identified in this extract were quercetin, luteolin, ellagic acid, caffeic acid, chlorogenic acid, apigenin, and *p*-coumaric acid.

A very interesting factor that can influence the composition of Propolis is the season in which it is collected. Regueira-Neto et al. (2017) showed that the hydroethanol extract of Red Propolis from the state of Pernambuco (Brazil) collected in the rainy season (RP-PER) did not contain the compound *p*-coumaric acid, while in dry season (RP-PED), it did not contain vitexin and rutin compounds. Using the same strains as in the previous study, RP-PER showed MIC of 128 μg/mL and RP-PED of 512 μg/mL against *E. coli*06; 512 μg/mL and 512 μg/mL against *P. aeruginosa* 03; 512 μg/m and 512 μg/mL against *P. aeruginosa* 24; 64 μg/mL and 101.6 μg/mL against *S. aureus*10. RP-PED was synergistic with the antibiotics imipenem and gentamicin for *P. aeruginosa* 03 and *P. aeruginosa* 24; with gentamicin for *S. aureus*10. RP-PER was synergistic with the antibiotic gentamicin for *P. aeruginosa* 24.

The same RP-PED was also used in the study of Costa et al. (2021) against the following toxigenic strains of *Clostridioides difficile*: NAP1/ST01/RT027 (LIBA-5756); NAP4/RT014–020 (ICC14); SLO231/UK[CE]821 (ICC45), with MIC values of 625 μg/mL for all of them. Concentrations of 312.5 μg/mL (½MIC) inhibited 70% of growing biofilm and 625 μg/mL (MIC) 80%. 312.5 μg/mL (½MIC) and 625 μg/mL (MIC) were able to reduce 100% of matrix protein, while 625 μg/mL (MIC) were able to reduce 80% of matrix carbohydrates. Concentrations of 2,500 μg/mL (4 × MIC) inhibited 50% of mature biofilm, while 5,000 μg/mL

(8 × MIC) inhibited 70%. 625 μg/mL (MIC) was able to reduce 90% of matrix protein, while 2500 μg/mL (4 × MIC) and 5,000 μg/mL (8 × MIC) reduced 95%. According to scanning electron microscopy analysis, RP-PED at a concentration of 312.5 μg/mL caused small perforations in ICC45 and loss of cellular integrity in LIBA-5756, while at 625 μg/mL, perforations were more frequent and larger, and also caused cell fragmentation in ICC 45.

Another aspect to be considered in research is the methodology of extraction, which led Dos Santos et al. (2021) to develop an alcohol-free optimized process using 1-hexyl-3-methylimidazolium chloride ([C6mim]Cl) and Red Propolis (state of Alagoas/Brazil). The extract with no dilution obtained was assessed by diffusion technique and showed an inhibition halo of 23 mm against *S. aureus* (ATCC 19095), enterotoxin C producer, while the extract in a concentration of 25 mg/mL and 75 mg/mL presented a halo of 8 and 10 mm, respectively. The compounds identified by this new process were medicarpin, isoliquiritigenin, hesperetin derivative, formononetin, liquiritigenin, daidzein, biochanin A, and pinocembrin.

Machado et al. (2016) and Dantas Silva et al. (2017) used, in addition to conventional ethanol extraction, the CO_2 as a supercritical fluid technique. In the first study against *S. aureus* (ATCC 33591), MRSA, MIC/MBC values of ethanolic extract of Sergipe's Red Propolis, Brazil (RSE_EtOH) were 400–100/1,600–800 μg/mL; ethanolic extract of Alagoas' Red Propolis, Brazil (RAL_EtOH) was 400–100/1,600–800 μg/mL; supercritical fluid extract of Sergipe's Red Propolis (RSE_SCO_2) 600–200/1,600 μg/mL and supercritical fluid extract of Alagoas' Red Propolis (RAL_SCO_2) 800–400/1,600 μg/mL (Machado et al. 2016). In the second study against *Klebsiella*sp (ATCC 1706/700603), *β*-lactamase producer, MIC/MBC values of RSE_EtOH were 62.5 μg/mL; RAL_EtOH 31.3 μg/mL; RSE_SCO_2 250 μg/mL; RAL_SCO_2 62.5 μg/mL (Dantas Silva et al. 2017).

In turn, Santos et al. (2020) innovated in the production of a yogurt incorporated with the ethanolic extract of Red Propolis (state of Bahia/Brazil). The yogurt presented MIC/Minimum Biocide Concentration (MBC) of 50–500/100–500 μg/mL against *S. aureus* (ATCC 43300), resistant to methicillin and oxacillin, and 500/500 μg/mL against *Candida albicans* (ATCC 10231), resistant to itraconazole, voriconazole, fluconazole, and anidulafungin.

Sokolonski et al. (2021) approached the issue of ultrasound in the preparation of the ethanolic extract of Red Propolis (state of Bahia/Brazil) against clinical isolates of *Candida* spp., among them: *C. dubliniensis*PAC 01, resistant to itraconazole; *C. tropicalis*PAC 05, resistant to itraconazole and fluconazole; *C. tropicalis*PAC 15, resistant to itraconazole; *C. albicans*PAC 17, resistant to itraconazole and fluconazole; *C. albicans*PAC 19, resistant to itraconazole and *C. albicans*PAC 20, resistant to itraconazole. MIC/Minimum Fungicidal Concentration (MFC) values were 1.0/2.0 mg/mL of Red Propolis extract without ultrasound pretreatment (RP_EtOH) and 2.0/2.0 mg/mL of Red Propolis extract with ultrasound pretreatment (RP_US) for PAC 05; 4.0/>8.0 mg/mL RP_EtOH and 2.0/2.0 mg/mL RP_US for PAC 15; 2.0/>8.0 mg/mL RP_EtOH and 1.0/2.0 mg/mL RP_US for PAC 17; 1.0/1.0 mg/mL RP_EtOH and 0.015/0.125 mg/mL for PAC 19; 1.0/2.0 mg/mL RP_EtOH and 1.0/2.0 mg/mL RP_US for PAC 20. Regarding the interference in biofilm formation,

RP_EtOH was tested in concentrations of 1.0, 2.0, 4.0, and 8.0 mg/mL against PAC 01 interfering 50.1%, 78%, 80.2%, and 84.3%, respectively, and against PAC 15 interfering 72.9%, 76.5%, 69.1%, and 89.8% in biofilm formation.

According to Santos et al. (2021), the ethanolic extract of Red Propolis (state of Alagoas/Brazil) showed a dose-effect response (50, 100,and 500 mg/mL) against the highly virulent *Paracoccidioides Brasiliensis* Pb18 strain, in which 500 mg/mL presented the best antifungal activity with the lowest number of viable *P. brasiliensis*. The compounds identified in this extract were formononetin, vestitol, neovestitol, biochanin A, quercetin, liquiritigenin, isoliquiritigenin, and daidzein.

In relation to viruses, MS2 and Av-08 bacteriophages, genetically and structurally similar to *Enterovirus surrogates*, were tested with the ethanolic extract of Red Propolis from the state of Alagoas (Brazil) with and without ultrasound pretreatment by Silva-Beltrán et al. (2020). The results demonstrated an antiviral activity in a concentration- and time-dependent manner. The ethanolic extract with ultrasound was the most effective with titer values varying between ~4.0-$\log_{10}$ PFU/mL for Av08 and ~4.2-$\log_{10}$ PFU/mL for MS2 during 45 min of contact time at 1,000 μg/mL.

In the study of Silva et al. (2022), the *n*-butanol fraction of hydroalcoholic extract of Red Propolis (state of Bahia/Brazil), in concentrations of 50 μg/mL, significantly inhibited 69% of CHIKV replication. The compounds medicarpin, neovestitol, and vestitol in concentrations of 11 μg/mL inhibited CHIKV replication in vitro in 86%, 94%, and 97%, respectively.

The antiviral activity of the ethanolic extract of Red Propolis from the state of Bahia (Brazil), as well as the compounds quercetin, caffeic acid, and rutin, were also evaluated by Silva-Beltrán et al. (2023) against the human coronavirus 229E. Quercetin showed a reduction greater than 90% of viral cytopathic effects in all concentrations of 50, 100, and 200 μg/mL. Caffeic acid resulted in a reduction between 80% and 90% of infection in concentrations of 75, 150, and 300 μg/mL. Rutin presented a reduction no greater than 75% in the highest concentration of 250 μg/mL, while the extract exhibited a reduction of less than 70% even at the highest concentration of 50 μg/mL.

29.4 Action Mechanism of Brazilian Red Propolis

Brazilian Red Propolis and its compounds can act in different ways on microorganisms, making the understanding of the mechanism of action a key to provide greater clarification regarding natural product capabilities. In the literature, the mechanism of action is generally tested indirectly (such as, for example, through molecular docking, assessment of cellular permeability, proteins, etc.), as the indirect study is simpler and more dynamic in effectively demonstrating the result of mechanisms correlated with the microorganism. However, certain methodologies for studying mechanisms indirectly lead to results that are not very precise, as they state the connection and interaction but do not demonstrate how this interaction affects the microorganism.

Ripari et al. (2023) analyzed, through molecular docking, the interaction factor of pure substances isolated from the Brazilian Red Propolis extract with proteins from *S. aureus* resistant to β-lactams. The Penicillin Binding Protein (PBP) is a membrane-forming protein and the binding site for β-lactam antibiotics. Binding of β-lactam antibiotics, such as penicillin receptors and PBPs, results in the inhibition of the synthesis of cell wall structures (Fishovitz et al. 2014). The β-lactam resistant bacterium *S. aureus* has the PBP2a protein instead of the common PBPs. PBP2a has a very low binding force with β-lactam drugs, which guarantees resistance to this strain of *S. aureus.* The study revealed that compounds isolated from Brazilian Red Propolis such as liquiritigenin, isoliquiritigenin, vestitol, neovestitol, medicarpine, *O*-methylvestitol, gutiferone, and oblongifolin interact more strongly with PBP2a. It is also suggested that the method of action of the isolated compounds is in association with PBP2a, it exposes other sensitive PBPs that are not resistant (Foster 2019).

In relation to xanthochymol and guttiferone E, present in the hexane and dichloromethane fractions of the extract, they inhibit the activity of the catalase enzyme. According to De Souza et al. (2021), who used antivirulence and molecular docking assays for *S. aureus*, this inhibition may occur due to the fact that these two compounds can bind to amino acids present in the structure of the catalase enzyme. The compounds were able to interact with the heme group of the protein, preventing the binding of the amino acid and its substrate. The catalase enzyme plays an important role in the virulence of *S. aureus* by inhibiting neutrophil death mechanisms, described by Cheung et al. (2021), who concluded that this enzyme transforms toxic Reactive Oxygen Species (ROS), normally secreted by neutrophils when they phagocytose, into a less toxic product (H_2O_2).

Another mechanism of action was reported by Ozbey et al. (2023), who isolated the compounds of a Propolis extract from Turkey. One of the isolated compounds was naringenin, which is also present in Brazilian Red Propolis. This compound can interact with proteins from different bacteria: 5LMM from *E. coli*, 5YHG from *S. aureus*, 5FXT from *Helicobacter pylori* and 4NX9 from *P. aeruginosa*. The 5LMM protein is a structure of the *E. coli* Hyd-1 hydrogenase enzyme, and its role in cell metabolism is the degradation of toxic hydroxyl radicals, which can cause damage to DNA, other proteins and lipids (Nie et al. 2012); the 5YHG protein of *S. aureus*is a membrane structure and the interaction may occasionally result in damage to membrane; the 5FXT protein from *H. pylori* plays an important role in DNA duplication, and its inhibition can cause damage to genetic material (Woo et al. 2020); the 4NX9 protein is a flagellin present in the flagella of *P. aeruginosa* (Totten and Lory 1990), and its damage can lead to difficulty in movement and decreased virulence.

In turn, the mechanism of action of Brazilian Red Propolis on fungi is explored by de Pippi et al. (2015), who carried out a molecular docking test with Brazilian Red Propolis and *Candida*sp and suggested that the extracts act on the structure of the yeast cell wall. The antivirulence mechanism of Brazilian Red Propolis was explored by D'Auria et al. (2003) and Packer and Luz (2007), who confirmed that propolis compounds have antivirulence potential by inhibiting the activity of the extracellular phospholipase enzyme, which makes the adhesion of these cells difficult to host epithelial cells.

Last, regarding viruses, one of the mechanisms of action of Brazilian Red Propolis is described by Guler et al. (2021) and focused on the SARS-CoV-2 virus. The mechanism studied does not act directly on the virus but on its ability to penetrate human cells. This virus uses angiotensin-converting enzyme to pass through the cell membrane and enter the intracellular space in host cells. Thus, flavonoids and isoflavonoids from Brazilian Red Propolis such as rutin (quercetin-3-rutinoside hydrate) and luteolin (2-(3,4-dihydroxyphenyl)-5,7-dihydroxy-4-chromenone) can bind to this enzyme in order to compete with the virus for the binding site, which prevents and delays the entry of the virus into cells. The test used molecular docking assays to analyze the possible interactions of Brazilian Red Propolis and the viral replication cycle.

29.5 Future Perspective

Several gaps still need to be explored when it comes to studies with Brazilian Red Propolis. More studies are needed so that researchers can advance the frontier of knowledge about this product. Future directions and perspectives regarding Brazilian Red Propolis imply a broad investigation into the chemical properties of each compound of the extract and their respective mechanisms of action, as well as a more complete approach regarding the synergistic capacity of Propolis and pharmaceuticals. In this way, the scientific community will leverage the limits of knowledge regarding Brazilian Red Propolis in addition to opening a door to the creation of more efficient drugs against bacteria, fungi, and viruses.

Several of these objectives can be achieved through the efficient use of Artificial Intelligence (AI), evaluation of synergistic activity against various synthetic antibiotics and other natural products, as well as the evaluation of their antimicrobial activity against a panel diversity of resistant microorganisms. The antifungal potential of Brazilian Red Propolis should be further explored given the scarcity of scientific articles published to date, especially against resistant fungi. The efficient use of AI can also be a possible methodology to be followed for better understanding and boosting knowledge, which are already used for different purposes. The use of AI for scientific research has already been used to discover new antimicrobial agents (Ali et al. 2023), therefore they could be used to study the mechanisms of action of natural products given the scarcity of studies in this area. In vivo tests in animals and clinical trials are another gap that must be filled in the study of Brazilian Red Propolis. There are few studies that come close to clinical reality and are in the advanced stages of becoming medicines. Furthermore, using unexplored methodologies, where the area of biotechnology and biomedical engineering could be allies in this type of study.

The Brazilian Health Regulatory Agency (ANVISA) published a technical note on the Registration of products containing propolis, as well as a resolution in 2011 authorizing the registration of propolis-based medicines. That encourages more study on this topic so that in the future several formulations of different propolis can be developed for use by the population, especially medicines that can be allies in infections caused by resistant microorganisms. More clinical studies are needed for propolis to be used safely and efficiently, but the results seen so far are encouraging.

29.6 Conclusion

Based on the literature, there are few studies approaching antimicrobial evaluation against resistant microorganisms. However, the results described in this chapter confirm that Brazilian Red Propolis has potential against bacteria, fungi, and viruses, especially its crude extract. Therefore, the scarcity of studies shows that it is necessary to further explore Brazilian Red Propolis against resistant microorganisms, as well as the use of new approaches to investigate its mechanism of action, synergistic activity in association with other natural products or synthetic antibiotics. One must advance the frontier of knowledge about Brazilian Red Propolis so that in the near future this product can be tested in clinical use and approved as a medicine against infections caused by resistant microorganisms.

References

Aldana-Mejía JA, Ccana-Ccapatinta GV, Squarisi IS, Nascimento S, Tanimoto MH, Ribeiro VP, Ribeiro AB, de Freitas KS (2021a) Nonclinical toxicological studies of Brazilian red propolis and its primary botanical source *Dalbergiaecastaphyllum*. J Clin Toxicol 34(4):1024–1033. https://doi.org/10.1021/acs.chemrestox.0c00356

Aldana-Mejía JA, Ccana-Ccapatinta GV, Ribeiro VP, Arruda C, Veneziani RCS, Ambrósio SR, Bastos JK (2021b) A validated HPLC-UV method for the analysis of phenolic compounds in Brazilian red propolis and *Dalbergiaecastaphyllum*. J Pharm Biomed Anal 198:114029. https://doi.org/10.1016/j.jpba.2021.114029

Alencar SM, Oldoni TL, Castro ML, Cabral IS, Costa-Neto CM, Cury JA, Ikegaki M (2007) Chemical composition and biological activity of a new type of Brazilian propolis: red propolis. J Ethnopharmacol 113(2):278–283. https://doi.org/10.1016/j.jep.2007.06.005

Ali T, Ahmed S, Aslam MJA (2023) Artificial intelligence for antimicrobial resistance prediction: challenges and opportunities towards practical implementation. Antibiotics 12(3):523. https://doi.org/10.3390/antibiotics12030523

Atanasov AG, Zotchev SB, Dirsch VM, Supuran CT, International Natural Product Sciences, T (2021) Natural products in drug discovery: advances and opportunities. Nat Rev Drug Discov 20(3):200–216. https://doi.org/10.1038/s41573-020-00114-z

Awale S, Li F, Onozuka H, Esumi H (2008) Constituents of Brazilian red propolis and their preferential cytotoxic activity against human pancreatic PANC-1 cancer cell line in nutrient-deprived condition. Bioorg Med Chem 16(1):181–189. https://doi.org/10.1016/j.bmc.2007.10.004

Banzato TP, Gubiani JR, Bernardi DI, Nogueira CR, Monteiro AF, Juliano FF, Berlink RGS (2020) Antiproliferative flavonoid dimers isolated from brazilian red própolis. J Nat Prod 83(6):1784–1793. https://doi.org/10.1021/acs.jnatprod.9b01136

Botteon CEA, Silva LB, Ccana-Ccapatinta GV, Silva TS, Ambrosio SR, Veneziani RCS, Marcato PD (2021) Biosynthesis and characterization of gold nanoparticles using Brazilian red propolis and evaluation of its antimicrobial and anticancer activities. Sci Rep 11(1):1974. https://doi.org/10.1038/s41598-021-81281-w

Bueno-Silva B, Alencar SM, Koo H, Ikegaki M, Silva GV, Napimoga MH, Rosalen PL (2013a) Anti-inflammatory and antimicrobial evaluation of neovestitol and vestitol isolated from Brazilian red propolis. J Agric Food Chem 61(19):4546–4550. https://doi.org/10.1021/jf305468f

Bueno-Silva B, Koo H, Falsetta ML, Alencar SM, Ikegaki M, Rosalen PL (2013b) Effect of neovestitol-vestitol containing Brazilian red propolis on accumulation of biofilm in vitro and development of dental caries in vivo. Biofouling 29(10):1233–1242. https://doi.org/10.1080/08927014.2013.834050

Ccana-Ccapatinta GV, Mejía JAA, Tanimoto MH, Groppo M, de Carvalho JCAS, Bastos JK (2020) *Dalbergiaecastaphyllum* (L.) Taub. And *Symphoniaglobulifera* L. f. and botanical sources of isoflavonoids and benzophenones in Brazilian red propolis. Molecules 25:2060. https://doi.org/10.3390/molecules25092060

Cheung GY, Bae JS, Otto M (2021) Pathogenicity and virulence of *Staphylococcus aureus*. Virulence 12(1):547–569. https://doi.org/10.1080/21505594.2021.1878688

Chopra B, Dhingra AK (2021) Natural products: A lead for drug discovery and development. Phytother Res 35(9):4660–4702. https://doi.org/10.1002/ptr.7099

Costa CL, Azevedo CP, Quesada-Gómez C, Brito GAC, Regueira-Neto MDS, Guedes GMM, Rocha MFG, Sidrim JJC, Cordeiro RA, Carvalho CBM, Castelo-Branco DSCM (2021) Inhibitory effect of Brazilian red propolis on planktonic and biofilm forms of Clostridioides difficile. Anaerobe 69:102322. https://doi.org/10.1016/j.anaerobe.2021.102322

D'Auria FD, Tecca M, Scazzocchio F, Renzini V, Strippoli V (2003) Effect of propolis on virulence factors of *Candida albicans*. J Chemother 15(5):454–460. https://doi.org/10.1179/joc.2003.15.5.454

Dantas Silva RP, Machado BA, Barreto GA, Costa SS, Andrade LN, Amaral RG, Umsza-Guez MA (2017) Antioxidant, antimicrobial, antiparasitic, and cytotoxic properties of various Brazilian propolis extracts. PLoS One 12(3):e0172585. https://doi.org/10.1371/journal.pone.0172585

Das Neves MV, Da Silva TM, Lima Ede O, Da Cunha EV, Oliveira Ede J (2016) Isoflavone formononetin from red propolis acts as a fungicide against Candida sp. Braz J Microbiol 47(1):159–166. https://doi.org/10.1016/j.bjm.2015.11.009

De Freitas MCD, De Miranda MB, De Oliveira DT, Vieira-Flho SA, Caligiorne RB, De Figueiredo SM (2017) Biological activities of red propolis: a review. Recent Pat Endocr Metab Immune Drug Discov 11(1):3–12. https://doi.org/10.2174/1872214812666180223120316

De Mendonça ICG, Porto ICCDM, Do Nascimento TG, De Souza NS, Oliveira JMDS, Arruda REDS, Mousinho KC, Dos Santos AF, Basílio-Júnior ID, Parolia A, Barreto FS (2015) Brazilian red propolis: phytochemical screening, antioxidant activity and effect against cancer cells. BMC Complement Altern Med 15(1):1–12. https://doi.org/10.1186/s12906-015-0888-9

De Souza ST, Silva JMB, Braun GH, Mejia JAA, Ccapatinta GVC, Santos MFC, Tanimoto MH, Bastos JK, Parreira RLT, Orenha RP, Borges A, Berretta AA, Veneziani RCS, Martins CHG, Ambrósio SR (2021) Green and red Brazilian propolis: antimicrobial potential and anti-virulence against ATCC and clinically isolated multidrug-resistant bacteria. Chem Biodivers 18(8):e2100307. https://doi.org/10.1002/cbdv.202100307

Devequi-Nunes D, Machado BAS, Barreto GA, Reboucas Silva J, da Silva DF, da Rocha JLC, Umsza-Guez MA (2018) Chemical characterization and biological activity of six different extracts of propolis through conventional methods and supercritical extraction. PLoS One 13(12):e0207676. https://doi.org/10.1371/journal.pone.0207676

Do Nascimento TG, Dos Santos Arruda RE, Almeida DC, Dos Santos Oliveira JM, Basílio-Júnior ID, De Moraes Porto ICC, Watson DG (2019) Comprehensive multivariate correlations between climatic effect, metabolite-profile, antioxidant capacity and antibacterial activity of Brazilian red propolis metabolites during seasonal study. Sci Rep 9(1):18293. https://doi.org/10.1038/s41598-019-54591-3

dos Santos DA, Munari FM, da Silva Frozza CO, Moura S, Barcellos T, Henriques JAP, Roesch-Ely M (2019) Brazilian red propolis extracts: study of chemical composition by ESI-MS/MS (ESI+) and cytotoxic profiles against colon cancer cell lines. Biotechnol Res Innovat 3(1):120–130. https://doi.org/10.1016/j.biori.2019.02.001

Dos Santos CM, De Souza Mesquita LM, Braga ARC, de Rosso VV (2021) Red propolis as a source of antimicrobial phytochemicals: extraction using high-performance alternative solvents. Front Microbiol 12:659911. https://doi.org/10.3389/fmicb.2021.659911

Dutra RP, De Sousa Júnior MM, Mignoni MSPM, De Oliveira KGM, Pereira EB, Figueredo AS, Lopes AJO (2023) Brazilian amazon red propolis: Leishmanicidal activity and chemical composition of a new variety of red própolis. Meta 13(9):1027. https://doi.org/10.3390/metabo13091027

Fishovitz J, Hermoso JA, Chang M, Mobashery S (2014) Penicillin-binding protein 2a of methicillin-resistant *Staphylococcus aureus*. IUBMB Life 66(8):572–577. https://doi.org/10.1002/iub.1289

Foster TJ (2019) Can β-lactam antibiotics be resurrected to combat MRSA? Trends Microbiol 27(1):26–38. https://doi.org/10.1016/j.tim.2018.06.005

Freires IA, de Alencar SM, Rosalen PL (2016) A pharmacological perspective on the use of Brazilian Red propolis and its isolated compounds against human diseases. Eur J Med Chem 110:267–279. https://doi.org/10.1016/j.ejmech.2016.01.033

Gomes Sá SH, Chalella Mazzocato M, Saliba ASMC, Alencar SM, Favaro-Trindade CS (2023) Evaluation of the release, stability and antioxidant activity of Brazilian red propolis extract encapsulated by spray-drying, spray-chilling and using the combination of both techniques. Food Res Int J 164:112423. https://doi.org/10.1016/j.foodres.2022.112423

Guler HI, Tatar G, Yildiz O, Belduz AO, Kolayli S (2021) Investigation of potential inhibitor properties of ethanolic propolis extracts against ACE-II receptors for COVID-19 treatment by molecular docking study. J Appl Microbiol 203(6):3557–3564. https://doi.org/10.1007/s00203-021-02351-1

Huemer M, MairpadyShambat S, Brugger SD, Zinkernagel AS (2020) Antibiotic resistance and persistence-implications for human health and treatment perspectives. EMBO Rep 21(12):e51034. https://doi.org/10.15252/embr.202051034

Inui S, Hatano A, Yoshino M, Hosoya T, Shimamura Y, Masuda S, Kumazawa S (2014) Identification of the phenolic compounds contributing to antibacterial activity in ethanol extracts of Brazilian red propolis. Nat Prod Res 28(16):1293–1296. https://doi.org/10.1080/14786419.2014.898146

Kausar S, Said Khan F, IshaqMujeeb Ur Rehman M, Akram M, Riaz M, Rasool G, Malik A (2021) A review: mechanism of action of antiviral drugs. Int J Immunopathol Pharmacol 35:20587384211002621. https://doi.org/10.1177/20587384211002621

Machado CS, Mokochinski JB, de Lira TO, de Oliveira FC, Cardoso MV, Ferreira RG, Sawaya AC, Ferreira AG, Pessoa C, Cuesta-Rubio O, Monteiro MC, de Campos MS, Torres YR (2016) Comparative study of chemical composition and biological activity of yellow, green, brown, and red Brazilian propolis. Evid Based Complement Alternat Med 2016:6057650. https://doi.org/10.1155/2016/6057650

Martorano-Fernandes L, Cavalcanti YW, de Almeida LFD (2020) Inhibitory effect of Brazilian red propolis on *Candida* biofilms developed on titanium surfaces. BMC Complement Med Ther 20(1):104. https://doi.org/10.1186/s12906-020-02893-9

Mayworm MAS, Lima CA, Tomba ACB, Fernandes-Silva CC, Salatino MLF, Salatino A (2014) Does própolis contain tannins? Evid Based Complament Alternat Med 2014:613647. https://doi.org/10.1155/2014/613647

Miranda SLF, Damasceno JT, Faveri M, Figueiredo L, da Silva HD, Alencar SMA, Bueno-Silva B (2019) Brazilian red propolis reduces orange-complex periodontopathogens growing in multispecies biofilms. Biofouling 35(3):308–319. https://doi.org/10.1080/08927014.2019.1598976

Morrison L, Zembower TR (2020) Antimicrobial Resistance. Gastrointest Endosc Clin N Am 30(4):619–635. https://doi.org/10.1016/j.giec.2020.06.004

Newman DJ, Cragg GM (2020) Natural products as sources of new drugs over the nearly four decades from 01/1981 to 09/2019. J Nat Prod 83(3):770–803. https://doi.org/10.1021/acs.jnatprod.9b01285

Nie W, Tang H, Fang Z, Chen J, Chen H, Xiu Q (2012) Hydrogenase: the next antibiotic target? Crit Rev Microbiol 122(12):575–580. https://doi.org/10.1042/CS20110396

Nunes JO, Domingues RAS, Barcellos RJ, Alves B, Carvalho I, Tavares NUL (2022) Policy and strategies addressing prevention and control of antimicrobial resistance in Brazil: a scoping review protocol. PLoS One 17(1):e0263305. https://doi.org/10.1371/journal.pone.0263305

Oladipo A, Enwemiwe V, Ejeromedoghene O, Adebayo A, Ogunyemi O, Fu F (2022) Production and functionalities of specialized metabolites from diferentes organic sources. Meta 12(6):534. https://doi.org/10.3390/metabo12060534

Oliveira AC, Shinobu CS, Longhini R, Franco SL, Svidzinski TI (2006) Antifungal activity of propolis extract against yeasts isolated from onychomycosis lesions. Mem Inst Oswaldo Cruz 101(5):493–497. https://doi.org/10.1590/S0074-02762006000500002

Ozbey G, Muz MN, Tanriverdi ES, Erkan S, Bulut N, Otlu B, Zigo F (2023) Chemical composition, antimicrobial activities, and molecular docking studies of Turkish propolis ethanol extract. J Chem 41(2):144–154. https://doi.org/10.17221/100/2022-CJFS

Packer JF, Luz MMS (2007) Evaluation and research methods for natural products inhibitory activity. Rev Bras 17(1):102–107. https://doi.org/10.1590/S0102695X2007000100019

Piccinelli AL, Lotti C, Campone L, Cuesta-Rubio O, Campo Fernandez M, Rastrelli L (2011) Cuban and Brazilian red propolis: botanical origin and comparative analysis by high-performance liquid chromatography-photodiode array detection/electrospray ionization tandem mass spectrometry. J Agric Food Chem 59(12):6484–6491. https://doi.org/10.1021/jf201280z

Pippi B, Lana AJ, Moraes RC, Guez CM, Machado M, de Oliveira LF, Fuentefria AM (2015) In vitro evaluation of the acquisition of resistance, antifungal activity and synergism of Brazilian red propolis with antifungal drugs on *Candida* spp. J Appl Microbiol 118(4):839–850. https://doi.org/10.1111/jam.12746

Rawat JM, Pandey S, Rawat B, Rai N, Preeti P, Thakur A, Wang C-H (2023) Traditional uses, active ingredients, and biological activities of Paris polyphylla smith: A comprehensive review of an important Himalayan medicinal plant. J Chem 2023:1–18. https://doi.org/10.1155/2023/7947224

Regueira-Neto MSN, Tintino SR, da Silva ARP, Costa MDS, Boligon AA, Matias EFF, de Queiroz Balbino V, Melo Coutinho HD (2017) Seasonal variation of Brazilian red propolis: antibacterial activity, synergistic effect and phytochemical screening. Food Chem Toxicol 107(Pt B):572–580. https://doi.org/10.1016/j.fct.2017.03.052

Regueira-Neto M, Relison Tintino S, Pereira da Silva AR, do Socorro Costa M, de Morais Oliveira-Tintino CD, Augusti Boligon A, Menezes IRA, de Queiroz Balbino V, Melo Coutinho HD (2019) Comparative analysis of the antibacterial activity and HPLC phytochemical screening of the Brazilian red Propolis and the resin of *Dalbergiaecastaphyllum*. Chem Biodivers 16(9):e1900344. https://doi.org/10.1002/cbdv.201900344

Reis JHO, Barreto GA, Cerqueira JC, Anjos JPD, Andrade LN, Padilha FF, Machado BAS (2019) Evaluation of the antioxidant profile and cytotoxic activity of red propolis extracts from different regions of northeastern Brazil obtained by conventional and ultrasound-assisted extraction. PLoS One 14(7):e0219063. https://doi.org/10.1371/journal.pone.0219063

Ripari N, Pereira AFM, Júnior AF, Rall VLM, Aldana-Mejía JA, Bastos JK, Sforcin JM (2023) Brazilian red propolis in combination with β-lactams exerts an efficient antibacterial action over methicillin-resistant *Staphylococcus aureus* (MRSA) strains. J Appl Microbiol 134(2):lxac080. https://doi.org/10.1093/jambio/lxac080

Rodrigues DM, De Souza MC, Arruda C, Pereira RAS, Bastos JK (2020) The role of *Baccharisdracunculifolia* and its chemical profile on green propolis production by *Apismellifera*. J Chem Ecol 46(2):150–162. https://doi.org/10.1007/s10886-019-01141-w

Rufatto LC, Luchtenberg P, Garcia C, Thomassigny C, Bouttier S, Henriques JAP, Roesch-Ely M, Dumas F, Moura S (2018) Brazilian red propolis: chemical composition and antibacterial activity determined using bioguided fractionation. Microbiol Res 214:74–82. https://doi.org/10.1016/j.micres.2018.05.003

Santos MS, Estevinho LM, de Carvalho CAL, da Silva Conceicao AL, de Castro Almeida RC (2020) Rheological and sensorial evaluation of yogurt incorporated with red propolis. J Food Sci Technol 57(3):1080–1089. https://doi.org/10.1007/s13197-019-04142-5

Santos LA, Rosalen PL, Dias NA, Grisolia JC, Nascimento Gomes BJ, Blosfeld-Lopes L, Ikegaki M, Alencar SM, Burger E (2021) Brazilian Red Propolis shows antifungal and immunomodulatory activities against *Paracoccidioidesbrasiliensis*. J Ethnopharmacol 15:114181. https://doi.org/10.1016/j.jep.2021.114181

Sawaya AC, Palma AM, Caetano FM, Marcucci MC, da Silva Cunha IB, Araujo CE, Shimizu MT (2002) Comparative study of in vitro methods used to analyse the activity of propolis extracts with different compositions against species of *Candida*. Lett Appl Microbiol 35(3):203–207. https://doi.org/10.1046/j.1472-765X.2002.01169.x

Silva LA, Dermody TS (2017) Chikungunya virus: epidemiology, replication, disease mechanisms, and prospective intervention strategies. J Clin Invest 127(3):737–749. https://doi.org/10.1172/JCI84417

Silva NBS, De Souza JH, Santiago MB, Da Silva Aguiar JR, Martins DOS, Da Silva RA, Dos Santos Pedroso RJ, Ambrósio SR, Veneziani RCS, Bastos JK, Pires RH, Martins CHG (2022) Potential in vitro anti-periodontopathogenic, anti-chikungunya activities and in vivo toxicity of Brazilian red propolis. Sci Rep 12(1):21165. https://doi.org/10.1038/s41598-022-24776-4

Silva-Beltrán NP et al (2020) Antiviral effects of Brazilian green and red propolis extracts on *Enterovirus surrogates*. Environ Sci Pollut Res 27(23):28510–28517. https://doi.org/10.1007/s11356-019-07458-z

Silva-Beltrán NP, Galvéz-Ruíz JC, Ikner LA, Umsza-Guez MA, de Paula Castro TL, Gerba CP (2023) *In vitro* antiviral effect of Mexican and Brazilian propolis and phenolic compounds against human coronavirus 229E. Int J Environ Health Res 33(12):1591–1603. https://doi.org/10.1080/09603123.2022.2110576

Siqueira AB, Gomes BS, Cambuim I, Maia R, Abreu S, Souza-Motta CM, Porto AL (2009) *Trichophyton* species susceptibility to green and red propolis from Brazil. Lett Appl Microbiol 48(1):90–96. https://doi.org/10.1111/j.1472-765X.2008.02494.x

Sokolonski AR et al (2021) Activity of antifungal drugs and Brazilian red and green propolis extracted with different methodologies against oral isolates of *Candida* spp. J Mycol 21(1):1–14. https://doi.org/10.1590/1807-3107BOR-2015.vol29.0083

Totten PA, Lory S (1990) Characterization of the type a flagellin gene from *Pseudomonas aeruginosa* PAK. J Bacteriol 172(12):7188–7199. https://doi.org/10.1128/jb.172.12.7188-7199.1990

Trusheva B, Popova M, Bankova V, Simova S, Marcucci MC, Miorin PL, Pasin FDR, Tsvetkova I (2006) Bioactive constituents of Brazilian red propolis. Evid Based Complement Alternat Med 3:249–254. https://doi.org/10.1093/ecam/nel006

Valli M, Russo HM, Bolzani VS (2018) The potential contribution of the natural products from Brazilian biodiversity to bioeconomy. Ann Braz Acad Sci 90:763–778. https://doi.org/10.1590/0001-3765201820170653

WHO (2019) New report calls for urgent action to avert antimicrobial resistance crisis. Retrieved April 29, 2019, from https://www.who.int/news/item/29-04-2019-new-report-calls-for-urgent-action-to-avert-antimicrobial-resistance-crisis

Woo HJ, Yang JY, Lee MH, Kim HW, Kwon HJ, Park M, Kim J-B (2020) Inhibitory effects of β-caryophyllene on *Helicobacter pylori* infection in vitro and in vivo. Int J Mol Sci 21(3):1008. https://doi.org/10.3390/ijms21031008

Zabaiou N, Fouache A, Trousson A, Baron S, Zellagui A, Lahouel M, Lobaccaro JA (2017) Biological properties of propolis extracts: something new from an ancient product. Chem Phys Lipids 207(Pt B):214–222. https://doi.org/10.1016/j.chemphyslip.2017.04.005

30 Phytochemicals as Source of Novel Antimicrobial Agents Against the Rapidly Evolving Antimicrobial Resistance

Ashapurna Sinha, Anupama Rani, Vaishali Mishra, Shiwa Chaubey, and Farina Mujeeb

Abstract

It has been widely accepted that plants possess inherent biological potential, fiber, and phenolic content, which collectively mount up to health promotion. Globally, the bioactive compounds of medicinal and therapeutic plants are an inestimable asset in the struggle against severe diseases. Antibiotics have been employed to treat microbial infections and these bioactive agents with low molecular weight have been found pertinent in various medicinal contexts. Abusive and excessive use of antibiotics over the last few decades has spur the proliferation of microbe's resistant to antibiotics, making medication treatments ineffectual. As an excellent source of antimicrobial compounds with potential applications, plants can supplement or alter antibiotics. Many secondary metabolites of different classes like terpenes, polyphenols, and alkaloids have been shown to showcase antimicrobial properties. Moreover, antimicrobial resistance (AMR) occurs when bacteria, viruses, fungi, and parasites deplete their responsiveness to antimicrobial medicines. The use of phytomolecules or natural products from medicinal plants as an alternative to antibiotics in combating microbial infections has gained attention due to the increasing concerns related to antibiotic resistance in microbes. The current state of knowledge regarding the utilization and therapeutic applications of naturally occurring bioactive is the subject of this chapter.

A. Sinha · A. Rani · F. Mujeeb (✉)
Department of Biosciences, Integral University, Lucknow, Uttar Pradesh, India
e-mail: farinam@iul.ac.in

V. Mishra · S. Chaubey
Phytochemistry Division, CSIR-National Botanical Research Institute, Lucknow, Uttar Pradesh, India

M. Gangwar, G. Nath (eds.), *Emerging Paradigms for Antibiotic-Resistant Infections: Beyond the Pill*, https://doi.org/10.1007/978-981-97-5272-0_30

Keywords

Natural products · Phytomolecules · Metabolites · AMR · Antibiotics

30.1 Introduction

Antimicrobial resistance (AMR), commonly known as drug resistance, occurs when bacteria, viruses, fungi, and parasites do not respond to drugs that were previously effective in treating the disease. This makes the disease difficult to treat and increases the risk of infection, serious illness, and death (Prestinaci et al. 2015).

Phytochemical compounds have potential against microbial antibiotics due to their ability to target many mechanisms that make bacteria resistant to antibiotics. Antibacterial agents refer to drugs used to prevent and treat infections caused by bacteria. Many peptides have been extracted from nature, such as bacteria, plants, fungi, and invertebrates. Some of these compounds have been shown to have significant antimicrobial activity in both preclinical and clinical studies (Youssef et al. 2019). The period from 1940 to 1966 is known as the Golden Age of Antibiotics, with the introduction of many new antibiotics that changed the definition of treating infections (Hutchings et al. 2019). Antimicrobial resistance has become a major problem in global health, characterized by bacteria and fungi that do not respond to existing antibiotics. This problem makes treatment options difficult and increases mortality (Prestinaci et al. 2015). The discovery of antibiotics represents a major advance in medicine and public health. It makes interventions such as surgery safer and results in fewer infant and maternal deaths. The fungus Penicillium, which produces penicillin and has bactericidal properties, is an example of how bacteria or fungi can produce antibiotics (Yadav et al. 2021).

Antibiotics are widely used in industry, mainly due to drug overuse in humans and the daily management of livestock (Davies and Davies 2010). The expansion in antibiotics causes a lot of selection for bacteria, resulting in the emergence of highly resistant bacteria (Patel et al. 2020). Plants may be an important component of various phytochemical compounds capable of protecting against multidrug-resistant microorganisms. Secondary metabolites, or phytochemicals, include alkaloids, flavonoids, quinones, coumarins, Carotenoids, polyphenolics, alkaloids, terpenoids, terpenes, sulfur-containing groups and others. These compounds use different strategies to wield their antibacterial effects on bacteria (Tiwari et al. 2021). The mechanism of action of phytochemicals against AML mainly includes changes in cellular morphology, disruption of pH balance in the cell, membrane disruption leading to ion leakage, membrane reduction ability, inhibition of ATPase activity and enzymatic reactions, transcriptome and protein groups. Mutations include a reduction in synthesis of nucleic acid, inhibition of toxin biosynthesis, biofilm inhibition, lessening of bacterial virulence, interference with quorum sensing, and pH control (Fleitas Martínez et al. 2019).

Therefore, plant bioactive compounds cause changes in various activities in cells, ultimately leading to the disruption of important cellular processes necessary for the survival of the organism (Grad et al. 2007). In this context, it has been determined that many phytochemicals in extract form have antimicrobial properties against microbial species. Medicinal plant extracts such as neem, lemongrass, aloe vera, rosemary, thyme, tulsi, and bryophytes obtained through various solvent methods have been shown to be effective against Gram-positive and Gram-positive antibacterial activity of C-negative bacteria (Khare et al. 2021).

This chapter discusses the molecular and bactericidal factors behind antibiotics, as well as the ability of phytochemicals to reverse antibiotics. Various mechanisms have been found in microbial organisms that make them resistant to antibiotics.

30.2 How Microbes Develop AB-Resistance?

Antibiotics should help improve the quality and longevity of human life, not only by combating diseases/fatal diseases but also by promoting advancements in medical procedures. The pre-antibiotic era with the discovery of the first antibiotic by Sir Alexander Fleming in 1928 (Tan and Tatsumura 2015). Florey and Chain took around 11 years to isolate and later commercialize the highly potent penicillin. Although antibiotics are effective in treating bacterial infections, research shows that bacteria can become resistant to antibiotics. There is always a risk that improper use of medication by individuals can worsen the disease and lead to drug resistance. There is always a risk that people who do not know how to use the right medicine will cause the disease to become lethal and lead to resistance] (Yadav et al. 2021). Genes can arise from spontaneous DNA mutations and be inherited through generations, or they can be transferred horizontally from other DNA sources such as phages, plasmids, and transposons (Davies and Davies 2010). Drug resistance is a complex and constantly competing process, and the commitment to developing new drugs has made it difficult for pharmaceutical companies to invest in developing drugs for new diseases. Multidrug resistance (MDR) is characterized by the simultaneous development of resistance to more than one antibiotic when bacteria have different strains (all bacteria are drug-resistant) (Davies and Davies 2010). Multidrug resistance (MDR), characterized by resistance to multiple classes of antibiotics simultaneously, can arise when a bacterial strain possesses various genes to resistance, each one confronting a diverse antibiotic, or when a sole mechanism of resistance confers resistance to numerous antibiotics. Such bacteria commonly stand as "superbugs" as they cause infections that stay resistant to conventional antibiotics (Yadav et al. 2023).

30.3 Mechanism of Antibiotic Resistance

Mechanisms of antimicrobial resistance can be classified into four primary groups:

- Enzymatic control/inhibition.
- Porin changes.
- Excessive activation of efflux pump.
- Target change.

Pressure exerted by microbe's antibiotics (AB) (Ray et al. 2017). It activates the survival instincts of bacteria and makes them resistant. Although this phenomenon is evolving, its value has increased in recent years due to the widespread and irresponsible use of antibiotics. Bacterial genomes are small, short, highly variable, and rapidly changing. The development of resistance in infectious diseases occurs in many ways, divided into internal resistance and external resistance (Reygaert 2018). In the case of bacteria, intrinsic resistance refers to the presence in their genomes of specific genes that encode traits compatible with antibiotics. In an environment where bacteria coexist, they may carry resistance markers that alter their genome so that they do not respond to antibiotics (AB) (Monteiro et al. 2020). Bacteria can acquire disease in the host through new transfer, upright and level transfer from further resistant bacteria, or a combination of processes. Gene transfer can occur through various means, such as conjugation (using conjugative and plasmid transposons), transformation (using phages), and mutation through integration into plasmids or chromosomal DNA (Fig. 30.1) (Bertani and Ruiz 2018).

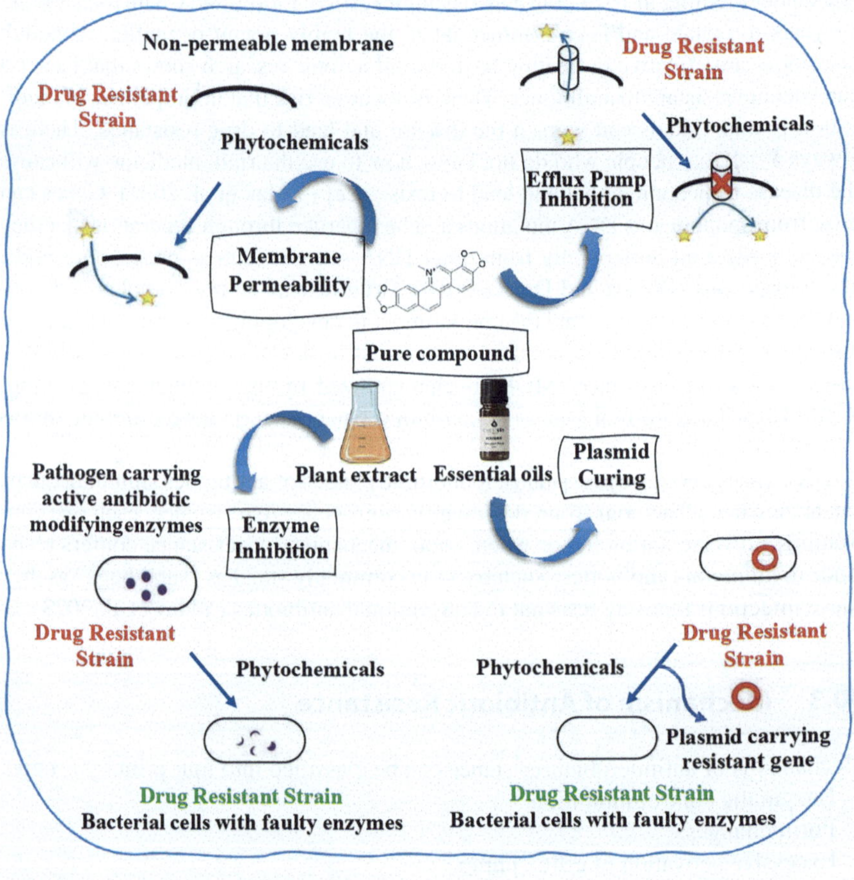

Fig. 30.1 Mechanism of antibiotic resistant

30.3.1 Enzymatic Control/Inhibition

Alteration of the intended location. One of the most prevalent mechanisms of antibiotic resistance in bacterial pathogens, impacting nearly all families of antimicrobial drugs, is the introduction of changes to the target site.

30.3.2 Antibiotic Modification

Many bacterial enzymes show activity against different types of antibiotics (AB), causing hydrolysis, group changes, and structural changes, reducing their benefits (Fig. 30.2). The main reason for the ineffectiveness of beta-lactam antibiotics is hydrolysis. The functional group's transfer such as acyl, phosphoryl, and nuclotide are facilitated by enzymes such as acyltransferase, thiol transferase, nucleotidyltransferase phosphatase in aminoglycosides. The existence of bacterial oxidoreductases and lyases is associated with antibiotics such as tetracyclines, rifampin, mycomycin, chloramphenicol, and streptogramins promoting group transfer and hydrolysis activity (Garneau-Tsodikova and Labby 2016).

30.3.3 Alteration of Porins

The gram-negative bacteria consist of the hydrophobic outer lipid bilayer which usually contains lipopolysaccharide as a barrier that limits the passage of hydrophilic substances such as antibiotics (Bertani and Ruiz 2018). The porins of bacteria such as Omp-E in Salmonella Enteritidis, Omp-C and Omp-F in E. coli exhibit strain-specific levels (Fig. 30.3) (Choi and Lee 2019; Valero-Pacheco et al. 2020).

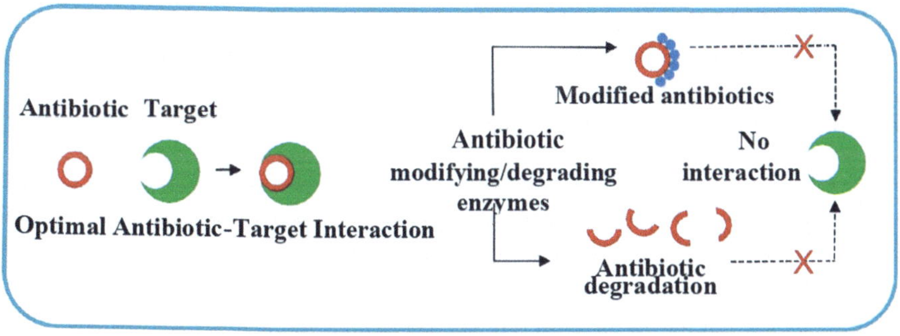

Fig. 30.2 Bacterial mechanism conferring antibiotic resistance

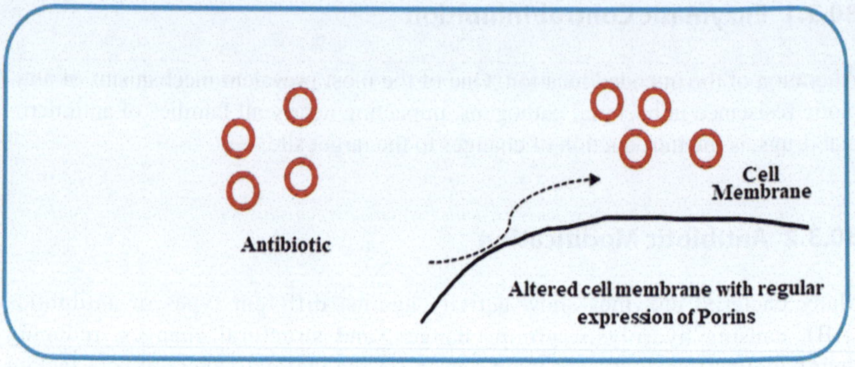

Fig. 30.3 Bacterial mechanism conferring antibiotic resistance through altered cell permeability

In Gram-negative bacteria, the outer hydrophobic lipid bilayer usually consists of lipopolysaccharide as a barrier that limits the passage of hydrophilic substances such as antibiotics (Bertani and Ruiz 2018). Bacterial porins such as Omp-E in *Salmonella enteritidis*, Omp-C, and Omp-F in *E. coli* exhibit strain-specific levels (Choi and Lee 2019). Studies have shown that resistant bacteria like *Escherichia coli, Pseudomonas aeruginosa, Enterobacter aerogenes, Neisseria gonorrhoeae,* and *Klebsiella pneumoniae* reduce porin production (Beceiro et al. 2013). In addition, in many species such as *Klebsiella pneumoniae*, replacing OmpK35, having a large channel diameter, with OmpK36, which has a small channel diameter, will reduce sensitivity to β-lactam antibiotics. Similarity, the Omp36 G112D mutation found in ***B. aerogenes*** also reduces cefepime uptake by disrupting porin channel function (Bornet et al. 2004).

30.3.4 Efflux Pump Hyperactivation

Bacterial efflux pumps are the prime defense mechanism of the disease. Overexpression of the pump in resistant strains results in antibiotic (AB) efflux from cells reduced intracellular concentration, and ineffectiveness. Bacteria can produce efflux pumps encoded on plasmids or chromosomes that are selective or able to eliminate different types of antibiotics, so they are called multidrug efflux pumps (Sharma et al. 2019). Multidrug resistance (MDR) efflux pumps are divided into five groups based on their energy requirements and properties: the major facilitator superfamily (MFS), the small multidrug resistance (SMR) family, the multidrug and toxic compound efflux (MATE) family, resistant to cell division.

Resistance-Nodulation-division family (RND) and Adenosine triphosphate binding cassette (ABC) family as shown in (Fig. 30.4) (Blanco et al. 2016; Sun et al. 2014). Efflux pumps in Gram-negative bacteria form a complex system. These generally include three parts of the protein in the membrane; these include membrane fusion proteins that cross the periplasm, outer membrane efflux proteins, and transport proteins in the cytoplasmic membrane (Nikaido and Takatsuka 2009; Fernández and Hancock 2012). Most Gram-positive bacteria have carriers for drug use in the

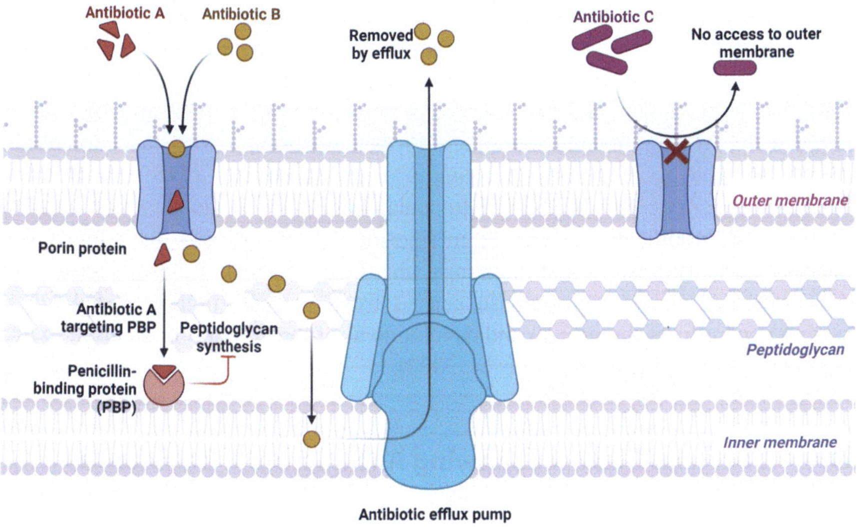

Fig. 30.4 Bacterial mechanism and its antibiotic resistance

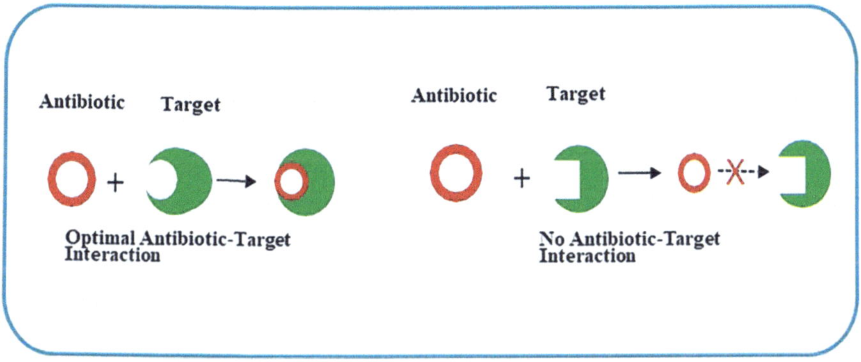

Fig. 30.5 Bacterial mechanism conferring antibiotic resistance through target modification

SMR, MFS, or ABC families. The best-known efflux pumps include NorA in *Staphylococcus aureus* and Bmr and Blt in *Bacillus subtilis* (Monteiro et al. 2020).

30.3.5 Alteration at Target Site

One typical route of resistance is the acquisition or natural variation of alterations in the target sites of antimicrobials that inhibit drug binding. Spontaneous mutations of bacterial genes on chromosomes frequently lead to alterations in target sites (Fig. 30.5).

Most antibiotics (AB) target ribosomes, thus disrupting the protein production process (Chukwudi 2016). Biochemical studies aimed at identifying the AB binding site have demonstrated direct interactions between ribosomal nucleotides and compounds such as lincosamides, streptogramins, and macrolides (Munita and Arias 2016). It has been shown that even a small change in the amino acid sequence in ribosomal proteins can cause changes in protein structure without affecting function, causing AB binding and subsequent further functioning. Resistance to antibiotics occurs due to changes in the amino acid sequence of ribosomal proteins (Podlesek and Žgur Bertok 2020). Molecular targets for drugs are often derived from genetic modification. Changes in ribosomal RNA (rRNA), including post-translational modifications and mutations, are associated with drug resistance. The SOS response is a stress response caused by abnormal single-stranded DNA (ssDNA) and affects the growth and development of the reaction in addition to the processes described above (Shen et al. 2022).

30.4 Bacterial Diseases showing Antimicrobial Resistance

30.4.1 Pyelonephritis

Acute pyelonephritis is a urinary trap infection that is clinically recognized as a tubulointerstitial infection. In these diseases bacteria urea causes inflammation of the renal pyelonephritis. It starts in the bladder, passes through the ureters and reaches the kidneys (Czaja et al. 2007). If treatment is inadequate or delayed, patients develop severe pyelonephritis. Due to pathophysiological reasons, this disease is caused by bacteria collected from the host of intestinal bacteria, *Escherichia coli, Klebsiella pneumoniae, Enterococcus faecalis,* and *Proteus mirabilis* (Kaur et al. 2022). Hormonal changes and some lifestyle factors can cause pyelonephritis. Menstruation may reduce uterine mortality possibly causing pyelonephritis. The use of oestrogen and birth control pills rapidly increases the risk of hormonal deficiencies (Yano et al. 2022).

30.4.1.1 Pathogens of Pyelonephritis

Several strains of *E. coli* are responsible for more than 90% of pyelonephritis cases in young, healthy women. However, *E.coli* infections are still more common in men, older women, and people with immune system problems or non-*E. coli* negative or *E. coli* strain bacilli, gram-positive bacteria, and Candida strains. The increase in antibiotic resistance is mainly due to the spread of potent *E. coli. E. coli* clones, especially the H30 Group in the ST131 system. The World Health Organisation (WHO) published the first list of immune-mediated diseases, identifying 12 disease families that pulls the greatest risk to human health.

This list divides drug-resistant bacteria into three levels based on the importance of developing new drugs. The first category includes important bacteria such as *Acinetobacter baumannii, Pseudomonas aeruginosa,* and several *Enterobacteriaceae (E. coli, Klebsiella, Serratia, and Proteus).* The second category includes important bacteria such as *Enterococcus faecium, Helicobacter pylori, Staphylococcus aureus, Campylobacter, Salmonella,* and *Neisseria gonorrhoeae*. The second category includes pathogens of moderate concern, including *Streptococcus pneumoniae, Haemophilus influenza,* and *Shigella* (Schwartz et al. 2023).

30.4.1.2 Pathogenic Microbes

Microbial organisms such as bacteria and fungi are responsible for many diseases. Despite much prevention, treatment, and control strategies implemented, microbial infections are still a major public health problem worldwide and cause many deaths every year. Human microbial infections, including bacteria and fungi, are proliferating rapidly. Research is needed to identify and implement molecular or genetic strategies to effectively control the global increase in human infectious diseases everywhere. They are important for protecting the ecology we live in. Diseases and illnesses are usually caused by a small percentage of bacteria. But these bacteria can affect people's health, bacterial infections are generally easier to control than infectious diseases because there are many antibiotics available (Sullivan and Ulett 2020).

30.4.2 Bacterial Infection

Pathogenic bacteria will produce a variety of antibodies to protect against the effects of different antibiotics. AMR processes generally fall into four categories:

- Drug resistance.
- Modification.
- Drug failure.
- Drug use.

Some types of diseases have immune mechanisms in the body, such as drug resistance, drug addiction, and drug abuse.

The resistance mechanisms of Gram-negative and Gram-positive bacteria differ according to their structures as explained. Acute pyelonephritis is a malignant disease and urinary tract infection (UTI). Other symptoms of urinary tract infection include febrile urinary tract infection (urinary tract infection is associated with fever or flank pain or discomfort), painful prostatitis and urogenic bacteremia. These conditions can lead to uncontrolled host response, resulting in some sepsis cases, and septic shock. Pyelonephritis is inflammation of the kidney and renal pelvis. The location of the treatment is usually determined by the presence of discomfort or tenderness. Bacteriurea or pyuria on urinalysis and significant levels of uropathogens (usually *E.coli* or other gram-negative bacteria) in urine cultures suggest pyelonephritis infection. Pyelonephritis usually occurs suddenly, symptoms of body inflammation (such as chills, fever, and malaise) and bladder inflammation such as urinary frequency urgency and dysuria). Urinary tract infection can be caused by fungi. Bacteria in urine cultures suggest pylon nephritis inflammation (such as urinary frequency, urgency, and dysuria).

Urinary tract infections can be caused by fungi, bacteria, Gram positive and Gram-negative bacteria.

Uropathogenic *Escherichia coli* (UPEC) is the cause of serious and controversial urinary tract infections. *Klebsiella pneumoniae, Enterococcus faecalis, Staphylococcus saprophyticus,* Group B Streptococcus (GBS), *Proteus mirabilis, Pseudomonas aeruginosa, Staphylococcus aureus and Candida species.* The most common bacteria that cause urinary tract infections are *Enterococcus, Klebsiella, Pseudomonas aeruginosa,* and GBS, UPEC is the second most common (Schwartz et al. 2023).

30.5 Antibacterial Phytochemicals

Plants are known for their medicinal properties first as raw materials or powders and later for their benefits as well. Made medicines containing the essence of active plants. This dates back to the publication in 1803 of the isolation of morphine from opium poppies. Phytochemicals have attracted the attention of the scientific community due to their many uses in medicine, herbal medicine, cosmetics and antibiotics. Many phytochemicals have been shown to have anti-inflammatory properties, even against different types of antibiotics although a variety of phytochemicals have been studied; only a few have been approved by US Food and Drug Administration (FDA), including codeine, capsaicin, reserpine, paclitaxel, and Colchicine. Compesterol has been shown to be effective against pyelonephritis (Kushaish et al. 2020).

30.5.1 ß-Sitosterol

Tribulus terrestris: Belongs to the Tyrannidae Family. Famous names are Kharkhasak and Gokhru. It has diuretic anti-cancer anthelmintic, antibacterial, and aphrodisiac properties. Its ingredients have medicinal properties and include gitogenin, chlorogenin, kaempferol, rhamnose, saponin, stigmasterol, beta-sitosterol, neotigogenin, hekogenin, tribloside, and cinamide (Fig. 30.6) (Fozia et al. 2022).

30.5.2 *Terminalia Chebula*

It is classified in the Combretaceae family. Popular name is Hareer. It has hypopidemic and antibacterial properties, these substances include chebulin, tannin acid, gallic acid, beta-sitosterol, and betulinic acid.

Holy basil: It belongs to the Lamiaceae family, and its name is Tulsi. It has antibacterial, antipyretic, anti-inflammatory effects and energetic properties. The compounds include flavonoids, polyphenols, flavonols, carnosic acid, beta-sitosterol, luteolin, metaldehyde, apigenin, rosmarinic acid, eugenol, vigenin, and orin Ting (Vijayalakshmi and Mohankumar 2023).

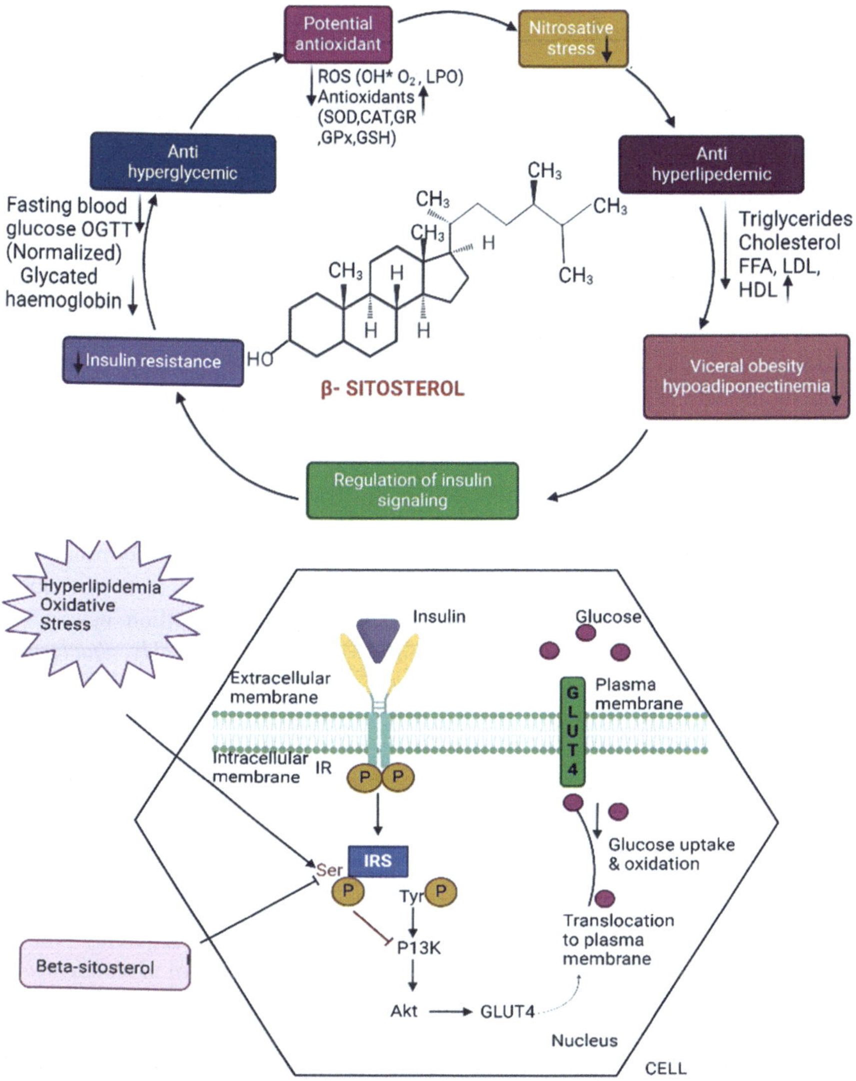

Fig. 30.6 Anti-diabetic and anti-oxidant activity of β-sitosterol

30.5.3 *Trachyspermum Copticum*

It belongs to the *Apiaceae* family. The plant is popularly known as Ajwain. It possesses antibacterial properties due to the presence of compounds such as terpinene, p-cymene, thymol, xylene, palmitic acid, oleic acid, beta-pinene (Wilson et al. 2022).

30.5.4 Juniper

It belongs to the Cupressaceae juniper family. A common name is Juniper. It has diuretic and antibacterial properties. The chemicals include oxygenated sesquiterpenes, monoterpenes, hydrocarbons, beta-pinene, limonene, sabinene, and myrcene (Iqbal et al. 2023; Fukui et al. 2022).

30.6 Tuberculosis

Human tuberculosis (TB), an in capacitating illness caused by the gram-positive, acid-resistant bacterium *Mycobacterium tuberculosis*, was designated a worldwide crisis by the World Health Organization in 1993. Tuberculosis remains one of the largest epidemics, causing the death of approximately one million 8000 people deaths to occur every year, mostly in developing regions (World Health Organization 2016). Although the incidence of tuberculosis has decreased in recent years, approximately 10.4 million new cases were reported in 2015; of these 480,000 have multidrug resistance (MDR), meaning they are resistant to the first-line drug rifampicin and isoniazid (World Health Organization 2013). Although the incidence of tuberculosis has decreased in recent years, approximately 10.4 million new cases occurred in 2015, of which 480,000 multidrugs were caused by *Mycobacterium tuberculosis*. The availability of anti-tuberculosis drugs poses a major challenge to global health because treatment of these diseases requires the use of second-line drugs that are less effective, more expensive and more toxic. In addition, it often requires clinical trials of immunosuppressive drugs for which limited resources are available. Effective treatment of tuberculosis caused by MDR/XDR (multidrug-resistant) mycobacterial variants. Streptomycin, a tuberculosis drug after its discovery in 1944, was immediately used in the treatment of tuberculosis pateints (Schatz et al. 1944). Many TB patients recover during the first few months of streptomycin treatment, but their health then deteriorates as treatment improves. It quickly became clear that this was due to the emergence of antibiotic resistance of *Mycobacterium tuberculosis* and the inactivation of streptomycin (Crofton and Mitchison 1948). TB treatment usually involves an initial period lasting 2 months, during which patients take the medications isoniazid, rifampin, ethambutol, and pyrazinamide. This is followed by a continuous phase for 4 months, in which patients receive isoniazid and rifampicin. However, noncompliance with treatment protocols and inadequacies in healthcare services have led to the emergence of multidrug-resistant TB (MDR-TB), which is characterized by at least the prevention of isoniazid and rifampicin, the two main drugs used in the treatment of TB (World Health Organization 2013). Treatment of many types of TB requires a 20-month course of second-line drugs such as capreomycin, kanamycin, amikacin, and fluoroquinolones. These drugs are toxic and ineffective, and recovery rates vary between 60–75% (Lienhardt et al. 2012). Recently, there have been many reports of the emergence of drug-resistant tuberculosis, which causes serious problems in the treatment of MTB diseases (Blanco et al. 2015). There is also an urgent need to reach consensus on the definition and especially the severity of these MTB diseases. Therefore, the search for new antibiotics is important to solve drug resistance and eventually eliminate tuberculosis. Furthermore, there is an immediate requirement to establish a consensus regarding the definition of these strains of MTB, particularly concerning their severity. Therefore, the pursuit of novel antitubercular medications is paramount to address the issue of drug resistance and ultimately eradicate TB.

30.7 Bioactive Compounds Effective Against *Mycobacterium Tuberculosis*

30.7.1 Allicin

Allicin belongs to the class of chemical known as organosulfur compounds. These compounds contain sulfur atoms bonded to organic molecules and are commonly found in garlic and other members of the Allium genus. Allicin is found abundant in freshly crushed garlic and is one of the most potent bioactive compounds in garlic. It has antibacterial, antiviral, and antiparasitic properties (Oommen et al. 2004). The strong efficacy of garlic extract was revealed in 1996, after previous research demonstrated its ability to inhibit the growth of both resistant and isoniazid-resistant strains of *Mycobacterium tuberculosis*, namely (H37Rv and TRC-C1193, respectively). Minimum inhibitory concentrations (MICs) range from 80 to 160 μg/mL for susceptible strains and from 100 to 200 μg/mL for the resistant strains. Additionally, aqueous garlic extract has been shown to block 14C glycine formation throughout the cell, suggesting that its main function is related to the inhibition of protein synthesis (Ratnakar and Suryanarayana Murthy 1996). Aqueous extracts of garlic are effective against two multidrug-resistant (MDR) strains of *Mycobacterium tuberculosis*, which are resistant to both rifampicin and isoniazid (Gupta et al. 2010). Abdul Hannan et al. (2011) conducted a study to evaluate the antibacterial activity of garlic against non-multidrug-resistant and multidrug-resistant *Mycobacterium tuberculosis* strains. 20 clinical isolates, including 15 MDR strains and 5 non-MDR strains, were included in this study. The preparation of ethanolic extract of garlic was done using the maceration method. The minimum inhibitory concentration of garlic is 1–3 mg/mL, indicating its inhibitory effects against non-multidrug and multidrug strains of *M. tuberculosis* (Abdul Hannan et al. 2011).

30.7.2 Artemisinin

Artemisinin belongs to the sesquiterpene lactone class. It is characterized by having a 15-carbon skeleton and a lactone ring, these compounds are often found in plants such as *Artemisia annua*, from which artemisinin is extracted. Research shows that Artemisinin has the ability to increase the effectiveness of traditional medicines in the treatment of tuberculosis. It acts as an anti-tuberculosis agent by preventing tuberculosis infection from becoming dormant. Dormancy is a state in which bacteria are difficult to get rid of, as they protect themselves from a low level of disease, usually created by the immune system to control the growth of bacteria and prevent infection. In this case, dormant bacteria become very resistant to drugs. Artemisinin prevents bacteria from sensing oxygen by targeting their heme molecules, thus preventing them from entering a dormant state, ultimately causing their death. This process can shorten treatment time and prevent the development of resistance (Zhang et al. 2012).

30.7.3 Linalool

Linalool belongs to the class of chemicals known as terpenoids, especially monoterpenoids. Terpenoids are a large and diverse class of organic compounds derived from five-carbon isoprene units, commonly found in essential oils of multiple plants. Linalool is a monoterpene alcohol and is commonly found in many essential oils, including lavender, rosewood, and basil. Linalool is a terpene alcohol compound found in many natural plants such as *Eucalyptus citriodora*. It has many commercial uses and is especially known for its pleasant scent. According to research, linalool has the ability to prevent the inhalation of *Mycobacterium tuberculosis*, thus contributing to the prevention of infections and various airborne, bacterial, and fungal infections (Gao and Li 2010).

30.7.4 (+)-Calanolide A

Calanolide A belongs to the class of secondary metabolites called coumarins. Coumarins are a group of natural compounds with the benzopyrone structure found in many plants. Calanolide A is isolated from the plant *Calophyllum lanigerum* in Malaysia. (+)-Calanolide A is a naturally occurring compound obtained from the *Calophyllum lanigerum* plant in Malaysia. It works as a non-nucleoside reverse transcriptase inhibitor (NNRTI) and is said to have anti-HIV-1 properties. Accordingly, in previous studies, the compound was shown to be potent against all strains of *Mycobacterium tuberculosis*, including resistance to antibiotics. (+)-Calanolide A acts by rapidly inhibiting DNA, RNA, and protein synthesis. Its effects are comparable to the standard tuberculosis drug rifampicin, which also inhibits RNA synthesis (Bueno et al. 2011).

30.8 Pneumocystis Pneumonia

Pneumocystis pneumonia (PCP), now known as Pneumocystis jirovecii pneumonia (PJP), is caused by the fungus *Pneumocystis jirovecii* (Chabé et al. 2011). It was first recorded in the early twentieth century but dates back to World War II. Until World War II, there was no serious disease in humans (Aliouat-Denis et al. 2008). Pneumocystis was first classified as a protozoan, and in 1988 advances in RNA sequencing revealed that Pneumocystis is an ascomycete. Since then, scientists have identified five types of

Pneumocystis jirovecii. Among them, Pneumocystis jiroveci pneumonia (PJP) has become a major threat, especially for people living with HIV (Kovacs et al. 2001).

30.9 Treatment and Outcomes

Primary treatment and prevention of PJP usually involves the combination use of trimethoprim and sulfamethoxazole (Helweg-Larsen et al. 1999). But there are other drugs such as the combination of primaquine and clindamycin, pentamidine, atovaquone, and the combination of dapsone and trimethoprim (Thomas Jr and Limper 2007). Pneumocystis pneumonia can inhibit the phagocytic activity of alveolar macrophages and cause their apoptosis. During PJP, when there is more polyamine in the lungs and alveolar macrophages, they activate caspase (Hardak et al. 2010). This initiates the apoptotic process. Although trimethoprim-sulfamethoxazole remains the mainstay of current treatment, its effectiveness in treating and preventing PCP-related mortality is limited (Krajicek et al. 2009). Hospitalized patients with PCP continue to have high mortality, indicating that current treatment is inadequate (Mahmood et al. 2013).

30.9.1 Daidzein

Classified as a phytoestrogen, daidzein is a diphenolic compound found in many plants such as *Pueraria mirifica*, *Pueraria lobata*, soybeans, and soy products. Recent studies have demonstrated its ability to prevent the growth of cancer cells, especially in breast cancer (Chen et al. 2015). Daidzein is structurally similar to estradiol, a female hormone, and has properties that support collagen production, suggesting it may aid skin and wound healing (Clarke et al. 2008). Similarly, genistein, another isoflavone found in soybeans, has been shown to increase collagen synthesis in fibroblasts damaged by oxidative stress (Kajta et al. 2013). Genistein achieves this regeneration by modulating ERK1/2 signaling, demonstrating its ability to prevent oxidative damage and promote skin regeneration (Song et al. 1999). Continuing the discussion of triterpenoids, ursolic acid, and asiaticoside, they all belong to this group and are expected to have the following benefits: They have many biological effects (Yoshida et al. 2001).

30.9.2 Glycitein

This drug is a methoxy isoflavone characterized by having a methoxy group at the 6-position and a hydroxyl group at the 7- and 4-position. It is obtained from the mycelium of the *Cordyceps sinensis* mushroom (Hu et al. 2015). The biological activity of daidzein, a little-known substance found in soy products, as well as the estrogen diethylstilbestrol (DES) Isoflavones, was discovered when compared to well-studied isoflavones such as daidzein and genistein (Gutierrez-Zepeda et al. 2005). The uterine weight of mice given daidzein, genistein, and DES significantly increased compared to the control group. Daidzein showed moderate estrogenic effects similar to those observed for daidzein and genistein but were weaker compared to DES and 17β-estradiol (Aboushanab et al. 2022).

30.9.3 Nepetin

Nepetin is obtained by O-methylation of the flavonoid 6-hydroxy luteolin (Chen et al. 2018). This study aims to evaluate the potential of nepetin, a flavonoid derived from O methylation of 6-hydroxy luteolin, widely found in plants and known for its diversity of pharmacological effects, especially its anti-inflammatory effects (Militão et al. 2004). However, treatment with nepetin prevented LPS-induced cell death and effectively suppressed the production of inflammatory mediators and cytokines in cultured keratinocytes. In the AD mouse model, nepetin treatment demonstrated a dose-dependent reduction in skin inflammation symptoms, leading to significant decreases in inflammation-related cytokines, skin lesions, and behavioral scores (Bogaevskii et al. 1970).

30.9.4 Glycyrrhizae Radix

Glycyrrhizae Radix, a traditional Chinese medicine commonly referred to as Gan Cao (GC), is often prescribed in clinical settings to treat pneumonia (Wang et al. 2019). Due to its pharmacological properties including antibacterial, anti-inflammatory, anti-tumor, and antiviral effects, Glycyrrhizae Radix et Rhizoma (GC) has been used clinically to treat various conditions such as cough, bronchitis, peptic ulcer, and dermatitis (Ramalingam et al. 2018). Existing studies on the treatment of pneumonia with Glycyrrhizae Radix et Rhizoma (GC) primarily concentrate on its general pharmacological effects or the mechanisms of action of its main constituent compounds. For instance, glycyrrhizin, a key component, has been shown to inhibit the replication of severe acute respiratory syndrome coronavirus 2 (SARS-CoV-2) due to its ability to target the angiotensin-converting enzyme 2 receptor. However, research investigating the specific compositions or active medicinal components of GC that are effective in treating respiratory diseases like pneumonia in vivo has not yet been conducted.

30.10 Phytochemicals Signaling Pathway

There are four ways that Artemisia compounds have anti-inflammatory effects. Phytochemicals inhibit pathogenic bacteria. This method focuses on different molecular components and components in the organism. Nine essential oils obtained from wormwood affect the cell membranes and/or cell walls of bacteria, changing their functioning and structure. This causes changes in membrane permeability and the breakdown of bacterial and/or cell walls. As a result, important cell products are released, eventually leading to cell death. Chemicals such as (+)-limonene, vulgarone B, and thymol can affect the structure and function of DNA, causing SOS and oxidative stress caused by the overproduction of reactive oxygen species (ROS). Excessive ROS production can cause bacterial cell death. Additionally, other compounds include (+)-α pinene, thymol, chrysoplenetin, chrysoeriol, penduletin, RA, jaceosidin, (Z)-2 (hexa-2,4-diyn-1-ylidene)-1,6 dioxaspiro gets (4.5) dec-3-ene and (E)-2-(hexa 2,4-diyn-1-ylidene)-1,6-dioxaspiro(4.5)dec-3-ene can be modified. Protein targets and enzymes activities can cause protein denaturation and reduce enzyme activity. Therefore, there is inhibition of oxidative stress, heat shock and virulence factor expression in bacteria. Additionally, four compounds target other pathways: thymol and eugenol disrupt the biofilm pathway by reducing and inactivating biofilms, thereby reducing bacterial virulence and survival. Artemisia and artesunate inhibit efflux pumps, causing antibiotic accumulation in bacteria (Fig. 30.7).

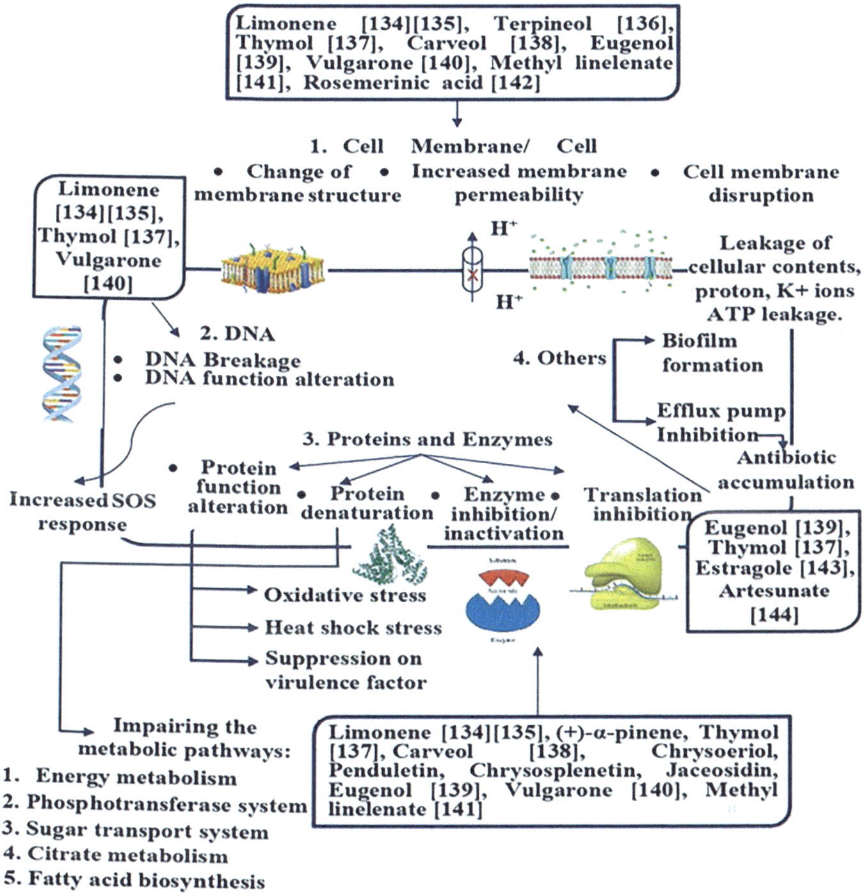

Fig. 30.7 Schematic representation of phytochemical signaling pathway

30.11 Fungal Infections

Fungi are widely distributed in the environment, and there are more than 1,00,000 species, out of which 300 can cause disease in animals and people. Infections in humans are now difficult to control (Pathakumari et al. 2020). Most of the mushrooms containing bioactive components are used by HIV, asthma, diabetic, and cancer patients, as well as people suffering from autoimmune diseases, etc. They are responsible for more than 90% of sexually transmitted fungal infections (Makhuvele et al. 2020). Fungal infections range from severe life-threatening to severe infections. Treatment of fungal infections is severely limited due to the limited availability of antibiotics such as polyenes and azoles. Fungal organisms have developed resistance mechanisms to antibiotics, such as efflux pump overexpression and biofilm formation and these mechanisms need to be understood and studied. Phytochemical compounds provide very toxic antibiotics, and they are widespread. Use leads to the rapid emergence of resistant bacteria, causing serious problems in agriculture and medical facilities More than 1000 microbial compounds have been discovered, many with potential therapeutic uses (Donnelly et al. 2020).

30.12 Diabetes Mellitus

Diabetes mellitus is a serious problem caused by insulin deficiency, Causing serious disorders in sugar fat and protein metabolism. Blood sugar is a major factor in causing fungal infections, leading to serious illness and death. Diabetic patients are more susceptible to disease due to the hyperglycemic environment resulting in cell-mediated and humoral resistance (Cole and Florez 2020). Blood sugar changes and the adverse effect of lack of oxygen can affect the ability of white blood cells to destroy harmful bacteria and viruses, increasing the risk of infection. People with diabetes are at a higher risk of foot conditions and complications if they suffer from diabetes-related neuropathy and/or poor circulation. People with diabetes are more likely to get fungal infections than healthy people, and they are more likely to develop complications such as fungal infection, food infections, and lower back pain (Lao et al. 2020). A study was conducted to determine the frequency of fungal infections in diabetic feet and to compare the results with previous studies. The majority of patients with diabetic foot ulcers in our study were men and similar results showed that 175% of cultures in our study had a positive diagnosis of diseases. This was consistent with previous results (Saud et al. 2020). In our study, *Candida* is common, and *Candida albicans* is the male species counting for 42.85%. *Candida tropicalis* predominates. According to reports 210.42% of the *Candida species* isolated were *Candida tropicalis* and 140.28% were *Candida glabrata* other studies have demonstrated that presence of *Candida parapsilosis, Candida tropicalis, Candida girimondale, Candida foliaceus, Candida glabrata,* and *Candida kefirensis* (Mohammed et al. 2021).

Plants have been used since ancient times as effective treatments for many ailments, including diabetic footlers. Recent research has shown that powerful substances in the body found in plant have pharmacological properties that can prevent many types of cell damage, such as chronic pain. Many herbs such as Aloe vera,

Custard apple, Papaya, Turmeric, Dragon fruit, Pomegranate and *Centella asiatica* have very few side effects due to their wound healing, anti-inflammatory, antioxidant, and medicinal properties (Kumar et al. 2021) Antibiotics and many bioactive compounds are found in medicinal plants such as onion, sugar beet, citrus, taro, and sorghum. Bioactive compounds such as isoquercitrin, quercitin, kaempferol, apigenin, tangeretin, luteolin, catechin, gallic acid, methyl gallate, naringenin, routinely exert anti-inflammatory effects through different mechanisms (Shanmugam et al. 2021). These include improving β-cell function and insulin secretion, improving insulin sensitivity increasing glut4 expression reducing gluconeogenesis and inhibition of α-amylase, α glucosidase, DPP 4 activity, and production of advanced glycation end products. These phytochemicals may improve diabetic foot pain and play a role in preventing diabetes. Scientists are studying the healing properties of the plant and isolating bioactive substances to evaluate its potential to heal pain. Some bioactive compounds isoquercitrin, quercitin, and gallic acid are effective against diabetes (Mondal et al. 2021).

Following are some phytochemical compounds that have been shown antidiabetic activity.

Isoquercetin: Found in *Acasia leguminosarum.* It's a flower healer. The tree is native to the Middle East, Africa, and the Indian subcontinent. It is used medicinally for many ailments, such as wounds, burns, diabetes, diabetes, foot, pneumonia, fever, cuff, and abdominal pain. It eliminates the oxidative stress and inhibits the effects of tumor necrosis. Factor Alpha, interleukin 1β promotes cell proliferation, wound epithelial and skin regeneration. Bioactive compounds found in Acacia are gallic acid, ellagic acid, isoquercitrin, albacin, and kaempferol (Jayachandran et al. 2020).

Rutin: Found in Acacia (family: Crimsonaceae), Annonaceae (family: Annonaceae), *Melia azedarach* (Meliaceae), Papaya (Cariaceae), *Morinda officinalis* (Rubiaceae), *Prosopis cineraria* (Fabaceae), Pomegranate (Punicaceae). This herb enhances wound healing, increases nitric oxide, DNA content, improves tissue repair, increases hydroxy proline and protein content (Maradesha et al. 2022).

Kaempferol: This bioactive molecule is found in Lawsonia alba, whose leaves have antibacterial properties and can improve wound healing and diabetic foot ulcers. Moreover, it can also reduce ulcers, stimulate fibroblast, collagen production and promote the formation of fibroblast. *Centella asiatica.* Diabetes mellitus is caused by insulin. Gluconeogenesis genes are over-expressed, resulting in increased hepatic glucose production A decrease in hepatic glucose synthesis and glucose absorption leads to an increase in glucose, leading to hyperglycemia. Kaempferol stimulates insulin production and estimates Akt activity by regulating mitochondrial calcium absorption. Kaempferol can directly reactivate Akt. Thus with Akt in activation, increased gluconeogenesis, decreased liver glucose, and glucose absorption are reversed. Additionally, the antioxidant Kaempferol can inhibit autophagy and apoptosis. IRS stands for insulin receptor substrate; PI3K represents inosine phosphate 3 kinase; AKT is a threonine protein kinase; PIP 3 stands for phosphoinositol. 3,4,5 Triphosphate, PDK1 represents phosphoinositide, hol 6 kinase, ribosomal Protein S6 kinase; GSK3β is glycogen synthesis kinase 3β; The FOXO bifurcate box represents 0; FOXK is a K-class fork box; TSC2 refers to tuberous sclerosis complex 2, ROS stands for Reactive Oxygen Species, Tbc1d4/AS160 is the 10 AKT

Receptor Bcl-2 stands for B cell leukaemia lymphoma-2,mTORC1 is rapamycin complex 1, PEPCK is phosphoenolpyruvate carboxylase. G6P stands for glucose 6 phosphatase (Sha et al. 2021).

30.13 Medicinal Herbs

Citrus sinensis: (family: Rutaceae): The peel and food protect against oxidative damage, and cell damage caused by hydrogen peroxide. It also supports blood flow cell proliferation, angiogenesis, keratinocyte, fibroblast migration and tissue wound healing (Asabi et al. 2021).

Euphorbia aleppica: (family: Euphorbiaceae): Is a flowering plant whose leaves and flowers promote wound healing by supporting the epidemic collagen tissue and fibrosis, increasing granulation tissue weight, reducing inflammation, cell damage and promoting re-epithelialization. Lipid peroxidation can also be reduced. The plant contains bioactive compounds, such as hepatocosane, gallic acid, leucoanthocyanins, kaempferol, and camphor (Onyeka et al. 2021).

Centella asiatica: (family: Apiaceae): It improves angiogenesis, collagen levels, fibroblasts, growth factor, and vesicular endothelial growth factor, inhibits TNF-α, IL-6, IL-1β, COX-2, pProstaglandin E-2, and lipoxygenase, leads to re -epithelization. Papaya (Papaya family) its fruit and leaves improve granulation tissue, weight, collagen, and hydroxyproline content, wound size and contraction, rapid epithelization, and resistance to microbial infection (Fig. 30.8) (Yang et al. 2022).

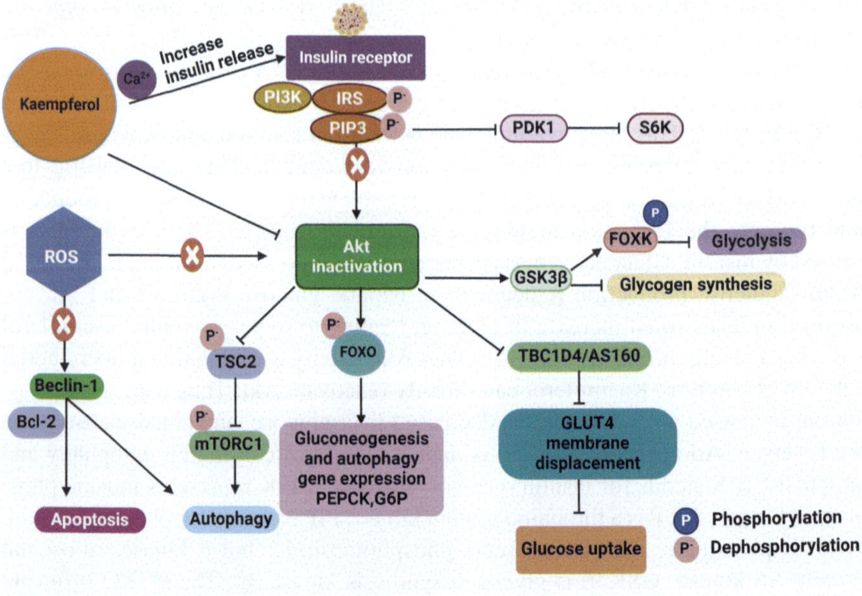

Fig. 30.8 Mechanism of kaempferol (diabetes)

30.14 Candidiasis

Pathogenic yeasts (Candida species) are opportunistic pathogens that cause candidiasis. Candida species are found in the skin, in the gastrointestinal and genital tracts of humans. *Candida auris* is a multi-drug resistant that is often misdiagnosed and has now become a serious medical health problem. From 2015 to 2018, there was a 318% increase in *Candida auris* cases (Salerno et al. 2011). According to the (Salerno et al. 2011) Centers for Disease Control and Prevention (CDC), 1700 people died from candidiasis in 2017 and nearly 34,000 people were reported to have the disease in the hospital (Bhattacharya et al. 2020). Candida infections are also commonly associated with medical devices such as central venous lines, cardiovascular devices, and urinary catheters (Kojic and Darouiche 2004). Fungal infections, in particular, are a serious human health problem. It is becoming increasingly difficult to treat fungal infections with antibiotic-resistant drugs.

In tropical regions, fungal diseases reach a very serious endemic level. It has become difficult for some immune system-boosting drugs to treat fungal infections, so we turn to herbal medicines to treat fungal infections. These natural products can treat fungal infections by inhibiting the growth of various bacteria (López et al. 2002).

There are many kinds of antifungal medications, such as polyenes azoles, and echinocandins, are used to treat fungal infections. These drugs target various specific metabolic pathways, particularly azoles that target the synthesis of ergosterol, a specific cell membrane component found in fungi and other methods (Veen et al. 2003). The quality, location, and strength of the host's immune system depend on the treatment of drugs (White et al. 1998). These mechanisms include overexpression of membrane transporters, cell wall defects, functional changes in transcription factors that regulates transporter processes and ergosterol biosynthesis (Popolo et al. 2001).

30.14.1 Pathogens

Candida tropicalis, *Candida glabrata*, *Candida parapsilosis*, *Candida krusei*, *Candida dubliniensis* and *Candida auris* are the various kinds of pathogens which causes Candidiasis infection. Approximately 50% of candidiasis are caused by *Candida albicans* pathogen while the remaining half of candidiasis infection is caused by non-albicans candida species (Salerno et al. 2011).

30.15 Antifungal Phytochemicals

30.15.1 Caffiec Acid

Caffeic acid, also known as 3,4-dihydroxy cinnamic acid, it is one of the most specific important phenolic compounds. It is widely found in different places, including natural substances in human foods, vegetables, alcohol, olive oil, and coffee. Caffeic acid has different organic properties, such as antithrombotic, antioxidant, antihypertensive, antifibrotic, antiviral properties acting as chelating agent for metallic particles (Rajendra Prasad et al. 2011; Pappas et al. 2009). It is one of the most widely used hydroxycinnamates esters and can be exist in different forms such as esters and amides (Vermes et al. 2000). This acid protects cells from UV light-induced cell damage (Rajendra Prasad et al. 2011).

30.15.2 Berberine

Berberine (BBR) is a quaternary ammonium salt of isoquinoline alkaloid (Cicero and Baggioni 2016) found in traditional Chinese herb, such as *Coptis chinensis* and *Mahonia aquifolium* belonging to the Ranunculaceae family (Sanglard et al. 2009). Previous studies have been shown to have multiple pharmacological effects, including anti-inflammatory, antibacterial, antiviral, and antifungal. It has been used to treat diarrhea, diabetes, and other fungal diseases. *HSF1* a transcription regulator protects cells from thermal shock by activating the HSPs (Heat Shock Proteins) expression which act as chaperones for rescue proteins (Nicholls et al. 2009). Researchers have shown in previous studies that HSF1 affects Candida cell sensitivity to berberine and vulnerability to various drugs with comparable targets. The effects of berberine and HSF1 on drug sensitivity in Candida cells have not been demonstrated in a dependent manner. Berberin analysis showed that HSF1 mutant strain was more susceptible to this antibiotic. The Calcineurin pathway, which leads to mitochondrial dysfunction and cell wall death, were altered by berberin treatment. Regarding calcineurin, the HSF1 phenotype does not affect calcineurin expression, ROS levels and mitochondrial dysfunction but HSF1 does not affect the cell wall integrity through its inhibitors (Fig. 30.9) (Dhamgaye et al. 2014).

30.15.3 Pinocembrin

Pinocembrin also known as 5,7-dihydroxyflavanone, is one of the main flavonoids secondary metabolites belongs to piperaceae family and have been obtained from many plant species, including Pine heartwood, Eucalyptus, Populus, Euphorbia, and white mustard (Danelutte et al. 2003). It is also extracted from multiple different plants belonging to Lauraceae and Asteraceae family (Feng et al. 2012). Pinocembrin, a emblematic flavonoid class of natural product found in variety of foods, fruits, vegetables, nuts, seeds, flowers, stems, bevarages (Rasul et al. 2013). Pinocembrin demonstrated antifungal activity by inhibiting hyphal growth and cell membrane damage, with a minimal inhibitory concentration value of 100microg/mL (López et al. 2002).

30.15.4 Magnolol

Magnolol, one of the main pharmacologically active compounds found in *Magnolia officinalis*, *Magnolia obovata*, and *Magnolia grandiflora* are the main traditional

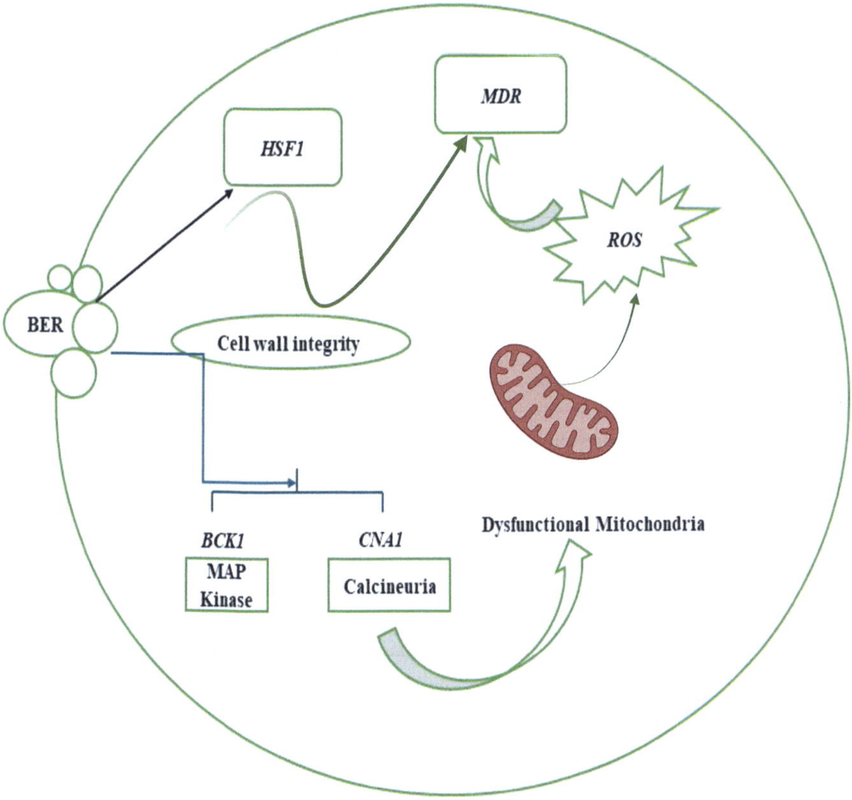

Fig. 30.9 Pathways affected by BER treatment in *C. albicans*

Chinese and Japanese natural plants (Bhattacharya et al. 2020). It is a type of neolignan obtained from parts of the plant such as root, stem, and bark of magnolia species. Magnolol is used to improve symptoms such as anxiety, asthma, neurological disorders, and digestive problems and may reduce the elemental messages of ergosterol commonly used in the *Candida albicans* SC5314. Magnolon treatment downregulates the Ras1cAMP-Efg1 pathway and inhibits hypha-specific gene expression level such as ECE1, HWP1, and ALS3 (Sun et al. 2015).

30.16 Conclusion

In conclusion, the utilization of phytochemicals from medicinal plants as novel antimicrobial agents presents a promising solution to the escalating threat of antimicrobial resistance (AMR). With the overuse of antibiotics leading to the proliferation of resistant microbes, alternative strategies are urgently needed. Phytochemicals, including terpenes, polyphenols, and alkaloids, exhibit potent antimicrobial properties, offering potential alternatives or supplements to conventional antibiotics. By harnessing the inherent biological potential of plants, we can address the challenges posed by AMR and pave the way for effective treatments against microbial infections.

References

Abdul Hannan AH, Ullah MI, Muhammad Usman MU, Shahid Hussain SH, Muhammad Absar MA, Khursheed Javed KJ (2011) Anti-mycobacterial activity of garlic (Allium sativum) against multi-drug resistant and non-multi-drug resistant mycobacterium tuberculosis. Pak J Pharm Sci 24(1):81–85. https://doi.org/10.5555/20113054770

Aboushanab S, Shevyrin V, Kamel M, Kambele J, Kovaleva E (2022) Phytochemical screening and properties of botanical crude extracts and ethyl acetate fractions isolated by deep eutectic solvent. Chimica Techno Acta 9(4):20229404

Aliouat-Denis C-M, Chabé M, Demanche C, Viscogliosi E, Guillot J, Delhaes L, Dei-Cas E (2008) Pneumocystis species, co-evolution and pathogenic power. Infect Genet Evol 8(5):708–726

Asabi OA, Oisemuzeimen JO, Abiodun OO, Blessing K (2021) Antioxidant and in-vitro anti-diabetic activities of fermented peels of citrus x Sinensis (L.) Osbeck (Rutaceae). Prog Chem Biochem Res 4:414–425

Beceiro A, Tomás M, Bou G (2013) Antimicrobial resistance and virulence: A successful or deleterious Association in the Bacterial World? Clin Microbiol Rev 26(2):185–230. https://doi.org/10.1128/CMR.00059-12

Bertani B, Ruiz N (2018) Function and biogenesis of lipopolysaccharides. EcoSal Plus 8(1):10.1128/ecosalplus.ESP-0001–2018. https://doi.org/10.1128/ecosalplus.esp-0001-2018

Bhattacharya S, Sae-Tia S, Fries BC (2020) Candidiasis and mechanisms of antifungal resistance. Antibiotics 9(6):312

Blanco D, Perez-Herran E, Cacho M, Ballell L, Castro J, González Del Río R, Lavandera JL, Remuiñán MJ, Richards C, Rullas J, Vázquez-Muñiz MJ, Woldu E, Zapatero-González MC, Angulo-Barturen I, Mendoza A, Barros D (2015) Mycobacterium tuberculosis gyrase inhibitors as a new class of antitubercular drugs. Antimicrob Agents Chemother 59(4):1868–1875. https://doi.org/10.1128/AAC.03913-14

Blanco P, Hernando-Amado S, Reales-Calderon JA, Corona F, Lira F, Alcalde-Rico M, Bernardini A, Sanchez MB, Martinez JL (2016) Bacterial multidrug efflux pumps: much more than antibiotic resistance determinants. Microorganisms 4(1):14

Bogaevskii AK, Dranik LI, Borisov MI (1970) The hydroxycinnamic acids of Galium fagetorum and G. pseudomollugo. Chem Nat Compd 6(6):761–762. https://doi.org/10.1007/BF00565353

Bornet C, Saint N, Fetnaci L, Dupont M, Davin-Régli A, Bollet C, Pagès J-M (2004) Omp35, a new *Enterobacter aerogenes* porin involved in selective susceptibility to cephalosporins. Antimicrob Agents Chemother 48(6):2153–2158. https://doi.org/10.1128/AAC.48.6.2153-2158.2004

Bueno J, Escobar P, Martínez JR, Leal SM, Stashenko EE (2011) Composition of three essential oils, and their mammalian cell toxicity and antimycobacterial activity against drug resistant-tuberculosis and nontuberculous mycobacteria strains. Nat Prod Commun 6(11):1934578X1100601. https://doi.org/10.1177/1934578X1100601143

Chabé M, Aliouat-Denis C-M, Delhaes L, Aliouat EM, Viscogliosi E, Dei-Cas E (2011) Pneumocystis: from a doubtful unique entity to a group of highly diversified fungal species. FEMS Yeast Res 11(1):2–17

Chen J, Duan Y, Zhang X, Ye Y, Ge B, Chen J (2015) Genistein induces apoptosis by the inactivation of the IGF-1R/p-Akt signaling pathway in MCF-7 human breast cancer cells. Food Funct 6(3):995–1000

Chen X, Han R, Hao P, Wang L, Liu M, Jin M, Kong D, Li X (2018) Nepetin inhibits IL-1β induced inflammation via NF-κB and MAPKs signaling pathways in ARPE-19 cells. Biomed Pharmacother 101:87–93

Choi U, Lee C-R (2019) Distinct roles of outer membrane porins in antibiotic resistance and membrane integrity in Escherichia coli. Front Microbiol 10:953

Chukwudi CU (2016) rRNA binding sites and the molecular mechanism of action of the tetracyclines. Antimicrob Agents Chemother 60(8):4433–4441. https://doi.org/10.1128/AAC.00594-16

Cicero AFG, Baggioni A (2016) Berberine and its role in chronic disease. In: Gupta SC, Prasad S, Aggarwal BB (eds) Anti-inflammatory nutraceuticals and chronic diseases, vol 928. Springer International Publishing, pp 27–45. https://doi.org/10.1007/978-3-319-41334-1_2

Clarke DB, Bailey V, Lloyd AS (2008) Determination of phytoestrogens in dietary supplements by LC-MS/MS. Food Addit Contam Part A 25(5):534–547. https://doi.org/10.1080/02652030701658340

Cole JB, Florez JC (2020) Genetics of diabetes mellitus and diabetes complications. Nat Rev Nephrol 16(7):377–390

Crofton J, Mitchison DA (1948) Streptomycin resistance in pulmonary tuberculosis. Br Med J 2(4588):1009

Czaja CA, Scholes D, Hooton TM, Stamm WE (2007) Population-based epidemiologic analysis of acute pyelonephritis. Clin Infect Dis 45(3):273–280

Danelutte AP, Lago JHG, Young MCM, Kato MJ (2003) Antifungal flavanones and prenylated hydroquinones from Piper crassinervium Kunth. Phytochemistry 64(2):555–559

Davies J, Davies D (2010) Origins and evolution of antibiotic resistance. Microbiol Mol Biol Rev 74(3):417–433. https://doi.org/10.1128/MMBR.00016-10

Dhamgaye S, Devaux F, Vandeputte P, Khandelwal NK, Sanglard D, Mukhopadhyay G, Prasad R (2014) Molecular mechanisms of action of herbal antifungal alkaloid Berberine, in Candida albicans. PLoS One 9(8):e104554. https://doi.org/10.1371/journal.pone.0104554

Donnelly JP, Chen SC, Kauffman CA, Steinbach WJ, Baddley JW, Verweij PE, Clancy CJ, Wingard JR, Lockhart SR, Groll AH (2020) Revision and update of the consensus definitions of invasive fungal disease from the European organization for research and treatment of cancer and the mycoses study group education and research consortium. Clin Infect Dis 71(6):1367–1376

Feng R, Guo ZK, Yan CM, Li EG, Tan RX, Ge HM (2012) Anti-inflammatory flavonoids from Cryptocarya chingii. Phytochemistry 76:98–105

Fernández L, Hancock REW (2012) Adaptive and mutational resistance: role of porins and efflux pumps in drug resistance. Clin Microbiol Rev 25(4):661–681. https://doi.org/10.1128/CMR.00043-12

Fleitas Martínez O, Cardoso MH, Ribeiro SM, Franco OL (2019) Recent advances in anti-virulence therapeutic strategies with a focus on dismantling bacterial membrane microdomains, toxin neutralization, quorum-sensing interference and biofilm inhibition. Front Cell Infect Microbiol 9:74

Fozia F, Ahmad I, Wadood A, Khan HU, Ahmed M, Jamila N, Ullah R, Alotaibi A, Sultan MA (2022) Urease and α-chymotrypsin inhibitory activities and molecular docking studies of alkaloids isolated from medicinal plant Isatis minima bunge. Evid Based Complement Alternat Med 2022:1904874. https://www.hindawi.com/journals/ecam/2022/1904874/

Fukui S, Inui A, Saita M, Kobayashi D, Naito T (2022) Clinical prediction rule for bacteremia with pyelonephritis and hospitalization judgment: Chi-square automatic interaction detector (CHAID) decision tree analysis model. J Int Med Res 50(1):030006052110656. https://doi.org/10.1177/03000605211065658

Gao Q, Li X (2010) Transmission of MDR tuberculosis. Drug Discov Today Dis Mechan 7(1):e61–e65

Garneau-Tsodikova S, Labby KJ (2016) Mechanisms of resistance to aminoglycoside antibiotics: overview and perspectives. Medchemcomm 7(1):11–27

Grad LI, Sayles LC, Lemire BD (2007) Isolation and functional analysis of mitochondria from the nematode Caenorhabditis elegans. In: Leister D, Herrmann JM (eds) Mitochondria, vol 372. Humana Press, pp 51–66. https://doi.org/10.1007/978-1-59745-365-3_4

Gupta R, Thakur B, Singh P, Singh HB, Sharma VD, Katoch VM, Chauhan SVS (2010) Anti-tuberculosis activity of selected medicinal plants against multi-drug resistant Mycobacterium tuberculosis isolates. Indian J Med Res 131(6):809–813

Gutierrez-Zepeda A, Santell R, Wu Z, Brown M, Wu Y, Khan I, Link CD, Zhao B, Luo Y (2005) Soy isoflavone glycitein protects against beta amyloid-induced toxicity and oxidative stress in transgenic Caenorhabditis elegans. BMC Neurosci 6(1):54. https://doi.org/10.1186/1471-2202-6-54

Hardak E, Brook O, Yigla M (2010) Radiological features of Pneumocystis jirovecii pneumonia in immunocompromised patients with and without AIDS. Lung 188(2):159–163. https://doi.org/10.1007/s00408-009-9214-y

Helweg-Larsen J, Benfield TL, Eugen-Olsen J, Lundgren JD, Lundgren B (1999) Effects of mutations in Pneumocystis carinii dihydropteroate synthase gene on outcome of AIDS-associated P carinii pneumonia. Lancet 354(9187):1347–1351

Hu YJ, Gao KG, Zheng CT, Wu ZJ, Yang XF, Wang L, Ma XY, Zhou AG, Jiang ZJ (2015) Effect of dietary supplementation with glycitein during late pregnancy and lactation on antioxidative indices and performance of primiparous sows. J Anim Sci 93(5):2246–2254

Hutchings MI, Truman AW, Wilkinson B (2019) Antibiotics: past, present and future. Curr Opin Microbiol 51:72–80

Iqbal A, Ibrahim M, Muhammad N (2023) Natural approach used for urinary tract infections. Phytopharmacol Res J 2(2):1–17

Jayachandran M, Zhang T, Wu Z, Liu Y, Xu B (2020) Isoquercetin regulates SREBP-1C via AMPK pathway in skeletal muscle to exert antihyperlipidemic and anti-inflammatory effects in STZ induced diabetic rats. Mol Biol Rep 47:593–602

Kajta M, Rzemieniec J, Litwa E, Lason W, Lenartowicz M, Krzeptowski W, Wojtowicz AK (2013) The key involvement of estrogen receptor β and G-protein-coupled receptor 30 in the neuroprotective action of daidzein. Neuroscience 238:345–360

Kaur P, Sachan RSK, Karnwal A, Devgon I (2022) A review on clinical manifestation and treatment regimens of UTI in diabetic patients. Iran J Med Microbiol 16(2):98–115

Khare T, Anand U, Dey A, Assaraf YG, Chen Z-S, Liu Z, Kumar V (2021) Exploring phytochemicals for combating antibiotic resistance in microbial pathogens. Front Pharmacol 12:720726

Kojic EM, Darouiche RO (2004) *Candida* infections of medical devices. Clin Microbiol Rev 17(2):255–267. https://doi.org/10.1128/CMR.17.2.255-267.2004

Kovacs JA, Gill VJ, Meshnick S, Masur H (2001) New insights into transmission, diagnosis, and drug treatment of Pneumocystis carinii pneumonia. JAMA 286(19):2450–2460

Krajicek BJ, Thomas CF, Limper AH (2009) Pneumocystis pneumonia: current concepts in pathogenesis, diagnosis, and treatment. Clin Chest Med 30(2):265–278

Kumar S, Mittal A, Babu D, Mittal A (2021) Herbal medicines for diabetes management and its secondary complications. Curr Diabetes Rev 17(4):437–456

Kushaish RA, AL-Salem BA, Hussein MA (2020) The effect of alkaloid extract of teucrium Polium L. against some pathogenic bacteria of urinary tracts and on pyelonephritis induced in rats. Indian J Public Health Res Dev 11(1):1450–1454. https://search.ebscohost.com/login.aspx?direct=true&profile=ehost&scope=site&authtype=crawler&jrnl=09760245&AN=142236904&h=DTm5jFt%2Fwbp9y44Gl%2FyJuKl1%2B3mjHrYYIARp%2BxqBIokVptX8%2Fy5sPv9gGYBcJMmrSlQhCwpOaMwAMKU5WtmvxQ%3D%3D&crl=c

Lao M, Li C, Li J, Chen D, Ding M, Gong Y (2020) Opportunistic invasive fungal disease in patients with type 2 diabetes mellitus from southern China: clinical features and associated factors. J Diabetes Investig 11(3):731–744. https://doi.org/10.1111/jdi.13183

Lienhardt C, Raviglione M, Spigelman M, Hafner R, Jaramillo E, Hoelscher M, Zumla A, Gheuens J (2012) New drugs for the treatment of tuberculosis: needs, challenges, promise, and prospects for the future. J Infect Dis 205(suppl_2):S241–S249

López A, Ming DS, Towers GHN (2002) Antifungal activity of benzoic acid derivatives from *Piperlanceaefolium*. J Nat Prod 65(1):62–64. https://doi.org/10.1021/np010410g

Mahmood A, Mahmood A, Malik RN, Shinwari ZK (2013) Indigenous knowledge of medicinal plants from Gujranwala district, Pakistan. J Ethnopharmacol 148(2):714–723

Makhuvele R, Naidu K, Gbashi S, Thipe VC, Adebo OA, Njobeh PB (2020) The use of plant extracts and their phytochemicals for control of toxigenic fungi and mycotoxins. Heliyon 6(10):e05291. https://www.cell.com/heliyon/pdf/S2405-8440(20)32134-4.pdf

Maradesha T, Patil SM, Phanindra B, Achar RR, Silina E, Stupin V, Ramu R (2022) Multiprotein inhibitory effect of dietary polyphenol rutin from whole green jackfruit flour targeting different stages of diabetes mellitus: defining a bio-computational stratagem. Separations 9(9):262

Militão GCG, Albuquerque M, Pessoa ODL, Pessoa C, Moraes MEA, De Moraes MO, Costa-Lotufo LV (2004) Cytotoxic activity of nepetin, a flavonoid from eupatorium ballotaefolium HBK. Pharmazie 59(12):965–966

Mohammed L, Jha G, Malasevskaia I, Goud HK, Hassan A (2021) The interplay between sugar and yeast infections: do diabetics have a greater predisposition to develop oral and vulvovaginal candidiasis? Cureus 13(2):e13407. https://www.cureus.com/articles/52092-the-interplay-between-sugar-and-yeast-infections-do-diabetics-have-a-greater-predisposition-to-develop-oral-and-vulvovaginal-candidiasis.pdf

Mondal S, Soumya NPP, Mini S, Sivan SK (2021) Bioactive compounds in functional food and their role as therapeutics. Bioact Comp Health Dis 4(3):24–39

Monteiro KL, de Aquino TM, Mendonça Junior FJB (2020) An update on Staphylococcus aureus NorA efflux pump inhibitors. Curr Top Med Chem 20(24):2168–2185

Munita JM, Arias CA (2016) Mechanisms of antibiotic resistance. In: Kudva IT, Cornick NA, Plummer PJ, Zhang Q, Nicholson TL, Bannantine JP, Bellaire BH (eds) Virulence mechanisms of bacterial pathogens. ASM Press, pp 481–511. https://doi.org/10.1128/9781555819286.ch17

Nicholls S, Leach MD, Priest CL, Brown AJP (2009) Role of the heat shock transcription factor, Hsf1, in a major fungal pathogen that is obligately associated with warm-blooded animals. Mol Microbiol 74(4):844–861. https://doi.org/10.1111/j.1365-2958.2009.06883.x

Nikaido H, Takatsuka Y (2009) Mechanisms of RND multidrug efflux pumps. Biochim Biophys Acta 1794(5):769–781

Onyeka IP, Aziakpono MO, Morikwe CU, Nwafor OH, Adione NM (2021) Physicochemical and anti-diabetic effect of the crude leaf extract of mallotus oppostifollous (Euphorbiaceae) in alloxan induced diabetic mice. J Bioanalys Biomed 13(8):279–286

Oommen S, Anto RJ, Srinivas G, Karunagaran D (2004) Allicin (from garlic) induces caspase-mediated apoptosis in cancer cells. Eur J Pharmacol 485(1–3):97–103

Pappas PG, Kauffman CA, Andes D, Benjamin DK Jr, Calandra TF, Edwards JE Jr, Filler SG, Fisher JF, Kullberg B-J, Ostrosky-Zeichner L (2009) Clinical practice guidelines for the management of candidiasis: 2009 update by the Infectious Diseases Society of America. Clin Infect Dis 48(5):503

Patel S, Pragati, Dwivedi SD, Yadav K, Kanwar JR, Singh MR, Singh D (2020) Pathogenesis and molecular targets in treatment of diabetic wounds. In: Faintuch J, Faintuch S (eds) Obesity and diabetes. Springer International Publishing, pp 747–758. https://doi.org/10.1007/978-3-030-53370-0_55

Pathakumari B, Liang G, Liu W (2020) Immune defence to invasive fungal infections: A comprehensive review. Biomed Pharmacother 130:110550

Podlesek Z, Žgur Bertok D (2020) The DNA damage inducible SOS response is a key player in the generation of bacterial persister cells and population wide tolerance. Front Microbiol 11:1785

Popolo L, Gualtieri T, Ragni E (2001) The yeast cell-wall salvage pathway. Sabouraudia 39(1):111–121

Prestinaci F, Pezzotti P, Pantosti A (2015) Antimicrobial resistance: a global multifaceted phenomenon. Pathog Glob Health 109(7):309–318. https://doi.org/10.1179/2047773215Y.0000000030

Rajendra Prasad N, Karthikeyan A, Karthikeyan S, Venkata Reddy B (2011) Inhibitory effect of caffeic acid on cancer cell proliferation by oxidative mechanism in human HT-1080 fibrosarcoma cell line. Mol Cell Biochem 349:11–19

Ramalingam M, Kim H, Lee Y, Lee Y-I (2018) Phytochemical and pharmacological role of liquiritigenin and isoliquiritigenin from radix glycyrrhizae in human health and disease models. Front Aging Neurosci 10:348

Rasul A, Millimouno FM, Ali Eltayb W, Ali M, Li J, Li X (2013) Pinocembrin: a novel natural compound with versatile pharmacological and biological activities. BioMed Res Int 2013:379850. https://www.hindawi.com/journals/bmri/2013/379850/abs/

Ratnakar P, Suryanarayana Murthy P (1996) Preliminary studies on the antitubercular activity and the mechanism of action of the water extract of garlic (Allium sativum) and its two partially purified proteins (Garlic defensins?). Indian J Clin Biochem 11(1):37–41. https://doi.org/10.1007/BF02868409

Ray S, Das S, Suar M (2017) Molecular mechanism of drug resistance. In: Arora G, Sajid A, Kalia VC (eds) Drug resistance in bacteria, fungi, malaria, and cancer. Springer International Publishing, pp 47–110. https://doi.org/10.1007/978-3-319-48683-3_3

Reygaert WC (2018) An overview of the antimicrobial resistance mechanisms of bacteria. AIMS Microbiol 4(3):482

Salerno C, Pascale M, Contaldo M, Esposito V, Busciolano, M., Milillo L, Guida A, Petruzzi M, Serpico R (2011) Candida-associated denture stomatitis. https://roderic.uv.es/handle/10550/60147

Sanglard D, Coste A, Ferrari S (2009) Antifungal drug resistance mechanisms in fungal pathogens from the perspective of transcriptional gene regulation. FEMS Yeast Res 9(7):1029–1050

Saud B, Bajgain P, Paudel G, Shrestha V, Bajracharya D, Adhikari S, Dhungana G, Awasthi MS (2020) Fungal infection among diabetic and nondiabetic individuals in Nepal. Interdiscip Perspect Infect Dis 2020:7949868. https://www.hindawi.com/journals/ipid/2020/7949868/

Schatz A, Bugle E, Waksman SA (1944) Streptomycin, a substance exhibiting antibiotic activity against gram-positive and gram-negative bacteria.*. Exp Biol Med 55(1):66–69. https://doi.org/10.3181/00379727-55-14461

Schwartz L, de Dios Ruiz-Rosado J, Stonebrook E, Becknell B, Spencer JD (2023) Uropathogen and host responses in pyelonephritis. Nat Rev Nephrol 19(10):658–671

Sha W, Hu F, Xi Y, Chu Y, Bu S (2021) Mechanism of ferroptosis and its role in type 2 diabetes mellitus. J Diabetes Res 2021:9999612. https://www.hindawi.com/journals/jdr/2021/9999612/

Shanmugam KR, Shanmugam B, Subbaiah GV, Ravi S, Reddy KS (2021) Medicinal plants and bioactive compounds for diabetes management: important advances in drug discovery. Curr Pharm Des 27(6):763–774

Sharma A, Gupta VK, Pathania R (2019) Efflux pump inhibitors for bacterial pathogens: from bench to bedside. Indian J Med Res 149(2):129

Shen Z, Tang CM, Liu G-Y (2022) Towards a better understanding of antimicrobial resistance dissemination: what can be learnt from studying model conjugative plasmids? Mil Med Res 9(1):3. https://doi.org/10.1186/s40779-021-00362-z

Song TT, Hendrich S, Murphy PA (1999) Estrogenic activity of Glycitein, a soy Isoflavone. J Agric Food Chem 47(4):1607–1610. https://doi.org/10.1021/jf981054j

Sullivan MJ, Ulett GC (2020) Evaluation of hematogenous spread and ascending infection in the pathogenesis of acute pyelonephritis due to group B streptococcus in mice. Microb Pathog 138:103796

Sun J, Deng Z, Yan A (2014) Bacterial multidrug efflux pumps: mechanisms, physiology and pharmacological exploitations. Biochem Biophys Res Commun 453(2):254–267

Sun L, Liao K, Wang D (2015) Effects of magnolol and honokiol on adhesion, yeast-hyphal transition, and formation of biofilm by Candida albicans. PLoS One 10(2):e0117695

Tan SY, Tatsumura Y (2015) Alexander Fleming (1881–1955): discoverer of penicillin. Singapore Med J 56(7):366

Thomas CF Jr, Limper AH (2007) Current insights into the biology and pathogenesis of pneumocystis pneumonia. Nat Rev Microbiol 5(4):298–308

Tiwari P, Khare T, Shriram V, Bae H, Kumar V (2021) Plant synthetic biology for producing potent phyto-antimicrobials to combat antimicrobial resistance. Biotechnol Adv 48:107729

Valero-Pacheco N, Blight J, Aldapa-Vega G, Kemlo P, Pérez-Toledo M, Wong-Baeza I, Kurioka A, Perez-Shibayama C, Gil-Cruz C, Sánchez-Torres LE (2020) Conservation of the OmpC porin among typhoidal and non-typhoidal Salmonella serovars. Front Immunol 10:2966

Veen M, Stahl U, Lang C (2003) Combined overexpression of genes of the ergosterol biosynthetic pathway leads to accumulation of sterols in Saccharomyces cerevisiae. FEMS Yeast Res 4(1):87–95

Vermes A, Guchelaar H-J, Dankert J (2000) Flucytosine: a review of its pharmacology, clinical indications, pharmacokinetics, toxicity and drug interactions. J Antimicrob Chemother 46(2):171–179

Vijayalakshmi S, Mohankumar A (2023) Bioactive ethnomedicinal plant extracts for the management of urinary tract infection in pregnancy women challenges in pandemic period. J Drug Deliv Therap 13(3):30–36

Wang Y, Zhang X, Ma X, Zhang K, Li S, Wang X, Liu X, Liu J, Fan W, Li Y (2019) Study on the kinetic model, thermodynamic and physicochemical properties of Glycyrrhiza polysaccharide by ultrasonic assisted extraction. Ultrason Sonochem 51:249–257

White TC, Marr KA, Bowden RA (1998) Clinical, cellular, and molecular factors that contribute to antifungal drug resistance. Clin Microbiol Rev 11(2):382–402. https://doi.org/10.1128/CMR.11.2.382

Wilson K, Kumar M, Sharma P, Nehra H, Verma I (2022) Herbal bioactives for treating urinary tract infections. In: Singh I, Sindhu RK, Shirkhedkar A, Panichayupakaranant P (eds) Herbal drugs for the management of infectious diseases, 1st edn. Wiley, pp 427–442. https://doi.org/10.1002/9781119818779.ch15

World Health Organization (2013) Global tuberculosis report 2013. World Health Organization https://books.google.com/books?hl=en&lr=&id=1rQXDAAAQBAJ&oi=fnd&pg=PP1&dq=3.+Organization.,+World+Health.+Global+Tuberculosis+Report+2016.+Geneva+:+s.n.,+2016&ots=laYdVn2r_-&sig=9AbSh32IZlQjKIB_rsgCQk6RCZM

Yadav K, Singh D, Singh MR (2021) Novel archetype in psoriasis management bridging molecular dynamics in exploring novel therapies. Eur J Pharmacol 907:174254

Yadav H, Mahalvar A, Pradhan M, Yadav K, Sahu KK, Yadav R (2023) Exploring the potential of phytochemicals and nanomaterial: a boon to antimicrobial treatment. Med Drug Discov 17:100151

Yang Y, Chen Z, Zhao X, Xie H, Du L, Gao H, Xie C (2022) Mechanisms of Kaempferol in the treatment of diabetes: A comprehensive and latest review. Front Endocrinol 13:990299

Yano T, Takada T, Fujiishi R, Fujii K, Honjo H, Miyajima M, Takeshima T, Hayashi M, Miyashita J, Azuma T, Fukuhara S (2022) Usefulness of computed tomography in the diagnosis of acute pyelonephritis in older patients suspected of infection with unknown focus. Acta Radiol 63(2):268–277. https://doi.org/10.1177/0284185120988817

YosHIDA H, Teramoto T, Ikeda K, YAMORI Y (2001) Glycitein effect on suppressing the proliferation and stimulating the differentiation of osteoblastic MC3T3-E1 cells. Biosci Biotechnol Biochem 65(5):1211–1213

Youssef FS, Ashour ML, Singab ANB, Wink M (2019) A comprehensive review of bioactive peptides from marine fungi and their biological significance. Mar Drugs 17(10):559

Zhang D, Lu Y, Liu K, Liu B, Wang J, Zhang G, Zhang H, Liu Y, Wang B, Zheng M, Fu L, Hou Y, Gong N, Lv Y, Li C, Cooper CB, Upton AM, Yin D, Ma Z, Huang H (2012) Identification of less lipophilic riminophenazine derivatives for the treatment of drug-resistant tuberculosis. J Med Chem 55(19):8409–8417. https://doi.org/10.1021/jm300828h

Natural Antimicrobial Compounds Are an Alternative to Synthetic Antibiotics

31

Varsha Rini Guria, Pavan Kumar Padarthi, P. Gayatri Devi, Vishal Chauhan, and K. S. Sridevi Sangeetha

Abstract

Natural antimicrobial substances are compounds having a natural origin, such as plants, microorganisms or even insects. These organisms use these types of compounds to defend themselves from pathogens and parasites or to compete with other organisms in their ecological niche. The use of natural antimicrobial compounds has been known for centuries to treat many infectious diseases. Here are a few examples of broad categories of these compounds: phenolic compounds, alkaloids, essential oils, enzymes, peptides, fatty acids, polyphenols and saponins. Antibiotic resistance has aroused the need to use antimicrobial solutions for the treatment of various diseases. It is a matter of global concern because there are a limited number of antibiotics at present that are suitable to consume and that aid in fighting various types of infectious diseases. These naturally obtained antimicrobial compounds are very sustainable and eco-friendly as compared to those antibiotics, which are synthetic. These antimicrobial compounds also have

V. R. Guria (✉)
Special Centre for Molecular Medicine, Jawaharlal Nehru University, New Delhi, India

KIIT School of Biotechnology, KIIT University, Bhubaneshwar, Odisha, India

P. K. Padarthi
Chebrolu Hanumaiah Institute of Pharmaceutical Sciences, Chowdavaram, Guntur, Andhra Pradesh, India

P. Gayatri Devi
NRI College of Pharmacy, Vijayawada, Andhra Pradesh, India

V. Chauhan
MAHSA University, Jenjarom, Selangor, Malaysia

K. S. S. Sangeetha
Faculty of Allied Health Sciences, Meenakshi Academy of Higher Education and Research, Chennai, Tamil Nadu, India

M. Gangwar, G. Nath (eds.), *Emerging Paradigms for Antibiotic-Resistant Infections: Beyond the Pill*, https://doi.org/10.1007/978-981-97-5272-0_31

diverse modes of action. Some may disrupt the cell membrane and its synthesis, some may inhibit protein and nucleic acid synthesis, and some may disrupt the metabolic pathways. As these compounds are present naturally, they exhibit very few side effects. Additionally, these compounds also aid in providing additional health benefits.

Keywords
Natural antimicrobial compounds · Synthetic antibiotics · Antimicrobial resistance · Bioactive compounds · Plant extracts

31.1 Introduction

31.1.1 Natural Antimicrobial Compounds and Importance in Therapeutics

Natural antibiotic compounds are chemical substances derived from plants, microorganisms, animals and insects. These chemicals can kill the pathogens or have the ability to inhibit their growth. The first antibiotic to be discovered was penicillin by Alexander Fleming, which was isolated from *Penicilliumnotatum*. He was also worried about the emergence of penicillin-resistant bacteria, and his prediction came true. (Tan and Tatsumura 2015) The emerging case of multidrug resistance against antibiotics is a major concern. In developing countries, the mortality of critical care unit patients is higher because of secondary hospital-acquired infections than primary infections. Multidrug resistance plays a major role in it. Hence, there is a need for the distribution of new antibiotics. However, this is not easy because the antibiotics should maintain high efficacy and low toxicity. The FDA will never approve a medication that is very effective at eradicating pathogens but is only mildly toxic. This increases the difficulty of clinical trials of new antibiotic discoveries. Natural products are a good source of antibiotics. Antibiotics derived from natural compounds like sulfa drugs, trimethoprim, u-oroquinolones and oxazolidinones are still in use. These products are the result of natural selection, and they can penetrate the microorganisms, bind effectively to the target, work efficiently, and show less toxicity. There are various categories of natural microbial compounds, which include phenolics, terpenoids, polyphenols, antimicrobial plants, polyenes, and cyclic peptides. All these compounds are derived from different sources. This chapter covers phytochemicals and their antimicrobial properties.

Phytochemicals are plant-derived secondary metabolites. They are bioactive compounds, and plants use them as defence mechanisms against various pathogens. They have been used traditionally for a long time. Many of the currently used antibiotics have been derived from plants. For example, In Peru, the bark of the Cinchona tree (*Cinchona officinalis*) was used for the treatment of malaria.

Later, studies confirmed that an alkaloid compound quinine present in the Cinchona can be used to treat malaria. Synthetic forms of quinine, like chloroquine and hydroxychloroquine, are now used for the treatment of malaria.

Phytochemicals can be categorized into various primary groups based on their chemical structures, sources, biosynthesis pathways or characteristics. The primary categorization of phytochemicals is done on chemical structures such as phenolics, alkaloids, saponins, terpenoids, limonoids, polyacetylenes, secoiridoids and others (Table 31.1).

31.1.2 *O. vulgare* as an Antimicrobial Compound

O. vulgare, commonly known as 'oregano', is a native plant of the Western and Southwestern Eurasia regions and the Mediterranean region (Kirschner 1981). It is an aromatic medicinal plant used for the treatment of rheumatoid arthritis, stomachaches, respiratory disorders, urinary problems and menstrual pain. It has remarkable antimicrobial activity, even against multidrug-resistant bacteria and fungi (Soltani et al. 2021). There are two types of phytochemicals present in *O. vulgare*: essential oil and phenolic compounds. The phenolic compounds include flavonoids and phenolic acids. *O. vulgare* also has compounds like sterols, tannins and terpenoids (Morshedloo et al. 2017). The essential oil is volatile in nature and has monocyclic monoterpenoids, acyclic monoterpenes, bicyclic monoterpenes and sesquiterpenoids. Rosmarinicacid, which is non-volatile in nature, is the major phenolic acid present in *O. vulgare*. Free flavonoids like flavones, flavonolsanddihydroflanonols are also present, along with flavonoid glycosides (Soltani et al. 2021).

31.1.2.1 Antibacterial Effects of *O. vulgare*

There are three mechanisms by which *O. vulgare* essential oil acts as an antibacterial compound. First, this essential oil inhibits bacterial enzymes. Second, it can inhibit the efflux pump. And third, it acts as an antibiofilm agent (Soltani et al. 2021). The essential oil of *O. vulgare* can inhibit the lipase and coagulase enzymes

Table 31.1 There is a variety of natural antimicrobial compounds. They have various sources of origin

Source	Category	Examples
Plant-derived	Essential oils	Tea tree oil, oregano oil, thyme oil
	Phenolic compounds	Curcumin, thymol, eugenol
	Flavonoids	Epigallocatechin gallate, quercetin
	Tannins	Ellagitannins
	Alkaloids	Berberin, Caffeine
Microbial-derived	Bacteriocins	Nisin, colicins
	Antifungal peptides	Caspofungin, echinocandins
	Lysostaphin	Enzyme from *Staphylococcus simulans*
Animal-derived	Defensins	Human β-defensin 1, cathelicidins
	Lysozyme	Found in tears, saliva, egg whites
	Lactoperoxidase system's Polyphenols	Hypothiocyanite, hypothiocyanous acid

of the bacteria. Bacteria use lipase to break down fats present in the host cell membrane into glycerol and fatty acids. This helps them invade the host tissues. Whereas coagulase causes the blood of the host to clot by converting fibrinogen to fibrin. This mechanism helps the bacteria form a protective clot around them and escape the host immune response.

A drug-resistant bacterium develops many mechanisms to survive in the presence of antibiotics. One such mechanism is to pump out the antibiotics with the help of efflux pumps. *O. vulgare* essential oil can inhibit these pumps and cause the death of bacteria in the presence of antibiotics (Langeveld et al. 2014). Antibiotics like ciprofloxacin and ethidium bromide, in synergy with this essential oil, are effective against fluoroquinolone-resistant *Streptococcus pneumoniae*. PmrA efflux pump gene, which makes *S. pneumonia* fluroquinolone resistant, is inhibited by this essential oil (Ghafari et al. 2018). Thymol, which is present in the essential oil of *O. vulgare*, and tetracycline work together to inhibit the TetKefflux protein of *S. aureus*.

This essential oil can inhibit the biofilm production of *S. pyogenes* with a minimum inhibitory concentration value of 0.25–0.5 mg/mL (Wijesundara and Rupasinghe 2018). This essential oil, along with thymol and carvacrol, decreases the production of fimbriae in *E. coli* at a sub-inhibitory concentration of less than 0.01%. *E. coli*'s hemagglutinating activity decreases as a result, and the human whole blood cells then kill it (Fig. 31.1).

31.1.2.2 Antifungal Effects of *O. vulgare* Essential Oil

The essential oil of *O. vulgare* is effective against fugal pathogens like *Aspergillus*, *Candida*, *Dermatophyte*, and *Cryptococcus*, with a minimum inhibitory concentration value of 16–2.5 μL/mL (Vale-Silva et al. 2012). This essential oil attacks the cell wall of these fungi and causes retraction of the mycelium and the death of the hyphae. This essential oil also inhibits the production of the phospholipase enzyme in *Candida*. The phospholipase enzyme is involved in membrane remodelling, signalling, pathogenesis and nutrient acquisition in fungi. This essential oil also inhibits cell wall synthesis and affects morphogenesis and fungal growth. (Souza et al. 2010).

31.1.3 Antimicrobial Properties of Garlic Extracts

Allium sativum, commonly known as garlic, is an edible plant. Its medicinal use has been known for ages. It shows antimicrobial properties against various gram-positive and gram-negative strains of bacteria, including some multidrug-resistant enterotoxigenic strains of *E. coli*. It also inhibits the growth of fungi, protozoa and viruses. It also has other medicinal properties. It acts as an anti-cancer compound and can also reduce blood lipid levels. Allicin is the main component in garlic, which imparts antimicrobial properties. It is an oxygenated sulphur compound that is poorly soluble in water and is volatile in nature. The enzyme alliinase converts alliin into the active form of allicin (Fig. 31.2) (Ankri and Mirelman 1999). Two molecules of allyl sulfenic acid condense spontaneously to one molecule of allicin.

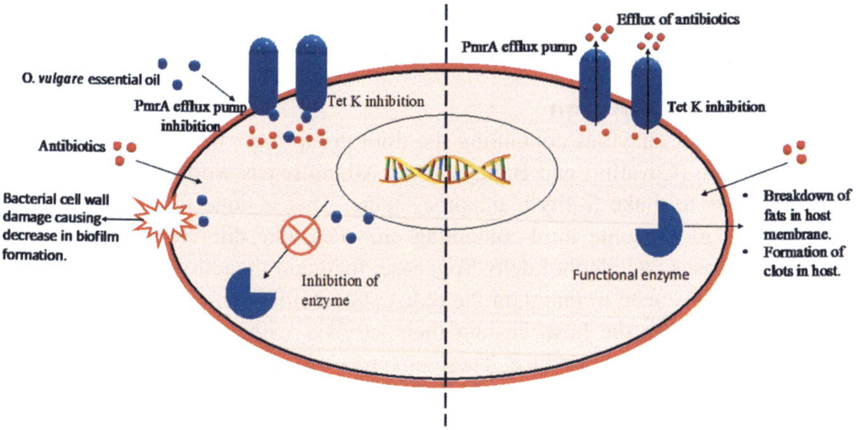

Fig. 31.1 The essential oil of O. vulgare has three modes of action against the bacterial. (1) It inhibits the antibiotics efflux pumps and antibiotics retain inside the bacteria which causes its death. (2) It inhibits the bacterial enzymes like lipase and coagulase. This inhibits the bacteria from invading host's tissues. (3) It destroys the cell wall and cell membrane and cause leakage of the cell content. Hence the bacteria fail to form biofilms

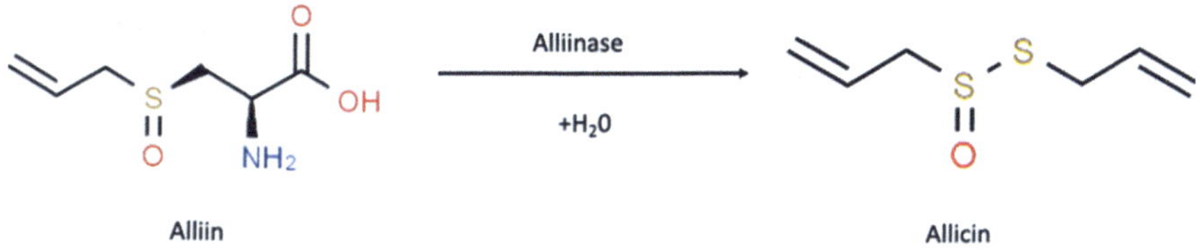

Fig. 31.2 Hydrolysed reaction of Alliin for the production of dehydroalanine and allyl sulfenic acid

Crushed garlic extracts show a broad spectral range of antibacterial activities for bacteria like *Bacillus*, *Clostridium*, *Escherichia*, *Klebsiella*, *Proteus*, *Staphylococcus*, *Salmonella* and *Streptococcus*. It also inhibits *Helicobacter pylori* and *Mycobacterium tuberculosis*, which causes gastric ulcers and tuberculosis, respectively (Uchida et al. 1975)(Cellini et al. 1996). It also inhibits methicillin-resistant *Staphylococcus aureus* and other multidrug-resistant bacteria like *S. sonnie*, *E. coli*, *S. flexneri*, *Shigelladysenteriae* and *Enterococcus*.

Allicin also acts as an antifungal compound against *Aspergillus parasiticus* by inhibiting mycotoxins like aflatoxin (Lawson 1996). *Cryptococcus neoformans* is inhibited by the concentrated garlic extracts, with a minimum inhibition concentration value of 6–12 μg/mL. *Candida* isolates are also sensitive to pure allicin with a minimum inhibitory concentration of 7 μg/mL (Hughes and Lawson 1991).

Allicin also acts as an antiparasitic compound against *Entamoebahistolytica*, a causative agent of dysentery, with a minimum inhibitory concentration of 30 μg/mL (Mirelman et al. 1987). Allicin at a concentration of 30 μg/mL also inhibits the growth of other protozoans such as *Leishmania major*, *Crithidiafasciculata*, *Giardia lamblia*, and *Leptomonascolosoma*.

31.1.3.1 Mode of Action

Several microbial enzymes containing the thiol group react with the thiosulfinate group of allicin (Cavallito and Bailey 1944). Allicin reacts with the L-cysteine in these enzymes to make S-allylmercaptocysteine. This is done through a process called S-thiolation. Some thiol-containing enzymes, like thioredoxin reductases, cysteine proteases and alcohol dehydrogenase, maintain the redox state within the parasite. Parasites need to maintain the redox state within them to adapt to diverse environments within the host, sustain their survival within the host, evade host immune responses and reproduce. These enzymes are sensitive to allicin at a concentration lower than 10 μg/mL. It has also been observed that mammalian cells are also sensitive to allicin at concentrations higher than 100 μg/mL (Ankri et al. 1997). Microorganisms are sensitive to allicin at a very low concentration as compared to mammalian cells.

Allicin also inhibits enzymes like acetate kinase and phosphotransacetyl-coA synthase noncovalently and reversibly. (Focke et al. 1990) These enzymes play a critical role in the formation of acetyl-coA. Acetyl-coA is used in cellular energy metabolism and is a precursor for the synthesis of fatty acids, cholestrols, keton bodies and certain amino acids. Hence, their production can inhibit the growth and reproduction of the microorganisms.

31.1.4 Antimicrobial Properties of Essential Oil of Tea Tree

Tea tree oil (TTO) is a volatile essential oil that is extracted from the plant *Melaleuca alternifolia*. This plant is a native of Australia, North America, and Europe. Tea tree oil has antimicrobial properties and has been used traditionally in Australia for 100 years. It is recently gaining popularity for the treatment of cutaneous infections. Terpene hydrocarbons, comprising monoterpenes and sesquiterpenes, along with some associated alcohols, are the main components of this essential oil. Terpenes with a molecular formula of C5H8 are considered polymers of isoprene. It has aromatic hydrocarbons and is volatile in nature. It is sparingly soluble in water and is miscible in nonpolar solutions. Six different varieties of *Melaleuca alternifolia* plants produce oil with different compositions. These varieties produce four types of 1,8-cineole, one type of terpinen-4-ol, and terpinolene (Carson et al. 2006). 30–40% of terpinen-4-ol produced by plants is used for the commercial production of TTO. The bioactivity of tea TTO increases when globulol, sabinene and viridiflorol are included along with the major constituents of TTO.

The antimicrobial activity is primarily shown by terpinen-4-ol. Studies of the antimicrobial properties of TTO were published in the 1920s and 1930s by Penfolds (Penfold and Grant 1926). In these reports, a comparative study of the disinfectant properties of tea tree oil was done with phenol by measuring the Rideal-Walker coefficient. It was reported that tea tree oil showed 11 times more disinfectant activity than that of phenol. The Rideal-Walker coefficients of other components of tea tree oil were also studied. The Rideal-Walker coefficients of cineole and cymene were found to be 3.5 and 8, respectively; for linalool, it was 13, and for terpinen-4-ol and terpineol, it was 13.5 and 16, respectively (Penfold and Grant 1923, 1926). Tea tree oil shows a broad spectrum of antimicrobial activity against viruses, bacteria, protozoa, and fungi. Since the constituents of TTO are sparingly soluble in water, they were incorporated in agar test media and broth to study the minimum inhibitory concentration value.

31.1.4.1 Tea Tree Oil Is an Antibacterial Compound, and Its Mode of Action

In a few studies, it was seen that a broad range of bacteria were susceptible to tea tree oil with a concentration of 1% or less. The minimum inhibitory concentration value for microorganisms like *Pseudomonas aeruginosa*, *Staphylococcus*, *Enterococcus faecalis* and *Micrococci* was not more than 2% (Banes-Marshall et al. 2001; Hammer et al. 1996). For 64 methicillin-resistant *Staphylococcus aureus* isolates from Australia and 33 mupirocin-resistant strains, the minimum inhibitory concentration value was 0.25% and the minimum bactericidal concentration value was 0.5%. For isolates from the UK, they were 0.312% and 0.625%, respectively. In addition to this, the vapours of tea TTO can restrict the growth of *Escherichia coli*, *Streptococcus pneumoniae*, *Streptococcus pyogenes*, *Haemophilus influenzae* and *Mycobacterium avium* ATCC4676 (Maruzzella and Sicurella 1960).

The mode of action of TTO is dependent on the hydrocarbon structure of the oil and the lipophilicity of the microorganisms. These hydrocarbons are partitioned preferentially in the cell membrane of the microorganisms because they have an affinity for the lipid bilayers. This can affect the membrane properties, like permeability and membrane fluidity, of the microorganisms (Sikkema et al. 1995). The lysis of the lipid membrane and the loss of its integrity cause the leakage of ions and disrupt respiration. Also, tea tree oil increases the sensitivity of *S. aureus* to sodium chloride, which changes the morphology of the cells. Whole-cell lysis because of ion imbalance is also observed in *E. coli* when treated with TTO (Gustafson et al. 1998).

31.1.4.2 Antifungal Properties of the Essential Oil of Tea Tree and Its Mode of Action

TTO inhibits a wide variety of fungi, with a minimum inhibitory concentration value ranging from 0.03% to 0.5%. Tea tree oil also acts as a fungicide, with concentrations ranging from 0.12% to 2%, except for *Aspergillus niger*, which requires a concentration of 8%. The mode of action against fungus is similar to that of bacteria; that is, tea tree oil alters the permeability of the cell membrane. The treatment also increases membrane fluidity significantly (Hammer et al. 2004). Depending on the dose, tea tree oil can also inhibit respiration and glucose-induced medium acidification, which is crucial for fungal growth. Medium acidification occurs by the efflux of protons from the plasma membrane or mitochondrial membrane using ATP as a source of energy. Since tea tree oil inhibits respiration, it also inhibits ATP production, which leads to the death of fungi (Cotmore et al. 1979). Tea tree oil also changes how fungi reproduce and grow by stopping the creation of germ tubes and mycelial conversion (D'Auria et al. 2001)(Hammer et al. 2000). It does this by messing up the integrity of the cell wall, cell wall synthesis, and cellular processes.

31.1.5 Curcumin as an Natural Antimicrobial Compound

Curcumin, or the diferuloyl methane molecule (1E,6E)-1,7-bis (4-hydroxy-3-methoxyphenyl)-1,6-heptadiene-3,5-dion is a polyphenolic compound and is one of the main phytochemicals of *Curcumin longa*, commonly known as turmeric. This plant belongs to the family *Zingiberaceae*. Curcumin and other extracts of the *Curcumin longa* rhizome show a broad spectrum of antimicrobial activity that prevents the growth of bacteria, fungi, parasites and viruses. In Asia, curcumin is traditionally used as a food colouring agent. It also has wound-healing properties and acts as an insect repellent.

31.1.5.1 Antibacterial Activity of Curcumin

In a study, the antibacterial activity of *C. longa* rhizome's aqueous extract was shown. The study found that the minimum inhibitory concentration value was around 4–16 g/L and the minimum bactericidal concentration value was around 16–32 g/L against the bacterial strains *Staphylococcus aureus* ATCC 25923, *Staphylococcus epidermis* ATCC12228, *E. coli* ATCC25922 and *Klebsiella pneumoniae* ATCC10031(Fig. 31.3) (Niamsa and Chaiyasit 2009). When the extraction of *C. longa* rhizome was done using methanol, the minimum inhibitory concentration value was found to be 16 g/L for *Bacillus subtilis* and 128 g/L for *Staphylococcus aureus*. Further, in a study where around 24 bacteria were isolated from chickens and shrimp, the minimum inhibitory concentration values of ethanol and hexane extracts were found to be 3.19–125 ppt and 125–1000 ppt against these pathogens (Lawhavinit et al. 2010). These extracts were able to inhibit the growth of 13 bacteria, which are *Staph. epidermis*, *Staph. intermedius*, *Vibrocholerae*, *V. harveyi*, *V.parahaemolyticus*, *V. alginolyticus*, *V. vulnificus*, *Edwardsiellatarda*, *Bacillus cereus*, *B.subtilis* and *Aeromonashydrophila*. It is also found that methicillin-resistant *Staph aureus* strains can also be inhibited by *curcumin,* with a minimum inhibitory concentration value of around 125–250 g/L. Curcumin suppresses the growth of *B. subtilis* by affecting the assembly of the FTsZprotofilament and its distability, which is crucial for the process of bacterial cytokinesis (Fig. 31.4). It inhibits the polymerization of FtsZ and, hence, cell division. It also suppresses the formation of cytokinetic Z ring formation by affecting the organization and segregation of nucleoids (Rai et al. 2008).

31.1.5.2 Antiviral Activity of Curcumin

There are very few antiviral compounds that are in use because of the lack of efficiency in viral diseases and the emergence of drug resistance. The new antiviral compounds may have allergic or toxic effects on patients. Hence, the

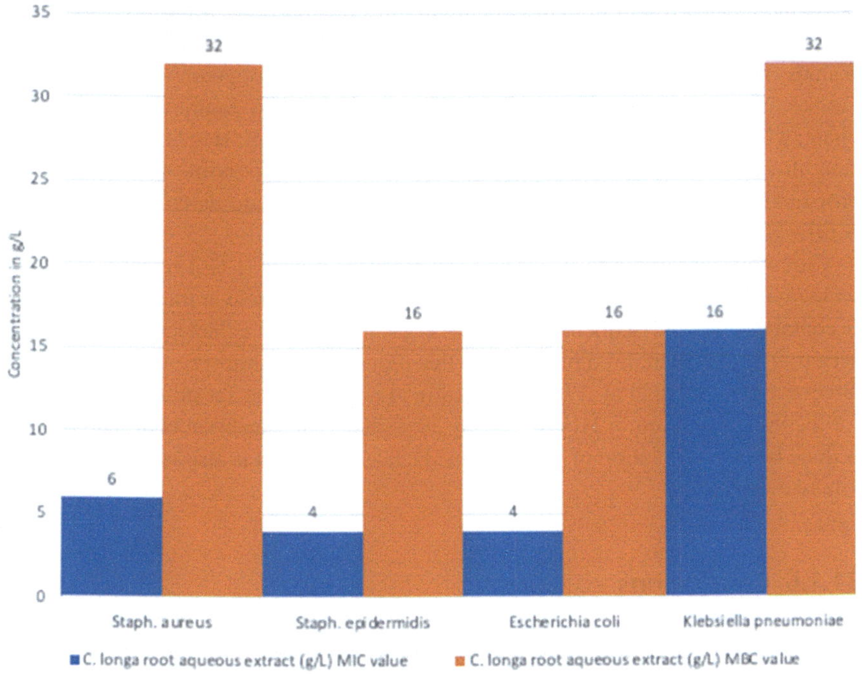

Fig. 31.3 Comparison of Minimum inhibitory concentration (MIC) and minimum bactericidal concentration (MIB) of *C. longa* aqueous extracts against different bacteria

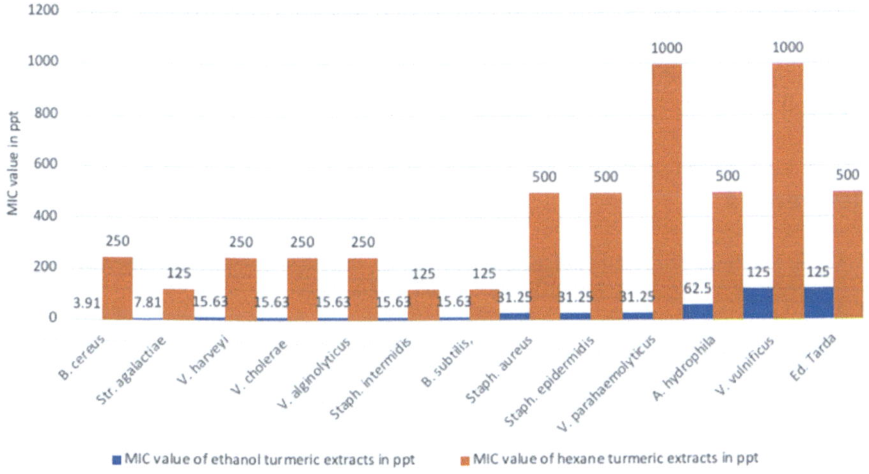

Fig. 31.4 Comparison of Minimum inhibitory concentration (MIC) and minimum bactericidal concentration (MIB) of *C. longa* methanol and hexane extracts against different bacteria

phytochemicals present in plants are studied for this purpose. It has been seen that inosine monophosphate dehydrogenase can be a potential target for antiviral compounds. It plays a crucial role in the de novo biosynthesis of guanine nucleotides in viruses. This enzyme converts inosine monophosphate to xanthosine monophosphate, which is a precursor for guanine nucleotide synthesis. Hence, inhibiting inosine monophosphate dehydrogenase activity can disrupt viral replication and propagation. Curcumin inhibits the inosine monophosphate dehydrogenase competitively and non-competitively.

Curcumin also inhibits the viral long terminal repeat (LTR) in *type 1 human immunodeficiency virus* (HIV-1). LTR plays an important role in the viral life cycle. It contains essential sequences that act as promoters and enhancers in viral transcription. It also contains a binding site that regulates viral transcription (Cullen and Greene 1989; Nabel et al. 1988). Curcumin also inhibits the Tat protein responsible for the transactivation of HIV-1 LTR by 70%–80%(Ungphaiboon et al. 2005). This is done by the acetylation of Tat proteins. Hence, curcumin is also a promising antiviral compound.

31.1.6 Conclusions

Phytochemicals, which are plant-derived secondary metabolites, show promising results against various microorganisms like bacteria, fungi, parasites and even viruses. They provide natural defense mechanisms for plants against pathogens. They all have various modes of action. And as they provide natural defense mechanisms to plants, they can easily penetrate the pathogen's cells. They also bind efficiently to the drug target and inhibit their growth. They are also less toxic to humans. Thus, their toxicity and efficiency are low. They work more efficiently in synergy with the present antibiotics available. Thus, there is a need to study them to know their full potential.

References

Ankri S, Mirelman D (1999) Antimicrobial properties of allicin from garlic. Microbes Infect 1(2):125–129. https://doi.org/10.1016/S1286-4579(99)80003-3

Ankri S, Miron T, Rabinkov A, Wilchek M, Mirelman D (1997) Allicin from garlic strongly inhibits cysteine proteinases and cytopathic effects of Entamoeba histolytica. Antimicrob Agents Chemother 41(10):2286–2288

Banes-Marshall L, Cawley P, Phillips CA (2001) In vitro activity of Melaleuca alternifolia (tea tree) oil against bacterial and Candida spp. isolates from clinical specimens. Br J Biomed Sci 58(3):139–145

Carson CF, Hammer KA, Riley TV (2006) Melaleuca alternifolia (tea tree) oil: a review of antimicrobial and other medicinal properties. Clin Microbiol Rev 19(1):50–62. https://doi.org/10.1128/CMR.19.1.50-62.2006

Cavallito CJ, Bailey JH (1944) Allicin, the antibacterial principle of Allium sativum. I. Isolation, physical properties and antibacterial action. J Am Chem Soc 66(11):1950–1951

Cellini L, Di Campli E, Masulli M, Di Bartolomeo S, Allocati N (1996) Inhibition of Helicobacter pylori by garlic extract (Allium sativum). FEMS Immunol Med Microbiol 13(4):273–277

Cotmore JM, Burke A, Lee NH, Shapiro IM (1979) Respiratory inhibition of isolated rat liver mitochondria by eugenol. Arch Oral Biol 24(8):565–568. https://doi.org/10.1016/0003-9969(79)90014-1

Cullen BR, Greene WC (1989) Regulatory pathways governing HIV-1 replication. Cell 58(3):423–426

D'Auria FD, Laino L, Strippoli V, Tecca M, Salvatore G, Battinelli L, Mazzanti G (2001) In vitro activity of tea tree oil against Candida albicans mycelial conversion and other pathogenic fungi. J Chemother (Florence, Italy) 13(4):377–383. https://doi.org/10.1179/joc.2001.13.4.377

Focke M, Feld A, Lichtenthaler HK (1990) Allicin, a naturally occurring antibiotic from garlic, specifically inhibits acetyl-CoA synthetase. FEBS Lett 261(1):106–108

Ghafari O, Sharifi A, Ahmadi A, Nayeri Fasaei B (2018) Antibacterial and anti-PmrA activity of plant essential oils against fluoroquinolone-resistant Streptococcus pneumoniae clinical isolates. Lett Appl Microbiol 67(6):564–569. https://doi.org/10.1111/lam.13050

Gustafson JE, Liew YC, Chew S, Markham J, Bell HC, Wyllie SG, Warmington JR (1998) Effects of tea tree oil on Escherichia coli. Lett Appl Microbiol 26(3):194–198. https://doi.org/10.1046/j.1472-765x.1998.00317.x

Hammer KA, Carson CF, Riley TV (1996) Susceptibility of transient and commensal skin flora to the essential oil of Melaleuca alternifolia (tea tree oil). Am J Infect Control 24(3):186–189. https://doi.org/10.1016/s0196-6553(96)90011-5

Hammer KA, Carson CF, Riley TV (2000) Melaleuca alternifolia (tea tree) oil inhibits germ tube formation by Candida albicans. Med Mycol 38(5):355–362

Hammer KA, Carson CF, Riley TV (2004) Antifungal effects of Melaleuca alternifolia (tea tree) oil and its components on Candida albicans, Candida glabrata and Saccharomyces cerevisiae. J Antimicrob Chemother 53(6):1081–1085. https://doi.org/10.1093/jac/dkh243

Hughes BG, Lawson LD (1991) Antimicrobial effects of Allium sativum L.(garlic), Allium ampeloprasum L.(elephant garlic), and Allium cepa L.(onion), garlic compounds and commercial garlic supplement products. Phytother Res 5(4):154–158

Kirschner J (1981) J. H. Ietswaart A taxonomic revision of the genus origanum (labiatae). Folia Geobotanica et Phytotaxonomica 16(4):390. https://doi.org/10.1007/BF02852419

Langeveld WT, Veldhuizen EJA, Burt SA (2014) Synergy between essential oil components and antibiotics: a review. Crit Rev Microbiol 40(1):76–94. https://doi.org/10.3109/1040841X.2013.763219

Lawhavinit O-A, Kongkathip N, Kongkathip B (2010) Antimicrobial activity of curcuminoids from Curcuma longa L. on pathogenic bacteria of shrimp and chicken. Kasetsart J Nat Sci 44

Lawson LD (1996) The composition and chemistry of garlic cloves and processed garlic. In: Koch HP, Lawson LD, editors. Garlic: the science and therapeutic application of Allium sativum L. and related species. Baltimore: Williams & Wilkins; pp. 37–108

Maruzzella JC, Sicurella NA (1960) Antibacterial activity of essential oil vapors. Received February 2, 1960, from the Department of Biology, Long Island University, Brooklyn, NY. J Am Pharm Assoc (Scientific Ed.) 49(11):692–694. https://doi.org/10.1002/jps.3030491103

Mirelman D, Monheit D, Varon S (1987) Inhibition of growth of Entamoeba histolytica by allicin, the active principle of garlic extract (Allium sativum). J Infect Dis 156:243–244

Morshedloo MR, Craker LE, Salami A, Nazeri V, Sang H, Maggi F (2017) Effect of prolonged water stress on essential oil content, compositions and gene expression patterns of mono- and sesquiterpene synthesis in two oregano (Origanum vulgare L.) subspecies. Plant Physiol Biochem 111:119–128. https://doi.org/10.1016/j.plaphy.2016.11.023

Nabel, G. J., Rice, S. A., Knipe, D. M., & Baltimore, D. (1988). Alternative mechanisms for activation of human immunodeficiency virus enhancer in T cells (Science (New York, NY), 239(4845), 1299–1302. doi: https://doi.org/10.1126/science.2830675

Niamsa N, Chaiyasit S (2009) Antimicrobial activity of Curcuma longa aqueous extract. J Pharmacol Toxicol 4:173–177. https://doi.org/10.3923/jpt.2009.173.177

Penfold ADRADRP, Grant JR (1923) The germicidal values of the principal commercial eucalyptus oils and their pure constituents, with observations on the value of concentrated disinfectants. J Proc R Soc NSW 57:80–89. https://doi.org/10.5962/p.359839

Penfold AR, Grant R (1926) The germicidal values of some Australian essential oils and their pure constituents: together with those for some essential oil isolates and synthetics (issue pt. 3). Government Printer, South Africa. https://books.google.co.in/books?id=lOU8NAAACAAJ

Rai D, Singh JK, Roy N, Panda D (2008) Curcumin inhibits FtsZ assembly: an attractive mechanism for its antibacterial activity. Biochem J 410(1):147–155. https://doi.org/10.1042/BJ20070891

Sikkema J, de Bont JA, Poolman B (1995) Mechanisms of membrane toxicity of hydrocarbons. Microbiol Rev 59(2):201–222. https://doi.org/10.1128/mr.59.2.201-222.1995

Soltani S, Shakeri A, Iranshahi M, Boozari M (2021) A review of the phytochemistry and antimicrobial properties of origanum vulgare l. and subspecies. Iran J Pharm Res 20(2):268–285. https://doi.org/10.22037/ijpr.2020.113874.14539

Souza NAB, de Oliveira Lima E, Guedes DN, de Oliveira Pereira F, de Souza EL, de Sousa FB (2010) Efficacy of Origanum essential oils for inhibition of potentially pathogenic fungi. Braz J Pharm Sci 46(3):499–508. https://doi.org/10.1590/s1984-82502010000300013

Tan SY, Tatsumura Y (2015) Alexander Fleming (1881-1955): discoverer of penicillin. Singapore Med J 56(7):366–367. https://doi.org/10.11622/smedj.2015105

Uchida Y, Takahashi T, Sato N (1975) The characteristics of the antibacterial activity of garlic (author's transl). Jpn J Antibiot 28(4):638–642

Ungphaiboon S, Tanomjit S, Pechnoi S, Supreedee S, Pranee R, Itharat A (2005) Study on antioxidant and antimicrobial activities of turmeric clear liquid soap for wound treatment of HIV patients. Songklanakarin J Sci Technol 27

Vale-Silva L, Silva M-J, Oliveira D, Gonçalves M-J, Cavaleiro C, Salgueiro L, Pinto E (2012) Correlation of the chemical composition of essential oils from Origanum vulgare subsp. virens with their in vitro activity against pathogenic yeasts and filamentous fungi. J Med Microbiol 61(Pt 2):252–260. https://doi.org/10.1099/jmm.0.036988-0

Wijesundara NM, Rupasinghe HPV (2018) Essential oils from Origanum vulgare and Salvia officinalis exhibit antibacterial and anti-biofilm activities against Streptococcus pyogenes. Microb Pathog 117:118–127. https://doi.org/10.1016/j.micpath.2018.02.026

Part VI

Bacteriophage-Derived Enzymes: Enlisting Nature's Allies

Phage Power: Harnessing Nature's Arsenal in the Battle Against Microbial Threats for Sustainable Healthcare

32

Jyotirmayee, Reshma Khanda, and Malkhey Verma

Abstract

In the fight against microbial threats, researchers are turning to natural antimicrobial solutions as a promising strategy to tackle the challenges facing microbial infections. Researchers are exploring new approaches to address the threats caused by microbial infections. Natural antimicrobials are compounds, derived from plants, fungi, and other organisms that exhibit potent antibacterial, antiviral, and antifungal activities. Plant-derived compounds such as essential oils, are particularly effective against a broad spectrum of pathogens. They can disrupt microbial membranes, impede essential enzymatic processes, and hinder viral replication, making them promising therapeutic agents in the medical field. One significant advantage of natural antimicrobials is their complex chemical profiles, which make it challenging for microbes to develop rapid resistance. This makes them a sustainable alternative to traditional antibiotics, especially in the face of escalating antibiotic resistance, which causes an expressive threat to global public health. Incorporating natural antimicrobial solutions into medical practices aligns with the imperative for sustainable and eco-conscious healthcare. The environmentally friendly extraction processes and reduced ecological footprint associated with these solutions present an attractive option for medical

Jyotirmayee and Reshma Khanda contributed equally and are joint first authors.

Jyotirmayee · M. Verma (✉)
School of Biotechnology, Institute of Science, Banaras Hindu University,
Varanasi, Uttar Pradesh, India
e-mail: malkhey@bhu.ac.in

R. Khanda
School of Life Sciences, Sambalpur University, Jyoti Vihar, Burla, Sambalpur, Odisha, India

M. Gangwar, G. Nath (eds.), *Emerging Paradigms for Antibiotic-Resistant Infections: Beyond the Pill*, https://doi.org/10.1007/978-981-97-5272-0_32

professionals striving to integrate green practices into patient care. Exploration and integration of natural antimicrobial solutions into medical protocols is a promising frontier in the fight against infectious diseases. The potential synergy between traditional pharmaceutical interventions and nature's weaponry may yield innovative and sustainable solutions to combat microbial infections, offering renewed hope for effective and environmentally conscious medical treatments.

Keywords

Bacteriophage enzymes · Endolysins · Biofilm · Horizontal gene transfer · Viral genome

32.1 Overview of Phage Enzymes

Bacteriophages, commonly referred to as phages, are viruses that infect and replicate within bacteria. They are estimated to outnumber bacteria by at least ten times and play an essential role in regulating bacterial populations in various environments, including soil, water, and the human body. Phage enzymes are enzymes produced by viruses called bacteriophages that infect and reproduce within bacteria (Sahu et al. 2023). These enzymes play a critical role in the viral life cycle by aiding in the attachment, penetration, replication, and release of phage particles from the bacterial host. In recent years, phage enzymes have gained significant attention due to their potential applications in various fields like biotechnology, medicine, and agriculture. They catalyse biochemical reactions that are crucial for phage replication, genome maintenance, and evasion of host defense mechanisms (Bhushan and Bhagyawant 2023). The history of phages is closely intertwined with that of bacteria, with both engaged in a constant arms race shaped by selective pressures and ecological dynamics. Phages have evolved strategies to overcome bacterial defaces, and their enzymatic warfare is crucial to their survival and proliferation. Phage enzymes target specific cellular components and metabolic pathways, modulating bacterial diversity and driving the flux of nutrients and energy within microbial communities (Burz et al. 2023). By exploring the interplay between phages, bacteria, and their enzymatic weaponry, we can understand the mechanisms that underpin microbial ecology and ecosystem functioning. The intricate dance of molecular interactions between phages and their bacterial hosts is mediated by various enzymes that catalyze viral replication, host lysis, and nucleic acid metabolism (Alexyuk et al. 2023). By deciphering the molecular choreography of phage–host interactions, we can illuminate the pivotal role of phage enzymes in orchestrating the infection process, hijacking host machinery, and ensuring the successful propagation of viral progeny. Through a molecular lens, we can understand the exquisite specificity and versatility of phage enzymes, shedding light on the fundamental principles governing virus–host interactions.

32.2 Enzymes Produced by Bacteriophages

Bacteriophages are divided into two categories, lytic and lysogenic phages, based on their morphology, nucleic acid content, and life cycles. In the lytic cycle, bacteriophages invade a bacterial cell, take control of its machinery to replicate, and then cause the cell to rupture, releasing new phages that are responsible for the destruction of the bacterial host. In the lysogenic cycle, phages integrate their genetic material into the bacterial chromosome, becoming prophages. Mathematical models by Sousa and Rocha 2019 suggest that phage particle diffusion is a significant factor in phage infectivity in structured environments, affecting population dynamics. Spatial structures, such as biofilms, can affect the direct interaction between phage and host. In a multispecies biofilm, phage-resistant *Escherichia coli* is protected from phage T7 by microenvironments, and the biofilm matrix changes and other environmental factors, such as temperature, UV light, hydrogen peroxide, salinity, nutrient availability, aeration, and pH, also impact phage infectivity and host range (Holtappels et al. 2023). Bacteriophages use a variety of enzymes to aid in their replication and propagation with host bacteria. In the initial phase, depolymerases help phages break down bacterial polysaccharides and biofilms, thus facilitating host attachment and penetration (Topka-Bielecka et al. 2021). Nucleases and DNA polymerases play a critical role in replicating the viral genome and synthesizing viral components after attachment and penetration. Nucleases assist in breaking down host DNA, freeing nucleotides for phage genome replication, while DNA polymerases catalyze the synthesis of new phage genome strands using host nucleotides as building blocks (Morcinek-Orłowska et al. 2022; Salas et al. 2016). Proteases play a crucial role in cleaving viral precursor proteins into functional components during the assembly phase, while lysozymes aid in cell lysis and virion release by breaking down the host cell wall (Sharma and Gupta 2017). The host bacterium continues to grow and reproduce without being immediately destroyed. Under certain circumstances, the prophage may switch to the lytic cycle, leading to the production of new phages and cell lysis.

32.3 Types of Bacteriophage-Derived Enzymes

Bacteriophages are viruses that infect bacteria. They contain various enzymes that are important for their lifecycle and interactions with bacterial hosts. These enzymes are diverse in their structure, function, and mode of action. Some enzymes are involved in viral replication and nucleic acid metabolism, while others mediate host cell lysis and immune evasion. These phage-derived enzymes have multifaceted implications across microbiology, biotechnology, and medicine. In this comprehensive overview, we aim to discuss the different types of bacteriophage-derived enzymes, their molecular characteristics, biological functions, and applications in various fields (Fig. 32.1).

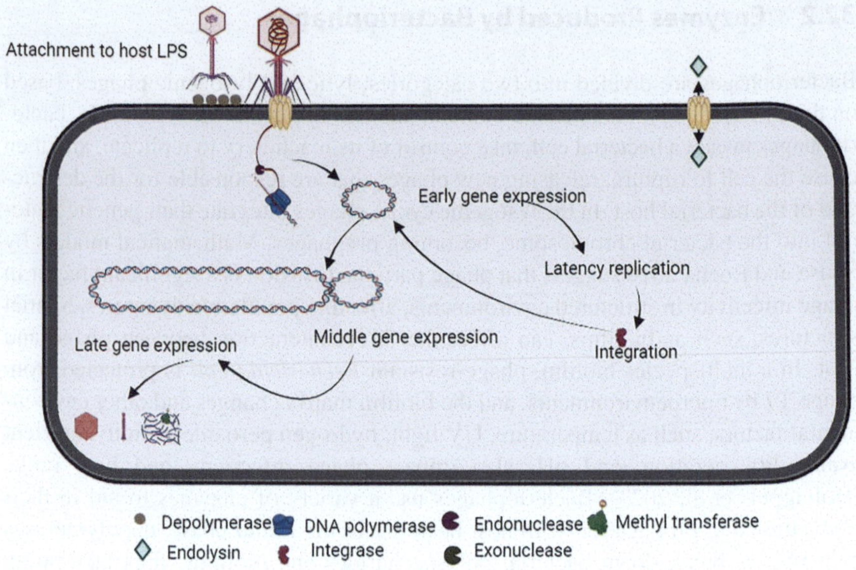

Fig. 32.1 Production of bacteriophage-derived enzymes in phage life cycle

32.3.1 DNA Polymerases and Replication

Phage DNA replication is a crucial area of study in virology and molecular biology, as these enzymes play a pivotal role in the intricate dance between phages and their hosts. They catalyze the synthesis of new DNA strands using existing templates, with some resembling the host while others resemble their structural activity specific to viral replication (Zhou et al. 2024; Tan et al. 2024). Phase DNA initiates DNA synthesis by binding to specific DNA sequences, known as origins of replication (ori), in the viral genome. Initiation factors and accessory proteins may facilitate the assembly of the replication complex and promote the loading of the polymerase onto the DNA template. The DNA polymerase enzyme begins adding various nucleotides complementary to the template strand, synthesizing a new DNA strand in the 5′ to 3′ direction by forming phosphodiester bonds between adjacent nucleotides (de Jong and Oosting 2024; Fenstermaker et al. 2024). Many of these enzymes possess proofreading activity, allowing them to detect and correct errors in DNA synthesis, ensuring high-fidelity replication of the viral genome. Understanding the structural and functional relationships of these enzymes provides valuable insights into molecular mechanisms underlying replication and polymerization. One of the best-studied examples is the T7 DNA polymerase, encoded by the T7 bacteriophage. It belongs to the family B DNA polymerases and features a modular architecture consisting of finger, palm, and thumb domains (Juarez-Quintero et al. 2021). A series of Bst DNA polymerase mutants have been developed with enhanced catalytic efficiency, enabling real-time LAMP at high temperatures (Li et al. 2023). Metabolic changes, such as metabolic reprogramming, downregulate surface receptors and nutrient transporters, suggesting a novel bacterial adaptation strategy to limit phage predation (Schroven et al. 2021).

T7 DNA polymerase eliminates leftover genomic DNA during circular DNA purification, whereas T4 DNA polymerase is used for cloning and synthesizing tagged DNA (Dangerfield and Johnson 2023). Moreover, T7 DNA polymerase can be employed for the in-situ detection of fragmented DNA resulting from apoptosis (Dronina et al. 2022) and is effective in expanding templates with a significant number of base pairs. They can label the 3′ ends of DNA, typically provided by primase (Soffer et al. 2021). The highly flexible φ29 DNAP is used for xenobiotic nucleic acids (XNAs) biosynthesis, substituting only one amino acid. It is used in a rolling circle and multiple displacement amplification and has limited reverse transcriptase activity, allowing it to replicate circular, RNA-containing templates (Morcinek-Orłowska et al. 2022). DNA polymerases are crucial in synthesizing DNA accurately and efficiently. However, they cannot pass through damaged DNA, leading to stalled replication forks and replisome disassembly (Berti et al. 2020). Translesion synthesis (TLS) polymerases take over from replicative polymerases to bypass DNA lesions but are error-prone and have low processivity (Anand et al. 2023). The balance between high fidelity, processivity, and damage tolerance is crucial in phage DNA replication, particularly in temperate bacteriophages that can accumulate DNA lesions during the lysogenic cycle (Domingo et al. 2021). Bam35, a phage-encoded DNA polymerase found in *Bacillus thuringiensis* phage, is a game-changer in this field (Lechuga et al. 2021). It can bypass basic sites during DNA replication, shares structural and functional similarities with ϕ29 DNA polymerase, and has efficient polymerase activity and 3′ → 5′ exonuclease activity (Han et al. 2023). Phi29 DNA polymerase is a bacteriophage polymerase known for its high fidelity and processivity and is widely used in biological techniques such as rolling circle amplification and DNA sequencing (Zhang et al. 2023a, b). Other bacteriophage-derived enzymes, such as PyroPhage 3173 DNA Pol polymerase and OmniAmp Pol, are commonly used in molecular biology, veterinary medicine, and One Health (Liu et al. 2023a, b).

32.3.2 Endonucleases: Tools for Precision Editing

Phage endonucleases facilitate the viral life cycle by cleaving phosphodiester bonds within nucleic acid molecules, thereby allowing for the manipulation of host nucleic acids. These enzymes can be grouped into families according to sequence homology, biochemical properties, and structural features. Certain endonucleases encoded by template phages may have modular architectures, including catalytic domains, DNA binding domains, and regulatory regions. At different phases of the viral life cycle, these architectures enable accurate control of enzymatic activity (Mayo-Muñoz et al. 2024; Grabowski et al. 2021). Since these enzymes frequently have strict sequence requirements for recognition and cleavage, genetic engineering and genome editing applications can benefit more from using them as tools. The enzymes specifically target the nucleic acids of bacteria, altering their diversity, abundance, and composition through a combination of genetic recombination, lysis of host cells, and horizontal gene transfer. T7 bacteriophage Holliday structures are resolved by endonuclease I, whereas DNA polymerase and exonuclease halt strand-displacement synthesis (Falkenberg and Gustafsson 2020). In phages, homing endonuclease genes (HEGs) are essential selfish components that cause double-stranded DNA breaks and aid in the evolution of genomes. Examples of engineered endonucleases that enable targeted modification of genomic DNA sequence with previously unheard-of precision and specificity include ZFNs, Crispr-Cas, and transcription affecter-effector nucleases (TALENs) (Chaudhary et al. 2021; Baptista et al. 2021). Each phage has a different distribution of ICP1-encoded HEGs, and as noted by Netter et al. (2021), HEGs are frequently found inside or close to critical genes.

32.3.3 Lysozymes: Cell Wall Degraders

According to Molina-Quiroz et al. (2023), Nikolai Gamaleya reported the first bacterio-lysin of Bacillus anthracis, a transmissible and dissolved bacterium. Louis Pasteur's work influenced D'Herelle's research, leading to frequent discussions with Jules Bordet, a phage and immunology pioneer awarded the Nobel Prize in 1919 for discovering the complement system (Letarov 2020). Research conducted between 1930 and 1940 found that phage-treated Vibrio cholerae colonies had changed morphologically. Molina-Quiroz et al. (2023) reported notable alterations in phenotypic and phage sensitivity due to this phage-mediated modification. These discoveries shed light on the co-evolutionary arms race between bacteria and phages. In the early 1900s, Clark and Clark isolated a "sludge phage" from a Milwaukee sewage treatment plant. This phage was discovered only to infect animals—not humans—and could multiply in streptococci. Evans used this phage to study "nascent lysis," or the strain's lytic activity on streptococci strains. In his follow-up research, Maxted discovered that the "lytic factor" eliminated the streptococci of groups A, C, and E without needing active phages (Ho et al. 2022). The C1 phage lysin, which was used to break down the cell walls of groups A and C streptococci, was partially purified by Krause, a medical doctor at Rockefeller University (King et al. 2021). Subsequent research revealed that the term "endolysin" can relate to several different types of enzymes that attack peptide or glycosidic bonds, which are responsible for the mechanical rigidity of peptidoglycan (Abdelrahman et al. 2021). In the early 2000s, endolysins were used to treat in-vivo bacterial infections (Roehrig et al. 2024).

Enzymes called endolysins and explosions cause the bacterial cell wall to rupture, lysis, and release fresh viral particles. Endolysins are classified into five types: acetylmuramidases, transglycosylases, glucosaminidases, amidases, and endopeptidases. During cell lysis, they take part in the degradation of peptidoglycan, which is regulated by holins and phage lytic enzymes (Vermassen et al. 2019; Grishin et al. 2020). The process of phage lysis in Gram-negative bacterial cells involves three different protein types: spanins, pinholins, and holins. The cytoplasmic membrane becomes porous due to accumulated holins, which endolysins use to break down the peptidoglycan cell wall. Pinholins create tiny channels that allow endolysins to hydrolyse peptidoglycan cell wall bonds by depolarising the membrane and collapsing the proton motive force (PMF). A spanin complex disrupts the outer membrane. According to Fernandes and São-José (2018), most phage-encoded holins release endolysin and disperse PMF. Phage depolymerase, either hydrolases or lyases, hydrolyses or degrades the carbohydrate polymers on bacterial surfaces. As lyases break down the glycosidic bond that unites monosaccharides with uronic acid, creating a double bond, hydrolases break down O-polysaccharides, CPSs, EPSs, and LPSs. Some unidentified phage depolymerase classes can break down or inhibit the growth of bacterial biofilms. According to Tan et al. (2022), the tail spike protein (TSP) of *Acinetobacter baumannii* strain φAB6 inhibited host cell colonization and increased zebrafish infected with *A. baumannii* 54149 survival rates. According to Guo et al. (2023), the *K. pneumoniae* phage P560 depolymerase P560dep prevented 90%–100% of infected mice from dying. This is because it prevented the formation of biofilms.

The antibacterial qualities of cationic dendrimers and how well they complement antibiotics are the subject of further investigation. By using polyethylene glycol (PEG), sugar moieties, or pyrrolidone, surface fictionalization has been used to reduce cytotoxicity and increase selectivity towards the bacterial membrane (Zeb et al. 2024; Dean et al. 2024). Endolysins are produced during the late stage of infection, while exolysins are made during the initial stages (Boroujeni et al. 2024). Ciepluch et al. (2019), evaluated the synergistic effect of poly-propyleneimine (PPI) cationic dendrimers or maltose-modified PPI glycodendrimers with KP27 lysin, finding that the secondary structure of KP27 changed slightly in the presence of both dendrimers, inhibiting its PG degradation activity. Endolysins derived from phages that infect The specificity and catalytic activity of phage lysins in Gram-positive bacteria are determined by their N-terminal catalytic domain and C-terminal cell wall-binding domain (CBD). Bacteriolytic activity typically requires cell wall binding, and endolysins can show near species- or serovar-specificity. Two *Pseudomonas aeruginosa* phage endolysins, KZ144 and EL188, have modular architectures and N-terminal CBD and C-terminal EAD. Endolysins have efficiently inhibited various Gram-positive bacteria, including PlyC, C1, ClyR, Cpl-1, ClyV, and ClyJ (Guo et al. 2017). The enzymatic activity of endolysins can be evaluated using a variety of assays, such as bacterial suspension clearing, native peptidoglycan degradation, a decrease in the number of viable cells, and observation of the bacterial suspension's decreased optical density. Zhang and Hu (2013) assert that phage, endolysin, and chlorine effectively manage and eradicate bacterial biofilm on surfaces. Lysando's Artilysin (Art-X01) prevented *P. aeruginosa*-induced bacteremia in 80% of the mice (Lai et al. 2020). Using a mouse bacteremia model, Kim et al. 2020 assessed the efficacy of the lysin LysSS and found a 40% survival rate. Heselpoth et al. (2019) evaluated the quantity of a modified PyS2-GN4 lysin needed to maintain a fatal mouse model of *P. aeruginosa*-induced bacteremia. The strong peptide P307SQ-8C has decreased bacterial burden in mice, similar to antibiotic therapy (Thandar et al. 2016). Additionally, it was demonstrated that PlyPa03 and

PlyPa91 were useful in managing *P. aeruginosa* PAO1 skin infections, with PlyPa03 demonstrating superior effectiveness (Raz et al. 2019). The intestinal bactericidal action of Colicin-Lysep3, a modified lysin, was investigated by Yan et al. (2017), proving the product's efficacy in treating G-ve bacterial GI tract infections. Blasco et al. 2020 confirmed the complementary actions of colistin and lysin ElyA1 in addressing certain MDR G-ve species, such as *A. baumannii*, *P. aeruginosa*, and *K. pneumonia*. Raz et al. (2019) examined the effectiveness of PlyPa91 lysin in treating *P. aeruginosa* PAO1-induced lung infections in mice. CF-370 lysin, an engineered lysin targeting *P. aeruginosa*, showed promising preclinical data in a rabbit pneumonia model, with a 100% survival rate compared to a 40% survival rate for the vehicle control group (Sauve et al. 2024).

Chimeric endolysins have been developed to alter their specificity and catalytic activity. These include combining the peptide chain endonuclease structural domain of streptococcal phage lysin λSA2 with the SH3b-binding structural domains of staphylococcal phage lysin LysK and staphylococcal endolysin to create a chimeric enzyme with higher lytic activity against *Staphylococcus aureus* and maintain its original activity (Liu et al. 2023a). Other fusion proteins, such as HydH5SH3b, CHAPSH3b, and HydH5Lyso, have shown higher cleavage capacity than HydH5 (Rodrıguez et al. 2013). The bactericidal function of endolysins can be enhanced by combining different lytic enzyme structural domains. For instance, the lytic enzyme ClyH, formed by fusing Ply187 with phiNM3, showed strong bactericidal activity in vitro and in vivo (Yang et al. 2014). The fusion protein SMAP-29-KZ144, formed by fusing SMAP-29 with KZ144, has shown in vivo killing effects against *P. aeruginosa* and its biofilms (Briers et al. 2014). The peptidase CHAPk produced by the truncated structural domain can act as a biocide to rapidly degrade *S. aureus* biofilms and prevent and treat biofilm-associated staphylococcal infections (Kranjec et al. 2021). Lysins have been found to have PG hydrolytic activity against G-ve bacteria, which can be enhanced by using OMPS (EDTA, organic acids, polycationic peptides, and colistin) (Chen 2021). PlyGVE2CpCWB endolysin, a bacteriophage-derived enzyme, is known for its ability to lyse foodborne pathogens like *Clostridium perfringens* and has antimicrobial activities against both Gram-negative and Gram-positive bacteria (Swift et al. 2019).

32.3.4 Methyl Transferases

Methyl transferases are a class of enzymes that catalyse the transfer of methyl groups from donor molecules like SAM (S-adenosylmethionine) to specific substrates, including DNA, RNA, proteins, and small molecules. Lefkowitz et al. (2024) reported that the accessibility of DNA-binding proteins (TFs and RNA polymerase) may be affected by the viral genome's methylation of specific nucleotide residues. By methylating specific DNA sequences, it is possible to prevent viral genome damage and deterioration and maintain a certain level of genomic stability. Furthermore, it provides a molecular shield against substances that break DNA, such as nucleases and mutagens, which could compromise the stability and integrity of the phage genome. To prevent their DNA from being recognized as foreign and cleaved by a bacterial restriction-modification (R-M) system, phages frequently encode methyltransferase (Agapov et al. 2024). By methylating the viral genome, phages can mimic the methylation patterns of the host DNA, rendering themselves resistant to recognition and degradation by host restriction enzymes. In *V. cholerae*, mobile genetic elements known as integrative and conjugative (ICEs) contain anti-phage RM systems (Horne et al. 2023). Variable genes within the SXT element include a DNase that inhibits natural transformation and RM systems linked to anti-phage defence against ICP1 and other phages (Molina-Quiroz et al. 2023). Phages must acquire the methylase or make inhibitory proteins to counteract RM systems. Due to selection pressures from the host immune system and environmental fluctuations, methyltransferases derived from horizontal gene transfer can rapidly evolve and diversify. Methyl transferase has variable degrees of sequence specificity; it targets specific motifs and sequences. Every phage lineage has a distinct evolutionary history and set of selective forces that have moulded the motif composition, size, and spatial distribution throughout the genome. Phases may need methyltransferase genes from their host bacteria or other phages through horizontal gene transfer events, which can spread and diversify these genes throughout different phage lineages (Shuwen and Kefeng 2022).

32.3.5 Integrase and Recombinases

The first descriptions of integrase-mediated trans-gene chromosomal integration systems were published for Actinobacteria, Staphylococcus aureus, and *Escherichia coli* in the early 1990s (Stegmüller et al. 2024). These techniques, according to Mitousis et al. (2020), used a two-plasmid strategy in which target cells with an attP are injected into those that already have an alternative integrase expression plasmid. An alternative would be to employ a single suicide plasmid that expresses either the attP or int selectable flag gene. The endogenous attB site facilitates site-specific integration and makes selecting for additional conjugation or transformation simple. The integrated plasmids are very stable if the xis gene is not included in the integrating vector. These methods were refined in Pseudomonas, where integrating vectors exploited the ΦCTX recombinase. Conjugation into *P. aeruginosa* produced efficient integration into the native chromosomal attB site (McKeithen-Mead and Grossman 2023). The FRT sites that surround the multiple cloning sites facilitate the removal of the entire plasmid backbone when Flp recombinase is expressed, rendering the final recombinant strain markerless (Peri et al. 2023). Similar single-copy integration vectors have been developed for use with Francisellatularensis and other species. Some phages integrate their genetic material into the host bacterial chromosome upon infection. Integrase facilitates integration by catalyzing site-specific recombination between the phage genome and the bacterial chromosome (Qi et al. 2023). Recombinases are enzymes involved in genetic recombination events during phage replication and integration. Site-specific recombination reactions involve the precise integration, excision, resolution, or inversion of

DNA. Recombinases bind to two substrates and use protein–protein interactions to cleave them; reorganizing DNA ends through a strand exchange reaction. The recombinase covalently linked to the phosphodiester backbone is an important intermediate during DNA breakage and rejoining (Sun et al. 2020). Two distinct families of site-specific recombinase proteins exist tyrosine and serine. Tyrosine integrases mediate integration, excision, or inversion. λ Int, a member of the tyrosine recombinase family, is a key player in DNA exchange. Its recombination mechanisms are similar to those of most tyrosine integrases (Badel et al. 2021). Recombination substrates consist of inverted repeats and a core sequence. Tyrosine recombinases bind to these sequences, forming a tetrameric complex, or synapse. The correct synaptic complex is necessary for recombinase activation. Two recombinase monomers attack the DNA backbone, producing a covalent 3′ phosphortyrosine enzyme intermediate and a free 5′ OH-terminated DNA strand (Das et al. 2021). The free OH groups displace the integrases, creating a recombinant joint resembling a Holliday junction.

32.3.6 Depolymerase

In 1929, Sertic found an enzyme that caused halo to grow again in bacterial cultures. This discovery revealed the mechanism of the enzyme and showed that it could eliminate surface antigens without causing cell death (Sertic 1929; Wang et al. 2023). In 1956, Adams and associates identified a novel enzyme that could break down bacterial capsular polysaccharides, which resulted in multiple kinds of depolymerase (Adams and Park 1956). The phage P22 tail spike protein (TSP) was crystallized by Steinbacher et al. in 1996, revealing the structural underpinnings of the interaction between TSP and its particular polysaccharide substrate. Swapping host-specific domains and changing active site residues observed evolutionary changes between these proteins. Phage-bacteria evolutionary machinery alters bacterial carbohydrate structures, resulting in the diversification of active depolymerases. Depolymerases are structural components of the phage adsorption apparatus, located within receptor binding proteins (RBPs), that initiate infection by recognizing and degrading bacterial capsules. Bacteria vary greatly in them, and they can be altered by deletions, mutations, or loci involved in the biosynthesis of carbohydrates (Vieira et al. 2023; Drobiazko et al. 2022). According to Topka-Bielecka et al. (2021), depolymerase can mediate biofilm degradation, decrease biomass and adhered cells in the preformed biofilm, and stop host bacterial biofilms from forming.

32.4 Isolation and Characterization

Phage amplification was performed using a Luria broth medium containing peptone, sodium chloride, and yeast extract. Preparing a phage culture medium involved incubating phage lysate with a bacterial host cultured in LB broth overnight. The culture was centrifuged at 8000 × *g* for 8 min to separate the bacteria from the proliferating bacteriophages. The supernatant was then filtered using a syringe filter. The phage activity was evaluated using the double-layer plate method. Phage lysates were prepared to achieve a MOI (multiplicity of infections) coefficient of 1.0. Optical density measurements were performed every 30 min using a broadband filter. The effectiveness of the phage application was assessed compared to a control culture. For 24–72 h, phage lysates were cultured in an extended manner to produce a solution that contained more viral particles and lytic enzymes. The suspension was centrifuged at 8000 × *g* for 10 min to extract the bacteria pellet. The supernatant was filtered using a syringe filter with a pore size of 0.45 μm. The purified phage lysate was diluted 1:1 with LB medium, and then 1 mL of overnight bacterial culture was added. The mixture was then cultured for an extra 24 h. Phage titers were determined using the two-layer plate method (Wiebe et al. 2024; Buzikov et al. 2023). This investigation employed two techniques for isolating enzymes: ultracentrifugation and enzyme precipitation with cold acetone. The first method involved adding ice-cold acetone to the lysate, stirring, and centrifuging for 24 h. The precipitate was then evaporated, and saline was added. The protein solution was tested for antimicrobial properties by applying it to a host. The second method used ultracentrifugation to decrease bacterial cell division and determine protein concentration using the Bradford method (Fig. 32.2).

Biochemical tests are essential in identifying the existence of different enzymes that are linked with bacteriophages or phages—viruses that invade and reproduce inside bacteria. These enzymes assist in the infection process by aiding the penetration of the host cell wall, injecting nucleic acid, and facilitating the replication of the phage genome (Table 32.1).

32.5 Role in Bacterial Evolution

Endosymbiotic events have greatly impacted eukaryotic evolution; the earliest known event most likely involved an α-proteobacterium and a primordial eukaryotic cell. According to Olszak et al. (2017), phages coexist dynamically with their bacterial hosts, making them the most prevalent biological entities in the biosphere. There are rare instances of a mutualistic relationship between phages and bacteria in which the phages are adept at carrying important bacterial genes, such as virulence factors (Rands and Brüssow 2021). Phages can continue to be integrated into the host genome as prophages (lysogens) or go through the lytic cycle. Prophages can transfer genes that confer advantageous traits to their host, which facilitates the maintenance of these traits within bacterial genomes by producing virulence determinants, antibiotic-resistance genes, metabolic pathways, or other favorable traits. When phages co-infect, the degraded phages' genes may re-enter the viral gene pool and remain inducible, lysing host cells. Anti-phage defense systems like Pycsar and CBASS rely on secondary messenger signals like cyclic nucleotides, and the cleavage of this is a key strategy of immune evasion in phages (Athukoralage and White 2022). Phages can overcome defense mechanisms by preventing activation,

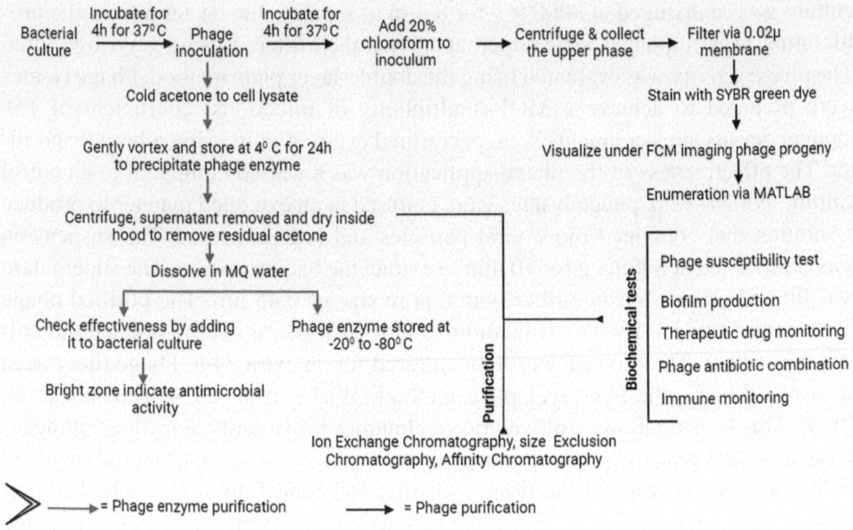

Fig. 32.2 Schematic of phage purification and phage enzyme purification and various biochemical tests for its identification and characterization

becoming resistant to the system, or acquiring or modifying a factor inhibiting the system. Examples include *E. coli* phage T5, which produces the lipoprotein Llp, and filamentous phages infecting *P. aeruginosa* (Dantas et al. 2023). For phages to finish their life cycle, phage satellites—virus-like elements—paralyze them using their genetic machinery. This gives the bacterial host an evolutionary advantage regarding virulence, biofilm formation, and toxins or protects against phage infections (Villarreal 2009).

The evolutionary history and adaptive significance of bacteriophage enzymes can be used to comprehend the coevolutionary dynamics between phages and their hosts. Phages have evolved strategies to overcome host defense and exploit cellular machinery for their replication. Phage enzymes may exhibit sequence variability and structural plasticity, allowing them to function in diverse host backgrounds and adapt to challenging environmental conditions (Bobay et al. 2013). Comparative genomics and phylogenetic analysis of genes producing respective enzymes reveal insights into the evolutionary relationships between different phage families and the molecular mechanisms underlying viral genome replication. Bacteria develop defense mechanisms against predatory phages, while phages counter-adapt by mutating or capturing new genes to maintain infectivity (Brockhurst et al. 2021). Through the acquisition of novel traits that boost bacterial or phage fitness, the ongoing struggle between phages and bacteria propels evolution. Mutations or loss of the receptor that a particular phage recognizes are common causes of resistance to phages (Hesse et al. 2020). *Given that surface receptors are essential virulence factors, V. cholerae is under pressure to develop phage-escape mechanisms without involving them* (Seed et al. 2013). Additional mechanisms of phage resistance

Table 32.1 Biochemical tests used in the healthcare system to detect the presence of phage enzymes

Tests	Specificity	Category	Use
PST (Phage susceptibility testing)	Highly specific	Plaque assays	Best testing technique: Examine the growth inhibition that results from cultivating combinations of bacteriophages on solid agar.
		Spot assays	Have limitations, including the inability to differentiate between different components of a phage cocktail and visibility restrictions, but they can be used to determine phage titres.
		Liquid-based assays	Assess the kinetic interactions of bacteriophages in real time; however, they cannot differentiate between viable and dead bacterial cells.
		Fluorescent tagging of phage DNA and hydrogel matrix	Rapid analysis and higher throughput.
Biofilm assay	Biofilm confirmatory	Isothermal microcalorimetry	Inhibition of heat production associated with growth.
		Standardized biofilm susceptibility assay	Model an in vivo biofilm environment and forecast medical results.
Therapeutic drug monitoring	Real-time decision, duration and dose adjustment of drugs	Drugs like aminoglycosides and vancomycin	To obtain sufficient serum phage concentrations.
Phage-antibiotic combining testing (PAS)	Mechanism of action	Synogram	Phage burst size, host environment, and antibiotic mode of action, which cause both synergistic and antagonistic effects.
Immune monitoring	Immunity check	ELISA	Block infectivity of neutralizing Ab, gene expression profile of both innate and adaptive immune response

include secreting outer membrane vesicles (OMVs) containing the phage receptor, making the receptor unavailable through biofilms, and the presence of auto-inducers promoting phage-resistant *V. cholerae* by producing hemagglutinin protease and downregulating the O1-antigen phage receptor (Molina-Quiroz et al. 2023). Coevolution plays a crucial role in ecological environments, particularly in phage resistance. Nevertheless, more is needed to know about these trade-off effects and how they affect community competitiveness. Subsequent investigations should focus on determining the significance of every evolutionary pressure within this dynamic structure.

32.6 Application

Phage enzymes have garnered significant interest in biotechnology and medicine due to their unique properties and enzymatic activities. These enzymes serve as valuable tools for various molecular biology techniques, including PCR, and DNA sequencing, enabling the amplification and analysis of nucleic acid sequences with high specificity and efficiency. PCR-based methods utilizing phage DNA polymerases have been developed for the detection and identification of viral pathogens, genetic disorders, and microbial contaminants in food and water samples. The robustness and versatility of phage enzymes make them valuable assets in the quest for novel diagnostic tools and therapeutic interventions. The structural diversity, enzymatic activities, and evolutionary significance underscore the complexity and adaptability of viral biology, developing novel biotechnological tools and combating infectious diseases in the era of molecular medicine (Table 32.2).

32.6.1 Pharmaceutical Industry

Integrase's site-specific integration techniques and cloning techniques are combined into a single process to transfer DNA of interest precisely into a native ATtB site. Since most of these integration vectors depend on host-encoded elements and endogenous attB sites, obtaining the integrase from a phage that infects the bacterial strain that will be genetically modified (Henson 2023) is customary. *Phage endolysins* are a naturally occurring bio-preservative that may help enhance food safety by quickly getting rid of bacterial pathogens. Based on phage endolysins, biotech companies have developed phage lytic proteins to treat skin infections, bloodstream infections, and chronic kidney disease. Companies such as GangaGen (Secco 2023), ContraFect (Fowler Jr et al. 2024), and Micreos (Roehrig et al. 2024) have developed phage lytic proteins for the treatment of *S. aureus*, demonstrating the additional applications of these endolysins in immune-based therapeutic and diagnostic approaches. Overall, phage endolysins are a promising approach for diagnosing and killing pathogenic populations in agriculture, human medicine, and veterinary medicine. Novel anti-biofilm techniques are needed to provide further control strategies. Because lytic phages harm bacterial hosts and do not contain the necessary enzymes for horizontal gene transfer, they are employed in phage-based therapeutics. Endolysins are useful in biofilm treatment, with studies showing significant results. LysPA26, a novel endolysin, was used to eliminate *P. aeruginosa* and other Gram-negative bacteria under a broad range of temperatures (Nazir et al. 2023). LySMP, a synthetic bacteriophage lysin, treated Streptococcus suis biofilm alone and with antibiotics (Xi et al. 2023). Engineered peptidoglycan hydrolases and glycoside hydrolases have increased bacterial colony lysis and biofilm removal. Persister

Table 32.2 Common phage enzymes and their uses in the healthcare system

Phage enzymes category	Enzymes	Use	References
DNA polymerases	PolI_G20c	Similar to klenow fragment of DNA pol 1 of *E coli*	Doss et al. (2023)
	3173 Pol	Has DNA-dependent polymerase activity but also has RNA-dependent reverse transcription activity so used in RT-PCR	Moser et al. (2012)
	PyroPhage 3173 DNA Pol polymerase	RT-PCR/PCR	Heller et al. (2019)
	OmniAmp Pol	Use the RT-LAMP assay to identify the following pathogens: *Bacillus atrophaeus*, swine influenza virus, *Edwardsiellaictaluri*, Ebola virus, porcine circovirus-2, and *Staphylococcus aureus*	Chander et al. (2014)
	MMLV	RNA to cDNA for PCR	Zajac et al. (2013)
	DNA polymerase from NrS-1 bacteriophage	Fresh insight into de novo DNA synthesis reveals the requirement for a helix bundle domain and a prim/pol domain for priming activity.	Zabrady et al. (2023)
Ligases	RM378 RNA ligase 1	sDNA and RNA molecules	Blondal et al. (2003)
	TS2126 ligase	Circulization of ssDNA	Lama and Ryan (2016)
	T4 DNA ligase	Stable at high temperatures (60–65 °C) and can circularize single-stranded DNA	Ma et al. (2023)
Endolysins	Holins from Shiga toxigenic *Escherichia coli*	Significantly decreased the MDR *S. aureus* isolates' established biofilm biomass and biofilm development.	Chaudhary et al. (2022)
	PlyG	Against *Bacillus anthracis*, *Streptococcus equi*	Fenton et al. (2010)
	PlyV12	Against antibiotic-resistant *Enterococcus faecium*, *Enterococcus faecalis s*& treating *L. monocytogenes* contamination	Stritt (2014), Dorval Courchesne et al. (2009)
	LysSA11	Antimicrobial activity against *S. aureus*	Chang et al. (2017)
	Lys1S-L9P	Killed colistin-resistant *E. coli* and had greater success removing *P. aeruginosa*, *A. baumannii*, and *E. coli* biofilms.	Son et al. (2023)
	LysSPN1S and KL-L9P	Lytic activity across a wider range of multidrug-resistant Gram-negative bacteria.	Son et al. (2023)

(continued)

Table 32.2 (continued)

Phage enzymes category	Enzymes	Use	References
	Lysep3	Moderately active against *E. coli*	Wang et al. (2017)
	PlyGVE2CpCWB	Lyses *Clostridium perfringens*	Swift et al. (2019)
	MMPphg	Bacteria susceptible to its antimicrobial activities include antibiotic-resistant strains, *Staphylococcus aureus*, *Salmonella enterica* serovar Typhi, *Salmonella enterica* serovar Enteritidis, *Salmonella enterica* serovar Paratyphi B, *Escherichia coli* O157, *Shigella dysenteriae*, and *Klebsiella pneumoniae*.	Liu et al. (2023a, b)
	Ply17	Reduce the number of viable Gram-positive bacteria including *S. epidermidis* and *S. aureus* as well as Gram-negative bacteria such as *P. aeruginosa* and *E. coli*	Yang et al. (2018), Nguyen et al. (2023)
Depolymerase	Bam35, primase polymerase	High fidelity, damage tolerant and de novo synthesis of DNA from dNTPs	Redrejo-Rodríguez et al. (2017)
	KP34p57	Reduce bacterial count of *Klebsiella pneumoniae*	Latka and Drulis-Kawa (2020)
	EC3a	Antibiofilm activity with honey against *E. coli* CECT 434	Oliveira et al. (2017)
	CF-301, P128	Drug-resistant *S. aureus* biofilms and human synovial fluid lysing biofilms.	Tian et al. (2021)
	LySMP	Removing biofilms produced by strains of *Streptococcus suis* SS2-4 and SS2-H.	Meng et al. (2011)
	HydH5 + LysH5	Anti-staphylococcal effects	Rodrıguez et al. (2013)
	Ply6A3	Saving mice from *A. baumannii*'s deadly bacteremia	Wu et al. (2019)
	KP34gp57	Against *K. pneumonia*	Maciejewska et al. (2023)
Endonucleases	Cap V	Degrade PE and PG layer	Severin et al. (2018)
	NucC	Degrade foreign DNA in the CRISPR system	Athukoralage and White (2021)
	Cap 5	Bind to the SAVED domain and execute DNA degradation	Lowey et al. (2020)

(continued)

Table 32.2 (continued)

Phage enzymes category	Enzymes	Use	References
	Cap 4	Recognize secondary messenger	Kato et al. (2017)
	ssRNAses, Csm6, Csx1	Cleave dsDNA/ssRNA invasion	Rostøl et al. (2021)
	Card1	REase activity on ssDNA/ssRNA causing cell death	Patel et al. (2022)
	Cap15	Membrane damage without secondary messenger	Duncan-Lowey et al. (2021)
	Cas13	Target sgRNA	Baptista et al. (2021)
Methyl transferase	5-methylcytosine	Shiga toxin-producing *E. coli* O157:H7	Gong et al. (2024)
	N6-methyladenine	MDR *Shigella* spp.	Ahamed et al. (2023)
Integrase	IntI	Causes lateral gene transfer	Giengkam et al. (2023)
	Dihydrofolate reductase	Inhibit the parasite enzyme, act directly on folate biosynthesis	Aragaw et al. (2024)
	Streptothricin acetyltransferase	Marker for methanosarcina transformation	Halder et al. (2023)
	Aminoglycoside adenyltransferase	Causes resistance to streptomycin and spectinomycin causing resistant *E. coli*	Zhang et al. (2023a, b)

cells, which show high antibiotic resistance, are a significant issue in biofilm removal. Endolysins, like PlyV12, selectively treat specific pathogenic bacteria without disrupting the surrounding microbiota (Khan et al. 2023). They induce the immune system's response without neutralizing or preventing antimicrobial activity, making them suitable for systemic infection treatments and rapid lysis of Gram-positive bacteria.

32.6.2 Biomedical Application

Scientists For the past 20 years, researchers have examined endolysins' antimicrobial capabilities. When endolysins are applied externally to susceptible Gram-positive bacterial cells, they can cause rapid bacterial lysis. They are impervious to metabolic pathways, penicillin-binding proteins, efflux pumps, and resistance mechanisms found in conventional antibiotics. Gram-negative (G-ve) bacteria, including *P. aeruginosa*, *A. baumannii*, *K. pneumonia* and *E. coli*, are susceptible to most synthetic and naturally occurring lysins. They can act bactericidal without being able to identify the bacteria because of their globular structure, which has one EAD and no CBD. A single gene produces endolysins, first found in the periplasmic membrane before being released to cause lysis in the host. As seen in Artilysins, endolysins were fused with outer membrane permeabilizing peptides to overcome their limited activity (Sisson et al. 2024). Lysogeny is associated with high viral particle decay rates and decreased host cell counts, such as VMR in the human gut lumen, which are inversely correlated with primary microbial productivity.

Human pathogen Group A Streptococcus (GAS) can cause superficial skin, nasopharyngeal infections, and potentially fatal infections like *Necrotizing fasciitis* or scarlet fever. PlyGBS and B30, two promising endolysins discovered and tested against *Streptococcus agalactiae*, have demonstrated protection in mouse models of *S. agalactiae* infection (Kajsikova et al. 2024). The majority of the tested streptococcal strains have been neutralized by the chimeric endolysin ClyR, which is a combination of PlyC CHAP EAD and PlySs2 CBD, in dental caries (Vander Elst et al. 2023). Pneumococcal phage endolysins, such as Pal and Cpl-1, have been studied in vivo for their efficacy and synergy (Khan et al. 2023). According to recent research, the creation of biofilms increases TCP and CT, which increases the infectiousness of *V. cholerae* (Kitts et al. 2023). Phages have been developing antibacterial agents from bacteriostatic and bacteriolytic proteins, such as endolysins, virion-associated peptidoglycan hydrolase (VAPGH), polysaccharide depolymerases, and holins, which may offer new therapeutics for treating bacterial infections and diseases in humans. Research indicates that the immune response may not always reduce lysin's antibacterial action in acute infections. Masking techniques such as PEGylation, glycosylation, and T-cell epitope depletion have been developed to prevent hypersensitive reactions or antibody inhibition. A vaccine to prevent streptococcal pharyngeal infection using streptococcal M protein characteristics (Walkinshaw et al. 2023) and developed a third therapeutic lysin, PlyG, for *Bacillus anthracis* (Chen et al. 2023). Enzymatic activity of endolysins can be detected through various assays, such as clearing bacterial suspension, native peptidoglycan degradation, and a decrease in viable cell count. Endolysins SAL200 and CF-301 have been developed against *S. aureus* for bloodstream and cardiac infections (Roehrig et al. 2024). Phage muralytic enzyme mutants, such as. PlySs2 and SAL-1 have also been immobilized on spider silk proteins and combined with Dispersin B can act as a target for biofilm (Efremenko et al. 2023). Another endolysin CF-301, directed at MRSA, has completed phase 1 clinical trials and is currently entering phase 2 trials to treat MRSA bacteraemia or endocarditis (Douglas et al. 2023).

Lysins work well against persister cells because their mode of action does not involve metabolic activity. Becker et al. (2016) combined 11 cell-penetrating peptides (CPPs) that target intracellular *S. aureus*. These PPPs contained triple-acting chimeric lysins, lysostaphin, and LysK. To further benefit from the transducing ability of CPPs, they merged with JDlys, an endolysin derived from phage JD007 that infects *S. aureus* at either its N- or C-terminus. De Maesschalck et al. (2020) have created advanced formulation techniques to increase the topical effects of lysins, especially those of bacteriocin lysostaphin and first-generation lysins. These tactics consider infection-related variables like inflammation while trying to boost lysin's antibacterial activity. Examples include the topical and invasive use of polyethylene glycol hydrogels, the application of lysostaphin-loaded chitosan-protein sponges as wound dressings, the co-formation of lysins with anti-inflammatory substances like apigenin, and the immobilization of antibacterial enzymes on spider silk proteins.

32.7 Phage Therapy for Bacterial Infections

Phages were used in the Soviet Union for preventative and therapeutic purposes during World War II. The German army treated soldiers' dysentery with a suspension called Polyfagin, which contained phages specific to various strains of Shigella (Reuter and Kruger 2020). Phage therapy for burn wounds infected with *Escherichia coli* and *Pseudomonas aeruginosa* in Europe was assessed in a multicenter trial called "PhagoBurn," but the trial was terminated for lack of efficacy. The Queen Astrid Military Hospital in Belgium, operating under the Magistral Framework, enabled the first phage treatments to be conducted regularly in the West since the 1980s (De Vos et al. 2019). Combining different antimicrobial treatment options, including antibiotics, may be more effective in combating multi-resistant bacterial pathogens.Recombinant depolymerases are advised for therapeutic applications to lower the danger of resistance because they hardly ever encourage bacterial mutations. According to studies, phage-resistant mutants are present when a depolymerase and its parent phage, PMK34, are present (Abdelkader et al. 2022). It is difficult to determine if bacteria challenged with Dpo71 and cultures challenged with K2 depolymerase have developed phage-resistant mutants. The combined therapy approach can stop or slow the emergence of phage-resistant variants by utilising an effective phage cocktail or phage/depolymerase combination (Wang et al. 2023). Pathogenic bacterium management strategies include recombinant depolymerase, engineered recombinant depolymerase, phage-equipped depolymerase, engineered phages with shield depolymerase, and engineered phages with depolymerase.

Czaplewski et al. 2016 reviewed potential antibiotic-resistant therapies, focusing on phage lysins, vaccines, antibodies, and probiotics. Enzymes produced from phages, like those that degrade peptidoglycans and polysaccharides, show great promise as therapeutic candidates for the prevention and elimination of biofilms. Phage endolysins are arranged modularly into discrete functional domains, which confers substrate selectivity and makes it easier to protein engineer new proteins with improved antibacterial properties. Clarithromycin's antibacterial efficacy against *P. aeruginosa* increased greatly by encasing the medication in positively charged liposomes (Alhajlan et al. 2013). Cellulose nanocrystals (CNC) are being used more often as drug delivery vehicles for the selective immobilization of enzymes due to their excellent stability, affordability, and biocompatibility. The goal of the developing field of phage engineering is to increase the antimicrobial activity of phages to treat pathogens. Phage engineering techniques have been successfully implemented, and businesses worldwide employ these approaches. Because prophage genes encode potential toxins such as zot, RTX, and zonula occludens in different Vibrio species, they can also affect host fitness.

32.8 Synergies with Immunotherapies

In order to maximize antimicrobial efficacy and minimize side effects, researchers are investigating potential synergies between phage therapy and immunotherapies. To maximize the integration of phage therapy with immunotherapies, it is imperative to comprehend the intricate relationships between phages and the host immune system. Characterizing phage immunomodulatory effects, including how they affect innate and adaptive immune responses, should be the main goal of research. It is possible to create immunomodulatory phage formulations specifically suited to maximize antimicrobial efficacy and minimize immunogenicity by thoroughly understanding how phages affect immune cell activation, cytokine production, and regulatory pathways. Identifying and removing foreign pathogens, such as phages, from the body largely depends on the innate immune system. It is important to investigate methods for using innate immune mechanisms to improve phage clearance and stop systemic dissemination. Phage recognition receptors, including Toll-like and scavenger receptors, can be targeted to promote quick phage clearance while lowering the possibility of immunotoxicity and unfavorable immune responses. The generation of antibodies and T-cell activation are examples of adaptive immune responses crucial in determining phage therapy's long-term safety and effectiveness. Epi-Seq, DeepAntigen, TIminer, Neopepsee, and EpiSweep are examples of T-cell epitope mapping tools that have made it possible to screen peptides for potential immunogenicity and digest them in silico (Schaap-Johansen et al. 2021). A proof-of-concept for deimmunization was discovered in Lysostaphin, which reduced immunogenic variants and antibacterial activity that was particular to the human MHC complex, which generates antibodies. Immunotherapeutic platforms offer innovative opportunities for targeted and controlled delivery of phages to infection sites while minimizing off-target effects. Further research is needed to elucidate the underlying mechanisms of synergy and optimize combinatorial treatment regimens for maximal therapeutic benefit.

32.9 Challenges and Future Directions

To lessen these risks, the One Health approach encourages antimicrobial resistance (AMR), sustainable agriculture, and the health of animals, plants, forests, and aquaculture. With an emphasis on AMR bacteria specifically, this strategy attempts to stop, detect, and manage infections that transfer from animals to people. Lysogenic and lytic phages are the two phages most useful for biocontrol applications because of their high specificity and capacity to eradicate target bacteria without harming the environment or the host's microbiota. Food-grade phage-based biocontrollers are approved for use and can be used with disinfectants or antibiotics for enhanced control. Due to publication bias and a dearth of thorough registries, the success rate of compassionate use cases still needs to be discovered. Factors contributing to this include bacterial resistance, immune neutralization, phage host range, and infection concentration challenges. Microbiological techniques deal with these issues.

In addition to being important for human health, ecosystems, and nutrient cycling, viruses also impact microbes by modifying microbial communities and host cell metabolism. Thorough mapping of the sequence space of viruses can aid in characterizing mechanisms for microbial manipulation encoded in the viruses. According to Kieft et al. (2020), the virus identification program VIBRANT performed better than other programs in recovering high-quality viruses and decreasing the false identification of non-viral genome fragments. From 120,834 sequences derived from the metagenome, 94% of the viruses were recovered, and additional viral sequences and proteins were identified. Viral detection rates were improved over viral RefSeq and IMG/VR by 182 and 2.6 times, respectively, with the creation of the human Gut Virome Database (GVD) utilizing 2697 viral particle or microbial metagenomes from 1986 individuals across 16 countries (Gregory et al. 2020). Phage defense systems, which use phages to protect hosts and facilitate phage transmission, are on mobile genetic elements (MGEs), such as phage satellites in *V. cholerae* (Netter et al. 2021).

The evolution of phage therapy necessitates ongoing surveillance and adaptation to maintain its efficacy. A robust regulatory framework is crucial for safety, efficacy, and quality. Production yields depend on host strain, culture condition, and purification techniques. Novel production platforms and process optimization strategies are essential for cost-effective manufacturing technologies. Phages exhibit varying kinetics within the host, and factors like host immune responses, biofilm formation, and physiological barriers influence their distribution and persistence. Characterizing phages' pharmacokinetic profiles and eradicating their interactions with target bacteria are critical for optimizing therapeutic outcomes and minimizing treatment failure. Tailored trial designs account for variability in disease severity, antimicrobial susceptibility profiles, and patient comorbidities. Ethical considerations, patient consent, and regulatory compliance add complexity to the clinical trial process. Implementing standardized endpoints, outcome measures, and surveillance protocols is essential for generating clinical evidence and guiding phage therapy. Economic factors like production costs, reimbursement mechanisms, and market competition influence adoption and investment decisions. Addressing affordability, distribution channels, and regulatory barriers is crucial for equitable access to phage-based interventions.

Optimizing the pharmacokinetics and formulation of phage therapeutics is crucial for their clinical efficacy and safety. This can be achieved by characterizing the pharmacokinetic profiles of phages in vivo, enhancing their stability and bioavailability. Innovative formulation strategies, such as encapsulation and controlled release systems, can also enhance the stability and bioavailability of phage preparations. To advance phage therapy from the laboratory to clinical practice, rigorous evaluation through well-designed clinical trials is necessary. Large-scale trials assessing safety, efficacy, and long-term outcomes are also essential. Regulatory frameworks and concerns regarding phage quality control, standardization, and approval processes are also critical for widespread adoption. Furthermore, integrating phage therapy with conventional antimicrobial treatments can maximize therapeutic outcomes and minimize the risk of treatment failure and antimicrobial resistance (AMR). A One Health approach considering human, animal, and environmental health is essential for addressing AMR and promoting sustainable healthcare practices (Nardulli et al. 2023).

32.10 Future Directions and Research Opportunities

Applying depolymerase in the presence of human serum or macrophages has been shown to inhibit the growth of bacteria dramatically. Bacteria are being explored as substitutes for traditional antibiotics for therapeutic purposes. However, their use may be restricted due to potential bacterial resistance, limitations in the immune system's response, and concerns about health safety. Phage endolysins show promise in the fight against antibiotic resistance; however, their acceptance is contingent upon several factors, including immunogenicity, safety profile, and the potential for resistance development in bacteria. To gain a deeper understanding of these concerns, endolysin clinical trials are required. Antibodies have a reputation for being ineffective at inactivating endolysin. Endolysins have a high safety profile and no side effects, making them safe to use in the future based on pre-clinical research on animal models. Endolysins have not been found to have any negative effects on dogs, monkeys, or humans, with notable clinical effects. According to a human study, the risk of resistance is much lower with SAL200 endolysins than with traditional antibiotics. There are still resistance strategies to be addressed, such as altering phage receptors, secreting chemicals, preventing phage DNA injection, and phage replication and release. The challenge of treating antibiotic-resistant bacterial infections is increasing, calling for novel approaches to antimicrobial therapy. Studies have shown that treating bacteria with an endolysin-antibiotic combination is synergistic.

Multi-drug resistant *Streptococcus pneumoniae* strains can be successfully treated with Cpl-711 endolysin and widely used antibiotics such as amoxicillin, levofloxacin, vancomycin, and cefotaxime, according to research by Letrado et al. (2018). Antibiotics were found to have a synergistic effect by breaking down the peptidoglycan cell wall, which made the organism more vulnerable to endolysins. Numerous studies support the idea that the synergy between endolysin and antibiotics improves therapeutic efficacy by lowering antibiotic minimum inhibitory concentrations and lowering antibiotic dosage administration. The therapeutic agents' route and delivery systems must be considered when administering them to target infection sites. Enzymes derived from phages can be delivered orally, vaginally, transnasally, or through topical treatments. The acidic pH and the harsh gastric proteolytic enzymes make it difficult for oral routes to maintain enzymatic activity.

32.11 Conclusion

Regulatory considerations and safety assessments are crucial when developing and commercializing bacteriophage-based enzymes across various industries. Bacteriophage enzymes have shown promise in numerous applications, including food safety, agriculture, and healthcare. Strategies include using outer membrane permeabilizers, fusion peptides, and molecular engineering techniques to enhance enzyme lytic activity. Phage therapies should be designed to minimize bacterial resistance and potentially contribute to septic shock in invasive infections. Using a combination of phages, minimizing bacterial lysis, comprehending the evolutionary arms race between phages and hosts, utilizing non-transducing phages and avirulent host strains, and knowing phage biology are important components.

Phages phage-based approaches are effective for treating extracellular pathogens, but knowledge of phage therapy for intracellular pathogens is limited. Intracellular bacteria can evade humoral immunity and antibiotics, making synthetic biology and genetic engineering an attractive strategy. Different protocols for phage isolation and the continuous development of sequencing tools will contribute to understanding phage biology and taxonomic classification, ultimately benefiting therapies. More trials are needed to understand endolysins' immunogenicity. Phages have become more popular as a biocontrol agent because of the rise of AMR bacteria and the decreased usage of antibiotics in crops and livestock. Nevertheless, factors like high specificity, resistance, and self-dosage constrain their broad application. Their phage specificity significantly hampers their effectiveness as antimicrobials in biocontrol. It could take a while and be expensive to build phage collections or biobanks to combat most pathogenic bacterial strains. Phage biocontrol can be achieved as a customized treatment, requiring prior knowledge of the bacterial host and phage hunting. Phage-based products can be embedded in materials like nanoparticles or freeze-drying to maintain stability and infectivity. Encapsulation is preferred for food protection, but further research is needed to understand its pharmacokinetics and pharmacodynamics. Phage–host interactions can predict treatment failure and phage cocktails can help reduce resistance. Compliance with legislation is crucial for phage use, but the complexity of phage morphologies and self-replication makes it challenging for regulatory agencies to regulate all phage-derived products. However, obstacles like the low potency of lysins, stability problems, extrapolation of results from animal models, safety concerns with OMPs and lysins with OM-disrupting properties, and the low potency of lysins impede their clinical application. One major obstacle to OMPs and lysins with OM-disrupting properties' clinical application is safety concerns.

References

Abdelkader K, Gutiérrez D, Latka A, Boeckaerts D, Drulis-Kawa Z, Criel B et al (2022) The specific capsule depolymerase of phage PMK34 sensitizes Acinetobacter baumannii to serum killing. Antibiotics 2022(11):677. https://doi.org/10.3390/antibiotics11050677

Abdelrahman F, Easwaran M, Daramola OI, Ragab S, Lynch S, Oduselu TJ, Khan FM et al (2021) Phage-encoded endolysins. Antibiotics 10(2):124. https://doi.org/10.3390/antibiotics10020124

Adams MH, Park BH (1956) An enzyme produced by a phage-host cell system: II. The properties of the polysaccharide depolymerase. Virology 2(6):719–736. https://doi.org/10.1016/0042-6822(56)90054-x

Agapov A, Baker KS, Bedekar P, Bhatia RP, Blower TR, Brockhurst MA et al (2024) Multi-layered genome defences in bacteria. Curr Opin Microbiol 78:102436. https://doi.org/10.1016/j.mib.2024.102436

Ahamed SK, Rai S, Guin C, Jameela RM, MuthuirulandiSethuvel DP, Balaji V, Giri N (2023) Characterizations of novel broad-spectrum lytic bacteriophages Sfin-2 and Sfin-6 infecting MDR Shigella spp. with their application on raw chicken to reduce the Shigella load. Front Microbiol 14:1240570. https://doi.org/10.3389/fmicb.2023.1240570

Alexyuk MS, Bukin YS, Butina TV, Alexyuk PG, Berezin VE, Bogoyavlenskiy AP (2023) Viromes of coastal waters of the North Caspian Sea: initial assessment of diversity and functional potential. Diversity 15(7):813

Alhajlan M, Alhariri M, Omri A (2013) Efficacy and safety of liposomal clarithromycin and its effect on Pseudomonas aeruginosa virulence factors. Antimicrob Agents Chemother 57(6):2694–2704. https://doi.org/10.1128/AAC.00235-13

Anand J, Chiou L, Sciandra C, Zhang X, Hong J, Wu D et al (2023) Roles of trans-lesion synthesis (TLS) DNA polymerases in tumorigenesis and cancer therapy. NAR Cancer 5(1):zcad005. https://doi.org/10.1093/narcan/zcad005

Aragaw WW, Negatu DA, Bungard CJ, Dartois VA, Marrouni AE, Nickbarg EB et al (2024) Pharmacological validation of dihydrofolate reductase as a drug target in Mycobacterium abscessus. Antimicrob Agents Chemother 68(1):e00717–e00723. https://doi.org/10.1128/aac.00717-23

Athukoralage JS, White MF (2021) Cyclic oligoadenylate signaling and regulation by ring nucleases during type III CRISPR defense. RNA 27(8):855–867. https://doi.org/10.1261/rna.078739.121

Athukoralage JS, White MF (2022) Cyclic nucleotide signaling in phage defense and counter-defense. Annu Rev Virol 9:451–468. https://doi.org/10.1146/annurev-virology-100120-010228

Badel C, Da Cunha V, Oberto J (2021) Archaeal tyrosine recombinases. FEMS Microbiol Rev 45(4):fuab004. https://doi.org/10.1093/femsre/fuab004

Baptista B, Carapito R, Laroui N, Pichon C, Sousa F (2021) mRNA, a revolution in biomedicine. Pharmaceutics 13(12):2090. https://doi.org/10.3390/pharmaceutics13122090

Becker SC, Roach DR, Chauhan VS, Shen Y, Foster-Frey J, Powell AM et al (2016) Triple-acting lytic enzyme treatment of drug-resistant and intracellular Staphylococcus aureus. Sci Rep 6(1):25063. https://doi.org/10.1038/srep25063

Berti M, Cortez D, Lopes M (2020) The plasticity of DNA replication forks in response to clinically relevant genotoxic stress. Nat Rev Mol Cell Biol 21(10):633–651. https://doi.org/10.1038/s41580-020-0257-5

Bhushan K, Bhagyawant SS (2023) The perpetual battle of bacteria and phages. In: Sustainable agriculture reviews 60: microbial processes in agriculture, pp 1–22

Blasco L, Ambroa A, Trastoy R, Bleriot I, Moscoso M, Fernández-Garcia L et al (2020) *In vitro* and *in vivo* efficacy of combinations of colistin and different endolysins against clinical strains of multi-drug resistant pathogens. Sci Rep 10(1):7163. https://doi.org/10.1038/s41598-020-64145-7

Blondal T, Hjorleifsdottir SH, Fridjonsson OF, Ævarsson A, Skirnisdottir S, Hermannsdottir AG et al (2003) Discovery and characterization of a thermostable bacteriophage RNA ligase homologous to T4 RNA ligase 1. Nucleic Acids Res 31(24):7247–7254. https://doi.org/10.1093/nar/gkg914

Bobay LM, Rocha EP, Touchon M (2013) The adaptation of temperate bacteriophages to their host genomes. Mol Biol Evol 30(4):737–751. https://doi.org/10.1093/molbev/mss279

Boroujeni MB, Mohebi S, Malekian A, Shahraeini SS, Gharagheizi Z, Shahkolahi S et al (2024) The therapeutic effect of engineered phage, derived protein and enzymes against superbug bacteria. Biotechnol Bioeng 121(1):82–99. https://doi.org/10.1002/bit.28581

Briers Y, Walmagh M, Grymonprez B, Biebl M, Pirnay JP, Defraine V et al (2014) Art-175 is a highly efficient antibacterial against multidrug resistant strains and persisters of Pseudomonas aeruginosa. Antimicrob Agents Chemother 58(7):3774–3784. https://doi.org/10.1128/AAC.02668-14

Brockhurst MA, Koskella B, Zhang QG (2021) Bacteria-phage antagonistic coevolution and the implications for phage therapy. In: Bacteriophages: biology, technology, therapy, pp 231–251

Burz SD, Causevic S, Dal Co A, Dmitrijeva M, Engel P, Garrido-Sanz D et al (2023) From microbiome composition to functional engineering, one step at a time. Microbiol Mol Biol Rev 87(4):e00063–e00023. https://doi.org/10.1128/mmbr.00063-23

Buzikov RM, Kazantseva OA, Piligrimova EG, Ryabova NA, Shadrin AM (2023) Bacteriolytic potential of enterococcus phage iF6 isolated from "Sextaphag®" therapeutic phage cocktail and properties of its endolysins, Gp82 and Gp84. Viruses 15(3):767. https://doi.org/10.3390/v15030767

Chander Y, Koelbl J, Puckett J, Moser MJ, Klingele AJ, Liles MR et al (2014) A novel thermostable polymerase for RNA and DNA loop-mediated isothermal amplification (LAMP). Front Microbiol 5:395. https://doi.org/10.3389/fmicb.2014.00395

Chang Y, Kim M, Ryu S (2017) Characterization of a novel endolysin LysSA11 and its utility as a potent biocontrol agent against Staphylococcus aureus on food and utensils. Food Microbiol 68:112–120. https://doi.org/10.1016/j.fm.2017.07.004

Chaudhary MF, Saleem U, Khan S, Khan R, Jehanzaib M (2021) Cancer therapy with CRISPR/Cas9: prospects and challenges. Int J Environ Agric Biotechnol 6:3

Chaudhary N, Mohan B, Mavuduru RS, Kumar Y, Taneja N (2022) Characterization, genome analysis and *in vitro* activity of a novel phage vB_EcoA_RDN8. 1 active against multi-drug resistant and extensively drug-resistant biofilm-forming uropathogenic Escherichia coli isolates, India. J Appl Microbiol 132(4):3387–3404. https://doi.org/10.1111/jam.15439

Chen X (2021) Strategies to endow phage-encoded enzymes against gram-negative bacteria (Doctoral dissertation, The Chinese University of Hong Kong (Hong Kong))

Chen Y, Ying Y, Lalsiamthara J, Zhao Y, Imani S, Li X et al (2023) From bacteria to biomedicine: developing therapies exploiting NAD+ metabolism. Bioorg Chem 142:106974. https://doi.org/10.1016/j.bioorg.2023.106974

Ciepluch K, Maciejewska B, Gałczyńska K, Kuc-Ciepluch D, Bryszewska M, Appelhans D et al (2019) The influence of cationic dendrimers on antibacterial activity of phage endolysin against P. aeruginosa cells. Bioorg Chem 91:103121. https://doi.org/10.1016/j.bioorg.2019.103121

Czaplewski L, Bax R, Clokie M, Dawson M, Fairhead H, Fischetti VA et al (2016) Alternatives to antibiotics—a pipeline portfolio review. Lancet Infect Dis 16(2):239–251. https://doi.org/10.1016/S1473-3099(15)00466-1

Dangerfield TL, Johnson KA (2023) Kinetics of DNA strand transfer between polymerase and proofreading exonuclease active sites regulates error correction during high-fidelity replication. J Biol Chem 299(1). https://doi.org/10.1016/j.jbc.2022.102744

Dantas R, Brocchi M, Pacheco Fill T (2023) Chemical-biology and metabolomics studies in phage-host interactions. In: Microbial natural products chemistry: a metabolomics approach. Springer International Publishing, Cham, pp 71–100. https://doi.org/10.1007/978-3-031-41741-2_4

Das BB, Ghosh A, Bhattacharjee S, Bhattacharyya A (2021) Trapped topoisomerase-DNA covalent complexes in the mitochondria and their role in human diseases. Mitochondrion 60:234–244. https://doi.org/10.1016/j.mito.2021.08.017

De Maesschalck V, Gutiérrez D, Paeshuyse J, Lavigne R, Briers Y (2020) Advanced engineering of third-generation lysins and formulation strategies for clinical applications. Crit Rev Microbiol 46(5):548–564. https://doi.org/10.1080/1040841X.2020.1809346

De Vos D, Verbeken G, Quintens J, Pirnay JP (2019) Phage therapy in Europe: regulatory and intellectual property protection issues. Phage therapy: a practical approach, pp 363–377

Dean TT, Jelú-Reyes J, Allen ALC, Moore TW (2024) Peptide–drug conjugates: an emerging direction for the next generation of peptide therapeutics. J Med Chem 67:1641. https://doi.org/10.1021/acs.jmedchem.3c01835

Domingo E, García-Crespo C, Lobo-Vega R, Perales C (2021) Mutation rates, mutation frequencies, and proofreading-repair activities in RNA virus genetics. Viruses 13(9):1882. https://doi.org/10.3390/v13091882

Dorval Courchesne NM, Parisien A, Lan CQ (2009) Production and application of bacteriophage and bacteriophage-encoded lysins. Recent Pat Biotechnol 3(1):37–45. https://doi.org/10.2174/187220809787172678

Doss RK, Palmer M, Mead DA, Hedlund BP (2023) Functional biology and biotechnology of thermophilic viruses. Essays Biochem 67:671–684. https://doi.org/10.1042/EBC20220209

Douglas EJ, Wulandari SW, Lovell SD, Laabei M (2023) Novel antimicrobial strategies to treat multi-drug resistant Staphylococcus aureus infections. Microb Biotechnol 16:1456. https://doi.org/10.1111/1751-7915.14268

Drobiazko AY, Kasimova AA, Evseev PV, Shneider MM, Klimuk EI, Shashkov AS et al (2022) Capsule-targeting depolymerases derived from Acinetobacter baumannii prophage regions. Int J Mol Sci 23(9):4971. https://doi.org/10.3390/ijms23094971

Dronina J, Plausinaitis D, Samukaite-Bubniene U, Ramanavicius A (2022) Real-time label-free assessment of T7 DNA polymerase immobilization. Mater Today Nano 19:100232

Duncan-Lowey B, McNamara-Bordewick NK, Tal N, Sorek R, Kranzusch PJ (2021) Effector-mediated membrane disruption controls cell death in CBASS antiphage defense. Mol Cell 81(24):5039–5051. https://doi.org/10.1016/j.molcel.2021.10.020

Efremenko E, Stepanov N, Aslanli A, Lyagin I, Senko O, Maslova O (2023) Combination of enzymes with materials to give them antimicrobial features: modern trends and perspectives. J Funct Biomater 14(2):64. https://doi.org/10.3390/jfb14020064

Falkenberg M, Gustafsson CM (2020) Mammalian mitochondrial DNA replication and mechanisms of deletion formation. Crit Rev Biochem Mol Biol 55(6):509–524. https://doi.org/10.1080/10409238.2020.1818684

Fenstermaker TK, Petruk S, Mazo A (2024) An emerging paradigm in epigenetic marking: coordination of transcription and replication. Transcription 15:22–37. https://doi.org/10.1080/21541264.2024.2316965

Fenton M, McAuliffe O, O'Mahony J, Coffey A (2010) Recombinant bacteriophage lysins as antibacterials. Bioeng Bugs 1(1):9–16. https://doi.org/10.4161/bbug.1.1.9818

Fernandes S, São-José C (2018) Enzymes and mechanisms employed by tailed bacteriophages to breach the bacterial cell barriers. Viruses 10(8):396. https://doi.org/10.3390/v10080396

Fowler VG Jr, Das AF, Lipka-Diamond J, Ambler JE, Schuch R, Pomerantz R et al (2024) Exebacase in addition to standard-of-care antibiotics for staphylococcus aureus bloodstream infections and right-sided infective endocarditis: a phase 3, superiority-design, placebo-controlled, randomized clinical trial (DISRUPT). Clin Infect Dis:ciae043. https://doi.org/10.1093/cid/ciae043

Giengkam S, Kullapanich C, Wongsantichon J, Adcox HE, Gillespie JJ, Salje J (2023) Orientiatsutsugamushi: comprehensive analysis of the mobilome of a highly fragmented and repetitive genome reveals the capacity for ongoing lateral gene transfer in an obligate intracellular bacterium. Msphere 8(6):e00268–e00223. https://doi.org/10.1128/msphere.00268-23

Gong C, Chakraborty D, Koudelka GB (2024) A prophage encoded ribosomal RNA methyltransferase regulates the virulence of Shiga-toxin-producing Escherichia coli (STEC). Nucleic Acids Res 52(2):856–871. https://doi.org/10.1093/nar/gkad1150

Grabowski Ł, Łepek K, Stasiłojć M, Kosznik-Kwaśnicka K, Zdrojewska K, Maciąg-Dorszyńska M et al (2021) Bacteriophage-encoded enzymes destroying bacterial cell membranes and walls, and their potential use as antimicrobial agents. Microbiol Res 248:126746. https://doi.org/10.1016/j.micres.2021.126746

Gregory AC, Zablocki O, Zayed AA, Howell A, Bolduc B, Sullivan MB (2020) The gut virome database reveals age-dependent patterns of virome diversity in the human gut. Cell Host Microbe 28(5):724–740. https://doi.org/10.1016/j.chom.2020.08.003

Grishin AV, Karyagina AS, Vasina DV, Vasina IV, Gushchin VA, Lunin VG (2020) Resistance to peptidoglycan-degrading enzymes. Crit Rev Microbiol 46(6):703–726. https://doi.org/10.1080/1040841X.2020.1825333

Guo M, Feng C, Ren J, Zhuang X, Zhang Y, Zhu Y et al (2017) A novel antimicrobial endolysin, LysPA26, against *Pseudomonas aeruginosa*. Front Microbiol 8:293. https://doi.org/10.3389/fmicb.2017.00293

Guo Z, Liu M, Zhang D (2023) Potential of phage depolymerase for the treatment of bacterial biofilms. Virulence 14(1):2273567. https://doi.org/10.1080/21505594.2023.2273567

Halder U, Biswas R, Shaw R, Chitikineni A, Varshney RK, Bandopadhyay R (2023) Genome-based characterization of the deep-sea psychrotolerant bacterium Bacillus altitudinis SORB11 isolated from the Indian Sector of the Southern Ocean. Polar Biol:1–19

Han P, Fan H, Tong Y (2023) Identification of a novel family B DNA polymerase from Enterococcus phage IME199 and its overproduction in Escherichia coli BL21 (DE3). Microb Cell Factories 22(1):217. https://doi.org/10.1186/s12934-023-02228-6

Heller RC, Chung S, Crissy K, Dumas K, Schuster D, Schoenfeld TW (2019) Engineering of a thermostable viral polymerase using metagenome-derived diversity for highly sensitive and specific RT-PCR. Nucleic Acids Res 47(7):3619–3630. https://doi.org/10.1093/nar/gkz104

Henson MJ (2023) Preliminary investigation of cluster M lysogeny and prophage-mediated defense mechanisms (Doctoral dissertation, Western Carolina University)

Heselpoth RD, Euler CW, Schuch R, Fischetti VA (2019) Lysocins: bioengineered antimicrobials that deliver lysins across the outer membrane of Gram-negative bacteria. Antimicrob Agents Chemother 63(6):10–1128. https://doi.org/10.1128/AAC.00342-19

Hesse S, Rajaure M, Wall E, Johnson J, Bliskovsky V, Gottesman S, Adhya S (2020) Phage resistance in multidrug-resistant *Klebsiella pneumoniae* ST258 evolves via diverse mutations that culminate in impaired adsorption. MBio 11(1):10–1128. https://doi.org/10.1128/mBio.02530-19

Ho MKY, Zhang P, Chen X, Xia J, Leung SSY (2022) Bacteriophage endolysins against gram-positive bacteria, an overview on the clinical development and recent advances on the delivery and formulation strategies. Crit Rev Microbiol 48(3):303–326. https://doi.org/10.1080/1040841X.2021.1962803

Holtappels D, Alfenas-Zerbini P, Koskella B (2023) Drivers and consequences of bacteriophage host range. FEMS Microbiol Rev 47(4):fuad038. https://doi.org/10.1093/femsre/fuad038

Horne T, Orr VT, Hall JP (2023) How do interactions between mobile genetic elements affect horizontal gene transfer? Curr Opin Microbiol 73:102282. https://doi.org/10.1016/j.mib.2023.102282

de Jong OG, Oosting RS (2024) Molecular biotechnology: from DNA sequence to therapeutic protein. In: Pharmaceutical biotechnology: fundamentals and applications. Springer International Publishing, Cham, pp 3–28

Juarez-Quintero V, Peralta-Castro A, Benítez Cardoza CG, Ellenberger T, Brieba LG (2021) Structure of an open conformation of T7 DNA polymerase reveals novel structural features regulating primer-template stabilization at the polymerization active site. Biochem J 478(13):2665–2679. https://doi.org/10.1042/BCJ20200922

Kajsikova M, Kajsik M, Bocanova L, Papayova K, Drahovska H, Bukovska G (2024) Endolysin EN572-5 as an alternative to treat urinary tract infection caused by Streptococcus agalactiae. Appl Microbiol Biotechnol 108(1):1–14. https://doi.org/10.1007/s00253-023-12949-8

Kato K, Omura H, Ishitani R, Nureki O (2017) Cyclic GMP–AMP as an endogenous second messenger in innate immune signaling by cytosolic DNA. Annu Rev Biochem 86:541–566. https://doi.org/10.1146/annurev-biochem-061516-044813

Khan FM, Chen JH, Zhang R, Liu B (2023) A comprehensive review of the applications of bacteriophage-derived endolysins for foodborne bacterial pathogens and food safety: recent advances, challenges, and future perspective. Front Microbiol 14:1259210. https://doi.org/10.3389/fmicb.2023.1259210

Kieft K, Zhou Z, Anantharaman K (2020) VIBRANT: automated recovery, annotation and curation of microbial viruses, and evaluation of viral community function from genomic sequences. Microbiome 8(1):1–23. https://doi.org/10.1186/s40168-020-00867-0

Kim S, Lee DW, Jin JS, Kim J (2020) Antimicrobial activity of LysSS, a novel phage endolysin, against Acinetobacter baumannii and Pseudomonas aeruginosa. J Glob Antimicrob Resist 22:32–39. https://doi.org/10.1016/j.jgar.2020.01.005

King H, Ajay Castro S, Pohane AA, Scholte CM, Fischetti VA, Korotkova N et al (2021) Molecular basis for recognition of the group A carbohydrate backbone by the PlyC streptococcal bacteriophage endolysin. Biochem J 478(12):2385–2397. https://doi.org/10.1042/BCJ20210158

Kitts G, Rogers A, Teschler JK, Park JH, Trebino MA, Chaudry I et al (2023) The Rvv two-component regulatory system regulates biofilm formation and colonization in Vibrio cholerae. PLoS Pathog 19(5):e1011415. https://doi.org/10.1371/journal.ppat.1011415

Kranjec C, Morales Angeles D, TorrissenMårli M, Fernández L, García P, Kjos M, Diep DB (2021) Staphylococcal biofilms: challenges and novel therapeutic perspectives. Antibiotics 10(2):131. https://doi.org/10.3390/antibiotics10020131

Lai WCB, Chen X, Ho MKY, Xia J, Leung SSY (2020) Bacteriophage-derived endolysins to target gram-negative bacteria. Int J Pharm 589:119833. https://doi.org/10.1016/j.ijpharm.2020.119833

Lama L, Ryan K (2016) Adenylylation of small RNA sequencing adapters using the TS2126 RNA ligase I. RNA 22(1):155–161. https://doi.org/10.1261/rna.054999.115

Latka A, Drulis-Kawa Z (2020) Advantages and limitations of microtiter biofilm assays in the model of antibiofilm activity of Klebsiella phage KP34 and its depolymerase. Sci Rep 10(1):20338. https://doi.org/10.1038/s41598-020-77198-5

Lechuga A, Lood C, Berjón-Otero M, Del Prado A, Wagemans J, van Noort V et al (2021) Unraveling protein interactions between the temperate virus Bam35 and its bacillus host using an integrative yeast two hybrid–high throughput sequencing approach. Int J Mol Sci 22(20):11105. https://doi.org/10.3390/ijms222011105

Lefkowitz RB, Miller CM, Martinez-Caballero JD, Ramos I (2024) Epigenetic control of innate immunity: consequences of acute respiratory virus infection. Viruses 16(2):197. https://doi.org/10.3390/v16020197

Letarov AV (2020) History of early bacteriophage research and emergence of key concepts in virology. Biochem Mosc 85:1093–1112. https://doi.org/10.1134/S0006297920090096

Letrado P, Corsini B, Díez-Martínez R, Bustamante N, Yuste JE, García P (2018) Bactericidal synergism between antibiotics and phage endolysin Cpl-711 to kill multidrug-resistant pneumococcus. Future Microbiol 13(11):1215–1223. https://doi.org/10.2217/fmb-2018-0077

Li J, Li Y, Li Y, Ma Y, Xu W, Wang J (2023) An enhanced activity and thermostability of chimeric Bst DNA polymerase for isothermal amplification applications. Appl Microbiol Biotechnol 107(21):6527–6540. https://doi.org/10.1007/s00253-023-12751-6

Liu B, Guo Q, Li Z, Guo X, Liu X (2023a) Bacteriophage endolysin: a powerful weapon to control bacterial biofilms. Protein J 42(5):463–476. https://doi.org/10.1007/s10930-023-10139-z

Liu H, Kheirvari M, Tumban E (2023b) Potential applications of thermophilic bacteriophages in one health. Int J Mol Sci 24(9):8222. https://doi.org/10.3390/ijms24098222

Lowey B, Whiteley AT, Keszei AF, Morehouse BR, Mathews IT, Antine SP et al (2020) CBASS immunity uses CARF-related effectors to sense 3′–5′-and 2′–5′-linked cyclic oligonucleotide signals and protect bacteria from phage infection. Cell 182(1):38–49. https://doi.org/10.1016/j.cell.2020.05.019

Ma Z, Chen H, Yang Y, Gao S, Yang J, Cui S et al (2023) Characterization of an ssDNA ligase and its application in aptamer circularization. Anal Biochem:115409. https://doi.org/10.1016/j.ab.2023.115409

Maciejewska B, Squeglia F, Latka A, Privitera M, Olejniczak S, Switala P et al (2023) Klebsiella phage KP34gp57 capsular depolymerase structure and function: from a serendipitous finding to the design of active mini-enzymes against K. pneumoniae. MBio 14(5):e01329–e01323. https://doi.org/10.1128/mbio.01329-23

Mayo-Muñoz D, Pinilla-Redondo R, Camara-Wilpert S, Birkholz N, Fineran PC (2024) Inhibitors of bacterial immune systems: discovery, mechanisms and applications. Nat Rev Genet 25:237–254. https://doi.org/10.1038/s41576-023-00676-9

McKeithen-Mead SA, Grossman AD (2023) Timing of integration into the chromosome is critical for the fitness of an integrative and conjugative element and its bacterial host. PLoS Genet 19(2):e1010524. https://doi.org/10.1371/journal.pgen.1010524

Meng X, Shi Y, Ji W, Meng X, Zhang J, Wang H et al (2011) Application of a bacteriophage lysin to disrupt biofilms formed by the animal pathogen Streptococcus suis. Appl Environ Microbiol 77(23):8272–8279. https://doi.org/10.1128/AEM.05151-11

Mitousis L, Thoma Y, Musiol-Kroll EM (2020) An update on molecular tools for genetic engineering of actinomycetes—the source of important antibiotics and other valuable compounds. Antibiotics 9(8):494. https://doi.org/10.3390/antibiotics9080494

Molina-Quiroz RC, Camilli A, Silva-Valenzuela CA (2023) Role of bacteriophages in the evolution of pathogenic Vibrios and lessons for phage therapy. In: Vibrio spp. infections. Springer International Publishing, Cham, pp 149–173. https://doi.org/10.1007/978-3-031-22997-8_8

Morcinek-Orłowska J, Zdrojewska K, Węgrzyn A (2022) Bacteriophage-encoded DNA polymerases—beyond the traditional view of polymerase activities. Int J Mol Sci 23(2):635. https://doi.org/10.3390/ijms23020635

Moser MJ, DiFrancesco RA, Gowda K, Klingele AJ, Sugar DR, Stocki S et al (2012) Thermostable DNA polymerase from a viral metagenome is a potent RT-PCR enzyme. PLoS One 7(6):e38371. https://doi.org/10.1371/journal.pone.0038371

Nardulli P, Ballini A, Zamparella M, De Vito D (2023) The role of stakeholders' understandings in emerging antimicrobial resistance: a one health approach. Microorganisms 11(11):2797. https://doi.org/10.3390/microorganisms11112797

Nazir A, Xu X, Liu Y, Chen Y (2023) Phage endolysins: advances in the world of food safety. Cells 12(17):2169. https://doi.org/10.3390/cells12172169

Netter Z, Boyd CM, Silvas TV, Seed KD (2021) A phage satellite tunes inducing phage gene expression using a domesticated endonuclease to balance inhibition and virion hijacking. Nucleic Acids Res 49(8):4386–4401. https://doi.org/10.1093/nar/gkab207

Nguyen HM, Watanabe S, Sharmin S, Kawaguchi T, Tan XE, Wannigama DL, Cui L (2023) RNA and single-stranded DNA phages: unveiling the promise from the underexplored world of viruses. Int J Mol Sci 24(23):17029. https://doi.org/10.3390/ijms242317029

Oliveira A, Ribeiro HG, Silva AC, Silva MD, Sousa JC, Rodrigues CF et al (2017) Synergistic antimicrobial interaction between honey and phage against Escherichia coli biofilms. Front Microbiol 8:2407. https://doi.org/10.3389/fmicb.2017.02407

Olszak T, Latka A, Roszniowski B, Valvano MA, Drulis-Kawa Z (2017) Phage life cycles behind bacterial biodiversity. Curr Med Chem 24(36):3987–4001. https://doi.org/10.2174/0929867324666170413100136

Patel DJ, Yu Y, Jia N (2022) Bacterial origins of cyclic nucleotide-activated antiviral immune signaling. Mol Cell 82:4591. https://doi.org/10.1016/j.molcel.2022.11.006

Peri KV, Faria-Oliveira F, Larsson A, Plovie A, Papon N, Geijer C (2023) Split-marker-mediated genome editing improves homologous recombination frequency in the CTG clade yeast Candida intermedia. FEMS Yeast Res 23:foad016. https://doi.org/10.1093/femsyr/foad016

Qi Q, Rajabal V, Ghaly TM, Tetu SG, Gillings MR (2023) Identification of integrons and gene cassette-associated recombination sites in bacteriophage genomes. Front Microbiol 14:58. https://doi.org/10.3389/fmicb.2023.1091391

Rands CM, Brüssow H (2021) The role of bacteriophages in bacterial evolution, academic press, 1:633–643. https://doi.org/10.1016/B978-0-12-809633-8.20971-0

Raz A, Serrano A, Hernandez A, Euler CW, Fischetti VA (2019) Isolation of phage lysins that effectively kill Pseudomonas aeruginosa in mouse models of lung and skin infection. Antimicrob Agents Chemother 63(7):10–1128. https://doi.org/10.1128/AAC.00024-19

Redrejo-Rodríguez M, Ordonez CD, Berjón-Otero M, Moreno-González J, Aparicio-Maldonado C, Forterre P et al (2017) Primer-independent DNA synthesis by a family B DNA polymerase from self-replicating mobile genetic elements. Cell Rep 21(6):1574–1587. https://doi.org/10.1016/j.celrep.2017.10.039

Reuter M, Kruger DH (2020) Approaches to optimize therapeutic bacteriophage and bacteriophage-derived products to combat bacterial infections. Virus Genes 56(2):136–149. https://doi.org/10.1007/s11262-020-01735-7

Rodrıguez-Rubio L, Martínez Fernández B, Donovan DM, García Suárez MP, Rodríguez González A (2013) Potential of the Virion-associated peptidoglycan hydrolase HydH5 and its derivative fusion proteins in milk biopreservation. PLoS One 8:e54828. https://doi.org/10.1371/journal.pone.0054828

Roehrig C, Huemer M, Lorgé D, Arn F, Heinrich N, Selvakumar L et al (2024) MEndoB, a chimeric lysin featuring a novel domain architecture and superior activity for the treatment of staphylococcal infections. MBio 15(2):e02540–e02523. https://doi.org/10.1128/mbio.02540-23

Rostøl JT, Xie W, Kuryavyi V, Maguin P, Kao K, Froom R et al (2021) The Card1 nuclease provides defence during type III CRISPR immunity. Nature 590(7847):624–629. https://doi.org/10.1038/s41586-021-03206-x

Sahu R, Singh AK, Kumar A, Singh K, Kumar P (2023) Bacteriophages concept and applications: a review on phage therapy. Curr Pharm Biotechnol 24(10):1245–1264. https://doi.org/10.2174/1389201024666221104142457

Salas M, Holguera I, Redrejo-Rodríguez M, De Vega M (2016) DNA-binding proteins essential for protein-primed bacteriophage Φ29 DNA replication. Front Mol Biosci 3:37. https://doi.org/10.3389/fmolb.2016.00037

Sauve K, Watson A, Oh JT, Swift S, Vila-Farres X, Abdelhady W et al (2024) The engineered Lysin, CF-370, is active against antibiotic-resistant gram-negative pathogens *in vitro* and synergizes with meropenem in experimental Pseudomonas aeruginosa pneumonia. J Infect Dis:jiae027. https://doi.org/10.1093/infdis/jiae027

Schaap-Johansen AL, Vujović M, Borch A, Hadrup SR, Marcatili P (2021) T cell epitope prediction and its application to immunotherapy. Front Immunol 12:712488. https://doi.org/10.3389/fimmu.2021.712488

Schroven K, Aertsen A, Lavigne R (2021) Bacteriophages as drivers of bacterial virulence and their potential for biotechnological exploitation. FEMS Microbiol Rev 45(1):fuaa041. https://doi.org/10.1093/femsre/fuaa041

Secco V (2023) Microbes and biocultural diversity in the Ganges: antibiotic modernity and the revival of phage therapy. In: Nurturing alternative futures. Routledge, pp 169–186

Seed KD, Lazinski DW, Calderwood SB, Camilli A (2013) A bacteriophage encodes its own CRISPR/Cas adaptive response to evade host innate immunity. Nature 494(7438):489–491. https://doi.org/10.1038/nature11927

Sertic VLADIMIR (1929) Origine de la lysine d'une race du bactériophage. Compt rend soc biol 100:477–479

Severin GB, Ramliden MS, Hawver LA, Wang K, Pell ME, Kieninger AK et al (2018) Direct activation of a phospholipase by cyclic GMP-AMP in El Tor Vibrio cholerae. Proc Natl Acad Sci 115(26):E6048–E6055. https://doi.org/10.1073/pnas.1801233115

Sharma A, Gupta SP (2017) Fundamentals of viruses and their proteases. In: Viral proteases and their inhibitors. Academic Press, pp 1–24

Shuwen H, Kefeng D (2022) Intestinal phages interact with bacteria and are involved in human diseases. Gut Microbes 14(1):2113717. https://doi.org/10.1080/19490976.2022.2113717

Sisson HM, Jackson SA, Fagerlund RD, Warring SL, Fineran PC (2024) Gram-negative endolysins: overcoming the outer membrane obstacle. Curr Opin Microbiol 78:102433. https://doi.org/10.1016/j.mib.2024.102433

Soffer A, Eisdorfer SA, Ifrach M, Ilic S, Afek A, Schussheim H et al (2021) Inferring primase-DNA specific recognition using a data driven approach. Nucleic Acids Res 49(20):11447–11458. https://doi.org/10.1093/nar/gkab956

Son SM, Kim J, Ryu S (2023) Development of sensitizer peptide-fused endolysin Lys1S-L9P acting against multidrug-resistant gram-negative bacteria. Front Microbiol 14:1296796. https://doi.org/10.3389/fmicb.2023.1296796

Sousa JAMD, Rocha EP (2019) Environmental structure drives resistance to phages and antibiotics during phage therapy and to invading lysogens during colonisation. Sci Rep 9(1):3149. https://doi.org/10.1038/s41598-019-39773-3

Stegmüller J, Estévez MR, Shu W, Gläser L, Myronovskyi M, Rückert-Reed C et al (2024) Systems metabolic engineering of the primary and secondary metabolism of *Streptomyces albidoflavus* enhances production of the reverse antibiotic nybomycin against multi-resistant Staphylococcus aureus. Metab Eng 81:123–143. https://doi.org/10.1016/j.ymben.2023.12.004

Steinbacher S, Baxa U, Miller S, Weintraub A, Seckler R, Huber R (1996) Crystal structure of phage P22 tail spike protein complexed with Salmonella sp. O-antigen receptors. Proc Natl Acad Sci 93(20):10584–10588. https://doi.org/10.1073/pnas.93.20.10584

Stritt K (2014) Cloning of bacteriophage lysins and evaluation of their antibacterial activity against Enterococcus faecalis (Doctoral dissertation, Université de Lausanne, Faculté de Biologie et Medecine)

Sun Y, Saha S, Wang W, Saha LK, Huang SYN, Pommier Y (2020) Excision repair of topoisomerase DNA-protein crosslinks (TOP-DPC). DNA Repair 89:102837. https://doi.org/10.1016/j.dnarep.2020.102837

Swift SM, Reid KP, Donovan DM, Ramsay TG (2019) Thermophile lytic enzyme fusion proteins that target Clostridium perfringens. Antibiotics 8(4):214. https://doi.org/10.3390/antibiotics8040214

Tan Y, Su J, Fu M, Zhang H, Zeng H (2022) Recent advances in phage-based therapeutics for multi-drug resistant Acinetobacter baumannii. Bioengineering 10(1):35. https://doi.org/10.3390/bioengineering10010035

Tan WS, Rong E, Dry I, Lillico S, Law A, Digard P et al (2024) Validation of candidate host cell entry factors for bovine herpes virus Type-1 based on a genome-wide CRISPR knockout screen. Viruses 16(2):297. https://doi.org/10.3390/v16020297

Thandar M, Lood R, Winer BY, Deutsch DR, Euler CW, Fischetti VA (2016) Novel engineered peptides of a phage lysin as effective antimicrobials against multidrug-resistant Acinetobacter baumannii. Antimicrob Agents Chemother 60(5):2671–2679. https://doi.org/10.1128/AAC.02972-15

Tian F, Li J, Nazir A, Tong Y (2021) Bacteriophage—a promising alternative measure for bacterial biofilm control. Infection and drug resistance 14:205–217. https://doi.org/10.2147/IDR.S290093

Topka-Bielecka G, Dydecka A, Necel A, Bloch S, Nejman-Faleńczyk B, Węgrzyn G, Węgrzyn A (2021) Bacteriophage-derived depolymerases against bacterial biofilm. Antibiotics 10(2):175. https://doi.org/10.3390/antibiotics10020175

Vander Elst N, Bert J, Favoreel H, Lavigne R, Meyer E, Briers Y (2023) Development of engineered endolysins with *in vitro* intracellular activity against streptococcal bovine mastitis-causing pathogens. Microb Biotechnol 16(12):2367–2386. https://doi.org/10.1111/1751-7915.14339

Vermassen A, Leroy S, Talon R, Provot C, Popowska M, Desvaux M (2019) Cell wall hydrolases in bacteria: insight on the diversity of cell wall amidases, glycosidases and peptidases toward peptidoglycan. Front Microbiol 10:331. https://doi.org/10.3389/fmicb.2019.00331

Vieira M, Duarte J, Domingues R, Oliveira H, Dias O (2023) PhageDPO: phage depolymerase finder. BioRxiv, 2023-02

Villarreal LP (2009) The prokaryotes: virus, hyperparasites and the origin of group identity. In: Origin of group identity: viruses, addiction and cooperation, pp 27–100

Walkinshaw DR, Wright ME, Mullin AE, Excler JL, Kim JH, Steer AC (2023) The *Streptococcus pyogenes* vaccine landscape. NPJ Vaccines 8(1):16. https://doi.org/10.1038/s41541-023-00609-x

Wang S, Gu J, Lv M, Guo Z, Yan G, Yu L et al (2017) The antibacterial activity of E. coli bacteriophage lysin lysep3 is enhanced by fusing the Bacillus amyloliquefaciens bacteriophage endolysin binding domain D8 to the C-terminal region. J Microbiol 55:403–408. https://doi.org/10.1007/s12275-017-6431-6

Wang H, Liu Y, Bai C, Leung SSY (2023) Translating bacteriophage-derived depolymerases into antibacterial therapeutics: challenges and prospects. Acta Pharm Sin B 14:155–169. https://doi.org/10.1016/j.apsb.2023.08.017

Wiebe KG, Cook BW, Lightly TJ, Court DA, Theriault SS (2024) Investigation into scalable and efficient enterotoxigenic Escherichia coli bacteriophage production. Sci Rep 14(1):3618. https://doi.org/10.1038/s41598-024-53276-w

Wu M, Hu K, Xie Y, Liu Y, Mu D, Guo H et al (2019) A novel phage PD-6A3, and its endolysin Ply6A3, with extended lytic activity against Acinetobacter baumannii. Front Microbiol 9:3302. https://doi.org/10.3389/fmicb.2018.03302

Xi H, Fu Y, Chen C, Feng X, Han W, Gu J, Ji Y (2023) Aerococcusviridans phage Lysin AVPL had lytic activity against Streptococcus suis in a mouse bacteremia model. Int J Mol Sci 24(23):16670. https://doi.org/10.3390/ijms242316670

Yan G, Liu J, Ma Q, Zhu R, Guo Z, Gao C et al (2017) The N-terminal and central domain of colicin A enables phage lysin to lyse Escherichia coli extracellularly. Antonie Van Leeuwenhoek 110:1627–1635. https://doi.org/10.1007/s10482-017-0912-9

Yang H, Zhang Y, Yu J, Huang Y, Zhang XE, Wei H (2014) Novel chimeric lysin with high-level antimicrobial activity against methicillin-resistant Staphylococcus aureus *in vitro* and *in vivo*. Antimicrob Agents Chemother 58(1):536–542. https://doi.org/10.1128/AAC.01793-13

Yang Y, Le S, Shen W, Chen Q, Huang Y, Lu S et al (2018) Antibacterial activity of a lytic enzyme encoded by Pseudomonas aeruginosa double stranded RNA bacteriophage phiYY. Front Microbiol 9:1778. https://doi.org/10.3389/fmicb.2018.01778

Zabrady K, Li AWH, Doherty AJ (2023) Mechanism of primer synthesis by primase-polymerases. Curr Opin Struct Biol 82:102652. https://doi.org/10.1016/j.sbi.2023.102652

Zajac P, Islam S, Hochgerner H, Lönnerberg P, Linnarsson S (2013) Base preferences in non-templated nucleotide incorporation by MMLV-derived reverse transcriptases. PLoS One 8(12):e85270. https://doi.org/10.1371/journal.pone.0085270

Zeb A, Khan N, Shah FA, Ullah A, Choi HI, Ryu JS et al (2024) A review on the nanotechnology-based approaches for managing sexually transmitted infections. J Pharm Investig 54:1–34

Zhang Y, Hu Z (2013) Combined treatment of *Pseudomonas aeruginosa* biofilms with bacteriophages and chlorine. Biotechnol Bioeng 110(1):286–295. https://doi.org/10.1002/bit.24630

Zhang J, Su X, Wang Y, Wang X, Zhou S, Jia H et al (2023a) Improved single-cell genome amplification by a high-efficiency phi29 DNA polymerase. Front Bioeng Biotechnol 11. https://doi.org/10.3389/fbioe.2023.1233856

Zhang S, Zhang Y, Liu R, Yuan S, Chen Y, Li W et al (2023b) Characterization and molecular mechanism of aminoglycoside-6-Adenyl transferase associated with aminoglycoside resistance from Elizabethkingiameningoseptica. Infect Drug Resist 16:5523–5534. https://doi.org/10.2147/IDR.S423418

Zhou Y, Shen W, Gao Y, Peng J, Li Q, Wei X et al (2024) Protein-templated ligand discovery via the selection of DNA-encoded dynamic libraries. Nat Chem 16:543–555. https://doi.org/10.1038/s41557-024-01442-y

Bacteriophage-Derived Enzymes as Potential Alternatives for the Treatment of MDR Bacterial Infections

33

Subhash Lal Karn, Satyanam Kumar Bhartiya, Sandeep Pokhrel, Mayank Gangwar, and Gopal Nath

Abstract

Bacteriophage lytic enzymes, also known as lysins, are highly evolved chemicals that are produced by bacteriophage to break down bacterial cell walls and release their progeny. These enzymes, known as "enzybiotics," have an antibacterial property due to their ability to degrade peptidoglycan, a crucial part of the bacterial cell wall. Phage lytic products kill the target bacteria when introduced exogenously as recombinant proteins. Gram-positive bacteria immediately lyse when small amounts of purified recombinant lysin are added, which causes the target bacterium to die log-fold. Lysins have been successfully employed in animal studies to combat pathogenic antibiotic-resistant bacteria. Their distinct mechanism of action is responsible for their quick action, potent bactericidal impact, absence of resistance development, and activity against persister cells. Thus, in a time when antibiotic resistance is increasing, lysins may be a useful anti-

S. L. Karn (✉) · S. Pokhrel
Department of Microbiology, Nepalgunj Medical College, Kohalpur, Banke, Nepal

S. K. Bhartiya
Department of General Surgery, Institute of Medical Sciences, Banaras Hindu University, Varanasi, India

M. Gangwar
Faculty of Medicine, Viral Research and Diagnostic Laboratory, Department of Microbiology, Institute of Medical Sciences, Banaras Hindu University, Varanasi, Uttar Pradesh, India

Clinical Studies and Trials Unit, Division of Development Research, Indian Council of Medical Research, Department of Health Research, Ministry of Health and Family Welfare, Govt. of India, V. Ramalingaswami Bhawan, New Delhi, India

G. Nath
Department of Microbiology, Institute of Medical Sciences, Banaras Hindu University, Varanasi, India

M. Gangwar, G. Nath (eds.), *Emerging Paradigms for Antibiotic-Resistant Infections: Beyond the Pill*, https://doi.org/10.1007/978-981-97-5272-0_33

infective. This chapter thoroughly overviews the therapeutic application and eventual development of phage-encoded lysins as novel anti-infectives. It also describes the limitations of both agents with regard to their composition, specificity, mechanism of action, emergence of resistance, method of product production, and how they interact with the immune system.

Keywords

Bacteriophages · Lytic enzymes · Enzybiotics · MDR

33.1 Introduction

Antibiotics have been overused and misused in humans and animals since they were first used in clinical practice in the 1940s. This resulted in the uncontrolled emergence and dissemination of antibiotic-resistance determinants in nearly every bacterial pathogen over time, some of which are now highly resistant to all available antibiotics (Ventola 2015) If effective solutions for tackling antibiotic resistance are not found quickly, several worldwide studies predict catastrophic global scenarios with tens of millions of deaths annually and expenditures rising to trillions of USD by 2050 (Jonas et al. 2017; O'Neill 2016). This leads to the resurrection of phage therapy.

There have been attempts to create antibacterial drugs from the bacteriostatic and bacteriolytic proteins that phages encode in parallel with phage therapy (Oliveira et al. 2013; Schmelcher et al. 2012). The most well-known ones are the endolysins and phage-encoded peptidoglycan hydrolases (PGH). Endolysins break down peptidoglycan, which weakens the cell wall and allows the host cell to rupture, contributing to the release of freshly produced viral particles. Certain phages also encode a second PGH type called virion-associated peptidoglycan hydrolase (VAPGH). During infection, or the initial cell wall penetration, VAPGH breaks down localized peptidoglycan. Both forms of PGH, when isolated, can be bacteriolytic when added exogenously to cells. Polysaccharide depolymerases, which break down the bacterial cell wall envelope's macromolecule polysaccharides, are another enzyme that phages produce. Extracellular polysaccharides can shield bacteria from the host immune system, desiccation, and antibiotics. The viral particle uses depolymerases to help it adsorb, invade, and break down the host bacterium. This process can go on even inside biofilms. The cytoplasmic membrane is disrupted by holins, which help endolysins lyse cells (Roach et al. 2013; Roach and Donovan 2015).

Since the peptidoglycan layer gives the cell its structural rigidity and integrity, its degradation causes the cell wall to become fragile and eventually rupture due to differences in the osmotic pressures inside and outside the cell (osmolysis). This is particularly applicable to bacteria that are Gram-positive and do not have an outer membrane. Gram-negative cells are significantly more susceptible to endolysin-induced lysis, which requires additional disruption of the outer membrane by spanin complexes, catalyzing a fusion of the inner and outer membranes (Rajaure et al. 2015; Young 2014). Even while lytic enzymes have been known to have phage lytic activity for almost a century, their application for in vivo bacterial control is relatively new (Loeffler et al. 2003; Nelson et al. 2012; Schuch et al. 2002).

33.2 Structure, Specificity, and Host Range

The endolysin derived from phages that infect Gram-positive hosts commonly possesses a modular domain structure consisting of an N-terminal enzymatically active domain (EAD) and a C-terminal cell wall binding domain (CBD), which are linked together by a short linker region (Fig. 33.1). Nevertheless, endolysins may not always share the same proximity or orientation of these domains; for example, a CBD may be present or lacking on the N-terminus, mid-protein, or C-terminus. While endolysins from Gram-positive phages exhibit this multi-domain organization, endolysins from Gram-negative phages are typically globular and devoid of CBDs (Lai et al. 2011; Lood et al. 2015; Walmagh et al. 2013).

According to various studies (Latka et al. 2017; Paul et al. 2011; Rodríguez et al. 2013), the specificity and host range of PG degrading lysins (endolysins and VALs) differ and depend on both the features of the protein as well as the species/genus of phage. Since their target is characterized by significant diversity in peptide composition, crosslinks, and glycan chain modification, lysins encoded by Gram-positive-specific phages have developed along their course (Schleifer and Kandler 1972). As a result, only specific bacterial species or even serotypes can utilize such enzymes. On the contrary hand, the PG of Gram-negative bacteria has a very conservative structure with substantial species-to-species similarities. Consequently, endolysins and VALs typically exhibit broad host range activity (Briers et al. 2007; Latka et al. 2017). Since bacteria may create a wide variety of glycans, including capsules (CPS, K-serotype), O-polysaccharide chains (LPS, O-serotype), and extracellular polysaccharides (EPS), the third class of antibacterial enzymes, known as depolymerases, exhibits significant substrate specificity.

Phage lysins result from millions of years of evolution by bacteriophage during their interaction with bacteria, much as antibiotics, which are utilized by bacteria to control the organisms surrounding them in the environment. It has now become possible to identify and purify certain lysins that, when given to Gram-positive bacteria, would kill them within seconds of contact(Loeffler et al. 2001; Nelson et al. 2001). Lysin in nanogram amounts, for instance, might reduce 10^7 *S. pyogenes* by >6 logs shortly after the addition of the enzyme. Besides chemical agents, no known biological substance can kill bacteria rapidly (Fischetti 2005).

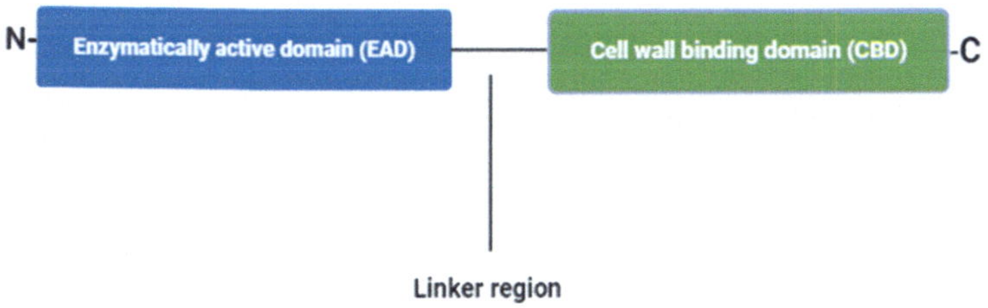

Fig. 33.1 Schematic diagram of endolysin structure

33.3 Mode of Action

Based on their mode of action, phage-encoded lysins can be divided into two main groups: (1) hydrolases that hydrolyze PG bonds and (2) lytic.

Transglycosylases cleave glycoside bonds in glycan chains to create 1,6-anhydro rings at the N-acetylmuramic acid residue. Glucosaminidases and lysozymes (muramidases) hydrolyze glycoside bonds in the glycan chain, while amidases hydrolyze amide bonds. These three groups of enzymes are distinguished according to the nature of the chemical bond degraded in PG (Nelson et al. 2012).

Lysin has been shown to be highly efficient against Gram-positive infections such as *Bacillus anthracis*, MRSA, and *Streptococcus pneumoniae* in numerous in vivo experiments. On the contrary, because of the outer membrane layer, exogenous enzymes have a limited impact on Gram-negative bacteria. Up until now, research has only demonstrated that a small number of endolysins, such as SPN9CC, PlyF307, and CfP1gp153, can act as external agents to penetrate the outer membrane and break down Gram-negative PG (Lim et al. 2014; Lood et al. 2015; Oliveira et al. 2016b). The lytic activity of phage lysins varies greatly, from 100 to 10^8 U/mg, yet they are still considered the strongest PG hydrolyzers. PlyC endolysin, which is obtained from the streptococcal C1 phage, is several orders of magnitude more active than any other endolysin and may quickly clear a bacterial culture (Schmelcher et al. 2012).

Phage depolymerases break down carbohydrate macromolecules in the bacterial cell membrane. Based on how they function, depolymerases are classified as either (1) hydrolases or (2) lyases that cleave a glycosidic bond through trans-β-elimination. The hydrolases include sialidases, which hydrolyze internal α-2,8-linkages in capsular polysialic acid; rhamnosidases, which hydrolyze α-1,3 O-glycosidic bonds between L-rhamnose and D-galactose in the *Salmonella* LPS O-antigen; levanases, which hydrolyze β-2,6-bonds between fructose monomers in levan; xylanases, which cut β-1,4 bonds within xylan; dextranases, which hydrolyze α-1,6-linkages between glucose units in dextran); and LPS deacetylases, which deacetylate the O-antigen instead of breaking the polysaccharide chain (Latka et al. 2017; Prokhorov et al. 2017). The lyases are alginate lyase (cutting α-1,4 bonds in polygalacturonic acid), pectate lyase (cleaving α-1,4 bonds in hyaluronic acid), and hyaluronate lyase (cleaving β-1,4 bonds in hyaluronic acid).

According to Latka et al. (2017) and Pires et al. (2016), depolymerases, which function as antimicrobials, can be effectively employed as exogenous agents to break down bacterial capsules, lipopolysaccharides, and exopolysaccharides. They work as anti-virulent agents, making bacteria more susceptible to antimicrobials, the immune system, and desiccation.

An outer membrane prevents endolysins from reaching the PG in Gram-negative bacteria (Fenton et al. 2010). The use of organic acids and EDTA, for example (Oliveira et al. 2014), as well as a recombinant fusion of endolysins with LPS-destabilizing peptides (engineered endolysins) (Briers et al. 2014), are two strategies that damage the outer membrane of Gram-negative pathogens. Because their positively charged portions interact with the LPS molecules, a few endolysins of Gram-negative pathogens exhibit an intrinsic outer membrane permeabilizing behavior (Oliveira et al. 2016a; Pires et al. 2016). Additional methods of cell permeabilization are not necessary for the antibacterial action of these endolysins.

Mycobacteria have a structurally distinct cell wall compared to Gram-positive and Gram-negative bacteria. Mycobacterium-infecting phages secrete two distinct lytic enzymes that break the mycolyl–arabinogalactan–PG complex that constitutes this cell wall. LysA carries out the degradation of the PG layer, whereas LysB breaks the bond between the arabinogalactan layer (AG), mycolic acids (MA), glycolipids (GL), and lipoarabinomannan (LA) mesh (Catalão and Pimentel 2018; Gerstmans et al. 2016).

Some pathogens, like *S. aureus*, live intracellularly in their host eukaryotic cells, where they avoid both the host immune system and medications (Hébert et al. 2000). Protein transport across eukaryotic cell membranes is naturally facilitated by cationic peptides called protein transduction domains (PTDs) (Torchilin 2008). It is also feasible for an endolysin to fuse with a PTD (Borysowski and Górski 2010; Schmelcher et al. 2012) Using PTD-staphylolytic endolysin hybrids, the intracellular *S. aureus* load in three distinct eukaryotic epithelial cell lines can be reduced by up to 95% when PTD-assisted uptake is compared to non-PTD enzymes (Roach and Donovan 2015).

33.4 Biofilm Eradication

Phage depolymerases are responses against the thick polysaccharides that envelop bacterial cells, covering the phage receptor necessary for the bacterium to attach to its host. One of these layers is the biofilm matrix, which is primarily made up of exopolysaccharides. Additionally, when LPS-containing outer membrane vesicles are anchored inside the matrix, LPS-degrading enzymes can also loosen the structure of biofilms (Olszak et al. 2017). Many studies (Bansal et al. 2014; Cornelissen et al. 2011; Gutiérrez et al. 2015; Mushtaq et al. 2004, 2005; Pan et al. 2015). confirm the effectiveness of depolymerases in eliminating biofilms produced by both Gram-positive and Gram-negative bacteria. Phage lysins, in addition to depolymerases, have also been effectively employed in eliminating bacterial biofilms.

The majority of the studies focused on treating streptococcal biofilms (Domenech et al. 2011; Rico-Lastres et al. 2015; Shen et al. 2013) and *S. aureus* biofilms (Fenton et al. 2013; Gutierrez et al. 2014; Singh et al. 2014; Yang et al. 2015). The endolysins Lys68 LysPA26 (Guo et al. 2017), and PlyF307 (Lood et al. 2015) have demonstrated their efficacy against *Salmonella Typhimurium*, *P. aeruginosa*, and *A. baumannii*, respectively.

33.5 Development of Resistance

Given that resistance to phage lysins is an exceptionally uncommon event, they are distinct from whole phages and antibiotics. According to Fischetti (2010), lysin can attach to and cleave highly conserved targets inside the cell wall. Furthermore, bacterial resistance is less likely to develop due to the high specificity of most endolysins (Fischetti 2005). Nonetheless, secondary alterations in bacterial cell walls, like D-alanylation in teichoic acids or O-acetylation and N deacetylation in PG, can be viewed as possible defense mechanisms against phage lysins, similar to what has been documented for human lysozyme (Guariglia-Oropeza and Helmann 2011; Vollmer et al. 2008). Studies have demonstrated that repeated treatment with low concentrations of the enzyme does not result in resistance phenotypes to native or engineered phage lysins (Fischetti 2005; Gilmer et al. 2017; Loeffler et al. 2001; Pastagia et al. 2011; Schuch et al. 2002). Resistance to phage depolymerases frequently arises from modifications or variations in the polysaccharide content of capsules, exopolysaccharides or LPS); when depolymerase was used, resistant mutants of *Klebsiella* O1:K20 and *Escherichia coli* O9:K30 quickly emerged (McCallum et al. 1989).

33.6 Product Modification

Phage lysins can be made more effective by utilizing modern synthetic biology techniques. Two ways to boost lytic activity are exchanging the cell wall binding domain (CBD) or enzymatic domain (EAD) by random mutagenesis. Adding a heterologous EAD to a full-length enzyme, duplicating CBD, fusing two full-length endolysins, adding a heterologous CBD to a truncated endolysin, or combining two heterologous CBDs were the five methods used in experiments to expand the spectrum of enzymes (Cheng and Fischetti 2007; Schmelcher et al. 2012). According to recent studies, PG in Gram-negative bacteria can be effectively destroyed by applying genetically engineered endolysins (Artilysins®) or enzymes in combination with membrane permeabilized (Briers et al. 2011; Yang et al. 2015).

To increase the antibacterial action against Gram-negative bacteria, in addition to structure-based alteration, lysin preparation could be combined with outer membrane destabilizing agents such as EDTA, weak organic acids (citric acid), and polycationic agents (Oliveira et al. 2014). Information regarding phage depolymerase modification is scarce. The fact that these enzymes are rather large, typically ~1000 aa—and create a complex spatial structure of trimers or occasionally tetramers is one of the primary causes. Despite being an essential component of phage particles, depolymerases are still poorly understood proteins in terms of their specificity and enzymatic activity (Latka et al. 2017).

33.7 Effect on the Immune System

Phage enzymes are proteinaceous, which causes them to quickly trigger an immunological response and produce neutralizing antibodies (Fischetti 2010). Several animal studies (Daniel et al. 2010; Jado et al. 2003; Jun et al. 2014; Rashel et al. 2007) showed the presence of antibodies against Cpl-1, Pal, MVL, ClyS, and SAL-1 endolysins.

Antibodies significantly shortened the systemic half-life of enzymes to around 20 min despite their weak efficacy in inactivating lysin. The antibacterial efficacy of lysins is slowed down but not eliminated by antibodies, according to in vitro and in vivo studies conducted on several endolysins and pathogens (Loeffler et al. 2003).

33.8 Product Preparation

There should not be any significant issues while preparing lysin/depolymerase formulations. Recombinant protein production and purification techniques and tactics are constantly being improved, and large-scale, ultrapure protein can be produced quickly under well-established circumstances (Wingfield 2015). The phage-based products that have been characterized thus far have the ideal formulation properties; they can be frozen or lyophilized and can withstand weeks or even months of stability at refrigerator storage temperature (4 °C) (Gilmer et al. 2013; Jun et al. 2013; Pastagia et al. 2011) Protein stability can be further enhanced through molecular engineering (Heselpoth et al. 2015) or by choosing the ideal conditions (protein concentration, buffer, pH, temperature, and extra stabilizers) (Jun et al. 2013).

Most of the phage enzymes that have been described thus far have demonstrated a considerable degree of stability across a broad pH range (Maciejewska et al. 2017; Olszak et al. 2017), indicating that they may continue to operate even after being administered orally. A number of phage-based products demonstrated a high degree of thermostability (above 80 °C), a characteristic that broadens their industrial application as, for example, food preservatives (Maciejewska et al. 2017; Majkowska-Skrobek et al. 2016; Olszak et al. 2017; Rodríguez-Rubio et al. 2016).

33.9 Clinical Trials of Endolysin

Stevens et al. (2023) reported the first successful treatment and resolution of a persistent *Enterococcus faecalis* prostate infection using endolysin, a muralytic enzyme generated from bacteriophages. *E. faecalis* was eliminated after a seven-day endolysin therapy course, and symptoms were reduced to almost normal levels. In addition, the patient had no adverse reactions while receiving endolysin. The study illustrates how an endolysin, a novel treatment option, can effectively manage a recalcitrant infection that conventional antibiotic therapy failed to control.

A single dose of CF-301 has a low potential to cause an inflammatory response, according to a clinical trial by Jandourek et al. (2017) examining the safety and acceptability of intravenous doses in healthy subjects. A 180-day immunological monitoring period could not find any correlation between the production of a particular antibody and hypersensitivity factors (basophils and IgE).

There have been two announcements of additional clinical phase II trials, and the clinical efficacy can be assessed soon (Abdelkader et al. 2019).

33.10 Conclusions

There is an increasing need to revitalize outdated antimicrobials and develop new ones to cure bacterial infections in an era where multi-drug resistant infections are emerging more frequently, and few new antibiotics are available. In this situation, endolysins and VAPGH are useful against various bacterial pathogens that are primarily Gram-positive, according to preclinical and clinical data. Additionally, endolysins capable of targeting Gram-negative pathogens are on the rise. It is anticipated that additional endolysin discovery and engineering of endolysin-based enzymes with outer membrane transporters would increase the potential use of endolysins as therapeutics for treating human and animal health problems. However, the recent safety and efficacy validation for two stapholytic endolysins may offer the required level of efficacy for this class of antibacterial agents to control MDR. Further development of phage-encoded lysins would be a challenge. For particular purposes, improved lytic enzymes may result from protein engineering, domain swapping, and gene shuffling. The estimated number of phages on Earth is 10^{31}. Consequently, there is an enormous chance to discover new lysins. Phage lysins might eventually become a vital weapon in our arsenal against pathogenic bacteria infections.

References

Abdelkader K, Gerstmans H, Saafan A, Dishisha T, Briers Y (2019) The preclinical and clinical progress of bacteriophages and their lytic enzymes: the parts are easier than the whole. Viruses 11(2):96

Bansal S, Soni SK, Harjai K, Chhibber S (2014) Aeromonas punctata derived depolymerase that disrupts the integrity of Klebsiella pneumoniae capsule: optimization of depolymerase production. J Basic Microbiol 54(7):711–720

Borysowski J, Górski A (2010) Fusion to cell-penetrating peptides will enable lytic enzymes to kill intracellular bacteria. Med Hypotheses 74(1):164–166

Briers Y, Volckaert G, Cornelissen A, Lagaert S, Michiels CW, Hertveldt K, Lavigne R (2007) Muralytic activity and modular structure of the endolysins of Pseudomonas aeruginosa bacteriophages φKZ and EL. Mol Microbiol 65(5):1334–1344

Briers Y, Walmagh M, Lavigne R (2011) Use of bacteriophage endolysin EL188 and outer membrane permeabilizers against Pseudomonas aeruginosa. J Appl Microbiol 110(3):778–785

Briers Y, Walmagh M, Van Puyenbroeck V, Cornelissen A, Cenens W, Aertsen A, Oliveira H, Azeredo J, Verween G, Pirnay J-P (2014) Engineered endolysin-based "Artilysins" to combat multidrug-resistant gram-negative pathogens. MBio 5(4):10–1128

Catalão MJ, Pimentel M (2018) Mycobacteriophage lysis enzymes: targeting the mycobacterial cell envelope. Viruses 10(8):428

Cheng Q, Fischetti VA (2007) Mutagenesis of a bacteriophage lytic enzyme PlyGBS significantly increases its antibacterial activity against group B streptococci. Appl Microbiol Biotechnol 74:1284–1291

Cornelissen A, Ceyssens P-J, T'syen J, Van Praet H, Noben J-P, Shaburova OV, Krylov VN, Volckaert G, Lavigne R (2011) The T7-related pseudomonas putida phage φ15 displays virion-associated biofilm degradation properties. PLoS One 6(4):e18597

Daniel A, Euler C, Collin M, Chahales P, Gorelick KJ, Fischetti VA (2010) Synergism between a novel chimeric lysin and oxacillin protects against infection by methicillin-resistant Staphylococcus aureus. Antimicrob Agents Chemother 54(4):1603–1612

Domenech M, García E, Moscoso M (2011) In vitro destruction of Streptococcus pneumoniae biofilms with bacterial and phage peptidoglycan hydrolases. Antimicrob Agents Chemother 55(9):4144–4148

Fenton M, McAuliffe O, O'Mahony J, Coffey A (2010) Recombinant bacteriophage lysins as antibacterials. Bioeng Bugs 1(1):9–16

Fenton M, Keary R, McAuliffe O, Ross RP, O'Mahony J, Coffey A (2013) Bacteriophage-derived peptidase eliminates and prevents Staphylococcal biofilms. Int J Microbiol 2013:625341

Fischetti VA (2005) Bacteriophage lytic enzymes: novel anti-infectives. Trends Microbiol 13(10):491–496

Fischetti VA (2010) Bacteriophage endolysins: a novel anti-infective to control gram-positive pathogens. Int J Med Microbiol 300(6):357–362

Gerstmans H, Rodríguez-Rubio L, Lavigne R, Briers Y (2016) From endolysins to Artilysin® s: novel enzyme-based approaches to kill drug-resistant bacteria. Biochem Soc Trans 44(1):123–128

Gilmer DB, Schmitz JE, Euler CW, Fischetti VA (2013) Novel bacteriophage lysin with broad lytic activity protects against mixed infection by Streptococcus pyogenes and methicillin-resistant Staphylococcus aureus. Antimicrob Agents Chemother 57(6):2743–2750

Gilmer DB, Schmitz JE, Thandar M, Euler CW, Fischetti VA (2017) The phage lysin PlySs2 decolonizes Streptococcus suis from murine intranasal mucosa. PLoS One 12(1):e0169180

Guariglia-Oropeza V, Helmann JD (2011) Bacillus subtilis σV confers lysozyme resistance by activation of two cell wall modification pathways, peptidoglycan O-acetylation and D-alanylation of teichoic acids. J Bacteriol 193(22):6223–6232

Guo M, Feng C, Ren J, Zhuang X, Zhang Y, Zhu Y, Dong K, He P, Guo X, Qin J (2017) A novel antimicrobial endolysin, LysPA26, against Pseudomonas aeruginosa. Front Microbiol 8:293

Gutierrez D, Ruas-Madiedo P, Martinez B, Rodríguez A, Garcia P (2014) Effective removal of staphylococcal biofilms by the endolysin LysH5. PLoS One 9(9):e107307

Gutiérrez D, Briers Y, Rodríguez-Rubio L, Martínez B, Rodríguez A, Lavigne R, García P (2015) Role of the pre-neck appendage protein (Dpo7) from phage vB_SepiS-phiIPLA7 as an anti-biofilm agent in staphylococcal species. Front Microbiol 6:1315

Hébert A, Sayasith K, Sénéchal S, Dubreuil P, Lagacé J (2000) Demonstration of intracellular Staphylococcus aureus in bovine mastitis alveolar cells and macrophages isolated from naturally infected cow milk. FEMS Microbiol Lett 193(1):57–62

Heselpoth RD, Yin Y, Moult J, Nelson DC (2015) Increasing the stability of the bacteriophage endolysin PlyC using rationale-based FoldX computational modeling. Protein Eng Des Sel 28(4):85–92

Jado I, López R, García E, Fenoll A, Casal J, García P (2003) Phage lytic enzymes as therapy for antibiotic-resistant Streptococcus pneumoniae infection in a murine sepsis model. J Antimicrob Chemother 52(6):967–973

Jandourek A, Boyle J, Cassino C, Wittekind M, Kirby H (2017) Long term immunology results of a phase 1 placebo controlled dose escalating study to examine the safety of CF-301 in human volunteers. Proceedings of the 27th ECCMID, Vienna, Austria, 22–25

Jonas OB, Irwin A, Berthe FCJ, Le Gall FG, Marquez PV (2017) Drug-resistant infections: a threat to our economic future (Vol. 2): final report. HNP/Agriculture Global Antimicrobial Resistance Initiative

Jun SY, Jung GM, Yoon SJ, Oh M-D, Choi Y-J, Lee WJ, Kong J-C, Seol JG, Kang SH (2013) Antibacterial properties of a pre-formulated recombinant phage endolysin, SAL-1. Int J Antimicrob Agents 41(2):156–161

Jun SY, Jung GM, Yoon SJ, Choi Y-J, Koh WS, Moon KS, Kang SH (2014) Preclinical safety evaluation of intravenously administered SAL200 containing the recombinant phage endolysin SAL-1 as a pharmaceutical ingredient. Antimicrob Agents Chemother 58(4):2084–2088

Lai M-J, Lin N-T, Hu A, Soo P-C, Chen L-K, Chen L-H, Chang K-C (2011) Antibacterial activity of Acinetobacter baumanniiphage ϕAB2 endolysin (LysAB2) against both gram-positive and gram-negative bacteria. Appl Microbiol Biotechnol 90:529–539

Latka A, Maciejewska B, Majkowska-Skrobek G, Briers Y, Drulis-Kawa Z (2017) Bacteriophage-encoded virion-associated enzymes to overcome the carbohydrate barriers during the infection process. Appl Microbiol Biotechnol 101:3103–3119

Lim J-A, Shin H, Heu S, Ryu S (2014) Exogenous lytic activity of SPN9CC endolysin against gram-negative bacteria. J Microbiol Biotechnol 24(6):803–811

Loeffler JM, Nelson D, Fischetti VA (2001) Rapid killing of Streptococcus pneumoniae with a bacteriophage cell wall hydrolase. Science 294(5549):2170–2172

Loeffler JM, Djurkovic S, Fischetti VA (2003) Phage lytic enzyme Cpl-1 as a novel antimicrobial for pneumococcal bacteremia. Infect Immun 71(11):6199–6204

Lood R, Winer BY, Pelzek AJ, Diez-Martinez R, Thandar M, Euler CW, Schuch R, Fischetti VA (2015) Novel phage lysin capable of killing the multidrug-resistant gram-negative bacterium Acinetobacter baumannii in a mouse bacteremia model. Antimicrob Agents Chemother 59(4):1983–1991

Maciejewska B, Roszniowski B, Espaillat A, Kęsik-Szeloch A, Majkowska-Skrobek G, Kropinski AM, Briers Y, Cava F, Lavigne R, Drulis-Kawa Z (2017) Klebsiella phages representing a novel clade of viruses with an unknown DNA modification and biotechnologically interesting enzymes. Appl Microbiol Biotechnol 101:673–684

Majkowska-Skrobek G, Łątka A, Berisio R, Maciejewska B, Squeglia F, Romano M, Lavigne R, Struve C, Drulis-Kawa Z (2016) Capsule-targeting depolymerase, derived from Klebsiella KP36 phage, as a tool for the development of anti-virulent strategy. Viruses 8(12):324

McCallum KL, Laakso DH, Whitfield C (1989) Use of a bacteriophage-encoded glycanase enzyme in the generation of lipopolysaccharide O side chain deficient mutants of Escherichia coli O9: K30 and Klebsiella O1: K20: role of O and K antigens in resistance to complement-mediated serum killing. Can J Microbiol 35(11):994–999

Mushtaq N, Redpath MB, Luzio JP, Taylor PW (2004) Prevention and cure of systemic Escherichia coli K1 infection by modification of the bacterial phenotype. Antimicrob Agents Chemother 48(5):1503–1508

Mushtaq N, Redpath MB, Luzio JP, Taylor PW (2005) Treatment of experimental Escherichia coli infection with recombinant bacteriophage-derived capsule depolymerase. J Antimicrob Chemother 56(1):160–165

Nelson D, Loomis L, Fischetti VA (2001) Prevention and elimination of upper respiratory colonization of mice by group A streptococci by using a bacteriophage lytic enzyme. Proc Natl Acad Sci 98(7):4107–4112

Nelson DC, Schmelcher M, Rodriguez-Rubio L, Klumpp J, Pritchard DG, Dong S, Donovan DM (2012) Endolysins as antimicrobials. Adv Virus Res 83:299–365

O'Neill J (2016) Tackling drug-resistant infections globally: final report and recommendations. Government of the United Kingdom

Oliveira H, Melo LDR, Santos SB, Nóbrega FL, Ferreira EC, Cerca N, Azeredo J, Kluskens LD (2013) Molecular aspects and comparative genomics of bacteriophage endolysins. J Virol 87(8):4558–4570

Oliveira H, Thiagarajan V, Walmagh M, Sillankorva S, Lavigne R, Neves-Petersen MT, Kluskens LD, Azeredo J (2014) A thermostable Salmonella phage endolysin, Lys68, with broad bactericidal properties against gram-negative pathogens in presence of weak acids. PLoS One 9(10):e108376

Oliveira H, Pinto G, Oliveira A, Oliveira C, Faustino MA, Briers Y, Domingues L, Azeredo J (2016a) Characterization and genome sequencing of a Citrobacter freundii phage CfP1 harboring a lysin active against multidrug-resistant isolates. Appl Microbiol Biotechnol 100:10,543–10,553

Oliveira H, Vilas Boas D, Mesnage S, Kluskens LD, Lavigne R, Sillankorva S, Secundo F, Azeredo J (2016b) Structural and enzymatic characterization of ABgp46, a novel phage endolysin with broad anti-gram-negative bacterial activity. Front Microbiol 7:208

Olszak T, Shneider MM, Latka A, Maciejewska B, Browning C, Sycheva LV, Cornelissen A, Danis-Wlodarczyk K, Senchenkova SN, Shashkov AS (2017) The O-specific polysaccharide lyase from the phage LKA1 tailspike reduces Pseudomonas virulence. Sci Rep 7(1):16302

Pan Y-J, Lin T-L, Lin Y-T, Su P-A, Chen C-T, Hsieh P-F, Hsu C-R, Chen C-C, Hsieh Y-C, Wang J-T (2015) Identification of capsular types in carbapenem-resistant Klebsiella pneumoniae strains by wzc sequencing and implications for capsule depolymerase treatment. Antimicrob Agents Chemother 59(2):1038–1047

Pastagia M, Euler C, Chahales P, Fuentes-Duculan J, Krueger JG, Fischetti VA (2011) A novel chimeric lysin shows superiority to mupirocin for skin decolonization of methicillin-resistant and-sensitive Staphylococcus aureus strains. Antimicrob Agents Chemother 55(2):738–744

Paul VD, Rajagopalan SS, Sundarrajan S, George SE, Asrani JY, Pillai R, Chikkamadaiah R, Durgaiah M, Sriram B, Padmanabhan S (2011) A novel bacteriophage Tail-Associated Muralytic Enzyme (TAME) from Phage K and its development into a potent antistaphylococcal protein. BMC Microbiol 11(1):1–11

Pires DP, Oliveira H, Melo LDR, Sillankorva S, Azeredo J (2016) Bacteriophage-encoded depolymerases: their diversity and biotechnological applications. Appl Microbiol Biotechnol 100:2141–2151

Prokhorov NS, Riccio C, Zdorovenko EL, Shneider MM, Browning C, Knirel YA, Leiman PG, Letarov AV (2017) Function of bacteriophage G7C esterase tailspike in host cell adsorption. Mol Microbiol 105(3):385–398

Rajaure M, Berry J, Kongari R, Cahill J, Young R (2015) Membrane fusion during phage lysis. Proc Natl Acad Sci 112(17):5497–5502

Rashel M, Uchiyama J, Ujihara T, Uehara Y, Kuramoto S, Sugihara S, Yagyu K-I, Muraoka A, Sugai M, Hiramatsu K (2007) Efficient elimination of multidrug-resistant Staphylococcus aureus by cloned lysin derived from bacteriophage ϕMR11. J Infect Dis 196(8):1237–1247

Rico-Lastres P, Díez-Martínez R, Iglesias-Bexiga M, Bustamante N, Aldridge C, Hesek D, Lee M, Mobashery S, Gray J, Vollmer W (2015) Substrate recognition and catalysis by LytB, a pneumococcal peptidoglycan hydrolase involved in virulence. Sci Rep 5(1):16,198

Roach DR, Donovan DM (2015) Antimicrobial bacteriophage-derived proteins and therapeutic applications. Bacteriophage 5(3):e1062590

Roach DR, Sjaarda DR, Castle AJ, Svircev AM (2013) Host exopolysaccharide quantity and composition impact Erwinia amylovora bacteriophage pathogenesis. Appl Environ Microbiol 79(10):3249–3256

Rodríguez L, Martínez Fernández B, Donovan DM, García Suárez MP, Rodríguez González A (2013) Potential of the virion-associated peptidoglycan hydrolase HydH5 and its derivative fusion proteins in milk biopreservation. PLoS One 8:e54828

Rodríguez-Rubio L, Gerstmans H, Thorpe S, Mesnage S, Lavigne R, Briers Y (2016) DUF3380 domain from a Salmonella phage endolysin shows potent N-acetylmuramidase activity. Appl Environ Microbiol 82(16):4975–4981

Schleifer KH, Kandler O (1972) Peptidoglycan types of bacterial cell walls and their taxonomic implications. Bacteriol Rev 36(4):407–477

Schmelcher M, Donovan DM, Loessner MJ (2012) Bacteriophage endolysins as novel antimicrobials. Future Microbiol 7(10):1147–1171

Schuch R, Nelson D, Fischetti VA (2002) A bacteriolytic agent that detects and kills Bacillus anthracis. Nature 418(6900):884–889

Shen Y, Köller T, Kreikemeyer B, Nelson DC (2013) Rapid degradation of Streptococcus pyogenes biofilms by PlyC, a bacteriophage-encoded endolysin. J Antimicrob Chemother 68(8):1818–1824

Singh PK, Donovan DM, Kumar A (2014) Intravitreal injection of the chimeric phage endolysin Ply187 protects mice from Staphylococcus aureus endophthalmitis. Antimicrob Agents Chemother 58(8):4621–4629

Stevens RH, Zhang H, Kajsik M, Płoski R, Rydzanicz M, Sabaka P, Šutovský S (2023) Successful use of a phage endolysin for treatment of chronic pelvic pain syndrome/chronic bacterial prostatitis. Front Med 10:1238147. https://doi.org/10.3389/FMED.2023.1238147/BIBTEX

Torchilin VP (2008) Cell penetrating peptide-modified pharmaceutical nanocarriers for intracellular drug and gene delivery. Pept Sci 90(5):604–610

Ventola CL (2015) The antibiotic resistance crisis: part 1: causes and threats. Pharm Ther 40(4):277

Vollmer W, Joris B, Charlier P, Foster S (2008) Bacterial peptidoglycan (murein) hydrolases. FEMS Microbiol Rev 32(2):259–286

Walmagh M, Boczkowska B, Grymonprez B, Briers Y, Drulis-Kawa Z, Lavigne R (2013) Characterization of five novel endolysins from Gram-negative infecting bacteriophages. Appl Microbiol Biotechnol 97:4369–4375

Wingfield PT (2015) Overview of the purification of recombinant proteins. Curr Protoc Protein Sci 80(1):1–6

Yang H, Wang M, Yu J, Wei H (2015) Antibacterial activity of a novel peptide-modified lysin against Acinetobacter baumannii and Pseudomonas aeruginosa. Front Microbiol 6:1471

Young R (2014) Phage lysis: three steps, three choices, one outcome. J Microbiol 52:243–258

Nature's Allies: Bacteriophage-Derived Enzymes to Combat Resistant Bacteria

34

Shraddha Manish Gupta, Dinesh D. Rishipathak, Hemant U. Chikhale, and Iram Khan

Abstract

Bacterial multidrug resistance is a critical issue in medicine and veterinary practice. Bacterial infections that are resistant to most or all known antibiotics generate severe diseases with a high morbidity and mortality rate. As a result, research of novel therapeutic techniques is seen as one of the objectives in the era of the "antibiotic resistance crisis" Phage therapy is a potential strategy for the treatment of bacterial illnesses that makes use of bacteriophages, which are viruses that particularly target and kill bacteria. Phages and phage-derived proteins, particularly enzymes, are being extensively researched as potential alternatives or supporting antibacterial treatments, either alone or in combination with traditional antibiotic regimens. Enzymes derived from bacteriophages, including endolysins and lysins, offer a more tailored strategy for fighting bacterial infections. This chapter investigates the potential of these enzymes in treating antibiotic-resistant bacteria, as well as their mechanism of action and uses in a variety of therapeutic contexts. The chapter also discusses the challenges and the prospects of these medicines.

S. M. Gupta · I. Khan
Department of Pharmaceutical Sciences, School of Health Sciences and Technology, UPES, Dehradun, Uttarakhand, India
e-mail: shraddham.gupta@ddn.upes.ac.in

D. D. Rishipathak (✉)
Department of Pharmaceutical Chemistry, MET's Institute of Pharmacy, Nashik, Maharashtra, India
e-mail: dineshr_iop@bkc.met.edu

H. U. Chikhale
Department of Pharmaceutical Chemistry, Sandip Institute of Pharmaceutical Sciences, Nashik, Maharashtra, India

M. Gangwar, G. Nath (eds.), *Emerging Paradigms for Antibiotic-Resistant Infections: Beyond the Pill*, https://doi.org/10.1007/978-981-97-5272-0_34

Keywords

Bacteriophage · Bacterial multidrug resistance · Bacteriophage-derived enzymes · Antibacterial treatment

34.1 Outline

Enzymes derived from bacteriophages such as endolysins and lysins, offer a targeted approach to combating bacterial infections. This chapter explores the potential of these enzymes in treating antibiotic-resistant pathogens, their mode of action, and their applications in various clinical settings. The chapter also addresses the challenges and future prospects for these therapies.

34.2 Introduction

34.2.1 Overview of Bacteriophages

Bacteriophages, or just "phages," are viruses that multiply and infect both bacteria and archaea. Phaganein, a Greek word meaning "to devour," is where the term "bacteria" originates. Phage viruses, also known as bacteriophages, are the essential parasites that proliferate in bacterial hosts. The Caudovirales family of dsDNA-tailed viruses, which is further subdivided into lytic and temperate phages, comprises the majority of phages discovered.

Phages are perhaps the most prevalent form of biological debris on a global scale and have a significant impact on the environment, according to studies by Wommack and Colwell (2000) and Brussaard et al. (2008). They are responsible for four main processes: (1) regulating and reducing the amount of dominating bacteria in populations; (2) lysogenic conversions by temperate phages; and transforming bacterial DNA that is released during cell lysis in order to (3) promote horizontal gene transfer (HGT) by transduction. Thus, as phages are an essential component of every biological niche, including the human body, they have a significant impact on microbial diversity. The co-evolution of both organisms is fueled by the constant flow of genetic material resulting from the incredibly dynamic phage-host interactions (Biswas et al. 2006; Borysowski and Gorksi 2010; Cotter et al. 2013; Knecht et al. 2020; Parisien et al. 2008; Pestrak et al. 2019; Porayath et al. 2018; Qin et al. 2007; Riley and Wertz 2002; Zhang et al. 2022).

34.2.2 Introduction to Bacteriophage-Derived Enzymes—"Enzybiotics"

The word "enzybiotics" has been used more recently to refer to phage-derived enzymes called polysaccharide depolymerases (PSDs) (Yang et al. 2014), which additionally possess the capability to degrade bacterial polysaccharides like lipopolysaccharide (LPS), biofilm matrix, capsules, and layers of slime extracellularly. Enzybiotics can be divided into six major groups. Phage-derived enzymes are among them. They are classified as (1) Phosphoglycan hydrolases (PGHs), including endolysins and peptidoglycan hydrolases associated with vesicles; (2) Biofilm-disrupting enzymes, specifically polysaccharide depolymerases (PSDs). The term "enzybiotics" also refers to substances that are derived from bacteria: (3) biofilm-disrupting enzymes (Oliver and Wells 2015; Veiga-Crespo et al. 2007), (4) autolysins (LaFee and Buschman 2017; Prescott et al. 1996; Suttle 2005; Wommack and Colwell 2000), which are bacterial enzymes that facilitate the segregation of daughter cells during cell division, and (5) bacteriocins (D'Herelle 1961; Hankin 1896; Keen 2012; Twort 1915) are antimicrobial proteins or peptides synthesized by bacteria that inhibit the proliferation of closely similar microorganisms. Additionally, Enzymes derived from animals such as (6) including the lysozymes that are found in naturally occurring body fluids such as tears, saliva, milk, and mucus, are classified as enzybiotics (Aswani and Shukla 2021; Knecht et al. 2020; Pestrak et al. 2019). Fungal endoglucanases are one type of enzybiotic that has antifungal properties, albeit this has not been shown (Parisien et al. 2008). Enzybiotic sources and their intended substrates are linked by arrows.

34.2.3 Importance of Bacteriophage-Derived Enzymes in Biotechnology

Bacteriophages' inherent characteristics broaden their potential applications across several scientific domains, such as agriculture, food safety, industry, health, and molecular typing, control, and detection of bacteria. Understanding the roles of the proteins encoded by phages and identifying them also pave the way for the possible use of bacteriophages as medication delivery and vaccine delivery vehicles. The abundance of excellent studies based on phage tests shows the enormous potential that phages exhibit and the possibility that they may lead to biotechnological uses that will benefit all of humanity. Furthermore, by utilizing bacteriophages in conjunction with a range of transducers, new biosensors and innovative biomarkers that are specifically designed to detect molecules like proteins or harmful microbes have been made possible. Only a small number of phage-based sensing devices are already on the market, despite the fact that several innovative technologies have been patented (Ex. Corvium, In.) (Zhang et al. 2022). The high variability and abundance of these organisms suggest that phage-based technologies have a bright future ahead of them. Additional research in this area is expected to yield new applications of phage-based biotechnology that will not only benefit human health but also advance commercial methods (Biswas et al. 2006).

34.3 Bacteriophage-Derived Enzymes: Types and Functions

The viruses known as bacteriophages exclusively infect bacteria. In aquatic settings, the proportion of bacteriophages to bacteria exceeds 10:1, indicating the widespread presence of bacteriophages. Bacteriophage reproduction is intimately associated with its bacterial host. The two types of bacteriophages that exist are lytic and lysogenic bacteriophages. Lytic bacteriophages have the ability to: (a) bind to bacterial receptors; (b) impart the genome to bacteria; (c) replicate the virus in the cytoplasm by means of bacterial transcription, translation, and replication; and (d) liberate the virus from the host cells. Subsequently, fresh lytic bacteriophage particles will carry out this procedure in fresh vulnerable hosts. Lysogenic bacteriophages are also often found in nature, much like lytic bacteriophages. In the process of binary division, offspring cells can inherit the genes that lysogenic bacteriophages integrate into the host genome after being consumed by host bacteria. The host genome has been used to extract the lysogenic phage genes in response to environmental disruption or other physiological stresses.

34.3.1 Bacteriophage-Derived Enzymes

Bacteriophage-derived enzymes are enzymes that are obtained or derived from bacteriophages, which are viruses that infect and replicate within bacterial cells. Bacteriophages are specialized viruses that specifically target bacteria, and mechanisms to commandeer bacterial replication machinery. Enzymes derived from bacteriophages often play crucial roles in the viral life cycle by facilitating processes such as host cell recognition, attachment, penetration, and release of new viral particles. These enzymes can have diverse functions and are often involved in manipulating bacterial cell components to aid in the viral replication process (Qin et al. 2007).

Examples of bacteriophage-derived enzymes include:

34.3.1.1 Endolysins

These enzymes are responsible for breaking down the cell wall of bacteria in the final phases of the bacteriophage infection cycle, facilitating the release of new viral particles.

34.3.1.2 Exonucleases

Enzymes that degrade nucleic acids from the ends, which may be involved in various aspects of viral DNA replication and processing.

34.3.1.3 DNA Polymerases

Bacteriophages encode their DNA polymerases to replicate their genetic material within the bacterial host.

34.3.1.4 Restriction Enzymes

Some bacteriophages produce restriction enzymes as a defense mechanism against bacterial host defenses. These enzymes can cleave bacterial DNA, providing a selective advantage for the phage.

34.3.1.5 Methyltransferases

Bacteriophages may produce enzymes that modify DNA bases, such as methylation, to evade bacterial host restriction systems.

Studying and utilizing bacteriophage-derived enzymes has applications in biotechnology, medicine, and molecular biology. Researchers may explore these enzymes for various purposes, including the development of antibacterial agents, gene editing tools, and other biotechnological applications.

The concept of utilizing phages as potential therapeutic agents for bacterial infections emerged shortly after the identification of bacterial viruses (bacteriophages or phages) exactly a century ago. The emergence of antibiotic-resistant bacteria on a global scale is primarily attributable to the overuse as well as misuse of antibiotics; this has led to increased morbidity, mortality, and healthcare costs. Bassett et al. concluded, based on a review of medical literature and expert opinion, that over the past decade, no infection prevention or control measures have been successful in halting the fast dissemination of microorganisms that are resistant to treatment. According to the authors' estimation, the global antibiotic resistance pandemic will result in ten million deaths annually by 2050 (Porayath et al. 2018). As a result, it is critical to develop alternative strategies for reversing this trend and developing treatments for these pathogens (Borysowski and Gorksi 2010; Cotter et al. 2013; Parisien et al. 2008; Riley and Wertz 2002; Yang et al. 2014). Phage therapy represents a potentially fruitful avenue.

One aspect of endolysins that is influenced by their origin is their structure (Fig. 34.1). In contrast, endolysins that effectively target Gram-negative bacteria possess a straightforward globular structure (Berchieri Jr et al. 1991), whereas the vast majority of endolysins (typically molecular mass:15–40 kDa) possess modular configurations. Modular endolysins are frequently distinguished by one or two enzymatically active domains (EADs) located at their N-termini, which are linked to a cell wall-binding domain (CBD) at the C-terminus by a short, flexible linker region. Mycobacteriophages and Gram-positive phages share this configuration (Oliver and Wells 2015). Modular endolysins have an N-terminal enzymatically active domain (EAD) which cleaves relevant peptidoglycan bonds found in the murein layer of the host bacterium. To ensure accurate fixation of the catalytic effect of the EAD, the C-terminal cell wall-binding domain (CBD) attaches to and identifies diverse epitopes within the cell wall (Veiga-Crespo et al. 2007). Endolysins derived from Gram-positive phages exhibit a structural resemblance to fungal cellulases, which are also composed of EADs and CBD that are connected by a flexible linkage (Garcia et al. 2008).

In essence, the bacterial host lysis model system consists of three steps. Three proteins—endolysin, holin, and spanin—make up this model (Fig. 34.2). These proteins act on the peptidoglycan, inner membrane, and outer membrane of bacteria,

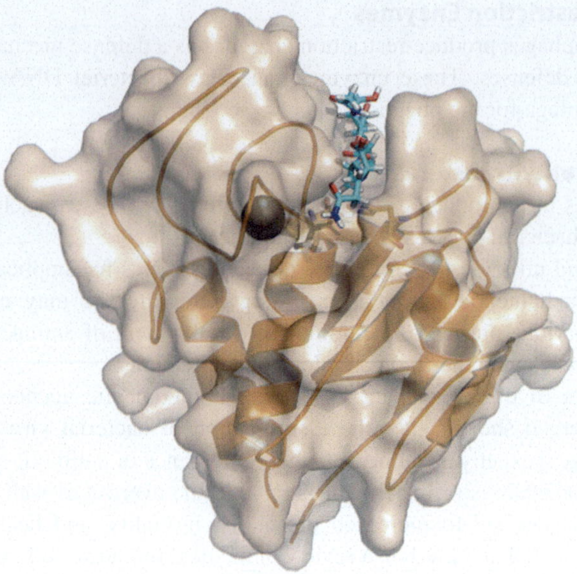

Fig. 34.1 Structure of endolysins

and they have the ability to lyse the host. In Caudovirales of Gram-negative hosts, the peptidoglycan is broken down by two different mechanisms: (1) Holinendolysin; and (2) Pinholin SAR endolysin. The lysis process is started by releasing endolysin inside the periplasm, which breaks down the peptidoglycan. In the previous phase, holin creates tiny holes in the host cell's inner membrane, which starts the peptidoglycan's breakdown. When pinholin can conduct membrane depolarization and trigger the secreted SAR endolysin, lysis begins in the final phase. It was previously believed that the lysis of the first two membrane barriers was adequate for the lysis of Gram-negative hosts. However, the latest research indicates that the outer membrane must also be broken down, requiring the lysis protein spanin. This mechanism leads to the destruction of the bacterial host cell (Fig. 34.3) (Berchieri Jr et al. 1991).

34.4 Applications

34.4.1 Medical Health Care

As a result of the diminished effectiveness of antibiotics, numerous infectious pathogens have evolved into agents that pose a risk to human life. As a consequence, scholarly investigations have been directed towards phage-derived endolysins as a potential treatment for systemic and topical infestations in humans (Loc Carrillo et al. 2005). Antimicrobial compounds derived from phages have been effectively utilized to combat the majority of Gram-positive pathogenic infections. Phage lysins have demonstrated efficacy against the *S. aureus* pathogen in recent studies,

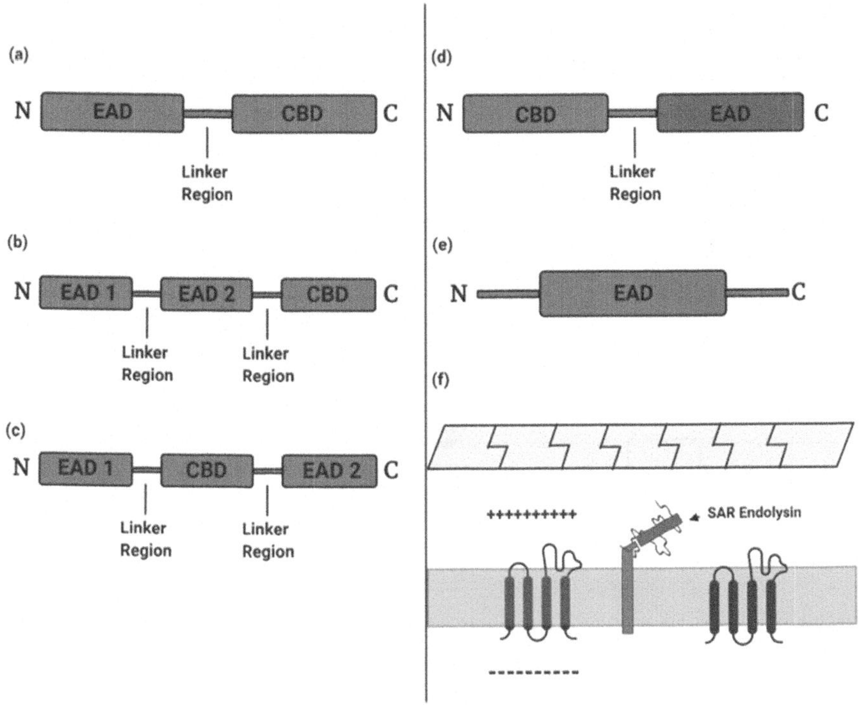

Fig. 34.2 Models of typical phage endolysins with modular configurations (Abdelrahman et al. 2021)

as evidenced by animal studies and human clinical studies (Fox 2000; Nakai and Park 2002; Park and Nakai 2003; Schnabel and Jones 2001). It is noteworthy that phage endolysins are presently being utilized in numerous human clinical trials. For example, endolysin SAL200 and CF-301, which target *S. aureus*, have demonstrated efficacy in the treatment of infections affecting the bloodstream and myocardium (Obradovic et al. 2004). Cheng and Fischetti investigated the relative lytic activity of the phage muralytic enzyme PlyGBS and its mutant PlyGBS90-1, which exhibited hyperthymic activity, a decade ago. A substantial reduction in the overall number of streptococci (twenty-eight-fold) was achieved through the utilization of a phage muralytic enzyme mutant, according to research. The most exemplary application using the phage-derived protein strategy is bacterial vaginosis. In addition, this research group observed that the mutant PlyGBS90-1 has a multitude of advantageous properties, including rabid and effective lysis activity against *streptococci* in a model of mouse vaginal colonization (Carlton et al. 2005). By employing this novel system comprised of phage-derived lysins, antimicrobial resistance could be effectively reduced in comparison to phage cocktail therapy.

In preclinical trials, the *pneumococcal* lysins Pal and Cpl-1 were effective in eliminating antibiotic-resistant *S. pneumoniae*, the pathogen responsible for acute otitis media (AOM), pneumonia, septicemia, bronchitis, and meningitis (Bassetti

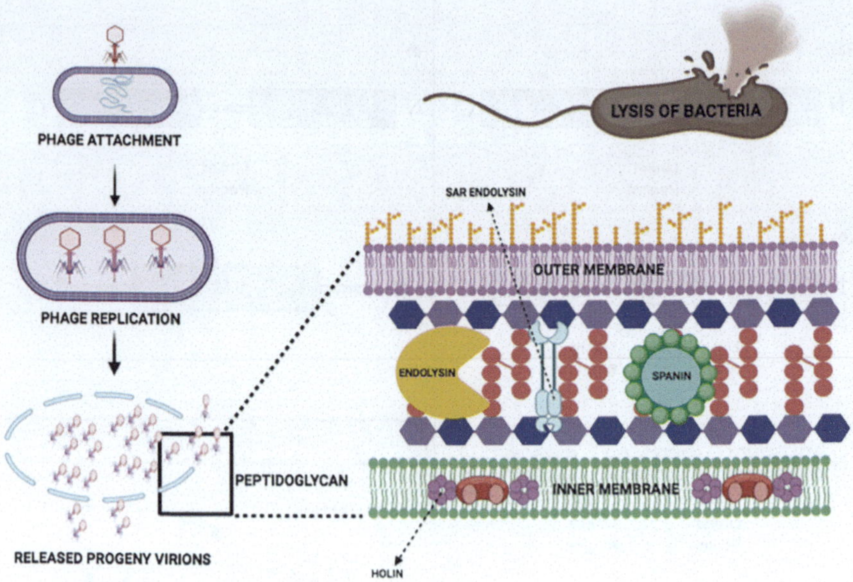

Fig. 34.3 Working mechanism of endolysin

et al. 2017; Holmes et al. 2016; Liu et al. 2004; Rios et al. 2016; Williams and Hergenrother 2008).

34.4.2 Veterinary Industry

In the veterinary sector, penicillins are said to be the most commonly administered antibiotics. Bacterial resistance has significantly evolved in the aquaculture industry, the agricultural sector, and veterinary medicine (Czaplewski et al. 2016; Oliveira et al. 2018) due to the excessive and inappropriate use of antibiotics. The transfer of antibiotic-resistance genes exists between contaminated humans and food which has been thoroughly researched (Matamp and Bhat 2019). Food animals including cattle, chicken, and swine are a significant source of bacteria resistant to antibiotics and their particular genes, which can be transmitted to humans through the food chain (Fischetti 2018). Consequently, the United States (Xu et al. 2004) as well as the European countries have outlawed the usage of antibiotics in animal feed. Hence, there is a substantial demand for antibiotic substitutes, including endolysins, along with preclinical research on endolysins for veterinary use (Cahill and Young 2019).

34.4.3 Food and Biotechnology

Food It is common knowledge that food products derived from animals such as piglets, chickens, and cattle are a reservoir of drug-resistant pathogens. Endolysins PlyV12 have the capacity to inhibit the development of resistance by exhibiting potent lytic activity against *Enterococcus faecium* and *Enterococcus faecalis*, both of which are resistant to antibiotics. It is noteworthy that plant-based milk can become contaminated with *L. monocytogenes*. Consequently, research has demonstrated that the administration of LysZ5 to soya milk results in exceptional sterilization efficacy (Love et al. 2018). Furthermore, the *L. monocytogenes* pathogen has been effectively treated in the context of hydrostatic pressure with alternative phage endolysins including PlyP825, PlyP40, and Ply511 (Álvarez et al. 2019; Cheng and Fischetti 2007; Jado et al. 2003; Jun et al. 2017; Loeffler et al. 2003).

The Biotechnology Sector Sophisticated biotechnological methodologies, including therapeutic protein (endolysin) development and genetic engineering, possess the capacity to enhance the food safety process by expeditiously eliminating bacterial pathogens by virtue of their enzymatic characteristics. As a means of preventing food-borne infections caused by antibiotic-resistant microorganisms, the food industry has acknowledged phage endolysins as natural bio-preservatives. It is imperative to maintain sanitary and secure surfaces throughout the food production process to avert foodborne outbreaks in the community. Peptoglycan hydrolases and endolysins have been effectively utilized to eradicate biofilms from food and clinical environments, thereby ensuring adequate disinfection of surfaces.

34.5 Challenges and Future Perspectives

34.5.1 Current Challenges in Utilizing Bacteriophage-Derived Enzymes

Bacteriophage-derived enzymes, known as phage enzymes or phage endolysins, face several challenges in their application. One significant obstacle is the potential development of bacterial resistance to these enzymes, as bacteria can evolve mechanisms to evade enzymatic attack. Additionally, the narrow host range of many bacteriophages limits the effectiveness of phage-derived enzymes against diverse bacterial species. Another challenge is the need for a thorough understanding of the enzyme's structure and function, as variations in bacterial cell wall structures may require tailored enzymes for optimal activity. Furthermore, issues related to stability, formulation, and delivery methods need to be addressed for practical applications, particularly in medical and food safety contexts. Despite these challenges, ongoing research and advancements in biotechnology hold promise for overcoming these obstacles and harnessing the potential of bacteriophage-derived enzymes for therapeutic and industrial purposes (Brown and Lerner 1998; Cabello 2006; Cantas et al. 2013; McEwen 2006; Silbergeld et al. 2008; Woolhouse and Ward 2013).

34.5.2 Future Directions and Emerging Trends in Bacteriophage-Based Biotechnology

The future scope of bacteriophage-derived enzymes holds great promise in various fields, including medicine, agriculture, and biotechnology. These enzymes, derived from bacteriophages, which are viruses that infect and replicate within bacteria, exhibit specificity in targeting bacterial pathogens. In medicine, bacteriophage-derived enzymes such as endolysins are being explored as prospective substitutes for antibiotics in the management of bacterial infections, specifically those induced by strains that are resistant to antibiotics. Additionally, these enzymes have applications in agriculture for controlling plant pathogens, thereby reducing the reliance on chemical pesticides. In biotechnology, the use of bacteriophage-derived enzymes in various industrial processes, such as food production and wastewater treatment, is being investigated for their efficiency and specificity. As the global concern over antibiotic resistance grows and the need for sustainable solutions increases, the development and utilization of bacteriophage-derived enzymes represent a promising avenue for addressing these challenges in diverse sectors.

34.5.3 Potential Ethical and Regulatory Considerations

Phage therapy (PT) is becoming more and more popular due in part to the issue of antibiotic resistance throughout the world. Antibiotics may be replaced by phage therapy. Phage therapy's increasing popularity is a result of the global problem of antimicrobial multidrug resistance. Nonetheless, there is still disagreement over the potential for both clinical safety and efficacy of the phage therapy strategies, which mostly include readily available preset cocktails or, on occasion, phage products made to order. Investigations are now being conducted to determine the safety and efficacy of phage therapy methods, the majority of which include commercial fixed cocktails or specially created phage products (Table 34.1).

The implementation of unique regulatory procedures in EU Member States and the subsequent treatment experience obtained in selected patients facilitate the acquisition of high-quality clinical trial data. Since regulatory surveillance will be a crucial component in determining the successful entry of bacteriophages into clinical practice, different countries have different ethical and regulatory rules for the compassionate use of PT (Casewell et al. 2003; Fernández et al. 2018; Marshall and Levy 2011). Currently, PT contains single phages with a variety of purposes, known as phage mixes. In addition, it makes it possible to quickly match a phage or phages to a bacterial isolate (from a patient with infection), enabling targeted and targeted therapy. Regulatory supervision is required for bacteriophages to be used in therapeutic settings. There is now just one endolysins-based medication available as a medicinal in Europe, Staphefekt SA.100. On the contrary, its regulatory approval indicates a bright outlook for the therapeutic implementation of enzybiotics in the treatment of prevalent bacterial diseases (Love et al. 2018; van Nassau et al. 2017; Yoong et al. 2004; Zhang et al. 2012).

Table 34.1 Phage Product approved to date by various regulatory body

Sr. no	Dosage form	Approved	Application	Ref
1	Phage spray (Phagyo®spray)	Approved	• Treatment and prophylaxis of bacterial purulent–inflammatory infections (multiple microorganism).	Khullar et al. (2022)
2	Phage tablet (Septaphage®table)			
3	Phage cocktail (Septaphage®, Phagyo+, PhageStaph)			
4	Phage capsule (Travelphag™)		• For bacterial infections, indigestion.	Sompolinsky et al. (1985)
5	Phage gel (Phagodent, Phagoderm, Phagogyn)	Approved	• Contains 72-phages complexes to normalize oral microflora, • Skin gel containing 64-Phages to prevent bacterial infection of the skin. • Gel containing 7-Phages that prevent bacterial diseases of the reproductive system.	Khullar et al. (2022)

References

Abdelrahman F, Easwaran M, Daramola OI, Ragab S, Lynch S, Oduselu TJ et al (2021) Phage-encoded endolysins. Antibiotics 10(2):124

Álvarez A, Fernández L, Gutiérrez D, Iglesias B, Rodríguez A, García P (2019) Methicillin-resistant Staphylococcus aureus in hospitals: latest trends and treatments based on bacteriophages. J Clin Microbiol 57(12):10–1128

Aswani VH, Shukla SK (2021) An early history of phage therapy in the United States: is it time to reconsider? Clin Med Res 19(2):82–89

Bassetti M, Poulakou G, Ruppe E, Bouza E, Van Hal SJ, Brink A (2017) Antimicrobial resistance in the next 30 years, humankind, bugs and drugs: a visionary approach. Intensive Care Med 43:1464–1475

Berchieri A Jr, Lovell MA, Barrow PA (1991) The activity in the chicken alimentary tract of bacteriophages lytic for Salmonella typhimurium. Res Microbiol 142(5):541–549

Biswas R, Voggu L, Simon UK, Hentschel P, Thumm G, Götz F (2006) Activity of the major staphylococcal autolysin Atl. FEMS Microbiol Lett 259(2):260–268

Borysowski J, Gorksi A (2010) Enzybiotics and their potential applications in medicine. In: Villa TG, Veiga-Crespo P (eds) Enzybiotics: antibiotic enzymes as drugs and therapeutics, pp 1–26

Brown PD, Lerner SA (1998) Community-acquired pneumonia. Lancet 352(9136):1295–1302

Brussaard CPD, Wilhelm SW, Thingstad F, Weinbauer MG, Bratbak G, Heldal M et al (2008) Global-scale processes with a nanoscale drive: the role of marine viruses. ISME Journal 2:575–578

Cabello FC (2006) Heavy use of prophylactic antibiotics in aquaculture: a growing problem for human and animal health and for the environment. Environ Microbiol 8(7):1137–1144

Cahill J, Young R (2019) Phage lysis: multiple genes for multiple barriers. Adv Virus Res 103:33–70

Cantas L, Shah SQA, Cavaco LM, Manaia CM, Walsh F, Popowska M et al (2013) A brief multidisciplinary review on antimicrobial resistance in medicine and its linkage to the global environmental microbiota. Front Microbiol 4:96

Carlton RM, Noordman WH, Biswas B, De Meester ED, Loessner MJ (2005) Bacteriophage P100 for control of Listeria monocytogenes in foods: genome sequence, bioinformatic analyses, oral toxicity study, and application. Regul Toxicol Pharmacol 43(3):301–312
Casewell M, Friis C, Marco E, McMullin P, Phillips I (2003) The European ban on growth-promoting antibiotics and emerging consequences for human and animal health. J Antimicrob Chemother 52(2):159–161
Cheng Q, Fischetti VA (2007) Mutagenesis of a bacteriophage lytic enzyme PlyGBS significantly increases its antibacterial activity against group B streptococci. Appl Microbiol Biotechnol 74:1284–1291
Cotter PD, Ross RP, Hill C (2013) Bacteriocins—a viable alternative to antibiotics? Nat Rev Microbiol 11(2):95–105
Czaplewski L, Bax R, Clokie M, Dawson M, Fairhead H, Fischetti VA et al (2016) Alternatives to antibiotics—a pipeline portfolio review. Lancet Infect Dis 16(2):239–251
D'Herelle MF (1961) Sur un microbe invisible antagoniste des bacilles dysenteriques. https://api.semanticscholar.org/CorpusID:222394659
Fernández L, Gutiérrez D, Rodríguez A, García P (2018) Application of bacteriophages in the agro-food sector: a long way toward approval. Front Cell Infect Microbiol 8:296
Fischetti VA (2018) Development of phage lysins as novel therapeutics: a historical perspective. Viruses 10(6):310
Fox J (2000) Phage treatments yield healthier tomato, pepper plants. Asm News 66(8):455–456
Garcia P, Martinez B, Obeso JM, Rodriguez A (2008) Bacteriophages and their application in food safety. Lett Appl Microbiol 47(6):479–485
Hankin E (1896) L'action bactericide des eaux de la Jumna et du Gange sur le vibrion du cholera. Ann Inst Pasteur 10:511
Holmes AH, Moore LSP, Sundsfjord A, Steinbakk M, Regmi S, Karkey A et al (2016) Understanding the mechanisms and drivers of antimicrobial resistance. Lancet 387(10014):176–187
Jado I, López R, García E, Fenoll A, Casal J, García P (2003) Phage lytic enzymes as therapy for antibiotic-resistant Streptococcus pneumoniae infection in a murine sepsis model. J Antimicrob Chemother 52(6):967–973
Jun SY, Jang IJ, Yoon S, Jang K, Yu K-S, Cho JY et al (2017) Pharmacokinetics and tolerance of the phage endolysin-based candidate drug SAL200 after a single intravenous administration among healthy volunteers. Antimicrob Agents Chemother 61(6):10–1128
Keen EC (2012) Felix d'Herelle and our microbial future. Future Microbiol 7(12):1337–1339
Khullar L, Harjai K, Chhibber S (2022) Exploring the therapeutic potential of staphylococcal phage formulations: current challenges and applications in phage therapy. J Appl Microbiol 132(5):3515–3532
Knecht LE, Veljkovic M, Fieseler L (2020) Diversity and function of phage encoded depolymerases. Front Microbiol 10:2949
LaFee S, Buschman H (2017) Novel phage therapy saves patient with multidrug-resistant bacterial infection. UC San Diego Today
Liu J, Dehbi M, Moeck G, Arhin F, Bauda P, Bergeron D et al (2004) Antimicrobial drug discovery through bacteriophage genomics. Nat Biotechnol 22(2):185–191
Loc Carrillo C, Atterbury RJ, El-Shibiny A, Connerton PL, Dillon E, Scott A, Connerton IF (2005) Bacteriophage therapy to reduce Campylobacter jejuni colonization of broiler chickens. Appl Environ Microbiol 71(11):6554–6563
Loeffler JM, Djurkovic S, Fischetti VA (2003) Phage lytic enzyme Cpl-1 as a novel antimicrobial for pneumococcal bacteremia. Infect Immun 71(11):6199–6204
Love MJ, Bhandari D, Dobson RCJ, Billington C (2018) Potential for bacteriophage endolysins to supplement or replace antibiotics in food production and clinical care. Antibiotics 7(1):17
Marshall BM, Levy SB (2011) Food animals and antimicrobials: impacts on human health. Clin Microbiol Rev 24(4):718–733

Matamp N, Bhat SG (2019) Phage endolysins as potential antimicrobials against multidrug resistant Vibrio alginolyticus and Vibrio parahaemolyticus: current status of research and challenges ahead. Microorganisms 7(3):84

McEwen SA (2006) Antibiotic use in animal agriculture: what have we learned and where are we going? Anim Biotechnol 17(2):239–250

Nakai T, Park SC (2002) Bacteriophage therapy of infectious diseases in aquaculture. Res Microbiol 153(1):13–18

van Nassau TJ, Lenz CA, Scherzinger AS, Vogel RF (2017) Combination of endolysins and high pressure to inactivate Listeria monocytogenes. Food Microbiol 68:81–88

Obradovic A, Jones JB, Momol MT, Balogh B, Olson SM (2004) Management of tomato bacterial spot in the field by foliar applications of bacteriophages and SAR inducers. Plant Dis 88(7):736–740

Oliveira H, São-José C, Azeredo J (2018) Phage-derived peptidoglycan degrading enzymes: challenges and future prospects for in vivo therapy. Viruses 10(6):292

Oliver WT, Wells JE (2015) Lysozyme as an alternative to growth promoting antibiotics in swine production. J Anim Sci Biotechnol 6(1):1–7

Parisien A, Allain B, Zhang J, Mandeville R, Lan CQ (2008) Novel alternatives to antibiotics: bacteriophages, bacterial cell wall hydrolases, and antimicrobial peptides. J Appl Microbiol 104(1):1–13

Park SC, Nakai T (2003) Bacteriophage control of Pseudomonas plecoglossicida infection in ayu Plecoglossus altivelis. Dis Aquat Org 53(1):33–39

Pestrak MJ, Baker P, Dellos-Nolan S, Hill PJ, Passos da Silva D, Silver H et al (2019) Treatment with the Pseudomonas aeruginosa glycoside hydrolase PslG combats wound infection by improving antibiotic efficacy and host innate immune activity. Antimicrob Agents Chemother 63(6):10–1128

Porayath C, Suresh MK, Biswas R, Nair BG, Mishra N, Pal S (2018) Autolysin mediated adherence of Staphylococcus aureus with fibronectin, gelatin and heparin. Int J Biol Macromol 110:179–184

Prescott LM, Harley JP, Klein DA (1996) Student study art notebook, Microbiology. Wm. C. Brown Publishers

Qin Z, Ou Y, Yang L, Zhu Y, Tolker-Nielsen T, Molin S, Qu D (2007) Role of autolysin-mediated DNA release in biofilm formation of Staphylococcus epidermidis. Microbiology 153(7):2083–2092

Riley MA, Wertz JE (2002) Bacteriocins: evolution, ecology, and application. Annu Rev Microbiol 56(1):117–137

Rios AC, Moutinho CG, Pinto FC, Del Fiol FS, Jozala A, Chaud MV et al (2016) Alternatives to overcoming bacterial resistances: state-of-the-art. Microbiol Res 191:51–80

Schnabel EL, Jones AL (2001) Isolation and characterization of five Erwinia amylovora bacteriophages and assessment of phage resistance in strains of Erwinia amylovora. Appl Environ Microbiol 67(1):59–64

Silbergeld EK, Graham J, Price LB (2008) Industrial food animal production, antimicrobial resistance, and human health. Annu Rev Public Health 29:151–169

Sompolinsky D, Samra Z, Karakawa WW, Vann WF, Schneerson R, Malik Z (1985) Encapsulation and capsular types in isolates of Staphylococcus aureus from different sources and relationship to phage types. J Clin Microbiol 22(5):828–834

Suttle CA (2005) Viruses in the sea. Nature 437(7057):356–361

Twort FW (1915) An investigation on the nature of ultra-microscopic viruses. Lancet 2:1241–1243

Veiga-Crespo P, Ageitos JM, Poza M, Villa TG (2007) Enzybiotics: a look to the future, recalling the past. J Pharm Sci 96(8):1917–1924

Williams JJ, Hergenrother PJ (2008) Exposing plasmids as the Achilles' heel of drug-resistant bacteria. Curr Opin Chem Biol 12(4):389–399

Wommack KE, Colwell RR (2000) Virioplankton: viruses in aquatic ecosystems. Microbiol Mol Biol Rev 64(1):69–114

Woolhouse MEJ, Ward MJ (2013) Sources of antimicrobial resistance. Science 341(6153):1460–1461

Xu M, Struck DK, Deaton J, Wang I-N, Young R (2004) A signal-arrest-release sequence mediates export and control of the phage P1 endolysin. Proc Natl Acad Sci 101(17):6415–6420

Yang S-C, Lin C-H, Sung CT, Fang J-Y (2014) Antibacterial activities of bacteriocins: application in foods and pharmaceuticals. Front Microbiol 5:241

Yoong P, Schuch R, Nelson D, Fischetti VA (2004) Identification of a broadly active phage lytic enzyme with lethal activity against antibiotic-resistant Enterococcus faecalis and Enterococcus faecium. J Bacteriol 186(14):4808–4812

Zhang H, Bao H, Billington C, Hudson JA, Wang R (2012) Isolation and lytic activity of the Listeria bacteriophage endolysin LysZ5 against Listeria monocytogenes in soya milk. Food Microbiol 31(1):133–136

Zhang J, Wu H, Wang D, Wang L, Cui Y, Zhang C et al (2022) Intracellular glycosyl hydrolase PslG shapes bacterial cell fate, signaling, and the biofilm development of Pseudomonas aeruginosa. Elife 11:e72778

Advancing Beyond Antibiotics: The Pioneering Use of Bacteriophage Enzymes

35

Nishu Sehrawat, Seema Sangwan, Babita, and Renu Lamba

Abstract

The increasing prevalence of antibiotic-resistant bacterial infections necessitates the development of novel therapeutic methods. Bacteriophages, which are viruses that target bacteria, encode a variety of enzymes with potent antibacterial properties, which present a promising substitute for conventional antibiotics. Among these, phage-encoded peptidoglycan hydrolases (PGHs), such as virion-associated PGHs (VAPGHs) and endolysins, as well as polysaccharide depolymerases and holins, have been recognized as putative antibacterial agents. Endolysins, which are synthesized during the final stages of the phage replication process, degrade the cell wall of bacteria, resulting in cell lysis. The VAPGHs aid in initial cell wall penetration, while polysaccharide depolymerases target the bacterial cell wall envelope, assisting phage adsorption and host bacteria disintegration. Holins cause cytoplasmic membrane disruption, allowing the access of endolysin. Recent investigations in preclinical animal models have shown that these phage-derived enzymes are effective against a variety of bacterial infections, including multidrug-resistant strains. Engineered recombinant versions of these enzymes, developed to improve their bacteriolytic and bactericidal activity, have produced promising findings, implying a possible role in the fight against bacterial infections. This chapter discusses the therapeutic potential of phage-encoded enzymes, particularly in light of developing antibiotic resistance, emphasizing the need for further research and clinical development.

N. Sehrawat · S. Sangwan (✉) · Babita
Department of Microbiology, CCS Haryana Agricultural University, Hisar, Haryana, India

R. Lamba
Department of Botany and Plant Physiology, CCS Haryana Agricultural University, Hisar, Haryana, India

M. Gangwar, G. Nath (eds.), *Emerging Paradigms for Antibiotic-Resistant Infections: Beyond the Pill*, https://doi.org/10.1007/978-981-97-5272-0_35

Keywords
Antibiotic Resistance · Bacteriophages · Endolysins · Polysaccharide Depolymerases · Holins

35.1 Introduction

The escalating crisis of resistance to antibiotics among pathogenic bacteria has propelled the quest for new antimicrobial methods. Bacteriophages and their encoded enzymes provide a novel way of treating infections caused by bacteria that have developed resistance to traditional antibiotics. These biological entities have evolved particular methods, such as the creation of lytic enzymes, to overcome bacterial defenses and trigger cell death. It includes the cell wall degrading enzymes such as VAPGHs (Rodríguez-Rubio et al. 2013) and endolysins (Schmelcher et al. 2012a) which break down the peptidoglycan layer during early phage entry and at lysis towards the end of the phage replication cycle, respectively. Furthermore, polysaccharide depolymerases (Yan et al. 2014) and holins (Young 2014), which raptures the extracellular polysaccharides of the bacterial cell wall and cytoplasmic membrane, respectively, are essential components of the phage life cycle. These enzymes support phages' bactericidal activity and provide models for the synthesis of new antimicrobials that can break through bacterial biofilms, a well-known obstacle to the successful use of antibiotics. The therapeutic potential of phage-derived enzymes goes beyond their direct bactericidal properties. By dissolving bacterial biofilms, these enzymes can improve the efficiency of traditional antibiotics thereby providing a synergistic approach for treating complex bacterial infections. Such characteristics are especially important in the context of chronic infections by biofilm-associated pathogens, which are naturally more resistant to antibiotic therapy.

Phage-encoded enzymes offer unique advantages over conventional antibiotics, including high selectivity towards target bacteria, no effect on the beneficial human microbiome, and the ability to penetrate and break bacterial biofilms (Haddad Kashani et al. 2018). Furthermore, by breaking through outer membrane barriers, these enzymes can be modified to broaden their antibacterial spectrum, increase their effectiveness against Gram-negative bacteria, and reduce the likelihood of resistance development. Phage-derived lytic enzymes are being actively investigated in preclinical and clinical settings due to their potential effectiveness against antibiotic-resistant bacteria (Danis-Wlodarczyk et al. 2021), such as those from the ESKAPE group ("*Enterococcus faecium*, *Staphylococcus aureus*, *Klebsiella pneumoniae*, *Acinetobacter baumannii*, *Pseudomonas aeruginosa*, and *Enterobacter* spp."), which currently pose a major challenge in the management of infectious diseases.

Despite having antimicrobials potential, the research and clinical applications of phage-derived enzymes still face several challenges. Although these enzymes are harmless and help preserve the microbiome, their specificity makes it necessary to use accurate diagnostic tools in order to determine selectively which enzyme is to be used to control the specific pathogen which is causing the infection. Further, in order to fully realize the therapeutic potential of enzyme-based medicines, logistic challenges related to their delivery and stability in the human body must also be addressed. Nevertheless, continued research and innovation, is leading phage-derived enzymes towards playing a pivotal role in antimicrobial arsenal, offering effective, targeted, and sustainable alternatives to traditional antibiotics.

35.2 Mechanisms of Bacterial Lysis by Phage-Encoded Enzymes

Polysaccharide depolymerases (PSDs) and peptidoglycan hydrolases (PGHs) are pivotal in the infection cycles of tailed phages (Fig. 35.1). Initially, PSDs (Knecht et al. 2020) facilitate phage entry by dissolving bacterial defense layers such as capsular polysaccharides (CPS), exopolysaccharides (EPS), and lipopolysaccharides (LPSs), enabling the phage to access specific receptors on the bacterial cell envelope (Harper et al. 2021). Following this, the phage tail mechanism penetrates the cell wall, aided by VAPGHs, specifically targeting the peptidoglycan layer for localized degradation without causing cell lysis, preserving cell integrity for the infection to proceed (Rodríguez-Rubio et al. 2013). In isolation, recombinantly produced VAPGHs can directly lyse bacteria upon contact. Once inside, phage DNA manipulates bacterial systems to replicate and assemble new phages. Holins then create membrane perforations (Wang et al. 2000), allowing endolysins and, in

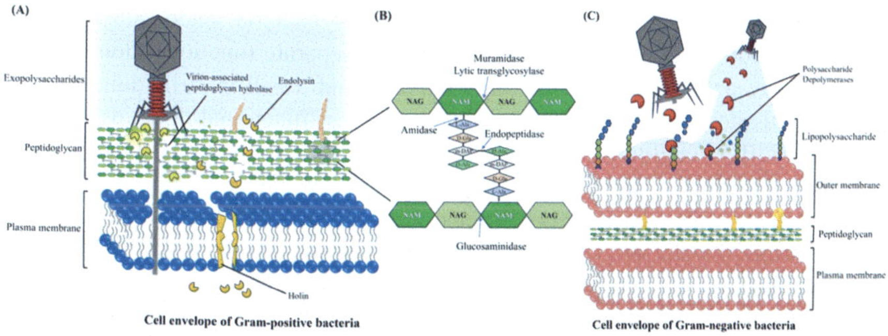

Fig. 35.1 Illustration of bacteriophage-derived enzymes attacking the cell envelope of bacteria during lytic replication. Bacteriophages deploy enzymes such as (**a**) Endolysins and holins to breach the bacterial cell wall, initiating cell rupture. Endolysins target specific bonds within the peptidoglycan layer, leading to its disintegration. This action, coupled with the cell's internal pressure, forces the plasma membrane outward, culminating in the cell bursting due to external forces. (**b**) A general peptidoglycan structure is shown, highlighting the specific sites targeted by various endolysin categories, including glucosaminidases, muramidases, endopeptidases, and amidases. Some phages employ virion-associated peptidoglycan hydrolases to penetrate the bacterial cell's protective envelope. (**c**) Depolymerases are utilized to break down the extracellular polysaccharides and lipopolysaccharides, which otherwise serve as obstacles to the viral attachment to the bacterial surface

Gram-negative bacteria, phages also produce spanins, which breach the bacterial outer membrane (Kongari et al. 2018), releasing new phages. PSDs may also act here to prevent newly liberated phages from biofilm entrapment (Yan et al. 2014).

35.3 Key Phage-Encoded Enzymes as Antimicrobial Agents

35.3.1 Endolysins

Endolysins, with their extensive lytic capabilities across both Gram-positive and Gram-negative bacterial types, are emerging as a potent alternative to antibiotics. These phage-derived enzymes break down the bacterial peptidoglycan layer at cell lysis, a critical step in phage proliferation. Their lytic spectrum includes acetylmuramidases, glucosaminidases, amidases, transglycosylases, and endopeptidases (Gondil et al. 2020). Despite the development of potential bacterial resistance, phage proteins like endolysins remain effective due to long-term phage-host genetic exchanges. However, bacterial defense strategies and human immune responses pose challenges to their application, highlighting the necessity of more in-vivo research and clinical studies to effectively employ endolysins in phage therapy.

35.3.1.1 Structure and Activity

Endolysins are peptidoglycan degrading enzymes encoded by almost all phages and are synthesized within a host bacterium towards the end of a replication cycle (Haddad Kashani et al. 2018). The structure of endolysins differs significantly when targeting Gram-positive versus Gram-negative bacteria, attributed to the distinct compositions of their cell walls. In Gram-positive bacteria, endolysin typically exhibits a modular architecture consisting of two separate functional domains: an enzymatically active (catalytic) domain (EADs) and a cell wall binding domain (CBDs). This design allows for the rearrangement of functional domains, notably EAD and CBD, facilitating the development of novel lysine enzymes (Binte Muhammad Jai et al. 2020). The EADs disrupt specific bonds within the bacterial peptidoglycan, while CBDs anchor the enzyme to the cell wall thereby dictating the lytic spectrum of lysine enzyme, preventing its diffusion, and facilitating the lysis of adjacent cells.

Conversely, the outer membrane of Gram-negative bacteria generally shields against such collateral damage, resulting in the absence of CBDs in most Gram-negative endolysins and instead featuring a singular globular catalytic domain (Loessner et al. 2002). However, exceptions exist, such as the Gram-negative endolysins KZ144 and EL188 from *Pseudomonas* phage, which include both CBD and EAD modules, exhibit high affinity and broad-spectrum efficacy towards Gram-negative bacterial pathogens (Briers et al. 2009, 2011).

The typical structure for Gram-positive endolysins includes an EAD at the N-terminus and a CBD at the C-terminus, though variations exist, as demonstrated by streptococcal lysins with differing arrangements of these domains. In streptococcal lysin, λSA2, centrally located CBD with two flanking EADs has been documented (Pritchard et al. 2007). Recently, a new endolysin from *Bacillus cereus*, named LysPBC2, was identified to possess an additional spore binding domain (SBD). This unique SBD selectively attaches to the spores of *B. cereus* without interacting with the vegetative cells of the bacterium (Kong et al. 2019). These structural variations among lysins underline their adaptability and specificity in targeting different bacterial cell wall architectures besides showcasing their potential as precise antibacterial agents.

Endolysins are usually encoded by a single gene, with the exception of the PlyC endolysin against *Streptococcus*, which comprises two genes: PlyCA for enzymatic activity and PlyCB for cell wall binding (Nelson et al. 2001, 2006). The EADs are categorized into four main types based on their action: acetylmuramidases, amidases, glucosaminidases, and endopeptidases, each targeting specific bonds within the peptidoglycan. Muramidases and glucosaminidases target the disaccharide backbone for cleavage; N-acetylmuramoyl-L-alanine amidases sever the amide linkage that connects MurNAc to peptide chains; and endopeptidases make cuts within the peptide cross-links (Schmelcher et al. 2012a). Amidases and muramidases are among the most common enzymes, targeting highly conserved bonds of polysaccharide chains. The net charge of EADs also influences the lytic activity, especially important in negatively charged membranes of Gram-positive bacteria (Low et al. 2011).

The CBDs are crucial in maintaining the specificity of the action of endolysin. It specifically targets the carbohydrate epitopes on the cell wall, with an affinity ranging from nano to picomolar level, comparable to the affinity of the antigen–antibody interactions (Schmelcher et al. 2012a). Some endolysins can retain or enhance their lytic activity without CBDs, while others see a reduction upon CBD removal (Haddad Kashani et al. 2018). The potential of endolysins extends beyond direct bacterial lysis to include disrupting pathogenesis and immune evasion mechanisms, with the O-acetylation of peptidoglycan emerging as a novel antivirulence target.

35.3.1.2 Role and Potential of Endolysins

Endolysins operate by degrading peptidoglycan during cell lysis, a process that involves additional phage lytic proteins and holins. Holins are tiny hydrophobic proteins that create holes (>100 nm in diameter) in the cell membrane (Wang et al. 2000; Young 2002; Donovan 2007). Remarkably, resistance to phage proteins like endolysins is uncommon among bacteria, likely due to the long history of horizontal gene transfer between phages and their hosts (Gondil et al. 2020). Yet, bacteria have evolved resistance mechanisms against endolysins, such as developing an outer membrane or exopolysaccharides like biofilms. In a similar realm, the host immune system such as pro-inflammatory cytokines and antibodies may have implications in the use of endolysin.

35.3.1.3 Endolysins as Therapeutic Agents

In the battle against escalating antibiotic resistance, endolysins are gaining prominence as potential substitutes for traditional antibiotics. Being effective towards both Gram-positive and Gram-negative pathogens, these enzymes have seen a skewed interest in targeting Gram-positive species due to their unique cell wall compositions. Research to date has unveiled a range of endolysins capable of combating lethal bacteria, such as *Streptococcus pneumoniae* (Silva et al. 2020), *S. pyogenes* (Shen et al. 2013; Lood et al. 2014) and *Staphylococcus aureus* (Son et al. 2021). Additionally, studies have demonstrated their effectiveness against *Mycobacterium* spp. (Payne and Hatfull 2012; Catalão and Pimentel 2018), *Pseudomonas aeruginosa* (Guo et al. 2017), *Escherichia coli* (Lv et al. 2015); *Acinetobacter baumannii* (Lood et al. 2015), and *Klebsiella pneumoniae* (Walmagh et al. 2013; Maciejewska et al. 2017) etc. showcasing the broad-spectrum potential of endolysins against antibiotic-resistant pathogens (Kim et al. 2020).

Endolysins Against Gram-Positive Bacterial Pathogens

The initial report on the application of purified endolysins as recombinant proteins for bacterial lysis dates back to 1959 (Freimer et al. 1959). Subsequently, numerous endolysins, especially those targeting Gram-positive bacteria, have been identified and characterized. However, as of now, endolysins have not received approval for human therapeutic applications. Contra Fect's Exebacase (formerly CF-301), the first and only FDA-approved endolysin therapy for *Staphylococcus aureus* bloodstream infections, encompassing right-sided endocarditis has reached Phase 3 trials of clinical development (ClinicalTrials.gov, Identifier NCT04160468). Despite its promising early results, the study was terminated due to futility following interim efficacy analysis, underscoring the challenges in advancing bacteriophage-derived therapeutics.

In a landmark study by Nelson et al. (2001), it was demonstrated for the first time that a purified recombinant endolysin could effectively diminish bacterial populations in an animal model of infection, specifically targeting *Streptococcus pyogenes*, a bacterium generally linked to pharyngitis and impetigo. The study revealed that oral administration of 24 ŋg of the group *C. streptococcal* phage endolysin PlyC could clear a significant streptococcal infection in the oral cavities of mice within just 2 h. Later, Gilmer et al. (2013) identified a novel bacteriophage lysin PlySs2, originating from a *Streptococcus suis*phage, to exhibit extensive lytic capabilities against a wide range of bacteria, including MRSA, vancomycin-intermediate *Listeria*, *Staphylococcus aureus*, *S. simulans*, *S. epidermidis Streptococcus suis*, *S. equi*,,*S. agalactiae*, *S. pneumoniae*, *S. pyogenes*, *S. sanguinis*, group G streptococci and group E streptococci. It has demonstrated remarkable stability for 30 min at 50 °C, for over 24 h at 37 °C, for 15 days at 4 °C and for over 7 months at −80 °C. Further it has retained full activity even after 10 freeze-thaw cycles. Under in-vitro conditions, PlySs2 significantly reduced the growth of MRSA and *S. pyogenes* by 5 logs and 3 logs within an hour, respectively, was found having a minimum inhibitory concentration (MIC) of 16 μg/ml against MRSA. In a bacteremia model involving a mixed infection of MRSA and *S. pyogenes* in mice, a single 2 mg dose of PlySs2 offered protection against 92% of the mice under test. On continual exposure of MRSA and *S. pyogenes* to PlySs2 or mupirocin, no resistance to PlySs2 was observed, whereas resistance had emerged against mupirocin. This level of broad-spectrum lytic activity, stability and effectiveness against multiple, deadliest human bacterial pathogens had not been documented for any other lysin before, crowing PlySs2 as a potentially powerful therapeutic agent.

In another significant contribution, Loeffler et al. (2001) demonstrated potent bactericidal activity of a purified pneumococcal bacteriophage lytic enzyme (Pal) which was originally identified as N-acetyl muramoyl l-alanine amidase, against 15 prevalent pneumococcal serotypes, including strains highly resistant to penicillin. In-vivo trials showed that the mice with prior pneumococcal colonization had undetectable levels of bacteria within 5 h following a single treatment with the enzyme. Notably, Pal exhibited minimal to no impact on the typical microflora of the human oropharynx and the exposure to enzymes did not result in the detection of Pal-resistant pneumococci after extensive testing. A significant breakthrough was made with the identification of Lys-phiSA012, an endolysin derived from phiSA012 and related to the Lys-phiK type, which demonstrated potent lytic activity against various Staphylococcal strains, including MRSA, *Staphylococcus pseudintermedius* and *S. haemolyticus* (Fujiki et al. 2018).

Endolysins Against Gram-Negative Bacterial Pathogens

Eliminating Gram-negative bacterial pathogens with externally applied PGHs is challenging as their outer membrane is exceptionally resistant to the penetration of large molecules, including the majority of endolysins (Morita et al. 2001). This barrier has resulted in a scarcity of successful endolysin-based animal studies targeting these pathogens and hence increased the concerns about escalating prevalence of multidrug-resistant Gram-negative bacteria (Briers et al. 2014). Despite these obstacles, there have been discoveries of phage-derived endolysins that inherently possess the ability to combat Gram-negative bacteria (Antonova et al. 2019). Recent research has highlighted a small subset of endolysins capable of exhibiting bactericidal actions against Gram-negative bacteria, pinpointing the importance of their terminal sequences (amino acids 1–20) that contain either positively charged or hydrophobic residues, facilitating their passage through the bacterial outer membrane (Plotka et al. 2019; Lai et al. 2020).

A study unveiled LysPA26, a novel endolysin with a domain similar to lysozyme, capable of targeting MDR *Pseudomonas aeruginosa*. Remarkably, LysPA26 could reduce bacterial populations by up to four logarithmic units in just 30 min without needing an outer-membrane permeabilizer. This study further revealed LysPA26's robust stability across diverse pH levels and temperatures, alongside its bactericidal effectiveness against other Gram-negative pathogens like *Klebsiella pneumoniae*, *Escherichia coli,* and *Acinetobacter baumannii,* though it was found ineffective against Gram-positive bacteria. Importantly, LysPA26 was also shown to disrupt biofilms formed by *P. aeruginosa*, positioning it as a potent new tool to fight against Gram-negative infections (Guo et al. 2017). Additionally, two endolysins LysG24 and LysCA from *K. pneumoniae* phage vB_KpnS_MK54, were found to introduce permeability in outer membrane in the presence of EDTA, displaying a strong antibacterial action against *K. pneumoniae* (Lu et al. 2022).

Crucial research (Gonçalves et al. 2024) tackled the public health menace posed by multidrug-resistant *K. pneumoniae* from Enterobacteriaceae family by the application of three specific lytic endolysins: Kp2948-Lys, Kaer26608-Lys, and Ps3418-Lys. In order to enhance the penetration of these endolysins through the outer membrane of Gram-negative bacteria, the liposomes were employed as a delivery method. The liposomes were composed of dimyristoyl phosphatidylcholine, cholesteryl hemisuccinate, and dioleoyl phosphatidyl ethanolamine. This method markedly boosted the endolysins' ability to penetrate and act upon their targets. The findings underscored the particular effectiveness of Kp2948-Lys against *K. pneumoniae*, while indicating that Kaer26608-Lys and Ps3418-Lys could also be effective against a wider range of bacterial pathogens. This innovative strategy of using liposome-encapsulated endolysins presents a promising avenue for addressing multidrug-resistant infections, consequently supporting the 'One Health' concept of seeking comprehensive solutions across human, animal, and environmental health sectors.

In another study, Wu et al. (2019a) examined the efficacy of Ply6A3 endolysin, derived from vB_AbaP_PD-6A3 phage, against MDR *Acinetobacter baumannii*. The research has demonstrated that Ply6A3 endolysin is not only effective in neutralizing MDR *A. baumannii* strains but also has the capacity to act against *E. coli* and methicillin-resistant *Staphylococcus aureus* (MRSA) strains. Additionally, the study provided evidence that intraperitoneal injections of Ply6A3 endolysin could successfully treat lethal sepsis in mice caused by *A. baumannii*, showcasing its potential as a therapeutic agent against multiple drug-resistant infections.

35.3.2 Virion-Associated Peptidoglycan Hydrolases (VAPGHs)

Phage-encoded VAPGHs play a pivotal role early in the infection process by degrading the bacterial peptidoglycan layer, which is crucial for the entry of phage genetic material into host cells. These enzymes are distinct from endolysins, which act later in the phage lifecycle to facilitate the release of newly formed phage particles by breaking down the host cell wall. The VAPGHs facilitate the initial phase of phage infection by generating a narrow aperture, enabling the phage's tail tube to breach the cell envelope and deliver its genetic material into the host. This underscores the essential function served by these enzymes in initiating phage infection. The antimicrobial properties of VAPGHs were unveiled through the observation of the 'lysis from without' event, in which the bacterial cell wall is compromised prior to the generation of new phages, due to the attachment of multiple phages to the bacterial exterior. The specificity, robustness at elevated temperatures, and structural modularity render VAPGHs, promising candidates for developing new antibacterial strategies. These enzymes hold potential as effective tools against bacteria resistant to antibiotics in both human and veterinary healthcare, as well as bio-preservatives for ensuring food safety and as means to manage detrimental bacteria in agricultural settings (Rodríguez-Rubio et al. 2013).

35.3.2.1 Diversity and Biochemical Characterization of VAPGHs

Virion-associated peptidoglycan hydrolases are found in phages that infect both Gram-negative and Gram-positive bacterial types, aiding in delivery of their genetic material into the host cell. Not all phages, however, depend on VAPGH for the transfer of their genetic material, as many lack genes coding for these enzymes. Despite the availability of limited biochemical characterizations of VAPGHs they are believed to exhibit significant similarities to endolysins, particularly in their likely mechanisms of action involving breaking down of specific peptidoglycan bonds. For instance, the temperate *S. aureus* phage DW2 produces both a potential endolysin (LysDW2) and a VAPGH (THDW2), each capable of lysing staphylococcal cells on their own when applied exogenously to cells. These enzymes share a CHAP domain at the N-terminus but differ in their EADs, with LysDW2 featuring an amidase and THDW2 a muramidase, and notably, THDW2 does not have a CBD (Keary et al. 2014). Similarly, the bacteriophage FMR11, which targets *S. aureus*, produces a potential VAPGH featuring a CHAP domain followed by a muramidase domain. These domains function independently to lyse staphylococcal cells and do not include any CBDs (Rashel et al. 2007). In another example, the *S. aureus* phage vB_SauS-phiIPLA88's VAPGH, named HydH5, functions as an enzyme with two EADs (Rodríguez et al. 2011). In contrast, the VAPGH, P68 from *S. aureus* phage, not only exhibits muralytic activity but also contains a C-terminal region for substrate recognition or binding, which is speculated to attach to a carbohydrate moiety linked to the peptidoglycan. This enzyme has shown effectiveness against various clinical strains of *S. aureus*, including MRSA (Takác and Bläsi 2005). The diversity observed among VAPGHs, despite the small limited studies, highlighted their potential for engineering due to their modular design, similar to endolysins. This was demonstrated by enhancing the lytic activity of HydH5 and its derivatives through fusion with the SH3b CBD from lysostaphin, achieving greater bacteriolytic efficacy than the original enzyme (Rodríguez-Rubio et al. 2012).

35.3.2.2 Therapeutic Application of VAPGH

The employment of VAPGHs in therapeutic contexts, particularly HydH5 and its SH3b CBD fused derivative, when combined with the endolysin LysH5 from the same phage, displayed synergistic effects due to their ability to target different sites in the peptidoglycan structure of staphylococcal cells. This combination proved to be more effective at breaking down the peptidoglycan than when the enzymes were used separately (Rodríguez-Rubio et al. 2012). This strategy of domain swapping was also successful in enhancing the bacteriolytic potency of the VAPGH within phage K's orf56. By removing the N-terminus of orf56, a smaller PGH (Lys16) with staphylolytic capabilities was identified. A fusion of Lys16 with the SH3b CBD from lysostaphin resulted in a chimeric enzyme (P128) that exhibited enhanced effectiveness against a range of *S. aureus*strains, MRSA included (Paul et al. 2011). Notably, P128 is the first VAPGH-based therapeutic to be tested in an animal model for bacterial infections. In a study involving rats with MRSA in their nares, treatment with P128 formulated in a hydrogel significantly reduced MRSA colonization compared to a mupirocin ointment, which was ineffective (Paul et al. 2011). The hydrogel formulation also demonstrated bacteriolytic activity against staphylococcal isolates from human nares without needing any preparatory steps, indicating potential for direct application (Vipra et al. 2012). Cornelissen et al. (2011) investigated how phage φ15, identified as a 'T7-like virus' harboring an exopolysaccharide (EPS) depolymerase, breaks down biofilms formed by *Pseudomonas putida*. It was found that φ15 created clear zones around plaques on several *P. putida* strains, indicating EPS breakdown, essential for phage infection. The degradation of biofilms showed a time and dose-dependent efficacy, peaking at 8 h with higher phage doses and showing resistance by 24 h. Biofilm age impacted the degradation effectiveness, with older PpG1 biofilms showing reduced susceptibility, unlike RD5PR2 biofilms. Planktonic cultures were killed more efficiently than biofilms were degraded, highlighting the importance of evaluating phage therapy in biofilm scenarios. The study confirmed the EPS degrading capability of the phage's recombinant tail spike protein, underscoring the potential of high-titer, EPS-targeting phages in phage therapy development. These findings suggest that, given their similar PGH properties, newly identified VAPGHs could offer therapeutic benefits against Gram-positive pathogens akin to endolysins in future bacterial infection and disease models.

35.3.3 Polysaccharide Depolymerases

Phage-encoded depolymerases are enzymes that specialize in breaking down bacterial polysaccharides such as exopolysaccharides (EPS), capsular polysaccharides (CPS), and lipopolysaccharides (LPS). These enzymes play an important role in bacteriophage infection by clearing the way for the bacteriophage to deliver its genetic material inside the bacterial cell (Knecht et al. 2020). Unlike other phage enzymes that directly lyse bacterial cells, depolymerases target and degrade the protective polysaccharide layers of bacteria. This degradation not only facilitates phage infection but also leaves bacteria vulnerable to immune system attacks and antibacterial agents, enhancing their effectiveness (Majkowska-Skrobek et al. 2018).

35.3.3.1 Structure

The evolutionary interplay between phages and bacteria leads to the variation in bacterial carbohydrate compositions, thereby influencing the diversity of active depolymerases. Despite their varied enzymatic targets, many depolymerases exhibit similar structural characteristics due to the evolving carbohydrate structures of their bacterial hosts (Mostowy and Holt 2018). A comprehensive review by Pires et al. (2016) identified 160 potential depolymerases across 143 phages, targeting 24 varied bacterial genera. These depolymerases, integral to the phage's binding machinery, are commonly found as components of tail fibers or spikes connected to the phage's baseplate, with some instances of their occurrence on the phage neck or as separate soluble entities (Drulis-Kawa et al. 2015). They are primarily located within receptor-binding proteins (RBPs), playing a crucial role in initiating phage infection by recognizing and dismantling bacterial capsules (Nobrega et al. 2018).

Structurally, RBPs with depolymerase functions are characterized by a modular design that includes a stable N-terminal domain forming a dome-like attachment site to the phage tail, a variable central domain exhibiting a parallel β-helical homotrimer configuration with a catalytic pocket essential for recognizing the host and facilitating enzymatic breakdown, and a domain at C-terminal end for receptor engagement, potentially aided by a chaperone for correct folding and activity (Latka et al. 2017, 2019; Latka and Drulis-Kawa 2020; Squeglia et al. 2020). This configuration facilitates rapid alterations in the host range through either horizontal gene transfer, which exchanges the host-specific central domain, or vertical gene transfer, which modifies the polysaccharide-depolymerizing domains through mutation accumulation (Latka et al. 2017, 2019; Latka and Drulis-Kawa 2020).

35.3.3.2 Mechanism of Action of Polysaccharide Depolymerases

Depolymerases, as characterized by their enzymatic functions, engage in the cleavage of O-glycosidic linkages within bacterial polysaccharides, yielding oligosaccharide fragments through either hydrolytic or lytic processes. This bifurcation in operational modes segregates depolymerases into two primary categories: hydrolases and lyases, a classification thoroughly reviewed by Pires et al. (2016). Hydrolases, which encompass enzymes such as sialidases, xylosidases, hyaluronidases, dextranases, levanases, and peptidases, facilitate bond cleavage via the incorporation of a water molecule. Conversely, lyases, which include enzymes like alginate lyases, hyaluronate lyases, K5 lyases, and pectin/pectate lyases, execute glycosidic bond cleavage through β-elimination, introducing a novel double bond in the absence of a water molecule (Yan et al. 2014; Drulis-Kawa et al. 2015; Knecht et al. 2020).

35.3.3.3 Therapeutic Potential of Depolymerases

A variety of bacteriophages synthesize specialized enzymes known as polysaccharide depolymerases, which are generally found in association with components such as tail spikes, base plates, tail fibers, or tail tubular proteins (Pires et al. 2016). These enzymes play a crucial role in dismantling the capsular polysaccharide (CPS) structures of bacteria, facilitating the phage's attachment to outer membrane receptors, enabling the transfer of phage DNA inside the target cell, and leading to the cell's eventual rupture. Consequently, bacteriophages and their CPS-depolymerizing enzymes are recognized as potent agents in combating polysaccharide-encased bacterial pathogens. Notable examples include endorhamnosidases targeting the LPS of Gram-negative bacteria (Waseh et al. 2010), endosialidases breaking down *E. coli* capsular polysaccharides (Mushtaq et al. 2005), alginate lysases targeting *P. aeruginosa* capsules (Glonti et al. 2010), and hyaluronidases that act on streptococcal species capsules (Mushtaq et al. 2005). To date, depolymerases targeting a wide range of capsular types in *K. pneumoniae* have been identified, showcasing their specificity across various strains (Lin et al. 2014; Hsieh et al. 2017; Majkowska-Skrobek et al. 2018; Solovieva et al. 2018; Wang et al. 2019; Volozhantsev et al. 2020; Pertics et al. 2021). As potential antimicrobial agents, depolymerases are harnessed in two primary formats: as phage-derived enzymes and as recombinant proteins. Both forms can be genetically modified or combined into mixtures to broaden their applicability across different bacterial hosts (Oliveira et al. 2022). Although depolymerases have not yet been documented to directly cause bacterial lysis or stasis, their therapeutic value primarily lies in their ability to remove extracellular polysaccharides. These polysaccharides are exploited by numerous bacterial pathogens for biofilm formation, virulence enhancement, and evasion of host immune responses, as well as resistance to antimicrobials and phages. Thus, depolymerases hold promise as agents against biofilms, as adjuncts to phage and antibiotic therapies, in curtailing resistance development, and as supplements to innate immune mechanisms (Wang et al. 2023).

In a pioneering study by Gutiérrez et al. (2012), a temperate phage named vB_SepiS-phiIPLA7, which infects *Staphylococcus epidermidis* strain *F12*, was found to carry a putative EPS depolymerase. This research demonstrated the phage's capacity to dismantle biofilms, significantly reducing the viability of bacteria within the biofilm, with survival rates plummeting to 5% after 3 h. A follow-up investigation by Gutiérrez et al. (2015) successfully isolated and cloned the gene responsible for encoding the depolymerase, leading to the production and purification of a recombinant protein dubbed Dpo7. This enzyme showed a potent ability to diminish both the biomass and the number of adhered cells within pre-existing biofilms in a manner that was dependent on the dosage but not the duration of treatment. Furthermore, Dpo7 proved effective in hindering the biofilm formation of the host bacteria and was capable of dispersing biofilms formed by various strains of *S. epidermidis* and *Staphylococcus aureus*, though it did not achieve complete eradication of preformed biofilms or fully prevent new biofilm formation. The findings also suggested that deploying depolymerases in conjunction with antibiotics or other antimicrobial strategies could significantly bolster the efficacy of biofilm disruption. In a related exploration, Chen et al. (2022) examined the impact of another depolymerase, Dpo71, when used together with colistin against biofilms formed by *Acinetobacter baumannii*. This synergistic approach markedly improved the disruption of biofilms beyond what was achievable with either agent alone.

Studies have highlighted the effectiveness of combining depolymerases with antimicrobial agents in tackling infections caused by multi-drug resistant (MDR) organisms. For instance, the collaborative impact of depolymerase and gentamicin on the biofilms of *Klebsiella pneumoniae* was explored by Bansal et al. (2015), demonstrating that depolymerases could break down the biofilm structure, enhancing gentamicin's ability to penetrate and significantly lower bacterial populations in comparison to treatments using either agent alone. This cooperative effect was similarly noted when combining Dep42 depolymerase with polymyxin in targeting *K. pneumoniae* biofilms (Wu et al. 2019b), and when Dpo71 depolymerase (Chen et al. 2022) was used alongside colistin against biofilms formed by *A. baumannii*.

Recently, Volozhantsev et al. (2022) investigated the antimicrobial capabilities of the bacteriophage KpV74 and its depolymerase enzyme, Dep_kpv74, targeting the K2 capsular type of *Klebsiella pneumoniae*. The enzyme Dep_kpv74, functioning as a glucosidase, specifically attacked and broke down the capsular polysaccharides of the K2 type *K. pneumoniae*, proving to be effective in treating soft tissue infections in the thighs of mice without any noticeable side effects or toxicity. The study noted that while the phage KpV74 was only effective against its host strain, Dep_kpv74 showed effectiveness across all tested *K. pneumoniae* strains. Resistance to both the phage and the depolymerase treatment was observed, linked to a single base deletion in the wzx gene, which affects the capsule structure and diminishes virulence. The exploration of recombinant depolymerases for the treatment of various MDR bacterial infections in vivo is growing, alongside advancements in formulation strategies for efficient delivery of these protein-based agents, enhancing the potential for their clinical application.

The validation of depolymerases' therapeutic potential extends beyond in vitro studies, necessitating in vivo demonstrations of efficacy. So far, the in vivo assessment of depolymerase effectiveness predominantly involves animal testing (Danis-Wlodarczyk et al. 2021). A range of preclinical investigations has highlighted the efficacy of recombinant depolymerases against multiple MDR pathogens using animal infection models (Olszak et al. 2017; Liu et al. 2019; Chen et al. 2020, 2022; Wang et al. 2020; Gorodnichev et al. 2021; Li et al. 2022; Volozhantsev et al. 2022). These studies collectively affirm that depolymerases, when administered in appropriate doses, are both safe and efficacious for in vivo antimicrobial applications.

35.3.4 Holins

Both types of bacteria, Gram-positive and Gram-negative, feature an inner membrane in their cellular architecture which acts as a barrier, preventing the cell's internal substances from interacting with the peptidoglycan layer of the cell wall. This structural arrangement usually restricts endolysins, which accumulate during the replication of phages to facilitate cell lysis and the release of new phage particles, from directly accessing their target. This is where the role of holins becomes crucial. Holins are small hydrophobic proteins that integrate into the membrane. Once they reach a specific, genetically determined concentration, they activate to create openings. These "scheduled" openings are typically large enough to permit the passage of fully formed endolysins to reach and interact with the peptidoglycan. This interaction leads to the breakdown of the peptidoglycan, causing the cell to rupture and subsequently release the newly formed phage particles (Park et al. 2006; Fernandes and São-José 2018).

35.3.4.1 Structure of Holin

A diverse array of holin proteins, encompassing 52 known families, has been characterized, playing a crucial role in phage biology. These proteins, which can embed themselves across the cytoplasmic membrane through one to four α-helical transmembrane segments, exhibit a remarkable consistency in size and structure within each family. Holins have been detected in viruses that target both Gram-positive and Gram-negative bacteria, underscoring their widespread relevance (Reddy and Saier Jr 2013). Despite the challenge in pinpointing holins due to their sequence variability, they share notable features: transmembrane domains, a charged, hydrophilic

C-terminal region, and a gene location near the endolysin gene. The precise mechanism through which holins facilitate endolysins' access to the bacterial cell wall—whether by secretion, causing leakage, or lysing the membrane—remains an area for further clarification, potentially varying across different holin types (Saier Jr and Reddy 2015).

35.3.4.2 Holin as Antimicrobial Compound

Holins, beyond facilitating bacterial lysis, exhibit a capability for inducing cell death on their own, demonstrating a broad-spectrum antibacterial effect towards Gram-positive as well as Gram-negative bacteria, distinct from the more targeted action of endolysins. The holin HolGH15, derived from *S. aureus* phage GH15, has been shown to possess potent antimicrobial properties against a range of bacteria including *S. aureus*, *B. subtilis*, *L. monocytogenes*, *P. aeruginosa*, and *K. pneumoniae*, particularly demonstrating significant activity against *Listeria* and *Staphylococcus*, with bacterial reductions nearing 5 log units. Notably, HolGH15's action on cellular membranes does not lead to hemolysis, as evidenced by the lack of erythrocyte rupture in tests using defibrinated rabbit blood, highlighting a selective mechanism of action (Song et al. 2016). Furthermore, a metagenomic exploration on goat skin uncovered a holin-like peptide, Tmp1, which compensated for the lysis function in a lysis-deficient lambda bacteriophage. This 34-amino-acid peptide showed efficacy against Gram-positive bacteria but not Gram-negative ones. Through error-prone PCR, variants of Tmp1 with expanded antibacterial activity, including against Gram-negative bacteria, were developed, achieving up to 4 log units reduction in bacterial counts. This enhancement in antibacterial potency is attributed to an increase in hydrophobicity, without affecting the protein's net charge or inducing hemolytic activity. Such findings not only underscore holins' antimicrobial potential but also their capacity for optimization via protein engineering (Rajesh et al. 2011). The lysis cassette of the *S. suis* phage SMP, which includes holinHolSMP and endolysin LySMP, was examined (Shi et al. 2012). It was found that the exogenous application of LySMP was capable of lysing various strains of *S. suis* and *S. aureus*, demonstrating a wider range of lytic activity compared to the complete phage. Additionally, LySMP proved effective in disrupting biofilms formed by *S. suis*. Interestingly, when applied by itself, HolSMP induced a mild lysis effect on *S. aureus* and *B. subtilis*, even in strains not affected by endolysin. The combination of holins and endolysins enhanced the spectrum of activity of LySMP, including additional strains of multidrug-resistant *S. suis* and *S. aureus*. These findings suggested that leveraging both holins and endolysins together might serve as an improved approach for enhancing lytic efficiency. In a previous study, the roles of a holin-endolysin pair derived from bacteriophage HH109 of *V. alginolyticus* were explored by analyzing their overexpression effects on *E. coli* and *V. alginolyticus*. The findings indicated that phage HH109's endolysin exhibited more potent bactericidal activity than its holin, leading to increased cell death and significant structural damage in endolysin-expressing *E. coli*. A synergistic effect was observed when holA and lysin were co-expressed, despite no direct interaction being found. This system also showed effectiveness against *V. alginolyticus*, suggesting potential for treating vibriosis (Li et al. 2021). In a recent study, a phage holin gene was successfully cloned and expressed, revealing the recombinant holin pEF191 had superior antibacterial and biofilm removal capabilities against *E. faecalis* compared to its parental phage PEf771. In Sprague-Dawley rats, pEF191 proved safe and enhanced protection against *E. faecalis* without impacting the intestinal microflora diversity, even increasing beneficial bacteria. This positions pEF191 as a promising antibiotic alternative for *E. faecalis* infections (Xiang et al. 2024).

Recent investigations have explored the application of holins in development of potential vaccines. By incorporating the genes responsible for holin and endolysin production (genes S and R, respectively) from the λ phage, alongside the gene for the lysis protein E from the φX174 phage into *Salmonella* strains, researchers succeeded in generating bacterial ghosts (BGs). These BGs are capable of provoking both an antibody-mediated and cell-mediated immune response (Won et al. 2017a, 2017b; Won and Lee 2017), offering a promising platform for vaccine vector development (Won et al. 2020). Additionally, the use of the nonpathogenic *Lactobacillus casei* as a carrier for DNA vaccine delivery has been explored. In this scenario, the activation of holin gene expression alone was adequate for the formation of BGs, which were then efficiently and securely loaded with DNA, showcasing a novel approach for vaccine delivery (Hou et al. 2018).

35.4 Development of Resistance

The mechanism of action of phage-derived enzymes, focusing on highly conserved elements of the cell wall, significantly lowers the likelihood of bacterial resistance development (Fischetti 2005, 2010). This is in stark contrast to whole phages and conventional antibiotics, where resistance is a more common occurrence. For instance, the specificity of endolysins towards their bacterial targets further minimizes the risk of resistance emergence. Studies have consistently shown that repeated exposure to these enzymes does not lead to resistance, underlining their potential as therapeutic agents (Schuch et al. 2002; Pastagia et al. 2011; Gilmer et al. 2013). However, certain cell wall modifications in bacteria, such as O-acetylation, D-alanylation, and N-deacetylation, could theoretically contribute to resistance against phage lysins, drawing parallels to resistance mechanisms against human lysozyme (Guariglia-Oropeza and Helmann 2011; Vollmer et al. 2008). Despite these potential resistance pathways, real-world evidence of lysin resistance remains scarce. This is attributed to the external application of endolysins and their action on the peptidoglycan layer outside the bacterial cell, which are factors limiting the development of resistance (Spratt 1994).

Conversely, resistance to phage depolymerases, enzymes that degrade protective bacterial capsules, has been noted due to alterations in the polysaccharide composition of bacterial capsules or surface structures. This form of resistance suggests a more frequent adaptation (McCallum et al. 1989). Recent studies advocate for the therapeutic use of recombinant depolymerases and combination treatments involving both depolymerases and other antimicrobials to counteract bacterial resistance effectively (Oliveira et al. 2019; Majkowska-Skrobek et al. 2021; Abdelkader et al. 2022). Such strategies have shown promise in preventing the emergence of resistant bacterial strains, indicating a synergistic potential that could be leveraged for future therapeutic interventions. In summary, phage-derived enzymes present a viable alternative to traditional antibiotics, with a significantly reduced risk of inducing bacterial resistance. Their specific mechanisms of action, combined with strategic application methods, such as combination therapies, hold great promise for overcoming the challenges posed by antibiotic-resistant bacterial infections.

35.5 Engineering Phage Enzymes for Enhanced Therapeutic Efficacy

Historically, native phage-encoded enzymes especially those targeting Gram-positive bacteria, have demonstrated effective antibacterial activity. Recent studies have expanded on this foundation, reporting engineered bacteriophage lytic enzymes with improved characteristics (São-José 2018). These advancements include increased antibacterial activity across various growth conditions, a broader bactericidal spectrum, and extended applicability against Gram-negative bacteria. Efforts to enhance the solubility as well as stability of these enzybiotics have also been made when produced heterologously, typically in *E. coli*.

A key strategy in phage lytic enzyme (PLE) engineering involves the shuffling and fusion of catalytic domains (CDs) and cell wall-binding domains (CBDs) from different sources to create chimeric lytic enzymes or chimeolysins. The growing availability of PLE structural data and bioinformatics tools has facilitated this process by allowing for the precise identification and manipulation of enzyme functional domains. Interestingly, the inherent properties of these domains are generally maintained in the engineered chimeolysins, enabling the production of enzybiotics with tailored characteristics. Other engineering approaches have included the fusion of complete lytic enzymes, modifications of domain structures through deletion, addition, or duplication, and various mutagenesis techniques. These methods, along with the fusion of specific peptides and the combination of multiple strategies, have proven effective in enhancing the antibacterial efficacy of PLEs (Yang et al. 2014; Roach and Donovan 2015; Gerstmans et al. 2018). As we continue to face the challenges of antibiotic resistance, the engineering of phage-encoded enzymes offers a promising pathway to develop novel antimicrobial agents. The ongoing advancements in synthetic biology and structural biology provide a robust framework for the rational design and optimization of these enzymes, paving the way for the development of novel antimicrobial agents capable of addressing the growing challenge of antibiotic-resistant infections (Chen et al. 2019; Kilcher and Loessner 2019).

35.6 Phage-Encoded Enzymes in Combination Therapy

The potential of phage-derived enzymes to act in synergy with other antimicrobial agents has opened new avenues in combating antibiotic-resistant infections. Research has demonstrated the enhanced antimicrobial efficacy of these enzymes when combined with various classes of antibiotics and antimicrobial peptides. Notably, the collaborative action of glycopeptides and β-lactams with endolysins like MV-L (Rashel et al. 2007) and ClyS (Daniel et al. 2010) has shown promising results in vivo in treating systemic infections caused by MRSA. Further illustrating this synergy, the chimeric endolysin λSA2-ELysK-SH3b and lysostaphin combination effectively combated infections in a mouse model (Schmelcher et al. 2012b), mirroring the cooperative impact of Cpl-1 with Pal endolysins (Jado et al. 2003).

A notable strategy against multi-drug resistant (MDR) *S. Typhimurium* involves the combination of endolysin LysPB32 with the membrane-disrupting antibiotic polymyxin B, significantly improving the susceptibility of both antibiotic-sensitive and -resistant strains to treatment (Kim et al. 2022). This synergy not only broadens the antimicrobial spectrum but also enhances the efficacy of treatment protocols, presenting a viable method to tackle MDR bacterial challenges. Smug et al. (2022) suggested that using a combination of various phages or a blend of phages with depolymerases could potentially slow down or reduce the emergence of bacteria resistant to phages. By substituting the KP34 phage with its corresponding capsule depolymerase (KP34p57) in a treatment mix alongside KP15 and KP27, they observed a decrease in bacterial resistance, highlighting the efficacy of this strategy. In a previous investigation, the combined use of LysGH15, a bacteriophage-derived lysin, with apigenin, was evaluated for treating MRSA-induced pneumonia in mice. This combination therapy significantly outperformed the monotherapies of LysGH15 and apigenin in reducing bacterial lung load and improving overall lung health. Results showed a marked decrease in bacterial presence, reduced lung tissue damage, and lower inflammation levels compared to either treatment alone, suggesting a promising therapeutic strategy against MRSA pneumonia (Xia et al. 2016). The utility of phage-derived enzymes extends to the disruption of biofilms, where their combination with traditional antibiotics has shown potential in enhancing antibiofilm activity (Chopra et al. 2015; Latka and Drulis-Kawa 2020). The integration of phages and enzymes into antimicrobial cocktails, including antibiotics and other antimicrobial agents, offers a multifaceted approach to infection control. These findings underscore the immense potential of leveraging phage-encoded enzymes in combination therapies as a novel and effective strategy to overcome the hurdles posed by antibiotic resistance.

35.7 Safety, Immunogenicity, and Future Approaches

Phage-derived enzymatic therapies, encompassing PGHs, depolymerases, holins, and other bacteriophage-encoded enzymes, represent a promising avenue in the fight against antibiotic-resistant bacterial infections. These enzymes offer a targeted approach to bacterial eradication, posing minimal risk to human cells and the beneficial microbiota. The inherent specificity of these enzymes to bacterial components, such as peptidoglycan and polysaccharides, underscores their potential as non-toxic and biodegradable therapeutic agents. Integral to the natural microbiota in humans, phages continuously secrete lysins and depolymerases, a process that is harmless to human health. (Navarro and Muniesa 2017). This safety has been corroborated through various animal model trials and initial human clinical trials, such as the intravenous administration of SAL200, which demonstrated tolerability in healthy volunteers (Jun et al. 2017). Despite their protein nature, which could theoretically provoke allergic reactions or adverse side effects, such incidents have not been substantiated in practical observations (Jado et al. 2003; Gu et al. 2011; Gupta and Prasad 2011; Jun et al. 2014; Pan et al. 2015). Their specificity towards bacterial polysaccharides or peptidoglycan, absent in mammalian cells, contributes to their non-toxic profile. Their biodegradability further enhances their safety (Nelson et al. 2012). However, the rapid bacterial cell lysis induced by these enzymes can lead to the release of endotoxins and bacterial debris, potentially triggering severe immune responses in cases of massive infections, a phenomenon similar to that observed with lytic phages and bactericidal drugs (Ramachandran 2014).

The specificity of Bacteriophage-encoded enzymes for bacterial targets theoretically minimizes disturbance to the host's normal microflora compared to broad-spectrum antibiotics. Nevertheless, in vivo administration of endolysin has been linked with the production of pro-inflammatory cytokines in response to host cell debris, potentially leading to septic shock (Entenza et al. 2005). Yet, some studies have reported reduced cytokine levels following endolysin treatment (Witzenrath et al. 2009), and no significant immune system activation or adverse effects on human cell lines have been observed in preclinical studies (Harhala et al. 2018). Despite these promising attributes, challenges such as the potential for the production of neutralizing antibodies and inflammatory cytokines, and short in vivo half-life, persist (Nau and Eiffert 2002; Jado et al. 2003).

Depolymerases, another class of phage-derived enzymes, have shown potential in modulating immune responses, reducing virulence, and minimizing tissue damage in animal models (Bansal et al. 2014; Oliveira et al. 2019). Although systemic administration at low doses has not led to adverse effects or diminished antibacterial activity, the possibility of immune responses at higher doses remains an area for further investigation. Comprehensive assessments, including acute toxicity studies, have not indicated significant toxicity, supporting their potential safety (Pan et al. 2015; Lin et al. 2017; Shahed-Al-Mahmud et al. 2021). Limited data from clinical trials further constrain their widespread clinical adoption. Future approaches to enhancing the therapeutic potential of phage-derived enzymes involve protein engineering and the development of novel delivery systems. Tailoring these enzymes through genetic modifications can improve their stability, specificity, and reduce potential immunogenicity. Moreover, targeted delivery systems can ensure that these therapeutic agents reach the site of infection efficiently, mitigating systemic side effects and maximizing bacterial eradication.

35.8 Conclusion

In conclusion, bacteriophage-encoded enzymes; such as endolysins, VAPGHs, polysaccharide depolymerases, and holins stand at the forefront of a promising revolution against antibiotic-resistant pathogens. Their high specificity, ability to disrupt both planktonic and biofilm bacterial structures, and compatibility with existing antimicrobials, highlight their potential as superior therapeutic agents. While endolysins have demonstrated significant efficacy in preclinical models, the journey towards clinical application necessitates further exploration into scalable production, cost-effectiveness, and regulatory approval. Depolymerases, with their unique biofilm-disrupting capabilities and potential for rapid bacterial typing, offer an innovative approach to diagnosing and treating biofilm-associated infections. The road to clinical application is paved with challenges, including the need for comprehensive preclinical data and the development of stable, effective delivery systems. However, the promise of these phage-derived enzymes in combating multidrug-resistant bacteria and biofilms is undeniable. Future research, embracing combination therapies, advanced genetic engineering, and leveraging cutting-edge technologies, will undoubtedly unlock the full therapeutic potential of these enzymes, offering hope in the battle against antibiotic-resistant pathogens.

References

Abdelkader K, Gutiérrez D, Latka A, Boeckaerts D, Drulis-Kawa Z, Criel B et al (2022) The specific capsule depolymerase of phage PMK34 sensitizes *Acinetobacter baumannii* to serum killing. Antibiotics 11(5):677

Antonova NP, Vasina DV, Lendel AM, Usachev EV, Makarov VV, Gintsburg AL et al (2019) Broad bactericidal activity of the *Myoviridae* bacteriophage lysins LysAm24, LysECD7, and LysSi3 against Gram-negative ESKAPE pathogens. Viruses 11(3):284

Bansal S, Harjai K, Chhibber S (2014) Depolymerase improves gentamicin efficacy during *Klebsiella pneumoniae* induced murine infection. BMC Infect Dis 14:1–11

Bansal S, Harjai K, Chhibber S (2015) *Aeromonas punctata* derived depolymerase improves susceptibility of *Klebsiella pneumoniae* biofilm to gentamicin. BMC Microbiol 15(1):1–10

Binte Muhammad Jai HS, Dam LC, Tay LS, Koh JJW, Loo HL, Kline KA, Goh BC (2020) Engineered lysins with customized lytic activities against enterococci and staphylococci. Front Microbiol 11:574739

Briers Y, Schmelcher M, Loessner MJ, Hendrix J, Engelborghs Y, Volckaert G, Lavigne R (2009) The high-affinity peptidoglycan binding domain of *Pseudomonas* phage endolysin KZ144. Biochem Biophys Res Commun 383(2):187–191

Briers Y, Walmagh M, Lavigne RJMM (2011) Use of bacteriophage endolysin EL188 and outer membrane permeabilizers against *Pseudomonas aeruginosa*. J Appl Microbiol 110(3):778–785

Briers Y, Walmagh M, Van Puyenbroeck V, Cornelissen A, Cenens W, Aertsen A et al (2014) Engineered endolysin-based "Artilysins" to combat multidrug-resistant gram-negative pathogens. MBio 5(4):10–1128

Catalão MJ, Pimentel M (2018) Mycobacteriophage lysis enzymes: targeting the mycobacterial cell envelope. Viruses 10(8):428

Chen X, Liu M, Zhang P, Xu M, Yuan W, Bian L et al (2022) Phage-derived depolymerase as an antibiotic adjuvant against multidrug-resistant *Acinetobacter baumannii*. Front Microbiol 13:845500

Chen Y, Batra H, Dong J, Chen C, Rao VB, Tao P (2019) Genetic engineering of bacteriophages against infectious diseases. Front Microbiol 10:954

Chen Y, Li X, Wang S, Guan L, Li X, Hu D et al (2020) A novel tail-associated O91-specific polysaccharide depolymerase from a podophage reveals lytic efficacy of Shiga toxin-producing *Escherichia coli*. Appl Environ Microbiol 86(9):e00145–e00120

Chopra S, Harjai K, Chhibber S (2015) Potential of sequential treatment with minocycline and *S. aureus* specific phage lysin in eradication of MRSA biofilms: an *in vitro* study. Appl Microbiol Biotechnol 99:3201–3210

Cornelissen A, Ceyssens PJ, T'syen J, Van Praet H, Noben JP, Shaburova OV et al (2011) The T7-related *Pseudomonas putida* phage φ15 displays virion-associated biofilm degradation properties. PLoS One 6(4):e18597

Daniel A, Euler C, Collin M, Chahales P, Gorelick KJ, Fischetti VA (2010) Synergism between a novel chimeric lysin and oxacillin protects against infection by methicillin-resistant *Staphylococcus aureus*. Antimicrob Agents Chemother 54(4):1603–1612

Danis-Wlodarczyk KM, Wozniak DJ, Abedon ST (2021) Treating bacterial infections with bacteriophage-based enzybiotics: *in vitro*, *in vivo* and clinical application. Antibiotics 10(12):1497

Donovan DM (2007) Bacteriophage and peptidoglycan degrading enzymes with antimicrobial applications. Recent Pat Biotechnol 1(2):113–122

Drulis-Kawa Z, Majkowska-Skrobek G, Maciejewska B (2015) Bacteriophages and phage-derived proteins–application approaches. Curr Med Chem 22(14):1757–1773

Entenza JM, Loeffler JM, Grandgirard D, Fischetti VA, Moreillon P (2005) Therapeutic effects of bacteriophage Cpl-1 lysin against *Streptococcus pneumoniae* endocarditis in rats. Antimicrob Agents Chemother 49(11):4789–4792

Fernandes S, São-José C (2018) Enzymes and mechanisms employed by tailed bacteriophages to breach the bacterial cell barriers. Viruses 10(8):396

Fischetti VA (2005) Bacteriophage lytic enzymes: novel anti-infectives. Trends Microbiol 13(10):491–496

Fischetti VA (2010) Bacteriophage endolysins: a novel anti-infective to control Gram-positive pathogens. Int J Med Microbiol 300(6):357–362

Freimer EH, Krause RM, McCarty M (1959) Studies of L forms and protoplasts of group A streptococci: I. Isolation, growth, and bacteriologic characteristics. J Exp Med 110(6):853–874

Fujiki J, Nakamura T, Furusawa T, Ohno H, Takahashi H, Kitana J et al (2018) Characterization of the lytic capability of a LysK-like endolysin, Lys-phiSA012, derived from a polyvalent *Staphylococcus aureus* bacteriophage. Pharmaceuticals 11(1):25

Gerstmans H, Criel B, Briers Y (2018) Synthetic biology of modular endolysins. Biotechnol Adv 36(3):624–640

Gilmer DB, Schmitz JE, Euler CW, Fischetti VA (2013) Novel bacteriophage lysin with broad lytic activity protects against mixed infection by *Streptococcus pyogenes* and methicillin-resistant *Staphylococcus aureus*. Antimicrob Agents Chemother 57(6):2743–2750

Glonti T, Chanishvili N, Taylor PW (2010) Bacteriophage-derived enzyme that depolymerizes the alginic acid capsule associated with cystic fibrosis isolates of *Pseudomonas aeruginosa*. J Appl Microbiol 108(2):695–702

Gonçalves T, Marques AT, Manageiro V, Tanoeiro L, Vital JS, Duarte A et al (2024) Antimicrobial activity of prophage endolysins against critical Enterobacteriaceae antibiotic-resistant bacteria. Int J Pharm 651:123758

Gondil VS, Harjai K, Chhibber S (2020) Endolysins as emerging alternative therapeutic agents to counter drug-resistant infections. Int J Antimicrob Agents 55(2):105844

Gorodnichev RB, Volozhantsev NV, Krasilnikova VM, Bodoev IN, Kornienko MA, Kuptsov NS et al (2021) Novel *Klebsiella pneumoniae* K23-specific bacteriophages from different families: similarity of depolymerases and their therapeutic potential. Front Microbiol 12:669618

Gu J, Xu W, Lei L, Huang J, Feng X, Sun C et al (2011) LysGH15, a novel bacteriophage lysin, protects a murine bacteremia model efficiently against lethal methicillin-resistant *Staphylococcus aureus* infection. J Clin Microbiol 49(1):111–117

Guariglia-Oropeza V, Helmann JD (2011) Bacillus subtilis σV confers lysozyme resistance by activation of two cell wall modification pathways, peptidoglycan O-acetylation and D-alanylation of teichoic acids. J Bacteriol 193(22):6223–6232

Guo M, Feng C, Ren J, Zhuang X, Zhang Y, Zhu Y et al (2017) A novel antimicrobial endolysin, LysPA26, against *Pseudomonas aeruginosa*. Front Microbiol 8:293

Gupta R, Prasad Y (2011) P-27/HP endolysin as antibacterial agent for antibiotic resistant *Staphylococcus aureus* of human infections. Curr Microbiol 63:39–45

Gutiérrez D, Briers Y, Rodríguez-Rubio L, Martínez B, Rodríguez A, Lavigne R, García P (2015) Role of the pre-neck appendage protein (Dpo7) from phage vB_SepiS-phiIPLA7 as an anti-biofilm agent in staphylococcal species. Front Microbiol 6:1315

Gutiérrez D, Martínez B, Rodríguez A, García P (2012) Genomic characterization of two *Staphylococcus epidermidis* bacteriophages with anti-biofilm potential. BMC Genomics 13(1):1–10

Haddad Kashani H, Schmelcher M, Sabzalipoor H, Seyed Hosseini E, Moniri R (2018) Recombinant endolysins as potential therapeutics against antibiotic-resistant *Staphylococcus aureus*: current status of research and novel delivery strategies. Clin Microbiol Rev 31(1):10–1128

Harhala M, Nelson DC, Miernikiewicz P, Heselpoth RD, Brzezicka B, Majewska J et al (2018) Safety studies of pneumococcal endolysins Cpl-1 and Pal. Viruses 10(11):638

Harper DR, Abedon ST, Burrowes BH, McConville ML (eds) (2021) Bacteriophages: biology, technology, therapy. Springer Nature

Hou R, Li M, Tang T, Wang R, Li Y, Xu Y et al (2018) Construction of *Lactobacillus casei* ghosts by Holin-mediated inactivation and the potential as a safe and effective vehicle for the delivery of DNA vaccines. BMC Microbiol 18:1–9

Hsieh PF, Lin HH, Lin TL, Chen YY, Wang JT (2017) Two T7-like bacteriophages, K5-2 and K5-4, each encodes two capsule depolymerases: isolation and functional characterization. Sci Rep 7(1):4624

Jado I, López R, García E, Fenoll A, Casal J, García P (2003) Phage lytic enzymes as therapy for antibiotic-resistant *Streptococcus pneumoniae* infection in a murine sepsis model. J Antimicrob Chemother 52(6):967–973

Jun SY, Jang IJ, Yoon S, Jang K, Yu KS, Cho JY et al (2017) Pharmacokinetics and tolerance of the phage endolysin-based candidate drug SAL200 after a single intravenous administration among healthy volunteers. Antimicrob Agents Chemother 61(6):10–1128

Jun SY, Jung GM, Yoon SJ, Choi YJ, Koh WS, Moon KS, Kang SH (2014) Preclinical safety evaluation of intravenously administered SAL200 containing the recombinant phage endolysin SAL-1 as a pharmaceutical ingredient. Antimicrob Agents Chemother 58(4): 2084–2088

Keary R, McAuliffe O, Ross RP, Hill C, O'Mahony J, Coffey A (2014) Genome analysis of the staphylococcal temperate phage DW2 and functional studies on the endolysin and tail hydrolase. Bacteriophage 4(1):e28451

Kilcher S, Loessner MJ (2019) Engineering bacteriophages as versatile biologics. Trends Microbiol 27(4):355–367

Kim J, Kim JC, Ahn J (2022) Assessment of bacteriophage-encoded endolysin as a potent antimicrobial agent against antibiotic-resistant *Salmonella Typhimurium*. Microb Pathog 168: 105576

Kim S, Jin JS, Choi YJ, Kim J (2020) LysSAP26, a new recombinant phage endolysin with a broad spectrum antibacterial activity. Viruses 12(11):1340

Knecht LE, Veljkovic M, Fieseler L (2020) Diversity and function of phage encoded depolymerases. Front Microbiol 10:2949

Kong M, Na H, Ha NC, Ryu S (2019) LysPBC2, a novel endolysin harboring a *Bacillus cereus* spore binding domain. Appl Environ Microbiol 85(5):e02462–e02418

Kongari R, Rajaure M, Cahill J, Rasche E, Mijalis E, Berry J, Young R (2018) Phage spanins: diversity, topological dynamics and gene convergence. BMC Bioinform 19:1–26

Lai WCB, Chen X, Ho MKY, Xia J, Leung SSY (2020) Bacteriophage-derived endolysins to target gram-negative bacteria. Int J Pharm 589:119833

Latka A, Drulis-Kawa Z (2020) Advantages and limitations of microtiter biofilm assays in the model of antibiofilm activity of *Klebsiella* phage KP34 and its depolymerase. Sci Rep 10(1):20338

Latka A, Leiman PG, Drulis-Kawa Z, Briers Y (2019) Modeling the architecture of depolymerase-containing receptor binding proteins in *Klebsiella* phages. Front Microbiol 10:2649

Latka A, Maciejewska B, Majkowska-Skrobek G, Briers Y, Drulis-Kawa Z (2017) Bacteriophage-encoded virion-associated enzymes to overcome the carbohydrate barriers during the infection process. Appl Microbiol Biotechnol 101:3103–3119

Li M, Wang H, Chen L, Guo G, Li P, Ma J et al (2022) Identification of a phage-derived depolymerase specific for KL47 capsule of *Klebsiella pneumoniae* and its therapeutic potential in mice. Virol Sin 37(4):538–546

Li X, Zhang C, Wei F, Yu F, Zhao Z (2021) Bactericidal activity of a holin-endolysin system derived from *Vibrio alginolyticus* phage HH109. Microb Pathog 159:105135

Lin H, Paff ML, Molineux IJ, Bull JJ (2017) Therapeutic application of phage capsule depolymerases against K1, K5, and K30 capsulated *E. coli* in mice. Front Microbiol 8:2257

Lin TL, Hsieh PF, Huang YT, Lee WC, Tsai YT, Su PA et al (2014) Isolation of a bacteriophage and its depolymerase specific for K1 capsule of *Klebsiella pneumoniae*: implication in typing and treatment. J Infect Dis 210(11):1734–1744

Liu Y, Leung SSY, Guo Y, Zhao L, Jiang N, Mi L et al (2019) The capsule depolymerase Dpo48 rescues *Galleria mellonella* and mice from *Acinetobacter baumannii*systemic infections. Front Microbiol 10:545

Loeffler JM, Nelson D, Fischetti VA (2001) Rapid killing of *Streptococcus pneumoniae* with a bacteriophage cell wall hydrolase. Science 294(5549):2170–2172

Loessner MJ, Kramer K, Ebel F, Scherer S (2002) C-terminal domains of Listeria monocytogenes bacteriophage murein hydrolases determine specific recognition and high-affinity binding to bacterial cell wall carbohydrates. Mol Microbiol 44(2):335–349

Lood R, Raz A, Molina H, Euler CW, Fischetti VA (2014) A highly active and negatively charged *Streptococcus pyogenes* lysin with a rare D-alanyl-L-alanine endopeptidase activity protects mice against streptococcal bacteremia. Antimicrob Agents Chemother 58(6):3073–3084

Lood R, Winer BY, Pelzek AJ, Diez-Martinez R, Thandar M, Euler CW et al (2015) Novel phage lysin capable of killing the multidrug-resistant gram-negative bacterium *Acinetobacter baumannii* in a mouse bacteremia model. Antimicrob Agents Chemother 59(4):1983–1991

Low LY, Yang C, Perego M, Osterman A, Liddington R (2011) Role of net charge on catalytic domain and influence of cell wall binding domain on bactericidal activity, specificity, and host range of phage lysins. J Biol Chem 286(39):34391–34403

Lu B, Yao X, Han G, Luo Z, Zhang J, Yong K et al (2022) Isolation of *Klebsiella pneumoniae* phage vB_KpnS_MK54 and pathological assessment of Endolysin in the treatment of pneumonia mice model. Front Microbiol 13:854908

Lv M, Wang S, Yan G, Sun C, Feng X, Gu J et al (2015) Genome sequencing and analysis of an *Escherichia coli* phage vB_EcoM-ep3 with a novel lysin, Lysep3. Virus Genes 50:487–497

Maciejewska B, Roszniowski B, Espaillat A, Kęsik-Szeloch A, Majkowska-Skrobek G, Kropinski AM et al (2017) *Klebsiella* phages representing a novel clade of viruses with an unknown DNA modification and biotechnologically interesting enzymes. Appl Microbiol Biotechnol 101:673–684

Majkowska-Skrobek G, Latka A, Berisio R, Squeglia F, Maciejewska B, Briers Y, Drulis-Kawa Z (2018) Phage-borne depolymerases decrease *Klebsiella pneumoniae* resistance to innate defense mechanisms. Front Microbiol 9:2517

Majkowska-Skrobek G, Markwitz P, Sosnowska E, Lood C, Lavigne R, Drulis-Kawa Z (2021) The evolutionary trade-offs in phage-resistant *Klebsiella pneumoniae* entail cross-phage sensitization and loss of multidrug resistance. Environ Microbiol 23(12):7723–7740

McCallum KL, Laakso DH, Whitfield C (1989) Use of a bacteriophage-encoded glycanase enzyme in the generation of lipopolysaccharide O side chain deficient mutants of *Escherichia coli* O9: K30 and *Klebsiella* O1: K20: role of O and K antigens in resistance to complement-mediated serum killing. Can J Microbiol 35(11):994–999

Morita M, Tanji Y, Orito Y, Mizoguchi K, Soejima A, Unno H (2001) Functional analysis of antibacterial activity of *Bacillus amyloliquefaciens* phage endolysin against Gram-negative bacteria. FEBS Lett 500(1–2):56–59

Mostowy RJ, Holt KE (2018) Diversity-generating machines: genetics of bacterial sugar-coating. Trends Microbiol 26(12):1008–1021

Mushtaq N, Redpath MB, Luzio JP, Taylor PW (2005) Treatment of experimental *Escherichia coli* infection with recombinant bacteriophage-derived capsule depolymerase. J Antimicrob Chemother 56(1):160–165

Nau R, Eiffert H (2002) Modulation of release of proinflammatory bacterial compounds by antibacterials: potential impact on course of inflammation and outcome in sepsis and meningitis. Clin Microbiol Rev 15(1):95–110

Navarro F, Muniesa M (2017) Phages in the human body. Front Microbiol 8:566

Nelson DC, Schmelcher M, Rodriguez-Rubio L, Klumpp J, Pritchard DG, Dong S, Donovan DM (2012) Endolysins as antimicrobials. Adv Virus Res 83:299–365

Nelson D, Loomis L, Fischetti VA (2001) Prevention and elimination of upper respiratory colonization of mice by group A streptococci by using a bacteriophage lytic enzyme. Proc Natl Acad Sci 98(7):4107–4112

Nelson D, Schuch R, Chahales P, Zhu S, Fischetti VA (2006) PlyC: a multimeric bacteriophage lysin. Proc Natl Acad Sci 103(28):10765–10770

Nobrega FL, Vlot M, de Jonge PA, Dreesens LL, Beaumont HJ, Lavigne R et al (2018) Targeting mechanisms of tailed bacteriophages. Nat Rev Microbiol 16(12):760–773

Oliveira H, Drulis-Kawa Z, Azeredo J (2022) Exploiting phage-derived carbohydrate depolymerases for combating infectious diseases. Trends Microbiol 30:707

Oliveira H, Mendes A, Fraga AG, Ferreira A, Pimenta AI, Mil-Homens D et al (2019) K2 capsule depolymerase is highly stable, is refractory to resistance, and protects larvae and mice from *Acinetobacter baumannii* sepsis. Appl Environ Microbiol 85(17):e00934–e00919

Olszak T, Shneider MM, Latka A, Maciejewska B, Browning C, Sycheva LV et al (2017) The O-specific polysaccharide lyase from the phage LKA1 tailspike reduces *Pseudomonas* virulence. Sci Rep 7(1):16302

Pan YJ, Lin TL, Lin YT, Su PA, Chen CT, Hsieh PF et al (2015) Identification of capsular types in carbapenem-resistant *Klebsiella pneumoniae* strains by wzc sequencing and implications for capsule depolymerase treatment. Antimicrob Agents Chemother 59(2):1038–1047

Park T, Struck DK, Deaton JF, Young R (2006) Topological dynamics of holins in programmed bacterial lysis. Proc Natl Acad Sci 103(52):19,713–19,718

Pastagia M, Euler C, Chahales P, Fuentes-Duculan J, Krueger JG, Fischetti VA (2011) A novel chimeric lysin shows superiority to mupirocin for skin decolonization of methicillin-resistant and-sensitive *Staphylococcus aureus* strains. Antimicrob Agents Chemother 55(2):738–744

Paul VD, Rajagopalan SS, Sundarrajan S, George SE, Asrani JY, Pillai R et al (2011) A novel bacteriophage Tail-Associated Muralytic Enzyme (TAME) from Phage K and its development into a potent antistaphylococcal protein. BMC Microbiol 11(1):1–11

Payne KM, Hatfull GF (2012) Mycobacteriophage endolysins: diverse and modular enzymes with multiple catalytic activities. PLoS One 7(3):e34052

Pertics BZ, Cox A, Nyúl A, Szamek N, Kovács T, Schneider G (2021) Isolation and characterization of a novel lytic bacteriophage against the K2 capsule-expressing hypervirulent *Klebsiella pneumoniae* strain 52145, and identification of its functional depolymerase. Microorganisms 9(3):650

Pires DP, Oliveira H, Melo LD, Sillankorva S, Azeredo J (2016) Bacteriophage-encoded depolymerases: their diversity and biotechnological applications. Appl Microbiol Biotechnol 100:2141–2151

Plotka M, Kapusta M, Dorawa S, Kaczorowska AK, Kaczorowski T (2019) Ts2631 endolysin from the extremophilic *Thermus scotoductus* bacteriophage vB_Tsc2631 as an antimicrobial agent against gram-negative multidrug-resistant bacteria. Viruses 11(7):657

Pritchard DG, Dong S, Kirk MC, Cartee RT, Baker JR (2007) LambdaSa1 and LambdaSa2 prophage lysins of *Streptococcus agalactiae*. Appl Environ Microbiol 73(22):7150–7154

Rajesh T, Anthony T, Saranya S, Pushpam PL, Gunasekaran P (2011) Functional characterization of a new holin-like antibacterial protein coding gene tmp1 from goat skin surface metagenome. Appl Microbiol Biotechnol 89:1061–1073

Ramachandran G (2014) Gram-positive and gram-negative bacterial toxins in sepsis: a brief review. Virulence 5(1):213–218

Rashel M, Uchiyama J, Ujihara T, Uehara Y, Kuramoto S, Sugihara S et al (2007) Efficient elimination of multidrug-resistant *Staphylococcus aureus* by cloned lysin derived from bacteriophage ϕMR11. J Infect Dis 196(8):1237–1247

Reddy BL, Saier MH Jr (2013) Topological and phylogenetic analyses of bacterial holin families and superfamilies. Biochim Biophys Acta 1828(11):2654–2671

Roach DR, Donovan DM (2015) Antimicrobial bacteriophage-derived proteins and therapeutic applications. Bacteriophage 5(3):e1062590

Rodríguez L, Martínez B, Zhou Y, Rodríguez A, Donovan DM, García P (2011) Lytic activity of the virion-associated peptidoglycan hydrolase HydH5 of *Staphylococcus aureus* bacteriophage vB_SauS-phiIPLA88. BMC Microbiol 11:1–11

Rodríguez-Rubio L, Martínez B, Donovan DM, Rodríguez A, García P (2013) Bacteriophage virion-associated peptidoglycan hydrolases: potential new enzybiotics. Crit Rev Microbiol 39(4):427–434

Rodríguez-Rubio L, Martínez B, Rodríguez A, Donovan DM, García P (2012) Enhanced staphylolytic activity of the *Staphylococcus aureus* bacteriophage vB_SauS-phiIPLA88 HydH5 virion-associated peptidoglycan hydrolase: fusions, deletions, and synergy with LysH5. Appl Environ Microbiol 78(7):2241–2248

Saier MH Jr, Reddy BL (2015) Holins in bacteria, eukaryotes, and archaea: multifunctional xenologues with potential biotechnological and biomedical applications. J Bacteriol 197(1): 7–17

São-José C (2018) Engineering of phage-derived lytic enzymes: improving their potential as antimicrobials. Antibiotics 7(2):29

Schmelcher M, Donovan DM, Loessner MJ (2012a) Bacteriophage endolysins as novel antimicrobials. Future Microbiol 7(10):1147–1171

Schmelcher M, Powell AM, Becker SC, Camp MJ, Donovan DM (2012b) Chimeric phage lysins act synergistically with lysostaphin to kill mastitis-causing *Staphylococcus aureus* in murine mammary glands. Appl Environ Microbiol 78(7):2297–2305

Schuch R, Nelson D, Fischetti VA (2002) A bacteriolytic agent that detects and kills *Bacillus anthracis*. Nature 418(6900):884–889

Shahed-Al-Mahmud M, Roy R, Sugiokto FG, Islam MN, Lin MD, Lin LC, Lin NT (2021) Phage φAB6-borne depolymerase combats *Acinetobacter baumannii* biofilm formation and infection. Antibiotics 10(3):279

Shen Y, Köller T, Kreikemeyer B, Nelson DC (2013) Rapid degradation of *Streptococcus pyogenes* biofilms by PlyC, a bacteriophage-encoded endolysin. J Antimicrob Chemother 68(8):1818–1824

Shi Y, Li N, Yan Y, Wang H, Li Y, Lu C, Sun J (2012) Combined antibacterial activity of phage lytic proteins holin and lysin from *Streptococcus suis* bacteriophage SMP. Curr Microbiol 65:28–34

Silva MD, Oliveira H, Faustino A, Sillankorva S (2020) Characterization of MSlys, the endolysin of *Streptococcus pneumoniae* phage MS1. Biotechnol Rep 28:e00547

Smug BJ, Majkowska-Skrobek G, Drulis-Kawa Z (2022) PhREEPred: phage resistance emergence prediction web tool to foresee encapsulated bacterial escape from phage cocktail treatment. J Mol Biol 434(14):167670

Solovieva EV, Myakinina VP, Kislichkina AA, Krasilnikova VM, Verevkin VV, Mochalov VV et al (2018) Comparative genome analysis of novel Podoviruses lytic for hypermucoviscous*Klebsiella pneumoniae* of K1, K2, and K57 capsular types. Virus Res 243:10–18

Son B, Kong M, Lee Y, Ryu S (2021) Development of a novel chimeric endolysin, Lys109 with enhanced lytic activity against *Staphylococcus aureus*. Front Microbiol 11:615887

Song J, Xia F, Jiang H, Li X, Hu L, Gong P et al (2016) Identification and characterization of HolGH15: the holin of *Staphylococcus aureus* bacteriophage GH15. J Gen Virol 97(5):1272–1281

Spratt BG (1994) Resistance to antibiotics mediated by target alterations. Science 264(5157):388–393

Squeglia F, Maciejewska B, Łątka A, Ruggiero A, Briers Y, Drulis-Kawa Z, Berisio R (2020) Structural and functional studies of a *Klebsiella* phage capsule depolymerase tailspike: mechanistic insights into capsular degradation. Structure 28(6):613–624

Takác M, Bläsi U (2005) Phage P68 virion-associated protein 17 displays activity against clinical isolates of *Staphylococcus aureus*. Antimicrob Agents Chemother 49(7):2934–2940

Volozhantsev NV, Shpirt AM, Borzilov AI, Komisarova EV, Krasilnikova VM, Shashkov AS et al (2020) Characterization and therapeutic potential of bacteriophage-encoded polysaccharide depolymerases with β galactosidase activity against *Klebsiella pneumoniae* K57 capsular type. Antibiotics 9(11):732

Vipra AA, Desai SN, Roy P, Patil R, Raj JM, Narasimhaswamy N et al (2012) Antistaphylococcal activity of bacteriophage derived chimeric protein P128. BMC Microbiol 12:1–9

Vollmer W, Joris B, Charlier P, Foster S (2008) Bacterial peptidoglycan (murein) hydrolases. FEMS Microbiol Rev 32(2):259–286

Volozhantsev NV, Borzilov AI, Shpirt AM, Krasilnikova VM, Verevkin VV, Denisenko EA et al (2022) Comparison of the therapeutic potential of bacteriophage KpV74 and phage-derived depolymerase (β-glucosidase) against *Klebsiella pneumoniae* capsular type K2. Virus Res 322:198951

Walmagh M, Boczkowska B, Grymonprez B, Briers Y, Drulis-Kawa Z, Lavigne R (2013) Characterization of five novel endolysins from Gram-negative infecting bacteriophages. Appl Microbiol Biotechnol 97:4369–4375

Wang C, Li P, Niu W, Yuan X, Liu H, Huang Y et al (2019) Protective and therapeutic application of the depolymerase derived from a novel KN1 genotype of *Klebsiella pneumoniae* bacteriophage in mice. Res Microbiol 170(3):156–164

Wang C, Li P, Zhu Y, Huang Y, Gao M, Yuan X et al (2020) Identification of a novel *Acinetobacter baumannii* phage-derived depolymerase and its therapeutic application in mice. Front Microbiol 11:1407

Wang H, Liu Y, Bai C, Leung SSY (2023) Translating bacteriophage-derived depolymerases into antibacterial therapeutics: challenges and prospects, vol 14. Acta Pharm Sin B, p 155

Wang IN, Smith DL, Young R (2000) Holins: the protein clocks of bacteriophage infections. Annu Rev Microbiol 54(1):799–825

Waseh S, Hanifi-Moghaddam P, Coleman R, Masotti M, Ryan S, Foss M et al (2010) Orally administered P22 phage tailspike protein reduces *Salmonella* colonization in chickens: prospects of a novel therapy against bacterial infections. PLoS One 5(11):e13904

Witzenrath M, Schmeck B, Doehn JM, Tschernig T, Zahlten J, Loeffler JM et al (2009) Systemic use of the endolysin Cpl-1 rescues mice with fatal pneumococcal pneumonia. Crit Care Med 37(2):642–649

Won G, Lee JH (2017) *Salmonella Typhimurium*, the major causative agent of foodborne illness inactivated by a phage lysis system provides effective protection against lethal challenge by induction of robust cell-mediated immune responses and activation of dendritic cells. Vet Res 48:1–12

Won G, Hajam IA, Lee JH (2017a) Improved lysis efficiency and immunogenicity of *Salmonella* ghosts mediated by co-expression of λ phage holin-endolysin and ϕX174 gene E. Sci Rep 7(1):45139

Won G, Kim B, Lee JH (2017b) A novel method to generate *Salmonella Typhi* Ty21a ghosts exploiting the λ phage holin-endolysin system. Oncotarget 8(29):48,186

Won G, Senevirathne A, Lee JH (2020) *Salmonella Enteritidis* ghost vaccine carrying the hemagglutinin globular head (HA1) domain from H1N1 virus protects against salmonellosis and influenza in chickens. Vaccine 38(28):4387–4394

Wu M, Hu K, Xie Y, Liu Y, Mu D, Guo H et al (2019a) A novel phage PD-6A3, and its endolysin Ply6A3, with extended lytic activity against *Acinetobacter baumannii*. Front Microbiol 9:3302

Wu Y, Wang R, Xu M, Liu Y, Zhu X, Qiu J et al (2019b) A novel polysaccharide depolymerase encoded by the phage SH-KP152226 confers specific activity against multidrug-resistant *Klebsiella pneumoniae* via biofilm degradation. Front Microbiol 10:2768

Xia F, Li X, Wang B, Gong P, Xiao F, Yang M et al (2016) Combination therapy of LysGH15 and apigenin as a new strategy for treating pneumonia caused by *Staphylococcus aureus*. Appl Environ Microbiol 82(1):87–94

Xiang Y, Wang S, Huang H, Li X, Li H, Tu Y et al (2024) A novel holin from an *Enterococcus faecalis* phage and application *in vitro* and *in vivo*. Microb Pathog 186:106471

Yan J, Mao J, Xie J (2014) Bacteriophage polysaccharide depolymerases and biomedical applications. BioDrugs 28(3):265–274

Yang H, Yu J, Wei H (2014) Engineered bacteriophage lysins as novel anti-infectives. Front Microbiol 5:542

Young R (2014) Phage lysis: three steps, three choices, one outcome. J Microbiol 52:243–258

Young RY (2002) Bacteriophage holins: deadly diversity. J Mol Microbiol Biotechnol 4(1):21–36

Part VII

Overcoming Challenges and Regulatory Hurdles

Overcoming Challenges and Regulatory Hurdles for Sustainable Bacteriophage Therapy Products

36

Rajal Dave and Sangeeta Ahiwale

Abstract

Bacteriophages are naturally occurring bacteria eaters that instantly kill a bacterial cell or integrate its genome into a bacterial chromosome. The potential use of lytic bacteriophage emerged as a new promising alternative to combat antibiotic resistance among bacterial pathogens and eliminate targeted nuisance bacteria. Therefore, it is important to understand their mode of action, application, dosage and potential side effects which is relatively complex. To maximize the potential of phage therapy, certain biological and regulatory challenges should be addressed including bacterial resistance to phage, quorum sensing, large-scale production, and receptor adaptation. In this chapter, we discuss the role of phage cocktail, phage product deployment, phage product modification, phage encapsulation to enhance shelf life of phage to allayed challenges. Also Engineered Phage with antibacterial CRISPR-Cas to allayed challenges. Safety and quality are other key concerns for prospective phage therapeutics. The production and purification process of phages requires rigorous protocols to ensure the final product contains a pure and effective cocktail. Regulators must establish guidelines and benchmarks to assess the quality, safety, and consistency ensuring to meet necessary standards for clinical settings. We also discuss here the National regulatory framework, development of Good Laboratory Practices based processes to meet international quality standards, idea of Emergency use of Investigational New Drug (IND) protocols in accordance with FDA and clinical laboratory standards for phage susceptibility testing. Information discussed in

R. Dave
Department of Microbiology, Faculty of Science, Atmiya University, "Yogidham Gurukul", Rajkot, Gujarat, India

S. Ahiwale (✉)
Department of Microbiology, Mahatma Phule Mahavidyalaya, Pimpri, Pune, Maharashtra, India

M. Gangwar, G. Nath (eds.), *Emerging Paradigms for Antibiotic-Resistant Infections: Beyond the Pill*, https://doi.org/10.1007/978-981-97-5272-0_36

this chapter will serve as basic guidelines for addressing these challenges and bacteriophage therapy can contribute significantly to the fight against Multi-drug resistant bacterial infection, offering new hope against antibiotic resistant.

Keywords
Bacteriophage therapy · Phage encoded lytic enzymes · CRISPR-Cas antibacterial system · Regulatory framework

Abbreviation

AMR	Antimicrobial resistant
ARGs	Antibiotic resistant genes
cPT	Compassionate phage therapy
CRISPR	Clustered Regulatory Interspaced Short Palindromic Report
DNA	Deoxyribonucleic acid
EIBMV	Eliava Institute of Bacteriophage, Microbiology and Virology
FDA	Food and Drug Administration
GMP	Good Manufacturing Practices
HIIET	The Hirszfeid Institute of Immunology and Experimental Therapy
MDR	Multi drug resistant
PFU	Plaque forming unit
RNA	Ribonucleic acid
USA	United State of America

36.1 Introduction

The global rise of antibiotic resistance resulted due to transmission of Antibiotic Resistance Genes (ARGs) is interlinked in animal–human—and environmental sectors in food chains. This rise of antibiotic resistance ultimately led to insecurity in public health because of the resistance exhibited by bacteria to the last resort antibiotics. In such situations, Bacteriophage, a traditionally known weapon against nuisance bacterial pathogens can be a possible alternative to antibiotics.

Bacteriophages are viruses which require bacteria as a host to replicate themselves. Bacteriophages are one of the most abundant biological forms in the earth. It is believed that approximately 10^{31} bacteriophages are found on Earth which surpasses the total count of all other living organisms, including bacteria.

Bacteriophages are also present in soil, water bodies, and in human body too. Human blood, muco-saliva, feces, and urine a normal flora (Górski et al. 2020). Human colon also contains a large portion of bacteriophages. Human gut contains a large proportion of double-stranded DNA phages belongs to order Caudovirales which includes three families viz., *Siphoviridae*, *Myoviridae* and *Podoviridae*family bacteriophages present also Single-stranded DNA phages are also found in the human gut which belongs to *Microviridae and Inoviridae* families (Lusiak-Szelachowska et al. 2017). It is found that the human intestine contains 10^8–10^{10} bacteriophages per 1 g of stool excretion (Kim et al. 2011). A recent study shows that bacteriophage also plays an important role in maintaining homeostasis, in the regulation of microbial diversity of normal flora, in the permeability of the gut and also in the immune responses (Tetz et al. 2017). Recent research study shows that Caudovirales, double-stranded DNA, *Myoviridae*phages play a vital role in the regulation of gut virome (Sutton and Hill 2019).

36.1.1 Mechanism of Bacteriophage and Basics of Phage Therapy

Phage genome is packed into the head called by the headful mechanism of packaging. Capsid is connected to the elongated neck in the form of sheath. The phage sheath contains a hollow tube-like structure for injecting their genome into the host cells. At the bottom of the sheath, there is base plate. It contains minimum six contractile or non-contractile tail fibers as per the phage type. These tail fibers allow phages to attach to the bacterial cell surface. On the basis of their infection, phages can be classified into two types. Lytic or virulent phage, and Lysogenic Phages or non-virulent phage. Phage takes approximately 40 min to complete whole lytic cycles and at the end, it lyses the host cell and releases mature phages into environment.

While in the case of Lysogenic phage, Phage integrates its genome with bacterial chromosomes in a form of prophage. It can transit many generations of bacteria. Because of their unique characteristics, bacteriophage possess novel antimicrobial properties and therefore can be explored as a research tool in many biotechnological aspects like phage display, biosensor, and clustered regularly interspaced short palindromic repeats (CRISPR)-Cas systems. So, for successful phage therapy, Lytic phages have been known as an excellent solution against antibiotics.

36.1.2 Advantages of Phage Therapy

In pandemic of antibiotic resistance with a barrier to discovery of new antibiotics has strongly urged researchers to discover alternatives to antibiotics in medicine, food industries, and in field of agriculture. Bacteriophage therapy utilizes lytic phages which infect only target bacteria. So, it has many potential advantages compared to antibiotics and already shown a promising result in addressing rise of antimicrobial resistance.

- Phages have high specificity for their host bacteria; therefore, can infect only target bacteria without disturbing normal flora.
- Phage has the ability to emerge by adopting in some situations to express bacterial antiviral mechanism. So, by continuous adaptation mechanisms, phage alleviates themselves from the risk of arising anti-phage resistance in bacterial cells.

Applications of phage cocktail can also decrease the risk of developing phage-resistant mutants. Phage-sensitive bacterium which is sensitive to one phage may become susceptible to other phages.
- Phages have the ability to multiply and destroy the bacterial biofilm with the help of depolymerase enzymes. All these properties potentiate phage therapy in terms of safety with economic and environmental sustainability. In this chapter, we discuss the current situation of phage therapy, important aspect of biological and regulatory challenges and the strategies for successful bacteriophage therapy in the era of emerging antimicrobial resistant.
- Phage therapy can prevent antibiotic-induced dysbiosis and secondary infections.

36.2 Current Situation of Bacteriophage Therapy

A century ago, the first phage therapy was discovered by Frederick Twort and Felix d'Herelle but this discovery was quickly surpassed by first discovered antibiotics, penicillin. In certain institutes like The Eliava Institute of Bacteriophages, Microbiology, and Virology, Tblilisi (EIBMV), Georgia, and The Ludwik Hirszfeld Institute of Immunology and Experimental Therapy (HIIET), Wroclaw, Poland Institute is still functioning. Moreover, the Eliava Institute has developed Phage Bank which contains approximately 1000 phages and the second is in Poland which also has approximately 850 phages. Tentatively ten thousand patients have acquired phage treatment from these institutes (Żaczek et al. 2020). In spite of many successful treated cases of bacteriophage include their documents and history of western countries, phage therapy is still facing certain challenges in terms of biological and regulatory.

36.3 Biological Challenges in Phage Therapy

Despite of successful history of phage research, there are still many gaps which should be identified and implemented in future for clinical trials. Biological challenges are discussed in (Fig. 36.1),

36.3.1 Limitation of Host Range

Proper phage and host interaction determines the spectrum of host range in successful phage therapy. Selection of host range is a very critical step for identifying the application of phage candidate (de Jonge et al. 2018). There is no any standard protocol that can decide minimum host titer that should be infected to consider it as "Broad" host range. There are various methods for host range determination. First is Spot assay or direct inoculation of phage in the agar lawn of bacteria needs to determine turbidity or clarity of plaque. There is a scaling technique where rating has been assigned like for +4 is for clear plaque while +1 for turbid plaque. For good

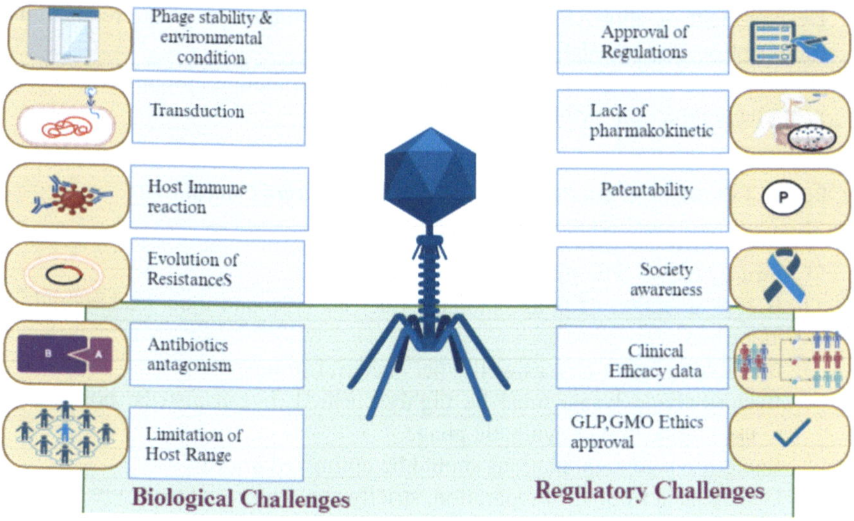

Fig. 36.1 Major challenges for successful phage therapy

results, bacteria should be in the exponential phase of growth which is in lawn culture. Second is Efficiency of Plating technique, where phage infection rate is determined from number of plaques. It is be calculated by average plaque forming unit by target bacteria/average plaque forming unit (PFU) formed by host bacteria (Mirzaei and Nilsson 2015). Broad host range phage must be used carefully while narrow host range phage used in pathogen identification as a specificity that can disappear by increase host range. So, while selection of host range, special attention should be given to host evolutionary feature. It may decrease the host range in future.

36.3.2 Bacterial Resistance to Phage

One of the important issues is possibility to develop bacteriophage-resistant bacteria that could make phage therapy unsuccessful, It has been found that the emergence of bacterial resistant to phage is possible at any time and is not avoidable (Oechslin 2018). There are many possible mechanisms of bacteria to develop resistance.

36.3.2.1 Bacterial Resistance Mechanism

- Modification of the receptor as a result of failure of phage adsorption.
- Adaptation of superinfection exclusion systems in bacteria as a result of failure to insert their nucleic acid.
- Development of systems like restriction-modification systems, CRISPR-Cas system in bacteria results in degradation of phage DNA.

- Development of abortive infection system that inhibits the phage replication, transcription, and translation processes.
- Development of cyclic oligonuclotide-dependent signaling systems which prevent phage infection (Aslam et al. 2020).
-

Box 36.1 Checklist of Successful Phage Therapy in Future Clinical Trial

Pre-requisitions

1. Strictly use of lytic phage
2. Before selection of phage therapy, Bacteria type and their load should identified.
3. It is able to infect host as well other bacterial strains.
4. Bacteriophage lysate must be highly purified. It can also be possible to use engineered or synthetic phage.
5. Bacteriophage combinations should be optimized prior to use.
6. During bacteriophage preparation, strictly avoid bacterial exposure.
7. Selected Bacteriophage should tested under certain condition to measure its efficacy like phage susceptibility testing, pH and temperature tolerance check, Stability check, their genetic analysis to determine virulence or prophage genes like repressor, integrases, excises, terminases and recombinases, measure their level of ability to retain maximum pathogenicity (Kumari et al. 2009)
8. Determine Various routes of administration to measure their spectrum of affectivity (Viazis et al. 2011).
9. Phage titre should be determined prior to application (Carvalho et al. 2010)
10. Determine phage toxicity in vitro.
11. In absence of host, Bacteriophage persistency should be monitored (Hall et al. 2012).
12. Use only GMP—phage
13. Prefer the selection of personalized phage rather than polyphage preparations.
14. During intravenous route of administration, monitor phage concentration at serum level.
15. Phage titre should have $>10^6$/ml of the lysate for patients
16. Quality control criteria of phage stock solution is that it should be periodically assessed by checking their sheft life, cytotoxicity, sterility, pH temperature tolerance test.

36.3.2.2 Possible Strategies to Overcome Bacterial Resistance to Phage

- Use of polyphage, a combination of different lytic phages (McCallin and Oechslin 2019).
- A single phage preparation

- Customized phage for specific patients (Pirnay et al. 2011).
- Combinational therapy of phage and antibiotics (Torres-Barcelo and Hochberg 2016).
- Replacement of resistant phage with another susceptible phage should be done on priority basis to provide a different cocktail or to adapt repetitive phage susceptibility testing.
- Use of genetically engineered phages or synthetic phages to increase the phage purity and overcome the irresponsive phage

36.3.3 Quality and Effective Titer of Phage

Major important challenge is to determine the efficacy of phage for antibacterial activity. It is also very important to maintain the quality of selected phage for clinical applications. There is a linear relationship between concentrations of phage against bacterial inhibition. Some phages are also active even at very low titre i.e. 10^5/mL (Worley-Morse et al. 2014). In a clinical trial which was under control of Good Manufacturing practices (GMP), PhagoBurn, initially was with very low titer for application (10^6/mL) after some time period, the titer was decreased and diluted phage was given by keeping endo-toxicity level for daily dose administration. The possible daily dose could be 10–100 PFU/mL instead of standard concentration (10^6/mL) (Jault et al. 2019).

Successful phage therapy depends on safety phage preparations as this is a big challenge during the time of formulation and production. Since lack of phage regulation, phage preparation must be under strict standard regulation. Many phage researchers have defined certain requirements that ensure sustainable phage preparation with safety and quality (Box 36.1).

In contrast to PhagoBurn, Clinical trial in California used 10^{10} pfu/mL of the phage lysate was administered through IV (Intravenously). It has been shown improved results with no side effects. But some side effects have been observed at 10^{11}/mL phage titre (Aslam et al. 2020). In Some cases, significant results have been observed at 10^9/ mL phage titre in the treatment of empyema and ventilator-associated pneumoniae infections (Maddocks et al. 2019). Dedrick et al. studied that when phage titre was used at 10^9/mL every 12 h, serum concentration was achieved as high as 10^9/mL (Dedrick et al. 2019). Since we don't have enough records of clinical trials to monitor the phage concentration at serum level to determine the effective phage titre.

36.3.4 Optimization of Route of Administration

There is a very major issue in optimization of the dosage administration. Bacteriophage enters our body and interact with different organs; its efficiency and survival rate should be same as in vitro testing. If the patient having a local infection development through systemic infection then phage candidate should also survive much enough to achieve their action at the site of infection. Our candidate phage must survive in blood stream, if the phage is administered in intestine. Research data indicates phage can be inactivated at low pH, so in such cases phage should be delivered through protected shield. Phage can be embedded in a natural polymer matrix, microencapsulation form. The encapsulated bacteriophage can survive at low pH, protected in bile salt, and easily clear through excretion system (Otero et al. 2019).

36.3.5 Stability and Storage of Phage

The stability of phage is the essential requirement for regulation of bacteriophage as pharmaceutical entity and for successful phage treatment. For potential bacteriophage candidate, it should be prepared and maintained in such a way that not even a minor drop of their titer is wasted. So, each stored phage candidate should be named with its titer. Strategies should be developed to increase phage stability are, (Malik et al. 2017).

- Freeze dried,
- Spray dried,
- Polymerization technique,
- Emulsion technique and.
- Dripping technique.
- Encapsulation on matrix such as cellulose, alginate, and liposome—Important to increase phage stability and phage delivery during therapy (Malik et al. 2017).

However, clearance and inactivation of phages also depend on medium pH, in some cases low pH can also easily inactivate the phage has seen in human body. So, phage stability and efficacy totally depend on phage concentration and protection of phage from harsh environments such as the human body.

Furthermore, studies are needed to maximize phage stability which in turn protects phage genome from mutation). Phage can be stored from −4 °C to −80 °C. For long-term storage, liquid nitrogen, −196 °C is the best solution to maintain its efficacy (Tanir et al. 2021). Many additives and cryopreservatives can be added to maintain efficacy. Calcium chloride and magnesium sulphate with combination of cryopreservants like gelatine, glycerol, disaccharides sugars, and polyethylene glycol (Vázquez et al. 2022).

36.3.6 Phage Neutralizing Antibodies

A century year ago, Jerne et al. indicated that phages are able to produce immunological responses by generating antibodies which can affect the functioning of phage (Jerne 1956). However, there is very few reports of such kinds or are not considered when patient's outcome was analysed. In some case studies, patients have received phages but immune responses have not been observed (Aslam et al.

2020; Fabijan et al. 2020) A study by Dedrick et al. indicated that no phage-neutralizing antibodies generated even though patient has received phage intravenously till 32 weeks (Dedrick et al. 2019). Therefore we can conclude that challenge of phage neutralizing antibody is dependent on type and dose of bacteriophage, duration of therapy received and route of administration. Moreover, there are no clear evidences of lever of phage neutralizing antibodies response with outcome of phage therapy.

Hence further research data of clinical trial should include this factor to this clear concept.

36.3.7 Immunological Response

Recent research has shown that phage is able to interact with human cells and through this interaction; phage can enhance their antimicrobial activity. Experimental study by Shan et al. indicates that phage can able to lyse bacteria effectively in presence of human cells. This study indicates that phage is host specific and this is due to presence of a specific phage receptor which has affinity with human cell receptor (Shan et al. 2018). If this phenomenon is relevant then in-vivo phage therapy generates good output. However, currently no much data is available to support this phenomenon. Further clinical studies on phage-eukaryotic cell interactions especially with organ-specific cells and tissues are suggested for successful phage therapy.

36.3.8 Environmental Challenges

Bacteriophage is made up of nucleic acid and proteins. There are possibilities that under any physiological and chemical condition, it may alter or lose their infectivity and efficiency. Apart from this, Phage preparation and storage conditions also change the actual Phage titer. So, there are many parameters that ensure quality for improved phage production process, storage & stability for the successful execution of clinical trials (Fig. 36.2).

36.3.9 Phage—Antibiotics Synergy

Another challenge is that phage therapy is in combination with antibiotics. There are many research studies that support combination therapy (Jeon and Ahn 2020; Luscher et al. 2020. Phage can be used alone as a single therapeutic agent. In an antibiotic resistance scenario, it is not possible for combinational therapy and on other side, it is not possible to exclude antibiotics to treat bacterial infections. So clinical trials also include one group where patients treated with combination. However, these combinational strategies may not be recommended with all groups of antibiotics and all types of phages. Recent study indicated that aminoglycosides

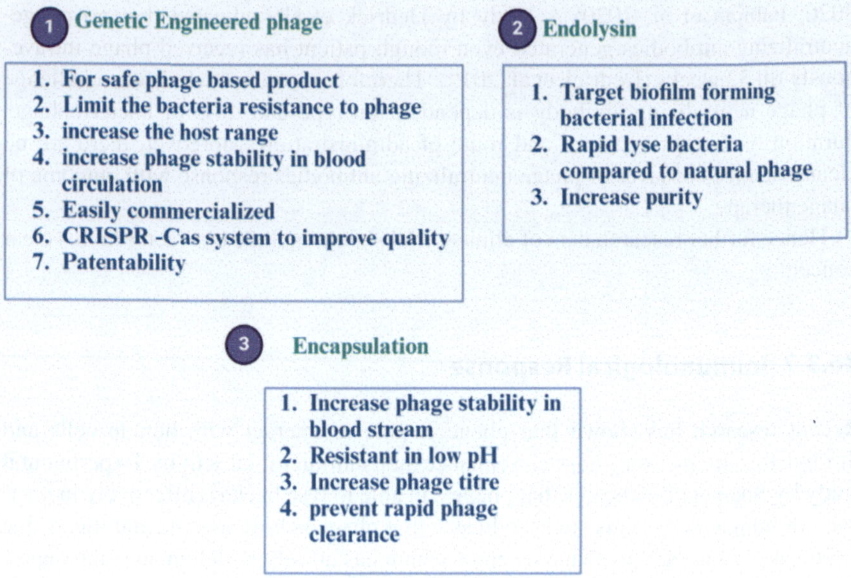

Fig. 36.2 Advanced solution to overcome existing phage therapy challenges

group of antibiotics inhibits the multiplication of mycobacteriophage by interfering at the level of phage DNA synthesis (Jiang et al. 2020).

36.3.10 External Elements

From some of the studies, it is also found that phage alters their function in presence of some external elements (Jończyk-Matysiak et al. 2019). Phages can be inactivated in presence of tea (Liu et al. 2015). Moreover, susceptibility of phage affects in presence of T-even phage while T-odd phage shows activity in presence of alcohol (Yamashita et al. 2000). Oral administration of phage also neutralizes gastric acid e.g., *S. aureus* specific phage A5/80 is affected by ranitidine or aluminium-containing drugs. Research studies have been shown that in presence of antacid 2% phage titre reduced (Międzybrodzki et al. 2017).

36.4 Regulatory Challenges

During the time of former Soviet Union, phages were produced in mass and been available as the first clinical trial was back at the 1920s (Myelnikov 2018). Today, issues are still to develop regulatory guidelines because phage production according to Good Manufacturing Practices (GMP) requires a controlled manufacturing process. Phages are not suitable for this because its complex characteristics seem time consuming, costly for pharmaceutical companies where limited also issue of limited patentability.

36.4.1 Lack of Clinical Trials

One of the main and major challenges in phage therapy to fit under GMP and regulatory standards is very few clinical trial data available to evidence phage as a medicinal product. Good clinical trials can be used with placebo as control group. Still many researchers and biotechnological companies have already launched clinical trials to highlight the potential of bacteriophage therapy. To provide effective treatment, Bacteriophage therapy with many solutions against challenges can reduce the risk to public health. In just a few years, many companies, researchers have come forward across the global with innovative ideas which will pave phage therapy towards the market (UK Parliament 2024).

Currently, the increasing clinical trials provide strength to the base of evidence. Some of the issues with clinical trial are (UK Parliament 2024),

- Phage is no single compound like "Paracetamol."
- Lack of consistent high specified phages for placebo control and double-blind studies.
- Possibly one phage might work efficiently for one patient but not for other patients.

Phage is "biological entity," it can be used as cocktail, combination with antibiotics or by genetically modification. These characteristics conclude into fact that phage treatment can be tailored as a personalized way for each patient. So, in future it can also possible that provision of personalized used medicine license of such a unique regime allows phage use with combination of antibiotics, engineered phage or natural generic phage (Anomaly 2020).

36.4.2 Adapted Regulatory Models for Non-GMP Phages

36.4.2.1 Emergency Use Investigational New Drug (eIND)—United States of America (USA)

In USA, phage is used to treat patients through Food and Drug Administration (FDA)—Expanded pathway, emergency investigational new drug (eIND) (FDA 2020). It requires one document which should be a clear mention of fail to antibiotic therapy.

36.4.2.2 Magistral Phage—Belgium

Belgium has adapted phage therapy as a "three-page—prescription." It is defined as "any medicine a product which is prepared in pharmacy is according to medical prescription for each individual patient". Means that the product requires manufacturing in the pharmacy, second it must be prepared according to patient's prescription and third is prescription is only for one individual patient. This procedure includes dictation of all the details in three different documents. Details need to mention are how the phage prepared, bacterial identification with load of infection, level of bacterial endotoxin, what equipment used to prepare, what are the quality standard followed for the procedure, water content, declaration of assure chloroform removal, phage shelf life etc. (Żaczek et al. 2020).

36.4.3 Intellectual Properly Issues with Phage Therapy

Last, major hurdle associated to further implementation of phage therapy is in the economic market. Pharmaceutical companies can achieve success by translating phage research into profitable scale. Many concerns are associated in the investment of bacteriophage therapy at pharmaceutical level. One of the major issues is limited options on patentability. As phage is considered under "Natural products," many phage cocktails are life saving for patients while its patent considered as frail. In such cases, genetically engineered phage with novel modifications which can enhance its antibacterial activity can possibly to patent (Pires et al. 2020). Natural phages which can be used for research and clinical trials represents frail and hurdle for financial investment to achieve regulatory standards and patent (Henein 2013).

36.5 Awareness about Phage Therapy and Linking of Phage Experts

Another key challenge is more use of phage and less awareness among consultants, doctors, and many clinicians. However, more phage therapy centers were found in the United Kingdom, one survey published in 2023 showed that very few clinicians known about phage and their use in multi-drug resistant bacterial infection treatment. Lack of proper understanding, their routes and challenges among many health care workers, Provision of medical training is way to address this challenge (UK Parliament 2024).

36.6 Conclusions

In the era of antibiotic resistance, we need to ensure that even if phage therapy recommended a safe and definitive tool, it should be developed to overcome this issue. To confirm phage as a medicine, more clinical trials data is required for launching its practicality and ethics. In this chapter, we tried to serve a few challenging factors and their solutions to implement clinical trials of bacteriophage therapy in future.

36.7 Summaries

Challenges	Solution
Determine effective titer of phage	Monitor the phage concentration at serum level
Phage—Antibiotic synergy	Not recommended with every phage and every antibiotic. Certain antibiotics can inhibit phage multiplication.
Biofilm and capsule producer bacterial infection	Application of Endolysin. Lysin and depolymerase have ability to destroy biofilm.
Development of Bacterial resistant to phage	Application of Polyphage or cocktail of single phage which will reduce the chance of development of bacterial resistance to phage.
Phage Neutralizing antibodies	Phage neutralizing antibody is depending on type and dose of bacteriophage, Duration of therapy received and route of administration. However clinical trial should involve this point for enough supporting data to finalize conclusion.
Phage purity	Use of genetic engineered phage or synthetic phage, Purify with Caesium chloride before application in human.
Stability of phage	Freeze dried, Spray dried, Polymerisation technique, Emulsion technique and Dripping technique, Encapsulation
Storage of phage	From 4 to −80 °C, Long term storage by −196 °C
Loss of efficacy	Addition of calcium chloride with magnesium sulphate + cryopreservants like gelatine, disaccharides, glycerol or polyethylene glycol.
Contamination	Isolation and separation of phage genomes and enzymes which reduce the risk of contamination.
Lysogenic phage	Strictly use lytic phage
Lack of regulatory standard	Design the regulation and policies for improvement
Administration and dosage decision	Explore the pros and cons of different route of administration
Endotoxin release	Define the treatment plan and phage titre on the basis of type of disease and its severity
Patentability	Genetically engineered phage can be patented

References

Anomaly J (2020) The future of phage: ethical challenges of using phage therapy to treat bacterial infections. Public Health Ethics 13(1):82–88. https://doi.org/10.1093/phe/phaa003

Aslam S, Lampley E, Wooten D, Karris M, Benson C, Strathdee S, Schooley RT (2020) Lessons learned from the first 10 consecutive cases of intravenous bacteriophage therapy to treat multidrug-resistant bacterial infections at a single Center in the United States. Open Forum Infect Dis 7:ofaa389. https://doi.org/10.1093/ofid/ofaa389

Carvalho CM, Gannon BW, Halfhide DE et al (2010) The *in vivo* efficacy of two administration routes of a phage cocktail to reduce numbers of Campylobacter coli and Campylobacter jejuni in chickens. BMC Microbiol 10:232. https://doi.org/10.1186/1471-2180-10-232

Dedrick RM, Guerrero-Bustamante CA, Garlena RA, Russell DA, Ford K, Harris K, Gilmour KC, Soothill J, Jacobs-Sera D, Schooley RT et al (2019) Engineered bacteriophages for treatment of a patient with a disseminated drug-resistant mycobacterium abscessus. Nat Med 25:730–733. https://doi.org/10.1038/s41591-019-0437-z

Fabijan AP, Lin RCY, Ho J, Maddocks S, Ben Zakour NL, Iredell JR (2020) Westmead Bacteriophage Therapy Team. Safety of bacteriophage therapy in severe Staphylococcus aureus infection. Nat Microbiol 5(3):465–472. https://doi.org/10.1038/s41564-019-0634-z. Epub 2020 Feb 17. Erratum in: Nat Microbiol. 2020 Apr; 5(4):652

FDA (2020). https://www.fda.gov/drugs/investigational-new-drug-ind-application/emergencyinvestigational-new-drug-eind-applications-antiviral-products. Accessed 4 Mar 2020

Górski A, Borysowski J, Międzybrodzki R (2020) Phage therapy: towards a successful clinical trial. Antibiotics (Basel) 9(11):827. https://doi.org/10.3390/antibiotics9110827

Hall AR, De VD, Friman VP, Pirnay JP, Buckling A (2012) Effects of sequential and simultaneous application of bacteriophages on populations of Pseudomonas aeruginosa *in vitro* and in wax moth larvae. Appl Environ Microbiol 78(16):5646–5652

Henein A (2013) What are the limitations on the wider therapeutic use of phage? Bacteriophage 3(2):e24872.2012BACTERIOPHAGE0028R. https://doi.org/10.4161/bact.24872

Jault P, Leclerc T, Jennes S, Pirnay J-P, Que Y-A, Resch G, Rousseau AF, Ravat F, Carsin H, Le Floch R et al (2019) Efficacy and tolerability of a cocktail of bacteriophages to treat burn wounds infected by Pseudomonas aeruginosa (PhagoBurn): a randomised, controlled, double-blind phase 1/2 trial. Lancet Infect Dis 19:35–45. https://doi.org/10.1016/S1473-3099(18)30482-1

Jeon G, Ahn J (2020) Assessment of phage-mediated inhibition of salmonella typhimurium treated with sublethal concentrations of ceftriaxone and ciprofloxacin. FEMS Microbiol Lett 367:fnaa159. https://doi.org/10.1093/femsle/fnaa159

Jerne NK (1956) The presence in normal serum of specific antibody against bacteriophage T4 and its increase during the earliest stages of immunization. J Immunol 76:209–216

Jiang Z, Wei J, Liang Y, Peng N, Li Y (2020) Aminoglycoside antibiotics inhibit mycobacteriophage infection. Antibiotics 2020(9):714. https://doi.org/10.3390/antibiotics9100714

Jończyk-Matysiak E, Łodej N, Kula D, Owczarek B, Orwat F, Międzybrodzki R, Neuberg J, Bagińska N, Weber-Dąbrowska B, Górski A (2019) Factors determining phage stability/activity: challenges in practical phage application. Expert Rev Anti-Infect Ther 17(8):583–606. https://doi.org/10.1080/14787210.2019.1646126

de Jonge PA, Nobrega FL, Brouns SJJ, Dutilh BE (2018) Molecular and evolutionary determinants of bacteriophage host range. Trends Microbiol 27(1):51–63. https://doi.org/10.1016/j.tim.2018.08.006

Kim MS, Park EJ, Roh SW, Bae JW (2011) Diversity and abundance of single stranded DNA viruses in human feces. Appl Environ Microbiol 77:8062–8070. https://doi.org/10.1128/AEM.06331-11

Kumari S, Harjai K, Chhibber S (2009) Efficacy of bacteriophage treatment in murine burn wound infection induced by Klebsiella pneumoniae. J Microbiol Biotechnol 19(6):622–628

Liu H, Meng R, Wang J, Niu YD, Li J, Stanford K, Mcallister TA (2015) Inactivation of Escherichia coli O157 bacteriophages by using a mixture of ferrous sulfate and tea extract. J Food Prot 78(12):2220–2226. ISSN: 0362-028X. https://doi.org/10.4315/0362-028X.JFP-15-239

Luscher A, Simonin J, Falconnet L, Valot B, Hocquet D, Chanson M, Resch G, Köhler T, Van Delden C (2020) Combined bacteriophage and antibiotic treatment prevents Pseudomonas aeruginosa infection of wild type and cftr-epithelial cells. Front Microbiol 11. https://doi.org/10.3389/fmicb.2020.01947

Lusiak-Szelachowska M, Weber-Dabrowska B, Jonczyk-Matysiak E, Wojciechowska R, Gorski A (2017) Bacteriophages in the Gastointestinal tract and their implications. Gut Pathog 9:44. https://doi.org/10.1186/s13099-017-0196-7

Maddocks S, Fabijan AP, Ho J, Lin RCY, Ben Zakour NL, Dugan C, Kliman I, Branston S, Morales S, Iredell JR (2019) Bacteriophage therapy of ventilator-associated pneumonia and empyema caused by *Pseudomonas aeruginosa*. Am J Respir Crit Care Med 200:1179–1181. https://doi.org/10.1164/rccm.201904-0839LE

Malik DJ, Sokolov IJ, Vinner GK, Mancuso F, Cinquerrui S, Vladisavljevic GT, Clokie MRJ, Garton NJ, Stapley AGF, Kirpichnikova A (2017) Formulation, stabilisation and encapsulation of bacteriophage for phage therapy. Adv Colloid Interface Sci 249:100–133, ISSN: 0001-8686. https://doi.org/10.1016/j.cis.2017.05.014

McCallin S, Oechslin F (2019) Bacterial resistance to phage and its impact on clinical therapy. Phage therapy: a practical approach. Switzerland AG: Springer International Publishing, pp 59–88

Międzybrodzki R, Kłak M, Jończyk-Matysiak E, Bubak B, Wójcik A, Kaszowska M, Weber-Dąbrowska B, Łobocka M, Górski A (2017) Means to facilitate the overcoming of gastric juice barrier by a therapeutic staphylococcal bacteriophage A5/80. Front Microbiol 8:467. https://doi.org/10.3389/fmicb.2017.00467

Mirzaei MK, Nilsson AS (2015) Isolation of phages for phage therapy: a comparison of spot tests and efficiency of plating analyses for detection of host range and efficacy. PLoS One 10(5):e0127606

Myelnikov D (2018) An alternative cure: the adoption and survival of bacteriophage therapy in the USSR, 1922–1955. J Hist Med Allied Sci 73(4):385–411. https://doi.org/10.1093/jhmas/jry024

Oechslin F (2018) Resistance development to bacteriophages occurring during bacteriophage therapy. Viruses 10(7):351. https://doi.org/10.3390/v10070351

Otero J, García-Rodríguez A, Cano-Sarabia M, Maspoch D, Marcos R, Cortés P, Llagostera M (2019) Biodistribution of liposome-encapsulated bacteriophages and their transcytosis during Oral phage therapy. Front Microbiol 10:689. https://doi.org/10.3389/fmicb.2019.00689

Pires DP, Costa AR, Pinto G, Meneses L, Azeredo J (2020) Current challenges and future opportunities of phage therapy. FEMS Microbiol Rev 44(6):684–700. https://doi.org/10.1093/femsre/fuaa017

Pirnay JP, De Vos D, Verbeken G et al (2011) The phage therapy paradigm: prêt-à-porter or sur-mesure? Pharm Res 28:934–937. https://doi.org/10.1007/s11095-010-0313-5

Shan J, Ramachandran A, Thanki AM, Vukusic FBI, Barylski J, Clokie MRJ (2018) Bacteriophages are more virulent to bacteria with human cells than they are in bacterial culture; insights from HT-29 cells. Sci Rep 8:1–8. https://doi.org/10.1038/s41598-018-23418-y

Sutton TD, Hill C (2019) Gut bacteriophage: current understanding and challenges. Front Endocrinol 10:784. https://doi.org/10.3389/fendo.2019.00784

Tanir T, Orellana M, Escalante A, Moraes de Souza C, Koeris MS (2021) Manufacturing bacteriophages (part 1 of 2): cell line development, upstream, and downstream considerations. Pharmaceuticals (Basel) 14(9):934. https://doi.org/10.3390/ph14090934

Tetz GV, Ruggles KV, Zhou H, Heguy A, Tsirigos A, Tetz V (2017) Bacteriophage as potential new Mannalian pathogens. Sci Rep 7:7043. https://doi.org/10.1038/s41598-017-07278-6

Torres-Barcelo C, Hochberg ME (2016) Evolutionary rationale for phages as complements of antibiotics. Trends Microbiol 24:249–256. https://doi.org/10.1016/j.tim.2015.12.001

UK Parliament (2024) Äntimicrobial potential of bacteriophage. Science, Innovation and Technology Committee. https://publications.parliament.uk/pa/cm5804/cmselect/cmsctech/328/report.html#

Vázquez R, Díez-Martínez R, Domingo-Calap P, García P, Gutiérrez D, MaiteMuniesa MR-R, Sanjuán R, Tomás M, ÁngelesTormo-Mas M et al (2022) Essential topics for the regulatory consideration of phages as clinically valuable therapeutic agents: a perspective from Spain. Microorganisms 10(4):717. https://doi.org/10.3390/microorganisms10040717

Viazis S, Akhtar M, Feirtag J, Diez-Gonzalez F (2011) Reduction of Escherichia coli O157:H7 viability on hard surfaces by treatment with a bacteriophage mixture. Int J Food Microbiol 145(1):37–42. https://doi.org/10.1016/j.ijfoodmicro.2010.11.021

Worley-Morse TO, Zhang L, Gunsch CK (2014) The long term effect of phage concentration on the inhibition of planktonik bacterial culture. Environ Sci Processs Impact 16:81–87. https://doi.org/10.1039/C3EM00427A

Yamashita M, Murahashi H, Tomita T, Hirata A (2000) Effect of alcohols on Escherichia coli phages. Biocontrol Sci 5:9–16. https://doi.org/10.4265/bio.5.9. https://www.researchgate.net/publication/271799663_Effect_of_Alcohols_on_Escherichia_coli_Phages
Żaczek M, Weber-Dąbrowska B, Międzybrodzki R, Łusiak-Szelachowska M, Górski A (2020) Phage therapy in Poland—a centennial journey to the first ethically approved treatment facility in Europe. Front Microbiol 11. https://doi.org/10.3389/fmicb.2020.01056

Non-Traditional Antibacterial Therapeutic: Focus on Safety, Manufacturing, and Regulatory Dynamics

37

Deepa Dehari, Aiswarya Chaudhuri, Dulla Naveen Kumar, and Ashish Kumar Agrawal

Abstract

The escalating global menace posed by drug-resistant bacterial infections has fueled a surge in dedicated research aimed at uncovering revolutionary treatments. This chapter conducts a thorough examination of non-traditional approaches designed to navigate the complexities of bacterial drug resistance. Exploring a spectrum of strategies, the chapter investigates the regulatory landscape, manufacturing intricacies, and safety advancements associated with these innovative interventions. The array of non-traditional approaches discussed encompasses targeting virulence mechanisms, such as toxin production and virulence factor secretion, alongside inhibiting bacterial adhesion, biofilm formation, and disrupting communication pathways. Furthermore, the chapter delves into immune evasion, microbiome-modifying therapies, and the utilization of bacteriophages, both as treatments and carriers. Additional exploration is provided on the potential of nanoparticles, immunotherapy, antisense RNA, and drug-resistance-modulation approaches. Despite the promising potential offered by these non-antibiotic therapies, their development faces a multitude of obstacles and regulatory challenges. This chapter critically analyzes these hurdles, encompassing safety considerations, manufacturing complexities, and regulatory frameworks, shedding light on the collaborative endeavors among researchers, clinicians, and regulatory bodies. Through a comprehensive discussion, the chapter underscores the shared commitment to surmounting these challenges and propelling non-traditional approaches forward as viable solutions to counteract the escalating threat of drug-resistant bacterial infections.

D. Dehari · A. Chaudhuri · D. N. Kumar · A. K. Agrawal (✉)
Department of Pharmaceutical Engineering and Technology, Indian Institute of Technology (BHU), Varanasi, Uttar Pradesh, India
e-mail: ashish.phe@iitbhu.ac.in

M. Gangwar, G. Nath (eds.), *Emerging Paradigms for Antibiotic-Resistant Infections: Beyond the Pill*, https://doi.org/10.1007/978-981-97-5272-0_37

Keywords

Nontraditional antibacterial therapy · Antimicrobial resistance · Drug development · Regulatory approval · Bacteriophage therapy

37.1 Introduction

The improper and excessive use of antibiotics has resulted in the emergence of antibiotic-resistant microorganisms, highlighting the pressing need to explore alternative treatment modalities. While traditional therapies typically involve small-molecule agents that directly target bacterial components, inhibiting growth or inducing cell death (Mancuso et al. 2021), non-traditional treatments employ alternative mechanisms and substances beyond small molecules, targeting non-conventional bacterial elements. These non-traditional agents encompass a range of substances, including microbiome modulators, chelators that deactivate essential bacterial metal ions, nucleic acids disrupting bacterial DNA, and antibodies or bacteriophages specifically targeting microorganisms or their toxins (Munita and Arias 2016). The diverse array of non-traditional agents offers distinct mechanisms and approaches that could broaden treatment options and potentially mitigate the emergence of resistance (Salam et al. 2023).

Non-antibiotic medicines, including bacteriophage therapy, monoclonal antibody therapy, peptides inhibiting microbes, microbiome-based therapies, antimicrobial potentiators, antisense methods, and immune system modulation, have gained attention (Lin et al. 2017; Murugaiyan et al. 2022). While demonstrating the clinical impact of these non-traditional therapies, adequate resources will be crucial, despite potential challenges in meeting translational criteria associated with direct-acting small molecule therapeutics (Theuretzbacher et al. 2020). It is anticipated that non-traditional medicines will complement antibiotics rather than replace them, which may complicate determining clinical objectives for regulatory approval. Significant effort is still needed to manage the complex process of obtaining regulatory approval and facilitating the future adoption of unconventional medicines in clinical practice(Muteeb et al. 2023).

The utility and future potential of antibodies for treating bacterial infections are now being recognized, marking a shift from their successful history in other therapeutic domains. While bacteriophage therapy has long been employed in Eastern European countries, it is currently under investigation in Western countries (Ling et al. 2022). This chapter provides a contemporary perspective on drug development, exploring past achievements, setbacks, insights gained, and future projections. Bacteriophage endolysins, extensively studied and mainly targeting Gram-positive species due to Gram-negative impermeability barriers (Schmelcher et al. 2012), are discussed, with a detailed overview of newly discovered lysins with Gram-negative activity in early development phases. This insight holds significant interest for professionals in the field. The chapter also explores the unexplored potential of metal complexes as antibiotics, challenging the notion that chemical lessons useful for specific purposes should be disregarded as a source of new chemotypes to combat bacterial infections (Hansford 2020). Additionally, it emphasizes the importance of medical ethnobotany in the quest for new natural products to cure infections, particularly antifungals. Finally, the communication highlights the significance of drug repositioning in developing novel pathogenicity inhibitors. Testing the efficacy of various pharmaceuticals as quorum sensing (QS)-inhibitors identified 14 agents with dose-dependent QS-inhibitory properties in vitro, indicating that QS-based control of bacterial virulence is a promising approach warranting further study (Mekonnen et al. 2022).

By synthesizing existing research, this book chapter illuminates advancements, regulatory limitations, and potential strategies to overcome obstacles in the field of non-traditional antibacterial therapeutics.

37.2 Challenges in Non-Antibiotic Therapy Development

The exploration of non-antibiotic therapies emerges as a potential solution to address the growing concern of antimicrobial resistance. These alternative therapies, including monoclonal antibodies, virulence disruptors, and phage therapies, operate through nontraditional mechanisms and could potentially exhibit lower susceptibility to resistance compared to conventional antibiotics. Nonetheless, the development of non-antibiotic therapies encounters challenges related to achieving target specificity and broad applicability, as highlighted previously (Liu et al. 2024).

37.2.1 Target Specificity and Broad Applicability

Nontraditional therapies, characterized by their atypical features and targeting nontraditional elements, face challenges in ensuring target specificity. Moreover, the effectiveness of these therapies may exhibit variability across different bacterial species and patient characteristics. Despite these challenges, the exploration of nontraditional antibiotics is essential in adapting to the evolving landscape of bacterial pathogenesis and mitigating the rising incidence of antimicrobial-resistant infections. Continued research and development efforts are crucial to optimize the target specificity and enhance the broad applicability of non-antibiotic therapies for the effective management of bacterial infections (Tse et al. 2017).

37.2.1.1 Bacteriophages

Bacteriophages, while holding promise as antibiotic alternatives, encounter several challenges in both research and implementation. Their inherent specificity, targeting particular bacterial species or strains, proves advantageous in minimizing harm to beneficial bacteria. However, this selectivity necessitates the use of multiple bacteriophages for treating diverse bacterial infections, extending treatment durations (Taati Moghadam et al. 2020). The broad application and species-specific nature of bacteriophages present significant hurdles, limiting their efficacy against diverse bacterial populations. Challenges include rapid clearance by the immune system, intracellular delivery difficulties, and unclear regulatory mechanisms, impeding clinical translation. The adjunctive usage paradigm and the need for diagnostic specificity further complicate matters (Maciejewska et al. 2018). Concerns about bacteriophage resistance emphasize the importance of rigorous evaluation in pharmacodynamics, pharmacokinetics, and combination products, despite their potential in combating antibiotic resistance. To fully unlock the therapeutic potential of bacteriophages against bacterial infections, creative thinking, and meticulous research are imperative. Overcoming these challenges requires innovative approaches and thorough investigations (Jo et al. 2023).

37.2.1.2 Lysin

The development and utilization of non-antibiotic treatments, such as lysins, face several challenges that need to be overcome. One major obstacle is the broad application and limited target specificity of lysins, primarily focusing on gram-positive bacteria. This restricts their use and highlights the need for more attention to gram-negative infections (Murray et al. 2021). Additionally, the regulatory pathway for lysins is unclear, hindering their integration into therapeutic settings. Introducing supplementary usage paradigms into current therapeutic protocols presents further challenges. Similar to bacteriophages, lysins require diagnostic specificity for effective implementation. To unlock the potential of lysins as non-antibiotic alternatives in the fight against bacterial infections, collaborative efforts are essential to elucidate regulatory mechanisms, refine treatment strategies, and enhance diagnostic capabilities (Fischetti 2008).

37.2.1.3 Immunomodulators

The development of non-antibiotic medicines encounters various challenges, with target specificity and broad application proving particularly formidable. Immunomodulators, for instance, face hurdles arising from the diverse response cascades triggered by different infections, such as those caused by gram-positive versus gram-negative bacteria. This diversity necessitates tailored techniques for effective treatment, complicating the development of therapies in this domain (Chinemerem Nwobodo et al. 2022).

Incorporating immunomodulators into current treatment regimens is further complicated by the adjunctive usage paradigm, demanding meticulous consideration of timing and compatibility with traditional antibiotics (Pirofski and Casadevall 2006). The notable failure rate observed in clinical studies of immunomodulatory therapy presents another substantial hurdle, emphasizing the importance of comprehensive preclinical testing, refined treatment strategies, and an improved understanding of the immune system's response to infections. To fully unlock the potential of immunomodulators in combating bacterial infections, it is imperative to foster interdisciplinary collaboration, adopt innovative research approaches, and take a nuanced approach to therapeutic development (Buchwald and Pirofski 2003).

37.2.1.4 Virulence Factor Disruptors

Developing non-antibiotic medicines encounters the challenge of creating virulence factor disruptors that are both broadly applicable and specific to their targets. Virulence factor disruptors signify a paradigm shift in medicine by targeting the processes that render bacteria harmful rather than merely eliminating them. This strategy necessitates a high degree of target specificity to ensure efficacy while minimizing off-target effects (Kane et al. 2018). A significant obstacle lies in ensuring the wide applicability of virulence factor disruptors across various bacterial strains and species, requiring a comprehensive understanding and characterization of bacterial virulence pathways. Additionally, the diagnostic specificity of these treatments determines their effectiveness in identifying diseases suitable for virulence factor disruptors (Qin et al. 2022). To fully realize the potential of virulence factor disruptors as effective non-antibiotic therapies for bacterial infections, overcoming these obstacles mandates a comprehensive strategy that integrates advancements in diagnostics, therapeutic design, and molecular understanding of bacterial virulence (Dehbanipour and Ghalavand 2022)

37.2.1.5 Microbiome Therapies

Two significant challenges in the development of non-antibiotic medicines, particularly within the realm of microbiome therapies, are target specificity and broad application. Microbiome therapeutics aim to modify the microbial populations within the human body to treat or enhance health. However, a lack of understanding regarding the optimal composition and function of the microbiome for various health conditions poses a major barrier (Mimee et al. 2016). The complexity of the microbiome and its interactions with the host makes it challenging to identify precise targets for treatment. Moreover, the regulatory landscape, particularly regarding recurring Clostridium difficile infections, remains uncertain, adding complexity to the approval process for microbiome medicine (Seekatz et al. 2022). The absence of well-defined regulatory pathways further complicates the creation and approval of such therapies. To overcome these obstacles, further research is necessary to elucidate the intricacies of the microbiome. Additionally, collaboration between regulatory bodies, researchers, and healthcare professionals is essential to establish standards and procedures for the development and authorization of microbiome-based therapies (Kho and Lal 2018).

37.2.1.6 Monoclonal Antibodies

Monoclonal antibodies offer a promising avenue in the development of non-antibiotic therapies, but they encounter numerous obstacles related to target selectivity and broad application. Cost presents a significant barrier, as the production and administration of monoclonal antibodies can be expensive, potentially limiting patient and healthcare system access (Chames et al. 2009). The adjunctive therapy paradigm adds complexity to treatment plans, requiring careful coordination with other medications regarding timing and combination. Infusion reactions pose another challenge, necessitating close monitoring and management when administering monoclonal antibodies (Cáceres et al. 2019). Additionally, if monoclonal antibodies consist of multiple components, they may face restrictions as combination products, complicating the regulatory approval process. To overcome these obstacles and unlock the full therapeutic potential of monoclonal antibodies in non-antibiotic therapy development, innovative solutions are required to reduce production costs, enhance treatment protocols, mitigate side effects, and navigate regulatory requirements (Struble et al. 2023).

37.2.1.7 Metallic Nanoparticles

The utilization of metallic nanoparticles, like silver and gold nanoparticles, in non-antibiotic therapy development offers promising prospects alongside significant challenges. Metallic nanoparticles inherently possess antimicrobial characteristics, interacting with bacterial cell membranes to induce cell death (Skłodowski et al. 2023). Yet, achieving target specificity while maintaining broad applicability remains challenging. While metallic nanoparticles show efficacy against diverse bacterial strains, ensuring selective targeting of pathogens while sparing beneficial microorganisms is critical. Strategies such as surface functionalization and nanoparticle conjugation with targeting ligands enhance specificity (Singh et al. 2024). Comprehensive characterization of metallic nanoparticles is essential for their broad applicability across various bacterial infections, considering factors like size, shape, and surface chemistry. Understanding potential cytotoxicity and long-term effects is crucial for safe usage. Thorough preclinical studies are necessary to evaluate biocompatibility and pharmacokinetics, elucidating their mechanisms against bacterial pathogens. Addressing challenges in target specificity and broad applicability is vital for successful integration into non-antibiotic therapies. Collaborative interdisciplinary research is key to unlocking metallic nanoparticles' full potential in combating antimicrobial resistance and improving patient outcomes (Burlec et al. 2023). The advantages and challenges of non-antibiotic therapy have been presented in Table 37.1.

37.2.2 Limited Spectrum of Activity

Non-antibiotic therapies might display a more limited spectrum of activity when compared to traditional antibiotics. This section delves into the constraints related to the variety of pathogens effectively targeted by non-antibiotic treatments. Narrow-spectrum antibiotics focus on a limited group of bacteria, either gram-positive or gram-negative. For instance, sodium fusidate exclusively acts against staphylococcal bacteria. In contrast, broad-spectrum antibiotics target both gram-positive and gram-negative bacteria, such as amoxicillin (Melander et al. 2018).

Table 37.1 Non-antibiotic therapy advantages and challenges

Name of therapy	Mechanism of action	Advantages	limitations	Ref
Microbiota-based therapy (microbiome therapies) • Natural microbiota strain • Fecal microbiota transplants • Gut toxin adsorber • Gut antibiotic inactivator • Genetically modified Probiotics	• Protect and restore the gut microbiome • Reduce survival of infectious bacteria • Reducing the load of antibiotic-resistance genes • Adsorb bacterial toxins	• Does not rely on diagnostic capabilities • Naturally derived • High rates of effectiveness.	• Difficulty during repeating gene sequence and response in genetic engineering • Lack of scientific data	Matzaras et al. (2022), Shim (2023)
Bacteriophage • Natural bacteriophage/cocktail • Modified bacteriophage • Bacteriophage isolated endolysin	• Replicate within bacterial cell and perform lysis	• Targeted therapy • Replicate on site of action • Lower investment risk • Ubiquitous • Economic	• Narrow host range, Lysogenic phenomenon • The lack of pharmacokinetic data	Coffey et al. (2016)
Antimicrobial peptides	• Inhibition of cell wall formation • Membrane disruption	• Broad range of action • Fight against multi-resistant bacteria	• Low stability • High production cost • Poor bioavailability	Lei et al. (2019), Xuan et al. (2023)
Monoclonal Antibodies	• Direct toxin neutralization • Inhibiting virulence factors • Complement depositions Enhanced opsonization	• Specifically targeted • Lower investment risk Batch to batch consistency	• Can interfere with diagnostic assay • Resistance • Expensive	Chen et al. (2024), Saylor et al. (2009)
Phytomedicine • Alkaloids • Flavonoids • Essential oils • Terpenes organic acids	• Plasma membrane disruption • Inhibiting nucleic acid and protein synthesis	• Safer than synthetic antibiotics • Broad spectrum antimicrobial activity • Rapid onset of action	• Lack of standardization during preparation • Short half life	Cowan (1999), Vaou et al. (2021)

(continued)

Table 37.1 (continued)

Name of therapy	Mechanism of action	Advantages	limitations	Ref
Antimicrobial metals	• Ions produce ROS, which may target lipids, DNA, RNA and proteins, and cause malfunctioning of membranes, proteins, and the DNA replication machinery	• Broad spectrum • Long term effect • Low bacterial resistance	• Toxicity needs to be control • Limited dispersibility	Sharma et al. (2022)
Immunomodulator	• Improve the host innate immune response	• Specific collateral damage not expected • Large pipeline with sepsis therapies • Repurposing of existing compounds possible	• Distinct response cascades for infections • Adjunctive use paradigm • High rate of failure in clinical trials	Kang et al. (2019), Pirofski and Casadevall (2006)

37.2.3 Dose Selection for Products Lacking Direct Antimicrobial Activity

The challenges faced by any of the four categories depend on the specific product. The development of antibiotics relies heavily on the dosage selection process, which is determined by the pharmacokinetics (PK) and pharmacodynamics (PD) of antibiotic exposures predicted to be effective based on animal models (Rodríguez-Gascón et al. 2021). The foundation of PK-PD's mathematical theory historically revolves around the concept of minimum inhibitory concentration (MIC). MIC refers to the lowest concentration of a medication in a Petri dish or test tube used in research, a critical component in determining the medication concentration levels required to elicit a meaningful response (Kowalska-Krochmal and Dudek-Wicher 2021). Efficacy proof at a specific drug concentration can facilitate the approval of an antibacterial agent based on a single clinical trial, leveraging the extensive experience in using MIC-based PK–PD to predict efficacy. However, certain novel products under development, like agents designed to absorb toxins generated by bacteria while sparing the bacteria themselves, lack a minimum inhibitory concentration (MIC) (Rizk et al. 2019). These medications may require more clinical data to establish a convincing case for their clinical activity compared to those with PK-PD evidence based on MIC. If a product is part of a combination that lowers the MIC of an existing medicine, it might pose a lesser hurdle (Zhou et al. 2022).

37.2.4 Industrial Manufacturing and Production

Industrial production and manufacturing of antibiotics play a vital role in modern healthcare, offering solutions for bacterial illnesses and saving countless lives worldwide. This process involves multiple stages, including fermentation, isolation, purification, formulation, and quality control. This comprehensive analysis will thoroughly examine each element of industrial antibiotic manufacture (Fedorenko et al. 2015). The development of microbiome-based therapeutics encounters numerous obstacles, primarily stemming from the composition of the bacterial species involved. A significant portion of the cocktail comprises anaerobic and/or spore-forming bacteria. Consequently, companies must allocate substantial resources to establish specialized manufacturing facilities capable of producing drug substances or products from these organisms (Yadav and Chauhan 2022).

37.2.4.1 Fermentation

Fermentation serves as the initial stage in the synthesis of antibiotics, involving the cultivation of microorganisms, particularly bacteria and fungi, in large-scale bioreactors or fermenters. The selected microbe serves as the primary source of the antibiotic. Critical fermentation parameters, such as temperature, pH, aeration, and agitation, are carefully controlled to optimize microbial growth and antibiotic synthesis. Nutrient-rich media, comprising carbohydrates, nitrogen sources, minerals, and growth factors, are utilized to support microbial growth and metabolite production (Kuila and Sharma 2018).

37.2.4.2 Isolation and Purification

Following fermentation, the antibiotic component is isolated and purified from the fermentation broth. The broth undergoes centrifugation to separate microbial cells from the supernatant containing the antibiotic. Various separation methods, including filtering, extraction, chromatography, and crystallization, are employed to isolate the antibiotic ingredient from other biological components and contaminants. Purification is a crucial step in ensuring that the final product meets regulatory standards for both efficacy and safety (Singh et al. 2018).

37.2.4.3 Formulation

Once the antibiotic component has been purified, it is formulated into dosage forms suitable for administration. The formulation process varies depending on the intended dosage form, which may include oral tablets, capsules, injectable solutions, topical creams, and suspensions. Various formulation processes are employed to enhance stability, bioavailability, and patient acceptability. Excipients, such as fillers, binders, disintegrants, lubricants, and preservatives, are added to the formulation to achieve these goals. Formulation scientists focus on optimizing the production process and composition to ensure consistent quality and performance in the final product (Adepu and Ramakrishna 2021).

37.2.4.4 Quality Control

Quality control plays a critical role in ensuring the safety, effectiveness, and uniformity of the finished antibiotic product. Throughout the manufacturing process, from obtaining raw materials to analyzing the final product, various analytical methods and tests are employed. Analytical techniques such as mass spectrometry (MS), gas chromatography (GC), high-performance liquid chromatography (HPLC), and microbiological tests are utilized to assess the identification, purity, potency, and stability of the antibiotic. Quality control laboratories adhere to Good Manufacturing Practices (GMP) and regulatory requirements to ensure compliance and maintain high levels of quality (Yu et al. 2014).

37.2.4.5 Challenges and Innovations

The industrial manufacture of antibiotics faces several challenges, including microbiological contamination, antibiotic resistance, environmental impact, and economic constraints. In response to these challenges, manufacturers continually strive to enhance and optimize their production processes. Advances in biotechnology, fermentation technology, downstream processing, and analytical techniques contribute to making antibiotic production more efficient, productive, and sustainable. Furthermore, research efforts are directed towards developing new antibiotics, alternative manufacturing platforms, and targeted delivery methods to address the growing issue of antibiotic-resistant organisms and enhance therapeutic outcomes (Kirwan et al. 2022).

37.3 Regulatory Hurdles in Non-antibiotic Therapy Approval

Current regulatory frameworks for non-traditional antibiotic treatments typically rely on established metrics of clinical efficacy demonstrated through non-inferiority or superiority trial designs. However, some products may require the expansion of existing paradigms, particularly those aiming for non-traditional outcomes with a focus on population-level benefits. Initiating discussions on these potential paradigms is imperative within the development community (Rex et al. 2019).

Decades ago, large international pharmaceutical corporations would have been instrumental in raising similar issues and advocating for effective approval criteria and research. However, many of these larger corporations have ceased antibiotic development efforts, including notable names like Novartis, AstraZeneca, Sanofi, Allergan, and The Medicines Company. As a result, the entities currently engaged in preclinical antibiotic programs are often small start-up businesses, many of which are part of the CARB-X portfolio (https://carb-x.org/), a non-profit initiative sponsored by the government and charitable foundations since 2016. CARB-X has a mandate to invest up to $500 million in early-stage antibiotic research globally(Klug et al. 2021).

The creation and management of CARB-X recognized the urgent need to address development challenges to ensure that the regulatory infrastructure is prepared to effectively evaluate these new technologies and advance them through development and approval processes. This is particularly crucial because smaller companies often lack the resources to effectively advocate for their collective needs (Simpkin et al. 2017). The regulatory status of non-antibiotic therapies is summarized in Table 37.2.

37.3.1 Lack of Standardized Regulatory Pathways

The regulatory approval process for non-antibiotic therapies often lacks the established framework found in antibiotics. This review critically assesses the current regulatory landscape, emphasizing the necessity for standardized pathways to facilitate the approval of these novel treatments. Initiatives aimed at consolidating technical standards, expediting approval procedures, and enhancing information-sharing systems across regulatory agencies are integral to standardizing regulatory routes for antibiotics (Bhavnani et al. 2020). Adopting uniform criteria and guidelines can improve the effectiveness and consistency of regulatory assessments while ensuring the safety, effectiveness, and quality of antibiotics globally. In low- and middle-income nations, where access to safe and effective antibiotics may be limited, greater cooperation and capacity-building initiatives are essential to enhance regulatory frameworks. This involves providing regulatory authorities with the tools, training, and technological support necessary to strengthen their capacity for evaluating and monitoring antibiotics (Bebell and Muiru 2014).

To summarize, various obstacles hinder ensuring the safety, effectiveness, and quality of these essential drugs due to the lack of established regulatory procedures

Table 37.2 Regulatory status of non-antibiotic therapy

Name of therapy	Regulatory approved	Clinical trial	Working Industries	Ref
Microbiome-based therapy	• REBYOTA™ by Ferring Pharmaceuticals • VOWST™ by Seres Therapeutics	• FMT product by MaaT Pharma under Phase III clinical trials • OxThera by Kibow Pharmaceuticals under Phase III clinical trials • EO2401 and EO2463 by Enterome Bioscience under Phase III clinical trials	• Second Genome Inc., USA • Enterome Bioscience, France • Yakult Honsha Company, Japan • Vedanta Biosciences, Inc, USA • Whole-biome, USA • Seres therapeutics, USA • RebiotixInc.USA	https://clinicaltrials.gov/
Bacteriophage therapy	• Phyla by Phi Therapeutics • • PhagoDermgel by Micromir • AgriPhage™ Citrus Canker by OmniLytics Inc. • agriPHIX™ by Fixed-Phage • PhageGuard S™ by Micreos • ShigActive™ by Intralytix • Various product by Eliava Bio Preparation	• AP-PA02 by Armata Pharmaceuticals completed Phase 1b/2a • BX001 by BiomX under Phase I/II clinical trial • PhagUTI by Pherecydes Pharma under Phase I/II clinical trial • AmpliPhage-004 by AmpliPhi under Pre-phase 1 clinical trial	• ACD Pharma • Eliava Bio Preparation (GEO) • IntraLytix Inc, USA • OmniLytics Inc. (USA) • Fixed-Phage (UK) • Pherecydes Pharma (FRA) • Micreos (NED) • BiomX (USA) • Phagelux (CHN)	Huang et al. (2022), Mboowa (2023), Shim (2023)

Antimicrobial metals	• Silverex by Silverex ionic & Sun Pharma • Sivel by Dattmediproducts limited • Mepilex® by Mölnlycke Health care • Dial® by		• Silverex ionic & Sun Pharma, India	Evans and Kavanagh (2021), Sharma et al. (2022)

for antibiotics. Strengthening international post-market surveillance methods, harmonizing regulatory standards, and streamlining approval procedures are crucial steps to address this challenge. International cooperation among regulatory bodies can enhance public health protection and promote the appropriate and responsible use of antibiotics (Morel et al. 2021).

Currently, firms and investors, especially in Europe, face confusion regarding the disparity between the stringent definition of a drug by the US FDA and the individual discretion wielded by EU national authorities in classifying microbiome-based medicines. However, additional clarity may arise from a recently proposed EU directive concerning pharmaceutical products, including microbiome-based therapeutics, introduced in April 2023. If ratified, this directive has the potential to encourage the EU, and indirectly other non-EU European nations such as the UK and Switzerland, to approve more microbiome-based treatments for commercial application (Hwang et al. 2020).

37.3.2 Safety and Efficacy Assessment

Demonstrating the safety and efficacy of non-antibiotic therapies presents unique challenges, and this section discusses the regulatory requirements for assessing the safety profile and efficacy of these therapies, emphasizing the need for robust preclinical and clinical studies (Kållberg et al. 2023).

To ensure the safety and efficacy of non-antibiotic therapies in treating bacterial infections, a comprehensive review of several parameters is necessary for their safety and efficacy assessment. This includes examining potential side effects and risks associated with their usage, evaluating toxicity levels, assessing medication interactions, and considering the likelihood of allergic responses. Preclinical research is conducted to evaluate the pharmacokinetics and pharmacodynamics of non-antibiotic therapies, understanding how the treatment is absorbed, distributed, metabolized, and eliminated from the body (Li et al. 2017). Animal studies are crucial in assessing the safety profile of these therapies before human testing, helping identify potential harmful effects such as carcinogenicity, organ toxicity, and reproductive toxicity. Animal models also aid in evaluating the risk of developing resistance and establishing appropriate dosage schedules. Clinical trials are then conducted to assess the safety and efficacy of non-antibiotic therapies in humans once safety information from preclinical research is available. These trials are typically conducted in stages and adhere to stringent ethical and legal requirements(Van Norman 2019).

In phase I studies, the primary objective is to evaluate the safety and tolerability of non-antibiotic therapies in a small sample of healthy volunteers, determining the ideal dosage range and potential side effects. Phase II studies involve a larger patient population with the target bacterial illness to assess safety and efficacy in a clinical context while treating the illness (Corey et al. 2014).

Phase III trials, on a larger scale, provide more information on effectiveness and safety in various patient populations by comparing the therapy with placebo or conventional treatment choices. Continuous monitoring of patients during clinical trials allows for observation of side effects, treatment outcomes, and the emergence of resistance. Adverse events are documented and examined to determine severity and impact on patient safety. Regulatory bodies evaluate safety and efficacy information following clinical studies to decide whether to authorize the therapy for commercialization and patient usage. The overall benefit-risk profile of the therapy is carefully assessed using data from preclinical and clinical research by regulatory bodies (Solomkin et al. 2019).

Post-market surveillance and monitoring follow approval to evaluate the long-term safety and efficacy of non-antibiotic therapies in real clinical settings. Healthcare professionals and regulatory bodies closely monitor adverse events, treatment outcomes, and the rise of resistance to ensure patient safety and improve treatment approaches. In summary, a comprehensive process of preclinical and clinical evaluation is involved in the safety and efficacy assessment of non-antibiotic therapies to guarantee both their effectiveness and safety in treating bacterial illnesses. Thorough research, testing, and monitoring enable healthcare professionals and regulatory bodies to make well-informed decisions on the use of non-antibiotic therapies, safeguarding public health and preventing antimicrobial resistance (Raj et al. 2019).

37.3.3 Market Access and Reimbursement

Market access and reimbursement issues pose significant challenges for non-antibiotic therapies, as discussed in the literature review exploring the economic and market-related obstacles associated with bringing these novel treatments to market and securing reimbursement from healthcare systems (Simoens et al. 2022).

The availability and affordability of non-antibiotic therapies for patients are profoundly influenced by market access and reimbursement processes, involving the management of legal constraints, securing payment from insurance companies, and ensuring widespread availability of these essential treatments. The market access pathway for non-antibiotic therapies requires regulatory clearance from government organizations, such as the European Medicines Agency (EMA) in Europe or the Food and Drug Administration (FDA) in the United States. To gain approval for marketing and distribution, these therapies must undergo extensive preclinical and clinical research to establish their safety, effectiveness, and quality. Once authorized, manufacturers face the challenge of negotiating payments with healthcare payers, including government-funded initiatives, commercial insurance providers, and healthcare systems (Saleem et al. 2021). Reimbursement decisions consider various factors, including budget impact, cost-effectiveness, and clinical efficacy. Manufacturers and payers engage in negotiations over prices and reimbursement schedules to ensure the company's profitability while providing proper coverage for patients. This may involve highlighting the advantages of non-antibiotic therapies over existing treatments, such as their ability to improve patient outcomes, reduce medical costs, and address unmet needs. In some cases, non-antibiotic therapies may qualify for accelerated review and approval processes, such as the Qualified Infectious Disease Product (QIDP) classification in the United States, designed to encourage the development of therapies for serious or life-threatening diseases. However, market dynamics present serious challenges for researchers in this field, including concerns about financial returns on investments due to low costs, brief treatment times, and considerations about antibiotic stewardship (Wettstein and Boes 2020). Policymakers, healthcare stakeholders, and business executives are exploring innovative reimbursement schemes and incentives to address these issues, promote research, and ensure the long-term availability of non-antibiotic therapies. Proposed tactics include decoupling reimbursement from sales volume, implementing subscription-based payment methods, and providing government-funded incentives for research and development in this field(Renwick et al. 2016).

The research and development of microbiome therapies heavily rely on sequencing and bioinformatics to engineer microorganisms with therapeutic traits. However, organizations proficient in both fields are rare, and production and quality control procedures in contract development and manufacturing organizations (CDMOs) often lack the specialized expertise and infrastructure required to ensure the survival of microorganisms throughout manufacturing processes (Yadav and Chauhan 2022).

37.4 Strategies for Overcoming Challenges and Regulatory Hurdles

37.4.1 Collaborative Research and Development

Encouraging collaborative efforts between academia, industry, and regulatory agencies is crucial for addressing challenges and streamlining the regulatory approval process for non-antibiotic therapies. Collaborative research and development (R&D) projects are seen as a strategic approach to overcoming obstacles and impediments related to regulations in the development of antibiotics (Metlay et al. 2006). These initiatives aim to combine resources, skills, and capacities, bringing together stakeholders from various sectors to address the growing problem of antimicrobial resistance and the depletion of the effective antibiotic pipeline. Partnerships involving pharmaceutical corporations, academic institutions, government agencies, and non-profit organizations are a crucial component of collaborative research and development (R&D) in the discovery of antibiotics (Miethke et al. 2021). Such collaborations leverage the unique abilities and assets of each partner to tackle the intricate problems related to antibiotic discovery, development, and regulatory approval. Academic institutions play a vital role in antibiotic research and development by conducting fundamental studies, identifying new drug targets, and exploring creative approaches to counteract antibiotic resistance. The discoveries made in academic settings can be effectively translated into viable drug candidates through partnerships with pharmaceutical firms, navigating through preclinical and clinical development phases (Wasan et al. 2023). Antibiotic R&D activities receive essential support in terms of financial backing, regulatory advice, and infrastructural support from government bodies. Programs like the Innovative Medicines Initiative (IMI) and the Combating Antibiotic-Resistant Bacteria Biopharmaceutical Accelerator (CARB-X) facilitate collaborative research and development efforts, providing funding to expedite the creation of novel antibiotics (Savic and Årdal 2018). Such collaborative endeavors enhance the efficiency of the antibiotic development process and contribute to the collective efforts aimed at addressing the challenges associated with regulatory approval.

37.4.2 Adaptive Regulatory Frameworks

The establishment of adaptive regulatory frameworks that can accommodate the unique characteristics of non-antibiotic therapies is essential. This section discusses potential strategies for adapting existing regulatory processes to better suit the approval of innovative treatments (Cooper et al. 2016).

The development of antibiotics poses continuous challenges and regulatory obstacles that demand flexible regulatory regimes. Traditional regulatory processes may not always be suitable for the approval of antibiotics due to the distinctive traits of antimicrobial resistance, the urgent need for novel therapies, and the complexities associated with designing clinical trials. One strategy to promote adaptive regulatory frameworks is the development of regulatory pathways specifically tailored to the creation of antibiotics. Adaptive licensing, also known as phased approval or progressive licensing, is gaining popularity as a flexible antibiotic regulatory strategy. This approach allows antibiotics to be conditionally approved based on early-stage safety and effectiveness data, with post-marketing surveillance and real-world data guiding subsequent stages of development and approval (Dutescu and Hillier 2021).

Regulatory bodies can contribute to adaptive frameworks by guiding novel trial designs, biomarkers, and surrogate endpoints that facilitate the effective evaluation of antibiotic candidates. Collaboration among industry, academics, and regulators can be fostered by regulatory bodies to establish consensus rules and standards for the development and assessment of antibiotics. Additionally, regulators can encourage the adoption of innovative incentives and regulatory mechanisms to stimulate antibiotic innovation, such as priority review vouchers, extended market exclusivity, and improved market access for antibiotics addressing unmet medical needs or serious public health threats (Brinkac et al. 2017).

Global cooperation and regulatory standardization can expedite the worldwide discovery and authorization of antibiotics. Initiatives like the International Council for Harmonization (ICH) of Technical Requirements for Pharmaceuticals for Human Use facilitate regulatory convergence and mutual recognition of data, enabling sponsors to navigate regulatory requirements more effectively across different countries (Brinkac et al. 2017).

To overcome regulatory obstacles, meticulously planned extensive clinical trials are crucial. These trials should generate robust, unbiased data on patient outcomes, complemented by an expanding body of real-world evidence. Such evidence is expected to convince stakeholders of a treatment's efficacy and facilitate the establishment of consistent regulatory guidelines across various markets. Furthermore, the increase in collaborations for microbiome research and development among businesses, academic institutions, and governments can enrich specialized knowledge and raise public awareness of microbiome therapeutics (Naidoo et al. 2021).

37.4.3 Public Awareness and Education

Raising public awareness and educating stakeholders about the benefits and challenges of non-antibiotic therapies are vital components of a successful implementation strategy. The literature review explores the importance of public engagement in fostering acceptance and understanding (Mathew et al. 2019).

Increasing public knowledge and educating the public are essential for addressing the complex issues and legal roadblocks related to the creation, distribution, and use of antibiotics. To tackle the growing danger of antimicrobial resistance (AMR), healthcare practitioners must receive thorough training on antibiotic stewardship concepts and best practices for prescription medications. With greater awareness of the negative effects of antibiotic overuse and abuse, healthcare providers will be better equipped to prescribe narrower-spectrum antibiotics when necessary and adhere to treatment recommendations. Additionally, programs designed to raise healthcare professionals' understanding of the value of diagnostic stewardship, which includes using fast diagnostic tests to inform treatment choices, can maximize the use of antibiotics and reduce avoidable exposure (Majumder et al. 2020).

Patients play a crucial role in antibiotic stewardship as well. Patient-focused public awareness campaigns and educational resources can increase knowledge of the dangers of antibiotic overuse, such as the emergence of antibiotic-resistant microorganisms and unfavorable medication responses. Stressing the significance of completing prescribed antibiotic courses as instructed by medical professionals is crucial to prevent the growth of resistant strains of bacteria and ensure the best possible treatment results. Moreover, programs aimed at improving communication between healthcare providers and patients about the reasons behind treatment decisions and antibiotic prescriptions might enable people to take an active role in their treatment plans and make informed decisions about their healthcare(Miller et al. 2020).

Additionally, outreach initiatives and community-based treatments play a critical role in reaching a variety of demographics and encouraging prudent antibiotic usage. Workshops, seminars, and social media campaigns can involve people from various age groups and socioeconomic backgrounds, promoting a shared understanding of the distinct roles that each person plays in the fight against antimicrobial resistance (Awad and Aboud 2015). Health literacy efforts and incorporating antibiotic education into school curricula can empower future generations to make informed decisions regarding antibiotic usage and infection prevention. Public awareness and education campaigns are vital in the global fight against antimicrobial resistance by fostering a culture of responsible antibiotic use and enabling people to take preventative action to protect public health. Stakeholders in non-traditional therapy, such as healthcare professionals, regulators, investors, and patients, may find it challenging to comprehend the intricacies of many innovative techniques due to their rapid development pace. To address this issue, companies should prioritize educating their target consumers and communicating a therapy's mechanism of action (MoA) and value proposition. This approach aims to facilitate its adoption and maximize its market potential (Kotwani et al. 2016).

37.5 Current Regulatory and Market Value

The commercial market for microbiome-based medicines is still in its early stages, and opinions on its global value vary. In 2021, the prescription-based industry was estimated to be worth $115 million, with projections suggesting growth to $1.3 billion by 2030, according to Strategic Industry Research. Another analysis by Research and Markets predicts an increase from $390 million in 2022 to $570 million by the end of 2023 in the global microbiome market (Daniel et al. 2023). Research and development in microbiomes are primarily led by small and medium-sized businesses. As of August 2023, over 100 companies were actively working on 243 potential microbiome modulator candidates, with only five of the top 20 pharmaceutical corporations represented. Regarding bacteriophage treatment, research has steadily increased since 2018, focusing on using bacteriophages to treat multidrug-resistant bacterial infections. The worldwide market for bacteriophages is anticipated to reach a value of USD 1441.3 million by 2028. In 2021, a minimum of 137 phage targets were reported worldwide, originating from 135 academic and commercial phage facilities, as well as 92 institutions involved in phage-related research and development, including phage banks and biotechnology companies (Federici et al. 2021). Adaptive Phage Therapeutics received a USD 12 million investment in June 2023 from funds controlled by Deerfield Management and the AMR Action Fund to develop its advanced phage treatment technology. Furthermore, national regulations and ethical standards concerning the humane application of bacteriophage treatment vary significantly.

37.6 Conclusion

In conclusion, this chapter offers a comprehensive exploration of the challenges and regulatory obstacles associated with the development of non-antibiotic therapies. By assessing the current state of research and regulatory frameworks, the chapter aims to contribute to the ongoing discussions on overcoming barriers in the pursuit of effective alternatives to traditional antibiotics. Future research directions and potential solutions are also deliberated upon to guide further progress in this crucial field. It is emphasized that the concept of a non-traditional product may not significantly impact regulatory decision-making. The majority of products falling into this category can and should be developed within the existing regulatory framework, employing assessment measures based on the impact of the therapeutic agent on the well-being, functionality, or survival of the treated individual. Successful navigation of these unconventional challenges requires coordinated efforts among businesses, government funders, and regulatory authorities.

References

Adepu S, Ramakrishna S (2021) Controlled drug delivery systems: current status and future directions. Molecules 26(19). https://doi.org/10.3390/molecules26195905

Awad AI, Aboud EA (2015) Knowledge, attitude and practice towards antibiotic use among the public in Kuwait. PLoS One 10(2):e0117910. https://doi.org/10.1371/journal.pone.0117910

Bebell LM, Muiru AN (2014) Antibiotic use and emerging resistance: how can resource-limited countries turn the tide? Glob Heart 9(3):347–358. https://doi.org/10.1016/j.gheart.2014.08.009

Bhavnani SM, Krause KM, Ambrose PG (2020) A broken antibiotic market: review of strategies to incentivize drug development. Open Forum Infect Dis 7(7):ofaa083. https://doi.org/10.1093/ofid/ofaa083

Brinkac L, Voorhies A, Gomez A, Nelson KE (2017) The threat of antimicrobial resistance on the human microbiome. Microb Ecol 74(4):1001–1008. https://doi.org/10.1007/s00248-017-0985-z

Buchwald UK, Pirofski L (2003) Immune therapy for infectious diseases at the dawn of the 21st century: the past, present and future role of antibody therapy, therapeutic vaccination and biological response modifiers. Curr Pharm Des 9(12):945–968. https://doi.org/10.2174/1381612033455189

Burlec AF, Corciova A, Boev M, Batir-Marin D, Mircea C, Cioanca O et al (2023) Current overview of metal nanoparticles' synthesis, characterization, and biomedical applications, with a focus on silver and gold nanoparticles. Pharmaceuticals (Basel) 16(10). https://doi.org/10.3390/ph16101410

Cáceres MC, Guerrero-Martín J, Pérez-Civantos D, Palomo-López P, Delgado-Mingorance JI, Durán-Gómez N (2019) The importance of early identification of infusion-related reactions to monoclonal antibodies. Ther Clin Risk Manag 15:965–977. https://doi.org/10.2147/tcrm.S204909

Chames P, Van Regenmortel M, Weiss E, Baty D (2009) Therapeutic antibodies: successes, limitations and hopes for the future. Br J Pharmacol 157(2):220–233. https://doi.org/10.1111/j.1476-5381.2009.00190.x

Chen H-C, Pan Y-L, Chen Y, Yang T-H, Hsu E-T, Huang Y-T, Chiang M-H (2024) Monoclonal antibodies as a therapeutic strategy against multidrug-resistant bacterial infections in a post-COVID-19 era. Life 14(2):246

Chinemerem Nwobodo D, Ugwu MC, Oliseloke Anie C, Al-Ouqaili MTS, Chinedu Ikem J, Victor Chigozie U, Saki M (2022) Antibiotic resistance: the challenges and some emerging strategies for tackling a global menace. J Clin Lab Anal 36(9):e24655. https://doi.org/10.1002/jcla.24655

Coffey A, O'Sullivan L, Buttimer C, McAuliffe O, Bolton D (2016) Bacteriophage-based tools: recent advances and novel applications. F1000Res 5: Faculty of 1000 Ltd

Cooper CJ, Khan Mirzaei M, Nilsson AS (2016) Adapting drug approval pathways for bacteriophage-based therapeutics. Front Microbiol 7:1209. https://doi.org/10.3389/fmicb.2016.01209

Corey R, Naderer OJ, O'Riordan WD, Dumont E, Jones LS, Kurtinecz M, Zhu JZ (2014) Safety, tolerability, and efficacy of GSK1322322 in the treatment of acute bacterial skin and skin structure infections. Antimicrob Agents Chemother 58(11):6518–6527. https://doi.org/10.1128/aac.03360-14

Cowan MM (1999) Plant products as antimicrobial agents. Clin Microbiol Rev 12(4):564–582. https://doi.org/10.1128/cmr.12.4.564

Daniel IK, Njue OM, Sanad YM (2023) Antimicrobial effects of plant-based supplements on gut microbial diversity in small ruminants. Pathogens 13(1). https://doi.org/10.3390/pathogens13010031

Dehbanipour R, Ghalavand Z (2022) Anti-virulence therapeutic strategies against bacterial infections: recent advances. Germs 12(2):262–275. https://doi.org/10.18683/germs.2022.1328

Dutescu IA, Hillier SA (2021) Encouraging the development of new antibiotics: are financial incentives the right way forward? A systematic review and case study. Infect Drug Resist 14:415–434. https://doi.org/10.2147/idr.S287792

Evans A, Kavanagh KA (2021) Evaluation of metal-based antimicrobial compounds for the treatment of bacterial pathogens. J Med Microbiol 70(5). https://doi.org/10.1099/jmm.0.001363

Federici S, Nobs SP, Elinav E (2021) Phages and their potential to modulate the microbiome and immunity. Cell Mol Immunol 18(4):889–904. https://doi.org/10.1038/s41423-020-00532-4

Fedorenko V, Genilloud O, Horbal L, Marcone GL, Marinelli F, Paitan Y, Ron EZ (2015) Antibacterial discovery and development: from gene to product and back. Biomed Res Int 2015:591349. https://doi.org/10.1155/2015/591349

Fischetti VA (2008) Bacteriophage lysins as effective antibacterials. Curr Opin Microbiol 11(5):393–400. https://doi.org/10.1016/j.mib.2008.09.012

Hansford KA (2020) Nontraditional antibiotics-challenges and triumphs. Antibiotics (Basel) 9(4). https://doi.org/10.3390/antibiotics9040169

Huang Y, Wang W, Zhang Z, Gu Y, Huang A, Wang J, Hao H (2022) Phage products for fighting antimicrobial resistance. Microorganisms 10(7). https://doi.org/10.3390/microorganisms10071324

Hwang TJ, Ross JS, Vokinger KN, Kesselheim AS (2020) Association between FDA and EMA expedited approval programs and therapeutic value of new medicines: retrospective cohort study. BMJ 371:m3434. https://doi.org/10.1136/bmj.m3434

Jo SJ, Kwon J, Kim SG, Lee SJ (2023) The biotechnological application of bacteriophages: what to do and where to go in the middle of the post-antibiotic era. Microorganisms 11(9). https://doi.org/10.3390/microorganisms11092311

Kållberg C, Mathiesen L, Gopinathan U, Salvesen Blix H (2023) The role of drug regulatory authorities and health technology assessment agencies in shaping incentives for antibiotic R&D: a qualitative study. J Pharm Policy Pract 16(1):53. https://doi.org/10.1186/s40545-023-00556-x

Kane TL, Carothers KE, Lee SW (2018) Virulence factor targeting of the bacterial pathogen Staphylococcus aureus for vaccine and therapeutics. Curr Drug Targets 19(2):111–127. https://doi.org/10.2174/1389450117666161128123536

Kang HK, Lee HH, Seo CH, Park Y (2019) Antimicrobial and immunomodulatory properties and applications of marine-derived proteins and peptides. Mar Drugs 17(6). https://doi.org/10.3390/md17060350

Kho ZY, Lal SK (2018) The human gut microbiome—a potential controller of wellness and disease. Front Microbiol 9:1835. https://doi.org/10.3389/fmicb.2018.01835

Kirwan JA, Gika H, Beger RD, Bearden D, Dunn WB, Goodacre R et al (2022) Quality assurance and quality control reporting in untargeted metabolic phenotyping: mQACC recommendations for analytical quality management. Metabolomics 18(9):70. https://doi.org/10.1007/s11306-022-01926-3

Klug DM, Idiris FIM, Blaskovich MAT, von Delft F, Dowson CG, Kirchhelle C et al (2021) There is no market for new antibiotics: this allows an open approach to research and development. Wellcome Open Res 6:146. https://doi.org/10.12688/wellcomeopenres.16847.1

Kotwani A, Wattal C, Joshi PC, Holloway K (2016) Knowledge and perceptions on antibiotic use and resistance among high school students and teachers in New Delhi, India: a qualitative study. Indian J Pharmacol 48(4):365–371. https://doi.org/10.4103/0253-7613.186208

Kowalska-Krochmal B, Dudek-Wicher R (2021) The minimum inhibitory concentration of antibiotics: methods, interpretation, clinical relevance. Pathogens 10(2). https://doi.org/10.3390/pathogens10020165

Kuila A, Sharma V (2018) Principles and applications of fermentation technology. John Wiley & Sons

Lei J, Sun L, Huang S, Zhu C, Li P, He J et al (2019) The antimicrobial peptides and their potential clinical applications. Am J Transl Res 11(7):3919–3931
Li J, Xie S, Ahmed S, Wang F, Gu Y, Zhang C et al (2017) Antimicrobial activity and resistance: influencing factors. Front Pharmacol 8:364. https://doi.org/10.3389/fphar.2017.00364
Lin DM, Koskella B, Lin HC (2017) Phage therapy: an alternative to antibiotics in the age of multi-drug resistance. World J Gastrointest Pharmacol Ther 8(3):162–173. https://doi.org/10.4292/wjgpt.v8.i3.162
Ling H, Lou X, Luo Q, He Z, Sun M, Sun J (2022) Recent advances in bacteriophage-based therapeutics: insight into the post-antibiotic era. Acta Pharm Sin B 12(12):4348–4364. https://doi.org/10.1016/j.apsb.2022.05.007
Liu GY, Yu D, Fan MM, Zhang X, Jin ZY, Tang C, Liu XF (2024) Antimicrobial resistance crisis: could artificial intelligence be the solution? Mil Med Res 11(1):7. https://doi.org/10.1186/s40779-024-00510-1
Maciejewska B, Olszak T, Drulis-Kawa Z (2018) Applications of bacteriophages versus phage enzymes to combat and cure bacterial infections: an ambitious and also a realistic application? Appl Microbiol Biotechnol 102(6):2563–2581. https://doi.org/10.1007/s00253-018-8811-1
Majumder MAA, Rahman S, Cohall D, Bharatha A, Singh K, Haque M, Gittens-St Hilaire M (2020) Antimicrobial stewardship: fighting antimicrobial resistance and protecting global public health. Infect Drug Resist 13:4713–4738. https://doi.org/10.2147/idr.S290835
Mancuso G, Midiri A, Gerace E, Biondo C (2021) Bacterial antibiotic resistance: the Most critical pathogens. Pathogens 10(10). https://doi.org/10.3390/pathogens10101310
Mathew P, Sivaraman S, Chandy S (2019) Communication strategies for improving public awareness on appropriate antibiotic use: bridging a vital gap for action on antibiotic resistance. J Family Med Prim Care 8(6):1867–1871. https://doi.org/10.4103/jfmpc.jfmpc_263_19
Matzaras R, Nikopoulou A, Protonotariou E, Christaki E (2022) Gut microbiota modulation and prevention of dysbiosis as an alternative approach to antimicrobial resistance: a narrative review. Yale J Biol Med 95(4):479–494
Mboowa G (2023) Reviewing the journey to the clinical application of bacteriophages to treat multi-drug-resistant bacteria. BMC Infect Dis 23(1):654. https://doi.org/10.1186/s12879-023-08621-1
Mekonnen AB, Mohammed AS, Tefera AK (2022) Ethnobotanical study of traditional medicinal plants used to treat human and animal diseases in Sedie Muja District, South Gondar, Ethiopia. Evid Based Complement Alternat Med 2022:7328613. https://doi.org/10.1155/2022/7328613
Melander RJ, Zurawski DV, Melander C (2018) Narrow-spectrum antibacterial agents. Medchemcomm 9(1):12–21. https://doi.org/10.1039/c7md00528h
Metlay JP, Powers JH, Dudley MN, Christiansen K, Finch RG (2006) Antimicrobial drug resistance, regulation, and research. Emerg Infect Dis 12(2):183–190. https://doi.org/10.3201/eid1202.050078
Miethke M, Pieroni M, Weber T, Brönstrup M, Hammann P, Halby L et al (2021) Towards the sustainable discovery and development of new antibiotics. Nat Rev Chem 5(10):726–749. https://doi.org/10.1038/s41570-021-00313-1
Miller BJ, Carson KA, Keller S (2020) Educating patients on unnecessary antibiotics: personalizing potential harm aids patient understanding. J Am Board Fam Med 33(6):969–977. https://doi.org/10.3122/jabfm.2020.06.200210
Mimee M, Citorik RJ, Lu TK (2016) Microbiome therapeutics—advances and challenges. Adv Drug Deliv Rev 105(Pt A):44–54. https://doi.org/10.1016/j.addr.2016.04.032
Morel CM, de Kraker MEA, Harbarth S (2021) Surveillance of resistance to new antibiotics in an era of limited treatment options. Front Med (Lausanne) 8:652638. https://doi.org/10.3389/fmed.2021.652638
Munita JM, Arias CA (2016) Mechanisms of antibiotic resistance. Microbiol Spectr 4(2). https://doi.org/10.1128/microbiolspec.VMBF-0016-2015
Murray E, Draper LA, Ross RP, Hill C (2021) The advantages and challenges of using endolysins in a clinical setting. Viruses 13(4). https://doi.org/10.3390/v13040680

Murugaiyan J, Kumar PA, Rao GS, Iskandar K, Hawser S, Hays JP et al (2022) Progress in alternative strategies to combat antimicrobial resistance: focus on antibiotics. Antibiotics (Basel) 11(2). https://doi.org/10.3390/antibiotics11020200

Muteeb G, Rehman MT, Shahwan M, Aatif M (2023) Origin of antibiotics and antibiotic resistance, and their impacts on drug development: a narrative Review. Pharmaceuticals (Basel) 16(11). https://doi.org/10.3390/ph16111615

Naidoo P, Bouharati C, Rambiritch V, Jose N, Karamchand S, Chilton R, Leisegang R (2021) Real-world evidence and product development: opportunities, challenges and risk mitigation. Wien Klin Wochenschr 133(15–16):840–846. https://doi.org/10.1007/s00508-021-01851-w

Pirofski LA, Casadevall A (2006) Immunomodulators as an antimicrobial tool. Curr Opin Microbiol 9(5):489–495. https://doi.org/10.1016/j.mib.2006.08.004

Qin S, Xiao W, Zhou C, Pu Q, Deng X, Lan L et al (2022) Pseudomonas aeruginosa: pathogenesis, virulence factors, antibiotic resistance, interaction with host, technology advances and emerging therapeutics. Signal Transduct Target Ther 7(1):199. https://doi.org/10.1038/s41392-022-01056-1

Raj N, Fernandes S, Charyulu NR, Dubey A, Ravi GS, Hebbar S (2019) Postmarket surveillance: a review on key aspects and measures on the effective functioning in the context of the United Kingdom and Canada. Ther Adv Drug Saf 10:2042098619865413. https://doi.org/10.1177/2042098619865413

Renwick MJ, Brogan DM, Mossialos E (2016) A systematic review and critical assessment of incentive strategies for discovery and development of novel antibiotics. J Antibiot (Tokyo) 69(2):73–88. https://doi.org/10.1038/ja.2015.98

Rex JH, Fernandez Lynch H, Cohen IG, Darrow JJ, Outterson K (2019) Designing development programs for non-traditional antibacterial agents. Nat Commun 10(1):3416. https://doi.org/10.1038/s41467-019-11303-9

Rizk ML, Bhavnani SM, Drusano G, Dane A, Eakin AE, Guina T et al (2019) Considerations for dose selection and clinical pharmacokinetics/pharmacodynamics for the development of antibacterial agents. Antimicrob Agents Chemother 63(5). https://doi.org/10.1128/aac.02309-18

Rodríguez-Gascón A, Solinís M, Isla A (2021) The role of PK/PD analysis in the development and evaluation of antimicrobials. Pharmaceutics 13(6). https://doi.org/10.3390/pharmaceutics13060833

Salam MA, Al-Amin MY, Salam MT, Pawar JS, Akhter N, Rabaan AA, Alqumber MAA (2023) Antimicrobial resistance: a growing serious threat for global public health. Healthcare (Basel) 11(13). https://doi.org/10.3390/healthcare11131946

Saleem Z, Saeed H, Akbar Z, Saeed A, Khalid S, Farrukh L et al (2021) WHO key access antibiotics price, availability and affordability in private sector pharmacies in Pakistan. Cost Eff Resour Alloc 19(1):10. https://doi.org/10.1186/s12962-021-00263-x

Savic M, Årdal CJJ (2018) A grant framework as a push incentive to stimulate research and development of new antibiotics. J Law Med Ethics 46(S1):9–24

Saylor C, Dadachova E, Casadevall A (2009) Monoclonal antibody-based therapies for microbial diseases. Vaccine 27(Suppl 6):G38–G46. https://doi.org/10.1016/j.vaccine.2009.09.105

Schmelcher M, Donovan DM, Loessner MJ (2012) Bacteriophage endolysins as novel antimicrobials. Future Microbiol 7(10):1147–1171. https://doi.org/10.2217/fmb.12.97

Seekatz AM, Safdar N, Khanna S (2022) The role of the gut microbiome in colonization resistance and recurrent Clostridioides difficile infection. Ther Adv Gastroenterol 15:17562848221134396. https://doi.org/10.1177/17562848221134396

Sharma B, Shukla S, Rattan R, Fatima M, Goel M, Bhat M et al (2022) Antimicrobial agents based on metal complexes: present situation and future prospects. Int J Biomater 2022:6819080. https://doi.org/10.1155/2022/6819080

Shim H (2023) Three innovations of next-generation antibiotics: evolvability, specificity, and non-immunogenicity. Antibiotics (Basel) 12(2). https://doi.org/10.3390/antibiotics12020204

Simoens S, De Groote K, Boersma C (2022) Critical reflections on reimbursement and access of advanced therapies. Front Pharmacol 13:771966. https://doi.org/10.3389/fphar.2022.771966

Simpkin VL, Renwick MJ, Kelly R, Mossialos E (2017) Incentivising innovation in antibiotic drug discovery and development: progress, challenges and next steps. J Antibiot (Tokyo) 70(12):1087–1096. https://doi.org/10.1038/ja.2017.124

Singh V, Haque S, Khare S, Tiwari AK, Katiyar D, Banerjee B et al (2018) Isolation and purification of antibacterial compound from Streptomyces levis collected from soil sample of north India. PLoS One 13(7):e0200500. https://doi.org/10.1371/journal.pone.0200500

Singh C, Mehata AK, Viswanadh MK, Tiwari P, Saini R, Singh SK et al (2024) Chitosan film of thiolated TPGS-modified Au-Ag nanoparticles for combating multidrug-resistant bacteria. Colloids Surf A: Physicochem Eng 686:133287

Skłodowski K, Chmielewska-Deptuła SJ, Piktel E, Wolak P, Wollny T, Bucki R (2023) Metallic nanosystems in the development of antimicrobial strategies with high antimicrobial activity and high biocompatibility. Int J Mol Sci 24(3). https://doi.org/10.3390/ijms24032104

Solomkin JS, Gardovskis J, Lawrence K, Montravers P, Sway A, Evans D, Tsai L (2019) IGNITE4: results of a phase 3, randomized, multicenter, prospective trial of eravacycline vs meropenem in the treatment of complicated intraabdominal infections. Clin Infect Dis 69(6):921–929. https://doi.org/10.1093/cid/ciy1029

Struble EB, Rawson JMO, Stantchev T, Scott D, Shapiro MA (2023) Uses and challenges of antiviral polyclonal and monoclonal antibody therapies. Pharmaceutics 15(5). https://doi.org/10.3390/pharmaceutics15051538

Taati Moghadam M, Amirmozafari N, Shariati A, Hallajzadeh M, Mirkalantari S, Khoshbayan A, Masjedian Jazi F (2020) How phages overcome the challenges of drug resistant bacteria in clinical infections. Infect Drug Resist 13:45–61. https://doi.org/10.2147/idr.S234353

Theuretzbacher U, Outterson K, Engel A, Karlén A (2020) The global preclinical antibacterial pipeline. Nat Rev Microbiol 18(5):275–285. https://doi.org/10.1038/s41579-019-0288-0

Tse BN, Adalja AA, Houchens C, Larsen J, Inglesby TV, Hatchett R (2017) Challenges and opportunities of nontraditional approaches to treating bacterial infections. Clin Infect Dis 65(3):495–500. https://doi.org/10.1093/cid/cix320

Van Norman GA (2019) Limitations of animal studies for predicting toxicity in clinical trials: is it time to rethink our current approach? JACC Basic Transl Sci 4(7):845–854. https://doi.org/10.1016/j.jacbts.2019.10.008

Vaou N, Stavropoulou E, Voidarou C, Tsigalou C, Bezirtzoglou E (2021) Towards advances in medicinal plant antimicrobial activity: a review study on challenges and future perspectives. Microorganisms 9(10). https://doi.org/10.3390/microorganisms9102041

Wasan H, Singh D, Reeta KH, Gupta YK (2023) Landscape of push funding in antibiotic research: current status and way forward. Biology (Basel) 12(1). https://doi.org/10.3390/biology12010101

Wettstein DJ, Boes S (2020) The impact of reimbursement negotiations on cost and availability of new pharmaceuticals: evidence from an online experiment. Health Econ Rev 10(1):13. https://doi.org/10.1186/s13561-020-00267-y

Xuan J, Feng W, Wang J, Wang R, Zhang B, Bo L et al (2023) Antimicrobial peptides for combating drug-resistant bacterial infections. Drug Resist Updat 68:100954. https://doi.org/10.1016/j.drup.2023.100954

Yadav M, Chauhan NS (2022) Microbiome therapeutics: exploring the present scenario and challenges. Gastroenterol Rep (Oxf) 10:goab046. https://doi.org/10.1093/gastro/goab046

Yu LX, Amidon G, Khan MA, Hoag SW, Polli J, Raju GK, Woodcock J (2014) Understanding pharmaceutical quality by design. AAPS J 16(4):771–783. https://doi.org/10.1208/s12248-014-9598-3

Zhou Y, Zhang J, Chen Y, Wu J, Guo B, Wu X et al (2022) Combined PK/PD index may be a more appropriate PK/PD index for Cefoperazone/Sulbactam against Acinetobacter baumannii in patients with hospital-acquired pneumonia. Antibiotics (Basel) 11(5). https://doi.org/10.3390/antibiotics11050703

Part VIII

Future of Non-antibiotics Therapies

38 Harnessing Artificial Intelligence and Machine Learning in the Battle Against Antimicrobial-Resistant Infections

Arup Acharjee, Sauparnika Prasanna Kumar, Papia Acharjee, and Graham Ball

Abstract

The escalating threat of antimicrobial-resistant infections presents a significant challenge to global healthcare systems. This chapter delves into the role of Artificial Intelligence (AI) and Machine Learning (ML) as critical tools in addressing this crisis. It emphasizes AI's utility in discovering novel antimicrobial agents, particularly its capability to predict resistance patterns and unearth new therapeutic compounds. The chapter also examines the use of ML in forecasting antibiotic resistance, assessing the accuracy and dependability of ML models that utilize genomic data to predict emerging resistance trends. This predictive ability is crucial for proactively managing resistance challenges. A key emphasis is placed on AI's role in personalized medicine and antibiotic stewardship, exploring AI-driven strategies for devising personalized treatment plans and optimizing antibiotic usage. Furthermore, the chapter integrates a discussion on predicting the origin of antimicrobial resistance genes in various environments to understand and mitigate the global spread of antimicrobial resistance. Lastly, the chapter acknowledges existing limitations and outlines potential future advancements, underscoring the ongoing need for research and innovation in AI and ML to combat antibiotic-resistant infections effectively.

A. Acharjee (✉)
Department of Zoology, University of Allahabad, Prayagraj, India
e-mail: arup@allduniv.ac.in

S. P. Kumar
SIES College of Arts Science and Commerce, Mumbai, India

P. Acharjee
Department of Zoology, Banaras Hindu University, Varanasi, India

G. Ball
Medical Technology Research Centre, Anglia Ruskin University, Cambridge, UK

M. Gangwar, G. Nath (eds.), *Emerging Paradigms for Antibiotic-Resistant Infections: Beyond the Pill*, https://doi.org/10.1007/978-981-97-5272-0_38

Keywords
Antimicrobial resistance · Artificial Intelligence · Machine Learning · Antibiotic Stewardship

38.1 Introduction

The increasing prevalence of antimicrobial-resistant (AMR) infections poses a critical threat to global health, with projections suggesting up to 300 million deaths by 2050 due to resistant pathogens (O'Neill 2016). The development of antimicrobial resistance, a consequence of bacteria adapting to antibiotics, undermines standard treatments, leading to persistent infections and heightened healthcare challenges. This narrative of AMR is a complex tale of microbial adaptability and scientific response that began in the 1940s. Following the success of penicillin, researchers discovered that *Escherichia coli* could produce penicillinase, an enzyme capable of neutralizing the antibiotic's effects (Abraham and Chain 1940). This early sign of bacterial resistance was a precursor to a broader trend; by the 1960s, approximately 80% of *Staphylococcus aureus* strains had developed resistance to penicillin, marking a significant milestone in the evolution of multi-drug resistant (MDR) pathogens (Fig. 38.1) (Lobanovska and Pilla 2017).

While antibiotic misuse and overuse have shared the majority of the blame (Malik and Bhattacharyya 2019), AMR develops even in the most unlikely corners. Hospital settings contribute significantly to AMR, with pathogens like *Pseudomonas aeruginosa* developing resistance through efflux pumps, antibiotic-binding proteins, biofilm formation, and horizontal gene transfer via plasmid transfer (Pachori et al. 2019). Studies indicate that hospitalization increases the risk of drug-resistant *Klebsiella pneumoniae* infections by 14% (Agarwal and Larson 2018), while research in Greece found high colonization rates of *Acinetobacter baumannii* and *Staphylococcus* spp. among healthcare workers in ICUs and surgical units (Tselebonis et al. 2016). Further complicating the scenario is the phenomenon of heteroresistance, where an isolate of bacterium has subpopulations showing resistance to different antibiotics. First observed in Gram-negative bacterium *Haemophilus influenzae* in 1947, this has since been observed in gram-positive *Staphylococcus* spp. (El-Halfawy and Valvano 2015) as well as *Acinetobacter baumannii* (Li et al. 2006). AMR also affects infants. Over the years, there have been significant increases in the incidence of community-based Methicillin-resistant *Staphylococcus aureus* (CA-MRSA) and HA-MRSA (Hospital acquired-MRSA) with a recent dominance of the CA-MRSA (Medernach and Logan 2018). Additionally, prolonged antibiotic exposure and extended hospital stay increase children's risk of vancomycin-resistant *Enterococcus* infections (Adams et al. 2016). Recently, in 2017, the World Health Organization (WHO) and the Centre for Disease Control (CDC) published a list of pathogens that require new antibiotic discovery as they have become resistant to most. Among them, ESKAPE (*Enterococcus faecium*, *Staphylococcus aureus*, *Klebsiella pneumoniae*,

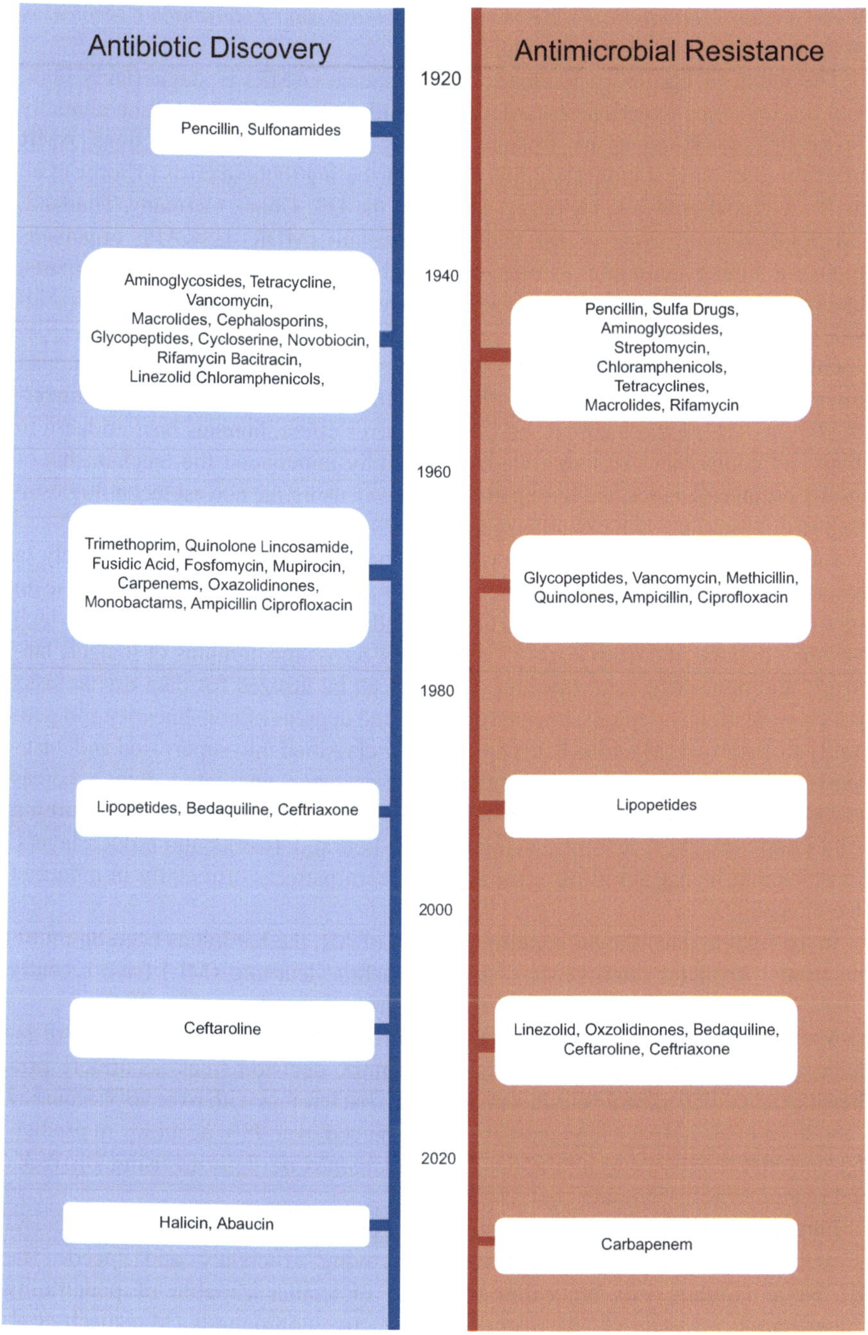

Fig. 38.1 Timeline depicting antibiotic discovery along with emerging antimicrobial resistance (AMR)

Acinetobacter baumannii, Pseudomonas aeruginosa, and Enterobacter species) is on the priority list (De Oliveira et al. 2020).

The financial burden of antimicrobial resistance (AMR) is staggeringly high, with the US alone experiencing a national healthcare cost of $2.2 billion annually for treating AMR infections. Despite a steady rate of bacterial infections, AMR infection cases have doubled, significantly contributing to these costs (Thorpe et al. 2018). A global analysis, including data from the US, China, Germany, Thailand, and South Korea, revealed that multidrug-resistant (MDR) ESKAPE organisms impose a higher economic burden compared to their susceptible counterparts, underscoring the substantial financial challenges posed by AMR infections worldwide (Zhen et al. 2019).

Given the above, it is imperative to investigate novel approaches to counteract this global-level problem. The solution to antimicrobial resistance lies in understanding its varied mechanisms, and, like a mirror effect, humans have to learn to adapt and counteract. Although we have not fully understood the mechanisms of antimicrobial resistance, we have come a long way using the newest technologies to use the information we have collected all these years.

Currently, Artificial Intelligence (AI) and machine learning are used widely in generating evidential-based hypotheses. Unsurprisingly, AI has found its use in fighting antimicrobial resistance. Artificial intelligence is an umbrella under which machine learning and deep learning are defined across the domains of images, language, and numerical data. Machine learning can be utilized for data-driven tasks like classification and regression analysis and lend degrees of non-linearity and generality to discovery. Machine learning is mainly classified into supervised and unsupervised learning. Supervised learning involves using annotated data, whereas unsupervised uses unannotated data that has not been labeled already. Deep learning comes under machine learning, using multiple neural networks and hidden layers. These hidden layers minimize error and can learn patterns efficiently in a dataset (Kufel et al. 2023).

In the fight against the burgeoning problem of AR, the battle has been on multiple fronts. Artificial intelligence (AI) and machine learning (ML) have recently offered promising avenues for tackling antimicrobial resistance (Fig. 38.2).

Machine learning has been effectively utilized to predict antimicrobial resistance, demonstrating high accuracy. For example, decision trees accurately predicted extended-spectrum beta-lactamase ESBL bacteremia with over 90% accuracy (Goodman et al. 2016), while another study achieved over 98% accuracy in predicting bacterial resistance to antibiotics in *Pseudomonas aeruginosa* using BioWeka and random forests (Noman et al. 2023).

Further, AI and ML have been instrumental in discovering new therapeutics. AI is revolutionizing antibiotic discovery, improving efficiency and speed. The ML-based Hunting Fox algorithm identified molecules capable of penetrating *Pseudomonas aeruginosa*, discovering five effective compounds (Mansbach et al. 2020). ML combined with high-throughput docking found seven YpkA inhibitors for *Yersinia* spp., completely inhibiting its growth (Hu et al. 2007). Moreover, deep learning identified halicin from drug repurposing data, showing strong action

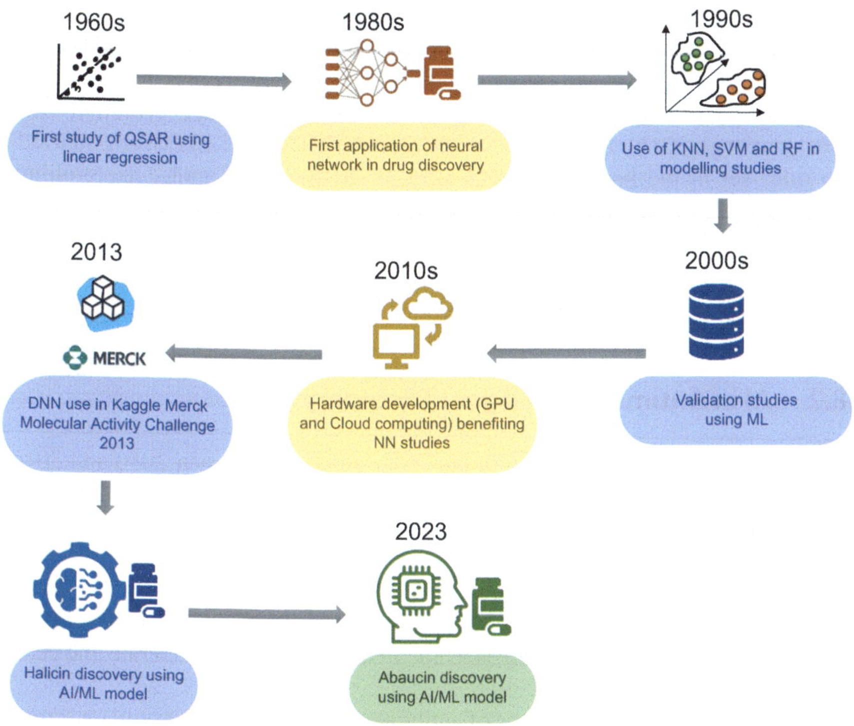

Fig. 38.2 The first study of Quantitative structure-activity relationships (QSAR) using a linear regression model was presented in the 1960s. Around the late 1980S, the first study of Quantitative structure-activity relationships (QSAR) using a linear regression model was presented in the rail network (NN) model. It was used by Tomoo Aoyama et al. to do QSAR analysis. The use of non-linear regression algorithms such as KNN, Support vector machine (SVM), and Random Forest (RF) algorithms in drug modeling started around the 1990s. Validation studies of potential antibiotic molecules were done using ML models around the 2000s (along with experimental validation). Hardware development (cloud computing and use of GPUs) in 2010 directly benefitted NN studies. After the Kaggle Merck activity conducted in 2013, which focused on predicting antibiotic molecules using data mining and machine learning models, Deep Neural network models became frequent in prediction studies. In 2020, MIT researchers recently found Halicin, a broad-spectrum antibiotic, using an ML algorithm. The latest discovery of abaucin reported in 2023, an antibiotic capable of treating superbug infection caused by *Acinetobacter baumannii*, marks the true potential of AI/ML algorithms in medicine

against resistant pathogens like *E. coli*, *Acinetobacter baumanii*, *M. tuberculosis*, and carbapenem-resistant Enterobacteriaceae (Stokes et al. 2020). Also, researchers have used ML to analyze antibiotics' effects with metabolite supplementation, aiding the understanding of bacterial responses and discovery pathways (Yang et al. 2019).

Not only have AI and ML shown promise in finding therapeutics, but they have also been instrumental in preventing the rise of AR. These measures include optimal selection, dosing, and treatment duration through the development of dose–response models. All the current advancements in antimicrobial resistance align with the idea of stewardship. Machine learning aids in clinical decision making, such as using Partially Observable Markov Decision Process (POMDP) models for predicting outcomes and antibiotic choices in sepsis, reducing resistance evolution risks (Tsoukalas et al. 2015). Also, studies have used the ML model on existing stewardship rules set in the hospital (Beaudoin et al. 2016). In the molecular profiling domain, ML has been instrumental in utilizing metagenomic datasets of different ecotypes to identify the source of AMR pollution (Li et al. 2018).

In the following sections, we shall look into the evolving role of AI and ML in understanding the different facets of the war against AMR.

38.2 AI in Antimicrobial Agent Discovery

The scarcity of new antibiotics, with only four approved between 2010 and 2014, underscores a critical challenge in combating antimicrobial resistance (Liu et al. 2024). Driven chiefly by economic factors such as delayed profitability, it has led to a significant decline in investment in novel antibiotic discovery since the early 1980s and has hindered the development of a new generation of antibiotics (Klug et al. 2021). This alarming trend highlights the urgent need for innovative solutions like Artificial Intelligence (AI) and Machine Learning (ML) to accelerate and economize antibiotic discovery.

In the mid-1950s, the golden age of antibiotic discovery relied heavily on screening soil microbes' secondary metabolites, a method so impactful that Nobel Prize laureate Selman Waksman defined antibiotics as compounds microbes produce to destroy other microbes (Hutchings et al. 2019). However, this approach has stalled due to dereplication issues, with researchers repeatedly discovering the same antibiotics. While high throughput screening (HTS) has identified novel antimicrobial molecules, it has not led to new antibiotics, likely due to its high costs and time demands, unlike AI/ML models, which bypass the need for in-vitro sample preparation (Table 38.1) (Liu and Stokes 2022).

AI/ML-based approach to antibiotic discovery can be delineated into four key steps. Initially, it involves gathering experimental data, which includes structural, dynamic, and physicochemical properties of known antimicrobial compounds. The second step transforms this data into a format suitable for AI, such as one-hot encoding. Following this, the processed data is utilized to train various AI/ML models, including neural networks, support vector machines, and random forests, which are among the most commonly employed. The final step is diversified into several analyses, notably the discovery of novel molecules, where the training data helps identify new putative compounds from the raw data (Melo et al. 2021).

A similar workflow was applied by Ma et al. where they discovered 11 highly effective antimicrobial peptides against antibiotic-resistant Gram-negative bacteria. They leveraged neural network models, specifically Long ShortTerm Memory (LSTM), attention mechanisms, and Bidirectional Encoder Representations from Transformers (BERT), for mining functional peptides. The efficacy of these

Table 38.1 The comparative landscape between traditional and AI-driven antibiotic discovery

Aspect	Traditional antibiotic discovery	AI-driven antibiotic discovery
Methodologies	Involves the screening of environmental samples or chemical libraries against pathogens, followed by optimization and testing.	Uses machine learning to predict antibiotic properties from large datasets, employing models for virtual screening and de novo design.
Timeframes	Typically spans 10–15 years from discovery to market approval.	Significantly reduces the discovery phase to months, though clinical validation still requires years.
Success Rates	Low success rate (<1%) from discovery to market due to high attrition.	Promises higher initial success rates through informed selection, pending real-world validation.
Costs	Estimated at $1 billion to $2.5 billion, influenced by long development times and high attrition rates.	Aims to reduce costs by streamlining the discovery process, with initial investments in computational resources required.
Scalability	Limited by extensive laboratory requirements.	Enhanced by the ability to process large datasets and explore vast chemical spaces digitally.
Adaptability to Emerging Resistance	Slow to adapt due to the reliance on discovering new compounds or classes.	Quickly adjusts to emerging resistance patterns through model updates, offering timely responses.
Reliance on Existing Knowledge	Dependent on known chemical libraries and natural product sources.	Leverages existing data and generative models to explore beyond current knowledge.
Collaborative Potential	Collaboration mainly within academic and industry for sharing resources.	Encourages open science and global collaboration across disciplines to accelerate discovery.

peptides was further confirmed in a mouse lung infection model (Ma et al. 2022). Wang et al. also used an LSTM model to design and classify short antimicrobial peptides with antibacterial activity against *Escherichia coli*, achieving about 88.9% validation accuracy (Wang et al. 2021). An intriguing study by Tucs et al. introduced a deep generative model, specifically a Generative Adversarial Network (GAN), for designing novel antimicrobial peptides. The study spotlighted PepGAN, a protein-specialized GAN model noted for its high statistical fidelity. Among the top six peptides synthesized, one exhibited a minimum inhibitory concentration (MIC) of 3.1 μg/mL, demonstrating twice the potency of ampicillin (Tucs et al. 2020).

In 2018, researchers leveraged genome mining coupled with deep learning models to bypass traditional culturing methods, facilitating the discovery of Malacidin. This novel metabolite exhibits potent antibacterial activity against Gram-positive bacteria, showcasing the potential of AI in uncovering previously unexplored antimicrobial compounds (Hover et al. 2018). Thereafter, in 2020, a team at MIT using deep learning identified the powerful antibiotic potential compound, Halicin, named after the fictional AI system HAL from the Hollywood blockbuster "2001: A Space Odyssey" in a mere 3 days' time. Initially recognized as an anti-diabetic drug, halicin was repurposed for its broad-spectrum antibacterial efficacy like that of *Clostridium difficile*, *Mycobacterium tuberculosis*, including activity against resistant strains of *Acinetobacter baumannii*. This discovery was facilitated by a machine learning algorithm analyzing the ZINC15 database, which contains nearly 1.5 billion compounds, illustrating the profound impact of AI on drug repurposing and antimicrobial discovery (Stokes et al. 2020).

One study combined an ML algorithm and biophysical-based minimal region determination to discover potent bacteriocin candidates. Support vector machine was found to be the best method as 676 putative bacteriocins were found using the tool. Out of these, 16 sequences were further tested for antimicrobial activity against *Escherichia coli* and *Pseudomonas aeruginosa* (Fields et al. 2020).

A novel family of medicines effective against MRSA and VRE infections was discovered in 2023 by a team led by James Collins at the Broad Institute of MIT and Harvard. This discovery was the result of an innovative effort. The group found and validated 283 compounds with antibacterial qualities by using a graph neural network model to predict antibiotic activity and cytotoxicity from over 12 million molecules. This study highlights the precision and scalability of AI in drug discovery and its critical role in addressing the challenge of antimicrobial resistance (Wong et al. 2024).

In 2024, Tao Shen et al. introduced AutoMolDesigner, adding even more functionality to the toolkit for AI-driven antibiotic discovery. The Qt-based graphical user interface of this open-source computational modeling program makes the process of designing small molecule antibiotics more accessible. AutoMolDesigner is an example of the growing field of AI applications in the hunt for new antimicrobial medicines, as it provides researchers worldwide with access to advanced computational tools (Shen et al. 2024).

Together, these achievements highlight the revolutionary influence of AI and machine learning technologies on the discovery of antibiotics, bringing in a new era of quick, effective, and creative ways to tackle the growing antimicrobial resistance issue.

However, despite such successes, the journey from discovery to market for AI-discovered antibiotics faces significant hurdles. A primary challenge is the dearth of investment in advancing AI-driven projects to the clinical trial phase. Currently, only a few major pharmaceutical companies are actively engaged in antimicrobial resistance research, with the majority of funding directed towards broad-spectrum antibiotics. This focus overlooks narrow-spectrum drugs, which, despite their potential, are often sidelined due to longer timelines for profitability. An illustrative case is Abaucin, a compound with potent antibacterial activity against *A. baumannii*, which has yet to progress to clinical trials (G. Liu et al. 2023). This phenomenon, known as the "valley of death," highlights the stark reality that many drugs do not proceed beyond preclinical success (Renwick et al. 2016).

However, given the escalating threat of antimicrobial resistance, there remains a pressing global imperative to persist in the quest for new antimicrobial agents. The innovative use of AI and ML in this domain not only offers a pathway to expedite the discovery of novel antibiotics but also emphasizes the necessity for sustained investment and commitment to overcoming the challenges inherent in bringing these discoveries to the market.

38.3 Machine Learning for Predicting Antimicrobial Resistance

The rising threat of antimicrobial resistance (AMR) demands innovative strategies to address its impact on global health pre-emptively. Artificial Intelligence (AI) and Machine Learning (ML) represent a paradigm shift in tackling the complexities of AMR, offering a more efficient and accurate approach than traditional detection methods. Such methods excel in dissecting extensive datasets encompassing microbial genetics, pharmacological interactions, and clinical data, enhancing our capacity to detect, predict, and decipher the underlying mechanisms of AMR (Masud et al. 2023).

The flexibility to engineer and amalgamate numerous ML models into an ensemble on a unified platform has revolutionized the potential for achieving unparalleled sensitivity and precision in AMR prediction. This shift towards integrating AI and ML in microbial resistance research accelerates the identification of resistance patterns and propels the development of novel intervention strategies, marking a significant advancement in the ongoing battle against AMR.

AI/ML integration provides a new perspective that helps to understand and manage the complexity of AMR. To capture the complex nature of resistance, these techniques use a wide range of raw data, including whole genome sequences, electronic health records (EHRs), results from wet laboratory studies, and raw imaging (Khaledi et al. 2020). A critical component of this methodology is the meticulous process of feature extraction and data cleansing, where predictive markers such as resistance genes, mutations, and novel genetic markers are identified and refined for analysis (Wiens et al. 2019). The development and training of AI/ML models are then based on a parsimonious set of these features. ML models can be trained in three different ways: supervised learning (using labeled data sets to identify specific features), unsupervised learning (letting the model form correlations on its own), and reinforcement learning (using an "agent" to optimize decisions based on interactions with the environment) (Alpaydın 2004). These models are put through a rigorous evaluation process using measures such as accuracy, sensitivity, specificity, and area under the receiver operating characteristic curve (AUC-ROC). To guarantee generalizability and minimize overfitting, cross-validation, and independent test sets are used. The quality and variety of the training data, which must include a broad range of bacterial species, drugs, and resistance mechanisms, are necessary for ML models to be reliable in predicting AMR (Liu et al. 2021). Interestingly, certain machine learning models have achieved an overall accuracy of 89.74%, especially when it comes to detecting AMR in ESKAPE infections. This emphasizes the significant influence and possibility of machine learning in transforming our strategy for addressing AMR (Yu et al. 2023).

As delineated in the preceding sections, machine and deep learning paradigms leverage a myriad of raw data types. Contemporary utilization of OMICS data, encompassing metagenomics, genomics, and transcriptomics, has become increasingly prevalent in the predictive modeling of antimicrobial resistance (AMR) by such algorithms. Arango-Argoty et al. introduced a pioneering deep learning model, specifically DeepAMRG-SS and DeepAMRG-LS, for the precise prediction of antibiotic-resistant genes (AMRG) with a high precision of 0.97. This model catered to both short-read sequences and full gene-length sequences, marking a notable advancement in the field (Arango-Argoty et al. 2018). Another study illustrated the capacity to forecast resistance to 11 compounds spanning four antibiotic classes through Whole Genome Sequencing (WGS) data, employing a gradient-boosted decision tree technique with an accuracy of 0.91. (Moradigaravand et al. 2018).

Other techniques for AMR prediction from whole-genome sequencing (WGS) data utilized genome distance-aware cross-validation combined with stacked generalization for more robust and accurate predictions. They emphasized the value of interpretable models like the Set Covering Machine (SCM) in contexts where understanding the model's decision-making process, such as biomarker discovery, is crucial (Lüftinger et al. 2021).

Further studies have described the employment of partial genome sequence data in differentiating susceptible and resistant bacterial phenotypes. Although these small core gene models have higher error rates, they do have the ability to predict AMR as the error rates are stable (Nguyen et al. 2020).

One group sequenced genomes and transcriptomes of 414 drug-resistant *P. aeruginosa* isolates. ML classification model predicted AMR by the presence or absence of particular genes, expression profiles, and sequence variation chosen as features. This ML model had a high sensitivity of almost 0.90, proving its potential use in routine lab susceptibility tests (Khaledi et al. 2020). Another recent study used ML, specifically RF, and BioWeka, to predict resistance in *Pseudomonas aeruginosa* strains to an array of 12 antibiotic classes. The models' accuracy exceeded 98% for BioWeka and 96% for RF, thus reinforcing the growing body of evidence supporting the use of ML in the precise prediction of AMR, thereby continuing the evolution of computational approaches in the fight against antimicrobial resistance (Noman et al. 2023).

An innovative approach to AMR prediction came through the use of deep transfer learning, which is distinct from traditional ML methods, including deep learning. This approach is especially effective in scenarios involving limited datasets, a common challenge when dealing with novel antibiotics. Utilizing two datasets of *Escherichia coli* (*E. coli*) that included WGS and resistance information for four antibiotics—ciprofloxacin, cefotaxime, ceftazidime, and gentamicin—the study demonstrated the model's efficacy, even when one dataset was incomplete, highlighting the versatility and potential of transfer learning in AMR prediction (Ren et al., 2022b). Complementing these findings, the same group conducted a comprehensive evaluation of various ML models, including logistic regression (LR), support vector machine (SVM), RF, and convolutional neural network (CNN), for predicting AMR against antibiotics such as ciprofloxacin cefotaxime, ceftazidime, and gentamicin. This evaluation found that RFs and CNNs outperformed LR and SVM, with Area under the Curve (AUC) values reaching up to 0.96, using WGS data. This study not only corroborated the efficacy of ML models in AMR prediction but also highlighted the superior performance of RFs and CNNs in leveraging genomic data for AMR prediction (Ren et al., 2022a).

In another approach to enhance ML models for predicting AMR, researchers ingeniously incorporated genome-wide mutation scoring, focusing on the functional impact of mutations across the proteomes of organisms such as *Pseudomonas aeruginosa, Citrobacter freundii*, and *Escherichia coli*. By utilizing the PROVEAN tool to predict mutation impacts as new features, they significantly improved the models' predictive accuracy. This methodological advancement was validated across a substantial dataset of 19,521 isolates, marking a pivotal enhancement in the accuracy and biological relevance of ML-based AMR prediction (Májek et al. 2021).

A comprehensive analysis of various parameters influencing ML predictions of AMR revealed that model performance fluctuated across different drugs, datasets, resistance metrics, and species, underscoring the complexity of AMR prediction and the challenges in achieving high accuracy across various conditions (Hicks et al. 2019).

Further, a recent study compared three machine learning architectures previously utilized for the WGS-AST (antimicrobial susceptibility testing) challenge against a comprehensive dataset of 8704 genome assemblies of five pathogens of clinical significance across 77 species-compound combinations (Lüftinger et al. 2021). They demonstrated that individual models could be effectively combined through ensemble techniques to enhance overall model performance. By applying stacked generalization alongside cross-validation, a strategy well-suited for small datasets, the average sensitivity and specificity of the models were increased by 1.77% and 3.20%, respectively. Moreover, these stacked models showed greater robustness, reducing the likelihood of outlier performance drops compared to their individual counterparts. This investigation underscored best practice methods for predicting antimicrobial resistance from WGS data, showcasing the integration of genome distance-aware cross-validation with stacked generalization as a means to achieve robust and precise WGS-AST predictions.

In another study, ML models were assessed for their capacity to predict AMR using WGS data from *Actinobacillus pleuropneumoniae*. They employed Support Vector Machine (SVM) and Set Covering Machine (SCM) algorithms to determine the resistance against a panel of five antibiotics: Tetracycline, Ampicillin, Sulfisoxazole, Trimethoprim, and Enrofloxacin. The SVM approach relied on identifying common k-mers shared between the bacterial isolates and reference genes, whereas the SCM strategy adopted a more aggressive tactic to assemble conjunctions or disjunctions of Boolean functions, pinpointing the most critical k-mers for precise predictions. By correlating the observed phenotypes with the predictions made for these drugs, it was discovered that both the SCM and SVM models demonstrated diagnostic potential, achieving testing accuracies exceeding 90%. This finding underscores the viability of SCM and SVM models as effective diagnostic tools in the identification of AMR (Liu et al. 2020).

Not only omics data researchers have also used ML in developing a decision algorithm based on electronic health record (EHR) data of a cohort of 3629 patients for predicting antibiotic susceptibility in urinary tract infection (UTI) patients. The implementation of this algorithm led to a significant 67% reduction in the use of second-line antibiotics, showcasing the role of ML in enhancing antimicrobial stewardship (Kanjilal et al. 2020).

Collectively, these studies highlight the advances and challenges in the application of ML for predicting AMR, offering insights into improving accuracy, interpretability, and clinical utility in the battle against antimicrobial resistance.

The field of AI and ML is relatively new and is constantly evolving. Its use in the medical field has helped tremendously with clinical decision-making and rapid diagnosis. Although its use in predicting and understanding AMR is growing, one cannot ignore its shortcomings. A systematic review evaluated the use of machine learning (ML) in analyzing MALDI-TOF mass spectra for microbial identification and antimicrobial resistance testing. They found that while ML techniques, such as Support Vector Machines and Artificial Neural Networks, are increasingly used, only a small fraction of 11.11% of studies validated their algorithms on external datasets, indicating a need for improved robustness and applicability in clinical settings (Weis et al. 2020).

One example is the challenges in using EHR data. EHR provides crucial information to clinicians used for decision-making. This dataset is highly prone to erroneous data due to simple entry errors. EHR datasets are often very specific to a particular place, which is a barrier to the generalizability of such models (Anahtar et al. 2024). Another challenge is the sheer complexity of AMR, as it depends on many factors. AMR pollution is a global threat, and understanding source–sink relations poses a big challenge. In another study, broad spectrum Antimicrobial resistant genes (AMRG) profiling was combined with the machine learning classification model Source Tracker and used to predict AMR pollution in influent and effluent wastewater sources from broad geographical regions (China, USA, Europe, Peru, etc.) (Li et al. 2018). Studies like these are paving the way for better AMR management.

Along with challenges come efforts from people all around the world trying to solve and disentangle the issues. One can expect the growing use of AI and ML in the fight against AMR as its evolution holds the potential to transform antibiotic stewardship and patient care, paving the way for a more targeted and effective approach to infectious disease treatment.

38.4 AI in Personalized Medicine and Antibiotic Stewardship

In the current scenario, Antibiotic Stewardship Programmes (ASP) have become crucial tactics in the fight against AMR, maximizing the utilization of antibiotics, and enhancing the quality of life for patients. These programmes will be more successful because to the incorporation of Artificial Intelligence (AI), which will offer data-driven insights, predictive analytics, and decision support tools. In order to solve the issues of AMR through a variety of cutting-edge ways, AI-driven solutions are being developed and used in ASP. Researchers have created a sophisticated automated analytic method to make the examination of antibiotic prescriptions easier by utilizing artificial intelligence (AI) algorithms. The system integrated a centralized prescription database with local recommendations for automatic analysis, thereby optimizing ASP technologies. It improves the responsible administration of antibiotics in resource-constrained situations by providing prescribers with feedback dashboards and real-time ASP analysis visualization (Gutierrez-Tobar et al. 2023).

Integration of AI has led to the generation of new systems, such as the Wise Antimicrobial Stewardship Program Support System (WASPSS). It is a Clinical Decision Support System (CDSS) designed to enhance antimicrobial stewardship activities in healthcare settings. It integrates into the Electronic Health Record (EHR) system, facilitating a multidisciplinary approach by incorporating knowledge from various experts and providing a customized view of information. WASPSS is characterized by its continuous, modular, shareable, adaptive, accurate, communicative, documental, and educational capabilities. These features support the ASP team by promoting bidirectional communication, enabling knowledge sharing across different hospitals, and providing educational resources for hospital staff (Palacios et al. 2016).

A second-generation AI system based on chronobiology to combat antimicrobial resistance in chronic infections is also under development. This approach utilizes algorithm-controlled regimens that adapt dosing and administration times to the body's circadian rhythms, aiming to enhance the long-term effectiveness of antimicrobials. By personalizing treatment schedules, the system seeks to prevent resistance development and improve patient outcomes. Ongoing clinical trials are expected to offer insights into resistance mechanisms and novel strategies for overcoming them (Kolben et al. 2023).

Envisioning the future of ASPs, AI could offer personalized clinical decision support tools to optimize antibiotic prescribing. Such tools would provide individualized, real-time recommendations to healthcare providers on appropriate, narrower spectrum antibiotic options, addressing the urgent need for targeted antibiotic stewardship interventions (Chang and Chen 2022).

However, implementing AI/ML-based technologies in ASPs presents several challenges, including ensuring data privacy and security, integrating AI tools into existing healthcare infrastructures, and addressing the need for interdisciplinary collaboration among healthcare professionals. However, the opportunities for improving antibiotic use and combating AMR through AI-driven ASPs are significant. By leveraging AI's predictive analytics, healthcare providers can make more informed decisions regarding antibiotic prescribing, ultimately leading to reduced antimicrobial resistance and better patient outcomes.

Nonetheless, AI represents a transformative approach to enhancing Antibiotic Stewardship Programs, offering innovative solutions to optimize antibiotic use and combat antimicrobial resistance effectively. By integrating AI-driven tools and strategies, ASPs can achieve their goals more efficiently, ensuring the judicious use of antibiotics and safeguarding their effectiveness for future generations. Continued research and development in AI technologies, along with collaborative efforts among healthcare professionals, are essential to realize the full potential of AI in antibiotic stewardship.

38.5 Challenges and Future Directions

38.5.1 Limitations of Current AI and Machine Learning Approaches in This Field

The limitations of current AI and ML approaches in addressing antimicrobial resistance stem from several key factors. First, data quality and availability are significant constraints. This runs the risk of finding only iterations of existing solutions. More comprehensive, standardized, and high-quality high-resolution datasets are necessary for training robust AI models. The effectiveness of AI in identifying and predicting antimicrobial resistance patterns relies heavily on diverse, accurate, and expansive datasets, which are often lacking due to inconsistent data collection and sharing practices across healthcare institutions and countries (Laws et al. 2019). The scarcity of such data, especially from low-resource settings, hinders model accuracy and applicability.

Second, the complexity of resistance mechanisms presents a challenge. AMR involves intricate biological processes and interactions that are difficult to capture and predict with current AI models. The evolutionary dynamics of bacterial resistance are influenced by a multitude of genetic, environmental, and social factors, making the prediction and management of resistance highly complex. Current AI models may not fully capture the intricate interplay of factors driving the emergence and spread of antimicrobial resistance, limiting their predictive accuracy and applicability (MacLean and San Millan 2019).

Lastly, interpretability, explainability, and trust in AI decisions are crucial for clinical adoption. Many AI models, particularly deep learning algorithms, act as "black boxes," offering little insight into how they derive their predictions. This lack of transparency can lead to scepticism and resistance from healthcare professionals, who must understand and trust AI recommendations to incorporate them into clinical practice (Popa et al. 2022).

38.5.2 Potential Future Developments and Research Directions

The future of AI and machine learning in combating antimicrobial resistance (AMR) looks promising, with several potential developments and research directions on the horizon. Enhanced data collection and sharing initiatives are crucial to overcoming current limitations and providing AI models with access to high-quality, diverse datasets. Efforts to improve the quality, standardization, and accessibility of antimicrobial resistance data through global collaborations and data-sharing platforms are essential. This includes the development of international databases that integrate genomic, phenotypic, and clinical data to facilitate the training of more accurate and generalizable AI models (Avershina et al. 2021).

The integration of multi-omics data promises a more holistic view of microbial life and resistance mechanisms, enriching the training data for AI models. Future research could focus on integrating multi-omics data (genomics, transcriptomics, proteomics, metabolomics) into AI models to capture the multifaceted nature of antimicrobial resistance. This approach could enhance the predictive power of AI models and provide deeper insights into the mechanisms of resistance and potential targets for novel antimicrobials (Zhang et al. 2021).

Developing interpretable AI models, such as Explainable AI (XAI), aims to build trust among healthcare professionals by making AI decisions transparent and understandable. This should not be confused with explainability in the context of existing hypotheses. Just because a solution is not explainable in the context of our current understanding doesn't mean it should be rejected as our current knowledge is incomplete. A growing emphasis is on creating more interpretable and explainable AI models that can provide understandable and actionable insights into their predictions. This would increase trust among healthcare professionals and facilitate the identification of novel resistance mechanisms and therapeutic targets (Cavallaro et al. 2023; Wong et al. 2024). Lastly, AI-driven antibiotic discovery is an exciting avenue, leveraging AI's pattern recognition capabilities to identify novel compounds and targets faster and more efficiently than traditional methods, potentially revolutionizing the field of antimicrobial research and providing new solutions to combat AMR. Leveraging AI to accelerate the discovery of new antibiotics and antimicrobial resistance breakers represents a promising research direction. AI can streamline the drug discovery process by predicting the antimicrobial activity of compounds, optimizing drug design, and identifying novel targets for antimicrobial action (Melo et al. 2021; Wong et al. 2024). Moreover, ASPs could leverage AI/ML to develop personalized antimicrobial therapy plans based on individual patient data, including genomic data and resistance profiles, which could significantly enhance treatment efficacy and reduce the emergence of resistance. Personalized therapy, guided by AI, could ensure the use of the most effective antibiotics while minimizing unnecessary exposure to antimicrobials (Tu et al. 2021).

The challenges faced by current AI and ML approaches underscore the need for interdisciplinary collaboration, improved data practices, and innovative research to harness the full potential of AI in combating antimicrobial resistance. Integrating advanced computational techniques with biological insights will be crucial in developing effective strategies to address this global health crisis as we move forward.

38.6 Conclusion

Artificial intelligence and machine learning are a disruptive leap toward the continuous fight with antibiotic-resistant infections. They have begun to mould approaches to infectious disease management and antimicrobial stewardship, for instance, in the invention of new creative ways of diagnosis and treatment optimization, or the revelation of original therapeutic agents. The area where AI is supportive of machine learning has an effect on several success elements in combating antibiotic-resistant infections. For the first time, AI-enabled analytics have greater predictive power on resistant pathogens and, therefore, can support intervening strategies much more effectively than ever before. Bacterial genomes predict antibiotic-resistant genes in them, and at the same time, they monitor phenotypic resistance patterns simultaneously, with lots of promise from machine learning algorithms as tools in finding possible potential mechanisms of resistance before they become full-blown clinical problems. In addition, they have developed rapid diagnostic tools that have equally identified infections that have not responded timely to give treatments.

And in such findings and development of the drug, so is the role of AI. That has the potential to speed up the process of exploration and therefore pave the way for new compounds in antimicrobials or improve on present treatments. These allow the practical designing of new forms of antibiotic molecules for effective action and hence predict its clinical effects even more efficiently and faster than by conventional methods.

In this regard, investment in further research and development in AI and machine learning cannot be overemphasized. Just like the nature of antimicrobial resistance itself, so must be our tools against it—ever-changing. New breakthroughs into AI research give hope not only for a deeper understanding of these mechanisms of resistance but also for breaking new ground in next-generation antibiotics and treatment strategies. Last but not least, interdisciplinary collaborations among computer scientists, microbiologists, and clinicians may be helpful to translate AI-driven innovations into practice in a useful way. In this line, a review of the available evidence indicates that AI and machine learning would be critical allies to fight against antimicrobial resistance. Consequently, its broader development and integration into healthcare systems could drive new hopes of reversing the trend of AMR. But to fully actualize this space with AI, sustained research to break silos and foster collaborations across faculties is very crucial, and push for innovation on a global stage needs to be created. The battle against antimicrobial resistance is surely not over, but we might be even better armed now with advanced artificial intelligence and machine learning tools to march ahead.

References

Abraham EP, Chain E (1940) An enzyme from bacteria able to destroy penicillin. Nature 146(3713):3713. https://doi.org/10.1038/146837a0

Adams DJ, Eberly MD, Goudie A, Nylund CM (2016) Rising vancomycin-resistant enterococcus infections in hospitalized children in the United States. Hosp Pediatr 6(7):404–411. https://doi.org/10.1542/hpeds.2015-0196

Agarwal M, Larson EL (2018) Risk of drug resistance in repeat gram-negative infections among patients with multiple hospitalizations. J Crit Care 43:260–264. https://doi.org/10.1016/j.jcrc.2017.09.033

Alpaydın E (2004) Introduction to machine learning. In: Machine learning, Vol 56
Anahtar MN, Yang JH, Kanjilal S (2024) Applications of machine learning to the problem of antimicrobial resistance: an emerging model for translational research. J Clin Microbiol 59(7):Article 7. https://doi.org/10.1128/JCM.01260-20
Arango-Argoty G, Garner E, Pruden A, Heath LS, Vikesland P, Zhang L (2018) DeepARG: a deep learning approach for predicting antibiotic resistance genes from metagenomic data. Microbiome 6(1):23. https://doi.org/10.1186/s40168-018-0401-z
Avershina E, Shapovalova V, Shipulin G (2021) Fighting antibiotic resistance in hospital-acquired infections: current state and emerging technologies in disease prevention, diagnostics and therapy. Front Microbiol 12. https://www.frontiersin.org/journals/microbiology/articles/10.3389/fmicb.2021.707330
Beaudoin M, Kabanza F, Nault V, Valiquette L (2016) Evaluation of a machine learning capability for a clinical decision support system to enhance antimicrobial stewardship programs. Artif Intell Med 68:29–36. https://doi.org/10.1016/j.artmed.2016.02.001
Cavallaro M, Moran E, Collyer B, McCarthy ND, Green C, Keeling MJ (2023) Informing antimicrobial stewardship with explainable AI. PLoS Digital Health 2(1):e0000162. https://doi.org/10.1371/journal.pdig.0000162
Chang A, Chen JH (2022) BSAC vanguard series: artificial intelligence and antibiotic stewardship. J Antimicrob Chemother 77(5):1216–1217. https://doi.org/10.1093/jac/dkac096
De Oliveira DMP, Forde BM, Kidd TJ, Harris PNA, Schembri MA, Beatson SA, Paterson DL, Walker MJ (2020) Antimicrobial resistance in ESKAPE pathogens. Clin Microbiol Rev 33(3):e00181–e00119. https://doi.org/10.1128/CMR.00181-19
El-Halfawy OM, Valvano MA (2015) Antimicrobial heteroresistance: an emerging field in need of clarity. Clin Microbiol Rev 28(1):191–207. https://doi.org/10.1128/CMR.00058-14
Fields FR, Freed SD, Carothers KE, Hamid MN, Hammers DE, Ross JN, Kalwajtys VR, Gonzalez AJ, Hildreth AD, Friedberg I, Lee SW (2020) Novel antimicrobial peptide discovery using machine learning and biophysical selection of minimal bacteriocin domains. Drug Dev Res 81(1):43–51. https://doi.org/10.1002/ddr.21601
Goodman KE, Lessler J, Cosgrove SE, Harris AD, Lautenbach E, Han JH, Milstone AM, Massey CJ, Tamma PD, Antibacterial Resistance Leadership Group (2016) A clinical decision tree to predict whether a Bacteremic patient is infected with an extended-Spectrum β-lactamase-producing organism. Clin Infect Dis 63(7):896–903. https://doi.org/10.1093/cid/ciw425
Gutierrez-Tobar IF, Bravo J, Hernandez M, Turmeque N, Londono-Ruiz JP, Alvarez C (2023) 1242. Advanced automated analysis system for antimicrobial stewardship programs including clinician-focused features developed with artificial intelligence tools in a resource-limited setting. Open Forum Infect Dis 10(Suppl_2):ofad500.1082. https://doi.org/10.1093/ofid/ofad500.1082
Hicks AL, Wheeler N, Sánchez-Busó L, Rakeman JL, Harris SR, Grad YH (2019) Evaluation of parameters affecting performance and reliability of machine learning-based antibiotic susceptibility testing from whole genome sequencing data. PLoS Comput Biol 15(9):9. https://doi.org/10.1371/journal.pcbi.1007349
Hover BM, Kim S-H, Katz M, Charlop-Powers Z, Owen JG, Ternei MA, Maniko J, Estrela AB, Molina H, Park S, Perlin DS, Brady SF (2018) Culture-independent discovery of the malacidins as calcium-dependent antibiotics with activity against multidrug-resistant Gram-positive pathogens. Nat Microbiol 3(4):415–422. https://doi.org/10.1038/s41564-018-0110-1
Hu X, Prehna G, Stebbins CE (2007) Targeting plague virulence factors: a combined machine learning method and multiple conformational virtual screening for the discovery of Yersinia protein kinase a inhibitors. J Med Chem 50(17):3980–3983. https://doi.org/10.1021/jm070645a
Hutchings MI, Truman AW, Wilkinson B (2019) Antibiotics: past, present and future. Curr Opin Microbiol 51:72–80. https://doi.org/10.1016/j.mib.2019.10.008
Kanjilal S, Oberst M, Boominathan S, Zhou H, Hooper DC, Sontag D (2020) A decision algorithm to promote outpatient antimicrobial stewardship for uncomplicated urinary tract infection. Sci Transl Med 12(568):eaay5067. https://doi.org/10.1126/scitranslmed.aay5067

Khaledi A, Weimann A, Schniederjans M, Asgari E, Kuo T-H, Oliver A, Cabot G, Kola A, Gastmeier P, Hogardt M, Jonas D, Mofrad MR, Bremges A, McHardy AC, Häussler S (2020) Predicting antimicrobial resistance in Pseudomonas aeruginosa with machine learning-enabled molecular diagnostics. EMBO Mol Med 12(3):e10264. https://doi.org/10.15252/emmm.201910264

Klug DM, Idiris FIM, Blaskovich MAT, von Delft F, Dowson CG, Kirchhelle C, Roberts AP, Singer AC, Todd MH (2021) There is no market for new antibiotics: this allows an open approach to research and development. Wellcome Open Res 6:146. https://doi.org/10.12688/wellcomeopenres.16847.1

Kolben Y, Azmanov H, Gelman R, Dror D, Ilan Y (2023) Using chronobiology-based second-generation artificial intelligence digital system for overcoming antimicrobial drug resistance in chronic infections. Ann Med 55(1):311–318. https://doi.org/10.1080/07853890.2022.2163053

Kufel J, Bargieł-Łączek K, Kocot S, Koźlik M, Bartnikowska W, Janik M, Czogalik Ł, Dudek P, Magiera M, Lis A, Paszkiewicz I, Nawrat Z, Cebula M, Gruszczyńska K (2023) What is machine learning, artificial neural networks and deep learning?-examples of practical applications in medicine. Diagnostics (Basel, Switzerland) 13(15):2582. https://doi.org/10.3390/diagnostics13152582

Laws M, Shaaban A, Rahman KM (2019) Antibiotic resistance breakers: current approaches and future directions. FEMS Microbiol Rev 43(5):490–516. https://doi.org/10.1093/femsre/fuz014

Li J, Rayner CR, Nation RL, Owen RJ, Spelman D, Tan KE, Liolios L (2006) Heteroresistance to colistin in multidrug-resistant Acinetobacter baumannii. Antimicrob Agents Chemother 50(9):2946–2950. https://doi.org/10.1128/AAC.00103-06

Li L-G, Yin X, Zhang T (2018) Tracking antibiotic resistance gene pollution from different sources using machine-learning classification. Microbiome 6(1):93. https://doi.org/10.1186/s40168-018-0480-x

Liu G, Stokes JM (2022) A brief guide to machine learning for antibiotic discovery. Curr Opin Microbiol 69:102190. https://doi.org/10.1016/j.mib.2022.102190

Liu Z, Deng D, Lu H, Sun J, Lv L, Li S, Peng G, Ma X, Li J, Li Z, Rong T, Wang G (2020) Evaluation of machine learning models for predicting antimicrobial resistance of Actinobacillus pleuropneumoniae from whole genome sequences. Front Microbiol 11. https://www.frontiersin.org/journals/microbiology/articles/10.3389/fmicb.2020.00048

Liu W, Ying N, Mo Q, Li S, Shao M, Sun L, Zhu L (2021) Machine learning for identifying resistance features of Klebsiella pneumoniae using whole-genome sequence single nucleotide polymorphisms. J Med Microbiol 70(11):001474. https://doi.org/10.1099/jmm.0.001474

Liu G, Catacutan DB, Rathod K, Swanson K, Jin W, Mohammed JC, Chiappino-Pepe A, Syed SA, Fragis M, Rachwalski K, Magolan J, Surette MG, Coombes BK, Jaakkola T, Barzilay R, Collins JJ, Stokes JM (2023) Deep learning-guided discovery of an antibiotic targeting Acinetobacter baumannii. Nat Chem Biol 19(11):1342–1350. https://doi.org/10.1038/s41589-023-01349-8

Liu G-Y, Yu D, Fan M-M, Zhang X, Jin Z-Y, Tang C, Liu X-F (2024) Antimicrobial resistance crisis: could artificial intelligence be the solution? Mil Med Res 11(1):7. https://doi.org/10.1186/s40779-024-00510-1

Lobanovska M, Pilla G (2017) Penicillin's discovery and antibiotic resistance: lessons for the future? Yale J Biol Med 90(1):135–145

Lüftinger L, Májek P, Beisken S, Rattei T, Posch AE (2021) Learning from limited data: towards best practice techniques for antimicrobial resistance prediction from whole genome sequencing data. Front Cell Infect Microbiol 11:610348. https://doi.org/10.3389/fcimb.2021.610348

Ma Y, Guo Z, Xia B, Zhang Y, Liu X, Yu Y, Tang N, Tong X, Wang M, Ye X, Feng J, Chen Y, Wang J (2022) Identification of antimicrobial peptides from the human gut microbiome using deep learning. Nat Biotechnol 40(6):921–931. https://doi.org/10.1038/s41587-022-01226-0

MacLean RC, San Millan A (2019) The evolution of antibiotic resistance. Science 365(6458):1082–1083. https://doi.org/10.1126/science.aax3879

Májek P, Lüftinger L, Beisken S, Rattei T, Materna A (2021) Genome-wide mutation scoring for machine-learning-based antimicrobial resistance prediction. Int J Mol Sci 22(23):13049. https://doi.org/10.3390/ijms222313049

Malik B, Bhattacharyya S (2019) Antibiotic drug-resistance as a complex system driven by socio-economic growth and antibiotic misuse. Sci Rep 9(1):9788. https://doi.org/10.1038/s41598-019-46078-y

Mansbach RA, Leus IV, Mehla J, Lopez CA, Walker JK, Rybenkov VV, Hengartner NW, Zgurskaya HI, Gnanakaran S (2020) Machine learning algorithm identifies an antibiotic vocabulary for permeating gram-negative bacteria. J Chem Inf Model 60(6):2838–2847. https://doi.org/10.1021/acs.jcim.0c00352

Masud R, Fahim N, Rana M, Islam M, Rahman M (2023) Artificial intelligence, a powerful tool to combat antimicrobial resistance: an update. J Adv Biotechnol Exp Ther 6(3):711. Crossref. https://doi.org/10.5455/jabet.2023.d161

Medernach RL, Logan LK (2018) The growing threat of antibiotic resistance in children. Infect Dis Clin North Am 32(1):1. https://doi.org/10.1016/j.idc.2017.11.001

Melo MCR, Maasch JRMA, de la Fuente-Nunez C (2021) Accelerating antibiotic discovery through artificial intelligence. Commun Biol 4(1):1–13. https://doi.org/10.1038/s42003-021-02586-0

Moradigaravand D, Palm M, Farewell A, Mustonen V, Warringer J, Parts L (2018) Prediction of antibiotic resistance in Escherichia coli from large-scale pan-genome data. PLoS Comput Biol 14(12):e1006258. https://doi.org/10.1371/journal.pcbi.1006258

Nguyen M, Olson R, Shukla M, VanOeffelen M, Davis JJ (2020) Predicting antimicrobial resistance using conserved genes. PLoS Comput Biol 16(10):e1008319. https://doi.org/10.1371/journal.pcbi.1008319

Noman SM, Zeeshan M, Arshad J, DeressaAmentie M, Shafiq M, Yuan Y, Zeng M, Li X, Xie Q, Jiao X (2023) Machine learning techniques for antimicrobial resistance prediction of *Pseudomonas aeruginosa* from whole genome sequence data. Comput Intell Neurosci 2023:e5236168. https://doi.org/10.1155/2023/5236168

O'Neill J (2016) Tackling drug-resistant infections globally: final report and recommendations. https://apo.org.au/node/63983

Pachori P, Gothalwal R, Gandhi P (2019) Emergence of antibiotic resistance Pseudomonas aeruginosa in intensive care unit; a critical review. Genes Dis 6(2):Article 2. https://doi.org/10.1016/j.gendis.2019.04.001

Palacios F, Campos M, Juarez JM, Cosgrove SE, Avdic E, Canovas-Segura B, Morales A, Martínez-Nuñez ME, Molina-García T, García-Hierro P, Cacho-Calvo J (2016) A clinical decision support system for an antimicrobial stewardship program. HEALTHINF 2016—9th international conference on health informatics, proceedings; part of 9th international joint conference on biomedical engineering systems and technologies, BIOSTEC 2016, pp 496–501. doi: https://doi.org/10.5220/0005824904960501

Popa SL, Pop C, Dita MO, Brata VD, Bolchis R, Czako Z, Saadani MM, Ismaiel A, Dumitrascu DI, Grad S, David L, Cismaru G, Padureanu AM (2022) Deep learning and antibiotic resistance. Antibiotics 11(11):Article 11. https://doi.org/10.3390/antibiotics11111674

Ren Y, Chakraborty T, Doijad S, Falgenhauer L, Falgenhauer J, Goesmann A, Hauschild A-C, Schwengers O, Heider D (2022a) Prediction of antimicrobial resistance based on whole-genome sequencing and machine learning. Bioinformatics (Oxford, England) 38(2):325–334. https://doi.org/10.1093/bioinformatics/btab681

Ren Y, Chakraborty T, Doijad S, Falgenhauer L, Falgenhauer J, Goesmann A, Schwengers O, Heider D (2022b) Deep transfer learning enables robust prediction of antimicrobial resistance for novel antibiotics. Antibiotics 11(11):Article 11. https://doi.org/10.3390/antibiotics11111611

Renwick MJ, Simpkin V, Mossialos E (2016) Background. In Targeting innovation in antibiotic drug discovery and development: the need for a one health—one Europe—one world framework [internet]. European Observatory on Health Systems and Policies. https://www.ncbi.nlm.nih.gov/books/NBK447326/

Shen T, Guo J, Han Z, Zhang G, Liu Q, Si X, Wang D, Wu S, Xia J (2024) AutoMolDesigner for antibiotic discovery: an AI-based open-source software for automated design of small-molecule antibiotics. J Chem Inf Model 64:575. https://doi.org/10.1021/acs.jcim.3c01562

Stokes JM, Yang K, Swanson K, Jin W, Cubillos-Ruiz A, Donghia NM, MacNair CR, French S, Carfrae LA, Bloom-Ackermann Z, Tran VM, Chiappino-Pepe A, Badran AH, Andrews

IW, Chory EJ, Church GM, Brown ED, Jaakkola TS, Barzilay R, Collins JJ (2020) A deep learning approach to antibiotic discovery. Cell 180(4):688–702.e13. https://doi.org/10.1016/j.cell.2020.01.021

Thorpe KE, Joski P, Johnston KJ (2018) Antibiotic-resistant infection treatment costs have doubled since 2002, now exceeding $2 billion annually. Health Aff 37(4):Article 4. https://doi.org/10.1377/hlthaff.2017.1153

Tselebonis A, Nena E, Nikolaidis C, Konstantinidis T, Kontogiorgis C, Panopoulou M, Constantinidis TC (2016) Monitoring of frequency and antimicrobial susceptibility of pathogens on the hands of healthcare workers in a tertiary hospital. Folia Med 58(3):200–205. https://doi.org/10.1515/folmed-2016-0028

Tsoukalas A, Albertson T, Tagkopoulos I (2015) From data to optimal decision making: a data-driven, probabilistic machine learning approach to decision support for patients with sepsis. JMIR Med Inform 3(1):e11. https://doi.org/10.2196/medinform.3445

Tu Q, Cotta M, Raman S, Graham N, Schlapbach L, Roberts JA (2021) Individualized precision dosing approaches to optimize antimicrobial therapy in pediatric populations. Expert Rev Clin Pharmacol 14(11):1383–1399. https://doi.org/10.1080/17512433.2021.1961578

Tucs A, Tran DP, Yumoto A, Ito Y, Uzawa T, Tsuda K (2020) Generating ampicillin-level antimicrobial peptides with activity-aware generative adversarial networks. ACS Omega 5(36):22847–22851. https://doi.org/10.1021/acsomega.0c02088

Wang C, Garlick S, Zloh M (2021) Deep learning for novel antimicrobial peptide design. Biomol Ther 11(3):471. https://doi.org/10.3390/biom11030471

Weis CV, Jutzeler CR, Borgwardt K (2020) Machine learning for microbial identification and antimicrobial susceptibility testing on MALDI-TOF mass spectra: a systematic review. Clin Microbiol Infect 26(10):1310–1317. https://doi.org/10.1016/j.cmi.2020.03.014

Wiens J, Saria S, Sendak M, Ghassemi M, Liu VX, Doshi-Velez F, Jung K, Heller K, Kale D, Saeed M, Ossorio PN, Thadaney-Israni S, Goldenberg A (2019) Do no harm: a roadmap for responsible machine learning for health care. Nat Med 25(9):1337–1340. https://doi.org/10.1038/s41591-019-0548-6

Wong F, Zheng EJ, Valeri JA, Donghia NM, Anahtar MN, Omori S, Li A, Cubillos-Ruiz A, Krishnan A, Jin W, Manson AL, Friedrichs J, Helbig R, Hajian B, Fiejtek DK, Wagner FF, Soutter HH, Earl AM, Stokes JM et al (2024) Discovery of a structural class of antibiotics with explainable deep learning. Nature 626(7997):177–185. https://doi.org/10.1038/s41586-023-06887-8

Yang JH, Wright SN, Hamblin M, McCloskey D, Alcantar MA, Schrübbers L, Lopatkin AJ, Satish S, Nili A, Palsson BO, Walker GC, Collins JJ (2019) A white-box machine learning approach for revealing antibiotic mechanisms of action. Cell 177(6):1649–1661.e9. https://doi.org/10.1016/j.cell.2019.04.016

Yu T, Fu Y, He J, Zhang J, Xianyu Y (2023) Identification of antibiotic resistance in ESKAPE pathogens through plasmonic nanosensors and machine learning. ACS Nano 17(5):4551–4563. https://doi.org/10.1021/acsnano.2c10584

Zhang A-N, Gaston JM, Dai CL, Zhao S, Poyet M, Groussin M, Yin X, Li L-G, van Loosdrecht MCM, Topp E, Gillings MR, Hanage WP, Tiedje JM, Moniz K, Alm EJ, Zhang T (2021) An omics-based framework for assessing the health risk of antimicrobial resistance genes. Nat Commun 12(1):4765. https://doi.org/10.1038/s41467-021-25096-3

Zhen X, Lundborg CS, Sun X, Hu X, Dong H (2019) Economic burden of antibiotic resistance in ESKAPE organisms: a systematic review. Antimicrob Resist Infect Control 8:137. https://doi.org/10.1186/s13756-019-0590-7

Correction to: The Art and Science of Defeating Superbugs: Present and Future of Phage Therapy

Sunil Kumar Raman, Chunna Yadav, Himanshu Bansode,
Shilpi Mishra, Unnati Sharma, and Niharika Mishra

Correction To:
Chapter 17 in: M. Gangwar, G. Nath (eds.) *Emerging Paradigms for Antibiotic-Resistant Infections: Beyond the Pill*,
https://doi.org/10.1007/978-981-97-5272-0_17

The original version of this book was inadvertently published with the below error:

Chapter 17 was originally published without including the chapter author "Shilpi Mishra." This has now been corrected, and the updated author sequence is as follows: "Sunil Kumar Raman, Chunna Yadav, Himanshu Bansode, Shilpi Mishra, Unnati Sharma, and Niharika Mishra."

The updated version of this chapter can be found at
https://doi.org/10.1007/978-981-97-5272-0_17

M. Gangwar, G. Nath (eds.), *Emerging Paradigms for Antibiotic-Resistant Infections: Beyond the Pill*, https://doi.org/10.1007/978-981-97-5272-0_39

The manufacturer's authorised representative in the EU is Springer Nature Customer Service Centre GmbH, Europaplatz 3, 69115 Heidelberg, Germany. If you have any concerns regarding our products, please contact ProductSafety@springernature.com

Printed and bound by CPI Group (UK) Ltd, Croydon, CR0 4YY

01/10/2026

02225941-0020